HANDBOOK OF TRIBOLOGY

Other McGraw-Hill Books of Interest

Avallone & Baumeister • MARKS' STANDARD HANDBOOK FOR MECHANICAL ENGINEERS
Brady & Clauser • MATERIALS HANDBOOK
Bralla • HANDBOOK OF PRODUCT DESIGN FOR MANUFACTURING
Brunner • HANDBOOK OF INCINERATION SYSTEMS
Corbitt • STANDARD HANDBOOK OF ENVIRONMENTAL ENGINEERING
Elliot • STANDARD HANDBOOK OF POWERPLANT ENGINEERING
Freeman • STANDARD HANDBOOK OF HAZARDOUS WASTE TREATMENT AND DISPOSAL
Gieck • ENGINEERING FORMULAS
Harris • HANDBOOK OF ACOUSTICAL MEASUREMENTS AND NOISE CONTROL
Harris & Crede • SHOCK AND VIBRATION HANDBOOK
Hicks • STANDARD HANDBOOK OF ENGINEERING CALCULATIONS
Juran and Gryna • JURAN'S QUALITY CONTROL HANDBOOK
Karassik et al. • PUMP HANDBOOK
Parmley • STANDARD HANDBOOK OF FASTENING AND JOINING
Rohsenow, Hartnett, & Ganić • HANDBOOK OF HEAT TRANSFER FUNDAMENTALS
Rohsenow, Hartnett, & Ganić • HANDBOOK OF HEAT TRANSFER APPLICATIONS
Rosaler & Rice • STANDARD HANDBOOK OF PLANT ENGINEERING
Rothbart • MECHANICAL DESIGN AND SYSTEMS HANDBOOK
Shigley & Mischke • STANDARD HANDBOOK OF MACHINE DESIGN
Tuma • ENGINEERING MATHEMATICS HANDBOOK
Tuma • HANDBOOK OF NUMERICAL CALCULATIONS IN ENGINEERING
Young • ROARK'S FORMULAS FOR STRESS AND STRAIN
Grimm & Rosaler • HANDBOOK OF HVAC DESIGN
Wadsworth • HANDBOOK OF STATISTICAL METHODS FOR ENGINEERS AND SCIENTISTS
Maynard • INDUSTRIAL ENGINEERING HANDBOOK

HANDBOOK OF TRIBOLOGY

Materials, Coatings, and Surface Treatments

Bharat Bhushan

IBM Research Division, Almaden Research Center
San Jose, California

B. K. Gupta

Industrial Tribology, Machine Dynamics,
and Maintenance Engineering Center
Indian Institute of Technology
New Delhi, India

McGRAW-HILL, INC.

New York St. Louis San Francisco Auckland Bogotá
Caracas Hamburg Lisbon London Madrid
Mexico Milan Montreal New Delhi Paris
San Juan São Paulo Singapore
Sydney Tokyo Toronto

Library of Congress Cataloging-in-Publication Data

Bhushan, Bharat, date.
Handbook of tribology : materials, coatings, and surface treatments / Bharat Bhushan, B.K. Gupta.
p. cm.
Includes bibliographical references (p.) and indexes.
ISBN 0-07-005289-2
1. Tribology—Handbooks, manuals, etc. I. Gupta, B. K. (Bal Kishan) II. Title.
TJ1075.B47 1991
621.8'9—dc20 91-7155
CIP

1 2 3 4 5 6 7 8 9 0 DOC/DOC 9 8 7 6 5 4 3 2 1

ISBN 0-07-005249-2

The sponsoring editor for this book was Robert Hauserman, the editing supervisor was Peggy Lamb, and the production supervisor was Pamela A. Pelton. This book was set in Times Roman. It was composed by McGraw-Hill's Professional Book Group composition unit.

Printed and bound by R. R. Donnelley & Sons, Inc.

Dedicated to our wives
Sudha Bhushan
and
Sarita Gupta

CONTENTS

Chapter 3. Physics of Tribological Materials 3.1

Chapter 4. Metals and Ceramics 4.1

Chapter 6. Coating Deposition and Surface Treatment Techniques: Classification 6.1

Chapter 7. Surface Preparation for Coating Deposition 7.1

Chapter 10. Coating Deposition by Miscellaneous Techniques 10.1

PREFACE

Effective lubrication between moving surfaces considerably reduces friction and wear. New designs of machines impose restrictions on the structural and surface characteristics of sliding materials and lubrication at the interface. Conventional liquid lubricants fail under extreme conditions, such as low pressure, corrosive or oxidative environment, high load, and high speed. Therefore, there is a need to develop solid lubricants, which can sustain extreme conditions without any failure, with prolonged life. A variety of bulk materials, such as ferrous and nonferrous metals and alloys, ceramics and cermets, and polymers, can be modified by alloying, mixing, compositing, and surface hardening, to achieve adequate resistance to wear and corrosion and low friction. Ceramics and cermets appear to be ideal wear-resistant materials that suit many tribological applications provided that their strength and toughness are acceptable. Self-lubricating solids, such as molybdenum disulfide (MoS_2), graphite, and polymers exhibit low friction and/or wear during sliding in the absence of an external supply of lubricant. These materials can be used in their dry state, combined with a carrier (e.g., oil and grease) or binder, or mixed with metals and polymers to achieve self-lubricating characteristics.

Many times, bulk materials are either incapable of satisfying design requirements or too expensive. Surface coatings and treatments can satisfy requirements, such as hard surface with a ductile core, at minimal costs. Soft coatings of graphite, MoS_2, and polytetrafluoroethylene (PTFE) that provide low friction and wear have been successfully used under extreme conditions. Hard coatings of oxides, carbides, nitrides, borides, and silicides can sustain heavy loads, high speeds, and high temperatures for extended periods without any deterioration and improve the wear performance of various tribological elements by several times. Surface treatment processes do not produce coating-to-substrate interface as found in coatings. The interface produced in coatings so often is subjected to mechanical weakness. Many surface treatment techniques can be applied very economically on large and intricate components and are preferred over coatings to reduce costs. The surface treatments are widely used, mainly for ferrous-based metals. Ion implantation can be used to incorporate virtually any kind of ion species into any material. The soft and hard coatings and surface treatments are finding increasing use in almost every industrial sector and provide means to conserve energy and materials.

This book presents a state-of-the-art review of bulk materials, coatings, and surface treatments selected to combat wear and reduce friction. It is a comprehensive source of an enormous amount of information on the production and properties of materials, coatings, and surface treatments that are used under a variety of tribological situations. Chapter 1 gives an overview of liquid and solid lubricants. The mechanisms of friction, wear, and lubrication are briefly reviewed in Chapter 2. Chapter 3 deals with the physics of materials, i.e., shape and properties of solid surfaces, the effect of interaction with operating environments, and phase equilibria, and microstructure's role in friction and wear characteristics of materials. Various bulk materials, such as ferrous and nonferrous metals and alloys, ceramics, cermets, carbon, graphite, molybdenum disulfide, polymers, and polymer composites employed for tribological applications are discussed in Chapters 4 and 5 with more emphasis on recently developed exotic materials, i.e., ceramics and polymer composites.

The classification and comparison of various coating deposition techniques and surface treatments are given in Chapter 6. The surface preparation techniques, i.e., cleaning and roughening, and requirements necessary to get good quality coatings are explained in Chapter 7. Chapters 8 to 10 provide the details of various technologies of coating deposition. These chapters cover all the commonly used techniques for depositing coatings in industrial applications. Chapters 11 and 12 deal with the surface treatments by thermal and chemical processes and ion beams employed for modifying the surfaces of metals, ceramics,and polymers to obtain low friction and wear. Chapters 13 and 14 describe the tribological characteristics of soft and hard coatings. These chapters cover different kinds of coatings commonly used for the control of friction and wear. The coating screening methodology that deals with physical examinations—thickness, density, porosity, electrical resistivity, surface roughness, hardness, residual stresses, adhesion, and friction and wear measurements—is covered in Chapter 15. Chapter 16 gives examples of the application of materials, coatings, and surface treatments to various tribological elements, such as bearings, seals, gears, cams and tappets, piston rings, electrical brushes, and tools, and some industrial applications such as gas turbines, fusion reactors, and magnetic storage devices.

This book is intended for graduate students of tribology, mechanical engineering, materials science, physics, and chemistry, for research workers who are active or intend to become active in materials aspects of tribology and coating technology, and for industrialists and practicing engineers who need state-of-the-art information on materials, coatings, and surface treatments that are required to improve tribological performance. The chapters that present coating deposition techniques and surface treatments in detail should prove useful to anyone interested in coatings for applications such as microelectronics, solar power, automotive engineering, oil and gas exploration and pipelines, earthmoving equipment, pumps, magnetic and optical peripheral devices, and printers for computers. Presenting the properties of materials, salient features of various coating deposition, and surface treatment techniques in tabular form provides a quick comparison of the data. This book can serve as an excellent reference for researchers and people in different industries with very little knowledge of this field who would like to select a material, coating, or surface treatment for their special applications.

We wish to thank all our colleagues at IBM Almaden Research Center, San Jose, California, the Indian Institute of Technology, New Delhi, India, and the Institute of Physics, University of Aarhus, Aarhus, Denmark, and our family members for their help and cooperation at various stages of preparation of this book.

Bharat Bhushan
Los Gatos, California, U.S.A.

B. K. Gupta
New Delhi, India

CHAPTER 1
INTRODUCTION

Lubrication is used to reduce friction and wear between solid surfaces having relative motion. Lubricants serve to (1) reduce the friction required to initiate or maintain the motion between contacting surfaces, (2) minimize wear of contacting surfaces, and (3) control, within reasonable limits, developed temperatures that arise due to friction (Anonymous 1968, 1971, 1978, 1984; Bartenev and Lavrentev, 1981; Benzing, 1964; Bhushan, 1980, 1987a; Bisson and Anderson, 1964; Booser, 1983, 1984; Bowden and Tabor, 1950, 1964; Braithwaite, 1964, 1967; Cameron, 1966; Campbell, 1972; Campbell et al., 1966; Clauss, 1972; Dowson, 1979; Gunther, 1971; Kragelsky and Alision, 1981; Lipp, 1976; Neale, 1973; Peterson and Winer, 1980; Sliney, 1982). Friction and wear losses cost the world economy hundreds of millions of U.S. dollars every year in lost time, lost production, lost material, and wasted energy (Davison, 1957).

Materials employed to reduce friction and to control wear may be in the forms of gases, liquids, or solids. In addition, surfaces can be treated to alter the properties of the skin of the surfaces to have desirable tribological properties. Use of coatings and surface treatments provides flexibility in selection of the substrate material. This helps in meeting conflicting requirements of the structural and tribological properties of the components at a fraction of the cost. Many of the surface-treatment techniques can be applied very economically on large and intricate components and are preferred over coatings for cost reasons.

Figures 1.1 and 1.2 represent a classification of various liquids, solid materials, coatings, and surface treatments used to lubricate the contacting surfaces (Bhushan, 1980, 1987a, 1987b). In this chapter we mention various types of liquid lubricants and give more emphasis to solid lubricant materials, coatings, and surface treatments. Table 1.1 lists the important properties, principal attributes, and applications of some of the commonly used liquid lubricants, coatings, and surface treatments. The values given in Table 1.1 are only approximate, to provide a guideline for selection of various types of liquid and solid lubricants.

1.1 LIQUID LUBRICATION

Liquid lubrication is obtained either by applying a very thin film of liquid at the interface of contacting surfaces or by maintaining a thick film of liquid between the two moving surfaces (Booser, 1983, 1984; Braithwaite, 1967; Cameron, 1966; Evans et al., 1972; Gunther, 1971). The former type leads to boundary lubrication phenomenon, in which the chemical properties of the liquid, the surface topog-

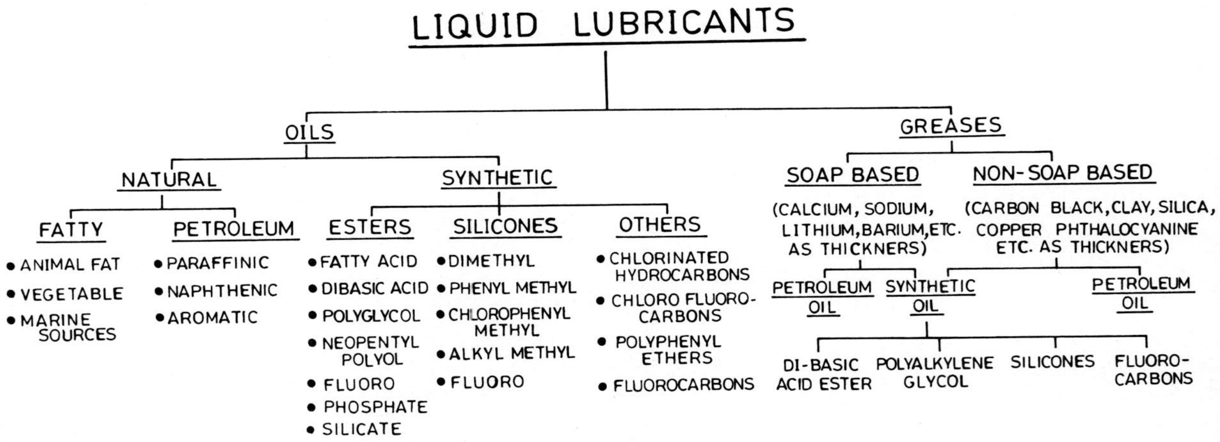

LIQUID LUBRICANTS

- OILS
  - NATURAL
    - FATTY
      - ANIMAL FAT
      - VEGETABLE
      - MARINE SOURCES
    - PETROLEUM
      - PARAFFINIC
      - NAPHTHENIC
      - AROMATIC
  - SYNTHETIC
    - ESTERS
      - FATTY ACID
      - DIBASIC ACID
      - POLYGLYCOL
      - NEOPENTYL POLYOL
      - FLUORO
      - PHOSPHATE
      - SILICATE
    - SILICONES
      - DIMETHYL
      - PHENYL METHYL
      - CHLOROPHENYL METHYL
      - ALKYL METHYL
      - FLUORO
    - OTHERS
      - CHLORINATED HYDROCARBONS
      - CHLORO FLUORO-CARBONS
      - POLYPHENYL ETHERS
      - FLUOROCARBONS
- GREASES
  - SOAP BASED (CALCIUM, SODIUM, LITHIUM, BARIUM, ETC. AS THICKNERS)
    - PETROLEUM OIL
    - SYNTHETIC OIL
      - DI-BASIC ACID ESTER
      - POLYALKYLENE GLYCOL
      - SILICONES
      - FLUORO-CARBONS
  - NON-SOAP BASED (CARBON BLACK, CLAY, SILICA, COPPER PHTHALOCYANINE ETC. AS THICKNERS)
    - SYNTHETIC OIL
    - PETROLEUM OIL

FIGURE 1.1 Classification of various liquid lubricants.

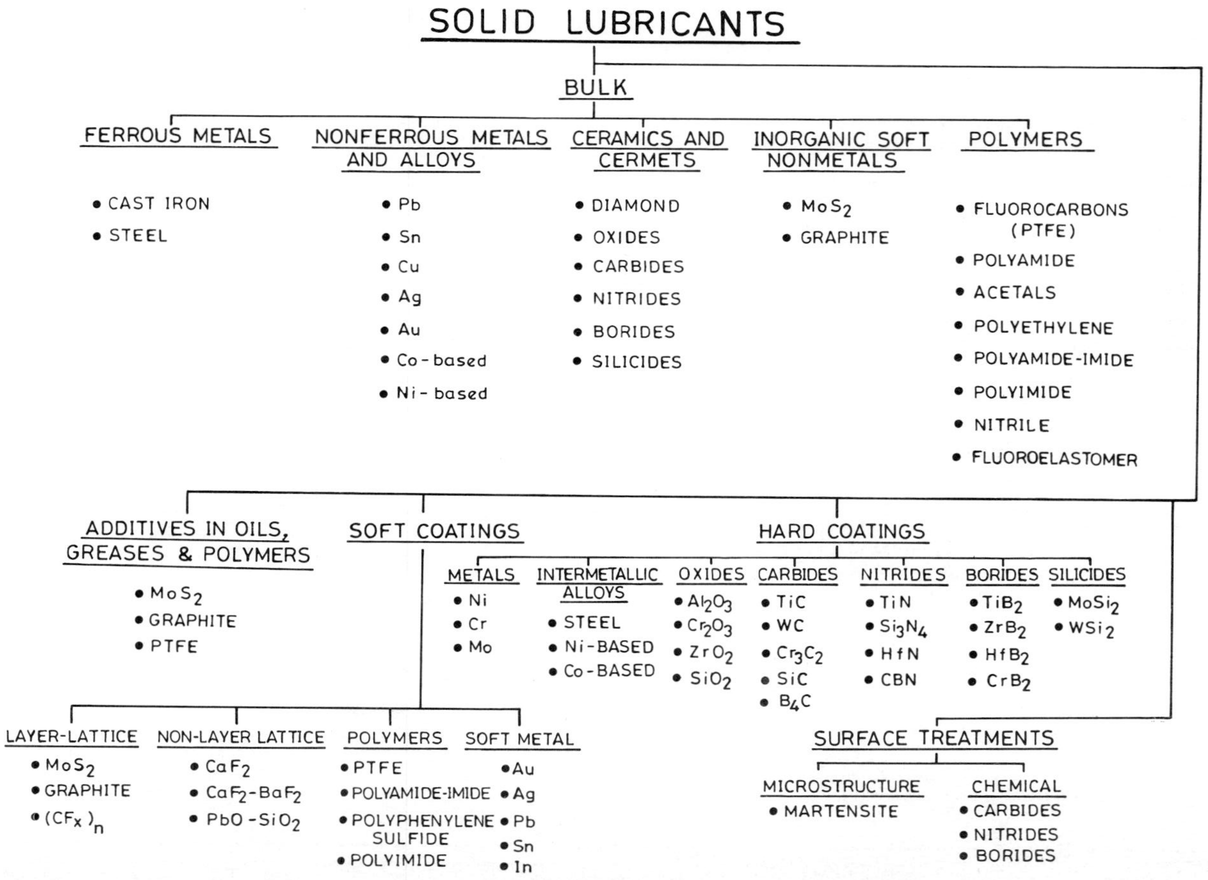

SOLID LUBRICANTS

BULK

FERROUS METALS
- CAST IRON
- STEEL

NONFERROUS METALS AND ALLOYS
- Pb
- Sn
- Cu
- Ag
- Au
- Co-based
- Ni-based

CERAMICS AND CERMETS
- DIAMOND
- OXIDES
- CARBIDES
- NITRIDES
- BORIDES
- SILICIDES

INORGANIC SOFT NONMETALS
- MoS_2
- GRAPHITE

POLYMERS
- FLUOROCARBONS (PTFE)
- POLYAMIDE
- ACETALS
- POLYETHYLENE
- POLYAMIDE-IMIDE
- POLYIMIDE
- NITRILE
- FLUOROELASTOMER

ADDITIVES IN OILS, GREASES & POLYMERS
- MoS_2
- GRAPHITE
- PTFE

SOFT COATINGS

LAYER-LATTICE
- MoS_2
- GRAPHITE
- $(CF_x)_n$

NON-LAYER LATTICE
- CaF_2
- CaF_2-BaF_2
- PbO-SiO_2

POLYMERS
- PTFE
- POLYAMIDE-IMIDE
- POLYPHENYLENE SULFIDE
- POLYIMIDE

SOFT METAL
- Au
- Ag
- Pb
- Sn
- In

HARD COATINGS

METALS
- Ni
- Cr
- Mo

INTERMETALLIC ALLOYS
- STEEL
- Ni-BASED
- Co-BASED

OXIDES
- Al_2O_3
- Cr_2O_3
- ZrO_2
- SiO_2

CARBIDES
- TiC
- WC
- Cr_3C_2
- SiC
- B_4C

NITRIDES
- TiN
- Si_3N_4
- HfN
- CBN

BORIDES
- TiB_2
- ZrB_2
- HfB_2
- CrB_2

SILICIDES
- $MoSi_2$
- WSi_2

SURFACE TREATMENTS

MICROSTRUCTURE
- MARTENSITE

CHEMICAL
- CARBIDES
- NITRIDES
- BORIDES

FIGURE 1.2 Classification of solid materials, coatings, and surface treatments for lubrication.

TABLE 1.1 Common Properties and Applications of Liquid Lubricants, Coatings, and Surface Treatments

Type of lubricants	Maximum usable temperature and other important properties	Properties and typical applications
	Liquid lubricants	
Natural oils		
Animal fats and vegetables	Up to 120°C η = 30–250 cSt at 38°C μ = 0.02	Old conventional lubricants, less stable than petroleum oils; not used as such, but suitable as additives in petroleum oils due to their good lubricity and emulsifying characteristics
Petroleum based	Up to 130°C Up to 200°C (superrefined) η = 15 cSt at 38°C	Good boundary lubricants; wide range of viscosity of oils is available; with additives these can be employed for internal combustion engines, diesel engines, steam turbines, compressors, gears, machine tool slideways, hydraulic systems
Synthetic oils		
Dibasic acid esters	Up to 230°C η = 15 cSt at 38°C η = 3 cSt at 100°C μ = 0.02	High viscosity index, low volatility, heat transfer properties not much compatible with rubbers and plastics, used in gas turbine engines and magnetic tapes
Polyglycol ether	Up to 230°C	Soluble in water, decompose without producing solid degradation products, sometimes replace conventional oils in many applications, such as bearings, fire-resistant hydraulic fluids in naval applications, wire drawing, and metalworking lubricant
Phosphate ester	Up to 240°C η = 11 cSt at 38°C η = 4 cSt at 100°C	Good boundary lubricant, not compatible with rubber and plastic components, good fire resistance; used as aircraft and industrial hydraulic fluid, in gas turbines, and in compressors
Silicones	Up to 290°C η = 65 cSt at 38°C η = 30 cSt at 100°C	Good chemical resistance, low vapor pressure, poor boundary lubrication for steel versus steel; used as lubricants for plastics and rubbers, medical syringes, photographic films, refrigerator and compressor pumps, oven conveyor belts, and as hydraulic fluid in fluid power transmission systems

Fluorocarbons	Up to 370°C η = 150–500 cSt at 38°C η = 35–40 cSt at 100°C	Good chemical resistance, extremely low vapor pressure, high oxidative and thermal stability, poor viscosity index; used to lubricate oxygen compressors and as lubricants in equipments for handling of strongly oxidizing materials and other reactive chemicals, magnetic disks
Greases		
Soap-based		
Calcium(lime)-based (petroleum oil)	Up to 80°C μ = 0.02–0.04	Water repellent and buttery in texture; used in plain bearings, antifriction bearings of moderate speed and temperature
Sodium(soda)-based (petroleum oil)	Up to 120°C	Emulsify with water; smooth, fibrous, or spongelike in structure; used in antifriction bearings of high or moderate speeds
Lithium-based (petroleum oil)	Up to 150°C	Water repellent, buttery to stringy in structure; used for general-purpose service
Non-soap-based		
Clay-based (synthetic oil)	Up to 300°C	High temperature range, insoluble both in hot and cold water, acid resistant to some extent, inhibit rust; used for aircraft and spacecraft
	Soft coatings	
Layer-lattice solids		
MoS_2	Up to 315°C in air Up to 500°C in vacuum μ = 0.05–0.1	Burnished and bonded coatings; used extensively for various sliding applications; sputtered coatings are used in high vacuum, such as in outer space; these are invaluable for high-precision components having close tolerances
Graphite	Up to 430°C μ = 0.1–0.2	Very effective in humid atmosphere; used extensively for various sliding applications, e.g., metalworking processes, bearings, and seals
Polyimide-bonded graphite fluoride	Up to 430°C μ = 0.05–0.2	Burnished coating gives longer wear life and higher load capacity than graphite

TABLE 1.1 Common Properties and Applications of Liquid Lubricants, Coatings, and Surface Treatments (*Continued*)

Type of lubricants	Maximum usable temperature and other important properties	Properties and typical applications
	Soft coatings (*Continued*)	
Nonlayer-lattice solids		
CaF_2/CaF_2-BaF_2	Up to 260–900°C μ = 0.3 at 25°C μ = 0.2 at 650°C	Lubricates effectively from 260 to 900°C; used in slow-speed applications in extreme environments, such as the space shuttle, and in the hot section of small jet engines
PbO-SiO_2	Up to 260–650°C μ = 0.2 at 650°C	Lubricates effectively from 260 to 650°C; at high speeds, friction is low over a wide range of temperature; used at high speeds and high temperatures, such as lubrication of dies for high-speed wire drawing
Polymers		
PTFE	Up to 150°C μ = 0.03–0.1	Excellent nonstick, extremely low friction and wear resistance; used in saw blades, hoppers and chutes for handling wet and dry materials, bearings, and several other automobile parts
Polyphenylene sulfide	Up to 200°C μ = 0.05–0.15	High-temperature polymeric coatings; used extensively as binder to produce solid lubricant coatings by mixing MoS_2 and metal oxides
Polyimides	Up to 300°C μ = 0.1–0.2	High thermal stability; used in bearings, gears, seals, and prosthetic human joints. Filling with graphite increases load-carrying capacity
Soft metals		
Gold, silver	Up to 600°C μ = 0.1–0.25	Successfully used to lubricate in vacuum, such as space bearings of satellite mechanisms, solar array drives, despin assemblies, and gimbals
Lead and lead-tin	Up to 200°C μ = 0.1–0.25	Used extensively as overlays in plain bearings; also finds applications in electrical contacts and vacuum (space bearings)

Hard coatings		
Metals		
Chromium	μ = 0.15–0.4 H = 900–1100 HV	Used on piston rings, cylinder liner and cross-head pins, etc. in internal combustion engines and compressors and cannon and gun tubes
Nickel	μ = 0.15–0.4 H = 125–450 HV (electroplated) H = 500–700 HV (electroless)	Used for decorative purposes, environmental corrosion resistance, on piston rings, for the electroforming of gramophone disks and tape path components of computer tape drives
Molybdenum	μ = 0.15–0.4 H = 500 HV	Used on piston rings
Intermetallic alloys		
Co-based alloy (Stellite)	μ = 0.15–0.4 H = 400–600 HV	Used for applications requiring resistance to metal-to-metal wear and low-stress abrasion, e.g., bearings, slurry pumps, rock bits, diesel engine valves
Co-based alloy (Tribaloy 800)	μ = 0.15–0.4 H = 630 HV	Used on journal bearings, ball and roller bearings, piston rings, valves, cams
Ni-based alloy (Haynes alloy 40)	μ = 0.2–0.4 H = 615 HV	Used for corrosion- and wear-resistant applications
Ceramics		
Aluminum oxide (Al_2O_3)	Up to 1800°C μ = 0.15–0.5 H = 1900–2100 HV	Sintered carbide cutting tips, bearings, gas turbine engine shafts, magnetic heads for tape and disk drives
Chromium oxide (Cr_2O_3)	Up to 1000°C μ = 0.1–0.4 H = 1200–1300 HV	Used in foil bearings, drill bits
Titanium carbide (TiC)	Up to 1100°C μ = 0.15–0.6 H = 2200–2800 HV	Very hard; does not yield under high loads; extremely stable; used on sintered carbide cutting tips, steel press tools, punches and dies, wear parts

TABLE 1.1 Common Properties and Applications of Liquid Lubricants, Coatings, and Surface Treatments (*Continued*)

Type of lubricants	Maximum usable temperature and other important properties	Properties and typical applications
	Hard coatings (*Continued*)	
Ceramics (*Continued*)		
Tungsten carbide (WC)	Up to 500°C μ = 0.15–0.6 H = 1800–2100 HV	Extensively used in carbide cutting tools, punches and dies, wear parts
Titanium nitride (TiN)	Up to 500°C μ = 0.15–0.5 H = 1500–1700 HV	Extensively used on carbide cutting tools, drill bits, bearings, and decorative coatings
	Surface treatments	
Microstructural treatments		
Induction/flame hardening (formation of martensite phase)	Up to 500°C μ = 0.2–0.6 Case depth = 250–5000 μm H = 500–750 HV	Hard, highly wear-resistant surface (deep case depth), good capacity for contact load, low-cost steels required
Chemical diffusion		
Nitride formation/nitriding	Up to 500°C μ = 0.15–0.6 Case depth = 5–250 μm H = 900–1150 HV	Hard, highly wear-resistant surface (medium case depth), excellent capacity for contact load, low- to medium-cost steels required
Boride formation/boriding	Up to 800°C μ = 0.15–0.6 Case depth = 5–100 μm H = 1600–2300 HV	Extremely hard, properties are dependent on steel chemistry, shallow case depth, good oxidation resistance
Ion implantation (e.g., formation of nitrides in case of N ion implantation)	μ = 0.15–0.6 Case depth = 0.01–0.5 μm H = 600–2000 HV	Hard, wear-resistant surface (extremely shallow depth), low-temperature process, finished products can be implanted

raphy of the contacting surfaces, and the operating temperatures are of major importance. The later type involves the complete separation of contacting surfaces known as hydrodynamic or elastohydrodynamic lubrication, in which the friction and load-carrying capacity depend on the liquid viscosity and not on its chemical constitution. In hydrodynamic lubrication, the coefficient of friction is very low (~0.02) and wear is almost negligible. Reynolds (1886) established the relationships between the film thickness, viscosity, speed, and load-carrying capacity of a liquid film. This theory has been utilized to design bearings and to predict their performance theoretically. Under extreme conditions, i.e., at high speeds and high loads, various factors such as pressure dependence of the liquid's viscosity, elastic deformation of the contacting surfaces, etc. play a key role in lubrication mechanisms.

The liquid lubricants are based on natural fatty oils, petroleum oils, and synthetic liquids (Bisson and Anderson, 1964; Booser, 1984; Bowden and Tabor, 1950, 1964; Braithwaite, 1967; Gunderson and Hart, 1962; Lansdown, 1982; McConnell, 1972). The performance of such oils can be considerably influenced by additives. A brief description of each generic class of liquid lubricants is given in the following subsections.

1.1.1 Fatty Oils

Early liquid lubricants were water-insoluble fatty substances extracted from plant, animal, and marine sources. Fatty oils decompose to fatty acids at slightly elevated temperatures and as a result of certain chemical reactions. Castor oil, lard, tallow, rapeseed oil, and sperm oil represent the important types of lubricating oils. Now these oils are rarely used alone but are usually employed as an additive to increase the lubricity of petroleum-based oils.

1.1.2 Petroleum-Based Oils

The petroleum-based oils, also known as *mineral oils*, are the most common liquid lubricants, produced from petroleum crudes by distillation and several refining processes. These oils are composed mostly of the following hydrocarbon groups, linked in a mixture of molecules:

- Paraffins: C_nH_{2n+2}
- Naphthenes: C_nH_{2n}
- Aromatics: C_nH_n

Most lubricating oils are paraffins, in which the carbon atoms are in straight or branched chains, but not rings. Naphthenes are second most common oils, in which the carbon atoms form rings. Finally, in aromatics, usually available in small percentage, about 2 percent carbon rings are present, but the proportion of hydrogen is reduced. Usually, in practice, the mineral oils contain mixtures of all the three types, i.e., paraffins, naphthenes, and aromatics. Some of these hydrocarbons contain more reactive heteroatoms of oxygen, nitrogen, or sulfur in concentrations of less than 10 percent by weight. Additionally, some oils contain small amounts of dissolved gas and water from the atmosphere. Aromatic hydrocarbons give better wear protection in air than paraffinic and naphthenic hydrocarbons of comparable viscosity.

Additives are added to petroleum-based oils in order (1) to modify some physical property such as pour point, foaming, or viscosity-temperature behavior, (2) to modify chemical behavior such as detergency, oxidation, or corrosion, and (3) to improve wear and extreme pressure resistance. Typical examples of additives dispersed in petroleum-based oils are tricresyl phosphate, lead naphthenate (for wear reduction), polyacrylates (viscosity-index improver), chlorine, phosphorous, and sulfur-containing oils (film-strength improver). The selection of an additive is made on the basis of operating conditions, environment, and function of the lubricant (Bisson and Anderson, 1964; Booser, 1984; Braithwaite, 1967; McConnell, 1972).

1.1.3 Synthetic Oils

Synthetic lubricants were developed to meet extraordinary requirements, such as low vapor pressure and fire resistance. Environments requiring such lubricants include those with extremes of temperatures and pressures, high vacuum, nuclear radiation, and chemical reducing effects. The lubricating properties of synthetic oils can be tailored to specific needs by changes in the size of certain molecules and by substitution of one group of atoms with another. The synthetic oils include esters, silicones, ethers, and chlorinated and fluorinated compounds.

The synthetic oils exhibit excellent thermal stability over a wide operating temperature range, i.e., up to 430°C. Because of the high cost and the toxicologic and environmental problems that exist, synthetic oils should be selected only after careful thought and adequate planning (Booser, 1984; Braithwaite, 1967; Gunderson and Hart, 1962).

1.1.4 Greases

Greases are used where circulating liquid lubricant cannot be contained because of space and cost and where cooling by the oil is not required or the application of a liquid lubricant is not feasible. A grease is a semisolid lubricant produced by the dispersion of a thickening agent in a liquid lubricant that may contain other ingredients that impart special properties (Boner, 1976; Braithwaite, 1967). The majority of greases are composed of petroleum and synthetic oils thickened with metal soaps and other agents such as clay, silica, carbon black, and polytetrafluoroethylene (PTFE). Most lubricant greases used in industry have petroleum oils as their liquid base. However, almost every lubricating liquid can serve with a suitable thickener or gelling agent to make a grease. All synthetic lubricants are eligible, but in practice, the cost of such materials restricts their use to applications having special requirements.

The petroleum oil–based greases can be used at temperatures up to 175°C depending on the thickener; e.g., barium metal soap gives the maximum usable temperature, i.e., 175°C. The synthetic oil–based greases can be used at higher temperatures. The best materials in these greases are silicones thickened with ammetine and silica, and they can be used at temperatures up to 300°C. However, there is some evidence that diester-based greases are superior in performance to silicon-based greases.

1.2 SOLID LUBRICATION

Solid lubrication is achieved by self-lubricating solids or by imposing a solid material having low shear strength and/or high wear resistance between the contacting surfaces (Anonymous, 1971, 1978, 1984; Benzing, 1964; Bhushan, 1980, 1987a; Bisson and Anderson, 1964; Booser, 1983, 1984; Bowden and Tabor, 1950, 1964; Braithwaite, 1964, 1967; Campbell, 1972; Campbell et al., 1966; Clauss, 1972; Jones and Scott, 1983; Lipp, 1976; McMurtrey, 1988; Sliney, 1982). The solid material may be a dispersion in oils and greases, loose powder, or coatings.

Solid lubricants are used when liquid lubricants do not meet the requirements. These lubricants do not displace the conventional oils and greases but provide successful lubrication in unusual situations, such as at very high temperatures, in cryogenics, under very high vacuum, in nuclear applications, under extreme loads, and in chemically reactive environments.

The maximum useful temperatures of solid lubricants depend strongly on the constituents of the environment, the required life at temperature, and such factors as oxygen availability at the lubricated surface, airflow rates, lubricant particle size, and addition of adjuvants and binders.

1.2.1 Bulk Solids

A number of bulk materials, both inorganic and organic, are used for low friction and/or low wear. They include ferrous metals, nonferrous metals and alloys, ceramics and cermets, inorganic soft nonmetals, and polymers. Ferrous metals and the large family of ferrous alloys called steels and cast irons are the most common of the commercial metals. The ferrous alloys have had a unique position as structural materials within a wide range of applications from the simple binary Fe-C alloys of the iron age to the sophisticated stainless steels and high-speed tool steels produced by powder technology. A unique property of steel, compared with many other materials, is the wide range within which hardness and toughness can be matched.

The nonferrous metals and alloys offer a wide variety of characteristics and mechanical properties. Several nonferrous metals (such as indium, thallium, lead, tin, cadmium, barium, and the precious metals) and alloys based on tin, lead, copper, and aluminum are relatively soft and exhibit low shear strength, good embeddability, and excellent conformability. Another class of nonferrous metals and alloys includes nickel, cobalt, molybdenum, and nickel- and cobalt-based alloys that are hard and can be used at elevated temperatures. These materials offer good wear resistance under corrosive conditions.

Ceramics are inorganic nonmetals. Cermets are composite materials composed of ceramics and a metal intimately bonded together. Ceramics such as oxides, carbides, nitrides, borides, and silicides of mostly refractory metals, with very high melting points, appear to be ideal wear-resistant materials in a variety of tribological applications, provided that their strength and toughness are acceptable for the particular application. Ceramics are chemically inert (e.g., oxide ceramics are chemically stable in air up to their melting points, say above 2000°C), and they retain their properties at high temperatures. However, they are expensive to fabricate to good dimensional tolerances.

The inorganic soft nonmetals include molybdenum disulfide (MoS_2), graphite, talc, and other inorganic salts (Braithwaite, 1964, 1967; Clauss, 1972; Hausner et al., 1970). MoS_2 and graphite have lubricating abilities because of their layered lattice structure. In MoS_2, low shear strength is an intrinsic property, whereas in others, notably graphite, the presence of adsorbed gases or intercalated impurities between the basal planes appears to be necessary to develop desirable friction properties. The maximum temperature of lubrication with MoS_2 and graphite in air atmosphere is limited by oxidation to about 315°C and 430°C, respectively.

Some polymers have friction and wear properties that cannot be obtained by any other group of materials (Bartenev and Lavrentev, 1981; Bhushan and Dashnaw, 1981; Bhushan and Wilcock, 1980, 1982; Bhushan and Winn, 1981; Bhushan et al., 1982; Lee, 1985; Moore, 1972; Sliney and Jacobson, 1975; Sliney and Johnson, 1972). In addition, the high chemical inertness does not allow any considerable change by reaction with the environment, such as by oxidation. The other properties of polymers are their high capacity for damping vibrations and their corrosion resistance, but they also show poor mechanical strength and thermal stability. The polymers (both plastics and elastomers) used as self-lubricating materials are polytetrafluoroethylene (PTFE), polyamide, acetals, polyethylene, polyamide-imide, polyimide, phenolic, nitrile, fluoroelastomer, etc. PTFE is outstanding in this group and exhibits a lower coefficient of friction (~0.04 to 0.1) than any other known polymer. Reinforcing fibers, fillers, and additives are commonly incorporated into polymers to improve mechanical properties. Some polymers, such as polyimides and phenolics, are used successfully at temperatures up to 300°C under dry conditions.

Soft solid lubricants (such as MoS_2, graphite, and PTFE) are also used as additives in oils, greases, polymers, and other solid materials.

1.2.2 Soft Coatings

The typical coatings used for solid lubrication are produced by bonding loose powder in a binder material, electrochemical deposition, and various physical vapor deposition processes (Anonymous, 1971, 1978, 1984; Campbell et al., 1966; Clauss, 1972; Lipp, 1976; McConnell, 1972; Spalvins, 1978). Resin-bonded coatings are extensively used as solid lubricants for many industrial applications. Solid powders such as MoS_2, graphite, and PTFE are bonded with polymers and applied to clean or pretreated surfaces by spraying, dipping, or burnishing (Bhushan, 1980, 1987a; Lipp, 1976; Sliney, 1982). The thickness of these coatings ranges from a fraction of a micrometer to about 50 μm. Binders used in air drying solid lubricants are thermoplastic polymers such as celluloses and acrylics and thermosetting polymers such as alkyds, phenolics, epoxies, silicones, and polyimides. Generally, the bonded solid lubricant coatings are stable at temperatures up to 200°C, but polyimides extend this range up to 300°C.

Ceramic-bonded and plasma-sprayed coatings of CaF_2-BaF_2, CaF_2, and PbO-SiO_2 exhibit flow characteristics above their softening point and are thus used for solid lubrication at high temperature (Sliney, 1979, 1982). These coatings are soft, nonabrasive, and chemically inert in strong acids and reducing environments. Some CaF_2-BaF_2 and CaF_2 coatings provide effective lubrication at temperatures up to 650 and 900°C in air atmosphere, respectively.

Many polymer coatings are commercially available. These coatings are a blend of PTFE, polyphenylene sulfide, polyamide-imide, and/or polyimide in an organic resin binder and solvent system. They can be purchased in liquid form, can be sprayed, baked, and burnished to 7.5 to 50 μm thicknesses, and can perform in operating temperatures generally up to about 200°C (Bhushan, 1980, 1987a). A few organic resin binders are oxidatively stable and have glass transition temperatures above 300°C, such as polyquinoxilines, polybenzimidazoles, and polyimides. Of these, polyimides are by far the most readily available and have been used as self-lubricated varnish and as resin binders for inorganic solid lubricants.

The soft metal coatings, such as Ag, Au, Pb, Sn, and In, are applied by electrodeposition and various physical vapor deposition processes (Gerkema, 1985; Gupta et al., 1987; Sherbiney and Halling, 1977; Spalvins, 1978). These coatings are particularly valuable at very high temperatures and under severe environmental conditions, as in spacecraft. Pb-Sn-Cu in the form of a plating on steel has been used as a lubricant for many years. Gold and silver coatings are used as lubricants in space capsules, high-performance jet engines, and high-speed machines operating under lightly loaded conditions.

1.2.3 Hard Coatings

Hard coatings have found extensive use in highly loaded and/or high-temperature (up to 1000°C or even higher) applications to reduce wear (Bair et al., 1980; Bhushan, 1980, 1981a, 1981b, 1987a; Bhushan et al., 1978; Brainard, 1978; Hintermann, 1984; Hintermann et al., 1978; Holleck, 1986; Peterson and Ramalingam, 1981; Peterson and Winer, 1980; Sundgren and Hentzell, 1986). Hard coatings primarily reduce wear and may not exhibit very low friction. Coatings of ferrous and nonferrous metals, intermetallic alloys, ceramics, and cermets provide good wear resistance owing to their inherent high hardness. These coatings, ranging in thickness from a fraction of a micrometer to several millimeters, can be applied by a variety of deposition techniques, such as electrochemical deposition, thermal spraying, physical vapor deposition, and chemical vapor deposition.

Among metals, nickel is the most widely used coating after hard chromium. Compared with chromium, as-plated hardness of Ni is relatively low, but it has good mechanical strength and ductility, is a good conductor of heat, and has good resistance to corrosion and oxidation.

Coatings of ferrous-based alloys (steels and cast irons) are used where heavy wear is encountered, under conditions that impose mechanical and thermal shock. They are used for corrosion and wear resistance at temperatures up to 650°C. Cobalt- and nickel-based alloy coatings are superior in hardness (wear resistance) at elevated temperatures and are corrosion- and oxidation-resistant. Coatings of Stellites, Tribaloys, Haynes alloys, Hastelloys, and MCrAlY (M = Ni, Co, or Fe) alloys are commonly used for applications at elevated temperatures (up to 1000°C or even higher).

The ceramic coatings that are normally recommended and used depend on the application. The commonly used materials are oxides, carbides, nitrides, borides and silicides of the refractory transition metals, nonmetallic oxides, and nonmetallic nonoxides such as aluminum oxide (Al_2O_3), chromium oxide (Cr_2O_3), zirconium oxide (ZrO_2), silicon dioxide (SiO_2), titanium carbide (TiC), tungsten carbide (WC), chromium carbide (Cr_3C_2), silicon carbide (SiC), boron carbide (B_4C), titanium nitride (TiN), silicon nitride (Si_3N_4), hafnium nitride (HfN), cubic boron nitride (CBN), titanium diboride (TiB_2), zirconium diboride (ZrB_2), hafnium diboride (HfB_2), chrome diboride (CrB_2), molybdenum disilicide ($MoSi_2$), and tungsten disilicide (WSi_2). The Vickers hardness of these coatings varies from 600 to 3500 kg mm^{-2} or higher.

1.3 SURFACE TREATMENTS

Surface treatments provide another approach to tailoring the surface characteristics of bulk materials (Anonymous, 1981; Bhushan, 1980, 1987a; Child, 1983; Elliott, 1978; Gregory, 1978; Smart, 1978; Thelning, 1984). These processes involve heating materials in reactive or nonreactive atmospheres and result into microstructural changes and/or the formation of chemical compounds only on the top layer of thicknesses ranging from 5 to 3000 μm. In addition, chemical changes also can be realized through ion implantation (Dearnaley, 1980; Dearnaley and Hartley, 1978; Hartley, 1979; Hirvonen, 1978).

Surface treatments that bring about microstructural changes on the case of bulk metallic materials involve heating and cooling/quenching through induction, flame, laser, and electron beam and such mechanical treatments as cold working. Surface treatments that alter the chemistry of the surface include carburizing, nitriding, carbonitriding, nitrocarburizing, boriding, siliconizing, chromizing, and aluminizing. These processes produce a diffusion layer of, for example, carbides and nitrides in the carburizing and nitriding processes, respectively.

Ion implantation is another surface treatment in which the metallic or nonmetallic ionized atomic species are injected into the surface through the bombardment/impingement of an ion beam having a very high energy, i.e., 2 to 200 keV. The ion penetration is limited by energy constraints and is typically a few hundred nanometers and rarely more than 0.5 μm, which is very shallow compared with conventional treatments.

1.4 REFERENCES

Anonymous (1968), *Lubrication and Wear: Fundamentals and Applications to Design*, 182, Part 3A, Institute of Mechanical Engineers, London.

——— (1971), *Proc. 1st Int. Conf. on Solid Lubrication*, SP-3, ASLE, Park Ridge, Ill.

——— (1978), *Proc. 2d Int. Conf. on Solid Lubrication*, SP-6, ASLE, Park Ridge, Ill.

——— (1981), *Metals Handbook*, Vol. 4: *Heat Treating*, Ninth Ed., American Society for Metals, Metals Park, Ohio.

——— (1984), *Proc. 3rd Int. Conf. on Solid Lubrication*, SP-14, ASLE, Park Ridge, Ill.

Bair, S., Ramalingam, S., and Winer, W. O. (1980), "Tribological Experience with Hard Coats on Soft Metallic Substrates," *Wear*, Vol. 60, pp. 413–419.

Bartenev, G. M., and Lavrentev, V. V. (1981), *Friction and Wear of Polymers*, Elsevier, Amsterdam.

Benzing, R. J. (1964), "Solid Lubricants," in *Modern Materials*, Vol. 4 (B. W. Gonser and H. H. Hausner, eds.), Academic, New York.

Bhushan, B. (1980), "High Temperature Self-Lubricating Coatings and Treatments, A Review," *Metal Finish*. Vol. 78, May, pp. 83–88, June, pp. 71–75.

——— (1981a), "Structural and Compositional Characterization of RF Sputter-Deposited Cr_2O_3 + Ni-Cr Films," *J. Lub. Technol.* (*Trans. ASME*), Vol. 103, pp. 211–217.

——— (1981b), "Friction and Wear Results from Sputter-Deposited Chrome Oxide with and without Nichrome Metallic Binders and Interlayers," *J. Lub. Technol.* (*Trans* ASME), Vol. 103, pp. 218–227.

——— (1987a), "Overview of Coating Materials, Surface Treatments, and Screening Techniques for Tribological Applications, Part 1: Coating Materials and Surface Treatments," in *Testing of Metallic and Inorganic Coatings* (W. B. Harding and G. A. DiBari, eds.), pp. 289–309, STP-947, ASTM, Philadelphia.

——— (1987b), "Overview of Coating Materials, Surface Treatments, and Screening Techniques for Tribological Applications, Part 2: Screening Techniques," in *Testing of Metallic and Inorganic Coatings* (W. B. Harding and G. A. DiBari, eds.), pp. 310–319, STP-947, ASTM, Philadelphia.

———, and Dashnaw, F. (1981), "Material Study for Advanced Stern-Tube Bearings and Face Seals," *ASLE Trans.*, Vol. 24, pp. 398–409.

———, and Wilcock, D. F. (1980), "Friction and Wear of Polymeric Composites for Reciprocating Stirling Engine Seals, Part 1: Initial Screening Tests," NASA/DOE Tech. Rep. No. MTI-80ASE141ER9, Mechanical Technology Inc., Latham, N.Y.

———, and ——— (1982), "Wear Behavior of Polymeric Compositions in Dry Reciprocating Sliding," *Wear*, Vol. 75, pp. 41–70.

———, and Winn, L. W. (1981), "Material Study for Advanced Stern-Tube Lip Seals," *ASLE Trans.*, Vol. 24, pp. 410–422.

———, Gray, S., and Graham, R. W. (1982), "Development of Low-Friction Elastomers for Bearing and Seals," *Lub. Eng.*, Vol. 38, pp. 626–634.

———, Ruscitto, D., and Gray, S. (1978), "Hydrodynamic Air-Lubricated Complaint Surface Bearing for an Automotive Gas Turbine Engine, Part 2: Materials and Coatings," Tech. Rep. No. CR-135402, NASA, Cleveland, Ohio.

Bisson, E. E., and Anderson, W. J. (1964), "Advanced Bearing Technology," Special Publication SP-38, NASA, Washington, D.C.

Booser, E. R. (1983), *Handbook of Lubrication: Theory and Practice of Tribology*, Vol. 1: *Application and Maintenance*, CRC Press, Boca Raton, Fla.

——— (1984), *Handbook of Lubrication: Theory and Practice of Tribology*, Vol. 2: *Theory and Design*, CRC Press, Boca Raton, Fla.

Boner, C. J. (1976), *Modern Lubricating Greases*, Scientific Publications, Broseley, Shropshire, U.K.

Bowden, F. P., and Tabor, D. (1950), *Friction and Lubrication of Solids*, Part I, Clarendon Press, Oxford, England.

———, and ——— (1964), *Friction and Lubrication of Solids*, Part II, Clarendon Press, Oxford, England.

Brainard, W. A. (1978), "The Friction and Wear Properties of Sputtered Hard Refractory Compounds," *Proc. 2nd Int. Conf. on Solid Lubrication*, pp. 139–147, ASLE, Park Ridge, Ill.

Braithwaite, E. R. (1964), *Solid Lubricants and Surfaces*, Pergamon, Oxford, England.

——— (1967), *Lubrication and Lubricants*, Elsevier, Amsterdam.

Cameron, A. (1966), *The Principles of Lubrication*, Longman, London.

Campbell, M. E. (1972), "Solid Lubricants: A Survey," Special Publication SP-5059(01), NASA, Washington, D.C.

———, Loser, J. B., and Sneegas, E. (1966), "Solid Lubricants," Special Publication SP-5059, NASA, Washington, D.C.

Child, T. H. C. (1983), "Surface Treatments for Tribology Problems," in *Coatings for High Temperature Applications* (E. Lang, ed.), pp. 395–433, Elsevier Science Publishers, London.

Clauss, F. J. (1972), *Solid Lubricant and Self-Lubricating Solids*, Academic, New York.

Davison, C. St.C. (1957), "Wear Prevention in Early History," *Wear*, Vol. 1, pp. 155–159.

Dearnaley, G. (1980), "Practical Applications of Ion-Implantation," in *Ion-Implantation Metallurgy* (C. M. Preece and J. K. Hirvonen, eds.), pp. 1–20, The Metallurgical Society of AIME, Warrendale, Pa.

———, and Hartley, N. E. W. (1978), "Ion Implantation into Metals and Carbides," *Thin Solid Films*, Vol. 54, pp. 215–232.

Dowson, D. (1979), *History of Tribology*, Longman, London.

Elliott, T. L. (1978), "Surface Hardening," *Tribol. Int.*, Vol. 11, pp. 121–128.

Evans, G. G., Galvin, V. M., Robertson, W. S., and Waller, W. F. (1972), *Lubrication in Practice*, Macmillan, Hampshire, United Kingdom.

Gerkema, J. (1985), "Lead Thin Film Lubrication," *Wear*, Vol. 102, pp. 241–252.

Gregory, J. C. (1978), "Chemical Conversion Coatings on Metals to Resist Scuffing and Wear," *Tribol. Int.*, Vol. 11, pp. 105–112.

Gunderson, R. C., and Hart, A. W. (1962), *Synthetic Lubricants*, Reinhold, New York.

Gunther, R. C. (1971), *Lubrication*, Bailey Brothers and Swinfen, Ltd., Folkestone, United Kingdom.

Gupta, B. K., Kulshrestha, S., and Agarwal, A. K. (1987), "Friction and Wear Behavior of Ion-Plated Lead-Tin Coatings," *J. Vac. Sci. Technol.*, Vol. A5, pp. 358–364.

Hartley, N. E. W. (1979), "Friction and Wear of Ion-Implanted Metals: A Review," *Thin Solid Films*, Vol. 64, pp. 177–190.

Hausner, H. H., Roll, K. H., and Johnson, P. K. (eds.) (1970), *Friction and Antifriction Materials*, Plenum, New York.

Hintermann, H. E. (1984), "Adhesion, Friction, and Wear of Thin Hard Coatings," *Wear*, Vol. 100, pp. 381–397.

———, Boving, H., and Hanni, W. (1978), "Wear-Resistant Coatings for Bearing Applications," *Wear*, Vol. 48, pp. 225–236.

Hirvonen, J. K. (1978), "Ion Implantation and Corrosion Science," *J. Vac. Sci. Technol.*, Vol. 15, pp. 1662–1668.

Holleck, H. (1986), "Material Selection for Hard Coatings," *J. Vac. Sci. Technol.*, Vol. A4, pp. 2661–2669.

Jones, M. H., and Scott, D. (eds.) (1983), *Industrial Tribology: The Practical Aspects of Friction, Lubrication, and Wear*, Elsevier, Amsterdam.

Kragelsky, I. V., and Alision, V. V. (1981), *Friction, Wear, and Lubrication: Tribology Handbook*, Vols. 1, 2, and 3, Mir Publishers, Moscow.

Lansdown, A. R. (1982), *Lubrication: A Practical Guide to Lubricant Selection*, Pergamon, Oxford, England.

Lee, L. H. (1985), *Polymer Wear and Its Control*, ACS Symp. Series 287, American Chemical Society, Washington, D.C.

Lipp, L. C. (1976), "Solid Lubricants: Their Advantages and Limitations," *Lub. Eng.*, Vol. 32, pp. 574–584.

McConnell, B. D. (1972), *Assessment of Lubricant Technology*, ASME, New York.

McMurtrey, E. L. (1988), *High Performance Solid and Liquid Lubricants: An Industrial Guide*, Noyes, Park Ridge, N.J.

Moore, D. F. (1972), *The Friction and Lubrication of Elastomers*, Pergamon, Oxford, England.

Neale, M. J. (ed.) (1973), *Tribology Handbook*, Newnes Butterworth, London.

Peterson, M. B., and Ramalingam, S. (1981), "Coatings for Tribological Applications," in *Fundamentals of Friction and Wear of Materials* (D. A. Rigney, ed.), pp. 331–372, American Society for Metals, Metals Park, Ohio.

———, and Winer, W. O. (1980), *Wear Control Handbook*, ASME, New York.

Reynolds, O. (1886), "On the Theory of Lubrication and Its Application to Mr. Beauchamp Tower's Experiments, Including an Experimental Determination of the Viscosity of Olive Oil," *Philos. Trans. R. Soc. Lond.*, Vol. 177, pp. 157–234.

Sherbiney, M. A., and Halling, J. (1977), "Friction and Wear of Ion-Plated Soft Metallic Films," *Wear*, Vol. 45, pp. 211–220.

Sliney, H. E. (1979), "Wide Temperature Spectrum Self-lubricating Coatings Prepared by Plasma Spraying," *Thin Solid Films*, Vol. 64, pp. 211–217.

——— (1982), "Solid Lubricant Materials for High Temperatures: A Review," *Tribol. Int.*, Vol. 15, pp. 303–314.

———, and Jacobson, T. P. (1975), "Performance of Graphite Fiber-Reinforced Polyimide Composites in Self-Aligning Plain Bearings to 315°C," *Lub. Eng.*, Vol. 31, pp. 609–613.

———, and Johnson, R. L. (1972), "Graphite Fiber-Polyimide Composites for Spherical Bearings to 340°C," Tech. Rep. No. TN-D-7078, NASA, Washington, D.C.

Smart, R. F. (1978), "Selection of Surfacing Treatments," *Tribol. Int.*, Vol. 11, pp. 97–104.

Spalvins, T. (1978), "Coatings for Wear and Lubrication," *Thin Solid Films*, Vol. 53, pp. 285–300.

Sundgren, J.-E., and Hentzell, H. T. G. (1986), "A Review of the Present State of the Art in Hard Coatings Grown from the Vapor Phase," *J. Vac. Sci. Technol.*, Vol. A4, pp. 2259–2279.

Thelning, K. E. (1984), *Steel and Its Heat Treatment*, Butterworth, London.

CHAPTER 2
FRICTION, WEAR, AND LUBRICATION

Friction and wear processes are inevitable when two surfaces undergo sliding or rolling under load. Friction is a serious cause of energy dissipation, and wear is the main cause of material wastage. Considerable savings can be made by reducing the friction and controlling the wear. Suitable materials are selected for mating bodies and/or solid/liquid lubrication is used to control friction and wear to an acceptable level.

An understanding of various friction and wear mechanisms is necessary for one to make the right selection of materials, coatings, surface treatments, liquid lubricants, and operating conditions for a given application. This chapter provides a brief review of various theories of friction and wear, discusses some of the methods commonly used to prevent wear, and presents various lubrication modes.

2.1 *FRICTION*

Friction is the resistance to relative motion of contacting bodies. The degree of friction is expressed as a coefficient of friction μ, which is the ratio of the force F_T required to initiate or sustain relative motion to the normal force F_N that presses the two surfaces together. Friction experienced during a sliding condition is known as *sliding friction*, and friction experienced during a rolling condition is known as *rolling friction*.

Despite the considerable amount of work that has been devoted to the study of friction since the early investigations of Leonardo da Vinci, Amontons, Coulomb, and Euler (Bowden and Tabor, 1950, 1964; Dowson, 1979), there is no simple model to predict or calculate the coefficient of friction of a given pair of materials. From the results of various studies, it is obvious that friction originates from complicated molecular-mechanical interactions between contacting bodies and that these interactions differ from one application to another.

2.1.1 Laws of Friction

The earliest law of friction was put forward by Leonardo da Vinci, and it states that the force required to initiate or sustain sliding F_T is proportional to F_N, the normal force. Thus

$$F_T \propto F_N \qquad \text{or} \qquad F_T = \mu F_N \tag{2.1}$$

If force F_S is applied on the body to initiate sliding and force F_K is applied to sustain sliding of the body, the starting (or static) coefficient of friction is given as F_S/F_N and the kinetic (or dynamic) coefficient of friction is given as F_K/F_N.

For a situation in which a body is allowed to slide on an inclined surface with an angle of inclination of θ (Fig. 2.1), the preceding law may be stated in terms of constant angle of repose or friction angle θ as

$$\tan \theta = \mu \tag{2.2}$$

The second law of friction states that the friction force F_T is independent of the apparent area of contact A_a. Thus a brick can slide as easily on its side as on its end.

Coulomb introduced a third law, which states that friction is independent of sliding speed. This implies that the force required to initiate sliding will be the same as the force needed to maintain sliding at any specified velocity, which is generally not true. It has been found experimentally that the first two laws of friction are obeyed over ranges of conditions. However, a number of notable exceptions exist.

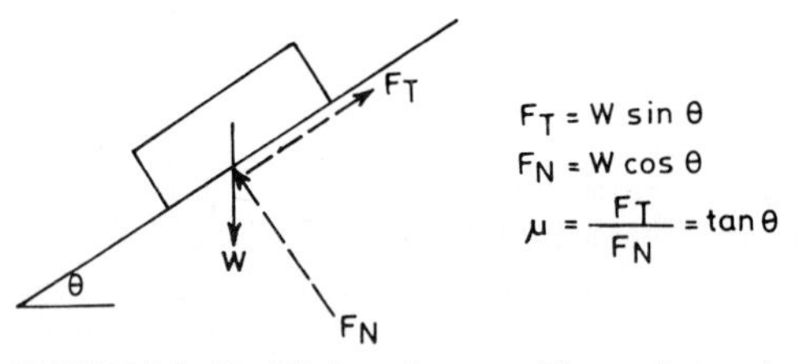

FIGURE 2.1 Equilibrium diagram of forces for an object sliding down the plane.

2.1.2 Sliding Friction

The friction between sliding surfaces is due to the various combined effects of adhesion between the flat surfaces, ploughing by wear particles and hard surface asperity, and asperity deformation. The relative contribution of these components depends on the specific material used, the surface topography, the conditions of sliding interface, and the environment.

2.1.2.1 Adhesion Component of Friction. The adhesion component of friction is due to the formation and rupture of interfacial bonds (Bowden and Tabor, 1950, 1964). These bonds are the result of interfacial interatomic forces that depend on the degree of interpenetration of asperities and the surface composition. If the sliding is to take place, the friction force is needed to shear the weakest tangential planes at the areas of real (actual) contact.

The surfaces are covered with asperities of a certain height distribution that deform elastically or plastically under the given load during an interaction. The summation of individual contact spots gives the real area of contact. There are two classes of properties, namely, mechanical properties, i.e., elastic modulus E', yield pressure (or strength) P_Y, and hardness H and surface topography properties, i.e., asperity distribution (asperity tip radius of curvature β and standard deviation of asperity height σ_p), that are relevant to the real area of contact. Whether the elastic or plastic deformation occurs [Fig. 2.2(*a*)] depends on the

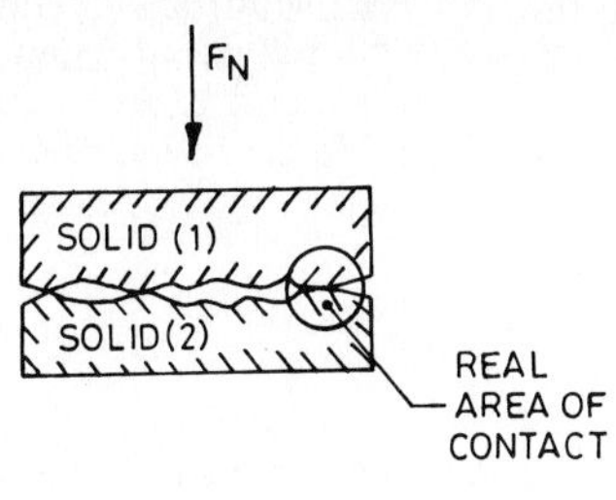

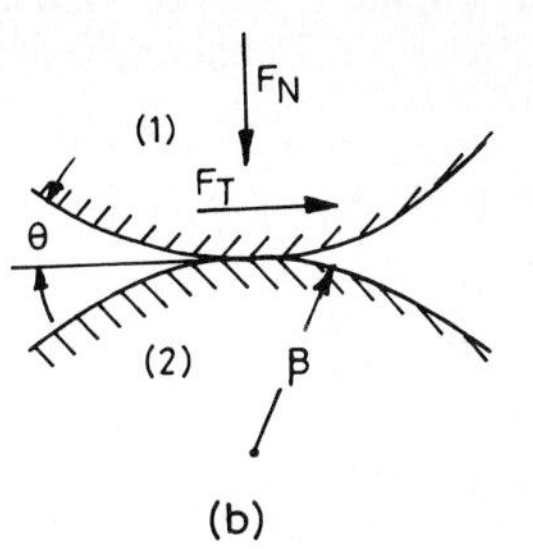

(a)

(b)

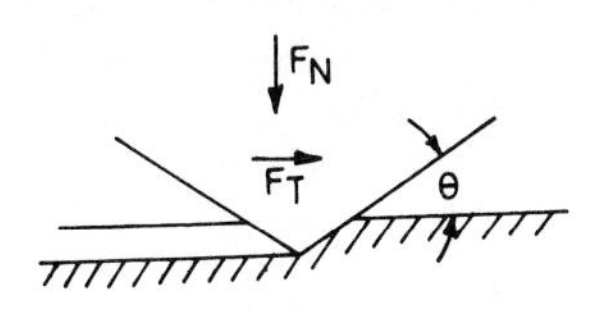

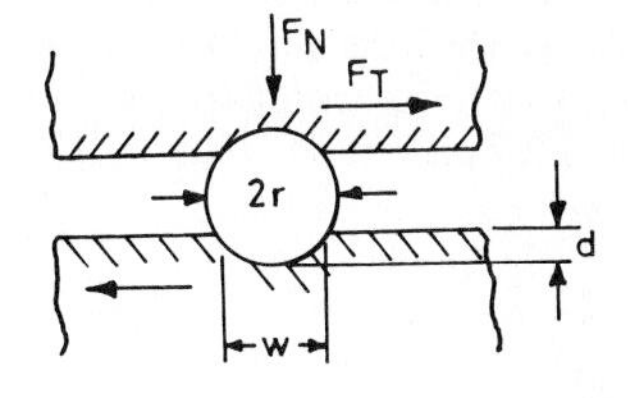

(c)

(d)

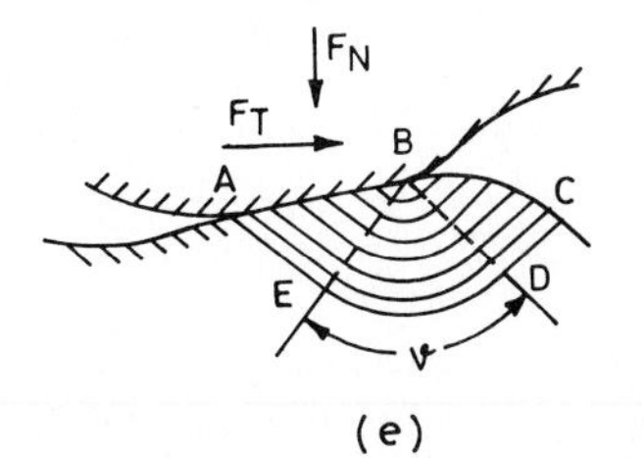

(e)

FIGURE 2.2 (*a*) Contact between two mating surfaces, (*b*) adhesion between contacting asperities, (*c*) ploughing of softer surface by an asperity of harder surface, (*d*) ploughing by foreign particle entrapped between moving surfaces, and (*e*) slip line deformation model of friction.

value of plasticity index, which can be written as (Greenwood and Williamson, 1966)

$$\psi = \frac{E'}{H}\left(\frac{\sigma_p}{\beta}\right)^{1/2} \tag{2.3}$$

where $\psi < 0.6$ for elastic contact
$\psi > 1.0$ for plastic contact
E' = composite elastic modulus = $1/[(1 - \nu_1^2)/E_1 + (1 - \nu_2^2)/E_2]$
H = hardness of the softer material
σ_p = composite standard deviation of asperity peak heights = $(\sigma_1^2 + \sigma_2^2)^{1/2}$
β = composite mean radius of curvature of asperity tips = $1/(1/\beta_1 + 1/\beta_2)$
ν = Poisson's ratio

The numbers 1 and 2 refer to values of the two mating surfaces. The real area of contact is given by (Bhushan, 1984, 1985, 1990) the following equations.

For elastic contact,

$$A_{or} \approx \frac{3.2F_N}{E'(\sigma_p/\beta)^{1/2}} \tag{2.4}$$

For plastic contact,

$$A_{or} = C\frac{F_N}{P_Y} \approx \frac{F_N}{H} \tag{2.5}$$

The term C is a constant that depends on the plastic behavior of materials. If in addition to the normal load a tangential force is also introduced, a junction growth of asperity contacts may occur, leading to a considerably larger area of contact (McFarlane and Tabor, 1950), which is given by

$$A_r = A_{or}\left[1 + \alpha\left(\frac{F_T}{F_N}\right)^2\right]^{1/2} \tag{2.6}$$

where α = a constant

If the shear strength at the interface is τ_S, the force required to break the junction will be given by $\tau_S A_r$, that is, F_T, and the adhesion component of friction is given by [Fig. 2.2(*b*)]

$$\mu_a = \frac{F_T}{F_N} = \frac{\tau_S}{H} \tag{2.7}$$

This equation satisfies Amontons' law and suggests that intentional and unintentional contaminants that change the effective shear strength τ_S at the interface affect the coefficient of friction.

Rabinowicz (1965) extended Bowden and Tabor's model by taking into account the surface energy of contacting bodies, which gives

$$\mu_a = \frac{\tau_S}{H}\left[1 - 2\left(\frac{W_{12}\cot\theta}{\beta H}\right)\right]^{-1} \tag{2.8}$$

where $W_{12} = \gamma_1 + \gamma_2 - \gamma_{12}$

θ = average slope of asperities

The term W_{12} represents the energy that must be applied to separate a unit area of the interface between bodies 1 and 2, involving the need to create two surfaces of energy γ_1 and γ_2, but destroying an interface that had interfacial energy of amount γ_{12}. W_{12} is generally referred to as the *work of adhesion* of the contacting materials.

Marx and Feller (1978, 1979) calculated the adhesion component of friction on the basis of the fracture mechanics model, which considers the fracture of an adhesive junction and introduces as influencing parameters a critical crack opening factor and a work hardening factor. The adhesion component of friction in this case is given by

$$\mu_a = c\,\frac{\sigma_{12}\delta_c}{N^2(F_N H)^{1/2}} \tag{2.9}$$

where σ_{12} = interfacial tensile strength
δ_c = critical crack opening factor
N = work hardening factor
c = a constant

It is obvious from the preceding formulas that the adhesion component of friction depends on the interfacial shear strength and surface energy characteristics of the contacting pair of materials rather than on the properties of single component involved.

2.1.2.2 Ploughing Component of Friction. If one of the sliding surfaces is harder than the other, the asperities of the harder surface may penetrate and plough into the softer surface. Ploughing into the softer surface also may occur as a result of impacted wear particles. In tangential motion, the ploughing resistance is an addition to the friction force.

Rabinowicz (1965) has shown that for a hard surface with conical asperity slid over a soft surface [Fig. 2.2(*c*)], the ploughing component of friction is related to the tangents of the slope of the ploughing asperity, that is,

$$\mu_p = \frac{\tan\theta}{\pi} \tag{2.10}$$

Typical surface asperities seldom have an effective slope exceeding 5 or 6°. It follows, therefore, that the contribution to the coefficient of friction in this case is on the order of 0.05. This value may be regarded, however, only as a lower bound of the friction because of the experimentally observed fact that a pileup of material ahead of the grooving path occurs in most cases of ploughing during sliding that has not been taken into account.

Sin et al. (1979) calculated the ploughing component of friction by considering the ploughing due to penetrated wear particles [Figure 2.2(*d*)].

$$\mu_p = \frac{2}{\pi}\left\{\left(\frac{2r}{w}\right)^2 \sin^{-1}\left(\frac{w}{2r}\right) - \left[\left(\frac{2r}{w}\right)^2 - 1\right]^{1/2}\right\} \tag{2.11}$$

where r = radius of wear particles
w = groove width

This equation indicates that in addition to the material properties, the geometric properties of the asperities or penetrated wear particles may significantly influence the ploughing component of friction.

2.1.2.3 Deformation Component of Friction. When the asperities of two sliding surfaces come into contact with each other, they have to deform in such a way that the resulting displacement field is compatible with the sliding direction and that the sum of the vertical components of surface traction at the contacting asperities are equal to F_N, the normal load. Plastic deformation is always accompanied by a loss of energy, and it is this energy dissipation that accounts for the major part of the friction of metals. Although energy is required to deform a metal elastically, most of this energy is recoverable, and elastic energy losses are therefore negligible compared with the energy losses associated with plastic deformation. Drescher (1959) has worked out a slip-line deformation model of friction [Fig. 2.2(*e*)]. The deformation component of friction is given by the following equations.

For plastic deformation,

$$\mu_d = \frac{F_T}{F_N} = \lambda \tan \text{arc} \left\{ \sin \left[\frac{\sqrt{2}}{4} \frac{(2 + \nu)}{(1 + \nu)} \right] \right\} \tag{2.12}$$

For elastic deformation,

$$\mu_d = 0$$

where $\lambda = \lambda(E', H)$, which defines the proportion of plastically supported load, and ν is the slip angle shown in Fig. 2.2(*e*).

In this model, it is assumed that under an asperity contact [*AB* in Fig. 2.2(*e*)], three regions of plastically deformed material may develop by the regions *ABE*, *BED*, and *BDC*. The maximum shear stress in these areas is equal to the flow shear stress of the pertinent material. The important parameter here is λ, which defines the proportion of the plastically supported load and is related to ratio of hardness and elastic modulus. If the asperity contact is completely plastic and the asperity slope is 45°, a friction coefficient of 1 results. This value goes down to 0.55 if the asperity slope approaches zero.

Heilmann and Rigney (1981) suggested another model for the deformation component of friction that relates friction to plastic deformation. The main assumption is that the frictional work performed is equal to the work of plastic deformation during steady-state sliding. The deformation component of friction is given by

$$\mu_d = \frac{A_r}{F_N} \tau_{\max} F\left(\frac{\tau_s}{\tau_{\max}}\right) \tag{2.13}$$

where $F\left(\frac{\tau_s}{\tau_{\max}}\right) = 1 - 2 \frac{\ln\left[1 + (\tau_s/\tau_{\max})\right] - (\tau_s/\tau_{\max})}{\ln\left[1 - (\tau_s/\tau_{\max})^2\right]}$

τ_s = average interfacial shear strength

$\tau_{\max}$ = ultimate shear strength of material

The shear strength of materials actually achieved at interface may depend on the experimental conditions, such as load, sliding velocity, temperature, and such material characteristics as crystal structure, microstructure, work hardening rate, and recovery rate.

2.1.2.4 Other Components of Friction. In addition to components of friction just described, there can be other contributions to friction. For example, when viscoelastic surfaces with high hysteresis loss (i.e., high internal damping) slide against each other, external work must be done by the tangential component of surface friction to overcome the cyclic energy loss due to the hysteresis loss. This is known as *friction due to elastic hysteresis* and is generally significant in the sliding of viscoelastic materials, e.g., polymers (Moore, 1972).

Friction force also can arise when the wear debris is a viscoelastic or plastic substance that sticks to the sliding interface and undergoes repeated deformation resulting in consumption of energy. Indeed, friction is a result of work done at the interface, and when any energy-consuming mechanism exists between and near the interface, friction force is consumed. This can contribute to friction.

2.1.3 Rolling Friction

Rolling friction, although in general much smaller (typically 5×10^{-3} to 10^{-5}) than sliding friction, is also a very complex phenomenon because of its dependence on so many factors, such as varying amounts of sliding (or commonly referred to as *slip*) during rolling and energy losses during mixed elastic and plastic deformations. Rolling friction may be classified into two types, one in which large tangential forces are transmitted (e.g., the traction drives and driving wheels of a locomotive) and another in which small tangential forces are transmitted, often called *free rolling*. If a torque is introduced that causes a rolling motion, then in addition to the hertzian normal stresses, tangential stresses occur. This results in the division of the contact area into a region of microsliding and a region of adhesion within which the surfaces roll without relative (sliding) motion. The following factors mainly contribute to the friction in a rolling contact:

- Adhesion
- Microslip (i.e., microsliding)
- Elastic hysteresis
- Plastic deformation

2.1.3.1 Adhesion. The joining and separation of surface elements under rolling contacts are little different from those of sliding contacts. Owing to differences in kinematics, the surface elements approach and separate in a direction normal to the interface rather than in a tangential direction. Therefore, junction growth is unlikely in the main part of the contact area. Consequently, at the regions within the rolling contact interface where no relative motion in tangential direction occurs, adhesion forces may be mainly of the van der Waals type. Short-range forces such as strong metallic bonds may act only in microcontacts within the microslip area. If adhesion bonds are formed, they are separated at the trailing end of the rolling contact in tension rather than in shear, as in a sliding contact. Therefore, the adhesion component of friction may be only a small portion of the friction resistance. For rolling of metals under specific situations, an adhesion component can nevertheless be the dominant factor in determining the order of coefficient of rolling friction for different metal pairs (Czichos, 1978).

2.1.3.2 Microslip. If the contact of the two bodies (say a sphere on a flat surface) were a point, we would consider pure rolling conditions to prevail. In practice, however, the region of contact is elastically or plastically deformed in extreme cases so that contact is made over an area of some size, the points within it lying in different planes (Fig. 2.3). As a consequence, it is not possible for pure rolling action to take place except at a very small number of points, but rather at all other points we have a combination of rolling and a small degree of sliding or

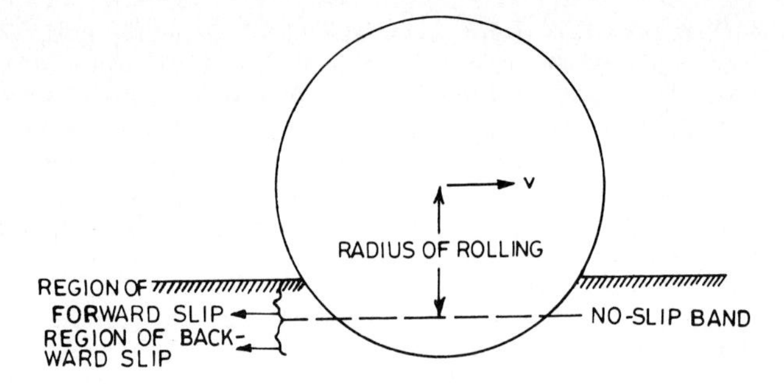

FIGURE 2.3 Forward and backward slip during rolling of a sphere on a flat surface. [*Adapted from Rabinowicz (1965).*]

slip (Rabinowicz, 1965). The literature identifies different cases of microslip, and these are discussed briefly in the following paragraphs.

Let us consider a hertzian contact of two bodies having different elastic constants. If the two bodies roll freely together, the pressure that acts on each gives rise to unequal tangential displacements of the surfaces, leading to interfacial slip processes. This is called *Reynolds slip* (Reynolds, 1876). Consider a rolling contact in which the rolling element is a ball that rolls inside a grooved track. In this case, the profile may conform closely in the transverse direction to the track on which the balls roll. Because the surface points lie at appreciably different distances from the axis of rotation, tangential tractions are introduced and microslip effects occur. This type of slip is called *Heathcote slip* (Heathcote, 1921). For the two-dimensional case (e.g., two cylinders rolling on each other) with a tangential force in the direction of rolling, the microslip area could be calculated. This work led to the important conclusion that, in rolling, the adhesion zone is adjacent to the leading edge of the contact area, in contrast to the static case, in which the adhesion zone is centrally placed (Fig. 2.4). This slip is termed as *Carter-Poritsky-Foppl slip* (Poritsky, 1950).

Tabor (1955) studied experimentally the slip effects of the Reynolds and Heathcote types using steel balls rolling in grooved tracks. He has shown that unless there is a high degree of conformity, both microslip effects contribute only very little to the observed rolling friction force.

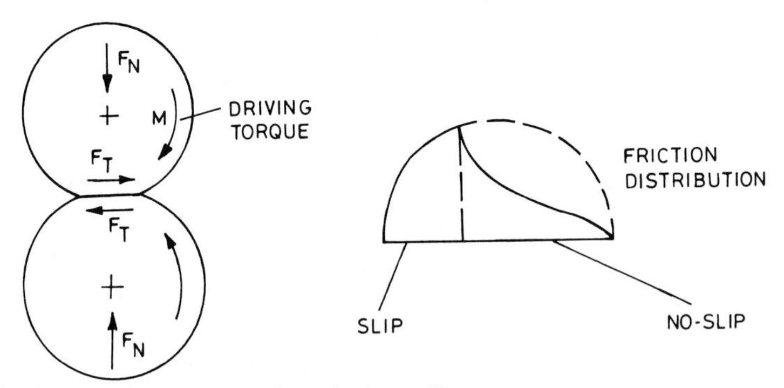

FIGURE 2.4 Slip and no-slip region in a rolling contact.

2.1.3.3 ***Elastic Hysteresis.*** During rolling, different regions on the ball and the flat surfaces are first stressed, and then the stress is released as rolling continues, and the point of contact moves on. Each time an element of volume in either body is stressed, elastic energy is taken up by it. Most of this energy is later released as the stress is removed from the element of the body, but a small part is lost owing to elastic hysteresis of the material of which the contacting bodies are made. This continued drain of energy is compensated for by the rolling force of friction (Rabinowicz, 1965).

The elastic hysteresis losses of a material are related to the damping and relaxation properties of the material and are more pronounced for viscoelastic materials than for metals. It has been observed for viscoelastic materials that the rolling friction can be related to the relaxation time of the material (Moore, 1972).

2.1.3.4 ***Plastic Deformation.*** If in a rolling contact of metallic bodies the contact pressure exceeds a certain value, gross yielding will occur. For freely rolling metallic cylinders, yield occurs first at a point beneath the surface when

$$P_H \approx 3P_Y \tag{2.14}$$

where P_H = maximum hertzian pressure
P_Y = yield stress of the softer material

Eldredge and Tabor (1955) analyzed the plastic deformation processes in the rolling of a ball over a plane. They observed that in the first traversal, a plastically deformed rolling track is created. They found for the rolling friction force an empirical expression of the form

$$F_T \propto \frac{F_N^{2/3}}{r} \tag{2.15}$$

where F_N = normal load
r = ball radius

It is assumed that rolling resistance is mainly due to the plastic deformation in front of the rolling ball.

In repeated rolling-contact cycles, however, the preceding yield criterion does not hold (Johnson, 1966, 1985). During the first contact cycle, the surface material is plastically compressed, and residual compressive stresses acting parallel to the surface are introduced. During subsequent rolling cycles, the material is subjected to the combined action of residual and contact stresses. Further yielding is less likely, and a steady state may be reached in which the material is no longer stressed beyond the elastic limit. This process is called *shakedown*, and the maximum load for which it occurs is called the *shakedown limit*, which is roughly equal to $4P_Y$.

If rolling cylinders are subjected to loads in excess of the shakedown limit, a new type of plastic deformation occurs. This particular mode of deformation in rolling contact was first observed by Crook (1957) and studied experimentally by Hamilton (1963). It consists of forward shearing of the surface of each cylinder relative to its core. The deformation is cumulative in the sense that an equal increment of plastic strain is acquired with each revolution. If follows that at loads above the shakedown limit, continuous and cumulative plastic deformation is observed, whereas at loads below it, even though some yielding is caused initially, after a few traversals the system shakes down to an elastic cycle of stress. Below the shakedown limit, no plastic deformation is observed, and stress is caused by elastic deformation only.

2.1.4 Coefficients of Friction for Various Material Combinations

We have seen that the coefficient of friction depends on several factors, such as intrinsic properties of the mating materials, operating conditions, sliding speed, load, and lubricant. It is very difficult to provide the exact data on coefficient of friction for a particular material combination for design applications. For reference, some typical data are presented in Table 2.1 for static and kinetic coefficients of friction measured under dry and lubricated conditions for various material combinations.

2.2 WEAR

Wear is a process of removal of material from one or both of two solid surfaces in solid-state contact. It occurs when solid surfaces are in sliding or rolling motion relative to each other. In well-designed tribological systems, the removal of material is usually a very slow process, but it is very steady and continuous.

Similar to friction, the wear behavior of a material is also a very complicated phenomenon in which various mechanisms and factors are involved (Booser, 1983, 1984; Bowden and Tabor, 1950, 1964; Burwell, 1957/58; Kragelsky et al., 1982; Lipson, 1960, 1967; Lipson and Colwell, 1961; Peterson and Winer, 1980; Rigney et al., 1984; Sarkar, 1976; Scott, 1979). Several types of wear phenomena occur, e.g., adhesive, abrasive, fatigue, corrosive, fretting, erosive wear by solid particles, fluids, and cavitation, and electrical arc–induced wear, which often mean different things to different people. Moreover, definition of wear processes is imprecise and ambiguous. For example, wear of an unlubricated metal pair sliding in a dusty atmosphere may be termed *dry wear, metallic wear, sliding wear, scratching wear, abrasive wear* depending on the emphasis intended. Since most of the imprecise wear terms in common use are connected with practical wear problems, a wear classification scheme is needed that maintains the common terms but avoids ambiguity. In the literature, the classification of wear is based on the type of relative motion, as suggested by Siebel (1938), and on the type of wear mechanism involved, as suggested by Burwell (1957/58) and Bhushan et al. (1985a, 1985b).

The classification of wear processes based on the type of wearing contacts, such as single phase and multiple phase, is shown in Fig. 2.5. In *single-phase wear*, a solid, liquid, or gas moving relative to a sliding surface causes material to be removed from the surface. The relative motion may be sliding (unidirectional or reciprocating) or rolling. In *multiple-phase wear*, wear also results from a solid, liquid, or gas moving across a surface, but in this case, solid, liquid, or gas act as a carrier for a second phase (particle, asperity, liquid drop, and gas bubble) that actually produce the wear. Each specific wear process in Fig. 2.5 is described in detail in Table 2.2.

In the present discussion, wear is classified into six categories, which are based on quite distinct and independent phenomena, as follows: (1) adhesive,

TABLE 2.1 Coefficient of Friction μ for Various Material Combinations

	μ, static		μ, sliding (kinetic)	
Materials	Dry	Greasy	Dry	Greasy
Hard steel on hard steel	0.78(1)	0.11(1,*a*)	0.42(2)	0.029(5,*h*)
		0.23(1,*b*)		0.081(5,*c*)
		0.15(1,*c*)		0.080(5,*i*)
		0.11(1,*d*)		0.058(5,*j*)
		0.0075(18,*p*)		0.084(5,*d*)
		0.0052(18,*h*)		0.105(5,*k*)
				0.096(5,*l*)
				0.108(5,*m*)
				0.12(5,*a*)
Mild steel on mild steel	0.74(19)		0.57(3)	0.09(3,*a*)
				0.19(3,*u*)
Hard steel on graphite	0.21(1)	0.09(1,*a*)		
Hard steel on babbitt (ASTM 1)	0.70(11)	0.23(1,*b*)	0.33(6)	0.16(1,*b*)
		0.15(1,*c*)		0.06(1,*c*)
		0.08(1,*d*)		0.11(1,*d*)
		0.085(1,*e*)		
Hard steel on babbitt (ASTM 8)	0.42(11)	0.17(1,*b*)	0.35(11)	0.14(1,*b*)
		0.11(1,*c*)		0.065(1,*c*)
		0.09(1,*d*)		0.07(1,*d*)
		0.08(1,*e*)		0.08(11,*h*)
Hard steel on babbitt (ASTM 10)		0.25(1,*b*)		0.13(1,*b*)
		0.12(1,*c*)		0.06(1,*c*)
		0.10(1,*d*)		0.055(1,*d*)
		0.11(1,*e*)		
Mild steel on cadmium silver				0.097(2,*f*)
Mild steel on phosphor bronze			0.34(3)	0.173(2,*f*)
Mild steel on copper lead				0.145(2,*f*)
Mild steel on cast iron		0.183(15,*c*)	0.23(6)	0.133(2,*f*)
Mild steel on lead	0.95(11)	0.5(1,*f*)	0.95(11)	0.3(11,*f*)
Nickel on mild steel			0.64(3)	0.178(3,*x*)
Aluminum on mild steel	0.61(8)		0.47(3)	
Magnesium on mild steel			0.42(3)	
Magnesium on magnesium	0.6(22)	0.08(22,*y*)		
Teflon on Teflon	0.04(22)			0.04(22,*f*)
Teflon on steel	0.04(22)			0.04(22,*f*)
Tungsten carbide on tungsten carbide	0.2(22)	0.12(22,*a*)		
Tungsten carbide on steel	0.5(22)	0.08(22,*a*)		
Tungsten carbide on copper	0.35(23)			
Tungsten carbide on iron	0.8(23)			
Bonded carbide on copper	0.35(23)			
Bonded carbide on iron	0.8(23)			
Cadmium on mild steel			0.46(3)	
Copper on mild steel	0.53(8)		0.36(3)	0.18(17,*a*)
Nickel on nickel	1.10(16)		0.53(3)	0.12(3,*w*)
Brass on mild steel	0.51(8)		0.44(6)	
Brass on cast iron			0.30(6)	
Zinc on cast iron	0.85(16)		0.21(7)	

TABLE 2.1 Coefficient of Friction μ for Various Material Combinations (*Continued*)

Materials	μ, static		μ, sliding (kinetic)	
	Dry	Greasy	Dry	Greasy
Magnesium on cast iron			0.25(7)	
Copper on cast iron	1.05(16)		0.29(7)	
Tin on cast iron			0.32(7)	
Lead on cast iron			0.43(7)	
Aluminum on aluminum	1.05(16)		1.4(3)	
Glass on glass	0.94(8)	0.01(10,*p*)	0.40(3)	0.09(3,*a*)
		0.005(10,*q*)		0.116(3,*v*)
Carbon on glass			0.18(3)	
Garnet on mild steel			0.39(3)	
Glass on nickel	0.78(8)		0.56(3)	
Copper on glass	0.68(8)		0.53(3)	
Cast iron on cast iron	1.10(16)		0.15(9)	0.070(9,*d*)
				0.064(9,*n*)
Bronze on cast iron			0.22(9)	0.077(9,*n*)
Oak on oak (parallel to grain)	0.62(9)		0.48(9)	0.164(9,*r*)
				0.067(9,*s*)
Oak on oak (perpendicular)	0.54(9)		0.32(9)	0.072(9,*s*)
Leather on oak (parallel)	0.61(9)		0.52(9)	
Cast iron on oak			0.49(9)	0.075(9,*n*)
Leather on cast iron			0.56(9)	0.36(9,*t*)
				0.13(9,*n*)
Laminated plastic on steel			0.35(12)	0.05(12,*t*)
Fluted rubber bearing on steel				0.05(13,*t*)

Note: Reference letters indicate the lubricant used; numbers in parentheses give sources (see References).

Key to Lubricants Used:

a = oleic acid
b = Atlantic spindle oil (light mineral)
c = castor oil
d = lard oil
e = Atlantic spindle oil plus 2% oleic acid
f = medium mineral oil
g = medium mineral oil plus ½% oleic acid
h = stearic acid
i = grease (zinc oxide base)
j = graphite
k = turbine oil plus 1% graphite
l = turbine oil plus 1% stearic acid
m = turbine oil (medium mineral)
n = olive oil
p = palmitic acid
q = ricinoleic acid
r = dry soap
s = lard
t = water
u = rape oil
v = 3-in-1 oil
w = octyl alcohol
x = triolein
y = 1% lauric acid in paraffin oil

References: (1) Campbell, *Trans. ASME*, 1939; (2) Clarke, Lincoln, and Sterrett, *Proc. API*, 1935; (3) Beare and Bowden, *Philos. Trans. R. Soc.*, 1935; (4) Dokos, *Trans. ASME*, 1946; (5) Boyd and Robertson, *Trans. ASME*, 1945; (6) Sachs, *Zeit. Angew. Math. Mech.*, 1924; (7) Honda and Yama la, *Jour. I. of M*, 1925; (8) Tomlinson, *Phil. Mag.*, 1929; (9) Morin, *Acad. Roy. Sci.*, 1838; (10) Claypoole, *Trans. ASME*, 1943; (11) Tabor, *J. Appl. Phys.*, 1945; (12) Eyssen, *General Discussion on Lubrication*, ASME, 1937; (13) Brazier and Holland-Bowyer, *General Discussion on Lubrication*, ASME, 1937; (14) Burwell, *J. SAE*, 1942; (15) Stanton, *Friction*, Longmans; (16) Ernst and Merchant, Conference on Friction and Surface Finish, M.I.T., 1940; (17) Gongwer, Conference on Friction and Surface Finish, M.I.T., 1940; (18) Hardy and Bircumshaw, *Proc. R. Soc. Lond.*, 1925; (19) Hardy and Hardy, *Phil. Mag.*, 1919; (20) Bowden and Young, *Proc. R. Soc. Lond.*, 1951; (21) Hardy and Doubleday, *Proc. R. Soc. Lond.*, 1923; (22) Bowden and Tabor, *The Friction and Lubrication of Solids*, Oxford; (23) Shooter, *Research*, 4, 1951.

Source: Adapted from Avallone and Baumeister (1987, p. 3-26).

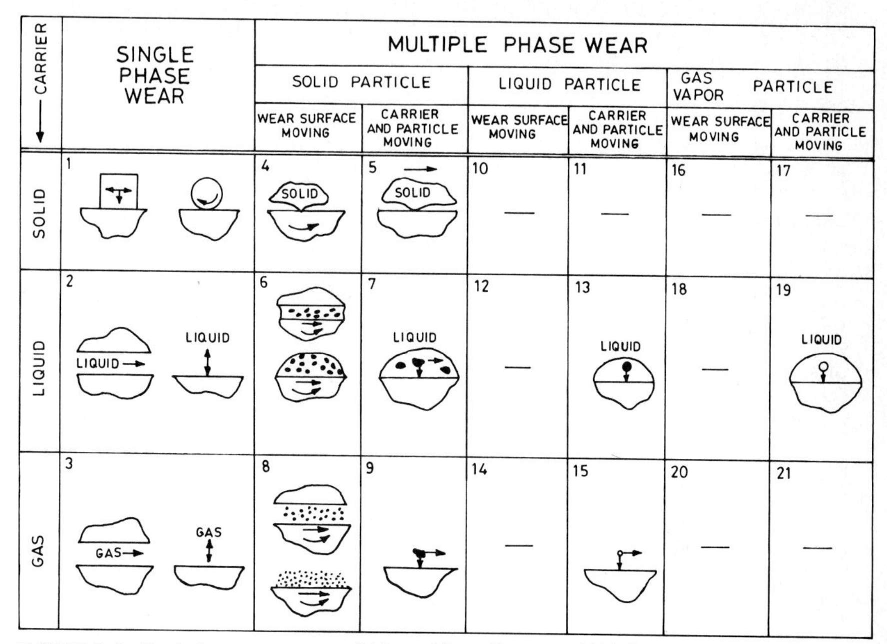

FIGURE 2.5 Typical wear processes. [*Adapted from Peterson (1980).*]

galling, or scuffing wear, (2) abrasive and cutting wear, (3) fatigue wear, (4) corrosive wear, (5) erosive wear by (*a*) solid particles, (*b*) fluid, and (*c*) cavitation, and (6) electrical arc–induced wear (Bhushan, 1990; Bhushan and Davis, 1983; Bhushan et al., 1985a, 1985b). Another commonly encountered wear mode is fretting and fretting corrosion. This is not a distinct mechanism, but rather a combination of adhesive, corrosive, and abrasive forms of wear. Impact wear is a subset of fatigue and erosive wear.

2.2.1 Adhesive Wear

Adhesive wear is often called *galling* or *scuffing*, although these terms are sometimes used loosely to describe other types of wear. Adhesive wear processes are initiated by the interfacial adhesive junctions that form if solid materials are in contact on an atomic scale (Keller, 1963). As a normal load is applied, local pressure at the asperities becomes extremely high. In some cases, the yield stress is exceeded, and the asperities deform plastically until the real area of contact has increased sufficiently to support the applied load. In the absence of surface films, the surfaces would adhere together, but very small amounts of contaminant minimize or even prevent adhesion under purely normal loading (Gane et al., 1974; Kinloch, 1980; Sargent, 1978; Sikorski, 1964). However, relative tangential motion at the interface acts to disperse the contaminant films at the point of contact, and cold welding of the junctions can take place. Continued sliding causes the junctions to be sheared and new junctions to be formed. The chain of events that leads to the generation of wear particles includes the adhesion and fracture of the

TABLE 2.2 Designation of Wear Processes

No.*	Contact	Particle	Motion	Common names	Typical examples
1	Solid†	None	Sliding	Dry sliding wear Metal-metal wear	Bushings, seals, brakes
			Reciprocating	Fretting	Clearance fits, fasteners
			Impact	Impact wear	Electrical contacts, hammers
			Rolling	Rolling wear	Rolling contact bearings, gears
2/3	Gas‡ or liquid§	None	Flow	Wire-drawing Erosion	Water erosion, valve wear
			Impact	—	Pipe bends, deflectors
			Hammer	—	—
4	Solid†	Hard rough surface	Wear surface rolling	—	Tire wear
5	Solid†	Hard rough surface	Particle sliding	Abrasion Two-body wear	Abrasive papers, files
6	Liquid¶	Solid	Wear surface sliding	Three-body wear Dirt abrasion	Dirt in machinery components, lapping
			Wear surface rolling	Three-body wear Dirt abrasion	Dirt in lubricants used for rolling contacts, cams
			Wear surface sliding	Erosion	Mixing, slurry pumping, blades
			Wear surface rolling	—	—
7/9	Liquid or gas	Solid	Particle sliding	Low-angle particle erosion	Pumping and transporting liquids or gases containing solid particles
			Particle impact	90° particle erosion	Shot and sand blasting
8	Gas	Solid	Wear surface sliding	Three-body wear	Soles of shoes
			Wear surface rolling	Three-body wear	Rock crushing
			Wear surface sliding	Two-body wear Low-stress abrasion Soil abrasion	Earthmoving equipment, digging

13	Liquid	Liquid	Particle impact	—	Slurry pumping
15	Gas	Liquid	Particle impact	Drop erosion Rain erosion	—
19	Liquid	Gas or vapor particle	Particle impact Particle flow	Cavitation Wire-drawing erosion	Valves, bearings, pipes Valves

*Refers to numbers in Fig. 2.5.
†Refers to a solid being worn by another solid.
‡Refers to a solid being worn by a gas.
§Refers to a solid being worn by a liquid.
¶Refers to two surfaces in contact being worn by particles in a liquid trapped between them.
Source: Adapted from Peterson (1980).

mating surfaces. Since both adhesion and fracture are influenced by surface contaminants and the environment, it is quite difficult to relate the adhesive wear process only with the bulk properties of solid surfaces.

The following observations are generally made on adhesive wear mechanisms:

- Interfacial metallic adhesion bonding depends on the electronic structure of the contacting bodies. It has been suggested that strong adhesion will occur if one metal can act as an electron donor and the other as an electron acceptor (Buckley, 1972; Czichos, 1972; Ohmae et al., 1980).
- Adhesive wear is directly related to the tendency of different mating materials to form solid solutions or intermetallic compounds with each other. The metallurgical compatibility indicated by the mutual solubility represents the degree of intrinsic attraction of the atoms of the contacting metals for each other (Rabinowicz, 1980, 1984).
- Crystal structure exerts an influence on adhesive wear processes. Hexagonal metals in general exhibit lower adhesive wear characteristics than either body-centered cubic or face-centered cubic metals. This difference is believed to be related to different asperity contact deformation modes and the number of operable slip systems in the crystals (Buckley and Johnson, 1968; Habig, 1968).
- Crystal orientation influences the adhesive wear behavior. In general, lower adhesive wear is observed for the high atomic density, lower surface energy planes (Buckley, 1981).
- When dissimilar metals are in contact, the adhesive wear process will generally result in the transfer of particles of the cohesively weaker of the two materials to the cohesively stronger material (Buckley, 1981).

The actual process of formation of wear particles has been studied by Sasada (1985) and Sasada et al. (1978, 1981) with a pin on disk machine. They concluded that an initial metal-to-metal junction is sheared by the frictional motion, and a small fragment of either surface becomes attached to the other surface. As sliding continues, this fragment constitutes a new asperity that becomes attached once more to the original surface. This transfer element is repeatedly passed from one surface to the other and grows quickly to a large size, absorbing many of the transfer elements so as to form a flakelike particle from materials of both rubbing elements (Eyre, 1976). Unstable thermal and dynamic conditions brought about by rapid growth of this transfer element finally account for its removal as a wear particle (Fig. 2.6).

The wear resulting from adhesive wear process has been described phenomenologically by the well-known Archard equation (Archard, 1953, 1980):

$$W_{ad} = \frac{V}{L} = K\frac{F_N}{H} \tag{2.16}$$

where W_{ad} = wear rate, i.e., worn volume per unit sliding distance
K = wear coefficient
V = wear volume
L = sliding distance
F_N = normal load
H = hardness of the softer material

This equation employs hardness as the only material property, but it is obvious from preceding discussions that the wear coefficient K depends on various

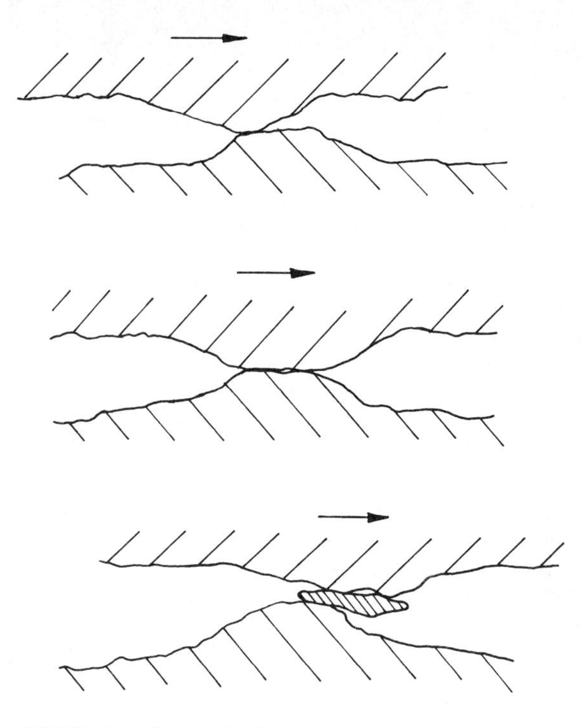

FIGURE 2.6 Schematic of generation of wear particle as a result of adhesive wear process.

properties of the mating materials. However, it is not possible to describe quantitatively the influence of material properties on wear coefficient K by a simple relationship.

Equation (2.16) is relevant for plastic contacts. This equation implies that wear rate is proportional to real area of contact in plastic contacts. Therefore, this may not be applicable to cases involving elastic contacts, such as magnetic storage devices (Bhushan, 1990).

The Archard equation, which employs only hardness as a material property, has been extended by Hornbogen (1975). He proposed a model to explain increasing relative wear rates with decreasing toughness of metallic materials. This model is based on a comparison of the strain that occurs during asperity interactions with the critical strain at which crack growth is initiated. If the applied strain is smaller than the critical strain, the wear rate is independent of toughness and Archard's law is followed. Applied strain larger than the critical strain of the material leads to an increased probability of crack growth and therefore to a higher wear. The modified wear equation is

$$W_{ad} = N^2 \frac{P_Y E' F_N^{3/2}}{K_{IC}^2 H^{3/2}} \tag{2.17}$$

where K_{IC} = fracture toughness

2.2.2 Abrasive Wear

Abrasive wear may be described as damage to a surface by a harder material. It is also sometimes called *scratching, scoring*, or *gouging* depending on the degree of severity. There are two general situations in which this type of wear occurs. In the first case, the hard surface is the harder of two rubbing surfaces (*two-body abrasion*), e.g., in mechanical operations such as grinding, cutting, and machining. In the second case, the hard surface is a third body, generally a small particle of grit or abrasive, caught between the two other surfaces and sufficiently harder than they are to abrade either one or both of them (*three-body abrasion*).

In the abrasive wear process, asperities of the harder surface press into the softer surface, with plastic flow of the softer surface occurring around the asperities from the harder surface. When a tangential motion is imposed, the harder surface removes the softer material by combined effects of microploughing, microcutting, and microcracking. Khruschov (1974) has studied the influence of material properties on abrasive wear and has shown that in the case of pure metals and nonmetallic hard materials, a direct proportionality exists between the relative wear resistance and the hardness. Moreover, he has also shown that heat treatment of structural steels (normal hardening and tempering) improves their abrasive wear resistance.

Rabinowicz (1965) derived an expression for abrasive wear by assuming that the asperities of the harder surface are conical [Fig. 2.2(*c*)]. The wear volume per unit sliding distance is given by

$$W_{ab} = \frac{V}{L} = \frac{KF_N}{\pi H} \overline{\tan \theta} \tag{2.18}$$

This expression relates the wear volume per unit sliding distance to the asperity slope ($\overline{\tan \theta}$, which is weighted average of the tan θ values of all the individual cones) of the penetrating abrasive particle and the hardness of the abraded material.

Another model was proposed by Zum Gahr (1982), who considered the detailed processes of microcutting, microploughing, and microcracking in the abrasive wear of ductile metals. This model includes other microstructural properties of the materials worn beside the flow pressure or hardness. The magnitude of the work hardening, the ductility, the homogeneity of the strain distribution, the crystal anisotropy, and the mechanical instability have been shown to influence abrasive wear mechanisms. The expression is

$$W_{ab} = \frac{f_{ab}}{K_1 K_2 \tau_c} \frac{\cos \rho \sin \theta}{[\cos (\theta/2)]^{1/2} \cos (\theta - \rho)} F_N \tag{2.19}$$

where f_{ab} = model factor (1 for microcutting)
K_1 = relation of normal and shear stress
K_2 = texture factor (1 for fcc metals)
τ_c = shear stress for dislocation movement
ρ = friction angle at abrasive-material interface

2.2.3 Fatigue Wear

In practice, all machines involve the periodic variations of stress. An element of metal at the surface of a rotating shaft will be subjected to reversal of bending stresses, the race of a rolling contact bearing will experience continual application and release of hertzian stress, and the surface of a conformal bearing will experience repeating stresses on a microscale due to the passage of asperities on the rotating surface. All these repeating stresses in a rolling or sliding contact can give rise to fatigue failure. These effects are mainly based on the action of stresses in or below the surfaces without needing a direct physical contact of the surfaces under consideration. This follows from the observation that surface fatigue failure can occur in journal bearings in which the interacting surfaces are fully seperated by a thick lubricant film (Lang, 1977). Subsurface- and surface-fatigue wear are the dominant failure modes in rolling-element bearings.

Since the shear stress is maximum some distance below the surface in a pure rolling contact (zero tangential stress at the surface), the failure initiates at the subsurface. If, superimposed on the rolling contact, there is also some sliding contact, then the location at which the shear stress is maximum, and therefore the position of failure, moves nearer to the surface. However, materials are rarely perfect, and the exact position of ultimate failure will be influenced by inclusions, porosity, microcracks, and other factors (Fig. 2.7). The fatigue failure requires a given number of stress cycles and often predominates after a component has been in service for a long period of time (Vingsbo and Hogmark, 1981).

Kuhlmann-Wilsdorf (1981) suggested several possible mechanisms of crack generation and propagation on the basis of dislocation theory. Halling (1975) derived an expression for wear rate by using a model that incorporates the concept of fatigue failure as well as simple plastic deformation failure. The expression for volume of wear per unit sliding distance is

$$W_{\mathrm{fa}} = K \frac{\eta\gamma}{\bar{\varepsilon}_1^2 H} F_N \tag{2.20}$$

where η = line distribution of asperities
γ = constant defining particle size
$\bar{\varepsilon}_1$ = strain to failure in one loading cycle
H = hardness of the softer material
K = wear coefficient

In studying the elastic-plastic stress fields in the subsurface regions of sliding asperity contacts and the possible dislocation interaction, Suh (1977) has suggested the following details of the events leading to subsurface fatigue wear:

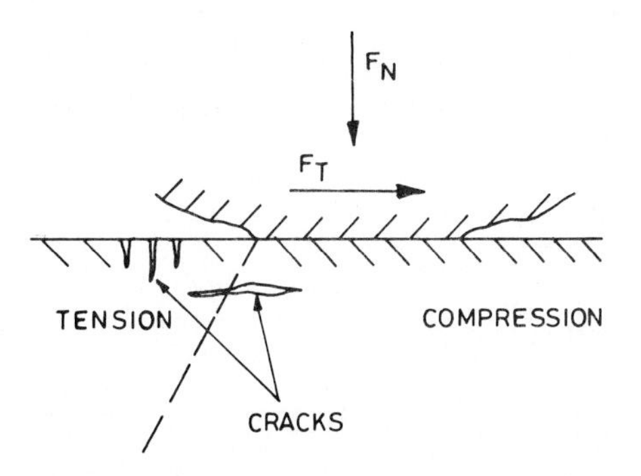

FIGURE 2.7 Schematic of fatigue wear due to formation of surface and subsurface cracks.

- When two sliding surfaces come into contact, asperities on the softer surface are deformed by repeated loading to generate a relatively smooth surface. Eventually, with asperity-plane contacts, the softer surface experiences cyclic loading as the asperities of the harder surface plough through it.
- Surface friction by the harder asperities on the softer surface induces plastic shear deformation that accumulates with repeated loading.
- As the subsurface deformation continues, cracks are nucleated below the surface. Crack nucleation very near to the surface is inhibited by the triaxial compressive stress existing just below the contact region.
- Further loading causes the cracks to propagate parallel to the surface.
- When these cracks finally intercept the surface, long, thin wear sheets delaminate, giving rise to platelike particles.

2.2.4 Corrosive Wear

The wear due to adhesion, abrasion, and fatigue can be explained in terms of stress interactions and deformation properties of the mating surfaces, but in corrosive wear, the dynamic interaction between environment and mating material surfaces plays a significant role (Fischer, 1988; Quinn, 1978, 1983). This interaction gives rise to a cyclic stepwise process:

- In the first step, the contacting surfaces react with environment, and reaction products are formed on the surface.
- In the second step, attrition of the reaction products occurs as a result of crack formation and/or abrasion in the contact interactions of the materials.

This process results in increased reactivity of the asperities because of increased temperature and changes in the mechanical properties of asperities.

Quinn and coworkers (Quinn, 1962, 1978, 1983; Quinn et al., 1980) have proposed a theory to explain wear in steels by assuming that tribochemically formed surface asperity layers are detached at a certain critical thickness. The expression derived by Quinn (1983) is given by (Fig. 2.8)

$$W_{\mathrm{corr}} = \frac{dA_c \exp\left[-Q/(R_c T_c)\right]}{3\varepsilon^2 \rho^2 v H} F_N \tag{2.21}$$

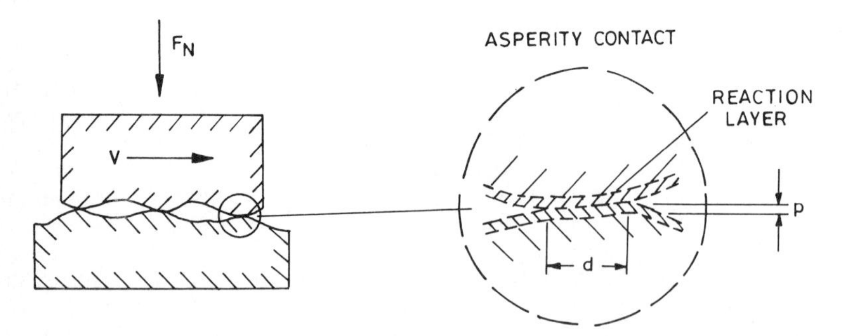

FIGURE 2.8 Model of corrosive wear.

where ρ = density of oxide layer
A_c = Arrhenius constant
Q = activation energy
R_c = gas constant
T_c = contact temperature
d = asperity contact diameter
v = sliding velocity
ε = critical thickness of reaction layer

This expression indicates the dependence of wear rate on the following group of parameters of the contacting surfaces:

- Operating parameters (F_N, v, T_c)
- General constant and material parameters (A_c, Q, R_c, ρ, H)
- Interaction features (d, T_c, ε)

Razavizadeh and Eyre (1982) investigated the oxidation behavior of aluminum alloys and concluded that corrosive wear occurs by a combined process of oxidation, deformation, and fracture to produce layers that are compacted into grooves in the metal surface. Oxidative surfaces become smoother with time as the troughs become filled with oxide, and oxidative wear then occurs by the fracture of platelike debris.

2.2.5 Fretting Wear

When components are subjected to very small relative vibratory movements at high frequency, an interactive form of wear, called *fretting*, takes place that is initiated by adhesion, is amplified by corrosion, and has its main effect by abrasion (Anonymous, 1955; Sproles and Duquette, 1978; Waterhouse, 1984). Examples of vulnerable components are shrink fits, bolted parts, and splines. Fretting wear frequently occurs between components that are not intended to move, e.g., press fits. Surfaces subjected to fretting have a characteristic appearance, with red-brown patches on ferrous metals and adjacent areas that are highly polished as a result of lapping by the hard iron oxide debris. It is observed that the environment plays a strong role in the wear of surfaces that undergo fretting. For example, the amount of fretting wear in moist atmosphere is considerably higher than that observed in dry atmosphere.

2.2.6 Erosive Wear by Solid Particles and Fluids

Erosion of materials and components caused by the impingement of solid particles or small drops of liquid or gas can be a life-limiting phenomenon for systems in erosive environments. Solid-particle impact erosion has been receiving increasing attention in recent years because of the development of coal conversion plants and their need for the movement and flow of solid particles into various parts of these plants. The impact of these particles on moving blades, valve constructions, pipe joints and bends, and other surfaces has resulted in severe erosion. Solid-particle erosion also has been a concern for aerospace systems, including sand erosion of helicopter blades and ingestion of sand and erosion of jet engine blades and vanes.

The basic mechanism involved is depicted schematically in Fig. 2.9. As a re-

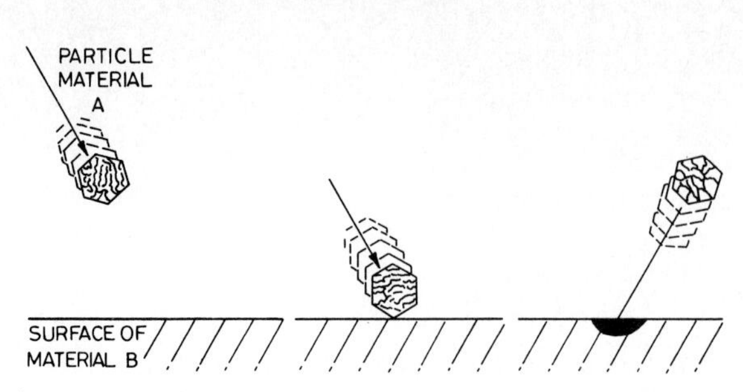

FIGURE 2.9 Typical events in erosive wear.

sult of the impact in solid-state contact of a particle of material *A* with the solid surface, part of the surface of material *B* is removed. The particle itself can vary in composition as well as in form. For example, it may be an atom or a molecule of a gaseous species. Basically, the sputtering process for the deposition of thin coatings is an erosive wear process on an atomic scale. In the sputtering process, an ionized argon ion, which is in gaseous form, is the incoming particle that strikes the solid target surface and removes material from that surface.

The response of engineering materials to the impingement of solid particles or liquid drops varies greatly depending on the class of material, the state of materials to which those materials have been exposed (i.e., thermal history, previous stresses, surface treatments), and the environmental parameters associated with the erosion process, such as impact velocity, impact angle, and particle type and size. A schematic presentation of the features of erosion on ductile and brittle materials as well as a comparison of cast iron and elastomer as a function of impact angle is given in Fig. 2.10 and Table 2.3. The mechanisms have been discussed by Bitter (1963), Finnie (1960), Finnie et al. (1967, 1979), Hutchings (1979), Levy (1981), Maji and Sheldon (1979), Preece (1979), Schmitt (1980), and Tilly (1969, 1973).

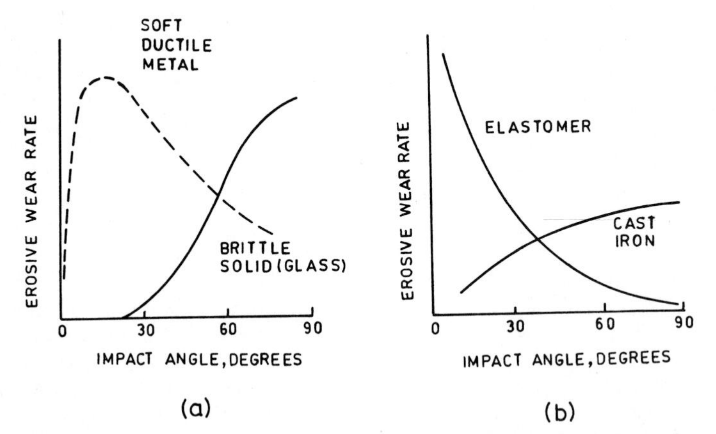

FIGURE 2.10(*a*) Dependence of solid-particle erosion rate on impact angle for ductile and brittle materials. [*Adapted from Wolfe (1963).*] (*b*) Dependence of solid-particle erosion rate on impact angle for cast iron and elastomer. [*Adapted from Eyre (1978).*]

TABLE 2.3 Effects of Impact Angle on Material Response to Liquid Impact

Type of material	Angle of impingement		
	<20°	45°	90°
Elastic	No effect	No observable effect	Surface deformed
Ductile	"Cutting" wear maximum	Mixed regime	Deformation wear maximum
Brittle	Very little erosion	Intermediate erosion	Maximum erosion

For ductile metals, the maximum erosion by solid-particle impingement occurs at an impingement angle of approximately 20°. This behavior was originally modeled by Finnie et al. (1979) by considering the abrasive cutting by a rigid angular particle into the surface of a ductile metal. The volume V removed from the surface by a mass M of eroding particles was predicted to be

$$V \approx \frac{Mv^2 f(\alpha)}{P} \tag{2.22}$$

where v = particle velocity
$f(\alpha)$ = function of angle, measured from plane of the surface to the particle velocity vector
P = horizontal component of flow pressure between particle and surface

This approach is not able to predict erosive behavior accurately at impingement angles greater than 45°. Later on, Bitter (1963) described the erosion process as consisting of two simultaneous processes: cutting wear, which dominates at low impact angles, and deformation wear, which dominates at high impact angles.

Tilly (1969, 1973) described a two-stage process whereby particles produce erosion by impact and then fragment to produce additional erosion. The fragmentation and outward flow of particle fragments causes the erosion at impingement angles of 90°.

In contrast to the erosion of ductile materials, in which erosion is maximum at impact angles of 20 or 30°, the erosion of brittle materials is maximum at 90° (see Fig. 2.10). The erosion by solid impact of ceramics, glass, and thermosetting polymer matrices in composite materials is becoming important because ceramics and composites are being employed more often for engine applications where dust ingestion is a concern and ceramics are being used as refractory liners in energy-conservation equipment. In the case of elastomers, the wear rate is minimum at high impact angles (see Fig. 2.10).

Vijh (1975, 1976) has derived expressions for erosive wear rate with various impacting particles and metal-metal bond energies or cohesive energies for the metals. For a variety of different impacting particles, there is a basic or fundamental relationship between material loss from the impact surface and the cohesive bonding energy of the metal. The cohesively stronger metals exhibit lower erosive wear than do the cohesively weaker metals when bombarded by quartz sand.

2.2.7 Erosive Wear by Cavitation

Cavitation erosion arises when a solid and a fluid are in relative motion and bubbles formed in the fluid become unstable and implode against the surface of the

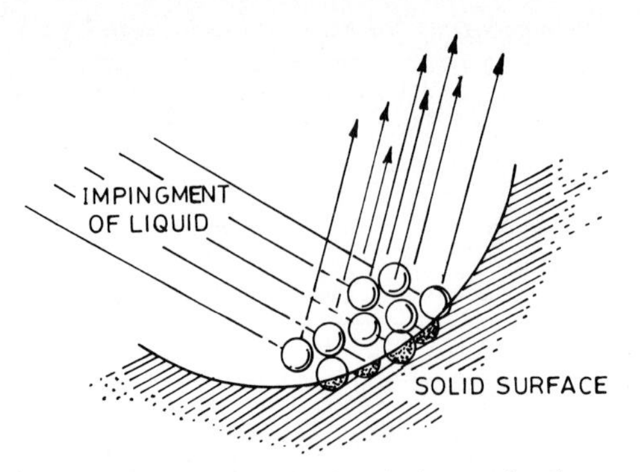

FIGURE 2.11 Schematic of cavitation erosion due to impingement of liquid bubbles.

solid (Anonymous, 1974; Hammitt, 1980). Cavitation damage generally occurs in such fluid-handling machines as marine propellors, hydrofoils, dam slipways, gates, and all other hydraulic turbines. In cavitation erosion, the impinging liquid brings the gaseous bubbles to the surface at relatively high velocities. The bubbles strike the solid surface and collapse on impact, and their collapse imposes shock waves on the solid surface and results in liberation of material from the surface, as shown in Fig. 2.11. Cavitation erosion roughens a surface much like an etchant would.

Lord Rayleigh (1917) related the instantaneous pressure developed in a liquid owing to collapse of a cavity (bubble) to the compressibility and density of the liquid and the speed of the collapse. Wilson and Graham (1957) reported that the weight loss of silver correlated well with this concept and that erosion damage may be related to the product of density and the speed of sound in a liquid. Cavitation erosion has been shown to be characterized by a delay period in which little or no damage occurs followed by a period of wear at a constant rate (Mathuson and Hobbs, 1960). The duration of the delay period is determined by the initial surface state and is closely related to the endurance limit in mechanical fatigue tests. Once cavitation has commenced, the rate of material removal is broadly related to its strength, as measured by hardness or ultimate tensile strength. The effect may not be entirely mechanical, because the nature of liquid, i.e., whether or not it is an electrolyte, markedly affects test results.

2.2.8 Electrical Arc–Induced Wear

When a high electrical potential is present across a thin air film in a sliding process, a dielectric breakdown results that leads to arcing. During arcing, a relatively high power density (on the order of 1000 W mm^{-2}) occurs over a very short period of time. This results in considerable melting, corrosion, hardness changes, and other phase changes and even in direct ablation of material. Arcing causes large craters, and any sliding or oscillation after an arc either shears or fractures the lips, leading to abrasion, corrosion, surface fatigue, and fretting. Arcing can thus initiate several modes of wear, resulting in catastrophic failures in electrical machinery (Bhushan and Davis, 1983; Bhushan et al., 1985a).

When contacts are rubbing, as in the case of a commutator, the sparking damage

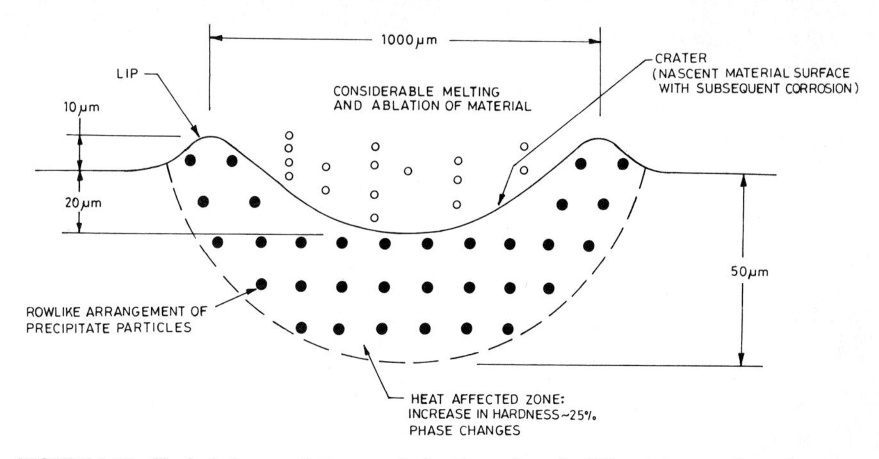

FIGURE 2.12 Typical changes that occur during the arcing of a 303 stainless steel specimen at 90 V (peak power 1300 W, duration 160 μs). [*Adapted from Bhushan and Davis (1983).*]

on the copper commutator can cause excessive wear of the brush by abrasion. This wear mode is also present in electroerosion printing (Bayer, 1983). In certain bearing applications in electrical machinery, there is the possibility that an electric current will pass through a bearing. When the current is broken at the contact surfaces between rolling elements and raceways, inner-race and shaft arcing or outer-race and housing arcing results. Both surfaces should be in the path of least resistance to a potential difference. Electrical arc–induced wear has been used productively as a method of removing metal in electrodischarge machining (EDM).

Very little has been published on the electrical arc–induced wear in bearing applications. Tallian et al. (1974) have briefly covered the subject. Bhushan and Davis (1983) conducted a study to examine the effect of arcing between two steel surfaces commonly used in bearing systems. Arcs were drawn at various voltages between 303 and 52,100 steel surfaces. A schematic of the changes that occur during arcing is presented in Fig. 2.12. The following two changes are significant from a wear standpoint: (1) the formation of lips and crater surfaces and (2) subsurface melting and resolidification, resulting in a material with increased hardness, a high density of crystalline-precipitate particles (nonmetallic carbides), and other phase changes. When a sliding, rolling, or oscillating member contacts the arced surface, the lips of the crater either shear or fracture the lips of the wear track (probably in the brittle mode caused by local phase changes), exposing nascent material, which promotes corrosion. The corrosion layer wears easily and enhances arcing. The debris generated by surface cracking and shearing of lips could lead to abrasion, fretting, and surface fatigue. In many cases it may be difficult to assess the wear modes in worn parts because wear could have been initiated by arcing that led to other modes. Incidentally, the wear induced by the electrical arc shall be more severe in the presence of surface films because of charge buildup.

2.3 WEAR PREVENTION

It is evident from Sec. 2.2 that wear is a complex process of material removal in which a variety of events take place independently or simultaneously. In order to

TABLE 2.4 Properties of a Material Preferred for the Control of a Particular Wear Process

Wear mode	Material property imparting wear resistance
Adhesive	Low solubility in mating surface material Resistance to thermal softening at interface temperature during sliding Low surface energy
Abrasive	Higher surface hardness than the abrasive medium Low work hardening coefficient
Fatigue	Resistance to subsurface deformation and crack nucleation and growth rate, i.e., high hardness and toughness, high flow strength
Corrosive	Resistance to corrosive medium when unpassivated
Fretting	Corrosion resistance to environment Capable of forming soft corrosion product Total immiscibility with mating surface High abrasion resistance
Solid-particle erosive	High hardness for low-angle impingement High toughness for high-angle impingement Heat treatment has no influence on erosion resistance

Source: Adapted from Glaeser (1980).

prevent or reduce wear, one must first understand the factors that contribute to wear.

The properties that are relevant to controlling a particular wear mode are summarized in Table 2.4.

2.3.1 Reduction of Adhesive Wear

Adhesive wear processes are initiated by the interfacial adhesive junctions that form if solid materials are in contact on an atomic scale. The relative tangential motion leads to the generation of wear particles. The increasing hardness of the surface reduces wear due to adhesion. The hardnesses of the mating members should be comparable in an unlubricated contact in order to distribute wear unless one of the members can be sacrificed. Adhesion tends to be favored by the presence of clean surfaces and chemical and structural similarities between the sliding metal couples. To prevent adhesive wear of metals, the sliding metal couples should be chosen to have a minimum tendency to form solid solutions, and this could be achieved by metals having different crystal structures and chemical properties (Rabinowicz, 1971, 1980, 1984).

Adhesive wear usually results from an inadequate supply of lubricant to the contact area. If the wear debris shows large metallic particles and the surfaces are damaged with material transferred from one surface to the other, then the amount of lubrication reaching the contact area should be investigated. Often with large area contacts, sufficient lubrication passages are not provided, and unlubricated regions result. This initiates the adhesive wear. If this does not appear to be the cause of the wear, then improved material combinations should be considered.

The role of microstructure in improving adhesive wear resistance is significant. First, the microstructure influences bulk properties, including the initiation of cracks. This is generally the case when the size of microstructural features is smaller than the depth of indentation of abrasive particles. Second, the microstructural features become effective as discrete components, and their individual properties assume increasing importance. It has been shown that the presence of hard phases (particularly carbides) has beneficial effects on wear resistance. Composition, amount, and morphology all affect the precise degree of improvement; chromium and vanadium carbides are superior to molybdenum and tungsten carbides and give optimal resistance to wear when used in amounts approaching 30 percent.

Adhesive wear is generally prevented by the interposition of dissimilar metal or nonmetal (can be similar) coatings. These may be soft deposits (e.g., Pb or Ag) or hard coatings of metals, alloys, and ceramics. Many surface treatments are also employed to reduce adhesive wear. In severe adhesive wear (i.e., scuffing) observed during running-in wear, the surfaces are relatively clean and require only transient protection. In such instances, short-life surface treatments such as phosphating and nitrocarburizing are useful.

Where there is no lubrication, harder surfaces are required, and ceramics and cermets or their coatings are normally used. In the case of nonmetals, it is acceptable to use identical material pairs. For low wear, the hardnesses of the mating members should be high and should be comparable with each other.

2.3.2 Reduction of Abrasive Wear

Abrasive wear results from a cutting action by (1) a rough, hard surface sliding against a softer surface or (2) contaminant hard particles trapped between the sliding surfaces. Hydrodynamic or elastohydrodynamic lubricant films either sufficiently thick to separate the surfaces in the former or thicker than the largest hard particles in the latter will greatly reduce this kind of abrasive wear.

As previously shown, the most important factor is the hardness of the cutting agent. To cut, it must have a hardness greater than about 1.2 times that of the material being cut. A contaminant commonly found in the environment is silica, or SiO_2, which has a hardness of approximately 1000 HV; therefore, a wear-resistant surface for use in an environment with silica should have a hardness of 1200 HV or more. Very few metals and surface treatments can reach this level of hardness. Under these circumstances, there is a need for enough strength and hardness in the surface layers to resist indentation from hard asperities and enough toughness to prevent fractures from surface imperfections or incipient fatigue cracks.

Various correlations have been attempted between abrasive wear and material properties such as hardness, modulus, hardness-to-modulus ratio, strain energy, and plastic work without much success when a large variety of metals and alloys are considered. However, for pure (homogeneous) materials (ceramics, plastics, metals), abrasive wear resistance is directly proportional to hardness (Khruschov, 1957). Hardness becomes less significant with duplex structures because hardness is only resistant to indentation, whereas abrasion requires independent penetration of both the hard and soft phases.

In addition to the hardness, the toughness (ductility) of the mating materials is also important (Khruschov and Babichev, 1965; Moore, 1974, 1981; Richardson, 1967). High hardness and toughness can be achieved simultaneously by depositing hard coatings onto ductile metals or by surface treating them. The hard-facing processes are widely used to provide abrasive wear-resistant surfaces.

2.3.3 Reduction of Fatigue Wear

Fatigue wear is the result of repeated stress on the surface of a body. Microcracks form, usually at inclusions, and are transmitted throughout the surface. Eventually these cracks coalesce, and wear particles can be formed. Although the overall stress is within the elastic range, the stresses at the cracktips are intensified, and fracture can take place.

Fatigue wear of materials can be minimized by slowing down or suppressing this microcrack formation. Raising the hardness of a material reduces the subsurface deformation and hence the crack nucleation rate. Raising the toughness decreases the crack growth rates. Therefore, for effective wear control, all that is necessary is to design the microstructure of the materials with high hardness and toughness.

Unfortunately, it is not possible to achieve both high hardness and toughness with a single microstructure. Normally, any manipulation of the microstructure to raise the hardness reduces the toughness of the material, and therefore, a careful optimization is necessary. The flow strength of materials should be made as high as possible without introducing any undesirable effects such as internal crack nucleation sites.

In many metals, hardening is often accomplished by the generation or introduction of second phases, since solution hardening may not produce an adequate increase in hardness. As soon as second-phase particles are introduced, crack nucleation occurs quite readily, and the crack propagation rate tends to become the wear rate-determining process. One interesting investigation would be to make two-phase materials in which the second phase is so small that it only contributes to hardness without providing crack nucleation sites. Another possibility is to increase the volume fraction of hard particles to the point where the matrix metal constitutes only a small fraction of the volume, such as, for example, in cutting tools made of tungsten carbide bonded with cobalt.

Another way of increasing wear resistance is to put a very hard coating on a substrate or to subject a substrate to surface treatment such that plastic deformation cannot take place in the substrate. Since a plastic deformation zone has not been observed deeper than 200 μm, the thickness of the hard layer should be of this order of magnitude, although in many situations it may be much less than this value. The coatings must be coherent and free of microcracks, since crack propagation is possible even in the absence of plastic deformation if there are preexisting cracks.

2.3.4 Reduction of Erosive Wear

This type of wear involves the erosion of surfaces as a result of solids, liquids, or gases impinging on them. The erosive wear under specific conditions depends on

- The kinetic energy of the erosive medium
- The impact angle (see Figure 2.10)

- The ultimate resilience of the surface (half the tensile strength divided by the elastic modulus)
- The metal-metal bond energy of the surface

The ultimate resilience is the amount of energy that can be absorbed before deformation or cracking occurs. Thus, for materials to withstand cavitation erosion, a high value for ultimate resilience should be preferred.

High metal-metal bond-energy values mean that the atoms of a metal are held together quite strongly in the crystal lattice and will not leave the lattice easily in any attempt to dislodge them, e.g., during impact erosion, and that it is the cluster of atoms that will be dislodged as particles. Thus materials having high values of metal-metal bond energy exhibit better erosive wear resistance (Vijh, 1975).

In recent reviews, attempts have been made to correlate the erosion rates of metals with their melting point, elastic moduli, and their indentation hardness values. The most attractive proposal was made by Hutchings (1979), who showed that erosion wear rates are inversely correlated with the product $\rho C_p \Delta T$, where ρ is the density of metal, C_p is the specific heat, and ΔT is the difference between the temperature of the metal and its melting point (a high ΔT value generally indicates low thermal softening). Hutchings' formula indicates that erosive wear resistance will be higher for materials having high melting points (i.e., high ΔT values).

2.3.5 Reduction of Electrical Arc–Induced Wear

Electrical arc–induced wear occurs when a high potential is present over a thin air film, resulting in a dielectric breakdown that leads to arcing. Methods to minimize electrical arc–induced wear (Bhushan and Davis, 1983; Bhushan et al., 1985a) would be to

- Eliminate the gap between the two surfaces that have a potential difference
- Provide an insulator with adequate dielectric strength (e.g., an elastomer or oxide ceramic coating) between the two surfaces
- Provide a low-impedance connection between the two surfaces to eliminate potential differences
- Have one of the surfaces not grounded

Bearing manufacturers suggest that bearings should be press-fitted to the shaft and that a conducting grease should be used to eliminate arcing between the shaft and the inner race and between the inner and outer races and the rolling elements.

2.4 *LUBRICATION*

The friction and wear mechanisms discussed so far are based on the physical interactions between two surfaces moving relative to each other. Lubrication is a process by which the friction and wear between two solid surfaces in relative motion is reduced significantly by interposing a lubricant between them. The role of lubrication is to separate the moving surfaces with a film of solid, liquid, or gaseous material that can be shared with low resistance without causing any damage to the surfaces (Booser, 1983, 1984; Cameron, 1966, 1981; Dorinson and Ludema, 1985; Fuller, 1984; Gunther, 1971; Summers-Smith, 1969).

Figure 2.13 shows a detailed classification of lubrication modes. In hydrostatic lubrication, the lubricant is fed under pressure to the interface, and this inhibits contact between the solids. In hydrodynamic, elastohydrodynamic, and squeeze-film lubrication, pressure in the lubricant is automatically created by the relative motion, and this inhibits contact between the solids. In boundary lubrication and in lubrication with solids, there is a considerable asperity interaction between the contacting solid surfaces. The friction under boundary lubrication is due partly to shear in the lubricant and partly to shear in the asperity contacts.

The differentiation of various lubrication modes other than hydrostatic can be done on the basis of the Stribeck curve, which provides the frictional behavior of different lubrication regimes. The Stribeck curve illustrates (Fig. 2.14) change in the coefficient of friction as a function of the lubrication parameter $\eta V P^{-1}$, where η is the lubricant viscosity, V is the sliding velocity, and P is the pressure. Therefore, depending on the geometry, the properties of the lubricant, and the operating conditions, three main lubrication modes may be distinguished:

1. Hydrodynamic and elastohydrodynamic (EHD) lubrication ($h \gg \sigma$)
2. Mixed lubrication ($h \approx \sigma$)
3. Boundary lubrication ($h \ll \sigma$)

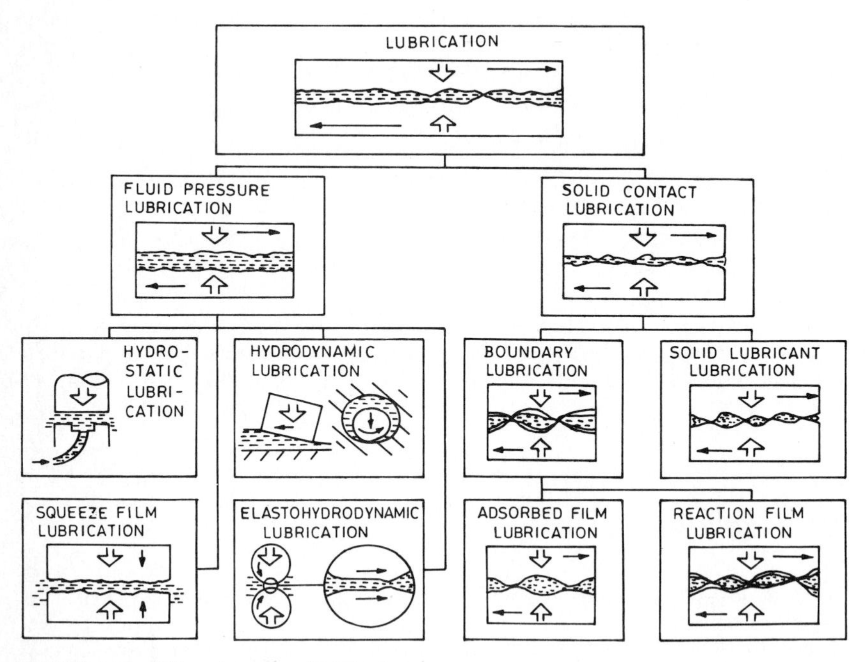

FIGURE 2.13 Different modes of lubrication.

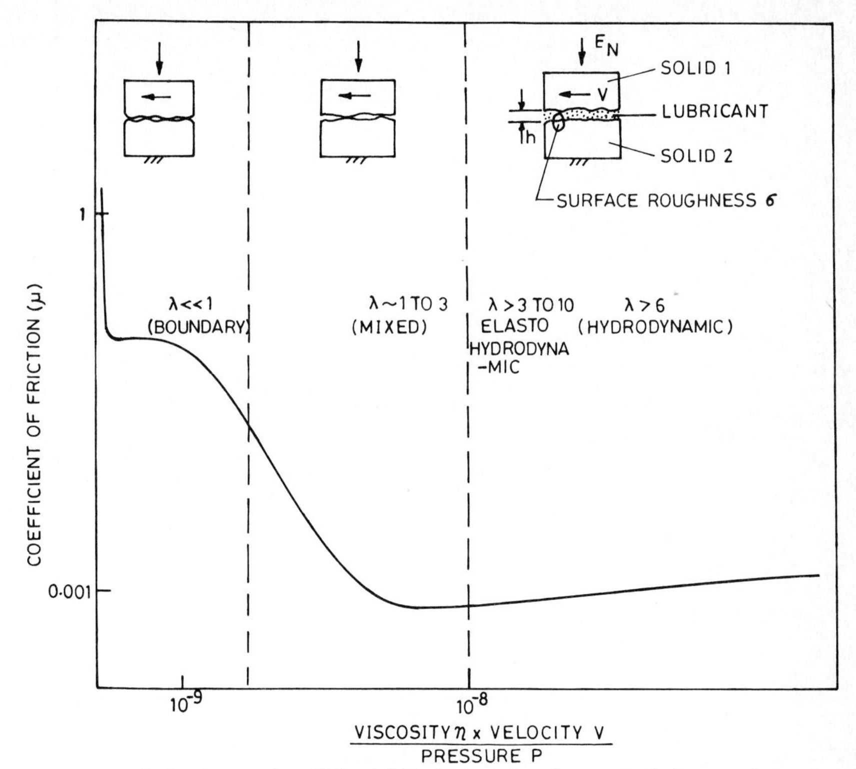

FIGURE 2.14 Stribeck curve (coefficient of friction vs. viscosity × velocity/pressure).

In these expressions, h is thickness of lubricant film, and σ is composite rms surface roughness of the mating surfaces, that is,

$$\sigma = (\sigma_1^2 + \sigma_2^2)^{1/2}$$

A better picture is obtained by considering the lubricant film thickness in relation to the surface roughness. This relationship is called *specific film thickness* λ and is defined as

$$\lambda = h_{\min}\sigma^{-1} \tag{2.23}$$

where $h_{\min}$ = minimum film thickness. The frictional behavior in relation to λ is shown in Fig. 2.15. Typical values of $h_{\min}$, λ, and μ are shown in Table 2.5.

2.4.1 Hydrostatic Lubrication

Hydrostatic bearings support load on a thick film of fluid supplied from an external pressure source, a pump, that feeds pressurized fluid to the film. For this rea-

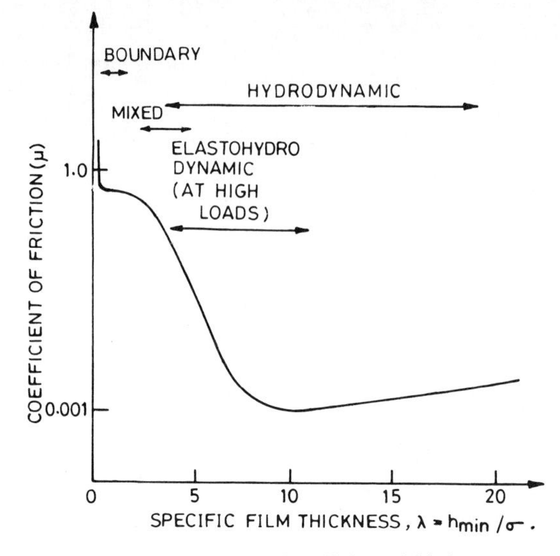

FIGURE 2.15 Dependence of coefficient of friction on specific film thickness.

TABLE 2.5 Typical Lubricant Film Properties in Machine Element Applications

	Lubricant film properties		
Mechanisms of lubrication	Lubricant film thickness, h_{min} (μm)	Specific film thickness, λ	Coefficient of friction, μ
Boundary lubrication	0.005–0.1	⩽1	0.03–1
Mixed lubrication	0.01–1	1–3	0.02–0.15
Elastohydrodynamic lubrication	0.01–10	3–10	0.01–0.1
Hydrodynamic lubrication	1–100	6–100	0.001–0.01

son, these bearings are often called *externally pressurized bearings*. Hydrostatic bearings are designed for use with both incompressible and compressible fluids. The film thickness and fluid pressure profile are fairly uniform across the interface (Fig. 2.16).

Since hydrostatic bearings do not require relative motion of the bearing surfaces to build up the load-supporting pressures, such as is necessary in hydrodynamic bearings, hydrostatic bearings are used in applications with little or no relative motion between the surfaces. Hydrostatic bearings also may be required in applications where, for one reason or another, touching or rubbing of the bearing surfaces cannot be permitted at startup and shutdown. In addition, hydrostatic bearings provide high stiffness. Hydrostatic bearings, however, do have the disadvantage of requiring high-pressure pumps and equipment for fluid cleaning, which adds to space and cost.

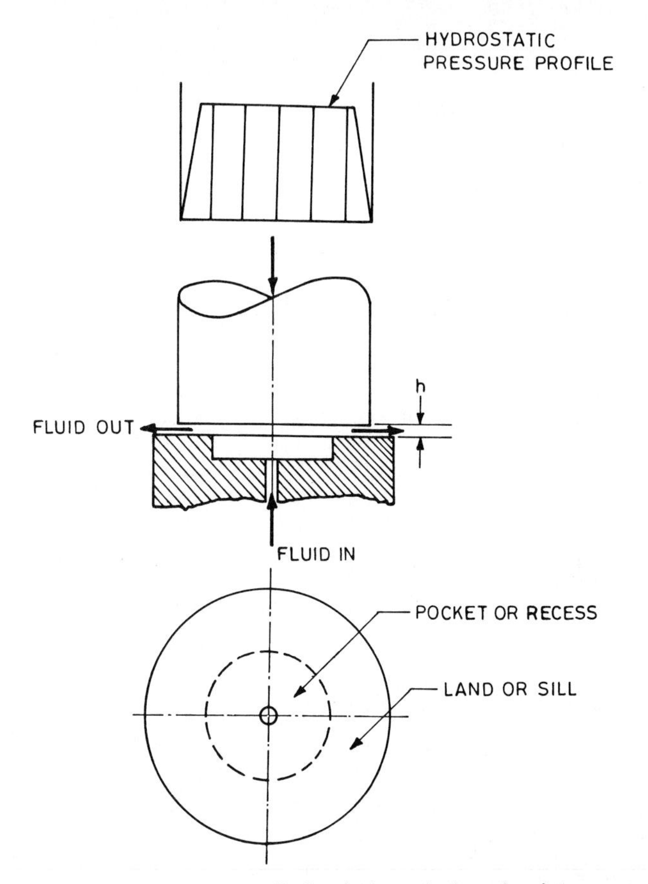

FIGURE 2.16 Pressure profile in a hydrostatic thrust bearing.

2.4.2 Hydrodynamic Lubrication

Hydrodynamic lubrication is based on the formation of a thick lubricant film of characteristic geometric profile that develops automatically between opposing nonparallel solid surfaces having relative motion to each other. The characteristic geometric profile resembles a converging wedge (Fig. 2.17) having maximum thickness at its leading edge and minimum thickness at its trailing edge. The leading edge of the converging wedge represents the point of lowest pressure within the film, while the trailing edge represents the point of maximal pressure. In between, the pressure increases from the leading edge to the trailing edge. The hydrodynamic pressure is low compared with the strength of the mating materials and will not cause appreciable local deformation. The condition of hydrodynamic lubrication is governed by the bulk physical properties of the lubricant, mainly viscosity, and the relative speed of the moving surfaces (Bisson and Anderson, 1964; Booser, 1984; Fuller, 1984).

Fluid film also can be generated only by a reciprocating or oscillating motion in the normal direction (squeeze) that may be fixed or variable in magnitude

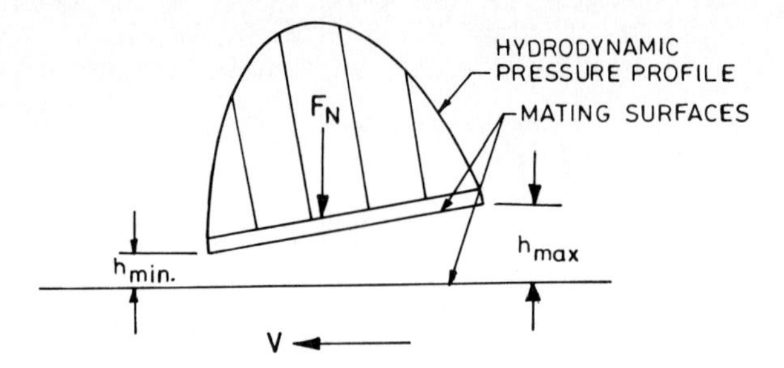

FIGURE 2.17 Generation of pressure due to hydrodynamic action.

(transient or steady state). This load-carrying phenomenon arises from the fact that a viscous fluid cannot be instantaneously squeezed out from between two surfaces that are approaching each other. It takes time for these surfaces to meet, and during this interval, because of the fluid's resistance to extrusion, pressure is built up and the load is actually supported by the fluid film. When the load is relieved or is reversed, the fluid is sucked in and the film often can recover its thickness in time for the next application. The squeeze phenomenon controls the buildup of a water film under the tires of automobiles and airplanes on wet roadways or landing strips (commonly known as *hydroplaning*) so that they have virtually no relative sliding motion (Booser, 1984; Fuller, 1984).

2.4.3 Elastohydrodynamic Lubrication

Elastohydrodynamic lubrication (EHD lubrication or EHL) applies to hydrodynamic conditions where surface deformation is comparable with the hydrodynamic film thickness and surface deformation affects the hydrodynamic behavior of the interface (Bhushan, 1980, 1981; Bisson and Anderson, 1964; Booser, 1984; Dowson and Higginson, 1966). The most prominent example of EHD lubrication is the hertzian contact in rolling-element bearings, gears, and cams, in which the contact areas are typically one-thousandth those occurring in journal bearings or pressures are 1000 times greater. The heavy loads cause local elastic deformation of the mating surfaces, which provides a coherent hydrodynamic film and avoids asperity interaction. Under the high pressures generated, the lubricant properties are markedly different from those of the bulk liquid. In particular, an enormous increase in viscosity may occur so that the lubricant behaves more like a solid than a liquid. It has been shown experimentally that under EHD lubrication conditions, a plane contact zone is formed with a length in the sliding direction of on the order of 100 μm and a film thickness of on the order of 1 μm. The pressure generated in the film reaches on the order of 1 GPa under a typical condition (Fig. 2.18).

2.4.4 Mixed Lubrication

In mixed lubrication, the contact behavior is mainly governed by a mixture of EHD or hydrodynamic and boundary lubrication. Although the contacting surfaces are separated by a thin lubricant film, asperity contact also may take place. The total applied load is thought to be partly carried by asperity contacts and

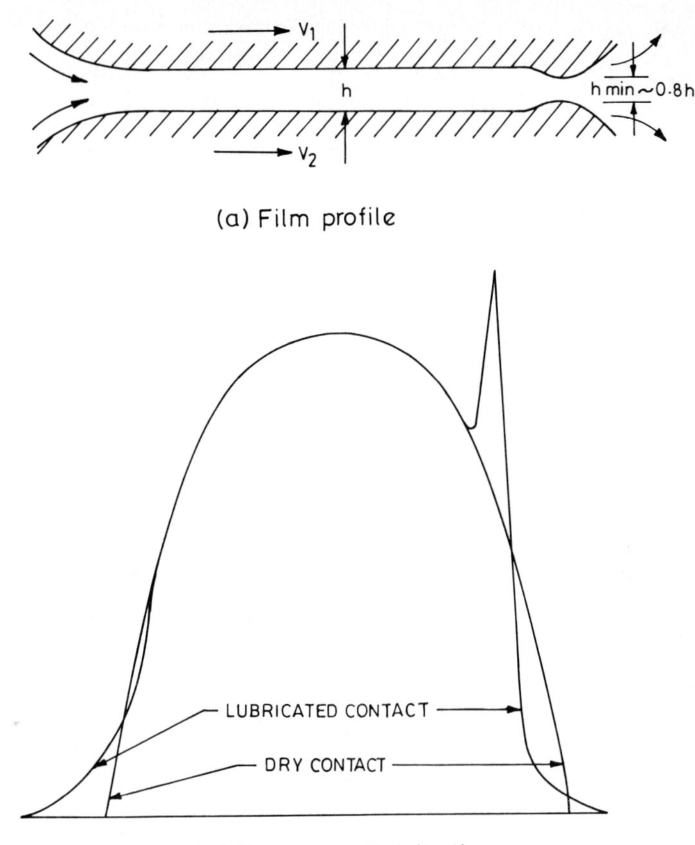

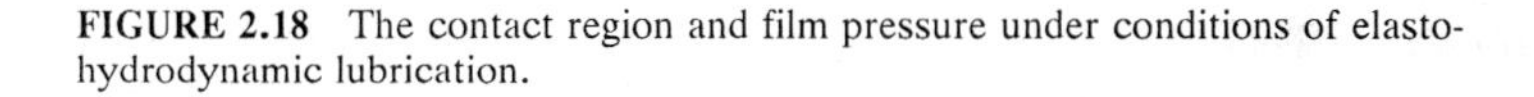

FIGURE 2.18 The contact region and film pressure under conditions of elastohydrodynamic lubrication.

partly by hydrodynamic action. The total friction results partly from asperity contacts and partly from viscous friction of lubrication.

Mixed lubrication exists when the specific film thickness λ ranges from 1 to 3.

2.4.5 Boundary Lubrication

In boundary lubrication, the contacting solid surfaces move so close to each other that there is a considerable asperity interaction. A thin film of lubricant is adsorbed to the solid surface and significantly inhibits asperity welding. Wear is reduced because of the less direct asperity contact, and the coefficient of friction is reduced because the adsorbed lubricant film has lower shear strength than the solid material (Bisson and Anderson, 1964; Booser, 1984; Georges, 1983; Ku, 1970; Ling et al., 1969).

In general, the solid-solid interactions are modified through the action of the boundary lubricant, so the friction and wear behavior of a boundary-lubricated contact is determined by the solid-lubricant-solid interface, which is significantly influenced by environmental conditions. The various type of solid-lubricant interactions, such as physisorption, chemisorption, and chemical reactions, are discussed at length in Chapter 3.

In practice, compared with other lubrication regimes of the Stribeck curve (see Figure 2.15), boundary lubrication is a complex phenomenon in which the solid-lubricant interactions are mixed with the effects of metallurgy, surface roughness, corrosion, temperature, pressure, and time of reaction.

2.4.6 Solid-Film Lubrication

The tendency to subject sliding surfaces to an ever-increasing range of thermal, mechanical, and environmental stresses has inevitably posed new problems in friction and wear behavior. Conventional lubricants are unable to lubricate effectively at temperatures higher than 300 to 450°C. The alternative is to use thin films of solids such as graphite and molybdenum disulfide interposed between surfaces of conventional materials in sliding/rolling contact or to develop new composite materials that intrinsically exhibit low friction and wear.

Solid-film lubrication arises when a soft or hard solid film of substantial thickness is interposed between the sliding surfaces. The lubrication action is achieved by a solid film that separates the surfaces completely or acts over a part of the junction or is present as an incomplete layer over the entire surface. The complete separation of the sliding surfaces provides most effective lubrication.

The solid film for lubrication can either be applied to the sliding surfaces or be formed by reaction of the metal with the environment. The interposed film in sliding or rolling contact influences the friction and wear behavior because shear takes place within and across the film or between the film and the sliding surfaces.

2.5 REFERENCES

Anonymous (1955), "Fretting and Fretting Corrosion," *Lubrication*, Vol. 41, pp. 85–96.

——— (1974), "Erosion, Wear, and Interfaces with Corrosion," Special Publication STP-567, ASTM, Philadelphia.

Archard, J. F. (1953), "Contact and Rubbing of Flat Surfaces," *J. Appl. Phys.*, Vol. 24, pp. 981–988.

——— (1980), "Wear Theory and Mechanisms," in *Wear Control Handbook* (M. B. Peterson and W. O. Winer, eds.), pp. 35–80, ASME, New York.

Avallone, E. A., and Baumeister, T., III (1987), *Marks' Standard Handbook for Mechanical Engineers*, 9th Ed., McGraw-Hill, New York.

Bayer, R. G. (1983), "Wear in Electroerosion Printing." *Wear*, Vol. 92, pp. 197–212.

Bhushan, B. (1980), "Review of Experimental Techniques in EHD-Lubricated Contacts," DOE Tech. Rep. No. AT80D069, SKF Industries, Inc., King of Prussia, Pa.

——— (1981), "Film Thickness Models in EHD-Lubricated Contacts," Tech. Rep. No. AT81D060, SKF Industries, Inc., King of Prussia, Pa.

——— (1984), "Analysis of the Real Area of Contact Between a Polymeric Magnetic Medium and a Rigid Surface," *J. Tribol.* (*Trans. ASME*), Vol. 106, pp. 26–34.

——— (1985), "The Real Area of Contact in Polymeric Magnetic Media: II. Experimental Data and Analysis," *ASLE Trans.*, Vol. 28, pp. 181–197.

——— (1990), *Tribology and Mechanics of Magnetic Storage Devices*, Spinger-Verlag, New York.

———, and Davis, R. E. (1983), "Surface Analysis Study of Electric-Arc Induced Wear," *Thin Solid Films*, Vol. 108, pp. 135–156.

———, ——— and Gordon, M. (1985a), "Metallurgical Reexamination of Wear Modes: I. Erosive, Electrical Arcing, and Fretting," *Thin Solid Films*, Vol. 123, pp. 93–112.

———, ———, and Kolar, H. R. (1985b), "Metallurgical Reexamination of Wear Modes: II. Adhesive and Abrasive," *Thin Solid Films*, Vol. 123, pp. 113–126.

Bisson, E. E., and Anderson, W. J. (1964), "Advanced Bearing Technology," Special Publication SP-38, NASA, Washington, D.C.

Bitter, J. G. A. (1963), "A Study of Erosion Phenomena, Parts I and II," *Wear*, Vol. 6, pp. 163–191.

Booser, E. R. (1983), *Handbook of Lubrication: Theory and Practice of Tribology*, Vol. 1: *Application and Maintenance*, CRC Press, Boca Raton, Fla.

——— (1984), *Handbook of Lubrication: Theory and Practice of Tribology*, Vol. 2: *Theory and Design,* CRC Press, Boca Raton, Fla.

Bowden, F. P., and Tabor, D. (1950), *Friction and Lubrication of Solids*, Part I, Clarendon Press, Oxford, England.

———, and ——— (1964), *Friction and Lubrication of Solids*, Part II, Clarendon Press, Oxford, England.

Buckley, D. H. (1972), "Adhesion of Metals to a Clean Iron Surface Studied with LEED and Auger Emission Spectroscopy," *Wear*, Vol. 20, pp. 89–103.

——— (1981), *Surface Effects in Adhesion, Friction, Wear, and Lubrication*, Elsevier, Amsterdam.

———, and Johnson, R. L. (1968), "The Influence of Crystal Structure and Some Properties of Hexagonal Metals on Friction and Adhesion," *Wear*, Vol. 11, pp. 405–419.

Burwell, J. T. (1957/58), "Survey of Possible Wear Mechanisms," *Wear*, Vol. 1, pp. 119–141.

Cameron, A. (1966), *The Principles of Lubrication*, Longman, London.

——— (1981), *Basic Lubrication Theory*, Ellis Harwood, London.

Crook, A. W. (1957), "Simulated Gear-Tooth Contacts: Some Experiments upon Their Lubrication and Subsurface Deformations," *Proc. Conf. Lub. Wear*, p. 701, Institute of Mechanical Engineers, London.

Czichos, H. (1972), "The Mechanism of the Metallic Adhesion Bond," *J. Phys. [D] Appl. Phys.*, Vol. 5, pp. 1890–1897.

——— (1978), *Tribology: A Systems Approach to the Science and Technology of Friction, Lubrication, and Wear*, Elsevier, Amsterdam.

Dorinson, A., and Ludema, K. C. (1985), *Mechanics and Chemistry in Lubrication*, Elsevier, Amsterdam.

Dowson, D. (1979), *History of Tribology*, Longman, London.

———, and Higginson, G. R. (1966), *The Fundamentals of Roller Gear Lubrication: Elastohydrodynamic Lubrication*, Pergamon Press, London.

Drescher, H. (1959), "The Mechanics of Friction Between Solid Bodies," *VDI-Z*, Vol. 101, p. 697 (in German).

Eldredge, K. R., and Tabor, D. (1955), "The Mechanisms of Rolling Friction: I. The Plastic Range," *Proc. R. Soc. Lond.*, Vol. A229, pp. 181–198.

Endo, K., and Goto, H. (1978), "Effect of Environment on Fretting Fatigue," *Wear*, Vol. 48, pp. 347–367.

Eyre, T. S. (1976), "Wear Characteristics of Metals," *Tribol. Int.*, Vol. 9, pp. 203–212.

——— (1978), "The Mechanisms of Wear," *Tribol. Int.*, Vol. 11, pp. 91–96.

Finnie, I. (1960), "Erosion of Surfaces by Solid Particles," *Wear*, Vol. 3, pp. 87–103.

———, Levy, A. V., and McFadden, D. H. (1979), "Fundamental Mechanisms of Erosive Wear of Ductile Metals by Solid Particles," in *Erosion, Prevention, and Useful Applications* (W. F. Adler, ed.), Special Publication STP-664, pp. 36–58, ASTM, Philadelphia.

———, Wolak, J., and Kabil, Y. H. (1967), "Erosion of Metals by Solid Particles," *J. Mater. Sci.*, Vol. 2, pp. 682–702.

Fischer, T. E. (1988), "Tribochemistry," *Ann. Rev. Mater. Sci.*, Vol. 18, pp. 303–323.

Fuller, D. D. (1984), *Theory and Practice of Lubrication for Engineers*, 2d Ed., Wiley, New York.

Gane, N., Pfaelzer, P. F., and Tabor, D. (1974), "Adhesion Between Clean Surfaces at Light Loads," *Proc. R. Soc. Lond.*, Vol. A340, pp. 495–517.

Georges, J. M. (ed.) (1983), *Microscopic Aspects of Adhesion and Lubrication*, Elsevier, Amsterdam.

Glaeser, W. M. (1980), "Wear-Resistant Materials," in *Wear Control Handbook* (M. B. Peterson and W. O. Winer, eds.), pp. 313–326, ASME, New York.

Greenwood, J. A., and Williamson, J. B. P. (1966), "Contact of Nominally Flat Surfaces," *Proc. R. Soc. Lond.*, Vol. A295, pp. 300–319.

Gunther, R. C. (1971), *Lubrication*, Bailey Brothers and Swinfen, Ltd., Folkestone, U.K.

Habig, K. H. (1968), "On the Dependence of Adhesion and Dry Friction of Metals on Structure and Orientation," *Mater. Prufung.*, Vol. 10, p. 417 (in German).

Halling, J. (1975), "A Contribution to the Theory of Mechanical Wear," *Wear*, Vol. 34, pp. 239–249.

Hamilton, G. M. (1963), "Plastic Flow in Rollers Loaded above the Yield Point," *Proc. Inst. Mech. Eng.*, Vol. 177, p. 667.

Hammitt, F. G. (1980), "Cavitation and Liquid Impact Erosion," in *Wear Control Handbook* (M. B. Peterson and W. O. Winer, eds.), pp. 161–230, ASME, New York.

Heathcote, H. L. (1921), "The Ball Bearings," *Proc. Inst. Auto. Eng.*, Vol. 15, p. 1569.

Heilmann, P., and Rigney, D. A. (1981), "An Energy-Based Model of Friction and Its Application to Coated Systems," *Wear*, Vol. 72, pp. 195–217.

Hornbogen, E. (1975), "The Role of Fracture Toughness in the Wear of Metals," *Wear*, Vol. 33, pp. 251–259.

Hutchings, I. M. (1979), "Mechanism of Erosion of Metals by Solid Particles," in *Erosion, Prevention, and Useful Applications* (W. F. Adler, ed.), Special Publication STP-664, pp. 59–76, ASTM, Philadelphia.

Johnson, K. L. (1966), "A Review of the Theory of Rolling Contact Stresses," *Wear*, Vol. 9, pp. 4–19.

——— (1985), *Contact Mechanics*, Cambridge University Press, Cambridge, Mass.

Keller, D. V. (1963), "Adhesion Between Solid Metals," *Wear*, Vol. 6, pp. 353–365.

Khruschov, M. M. (1957), "Resistance of Metals to Wear by Abrasion, as Related to Hardness," pp. 655–659, *Proc. Conf. on Lubrication and Wear*, Institute of Mechanical Engineers, London.

——— (1974), "Principles of Abrasive Wear," *Wear*, Vol. 28, pp. 69–88.

———, and Babichev, M. A. (1965), "The Effect of Heat Treatment and Work Hardening on the Resistance of Abrasive Wear of Some Alloy Steels," *Friction Wear Mach.*, Vol. 19, pp. 1–15.

Kinloch, A. J. (1980), "The Science of Adhesion," *J. Mater. Sci.*, Vol. 15, pp. 2141–2166.

Kragelsky, I. V., Dobychin, M. N., and Kombalov, V. S. (1982), *Friction and Wear: Calculation Methods*, Pergamon, Oxford, England.

Ku, P. M. (1970), "Interdisciplinary Approach to Friction and Wear," Special Publication SP-181, NASA, Washington, D.C.

Kuhlmann-Wilsdorf, D. (1981), "Dislocation Concepts in Friction and Wear," in *Funda-*

mentals of Friction and Wear of Materials (D. A. Rigney, ed.), pp. 119–186, American Society for Metals, Metals Park, Ohio.

Lang, O. R. (1977), "Surface Fatigue of Plain Bearings," *Wear*, Vol. 43, pp. 25–30.

Levy, A. V. (1981), "The Solid Particle Erosion Behavior of Steel as a Function of Microstructure," *Wear*, Vol. 68, pp. 269–288.

Ling, F. F., Klaus, E. E., and Fein, R. S. (1969), *Boundary Lubrication: An Appraisal of World Literature*, ASME, New York.

Lipson, C. (1960), *Wear Considerations in Design*, Univ. of Michigan Press, Ann Arbor, Mich.

——— (1967), *Wear Considerations in Design*, Prentice-Hall, Englewood Cliffs, N.J.

———, and Colwell, L. V. (eds.) (1961), *Handbook of Mechanical Wear*, Univ. of Michigan Press, Ann Arbor, Mich.

Maji, J., and Sheldon, G. L. (1979), "Mechanism of Erosion of a Ductile Material by Solid Particles," in *Erosion, Prevention, and Useful Applications* (W. F. Adler, ed.), Special Publication STP-664, pp. 136–147, ASTM, Philadelphia.

Marx, U., and Feller, H. G. (1978), "Correlation of Tribological and Mechanical Parameters Taking Gold Alloys, Gold-Tantalum Alloys, and Nickel as Examples: I. Determination of Tribological and Other Mechanical Parameters, *Metallurgy*, Vol. 32, pp. 1214–1218 (in German).

———, and ——— (1979), "Correlation of Tribological and Mechanical Parameters Taking Gold, Gold-Tantalum Alloys, and Nickel as Examples: III. Tribological Theories and Their Range of Validity," *Metallurgy*, Vol. 33, pp. 380–383 (in German).

Mathuson, R., and Hobbs, J. M. (1960), "Cavitation Erosion Comparison Tests," *Engineering*, Vol. 189, pp. 136–137.

McFarlane, J. S., and Tabor, D. (1950), "Relation Between Friction and Adhesion," *Proc. R. Soc. Lond.*, Vol. A202, pp. 244–253.

Moore, D. F. (1972), *The Friction and Lubrication of Elastomers*, Pergamon, Oxford, England.

Moore, M. A. (1974), "The Relationship Between Abrasive Wear Resistance, Hardness, and Microstructure of Ferritic Materials," *Wear*, Vol. 28, pp. 59–68.

——— (1981), "Abrasive Wear," in *Fundamentals of Friction and Wear of Materials* (D. A. Rigney, ed.), pp. 73–118, American Society for Metals, Metals Park, Ohio.

Ohmae, N., Okuyama, T., and Tsukizoe, T. (1980), "Influence of Electronic Structure on the Friction in Vacuum of 3d Transition Metals in Contact with Copper," *Tribol. Int.*, Vol. 13, pp. 177–180.

Peterson, M. B. (1980), "Classification of Wear Processes," in *Wear Control Handbook* (M. B. Peterson and W. O. Winer, eds.), pp. 1–8, ASME, New York.

———, and Winer, W. O., eds. (1980), *Wear Control Handbook*, ASME, New York.

Poritsky, H. (1950), "Stresses and Deflections of Cylindrical Bodies in Contact with Application to Contact of Gears and of Locomotive Wheels," *J. Appl. Mech.* (*Trans. ASME*), Vol. 17, pp. 191–201.

Preece, C. M. (1979), *Treatise on Materials Science and Technology*, Vol. 16: *Erosion*, Academic, New York.

Quinn, T. F. J. (1962), "The Role of Oxidation in the Mild Wear of Steel," *Br. J. Appl. Phys.*, Vol. 13, pp. 33–37.

——— (1978), "The Division of Heat and Surface Temperatures at Sliding Steel Interfaces and Their Relation to Oxidational Wear," *ASLE Trans.*, Vol. 21, pp. 78–86.

——— (1983), "A Review of Oxidational Wear," Parts I and II, *Tribol. Int.*, Vol. 16, pp. 257–271, 305–315.

———, Rowson, D. M., and Sullivan, J. L. (1980), "Application of the Oxidational Theory of Mild Wear to the Sliding Wear of Low-Alloy Steel," *Wear*, Vol. 65, pp. 1–20.

Rabinowicz, E. (1965), *Friction and Wear of Materials*, Wiley, New York.

——— (1971), "The Determination of the Compatibility of Metals Through Static Friction Tests," *ASLE Trans.*, Vol. 14, pp. 198–205.

——— (1980), "Wear Coefficients—Metals," in *Wear Control Handbook* (M. B. Peterson and W. O. Winer, eds.), pp. 475–506, ASME, New York.

——— (1984), "Wear Coefficients," in *Handbook of Lubrication: Theory and Practice of Tribology*, Vol. 2: *Theory and Design* (E. R. Booser, ed.), pp. 201–208, CRC Press, Boca Raton, Fla.

Rayleigh, L. (1917), "On the Pressure Developed in a Liquid During the Collapse of a Spherical Cavity," *Phil. Mag.*, Vol. 34, pp. 94–98.

Razavizadeh, K., and Eyre, T. S. (1982), "Oxidative Wear of Aluminum Alloys," *Wear*, Vol. 79, pp. 325–333.

Reynolds, O. (1876), "On Rolling Friction," *Philos. Trans. R. Soc. Lond.*, Vol. 166, pp. 155–174.

Richardson, R. C. D. (1967), "The Wear of Metals by Hard Abrasives," *Wear*, Vol. 10, pp. 291–309.

Rigney, D. A., Chen, L. H., Naylor, M. G. S., and Rosenfield, A. R. (1984), "Wear Processes in Sliding Systems," *Wear*, Vol. 100, pp. 195–219.

Sargent, L. B., Jr. (1978), "On the Fundamental Nature of Metal-Metal Adhesion," *ASLE Trans.*, Vol. 21, pp. 285–290.

Sarkar, A. D. (1976), *Wear of Metals*, Pergamon, Oxford, England.

Sasada, T. (1985), "Future Direction on Research in Wear and Wear-Resistant Materials: Status of Understanding," in *New Directions in Lubrication, Materials, Wear, and Surface Interactions: Tribology in the 80's* (W. R. Loomis, ed.), pp. 192–213, Noyes, Park Ridge, N.J.

———, Norose, S., and Nagai, T. (1978), "Wear in Different Metal Combinations in Vacuum," *Proc. Japan Congr. Mater. Res.*, p. 112, The Society of Material Science, Japan.

———, ———, and Mishina, H. (1981), "The Behavior of Adhered Fragments Interposed Between Sliding Surfaces and the Formation Process of Wear Particles," *J. Lub. Technol. (Trans. ASME)*, Vol. 103, pp. 195–202.

Schmitt, G. F., Jr. (1980), "Liquid and Solid Particle Impact Erosion," in *Wear Control Handbook* (M. B. Peterson and W. O. Winer, eds.), pp. 231–282, ASME, New York.

Scott, D. (1979), *Treatise on Material Science*, Vol. 13: *Wear*, Academic, New York.

Siebel, E. (1938), *Über die praktische Bewahrung dermit verşchlebiver suchen gewonnenen Ergebnisse*, S-4, J. T., Tagunsband VDI-Verschleibtagung, Stuttgart.

Sikorski, M. F. (1964), "The Adhesion of Metals and Factors that Influence It," *Wear*, Vol. 7, pp. 114–162.

Sin, H. C., Saka, N., and Suh, N. P. (1979), "Abrasive Wear Mechanisms and the Grit Size Effect," *Wear*, Vol. 55, pp. 163–190.

Sproles, E. S., and Duquette, D. J. (1978), "The Mechanism of Material Removal in Fretting," *Wear*, Vol. 49, pp. 339–352.

Suh, N. P. (1977), "An Overview of the Delamination Theory of Wear," *Wear*, Vol. 44, pp. 1–16.

Summers-Smith, D. (1969), *An Introduction to Tribology in Industry*, Machinery Publishing Company, London.

Tabor, D. (1955), "The Mechanism of Rolling Friction: II. The Elastic Range," *Proc. R. Soc. Lond.*, Vol. A229, pp. 198–220.

Tallian, T. E., Baile, G. H., Dalal, H., and Gusttafssen, O. (1974), *Rolling Element Damage Atlas*, SKF Industries, Inc., King of Prussia, Pa.

Tilly, G. P. (1969), "Erosion Caused by Air-Borne Particles," *Wear*, Vol. 14, pp. 63–79.

——— (1973), "A Two-Stage Mechanism of Ductile Erosion," *Wear*, Vol. 23, pp. 87–96.

Vijh, A. K. (1975), "The Influence of Metal-Metal Bond Energies on the Adhesion, Hardness, Friction, and Wear of Metals," *J. Mater. Sci.*, Vol. 10, pp. 998–1004.

——— (1976), "Resistance of Metals to Erosion by Solid Particles in Relation to the Solid-State Cohesion of Metals," *Wear*, Vol. 39, pp. 173–175.

Vingsbo, O., and Hogmark, S. (1981), "Wear of Steels," in *Fundamentals of Friction and Wear of Materials* (D. A. Rigney, ed.), pp. 373–408, American Society for Metals, Metals Park, Ohio.

Waterhouse, R. B. (1984), "Fretting Wear," *Wear*, Vol. 100, pp. 107–118.

Wilson, R. W., and Graham, R. (1957), "Cavitation of Metal Surfaces in Contact with Lubricants," *Proc. Conf. on Lubrication and Wear*, pp. 707–712, Institute of Mechanical Engineers, London.

Wolfe, G. F. (1963), "Effect of Surface Coatings on the Load-Carrying Capacity of Steel," *Lub. Eng.*, Vol. 19, p. 28.

Zum Gahr, K.-H. (1982), "Abrasive Wear of Ductile Materials," *Z. F. Metallkd.*, Vol. 73, pp. 267–276 (in German).

CHAPTER 3
PHYSICS OF TRIBOLOGICAL MATERIALS

The friction and wear behavior of materials is greatly dependent upon the surface material, the shape of mating surfaces, environment, and operating conditions. Various surface and bulk properties of materials that play a key role in determining the tribological behavior are listed in Fig. 3.1.

It is well established that clean metal surfaces adhere very strongly when pressed together, but a small amount of oxygen or other contamination greatly reduces that adherence. A surface film may change the mechanical behavior of the first few atomic layers of metal adjacent to the film through interaction with environment or diffusion. Thus some knowledge of the nature of surfaces and their interaction with the environment is necessary to understand the phenomena that take place during friction and wear processes.

The surface energy, which ultimately determines the adhesion energy between moving surfaces at the interface, depends on crystal orientation, bonding, etc. The mechanical properties such as shear strength, yield strength, elastic modulus, hardness, and toughness are greatly influenced by alloying one material with another. For instance, the addition of carbon in steel changes the hardness quite significantly (Chap. 4). The microstructure, grain size, dislocation, etc., also change during the alloying process.

In this chapter, the basic understanding of surface phenomena and some of the bulk processes and their relevance to friction and wear processes are briefly discussed. For further details, readers may refer to books on solid-state physics and metallurgy.

3.1 NATURE OF SOLID SURFACES

The tribological behavior of a solid surface is directly influenced by its chemical and physical states. The most widely used materials for tribological systems are metals and alloys. The surface properties of materials change markedly in different environments; for instance, if a metal is placed in a vacuum and heated, the surface always liberates water. The top surface of bulk material is known to consist of several zones having different physicochemical characteristics particular to the bulk material itself.

Figure 3.2 shows a schematic of a metal surface. At the base of surface layer, there is a zone of work-hardened material on the top of which is a region of amor-

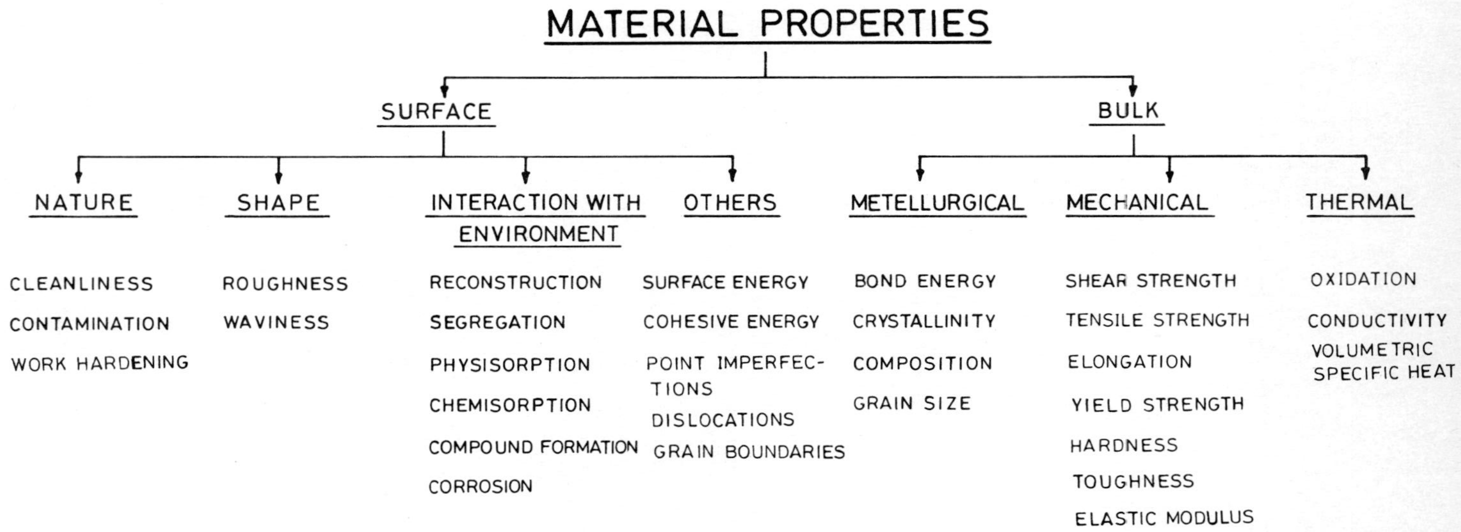

MATERIAL PROPERTIES

SURFACE

NATURE
CLEANLINESS
CONTAMINATION
WORK HARDENING

SHAPE
ROUGHNESS
WAVINESS

INTERACTION WITH ENVIRONMENT
RECONSTRUCTION
SEGREGATION
PHYSISORPTION
CHEMISORPTION
COMPOUND FORMATION
CORROSION

OTHERS
SURFACE ENERGY
COHESIVE ENERGY
POINT IMPERFEC- TIONS
DISLOCATIONS
GRAIN BOUNDARIES

BULK

METELLURGICAL
BOND ENERGY
CRYSTALLINITY
COMPOSITION
GRAIN SIZE

MECHANICAL
SHEAR STRENGTH
TENSILE STRENGTH
ELONGATION
YIELD STRENGTH
HARDNESS
TOUGHNESS
ELASTIC MODULUS

THERMAL
OXIDATION
CONDUCTIVITY
VOLUMETRIC SPECIFIC HEAT

FIGURE 3.1 Various surface and bulk properties that are relevant to tribological behavior of materials.

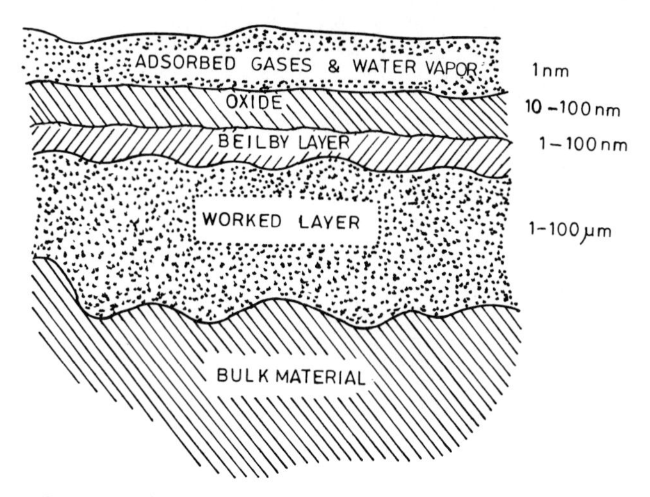

FIGURE 3.2 Schematic representation of a metal surface.

phous or microcrystalline structure. This is the so-called Beilby layer, which results because of melting and surface flow during machining of molecular layers which are subsequently hardened by quenching. On the top of the Beilby layer is an oxide layer, the formation of which depends on the environment, the amount of oxygen available to the surface, and surface oxidation mechanisms. For instance on heating a copper surface, the top layer is black, i.e., CuO, while an inner rose-colored layer is Cu_2O. Similarly oxides on an iron surface can be Fe_2O_3, Fe_3O_4, FeO, or their combinations.

The metallurgical effect of the work-hardened zone or worked layer (Fig. 3.3) shows a tapered section of a ground zinc metal surface (Samuels, 1960). At the base, the grain size is large, but near the surface, the grains are very small. The small grains result from recrystallization of the grains near the top surface. Furthermore, beneath the recrystallization zone, there is sufficient energy to bring about twinning in the individual grains of zinc. In a way, the energy produced during the grinding process is responsible for recrystallization near the top surface, where energy is maximum, and twinning the grains of the bulk material, where the energy is lower. The mechanical behavior of zones having grains of different sizes is significantly different.

On the top of the oxide layer, the whole multizone stack of surface contains a layer of adsorbate, which is generally water vapor or hydrocarbons from the environment that may have condensed and become physically or chemically adsorbed to the surface. In addition, the whole texture of the surface layer has a series of irregularities with different amplitudes and frequencies of occurrence.

3.2 THE SHAPE OF SURFACES

The surfaces used for machine components vary in initial surface topography, which depends upon the process of forming, e.g., casting, molding, cutting, and abrading. The geometrical texture of the surface is controlled by the characteris-

FIGURE 3.3 The section of ground zinc surface showing recrystallized surface layer and zone containing deformation twins, taper ratio 16.2. [*Adapted from Samuels (1960).*]

tics of the finishing process by which they are produced. A metal surface polished by fine aluminum oxide powder (particle size of 1 μm or less) looks *macroscopically* almost like a mirror. The same surface viewed *microscopically* is not smooth and contains surface irregularities called *asperities*. These surface irregularities (referred to as microroughness) are shown schematically in Fig. 3.4. A

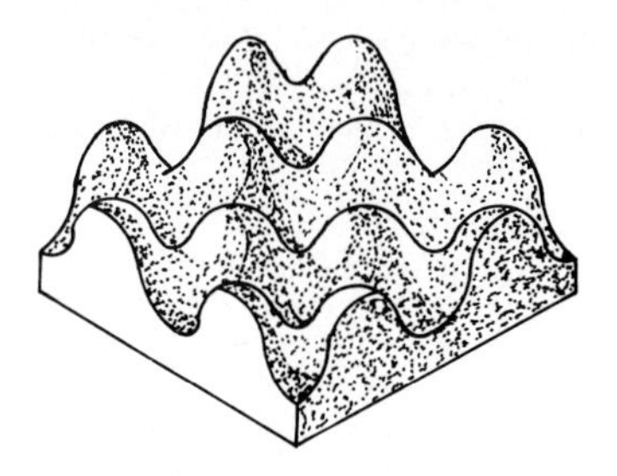

FIGURE 3.4 Surface asperities of a nominal smooth surface.

surface profile (Fig. 3.5) results mainly because of three different components of surface texture:

- *Roughness (or microroughness):* Surface irregularities of short wavelength, characterized by hills (or asperities) and valleys of varying amplitude and spacings that are larger than molecular dimensions. These create the predominant surface pattern, which exhibits those surface features intrinsic to the manufacturing process.
- *Waviness (or macroroughness):* Surface irregularities having much longer wavelength than microroughness. They often result from heat treatment, machine or workpiece deflections, vibration, or warping strains.
- *Errors of form:* Gross deviations from nominal shape. In general these deviations are not considered a part of the surface texture.

Numerous techniques to quantitatively measure surface roughness include optical methods using optical interference and light scattering and mechanical methods, such as oblique sectioning and stylus profilometry (Thomas, 1982). The op-

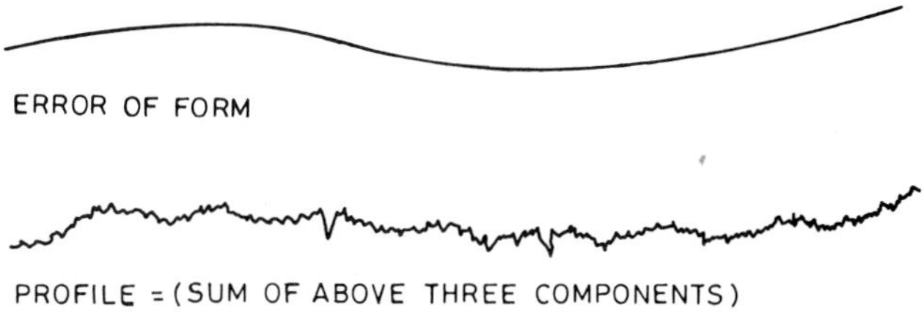

ROUGHNESS

WAVINESS

ERROR OF FORM

PROFILE = (SUM OF ABOVE THREE COMPONENTS)

FIGURE 3.5 The three components of surface texture.

tical techniques are noncontact and some of them provide three-dimensional measurements of the surface roughness (Bhushan, 1990). The various techniques for surface measurements are discussed in detail in Chap. 15.

3.3 PROPERTIES OF SURFACES

The rigidity of a solid arises because the atoms in the solid are held together by interatomic bonds. There are two types of solids: crystalline and amorphous. Amorphous solids lack any sort of regular pattern or array of atoms in the solids. In crystalline solids, however, there is a regular order or array of the atoms in the solids. Crystalline solids include most metals and alloys as well as inorganic solids, and even some lubricating species such as solid lubricants. We shall limit the present discussion to crystalline solids. The spatial arrangement of atoms in a crystalline solid is strongly influenced by the nature of the interatomic bonds, which in turn is influenced by the electronic structure of atoms.

3.3.1 Solid-State Bonding

Bonding in crystalline solids can be of four types, as shown in Fig. 3.6: (*a*) Van der Waals, (*b*) ionic, (*c*) metallic, and (*d*) covalent.

The *Van der Waals bond* is the weakest bond holding solids together. Van der Waals forces are attributed to nothing more than fluctuations in the charge distribution within the atoms or molecules. These forces can be represented in bonding the atoms of an inert gas, e.g., argon together when solidified. A very small amount of energy is required for sublimation. Van der Waals bonding is frequently encountered in tribological systems with the physisorption of liquid lubricants and gases to a solid surface.

Ionic bonding occurs in solids that are not elements but rather compounds of one element with one or more other elements such as NaCl (ordinary salt) as in Fig. 3.6(*b*). Ionic bonding forms between two oppositely charged ions, which are produced by the transfer of electrons from one atom to another. Ionic bonding can be very strong because there is an electron charge transfer. A classic example of this is aluminum oxide, a tribological material which has inherently high strength.

In case of *metallic bonds*, the sharing of electrons between neighboring atoms becomes delocalized, as there are not enough electrons to produce the inert gas configuration around each atom. The metallic state can be visualized as an array of positive ions, with a common pool of electrons, the so-called free electron cloud. The electrons move freely between the cores of ions allowing for, and contributing to, many of the metallic materials properties. This freedom is responsible for good thermal and electrical conduction properties of metals. With metals, a wide range of strength can be achieved: for instance, the weak binding energies associated with indium and thallium and the strong binding energies associated with osmium and tungsten.

In the fourth type, *covalent bonding*, sharing of electrons between neighboring atoms takes place. In Fig. 3.6(*d*), neutral atoms appear to be bound together by overlapping their electron distribution. Covalent bonding produces an extremely strong bond as evidenced by the high melting points and the high hardness of the material, such as the diamond. At the same time, the covalent bond found in organic molecules in polymers and lubricants is relatively weak.

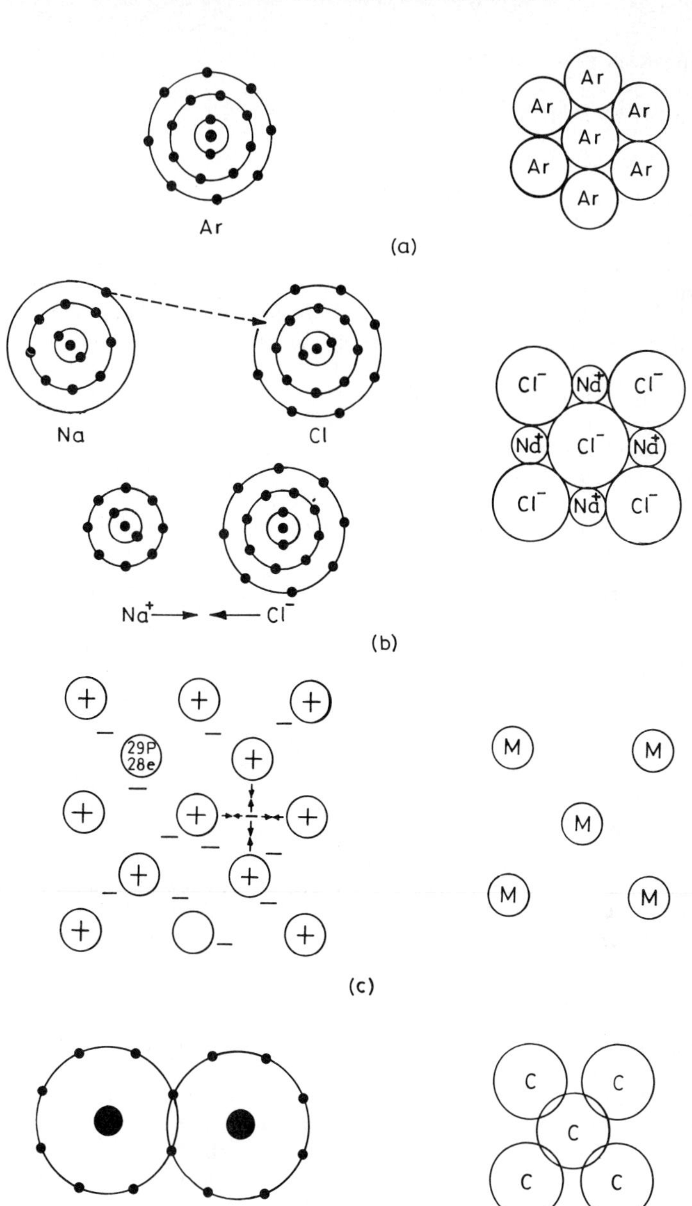

FIGURE 3.6 Main types of crystalline bonding: (*a*) Van der Waals (crystalline Ar), (*b*) ionic (NaCl), (*c*) metallic, and (*d*) covalent (diamond).

3.3.2 Crystalline Structure

An important aspect of any tribological material is its structure, because its properties are closely related to this feature. A crystal is defined as an orderly array of atoms in space. Generally the materials consist of an aggregate of many very small single crystals. Therefore, such materials are known as polycrystalline and the crystals in these materials are referred to as *grains*.

In crystals, the atoms tend to bond with one another in a particular arrangement or array which is repeated throughout the crystal. The atomic density varies in different directions, called *atomic planes*, through the crystal. In other words, the solid is made up of the individual atoms bonded within the crystalline planes, and the planes are bonded to each other throughout the solids. Thus the structure of a crystal can be represented by a unit cell because the entire lattice can be derived by repeating this cell as a unit.

Figure 3.7 shows the atomic arrangements in simple cubic, body-centered cubic (bcc), face-centered cubic (fcc), and hexagonal close-packed (hcp) structures. Materials such as metals, alloys, ceramics, and solid lubricants are crystalline with their atoms or molecules arranged or packed in accordance with a particular structure. For instance, metals such as copper, nickel, silver, gold, platinum, and aluminum have a face-centered cubic (fcc) structure; iron, tantalum, niobium, vanadium, and tungsten have a body-centered cubic (bcc) structure; and zinc, cobalt, rhenium, zirconium, and titanium have a hexagonal close-packed (hcp) structure. The atomic positions, characteristics, and typical examples of simple cubic, bcc, fcc, hcp, and diamond single crystals are listed in Table 3.1.

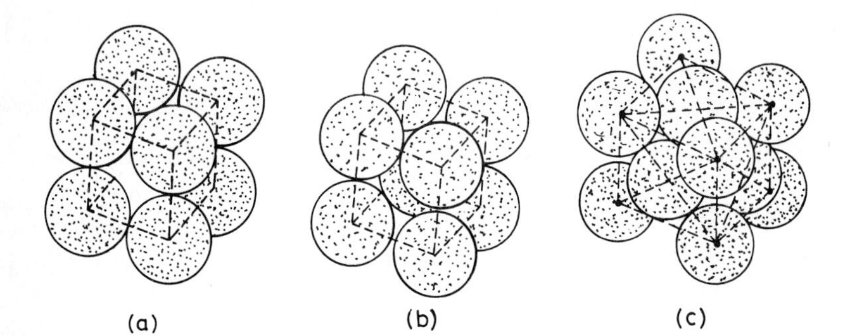

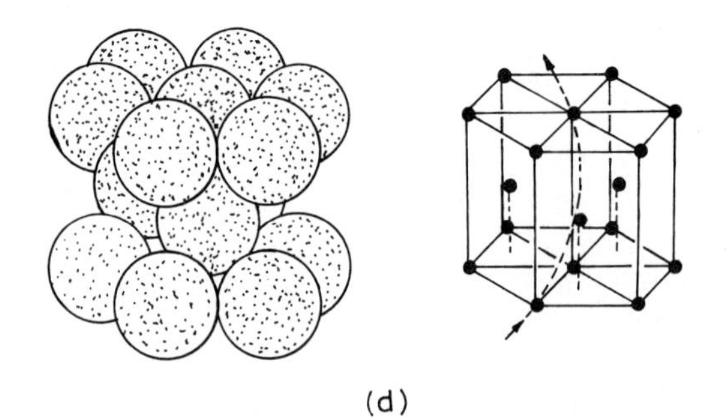

FIGURE 3.7 Atomic arrangements in (*a*) simple cubic, (*b*) bcc, (*c*) fcc, and (*d*) hcp unit cell structures.

TABLE 3.1 Crystal Structures, Atomic Positions and Some Characteristics, and Examples

Crystal structure	Atomic positions and characteristics	Examples
Simple cubic	(i) Only corner positions are occupied. (ii) There is 1 atom per unit cell. (iii) Each atom is surrounded by 6 nearest neighbors.	NaCl, MgO
bcc	(i) The atoms occupy corner (0 0 0) and body-centered (½ ½ ½) positions. (ii) There are 2 atoms per unit cell. (iii) Each atom is surrounded by 8 nearest neighbors.	Fe, Cr, Mn, W, Mo, V, Nb, Ta
fcc	(i) The corner atoms are (0 0 0) and the face-centered positions are (½ ½ 0), etc. (ii) There are 4 atoms per unit cell. (iii) Each atom is surrounded by 12 nearest neighbors.	Pb, Cu, Ag, Al, Pd, Pt, Sr, Ca
hcp	(i) The space lattice is simple hexagon with 2 atoms (0 0 0, ⅔ ⅓ ½) associated with lattice point. (ii) Hexagonal indices with the four axes (a_1 a_2 a_3 c) are used: (0 0 0 1) indicates the c-axis. (iii) The c/a ratio for close packing of spheres is $(8/3)^{1/2} = 1.633$. (iv) Each atom is surrounded by 12 nearest neighbors.	Ti, Zr, Hf, Cd, Zn, Y Co, Te, Se
Diamond	(i) The unit cell has cell constant a and comprises octants, four of which have the atoms arranged as shown here. The space lattice is fcc with a basis of two atoms at (0 0 0) and (¼ ¼ ¼) associated with each lattice point. (ii) The bonding is tetrahedral and each atom is surrounded by 4 nearest neighbors. (iii) There are 8 atoms per unit cell.	Diamond

In both fcc and bcc systems, there are three principal planes, i.e., (100), (110), and (111) which are referred to when discussing the crystalline nature of solids. As shown schematically in Fig. 3.8 these atomic planes move relative to each other when the crystal is deformed plastically and are therefore referred to as slip planes. Under applied stresses, the (110) set of planes is commonly observed to slip over one another in bcc structure. The (111) planes are those upon which slip and cleavage in fcc materials is most frequent (Buckley, 1981).

Figure 3.9(a) shows the packing arrangements for three atomic planes in the face-centered cubic system shown in Fig. 3.8. The close packing is a way of arranging equidimensional objects in space so that the available space is filled very efficiently. The notation (110) in Fig. 3.9(a) shows the direction in which atomic planes move relative to each other. From Fig. 3.9(a) it is obvious that (111)

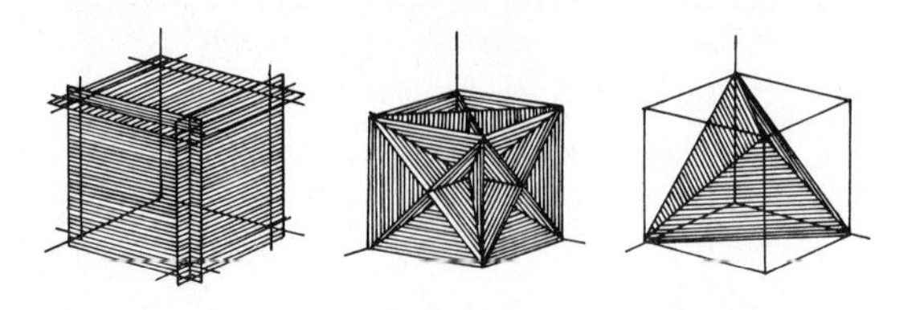

FIGURE 3.8 Planes of possible slip in bcc and fcc crystals: (*a*) three (100) planes, (*b*) six (110) planes, and (*c*) four (111) planes.

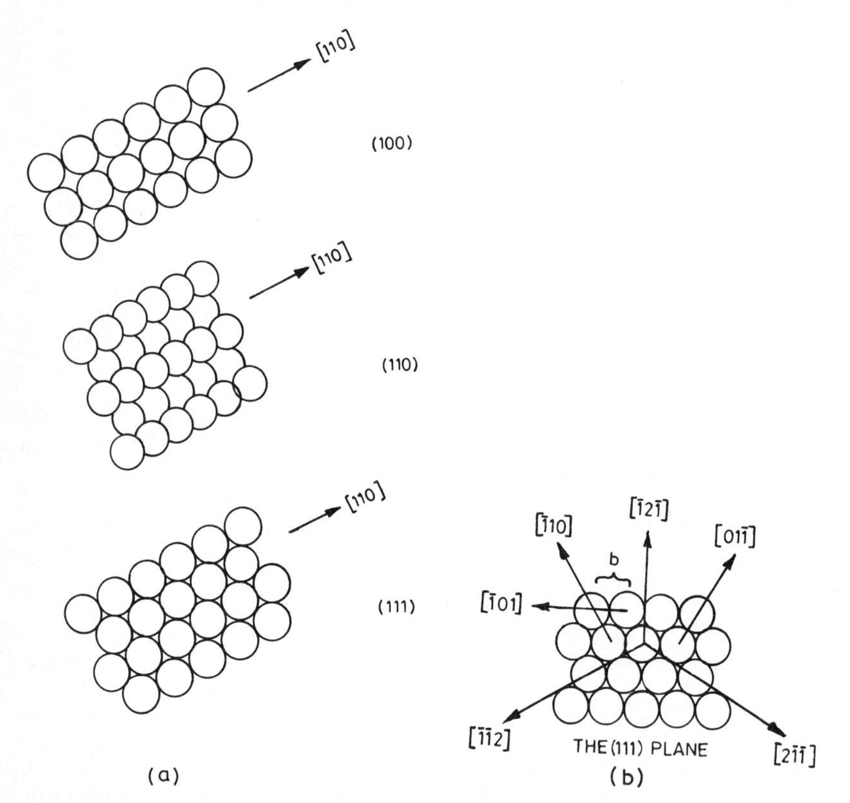

FIGURE 3.9 Atomic packing on various crystallographic planes and in various directions for fcc materials: (*a*) appearance of several crystal planes, and (*b*) the (111) planes.

planes have the closest atomic packing, which results in the lowest surface energy in the fcc system; in turn (111) planes are least likely to interact chemically with environmental constituents. On the other hand, (110) planes in the fcc system are least densely packed, exhibit higher surface energy, and therefore are more reactive to their environment. The elastic modulus and microhardness of

(110) planes will also be lower than (111) planes (Buckley, 1981). It is also observed that the packing density of a particular plane, say (111), also varies significantly with direction, shown in Fig. 3.9(*b*), which affects the surface energy and other properties. For further details, refer to standard texts (Azaroff, 1960; Barrett, 1943; Kittel, 1966; Phillips, 1956; and Van Vlack, 1985).

3.3.3 Surface Defects

In Sec. 3.3.2, it was implied that the regular crystalline order described by a unit cell exists throughout an entire crystal. In practice, numerous slight deviations from the ideal atomic array occur in most crystalline materials. The actual solids have grain boundaries which develop during solidifications as large defects which exist in the solids and extend to the surface. The grain boundaries do not exhibit a regular structure; they are highly active regions and on the surface are very energetic.

Crystalline imperfections can be classified on the basis of their geometry as follows: point imperfections, line imperfections, surface imperfections, and volume imperfections. Volume imperfections can be foreign-particle inclusions, large voids or pores, or noncrystalline regions which have the dimensions of at least a few nanometers.

3.3.3.1 Point Imperfections. Point imperfections are referred to as zero-dimensional imperfections. Figure 3.10 shows different types of point imperfections that can occur in crystals.

A vacancy refers to an atomic site from which the atom is missing, Fig. 3.10(*a*). The substitutional impurity refers to a foreign atom that substitutes for or replaces a parent atom in the crystal, Fig. 3.10(*b*). In an interstitial impurity a small sized atom occupies a void space in the crystal lattice, without dislodging any of the parent atoms from their sites, Fig. 3.10(*c*).

The presence of a point imperfection introduces distortions in the crystal. For example, if the imperfection is a vacancy, the bonds that the missing atom would have formed with its neighbors are not present. In a substitutional imperfection, a larger atom introduces compressive stresses and corresponding strains around it, while a smaller atom creates a tensile stress-strain field. All these imperfections tend to increase the enthalpy of the crystals.

3.3.3.2 Line Imperfections. Line imperfections are called *dislocations*. Since the concept of dislocations was put forward in the 1930s, the static and dynamic properties of dislocations have been studied in great detail, because of the vital role they play in determining the structure-sensitive properties of crystalline materials.

Figure 3.11 shows two types of dislocations, i.e., the edge dislocation and the screw dislocation. An edge dislocation viewed end on, as in Fig. 3.11(*a*), corresponds to the edge of an atomic plane that terminates within the crystal rather than passing all the way through. In the imperfect crystal just on the right, just above the edge of the incomplete plane the atoms are squeezed together and are in a state of compression. The bond lengths have been compressed to smaller than the equilibrium value. Just below the edge, the atoms are pulled apart and are in a state of tension. The bond lengths have been stretched to above normal values.

When the imperfect crystal shown in Fig. 3.11(*a*) is subjected to sufficiently large shear forces directed perpendicular to the dislocations, the dislocation will move as shown in Fig. 3.12. Thus, because of the applied force, the edge dislo-

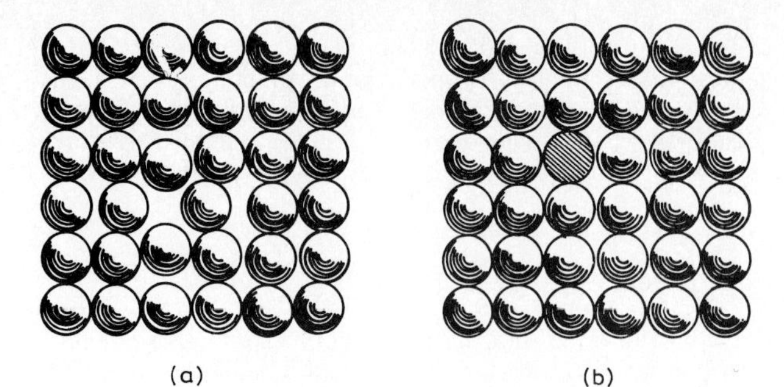

FIGURE 3.10 Point imperfections in a crystal: (*a*) vacancy, (*b*) substitutional impurity, and (*c*) interstitial impurity.

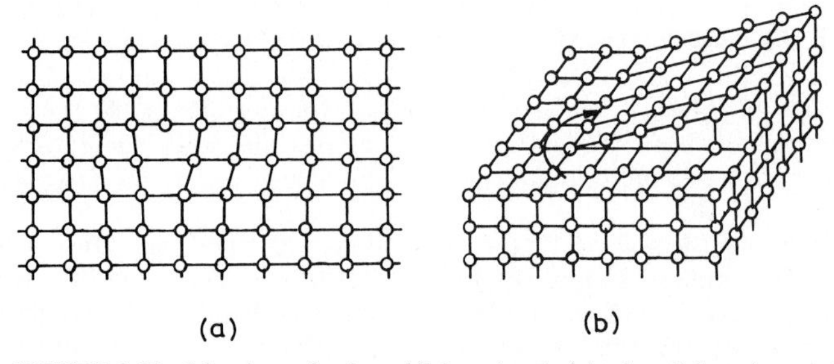

FIGURE 3.11 Line imperfections (dislocations): (*a*) edge dislocation, and (*b*) screw dislocation.

cation has moved, and through repetition of this process can continue to move or slip. If the dislocation moves through an entire crystal [Fig. 3.12(*c*)], then the upper half of the crystal shall be shifted by one slip vector (Burger vector $\vec{b}$) or interatomic spacing relative to the lower half. Plastic deformation occurs through the motion of dislocations, which requires much smaller forces than those necessary to permanently deform a perfect crystal.

The screw dislocation is shown in Fig. 3.11(*b*) as a step in the surface generated

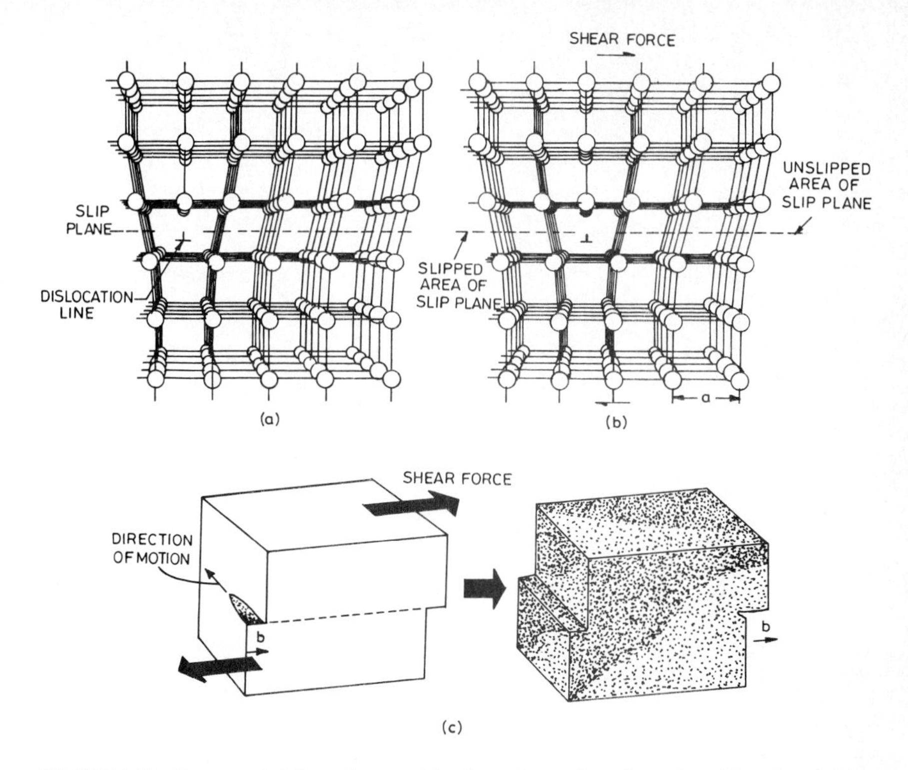

FIGURE 3.12 Permanent deformation resulting from the motion of an edge dislocation: (*a*) before, and (*b*) after slip by an applied shear force, and (*c*) permanent offset to the top half of the crystal with respect to the bottom half.

by the emergence of both a screw dislocation and a line defect perpendicular to the solid surface. The atomic bond in the region immediately surrounding the dislocation line has undergone a shear distortion. These line defects are readily discernible in a solid surface by examination in the field ion microscope (Buckley, 1981).

3.3.3.3 Surface Imperfections. Surface imperfections are two dimensional in the mathematical sense. They refer to regions or distortions that lie about a surface having a thickness of a few atomic diameters. The external surface of a crystal is an imperfection in itself, as the atomic bonds do not extend beyond the surface. This directly affects the surface energy. In addition to the external surface, crystals may have surface imperfections inside. Usually most materials consist of a number of crystals and are said to be polycrystalline. During solidification or recrystallization, new crystals that are randomly oriented with respect to one another form in different parts of materials. They grow by the addition of atoms from the adjacent regions and eventually impinge on each other. When two crystals impinge in this manner, the atoms that are caught in the two are being pulled by each to join its own configuration. Because of the opposing forces, they can join *neither* crystal and therefore take up a compromised position. These positions at the boundary region between two crystals are so distorted and unrelated to one another that we can compare the boundary region to an amor-

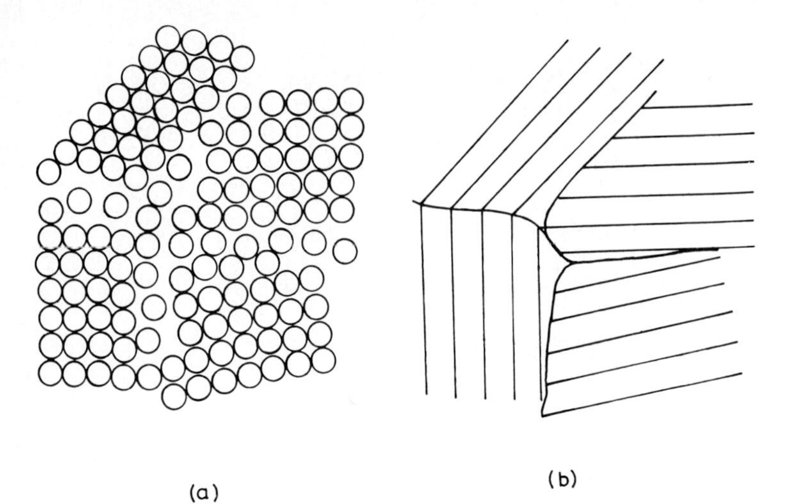

FIGURE 3.13 The atomic arrangements at grain boundaries are distorted and unrelated.

phous material. The boundary region is called a crystal boundary or *grain boundary* (Fig. 3.13).

The crystal orientation changes sharply at the grain boundary. The average number of nearest neighbors for an atom in the boundary or close-packed crystal is eleven, as compared to twelve in the interior of the crystal. Microhardness is generally higher at grain boundaries.

The grain boundary between two crystals that have different crystalline arrangements or differ in composition is given a different name—an *interphase boundary* or an *interface*. When the orientation difference between two crystals is less than 10°, the distortion in the boundary is not so drastic as to be compared with a noncrystalline material. Such boundaries have a structure that can be described by means of arrays of dislocations and are called low-angle boundaries. Figure 3.14 shows a low-angle tilt boundary, where neighboring crystalline regions are tilted with respect to each other only by small angles.

3.4 SURFACE INTERACTIONS WITH ENVIRONMENT

The mechanism of lubrication is partly dictated by the nature of interaction between the lubricant and the solid surface. The many interactions (Fig. 3.15) include:

- *Reconstruction:* This takes place when the outermost layers of atoms of the solid surface undergo a structural change [Fig. 3.15(*a*)].
- *Segregation:* In solids containing more than a single element, atoms from bulk can diffuse to the surface and segregate there [Fig. 3.15(*b*)].

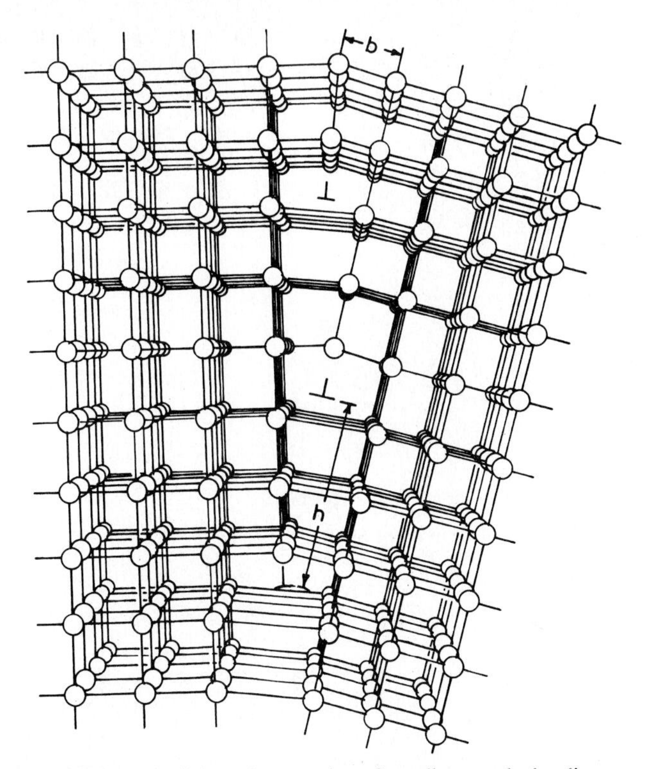

FIGURE 3.14 A tilt boundary consists of equally spaced edge dislocation of the same sign one above the other. $\vec{b}$ is the Burger vector and h is the vertical spacing between two neighboring edge dislocations.

- *Physisorption:* This attracts long-chain molecules in the liquid to the surface because of van der Waals-type electrostatic force [Fig. 3.15(*c*)].
- *Chemisorption:* This interaction involves strong chemical bonds between the liquid components and the solid surface [Fig. 3.15(*d*) and (*e*)].
- *Chemical reactions:* These form new compounds on the surface [Fig. 3.15(*f*)].

3.4.1 Surface Reconstruction

In this interaction, the top layer of solid surface undergoes a change in its structure. Miyoshi and Buckley (1982) observed this effect for the surface of silicon carbide. On heating silicon carbide above 800°C, silicon-to-carbon bond scission occurred and the silicon evaporated from the surface leaving the carbon behind. The carbon in the outermost layers collapsed into a pseudo-graphitic structure. The surface reconstruction resulted in a remarkable change in the coefficient of friction for the silicon carbide—from approximately 0.8 before reconstruction to 0.2 thereafter.

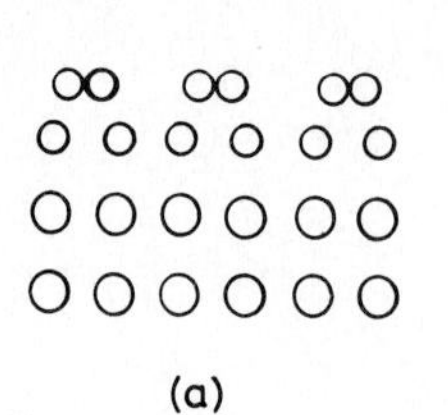

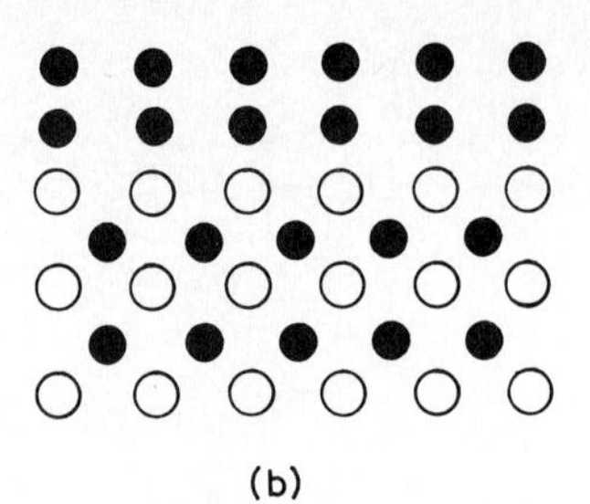

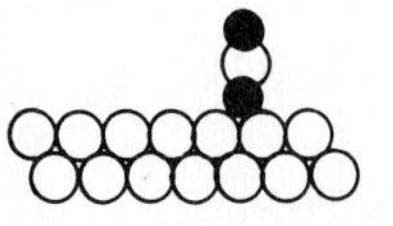

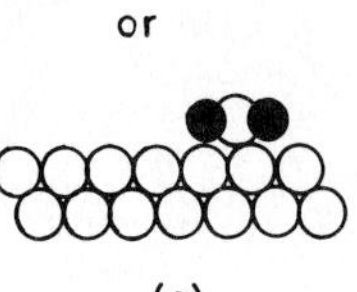

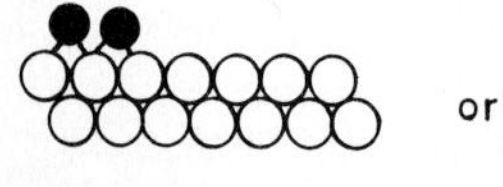

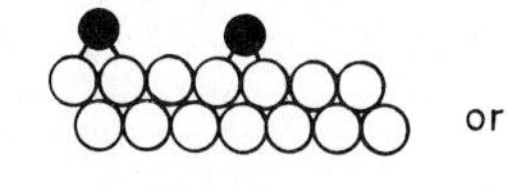

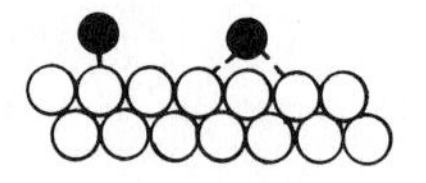

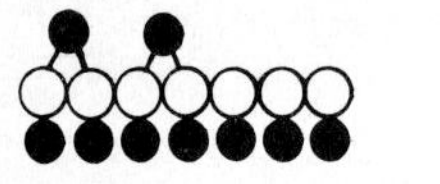

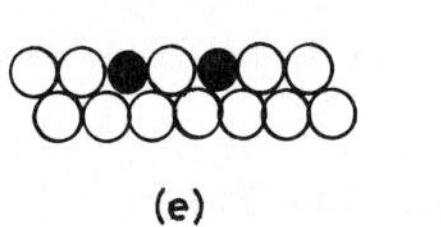

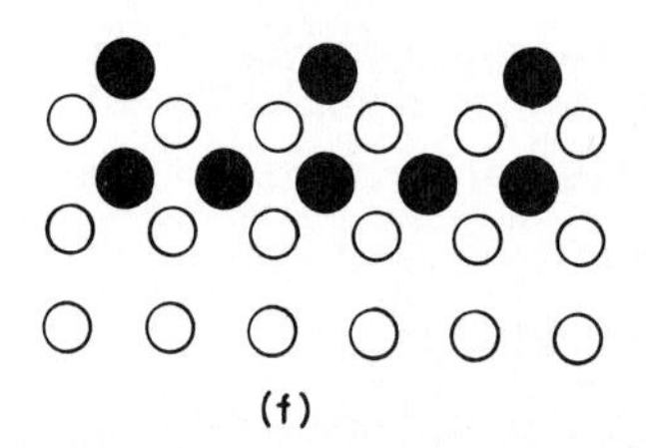

FIGURE 3.15 Schematic diagrams of various surface interactions: (*a*) reconstruction, (*b*) segregation, (*c*) physisorption, (*d*) chemisorption, (*e*) chemisorption with reorganization, and (*f*) compound formation.

3.4.2 Segregation

The segregation of alloy species to grain boundaries in alloys and its effect on mechanical properties have been known to metallurgists for some time (McLean, 1957). The same segregation occurs on solid surfaces, and segregation has been observed to exert considerable influence on adhesion, friction, and wear. In a simple binary alloy, solute atoms can diffuse from near surface regions to completely cover the surface of the solvent. This has been observed for many binary systems including aluminum in copper, tin in copper, indium in copper, aluminum in iron, and silicon in iron.

One hypothesis for the segregation mechanism is that the solute segregates on the surface because it reduces the surface energy. A second hypothesis is that solid produces a strain in the crystal lattice of the solvent, and this unusual lattice state ejects solute atoms from the bulk. The segregation of alloying elements towards the grain boundaries influence the surface energy of an interface which has a direct effect on the energy of adhesion. Segregation effects in tribology have been studied for copper-and-iron-based alloys within some instances, because of segregation, as small as 1 at. % of alloying element completely modify adhesion behavior (Buckley, 1969; Ferrante and Buckley, 1972). The segregation process is found to be irreversible; i.e., once alloying elements segregate to the surface, they remain there for some alloys such as carbon-to-iron surface, sulfur-to-iron surface, aluminum to the surfaces of both iron and copper, and indium and tin to copper surfaces. Miyoshi and Buckley (1982) observed segregation of boron to ferrous alloys surface.

3.4.3 Physisorption

Clean surfaces are extremely active chemically. The potential energy of an atom at the surface of a solid in vacuum is greater than that of a similar atom within the body, because an equilibrium interaction is established between all the atoms in a three-dimensional crystal which cannot be achieved at the surface. There is a smaller number of neighboring atoms at the surface, so each atom is less strongly bound and an excess free energy is stored. If a gas is admitted to the vacuum, the free energy can be lowered by adsorption. This occurs spontaneously, and the adsorbed particles are either absorbed at localized sites or retain mobility over the surface. Because surface atoms have this excess energy, the atoms can interact with each other, with other atoms from the bulk, and with species from the environment.

One of the most common types of surface interactions that can take place with a clean surface is the physical adsorption (physisorption) of species on that solid surface. The physical adsorption process, a relatively weak process, is shown in Fig. 3.15(*c*). The molecule shown in Fig. 3.15(*c*), for physisorption, bonding itself to the surface, is shown as a diatomic molecule such as might occur in oxygen (O_2). In such a case both oxygen atoms of diatomic molecules can bond to the already contaminated surface. However, a very small amount of energy is required to remove the physisorbed atoms. This process typically involves Van der Waals forces. It is observed that, if the interaction involves less than 10 kilocalories per mole, the process is one of physisorption. If, however, the energy involves in excess of 10 kilocalories per mole, the adsorption may be due to chemisorption.

The strength of physisorption depends on the electronic structure. Polar molecules (those in which there is a variation in electronic charge along the length of the molecule) tend to adsorb with their molecules perpendicular to the surface. A useful byproduct of physisorption is the so-called Rehbinder effect—the reduction of the modulus and yield stress of metals as well as nonmetals in the presence of an adsorbed film (Rehbinder and Likhtman, 1957; Ciftan and Saibel, 1979). As a result of this effect, lower stress is developed when asperities collide. In particular, during the running-in of new bearing surfaces, the removal of excessive asperities will take place more mildly.

3.4.4 Chemisorption

Chemisorption is a monolayer process (Hayward and Trapnell, 1964). When the individual gaseous atoms interact with a solid surface, chemisorption occurs and the atomic species become bonded to the solid surface [Fig. 3.15(*d*)]. Chemisorption is a much stronger bonding than that associated with physisorption. Bond strengths are a function of chemical activity of the solid surface, i.e., surface energy, degree of surface coverage of that adsorbate or another adsorbate, reactivity of the adsorbing species and its structure. The higher the surface energy of the solid surface, the stronger the tendency to chemisorb. In general, the high energy, low-atomic density crystallographic planes will chemisorb much more rapidly than will the high-atomic density, low-surface energy planes. Further, in case of copper, silver, and gold, it is observed that oxygen will chemisorb relatively strongly to copper, weakly to silver, and not at all to gold. The reactivity of adsorbent is also important; for instance, fluorine will adsorb more strongly than chlorine; chlorine more strongly than bromine; and bromine more strongly than iodine.

The adhesion of contacting solid surfaces is extremely sensitive to minor nuisance in the structure of adsorbed species. For instance, minor modifications in the hydrocarbon structure can produce considerable differences in measured adhesive forces. This is demonstrated in Table 3.2. Simply increasing the number of carbon atoms from one to two produces a significant change in adhesive behavior for a clean iron surface containing a chemisorbed layer of each of the hydrocarbons. Further, changing the degree of bond saturation has an effect, clearly obvious from Table 3.2, from the data of ethane, ethylene, and acetylene. In addition changing the surface active species in the two-carbon-atom structure can produce significant differences. This effect is demonstrated with the comparison results for vinyl chloride and ethylene oxide. We also note that both the quality of the adsorbed species on a solid surface and its concentration change adhesion.

TABLE 3.2 Effect of Various Hydrocarbons on Adhesion of Clean Iron

Hydrocarbon chemisorbed to iron	Adhesive force, dynes*
Clean Fe	>400
Ethane C_2H_6	280
Ethylene $H_2C=CH_2$	170
Acetylene $HC\equiv CH$	80
Vinyl chloride $H_2C=CHCl$	30
Ethylene oxide $H_2C\overset{O}{—}CH_2$	<10

*Load = 20 dynes, 10^{-10} torr, 20°C, both surfaces (001) planes.
Source: Adapted from Buckley (1985).

3.4.5 Compound Formation

Compound formation on solid surfaces plays an important role in tribological systems [Fig. 3.15(*f*)]. Physisorbed and chemisorbed films are very effective in reducing friction and mild wear under moderate rubbing. They fail fairly easily under severe rubbing conditions and therefore are not very effective in preventing wear or seizure. The naturally occurring oxides present on metals prevent their destruction during rubbing, but once they are removed, the reoxidation of the surface may be too slow to be effective.

With metals in contact with both metals and nonmetals, compound formation by chemical reaction has been observed to occur on the solid surface. Examples are gold on tungsten (Brainard and Buckley, 1971), gold on silicon (Mouttet et al., 1981), and phosphorus on iron (Egert and Panzner, 1982). The compound formation produces strong interfacial bonds at the contacting surfaces and influences adhesion behavior. Similarly more active chemicals can be added to the lubricants to react with rubbing surfaces and produce protective films. Suitable films include chlorides, iodides, sulfides, phosphides and phosphates, and any other chemical which will react with the rubbing surface to produce such substance will be effective in producing protective films.

3.5 PHASE EQUILIBRIA

An understanding of phase equilibria and the ways to represent them is very important for comprehending many tribological material systems. Various metals and nonmetals are mixed in different ratios to tailor properties such as hardness, toughness, elastic modulus, flow pressure, microstructure, and surface energy required for specific tribological applications. *Phase diagrams* provide a convenient way of representing which state of aggregation is stable for a particular set of conditions. Moreover, a phase diagram is a map that shows which phase is stable and not just the state of aggregation (Cottrell, 1975; Gulyaev, 1980; Guy, 1959; Lakhtin, 1977; Neely, 1984; Pascoe, 1978; Reed-Hill, 1973; Rhines, 1956; Rollason, 1973).

The components of an alloy either may be dissolved one in the other in the solid state to form solid solutions or chemical compounds, or after solidifying they may produce mechanical mixtures of the two components A and B solid solutions, etc. In the formation of a solid solution, atoms of the dissolved component (the solute) will substitute for a part of the atoms in the lattice of the solvent or they will be accommodated in the interstices (Goldschmidt, 1967). In most cases, the components of an alloy have limited mutual solubility in the solid state. Many metals, however, are mutually soluble to an unlimited extent. If the components are present in an amount exceeding the solubility limit, the surplus atoms form an independent phase comprising either a saturated solid solution or a chemical compound or finally crystals of one of the components.

A chemical compound differs from a solid solution in that chemical compound often has its own crystal lattice and in many cases a definite quantitative relation of the components is required for its formation. The transition of an alloy from the liquid to the solid state is associated with the conversion of a system to a state of less free energy. As in the solidification of pure metals, the transition will be initiated only if the alloy is supercooled to some extent and will then proceed by nucleation and subsequent growth of the nuclei as dendrites. In actual conditions, the crystals produced in the solidification of metal alloys are usually dendritic. When alloys solidify, the newly formed phases differ substantially in composition from the initial liquid solution. Therefore to form a stable nucleus or embryo, it will not be sufficient to have fluctuations of energy, as in a single-component system; there must also be a fluctuation of concentration which corresponds to zones not less than critical in size.

The solidification of metal alloys is clearly demonstrated by means of equilibrium diagrams which are convenient graphic representations of change in state due to variations in temperature and concentration. Equilibrium diagrams are also called constitutional and phase diagrams which enable the phase content of the alloy to be determined at any temperature and composition. A phase is a physically or chemically homogeneous portion of a system, separated from the other portion by a surface, the interface. All changes which take place in a system consisting of several phases, in accordance with external conditions (temperature and pressure), conform to the phase rule, which establishes the relationship between the number of degrees of freedom, the number of components, and the number of phases. It can be expressed as

$$F = C + (n - P) \tag{3.1}$$

where F = degrees of freedom in system (number of variable factors, e.g., temperature, pressure, and concentration)
C = number of components in system
P = number of phases in equilibrium
n = number of external factors (e.g., temperature and pressure)

Two-component systems have binary phase diagrams. Apart from temperature and pressure, we have one composition variable for each of the phases in equilibrium. We then need a three-dimensional diagram to plot the variations in pressure, temperature, and composition.

To simplify the representation of the phase relationships on paper, binary phase diagrams are usually drawn at atmospheric pressure, showing variations in temperature and composition only. The simplest binary phase diagram is obtained for a system exhibiting complete liquid solubility as well as solid solubility. The two components dissolve in each other in all proportions both in the liquid and solid states. Clearly the two components must have the same crystal structure, besides satisfying the other Hume Rothery's conditions and extensive solid solubility. Cu-Ni, Ag-Au, and Al_2O_3-Cr_2O_3 are examples of such systems. Figure 3.16 shows the phase diagram of Al_2O_3-Cr_2O_3, which has only two phases, the liquid and the solid. The single-phase regions are separated by a two-phase region (L + S) where both liquid and solid coexist. When only one phase is present, the composition axis gives the composition of that phase directly. When two phases are present, the composition of the phases is not the same.

When the solubility is limited and the melting point of the components is not vastly different, a eutectic phase diagram usually results. For instance, a Pb-Sn phase diagram is shown in Fig. 3.17. As there is complete liquid solubility, the liquid phase extends over all compositions above the melting temperatures of the components. The solid phase at the left end is the lead-rich α, which dissolves only a limited amount of tin. The solubility decreases with decreasing temperature. This limit is the solid solubility and is indicated by the phase boundary between α and $\alpha + \beta$, called *solvus*. The solid solution phase at the right end

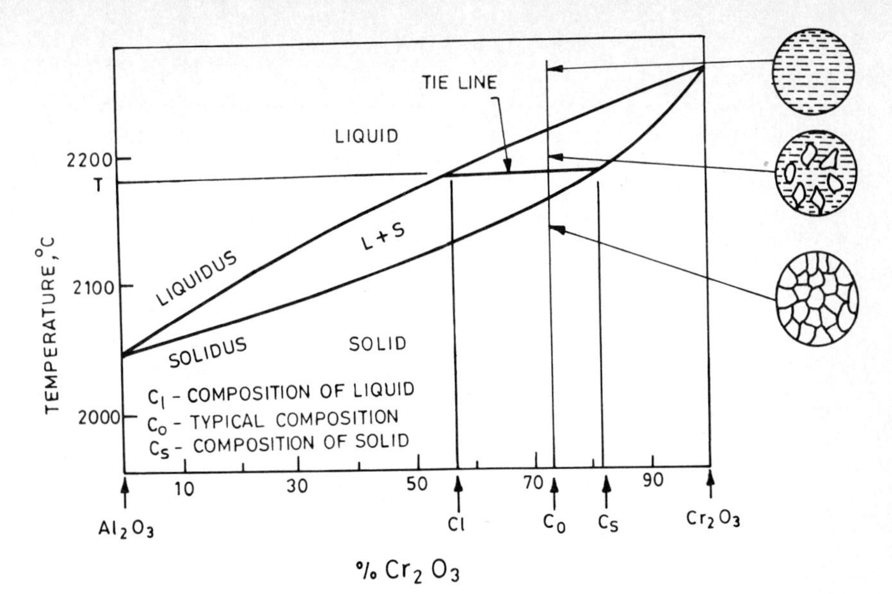

FIGURE 3.16 The Al_2O_3-Cr_2O_3 phase diagram. Microstructure changes shown for a composition compound with 28% Al_2O_3 and 72% Cr_2O_3.

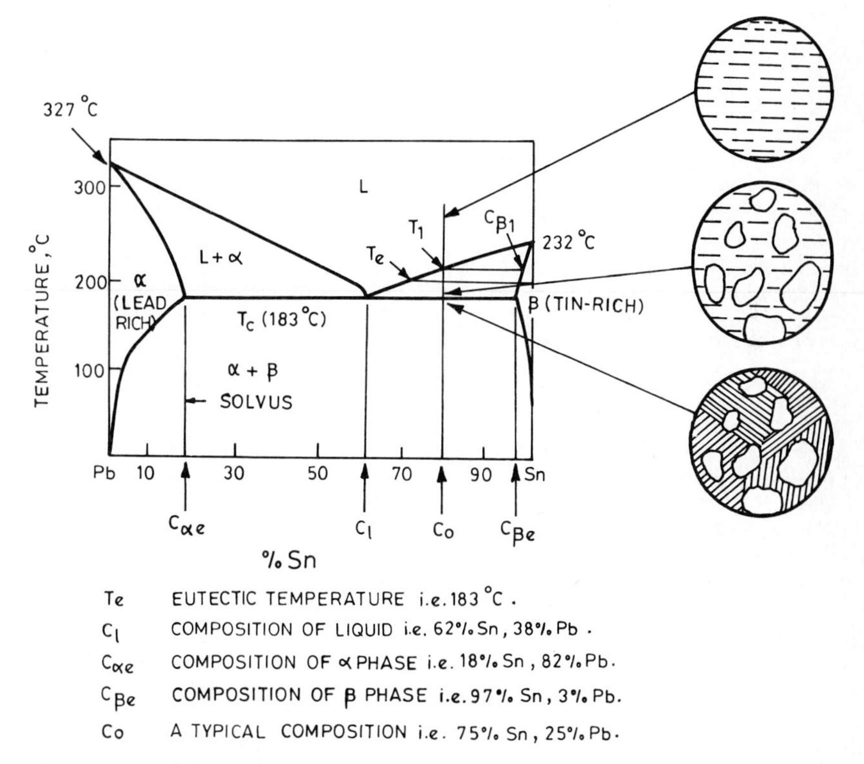

FIGURE 3.17 The Pb-Sn phase diagram. Microstructure changes shown for a typical composition compound, i.e., 75% Sn, 25% Pb.

is the tin-rich β, with only a very small quantity of lead dissolved in it. The three two-phase regions are separated by a horizontal line corresponding to the temperature T_e called the eutectic temperature, below which the material is fully solid for all compositions.

A phase diagram may look somewhat complicated if the system has more than one invariant reaction, as for example, the iron-carbon phase diagram presented in Chap. 4.

3.6 MICROSTRUCTURAL EFFECTS ON FRICTION AND WEAR

The microstructure of a material has an important role in physical properties, e.g., elastic modulus, hardness and toughness, and chemical properties, which in turn affect the friction and wear properties. It has been well established that the elastic modulus, hardness and the topography of the surface affect the number of asperity contacts and the size of each individual contact (Bhushan, 1984, 1985, 1990; Greenwood and Williamson, 1966), in addition to the resistance to deformation of the surface layer. In the case of metals, the crack nucleation and crack propagation rate are considerably affected by the toughness, which also affects the wear rate. Physical and chemical properties, such as surface energy, affect the adhesion at the interface which in turn affect friction and wear (Buckley, 1981).

This discussion attempts to explain how the microstructure affects certain basic properties and how these basic properties in turn affect the friction and wear behavior of materials (Latanision, 1980; Rabinowicz, 1980, 1984; Saka, 1980; Zum Gahr, 1987). Figure 3.18 presents a classification of metals based on binary phase diagrams. Metals can be classified into pure metals, solid solutions, two-phase alloys, and composites. Pure metals can be classified further in terms of crystal structure, substructure, grain structure, etc. Similarly solid solutions can be classified as ordered and disordered. Two-phase metals can be classified in terms of precipitates, size, shape, and particle-matrix interface. Composites can be classified in terms of size, shape, orientation, and connectivity of various phases. In composites, volume fractions of various phases are comparable. Eutectics and eutectoids belong to composites.

Since friction and wear phenomena involve two bodies in contact, they are functions of the microstructure of both the materials. As the friction and wear are not reversible processes, the initial microstructure is altered substantially during sliding to yield a completely different steady-state microstructure. The steady-state microstructure is a function of not only the initial microstructure but also the friction and wear behavior of materials in the transition period. Thus one must follow the microstructural changes from the time the materials are brought into contact until a steady-state is reached. Further, as friction and wear are surface phenomena, large gradients in microstructure are possible. Such gradients may affect some basic properties during sliding. For instance, migration of solute can take place due to substructure gradients which may affect the surface energy which in turn affect the friction coefficient and wear rate. It is not straightforward to correlate the friction and wear of materials to microstructure, but an attempt is made here.

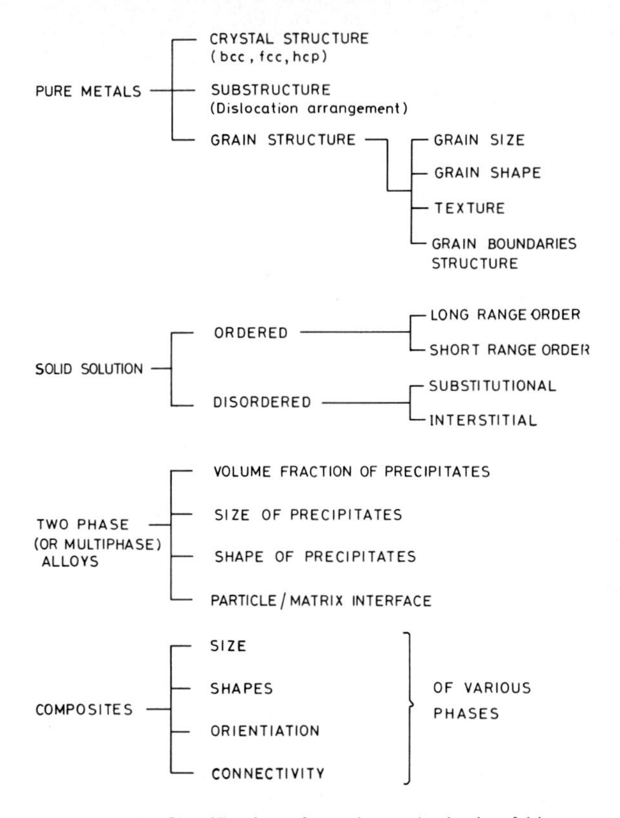

FIGURE 3.18 Classification of metals on the basis of binary phase diagrams.

3.6.1 Effect of Microstructure on Friction

According to Rabinowicz (1971), the effect of microstructure on friction coefficient is essentially through its influence on work of adhesion of the contacting materials W_{12} and hardness' H ratio. Equation (2.8) proposed by Rabinowicz (1965), which relates coefficient of friction to W_{12}/H, can be rewritten, based on experimental evidence, as

$$\mu_a \sim 0.3 + C_1 \frac{W_{12}}{H} \tag{3.2}$$

where C_1 is a constant for constant geometry. Since the maximum value of W_{12} is $(\gamma_1 + \gamma_2)$ and the minimum value is zero, a formula which is convenient to use is in the form

$$W_{12} = C_2(\gamma_1 + \gamma_2) \tag{3.3}$$

Substituting Eq. (3.3) into (3.2), we get

$$\mu_a \sim 0.3 + C_1C_2(\gamma_1 + \gamma_2)/H \tag{3.4}$$

We note that C_2 is expected to be a function of compatibility (referred to as compatibility parameter) having a high value where the metallurgical compatibility (metals with a high degree of mutual solubility) is high (compatibility = 1 for identical pairs), but a low value where the surfaces are metallurgically incompatible (metals being those which are insoluble in each other). Such compatibility ratings were obtained by consideration of binary phase diagrams. The compatibility diagram of a large number of metal pairs is shown in Fig. 3.19 (Rabinowicz, 1984).

Rabinowicz and his colleagues (Rabinowicz, 1971, 1977; Ohmae and Rabinowicz, 1980) have conducted numerous friction tests on commercial pure metals and reported the coefficient for hundreds of metal pairs (Table 3.3). Rabinowicz found that both bcc and fcc metals essentially followed Eq. (3.4) and

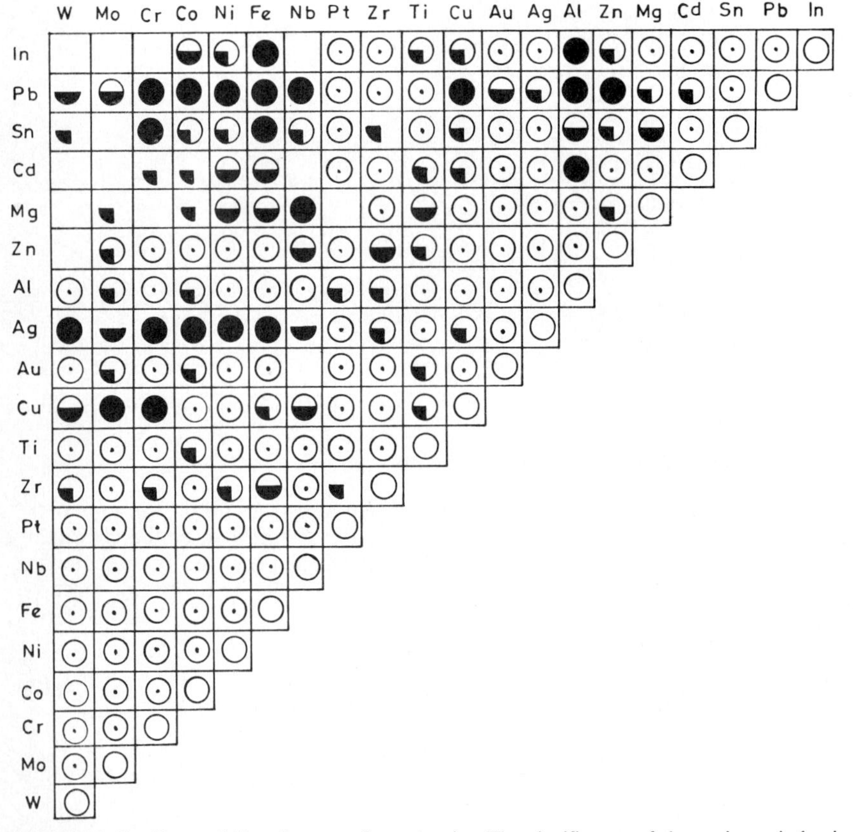

FIGURE 3.19 Compatibility diagram of metal pairs. The significance of the various circles is shown in Tables 3.4 and 3.5. Blank circles are due to insufficient information. [*Adapted from Rabinowicz (1984).*]

values for C_2 obtained are given in Table 3.4. He found that hcp metals had markedly different behavior and the coefficient of friction is generally lower. In this case friction coefficient was found to be lower for *c/a* values larger than the ideal value. It is argued that when *c/a* is larger than the ideal value, plastic deformation takes place at stresses below that given by the isotropic yield locus when the slip planes are favorably oriented.

Except for the deviation in hexagonal metals, we note that when the compatibility parameter C_2 is close to unity, the metals are compatible and the friction coefficients are high. Generally, mutually soluble metals are compatible and, therefore, exhibit large friction coefficients. The substructure effects do not seem to be important in pure metals. When metals are cold worked, the penetration hardness increases and hence W_{12}/H should decrease. Accordingly the friction coefficient should also decrease (cold work should not affect the W_{12} term appreciably). Although no systematic investigations were carried out on the effect of cold work, it appears that the friction coefficient does not depend on dislocation density, arrangement, etc. In fact, the friction coefficient values obtained on well-annealed metals after a substantial amount of sliding should be strictly interpreted as that for cold-worked metals because the material deforms substantially (perhaps reaching full hardness in a short time) during wear.

The presence of grain boundaries in polycrystalline materials influences friction behavior. In the sliding process, surface dislocations are blocked in their movement by a grain boundary. They accumulate at the grain boundary and produce strain hardening in the surfacial layers. This strain hardening makes sliding more difficult and increases the friction force for materials in sliding contact (Buckley, 1968, 1981; Buckley and Johnson, 1968). Strained metal—that is, metal which contains a high concentration of dislocations—is chemically more active, because the presence of defects increases the energy in the material.

Sliding friction experiments have been conducted across the surface of grain boundaries of single crystals and bicrystals of copper to measure the influence of the grain boundary on friction by Buckley (1981). The bicrystals, approximately the size of a dime, had a grain boundary running approximately through the center of the material. One grain had a (111) orientation; the other, a (210) orientation. A polycrystalline copper slider slid across the surface, and friction was measured (Fig. 3.20). We note that there is a marked decrease in coefficient of friction when moving from the (210) orientation to the (110) orientation because the (210) surface is a lower atomic energy surface than the (111) surface. A peak appears in the grain boundary region in the friction coefficient. The grain boundary indicates its own particular characteristic friction.

When alloying elements are added to pure metals, two effects are encountered: (1) the hardness of the metal increases by the substitutional or interstitial hardening mechanisms (Fleischer, 1964), and (2) the surface energy is decreased as the alloying content is increased (Murr, 1975). Since it is known that many impurities segregate to the grain boundaries and interfaces, the interfacial energy is also expected to decrease with alloying content. Although it is not clear whether W_{12} remains the same or decreases, W_{12}/H is expected to decrease and hence the friction coefficient should also decrease. Indeed, experimental results on substitutional solid solutions obtained by Pamies-Teixeira et al. (1977) and Saka and Suh (1977) show that the friction coefficient decreases when the concentration of the alloying elements is increased (Fig. 3.21). Recent data obtained by Tohkai (1978) on a variety of steels and some results on oxide dispersion strengthened alloys also indicate that the friction coefficient decreases when the solute content is increased. The solute effect should be much more predominant

TABLE 3.3 Coefficient of Friction for Commercially Pure Metals

Metals	Crystal structure	Measured hardness, kg mm^{-2}	Estimated surface energy, erg cm^{-2}	W	Mo	Cr	Co
In	Tetragonal	0.9	600	1.06	.73	.70	.68
Pb	fcc	4	450	.41	.65	.53	.55
Sn	Tetragonal	6	570	.43	.61	.52	.51
Cd	hcp	22	620	.44	.58	.58	.52
Mg	Hexagonal	40	560	.58	.51	.52	.54
Zn	hcp	38	790	.51	.53	.55	.47
Al	fcc	27	900	.56	.50	.55	.43
Ag	fcc	70	920	.47	.46	.45	.40
Au	fcc	70	1100	.46	.42	.50	.42
Cu	fcc	80	1100	.41	.48	.46	.44
Ti	hcp	250	1200	.56	.44	.54	.41
Zr	hcp	240	1400	.47	.44	.43	.40
Pt	fcc	140	1800	.57	.59	.53	.54
Nb	bcc	160	2000	.46	.47	.54	.42
Fe	bcc	200	1700	.47	.46	.48	.41
Ni	fcc	210	1700	.45	.50	.59	.43
Co	Hexagonal	230	1800	.48	.40	.41	.56
Cr	bcc	600	1400	.49	.44	.46	
Mo	bcc	300	2300	.51	.44		
W	bcc	500	2300	.51			

Source: Adapted from Rabinowicz (1971).

TABLE 3.4 Values of Compatibility Parameter C_2

Condition	Symbol	Measured value of C_2
Identical	○	1.00
Compatible	⦿	0.50
Limited compatibility	◕	0.32
Very limited compatibility	◒	0.20
Incompatible	●	0.125

Source: Adapted from Rabinowicz (1971).

in the interstitial alloys because of the high hardening rates obtained in interstitial solid solutions.

While the hardness increases enormously with the amount of second-phase particles, the surface energy should remain constant when the solubility limit is exceeded. Thus, even though the surface and interfacial energies of two-phase alloys are not substantially different from the corresponding values of saturated solid solutions, according to Eq. (3.2) the friction coefficient should decrease as the volume fraction of the second-phase particle is increased. Further, for a given volume fraction, the hardness can be increased by decreasing the particle size and interparticle spacing. This process again should not change either τ_s/H or

Ni	Fe	Nb	Pt	Zr	Ti	Cu	Au	Ag	Al	Zn	Mg	Cd	Sn	Pb	In
.59	.64	.67	.79	.70	.60	.67	.67	.82	.90	1.17	1.52	.74	.81	.93	1.46
.60	.54	.51	.58	.76	.88	.64	.61	.73	.68	.70	.53	.66	.84	.90	
.55	.55	.55	.72	.55	.56	.53	.54	.62	.60	.63	.52	.67	.74		
.47	.52	.56	.59	.50	.55	.49	.49	.59	.48	.58	.55	.79			
.52	.51	.49	.51	.57	.55	.55	.53	.55	.55	.49	.69				
.56	.55	.50	.64	.44	.56	.56	.47	.58	.58	.75					
.52	.54	.50	.62	.52	.54	.53	.54	.57	.57						
.46	.49	.52	.58	.45	.54	.48	.53	.50							
.54	.47	.50	.50	.46	.52	.54	.49								
.49	.50	.49	.59	.51	.47	.55									
.51	.49	.51	.66	.57	.55										
.44	.52	.56	.52	.63											
.64	.51	.57	.55												
.47	.46	.46													
.47	.51														
.50															

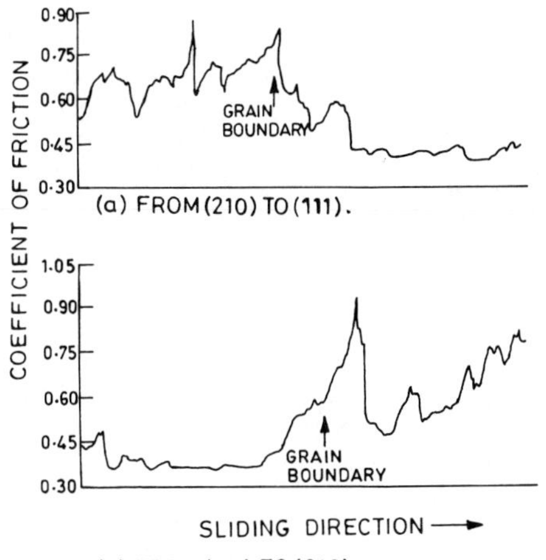

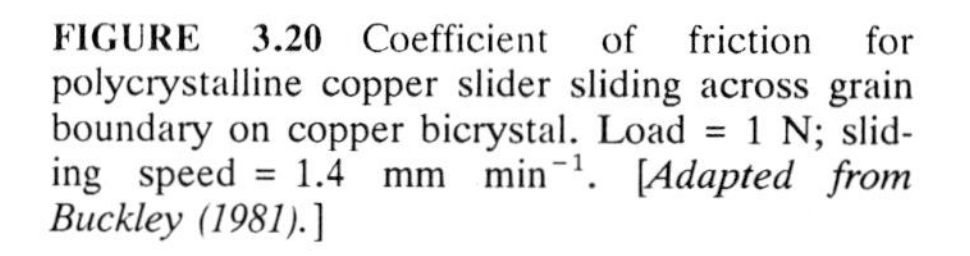

FIGURE 3.20 Coefficient of friction for polycrystalline copper slider sliding across grain boundary on copper bicrystal. Load = 1 N; sliding speed = 1.4 mm min^{-1}. [*Adapted from Buckley (1981).*]

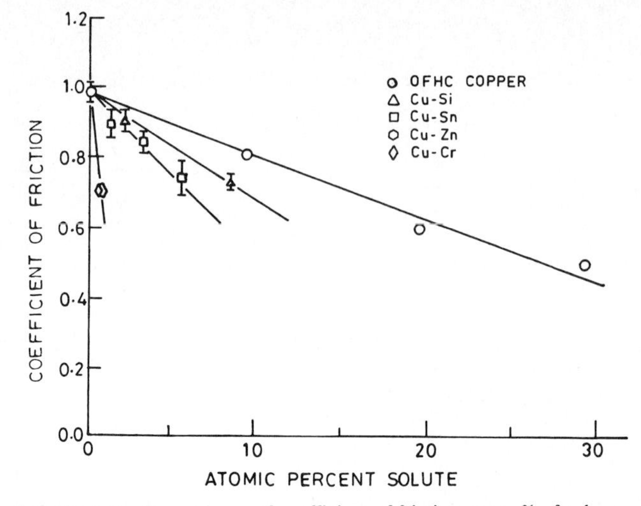

FIGURE 3.21 Dependence of coefficient of friction on at. % of solute of copper-based alloys. [*Adapted from Pamies-Teixeira et al. (1977).*]

W_{12}. Thus, the friction coefficient should decrease substantially in two-phase alloys when the particle size and spacing are decreased for a given volume fraction of the second phase.

Unfortunately, the experimental results on a variety of commercial steels (Tohkai, 1978), precipitation hardened alloys (Saka et al., 1977), and oxide dispersion strengthened alloys do not confirm this hypothesis. Thus Rabinowicz's theory, which is so successful in explaining the friction of pure metals, breaks down in the case of two-phase alloys. At present, there is no theory that can explain this anomaly. Eutectics and eutectoids have not been studied extensively because of their brittleness.

Substantial information exists on fiber-reinforced polymer-based composites slid against metals. The friction coefficient is *minimum* when the sliding direction coincides with the fiber direction and *maximum* when the fibers are normal to the sliding surface (Bhushan and Wilcock, 1980; Lancaster, 1972; and Sung and Suh, 1979). Further, the friction coefficient depends largely on the nature of the reinforcing phase. Surface energy considerations may be used for the fiber itself as the fiber generally gets exposed during sliding. However, as the hardness of the fiber is much larger than the matrix phase, Eq. (3.2) cannot be used for interpreting the results. In fact, ploughing and elastic deformations may have to be considered depending on whether the mating surface is soft or hard.

3.6.2 Effect of Microstructure on Wear

The adhesive wear is directly related to the tendency for different counterface materials to form solid solutions or intermetallic compounds with one another. The metallurgical compatibility, as indicated by the mutual solubility, represents the degree of intrinsic attraction of the atoms of the contacting metals for each other. Such compatibility is best determined from binary metal phase diagrams, which show the extent of mutual solubility or insolubility in the liquid or solid states (Fig. 3.19).

The significance of the various circles, in terms of metallurgical solubility at room temperature, metallurgical compatibility, sliding compatibility, and anticipated wear are shown in Table 3.5 (Rabinowicz, 1977, 1980, 1984). The general rule is that the blacker the circle, the better the sliding characteristics or the lower the adhesive wear coefficients (Table 3.6).

Experimental evidence indicates that the wear rate W [Eq. (2.16)] of metals is approximately proportional to the normal load F_N. Figure 3.22 indicates that inverse dependence of adhesive wear rate of metals on hardness is approximately obeyed. (Similar dependence for abrasive wear has been reported by Khruschov, 1957.) However, in case of solid solutions, the wear rate was not found to correlate well with hardness, as can be seen in Fig. 3.23. The disagreement is more pronounced in case of precipitation-hardened and dispersion-strengthened alloys. The key to an understanding of the microstructural effects on wear rate, in many cases, lies in the understanding of the subsurface deformation, crack nucleation, and propagation aspects as a function of microstructure (Sec. 2.2).

Since bcc metals are in general harder than fcc and hcp metals, the wear rate exhibited by bcc metals is generally lower than the other two crystal classes. An important effect in pure metals, at least in fcc metals, is the effect of substructure on wear. The development of substructure depends on the stacking-fault energy of metals. Low stacking-fault energy metals exhibit planar slip whereas high stacking-fault energy metals exhibit cross-slip and cellular dislocation structure. It has been speculated by Hirth and Rigney (1976) that the cell boundaries would act as potential crack nucleation sites and propagation paths. Although these speculations seem reasonable, no direct proof confirms them. Nevertheless, it has been shown experimentally by Suh and Saka (1977) that high stacking-fault energy materials indeed exhibit high wear rates. Grain size and grain shape also influence the wear rate. The texture also may have some influence on the wear rates of bcc metals when the cleavage planes are favorably oriented to promote crack propagation.

The addition of the solute atoms to pure metals generally increases the hardness without enhancing the tendency of crack nucleation. Pamies-Teixeira et al. (1977) have shown that the wear rate decreases as the solute content is increased (Fig. 3.24). Both interstitial and substitutional solid solutions are found to exhibit

TABLE 3.5 Compatibility Relationships for Metals

Symbol	Metallurgical solubility	Metallurgical compatibility	Sliding compatibility	Anticipated wear
○	100%	Identical	Very poor	Very high
⦿	Above 1%	Soluble	Poor	High
◔	0.1–1%	Intermediate soluble	Intermediate	Intermediate
◒	Below 0.1%	Intermediate insoluble	Intermediate or good	Intermediate or low
●	Two liquid phases	Insoluble	Very good	Very low

Source: Adapted from Rabinowicz (1984).

TABLE 3.6 Wear Coefficients of Materials Having Different Solubilities

Metal pair	Solubility	Wear coefficient, $\times 10^{-4}$
Cu-Pb	No	0.10
Ni-Pb	No	0.21
Fe-Ag	No	0.68
Fe-Pb	No	0.69
Al-Pb	No	1.4
Al-Zn	Yes	3.9
Al-Fe	Yes	1.0
Fe-Cu	Yes (Min)	19.0
Al-Al	Yes	30
Mg-Mg	Yes	36
Fe-Mg	Yes (Min)	38
Fe-Fe	Yes	77
Cu-Ni	Yes	81

Source: Adapted from Rabinowicz (1984).

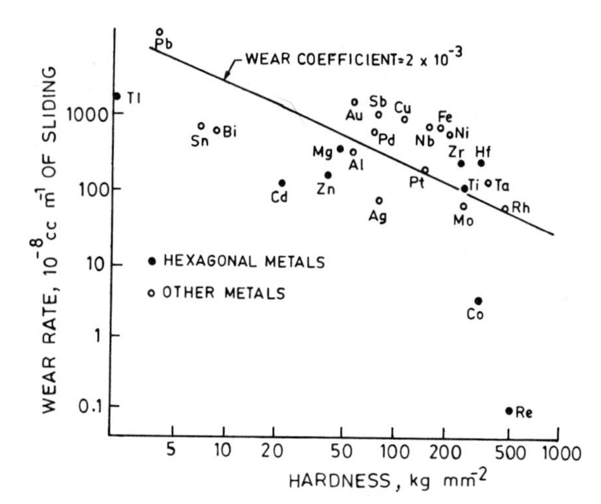

FIGURE 3.22 Wear rate as a function of hardness for pure metals in a pin-on-disk test at a load of 2 N. [*Adapted from Rabinowicz (1977).*]

superior wear properties. (We have seen earlier that the friction coefficient decreases with the increase in solute content.)

In general, abrasive wear is reduced by high interstitial content and high hardness—a relationship well known from practical experience. For instance, the popular bearing steel SAE 52100 is a very hard steel containing a high carbon content of 1% C. It is generally known that interstitially dissolved carbon atoms, carbon atmospheres (coherent precipitate and carbides incoherent precipitates), interfere with the passage of dislocation, and thus impede the process of plastic

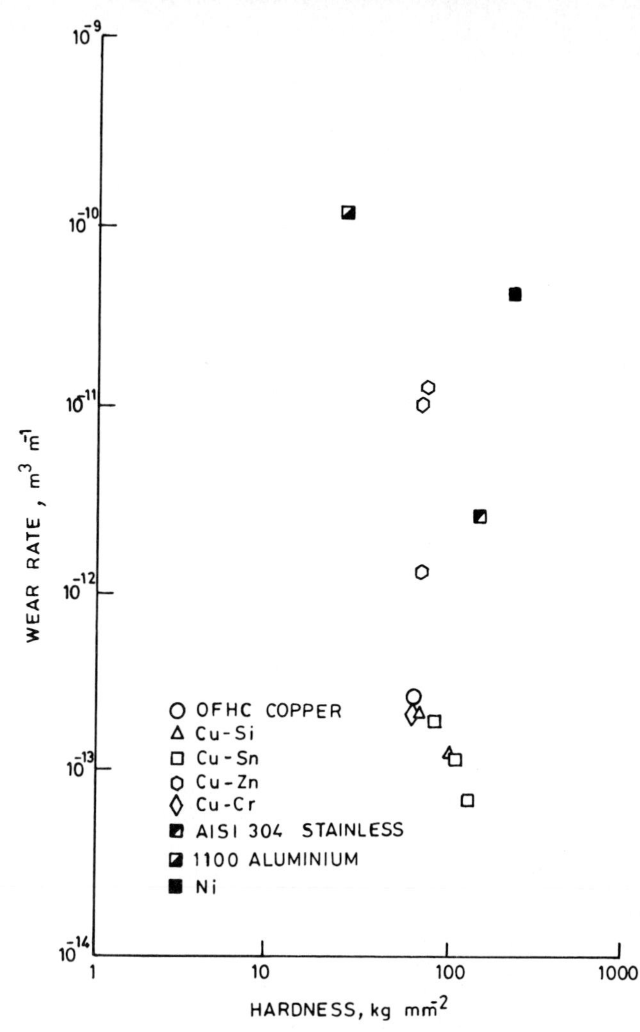

FIGURE 3.23 Wear rate versus hardness of commercially available pure metals and metal alloys. [*Adapted from Suh and Saka (1977).*]

flow. This impediment increases not only the yield strength (and hardness) but also intensifies the work hardening process. The wear resistance achieved with the aid of hard carbide microconstituents is retained at high temperatures as most carbides are stable (Borik, 1972).

Data for steels worn by hard abrasives (Fig. 3.25) show that wear resistance increases with increasing hardness, but the pearlitic and bainitic structures are superior to ferritic and martensitic structures of the same hardness (Borik, 1972; Khruschov and Babichev, 1965; Moore, 1974, 1981; Richardson, 1967; Tipnis, 1984; Zum Gahr, 1987). This is attributed to the higher strain hardening capacity and ductility of pearlite and bainite.

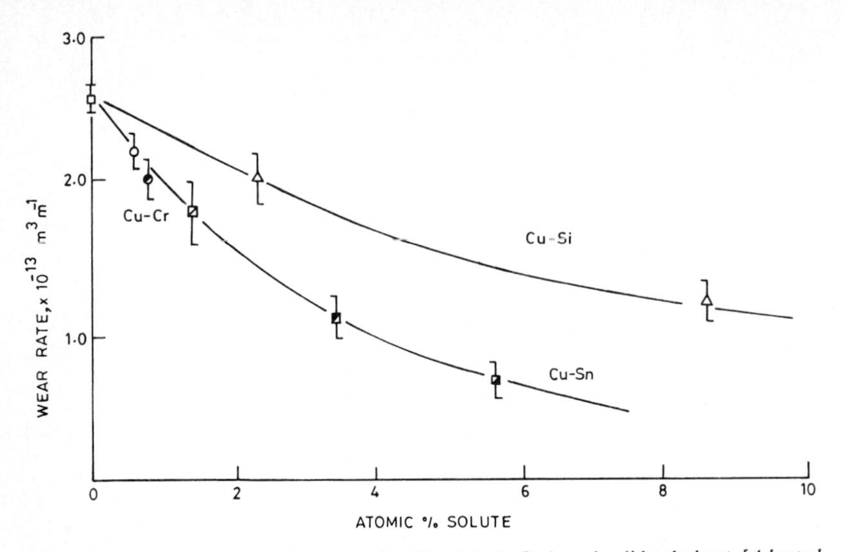

FIGURE 3.24 Wear rate as a function of at.% solute in Cu-based solid solutions. [*Adapted from Pamies-Teixeira et al. (1977).*]

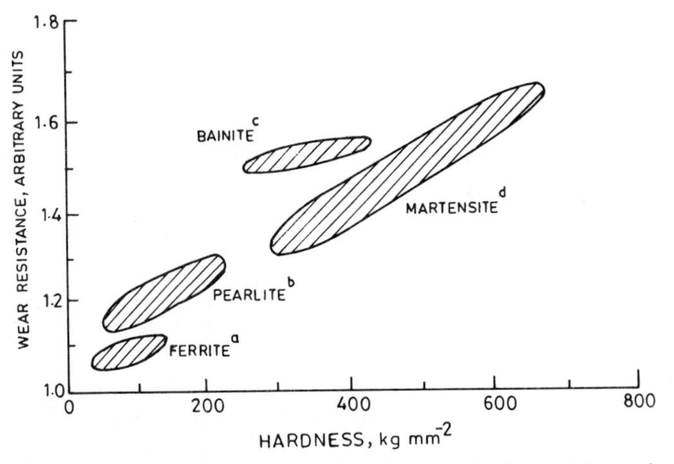

FIGURE 3.25 Effect of structure and hardness on abrasion resistance of steels: (*a*) low-carbon steels annealed or normalized; (*b*) medium-to-high carbon steels annealed or normalized; (*c*) austempered medium-to-high carbon steels; and (*d*) quenched and tempered medium-to-high carbon steels and carburized steels. [*Adapted from Zum Gahr (1987).*]

The mechanics of crack nucleation in two-phase alloys has been worked out by Jahanmir and Suh (1977). Basically from a microstructural point of view, we need to describe only the effects of particle morphology. When the mean free path of the particles (the distance traveled by a particle between two consecutive collisions without any disturbance) is decreased, the hardness increases and the

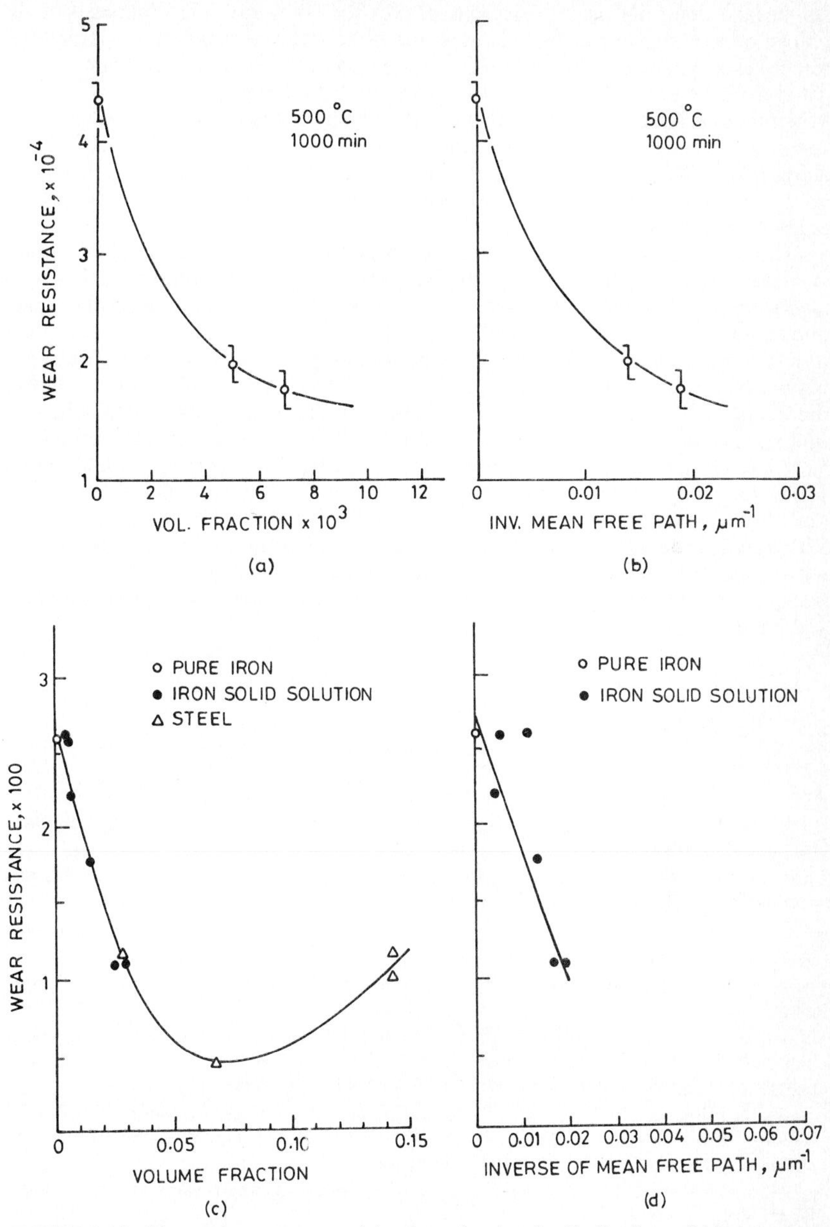

FIGURE 3.26 Wear resistance versus: (*a*) volume fraction for Cu-Cr alloys; (*b*) inverse mean free path for Cu-Cr alloys; (*c*) volume fraction for iron and spherodized steels; (*d*) inverse mean free path for iron and spherodized steels. [*Adapted from Saka et al. (1977) and Jahanmir et al. (1974).*]

subsurface deformation rate decreases. But the wear rate may increase or decrease depending on the particle size and coherency because these parameters affect crack nucleation rate. The coherency and size effects have been investigated by Saka et al. (1977) in Cu-Cr alloys and by Jahanmir et al. (1974) in steels. When the particles are large (>100 nm), the coherency is lost and the wear rate of two-phase metals increases even if the hardness increases. Age-hardenable Cu-Cr alloys [Fig. 3.26(*a*), (*b*)] and spherodized steels [Fig. 3.26(*c*), (*d*)] indicate that as the volume fraction of the second-phase particles is increased, the wear resistance decreases exponentially.

Although much work has been done on polymeric composites (Bhushan and Wilcock, 1982; Lancaster, 1972; and Sung and Suh, 1979), experimental data on metal matrix composites is scarce. In the case of particulate composites crack nucleation should take place readily, but the wear coefficient can either be high or low depending on the hardness. Experimental results of boron-reinforced aluminum composites indicate that the wear rate is dependent on the orientation of the fibers with respect to the sliding direction. In polymeric composites, when the fibers are oriented normal to the contact interface, the wear rates are found to be very low (Bhushan and Winn, 1979; Bhushan and Wilcock, 1980, 1982). When sliding takes place in the plane of fibers and normal to the fiber axis, the fibers are pulled out by debonding. As debonding, or fiber pull-out, are crack nucleation and growth processes, the fiber-matrix interface and the orientation of the fiber with respect to sliding direction are important aspects. Abrasion may also take place during the sliding wear of composites because the fibers are generally hard and have a high aspect ratio.

3.7 REFERENCES

Azaroff, L. V. (1960), *Introduction to Solids*, McGraw-Hill, New York.

Barrett, C. S. (1943), *Structure of Metals: Crystallographic Methods, Principles and Data*, McGraw-Hill, New York.

Bhushan, B. (1984), "Analysis of the Real Area of Contact Between a Polymeric Magnetic Medium and a Rigid Surface," *J. Tribol.* (*Trans. ASME*), Vol. 106, pp. 26–34.

——— (1985), "The Real Area of Contact in Polymeric Magnetic Media. II: Experimental Data and Analysis," *ASLE Trans.*, Vol. 28, pp. 181–197.

——— (1990), *Tribology and Mechanics of Magnetic Storage Devices*, Springer Verlag, New York.

———, and Winn, L. W. (1979), "Material Study for Advanced Sterntube Seals and Bearings for High Performance Merchant Ships," *Tech. Rep. No.* MA-920-79069, Office of Commercial Development, U.S. Maritime Administration, Washington, D.C.

———, and Wilcock, D. F. (1980), "Frictional Behavior of Polymeric Compositions in Dry Sliding," *Proc. 7th Leeds/Lyon Symp. on Tribology*, pp. 103–113, Westbury House, IPC Business Press Ltd., Guildford, United Kingdom.

———, and ——— (1982), "Wear Behavior of Polymeric Compositions in Dry Sliding," *Wear*, Vol. 75, pp. 41–70.

Borik, F., (1972), "Using Wear Tests to Determine the Influence of Metallurgical Variables on Abrasion," *Metals Eng. Q.*, Vol. 12, pp. 33–39.

Brainard, W. A., and Buckley, D. H. (1971), "Preliminary Studies by Field Ion Microscopy of Adhesion of Platinum and Gold to Tungsten and Iridium," *Report No. TND-6492*, NASA, Washington, D.C.

Buckley, D. H. (1968), "The Influence of the Atomic Nature of Crystalline Materials on Friction," *ASLE Trans.*, Vol. 11, pp. 89–100.

——— (1969), "A LEED Study of the Adhesion of Gold to Copper and Copper Aluminum Alloys," *Report No. TND-5351*, NASA, Washington, D.C.

——— (1981), *Surface Effects in Adhesion, Friction, Wear, and Lubrication*, Elsevier, New York.

——— (1985), "Importance and Definition of Materials in Tribology: Status of Understanding," *New Directions in Lubrication, Materials, Wear, and Surface Interactions: Tribology in the 80's* (W. R. Loomis, ed.), pp. 18–42, Noyes, Park Ridge, N.J.

———, and Johnson, R. L. (1968), "The Influence of Crystal Structure and Some Properties of Hexagonal Metals on Friction and Adhesion," *Wear*, Vol. 11, pp. 405–419.

Ciftan, M., and Saibel, E. (1979), "Rehbinder Effect and Wear," *Proc. Int. Conf. on Wear of Materials* (K. C. Ludema et al., eds.), pp. 659–664, ASME, New York.

Cottrell, A. (1975), *An Introduction to Metallurgy*, 2d ed., Edward Arnold, London.

Egert, B., and Panzner, G. (1982), "Electron Spectroscopic Study of Phosphorus Segregated to α-Iron Surfaces," *Surf. Sci.*, Vol. 118, pp. 345–368.

Ferrante, J., and Buckley, D. H. (1972), "Review of Surface Segregation, Adhesion, and Friction Studies Performed on Copper-Aluminum, Copper-Tin, and Iron-Aluminum Alloys," *ASLE Trans.*, Vol. 15, pp. 18–24.

Fleischer, R. L. (1964), in *The Strengthening of Metals* (D. Peckner, ed.), Reinhold, New York.

Goldschmidt, H. J. (1967), *Interstitial Alloys*, Butterworth, London.

Greenwood, J. A., and Williamson, J. B. P. (1966), "Contact of Nominally Flat Surfaces," *Proc. R. Soc. (Lond.)*, Vol. A295, pp. 300–319.

Gulyaev, A. (1980), *Physical Metallurgy*, Vols. 1 & 2, Mir Publishers, Moscow.

Guy, A. C. (1959), *Elements of Physical Metallurgy*, Addison-Wesley, Reading, Mass.

Hayward, D. O., and Trapnell, B. M. W. (1964), *Chemisorption*, 3d ed., Butterworth, Washington, D.C.

Hirth, J. P., and Rigney, D. A. (1976), "Crystal Plasticity and the Delamination Theory of Wear," *Wear*, Vol. 39, pp. 133–141.

Jahanmir, S., and Suh, N. P. (1977), "Mechanics of Subsurface Void Nucleation in Delamination Wear," *Wear*, Vol. 44, pp. 17–38.

———, ———, and Abrahamson, E. P. (1974), "Microscopic Observations of the Wear Sheet Formation by Delamination," *Wear*, Vol. 28, pp. 235–249.

Kittel, C. (1966), *Introduction to Solid State Physics*, 3d ed., Wiley, New York.

Khruschov, M. M. (1957), "Resistance to Metals to Wear by Abrasion, as Related to Hardness," *Proc. Conf. on Lub. and Wear*, Institute of Mechanical Engineers (London), pp. 655–659.

———, and Babichev, M. A. (1965), "The Effect of Heat Treatment and Work Hardening on the Resistance of Abrasive Wear of Some Alloy Steels," *Frict. Wear Mach.*, Vol. 19, pp. 1–15.

Lakhtin, Y. (1977), *Engineering Physical Metallurgy*, Mir Publishers, Moscow.

Lancaster, J. K. (1972), "Polymer-Based Bearing Materials: The Role of Fillers and Fiber Reinforcement," *Tribol. Int.*, Vol. 5, pp. 249–255.

Latanision, R. M. (1980), "Surface Effects in Crystal Plasticity," in *Fundamentals of Tribology* (N. P. Suh and N. Saka, eds.), pp. 255–294, M.I.T. Press, Cambridge, Mass.

McLean, D. H. (1957), *Grain Boundaries in Metals*, Clarendon Press, Oxford, England.

Miyoshi, K., and Buckley, D. H. (1982), "Friction and Surface Chemistry of Some Ferrous-Base Metallic Glasses," *Tech. Rep. No. TP-1991*, NASA, Washington, D.C.

Moore, M. A. (1974), "The Relationship Between Abrasive Wear Resistance, Hardness, and Microstructure of Ferritic Materials," *Wear*, Vol. 28, pp. 59–68.

——— (1981), "Abrasive Wear," in *Fundamentals of Friction and Wear of Materials* (D. A. Rigney, ed.), pp. 73–118, American Society for Metals, Metals Park, Ohio.

Mouttet, C., Gaspard, J. P., and Lambin, P. (1981), "Electronic Structure of the Si-Au Surface," *Surf. Sci.*, Vol. 111, pp. L755–L758.

Murr, L. E. (1975), *Interfacial Phenomena in Metals and Alloys*, Addison-Wesley, Reading, Mass.

Neely, J. E. (1984), *Practical Metallurgy and Materials of Industry*, Wiley, New York.

Ohmae, N., and Rabinowicz, E. (1980), "The Wear of Noble Metals," *ASLE Trans.*, Vol. 23, pp. 86–92.

Pamies-Teixeira, J. J., Saka, N., and Suh, N. P. (1977), "Wear of Copper-Based Solid Solutions," *Wear*, Vol. 44, pp. 65–76.

Pascoe, K. J. (1978), *An Introduction to the Properties of Engineering Materials*, Van Nostrand, New York.

Philips, F. C. (1956), *An Introduction to Crystallography*, Longman, London.

Rabinowicz, E. (1965), *Friction and Wear of Materials*, Wiley, New York.

——— (1971), "The Determination of the Compatibility of Metals Through Static Friction Tests," *ASLE Trans.*, Vol. 14, pp. 198–205.

——— (1977), "Dependence of the Adhesive Wear Coefficient on the Surface Energy of Adhesion," in *Proc. Int. Conf. on Wear of Materials* (W. A. Glaeser, K. C. Ludema, and S. K. Rhee, eds.), pp. 36–46, ASME, New York.

——— (1980), "Wear Coefficients—Metals," in *Wear Control Handbook* (M. B. Peterson and W. O. Winer, eds.), pp. 475–506, ASME, New York.

——— (1984), "Wear Coefficients," in *Handbook of Lubrication: Theory and Practice of Tribology*, Vol. 2, *Theory and Design*, (E. R. Booser, ed.), pp. 201–225, CRC Press, Boca Raton, Fla.

Reed-Hill, R. E. (1973), *Physical Metallurgy Principles*, 2d ed., Van Nostrand, New York.

Rehbinder, P. A., and Likhtman, V. I. (1957), "Effects of Surface Active Media on Strains and Rupture in Solids," *Proc. 2nd Int. Congress on Surface Activity*, No. 3, pp. 563–580, London.

Rhines, F. W. (1956), *Phase Diagram in Metallurgy*, McGraw-Hill, New York.

Richardson, R. C. D. (1967), "The Wear of Metals by Hard Abrasives," *Wear*, Vol. 10, pp. 291–309.

Rollason, E. C. (1973), *Metallurgy for Engineers*, Edward Arnold, London.

Saka, N. (1980), "Effect of Microstructure on Friction and Wear of Metals," in *Fundamental of Tribology* (N. P. Suh and N. Saka, eds.), pp. 135–170, M.I.T. Press, Cambridge, Mass.

———, and Suh, N. P. (1977), "Delamination Wear of Dispersion—Hardened Alloys," *J. Eng. Ind.* (*Trans. ASME*), Vol. 99, pp. 289–294.

———, Pamies-Teixeira, J. J., and Suh, N. P. (1977), "Wear of Two Phase Metals," *Wear*, Vol. 44, pp. 77–86.

Samuels, L. E. (1960), "Damaged Surface Layers: Metals," in *The Surface Chemistry of Metals and Semiconductors* (H. C. Gatos, ed.), pp. 82–103, Wiley, New York.

Suh, N. P., and Saka, N. (1977), "The Staking Fault Energy and Delamination Wear of Single Phase fcc Metals," *Wear*, Vol. 44, pp. 135–143.

Sung, N.-H., and Suh, N. P. (1979), "Effect of Fiber Orientation of Friction and Wear of Fiber Reinforced Polymeric Composition," *Wear*, Vol. 53, pp. 129–141.

Thomas, T. R. (1982), *Rough Surfaces*, Longman, London.

Tipnis, V. A. (1984), "Cutting Tool Wear," in *Wear Control Handbook* (M. B. Peterson and W. O. Winer, eds.), pp. 891–930, ASME, New York.

Tohkai, M. (1978), *Microstructural Aspects of Friction*, S. M. Thesis, Department of Mechanical Engineering, M.I.T., Cambridge, Mass.

Van Vlack, L. H. (1985), *Elements of Materials Science and Engineering*, Addison-Wesley, Reading, Mass.

Zum Gahr, K.-H. (1987), *Microstructure and Wear of Materials*, Elsevier, Amsterdam.

CHAPTER 4
METALS AND CERAMICS

In selecting materials for an application, one studies the wear modes and selects materials that satisfy the requirements under all operating conditions. Wide ranges of materials are available today for tribological applications. The emergence of new design concepts that quite often demand several conflicting properties in a material, e.g., high mechanical strength along with high conformability, high hardness along with high fracture toughness and a high strength-to-weight ratio, has required a tailoring of the properties of bulk materials to specific needs. Components employed under extreme conditions, i.e., high temperatures, high speeds, and heavy loads, also require materials with unusual characteristics. A determining factor for the extensive use of any material is its amenability to manipulation and the extent to which one can control and vary its properties, such as strength, hardness, ductility, and toughness, within the range of operating requirements (Amateau and Glaeser, 1964; Anonymous, 1974, 1980a; Booser, 1984; Clauss, 1972; Eyre, 1979; Glaeser, 1980; Peterson and Winer, 1980; Rigney, 1981; Scott, 1983).

High strength and hardness (but reduced ductility and toughness) of metals is obtained by cold or hot working or by heat treatment. Cold working (at ambient or slightly elevated temperature) strain-hardens metals. The grains elongate, distort, and generally increase the metal's resistance to plastic flow as deformation continues. Hot working is done at a high enough temperature to permit spontaneous formation of new, unstrained, undistorted grains within the metal. The heat treatment consists of heating the metal, holding it at an elevated temperature to bring the hardening constituents into solution, and then cooling it rapidly (quenching) to retain those constituents in solution. Some metals strain-harden severely (or can be cold worked), and some do not. Similarly, some metals are heat treatable, and some are not.

Metals are partially or fully annealed to increase ductility and toughness (and to reduce hardness), to relieve residual stresses, to improve machinability, to facilitate cold working, or to produce a desired microstructure. The annealing consists of heating the metal, holding it at an elevated temperature to produce the desired structure, and then cooling it slowly through the transformation range. Tempering of metals (mostly steels) is sometimes done to increase ductility and toughness. It is accomplished by heating previously hardened metal to a temperature below the transformation range and then cooling it at a suitable rate.

Materials can be cast to shape or wrought into various mill forms from which finished parts are machined, forged, formed, stamped, or otherwise shaped. Although the mechanical properties of cast parts are generally not as good as those of wrought products, cast parts remain in wide use today because casting is often the simplest, cheapest, or only way to produce a part with a particular combination of properties.

Table 4.1 lists various bulk materials commonly used to reduce friction and wear in different tribological applications. Surface treatments and coatings are also included for comparisons. The table indicates that the commonly used materials in tribology range from cast irons, stainless steels, high-temperature alloys, and lead- and tin-based alloys to ceramics, cemented carbides, and polymers, literally covering all technologically important materials. Many of these materials are used as a substrate for tribological applications requiring coatings and surface treatments for optimal performance.

For convenience, we have categorized various bulk materials into five types: ferrous metals; nonferrous metals and alloys; ceramics and cermets; carbon, graphite, and MoS_2; and polymers. Carbon, graphite, and MoS_2 and polymer materials are discussed in Chapter 5. Descriptions of metals and ceramic materials and their limitations and special attributes are presented in this chapter. The tables of important properties of different materials given in this chapter should only be used as a guide; for exact values, one should refer to the indicated literature or the manufacturer.

4.1 FERROUS METALS

Iron and the large family of iron alloy called *steels* and *cast irons* are the most common of the commercial metals. The iron alloys have had a unique position as structural materials within a wide range of applications from the simple binary Fe-C alloys of the iron age to the sophisticated stainless steels and high-speed tool steels produced by powder technology. A unique property of steel, compared with many other materials, is the wide range within which hardness and toughness can be matched.

Figure 4.1 presents an iron-carbon equilibrium diagram showing some of the microstructures that occur in carbon and alloy steels and cast irons. By definition, steels are iron alloys containing up to 2 percent carbon, and cast irons are iron alloys containing from 2 to 4 percent carbon. Important features of various phases are presented in Table 4.2. The performance of ferrous materials is mainly controlled by their internal structure. The structure, in turn, is the combined result of the chemical composition, percentage of alloying contents, heat treatment, and method of forming. Ferrous materials that have various alloying elements are used for specialty materials, which are discussed in the following sections. For further details, readers may refer to metals handbooks and metallurgy and engineering materials books (Anonymous, 1974, 1978a, 1978c, 1980a; Bain and Paxton, 1961; Budinski, 1980, 1983; Gulyaev, 1980; Lakhtin, 1977; Neely, 1984; Pascoe, 1978; Reed-Hill, 1973; Rhines, 1956; Thelning, 1984; Van Vlack, 1985).

4.1.1 Cast Irons

Cast irons are widely used because of their good mechanical properties, excellent machinability, and the ease with which they can be cast into complex shapes. Cast irons primarily are alloys of iron that contain from 2 to 4 percent carbon and from 1 to 3 percent silicon. Some types contain additional alloying contents. The

TABLE 4.1 Materials Used for Specific Tribological Applications

Materials	Tribological action							
	Sliding		Rolling					
	Unlubricated	Lubricated	Unlubricated	Lubricated	Impact	Abrasion	Particle erosion	Cavitation
Cast irons:								
Graphitic	X	X	X			X		
White	X					X		
Alloy steels		X	X	X	X	X	X	
Stainless steels		X	X	X		X		
Tool steels	X	X	X	X	X	X	X	X
High-temperature superalloys		X		X				
Soft bearing alloys:								
Babbits	X	X						
Copper-based alloys:								
Bronzes	X	X						
Be copper	X	X			X			
Ceramics	X		X			X	X	
Cermets	X		X		X	X	X	
Carbon, graphite, and MoS_2	X					X	X	X
Polymers:								
Thermoplastic	X	X						
Thermoset	X							
Elastomers	X	X				X	X	X
Polymer composites	X	X						
Surface treatments	X	X	X	X		X		
Soft coatings	X	X						
Hard coatings	X	X	X	X	X	X	X	

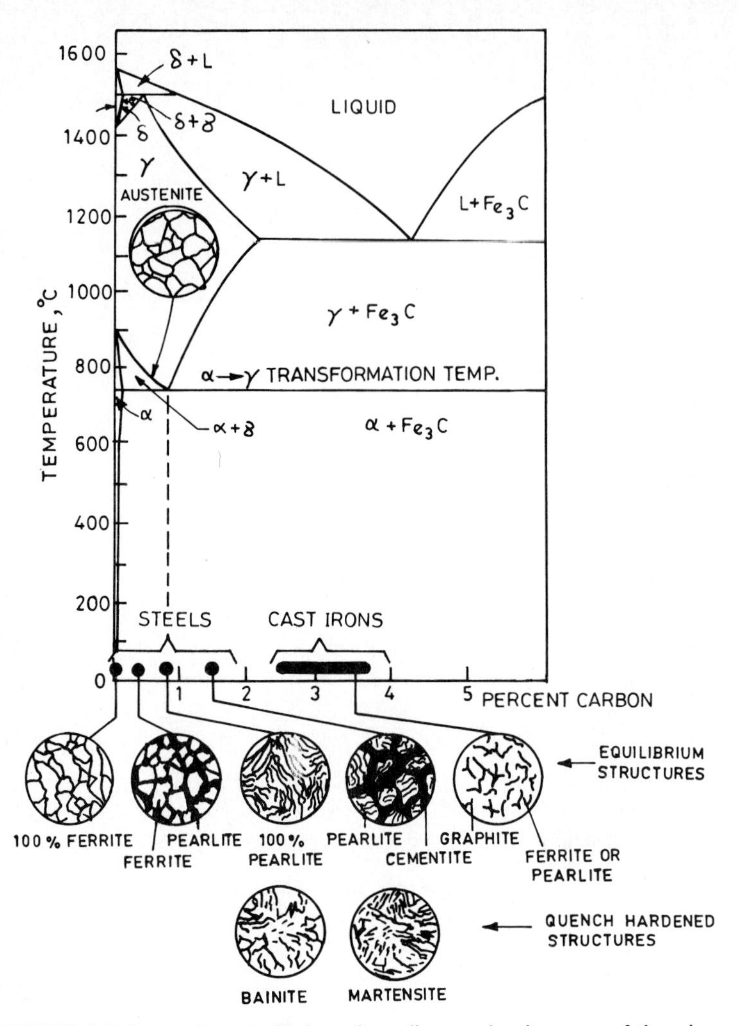

FIGURE 4.1 Iron-carbon equilibrium phase diagram showing some of the microstructures that occur in carbon and alloy steels and cast irons.

carbon content in iron is the principal strengthening and hardening element and is the key to its distinctive properties. The high carbon content makes molten iron very fluid, enabling it to be poured (cast) into intricate shapes. The carbon can be present as free carbon and combined carbon, which affects the lubrication and wear resistance of the cast irons. The factors that determine the microstructure are carbon content, alloy and impurity content, cooling rates after the casting, and the methods of heat treatment employed. The nature of carbon and the structure of the cast irons lead to quite distinct properties and types (Anonymous, 1978a).

Figure 4.1 illustrates the formation of several phases in iron-carbon systems. In some cases, the microstructure of the iron may be all ferrite—the same con-

TABLE 4.2 Description and Important Features of Various Phases of Iron-Carbon Systems

Structure	Description	Microhardness, kg mm^{-2}	Important features
Ferrite or α-iron	Pure iron with bcc structure	70–200	Quite soft and ductile, solubility of carbon is very low.
Cementite or iron carbide (Fe_3C)	A compound of iron and carbon, i.e., Fe_3C, with orthorhombic crystal structure	840–1100	Cementite is very hard compared with austenite, pearlite, and ferrite; provides excellent abrasion resistance.
Pearlite	A lamellar arrangement of ferrite and cementite formed by transforming austenite of eutectoid composition	250–320	Better abrasive wear resistance than ferrite but poorer than martensite.
δ-iron	Iron with bcc structure	—	Solubility of carbon is small but better than α-ferrite.
Austenite or γ-iron	A high-temperature phase of iron with an fcc crystal structure	—	At stable temperatures it is soft and ductile; stable at temperatures between 910 to 1400°C; better solubility (up to 2%) of carbon than ferrite.
Bainite	Metastable aggregate of ferrite and cementite obtained by special quench-hardening techniques	—	Structure and properties are similar to tempered martensite; hardening at high temperatures produces larger carbides and therefore softer, more ductile steels.
Martensite	Supersaturated solution of carbon in α-iron, body-centered tetragonal crystal structure	500–1100	Martensite is very hard, strong, and brittle; provides excellent abrasion resistance.
Graphite	Precipitate of carbon	—	Provides excellent machinability, damps vibrations, and aids lubrication on wearing surfaces.

stituents that make low-carbon steels soft and easily machined. Some cast iron grades consist of alternating layers of soft ferrite and hard iron carbide (cementite). This laminated structure—called *pearlite*—is strong and wear resistant but still quite machinable [Fig. 4.2(*a*) to (*c*)]. Cast irons that are flame hardened, induction hardened, or conventional flame furnace heated and oil quenched contain a martensite structure which (when tempered) provides machinability with maximum strength and good wear resistance.

In cast irons, more carbon is present than can be retained in solid solution in austenite at the eutectic temperature. The carbon that exceeds the solubility in austenite precipitates as graphite. The presence of graphite in the metal provides machinability (even at wear-resisting hardness levels), damps vibration, and aids lubrication on wearing surfaces (even under borderline lubrication conditions).

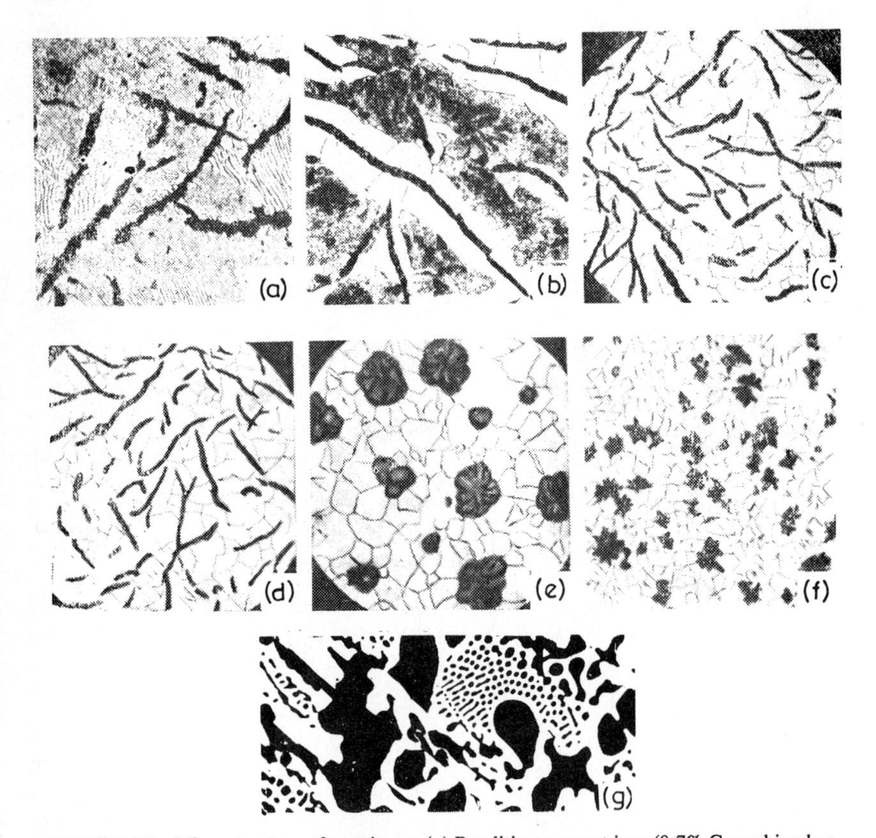

FIGURE 4.2 Microstructure of cast irons: (*a*) Pearlitic gray cast iron (0.7% C combined as Fe_3C) with structural components of pearlite and graphite; (*b*) ferritic-pearlitic gray cast iron (0.1% C combined as Fe_3C) with structural components of pearlite, ferrite, and graphite; (*c*) ferritic gray cast irons (no combined carbon) with structural components of ferrite and graphite; (*d*) gray cast iron with flaky graphite; (*e*) nodular cast iron with spheroidal graphite; (*f*) malleable cast iron with tempered carbon nodules; and (*g*) white cast iron (all carbon combined as Fe_3C) with structural components of ledeburite and pearlite. [*Adapted from Lakhtin (1977) and Gulyaev (1978).*]

Graphite can be present in cast irons mainly in three forms: flakes, true carbon nodules (spheroids), and tempered carbon nodules. In general practice, cast irons that have graphite in the form of flakes, true carbon nodules, and tempered carbon nodules are called *gray cast irons* [Fig. 4.2(*d*)], *nodular* (or *ductile* or *spheroidal-graphite*) *cast irons* [Fig. 4.2(*e*)], and *malleable cast irons* [Fig. 4.2(*f*)], respectively. When most of the carbon remains combined with the iron, such as in white cast iron, the presence of hard iron carbides or cementite [Fig. 4.2(*g*)] provides good abrasion resistance. Malleable iron is cast as white cast iron and then made malleable, i.e., heat-treated to impart ductility to an otherwise exceedingly brittle material. Besides these four cast irons, another form of cast iron, called *chilled cast iron*, is white cast iron that is produced by cooling very rapidly through the solidification temperature range.

In cast irons, the ratio of cementite phase can be controlled or altered by adding graphite-forming elements such as Si, Al, Ni, and C that promote cementite decomposition and carbide-forming elements such as Mn, Cr, W, V, and Mo that dissolve in cementite and impede its decomposition. In actual practice, this is easily accomplished by varying the silicon content. The structure of cast irons depends on the rate of cooling of the casting.

The following subsections briefly describe different types of cast irons. The mechanical properties and applications of some commonly used cast irons are compared in Table 4.3.

4.1.1.1 Gray Cast Irons. Gray cast iron is a high-carbon iron-carbon-silicon alloy. Its chemical composition ranges from about 2 to 4 percent total carbon with 1 to 3 percent silicon. The excess carbon is present as flake graphite. The iron matrix of gray cast iron usually has a microstructure of pearlite, ferrite-pearlite, and ferrite. In pearlitic gray cast iron, the structure [Fig. 4.2(*a*)] is pearlitic with graphite inclusions. This pearlite phase usually contains about 0.7 percent carbon in combined state (as Fe_3C or cementite) and balance in free state, i.e., as graphite. In ferritic-pearlitic gray cast iron [Fig. 4.2(*b*)], the structure is ferrite and pearlite with graphite inclusions. In this form of cast iron, the quantity of combined carbon (cementite) is about 0.1 percent. In ferritic gray cast iron, the metallic matrix is ferrite [Fig. 4.2(*c*)], and all the carbon is present in the form of graphite.

It is observed that the metallic matrixes of gray cast irons are similar to those of eutectoid steel and hypoeutectoid steel or iron. Therefore, gray cast irons differ from steels in their structure only by the presence of graphite inclusions, which dictate the properties of cast irons.

A ferrite matrix yields a low-strength gray cast iron; a pearlite matrix yields higher strength; and a quench-hardened high-carbon matrix yields a high-strength martensitic matrix. The gray cast irons possess excellent machinability, low notch sensitivity, high compressive strength, high damping capacity, excellent wear resistance, and low cost. However, the impact strength of gray iron is lower than that of most other cast ferrous metals. The inclusions of graphite also result into excellent lubrication properties. Gray cast iron is used for wear resistance in both cast and hardened conditions.

4.1.1.2 Nodular Cast Irons. Nodular cast iron, also known as *ductile* or *spheroidal graphite cast iron*, is very similar to gray cast iron in composition, but during casting of nodular iron, the graphite is caused to nucleate as spherical particles rather than as flakes. This is accomplished through the addition of a very

TABLE 4.3 Selected Physical Properties and Applications of Common Cast Irons (Density ≈ 7000 to 7800 kg m^{-3})

Types of cast irons/grades (ASTM Standards)	Yield strength, MPa	Tensile strength, MPa	Modulus of elasticity, GPa	Hardness, BHN (HV, kg mm^{-2})	Thermal conductivity, W $m^{-1}K^{-1}$	Coefficient of thermal expansion, $\times 10^{-6}$ °C^{-1}	Typical properties	Typical applications
Gray cast irons:								
20	—	140–170	65–100	155 (165)	38–46	9–19	Excellent machinability, ability to damp vibrations, excellent wear resistance	Automotive engine blocks, gears, flywheels, brake disks and drums
40		275–330	110–140	235 (250)				
60		410–450	140–160	300 (320)				
Nodular cast irons:								
80-55-06	380	550	140–170	190 (200)	—	—	Best combination of stiffness, strength, shock resistance, wear resistance, more ductile than gray iron	Crankshafts, heavy-duty gears and rollers
100–70–03	483	680	150–170	200 (210)	—	—		
White cast irons	—	140–345	—	400 (425)	22	12	Hard, excellent wear resistance, but difficult to machine, very brittle	Mill liners, shot-blasting nozzles, rolling-mill rolls
Malleable cast irons:								
Ferritic 32510	224	345	170	155 max. (165)	—	—	Machinable and ductile.	Heavy-duty bearing surfaces, railroad rolling stock
Pearlitic and martensitic:								
40010	275	415	180	150–200	—	—	Stronger, harder, and more wear resistant	Rocker arms, wheel hubs, bearing caps
50005				(170–210)				
90001	345	480	180	180–230				
				(190–240)				
	620	725	180–195	270–320				
				(285–340)				

small but definite amount of magnesium to the molten iron in a process called *nodulizing*. The chief advantage of nodular cast iron is its combination of high strength and ductility with a high modulus of elasticity and high yield strength.

Nodular cast iron is used in such applications as crankshafts because of its good machinability, fatigue strength, and high modulus of elasticity, heavy-duty gears because of its high yield strength and wear resistance, and automotive door hinges because of its ductility.

4.1.1.3 Malleable Cast Irons. Malleable cast irons are white cast irons that have been converted to a malleable condition by a two-stage heat treatment that converts the large amount of cementite to graphite in the distinctive form called *tempered carbon nodules* [Fig. 4.2(*f*)], compared with graphite flakes in gray iron or true carbon spheroids in nodular iron. Long cooling cycles result in the production of ferritic malleable cast irons, while a more rapid cooling rate results in pearlitic malleable cast irons. Ferritic grades are more machinable and ductile, whereas the pearlitic grades are stronger and harder. Actually, there is a third type of malleable iron-martensitic, which is a pearlitic or ferritic grade that has been heat-treated and transformed to a martensitic structure.

Malleable cast irons, like ductile irons, possess considerable ductility and toughness because of their combination of nodule-shaped graphite in a low-carbon metallic matrix. Malleable cast irons are often used for heavy-duty cast bearing surfaces in automobiles, trucks, and railroad rolling stock. The carbon in all malleable cast irons helps retain and store lubricants and in extreme-wear service; the pearlitic malleable cast iron surfaces wear away in harmless particles with the size on the order of a micrometer that are less damaging than the other types of iron particles.

4.1.1.4 White Cast Irons. White cast irons do not have any graphite in their microstructure. Instead, all the undissolved carbon exists in the form of carbides, chiefly of the Fe_3C type (cementite). Although cementite is the hardest and strongest phase normally encountered in iron and carbon alloys, it is very brittle. The presence of a large amount of cementite in white cast iron as a continuous interdendritic network [Fig. 4.2(*g*)] makes the iron hard, strong, and wear resistant but extremely difficult to machine. For this reason, white cast iron, for the most part, is used as surface layers on chilled cast iron castings that will be subjected to impact and abrasive wear. The hardness and wear resistance of white cast irons can be improved by increasing the cooling rate after casting, heat-treating the casting, and including alloying elements such as Cr, Ni, and Mo.

4.1.1.5 Chilled Cast Irons. By using a chilled casting technique, the working surface of a casting can be made of abrasion-resistant white cast iron while the core is composed of a tougher and more easily machinable gray or nodular cast iron. During chilling, that portion of the casting which is to resist wear is cooled by a metal or graphite heat sink (chill) in the mold. When the molten iron comes into contact with the chill, it solidifies so rapidly that iron and carbon cannot become dissociated, so it becomes white cast iron. Chilled cast irons are used in railroad wheels, making use of the high wear resistance of white cast iron and the high damping capacity and machinability of gray cast iron. Other applications include stamp shoes and dies, crushing rolls, plowshares, and several other heavy machines.

4.1.1.6 High-Alloy Cast Irons. High-alloy cast irons are gray, nodular, or white cast irons that contain 3 to 30 percent alloy content. Properties of these irons, which are usually produced by specialized foundaries, are significantly different from those of unalloyed irons. Silicon is of primary importance in controlling the properties of alloy cast irons because silicon determines the relative proportions of graphite and combined carbon. Dissolved silicon also adds to the strength and wear resistance.

White high-alloy cast irons containing nickel and chromium develop a microstructure with a martensitic matrix around primarily chromium carbides. This structure yields a very high hardness with extreme wear and abrasion resistance. High-chromium irons (typically, containing about 16 percent chromium) combine wear and oxidation resistance with toughness. Irons containing from 14 to 24 percent nickel are austenitic; they provide excellent corrosion resistance for nonmagnetic applications. The 35 percent nickel irons have an extremely low coefficient of thermal expansion and are also nonmagnetic and corrosion resistant.

4.1.2 Steels

Steels are basically iron alloys containing carbon (maximum up to 2 percent) and several other alloying elements, such as Cr, Ni, Mn, Mo, Co, Si, Cu, Ti, B, Al, V, Zr, W, and Cb. The exact structure present in a particular steel depends on the composition (predominately carbon content) and heat treatment. Steels can be cast to shape or wrought into various mill forms. Steels are suitable for numerous applications in which good wear resistance and corrosion resistance are required. Factors that determine wear resistance include hardness and toughness (which depend on carbon content and other alloying elements), thermal treatments (conventional quenching and tempering and surface-hardening treatments), and processing (cold finishing and hot rolling).

Various microstructures are possible in steels, as illustrated in Fig. 4.1. At room temperature, the solubility of carbon in α-iron is very low, and therefore, the carbon atoms are found only very infrequently in between iron atoms. Instead, the carbon is combined with iron carbide, also called *cementite*, Fe_3C. The iron carbide may be present as lamellae alternating with lamellae of ferrite, which together form a constituent called *pearlite*, the mean carbon content of which is 0.80 percent. The proportion of pearlite in the structure increases with the carbon content of steel up to 0.80 percent. Carbon in excess of this amount separates as grain-boundary carbides. Steels containing less than 0.80, 0.80, and greater than 0.80 percent carbon are said to be *hypoeutectoid, eutectoid*, and *hypereutectoid steels*, respectively. Quench-hardening results in hard bainite or martensite structure.

The iron-carbon equilibrium diagram (Fig. 4.1) describes only the situation when equilibrium has been established between the components carbon and iron. In most heat treatments, the time parameter is one of the determining factors. Varying the rate of heating to the heating temperature and holding time will have an effect on the rate of transformation and dissolution of the constituents. The influence of time is best explained by means of the time-temperature transformation curves for steels with three different carbon compositions shown in Fig. 4.3. The structure created during cooling is dependent on the temperature of transformation and on the time taken for the transformation to start. Fast quenching inhibits the structural transformations possible under equilibrium conditions. Structural transformations resulting from various cooling programs are shown in Fig. 4.4. Formation of bainite and martensite is sought for hard and wear-resistant steels.

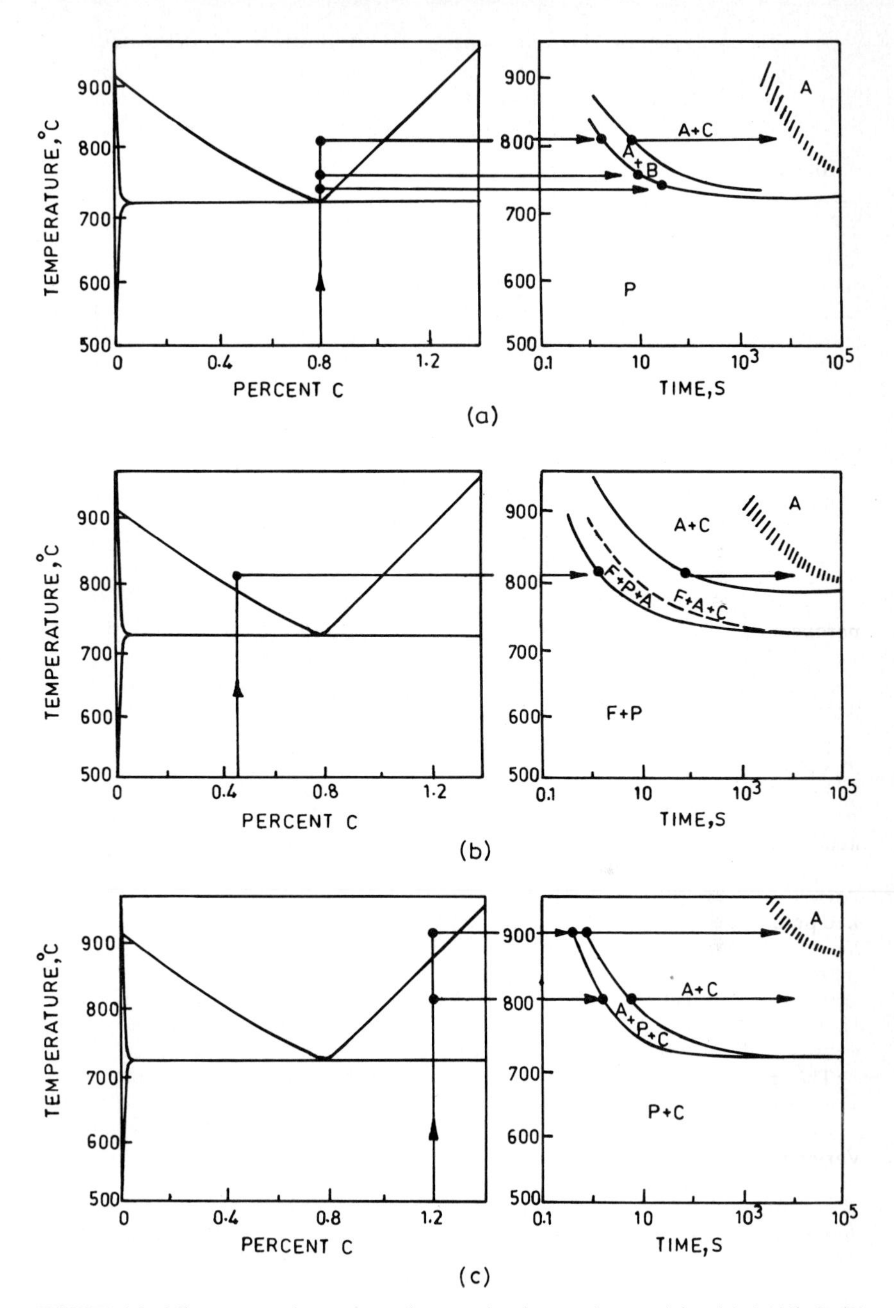

FIGURE 4.3 Microstructural transformations on heating steels containing (*a*) 0.80% C, (*b*) 0.45% C, and (*c*) 1.2% C. A = austenite, B = bainite, C = cementite, F = ferrite, P = pearlite.

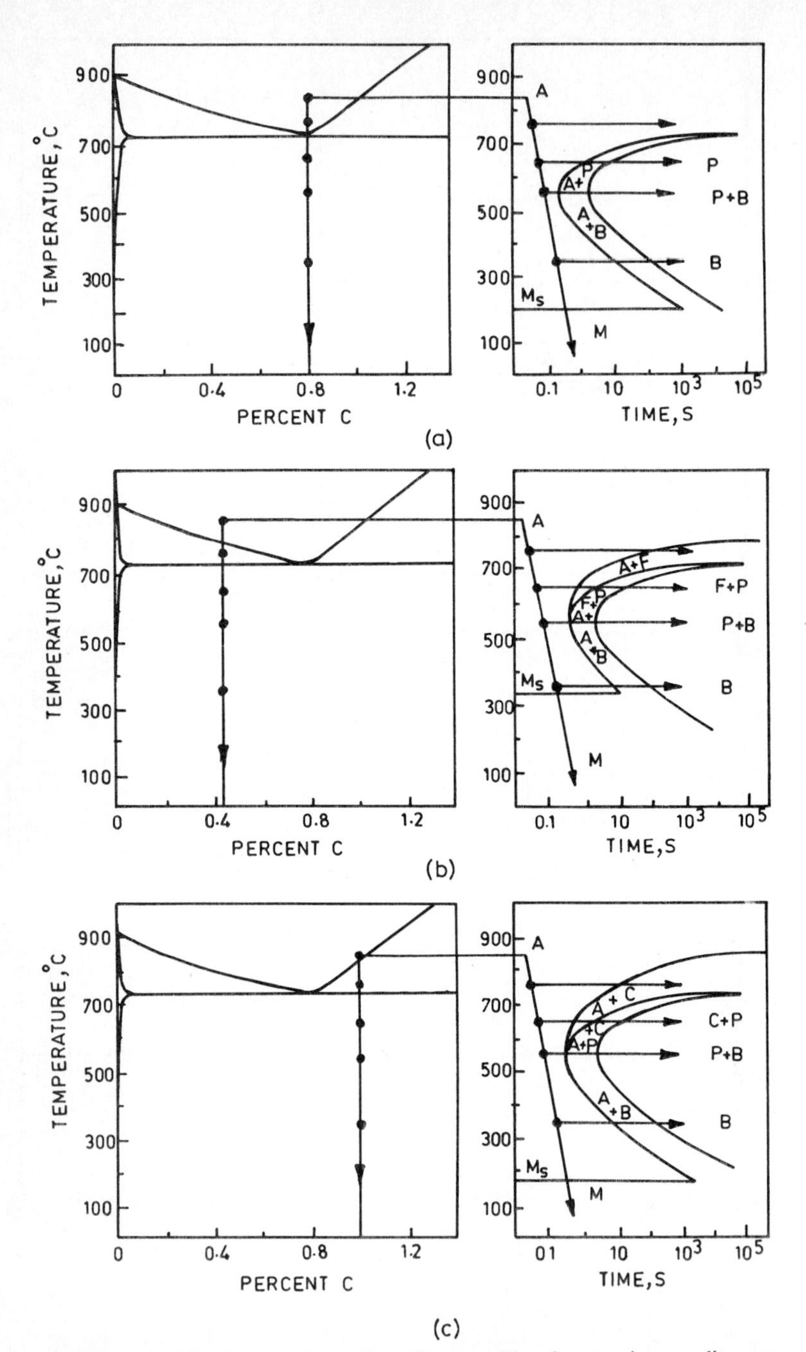

FIGURE 4.4 Microstructural transformations resulting from various cooling programs for steels containing (*a*) 0.80% C, (*b*) 0.45% C, and (*c*) 1.2% C. A = austenite, B = bainite, C = cementite, F = ferrite, P = pearlite, M = martensite, M_s = Start of martensite transformation.

Various heat treatments are also used for steels. The more important ones are described here (Fig. 4.5). Annealing, which involves heating until the austenitic structure is obtained and then cooling slowly, yields essentially the equilibrium structures shown in Fig. 4.1. Normalizing is like annealing, but a faster cooling rate (usually air cooling) is used. Normalizing yields steel of somewhat higher strength than annealing. Quench-hardening requires a fast quench from the austenizing temperatures. The resulting structure is martensite. Tempering is used to reduce the brittleness in quenched structures and to control working hardness. Stress relieving does not involve a transition in crystal structure and only serves to remove strains from the fabrication process. Spheroidizing involves long heat-treating times to cause microconstituents to agglomerate in spherical phases to aid machining or subsequent heat treatments. Austempering is simply an interrupted quench. It produces a semihard bainite structure. Martempering is similarly an interrupted quench, but the resulting structure is martensite. Martempering is used to minimize quenching distortion (Thelning, 1984).

Various alloying elements considerably improve strength, increase hardenability, refine the grain, and increase the resistance to softening on heating to moderate temperature (Anonymous, 1978a, 1978c; Thelning, 1984). Chromium forms a very stable, hard, wear-resistant carbide in the steel. The carbide is stable at high temperatures, making the alloy useful for elevated-temperature applications. Chromium also increases hardenability and promotes carburizing. Nickel-alloy steels exhibit improved toughness and ductility at low temperatures. Chromium and nickel also improve corrosion resistance. Manganese is a strengthening and hardening element that behaves in the same manner as carbon. Molybdenum increases high-temperature tensile strength and creep strength. Most important, however, molybdenum is second only to manganese in its ability to improve hardenability. Alloy steels specially formulated for nitriding contain from 0.90 to 1.3 percent aluminum. The aluminum combines with the nitrogen in the furnace atmosphere to form a hard, wear-resistant aluminum-nitride case.

Vanadium in concentrations of about 0.04 percent improves hardenability and resists grain growth during thermal treatment. Silicon increases oxidation resistance and also improves ductility and shock resistance in higher concentrations. Copper is added for corrosion resistance, and boron in concentrations as low as 0.0005 percent can significantly increase the hardenability of steel. When phosphorous and sulfur are present as impurities in high concentrations, these elements decrease ductility, although they are intentionally added to some steels to

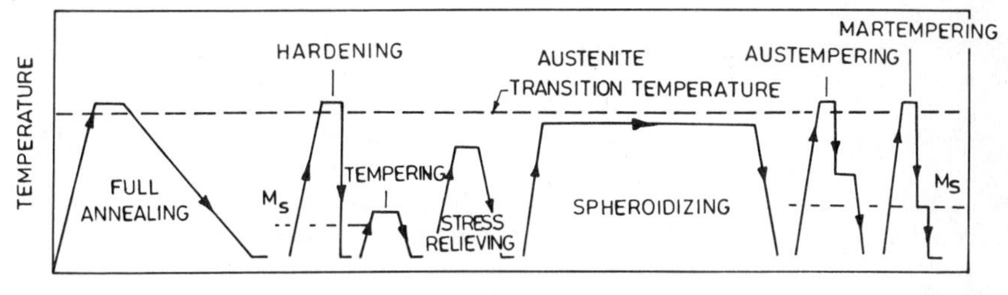

FIGURE 4.5 Heat treating thermal cycles for steels. M_s is the temperature for start of martensite formation.

improve machinability. Columbium and titanium are added to form stable carbides and reduce susceptibility to intergranular corrosion.

Table 4.4 lists selected mechanical and thermal properties of commonly used steels that belong to different families. Details of various steels follow.

4.1.2.1 ***Plain Carbon Steels.*** Plain carbon steels account for nearly 90 percent of total steel production. Plain carbon steel is a steel with up to 2 percent carbon and a trace amount of other elements (unavoidably present) such as Si, Mn, P, S, B, and Cu. Specifically, steels in this family are limited to a maximum of 1.65 percent Mn, 0.6 percent Si, and 0.6 percent Cu. There are three classifications: low-carbon steels, with less than 0.3 percent C, medium-carbon steels, with 0.3 to 0.6 percent C; and high-carbon steels, with a carbon content exceeding 0.6 percent. The microstructure of plain carbon steels depends on the carbon content and the heat treatment. Many of the carbon steels used commercially, whether wrought or cast, are primarily pearlitic in microstructure. Carbon steels are the familiar 1000 and 1100 series of AISI steels.

Carbon is the principal hardening and strengthening element in carbon steels. Plain carbon steels (10XX) are available with many different carbon contents. The carbon content of these alloys determines their hardening and strength characteristics. At least 0.4 percent carbon is required to develop a high hardness (600 BHN) through heat treatment; these steels are referred to as *direct-hardenable* (e.g., AISI 1060, 1080, 1090, and 1095). Low-carbon grades (e.g., AISI 1008, 1010, 1020, and 1025) cannot normally be direct-hardened. They can be hardened by cold working, but hardnesses seldom exceed 200 BHN. For high hardness, they can be case-hardened by carburizing and then quench-hardening or by other techniques such as carbonitriding.

Machinability in carbon steels can be improved in several ways, including (1) addition of such elements to the steel as lead, phosphorus, sulfur, tellurium, selenium, and bismuth, (2) cold finishing, (3) reducing the level of residual stress, and (4) adjusting the microstructure to minimize gumminess or hardness. A family of free-machining carbon steels with carbon contents from 0.1 to 0.5 percent (e.g., AISI 1110 and 1151) has been developed for fast and economical machining of parts. These steels are available as castings and as hot-rolled or cold-drawn bar stock and plate.

4.1.2.2 ***Low-Alloy Steels.*** Low-alloy steels are a low-cost material that can be processed in numerous ways to resist wear. A steel is considered to be an alloy when the content of the alloying element exceeds one or more of these limits: 1.65 percent Mn, 0.6 percent Si, or 0.6 percent Cu, or when a definite concentration of following elements is specified or required: Al, Cr (to 3.99 percent), Co, Mo, or other elements added to obtain an alloying effect. Technically, then, stainless and tool steels, for example, are alloy steels. Low-alloy steels are those which contain a small amount of alloying elements and depend on thermal treatment for the development of specific mechanical properties. Low-alloy steels are the AISI grades that start with digits 15XX, 31XX, 40XX, 41XX, 51XX, 86XX, etc.

The alloying elements make possible a large selection of microstructures and combinations of hardness and toughness. Several low-alloy steels provide a good substrate for surface treatments such as carburizing, nitriding and phosphating, hardfacing, chromium plating, and plasma spraying. Low-alloy steels react well to the conventional lubricants used for many tribological applications.

A large portion of low-alloy steels can be subdivided into (direct- or) through-

TABLE 4.4 Selected Physical Properties and Applications of Common Steels (Density ≈ 7600 to 8000 kg m^{-3}, melting point ≈ 1535°C)

Types of steels/grades	Yield strength, MPa	Tensile strength, MPa	Modulus of elasticity, GPa	Hardness (HV, kg mm^{-2})	Thermal conductivity, $Wm^{-1}K^{-1}$	Coefficient of thermal expansion, $\times 10^{-6}$ °C^{-1}	Typical applications
Plain-carbon steels (cast structural grades):							
Low strength*	240	435	190	130 BHN (140)	16	16.2	Used for high-production items such as automobiles and appliances; also used in machine design for baseplates, housings, structural members, etc.
High strength‡	520	725	200	210 BHN (220)	25	10.0	
Low-alloy steels:							
Carburizing grade							
AISI 4320*	420	580	—	600 BHN (640)	—	—	Used for structural components that are heat treated for wear, strength, and toughness; also used for axle shafts, gears, and hand tools; E 52100 used for rolling-element bearings, etc.
AISI 4320†	725–740	1020–1050	—		—	—	
Through-hardening grade							
AISI 4340*	470	745	—	525 BHN (560)	42	12.4	
AISI 4340‡	880–1080	960–1140	—	—			
AISI E52100‡	—	—	—	62–65 HRC (750–830)	—	—	
HSLA Steels:							
ASTM A607	310–345	445–485	—	—	—	—	Better strength-to-weight ratio compared with conventional low-carbon steels for automotive industry; HSLA bars are used in such equipment as power cranes, cement mixers, and power transmission towers
Stainless steels:							
Austenitic (Cr-Ni)							
AISI 303*	240	620	190	160 BHN (170)	16	17.3	Used where corrosion resistance and toughness are of primary concern, e.g., shaft pumps and fastners in the pumping industry
AISI 304*	240	560	190	150 BHN (160)	16	17.3	
AISI 316*	250	560	190	150 BHN (160)	16	16.0	

TABLE 4.4 Selected Physical Properties and Applications of Common Steels (Density ≈ 7600 to 8000 kg m^{-3}, melting point ≈ 1535°C) (*Continued*)

Types of steels/grades	Yield strength, MPa	Tensile strength, MPa	Modulus of elasticity, GPa	Hardness (HV, kg mm^{-2})	Thermal conductivity, $Wm^{-1}K^{-1}$	Coefficient of thermal expansion, $\times 10^{-6}$ °C^{-1}	Typical applications
Semiaustenitic (Cr-Ni, precipitation hardenable)							
631 (17-7 PH)*	580–1280	900–1400	195	160 BHN (170)	16	10.3	Used where high strength, moderate antigalling, and moderate corrosion resistance and good fabricability are required, e.g., foil bearings
631 (17-7 PH)§	1790	1825	—	49 HRC (520)	—	—	
Austenitic (Cr-Ni-Mn-N)							
AISI 201*	380	790	195	90 HRB (210)	—	15.7	Used where extreme corrosion resistance and high strength are required, e.g., seawater pumps, food industry, chemical processing applications
Nitronic 50*	410	830	190	98 HRB (225)	16	16.2	
Nitronic 60*	390	720	190	85 HRB (200)	—	15.8	
Ferritic (Cr)							
AISI 442*	345	550	200	170 BHN (180)	21.6	8.8	Used where moderate corrosion resistance is required and toughness is not a major need, e.g., automotive trim, exhaust systems, and heat-transfer equipment in the petrochemical industry
AISI 446*	345	550	220	170 BHN (180)	20.9	10.4	
Martensitic (Cr, precipitation hardenable)							
AISI 416*	275	515	200	155 BHN (165)	25	9.9	Used where hardness, strength, and wear and corrosion resistance are of concern, e.g., rolling-element bearings and shafts, aircraft structural parts, and turbine components
AISI 416§	—	—	—	42 HRC (430)	—	—	
AISI 440C*	450	760	200	230 BHN (240)	24	10.1	
AISI 440C§	—	—	—	60 HRC (635)	—	—	
Martensitic (Cr-Ni, precipitation hardenable)							
630 (17-4 PH)§	1170	1310	195	45 HRC (470)	17	10.8	Used where high strength, moderate corrosion resistance, and good fabricability are required, e.g., shafts, high-pressure pumps, high-temper springs, and aircraft components

Tool Steels:							
A-8*	450	710	—	100 HRB (225)	—	—	Specified for thin parts that are prone to crack or distort on hardening; used for applications such as machine ways and fuel-injector nozzles
A-8‡	1900	2350	—	60 HRC (635)	—	—	
S-1*	380	690	—	95 HRB (200)	—	—	High toughness, low hardness, and wear resistance; resist sudden and repeated loading; used for punches, shear blades, pneumatic tooling parts, springs
S-1‡	1890	2070	—	55 HRC (580)	—	—	
P-20*	525	700	—	97 HRB (215)	—	—	Used for plastic-molding and zinc die-casting dies
P-20‡	1435	1890	—	52 HRC (550)	—	—	
H-11 (AISI 610)*	365	690	210	95 HRB (210)	28	10.6	Moderate wear resistance, high toughness, low distortion during heat treatment; used for aircraft structural parts
H-11 (AISI 610)‡	1725	2035	—	55 HRC (580)	—	—	
M-2§	—	—	—	65–68 HRC (830–930)	21.3	10.1	High hardness and wear resistance, low toughness, high resistance to softening at low temperature; used for cutting tools, rolling-element bearings and shafts, etc.
M-50 (AISI 613)§	2300	2800	205	65–70 HRC (830–1000)	—	—	
L-2*	510	710	—	95 HRB (210)	—	—	Used where toughness and wear resistance are primary concern, e.g., cams, chucks, and collets
L-2‡	1790	2070	—	55 HRC (580)	—	—	
Iron-based superalloys:							
AISI 613 (M-50)	2300	2800	205	65–70 HRC (830–1000)	—	—	Used for high-temperature applications, e.g., high-temperature aircraft bearings and shafts, steam turbine blades
AISI 616	860–1200	1030–1650	200	—	—	—	
AISI 660 (ASTM A286)	650	1010	200	—	13	16.5	

*Annealed.
†Carburized.
‡Hardened and tempered.
§Hardened.

hardenable and carburizing grades (plus several specialty grades, such as nitriding steels). The medium-carbon (typically 0.4 to 0.6 percent) grade of alloy steel can be direct-hardened, and the purpose of various alloying elements is to control the hardenability—the ease with which a steel will harden. Through-hardenable grades (e.g., AISI 4140, 4340, 4615, and E52100), which are heat treated by quenching and tempering, are used when maximum hardness and strength must extend deep within a part. Type E52100 steel is a special grade of alloy widely used for rolling-element bearings. The high carbon (1 percent) and alloy content (0.35 percent Mn and 1.5 percent Cr) improves hardenability and promotes excess carbides in the structure.

If the carbon content is low (below 0.2 percent), the steels will not develop high surface hardness by direct-hardening. They must be subjected to some process such as carburizing to increase the alloy content at the surface. Carburizing grades (e.g., AISI 4320, 8620, and 9310) are used where a tough core and a relatively shallow, hard surface are needed. After surface-hardening treatments, these steels are suitable for parts that must withstand wear as well as high stresses.

4.1.2.3 HSLA Steels. The high-strength, low-alloy (HSLA) steels were developed to achieve high strength-to-weight ratios over those of conventional plain carbon steels. HSLA steels have a microstructure consisting mainly of ferrites, which imparts a fairly good toughness to the steel. An increase in carbon increases the amount of pearlite, which, in turn, causes a decrease in toughness. HSLA steels are low-carbon steels that are strengthened by the addition of small amounts of alloying elements, such as Mn, Cu, V, and Ti, and sometimes by special rolling and cooling techniques. Improved-formability HSLA steels contain proprietary additions and, in some cases, zirconium or rare earth elements such as cerium or lanthium for sulfide shape control. The improved grade has yield strengths up to 550 MPa, yet cost only about 25 percent more than a typical 240-MPa plain-carbon steel. HSLA steels were developed primarily for the automotive industry to replace low-carbon steels with steels with a thinner cross section for reduced weight without sacrificing strength. HSLA bar, with a minimum yield strength ranging from 345 to 480 MPa, is used in such equipment as power cranes, cement mixers, and power transmissions.

4.1.2.4 Stainless Steels. Effective resistance to atmospheric corrosion begins when iron-chromium alloys contain a minimum of about 10 percent chromium. Consequently, to be classed as stainless, steels must contain at least that much chromium. Larger amounts are added to improve the alloy's corrosion resistance and its resistance to oxidation at high temperatures. Chromium forms a nearly impervious, tightly adherent oxide-rich film on the surface of the steel as a result of oxidation that resists corrosion and further oxidation. If oxygen is present, any breaks in the film are repaired because the oxide film re-forms immediately. Nickel is added to chromium-iron alloys for a variety of reasons. In excess of about 6 percent, nickel increases corrosion resistance to some degree and greatly improves mechanical and fabricating properties. Manganese is sometimes added in conjunction with small amounts of nitrogen as a substitute for nickel.

Other elements are also added in small amounts. For example, molybdenum increases resistance to pitting-type corrosion and also improves high-temperature strength. Silicon increases oxidation resistance at high temperatures. Sulfur and selenium impart free-machining characteristics. Columbium and titanium are added to form stable carbides and reduce susceptibility to intergranular corrosion

(Anonymous, 1973). In general, the ability of each stainless steel to resist corrosion and oxidation is proportional to its chromium content and, within limits, its nickel and molybdenum content.

Stainless steels are classified on the basis of their metallurgical structure as austenitic, ferritic, or martensitic (Anonymous, 1978c, 1980a). Most of the Cr-Ni stainless steels and Cr-Ni-Mn nitrogen-strengthened steels are austenitic. However, some Cr-Ni steels are exceptions with semiaustenitic or martensitic structures, and these are seen more frequently as new classes of stainless such as precipitation-hardening types are developed. The stainless steels containing only chromium as their essential alloying element are classed as ferritic or martensitic. Martensitic steels, because of their carbon content and balance of alloying elements, can be precipitation-hardened. All other steels can be hardened by cold drawing (or cold working). The advantage of precipitation-hardening is that the components can be heat treated (hardened and tempered) after they have been machined in the annealed condition. Typical mechanical properties and composition of selected stainless steels are presented in Tables 4.4 and 4.5.

Austenitic Cr-Ni stainless steels (e.g., AISI 303, 304, 316, etc.) have a tendency to gall and cold weld when in dry sliding contact. These are used where toughness and corrosion resistance are of primary concern and the material should be nonmagnetic. Austenitic Cr-Ni-Mn nitrogen-strengthened stainless steels (e.g., Nitronic* 32, 50, and 60 and AISI 201, 202, etc.) provide a combination of exceptional corrosion resistance and strength.

Nitronic 50 steel provides the maximum corrosion resistance (Bhushan and Gray, 1978, 1979; Denhard and Espy, 1971). In these steels, the manganese enhances work-hardening characteristics, and the nitrogen is a solid solution-strengthening agent, acting like carbon. As a result, the galling and metal-to-metal wear characteristics of austenitic Cr-Ni-Mn nitrogen-strengthened steels are superior to austenitic Cr-Ni steels. The martensitic stainless steels (e.g., AISI 403, 410, 416, 420, 440C, 630, or 17-4 PH) and precipitation-hardenable semiaustenitic steels (e.g., AISI 631 or 17-7 PH) have high hardness and high abrasion resistance but low impact resistance (Schumacher, 1976). Ferritic steels (e.g., AISI 405, 430, 442, 446, etc.) fall some place in between austenitic Cr-Ni and martensitic steels as far as galling and wear resistance are concerned.

Precipitation-hardenable martensitic stainless steels, such as AISI 440C, behave like tool steels. These are widely used for dies, cutting devices, and rolling-element bearings that operate at temperatures up to 315°C and for instrument bearings where rusting from moist air will impair stringent performance requirements.

4.1.2.5 ***Tool Steels.*** Tool steels have been developed to obtain high strength at elevated temperatures, good corrosion resistance, high hardness, and good wear resistance. The tool steels are able to maintain hardness and cutting edges at the high temperatures generated during metal cutting. Tool steels are carbon or alloy steels that can be hardened and tempered. However, the real factor that differentiates tool steels from the carbon or alloy steels previously discussed is the manufacturing technique. Tool steels are usually metallurgically clean, high-alloy steels that are melted in electric furnaces and produced with careful attention to homogeneity. Tool steels are always heat-treated to develop their best properties. Tool steels are classified into seven subgroups on the basis of alloy composition, hardenability, and mechanical similarities (Budinski, 1980).

---

*Registered trademark of Armco Steel Company.

TABLE 4.5 Composition of Commonly Used Stainless Steels

Stainless Steel	Cr, %	Ni, %	C_{max}, %	Other significant elements, %
AISI 303	17–19	8–10	0.15	S, 0.15 minimum
AISI 304	18–20	8–10	0.08	—
AISI 316	16–18	10–14	0.08	Mo, 2–3
AISI 631 (17-7 PH)*	16–18	6.5–7.8	0.09	Al, 0.75–1.25
AISI 201	16–18	3.5–5.5	0.15	Mn, 5.5–7.5; N, 0.25 maximum
Nitronic 50 (22-13-5)	20.5–23.5	11.5–13.5	0.06	Mn, 4–6; Mo, 1.5–3.0; Si, 1 maximum; N, 0.2–0.4; V, 0.1–0.3; Cb, 0.1–0.3
Nitronic 60	16–18	8–9	0.10	Mn, 7–9; Si, 3.5–4.5; N, 0.08–0.18
AISI 442	18–23	—	0.25	—
AISI 446	23–27	—	0.20	—
AISI 416	12–14	—	0.15	S, 0.15 minimum
AISI 440 C	16–18	—	0.95–1.20	—
AISI 630 (17-4 PH)*	15.5–17.5	3–5	0.07	Cu, 3–5

*The numbers in the parentheses are more widely used designations. They correspond to the nominal chromium and nickel contents.

Water-hardening or *carbon tool steel* (designated by W by AISI) rely solely on carbon content for their useful properties. The surface hardness of these steels is usually in the range of 62 to 66 HRC, and they have a tough core. These characteristics make these steels candidates for wear components, such as knives and chopper blades, that are likely to be subjected to severe shocks. *Cold-work tool steels*, which include oil- and air-hardening types O, A, and D, are often more costly, but they can be quenched less drastically than water-hardening types and so are specified for thin parts. Type O steels (high carbon and low alloy) are *oil-hardening*; types A (medium alloy) and D (high carbon and high chromium) are *air-hardening*.

Shock-resistant tool steels (type S) are medium-carbon steels that are strong and tough but not as wear resistant as many other tool steels. These alloys are normally used for chisel-type applications. Type P mold steels are designed specifically for plastic-molding and zinc die-casting dies. These steels are seldom used for nontooling components.

Hot-work tool steels (type H) serve well at high temperatures. There are three categories of hot-worked tool steels containing either Cr, W, or Mo as the principal alloying element. The W and Mo alloy hot-work steels are heat and abrasion resistant even when they are subjected to 500 to 600°C. Modified hot-work steels (e.g., H11 or AISI 610) are used in landing gears and hydraulic and hinge parts because of their combined impact resistance, wear resistance, and high strength-to-weight ratio.

High-speed tool steels derive their name from their original intended application—to make tools that can machine other metals at high cutting speeds. They contain the highest percentage of alloying elements of any of the tool-steel classes. There are basically two classes of high-speed steels (HSS), the M and the T series. The M series contains molybdenum as a major alloying element, and the

T series contains tungsten. The M series is more widely used than the T series. Molybdenum in high-speed steels increases hardenability and promotes formation of complex alloy carbides similar to tungsten. Molybdenum's contribution to hot hardness is not as effective as that of tungsten. Thus many of the M series steels contain significant concentrations of both W and Mo. High-speed steels are extremely hard (65 to 70 HRC) and do not soften at elevated temperatures. They are used in cutting tools, dies, mandrels, screws, knives, and rolling-element bearings (for operation at temperatures up to 430°C). The most commonly used high-speed steels are M50 or AISI 613 and M2 (1 percent C, 6 percent W, 5 percent Mo, 4 percent Cr, 2 percent V, and the balance Fe).

Other grades, called *special-purpose tool steels*, include low-cost type L steels, which are low-alloy steels that are specified when wear resistance combined with toughness is important. However, these are not suitable for high temperatures or for shock service.

4.1.3 Ferrous Powder Metallurgy

Most metals can be fabricated into finished parts by the powder metallurgy (PM) process, but those most commonly fabricated are iron-based alloys. In the PM process, the metal powders are compacted at high pressure (100 to 200 MPa) to a high density by double pressing (coining), isostatic pressing, and cold and hot forging and are sintered at elevated temperatures, i.e., 1000 to 1200°C (Dreger, 1978).

Powder metallurgy processes are widely used for making parts that are difficult and more expensive to produce by other methods. However, the conventional powder-compacting system limits part geometry to shapes that can be removed from the die cavity in one direction. Low-density iron PM parts (5600 to 6000 kg m^{-3}) with a typical tensile strength of 110 MPa and modulus of elasticity of 70 GPa are usually employed in bearing applications. The addition of alloying elements to the iron-powder mix improves the mechanical properties. Copper is commonly added to improve both strength and bearing properties. The strength of iron PM or sintered iron-copper alloy may be varied by adjusting density, carbon content, and copper content (up to 15 percent) to satisfy specific design requirements. Low-density PM parts are used in bearing applications because they provide porosity for oil storage. Impregnating a sintered-metal bearing with oil usually eliminates the need of relubrication.

Carbon enables PM steel parts to be heat-treated to increase hardness, toughness, wear resistance, and strength. Ferrous PM parts containing 0.3 percent or higher combined carbon can be quenched-hardened for increased strength and wear resistance. Surface-hardness values of 500 to 650 kg mm^{-2} on a Knoop scale can be obtained by quenching.

4.1.4 Iron-Based Superalloys

Iron-based superalloys are defined as those alloys which have iron as the major constituent and which are hardened by a carbide or intermetallic precipitate. Superalloys are used primarily for elevated-temperature applications, such as in gas turbine engines. The iron-based grades (with high nickel and chromium content) are less expensive than cobalt- or nickel-based superalloys. Iron-based superalloys can be grouped into three types of alloys that (1) can be strengthened by a martensitic type of transformation, (2) are austenitic and are strengthened by a sequence of hot and cold working, and (3) are austenitic and are strengthend by precipitation-hardening. These alloys (AISI 600 series) consist of six subclasses of iron-based alloys: 601 through 604 (martensitic low-alloy steels), 610 through 613 (martensitic secondary-hardening steels), 614 through 619 (martensitic chromium steels), 630 through 635 (semiaustenitic and martensitic precipitation-hardening stainless steels); 650 through 653 (austenitic steels strengthened by hot or cold working), and 660 through 665 (austenitic superalloys; all but alloy 661 are strengthened by second-phase precipitation) (Anonymous, 1980a). Types H and M tool steels are often classified as iron-based superalloys. In general, the martensitic types are used at temperatures below 550°C, and the austenitic types are used above 550°C.

The significant properties of iron-based superalloys are high-temperature as well as room-temperature strength and resistance to creep, oxidation, corrosion, and wear. Wear resistance increases with carbon content. Maximum wear resistance is obtained in martensitic secondary-hardening steel alloys (AISI 610 through 613), which are used in high-temperature aircraft bearings and machinery parts subjected to sliding contact. Austenitic steel alloys (such as AISI 660 or ASTM A286), which can be strengthened by second-phase precipitation, are also used for high-temperature aircraft bearing journals because of their machinability, availability, and low cost. The composition of ASTM A286 is 26 percent Ni, 15 percent Cr, 2 percent Ti, 1.3 percent Mo, 1.3 percent Mn, 0.5 percent Si, 0.2 percent Al, 0.05 percent C, 0.015 percent B, and the balance Fe. Oxidation resistance increases with chromium content. The martensitic chromium steels, particularly AISI 616, are used in steam turbine blades.

4.2 NONFERROUS METALS AND ALLOYS

Although ferrous alloys are specified for more engineering applications than are nonferrous metals, the nonferrous metals and alloys offer a wide variety of characteristics and mechanical properties. Like ferrous metals and alloys, most nonferrous alloys can be cast to shape or wrought into various mill forms. Because of their structure, these metals can be cold worked and/or heat treated (Amateau and Glaeser, 1964; Anonymous, 1978b, 1978c). Availability, abundance, and the cost of converting the metal into useful forms all play important roles in the selection of a nonferrous metal or alloy. Another important aspect in material selection is how easily it can be shaped into a finished part and how its properties can be either inadvertently or intentionally altered in the process in order to meet the functional requirements.

Several nonferrous metals (such as indium, thallium, lead, tin, cadmium, barium, and precious metals) and alloys based on tin, lead, copper, and aluminum are relatively soft and exhibit low shear strength, good embeddability (ability to embed wear particles into the softer lining), and excellent conformability. These soft alloys have been used as linings for oil-lubricated plain bearings and other sliding applications (Dean and Evans, 1976; DeHart, 1984; Hausner et al., 1970; Kaufman, 1980; Kirkham et al., 1983; Pratt, 1973). Thin films of precious metals are used for lightly loaded unlubricated sliding applications. For operations in which some intermittent metallic contact occurs, attention must be paid to wear resistance and the compatibility of the mating surface. In most cases, plain bearings operate against a plain-carbon or low-alloy steel shaft that has been either normalized or heat treated. For example, a hardened shaft is recommended against a harder bronze bearing.

Another class of nonferrous alloys includes nickel- and cobalt-based alloys, which are hard and can be used at elevated temperatures. They offer good wear resistance under corrosive conditions (Anonymous, 1978c). Cobalt-based alloys have the unique ability to minimize corrosion, wear, and galling (a mild form of adhesive wear) simultaneously in many sliding systems. Nickel-based alloys offer better resistance to high temperatures than cobalt-based alloys, but their galling resistance is far inferior. The galling resistance of nickel-based alloys can be improved by adding various alloying elements such as Mo and/or Co.

Refractory metals, which have melting points above that of iron (1535°C), usually in the range of 1600 to 3340°C, are characterized by high-temperature strength and corrosion resistance. Although a number of metals meet this requirement, the least costly and most extensively used are tungsten, tantalum, molybdenum, and columbium. Another metal used for tribological application is titanium and its alloys. High-purity single-crystal silicon is used for semiconductor applications and sometimes as a smooth substrate for tribological coatings (Hampel, 1961).

Discussions of common nonferrous metals and their alloys used for tribological applications follow.

4.2.1 Precious Metals

The family of metals called *precious metals* consists of three subgroups: silver and silver alloys, gold and gold alloys, and the platinum metals, which are platinum, palladium, rhodium, ruthenium, iridium, and osmium. The six platinum metals are so grouped because they occupy the lower two rows of group VIII of the periodic table, together with silver and gold, which occupy adjacent positions in group I (Anonymous, 1978b; Hampel, 1961). These precious metals are soft metals that have low shear strengths, making them attractive as solid lubricants for special situations. They resist corrosion well, and to the extent that they tarnish in air, they form very thin oxide layers. Of all these precious metals, gold does not oxidize under normal atmospheric conditions. The precious metals also have high densities and high melting temperatures. Selected physical properties and typical applications of precious metals are presented in Table 4.6.

Silver is the least costly of the precious metal group. Silver, like gold and platinum metals, is very malleable, ductile, and corrosion resistant. Silver has the highest thermal and electrical conductivity of all metals. Gold is an extremely soft, ductile metal that undergoes very little work-hardening. Since pure gold is soft, it is often alloyed with other metals such as copper or nickel for greater strength and hardness. Platinum is inert and stable even at high temperatures. Palladium, rhodium, and iridium resemble and behave like platinum. Ruthenium and osmium are used as alloying elements to increase the hardness and electrical resistivity of platinum or other metals in this group. Rhodium has high hardness and exhibits high wear resistance.

Typical applications for precious metals include electrical contacts and wires for electronic products because of their chemical inertness and high melting points (Dillich and Kuhlmann-Wilsdorf, 1982; Ohmae and Rabinowicz, 1980; Steiger, 1982). Silver and gold are also used as additives in MoS_2- and graphite-based and calcium fluoride–based solid-lubricant coatings (see Chapter 13). Silver is added in a self-lubricating cage material for high-temperature rolling-

TABLE 4.6 Selected Physical Properties and Applications of Precious Metals

Material	Density, kg m^{-3}	Yield strength, MPa	Tensile strength, MPa	Modulus of elasticity, GPa	Hardness, Mohs*	Thermal conductivity, Wm^{-1} K^{-1}	Coefficient of thermal expansion, $\times 10^{-6}$ $°C^{-1}$	Melting point, °C	Electrical resistivity, $\mu\Omega$ cm^{-1}	Typical applications
Silver, annealed	10,500	55	150	75	2.5–3	420	19.6	960	1.47	Used for lightly loaded bearing applications, electrical contacts and wires for electronic products
50 percent cold worked	—	303	370	—	—	—	—	—	—	
Gold, annealed	19,300	0	135	83	2.5	300	14.2	1060	2.19	
50 percent cold worked	—	210	220	—	—	—	—	—	—	
Platinum, annealed	21,450	14–38	125–190	172	4.3	73	9.2	1770	9.83	
50 percent cold worked	—	190	195–220	—	—	—	—	—	—	
Palladium, annealed	12,000	35	140–190	125	—	70	11.7	1550	10.0	
Rhodium, annealed	12,380	—	500	325	4.5–5	86	7.9	1962	4.51	

*For comparisons, hardness of talc, gypsum, and calcite on the Mohs' scale is 1, 2, and 3, respectively.

element bearings. This cage material was initially developed by Westinghouse and is made of a silver-mercury-PTEF-molybdenum diselenide composite (Dayton et al., 1978).

4.2.2 Tin- and Lead-Based Alloys

Tin- and lead-based alloys are commonly known as *babbitt* or *white metals* and are among the most widely used bearing materials for lubricated conditions (Anonymous, 1978c; Bhushan and Winn, 1979; Dean and Evans, 1976; DeHart, 1984). These tin- or lead-based alloys are soft metals (~1.5 to 2 Mohs) that have excellent embeddability and conformability. They are unsurpassed in compatibility and thus prevent shaft scoring. They are normally cast onto a steel, bronze, or brass or directly into the bearing housing. In large rotating equipment, such as large motors, turbines, and generators, the finished machined babbitt can be as thick as 13 mm. This thickness provides a safety feature against shaft damage in case of loss of lubricant or other extreme conditions. Fatigue strength is increased by decreasing the thickness of the babbitt lining. The babbitt alloys will operate with hard or soft journal surfaces.

Tin-based alloys consist of a tin-rich matrix, with antimony up to 14 percent and copper up to 10 percent in solid solution; dispersed in the matrix are cuboids of Sb-Sn and needles of Cu_6Sn_5. The increase of Sb and Cu in tin matrix increases the hardness and tensile strength and decreases ductility. Lead-based alloys generally contain antimony and tin and have a very small quantity of copper. Antimony contents greater than 15 percent cause the material to become brittle. The tin content helps to improve the corrosion resistance. Addition of arsenic improves the mechanical properties at temperatures of 90 to 150°C (Table 4.7).

The choice between a tin- or lead-based alloy is a matter of convenience rather than a clear-cut advantage of one alloy over the other. Both alloys provide the best combination of properties for bearings, particularly their antiseizure characteristics. Tin-based alloys are more easily bonded, have better corrosion resistance, and are less likely to segregate, but lead-based alloys are cheaper. Today, many high-performance oil-lubricated bearings use some type of plated babbitt alloy of nominally 10 percent tin, with about 3 percent copper often added to confer additional hardness.

4.2.3 Copper-Based Alloys

A wide variety of copper-based alloys are used directly as bearing materials and also as backing materials for babbitt linings. Their use in the former category covers a range of physical properties, in terms of hardness and strength, that is wider than that of any other group of bearing alloys. Copper also has the highest thermal conductivity of any commercial metal, which allows for rapid dissipation of frictional heat. Copper and its alloys can be hot or cold worked, although they work-harden. The copper-based alloys can be cast or sintered (Anonymous, 1978b, 1978c).

The copper-based alloys used for tribological applications can be grouped according to composition into several general categories: brasses, bronzes, copper nickels, nickel silvers, and beryllium copper (see Table 4.7). Brasses and bronzes are very commonly used in both oil-lubricated and unlubricated applications, and sintered bronze or porous bronze (with 10 to 35 percent interconnected porosity) is also used in which the pores have been filled with oil.

TABLE 4.7 Selected Physical Properties and Applications of Soft Nonferrous Metals and Alloys

Materials/ compositions	Density, kg m^{-3}	Yield strength, MPa	Tensile strength, MPa	Modulus of elasticity, GPa	Hardness, BHN	Thermal conductivity, Wm^{-1} K^{-1}	Coefficient of thermal expansion, $\times 10^{-6}$ $°C^{-1}$	Melting point/ maximum operating temperature, °C	Maximum load, MPa	Typical applications
Tin-based alloys (babbitts):										
90–92% Sn, 0.35% Pb, 4–5% Sb, 4–5% Cu, 0.1% As	7,340	—	65	43	22 (~1.8 Mohs)	65	2.4	230/150	10 (cyclic)	Crankshaft bearings for petrol and diesel engines, gas turbine bearings, camshaft bushings, marine gearbox bearings, large plant and machinery bearings, general lubricated application
83–85% Sn, 0.35% Pb, 7.5–8.5% Sb, 7.5–8.5% Cu, 0.1% As	—	—	85	—	28	—	—	—	28 (unidirectional)	
Lead-based alloys (babbitts):										
9.3–10.7% Sn, 14–16% Sb, 0.5% Cu, 0.3–0.6% As, bal. Pb	11,350	—	75	20	22	—	—	330/95–150	10 (cyclic)	
5.5–6.5% Sn, 9.5–10.5% Sb, 0.5% Cu, 0.25% As, bal. Pb	—	—	70	—	19 (~1.5 Mohs)	—	—	—	28 (unidirectional)	
Copper-based alloys (brasses and bronzes):										
34% Zn, 11% Pb, 65% Cu (leaded brass)	—	210	340	105	70	115	20.3	920/240	27	Machined bushings, thrust washers, slides, etc, for light-duty applications, lining of thin- and medium-walled petrol engine, crankshaft bearings, gearbox bushings, gas turbine bearings
5% Zn, 5% Pb, 5% Sn, 85% Cu (leaded brass)	8,860	120	250	93	60	72	18	1,010/230	24	
35% Pb, 65% Cu (leaded bronze)	—	30	55	—	25	—	—	—/175	27	Bearings for pumps, automotive and aircraft internal combustion engines, and diesel engines
5% Sn, 25% Pb, 70% Cu (leaded bronze)	—	90	170	—	48	—	—	—/200	20	
10% Sn, 10% Pb, 80% Cu (leaded bronze)	—	100	240	—	63	—	—	—/230	27	

10% Sn, 2% Zn, 88% Cu (tin bronze)	—	160	310	—	65	—	—	—/260	27	Tin bronzes harder than leaded bronzes; used for high-load situations
3% Si, 97% Cu (silicon bronze)	8,530	220–360	500–700	—	180	—	—	—/260	29	Bushings for heavy-load and high-temperature applications
4% Fe, 11% Al, 85% Cu (aluminum bronze)	7,450	240–370	600–740	—	195	—	—	—/>260	31	Heavy-duty, high-temperature bearing and slide applications
Copper-based alloys (copper nickel and nickel silver):										
10% Ni, 90% Cu (copper nickel)	—	170	310	125	—	45	17.1	1,150/—	—	—
Ni-Zn-Cu (nickel silver)	—	120	240	110	—	29	16.2	1,040/—	—	—
Copper-based alloys (Be-Cu):										
2% Be, 0.6% Fe, 0.2% Co or Ni, 97.2% Cu (CA 172) (aged bar)		1,000–1,200	1,135–1,310	130	36 HRC	120	17.8	1,000/280	700	Bearings, cams, aircraft engine parts, connectors, electrical and electronic springs

Brasses contain zinc as the principal alloying element and may contain other elements, such as lead, tin, manganese, and silicon. Copper-zinc-lead alloys (leaded brasses) are more commonly used as bearing materials. Bronzes are composed of four main groups: copper-lead alloys (leaded bronzes), copper-tin alloys (tin bronzes), copper-silicon alloys (silicon bronzes), and copper-aluminum alloys (aluminum bronzes) (DeHart, 1984).

Leaded bronzes (Cu, Pb, Sn), which contain up to 25 percent Pb, provide a higher load-carrying capacity, fatigue strength, and temperature capability than babbitt alloys, but their antiseizure characteristics are poorer than those of the babbitt alloys. The increased lead content confers additional embeddability and eases their lubricant requirement, but it reduces strength, such as fatigue resistance. The lead content also means that these alloys are susceptible to corrosion. Up to 10 percent tin is added to impart corrosion and score resistance. The higher-lead bronzes (70 percent Cu, 5 percent Sn, 25 percent Pb) can be used with soft shafts, but harder shafts (300 BHN) are recommended with the harder (lower lead) bronzes, particularly under conditions of sparse lubrication. Leaded-bronze bearings are used in pumps, automotive and aircraft internal combustion engines, and diesel engines.

Tin bronzes are made up of copper, with tin being the major alloying constituent (typically 9 to 20 percent) and frequently small quantities of lead (usually less than 1 percent). Zinc is often added to improve castability, with some penalty for score resistance. The tin bronzes must rely on their lead content to obtain good score resistance at lower tin concentrations. Tin bronzes are harder than leaded bronzes. Consequently, they are best used for high-load situations in which low speeds are maintained, with ample lubricant and increased journal hardness.

Silicon bronzes (3 percent Si, 97 percent Cu) are used mainly for bushings, particularly little-end bushings, where heavy loads and high temperatures are encountered. The bearing properties of these materials are inferior to those of the tin bronzes, but selection is made largely on grounds of better corrosion resistance. The addition of 0.5 percent phosphorus has a marked effect on resistance to pounding; i.e., the alloy has outstanding resistance to wear when subjected to heavy loads at low sliding speeds.

Aluminum bronzes (85 percent Cu, 4 percent Fe, 11 percent Al) have the highest load-carrying capacity and can operate at higher temperatures ($>260°C$) than the other bronzes. They exhibit excellent shock resistance. They are very hard (195 BHN) bearing materials and thus can be used with lubricated hard shafts (350 to 440 BHN). The aluminum bronzes are popular for applications in which great strength is required without too much concern for score resistance and embeddability. They are used for rolling-mill bearings, for high-temperature bearings for coke-oven cars and heat-treat furnaces, and in heavy construction and road machinery industries.

Copper-nickel alloys contain nickel as the principal alloying element. Copper-nickel-zinc alloys are known as *nickel silvers* because of their color. All these alloys resist corrosion by freshwater and steam. Copper-nickel alloys and aluminum bronzes provide superior resistance to saltwater corrosion.

Beryllium*-copper (Be-Cu) alloys comprise another class of copper-based alloys (Anonymous, 1976b). The Be-Cu alloys offer a range of properties to meet

*Beryllium has several unique properties: low density 1830 kg m^{-3} (two-thirds that of aluminum), high elastic modulus (290 GPa), high modulus-to-weight ratio (five times that of the steels), high thermal conductivity (180 W m^{-1} K^{-1}), high specific heat (1930 J kg^{-1} K^{-1}), high strength/density, and excellent dimensional stability and transparency to x-rays. However, beryllium has a low impact strength compared with other metals and is very costly.

varying requirements of strength, conductivity (electrical and thermal), and resistance to corrosion and abrasion. Berylco 25 (CA 172), the standard high-performance alloy, is most widely used because of its corrosion resistance in air and salt spray, high strength, hardness, and excellent spring characteristics. It has good electrical conductivity and is highly resistant to stress relaxation, fatigue, abrasion, and corrosion. Be-Cu alloys can be precipitation heat treated, are capable of sustaining higher bearing loads than other copper-based alloys, and can be used in sliding contact with heat-treated steels. For example, age-hardened CA 172 (the hardest of the Be-Cu alloys) has a bearing yield strength of 2 GPa. Properly lubricated, this alloy can sustain operating bearing loads up to 700 MPa. Be-Cu bearings can be used with marginal lubrication against steel and other materials. Wikle et al. (1974) have reported that the wear performance of Be-Cu bearings is excellent when they are run against a bearing shaft of M-50 tool steel or 440C steel in a grease-lubricated bearing.

Special precautions are needed in machining beryllium because breathing air contaminated beyond safe limits with small particles of beryllium subjects a person to the risk of contracting lung diseases.

4.2.4 Aluminum-Based Alloys

Aluminum is lightweight (one-third the density of iron), strong, and readily formable (one-third the resistance of iron). Wrought aluminum alloys and alloy castings and foundry ingots are identified by two different four-digit number systems that designate the specific alloying-element combination for increasing the strength of the aluminum (Table 4.8). This number is followed by a temper designation that identifies thermal and/or mechanical treatments. Aluminum alloys can be strain-hardened or precipitation-hardened for increased strength. Some aluminum alloys can be anodized to convert the surface layer to alumina, which provides high hardness and wear resistance. Selected properties of commonly used wrought alloys are presented in Table 4.9 (Anonymous, 1978b).

Tin, cadmium, and lead are normally added to aluminum for sliding applications to upgrade its inherently poor compatibility. Lead-tin and lead-copper overlays are also used to improve compatibility. Aluminum-based alloys are used in certain medium- and heavy-duty engines for connecting-rod and main bearings.

TABLE 4.8 Aluminum Alloy Designations

Wrought alloy		Cast alloy	
Major alloying element	Designation	Major alloying element	Designation
Aluminum*	1XXX	Aluminum*	1XX.X
Copper	2XXX	Al-Cu alloys	2XX.X
Manganese	3XXX	Al-Si alloys†	3XX.X
Silicon	4XXX	Al-Si alloys	4XX.X
Magnesium	5XXX	Al-Mg alloys	5XX.X
Magnesium and silicon	6XXX	Unused series	6XX.X
Zinc	7XXX	Al-Zn alloys	7XX.X
Other	8XXX	Al-Sn alloys	8XX.X
Unused series	9XXX		

*99% or greater
†With Cu and/or Mg.

TABLE 4.9 Selected Physical Properties and Applications of Wrought Aluminum and Its Alloys

Composition	Density, kg m^{-3}	Yield strength, MPa	Tensile strength, MPa	Modulus of elasticity, GPa	Hardness, BHN	Thermal conductivity, W m^{-1} K^{-1}	Coefficient of thermal expansion, $\times 10^{-6}$ °C^{-1}	Melting point/maximum operating temperature, °C/ °C	Typical applications
AISI 1100 (99% min. Al, 0.12% Cu)									
Annealed	2715	35	90	69	23	220	23.6	650/200	Sheet-metal work, spun hollow arc fin stock, electrical conductor (alloy 1350)
H18 (cold worked)	—	150	165	—	44	215	—	—	
AISI 2024 (93.5% Al, 4.4% Cu, 0.6% Mn, 1.5% Mg)									
Annealed	2800	75	185	73	47	195	23.2	620/200	Truck wheels, screw-machine products, aircraft structures
T4 (heat treated)	—	325	470	—	120	120	—	—	
AISI 6061 (97.9% Al, 0.6% Si, 0.28% Cu, 1% Mg, 0.2% Cr)									
Annealed	2715	55	125	69	30	180	23.6	650/200	Heavy-duty struc tures requiring good corrosion resistance, truck and marine equipment, railroad cars, pipelines
T6 (heat treated)	—	275	310	—	95	165	—	—	
AISI 7075 (90% Al, 1.6% Cu, 2.5% Mg, 0.23% Cr, 5.6% Zn)									
Annealed	2800	105	225	71	60	—	23.6	620/200	Aircraft and other structures
T6 (heat treated)	—	505	570	—	150	130	—	—	

Alloys 852.0 (7 percent Sn, 1 percent Cu, 1.7 percent Ni, 0.6 percent Si, and 1 percent Mg) and 850.0 (6.25 percent Sn, 1 percent Cu, and 1 percent Ni) are normally permanent mold castings. Another composition, 828.0 (similar to 850.0 but containing a nominal 1.5 percent Si), is used in composite bearings with a steel backing and usually with a plated overlay.

4.2.5 Cobalt-Based Alloys

Cobalt has a hexagonal crystalline structure and is known to be wear resistant (Buckley and Johnson, 1964). Cobalt-based alloys are used extensively because of their inherent high strength, corrosion resistance, and ability to retain hardness at elevated temperatures (Table 4.10). Binary alloys of cobalt and chromium with more than 10 percent chromium exhibit excellent resistance to oxidation and corrosion plus outstanding hardness at temperatures up to about 1000°C, e.g., Haynes alloys 25 and 188 (Anonymous, 1970b, 1978c). Tungsten and molybdenum are added to increase strength and to improve frictional and wear properties, e.g., Haynes Stellite 6B and 6KC (Anonymous, 1976a, 1978c; Buckley, 1968).

Two types of cobalt-based alloys are available commercially: carbide-containing alloys and alloys containing Laves phase. The high carbon content of Co-Cr-W-C alloys (e.g., Stellite 6B and 6KC) promotes the formation of chromium, cobalt, and tungsten carbides during solidification that are dispersed in a cobalt-rich fcc matrix. The amount of carbide phase varies from 12 to 16 wt % for various Stellites. The best protection from abrasion is provided by alloys having large particle sizes and large volumes of the hard constituents (carbides) (Silence, 1978). Since hardness is usually decreased by increasing the particle size of the harder phase, at constant volume, hardness is not a good indicator of abrasive wear resistance. In metal-to-metal wear, the role of the hard phase (carbide) is found to be less critical, and the wear rate is less sensitive to structural features (Silence, 1978). The properties that distinguish Co-Cr-W-C alloys from other materials used in bulk or as hard-facing materials are their outstanding resistance to liquid impingement and cavitation, erosion, and galling.

The Co-Cr-W-C alloys made by Cabot Corporation are available in sheets, plates, bars, castings, and hard facing rods. Haynes alloys no. 25 and 188 meet critical high-temperature material requirements for hot sections of gas turbines as well as many applications in the chemical and nuclear fields. Hard facings of Stellite 6B and 6KC have better antigalling characteristics because of the W and Mo alloying elements and have been used in applications involving severe wear conditions in the unlubricated state and wear combined with corrosion at high temperatures. They are particularly suited for erosion shields in steam turbines, wear pads in gas turbines, shaft sleeves, and bearings.

There are also cobalt- and nickel-based alloys that during solidification form hard metallic compounds (Laves phase) rather than hard carbides. Tribaloys are free of carbide phase. Carbon is kept low so that carbide will not form in preference to the Laves phase. The wear resistance of Tribaloys is attributed to the hard Laves phase. These alloys are hypereutectic (with a eutectic temperature of about 1230°C) and structurally contain about 35 to 70 vol % Laves phase (nearly the same as by weight, with a melting point of about 1560°C). The composition is typically between CoMoSi and Co_3Mo_2Si inasmuch as the atom ratios are continuously variable between these two stoichiometric limits. Chromium partitions itself about one-third in the Laves phase and two-thirds in the solid solution, contributing significantly to the corrosion resistance of both phases. At higher chro-

TABLE 4.10 Selected Physical Properties and Applications of Common Cobalt-Based Alloys

Alloys	Density, kg m^{-3}	Yield strength, MPa	Tensile strength, MPa	Modulus of elasticity, GPa	Hardness, HRC (HV, kg mm^{-2})	Thermal conductivity, W m^{-1} K^{-1}	Coefficient of thermal expansion, $\times 10^{-6}$ $°C^{-1}$	Melting point, °C	Typical applications
Cobalt-chromium-tungsten-carbon alloys:									
20% Cr, 10% Ni, 15% W, 3% Fe, 1% Si, 1.5% Mn, 0.1% C, bal. Co (Haynes alloy no. 25)	8590	450	930	225	24 (260)* 57 (635)†	9.4	12.3	1330–1410	Used where extreme wear and corrosion resistant properties are required at high temperature; used in wrought form or as a hardfacing alloy
32% Cr, 3% Ni, 1.5% Mo, 3.5–5.5% W, 3% Fe, 2% Si, 2% Mn, 0.9–1.4% C, bal. Co (Haynes Stellite alloy no. 6B)	8390	600	920	215	41 (400)* 49 (510)†	14.8	13.9	1265–1355	Stellite 6B more common for dry metal-metal wear; used in gas turbine application, airframes, chemical and nuclear fields, bearings, wear plates, shaft sleeves, etc.
32% Cr, 3% Ni, 1.5% Mo, 3.5–5.5% W, 3% Fe, 2% Si, 2% Mn, 1.4–2.2% C, bal. Co (Haynes Stellite alloy no. 6KC)	8390	600	920	215	44 (440)* 51 (540)†	14.8	9.7	1265–1355	—
Cobalt-chromium-molybdenum-silicon alloys:									
8% Cr, 28% Mo, 2.5% Si, < 0.08% C, bal. Co (Tribaloy T-400)	9000	—	—	—	51–58 (540–610)	14.7	13.4	1230–1590	Used where outstanding wear- and corrosion-resistant properties are required; T-800 more wear resistant than T-400; used as a plasma-sprayed coating or hardfacing or hot-pressed compact; used in pumps, valves, bearings, and seals, especially in the chemical industry
17% Cr, 28% Mo, 3% Si, < 0.08% C, bal. Co (Tribaloy T-800)	8640	—	—	—	54–62 (570–660)	14.3	13.1	—	

*Heat treated.
†Cold worked.

mium concentrations, the corrosion resistance is excellent in most aggressive environments. Laves phase has a hexagonal structure similar to M_7C_3 carbides, with a hardness of about 1200 HV (lower than that of carbides and thus less resistant to abrasion than carbides), and a solid-solution matrix in the range of 300 to 800 HV, with overall hardness in the range of 450 to 700 HV.

Two commonly used cobalt-based alloys are Tribaloy T-400 and Tribaloy T-800. Carbon is not a significant alloying element in Tribaloy. Tribaloys are available in a powder form that may be plasma-sprayed or compacted by conventional powder metallurgy. Most Tribaloys are also castable, and welding rods are also available for application with oxyacetylene or inert-gas tungsten arc welding (Anonymous, 1975). These alloys have outstanding wear and corrosion resistance at temperatures up to 1000°C (Schmidt and Ferriss, 1975). At higher chromium concentrations, the corrosion resistance is excellent in most environments. These alloy coatings are particularly suited to severe unlubricated wear applications at high temperatures. They have been used in bearings, seals, piston rings, valves, vanes, and pump parts, especially in the chemical industry.

The cast Co-Cr-W-Nb-C alloys, in general, bridge the gap between high-speed tool steels and cemented carbides and are used for applications requiring wear resistance. These alloys are produced by electrical or induction melting under a protective atmosphere. The typical grades are Tantung G, which has a nominal composition of 47 percent Co, 30 percent Cr, 15 percent W, 3 percent Nb, 2.5 percent C, 2 percent Mo, and 0.5 percent B, and Tantung 144, which has a composition of 45 percent Co, 27 percent Cr, 18 percent W, 5 percent Nb, 2 percent Mn, 2.3 percent C, and 0.7 percent B. These grades are recommended for use where resistance to abrasion is paramount and there is no shock or impact. Both the grades tarnish on short heating at 400°C and lose considerable weight at 750°C. The typical wear applications for these alloys include wear strips for belt sanders, dies for extruding copper, dies for extruding Mo tubes, burnishing rolls, drill bushing, etc.

4.2.6 Nickel-Based Alloys

Nickel and its alloys are used in a variety of structural applications usually requiring moderate resistance to wear under corrosive environments or strength at high temperatures. Several nickel-based superalloys are specified for temperatures up to 1100°C. High-carbon nickel-based casting alloys are commonly used at moderate stresses above 1200°C (Anonymous, 1978c). From a structural standpoint, some nickel alloys are among the toughest known materials. The nickel alloys have ultrahigh strength and high moduli compared with steel. Nickel-based alloys are strong, tough, and ductile at cryogenic temperatures. However, nickel-based alloys in metal-to-metal sliding usually exhibit poor galling resistance (Foroulis, 1984; Schumacher, 1977, 1981). Mo and/or Co is normally added to obtain superior sliding characteristics. This is probably because of the formation of beneficial oxide films. Some important properties of commonly used nickel-based alloys are summarized in Table 4.11.

Most commercially available nickel-based alloys can be divided into four groups: nickel alloys, carbide-containing alloys, boride-containing alloys, and Laves phase–containing alloys. Nickel and Duranickel alloys have a nickel content over 94 percent. When Duranickel, a precipitation-hardenable metal alloy, is thermally treated, its strength is greatly enhanced by the precipitation of submicroscopic particles of Ni_3AlTi throughout the matrix. It has excellentspring

TABLE 4.11 Selected Physical Properties and Applications of Common Nickel-Based Alloys

Alloys		Density, kg m^{-3}	Yield strength, MPa	Tensile strength, MPa	Modulus of elasticity, GPa	Hardness, Rockwell (HV, kg mm^{-2})	Thermal conductivity, W m^{-1} K^{-1}	Coefficient of thermal expansion, X 10^{-6} °C^{-1}	Melting point, °C	Typical applications
Nickel alloys:										
Nickel 201 (99.3% Ni*, 0.2% Mn, 0.2% Fe, 0.2% Si, 0.1% Cu)	Annealed bar	8890	70–200	345–550	205	40–65 HRB (80–115)	75	15.3	1440	Chemical industry, production of caustic soda, nickel plating, and heat-resisting articles
	Hot-rolled bar		105–310	380–550	—	40–95 HRB (80–210)	—	—	—	
Duranickel 301 (95% Ni, 4.5% Al, 0.5% Ti)	Annealed bar	8250	200–400	620–830	207	8–31 HRC (75–310)	24	14.8	1440	Large sections of pump rods, springs, shafts, etc.
	Cold-drawn aged bar		860–1200	1170–1450	—	31–40 HRC (310–390)	—	—	—	
Nickel-copper alloys:										
Monel 400 (66% Ni*, 32% Cu, 1% Mn, 1% Fe)	Annealed bar	8840	175–345	480–620	180	60–75 HRB (100–140)	21.7	16.4	1320	Used where high strength and corrosion resistance are required, such as chemical, pharmaceutical, marine, and pulp and paper equipment
	Hot-rolled bar		275–690	550–760	—	15–24 HRC (150–255)	—	—	—	
Monel K-500 (66% Ni*, 29% Cu, 3% Al, 1% Fe, 1% Mn)	Annealed bar	8470	275–400	620–780	180	75–90 HRB (140–185)	17.4	15.7	1320	For industrial applications that demand high strength and corrosion resistance, nonmagnetic parts for aircraft, pumps, springs, shafts, valve stems
	Aged bar		690–1035	960–1170	—	27–33 HRC (280–330)	—	—	—	
Nickel-molybdenum-carbon alloys:										
Hastelloy B-2 (62% Ni, 28% Mo, 5% Fe, 0.05% C)	Aged sheet	9250	525	955	195	22 HRC (235)	10.4	10.1	1320	Used at up to 700°C in reducing atmospheres; resists HCl and other acids; available in the form of sheets, strips, plates, bars, welding electrodes, etc.
	Aged plate		390	840	—	92 HRB (195)	—	—	—	

Nickel-chromium-molybdenum-carbon alloys:									
Hastelloy X (47% Ni, 9% Mo, 22% Cr, 18% Fe, 0.1% C)	Annealed sheet	8225	360 (20°C) 55 (900°C)	785 (20°C) 90 (900°C)	195	90 HRB (185)	9.1	13.9	1300 —	Used in industrial furnaces, aircraft parts, e.g., jet engine tail pipes, afterburners, turbine blades, and vanes; available in the form of sheets, strips, plates, bars, welding electrodes, etc.
Hastelloy C (65.9% Ni, 17% Cr, 17% Mo, 0.12% C)	Annealed bar	8865	—	—	—	95 HRB (200)	—	13.7	—	—
Inconel 625 (62% Ni*, 22% Cr, 9% Mo, 4% Cb, 3% Fe, 0.05% C)	Annealed bar	8450	490 (20°C) 415 (900°C)	960 (20°C) 535 (900°C)	205	65–85 HRB (115–165)	9.8	14	1300	Used where good strength and high corrosion and oxidation resistance are needed at high temperatures; used in aerospace applications, combustion systems, fuel nozzles, etc.
Rene 41 (55.5% Ni, 19% Cr, 11% Co, 10% Mo, 3% Ti, 1.5% Al, 0.09% C)	Aged bar	8250	830 (20°C) 500 (900°C)	1100 (20°C) 545 (900°C)	220	—	8.9	11.9	1370	Good antigalling behavior because of Co and Mo alloying elements; used in numerous aerospace applications, e.g., gas turbine shafts
Waukesha 88 (73% Ni, 13% Cr, 4% Sn, 4% Bi, 3% Mo, 2% Fe, 1% Mn)		8590	260	310	185	80 HRB (160)	28.9	14.4	—	Good antigalling behavior and corrosion resistance; used for bearings in seawater and components for food-processing and chemical industries
Illium G (56% Ni, 22.5% Cr, 6.5% Mo, 6.5% Cu)	As cast	—	260	470	—	170 BHN (180)	—	—	—	Pumps, valves, and other process-industry components where corrosive environments are encountered

TABLE 4.11 Selected Physical Properties and Applications of Common Nickel-Based Alloys (*Continued*)

Alloys		Density, kg m^{-3}	Yield strength, MPa	Tensile strength, MPa	Modulus of elasticity, GPa	Hardness, Rockwell (HV, kg mm^{-2})	Thermal conductivity, W m^{-1} K^{-1}	Coefficient of thermal expansion, X 10^{-6} °C^{-1}	Melting point, °C	Typical applications
Nickel-chromium-iron-carbon alloys:										
Inconel X-750 (74.5% Ni*, 15% Cr, 7% Fe, 2.5% Ti, 1% Al, 0.04% C)	Aged bar	8255	630 (20°C) 570 (900°C)	1115 (20°C) 835 (900°C)	210	32–40 HRC (310–400)	12.0	14.6	1400	Typical applications are gas turbine parts, springs, foil bearings, nuclear reactors, aircraft sheet, etc.
Nickel-chromium-boron-silicon-carbon alloys:										
Wall Colmonoy 6 (72% Ni, 15% Cr, 4.5% Si, 4.5% Fe, 3.5% B, 0.5% C)		7800	—	—	220	56–61 HRC (590–645)	15.0	12	—	Hardfacings used for wear and corrosion resistance at high temperatures; used for bushings, screws, seal rings, shaft sleeves, cams, valve trims, etc.
Nickel-molybdenum-chromium-silicon alloys:										
Tribaloy T-700 (50% Ni, 32% Mo, 15% Cr, 3% Si)		8700	—	—	215	42–48 HRC (410–500)	11.4	12.4	—	For applications that require both wear and corrosion resistance; not as abrasion resistant as Tribaloy 800; used in pumps, valves, bearings, and seals, especially in chemical industry

*Includes cobalt.

properties up to 300°C. Typical applications are extrusion press parts, molds used in the glass industry, chips, diaphragms, and springs (Anonymous, 1970a).

Nickel-copper alloys (Monel series) are based on the mutual solubility of nickel and copper in all proportions. These are characterized by high strength, weldability, excellent corrosion resistance, and toughness over a wide temperature range. Monel alloys of the K-500 series are age-hardenable alloys that combine the excellent corrosion resistance of the Monel alloys with the added advantage of increased strength and hardness. Monel alloys give excellent service in seawater or brackish water under high-velocity conditions, such as in propellers, propeller shafts, pump shafts, impellers, and condensor tubes, where resistance to the effects of cavitation and erosion are important (Anonymous, 1970a; Bhushan and Winn, 1979).

Ni-Mo-C and Ni-Cr-Mo-C alloys (such as Hastelloy and Inconel alloys) are specified principally for corrosive, high-temperature, and high-strength service. These alloys may be used in oxidizing atmospheres up to 750°C. In reducing atmospheres, the alloy may be used at substantially higher temperatures. These alloys are available in sheets, strips, plates, bars, wire, welding electrodes, forging stock, and castings.

Hastelloy B series (Ni-Mo-C alloys) can be solution heat treated for maximum ductility, corrosion resistance, and machinability. Hastelloy alloy B-2 is suited for use in equipment handling HCl at all concentrations and temperatures, including the boiling point. It is also resistant to hydrogen chloride gas and sulfuric, acetic, and phosphoric acids. This alloy resists the formation of grain-boundary carbide precipitates in heat-affected zones of welds, making it suitable for most chemical-process uses in the as-welded condition. This alloy also has excellent resistance to pitting and stress-corrosion cracking (Anonymous, 1977).

Hastelloy X (Ni-Cr-Mo-C alloy) is exceptionally resistant to stress-corrosion cracking in petrochemical applications and has excellent forming and welding characteristics. Other applications include industrial furnace components and aircraft gas turbine tail pipes and afterburner parts.

Ni-Cr-Mo-C and Ni-Cr-Fe-C alloys (e.g., Inconel and Rene series) exhibit high corrosion and oxidation resistance and high strength up to 800 to 1100°C and are extensively used in aircraft gas turbine engines (Anonymous, 1970a). Inconel 600 series alloys present a desirable combination of high strength and workability and are hardened and strengthened by cold working. These alloys have excellent mechanical properties at cryogenic temperatures as well. Typical applications are furnace muffles, electronic components, chemical- and food-processing equipment, nuclear reactors, springs, etc.

Inconel 700 series alloys are age-hardenable and nonmagnetic. They are made age-hardenable by the addition of Al and Ti, which combined with Ni, following proper heat treatment, forms the intermetallic compound $Ni_3(Al, Ti)$. Typical applications for 700 series alloys are gas turbine parts, springs, nuclear reactors, aircraft sheet and foil bearings, etc. (Anonymous, 1970a; Bhushan et al., 1978). Rene 41 is another age-hardenable Ni-Cr alloy with additions of Co as well as Ti and Al. Co and Mo are added to improve the friction and wear (antigalling) properties of nickel-based alloys. Rene 41 is also used in numerous aerospace applications (Anonymous, 1970a).

Waukesha 88 (Waukesha Foundary Division, Abex Corp., Waukesha, Wisc.) is a nongalling alloy. It contains a second phase high in tin and bismuth, which acts as a solid lubricant and serves to reduce wear despite the high nickel content. It is normally used against stainless steels, chromium plating, Monel, and other sliding members to resist galling under both dry and lubricated conditions.

This alloy also offers excellent corrosion resistance. It is used in bearings operating in seawater and components for the food-processing and chemical industries. Illium G is a Ni-Cr–based alloy that is used for wear and corrosion resistance in pumps, valves, and other process-industry components.

Other Ni-Cr-Fe-C alloys, such as Haynes (or Nistelle) alloys no. 711 and 716 with a small amount of Co, designed for abrasion resistance at elevated temperatures and corrosive environments are used only as hard facings by the weld-deposition method. These will be discussed in Chapter 14.

The group of boride-containing alloys is primarily composed of Ni-Cr-B-Si-C alloys such as Wall Colmonoy 6 (Wall Colmonoy Corp., Detroit, Mich.), which is used for hard facing parts to resist wear, corrosion, heat, and galling. Hard facings of this material are used in bushings, screws, seal rings, shaft sleeves, cams, and valve trims. Another hard-facing boride-containing alloy is Haynes alloy no. 40 (or Deloro 60), which will be discussed in Chapter 14.

4.2.7 Refractory Metals

Refractory metals having melting points above that of iron (1535°C), usually above 1600°C, are characterized by high-temperature strength and corrosion resistance (Hampel, 1961; Samsonov and Vinitskii, 1980). Most-refractory-metal tungsten melts at 3340°C, and most-refractory-nonmetal carbon melts at about 3700°C. Although a number of metals meet this requirement, the least costly and most extensively used are tungsten, tantalum, molybdenum, and columbium. These four refractory metals are available in mill forms as well as in such products as screws, bolts, studs, and tubing.

The high melting points of these metals tend to be misleading because all four oxidize at temperatures much lower than their melting point. Protective coatings must be applied to use these metals at extremely high temperatures. The tensile and yield strengths of the refractory metals are substantially retained at high temperatures. These materials can be work-hardened for high hardness. Typical properties of the four refractory metals are shown in Table 4.12.

The principal applications for tantalum are in electrolytic capacitors, chemical processing, and high-temperature aerospace components. Columbium is used in superconductors and as an alloying element in steels and superalloys. The properties of molybdenum are similar to those of tungsten; however, "moly" is more ductile and easier to fabricate. Tungsten is the only refractory metal with an extremely high tensile strength that has excellent erosion resistance, good electrical and thermal conductivities, a low coefficient of thermal expansion, and high strength at elevated temperatures. The high melting point and low vapor pressure of pure tungsten make it ideal for filaments in incandescent lamps and cathodes in electronic tubes. In pure rod or wire form it is used in thermocouples for temperature measurements up to 3000°C. Tungsten and molybdenum are used as alloying elements in special steels. The tungsten alloy tungsten carbide is used in cutting tools, aerospace nozzles, and a variety of high-stress sliding conditions.

4.2.8 Titanium-Based Alloys

Titanium and its alloys are used in corrosive environments or in applications that take advantage of its light weight, high modulus, high strength-to-weight ratio, low coefficient of thermal expansion, and nonmagnetic properties (Anonymous,

TABLE 4.12 Selected Physical Properties and Applications of Some Refractory Metals

Material	Density, kg m^{-3}	Yield strength, MPa	Tensile strength, MPa	Modulus of elasticity, GPa	Hardness, HV	Thermal conductivity, W m^{-1} K^{-1}	Coefficient of thermal expansion, $\times 10^{-6}$ °C^{-1}	Melting point/ maximum operating temperature, °C/°C	Typical applications
Columbium	8,570	140 55 (980°C)	240 115 (980°C)	105	90* 150†	54.4	7.1	2468/425	Used as alloying element in special steels and superalloys
Molybdenum	10,220	565 205 (870°C)	655 345 (870°C)	325	210* 260†	146	5.4	2610/400	Used as alloying elements in special steels and superalloys; used for bolts, screws
Tantalum	16,620	240 165 (980°C)	310 185 (980°C)	185	95* 180†	54.4	6.5	3015/425	Used as alloying element in special steels and superalloys
Tungsten	19,310	1520 405 (870°C)	1520 455 (870°C)	350	320* 460†	130	4.5	3410/190	Exhibits very high tensile strength and hardness; used as alloying element in special steels and superalloys and in applications requiring erosion resistance

*Annealed.
†Cold worked.

1978c; Hampel, 1961). Titanium alloys have good resistance to oxidation at temperatures up to 500°C for short-term exposures and up to 350°C for long-term service. Ti alloys exhibit poor sliding (galling) characteristics.

Ti-6A1-4V is an alpha-beta titanium alloy with total alloy content controlled to produce an attractive annealed strength (Anonymous, 1972). It can be age-hardened to yield a material having mechanical properties in the range of those of many of the high-strength steels. This alloy has excellent toughness down to liquid hydrogen temperatures. Typical applications are tanks, pressure vessels, cryogenic storage vessels, fasteners, jet engine components, and marine parts. Selected properties of titanium and its alloys are presented in Table 4.13. The surface of titanium can be nitrided to increase its hardness (hardness of TiN ≈ 2000 kg mm^{-2}).

4.2.9 Silicon

Silicon is very abundant, is one of the cheapest elements (97 percent pure), is slightly lighter than aluminum, has high hardness and young modulus, and is corrosion- and heat-resistant. It is not widely used because it is not ductile from room temperature to at least 600°C. Because of its excellent electrical properties, it is widely used in semiconductor applications (such as computer chip substrate) (Hampel, 1961).

Very pure single-crystal silicon is used for semiconductor applications (semiconductor grade). Controlled amounts of impurities are added during a "doping" step, which can occur during production of the single crystal or in a subsequent operation. Silicon suppliers make wafers that are thin Si disks (usually 100 to 150 mm wide) with a very high surface smoothness. Semiconductor manufacturers etch millions of circuits onto the wafers and later cut them into individual chips. Since silicon can be finished to a high surface smoothness (comparable with highly polished optical-grade glass surfaces), it is also used as a substrate for coatings in applications that require a smooth-coated surface with a minimum number of asperities. Selected properties of silicon are presented in Table 4.13.

4.2.10 Wear and Galling Characteristics of Metallic Pairs

The Committee of Stainless Steel Producers (Anonymous, 1978d) commissioned several laboratory tests to evaluate the wear and galling properties of stainless steel and other metals. Wear tests were conducted under both water-lubricated and unlubricated conditions. To conduct tests under water-lubricated conditions, a piston/cylinder wear-test apparatus was used in which a piston moved in a reciprocation mode within a restrained cylinder. Temperature was maintained by conducting the test in an autoclave. These tests utilized components whose shape, relative motion, and load simulated the design and application of components used in water-cooled nuclear reactors (Anonymous, 1957). The tests were conducted in oxygenated water containing 10 to 30 cc of oxygen per kilogram of water. Wear factors were determined by measuring the weight loss in milligrams per kilogram of load per million cycles. Tests were conducted for 500,000 cycles at a loading of 55 kPa and a temperature of 260°C.

A commercial wear machine (Taber Met-Abrader, Model 500) was used to evaluate metal-to-metal sliding wear under unlubricated conditions. The machine was altered by replacing one grinding wheel with a fixture to hold a 12.7-mm-

TABLE 4.13 Selected Physical Properties and Applications of Titanium and Its Alloys and Silicon

Material	Density, kg m^{-3}	Yield strength, MPa	Tensile strength, MPa	Modulus of elasticity, GPa	Hardness, HV	Thermal conductivity, W m^{-1} K^{-1}	Coefficient of thermal expansion, $\times 10^{-6}$ °C^{-1}	Melting point/ maximum operating temperature, °C/°C	Typical applications
Commercially pure α-Ti (Ti-50A)	4515	275	350	105	220	17.3	9.5	1675/350	—
Alpha-beta alloy (Ti-6Al-4V)									
Annealed	4430	830	900	115	—	7.3	9.5	1675/350	Used in extreme environments, e.g., tanks, pressure vessels, cryogenic storage vessels, fastners, jet engine components, and marine applications
Heat treated							58 (540°C)		
	4430	1030	1100	115	30 HRC	7.3	—	—	
Silicon*	2330	—	—	107	800	84	42	1420/600	Very high purity and single-crystal silicon used in semiconductor applications (transistors, rectifiers, solar cells, computer chips); also used as a smooth substrate for various coating applications

*Electrical resistivity = 2.3×10^5 Ω cm^{-1}.

diameter specimen vertically while being pressed against a rotating 12.7-mm-diameter horizontal specimen under a 71-N load. Tests were conducted at 105 r/min for 10,000 cycles and at room temperature (Schumacher, 1977, 1981). Weight loss (milligrams per 1000 cycles) was measured.

In metal-to-metal wear tests, high stresses can result in catastrophic galling and eventual seizure even after a single cycle, so the stainless steel committee devised a button and block galling test. In this test, a small button specimen and a large block specimen were machined and polished to provide parallel contacting surfaces. The specimens were dead-weight loaded in a Brinell hardness tester, and the button was rotated 360° against the block. Specimens were then examined for galling at 10× magnification, with new specimens being tested at progressively higher stress levels until galling just began. This point was called the *unlubricated threshold galling stress*. Galling usually appears as a groove, or score mark, terminating in a mound of metal (Foroulis, 1984; Schumacher, 1977, 1981).

Numerous similar and dissimilar metal couples were tested. Martensitic stainless steels were tested in the hardened condition; the others were tested in the annealed condition, except the precipitation-hardening grades (17-4 PH). Surface finish of specimens prior to testing ranged from 0.2 to 0.4 μm rms. Results are presented in Tables 4.14 to 4.17 (Anonymous, 1978d; Bhansali, 1980; Foroulis, 1984; Schumacher, 1977, 1981).

Wear data show that among the various steels tested, types 201, 301, and hardened 440C and the proprietary Nitronic austenitic grades provide good wear resistance when mated to themselves under unlubricated conditions. High-nickel alloys generally rated intermediate between the austenitic and martensitic stainless steels. Cobalt-based alloys also did well. Considerable improvement in wear resistance can be achieved when dissimilar metals are coupled, and this is especially true for steels coupled with silicon bronze and Stellite alloys. The wear data further suggest that improvement in wear resistance can be achieved by altering the surface characteristics, such as by surface treatment or by adding a coating.

Galling data show that identical metal couples usually do poorly in terms of galling compared with dissimilar metal couples. When stainless steels are coupled with each other, with the exception of some Nitronic steels, they exhibit worse galling resistance than all other steels by a factor of 2 or more.

Cobalt-based alloys such as T-400 and Stellite 6B have, in general, good galling resistance. However, several nickel-based alloys exhibit a very low threshold galling stress when self-mated or coupled with other similar alloys (Bhansali, 1980). A nickel-based alloy such as Waukesha 88 can be modified specifically for galling resistance. Waukesha 88 exhibited extremely high galling resistance (350+ MPa) in combination with several stainless steels. It is also of note that a high nickel content in steels has a detrimental effect on galling resistance. When compared with steels, with the exception of 316 stainless steel, the remaining steels exhibit moderate threshold galling stress when coupled with the cobalt-based alloy Stellite 6B. Type 316 steel probably exhibits a low galling stress because it has higher nickel content than type 304 stainless steel.

4.3 CERAMICS AND CERMETS

Ceramics are inorganic nonmetallic materials as opposed to organic (polymers) or metallic materials. Cermets are composite materials composed of ceramics and a

TABLE 4.14 Summary of Piston/Cylinder Wear Test in Water Environment at 260°C

Piston material	Cylinder material	Wear factor*
	Self-mated couples	
Type 440C	Type 440C	35
Type 416	Type 416	365
Type 17-4 PH†	Type 17-4 PH†	970
Type 304	Type 304	7000
	Dissimilar couples	
Wall Colmonoy no. 6	Type 17-4 PH†	220
Type 304	Stellite no. 6	375
Stellite no. 6	Type 17-4 PH	410
Type 304	Haynes 25 (CW)‡	460
Wall Colmonoy no. 6	Type 440 C	600
Type 304	S-Monel	880
Silicon bronze	Type 304	1000
Type 17-4 PH†	Haynes 25 (CW)‡	1030
Type 17-4 PH†	Type 304	1050
Haynes 25 (CW)‡	Type 304	1780
Type 304	Type 17-4 PH†	105
	Dissimilar alloy/coating couples	
Type 347 nitrided	Type 347 nitrided	0
Honed chromium plating§	Type 17-4 PH†	45
Honed chromium plating§	Type 17-4 PH† (nitrided)	50
Type 17-4 PH	Honed chromium plating§	75
Honed chromium plating§	Type 304	140
Honed chromium plating§	Honed chromium plating§	300
As-plated chromium	Type 440C	330
Silicon bronze	Honed chromium plating§	1800

*Wear factor: mg/kg load/10^6 cycles.
†Precipitation hardened.
‡Cold worked.
§Honed chromium plating applied to type 17-4 PH stainless steel.
Note: The martensitic stainless steels were tested in the hardened condition; all others were tested in the solution-annealed condition unless otherwise noted; load, 55 kPa.
Source: Adapted from Anonymous (1978d).

metal intimately bonded together. The use of ceramics and cermets as structural materials offers many advantages over metals and alloys: (1) high strength-to-weight ratio, (2) high stiffness-to-modulus ratio, (3) high strength at elevated temperatures, and (4) resistance to corrosion. They are also economical and nonstrategic.

Some ceramics, such as carbides, nitrides, borides, silicides, and oxides of mostly refractory metals (substances melting at temperatures above that of iron, 1535°C), with very high melting points, appear to be ideal wear-resistant materials in a variety of tribological situations, provided that their strength and toughness are acceptable for the application (Amateau and Glaeser, 1964; Bersch, 1978; Bhushan et al., 1978; Burke et al., 1974; Chandrasekar and Bhushan, 1990; Evans and Marshall, 1981; Jack, 1980). Generally, ceramics and cermets are much harder than metals or alloys, so they are potentially abrasion-resistant. The coefficients of friction of ceramics are generally higher (~0.15 to 0.6) than those

TABLE 4.15 Summary of Taber Met-Abrader Wear Tests under Dry Conditions

Self-mated couples								
Austenitic stainless steels			Martensitic stainless steels			High-nickel alloys		
Alloy	Rockwell hardness	Rate	Alloy	Rockwell hardness	Rate	Alloy	Rockwell hardness	Rate
Nitronic 60	HRB 95	2.8	Type 440C	HRC 57	3.8	Waukesha 88	HRB 81	7.0
Type 201	HRB 90	4.9	Type 630 (17-4 PH*)	HRC 43	52.8	Inconel 718	HRC 38	9.4
Type 301	HRB 90	5.4	Type 416	HRC 39	58.1	Waspaloy	HRC 36	11.2
Nitronic 32	HRB 95	7.4	Type 420	HRC 46	169.7	Hastalloy C	HRB 95.5	13.8
Nitronic 40	HRB 93	8.9	Type 431	HRC 42	181.4	A-286	HRC 33	17.0
Nitronic 50	HRB 99	10.0				Inconel X750	HRC 36	18.7
Type 316	HRB 91	12.5				Monel K-500	HRC 34	30.6
Type 304	HRB 99	12.8				20Cr-80Ni	HRB 87	44.9
Type 303	HRB 98	386.1						

Self-mated couples (Miscellaneous alloys)					
Alloy	Rockwell hardness	Rate	Alloy	Rockwell hardness	Rate
Tufftrided PH* (nitride)	HRC 75	0.3	Haynes 25	HRC 28	1.7
D2 tool steel	HRC 61	0.4	Silicon bronze	HRB 93	5.5
AISI 4337 (low-alloy steel)	HRC 52	0.7	AISI 4130 (low-alloy steel)	HRC 47	9.4
Stellite 6B	HRC 48	1.0	Leaded brass	HRB 72	127.9
Hard chrome plating	—	1.6	AISI 1034 (mild steel)	HRB 95	134.0

Dissimilar stainless steel couples

Alloy (Rockwell hardness)	Type 304 (HRB 99)	Type 316 (HRB 91)	Type 630 (17-4 PH*) (HRC 43)	Nitronic 32 (HRB 95)	Nitronic 50 (HRB 99)	Nitronic 60 (HRB 95)	Type 440 C (HRC 57)
Type 304	12.8						
Type 316	10.5	12.5					
Type 630 (17-4 PH*)	24.7	18.5	52.8				
Nitronic 32	8.4	9.4	17.2	7.4			
Nitronic 50	9.0	9.5	15.7	8.3	10.0		
Nitronic 60	6.0	4.3	5.4	3.2	3.5	2.8	
Type 440C	4.1	3.9	11.4	3.1	4.3	2.4	3.8

Other dissimilar corrosion-resistant couples

Alloy (Rockwell hardness)		Silicon bronze (HRB 93)	Chrome plating	Stellite 6B (HRC 48)	Monel K-500 (HRC 34)	Waukesha 88 (HRB 81)
Type 304	(HRB 99)	2.1	2.3	3.1	—	8.1
Type 630 (17-4 PH*)	(HRC 43)	2.0	3.3	3.8	34.1	9.1
Nitronic 32	(HRB 95)	2.3	2.5	2.0	—	7.6
Nitronic 60	(HRB 95)	2.2	2.1	1.9	22.9	8.4
Type 316	(HRB 91)				33.8	9.6
Monel K-500	(HRC 34)				30.7	9.3

*Precipitation hardened.

Note: Weight loss corrected for density differences; all weight-loss rates in mg per 1000 cycles.

Source: Adapted from Anonymous (1978d).

TABLE 4.16 Galling Resistance of Stainless Steels

		Button material									
Block material	Condition and nominal hardness (BHN)	410	416	430	440C	303	304	316	630 (17-4 PH)	Nitronic 32	Nitronic 60
Type 410	Hardened and stress relieved (352)	21	28	21	21	28	14	14	21	320	350+
Type 416	Hardened and stress relieved (342)	28	90	21	145	60	165	290	14	310	350+
Type 430	Annealed (159)	21	21	14	14	14	14	14	21	21	250
Type 440C	Hardened and stress relieved (560)	21	145	14	75	35	21	250	21	350+	350+
Type 303	Annealed (153)	28	60	14	35	14	14	21	21	350+	350+
Type 304	Annealed (140)	14	165	14	21	14	14	14	14	210	350+
Type 316	Annealed (150)	14	290	14	255	21	14	14	14	21	260
Type 630 (17-4 PH)	H 950 (415)	21	14	21	21	14	14	14	14	350+	350+
Nitronic 32	Annealed (235)	315	310	55	350+	350+	210	21	350+	210	350+
Nitronic 60	Annealed (205)	350+	350+	250	350+	350+	350+	260	350+	350+	350+

Note: Values shown are threshold galling stress (MPa); condition and hardness apply to both the button and the block material; tests were terminated at 350 MPa, so values given as 350+ indicate the samples did not gall.

Source: Adapted from Anonymous (1978d).

TABLE 4.17 Galling Resistance of Alloys

Alloys in contact*	Threshold galling stress,† MPa
Silicon bronze (200) vs. silicon bronze (200)	28
Silicon bronze (200) vs. AISI 304 (140)	300
AISI 660 (A286) (270) vs. A286 (270)	21
AISI 4337 (484) vs. AISI 4337 (415)	14
AISI 1034 (415) vs. AISI 1034 (415)	14
Waukesha 88 (141) vs. AISI 303 (180)	350+
Waukesha 88 (141) vs. AISI 201 (202)	350+
Waukesha 88 (141) vs. AISI 316 (200)	350+
Waukesha 88 (141) vs. AISI 630 (405)	350+
Waukesha 88 (141) vs. 20Cr-80Ni (180)	350+
AISI 201 (202) vs. AISI 201 (202)	105
AISI 201 (202) vs. AISI 304 (140)	14
AISI 201 (202) vs. AISI 630 (17-4 PH) (382)	14
AISI 201 (202) vs. Nitronic 32 (231)	250
AISI 301 (169) vs. AISI 416 (342)	21
AISI 301 (169) vs. AISI 440C (560)	21
AISI 410 (322) vs. AISI 420 (472)	21
AISI 416 (342) vs. AISI 416 (372)	90
AISI 416 (372) vs. AISI 410 (322)	28
AISI 416 (342) vs. AISI 430 (190)	21
AISI 416 (342) vs. 20Cr-80Ni (180)	50
AISI 440C (560) vs. AISI 440C (604)	80
AISI 630 (17-4 PH) (311) vs. AISI 304 (140)	14
AISI 630 (17-4 PH) (380) vs. Nitronic 32 (401)	21
AISI 630 (435) vs. AISI 304 (140)	14
AISI 630 (400) vs. AISI 631 (400)	21
AISI 630 (435) vs. AISI 631 (435)	14
Nitronic 32 (235) vs. AISI 630 (380)	75
Nitronic 32 (401) vs. Nitronic 32 (401)	235
Nitronic 32 (235) vs. Nitronic 32 (401)	235
Nitronic 32 (235) vs. AISI 304 (140)	50
Nitronic 32 (401) vs. AISI 304 (140)	90
Nitronic 32 (205) vs. AISI 1034 (205)	14
Nitronic 50 (205) vs. Nitronic 50 (205)	14
Nitronic 50 (321) vs. Nitronic 50 (321)	14
Nitronic 50 (205) vs. Nitronic 32 (401)	90
Nitronic 50 (321) vs. Nitronic 32 (235)	55
Nitronic 50 (205) vs. AISI 304 (140)	28
Nitronic 60 (205) vs. AISI 301 (169)	350+
Nitronic 60 (205) vs. AISI 420 (472)	350+
Nitronic 60 (213) vs. AISI 630 (313)	350+
Nitronic 60 (205) vs. AISI 630 (332)	350+
Nitronic 60 (205) vs. Nitronic 50 (205)	350+
Nitronic 60 (205) vs. AISI 4337 (448)	350+
Nitronic 60 (205) vs. AISI 660 (A286) (270)	350+
Nitronic 60 (205) vs. 20Cr-80Ni (180)	250
Nitronic 60 (205) vs. Ti-6Al-4V (332)	350+
Nitronic 60 (205) vs. Stellite 6B (415)	350+
Stellite 6B (415) vs. AISI 304 (140)	240
Stellite 6B (415) vs. AISI 316 (140)	25
Stellite 6B (415) vs. Stellite 6B (415)	350+
Stellite 6B (415) vs. Tribaloy 400 (54 HRC)	350+
Stellite 6B (415) vs. Tribaloy 700 (47 HRC)	350+
Tribaloy 400 (54 HRC) vs. Tribaloy 400 (54 HRC)	350+
Tribaloy 700 (47 HRC) vs. Tribaloy 700 (47 HRC)	185

*Numbers in parentheses following alloy designations are nominal hardness (Brinell).
†Values given as 350+ indicate the samples did not gall.
Source: Adapted from Anonymous (1978d) and Foroulis (1984).

of solid lubricants (~0.05 to 0.2), and thus ceramics and cermets are used primarily in applications requiring high wear resistance under extreme operating conditions (high pressure, high sliding velocity, and/or high temperature).

Ceramics are chemically inert; e.g., oxide ceramics are chemically stable in air up to their melting points, say above 2000°C, and they retain these properties at high temperatures. However, they are brittle and fail readily under tension or as a result of mechanical or thermal shock. The ceramic materials fail without measurable flow from flaws that result from manufacturing defects or from surface damage in use. In addition, they are expensive to fabricate to good dimensional tolerances. Engineering uses of ceramics have tended to emphasize their refractory properties and low densities at the expense of poorer thermal shock resistance and low thermal conductivity.

The superior properties now attainable in ceramics arise from improved fabrication techniques for well-known materials, such as alumina, silicon carbide, and silicon nitride, as well as from the development of new materials, such as silicon aluminum oxynitride (Sialon) and partially stabilized zirconia (McColm and Clark, 1988).

Since raw materials and processing methods affect the microstructure and hence the properties of the ceramic materials, an understanding of their interdependence is vital (Chandrasekar and Bhushan, 1990; McColm and Clark, 1988). In applications where high strength, toughness, and oxidation resistance are required, high-purity starting materials and uniform packing are crucial. Generally, the processing method employed is a function of the properties and component configuration required. Cold pressing and sintering and hot pressing are better suited for forming simple parts, whereas slip casting and injection molding are used to fabricate parts in mass production.

In cold pressing and sintering, the powders are mixed, milled, and blended together and compacted at pressures of 35 to 100 MPa at room temperature and then are sintered at elevated temperatures (i.e., 850 to 1500°C) in a controlled atmosphere such as dry hydrogen, dissociated ammonia, or vacuum. In hot pressing, the pressure and temperature are applied simultaneously. Pressures are considerably lower (10 to 35 MPa) than for the cold-pressing method. Hot pressing is carried out in inert as well as reactive atmospheres. In hot isostatic pressing, the powder is held in a sealed flexible metallic or glass enclosure that is subjected to equal pressure from all directions at a temperature high enough to permit flow and sintering to take place. Hot isostatic pressing is becoming increasingly popular as a means of producing ceramics and cermets of very high density and superior surface quality.

In slip casting, ground material is mixed with water to form a creamy liquid that is poured into plaster molds, in which the surplus water is absorbed and a solid replica of the inside of the mold is obtained; then the part is further densified and strengthened by sintering at elevated temperatures. Injection molding is another promising method for mass production of ceramic components. The typical steps involved in injection molding are powder mixing, molding, debonding, and sintering. Accurate control of time/pressure and temperature is required during the forming process. This process is amenable to producing a component with thin and thick sections, such as bladed rotors, etc.

In the reaction-bonding method, also known as *reaction sintering*, the reactive components, produced by any of the previously described forming processes, react during sintering with each other or with the sintering atmosphere. For example, reaction-bonded Si_3N_4 is produced by forming the Si powder and then reacting it in a nitrogen atmosphere at elevated temperature. This method enables fabrication to near net shape, but the material is more porous and of lower strength than that produced by hot pressing. Table 4.18 compares the important features of various processing methods.

The material characteristics of a ceramic can be tailored to specific needs; i.e., the hardness, toughness, and impact strength can be optimized by mixing with other ceramic materials, varying the grain size, changing the manufacturing route, etc. Tribological applications for ceramics include bearings, seals, metal forming tools, cutting tools, wear-resistant parts, and several components for internal combustion and jet engines. Table 4.19 lists some of the engine components that have been considered for ceramic substitution, the material properties needed, the type of ceramic currently considered suitable, and the reasons why substitution is appropriate for better performance (Morrell, 1984).

Another class of ceramic materials is composed of cemented carbides, called *cermets*, which are a tungsten carbide or a titanium carbide mixed with a metallic binder such as tungsten, cobalt, nickel, or steel. This combination increases the toughness, ductility, and shock resistance of the composite material over wide limits according to the amount present, and the carbide provides the hardness and wear resistance. The main applications of these materials are in metalworking tools, mining tools, and wear-resistant parts.

Bulk ceramic and cermet materials are manufactured by various industries, e.g., (1) Norton Company, Industrial Ceramics Division, Worcester, Mass., (2) Carborundum Company, Niagara Falls, N.Y., (3) Kennametal, Inc., Latrobe, Pa., (4) General Electric, Schenectady, N.Y., (5) Coors Porcelain Company, Golden, Colo., (6) Corning Glass Works, Corning, N.Y., (7) Kawecki Berylco Industries, Inc., Reading, Pa., (8) Toshiba Corporation, Japan, (9) Kyocera International, Inc., Japan, and (10) Sumitomo Metals, Japan.

4.3.1 Classification and Properties of Ceramics and Cermets

The ceramic materials used for tribological applications are listed in Table 4.20. Ceramic materials can be divided into three groups on the basis of their chemical bonding: metallic, ionic, and covalent hard materials. Metallic hard materials include the binary compounds of the refractory transition metals of groups IVa to VIa with carbon, nitrogen, boron, and silicon. (Refractory substances melt at temperatures above that of iron, 1535°C. Transition metals refer to *d* and *f* elements having incomplete *d* and *f* shells.) Also in this group of materials can be included single- or multiple-phase systems that are derived from binary hard materials. The binary carbides and nitrides in most cases are completely miscible in the solid states, e.g., TiC and TiN systems.

Nonmetallic ceramics are either ionic or covalent. The ionic hard materials include oxides of such elements as Al, Cr, Zr, Hf, Ti, Si, and Mg (i.e., Al_2O_3, Cr_2O_3, ZrO_2, HfO_2, TiO_2, SiO_2, MgO, and others), their mixed crystals, and mixtures with each other or with other oxides (e.g., CaO, MgO, and Y_2O_3). Covalent hard materials include carbides, nitrides, and borides of Al, Si, and B (e.g., B_4C, SiC, BN, Si_3N_4, and AlN) as well as diamond. Cermets consist of two components, component A being a metal or alloy and component B being a ceramic, such as WC-Co or Ni-Cr-Al_2O_3.

Selected properties of various ceramic materials are presented in Table 4.21 (Alper, 1970a, 1970b, 1970c, 1971; Anonymous, 1976c, 1979, 1981; Holleck, 1986; Hove and Riley, 1965; Lynch et al., 1966; Samsonov and Vinitskii, 1980; Storms, 1967; Sundgren and Hentzell, 1986; Toth, 1971). Typical applications of selected ceramics are presented in Table 4.22. Melting point, density, thermal conductiv-

TABLE 4.18 Ceramic Forming Processes

Process	Size capability	Shape capability	Mold or die requirements	Production rate	Labor cost
Static cold pressing	Limited by press capability	Straight sided, no undercut	Hardened steel or carbide dies	High	Low to moderate
Cold hydrostatic pressing	Limited by pressure-vessel capacity	Intricate, limited dimensional accuracy	Rubber or plastic bags	Medium	Low
Static hot pressing	Limited by mold strength	Straight sided, no details	Graphite or alumina molds	Low	Moderate to high
Hot isostatic pressing	Limited by pressure-vessel capacity	Intricate, limited dimensional accuracy	High-temperature-resistant sheet metal containers	Medium	Low to moderate
Slip casting	Limited by green strength	Intricate, undercuts	Wood or metal patterns, plaster molds	Low	High
Injection molding	Limited by pressure capability	Intricate	—	High	Low

TABLE 4.19 Typical Applications of Ceramics for Engine Components

Component	Material properties needed	Material types being considered	Reasons for substituting for metals
Turbines:			
Flame can	1, 2, 3	A, B	Metal parts limited to 1100°C surface temperature; ceramic parts may not need parasitic losses of air cooling to cope with higher temperatures
Volute	1, 2	A, C	
Stator, nozzle rings	1, 2, 3	C, D, E, F	
Rotors	1, 2, 3	D, E, F	Lower mass, no cooling required
Bearings	1, 4	D, E, F	Bearings running hot are needed to cope with higher running temperatures
Diesels:			
Cylinder liners	2, 4, 5	A, G	Reduce heat losses through cylinder wall (adiabatic diesel)
Cylinder block, head	2, 4, 5	A, C	May no longer need cooling system, saving weight and space; may reduce emissions
Valve heads	1, 2	G	Reduce oxidation at higher running temperatures
Valve seats	1, 2, 3, 4	A, B, D, E, F	More corrosion resistant in hot exhaust stream
Pistons or piston caps, swirl chambers	2, 3, 5	C, D, G	Reduce heat losses through piston
Precombustion chambers	1, 2, 3	C, E	Cheaper than nickel-alloy types
Exhaust ports	1, 2, 5	C, G	Maintain high temperature to maximize turbocharger efficiency
Turbocharger rotors	1, 2, 3	D, E, F	Cheaper and lower inertia than nickel-alloy types
Diesel wear parts:			
Cam followers	3, 4	E, F, G	Reduce inertia and friction, giving less wear, less frequent adjustment, and smaller parasitic losses
Valve guides	3, 4	E, F, G	
Tappet faces	3, 4	E, F, G	
Gudgeon pins	3, 4	D, E, F	Lower thermal expansion to match that of ceramic piston
Petrol engines: For petrol engines that have temperature limitations on fuel injection, consideration is mostly being given to wear parts as above for the diesel.			

Key: 1, high-temperature capability (>1000°C); 2, thermal shock resistance; 3, high strength; 4, hard or wear resistant; 5, low thermal conductivity; coating on a metal: A, reaction-bonded silicon carbide; B, sintered silicon carbide; C, reaction-bonded silicon nitride; D, hot-pressed silicon nitride; E, sintered silicon nitride; F, sintered sialons; G, zirconia ceramics.

Source: Adapted from Morrell (1984).

TABLE 4.20 Selected Ceramic Materials for Tribological Applications

Metallic ceramics:		Refractory transition metals:		
Carbides	IVa	Ti	Zr	Hf
Nitrides	Va	V	Nb	Ta
Borides	VIa	Cr	Mo	W
Silicides				
Solid solutions				
Nonmetallic (ionic) oxides:				
Al_2O_3, Cr_2O_3, ZrO_2, HfO_2, TiO_2, SiO_2, BeO, MgO, ThO_2				
Nonmetallic (covalent) nonoxides:				
B_4C, SiC, BN, Si_3N_4, AlN, diamond				

ity, coefficient of thermal expansion, microhardness, and Young's modulus of elasticity of selected ceramics are compared in Fig. 4.6(*a*) to (*f*). Properties of selected refractory metals and nickel- and cobalt-based alloys are included for comparisons. Note that refractory metals and metallic and nonmetallic hard materials have high or very high melting points, as shown in Fig. 4.6(*a*). Silicides have the lowest melting points and hardnesses of all metallic hard materials. Figure 4.7 shows schematically the resistance to oxidation of metallic hard materials. To first order, the resistance to oxidation decreases in the order silicides, borides, carbides, and nitrides; however, for the metals of groups IVa, Va, and VIa, no such markedly decreasing sequence has been observed and there are many exceptions.

The changes from carbides of the transition metals of group IV to those of group VI of the periodic table also alter certain properties that are of great technological importance. The amount of metallic bonding increases from group IV to group VI transition-metal carbides. TiC and WC represent the limitations of this changing bonding character of refractory carbides (TiC → highest and WC → lowest amount of directed M–C bonding). As a consequence, depending on the group number of the carbide-forming transition metals, changes in the properties are observed. Figure 4.8 shows these changes in hardness, hot hardness, fracture toughness, solubility in metals of the iron groups, and angle of wetting with those metals. The hardness and wetting angle show a decreasing tendency, while the hot hardness, fracture toughness, and solubility in binder metals show an increasing tendency from TiC to WC (Holleck, 1986). These properties can be altered in a predetermined way by the formation of solid solutions. Such considerations lead to solid solutions of carbides, nitrides, borides, and silicides. TiC-TiN carbonitrides have gained technical importance. TiC and TiN are completely miscible.

Some general observations on each family of compounds follow.

4.3.1.1 Diamond. Diamond, the tetrahedral form of carbon, is the hardest material known. It is available in both single-crystal (natural and synthetic) and polycrystalline form (bonded). Diamond crystallizes in the modified face-centered cubic (fcc) structure with an interatomic distance of 0.154 nm. The diamond cubic lattice consists of two interpenetrating fcc lattices displaced by one-quarter of the cube diagonal (Fig. 4.9). Each carbon atom is tetrahedrally coordinated, making strong, directly σ (covalent) bonds to its four carbon neighbors using hybrid sp^3 atomic orbitals, which accounts for its chemical inertness,

TABLE 4.21(a) Selected Properties of Ceramic Carbides

Material	Crystalline structure	Density, kg m^{-3}	Knoop micro-hardness, GPa (kg mm^{-2})	Young's modulus, GPa	Poisson's ratio	Tensile strength, MPa	Compressive strength, MPa	Flexural strength, MPa	Fracture toughness, (K_{IC}) MPa · m$^{1/2}$	Thermal conductivity, W m^{-1} K^{-1}	Specific heat, J kg^{-1} K^{-1}	Coefficient of thermal expansion, × 10^{-6} °C^{-1}	Electrical resistivity, μΩ cm	Melting or decomposition temperature/ operating temperature, °C/°C
Covalent:														
Diamond	Cubic	3,515	78.4–102.0* (8000–10,400)	900–1050	0.20	—	—	1000–1050	—	900–2100	530	1.0	10^{13}–10^{26}	3800/1000
B_4C	Rhombohedral	2,520	31.4 (3200)	400	0.19	300	2850	350	—	26	920	4.3	5 × 10^5	2420/1100
SiC	β-Cubic, α-hexagonal at >2000°C	3,210	23.5–26.5* (2400–2700)	440	0.17	175	1400–3400	675	4.6	85–120	710	4.3	10^5	2830/1700
Metallic:														
TiC	Cubic	4,920	27.5 (2800)	450	0.19	250–475	1375–2950	860	4.0	27	520	7.2	60	3065/700
ZrC	Cubic	6,560	25.5 (2600)	410	0.19	>200	1650	250	—	22	250	6.3	42	3440/600
HfC	Cubic	12,670	26.5 (2700)	410	0.18	—	—	240	—	13	190	6.3	40–65	3930/600
VC	Cubic	5,480	26.5 (2700)	420	0.22	—	620	—	—	10	530	6.7	59	2730/600
NbC	Cubic	7,820	22.6 (2300)	450	0.22	250	2400	300	—	14	290	7.4	19	3500/650
TaC	Cubic	14,500	21.6 (2200)	510	0.24	300	—	300	—	23	190	6.7	35	3900/650
Cr_3C_2	Orthorhombic	6,680	17.7 (1800)	385	0.22	—	1050	325	—	19	530	9.9	75	1895/1100
Cr_7C_3	Hexagonal	—	—	—	—	—	—	—	—	—	—	—	—	—
$Cr_{23}C_6$	Cubic	—	—	—	—	—	—	—	—	—	—	—	—	—
Mo_2C	Hexagonal	9,120	16.7 (1700)	530	—	—	900	50	—	22	315	6.7	57	2500/500
WC	Hexagonal	15,800	20.6 (2100)	690	0.19	350	3500	700	8.0	29	200	5.2	20	2780/500
W_2C	Hexagonal	17,150	—	—	—	—	—	—	—	—	—	—	—	2860/—
WC-Co (94–6%)	Hexagonal	15,100	14.7 (1500)	640	0.26	—	5500	700–2000	—	90	210	5.4	—	—
High-speed steel: M-2 (for reference)	—	7,720 (860–960)	8.5–9.5	200	0.3	—	—	—	—	28	—	10.6	—	—

*Higher values of hardness are for single-crystal materials.

TABLE 4.21(b) Selected Properties of Ceramic Nitrides

Material	Crystalline structure	Density, kg m^{-3}	Knoop micro-hardness, GPa (kg mm^{-2})	Young's modulus, GPa	Poisson's ratio	Tensile strength, MPa	Compressive strength, MPa	Flexural strength, MPa	Fracture toughness (K_{IC}), MPa · $m^{1/2}$	Thermal conductivity, W m^{-1} K^{-1}	Specific heat, J kg^{-1} K^{-1}	Coefficient of thermal expansion, × 10^{-6} °C^{-1}	Electrical resistivity, μΩ cm^{-1}	Melting or decomposition temperature/maximum operating temperature, °C/°C
Covalent:														
β-BN	Cubic	3,480	34.3–46.6* (3500–4750)	660	0.15	65	300	110	—	300–600	670	5.0–9.0	10^{18}	3250/1100
Si_3N_4	Hexagonal	3,180	19.6 (2000)	310	0.22	350–425	700–2750	800–900	3.5–4.0	31	720	3.0	10^{18}	1900/1500
AlN	Hexagonal	3,260	12.0 (1225)	280	0.25	270	2000	950	—	36	710	5.8	10^{17}	2200/800
Metallic:														
TiN	Cubic	5,440	16.7 (1700)	600	—	—	1300	—	—	22	595	8.1	20–30	2950/500
ZrN	Cubic	7,350	14.7 (1500)	500	0.25	—	1000	—	—	10	400	5.9	21	2980/500
HfN	Cubic	13,940	15.7 (1600)	—	0.25	—	—	—	—	17	230	6.5	32	3300/500
VN	Cubic	6,080	13.7 (1400)	460	—	—	—	—	—	5.2	630	8.1	85	2170/450
NbN	Cubic	8,360	13.7 (1400)	480	—	—	—	—	—	4.3	420	3.2	58	2200/450
TaN	Hexagonal	16,360	10.3 (1050)	590	—	—	—	—	—	7.8	190	3.6	198	3090/750
CrN	Cubic	6,140	10.8 (1100)	400	—	—	—	—	—	12	710	0.7	640	1500/low
Mo_2N	Cubic	8,040	6.4 (650)	—	—	—	—	—	—	17	290	1.8	20	900/very low
W_2N	Cubic	1,220	—	—	—	—	—	—	—	—	—	—	—	840/very low

*Higher values of hardness are for single-crystal materials.

TABLE 4.21(c) Selected Properties of Ceramic Borides

Material	Crystalline structure	Density, kg m^{-3}	Knoop micro-hardness, GPa (kg mm^{-2})	Young's modulus, GPa	Poisson's ratio	Tensile strength, MPa	Compressive strength, MPa	Flexural strength, MPa	Thermal conductivity, W m^{-1} K^{-1}	Specific heat, J kg^{-1} K^{-1}	Coefficient of thermal expansion, $\times 10^{-6}$ °C^{-1}	Electrical resistivity, μΩ cm	Melting or decomposition temperature/maximum operating temperature, °C/°C
Covalent:													
B	Tetragonal	2,340	25.5 (2600)	440	—	2400	—	530	—	1280	1.1	7×10^{11}	2300/750
SiB_6	—	2,430	22.6 (2300)	330	—	—	—	—	—	—	5.4	10^7	1900/—
AlB_{12}	Orthorhombic/tetragonal	2,600	25.5 (2600)	430	—	—	—	—	—	—	—	2×10^{12}	2210/—
Metallic:													
TiB_2	Hexagonal	4,520	29.4 (3000)	410	—	—	1350	250	65	960	4.6	10–30	2790/1400
ZrB_2	Hexagonal	6,170	22.6 (2300)	340	—	200	1590	100	58	—	5.9	10–15	3245/1300
HfB_2	Hexagonal	11,200	23.5 (2400)	—	—	—	—	—	51	300	7.6	12	3250/1500
VB_2	Hexagonal	5,100	27.5 (2800)	260	—	—	—	—	42	670	7.6	13	2430/600
NbB_2	Hexagonal	7,210	24.5 (2500)	650	—	—	—	—	17	420	8.6	12	3000/850
TaB_2	Hexagonal	12,600	19.6 (2000)	250	—	—	—	—	16	250	8.5	14	3090/850
CrB_2	Hexagonal	5,600	21.1 (2150)	540	—	—	1280	620	32	690	11.1	18	2150/1000
Mo_2B_5	Rhombohedral/hexagonal	7,480	23.5 (2400)	685	—	—	—	350	27	530	5.0	18	2140/1000
W_2B_5	Hexagonal	13,100	24.5 (2500)	790	—	—	—	—	52	460	5.0	19	2365/1000

TABLE 4.21(d) Selected Properties of Ceramic Silicides

Material	Crystalline structure	Density, kg m^{-3}	Knoop micro-hardness, GPa (kg mm^{-2})	Young's modulus, GPa	Poisson's ratio	Tensile strength, MPa	Compressive strength, MPa	Flexural strength, MPa	Thermal conductivity, W m^{-1} K^{-1}	Specific heat, J kg^{-1} K^{-1}	Coefficient of thermal expansion, $\times 10^{-6}$ °C^{-1}	Electrical resistivity, μΩ cm	Melting or decomposition temperature/maximum operating temperature, °C/°C
Metallic:													
$TiSi_2$	Orthorhombic	4,150	8.8 (900)	270	—	—	1180	210	—	480	10.4	17	1500/1300
$ZrSi_2$	Orthorhombic	4,880	9.8 (1000)	270	—	—	—	—	—	—	8.3	76	1600/1150
$HfSi_2$	Orthorhombic	8,030	8.6 (875)	—	—	—	—	—	—	—	—	—	1700/1100
VSi_2	Hexagonal	5,100	9.8 (1000)	—	—	—	—	—	25	—	11.2	66	1700/1000
$NbSi_2$	Hexagonal	5,690	7.8 (800)	—	—	—	—	—	—	—	8.5	50	2050/900
$TaSi_2$	Hexagonal	9,140	13.7 (1400)	—	—	—	—	—	—	360	9.5	46	2300/800
Cr_3Si_2	Tetragonal	5,600	12.3 (1250)	—	—	—	—	380	—	—	5.9	80	1560/1150
$MoSi_2$	Tetragonal	6,240	14.2 (1450)	420	—	150	1130	380	49	550	8.3	20–30	2020/1700
WSi_2	Tetragonal	9,870	10.8 (1100)	300	—	—	1270	290	45	330	8.2	12	2110/1200

TABLE 4.21(e) Selected Properties of Ceramic (Ionic) Oxides

Material	Crystalline structure	Density, kg m^{-3}	Knoop microhardness, GPa (kg mm^{-2})	Young's modulus, GPa	Poisson's ratio	Tensile strength, MPa	Compressive strength, MPa	Flexural strength, MPa	Fracture toughness, (K_{IC}) $MPa \cdot m^{1/2}$	Thermal conductivity, $W\ m^{-1}\ K^{-1}$	Specific heat, $J\ kg^{-1}\ K^{-1}$	Coefficient of thermal expansion, $\times 10^{-6}\ °C^{-1}$	Electrical resistivity, $\mu\Omega$ cm	Melting or decomposition temperature/maximum operating temperature, °C/°C
Al_2O_3	Hexagonal	3,980	20.4* (2100)	390	0.23	300	3400	500–600	4.9	35	920	7.1	10^{20}	2050/1800
Cr_2O_3	Hexagonal	5,210	12.7 (1300)	—	—	—	—	95	—	7.5	730	5.4	—	2300/1000
TiO_2	Tetragonal	4,250	10.8 (1100)	280	—	70–100	800–1000	350	—	8.3	785	8.8	—	1840/700
SiO_2 (fused silica)	Amorphous	2,200	5.4–7.4† (550–750)	73	0.17	30–70	700–1400	50–120	—	1.2	670	0.9	10^{15}–10^{18}	1650/575
HfO_2-Y_2O_3	Tetragonal (partially stabilized)	10,010	7.4 (750)	—	—	—	—	350–2500	—	1.7	345	5.3	—	2840/1700
ZrO_2-Y_2O_3 (92–8%)	Tetragonal (partially stabilized)	6,100	12.7 (1300)	290	0.24	45	2000	500–700	8.5–9.5	1.8	630	11	10^{16}	2760/1700
BeO	Hexagonal	3,008	14.7 (1500)	380	0.34	105	2100	275	—	250	1000	7.4	10^{23}	2570/1100
MgO	Cubic	3,580	7.4 (750)	320	0.19	130	830	300	—	50	880	11.6	10^{12}	2850/1700
ThO_2	Cubic	10,010	6.4 (650)	240	0.28	100	1500	130	—	11	250	9.5	10^{16}	3220/1700

*Hardness value for single-crystal alumina (sapphire) is about 3000 kg mm^{-2} when measured in a direction 60° from the c axis, with about 2000 kg mm^{-2} when measured at 0 and 90°.

†Hardness value of single-crystal quartz glass (trigonal phase) is about 1100 kg mm^{-2}.

TABLE 4.22 Typical Applications of Selected Ceramics

Material	Typical applications
Diamond	Extremely hard and exhibit very high wear resistance; diamond tools are effective for cutting superalloys, cemented carbides, and ceramics at high speeds; also used for grinding and polishing
Silicon carbide (SiC)	High hardness and abrasive capacity; used for grinding and polishing, fillers for plastic composites for better wear resistance, and various internal combustion engine parts, such as flame can, valve seats, etc.
Titanium carbide (TiC)	High hardness and wear resistance; used extensively for wear-resistant parts
Tungsten carbide (WC)	High hardness and wear resistance; used in cemented-carbide tools, reinforcing elements for crowns used for drilling rocks; diamond substitute for dressing grinding wheels, etc.
Boron carbide (B_4C)	Applications that require high hardness and wear resistance and limited to somewhat lower temperature
Silicon nitride (Si_3N_4)	Excellent thermal shock resistance; used for bearings, rotors, valve seats, piston and piston caps, etc.
Sialon ($Si_{6-z}Al_zO_zN_{8-z}$)	High hardness, high toughness, and high wear resistance; used to machine alloys that cannot be machined by alumina tools; and used for several internal combustion engine parts, such as nozzle rings, bearings, turbochargers, rotors, cam followers, valve guides, etc.
Cubic boron nitride (CBN)	Retain high hardness at high temperature, high resistance to oxidation; effective tools to cut superalloys
Titanium nitride (TiN)	High wear resistance and very low affinity for metals; widely used in cutting tools and high-wear applications
Titanium diboride (TiB_2)	High hardness and wear resistance; cermet hard alloys for cutting metals and drilling rocks, wear-resistant coatings and linings, sand blasting nozzles
Zirconium diboride (ZrB_2)	Good heat resistance and scaling resistance; used in applications where high wear resistance is required at high temperatures, such as rocket nozzles, etc.
Molybdenum disilicide ($MoSi_2$)	High resistance to oxidation; heat-resistant alloys for gas turbines, combustion chambers, jet engines and guided missiles, sand blasting nozzles, etc.
Alumina (Al_2O_3)	Retain resistance to wear and deformation at elevated temperature; tools are used effectively for machining cast irons, low-carbon steels, and heat-treated steels at high speeds
Partially stabilized zirconia (10% Y_2O_3, 90% ZrO_2)	High strength, toughness, and mechanical reliability; good resistance to impact, abrasion, and cracking; used in mining industry applications, e.g., screw conveyers, bearings, and scraper blades; hot extrusion of copper and brass; and applications where good wear and corrosion resistance are required.
Fused silica (SiO_2)	Corrosion and wear resistance at elevated temperatures; also used as an abrasive for polishing

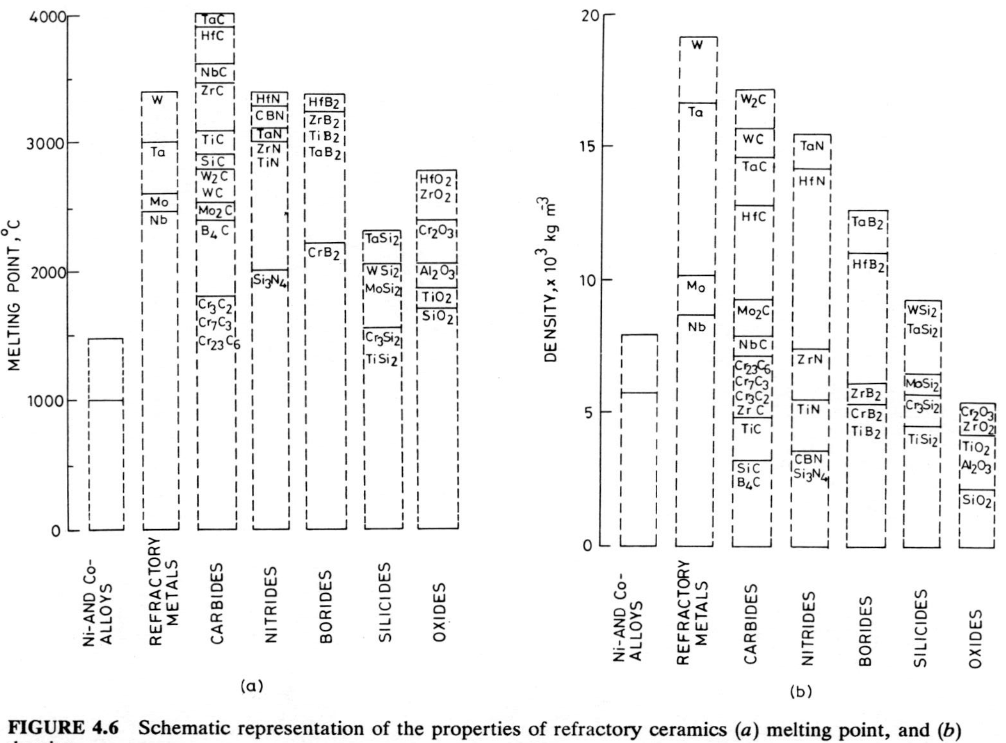

FIGURE 4.6 Schematic representation of the properties of refractory ceramics (*a*) melting point, and (*b*) density.

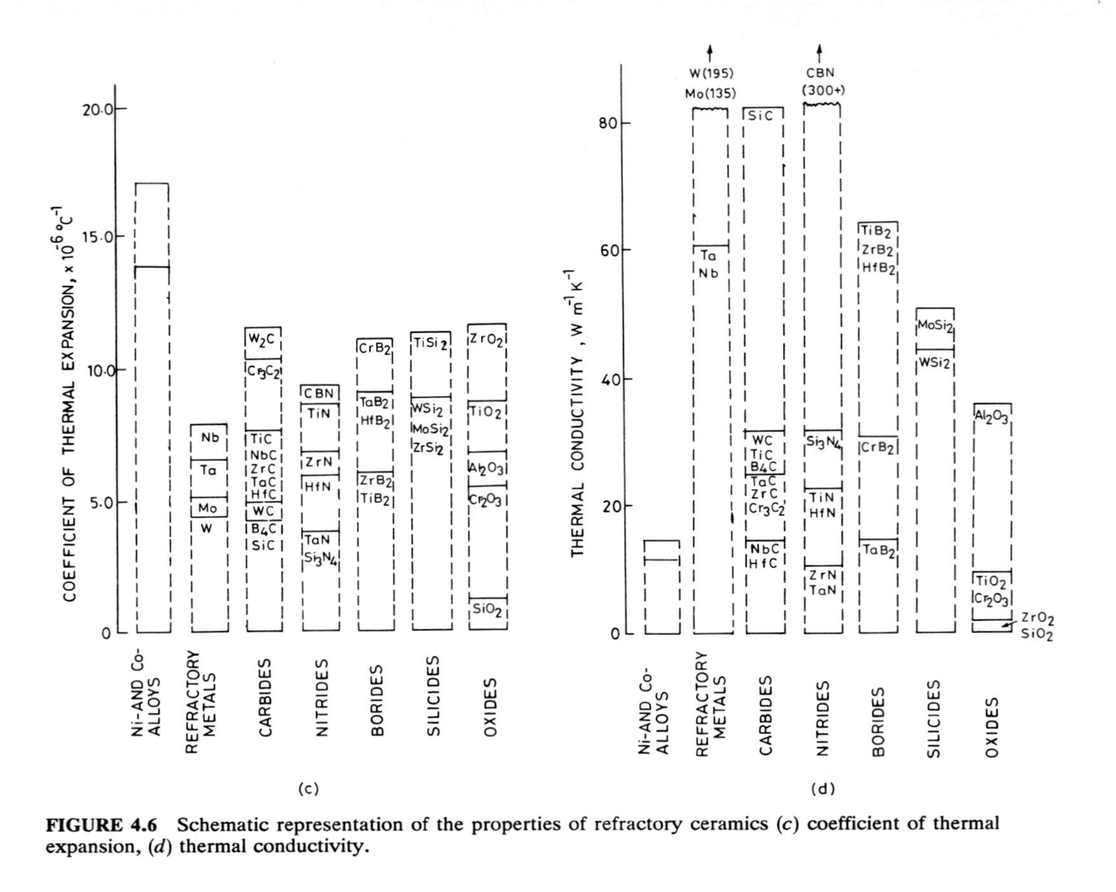

FIGURE 4.6 Schematic representation of the properties of refractory ceramics (*c*) coefficient of thermal expansion, (*d*) thermal conductivity.

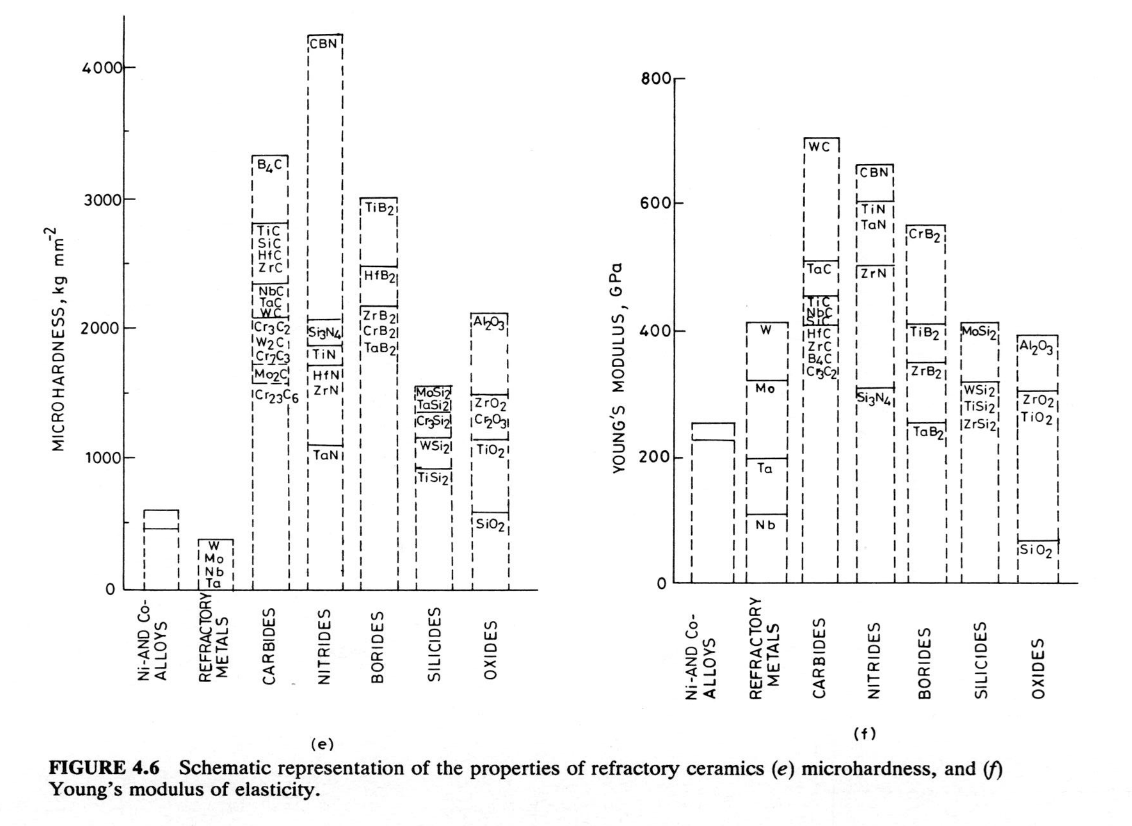

FIGURE 4.6 Schematic representation of the properties of refractory ceramics (*e*) microhardness, and (*f*) Young's modulus of elasticity.

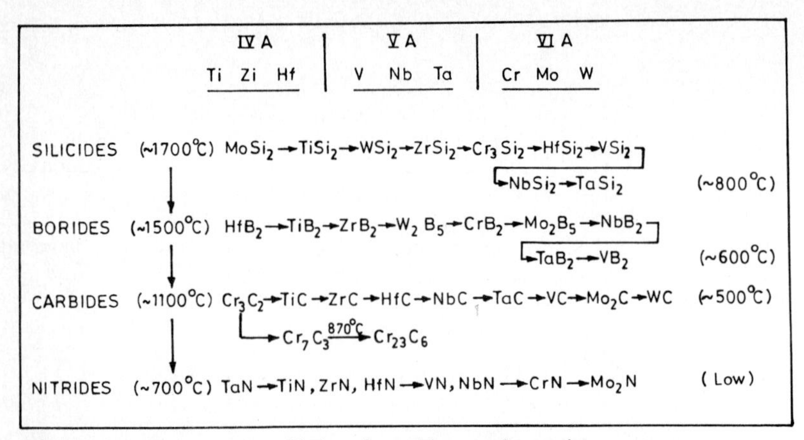

FIGURE 4.7 Resistance to oxidation of transition metal ceramics.

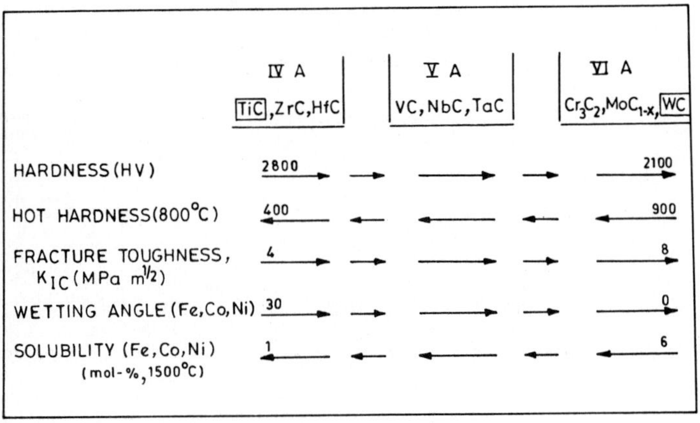

	IV A: TiC, ZrC, HfC	V A: VC, NbC, TaC	VI A: Cr_3C_2, MoC_{1-x}, WC	Trend
HARDNESS (HV)	2800		2100	→
HOT HARDNESS (800°C)	400		900	←
FRACTURE TOUGHNESS, K_{IC} (MPa $m^{1/2}$)	4		8	→
WETTING ANGLE (Fe,Co,Ni)	30		0	→
SOLUBILITY (Fe,Co,Ni) (mol-%, 1500°C)	1		6	←

FIGURE 4.8 Schematic representation of the properties of transition metal carbides.

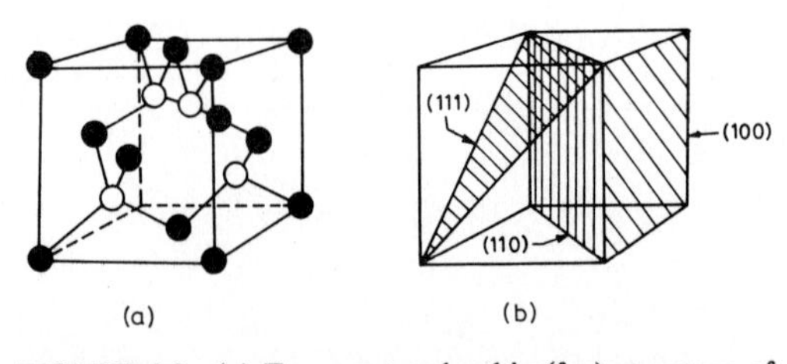

FIGURE 4.9 (*a*) Face-centered cubic (fcc) structure of diamond. For clarity, sites corresponding to one of the two interpenetrating fcc lattices are unshaded. The four nearest neighbors of each point form the vertices of a regular tetrahedron. (*b*) Various slip planes of diamond.

extreme hardness, high thermal conductivity, high electrical resistivity, and wide optical band gap (Field, 1979).

Single-crystal diamond can be classified into four categories: types Ia and Ib and types IIa and IIb. The classification is made primarily on the basis of the amount of nitrogen impurity the diamond contains. Type II natural diamond contains the least amount of nitrogen and therefore has a higher thermal conductivity than type I. Most natural diamonds (~98 percent) are type Ia, and only a few are type Ib. Synthetic diamonds are of types Ib, IIa, and IIb. Polycrystalline diamond is synthetic and is typically of two different types. One form consists of diamond particles embedded in some kind of tough matrix (usually metals such as nickel), forming a hard composite mass. The other form is a polycrystalline mass formed by direct diamond-to-diamond sintering without use of a binder (Chandrasekar and Bhushan, 1990; Field, 1979).

The hardness of diamond (~8000 to 10,400 kg mm^{-2}) is almost twice that of cubic boron nitride (CBN), the next hardest material known. The thermal conductivity of diamond is almost five times that of copper, which makes it quite resistant to thermal shock despite its brittle nature. In sliding asperity contacts, the high thermal conductivity of diamond would enable faster heat dissipation from the interface and hence lower interface temperatures.

Many of the mechanical properties of diamond single-crystal, such as hardness and strength, are a function of crystal orientation. Therefore, the coefficient of friction is sensitive to the orientation of the plane along which the sliding takes place. Usually, the coefficient of friction is the lowest when sliding occurs along the direction of the most closely packed atomic planes, which also exhibit higher hardness and low wear.

Diamond is the most attractive material for cutting tools. However, it is a very expensive material and is very brittle, cleaving easily along certain crystallographic planes. Diamond starts to graphitize at about 1000°C at atmospheric pressure and at about 1400°C in vacuum. Diamond tools are used effectively in machining high Si–cast aluminum alloys, copper-based alloys, sintered cemented carbides, glass fiber/plastic, carbon/plastic composites, and alumina ceramics. Diamond powders are used extensively in resin- and metal-bonded grinding wheels.

4.3.1.2 Carbides. The carbides are compounds of metallic elements with carbon (Me_xC_y) (Table 4.23). Each metal usually forms two or more carbides. Important exceptions are Si, Ti, Zr, and Hf, which form only monocarbides, and B, which forms only B_4C. Most of these compounds crystallize in either the sodium chloride (B1) structure (fcc array of one element superimposed on fcc array of another element) [Fig. 4.10(*a*)] or the nickel arsenide (B8) structure [Fig. 4.10(*b*)] or in a light modification of one of these. The B1 structure is described as cubic close-packed (ABC,... sequence) metal lattice with nonmetal atoms in the octahedral holes. All the monocarbides have this structure. In general, however, the nonmetal atom positions are not completely filled. This is particularly true for MoC, WC, and VC.

The B8 structure is described as a hexagonal close-packed (AB,... sequence) metal lattice with carbon atoms in the octahedral holes. WC crystallizes in a simple hexagonal structure with an A,A,... sequence of metal layers and with all the carbon atoms in the B positions [Figure 4.10(*c*)]. MoC also crystallizes in a hexagonal structure with an ABC,ACB,... sequence of metal layers and random filling by carbon of about two-thirds of the octahedral holes. This structure is stable at temperatures up to 1650°C. The hemicarbides of V, Nb, Ta, Mo, W, and Cr

TABLE 4.23 Crystal Structure and Hardness of Ceramic Carbides*

				Y_2C_3-C	YC_2-T
				La_2C_3-C	LaC_2-T
				Sm_2C_3-C	SmC_2-T
				Gd_2C_3-C	GdC_2-T
				Ho_2C_3-C	HoC_2-T
				Er_2C_3	ErC_2-T
				Lu_2C_3	LuC_2-T
			ThC-C		ThC_2-M
			UC-C (700)	U_2C_3-C	UC_2-C,T
			PuC-C	Pu_2C_3-C	PuC_2
			TiC-C (2000–3200)		
			ZrC-C (2360–2600)		
			HfC-C (2270–2650)		
	V_2C-H,R	V_3C_2	VC-C (2460–3150)		
	Nb_2C-H	Nb_3C_2	NbC-C (2300–2850)		
	Ta_2C-H,R	Ta_3C_2	TaC-C (1800–2450)		
$Cr_{23}C_6$-C	Cr_7C_3-H,R (1600)	Cr_3C_2-R,O (1800)	CrC-C		
	Mo_2C-H,R (1700)		MoC-C,H (1800)		
	W_2C-H,C (1450)		WC-H (2100–2200)		
			SiC-C, H (3500)		

*The values in parentheses are hardness values (in HV).
Key: C = cubic; H = hexagonal; T = tetragonal; R = rhombohedral; O = orthorhombic.

have a B8 structure with approximately 50 percent random vacancies in the carbon lattice. At lower temperatures, these compounds transform to structures with ordered filling of one-half the carbon sites (Hausner and Bowman, 1968; Storms, 1967; Toth, 1971).

HfC and TaC have the highest melting temperature (about 3900°C) known of materials that are extremely hard. Carbides are very hard, and in general, they are harder than the corresponding nitrides owing to a more pronounced covalent bonding. Boron carbide is the hardest carbide known, followed by SiC and TiC. WC has significantly higher moduli than other carbides. WC and W_2C have the highest densities, and B_4C has the lowest. At elevated temperatures, B_4C has the highest specific heat of this class of materials. The coefficients of thermal expansion of B_4C, SiC, and WC are the lowest among the carbides. An increase in thermal conductivity with an increase in temperature appears to be characteristic of the carbides.

Lack of oxidation resistance, particularly in a moist atmosphere, is a general limitation of carbides, silicon carbide being a notable exception. Consequently, a variety of silicon carbide products are commercially available and are used in high-temperature oxidizing environments. SiC has been widely used for a long time as an abrasive because of its extreme hardness, either in powdered form or in that of tools bonded by inorganic or organic binders.

Tungsten and titanium carbides are used in cutting tools or are applied as coatings for wear resistance. Boron carbide, with melting point 2350°C, is used because of its extreme hardness in various applications where high resistance to

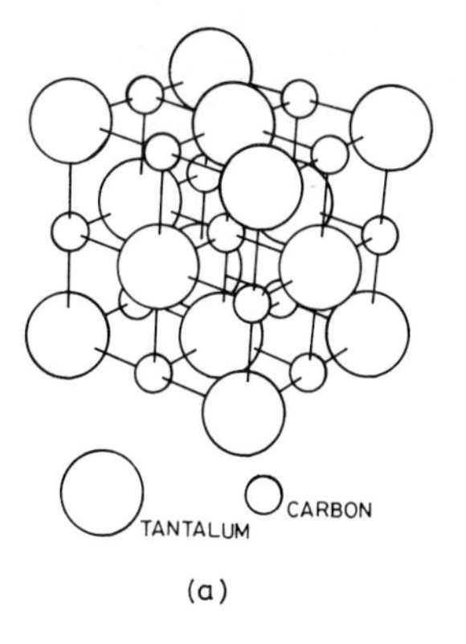

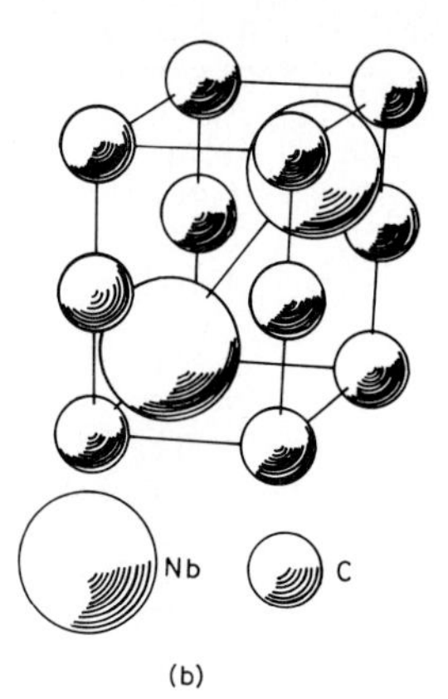

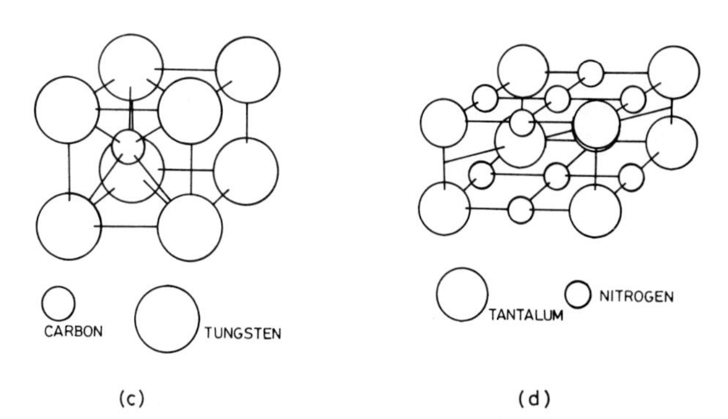

FIGURE 4.10 Unit cell structure of: (*a*) The B1 structure of TaC. (*b*) The B8 structure of β-Nb_2C. The carbon positions are half filled in a random manner. (*c*) Simple hexagonal structure of WC. (*d*) Hexagonal structure of TaN.

abrasion is required, e.g., the polishing of hard metals, pressure-blasting nozzles, and similar high-wear applications. However, boron carbide has poor toughness. Composites such as Al_2O_3/TiC are also used for metal cutting and other applications that require wear resistance (e.g., magnetic disk heads).

4.3.1.3 Nitrides. The nitrides are compounds of metallic elements with nitrogen (Me_xN_y) (Table 4.24). Most of the nitrides crystallize in either a B1 or B8 structure or in a light modification of one of these (Fig. 4.10). All mononitrides have a B1 structure except TaN, WN, and MoN. NbN crystallizes in hexagonal structure with an approximately AABB,... sequence of metal layers (Brauer and Kirner, 1964). The nitrogen atoms occupy both octahedral and trigonal prismatic holes. TaN crystallizes in a hexagonal structure with Ta_6 octahedra that share faces and edges [Fig. 4.10(*d*)]. The heminitrides of V, Nb, Ta, and Cr also have the B8 structure with random nitrogen vacancies (Hausner and Bowman, 1968; Toth, 1971).

TABLE 4.24 Crystal Structure and Hardness of Ceramic Nitrides*

	ScN-C		
	YN-C		
	LaN-C		
	LuN-C		
	ThN-C	Th_3N_4-H,R	
	UN-C	U_2N_3-H,C	UN_2-C
	PuN-C		
	TiN-C (1200–2200)		
	ZrN-C (1450)		
	HfN-C (1640)		
V_2N-H	VN-C (1400–1900)		
Nb_2N-H (1720)	NbN-H (1400)		
Ta_2N-H	TaN-H (1220)		
Cr_2N-H (2200)	CrN-H,C (1080)		
Mo_2N-C (650)	MoN-H		
W_2N-C (1200)	WN-H		
		Si_3N_4-H (3200)	

*Values in parentheses are hardness (in HV).
Key: C = cubic; H = hexagonal; R = rhombohedral.

The nitrides have the highest melting point next to the carbides. The mononitrides of boron, titanium, zirconium, hafnium, and tantalum are most refractory and thermally stable, with melting points of about 3000°C. Poor stability at elevated temperatures is characteristic of most of the other nitrides. Tantalum and hafnium nitrides have the highest theoretical densities. Nitrides of tungsten, boron, aluminum, and silicon have the lowest theoretical densities, with tungsten nitride having the lowest density. CBN, Si_3N_4, AlN, and CrN have very high specific heats compared with other nitrides. The thermal conductivities of nitrides of the transition metals increase with temperature. The coefficient of linear thermal expansion is moderate to low, silicon nitride and chromium nitride being exceptionally low.

Moderate strengths and high Young's moduli are reported for CBN, AlN, TiN, TaN, and ZrN. Si_3N_4, on the other hand, has low strength and a low Young's modulus. Microhardness of most nitrides is moderate, typically 1000 to 2000 kg mm^{-2}. Boron nitride is an exception, having a hardness of only about 200 kg mm^{-2} for the hexagonal form and a hardness second only to diamond for the cubic form. Nitrides are poor in oxidation resistance, exceptions being CBN and Si_3N_4, with Si_3N_4 being resistant at temperatures up to 1500°C. The nitrides of Ti, Zr, Hf, and Ta have moderate oxidation resistance, whereas the nitrides of Cr, V, and Nb are relatively poor. AlN is readily attacked by water vapor. Most nitrides are stable in reducing environment and, of course, in nitrogen.

Of this class of materials, Si_3N_4, CBN, and TiN are most readily available and useful. CBN exists both in hexagonal and cubic forms, which in many respects (e.g., hardness and electrical resistivity) are similar to the graphite and diamond phases of carbon. Si_3N_4 is less refractory than CBN but has greater strength at elevated temperatures. Si_3N_4 also is more resistant to oxidation. Si_3N_4 has a good thermal conductivity and low thermal expansion and thus has outstanding resistance to thermal shock. Silicon nitride ceramics find applications as components for gas turbine engines, rocket nozzles, and radomes.

Hot-pressed silicon nitride is an excellent material for many bearing applications because of its high fracture toughness, low density, good stability, high thermal shock resistance, and good hot hardness (Baumgartner, 1978; Bhushan and Sibley, 1982; Miner et al., 1981; Taylor et al., 1963). Si_3N_4 has two to three times the hardness (78 HRC) and one-third the dry coefficient of friction (~0.15) of rolling-element bearing steels and maintains its strength and oxidation resistance at temperatures up to 1200°C. The Si_3N_4 bearing surfaces are found to fail in rolling-contact fatigue cycling under high load for long periods by a spall-formation mechanism similar to that of bearing steels. This behavior is indifferent as compared with failure of alumina and silicon carbide, in which the material fails by rapid crushing wear beneath the loading disk. Impregnated graphite cages reduced the wear of dry Si_3N_4 rolling-spinning contacts to a wear coefficient of 2×10^{-7}, an order of magnitude lower than oil-lubricated steel (Bhushan and Sibley, 1982).

Sialons (Si-Al-O-N system) are a new group of solid-solution compositions that are synthesized by reacting together the β forms of silicon nitride (hexagonal structure), aluminum oxide, aluminum nitride, and silica. The formula of sialons can be represented as $Si_{6-z}Al_zO_zN_{8-z}$, where z denotes the number of nitrogen atoms substituted by oxygen atoms, ranging from 0 to 4.2 (Cother, 1987). The properties of sialons are similar to those of Si_3N_4, but sialons have superior resistance to oxidation at elevated temperatures. Sialon tools have good cutting characteristics that combine the high-speed capability of traditional hot-pressed ceramics with the impact resistance and high feed capabilities associated with carbides.

Cubic boron nitride (CBN) is the hardest known material after diamond. This material holds high hardness and resists oxidation at much higher temperatures than carbides. CBN oxidizes in air at around 450°C, forming a protective layer that prevents further oxidation up to about 1100°C. Conversion to hexagonal form occurs above 1500°C. Through the use of high-pressure and high-temperature processes, a layer of CBN can be bonded to a cemented carbide substrate. In many cases, CBN inserts can be brazed into a steel holder using the same procedure employed for cemented carbides. CBN tools are very effective for cutting superalloys at speeds several times higher than those possible with cemented carbides. The wear rate of CBN during machining of steels is markedly less than that for diamond owing to the remarkably low chemical affinity of CBN for steel.

Titanium nitride has very low affinity for metals and is widely used in cutting tools and other high-wear applications. It starts to oxidize around 425°C. Hafnium nitride is much harder than TiN and TiC at elevated temperatures and is a candidate for high-temperature applications.

4.3.1.4 Borides. Borides are compounds of metals with boron (Me_xB_y). Metal-boron compounds range from Me_3B to MeB_{12}. The diborides (MeB_2) are, in general, the most temperature-stable of these compounds, and they are characterized by a hexagonal structure. Of the borides, HfB_2 and ZrB_2 have the highest density. Borides have moderate coefficients of linear thermal expansion and thermal conductivity.

The hardness of borides is comparable with that of corresponding carbides, but unfortunately, borides are very brittle. Microhardness values for borides have a wide range from moderate to very high values for TiB_2. Resistance to oxidation is a main attraction of the borides, with HfB_2, ZrB_2, and TiB_2 being notable in this respect. They are very stable, only slightly volatile at temperatures up to about 2500°C. Their resistance to oxidation is superior to that of carbides at

TABLE 4.25 Hardness and Oxidation Properties of Borides Commonly Used for Tribological Applications*

Material	Hardness, HV	Weight change on oxidation $\times 10^{-5}$ g mm^{-2}	At temperature, °C
TiB_2	2200–3500	10 (2)*	1200
CrB_2	1800–2200	5.5 (1.5)	1240
HfB_2	2250–2900	—	—
ZrB_2	2250–2600	22 (20)	1100
TaB_2	2000–2900	3.34 (2)	900
VB	2070–2800	—	—
NbB_2	2100–2500	32.5 (1)	1000
WB	2400–2600	—	—
MoB	2350	—	—

*Values in parentheses indicate the oxidation time (in hours).
Source: Adapted from Samsonov and Vinitskii (1980).

temperatures up to 1300 to 1500°C. Water vapor increases the rate of oxidation of the borides at temperatures where liquid B_2O_3 would normally be present.

TiB_2, ZrB_2, HfB_2, and CrB_2 are most significant borides and are used as materials for rocket nozzles, cutting metals, and drilling rocks. Table 4.25 lists the hardness and oxidation properties of commonly used ceramic borides.

4.3.1.5 Silicides. Silicides are metalloid compounds of silicon and another metal (Me_xSi_y). Attractive properties found in this class of ceramics are good oxidation resistance and retention of strength at temperatures up to 1250°C. The resistance to oxidation results from the formation of a coating of silica or silicate on the surface on exposure. At low temperatures, below 1100°C, the surface is poorly protected. Preoxidation at higher temperatures alleviates this problem.

Among the transition-metal silicides, WSi_2 has the highest density. $MoSi_2$ and $TiSi_2$ exhibit relatively high specific heats. Silicides with low specific heat values include Ta and W. The silicides have moderate coefficients of linear thermal expansion. The thermal conductivity of $MoSi_2$ is the highest reported for the silicides. Microhardness values for silicides are moderate.

The disilicides of Mo, W, Cr, and Ta have the greatest potential for high-temperature applications. The bond strengths for the silicides of Mo, W, and Cr are relatively high (on the order of 275 MPa) and increase with temperature to values twice as high at about 1000°C.

4.3.1.6 Oxides. The most common or useful phases of the majority of the principal oxides crystallize in the hexagonal form. ThO_2, ZrO_2, and HfO_2 have the highest melting points of commonly used oxides, and TiO_2 has the lowest melting point. ThO_2 and HfO_2 have the highest theoretical density, and SiO_2 has the lowest. Oxides have very high electrical resistivity, moderate to high specific heat, moderate coefficients of thermal expansion, and low to moderate thermal conductivity.

Al_2O_3 has the highest hardness of the oxides, with single-crystal Al_2O_3 (sapphire) having a substantially higher hardness. Oxides are heavily used heat-resistant ceramics. One of their primary assets is greater resistance to oxidation at elevated temperatures than any other class of refractory materials. Oxides are also very stable in neutral environments but generally are reduced when subjected to elevated temperatures in reducing atmospheres. Alumina is chemically one of the most stable refractory compounds.

The oxide ceramics generally employed for tribological applications, such as cutting tools, bearings, seals, and other sliding/rolling applications, are single-crystal alumina (sapphire), polycrystalline Al_2O_3, ZrO_2, MgO, SiO_2, Cr_2O_3, and TiO_2. The two most important are polycrystalline alumina and zirconia. In these materials, oxygen ions generally form the closest packing in the arranged unit cell, and the cations are arranged in the octahedral interstices. These materials exhibit high hardness, high melting point, and, above all, a high thermodynamic stability.

Single-crystal alumina (sapphire) has a modified hexagonal close-packed structure, with the bond being a mixture of covalent and ionic type (Steijn, 1961). The thermal conductivity of sapphire is an order of magnitude higher than that of zirconia. Its hardness and Young's modulus are also significantly higher than those of zirconia. Polycrystalline alumina ceramics have good stability and hardness properties at high temperatures and hence are widely used as abrasives in grinding wheels and as cutting tools. The tool inserts may be made by cold pressing and sintering or by hot pressing. The retention of fine grain size is preferred for better tool performance, and this can be achieved by mixing other oxides or carbides of titanium, magnesium, molybdenum, nickel, and cobalt from 10 to 30 wt %. These materials retain their resistance to deformation and wear at much higher temperatures than the best-cemented carbides. Alumina tools are very effective for cutting low-carbon steels and cast irons at a cutting speed up to 10 m s^{-1}. An alumina–titanium carbide composite is used for magnetic-head sliders (Bhushan, 1990).

Zirconia is used primarily for its extreme inertness to most metals. Zirconia ceramics retain strength up to their melting point, well over 2000°C, the highest of all ceramics. It is known that if the particles of the transforming phase are kept below 1 μm in size, the strength and toughness improve dramatically. Ordinary zirconia has three different crystal structures: cubic above 2600 K, monoclinic below 1440 K, and tetragonal between these temperatures. The phase diagram of ZrO_2-Y_2O_3 (Figure 4.11) shows a decrease in temperature of the tetragonal-monoclinic transformation with increase in yttria content (Claussen et al., 1984; Green et al., 1989; Heuer and Hobbs, 1981). With increases in temperature above the monoclinic phase field, a narrow monoclinic-tetragonal field is encountered before a transformable tetragonal field is reached. The transformable tetragonal solid solution, i.e., a phase that will transform on cooling to the monoclinic structure, exists in a composition range of 0 to 5 mol % Y_2O_3. For compositions containing 6 to 12 mol % Y_2O_3, a nontransformable tetragonal solid solution exists at ambient temperature and is known as *partially stabilized zirconia* (PSZ). Finally, increasing the Y_2O_3 content above 17 mol % results in a homogeneous cubic solid solution that is stable from room temperature to the melting point and is known as *fully stabilized zirconia*. Small percentages (0.5 to 6.0 mol %) of MgO and CaO also can be added to stabilize the tetragonal structure at ambient temperature or to obtain partially stabilized zirconia (Kriven, 1988).

Partially stabilized zirconia is particularly popular for wear applications because if there are any cracks (or voids) in the material, the tetragonal structure changes to monoclinic and increases volume (density changes from 5560 to 6270 kg m^{-3}) and closes the crack. This phase-transformation toughening results in high fracture toughness, which plays an important role in influencing the tribological behavior of PSZ. Laser beam treatment of PSZ also can result in a

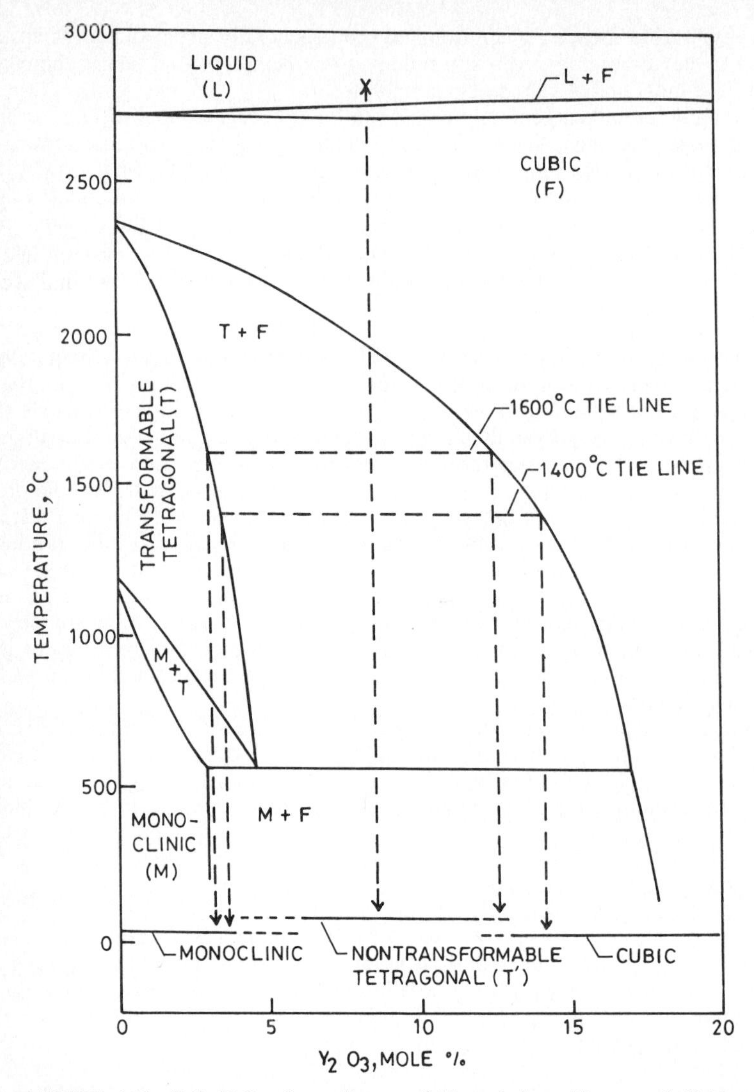

FIGURE 4.11 ZrO_2-Y_2O_3 phase diagram. [*Adapted from Heuer and Hobbs (1981).*]

phase change from tetragonal to monoclinic. The increase in the amount of the monoclinic phase as a result of this transformation results in the formation of compressive stresses on the surface. This compressive stress field causes reduction in the fracture component of the wear rate (Aronov and Benetatos, 1989). Sometimes a small percentage (up to 20 percent) TiO_2 is added to further increase wear resistance by providing high polish to mating surfaces.

Partially (yttria or MgO) stabilized zirconia has a high fracture toughness of around 10 MPa · $m^{1/2}$, which is almost twice that of other ceramics and an excellent tensile strength. Chemical stability of zirconia is also very good, with little or no oxidation and with resistance to attack by most acids and alkalis. Zirconia, however, has a low thermal conductivity of about 2 $Wm^{-1}K^{-1}$, which is an order of magnitude less than that of alumina. Heat dissipation from the sliding interface of friction couples involving zirconia can be a problem.

Magnesia ceramic has been used in internal combustion engines, in mining industry equipment, e.g., plain journal bearings in a screw conveyor system, in dies for hot extrusion, etc. Fused silica (SiO_2) is used for hot corrosion- and wear-resistant applications. Silica is also used as an abrasive.

*4.3.1.7 **Cermets: Cemented Carbides.*** Cemented carbides (cermets) are exceptionally hard materials that exhibit largely metallic characteristics (relatively high toughness). Cemented or sintered carbides are products of powder metallurgy (PM) and are made of finely divided, hard particles of carbide of a refractory metal that are sintered with one of the metals of the iron group (Fe, Ni, and Co), forming a structure of high hardness, compressive strength, and impact strength. These cemented carbides retain hardness at temperatures well above those at which high-speed steel begins to soften (Budinski, 1980, 1983).

The hard particles used for cermets are tungsten carbide and/or titanium carbide, usually in combination with a small amount of some other carbide, such as TaC and, under special circumstances, carbides of Mo, V, Cr, Zr, Cb, Nb, and Hf. Cemented carbides containing only tungsten carbides are commonly referred as *straight grades*, whereas those containing other metallic carbides in addition to tungsten carbide are referred to as *complex grades*. Straight grades are recommended for cutting tools, drawing dies, forming-die inserts, punches, and many other types of tools, whereas complex grades are recommended for cutting-tool inserts for machining plain carbon and low-alloy steels.

Cobalt-Bonded Carbides. The composition and properties of several grades of cobalt-bonded tungsten carbides are given in Table 4.26 (Anonymous, 1980b). Values for abrasion resistance are relative values and are based on 100 for the most abrasion-resistant grade. The properties of cemented tungsten carbides are optimized by changing the amount of metal used as the binder, changing the structure (grain size) of the carbide, and substituting other metallic carbides for part of the tungsten carbide. For instance, for tungsten carbides of comparable grain size, increasing cobalt content increases transverse strength and toughness and decreases hardness, compressive strength, elastic modulus, and abrasion resistance. The relative abrasion resistance is lowered as cobalt content or grain size is increased. However, abrasion resistance is lower for complex carbides than for straight WC grades having similar cobalt concentrations.

Nickel-Bonded Carbides. Cemented titanium carbides are produced by adding molybdenum carbide (Mo_2C) and nickel. This material exhibits good crater resistance, a low coefficient of friction, and a low thermal conductivity. Although penetration hardness is high, abrasion resistance is poorer than that of cobalt-bonded WC that has equal hardness. Typical cemented TiC contain 8 to 25 percent Ni, 8 to 15 percent Mo_2C, and 60 to 80 percent TiC. Occasionally, WC, Co, TiN, and other additives may be present in small amounts. The properties of nickel-bonded TiC are given in Table 4.27 (Anonymous, 1980b).

TABLE 4.26 Selected Mechanical Properties and Abrasion Resistance of Commonly Used Cobalt-Bonded Cemented Carbides

Composition	Hardness, HRA	Flexural strength, GPa	Compressive strength, GPa	Tensile strength, GPa	Impact strength, MPa	Abrasion resistance*
97% WC, 3% Co	92.5–93.2	1.59	5.86	—	1.13	100
94% WC, 6% Co						
Fine grain	92.5–93.1	1.79	5.93	—	1.02	100
Coarse grain	90.5–91.5	2.21	5.17	1520	1.36	25
90% WC, 10% Co						
Fine grain	90.7–91.3	3.10	5.17	—	1.69	22
Coarse grain	87.4–88.2	2.76	4.00	1340	2.03	7
75% WC, 25% Co	83–85	2.55	3.10	1380	3.05	3
71% WC, 12.5% TiC, 12% TaC, 4.5% Co	92.1–92.8	1.38	5.79	—	0.79	11
72% WC, 8% TiC, 11.5% TaC, 8.5% Co	90.7–91.5	1.72	5.17	—	0.90	13
28% WC, 64% TiC, 2% TaC, 2% Cr_2C_3, 4% Co	94.5–95.2	0.69	4.34	—	—	8
57% WC, 27% TaC, 16% Co	84.0–86.0	2.69	3.72	—	2.03	3

*Relative ranking; 100 is for most resistant cermet.
Source: Adapted from Anonymous (1980b).

TABLE 4.27 Selected Mechanical Properties of Nickel-Bonded Cemented Carbides

Composition*	Hardness, HRA (HV)	Flexural strength, GPa	Compressive strength, GPa	Tensile strength, GPa	Impact strength, MPa
High $TiC-Mo_2C$, low Ni (P01)	93.3 (1970)	1.17	3.58	0.97	0.79
High $TiC-Mo_2C$, low Ni (P10)	93.0 (1890)	1.38	3.45	1.10	0.90
TiC, Mo_2C, intermediate Ni (P20)	91.7 (1600)	1.72	3.27	1.17	1.02
Lower TiC, Mo_2C, high Ni (P30)	90.5 (1440)	1.89	2.96	1.24	1.24

*60–80% TiC, 8–15% Mo_2C, 8–25% Ni; P01, P10, P20, and P30 are ISO grades.
Source: Adapted from Anonymous (1980b).

Steel-Bonded Carbides. Steel-bonded carbides are a tool material that is intermediate in wear resistance between tool steel and cemented carbides. Steel-bonded carbides consist of 40 to 55 vol % TiC homogeneously dispersed in a steel matrix. Several proprietary machinable grades are commercially available with various combinations of steel and titanium carbide, e.g., Ferro-TiC (manufactured by Chromalloy Metal Tectonics Company, Sintercast Division, W. Nyack, N.Y.) for wear-resistant applications. The properties and recommended heat treatments for steel-bonded TiC are given in Table 4.28 (Anonymous, 1980b).

Steel-bonded carbides perform well in severe wear situations involving sliding friction because of their outstanding hardness and extremely fine and rounded grains of TiC. These carbides are not recommended for use in cutting tools employed to machine metals, which generate high interface temperatures because high interface temperatures drop the hardness significantly.

4.3.2 Friction and Wear Properties of Ceramics

There are numerous papers on friction and wear of ceramics in ambient and extreme environments (Anonymous, 1987). The ranking of the materials, however, depends on the processing of these materials and the test methods used. Therefore, no direct comparisons can be made between the data presented by various authors. This section presents limited data to provide readers with a feel for the typical friction and wear values of some ceramics.

Bhushan and Sibley (1982) reported friction and wear data for evaluation of a material suitable for use in rolling-element bearings that can be operated at high temperatures with or without liquid lubricants. Their measurements were made using a pin-on-disk method at a surface speed of 3 mm s^{-1} with a 20-N load on the pin for unlubricated and lubricated conditions (Dalal et al., 1975). The pin was 25 mm long and 6.3 mm in diameter with a truncated conical end. Various ceramic pairs were tested (Table 4.29). A commonly used bearing steel for high-performance applications, i.e., M-50, also was included for reference. Note that both Si_3N_4 on Si_3N_4 and Si_3N_4 on M-50 steel have low friction coefficients under both lubricated and unlubricated conditions. Bhushan and Sibley also reported friction measurements up to a temperature of 250°C and noted that friction did not increase appreciably under both lubricated and unlubricated conditions. They hypothesized that Si_3N_4 has a low dry coefficient of friction as a result of the formation of a very thin SiO_2 film on its surfaces, which is a good lubricant used in the metal-forming industry.

Dalal et al. (1975) reported wear data on the material combinations just discussed. Three-body abrasive wear experiments were conducted on a commercial lapping machine. A normal specimen load of 10 N, a commercial hydrocarbon-base lapping fluid, and 600-grit SiC were used, and the test was run for 1 to 2 hours. Wear volume was computed from the measured weight loss of the specimen. The abrasive wear coefficient (tan θ) was calculated using Eq. (2.18) for abrasive wear (see Chapter 2). A comparison of the values for tan θ is given in Table 4.30. The variations in the abrasive wear coefficients for the three silicon nitride samples is indicative of the effect of processing history and composition of the material.

Zeman and Coffin (1960) measured friction and wear of various ceramics at elevated temperatures. Tests incorporating crossed cylinders (19 mm in diameter) were conducted. The material couples selected for study consisted either of materials that had complete solid solubility or materials that had no solid solubility. Results are presented in Table 4.31. Note that the friction and wear of material

TABLE 4.28 Selected Mechanical Properties of Steel-Bonded Cemented Carbides

Composition	Hardness, HRC (HV)		Flexural strength, GPa	Compressive strength, GPa	Impact strength, MPa	Maximum service temperature, °C	Thermal shock resistance (quench cycle)
	Annealed	Hardened*					
45% TiC, 0.6% C, 3% Cr, 3% Mo, balance Fe (grade C)	40 (392)	70 (960)	2.0	3.2	2.2	200	4
45% TiC, 0.85% C, 10% Cr, 3% Mo, balance Fe (grade CM)	45 (446)	68 (940)	2.1	3.6	1.6	540	1
40% TiC, 0.4% C, 5% Cr, 0.5% Ni, 4% Mo, balance Fe (grade SK)	37 (363)	64 (800)	2.4	2.6	3.4	540	15
45% TiC, 10.5% Cr, 0.75% C, 0.5% Mo, balance Fe	48 (484)	68 (940)	1.4	3.2	1.3	—	1

*Grade C, austenized at 950°C, oil quenched and tempered at 190°C; Grade CM, austenized at 1100°C, oil quenched and oil tempered at 525°C; Grade SK, austenized at 1000°C, oil quenched and double tempered at 525°C.

Source: Adapted from Anonymous (1980b).

TABLE 4.29 Coefficients of Friction of Various Ceramic Pairs at Room Temperature

Lubricant	Si_3N_4* on Si_3N_4*	M-50 on Si_3N_4*	SiC on SiC	M-50 on SiC	B_4C on B_4C	M-50 on B_4C	WC on WC	M-50 on WC	M-50 on M-50
Unlubricated	0.17	0.15	0.52	0.29	0.53	0.29	0.34	0.19	0.54
Hydrocarbon base	0.13	0.11	0.14	0.13	0.25	0.25	0.18	0.13	0.12
Ester base (MIL-L-23699)	0.12	0.11	0.18	0.13	0.31	0.24	0.18	0.14	0.13
Silicone base	0.24	0.36	0.23	0.42	0.28	0.30	0.40	0.39	0.35

*Ground surface finish of 0.6 μm AA.

Note: AISI M-50 is a tool steel used in hardened and tempered condition (~HRC 60).

Source: Adapted from Bhushan and Sibley (1982).

TABLE 4.30 Abrasive Wear Coefficient Values (Tan θ) of Various Ceramic Pairs at Room Temperature (Self-Mated Couples)

Material	Tan θ, × 10^{-3}
SiC	1.1
WC	3.3
B_4C	5.5
Si_3N_4 (HS 110)	43
Si_3N_4 (HS 130)	46
Si_3N_4 (HS 132)	38

Source: Adapted from Dalal et al. (1975).

TABLE 4.31 Friction and Wear of Ceramic Pairs at Elevated Temperatures

Couple	Temperature, °C	Coefficient of friction	Wear scar diameter, mm
Couples having complete solid solubility			
TiB_2-TiB_2	590	2.40	1.1
SiC-SiC	650	0.78	2.0
Si_3N_4-Si_3N_4	650	0.90	2.7
Al_2O_3-Al_2O_3	310	1.08	2.6
Al_2O_3-Al_2O_3	650	0.80	1.2
Al_2O_3-Al_2O_3	760	0.82	1.6
MgO-MgO	980	0.39	1.6
$3Al_2O_3 \cdot 2SiO_2$-$3Al_2O_3 \cdot 2SiO_2$	650	1.05	3.6
$MgO \cdot Al_2O_3$-$MgO \cdot Al_2O_3$	650	0.88	2.7
$ZrO_2 \cdot SiO_2$-$ZrO_2 \cdot SiO_2$	650	1.00	1.5
$MoSi_2$-$MoSi_2$	650	0.83	1.6
$TiSi_2$-$TiSi_2$	650	0.82	2.5
VSi_2-$TaSi_2$	660	0.78	3.1
$MoSi_2$-$TiSi_2$	650	0.85	2.2
Couples having no solid solubility, known or predicted, and exhibiting interfacial sliding			
TiC-Al_2O_3	60	0.26	0.6
SiC-SiO_2	670	0.42	0.7
Couples having no solid solubility, known or predicted, and exhibiting bulk shearing			
SiO_2-$3Al_2O_3 \cdot 2SiO_2$	650	0.87	3.8
SiO_2-$3Al_2O_3 \cdot 2SiO_2$	650	0.72	3.8
SiO_2-$ZrO_2 \cdot 2SiO_2$	650	0.88	1.6
SiO_2-$ZrO_2 \cdot 2SiO_2$	650	0.69	3.4
Al_2O_3-$3Al_2O_3 \cdot 2SiO_2$	650	1.00	2.8
Al_2O_3-$MgO \cdot Al_2O_3$	650	0.88	1.2
Al_2O_3-$MgO \cdot Al_2O_3$	650	0.90	1.2
Al_2O_3-$ZrO_2 \cdot SiO_2$	650	1.02	1.7
$3Al_2O_3 \cdot 2SiO_2$-$MoSi_2$	650	0.82	2.6

Source: Adapted from Zeman and Coffin (1960).

couples with known or predicted solid solubility and of those with no solid solubility cannot be distinguished easily. Therefore, solid-solubility criteria cannot be used with confidence in the selection of ceramic materials.

Finkin et al. (1973) measured the friction and wear of various ceramics at 315 to 980°C. The tests were conducted using a flat-ended pin (7 mm diameter) against a flat plate in an oscillating wear mode. Results are presented in Tables 4.32 and 4.33. From the data it appears that many material combinations gave low friction and wear. The most outstanding material combination is nickel-molybdenum–bonded titanium carbide on either itself, cobalt-coated, or cobalt-coated thoria dispersion-hardened nichrome.

Bhushan and Gray (1978, 1979) reported wear data for several ceramics in the presence of synthetic seawater. The wear tests were conducted using a pin-on-flat plate apparatus in the reciprocating mode at a normal stress of about 5.5 MPa and 2000 strokes/min for about 6 hours. They found that the wear rate of high-purity alumina versus plasma-sprayed tungsten carbide coating under the test conditions was very low (~2 $\mu m\ h^{-1}$).

Hansen et al. (1979) evaluated a large variety of metals, alloys, and ceramics

TABLE 4.32 Mean Coefficient of Friction of Various Ceramic Parts after 30 Minutes of Sliding at Elevated Temperatures

	Load, MPa				
	3.5	3.5	3.5	10.5	21
	Temperature, °C				
Material combinations	310	650	980	980	980
Ceramics:					
Al_2O_3-TiC (70–30% by weight) on itself	0.90	0.70	0.48	0.70	(0.70)*
Si_3N_4 on itself	0.60	0.28	0.48	0.40	0.48
Porous ZrO_2 on itself	(—)	—	—	—	—
Cobalt-coated porous ZrO_2 on itself	(1.0)	(0.48)	—	—	—
Eutectics and alloys:					
Co-$Cr_{23}C_6$ eutectic on itself	0.25	0.22	0.28	0.13	—
Molybdenum ion-plated with Ag on itself	(0.5)	(0.23)	(0.33)	(0.22)	(0.17)
Thoria dispersion-hardened TD Nichrome on itself	(0.60)	(0.60)	—	—	—
Ni_3Al/Ni_3Nb eutectics	0.56	0.23	0.20	(0.34)	(0.43)
Ni_3Al/Ni_3Nb on NiAlNbCr	0.50	0.34	(0.19)	(0.35)	(0.47)
Tool steel (70% Fe, 15% Mo, 15% Co) on itself	—	0.30	0.26	0.33	—
Cermets and combinations:					
Nickel-molybdenum–bonded titanium carbide (NMBTC) on itself	0.40	0.30	0.17	0.13	0.27
Cobalt-coated NMBTC on itself	0.49	0.38	0.22	0.12	0.20
NMBTC on cobalt-coated TD Nichrome	0.48	0.23	0.23	0.16	—
NMBTC on TD Nichrome	0.42	0.38	0.18	0.15	—
NMBTC on dense ZrO_2	0.21	0.47	0.21	0.20	0.31
Carbon graphite on itself	0.18	(—)	—	—	—

*Parentheses indicate that the material failed.
Source: Adapted from Finkin et al. (1973).

TABLE 4.33 Wear of Flats of Various Material Pairs at Elevated Temperatures

Material combinations	Weight loss of flat, g
Ceramics:	
Al_2O_3-TiC (70–30% by weight) slid on itself	0.1563*
Si_3N_4 slid on itself	0.2958*
Porous ZrO_2 on itself	0.7150
Cobalt-coated porous ZrO_2 on itself	0.0753
Eutectics and alloys	
Co-$Cr_{23}C_6$ eutectic slid on itself	1.0077
Mo ion-plated with Ag on itself	35.05
TD Nichrome on itself	0.0805
Tool steel (70% Fe, 15% Mo, 15% Co)	(0.1454)†
Ni_3Al/Ni_3Nb eutectics	(0.0772)
Ni_3Al/Ni_3Nb on NiAlNbCr	(0.0723)
Cermets and combinations:	
Nickel-molybdenum–bonded titanium carbide (NMBTC) on itself	0.0116
	0.5192
Cobalt-coated NMBTC on itself	(0.0908)
NMBTC on cobalt-coated TD Nichrome	0.0041‡
NMBTC on TD Nichrome	0.0041‡
Carbon graphite on itself	2.3466

*Includes weight loss due to thermal stress, including chipping at cooldown.
†Parentheses indicate the weight gain.
‡The same specimen, one-half cobalt coated by sputtering.
Source: Adapted from Finkin et al. (1973).

in a sandblast type of test at 20 and 700°C with 90 and 20° angles of impingement. As shown in Table 4.34, some ceramics and carbides with a low metal binder content were more erosion resistant than metals or alloys. In the case of coatings, thicknesses of about 50 to 80 μm were necessary for adequate protection.

Finally, Bhansali and Mehrabian (1982) and Hosking et al. (1982) reported wear data for 2024 aluminum matrix composites containing Al_2O_3 and SiC powders (in sizes ranging from 16 to 142 μm). Based on dry sand/rubber wheel abrasion and pin-on-disk wear tests, they found that composites containing greater than 20 wt % of particulate additives exhibited excellent wear resistance compared with the matrix alloys. Particle size also played an important role; e.g., composites containing large (~142 μm) Al_2O_3 particles showed a weight loss about an order of magnitude less than composites containing small (~16 μm) particles.

4.4 CERAMIC-METAL COMPOSITES

4.4.1 Metal-Matrix Composites

The impetus for the development of reinforced metals with ceramic whiskers* and filaments (continuous or endless fibers) has been the continuing demand for

*A *whisker* is considered to be a filamentary single crystal of metal or nonmetal characterized by ultrahigh strength, micrometer-size width, and a high length-to-diameter ratio.

TABLE 4.34 Relative Erosion Resistance of Various Solid Materials and Coatings at a 90° Impact Angle

Material	Relative erosion factor*	
	20°C	700°C
Metals:		
Stellite 6B	1.0	1.0
Mild steel	0.75	—
Type 304 stainless steel	1.0	0.72
Molybdenum	0.46	—
Cemented carbides:		
Tungsten carbide + 25% cobalt	1.0	1.2
Tungsten carbide + 5.8% cobalt	0.41	0.81
Tungsten carbide + <1.5% cobalt	0.11	0.12
Pure ceramics:		
Silicon carbide	0.11	0.42
Cubic boron nitride	0†	0
Diamond	0	0
Hard coatings		
CVD silicon carbide	0.05	0
Electrodeposited TiB_2	0	0
Borided molybdenum	>0.1	—

*Relative erosion factor = mean specimen volume loss/mean volume loss of Stellite 6B control specimen.

†No measurable wear.

Source: Adapted from Hansen et al. (1979).

new and improved materials in the aerospace industry. The competitive advantages of reinforced metals over other metallic systems include the unique ability to (1) vary the metals' stiffness, (2) tailormake a composite to meet specific engineering strength and stiffness requirements, (3) improve the fatigue, creep, and stress-rupture properties of the metals, and (4) develop very high specific strengths (strength-to-density ratios) and specific stiffnesses (modulus-to-density ratios).

Early studies of reinforced metals centered on the incorporation of metal wires into such easily worked metals as copper, silver, and aluminum and later expanded to include the reinforcement of titanium and aluminum with beryllium wire and the reinforcement of nickel-based superalloys with refractory wire. More recently, with the general availability of high-strength and stiff whiskers, studies were initiated to explore their incorporation into metals. The sapphire (α-Al_2O_3) and silicon carbide (α-SiC and β-SiC) whisker–reinforced aluminum systems (e.g., 2024-T4, 6061-T6, 7075 T-6, Al–2.5 percent Si) have received the most attention. In the mid-1960s, high-strength, stiff, lightweight nonmetallic filaments were developed, most commonly boron, silicon carbide, single-crystal sapphire, and graphite. The most advanced of the filamentary systems are boron-reinforced aluminum and titanium (Hahn and Kershaw, 1970).

Metal-matrix composites containing particulate additives of hard particles also have been developed for wear-resistant applications. Hosking et al. (1982) developed 2024 aluminum-matrix composites containing Al_2O_3 or SiC particles ranging in size from 16 to 142 μm.

*4.4.1.1 **Ceramic Whisker–Reinforced Metals.*** Of all the reinforcing fibers for application in metal-matrix composites, whiskers are by far the strongest (McCreight et al., 1965). Their strength in the laboratory can approach the theoretical strength of matter, as calculated from the strength of the interatomic bonds. Typical diameters of whiskers used for reinforcement range from 0.1 to 10 μm. Whiskers are grown from solution or from the vapor phase. Alpha-alumina (sapphire) whiskers are prepared by a vapor-phase reaction in which moist hydrogen is passed over molten aluminum (Brenner, 1962a). The strength is approximately 7 GPa for whiskers having a cross section of approximately 1 μm, and their elastic modulus is approximately 500 GPa. These room-temperature properties, given the fact that the density (3950 kg m^{-3}) is close to that of aluminum, are considered suitable for reinforcing most metal matrices. The elevated-temperature properties are also excellent, and sapphire is inert to many structural metal matrices (Brenner, 1962b; Sutton and Chorne, 1965). The alpha form of silicon carbide is grown by a process in which organosilane is thermally decomposed in hydrogen at a temperature above 1400°C. Silicon nitride whiskers are grown by a vapor reaction between silicon and a silicate in nitrogen gas diluted with hydrogen at temperatures above 1400°C.

The properties of SiC whiskers are somewhat similar to those of sapphire, with the exception that their strength is less diameter-dependent. Their density (3200 kg m^{-3}) is lower than that of sapphire, and they are generally cheaper than sapphire whiskers. Unfortunately, SiC whiskers are reactive in Ni, Co, Cr, and other refractory metals and their alloys. The properties of Si_3N_4 whiskers are not as attractive as those of sapphire and SiC whiskers, and their compatibility with many matrices is poor. Table 4.35 lists the mechanical properties of a number of currently available whisker grades.

The choice of matrices with respect to sapphire whiskers is primarily dependent on the ability to achieve adequate whisker-matrix shear bond strength at elevated temperatures, while the choice of matrix with respect to SiC and Si_3N_4 is

TABLE 4.35 Selected Mechanical Properties of Commercially Available Whiskers and Filaments

	Density, kg m^{-3}	Tensile strength, GPa	Young's modulus, GPa	Specific strength, × 10^4 m	Specific stiffness, × 10^4 m	Flexural fatigue strength, MPa
Whiskers:						
Sapphire	3,990	3.5–20	485	52.90	1215	0.5–10
SiC	3,185	2.5–10	485	32.60	1520	0.1–1
Ceramic filaments:						
Boron	2,660	3.1	400	11.70	1510	—
SiC	3,460	2.5	365	7.00	1060	—
Sapphire	3,990	2.5–3.5	485	6.07	1215	—
Graphite	1,690	1.7	410	9.78	2450	—
Metal filaments:						
Beryllium	1,855	1.2	310	6.73	1680	—
Tungsten	19,300	4.0	405	2.07	213	—
Steel	8,030	4.2	200	5.18	250	—

Source: Adapted from Hahn and Kershaw (1970).

primarily limited by chemical incompatibility at elevated temperatures. Composites incorporating whiskers (discussed earlier) and various metals and alloys (e.g., aluminum and magnesium) have been made by various techniques, including liquid infiltration of aligned whisker bundles, casting, electroforming, high-energy-rate (i.e., Dynapac) consolidation, and a variety of PM techniques (Hahn and Kershaw, 1970). Typically, 15 to 30 vol % of whiskers is incorporated.

The whisker-reinforced metals have shown significant improvements in mechanical properties (Hahn and Kershaw, 1970). Selected mechanical properties of some composites are shown in Table 4.35. Whisker-reinforced composite materials will find applications in such areas as high-strength, high-modulus aircraft turbojet compressor blade material for intermediate temperature applications, high-strength fasteners, pressure vessels, and possibly armor.

4.4.1.2 Filament-Reinforced Metals. Continuous metallic and nonmetallic filaments are used for the reinforcement of metals (Rauch et al., 1968). Nonmetallic filaments have been developed with tensile strengths approaching 3.5 GPa and elastic moduli of nearly 420 GPa. Unreactive filament reinforcements (boron, silicon carbide, single-crystal sapphire, and graphite) have been developed for alloys of Al, Ti, and Ni. Surface coatings have been developed and applied to decrease the chemical interaction between filament and matrix.

Boron filaments are produced by chemical vapor deposition (CVD) of boron on a heated (1065°C) 12-μm tungsten wire substrate, typically according to

$$2BCl_3 + 3H_2 \rightarrow 6HCl + 3B$$

The boron deposits are extremely fine grained, bordering on being amorphous. Silicon carbide filaments are also produced by CVD on a heated 12-μm tungsten wire substrate. Single-crystal sapphire filaments are produced by drawing from molten alumina. Graphite filaments are produced by the graphitization at 2500 to 3200°C of polyacrylonitrile or rayon filamentary precursors. Single-crystal sapphire filaments are chemically unreactive in many alloys of interest. Because of the solubility of carbon in metals, the tendency to form metal carbides, and the potential for activated crystallization of the filament, efforts to reinforce metals with graphite have attained only limited success. The mechanical properties of various ceramic and metal filaments are in Table 4.35 (Hahn and Kershaw, 1970).

A spectrum of filament-matrix amalgamation techniques has been developed. The most successful techniques rely on diffusion bonding. Most extensively, titanium matrices have been reinforced with boron filaments. Other matrices that also have been reinforced are Ni and Co alloys (Vidoz et al., 1969). Filament-reinforced metals have shown significant improvement in mechanical properties. Selected mechanical properties of some composites are shown in Table 4.36. The fiber-reinforced composites (particularly boron-aluminum and boron-titanium) will find applications in many areas, such as compressor blades for gas turbine propulsion systems. These blades offer significant advantages over standard compressor blades, e.g., up to 40 percent weight savings, increased blade stiffness, and the potential for higher tip speeds and resulting higher engine efficiencies.

4.4.2 Ceramic-Matrix Composites.

Ceramics offer numerous advantages (high specific strength, high specific stiffness, high-temperature operation, corrosion resistance, and cost) over metals and

TABLE 4.36 Selected Mechanical Properties of Metal-Matrix Composites

Material	Tensile strength, GPa	Young's modulus, GPa	Specific strength, 10^4 m	Specific stiffness, 10^4 m	Flexural fatigue strength, MPa
2024-T4 Al	0.48	70	—	—	85
2024-T4 Al reinforced with 10 vol % β-SiC whiskers	0.58	—	—	—	175
2024-T4 Al reinforced with 25 vol % β-SiC whiskers	0.74	105	—	—	—
Aluminum unidirectionally reinforced with 50 vol % Boron filaments:					
0°*	1.15	230	4.45	890	—
90°†	0.11	145	0.40	560	—
Titanium unidirectionally reinforced with 50 vol % boron filaments:					
0°*	1.07	260	3.05	750	—
90°†	0.21	195	0.60	550	—

*Tested parallel to the direction of reinforcement.
†Tested perpendicular to the direction of reinforcement.
Source: Adapted from Hahn and Kershaw (1970).

alloys. However, their intrinsic flaw sensitivity and brittleness (poor thermal shock, impact, erosion, and fatigue resistance) continue to impede monolithic ceramic component development. A basic change in the material failure mechanism—from a flaw-sensitive, brittle, catastrophic failure to a tough, ductile, noncatastrophic failure—can be obtained through the use of particle- or fiber-reinforced* ceramic composite systems. Fibers act as barriers to crack propagation in ceramic-matrix composites.

Figure 4.12 shows tensile strength as a function of temperature for various materials. The selection of materials for operating temperatures above 1200°C is quite limited. Carbon and its composites exhibit equal or increased strength between room temperature and more than 2300°C. Carbon-carbon composites, which are a brittle fiber–brittle matrix ceramic system, are tough and somewhat foregiving materials. Data show that although microcracking does occur at load levels below ultimate failure, subsequent loading to equal or lesser values does not propagate these cracks. Fatigue life is very long (Warren, 1980). A typical application is in aircraft brakes, e.g., in F-14's, F-16's, the space shuttle, and the Concorde aircraft. The carbon-carbon composite provides friction, heat-sink, and structural capabilities.

Other ceramic-matrix composites that are under various stages of development are graphite fiber–reinforced carbon, silicon fiber–reinforced silicon carbide, glass fiber–reinforced glass, graphite or SiC fiber–reinforced glass, and tantalum fiber–reinforced Si_3N_4 composite. These composite systems have exhibited substantial improvements in toughness and lack of the flaw sensitivity so common in monolithic ceramics. For example, Warren (1980) has reported a sevenfold improvement in the Charpy impact strength of tantalum fiber–reinforced

*A *fiber* is a single, continuous material whose length is at least 200 times its width or diameter.

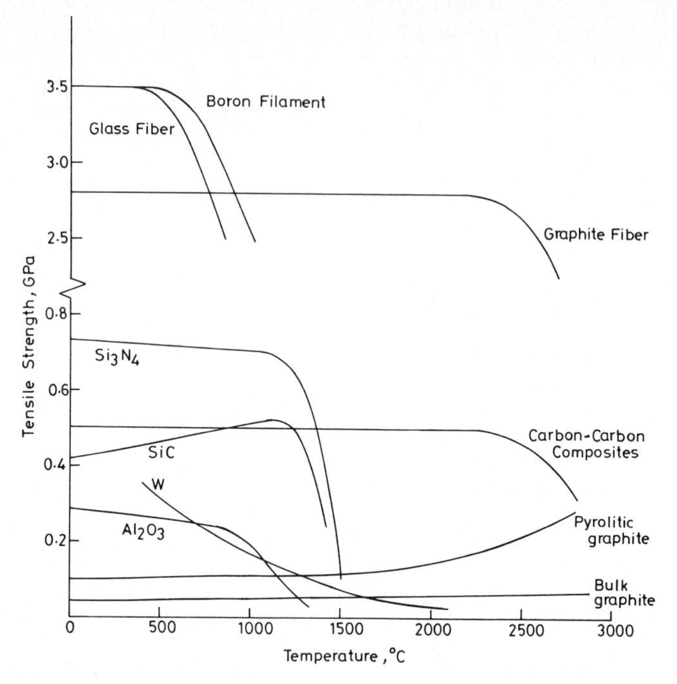

FIGURE 4.12 Effect of operating temperature on the tensile strength of various fibers and bulk materials.

Si_3N_4 (from 0.75 to 5.25 N · m). Flexural strengths of up to 800 MPa have been reported for carbon-Si-SiC materials. Many of the low-density, high-temperature ceramic-matrix composites will find applications in the rotating components (rotor and power train) of advanced automotive and aerospace gas turbines, which will operate at elevated temperatures.

4.5 REFERENCES

Alper, A. M. (1970a), *High-Temperature Oxides*, Part 1: *Magnesia, Lime, and Chrome Refractories*, Academic, New York.

——— (1970b), *High-Temperature Oxides*, Part 2: *Oxides of Rare Earths, Titanium, Zirconium, Hafnium, Niobium, and Tantalum*, Academic, New York.

——— (1970c), *High-Temperature Oxides*, Part 3: *Magnesia, Alumina, Beryllia Ceramics: Fabrication, Characterization, and Properties*, Academic, New York.

——— (1971), *High-Temperature Oxides*, Part 4: *Refractory Glasses, Glass-Ceramics, and Ceramics*, Academic, New York.

Amateau, M. J., and Glaeser, W. A. (1964), "Survey of Materials for High Temperature Bearing and Sliding Applications," *Wear*, Vol. 7, pp. 385–418.

Anonymous (1957), "Corrosion and Wear Handbook for Water-Cooled Reactors," TID 7006, United States Atomic Energy Commission, Oak Ridge, Tenn.

——— (1970a), *Huntington Alloys Handbook*, Huntington Alloys, Inc., Huntington, W.Va.

——— (1970b), "Haynes Alloy No. 25," Cabot Corporation, Stellite Division, Kokomo, Ind. Currently manufactured by Haynes International, Kokomo, Ind.

——— (1972), "Ti-6Al-4V Alloy Digest: Data on Worldwide Metals and Alloys," Eng. Alloy Digest, Inc., Upper Montclair, N.J.

——— (1973), "Armco Stainless Steels," Armco Steel Corporation, Middletown, Ohio.

——— (1974), "Special Issue on Metals," *Machine Design*, Vol. 46, Feb. 14.

——— (1975), "Tribaloy: Properties, Uses and Composition and Fabrication." DuPont Tribaloy Products, Wilmington Del.

——— (1976a), "Haynes Wrought Wear-Resistant Alloys," Cabot Corporation, Stellite Division, Kokomo, Ind. Presently manufactured by Deloro Stellite, Inc., Delleville, Ontario, Canada.

——— (1976b), "Berylco Beryllium Copper and Beryllium Nickel Alloys," Kawecki Berylco Industries, Inc., Reading, Pa.

——— (1976c), *Engineering Property Data on Selected Ceramics*, Vol. 1: *Nitrides, Metals, and Ceramics*, Information Center, Battelle Columbus Laboratory, Cleveland, Ohio.

——— (1977), "Hastelloy Alloy B-2," Cabot Corporation, Stellite Division, Kokomo, Ind.

——— (1978a), *Metals Handbook*, Vol. 1: *Properties and Selection: Irons and Steels*, 9th Ed., American Society for Metals, Metals Park, Ohio.

——— (1978b), *Metals Handbook*, Vol. 2: *Properties and Selection: Nonferrous Alloys and Pure Metals*, 9th Ed., American Society for Metals, Metals Park, Ohio.

——— (1978c), *Metals Handbook*, Vol. 3: *Properties and Selection: Stainless Steels, Tool Materials, and Special-Purpose Metals*, 9th Ed., American Society for Metals, Metals Park, Ohio.

——— (1978d), "Review of the Wear and Galling Characteristics of Stainless Steels," Committee of Stainless Steel Producers, American Iron and Steel Institute, Washington, D.C.

——— (1979), *Engineering Property Data on Selected Ceramics*, Vol. 2: *Carbides, Metals, and Ceramics*, Information Center, Battelle Columbus Laboratory, Cleveland, Ohio.

——— (1980a), "Materials Reference Issue," *Machine Design*, Vol. 52, April 17.

——— (1980b), "Superhard Tool Materials," in *Metals Handbook*, Vol. 3: *Properties and Selection: Stainless Steels, Tool Materials, and Special-Purpose Metals*, 9th Ed., pp. 448–465, American Society for Metals, Metals Park, Ohio.

——— (1981), *Engineering Property Data on Selected Ceramics*, Vol. 3: *Single Oxides, Metals, and Ceramics*, Information Center, Battelle Columbus Laboratory, Cleveland, Ohio.

——— (1987), "Tribology of Ceramics," Special Publications SP-23 and SP-24, STLE, Park Ridge, Ill.

Aronov, V., and Benetatos, M. (1989), "Wear Resistance of Laser Treated Partially Stabilized Zirconia," *J. Lub. Technol.* (*Trans. ASME*), Vol. 111, pp. 372–377.

Bain, E. C., and Paxton, H. W. (1961), *Alloying Elements in Steel*, 2d Ed., American Society for Metals, Metals Park, Ohio.

Baumgartner, H. R. (1978), "Ceramic Bearings for Turbine Applications," in *Ceramics for High Performance Applications*, Vol. II: *Proc. 5th Army Materials Technol. Conf.* (J. J. Burke, E. N. Lanoe, and R. N. Katz, eds.), pp. 423–443, Brook Hill, Chestnut Hill, Mass.

Bersch, C. F. (1978), "Overview of Ceramic Bearing Technology," in *Ceramics for High Performance Applications*, Vol. II: *Proc. 5th Army Materials Technol. Conf.* (J. J. Burke, E. N. Lanoe, and R. N. Katz, eds.), pp. 397–405, Brook Hill, Chestnut Hill, Mass.

Bhansali, K. J. (1980), "Wear Coefficients of Hard-Surfacing Materials," in *Wear Control Handbook* (M. B. Peterson and W. O. Winer, eds.), pp. 373–383, ASME, New York.

———, and Mehrabian, R. (1982), "Abrasive Wear of Aluminum Matrix Composites," *J. Metals*, Vol. 34, Sept., pp. 30–34.

Bhushan, B. (1980), "Coating Development and Friction Measurements for Solid Lubricated Rolling-Element Bearings," Hughes Aircraft/DARPA Report No. MTI-80TR30, Mechanical Technology, Inc., Latham, N.Y.

——— (1990), *Tribology and Mechanics of Magnetic Storage Devices*, Springer-Verlag, New York.

———, and Gray, S. (1978), "Materials Study for High Pressure Seawater Hydraulic Tool Motors," Tech. Rep. No. CR-78.012, Civil Engineering Lab, Naval Construction Batallion Center, Port Hueneme, Calif.

———, and ——— (1979), "Investigation of Material Combinations under High Load and Speed in Synthetic Seawater," *Lub. Eng.*, Vol. 35, pp. 628–639.

———, and Sibley, L. B. (1982), "Silicon Nitride Rolling Bearings for Extreme Operating Conditions," *ASLE Trans.*, Vol. 25, pp. 417–428.

———, and Winn, L. W. (1979), "Material Study for Advanced Sterntube Seals and Bearings for High Performance Merchant Ships," Tech. Rep. No. MA-920-79069, Office of Commercial Development, U.S. Maritime Administration, Washington, D.C.

———, Ruscitto, D., and Gray, S. (1978), "Hydrodynamic Air-Lubricated Compliant Surface Bearing for an Automotive Gas Turbine Engine, Part II: Materials and Coatings," Tech. Rep. No. CR-135402, NASA Lewis Research Center, Cleveland, Ohio.

Booser, E. R. (1984), *Handbook of Lubrication: Theory and Practice of Tribology*, Vol. 2: *Theory and Design*, CRC Press, Boca Raton, Fla.

Brauer, G., and Kirner, H. (1964), "High Pressure Synthesis of Niobium Nitride and the Constitution of δ-NbN," *Z. Anorg. Allgem. Chem.*, Vol. 328, pp. 34–43.

Brenner, S. S. (1962a), "The Case for Whisker-Reinforced Metals," *J. Metals*, Vol. 14, pp. 808–811.

——— (1962b), "Mechanical Behavior of Sapphire Whiskers at Elevated Temperatures," *J. Appl. Phys.*, Vol. 33, pp. 33–39.

Buckley, D. H. (1968), "Adhesion, Friction and Wear of Cobalt and Cobalt Base Alloys," *Cobalt*, Vol. 38, pp. 20–28.

———, and Johnson, R. L. (1964), "Marked Influence of Crystal Structure on the Friction and Wear Characteristics of Cobalt and Cobalt Base Alloys in Vacuum to 10^{-9} Millimeter of Mercury, I-Polycrystalline and Single-Crystal Cobalt," Tech. Rep. No. TN D-2523, NASA, Washington, D.C.

Budinski, K. G. (1980), "Tool Materials," in *Wear Control Handbook* (M. B. Peterson and W. O. Winer, eds.), pp. 931–985, ASME, New York.

——— (1983), *Engineering Materials: Properties and Selection*, Reston Publishing Co., Reston, Va.

Burke, J. J., Gorum, A. E., and Katz, R. N. (1974) (eds.), *Ceramics for High Performance Applications*, Vol. II: *Proc. 2nd Army Materials Technol. Conf.*, Brook Hill, Chestnut Hill, Mass.

Chandrasekar, S., and Bhushan, B. (1990), "Friction and Wear of Ceramics for Magnetic Recording Applications—Part I: A Review," *J. Tribol.* (*Trans. ASME*), Vol. 112, pp. 1–16.

Clauss, F. J. (1972), *Solid Lubricants and Self-Lubricating Solids*, Academic, New York.

Claussen, N., Ruhle, M., and Heuer, A. H. (eds.) (1984), *Advances in Ceramics*, Vol. 12: *Science and Technology of Zirconia II*, American Ceramic Society, Columbus, Ohio.

Cother, N. E. (1987), "Sialon Ceramics: Their Development and Engineering Applications," *Mater. Design*, Vol. 8, pp. 2–9.

Dalal, H. M., Chiu, Y. P., and Rabinowicz, E. (1975), "Evaluation of Hot-Pressed Silicon Nitride as a Rolling Bearing Material," *ASLE Trans.*, Vol. 18, pp. 211–221.

Dayton, R. D., Sheets, M. A., and Schrand, J. B. (1978), "Evaluation of Solid Lubricated Ball Bearing Performance," *ASLE Trans.*, Vol. 21, pp. 211–216.

Dean, R. R., and Evans, C. J. (1976), "Plain Bearing Materials: The Role of Tin," *Tribol. Int.*, Vol. 9, pp. 101–108.

DeHart, A. O. (1984), "Sliding Bearing Materials," in *Handbook of Lubrication: Theory and Practice of Tribology*, Vol. 2: *Theory and Design*, pp. 463–494, CRC Press, Boca Raton, Fla.

Denhard, E. E., and Espy, R. H. (1971), "Austenitic Stainless Steels with Unusual Mechanical and Corrosion Properties," presented at the ASME Petroleum Mechanical Engineering with Underwater Technology Conference, Houston, Texas, Sept. 19–23.

Dillich, S., and Kuhlmann-Wilsdorf, D. (1982), "Two Regimes of Current Conduction in Metal-Graphite Electrical Brushes and Resulting Instabilities," *Proc. IEEE 28th Holm Conf. on Electrical Contact Phenomena*, pp. 201–212, IEEE, Chicago, Ill.

Dreger, D. R. (1978), "Progress in Powder Metallurgy," *Machine Design*, Vol. 50, pp. 116–121.

Evans, A. G., and Marshall, D. B. (1981), "Wear Mechanisms in Ceramics," in *Fundamentals of Friction and Wear of Materials* (D. A. Rigney, ed.), pp. 439–450, American Society for Metals, Metals Park, Ohio.

Eyre, T. S. (1979), "Wear Resistance of Metals," in *Treatise on Materials Science and Technology*, Vol. 13: *Wear*, pp. 363–442, Academic, New York.

Field, J. E. (1979), *Properties of Diamond*, Academic, New York.

Finkin, E. F., Calabrese, S. J., and Peterson, M. B. (1973), "Evaluation of Materials for Sliding at 600F–1800F in Air," *ASLE Trans.*, Vol. 29, pp. 197–204.

Foroulis, Z. A. (1984), "Guidelines for the Selection of Hardfacing Alloys for Sliding Wear Resistant Applications," *Wear*, Vol. 96, pp. 203–218.

Glaeser, W. M. (1980), "Wear-Resistant Materials," in *Wear Control Handbook* (M. B. Peterson and W. O. Winer, eds.), pp. 313–326, ASME, New York.

Green, D. J., Hannink, R. H. J., and Swain, M. V. (1989), *Transformation Toughening of Ceramics*, CRC Press, Boca Raton, Fla.

Gulyaev, A. (1980), *Physical Metallurgy*, Vols. 1 and 2, Mir Publishers, Moscow.

Hahn, H., and Kershaw, J. P. (1970), "Development of Ceramic Fiber–Reinforced Metals," *Proc. 2nd Int. Am. Conf. Mater. Technol., Mexico City*, Part 1, pp. 559–582, ASME, New York.

Hampel, C. A. (1961), *Rare Metals Handbook*, 2d Ed., Reinhold, London.

Hansen, J. S., Kelly, J. E., and Wood, F. W. (1979), "Erosion Testing of Potential Valve Materials for Coal Gasification Systems," Rep. Invest. No. 8335, p. 26, Bureau of Mines, Washington, D.C.

Hausner, H. H., and Bowman, M. G. (eds.) (1968), *Fundamentals of Refractory Compounds*, Plenum, New York.

———, Roll, K. H., and Johnson, P. K. (1970), *Friction and Antifriction Materials*, Plenum, New York.

Heuer, A. H., and Hobbs, L. W. (eds.) (1981), *Advances in Ceramics*, Vol. 3: *Science and Technology of Zirconia*, American Ceramic Society, Columbus, Ohio.

Holleck, H. (1986), "Material Selection for Hard Coatings," *J. Vac. Sci. Technol.*, Vol. A4, pp. 2661–2669.

Hosking, F. M., Portillo, F. F., Wunderlin, R., and Mehrabian, R. (1982), "Composites of Aluminum Alloys: Fabrication and Wear Behavior," *J. Mater. Sci.*, Vol. 17, pp. 477–498.

Hove, J. E., and Riley, W. C. (1965), *Ceramics for Advanced Technology*, Wiley, New York.

Jack, K. H. (1980), "Energy and Ceramics," in *Material Science Monograph*, Vol. 6 (P. Vincenzini, ed.), p. 534, Elsevier, Amsterdam.

Kaufman, H. W. (1980), "Bearing Materials," in *Tribology: Friction, Lubrication, and Wear* (A. Szeri, ed.), McGraw-Hill, New York.

Kirkham, R. O., Park, G., Groves, W. S., and Stobo, J. J. (1983), "Plain Bearing Alloys," *Metals Forum* (*J. Aust. Inst. Metals*), Vol. 6, pp. 149–160.

Kriven, W. M. (1988), "Possible Alternative Transformation Tougheners to Zirconia: Crystallographic Aspects," *J. Am. Ceram. Soc.*, Vol. 71, pp. 1021–1030.

Lakhtin, Y. (1977), *Engineering Physical Metallurgy*, Mir Publishers, Moscow.

Lynch, J. F., Ruderer, C. G., and Duckworth, W. H. (1966), *Engineering Properties of Selected Ceramic Materials*, American Ceramic Society, Columbus, Ohio.

McColm, I. J., and Clark, N. J. (1988), *Forming, Shaping, and Working of High Performance Ceramics*, Chapman and Hall, London.

McCreight, L. R., Rauch, H. W., and Sutton, W. H. (1965), *Ceramic and Graphite Fibers and Whiskers*, Academic, New York.

Miner, J. R., Grace, W. A., and Valori, R. (1981), "A Demonstration of High-Speed Gas Turbine Bearings Using Silicon Nitride Rolling Elements," *Lub. Eng.*, Vol. 37, pp. 462–478.

Morrell, R. (1984), "Ceramics in Modern Engineering," *Phys. Technol.*, Vol. 15, pp. 252–261.

Neely, J. E. (1984), *Practical Metallurgy and Materials of Industry*, Wiley, New York.

Ohmae, N., and Rabinowicz, E. (1980), "The Wear of the Noble Metals," *ASLE Trans.*, Vol. 23, pp. 86–92.

Pascoe, K. J. (1978), *An Introduction of the Properties of Engineering Materials*, Van Nostrand–Reinhold, New York.

Peterson, M. B., and Winer, W. O. (1980), *Wear Control Handbook*, ASME, New York.

Pratt, G. C. (1973), "Materials for Plain Bearings," *Int. Metallurg. Rev.*, Vol. 18, pp. 2–28.

Rauch, H. W., Sutton, W. H., and McCreight, L. R. (1968), *Ceramic Fibers and Fibrous Composite Materials*, Academic, New York.

Reed-Hill, R. E. (1973), *Physical Metallurgy Principles*, 2d Ed, Van Nostrand, New York.

Rhines, F. W. (1956), *Phase Diagrams in Metallurgy*, McGraw-Hill, New York.

Rigney, D. A. (ed.) (1981), *Fundamentals of Friction and Wear of Materials*, American Society for Metals, Metals Park, Ohio.

Samsonov, G. V., and Vinitskii, I. M. (1980), *Handbook of Refractory Compounds*, IFI/Plenum, New York.

Schmidt, R. D., and Ferriss, D. P. (1975), "New Materials Resistant to Wear and Corrosion to 1000°C," *Wear*, Vol. 32, pp. 279–289.

Schumacher, W. J. (1976), "Metals for Nonlubricated Wear," *Machine Design*, March 11, Vol. 48, pp. 57–59.

——— (1977), "Wear and Galling Can Knock Out Equipment," *Chem. Eng.*, May 9, Vol. 84, pp. 155–160.

——— (1981), "A Stainless Steel Alternative to Cobalt Wear Alloys," *Chem. Eng.*, Sept. 21, Vol. 88, pp. 149–152.

Scott, D. (1983), "Materials for Tribological Applications," in *Industrial Tribology: The Practical Aspects of Friction, Lubrication, and Wear* (M. H. Jones and D. Scott, eds.), pp. 205–222, Elsevier, Amsterdam.

Silence, W. L. (1978), "Effect of Structure on Wear Resistance of Co-, Fe-, and Ni-Based Alloys," *J. Lub. Technol.* (*Trans. ASME*), Vol. 100, pp. 428–435.

Steiger, E. (1982), "Wear-Resistant Silver Combinations for Dynamic Electrical Contact," *Oberflache-surface*, Vol. 23, pp. 319–322.

Steijn, R. P. (1961), "On the Wear of Sapphire," *J. Appl. Phys.*, Vol. 32, pp. 1951–1958.

Storms, E. (1967), *Refractory Carbides*, Academic, New York.

Sundgren, J.-E., and Hentzell, H. T. G. (1986), "A Review of the Present State of the Art of Hard Coatings Grown from the Vapor Phase," *J. Vacuum Sci. Technol.*, Vol. A4, pp. 2259–2279.

Sutton, W. H., and Chorne, J. (1965), "Potential of Oxide Fiber Reinforced Metals," *Fiber Symp. Mater.* (S. H. Bush, ed.), p. 173, Ameican Society for Metals, Metals Park, Ohio.

Taylor, K. M., Sibley, L. B., and Lawrence, J. C. (1963), "Development of a Ceramic Rolling Contact Bearing for High-Temperature Use," *Wear*, Vol. 6, pp. 226–240.

Thelning, K. E. (1984), *Steel and Its Heat Treatment*, in Bofor's Handbook, Butterworth, London.

Toth, L. E. (1971), *Transition Metal Carbides and Nitrides*, Academic, New York.

Van Vlack, L. H. (1985), *Elements of Materials Science and Engineering*, Addison-Wesley, Reading, Mass.

Vidoz, A. E., Camahort, J. L., and Crossman, F. W. (1969), "Development of Nitrided Boron Reinforced Metal-Matrix Composites," *J. Comp. Mater.*, Vol. 3, pp. 254–261.

Warren, J. W. (1980), "Ceramic Matrix Composites for High-Performance Structural Applications," in *Advanced Fibers and Composites for Elevated Temperatures* (I. Ahmed and B. R. Noton, eds.), pp. 240–248, The Metallurgical Society, Warrendale, Pa.

Wikle, K. G., Fullerton-Batten, R. C., and Droegkamp, R. E. (1974), "Beryllium Copper in Wear Applications," *Machine Design*, Vol. 46, May, p. 66.

Zeman, K. P., and Coffin, L. F. (1960), "Friction and Wear of Refractory Compounds," *ASLE Trans.*, Vol. 3, pp. 191–202.

CHAPTER 5
SOLID LUBRICANTS AND SELF-LUBRICATING SOLIDS

Solid lubricants and self-lubricating solids are solid materials that exhibit low friction and/or wear during sliding in the absence of an external supply of lubricant. Solid lubricants and self-lubricating materials embrace a wide range of materials, of which the principal categories are carbon and graphite, molybdenum disulfide and other dichalcogenides, and polymers and polymer composites. These materials are used for providing lubrication at relatively low loads, low sliding speeds, and slightly elevated temperatures. Solid lubricants are used in their dry state as well as combined with a carrier (e.g., oils and greases) or binder that is applied on the mating surfaces. Sometimes these materials are mixed with sintered metals and polymers to achieve self-lubricating characteristics (Anonymous, 1988; Bhushan, 1987; Braithwaite, 1964, 1967; Campbell, 1972; Campbell et al., 1966; Clauss, 1972; Kanakia and Peterson, 1987; Lipp, 1976; Paxton, 1979).

The most commonly used self-lubricating solids are graphite, molybdenum disulfide (MoS_2), and polytetrafluoroethylene (PTFE). These materials, used singly or in various combinations with each other, make up the lubricant pigment in over three-quarters of the solid lubricants now available. Although they have been widely used, each of the materials has certain limitations. MoS_2 oxidizes rapidly at temperatures above 300°C; however, at lower temperatures it is a very stable material. Graphite requires the presence of adsorbed water vapor to help it lubricate; therefore, its use is limited to areas where some moisture is present and it exhibits high friction when used in high-temperature applications. PTFE is a soft, complex organic material. Because of its cold flow characteristics, its use is limited to light-load and moderate-temperature applications. These limitations of the common lubricating solids have stimulated the synthesis of other compounds and composites believed to have lubricating properties.

5.1 CARBON AND GRAPHITE

Carbon is an unusual material in that it exhibits both metallic and nonmetallic characteristics. It exists in three generic forms: diamond, graphite, and black, or amorphous, carbon. In its cubic structure, it is one of the hardest materials, i.e., diamond, while in hexagonal form, i.e., graphite, it is a relatively weak material that exhibits excellent lubricating characteristics. Amorphous carbon is a

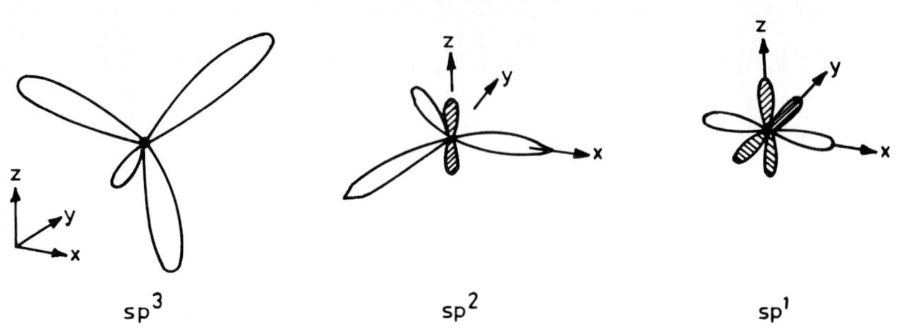

FIGURE 5.1 Schematic representation of sp^3 (diamond), sp^2 (graphite), and sp^1 (hydrocarbons and certain polymers) bonding types of carbon atoms.

noncrystalline material that exists as coke, charcoal, soot, and lampblack. Carbon is the most refractory nonmetal; it melts at 3700°C.

Each carbon atom has six electrons, with four electrons in its outer shell ($1s$ – 2; $2s$ – 2; $2p_x$ – 1; $2p_y$ – 1; $2p_z$ – none). A carbon atom can adapt three different bonding configurations (Fig. 5.1), that is,

- sp^3 Each of the carbon's four valence electrons is assigned to a tetrahedrally directed sp^3 hybrid orbital, which then forms a strong σ (covalent) bond with an adjacent atom. This structure exists in diamond, and strong bonding results in exceptionally high hardness and high thermal conductivity. The saturated bonding produces the wide 5.5-eV optical band gap and low electrical conductivity.
- sp^2 Three of the four electrons are assigned to trigonally directed sp^2 hybrid orbitals, which form σ bonds; the fourth electron lies in a p_z ($p\pi$) orbital lying normal to the σ bonding plane. The $p\pi$ orbital forms weaker π bonds (van der Waals type) with adjacent pπ orbitals. The weak π bond of the van der Waals type between planes accounts for the layered structure of graphite, which is responsible for its high electrical conductivity, softness, and lubricity.
- sp^1 Only two electrons form strong σ bonds along $\pm$Ox, and the remaining two electrons are left in orthogonal p_y and p_z orbitals to form weak π bonds. This structure exists in hydrocarbons and certain polymers.

The sp^3-type bonding in diamond is responsible for diamond's extreme hardness and excellent thermal conductivity. On the other hand, the sp^2-type bonding in graphite provides the ABAB stacking sequence along the c axis. The thermal conductivity and hardness are high along the basal plane (having strong σ bonds) and low along the c axis (having π bonds). Diamond, graphite, and carbon are used in a wide variety of materials whose structures and properties vary enormously depending on the other materials involved and the processing parameters (Badami and Wiggs, 1970; Bates, 1973; Hutcheon, 1970; Lancaster, 1979; Paxton, 1979). Selected physical properties of various forms of carbon are presented in Table 5.1.

5.1.1 Amorphous Carbon

Disordered carbon covers a wide range of materials and properties (Moore, 1973; Paxton, 1979; Robertson, 1986), e.g., coke, charcoal, soot, lampblack, amor-

TABLE 5.1 Selected Physical Properties of Various Forms of Carbon

Carbon form	Density, kg m^{-3}	Hardness (HV), kg mm^{-2}	Modulus of elasticity, GPa	Flexural strength, MPa	Thermal conductivity, W m^{-1} K^{-1}	Coefficient of thermal expansion, $\times 10^{-6}$ $°C^{-1}$	Electrical conductivity, $(\Omega \cdot cm)^{-1}$	Optical band gap, eV	Temperature limit in air, °C
Diamond	3515	8000–10,400	900–1050	1000–1050	900–2100	1.0	100^{-7}–10^{-20}	5.5	1000
Graphite	2267	Very soft	9–15	35–90	400,* 6†	−0.36,* +30†	2×10^4,* 250†	4×10^{-2}	425
Black or amorphous carbon	1600–2000	Soft	11–15	20–30	5–15	45	10^3–10^4	10^{-2}	260

*Parallel to layer planes.
†Perpendicular to layer planes.

phous (or glassy) carbon formed by heating certain organic polymers (Noda et al., 1969), carbon fibers of importance for their strength (Reynolds, 1973), and microcrystalline carbon (μ_C) produced by irradiating graphite (Kelly, 1981). Amorphous carbon is not soluble in any solvent, and it starts to oxidize at about 260°C. Therefore, for higher-temperature applications, carbon is graphitized/impregnated (see Table 5.1).

Carbon is often specified for low-friction applications. The coefficient of friction between carbon and another material depends on the grade of carbon, the mating material, surface finish, surface speed, load, and fluid environment. Even under the most severe conditions, however, galling and welding (common to metals) do not occur. Carbon black powder is used in printing inks, plastics, paints, paper, and other special applications. Carbon black is unique in that it has the smallest particle size (13 to 50 nm) and the highest oil absorption of the commonly used pigments (Anonymous, 1983). Because of its very small particle size, colloidal carbon black is easy to produce.

5.1.2 Graphite

Graphite, i.e., crystalline carbon, is a soft, shiny material (see Table 5.1). There are two broad classifications of graphite: natural and synthetic. Natural graphite is a mineral that occurs in flakes with varying degrees of crystallinity, ranging from amorphous to highly crystalline, and purity, ranging from 80 to 90 percent carbon (Kelly, 1981; Seeley, 1964). Synthetic or manufactured graphite is produced by graphitizing petroleum coke (a refinery by-product) or similar carbons in an electric furnace at temperatures in excess of 2000°C. Like most chemical and metallurgical processes, the conversion is not complete, and the graphite structure of artificial graphite hardly exceeds 85 percent. Natural graphite is silvery gray and synthetic graphite is dull black in appearance. Both natural and synthetic graphites are soft and slippery and are used as solid lubricants. Graphites are manufactured by many companies, e.g., Joseph Dixon Crucible Company, Jersey City, N.J., Pure Carbon Company, Pure Industries, Inc., St. Mary's, Pa., Union Poco Graphite, Inc., Decatur, Tex., and Wickes Engineering Materials, Saginaw, Mich.

*5.1.2.1 **Crystal Structure.*** Graphite has a hexagonal layered structure. The graphite crystal may be visualized as infinite parallel layers of hexagons stacked 0.34 nm apart. The distance between the layers is greater than the 0.1415-nm interatomic distance between the carbon atoms in the basal plane (Fig. 5.2). The bonds between the parallel layers are relatively weak (van der Waals type) in contrast with the strong covalent bonds in the basal plane (where each carbon atom is joined to three neighboring carbon atoms at 120° angles). This configuration of atoms is responsible for the extreme anisotropy displayed by aggregates of graphite crystal with preferred orientation. Tested parallel to the planes or layers of atoms, graphite is soft. However, at right angles to the planes, a graphite crystal is nearly as hard as diamond. Thermal conductivity in the planes is very high, the coefficient of thermal expansion is very low (in fact, negative), and electrical conductivity is very low. The opposite is true at right angles to the planes (Braithwaite, 1967; Clauss, 1972; Seeley, 1964).

As a rule, perfect crystals of graphite are as rare as diamonds of similar size. Crystallites as large as 1 mm in length are rare. Graphite is a birefringent crystal that appears to change color under polarized light as the microscope stage is ro-

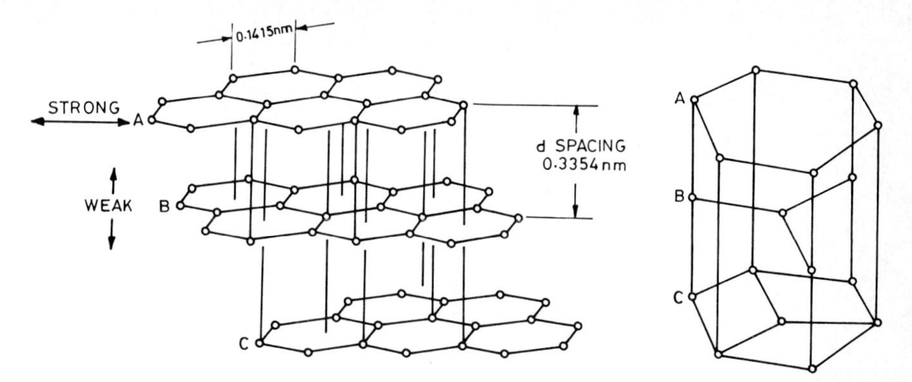

FIGURE 5.2 Crystal structure of graphite.

tated. The alignment of the crystallites in a filler grain or flake is often good enough to cause the entire grain to look like a single crystal. This type of microscopic examination is often used to estimate the concentration of larger graphite crystallites in a manufactured carbon part.

Graphite fluoride $(CF_x)_n$, prepared by the direct reaction of graphite with fluorine gas at controlled temperature and pressure, is a relatively new solid lubricant that can be loosely described as a layered-lattice intercalation compound of graphite (Ishikawa and Shimada, 1969). Graphite fluoride acts as an effective solid lubricant over a wide range of temperatures, and it will be discussed in greater detail in Chap. 13.

5.1.2.2 ***Lubrication Mechanism.*** The key to graphite's value as a self-lubricating solid lies in its layered-lattice structure and its ability to form strong chemical bonds with gases such as water vapor. Water vapor and other gases from the environment are adsorbed onto the crystalline edges, and this weakens the interlayer bonding forces, resulting in easy shear and transfer of the crystallite platelets to the mating surface (Bowden and Young, 1951; Bryant et al., 1964; Savage, 1948). Close examination of a metal mating surface that has slid against graphite in an ambient environment reveals a stained wear track. The buildup of solids in the wear track is generally 200 to 1000 nm thick (Buckley and Johnson, 1964; Paxton, 1979). Buckley and Johnson postulated that the wear track was coated with graphite platelets bonded to the metal surface through oxide linkages. This transfer film plays an important role in controlling friction and wear rates. Coefficients of friction in the range of 0.1 to 0.2 can be achieved with graphite in an ambient environment.

The coefficient of friction of graphite varies significantly with the gaseous composition of the environment. Figure 5.3 shows the dependence of the coefficient of friction of graphite on the vapor pressure of various gases, such as hydrogen, water vapor, oxygen, nitrogen, and heptane (Rowe, 1960). Figure 5.3 also shows that the quantity of water vapor necessary to reduce the coefficient of friction is very small. Buckley and Johnson (1964) noted a 1000-fold increase in the wear rate of graphite when a wear experiment with graphite sliding against a metal was conducted in a vacuum (13 μPa or 0.1 μtorr). Concurrent with this increase in wear was a disappearance of graphite from the wear track. Graphite starts to oxidize in air at about 425°C, but it can be used at higher temperatures

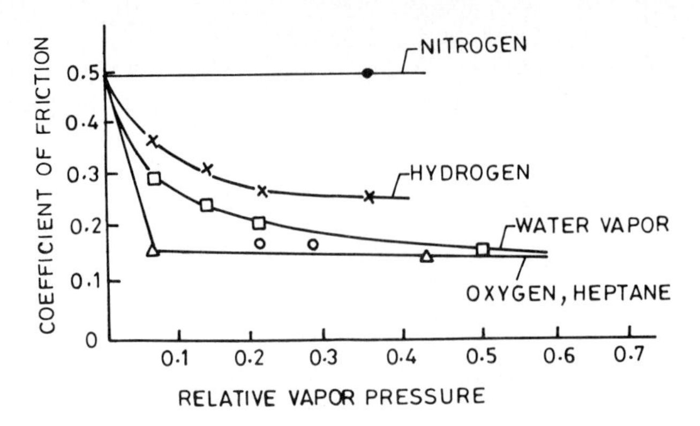

FIGURE 5.3 Effect of environment composition and its vapor pressure on coefficient of friction of graphite sliding against steel. [*Adapted from Rowe (1960).*]

(up to 3000°C) in a neutral atmosphere. For graphite to work effectively as a solid lubricant, however, the presence of moisture or other gaseous contaminants is necessary. Graphite also performs well under boundary lubrication conditions because of its good affinity for hydrocarbon lubricants.

Graphite also has excellent corrosion resistance, is an electrical conductor (1 to 1000 Ω^{-1} cm^{-1}), and is nonmagnetic. Graphite is an excellent thermal conductor and thus is highly resistant to thermal shock. This property is particularly important in jet engine seals, where temperatures can vary hundreds of degrees in a matter of seconds. Graphite closely approaches a true "black body" in its ability to radiate heat from its surface. This high radiating capacity, together with its good thermal conductivity, makes graphite very useful for dissipating heat in many applications, including tribocontacts—where frictional heat generated at the interface must be conducted away. Parts made of bulk graphite can be machined easily to close tolerances and polished to a flatness of a few light wavelengths.

5.1.2.3 ***Carbon-Graphite Solids.*** Solid carbon-graphite bodies are made by powder technology, in which coke and (synthetic or natural) graphite powders are mixed and bonded with a binder. These are sometimes called *manufactured carbon* (Paxton, 1979). The proportions (and types) of coke (carbon) and graphite are adjusted to obtain the desired lubricity and wear resistance. Generally, the more graphite in the composition, the softer is the material and the lower is the coefficient of friction. The binder is derived from coal-tar pitch (a petroleum-based product) with few exceptions. (Synthetic resins such as phenol formaldehyde are used in Pure Carbon's grade G-1 and metal binders such as silver and copper are used for higher strength, ductility, and high thermal conductivity for dry bearing applications.) A mixture of coke and graphite powder containing coal-tar pitch is formed by compression molding (under pressures ranging from 75 to 300 MPa). It is then sintered at elevated temperatures in the range of 1000 to 1400°C in oxygen-free furnaces. (A lower temperature is used for curing resin-bonded carbon-graphites.) Coal-tar pitch coats the coke and graphite particles and binds them during molding; then it is converted into carbon during the sintering process (Paxton, 1979).

For high-temperature applications, carbon-bonded carbon-graphite is converted to graphite-bonded graphite by graphitizing the carbon at temperatures in the range of 2700 to 3000°C in oxygen-free furnaces. Increased graphitization also increases the thermal conductivity (e.g., Pure Carbon's Purebon grades P03 and P003). High thermal conductivity is desirable for applications with high rubbing speeds, up to 150 m s^{-1} (e.g., encountered in aircraft jet engine seals), in order to conduct away the frictional heat generated at the interface.

For applications where strength, stiffness, and abrasion resistance are the prime requirements, e.g., vanes for dry air pumps, high-strength carbon grades (e.g., Pure Carbon's Purebon grades P-658 and P-5N and Wickes' grade 39) are used. Finely divided, calcinated petroleum coke is used in the filler to achieve these properties. Particle size of the coke filler controls the abrasiveness of the carbons. Calcinated petroleum coke is essentially 100 percent carbon, yet it is hard enough to scratch most metals. By mixing powdered coke and powdered graphite in the proper proportions, one can formulate a carbon with lubricating properties combined with extremely good wear resistance. The finest available filler carbon (coke), together with high pitch loading, is used to obtain high strength and abrasion resistance (Paxton, 1979).

Many carbon-graphites are impregnated with fillers to provide lower permeability, higher strength and hardness, lower friction, and wear, oxidation, and chemical resistance. Fillers include fine powders of various synthetic resins (e.g., epoxy, PTFE), metals (e.g., Ag, Cu, Cu-Pb alloys, babbitts, zinc), high-temperature chemical salts, ceramics, oil, wax, and various combinations of these materials. Basically, impregnation consists of saturating the porosity of carbon-graphite parts in the machined state with the molten resin and then thermal processing the part to develop the optimal properties of the impregnant. For example, in the case of a thermosetting plastic resin, the part is heat treated to polymerize the resin to a hard, tough state. Impregnation produces an integrally bonded two-phase material with a smooth, highly impermeable surface (Anonymous, 1964). Metal-rich composites of metals and graphite can be fabricated by powder metallurgy (PM) techniques, in which mixtures of the powders are either cold pressed and sintered in the solid state or hot pressed.

For maximum strength and wear resistance, graphite composites whose surfaces have been converted to silicon carbide are used, e.g., Purebide grades manufactured by Pure Carbon Company, St. Marys, Pa. (Anonymous, 1976a). To produce these materials, the graphite base material is subjected to a chemical vapor deposition process in the presence of a silicon monoxide atmosphere. A controlled chemical reaction takes place, and the graphite surface, about 0.5 mm thick, is converted to silicon carbide:

$$2C + SiO \rightarrow SiC + CO \tag{5.1}$$

These graphite-carbides combine the lubricity of graphite and the wear resistance of carbides. The surface hardness of these graphite-carbides is comparable with that of bulk SiC. They have a low coefficient of thermal expansion, low density (half that of Al_2O_3 and one-tenth that of WC), and high thermal conductivity (4 to 20 times that of Al_2O_3), and they can be used at elevated temperatures. However, they have a low elastic modulus and poor strength compared with ceramics.

Some carbon-graphite grades can be used up to the *PV* limit (a product of pressure and sliding velocity) of 0.70 MPa · m s^{-1} (or 20,000 psi × fpm) without lubrication and 17.5 MPa · m s^{-1} (or 500,000 psi × fpm) when fully lubricated. Graphite-carbides can be used at rather high stress levels and sliding velocities up to a *PV* limit of 52.5 MPa · m s^{-1} (or 1.5 × 10^6 psi × fpm) when fully lubricated. Carbon-graphites with high hardness values usually perform better at high loads with liquid cooling; those with lower hardness values perform better at high surface speeds and low loads. Some carbon-graphites and graphite-carbides can be used dry in air at temperatures up to 750°C or in nonlubricating fluids such as liquid nitrogen, gasoline, water, seawater, or other chemicals.

Selected physical properties of some carbon-graphites are shown in Table 5.2. Hardness values are given in Shore scleroscope C hardness (ANSI C-64.1), which is a measure of the rebound height of a falling diamond-tipped hammer. Hardened tool steel has a scleroscope hardness ranging up to 100, while annealed copper has a hardness of 5 to 20.

Ideal mating materials for all carbon-graphite grades include cast iron, chrome plating, stainless steel, hardened steel, iron-based superalloys (e.g., A286), nickel- and cobalt-based superalloys, carbides, and ceramic oxides finished to a surface roughness of 0.4 μm or less. Purebide grades can only be used with very hard materials (Rockwell hardness greater than 35 HRC), such as carbides, ceramic oxides, and various superalloys.

Figure 5.4 gives typical values for the coefficient of friction and wear rate of a variety of carbons and graphites and their composites during sliding against hardened steel (Lancaster, 1973). Carbon-graphite and graphite materials are commonly used in bearings, seals, and pump and valve parts in normal ambient and extreme operating conditions (high temperatures, high loads, and high surface velocities, such as jet engine seals; nonlubricated conditions, such as seawater pumps; and abrasive environment, such as sand slurry pumps) (Bhushan and Gray, 1978, 1979; Carson, 1965; Dillich and Kuhlmann-Wilsdorf, 1982; Paxton, 1967, 1979; Savage and Schaefer, 1956). Fibers and powders of carbon and graphite are commonly added to plastics and elastomers to increase strength and to reduce friction and wear (see next section). Metal salts (e.g., CdO) are normally added to graphite to improve its lubricating capability and to extend its operating temperature range to 540°C when used in the form of powder or a coating (see Chap. 13).

Carbon-graphite brushes are commonly used for electrical applications, e.g., commutating electric power in motors and generators. Electrical conductivity varies, for example, from a high 1000 to a low of 1 Ω^{-1} cm^{-1}. Aviation-grade brushes use additives that overcome the problem of rapid wear ("dusting") at high altitudes. A common additive used is MoS_2. Copper-graphite brushes (metal 40 to 95 percent) have very low resistivities and high current-carrying capacities. These are used for such applications as automotive starter brushes. In addition to copper in graphite, the brushes can contain lead and tin to help prevent excessive commutator wear. Silver-graphite brushes (metal 40 to 90 percent) provide very low noise level, low and stable contact resistance, low friction, and high conductivity. Silver-graphite brushes are used for slip rings and for commutators in low-voltage generators and motors. Silver-graphite brushes are generally used against such mating materials as silver, copper, bronze, silver-plated copper and bronze, gold, gold alloys, and other precious metal alloys. Typical properties of carbon-graphite brushes are shown in Table 5.3.

5.1.2.4 ***Dry Powders and Dispersions.*** Powdered graphite and stable suspensions of graphite in lubricating oils, as an additive to greases, are commonly used for a variety of industrial applications. In an experimental program, powdered lubricants entrained in a gaseous carrier were used for rolling-contact bearings and high-speed spur gears for operation at temperatures over 650°C in military aircraft. Natural graphite plus cadmium oxide (83:17) in air carrier and MoS_2 plus metal-free phthalocyanine (76:24) in nitrogen carrier were successfully used in a

TABLE 5.2 Selected Physical Properties of Some Manufactured (Molded) Carbon-Graphites

Identification/grade	Density, kg m^{-3}	Hardness, Shore scleroscope C	Tensile modulus, GPa	Tensile strength, MPa	Compressive strength, MPa	Flexural strength, MPa	Thermal conductivity, W m^{-1} K^{-1}	Coefficient of thermal expansion, $\times 10^{-6}$ $°C^{-1}$	Electrical conductivity, $(\Omega \cdot cm)^{-1}$	Temperature limit in air/neutral atmosphere, °C/°C	Typical applications
Resin-bonded carbon-graphite, Pure Carbon grade G-1 (Paxton, 1979)	1900	20	23	20	40	30	1.7	3.6	1.5×10^2	200	Used for light-duty seals and valve seats
Silver-bonded carbon-graphite, Pure Carbon grade S-95 (Paxton, 1979)	8050	10	14	55	210	100	140	2.0	7×10^5	370	Used for slow-moving bearings that carry heavy loads
Carbon-bonded carbon-graphite, Wickes grade 39 (Anonymous, 1976c)	1820	105	17	60	260	80	6.6	6.1	—	250	Used for high-speed, heavily loaded applications under lubricated or dry conditions
Carbon-bonded carbon-graphite, Purebon grade P-5N (Anonymous, 1975)	1850	90	21	60	280	85	8.6	4.3	—	815/315	Used for high-speed applications, e.g., jet engine main motor bearing seals
Carbon-bonded carbon-graphite, Purebon grade P-658 RCH (Anonymous, 1975)	1810	95	23	45	215	60	8.6	4.7	—	330/650	Used for high speeds and high loads in dry and liquid environments, e.g., compressor seals, chemical seals, bearings, and pump vanes
Graphite-bonded graphite, Purebon grade P-2003 (Anonymous, 1975)	2000	90	15	60	230	80	7.8	4.9	—	750/775	Used for high loads and high-temperature dry applications, e.g., jet engine bearing seals
Graphite-carbide, Purebide grade PE-7130 (Anonymous, 1976a)	1850	90*	16	14	85	28	52	4.3	—	460/1375	Used for high-temperature, high-load, high-speed applications requiring extreme wear resistance, e.g., sand slurry pump seals and mating rings and jet engine seals

*Rockwell 15 T.

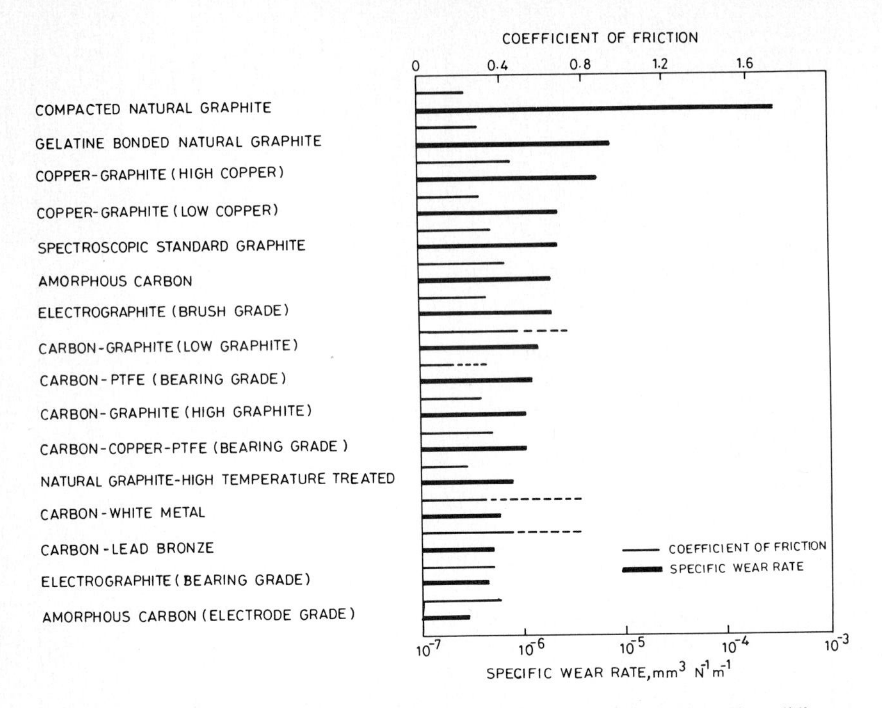

FIGURE 5.4 Friction and wear of various carbons, graphites, and their composites sliding on hardened 1% C steel (0.025 μm cla; load, 10 N; speed, 18 m s^{-1}). [*Adapted from Lancaster (1973).*]

TABLE 5.3 Selected Properties of Typical Carbon-Graphite Brushes

Grade	Hardness, Shore scleroscope C	Flexural strength, MPa	Specific electrical resistivity, Ω m^{-3}	Current capacity, A mm^{-2}	Coefficient of friction
Electrographitic	30–75	15–40	2–100	0.10–0.15	Low to medium
Carbon-graphite	50–80	20–30	7–35	0.08–0.10	Medium to high
Graphite	8–75	15–50	10–50	0.08–0.10	Medium to high
Copper-graphite	8–25	15–140	0.005–1	0.13–0.25	Very low
Silver-graphite	10–20	20–140	0.005–1	0.20–0.50	Very low

Source: Adapted from Clauss (1972).

laboratory test for a period of about 100 hours (Gray and Wilson, 1961; Schlosser, 1963).

Acheson (1907) produced a stable suspension of graphite in a fluid at the turn of the century. Now, colloidal suspensions of fine particles of high-purity natural or synthetic graphite (average particle size ranging from 40 to 0.3 to 0.5 μm or less) in petroleum oil, water, and various solvents are commercially available (for example, from Acheson Colloids Company, Port Huron, Mich., Joseph Dixon Crucible Company, Jersey City, N.J., and Union Carbide Corporation, Chicago, Ill.). Colloidal particles of graphite have an electronegative charge that assists in stabilizing the particles in suspension and in attaching the particles to metal surfaces of opposite polarity.

The addition of chemical antiwear and extreme pressure agents to petroleum and synthetic lubricants is common. The possible interrelations between these additives and solid lubricant suspensions should be understood. In some cases, chemical additives may prevent a solid lubricant film from forming on sliding metal surfaces, which is otherwise necessary to reduce friction and wear (Bartz, 1971; Bartz and Oppelt, 1978). Hisakado et al. (1983) studied the lubricating properties of petroleum machine oil no. 120 containing graphite, PTFE, and MoS_2 (Table 5.4). They found that MoS_2 dispersions gave the lowest wear, followed by graphite and then PTFE dispersions. However, graphite and PTFE dispersions gave a lower coefficient of friction than MoS_2 dispersions. It is also of note that the particle size of the solid lubricants, their concentrations, and the normal load have important effects on wear performance. Therefore, lubricant particle size and content should be optimized for a given sliding application (Bartz, 1971, 1972; Stock, 1966).

Graphite dispersions, along with MoS_2 and PTFE dispersions, are widely used in various industrial applications, e.g., mold release in the glass industry; die casting, chain extrusion, drawing lubricants, metal-cutting lubricants; heavily loaded bearings and gears; parting agents for screws, nuts, and bolts to prevent seizure of screw threads in joints; and automotive engine lubricants (Bartz, 1971; Bartz and Oppelt, 1978; Bennington et al., 1975; Clauss, 1972; Hisakado et al., 1983). Graphite particles are also used as an additive in greases. Table 5.5 shows the typical graphite dispersions available and their common applications.

5.1.2.5 ***Graphite Solids for Structural Applications.*** The strength of graphite bodies increases with temperature, whereas most materials lose strength at higher temperatures (Fig. 5.5). For this reason, impregnated graphite parts are often used in high-temperature structural applications such as missile-nozzle inserts,

TABLE 5.4 Coefficient of Friction and Wear Rate of a Copper Disk Slid Against a Bearing-Steel Pin (15 mm Diameter) in a Pin-on-Disk Sliding Test Under Lubricated Conditions*

Particle material	Mean particle size, μm	Solid content, %	Coefficient of friction	Wear rate,† $\times 10^{-12}$ m^3 $(N \cdot m)^{-1}$	
				At 75 mN	At 1.5 N
None	—	—	0.09	1.0	1.1
Graphite	1	1	0.05	0.5	0.55
Graphite	10	1	—	0.7	—
MoS_2	0.4	1	0.08	0.4	0.45
PTFE	5	1	0.05	0.8	—

*Test conditions: sliding velocity = 70 mm s^{-1}, ambient temperature, petroleum machine oil no. 120 (kinematic viscosity = 80 m^2 s^{-1}) with solid-lubricant additives.

†Volume of wear per unit sliding distance per unit normal load.

TABLE 5.5 Typical Graphite Dispersions

Carrier/diluent	Solid content, %	Average particle size, μm	Density, kg m^{-3}	Typical applications
Petroleum oil	1–10	0.5	980	General design and industrial uses; lubricant for dies, tools, and molds for metal working, etc.; additive for oils
Petroleum oil	40	40–60	1200	Formulated into forging lubricants, die-casting lubricants, ingot-mold coating
Water	20	0.5	1115	General design and industrial uses; lubricant for dies, tools, and molds for metal working, etc.; electrical-conductive coating
Water	30	40–60	1175	Mold wash for aluminum permanent molds and molds for mechanical rubber goods
Polyglycol (water insoluble)	1–10	2.5	1045	Extreme-temperature lubricant for conveyer chains and kiln car bearings
Organic vehicle	25	40–60	1610	Antiseize thread lubricant for oxygen system
Trichloroethylene (rapidly evaporating nonflammable carrier)	10	0.5	1440	Nonflammable dry-film lubricant

Source: Adapted from Clauss (1972).

heat shields, fuel-chamber liners, crucibles, furnace liners, and heating elements. Graphite parts are not suited for applications in which mechanical shock or loading is relatively high.

5.2 MOLYBDENUM DISULFIDE (MoS_2) AND OTHER DICHALCOGENIDES

Dichalcogenides of interest as self-lubricating solids include sulfides, diselenides, and ditellurides of the metals molybdenum, tungsten, niobium, and tantalum. These are crystallographically similar lamellar-type compounds between metals from either groups VB or VIB and nonmetals from group VIA of the periodic table. Whereas MoS_2 occurs naturally, the other dichalcogenides are produced synthetically by direct combination of the elements at elevated temperatures and are relatively expensive.

Various physical properties such as molecular weight, crystal structure, lattice

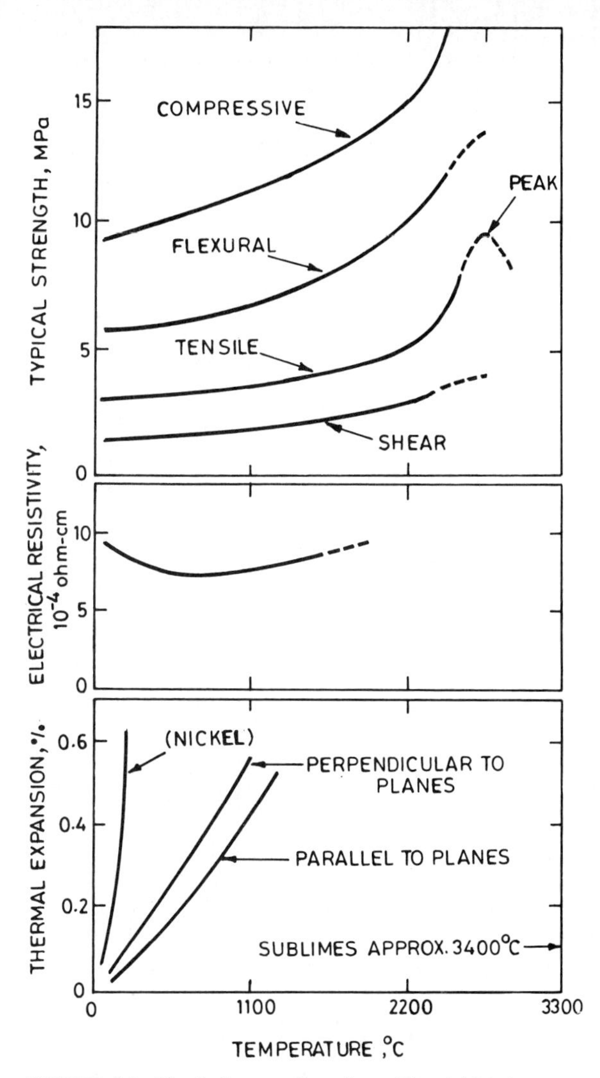

FIGURE 5.5 Physical properties of graphite at high temperatures.

constants, electrical resistivity, and coefficient of friction (at different loads and speeds) of several dichalcogenides are compared in Table 5.6 (Boes, 1964). The disulfides and diselenides of all four metals appear to exhibit low coefficients of friction. Ditellurides of all four metals have poor friction and wear. All compounds are relatively stable and inert. They are insoluble in water, oils, organic solvents, alkalis, and most acids. They are attacked by alkali metals. MoS_2 is the most commonly used of these compounds, and other dichalcogenides can be considered, whereas MoS_2 is limited by its poor oxidation resistance above 300°C or by its relatively high electrical resistivity (Clauss, 1972).

TABLE 5.6 Physical Properties and Friction Characteristics of the Dichalcogenides of the Group VB and VIB Metals[a]

Property	Disulfides				Diselenides				Ditellurides					
	MoS_2	WS_2	NbS_2[b]	TaS_2[b]	$MoSe_2$	WSe_2	$NbSe_2$[b,c,d]	$TaSe_2$[c,d]	$MoTe_2$[e]	WTe_2[f]	$NbTe_2$[f]	$TaTe_2$[f]	Graphite	PTFE
Molecular weight	160.08	248.05	157.04	245.01	253.87	341.84	250.83	238.80	351.17	439.14	348.13	436.10	—	—
Crystal structure[g]	Hex	Hex	Hex	Hex	Hex	Hex	Hex	Hex	Hex	Ortho	Trig	Trig	—	—
Lattice constants, nm														
a	0.316	0.329	0.331	0.332	0.329	0.329	0.345	0.343	0.352	1.403	1.090	1.090	—	—
c	1.229	1.297	1.189	1.230	1.280	1.295	1.300	1.274	1.397	0.627 (b = 0.349)	1.989	2.008	—	—
Volume resistivity,[h] $\Omega \cdot$ cm	851	14.4	0.0031	0.0033	0.018	144	0.00054	0.00223	869	0.0031	0.00057	0.00137	0.00264	—
Coefficient of friction														
At 0.55 MPa, 0.035 m s^{-1}	0.228	0.166	0.165	0.066	0.22	0.17	0.12	0.13	0.10	0.29	0.64	0.48	0.13	0.058
At 0.55 MPa, 0.35 m s^{-1}	0.170	0.136	0.098	0.068	0.20	0.10	0.15	0.10	0.34	0.38	0.61	0.55	0.16	0.21
At 1.7 MPa, 0.035 m s^{-1}	0.228	0.173	0.155	0.050	0.20	0.18	0.11	0.10	0.11	0.27	0.31	0.41	0.11	0.055
At 1.7 MPa, 0.35 m s^{-1}	0.166	0.148	0.074	0.062	0.17	0.10	0.17	0.095	0.19	0.34	0.74	0.59	0.14	0.25

[a]Compacts prepared by hot pressing 200-mesh powder of pure dichalcogenides in air at 200°C and 690 MPa. Friction measurements made by sliding hot-pressed buttons of dichalcogenides against flat disks of 440C stainless steel in air at room temperature.

[b]Friction values are those before increase due to heavy film buildup.

[c]Friction became erratic after ~5 minutes due to film buildup.

[d]Unannealed.

[e]Wore disk lightly under high-speed, high-load conditions.

[f]Wore disk under all conditions.

[g]Hex, hexagonal; Ortho, orthorhombic; Trig, trigonal.

[h]Measurements made on 25 × 6 × 3 mm bar hot pressed at 200°C and 205 MPa; measurements were made perpendicular to the pressing direction.

Source: Adapted from Boes (1964).

Natural MoS_2 is obtained by refining of the mineral molybdenite. The basic process consists of a series of alternating oil-floatation and grinding steps to reach the desired level of purity and particle size, followed by a deoiling step to remove the floatation oils still adhering (or vice versa) to the molybdenum disulfide. Oil floatation removes the impurities from the molybdenite ore (~0.4 percent MoS_2). The process involves aeration cells with selected floatation oils that have a high affinity for MoS_2 compared with impurities. Bubbling air through a series of such cells produces oil bubbles to which MoS_2 particles adhere. These bubbles are scraped from the top of the cells, and the MoS_2 particles are removed and ground.

Refined MoS_2 is a slippery, blue-gray powder with a molecular weight of about 160. Technical-grade MoS_2 consists of particles with an average size of 0.5 μm. Finer particle sizes are obtained by "air milling," in which particles are ground against themselves.

In addition to its natural occurrence, MoS_2 can be prepared synthetically by several methods, including direct union of the elements in pure nitrogen at 800°C, thermal decomposition of ammonium tetrathiomolybdate or molybdenum trisulfide, and reaction of MoO_3 with H_2S or H_2S/H_2 mixtures at 500°C (Killefer and Linz, 1952; Risdon, 1987). These techniques produce hexagonal crystalline MoS_2. Natural and synthetic MoS_2 both possess lubricating properties, but the natural hexagonal MoS_2 is preferred when cost and overall performance are considered and is primarily used. The bulk of MoS_2 consumed worldwide for tribological applications is mined and processed by Amax Mineral Sales, Greenwich, Conn. (formerly Climax Molybdenum Company). Dow-Corning Corporation, Midland, Mich., also is a major producer of various MoS_2 products, including colloidal MoS_2 (sold under the trade name Molykote).

5.2.1 Crystal Structure

The crystal structure of MoS_2 is a hexagonal layered structure similar to that of graphite. It consists essentially of planes of molybdenum atoms alternating with planes of sulfur atoms in the sequence S:Mo:S:S:Mo:S:···, as shown in Fig. 5.6. The atomic arrangement in each layer is hexagonal, and each Mo atom is surrounded by a trigonal prism of S atoms at a distance of 0.241 nm. Adjacent planes of S atoms are 0.308 nm apart. Thus the forces holding the atoms together in each group of S:Mo:S layers are relatively strong covalent bonds, whereas the forces between the adjacent planes of sulfur atoms are relatively weak van der Waals type bonds. As a result, the adjacent planes of S can slide readily over each other, and this is believed to be responsible for the low frictional resistance. At the same time, there is an immense resistance to penetration in the direction normal to the crystalline lamellae (Holinski and Gansheimer, 1972). Selected properties of natural MoS_2 are presented in Table 5.7 (Risdon, 1987).

5.2.2 Lubrication Properties

The hexagonal layered crystal structure of MoS_2, like that of graphite (discussed earlier), gives anisotropic shear properties with preferred planes for easy shear parallel to the basal planes of the crystallites. In graphite, we have seen that the presence of adsorbed gases or intercalated impurities on the edges of the basal planes is necessary to develop desirable frictional characteristics. However, in

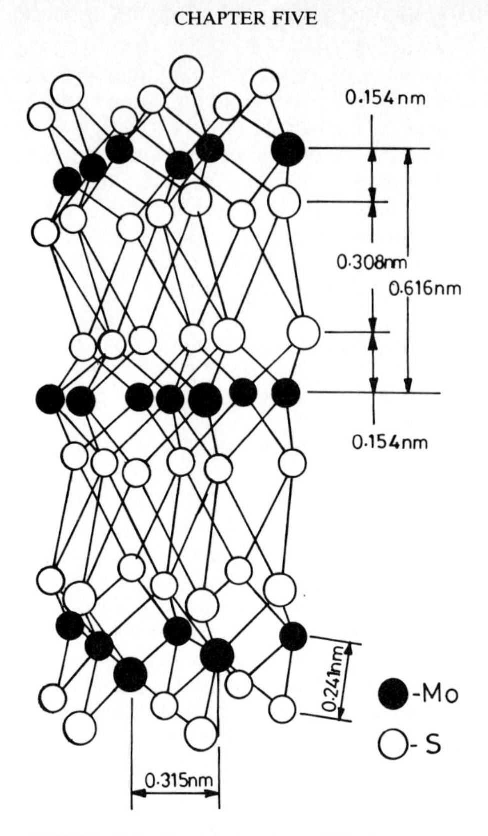

FIGURE 5.6 Crystal structure of MoS_2.

the dichalcogenides, notably MoS_2 and WS_2, a low shear strength is intrinsic to the structure of the pure material (Anonymous, 1973c; Braithwaite, 1966; Clauss, 1972; Farr, 1975; Feng, 1952; Holinski and Gansheimer, 1972; Lansdown, 1974, 1976, 1979; Lavik, 1959; Midgley, 1956; Winer, 1965).

MoS_2 readily reacts with the environment and promotes corrosion of metals and alloys in the presence of humid air by forming a surface layer of oxysulfide and sulfuric acid according to the following reactions (Kay, 1965):

$$MoS_2 + H_2O \rightarrow MoOS_2 + H_2 \tag{5.2a}$$

$$2MoS_2 + 7O_2 + 2H_2O \rightarrow 2MoO_2 \cdot SO_2 + 2H_2SO_4 \tag{5.2b}$$

Loss of adsorbed surface films, principally water vapor, actually enhances the lubricating properties of MoS_2, WS_2, and some of the other layered-lattice compounds. Peterson and Johnson (1953, 1955), using a low-speed friction apparatus (with a 50-mm diameter ring of SAE 4620 steel with a hardness of 62 HRC rotating on edge against the flat surface of a disk of SAE 1020 steel at a normal load of 178 N and sliding velocity of 30 mm s^{-1}) with an interface lubricated with powdered MoS_2, found that humidity had a very definite effect on the friction of MoS_2 [Fig. 5.7(*a*)]. The coefficient of friction increased with increasing relative humidity up to a relative humidity of approximately 60 percent and then dropped

TABLE 5.7 Selected Physical Properties of MoS_2 and MoS_2 Compacts

Identification	Density, kg m^{-3}	Hardness	Modulus of elasticity, GPa	Compressive strength, MPa	Flexural strength, MPa	Thermal conductivity, W m^{-1} K^{-1}	Coefficient of thermal expansion, $\times 10^{-6}$ °C^{-1}	Electrical conductivity, $(\Omega \cdot cm)^{-1}$	Temperature limit in air/neutral atmosphere, °C/°C
MoS_2	4900	60 HK (basal plane), 32 HK (001 plane), 900 HV (100 plane)	—	—	—	0.13	10.7	0.63 (basal plane), 1.58×10^{-4} (⊥ basal plane)	300/450+
MoS_2 compact containing refractory-metal matrix (Purebon Molalloy PM-107)	5900	Shore scleroscope C, 80	125	790	125	—	9.5	—	370/1150

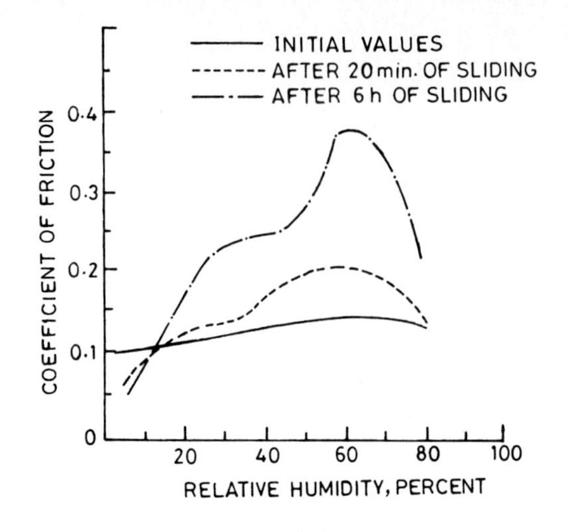

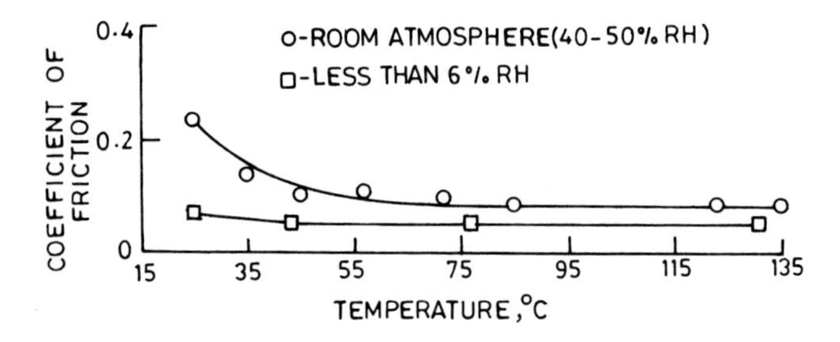

FIGURE 5.7 (*a*) Effect of various relative humidities on the coefficient of friction of steel specimens lubricated with powdered MoS_2 in air [load, 178 N; speed, 11 r/min (30 mm s^{-1})]. (*b*) Effect of temperature and relative humidity on the coefficient of friction of steel specimens lubricated with powdered MoS_2. [*Adapted from Peterson and Johnson (1953).*]

off again. The reason for this decrease at high humidity is not fully known, but it is believed that the MoS_2 coating was disrupted by moisture during the 30 to 60 percent humidity period and that sufficient powder accumulated in front of the slider on the disk to provide lubrication.

Peterson and Johnson (1953) further stated that an increase in temperature could drive volatile materials (most of which is adsorbed water) out of an MoS_2 film and restore lubrication at high humidity. This can be seen from Fig. 5.7(*b*), where the coefficient of friction in the 40 to 50 percent humidity range was lowered significantly to values approaching those in dry air (less than 6 percent relative humidity) as the temperature increased. The coefficient of friction in dry air is less than that in moist air, and it does not show as pronounced a drop with temperature. Peterson and Johnson (1953) also reported that as the relative hu-

TABLE 5.8 Friction Data for Some Layered-Lattice Compounds in Vacuum

Solid	Coefficient of friction	
	Air	Vacuum (pressure, torr)
MoS_2	0.18	0.07 (2×10^{-9})
WS_2	0.17	0.13 (3×10^{-9})
WS_2	0.17	0.08 (1×10^{-5})
BiI_3	0.34	0.39 (5×10^{-7})
LiOH	0.37	0.21 (2×10^{-9})
$CdCl_2$	0.35	0.16 (2×10^{-9})
CdI_2	0.24	0.18 (2×10^{-9})
$CdBr_2$	0.22	0.15 (2×10^{-9})
Phthalocyanine	0.35	0.33 (1×10^{-6})

Source: Adapted from Flom and Haltner (1968).

midity increased from 15 to 70 percent, the wear in a 6-hour period increased by 200 percent (along with an increase in friction from 0.1 to 0.38). Similar results also are reported by Wood (1961).

Layered-lattice compounds provide more effective boundary lubrication and lower friction in vacuum (~0.1) than in air (Table 5.8), probably owing to the removal of adsorbed water vapor, as discussed in the preceding paragraph. Many of these compounds, when used in air, are not only poorer lubricants, but also are actually corrosive to the surfaces they are lubricating because of the presence of adsorbed water vapor; e.g., MoS_2 in the presence of water vapor can form sulfuric acid, which is corrosive to steel surfaces (Bhushan, 1987; Buckley, 1971).

The MoS_2 can be used as a lubricant from cryogenic temperatures up to a maximum temperature of about 300°C (or about 400°C under favorable conditions) in air. The maximum-use temperature in air atmosphere is limited by oxidation:

$$2MoS_2 + 7O_2 \rightarrow 2MoO_3 + 4SO_2 \tag{5.3}$$

Some oxidation kinetics data for loosely compacted MoS_2 powders of 1-μm average particle size are given in Fig. 5.8(*a*) (Sliney, 1982). At a modest airflow rate over the compact, 50 percent of the MoS_2 was oxidized to molybdic oxide (MoO_3) in 1 hour at 400°C. At a six times higher airflow rate, the temperature for an oxidation half-life of 1 hour was reduced to 300°C. Figure 5.8(*b*) compares the oxidation kinetics of MoS_2 and WS_2 at the lower airflow rate. Note that the curves intersect at about 340°C, with MoS_2 oxidizing more rapidly above that temperature.

Tungsten disulfide has an advantage of about 50 to 100°C over MoS_2 with respect to oxidation resistance and thermal stability. The relatively slow rates of oxidation of WS_2 and WSe_2 (another dichalcogenide used for lubrication) are thought to be due to the formation of tungsten trioxide (WO_3), which is somewhat more protective against further oxidation and provides a lower coefficient of friction than molybdic trioxide (MoO_3). Kinetic coefficients of friction are about 0.2 to 0.3 for WO_3 and 0.5 to 0.6 for MoO_3. MoO_3 is generally believed to be a poor lubricant (i.e., it exhibits high friction); however, it is not abrasive (Grattan and Lancaster, 1967).

Comparison of data from friction experiments (Fig. 5.9) conducted with a pin-

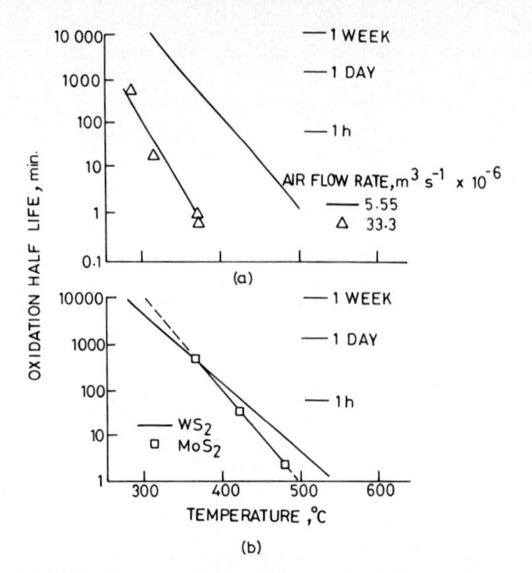

FIGURE 5.8 Oxidation kinetics of MoS_2 and WS_2 compacts with average particle size of 1 μm and compact density of 50 percent. (*a*) Oxidation characteristics of MoS_2 at two airflow rates. (*b*) Comparative oxidation of WS_2 and MoS_2 with an airflow rate of 0.05×10^{-3} $m^3 s^{-1}$. [*Adapted from Sliney (1982).*]

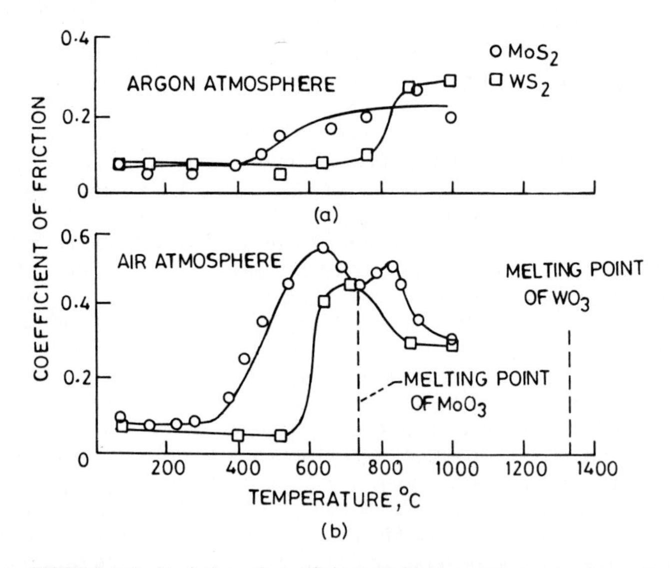

FIGURE 5.9 Variation of coefficient of friction with temperature of burnished MoS_2 and WS_2 disks slid against a pin in (*a*) argon and (*b*) air. [*Adapted from Sliney (1982).*]

on-disk apparatus using a hemispherically tipped pin in sliding contact with the flat surface of a rotating disk burnished with the solid lubricant with the oxidation data show that the loss of lubricating ability of MoS_2 and WS_2 in air coincides with the temperature at which rapid conversion to the oxide occurs. Figure 5.9 shows that both compounds lubricate at much higher temperatures in a nonreactive argon atmosphere (MoS_2, 500°C; WS_2, 750°C). Sliney (1982) has reported that based on the thermal dissociation rates, disulfides are the most stable, the diselenides are intermediate, and the ditellurides are the least stable.

MoS_2 performs well under a number of extreme environmental conditions. Its lubricating ability from cryogenic temperatures up to 300 to 400°C in air and even higher temperatures in nonoxidizing atmospheres makes it attractive for aerospace applications. Note that the low friction of graphite depends on the presence of water vapors and hydrocarbons, etc.; therefore, it is a poor lubricant in vacuum or nonoxidizing atmospheres. However, graphite in air can be used to a higher temperature than can MoS_2. PTFE (to be discussed later) does not perform well under high load because of its tendency to cold flow. Therefore, for high loads, MoS_2 and graphite are preferred.

5.2.3 Dry Powders and Dispersions

MoS_2 is used as a lubricant itself and as a lubricant additive to reduce both friction and wear. MoS_2 is often used in the form of dry powder, usually to facilitate assembly of parts or to provide lifetime lubrication. MoS_2 powder is frequently used in such applications as metal-forming dies, threaded parts, sleeve bearings, and electrical contacts in relays and switches.

Burnished coatings of MoS_2 (obtained by mechanically rubbing the solid particles onto the surface of a part) are commonly used in bearings and other sliding applications. Burnished coatings of MoS_2 adhere better to metals than does graphite, probably because of the strong sulfur-to-metal bonds. Sputtered and resin-bonded MoS_2 coatings are used in numerous applications, including aerospace (e.g., satellites and the space shuttle). These will be discussed in more detail in Chap. 13.

MoS_2 dispersions in fluids (typically petroleum oils) are commonly used for a variety of industrial applications (Banerjee, 1981; Bartz, 1980; Braithwaite, 1966; Clauss, 1972; Hisakado et al., 1983; McCabe, 1966; Stock, 1966). These dispersions are often a more convenient form than dry powders for applying uniform films. The effectiveness of solid lubricant additives, however, may be diminished by other chemical additives and detergents commonly added to the oils (Bartz, 1971; Bartz and Oppelt, 1978). Stock (1966) has shown that particle size, particle solid content, and normal load have an important effect on wear (Fig. 5.10). It should be noted that colloidal MoS_2 suspensions exhibit superior wear performance compared with graphite and WS_2 (also see Table 5.4). Within MoS_2 suspensions, finer particles and larger concentrations led to lower wear volume. In particular, the difference becomes obvious at higher loads. Therefore, MoS_2 particle size and content should be optimized for a given application (Bartz, 1971, 1972, 1980; Hisakado et al., 1983; Stock, 1966). Fine particles (micrometer or submicrometer size) are used more often than coarse ones, and the solid contents can cover a wide range, typically 1 to 3 percent by weight. Typical MoS_2 dispersions (manufactured by Acheson Colloids Company, Port Huron, Mich.) and their common industrial applications are presented in Table 5.9.

Incorporation of MoS_2 into greases is the most common means of utilizing its

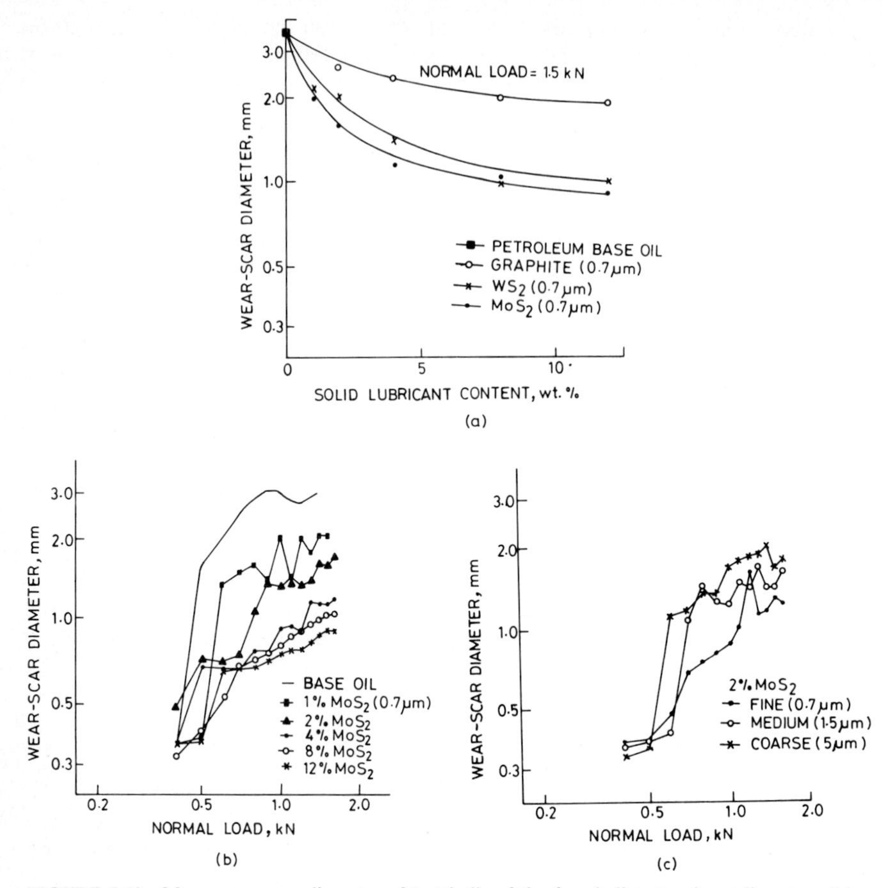

FIGURE 5.10 Mean wear-scar diameter of test balls of the four-ball tester depending on solid-lubricant content and its particle size within the SAE 90 petroleum oil for a running time of 1 minute at 1435 r/min. (*a*) Effect of solid lubricant content. (*b*) Effect of load and MoS_2 content. (*c*) Effect of load and MoS_2 particle size. [*Adapted from Stock (1966).*]

lubricant abilities, because oils and greases can be applied to many types of sliding surfaces and can be used in centralized lubrication systems (Bartz, 1971; Gansheimer and Holinski, 1973). MoS_2 lubricant additives for greases are commercially available from several companies, such as Hohman Plating and Manufacturing, Inc., Dayton, Ohio, and E/M Lubricants, Inc., W. Lafayette, Ind. Greases typically contain 5 to 10 percent MoS_2, but they may contain as much as 60 percent MoS_2. The heavily loaded greases have the consistency of paste and are known as *concentrates*. MoS_2-filled greases and concentrates are especially effective during the initial run-in of parts. They are commonly applied by brush or wipe methods during assembly of sliding surfaces, such as in gear teeth. They are also effective under heavy loads. Greases containing from 1 to 10 percent MoS_2 are used for automotive chassis lubrication. Trucks and buses use MoS_2 greases on steering joints, fifth wheels, wheel bearings, universal joints, etc.

TABLE 5.9 Typical MoS_2 Dispersions

Carrier/diluent	Solid content, %	Average particle size, μm	Density, kg m^{-3}	Typical applications
Petroleum oil	1–10	0.5	1008	Additives for specialized oils for general industrial and automotive uses in extreme pressure and high-temperature applications; drilling and tapping lubricant
Petroleum oil	60	40–60	1800	Formulated into specialty greases
Petroleum oil	35*	40–60	1200	Used as antiseize thread lubricant
Water	35	2.5	1320	Used in formulation of specialty greases
Polyglycol (water insoluble)	1–10	2.5	1068	Used as extreme-temperature bearing lubricant
Isopropanol	20	2.5	948	Used for dry coatings, press fittings, antiseize thread coatings, and lighty loaded bearing applications
Trichloroethylene (volatile carriers)	20	2.5	1464	

*Contains MoS_2+graphite.
Source: Adapted from Clauss (1972).

5.2.4 Solids Impregnated with MoS_2

Many polymers and PM compacts can be impregnated with MoS_2 particles. With such composites, the sliding process continues to feed a thin film of MoS_2 to the surfaces so that the friction remains low even after prolonged rubbing. Detailed information on plastics impregnated with MoS_2 is given in the next section. Impregnated polymers such as Nylon, PTFE, polyimide, phenolic, and epoxy resins are used in cams, thrust washers, ball-bearing retainers, compressor piston rings, and other automotive parts.

Metals such as copper, bronze, nickel, and copper-cobalt-beryllium and refractory metals such as molybdenum, tungsten, niobium, and tantalum are commonly impregnated with MoS_2 (Bowden, 1950; Johnson et al., 1950; Jost, 1956; Tsuya et al., 1976). The metal matrices impart their high strength at elevated temperatures to the lubricant. To manufacture compacts, the metal and the solid lubricant powders are mixed and cold pressed; then the mixed powder is molded into a specimen shape and sintered in vacuum at high temperature (Anonymous, 1973a; Johnson et al., 1950; Tsuya et al., 1976).

Tables 5.10 and 5.11 present friction and wear data illustrating the benefits of MoS_2 in metals. Impregnating MoS_2 into sintered bronze provides a greater improvement in friction and wear than impregnation with graphite. MoS_2-filled metals also appear to retain their low coefficient of friction even after repeated slid-

TABLE 5.10 Friction and Wear of Sintered Bronze Bearings (No Oil Lubrication) and Effects of Impregnation with Graphite and MoS_2

Bearing material	Coefficient of friction	Wear, μm	Test duration, h
Sintered bronze	0.67	60	70
With graphite	0.47	50	100
With MoS_2	0.34	25	400

Source: Adapted from Jost (1956).

TABLE 5.11 Friction of Copper and Copper Containing MoS_2 (No Oil Lubrication)

	Coefficient of friction		
Test conditions*	Solid copper	Sintered copper	Sintered copper containing MoS_2
At room temperature			
Initial	1.3	0.3	0.15
After repeated sliding	1.3	0.8	0.13
At 100°C	1.3	0.7	0.14
At 200°C	1.0	0.8	0.13
At 400°C	1.0	1.2	0.20

*Data was obtained with a steel slider and a load of 40 N.
Source: Adapted from Bowden (1950).

ing at elevated temperatures. Tsuya et al. (1976) studied the effect of solid lubricant content and particle size on the coefficient of friction and wear. The results for natural graphite and MoS_2 in copper-based compacts are shown in Fig. 5.11. Note that coarser particles gave lower friction and wear. In these measurements, graphite performed slightly better than MoS_2. Tsuya et al. (1976) also noted that the optimal concentration of solid lubricant varied with the sliding conditions.

A variety of sintered bronzes filled with MoS_2 (with a porosity on the order of 10 to 25 percent) is commercially available for lightly loaded journal bearings and other sliding applications. Because of their high porosity, these bronzes can be vacuum impregnated with a lubricating oil for further reduction of friction.

Refractory-based MoS_2 compacts are commercially available for use whenever high loads, high temperatures, or vacuum make conventional lubricants impossible. Selected physical properties of MoS_2 in a refractory metal matrix are presented in Table 5.7 (Anonymous, 1973a). The refractory metal–based composites can be used at loads up to 70 GPa at 500°C and 7 GPa at 800°C in vacuum. Such composites have many applications, e.g., bushings, seals, gears, clutch facings, and electric motor brushes. As an example, spherical bearings (37 mm in diameter) for the space shuttle door hinge are made of MoS_2 compacts in a refractory metal matrix.

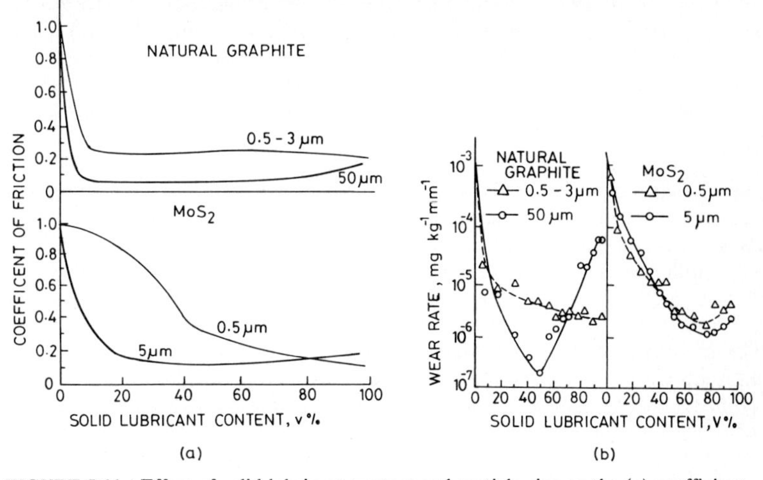

FIGURE 5.11 Effect of solid-lubricant content and particle size on the (*a*) coefficient of friction and (*b*) wear rate. Friction and wear were measured between a copper-based solid-lubricant compact and a copper surface at 50 N normal load and 500 r/min for 30 minutes. [*Adapted from Tsuya et al. (1976).*]

5.3 POLYMERS

Polymers and polymer composites are important tribological materials because of their self-lubricating properties, excellent wear and corrosion resistance, high chemical stability, high compliance, high capacity for damping vibrations or for shock resistance, low noise emission, light weight, and low cost (Anderson, 1982, 1986; Anderson and Davies, 1982; Anonymous, 1988; Bartenev and Lavrentev, 1981; Bhushan, 1978a, 1978b, 1980b; Bhushan and Dashnaw, 1981; Bhushan and Gray, 1978, 1979; Bhushan and Wilcock, 1981, 1982; Bhushan and Winn, 1979, 1981; Bhushan et al., 1982; Briscoe, 1981; Eiss and Czichos, 1986; Evans and Lancaster, 1979; Fusaro, 1988; Lancaster, 1968, 1972b, 1973; MacLaren and Tabor, 1963; Moore, 1972; Steijn, 1967; Tanaka, 1977; Wolverton et al., 1983). Polymer composites with self-lubricating properties can be used in unlubricated applications, unlike many metals and nonmetals. Polymer bearings rarely would seize like metals in the event of loss of lubricant, and thus polymer bearings have good tolerance to high stresses from edge loading due to shaft misalignment. During any misalignment, polymer bearings deflect and/or wear locally in the loaded zone and align under load so that there is no significant damage to the bearing and journal surfaces. Polymers are more tolerant to abrasive particles at the interface (like soft metals such as babbitts) because sharp particles generally would be embedded into polymer. Because of their high degree of resilience, elastomers are commonly used for sealing applications. Polymers are chemically inert and thus are preferred in corrosive environments (corrosive gases, marine applications).

Polymers are very light (lighter than any metal), and some polymer composites and reinforced polymers provide high strength. Therefore, they have a high strength-to-weight ratio. These composites are commonly used in the load-bearing structures of automobiles and airplanes to reduce weight in order to enhance fuel economy (Anonymous, 1979). Plastic structures are coated with metal films to give them a metal look. For example, the grills above the front bumpers of many automobiles are made by sputter deposition of chromium on plastic.

Polymers do suffer certain disadvantages, however, when compared with metals. The thermal conductivity of polymers is much lower than that of metals. Consequently, polymers are less able to dissipate frictional heat generated at rubbing interfaces. The coefficients of thermal expansion of polymers are 10 or more times less than those of metals. Polymers have been known to swell when immersed in various fluids; this phenomenon is accelerated at high temperatures. Bulk polymers soften with heat and cannot be used at as high temperatures as can metals. As a consequence of these limitations, polymers are restricted to lower-speed applications than metals. The strength and modulus of elasticity of polymers are much lower than those of hard metals so that the load-carrying capacity of polymer bearings is less than for a metal part of the same size. Adding fillers can compensate for the many disadvantages outlined here.

Polymers are a large group of materials consisting of combinations of carbon, oxygen, hydrogen, nitrogen, and other organic and inorganic elements (Anonymous, 1988; Brydson, 1982; Frados, 1977; Harper, 1970; Oleesky and Mohr, 1964). Polymers can be divided into plastics and elastomers. Many plastics are derived from fractions of petroleum or gas that are recovered during the refining process. Other sources are coal and agricultural oils, such as castor oil. From these basic sources come the feedstocks we call *monomers* (i.e., a small molecule). A monomer (one chemical unit) such as ethylene reacts with other monomer molecules forming a very long chain of repeating ethylene units known as *polyethylene*. Similarly, polypropylene is formed from propylene monomers, and other polymers are formed from their respective monomers. During the polymerization reaction, millions of separate polymer chains grow in length simultaneously until the monomer is exhausted. By adding predetermined amounts of hydrogen (or other chain stoppers), one can produce polymers with fairly consistent chain lengths. Chain length (molecular weight) is very important because it determines several polymer characteristics. Usually, with increasing chain length, toughness, creep resistance, stress-crack resistance, melt temperature, and viscosity increase. Another method for altering molecular symmetery is to combine two different monomers A and B in the polymerization reaction so that each polymer chain is composed partially of monomer A and partially of monomer B. Such polymers are known as *copolymers*. The final properties of copolymers depend on the percentages of monomers (Anonymous, 1988).

There are two main classes of plastics: *thermoplastics* and *thermosets*. The distinction is based on their elevated temperature characteristics and production processes (Harper, 1970; Oleesky and Mohr, 1964). At some stage in their conversion into finished products, both types are fluid enough to be formed and molded. Thermoplastics solidify by cooling and may be remelted repeatedly. Thermosets in their fluid condition are still chemically reactive, and they harden by the further reaction, called *cross-linking*, between groups on nearby chains, forming a three-dimensional network. Subsequent heating, though it may soften the structure somewhat, cannot restore the flowability that typifies the uncross-linked, uncured resin.

Thermoplastics generally can be produced by fast, continuous processes such as injection molding (only for melt-processable polymers), extrusion (direct forming), and compression molding. Injection molding is done by placing the polymer in powder or pellet form into the heated cavity of an injection molding machine, allowing it to melt, and then pushing it forward into a cooled cavity by a ram or reciprocating screw, where it remains until below its T_g or T_m to ensure that the part retains the desired form. Direct forming uses PM techniques to cold form a part, followed by sintering techniques at elevated temperatures. Direct forming also can be done in a single step by ramming the resin under pressure through a heated mold. Injection molding and extrusion can only be used to mold symmetrical and simple shapes, e.g., rods and tubes. Compression molding involves placing the material, usually in powder form, into the cavity of a heated mold and applying heat (several hundred °C) and pressure (tens of MPa). Injection molding followed by extrusion is preferred for mass production.

Thermoset plastics are produced by compression molding and extrusion; they cannot be produced by injection molding. Thermosetting plastics can be changed into a substantially infusible, insoluble product by curing with heat or chemicals. Polymerization (curing) of thermoset plastics is carried out in two steps, partly by the material supplier and partly by the molder. For instance, phenolic is first partially polymerized by reacting phenol with formaldehyde under heat and pressure. Unlike a monomer for a thermoplastic polymer, which has only two reactive ends for linear chain growth, a thermoset monomer must have three or more reactive ends so that its molecular chain cross-links in three dimensions. *Usually thermosets, because of their lightly cross-linked structure, resist higher temperatures and have better dimensional stability than thermoplastics.*

Elastomers are polymers that have rubberlike properties. More precisely, an elastomer is a polymer that has elongation generally greater than 200 percent before failure and a significant amount of resilience. Elastomers have low elastic moduli and exhibit great dimensional change when stressed but will return to, or almost to, their original dimensions immediately after the deforming stress is removed.

Of the very large number of polymers available commercially, only a few are in significant use for friction- and wear-related applications. These are listed in Table 5.12. The properties of a particular polymer can be tailored to specific requirements through fillers and reinforcement. The mechanical and self-lubricating properties of polymers can be further improved by blending them with fillers in the form of fibers, powders, and liquids. The addition of fillers allows the polymer to be used at high *PV* limits and higher operating temperatures. Selected polymers and polymer composites can be used up to a *PV* limit of 3.5 MPa · m s^{-1} under unlubricated conditions and up to 17.5 MPa · m s^{-1} under lubricated conditions, which is comparable with carbon-graphite compacts and soft metals. Many plastics are used as solid-lubricant coatings (see Chap. 13) and as colloidals in oils and other fluids for lubrication. Polymers and polymer composites are generally used in sliding contacts against hard and smooth surfaces, particularly metal surfaces (Evans and Lancaster, 1979; Lancaster, 1973). In some cases, polymers are also used against polymers (Jain and Bahadur, 1977; Steijn, 1967; Wolverton and Theberge, 1981).

Wear of polymers can be explained in terms of cohesive wear and interfacial wear. *Cohesive wear* results from the dissipation of frictional work and the damage it causes in relatively large volumes adjacent to the interface. Abrasion and fatigue wear induced by frictional forces are within this category (Sauer and Richardson, 1980). Cohesive wear is, by and large, controlled by the cohesive strength or toughness of the polymer. The level of deformation or work is a function of the details of interpenetration of the surface asperities and the magnitude of the frictional force. *Interfacial wear* results from the dissipation of frictional work in much thinner regions and at greater energy densities. Interfacial wear

TABLE 5.12 Typical Plastics, Elastomers, Fillers, and Reinforcements Used for Tribological Applications

Thermoplastic plastics	Fluorocarbons	Polytetrafluoroethylene (PTFE), fluorinated ethylene propylene (FEP), perfluoroalkoxy (PFA)
	Acetals	Homopolymer, copolymer
	Polyethylenes	High-density polyethylene (HDPE), ultrahigh-molecular-weight polyethylene (UHMWPE)
	Polyamides	Nylon 6, 6/6, 6/10, 6/11, 6/12
	Polycarbonate	
	High-temperature plastics	Polyphenylene sulfide (PPS), poly(amide-imide), polyimide,* polyester*
Thermosetting plastics		Phenolic, polyimide,* polyester,* epoxy
Elastomers		Natural and synthetic rubbers, styrene-butadiene (SBR), isobutylene isoprene (butyl), ethylene propylene, chloroprene (neoprene), butadiene-acrylonitrile (Buna-N, Nitrile, or NBR), polyacrylate, fluoroelastomer (vinylidene fluoride-hexafluoropropylene or Viton, perfluoroelastomer, phosphonitrilic fluoroelastomer), silicone, and thermoplastics (polyurethanes, copolyester, olefins)
Fillers and reinforcements (to improve mechanical properties and friction and wear characteristics)	Fibers	Asbestos, carbon, graphite, PTFE, glass, aromatic polyamide (Kevlar), polyamide (Nylon), polyester (Dacron or Terelene), Ornol, Dynel, and other textile fibers
	Powders	MoS_2, graphite, carbon, PTFE, polyimide, Ag, bronze, silica, alumina, mica
	Liquids	Mineral oil, fatty acids, silicones

*These can be either thermoplastic or thermosetting type.

includes adhesive or transfer wear and chemical wear. The frictional work is dissipated in a smaller volume, and the rates and extent of surface deformation are greater than those seen in cohesive wear (Briscoe, 1981).

When many polymers (more commonly, certain plastics) slide over clean (dry) and smooth counterfaces, the adhesion between the polymer and the counterface is of sufficient magnitude to inhibit sliding at the original interface. Instead, the asperity junctions rupture within the polymer itself, and a layer of polymer is deposited on the counterface in the form of a more or less coherent layer. Subsequent traversals of the polymer over this film results in lower friction and wear. Repeated sliding of the polymer over this film removes the transferred layer, which is ultimately displaced from the contact. A further layer is deposited, the process repeats, gradually wearing away the polymer surface. Formation of such a transfer film in many polymers, such as PTFE, polyethylene, and polyamide, is responsible for low friction and wear in dry applications. Generally, a thin (multimolecular), coherent film with strong adhesion to the counterface is desirable, and thick, patchy films with low adhesion to the counterface may lead to catastrophic failure (Briscoe, 1981; Lancaster et al., 1980; Low, 1979; Pooley and Tabor, 1972; Rhee and Ludema, 1978; Steijn, 1966; Tanaka and Uchiyama, 1974; Tanaka and Yamada, 1983). Surface roughness of the counterface plays an important role in the formation of suitable transfer films (see data to be presented later). Generally, composites containing abrasive fillers do not allow formation of the transfer films (Lancaster, 1968). However, these fillers improve the mechanical properties leading to improved load-carrying capabilities.

Various polymers that are commonly employed for tribological applications and listed in Table 5.12 are discussed in the following sections. Names and addresses of various manufacturers and typical properties of polymer resins and filled polymers can be found in Anonymous (1980, 1988).

5.3.1 Plastics

Selected properties of common plastics are presented in Table 5.13, and selected properties of high-temperature plastics are presented in Table 5.14. The wear coefficient K and PV limits for a number of plastics in the dry condition are presented in Table 5.15. The PV limit of the polymers in the lubricated condition (oils or water) can be up to an order of magnitude larger than that in the dry condition. Liquid medium removes frictional heat from the interface, thus allowing operation at high PV conditions. High-temperature polymers can be operated under lubricated conditions for a short period up to a PV of 35 MPa · m s^{-1} (1×10^6 psi × fpm), comparable to the PV limit of carbon-graphites (Bhushan and Winn, 1979, 1981). These data provide only a rough estimate of tribological behavior; absolute values, however, depend on operating conditions.

5.3.1.1 Thermoplastic Plastics. In thermoplastic polymers, the molecules slide past each other under shear force above a certain temperature such that the molecules have enough energy to overcome intermolecular bonding. In other words, above a certain temperature these materials are capable of flow, whereas below this temperature they are solid (Brydson, 1982; King and Tabor, 1953). Most thermoplastics are melt-processable; i.e., they can be produced by various processing methods, such as injection molding, extrusion, and extrusion and injection blow molding. These continuous processing methods are desirable for mass production. The thermoplastic polymers can be synthesized in amorphous or crystalline form. Usually amorphous polymers are relatively hard at room temperature and quite brittle, like glasses (Sharples, 1972). The majority of thermoplastics employed for tribological applications are crystalline. Of all the plastics, PTFE has been used most extensively at moderate PV conditions (especially under unlubricated conditions) in its pure form as well as in a large variety of filler

TABLE 5.13 Selected Properties of Common Plastics

Property	Fluorocarbons		Acetals		Polyethylene UHMWPE		Polyamide		Polycarbonate	
	PTFE	FEP	Unfilled	Filled with 30 wt % PTFE	Unfilled	Filled with 30 wt % glass fiber	Unfilled	Filled with 30 wt % glass fiber	Unfilled	Filled with 15 wt % PTFE and 20 wt % glass fiber
Specific gravity	2.13–2.3	2.12–2.17	1.41	1.51	0.94	1.18–1.28	1.13	1.40	1.20–1.22	1.43–1.50
Water absorption, 24 h (3 mm thick), %	<0.01	<0.01	0.2–0.25	—	<0.01	0.02–0.06	1.2–1.6	0.9	0.12–0.15	0.11
Hardness, Shore D	50–65	55	65	58	63–65	70	75	85	50–55	70
Tensile strength, MPa	9	8	4.8	45	15–20	52–62	81	192	52–59	85–103
Elongation at break, %	400	300	50	6	400–500	1.5–2.5	60–200	2.5	120–165	2
Tensile modulus, GPa	0.3–0.8	—	2.3	2.7	0.50–0.70	5–6.3	2.6–2.9	—	2.3–3.1	8.3
Flexural strength, MPa	No break	No break	—	70	7–9	77–82	—	280	75–86	125–160
Flexural modulus, GPa	0.6–0.7	0.65	—	2.60	0.69–1.17	5–5.6	2.8	8.9	1.92–2.24	5.9–6.3
Impact strength, Izod [(N · m) mm^{-1} of notch]	0.15	No break	0.04–0.08	0.05	No break	0.07–0.09	0.05	0.15	0.12–1.08	0.12–0.22
Thermal conductivity, W m^{-1} K^{-1}	0.25	0.25	0.23	0.23	0.48	0.37	0.24	—	0.18	—
Coefficient of thermal expansion, $\times 10^{-6}$ $°C^{-1}$	120–135	—	81	97	130	48	45	23	80–95	23
Injection moldable	No	Yes	Yes	Yes	No	No	Yes	Yes	Yes	Yes
Grade/manufacturer	Teflon/Du Pont	—/Du Pont	Milcon/ Tribol	Milcon RX-293/Tribol	Cadco 1900/ Cadillac	—/Cadillac	Nylon 6/6/ Du Pont	Nylon 6/6/ Du Pont	Lexan/GE	—/Polymer Resources

TABLE 5.14 Selected Properties of High-Temperature Plastics

Property	Polyphenylene sulfide			Poly(amide-imide)			Phenolic		Polyimide			Epoxy	
	Unfilled	Filled with 30 wt % carbon fiber, 15 wt % PTFE	Filled with 25 wt % glass fiber, 25 wt % PTFE	Unfilled	Filled with 3 wt % PTFE, 12 wt % graphite	Linear aromatic polyester filled with 20 wt % graphite	Unfilled	Filled with 40 wt % graphite fiber	Unfilled	Filled with 50 wt % glass fiber	Filled with 15 wt % graphite	Unfilled	Filled with glass fiber
Specific gravity	1.60	1.57	1.80	1.40	1.45	1.47	1.35	1.45	1.41	1.7	1.38	1.1–1.4	1.6–2.0
Water absorption, 24 h (3 mm thick), %	0.03	0.03	0.03	0.28	0.22	0.07	0.4–0.6	0.2	—	0.7	0.19	0.08–0.15	0.05–0.2
Hardness, Shore D	75	85	80	70	85	80	75	90	85	90	85	75	90
Tensile strength, MPa	65	175	105	185	135	58	62	117	117	186	66	30–90	69–138
Elongation at break, %	1.6	2–3	1.4	12	6	2.8	1.7	0.4	10	<1	4.5	3–6	4
Tensile modulus, GPa	—	—	—	5.0	4.7	2	4.0	—	1.3	—	1.8	2.4	21.0
Flexural strength, MPa	96	134–157	45–55	211	182	48	76	250	—	345	110	55–100	69–210
Flexural modulus, GPa	3.8	13.8	4.0	4.6	6.3	—	1.7	—	—	22.4	3.8	10–14	14–32
Impact strength, Izod [(N · m) mm^{-1} of notch]	0.03	0.06	0.04	0.15	0.08	—	0.03	0.58	0.15	0.36	0.13	0.05	0.1–0.5
Thermal conductivity, W m^{-1} K^{-1}	0.29	—	0.30	0.24	0.36	1.0	0.15	0.71	0.5	0.36	0.86	0.19	0.17–0.42
Coefficient of thermal expansion, $\times 10^{-6}$ °C^{-1}	49	—	63	36	27	79	68	3	51	13	49	45–65	11–50
Injection moldable	Yes	Yes	Yes	Yes	Yes	No	No	No	No	No	No	No	No
Grade/manufacturer	R-6/ Phillips	Thermo-comp OCL-4036/ LNP	Ryton NR09/ Phillips	Torlon 4203/ Amoco	Torlon 4301/ Amoco	Ekkcel CL-1340/ Carbo-rundum, Dart	SC 1008/ Mon-santo	—/Mon-santo	PI101/ Tribol	—	SP21/Du Pont	—/Hysol	—/Hysol

TABLE 5.15 The *PV* Limits, Wear Coefficients, and Coefficients of Friction of Various Unfilled and Filled Plastics Under Dry Conditions

Material	*PV* limits at (*V*) and 22°C, MPa · m s^{-1} (at m s^{-1})	Maximum operating temperature, °C	Wear coefficient, × 10^{-7} mm^3 $(N \cdot m)^{-1}$	Coefficient of friction
PTFE (unfilled)	0.06 (0.5)	110–150	4000	0.04–0.1
PTFE (glass fiber filled)	0.35 (0.05–5.0)	200	1.19	0.1–0.25
PTFE (graphite fiber filled)	1.05 (5.0)	200	—	0.1
Acetal	0.14 (0.5)	85–105	9.5	0.2–0.3
Acetal (PTFE filled)	0.19 (0.5)	—	3.8	0.15–0.27
	0.09 (5.0)	—	—	—
UHMW polyethylene	0.10 (0.5)	105	—	0.15–0.3
UHMW polyethylene (glass fiber filled)	0.19 (0.5)	105	—	0.15–0.3
Polyamide	0.14 (0.5)	110	38.0	0.2–0.4
Polyamide (graphite filled)	0.14 (0.5)	150	3.0	0.1–0.25
Polycarbonate	0.03 (0.05)	135	480	0.35
	0.01 (0.5)			
Polycarbonate (PTFE filled)	0.06 (0.5)	135	—	0.15
Polycarbonate (PTFE, glass fiber)	1.05 (0.5)	135	5.8	0.2
Polyphenylene sulfide	3.50 (0.5)	260–315	—	0.15–0.3
Polyphenylene sulfide (PTFE, carbon fibers)	3.50 (0.5)	260–315	—	0.1–0.3
Poly(amide-imide)	3.50 (0.5)	260	—	0.15–0.3
Poly(amide-imide) (PTFE, graphite)	1.75 (0.5)	260	—	0.08–0.3
Linear aromatic polyester (graphite filled)	1.75 (0.5)	260–315	—	0.2–0.4
Phenolic	0.17 (0.05)	260	—	0.9–1.1
Phenolic (PTFE filled)	1.38 (0.5)	—	—	0.1–0.45
Polyimide	3.50 (0.5)	315	30.0	0.15–0.3
Polyimide (graphite filled)	3.50 (0.5)	315	5.0	0.1–0.3
Epoxy (glass filled)	1.75 (0.5)	260	—	0.3–0.5

Note: The coefficient of friction and wear coefficient were measured with polymers sliding on steel. These are approximate values taken from various publications.

and reinforcement matrices. Acetals, polyethylenes, and polyamides are used for low *PV* applications, and polyphenylene sulfides, polyimides, and poly(amide-imides) are commonly used for high-temperature and/or high *PV* applications.

Fluorocarbons. Fluorocarbons are a family of polymers based on a fluorine atom substitute in olefins. Fluorocarbon thermoplastics are probably the most popular polymers for friction and wear applications. They combine outstanding chemical inertness, weatherability, dielectric properties, and heat resistance with low friction coefficients. This family of fluorocarbon polymers consists of polytetrafluoroethylene (PTFE), fluorinated ethylene propylene (FEP), perfluoroalkoxy (PFA), polychlorotrifluoroethylene (PCTFE), polyvinylidene fluoride (PVDF), and polyvinylfluoride (PVF).

PTFE monomer has one fluorine atom replacing each hydrogen atom on an ethylene molecule.

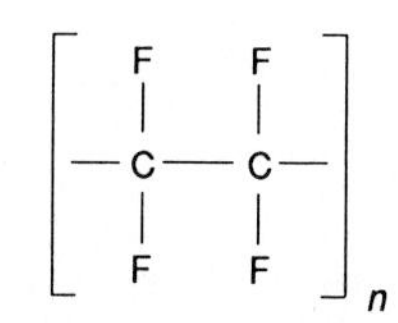

Polytetrafluoroethylene (PTFE)

PTFE is a crystalline polymer with a melting point of 325°C. In addition, PTFE has an exceptionally high melt viscosity, which precludes processing by conventional melt extrusion or molding techniques. PTFE is normally processed by compression molding. In the United States, PTFE is sold as Teflon by E. I. Du Pont Company, Wilmington, Del., as Fluon by ICI America, Inc., Wilmington, Del., as Hostaflon by Hoechst Celanese Corporation, Somerville, N.J., as Polyflon by Daikan, as Halon and Algoflon by Ausimont, USA, Inc., Morristown, N.J., and as Rulon by Dixon Corporation, Bristol, R.I. Rulon grade, by Dixon, is an injection-moldable resin.

Fluorinated ethylene propylene (FEP) is a copolymer of PTFE and hexafluoropropylene that closely approaches the properties of PTFE. FEP has a crystalline melting point of 290°C.

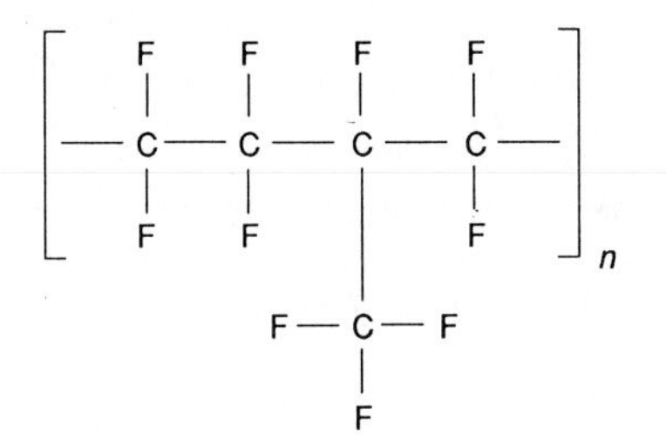

Fluorinated ethylene propylene (FEP)

Perfluoroalkoxy (PFA) is another fluorocarbon that has better temperature stability (260°C) than FEP and better creep resistance than PTFE.

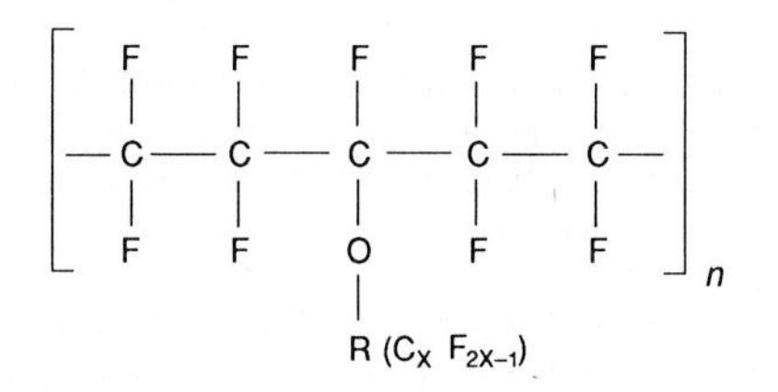

Perfluoroalkoxy (PFA)

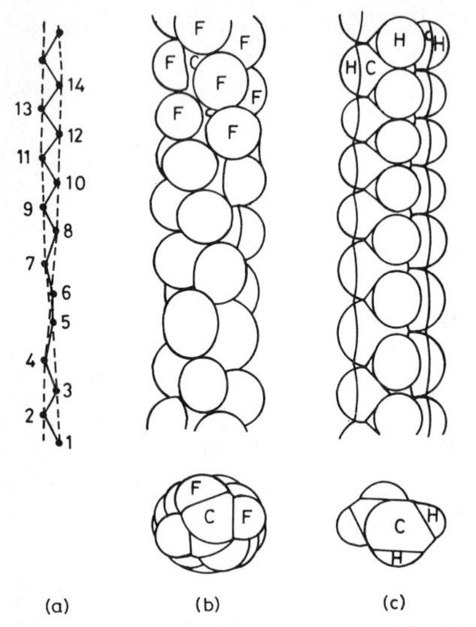

FIGURE 5.12 Molecular structure of PTFE: (*a*) twisted zigzag chain with 13 carbon atoms per 180° twist and repeat distance of 1.69 nm, (*b*) side and end views of a rodlike molecule of PTFE, and (*c*) ribbonlike hydrocarbon (e.g., polyethylene) molecule for comparison. [*Adapted from Bunn and Howells (1954).*]

FEP and PFA have better moldability than PTFE and can be processed by melt-processing techniques. The coefficients of friction of FEP (~0.1) and PFA are higher than that of PTFE (~0.05). Therefore, PTFE is used primarily for tribological applications.

Owing to the presence of the strong carbon-fluorine covalent bonds of extreme stability, PTFE is completely inert chemically. PTFE is highly crystalline below 20°C, with a molecular weight as high as 100 million. The melting point—the first-order transition from a partly crystalline to a completely amorphous structure—is about 330°C, very much higher than that of the corresponding hydrocarbon polymers. The molecular structure of PTFE is shown in Figure 5.12. Bunn and Howells (1954) have shown that the simple zigzag backbone of —CF_2—CF_2— groups that gradually twists through 180° over a distance corresponding to 13 CF_2 groups. The lateral packing of these rodlike molecules is hexagonal, with a_0 = 0.562 nm. On a larger scale, the individual units in the structure of PTFE are believed to consist of thin, crystalline bands that are separated from each other by amorphous or disordered regions. It has been suggested that the smooth profile of rodlike molecules permits easy slippage of the molecules with respect to each other in planes parallel to the C axis (Figure 5.12). This probably accounts for the low coefficient of friction (0.04 and up) of PTFE and easy transfer of PTFE material onto the sliding partner (Lancaster, 1973; Steijn, 1966, 1967; Tanaka et al., 1973).

The fundamental reasons for the low coefficient of friction of PTFE are

- The low adhesion of the PTFE surface
- The strong adhesions that are formed across the interface and then are sheared at quite low stresses
- The transfer of PTFE onto the bare surface, which results in subsequent PTFE-PTFE contact

The formation of a thin, coherent transfer film of PTFE on the counterface seems indeed to be one of the essential prerequisites for easy sliding and low friction. The static coefficient of friction for PTFE decreases with increasing load. Sliding speed and temperature have a marked effect on the friction characteristics of PTFE (Steijn, 1966, 1967; Tanaka et al., 1973; Uchiyama and Tanaka, 1980). Figure 5.13 illustrates the dependence of the coefficient of friction of PTFE on speed, load, and temperature. The friction coefficient of FEP is higher than that of PTFE. Comparable sliding tests of a steel ball against

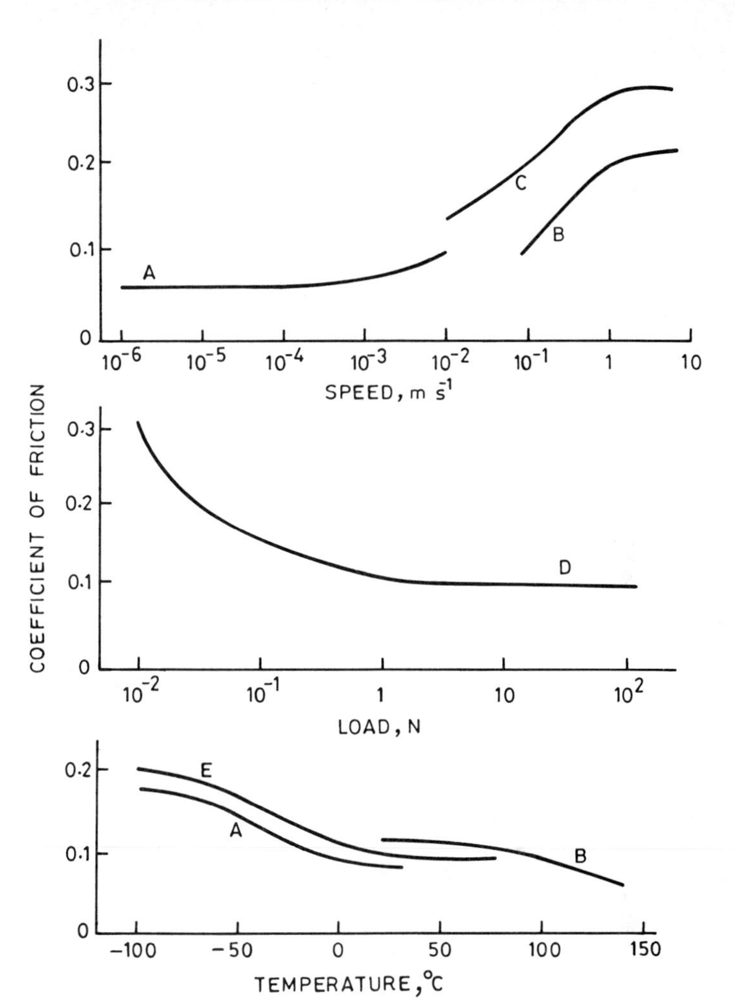

FIGURE 5.13 Effect of operating conditions such as speed, load, and temperature on the coefficient of friction of PTFE. [*Curve A adapted from Ludema and Tabor (1966); curve B, MacLaren and Tabor (1963); curve C, Tanaka et al. (1973); curve D, Shooter and Tabor (1952); and curve E, King and Tabor (1953).*]

these polymers yielded a friction coefficient of 0.05 for PTFE and of 0.18 for FEP (Bowers et al., 1954).

PTFE flows readily at modest pressures (~2 to 4 MPa) and modest temperatures (~110°C). Therefore, fillers and reinforcements are commonly added to improve its wear performance, *PV* limit, and temperature limit. PTFE is also used as filler for thermoplastic and other materials with very high strength and high temperature capabilities (Bhushan and Wilcock, 1982; Bhushan and Winn, 1979; Braithwaite, 1967; Lomax and O'Rourke, 1966). A prefinished bearing with PTFE-lead in a metallic matrix is commercially available from Garlock Bearing, Inc., under the trademark of DU bearing. DU is a composite material consisting of a steel backing strip that has a 0.25-mm-thick sintered porous bronze surface layer into which is impregnated a homogeneous PTFE-Pb (80:20) mixture with a thin overlay of PTFE-Pb (80:20) up to 25 μm thick (Anonymous, 1976b). DU bearing material is used for bushings, thrust washers, and slides in dry applications at high loads and high temperatures.

Acetals. Acetals are polymers of formaldehyde that technically are called *polyoxymethylene* (*POM*). Their linear polymer structures have oxymethylene (—CH_2O—) units repeating in their backbone.

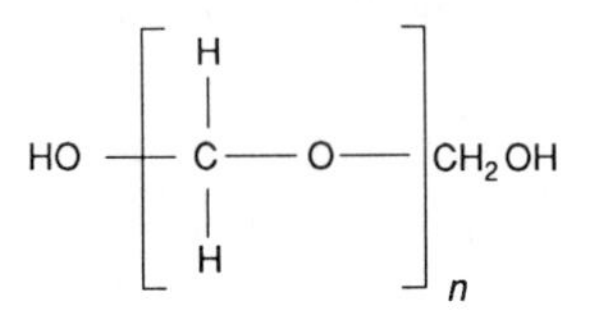

Polyoxymethylene (POM)

Acetals can be processed by all traditional thermoplastic processing methods. Acetal thermoplastics are strong, rigid, and highly crystalline resins. They have good resistance to moisture, heat, and solvents. Commercially produced acetals include homopolymers (Delrin, trade name of Dupont) and copolymers (Celcon, trade name of Celanese Plastics Company, Chatham, N.J., and Milcon, trade name of Fluorocarbon Tribol Division, Santa Ana, Calif.). Commercially produced homopolymer acetals have higher melting points, are harder, have higher resistance to fatigue, are more rigid, and have higher tensile and flexural strength with generally lower elongation. The homopolymers have higher (short term) heat-deflection temperatures than copolymers, but their continuous-use temperature is 80°C, compared with 105°C for copolymers. Both homopolymers and copolymers are available in unmodified and modified grades. The addition of PTFE in the form of fine fibers reduces the friction coefficient and increases the *PV* limit. Glass and carbon fibers are used for improving wear characteristics.

Polyethylene (PE). Polyethylene, the most widely used thermoplastic today, is available in a wide variety of grade formulations and properties. Polyethylene is made from ethylene monomers and varies in molecular weight (chain length), crystallinity (chain orientation), and branching characteristics.

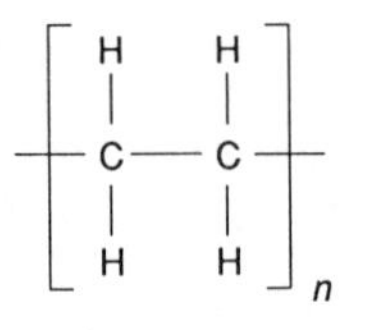

Polyethylene (PE)

Ultrahigh-molecular-weight polyethylene (UHMWPE) is a linear polyethylene with a molecular weight of from 2 to 5 million that exhibits excellent resistance to both abrasion and impact. Because the resin neither melts nor exhibits a measurable melt-flow index, it is processed by compression molding or ram extrusion. UHMWPE is produced by many companies, such as by Hoechst Celanese Corporation, Houston, Texas, H. Heller and Company, White Plains, N.Y., and Cadillac Plastic and Chemical Company, Detroit, Mich.

Typical coefficients of friction for UHMWPE range from 0.15 to 0.3. The outstanding abrasion resistance and high impact strength make it useful for bearings, gears, bushings, and many other sliding components. One of the important applications of UHMWPE is in prosthetic joints. Dumbleton and Shen (1976) achieved minimum wear of UHMWPE among the thermoplastics for applications in prosthetic joints. Other applications of UHMWPE are chemical pump parts, feed screws, and textile and other machine parts.

Polyamides (PA). The polyamides (Nylons, produced by Du Pont) are produced by condensation polymerization reactions between an organic diacid and a diamine. The basic reaction involved in forming a polyamide is

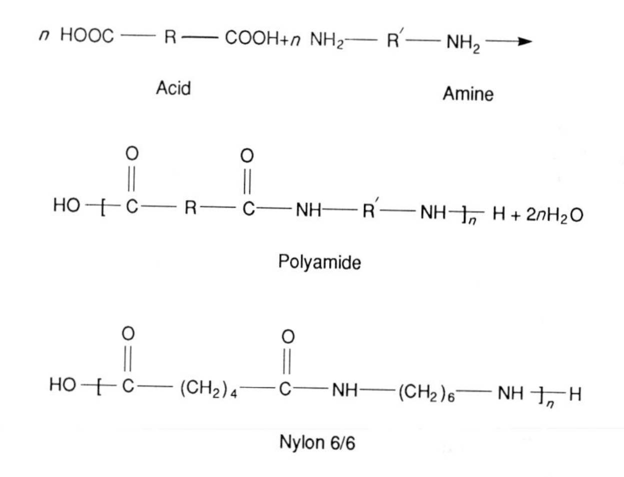

There are a number of common polyamides; the various types are usually designated as Nylon 6, Nylon 6/6, Nylon 6/10, Nylon 6/12, Nylon 11, and Nylon 12. The suffixes refer to the number of carbon atoms in each of the reacting substances involved in the polymerization process. The molecular weights of Nylons range from 11,000 to 34,000. They usually are semicrystalline polymers with melting points in the range of 175 to 300°C. The more C atoms, i.e., the lower the concentration of amide groups, the lower is the melting point. Nylon 6/6 and Nylon 6 are the most important commercial products. Their melting points are 269 and 228°C, respectively.

The polyamides were the first thermoplastic resins used on a large scale for manufacturing parts subjected to friction and wear. Nylons are available in a number of formulations for molding and extrusion. Although the coefficient of friction against steel is relatively high, Nylons have good abrasion resistance to wear within their *PV* limit. They have been used widely for gears, bearings, and other parts because of their excellent wear resistance, toughness, and ability to dampen vibration and reduce noise.

The friction and wear properties of Nylons are improved by the addition of several fillers, such as fiberglass, PTFE, graphite, and MoS_2. MoS_2-filled Nylon (Nylatron-GS) and glass fiber–reinforced Nylon are commercially available.

Polycarbonate (PC). Polycarbonate is really a polyester, since it is composed of esters of carbonic acid and an aromatic bisphenol.

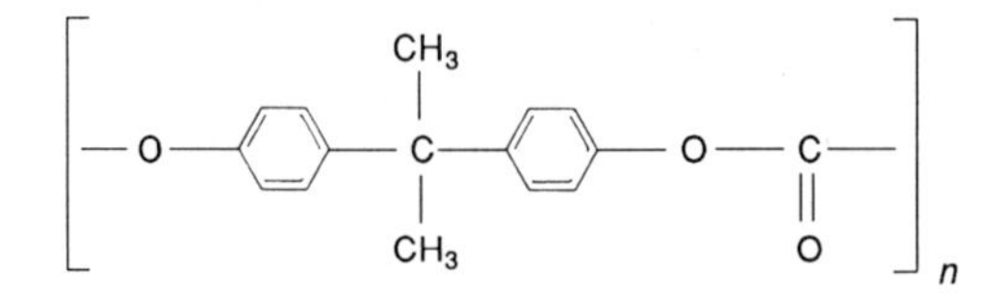

Polycarbonate (PC)

Polycarbonates are high-molecular-weight, low-crystalline thermoplastic polymers that have several outstanding physical and mechanical properties, such as very high impact strength, high transparency, low thermal expansion, and high heat-distortion temperature. Polycarbonates can be processed by all traditional thermoplastic processing methods (including injection molding). Polycarbonate resins are produced by General Electric Company, Pittsfield, Mass. (under the trade name Lexan) and by Mobay Chemical Corporation, Pittsburgh, Pa. (under the trade name Merlon). Lubricated polycarbonates are processed by several companies, such as Polymer Resources, Ltd., Berlin, Conn., and Wilson-Fiberfil International, Neshanic Station, N.J. Because of very low mold shrinkage of polycarbonates, gears and sliding parts demanding close tolerances have been made from this polymer for unlubricated service.

Unfortunately, friction and wear of unfilled polycarbonates are relatively high. However, significant improvements have been made by reinforcing polycarbonates with glass and by internal lubrication with PTFE.

Polyphenylene Sulfide (PPS). Polyphenylene sulfide (PPS) is a crystalline aromatic polymer in which recurring benzene rings are *para*-substituted with atom links.

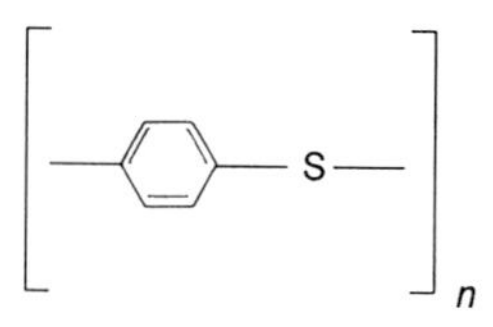

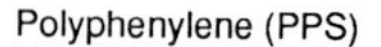

Polyphenylene (PPS)

PPS is a high-temperature plastic with continuous-use temperatures as high as 240°C. It is available commercially as Ryton (trade name of Phillips Chemical Company, Bartlesville, Okla.) and Thermocomp OCL (product code of LNP Corporation, Malvern, Pa.) in various glass fiber and mineral/glass fiber-reinforced grades for injection or compression molding and as powders for slurry coating and spraying.

Short and Hill (1972) and Boeke (1974) have proposed using PPS as a high-temperature matrix for solid lubricants, since PPS has a great affinity for fillers and fibers. Unfilled PPS exhibits friction coefficients from 0.2 to 0.6. The addition of MoS_2-Sb_2O_3 and glass fibers improves its friction and wear characteristics. Hopkins and Campbell (1977) used PPS as a binder for MoS_2- and Sb_2O_3-bonded coatings.

Poly(amide-imide) (PAI). Poly(amide-imide) polymers have a chemical structure that is, as the name implies, a combination of the typical nitrogen bond of polyamide and the ring structure of polyimide.

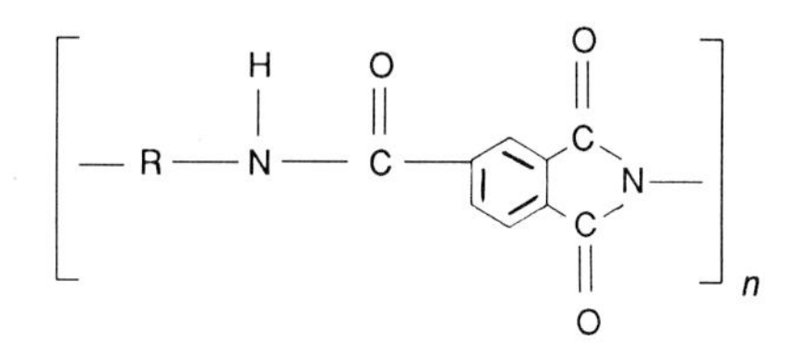

Poly(amide-imide) (PAI)

Poly(amide-imide) polymers are amorphous, high-temperature thermoplastics produced primarily by the condensation of trimellitic anhydride and various aromatic diamines. Poly(amide-imide), produced by Amoco Chemical Corporation, Chicago, Ill., and marketed under the trade name Torlon, is available in several grades, such as compression molding grades, injection molding and extrusion grade, and lubricated grade for bearing applications.

Of the commercially available amorphous thermoplastics, poly(amide-imide) polymers have among the highest glass-transition temperatures, from 270 to 285°C. Their thermal stability is similar to that of the polyimides, whereas their processability and mechanical properties are superior. Compared with most engineering polymers, the poly(amide-imide) polymers are higher in strength. Room-temperature tensile strength of PAI polymers is about 180 MPa, and compressive strength is about 207 MPa. At 230°C, tensile strength is about 65 MPa, and continued exposure at 260°C for up to 8000 hours produces no significant decline in tensile strength. In addition, PAI polymers exhibit a low coefficient of thermal expansion and good chemical resistance.

Their high mechanical strength and good thermal stability make poly(amide-imide) polymers suitable for structural and mechanical components, such as bearings, seals, and gears in such equipment as valves, pumps, compressors, and electric motors.

Polyester. Polyester resins are produced by the reaction of an organic alcohol with an organic acid, and a wide range of thermoplastic and thermosetting polymers can be produced. Based on their mechanical properties, they can be classified as plastics or elastomers. A variety of monomers and peroxide catalysts are used to cross-link with the polyester resins. A melt-processable (injection-moldable) liquid crystal polymer (aromatic copolyester) sold under the trade names of Vectra (from Hoechst Celanese) and Xyder (from Amoco) is based on *p*-hydroxybenzoic acid and hydroxynaphthoic acid. The structure of an aromatic polyester that is important for tribological applications is based on *p*-oxybenzoyl repeat units and is a linear thermoplastic with high crystallinity.

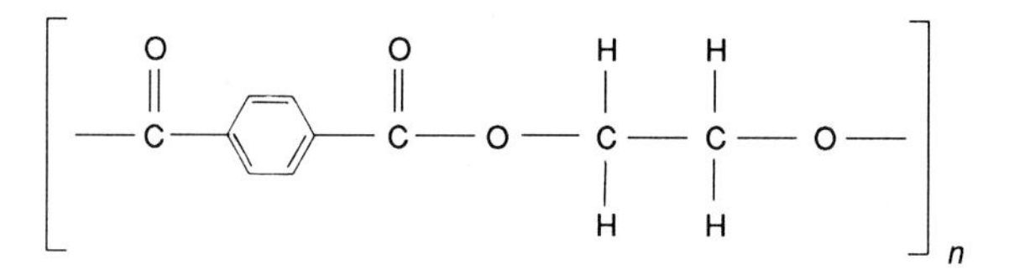

Polyester - polyethylene terephthalate (PET)

It is sold under the trade name Ekonol by Carborundum Company, Plastics and Adhesives Department, Atlanta, Ga. Another linear aromatic polyester is sold under the trade name Ekkcel by Dart Industries, Los Angeles, Calif. (previously sold by Carborundum Company, Niagara Falls, N.Y.). Although both these polyesters are thermoplastic, no melting point has been observed. These can be processed by compression molding.

Linear aromatic polyesters show excellent stability in air at temperatures over 315°C with inherent self-lubricating properties. They have high stiffness, high thermal conductivity, low thermal expansion, and high solvent resistance to any available polymer. They are usually combined with PTFE, polyimide, and graphite to produce the best mechanical properties and friction and wear performance. They are used in O-rings, seals, piston rings, and self-lubricating bearings (Economy et al., 1970).

5.3.1.2 ***Thermosetting Plastics.*** In thermosetting plastics, under the influence of heat or catalysts, the molecules join together and subsequently cross-link. Heating or curing of thermosetting polymers produces a three-dimensional structure that is relatively hard and brittle. Normally, these materials decompose on heating, rather than softening or melting. Thermosetting plastics are normally processed by compression molding. The most commonly used thermosetting polymers for tribological applications are phenolics, polyimides, and epoxies.

Polyimides are commonly used for high *PV* limits and high temperatures under both unlubricated and lubricated conditions. Phenolics and epoxies are normally selected for high *PV* limits with some lubrication. Some thermosetting polymers are used as matrices for fillers and as binders for solid lubricant coatings.

Phenolics. Phenolic resin is the reaction product of phenol and formaldehyde (CH_2O). It is a heat-cured thermoset. Monomers bond through the three unsaturated carbon atom bonds, which leads to a highly cross-linked, rigid thermosetting polymer. The cross-linked structure leads directly to the resin's excellent properties. Phenolics are best suited for applications requiring heat resistance, dimensional stability, and creep resistance.

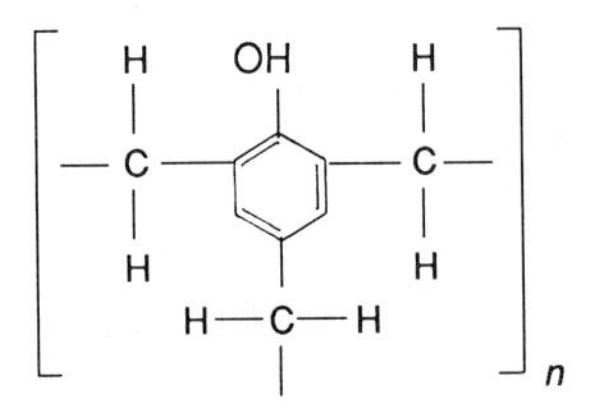

Phenol formaldehyde (Phenolic) (PF)

Phenolic resins are less costly than most thermoplastics, but fabrication costs tend to offset this advantage. The precursors are usually fluid, and the addition of hardeners and the catalyst induces cross-linking of the molecular chains, leading to an irreversible transformation to solids. Phenolic resins are produced by several companies, such as Occidental Chemical Corporation, Tonawanda, N.Y., Plastics Engineering Company, Sheboygan, N.Y., and Monsanto Company, St. Louis, Mo.

The phenolics are somewhat brittle and are almost always used in fiber-reinforced form. The simplest method of producing reinforced thermosetting bearing materials is by impregnation of fibrous material or cloth, cellulose, cotton, asbestos, glass, or carbon with liquid resins or a solution thereof, pressing into an appropriate shape, and curing at elevated temperatures. Phenolic laminates exhibit high load-carrying capabilities and high *PV* limits. Many automotive break linings consist of phenolic resins reinforced with asbestos, glass, metal powders, and friction modifiers. Compression-molded fabric-reinforced grades of phenolic resins have high compression strength (510 MPa), higher than any other plastics and several metals. They also have excellent dimensional stability, high-temperature creep resistance, and the ability to be molded to close tolerances. Notched Izod impact strength for general-purpose phenolics ranges up to 0.03 $N \cdot m\ mm^{-1}$ in impact-modified grades; engineering phenolics can go five times higher.

Stern-tube bearings and face seals lubricated with petroleum oil for commercial ships are made of asbestos-reinforced, filament-wound phenolic composites (produced by Railko Company, U.K.) (Bhushan and Winn, 1979; Richardson, 1957). Graphitized canvas-based phenolic laminate (Insurok T-768 by Spaulding Fiber Company, De Kalb, Ill.) is used for seal rings in U.S. ships and water-lubricated bearings. Gears, cams, and other power-transmission equipment are frequently machined or molded from the fabric-reinforced grades of phenolics.

Polyimides (PI). Polyimides are a class of long-chained polymers that have repeated imide groups as an integral part of the main chain. The polymeric chains consist of aromatic rings alternating with heterocyclic chains.

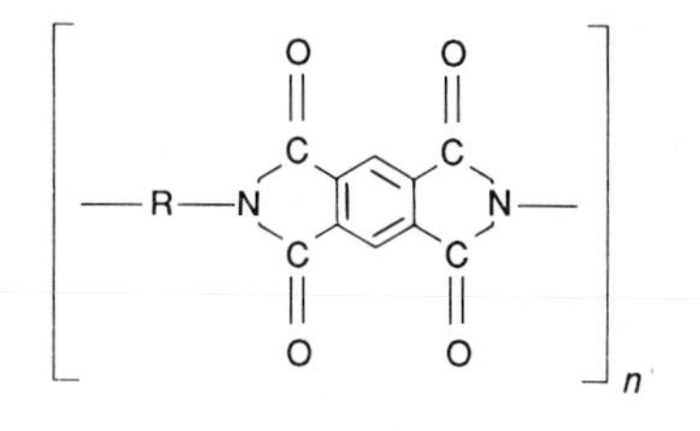

Polyimide (PI)

Because of the multiple bonds between these groups, the polyimides have high thermal stability. Polyimides are commercially produced by a poly-condensation reaction of aromatic dianhydride with aromatic diamine or aromatic diisocyanate.

Partially fluorinated polyimides, designated 100 PMDA, are formulated from the aromatic diamine 2,2-bis[4-(aminophenoxy)phenyl]hexafluoropane (4-BDAF) and the pyromellitic acid dianhydride (PMDA). The polyimide designated 80 PMDA/20 BTDA is a copolyimide made from the diamine 4-BDAF and 80 mol % of the dianhydride PMDA and 20 mol % of the benzophenonetetra-carboxilic acid dianhydride (BTDA) (Vaughan et al., 1976). The polyimide designated PMR-15 is a copolyimide made from dimethyl ester of 3,3′,4,4′-benzophenonetetra-carboxylic acid (BTDE), 4,4′-methylenediamine (MDA), and monomethyl ester of 5-norbornene-2,3-dicarboxylic acid (NE) (Serafini et al., 1972). These classes of polyimides possess great potential for long-term service in highly oxidative environments at temperatures up to 370°C.

Polyimides are formulated primarily as thermoset plastics. The thermosetting grades can be compression or transfer molded, sintered, or extruded into shapes or parts with some difficulty. A number of modified polyimides are currently available that are thermoplastic and can be more readily fabricated, but in each case the thermal resistance is somewhat compromised. Thermoplastic grades can be injection molded, compression molded, or direct formed using PM techniques. Films of polyimides can be cast by pouring liquid material onto rollers.

Thermoset-type polyimides are supplied by the Du Pont Company (Wilmington, Del.) under the trade names Vespel (sintered parts), Kapton (film), and Pyralin and Pyre ML (lacquers), by Fluorocarbon Tribol Industries Division (Santa Ana, Calif.) under the trade name Tribolon PI-500 series, by Dixon Corporation (Bristol, R.I.) under the trade name Meldin, and by TRW Bearing Division under the trade name Filimide. Thermoplastic polyimides include NR 150 from Du Pont, Matrimid 5218 from Ciba-Geigy (Hawthorne, N.Y.), Polyimide 280 from Dow Chemical (Midland, Wisc.), Meldin PI series from Dixon Corporation, Tribolon PI-100 series from Tribol, and Envex 1000 from Rogers Corporation (Atlanta, Ga.), all available in compression-molding powders and various solution forms.

Polyimides with excellent mechanical and physical properties at high temperatures and tensile strength in excess of 35 MPa at 325°C combine good wear and abrasion resistance with excellent sliding properties. Compared with other polymers, they exhibit low thermal expansion, excellent resistance to creep, even at temperatures as high as 315°C, and outstanding radiation resistance (Lewis, 1964). Buckley and Johnson (1963) and Fusaro (1988) have determined that even in the hostile environment of 10^{-4} to 10^{-8} Pa (10^{-6} to 10^{-10} torr) vacuum, the friction and wear performance of polyimides sliding against 440 C stainless steel was better than that of PTFE, poly(amide-imide), and polyphenylene sulfide. Further work on polyimides sliding in vacuum has indicated that the attainment of low friction depends on the thin film of transferred polyimide; without this film, the friction remains high (Buckley and Brainard, 1974).

While the sliding characteristics of unfilled polyimides are promising, substantial improvements in wear resistance can be made by adding graphite, MoS_2, and PTFE (Lewis, 1969). Devine and Kroll (1964) have shown the effectiveness of using graphite-filled polyimide retainer rings in ball-bearing assemblies at temperatures as high as 400°C and speeds of 10,000 r/min. Graphite fiber–reinforced polyimide (GFRPI) is used in oscillating plain spherical bearings (Sliney, 1979).

Epoxies. Epoxies are a group of polymers that are finding a growing use as engineering materials. Epoxies contain a reactive functional group (oxirane ring) in their molecular structure.

The most widely used epoxy resins are based on the reaction of bisphenol A and epiclorohydrin, producing the diglycidyl ether of bisphenol A. By controlling the degree of polymerization, the physical state of the resin produced can be varied from solid to liquid. In order to convert the epoxy resin into a rigid solid, a hardener is added that cross-links the relatively short linear chains (as small as 10 monomers). Several types of hardeners are available, but the most commonly used types are reagents possessing amino groups (—NH_2).

The equation showing cross-linking is as follows:

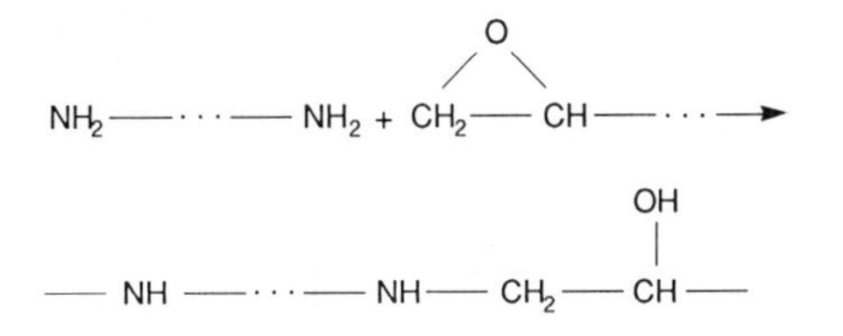

Both hydrogen atoms in every amino group can react with an epoxy group in this manner. Epoxy resins are available from Hysol AIP Division, Pittsburgh, Pa., Dow Chemical Company, Midland, Wisc., Fiberite Corporation, Winona, Minn., and Ciba-Geigy Corporation, Hawthorne, N.Y.

Epoxies are considerably more expensive than phenolics, but the mechanical properties of both are broadly similar. However, the fact that some epoxy formulations can be made less brittle than phenolics has led to their use in filled rather than reinforced forms, which partly offsets their cost. Epoxies filled with MoS_2 and graphite are available as solid bars for machine bearings or as two-component fluids for castings. Filled epoxies also can be sprayed as thin layers onto metal backings.

5.3.2 Elastomers

Elastomers usually have four ingredients: the base bum (polymer), the filler, plasticizers, and the curing or vulcanizing agent. The fillers commonly used (such as carbon black, graphite, and clay) are materials that reinforce the polymer to produce the degree of stiffness and hardness (abrasion resistance) needed for the intended application. Carbon black is used to achieve superior stiffness and low friction. Plasticizers are generally oils used to facilitate the mixing process and to establish the desired low-temperature characteristics of the material. Curing or vulcanizing results in the formation of chemical links between the molecular chains. Cross-linkage reduces the creepage and enhances the recovery of the rubber to its original dimensions. Vulcanization also reduces the stress relaxation. For each elastomer mix, there is an optimal degree of cross-linking to give the best overall properties in the finished vulcanisat (Bhushan and Winn, 1979, 1981).

Elastomers are selected for several tribological applications, such as soft-lined plain bearings, flexible thrust-pad bearings, seals, seal rings, tires, and abrasion-resistant parts, because of their ability to undergo reversible deformations as large as 1000 percent (Anonymous, 1973c, 1988). Elastomers exhibit the following common characteristics:

- They have glass-transition temperatures T_g well below room temperature.
- They have a high degree of elasticity and retract rapidly after stretching.
- They show a high modulus of elasticity when elongated.
- Most of the materials are amorphous.
- They have large molecular weights and are readily cross-linked.

Usually, elastomers have a high tolerance to abrasive particles, low wear, resilience in distributing the load under misaligned conditions (thus preventing seizure), low cost, and easy availability (Bartenev and Lavrentev, 1961; Bhushan and Winn, 1979, 1981; Bhushan et al., 1982; Harper, 1970; Ksieski, 1973; Ludema and Tabor, 1966; Reznikovskii, 1960; Schallamach, 1952, 1953, 1958, 1963; and Zaprivoda and Tenebaum, 1965). However, elastomers change chemically with lengthy exposures to high temperatures. As the temperature decreases, the hardness increases, resulting in loss of the compliance needed for some applications. Elastomers have been known to swell when immersed in various fluids. The swelling of elastomers is generally a physical phenomenon, although chemical changes also take place, particularly when swelling occurs at elevated temperatures. Swelling is often associated with a degradation in mechanical properties (Anonymous, 1973c; Bhushan and Winn, 1979; Blow, 1971).

Commonly used elastomers for various sealing devices (such as O-rings) include polyisoprene (synthetic rubber), butyl, ethylene propylene diamine monomer (EPDM), neoprene, Buna-N, polyacrylate, and olefins. Viton, silicone, and perfluoroelastomer (Kalrez) are used for high-temperature applications. Commonly used elastomers for bearing applications include Buna-N, polyacrylate, Viton, and polyurethane. A class of polyurethanes is also used to produce gears and rollers. Some important properties of common elastomers are listed in Table 5.16. Their descriptions follow.

5.3.2.1 ***Natural Rubber.*** The commercial base for natural rubber is latex, a milklike serum generated by the tropical tree *Hevea brasiliensis*. Many types and grades of natural rubber are available, having varying impurity levels, collection methods, and processing techniques. Natural rubber has a large deformability capacity, low compression, and low stress relaxation, which favor its application in sealing devices where maintenance of sealing forces and the surface conformability of high-quality soft stocks are important.

5.3.2.2 ***Synthetic Natural Rubber.*** The synthetic rubber that most closely duplicates the chemical composition of natural rubber is synthetic polyisoprene rubber (IR). It has the same chemical structure as natural rubber minus natural impurities.

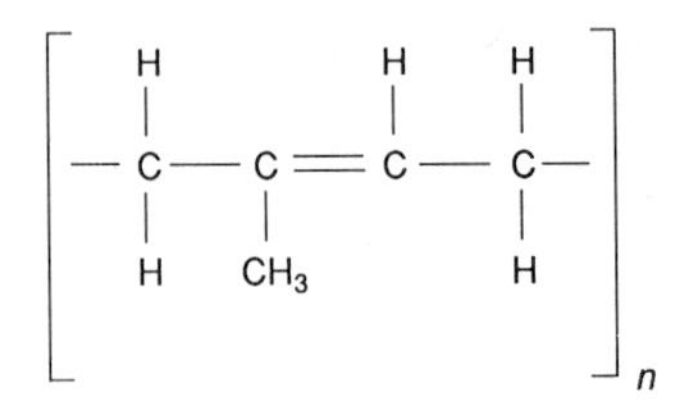

Polyisoprene rubber (IR)

Polyisoprene rubber shares the natural-rubber properties of good uncured tack, high reinforced strength, and good abrasion resistance. The green strength of synthetic natural rubber is significantly low. This material can be used interchangeably with natural rubber.

5.3.2.3 ***Styrene-Butadiene Rubber (SBR).*** Styrene-butadiene rubbers are a family of block copolymers containing a higher percentage of styrene than butadiene and having polymodal molecular weight distribution.

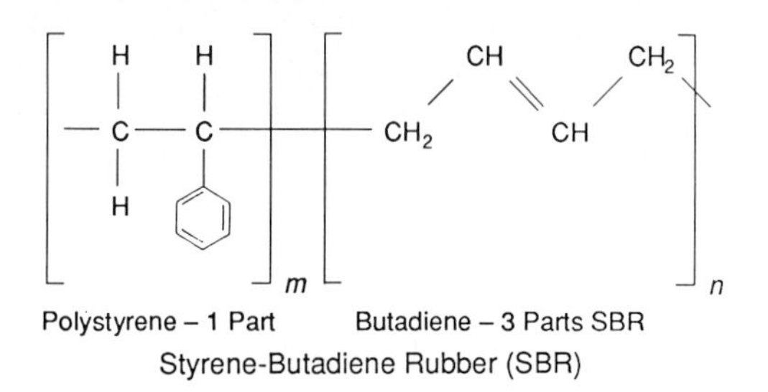

Polystyrene – 1 Part Butadiene – 3 Parts SBR

Styrene-Butadiene Rubber (SBR)

The butadiene mer can have various configurations (isomers) with the same number of carbon and hydrogen atoms. The SBR resin is amorphous, and formed parts made from SBR have high light transmission. These polymers are recognized for their transparency and high impact strength ($0.05\ N \cdot m\ mm^{-1}$). Phillips 66 Company (Pasadena, Texas) produces SBR under the trade name K-Resin. These grades are available for injection molding, blow molding, and sheet, profile, or film extrusion. They can be blended with other polymers or fillers to improve their mechanical properties.

SBR is used in a broad range of applications where clarity, good processability, toughness, and cost are significant factors. They are used in the disposable packaging market for lids, cups, and trays and in medical applications for centrifuge tubes and surgical devices.

5.3.2.4 ***Butadiene-Acrylonitrile Rubber (Butyl).*** Butyl rubber is a copolymer of isobutylene and isoprene (IIR).

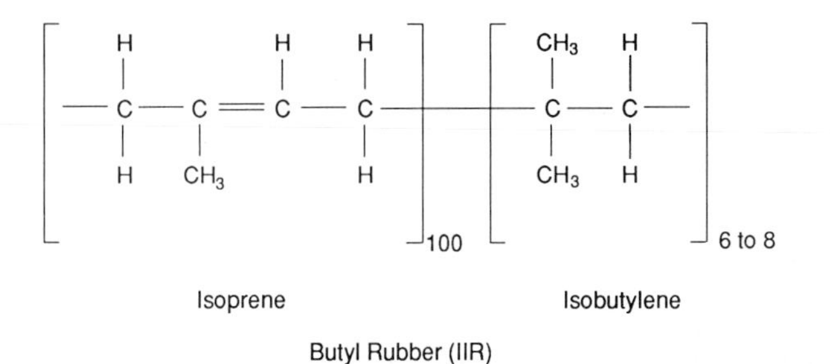

Butyl Rubber (IIR)

The two types of rubber in this category are both based on crude oil. The first is polyisobutylene with an occasional isoprene unit inserted into the polymer chain to enhance vulcanization characteristics. The second is the same, except that chlorine is added (up to 1.2 wt %), resulting in greater vulcanization flexibility and enhanced cure compatibility with general-purpose rubbers. Butyl rubber is used in automotive inner tubes and tubeless tire inner liners because of its excellent impermeability to air. The butyls are also used in O-rings and shock and vibration products.

5.3.2.5 ***Ethylene Propylene (EP).*** Like the butyls, the EP rubbers are of two types. One is a fully saturated (chemically inert) copolymer of ethylene and propylene (EPR); the other (EPDM) is the same, except that a third polymer building block (diene monomer) is attached to the side of the chain.

TABLE 5.16 Selected Properties of Common Elastomers

Material	Specific gravity	Water absorption in 24 h (3 mm thick), %	Hardness, Shore A	Tensile strength, MPa	Elongation at break, %	Tensile modulus (Graves' tear), MPa	Thermal conductivity, W m^{-1} K^{-1}	Coefficient of thermal expansion, $\times 10^{-6}$ $°C^{-1}$	Abrasion resistance*	Recommended maximum temperature, °C	Remarks
Thermosetting											
Natural polyisoprene/natural rubber	0.93	—	30–90	31	To 700	—	—	—	A	70	General purpose, poor resistance to hydrocarbon oils and oxidizing agents
Polyisoprene/synthetic natural rubber	0.93	—	40–80	27.5	700	—	—	—	A	70	As above
Styrene-butadiene rubber (SBR)	0.93	—	40–100	21	500	—	—	—	A	100	As above
Isobutylene isoprene (butyl)	0.93	—	40–90	17–20	800	—	—	—	B	100	Good resistance to oils
Ethylene propylene dienemonomer (EPDM)	0.86	—	30–90	20	600	—	—	—	B	125	Excellent weathering resistance
Chloroprene (neoprene)	1.24	—	40–95	28	600	—	—	—	B	100	Fair resistance to oils, poor low-temperature flexibility
Butadiene-acrylonitrile (Nitrile, Buna-N, or NBR)	1.00	—	40–90	21	150–350	9 at 100% elongation	—	—	A	125–150	Excellent resistance to hydrocarbons, but not resistant to oxidizing agents
Polyacrylate	1.10	—	40–90	17	400	—	—	—	C–B	150	Good heat resistance, good resistance to oil and aliphatic hydrocarbons, poor resistance to water and acids, poor tensile strength

Vinylidene fluoride–hexafluoropropylene (Viton, Du Pont)	1.80–1.86	—	60–90	17	150–300	18 at 100%	0.23	160	B	225–250	Excellent resistance to oil
Perfluoroelastomer (Kalrez 1050, Du Pont)	—	—	80–90	13	130	13 at 100%	—	—	B	260	Excellent resistance to oils and other chemicals; only used in applications requiring operation at extreme environments (very expensive)
Polysiloxane (silicone)	1.1–1.6	<1	25–90	11	120–700	5 at 50%	0.15–0.30	300–800	C–B	225–250	Excellent resistance over wide temperature range (−100 to 260°C), fair oil resistance, poor resistance to aromatic oils and abrasion
Fluorosilicone	1.4	0.9	40–80	8.3	—	—	—	—	D	225–250	Excellent aging properties over wide temperature range (−55 to 200°C), excellent oil and fuel resistance
					Thermoplastics						
Polyurethane	1.10–1.24	0.09	A65–D80	30–50	400–700	0.7	0.21	100–200	B	100–125	Good resistance to oils, fuels, and weather, good mechanical properties
Copolyester	1.17–1.25	0.3	A90–D70	35–45	390–800	—	—	—	B	100–125	As above
Olefin	0.86–0.90	0.01	55–95	5–13	180–250	—	0.21	130–170	B	135	As above

*A = excellent; B = good, C = fair, D = poor.

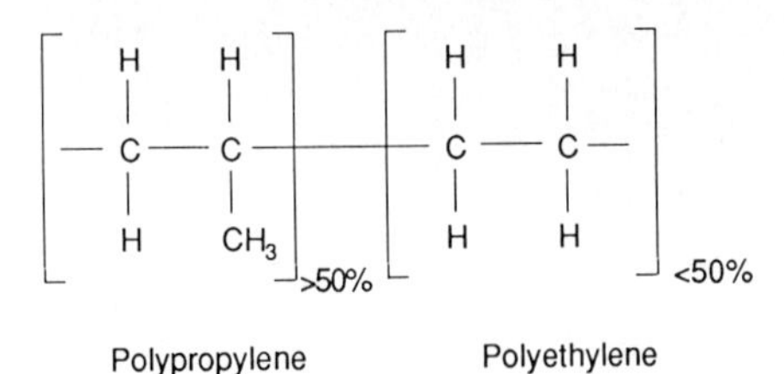

EPR

EPDM is chemically reactive and is capable of sulfur vulcanization. The polymer must be cured with peroxide. The physical properties of EPR and EPDM are not as good as those obtainable with natural rubber. However, property retention is better than that of natural rubber on exposure to heat, oxidation, or ozone. Typical applications are automotive hoses, body mounts, and pads, O-rings, conveyor belting, and other products requiring excellent weathering resistance.

5.3.2.6 ***Chloroprene (Neoprene).*** Except for polybutadiene and polyisoprene, chloroprene (or neoprene) rubber (CR) is perhaps the most rubberlike of all, particularly with regard to dynamic response. It is a carbon-chain polymer containing chlorine.

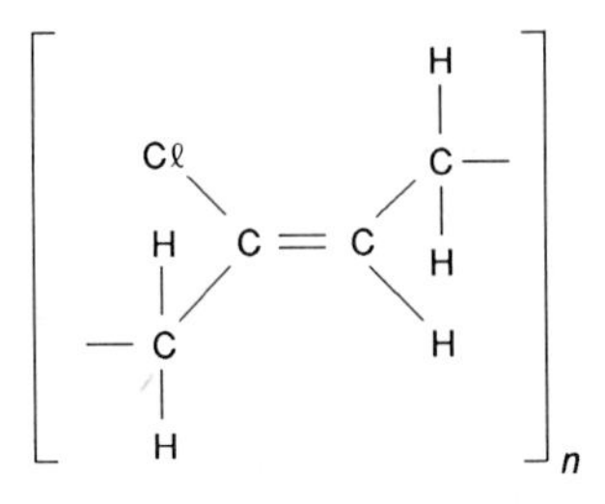

Polychloroprene (CR)

Neoprenes are a large family of rubbers that have a property profile approaching that of natural rubber and better resistance to oils, ozone, oxidation, and flame. They do not soften on heat exposure, although their high-temperature tensile strength may be lower than that of natural rubber. Neoprenes do not have the low-temperature flexibility of natural rubber, which detracts from their use in shock or impact applications. Typical applications include pump impellers, hoses, belting, wire and cable, footwear, tires, coated fabrics, bearing pads, O-rings, and seals.

5.3.2.7 ***Butadiene-Acrylonitrile (Buna-N, Nitrile, or NBR).*** The nitriles are copolymers of butadiene and acrylonitrile, also known as Buna-N, nitrile, or NBR.

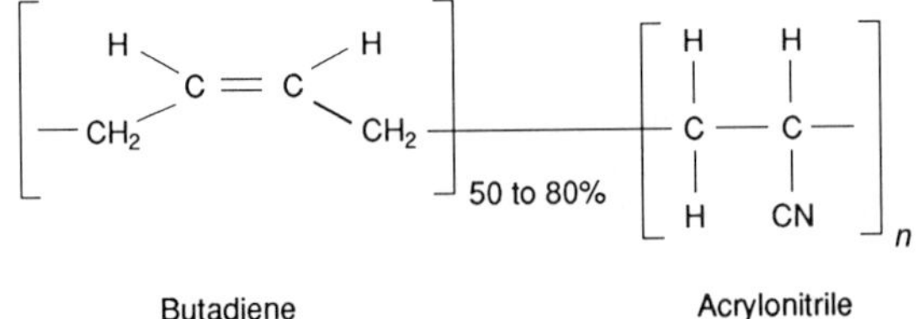

Nitrile rubber (NBR)

Acrylonitrile is added to increase oil resistance and hardness. Buna-N elastomers are used for applications requiring resistance to petroleum oil and gasoline.

Resistance to aromatic hydrocarbons is better than that of neoprene. With higher acrylonitrile content, the solvent resistance of nitrile compounds is increased, but low-temperature flexibility is decreased, and other low-temperature characteristics are inferior to those of natural rubber. These compounds have excellent resistance to oils. They are used in fuel-pump diaphragms, aircraft hoses, bearings operating in oil and seawater, and seals. Buna-N resins are manufactured by several companies, such as B. F. Goodrich Adhesive Systems, Inc., Akron, Ohio, and W. R. Grace and Company, Lexington, Mass.

5.3.2.8 ***Polyacrylate.*** Polyacrylates are special rubber-based on polymers of methyl, ethyl, or other alkyl acrylates and manufactured by Du Pont and W. R. Grace.

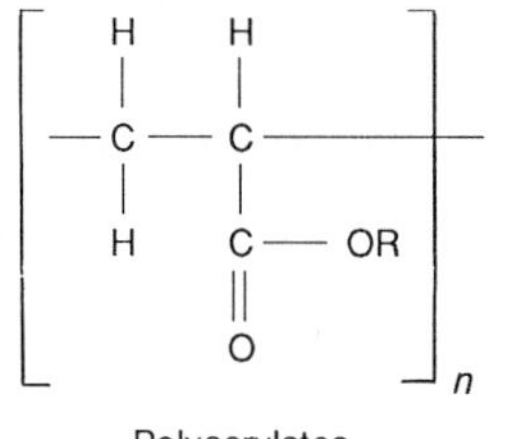

Polyacrylates

R is an alkyl group. They are highly resistant to oxygen, and their heat resistance is superior to that of all other commercial rubbers, except the silicones. Low-temperature flexibility is not good, and these rubbers decompose in alkaline solutions and do not swell by acids. Their low-temperature flexibility and water resistance can be improved, but only with a marked decrease in overall heat and oil resistance. These materials are used extensively for bearings, seals in transmissions, and O-rings and gaskets.

5.3.2.9 ***Fluoroelastomers.*** Fluoroelastomers are produced as a copolymer of vinylidene fluoride and hexafluoropropylene and are commercially available as Viton (trade name of Du Pont) and Fluorel (trade name of 3M).

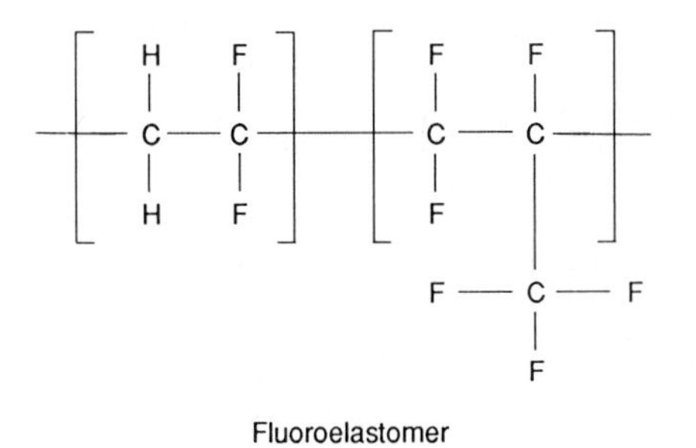

Fluoroelastomer

These rubbers have outstanding resistance to heat, as well as to many chemicals and oils, as compared with any other commercial rubber. The reinforced fluorocarbons offer moderate tensile strength but relatively low elongation characteristics. Several modifications of fluorocarbons are available, and conventional compounds within the hardness range of 60 to 98 Shore A are available. Typical applications are seals, gaskets, pump impellers, and high-vacuum components.

A new elastomer, perfluoroelastomer, with exceptional chemical and temperature durability was described by Oginiz (1978). It is marketed under the trade name Kalrez by Du Pont. This material can be used for sealing in many applications where other elastomers would fail because of chemical attack. Perfluoroelastomer's (Kalrez 1050) ability to retain sealing force at 200°C in air after 100 days was more than double that of fluoroelastomer (Viton 2282). In addition, perfluoroelastomer is serviceable continuously at temperatures up to 290°C and intermittently at 315°C. In other respects, perfluoroelastomer is comparable with fluoroelastomer. Perfluoroelastomers are extremely expensive and therefore are only used for applications requiring operation in extreme environments.

Another new class of elastomer, phosphonitrilic fluoroelastomer (PNF), with exceptionally low temperature capabilities (−51°C compared with −28°C for fluoroelastomer) and outstanding solvent resistance was described by Kyker and Antkowiak (1974). The poly(fluoroalkoxyphosphazene) elastomer is much more flexible than fluoroelastomer (Viton).

$$[\text{—}(CF_3CH_2O)_2PN\text{—}(HCF_2C_3F_6CH_2O)_2PN\text{—}]_n$$

The PNF elastomers have an effective range from −60 to 200°C. Their high-temperature range is comparable with that of Viton. Reynard et al. (1974) reported their physical and mechanical properties under fuel and fluid aging and extended exposure to high-humidity environments. Several grades of PNF fluoroelastomer (such as DG-27) are sold by Firestone Rubber Company, Akron, Ohio.

5.3.2.10 ***Silicone.*** Silicone rubbers are a versatile family of semiorganic synthetic rubbers that feel and look like synthetic rubber, yet they have a completely different type of structure from other rubbers. The backbone of silicone rubber is not a chain of carbon atoms, but an alternating chain of silicone and oxygen atoms.

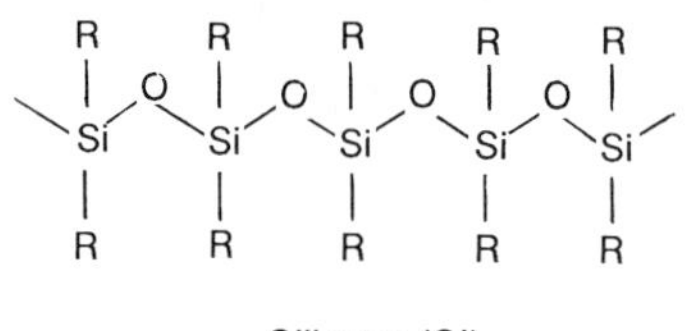

Silicone (Si)

Typically, the silicone atoms will have one or more organic side groups attached, generally phenyl, methyl, trifluoropropyl, or vinyl moieties. These groups impart such characteristics as solvent resistance, lubricity, compatibility, and reactivity with organic chemicals and polymers.

Silicones are thermosetting polymers that may be fluid, gel, elastomeric, or rigid in final form. Most elastomeric silicones are based on dimethyl siloxane polymers, but fuel- and solvent-resistant types, the fluorosilicones, contain polymers modified with trifluoropropyl groups. Silicone elastomers can be placed in two main categories: low- and high-consistency types. The low-consistency systems, e.g., RTV (a trade name of General Electric Company, Waterford, N.Y.), can be cured in place at room temperature or slightly elevated temperatures. These low-consistency elastomers are generally used as sealants. Two-component systems cure to form elastomers when the components are mixed, usually at room temperature. Cure rates depend on the catalyst type, concentration, and temperature. Two-part silicone rubber formulations are excellent for liquid injection molding.

In high-consistency, or rigid, silicone rubbers, the dimethyl polysiloxane polymer is basic. By replacing small quantities of methyl groups with phenyls and/or vinyls, significant variations can be achieved. The addition of phenyls improves low-temperature flexibility and resistance to gamma radiation. Vinyl side groups improve the vulcanization characteristics and the compression-set resistance of the cured rubber. For applications requiring resistance to oils and other fluids at high temperatures, for example, General Electric Company (Waterford, N.Y.) markets dimethyl vinyl polysiloxane (MeViSi) rubber with silica fillers under the trade name SE 3808U with a dicumyl peroxide curing agent.

The silicone rubbers can withstand temperatures as high as 260°C without deterioration. Although the strength of silicone elastomers is lower than that of other elastomers, these materials have outstanding fatigue and flex resistance. Typical applications include sealants, bonding agents, seals, diaphragms, and O-rings.

Another elastomer, fluorosilicone, provides most of the useful qualities of the regular silicone plus improved resistance to many fluids, except ketones and phosphate esters. Fluorosilicone elastomers have good dielectric properties, low compression set, and excellent resistance to weathering. They are expensive and definitely special-purpose elastomers. Typical applications include O-rings, seals, diaphragms, tank linings, gaskets, and protective boots in electrical equipment.

5.3.2.11 ***Polyurethanes.*** Polyurethane polymers available in various forms are classified either as plastics or elastomers and offer extremely wide variations in physical and mechanical properties. They can range in specific gravity from 0.008 to 1.1 in solid form and in hardness from rigid solids at 90 Shore D to soft, almost fluid elastomeric compounds. Urethane polymers can be either thermoplastic or thermosetting types. Thermoset polyurethanes are generally used for applications requiring maximum wear resistance; thermoplastics offer good toughness (elastomeric) and durability along with greater processing ease.

Urethanes are produced by the reaction of a polyisocyanate with long- and short-chain polyether or polyester or caprolactone glycol resins. The reactions are of the following type:

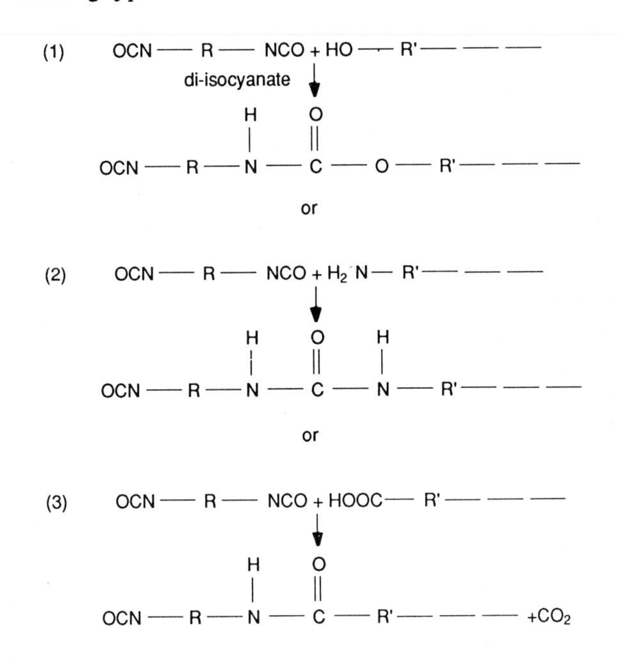

In general, polyether types have greater resilience and better hydrolytic stability, and polyester types have higher mechanical properties. Caprolactones offer a good compromise between polyester and polyether. Polyurethanes are manufactured by Dow Chemical USA Plastics Group, Midland, Mich., W. R. Grace and Company, Lexington, Mass., and Mobay Corporation.

The load-carrying capacity and abrasion resistance of polyurethanes are outstanding among elastomers. The addition of a filler can significantly reduce friction and improve dimensional stability and heat resistance. Polyurethanes are used to form a broad range of products, including flexible and rigid foams, flexible and rigid foam moldings, and solid elastomeric moldings and extrusions. Coatings and adhesives also can be made using polyurethanes. Typical applications of polyurethanes include gears, rollers, bumpers, pump impellers, shaft couplings, vibration-damping components, conveyor belts, etc. Since polyurethanes have poor compliance, they are not commonly used in sealing applications.

5.3.2.12 ***Copolyesters.*** Copolyester thermoplastic elastomers are generally tougher over a broader temperature range than the methanes. They are produced by Du Pont under the trade name Hytrel in four hardnesses ranging from 90 Shore A to 70 Shore D.

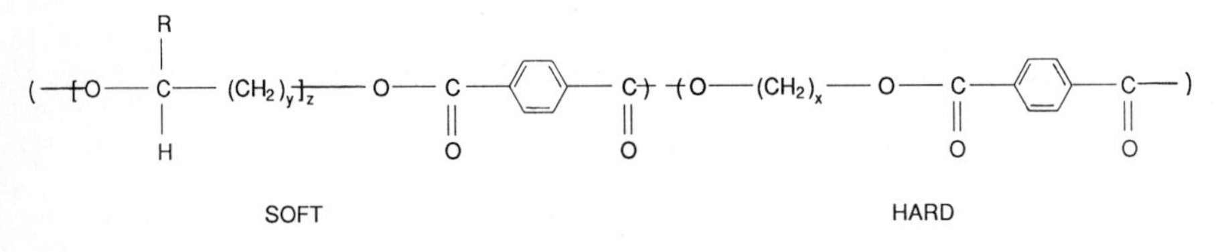

R = H, CH_3; y = 1, 3 z = 8-50 x = 2, 4

Copolyester - Hytrel

These materials are suitable for injection molding, extrusion, and thermoforming. Copolyesters, along with the methanes, are the highest-priced elastomers, have a high modulus of elasticity, good elongation and tear strength, and good resistance to flex fatigue at both low and high temperatures. Resistance to fluids is good to excellent, and weathering resistance is low unless antioxidants or carbon blacks are compounded with the resin. Applications include hoses, belts, couplings, tires, rolls, and wire and cable jacketing.

5.3.2.13 ***Olefins.*** Thermoplastic olefin elastomers are available in several grades, having room temperature hardnesses ranging from 55 to 95 Shore A.

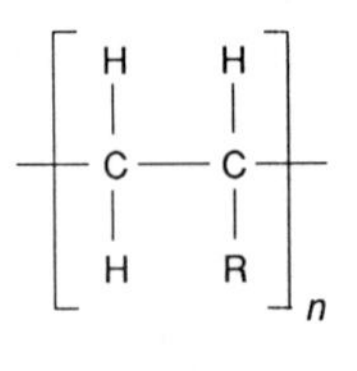

Olefins

R represents an atom or side group, e.g., H, Cl, CH_3. They are produced by several companies, including Du Pont, Exxon, B. F. Goodrich, Hercules, and Uniroyal. These materials have the lowest specific gravity of all elastomers. They remain flexible down to −50°C. They are autoclavable and can be used at temperatures up to 135°C in air. They have good resistance to most acids and bases. These elastomers are used for gaskets, seals, electrical components, and garden hoses.

5.4 POLYMER COMPOSITES

The bulk properties of a polymer can often be altered considerably by incorporating fillers or additives. Such changes in the bulk properties of composites generally result in improved friction and wear. Their friction and wear characteristics can be tailored to specific needs by mixing fillers in the form of solids and liquids in various shapes, sizes, and amounts (Anderson, 1986; Anonymous, 1978; Bhushan, 1978a, 1978b, 1980b, 1981; Bhushan and Dashnaw, 1981; Bhushan and Gray, 1978, 1979; Bhushan and Wilcock, 1981, 1982; Bhushan and Winn, 1979, 1981; Bhushan et al., 1982; Gardos and McConnell, 1981, 1982; Giltrow, 1973; Lancaster, 1968, 1972a, 1972b, 1973; Oleesky and Mohr, 1964; Seymour, 1990; Tanaka, 1977; Theberge and Arkles, 1974; Thorp, 1982; Voet, 1980). The typical benefits of fillers in polymer composites may be improvement in

- Mechanical properties such as impact strength, tensile strength, creep resistance, and load-carrying capacity
- Coefficient of thermal expansion and thermal conductivity
- Coefficient of friction, wear rate, and endurance life

The physical properties of some commonly used fibers are summarized in Table 5.17. Asbestos fibers are played down these days because of their carcinogenic properties. The other commonly used fibers are Kevlar 49 aramid (aromatic polyamide), glass, carbon (with various degrees of graphitization), Nylon, polyester (Dacron or Terelene), Oronol, Dynel, and other textile fibers. Carbon/graphite fibers are derived either from polyacrylonitrile (PAN) or a special petroleum pitch. The PAN-derived fibers have been available for several years, and these fibers are generally selected for their high strength. The pitch-based fibers are newer, and while they are not as strong as the low-modulus PAN fibers, the ease with which they can be processed to high moduli makes them attractive for stiffness-critical applications. PAN fibers are generally referred to as *high-strength* or *type II fibers*, and pitch fibers are referred to as *high-modulus* or *type I fibers*. To produce graphite fibers, the PAN or pitch precursors are processed at temperatures above 3000°C to promote a high degree of graphitization. This results in a fiber with a superior modulus and electrical/thermal properties and higher oxidation temperature. Glass fibers are very common because of their low cost, but they are abrasive. Kevlar, carbon, and graphite fibers are used for high-strength/stiffness/creep resistance and good tribological properties. Carbon fibers are slightly more abrasive than graphite fibers. It also should be noted that the addition of carbon and graphite fillers not only reduces the coefficient of friction and wear rate, but also reduces the thermal expansion coefficient and increases thermal strength.

The solid particulate fillers include MoS_2, graphite, carbon, PTFE, polyimide, silver, bronze, silica, alumina, etc., which are used to improve friction and wear.

TABLE 5.17 Selected Properties of Common Fibers Used in Polymer Composites

Fiber type/ designation	Shape and size	Density, kg m^{-3}	Tensile strength, GPa	Tensile modulus, GPa	Elongation at failure, % at 25°C	Coefficient of thermal expansion, $\times 10^{-6}$ °C^{-1}	Long-term use temperature, °C
			Glass				
Glass type E (alumina borosilicate) 739 AB (Owens Corning)	Round, 13 μm dia.	2590	3.45	72	4.8	5.1	600, softening point, 840°C
			Carbon				
Celanese 6000 (PAN)	Round, 7 μm dia.	1760	2.75	235	1.2	—	290
Thornel T-300 (PAN) (Amoco Performance Products)	Round, 7 μm dia.	1770	3.65	230	1.4	−0.54 (long.)	—
Thornel P-55 (Pitch) (Amoco Performance Products)	Round, 10 μm dia.	1990	1.90	380	0.5	−1.3 (long.)	—
			Graphite				
Celanese GY-70 (PAN)	Bilobal; aspect ratio 6:1, equiv. dia. 8 μm	1970	1.85	515	0.38	−1.3 (long.)	425
Thornel P-120 (Pitch) (Amoco Performance Products)	Round, 10 μm dia.	2190	2.25	830	0.27	−1.4 (long.)	—
Aromatic polyamide, Kevlar 49 (Du Pont)	Round, 12 μm dia.	1440	2.76	120	2.5	−2.0 (long.), +59.4 (radial)	160, decomposition temperature, 500°C
Polyamide (Nylon) type 728 (Du Pont)	Round, 25 μm dia.	1140	1.00	550	18.3	−300 (long.)	80
Dacron polyester type 68 (Du Pont)	Round, 20 μm dia.	1380	1.10	140	14.5	−1000 (long.)	120
Stainless steel (for reference)	—	7830	1.70	200	2.0	10–18	—

Lubricating oils have been added to polymers (e.g., polyethylene, butadiene-acrylonitrile copolymer) to improve their lubrication properties (Bhushan et al., 1982; Jamison, 1985).

Typically, 10 to 20 percent fiber content by weight is added to increase component stiffness, strength, and creep resistance. A higher fiber content makes the polymers hard to machine and less compliant. In some cases, where strength and creep resistance are crucial, fiber content by weight may be as high as 50 to 60 percent. The solid particulate fillers do not add much to stiffness and strength. Typically, 10 to 20 percent particulate fillers by weight is added for lubrication, but some polymers may lose strength with very large concentrations.

The fibers are typically 5 to 10 μm in diameter and can be long and continuous (known as *filaments*), milled, or chopped. In addition, they can be directionally or randomly oriented. Milled fibers are the shortest fibers; they range in length from 30 to 3000 μm, averaging approximately 300 μm. Chopped fibers range 5 to 50 mm in length. Continuous fibers are available in a number of forms, including yarns, tows containing 400 to 160,000 individual filaments, unidirectional impregnated tapes up to 1.2 m wide, multiple layers of tape with individual layers or plies at selected fiber orientations, and woven fabrics of many weights and weaves. In general, short fibers cost the least and fabrication costs are lowest, but continuous fibers provide the best strength and stiffness. Three-dimensional orthogonal weaves of fibers can be used for applications requiring extreme strength (e.g., solid-lubricating polymer cages for high-load, high-temperature rolling-element bearings).

The size and orientation of fibers in the polymer matrix are very important. For example, greater improvement in wear resistance can be achieved by orienting long, semicontinuous fibers normal to the surface. However, glass and other abrasive fibers should be oriented parallel to the sliding direction for reduced abrasion. The fibers also can be reinforced in chopped form so that their random orientations can produce enough resistance to the propagation of subsurface cracks.

5.4.1 Plastic Composites

5.4.1.1 Solid Fillers. Practically all plastics have been employed as matrices for polymer composites. PTFE has attracted the most attention because of its excellent frictional properties and chemical inertness. The wear rate of PTFE filled with many inorganic or organic fillers can be reduced by a factor of 100 to 1000 (Arkles et al., 1974; Mitchell and Pratt, 1957; Sung and Suh, 1979; Tanaka and Kawakami, 1982; and White, 1955). PTFE filled with polymeric fillers such as PPS and polyoxybenzoate and other fillers such as MoS_2 exhibits a low coefficient of friction (0.07) and a low wear rate (Arkles et al., 1974). Figure 5.14 shows the wear rate of PTFE composites worn against polished stainless steel (Ra = 0.15 μm cla) in a pin and ring test (Evans and Lancaster, 1979). Table 5.18 compares the physical properties and friction and wear performance of PTFE composites filled with inorganic and organic fillers. The filler additions to PTFE not only modify the friction and wear characteristics, but also modify the dependence of the wear rate on temperature (Evans, 1976). Figure 5.15 illustrates the influence of filler type on the dependence of the wear rate on temperature. This figure demonstrates that the wear rate of PTFE filled with inorganic fillers increases rapidly with temperatures over the range of 20 to 150°C, but for PTFE filled with organic fillers, the wear rate remains almost constant up to 150°C and

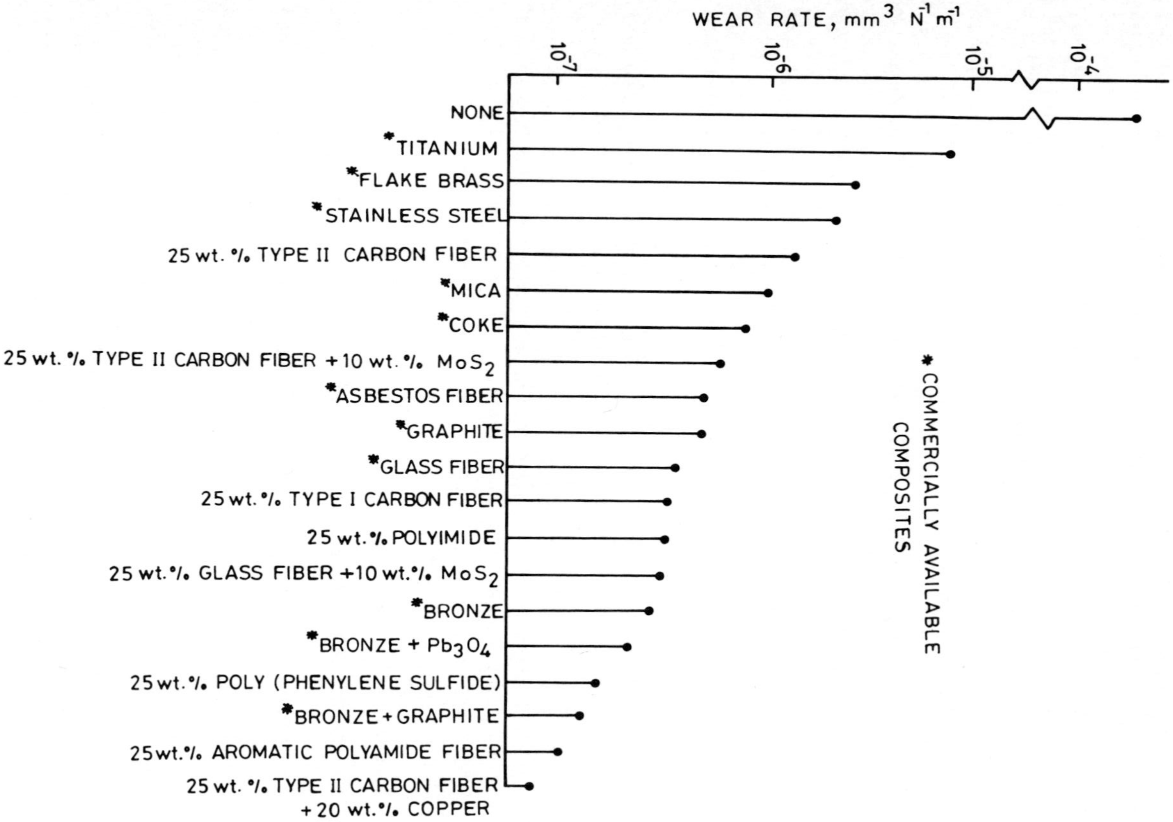

FIGURE 5.14 Effect of fillers on the wear rate of PTFE sliding against smooth steel (approximately 0.15 μm AA) based on data from pin-and-ring tests. [*Adapted from Evans and Lancaster (1979).*]

TABLE 5.18 Selected Properties of Filled and Unfilled PTFE

Fillers	Unfilled	15 wt % glass fiber	12.5 wt % glass fiber and 12.5 wt % MoS_2	15 wt % graphite fiber	20 wt % carbon and 5 wt % graphite fiber	55 wt % bronze and 5 wt % MoS_2*	25 wt % polyoxy-benzoate	40 wt % PPS	50 wt % PPS/carbon/MoS_2
Specific gravity	2.2	2.19	2.3	2.12	2.1	3.9	1.91	1.73	1.84
Hardness, Shore D	55	65–70	65	55	55–60	60–70	—	—	—
Tensile strength, MPa	9.0	17.5	13.0	9.5	11.6	13.0	12.4	11.0	8.3
Elongation, %	400	300	230	130	70	90	170	6	2
Flexural modulus, GPa	0.6	1.1	1.1	1.4	1.2	1.5	—	—	—
Thermal conductivity, W m^{-1} K^{-1}	0.25	0.43	0.51	0.45	0.44	0.72	—	—	—
Coefficient of thermal expansion, $\times 10^{-6}$ °C^{-1}	130	120	110	125	85	100	—	—	—
Wear coefficient, $\times 10^{-7}$ mm^3 $(N \cdot m)^{-1}$	4000	1.4	1.2	6.8	1.2	1.0	4.0	5.6	1.4
Coefficient of friction on steel at 0.01 m s^{-1}	0.1	0.09	0.09	0.12	0.12	0.13	0.05	0.07	0.07
PV (MPa · m s^{-1})									
At 0.05 m s^{-1}	0.04	0.33	0.50	0.35	0.53	0.44	0.52	0.38	0.52
At 5.0 m s^{-1}	0.06	0.5	0.62	0.95	0.42	0.44	0.43	0.26	0.26

*A commercial product of Crane Packing Company, Morton Grove, Ill.

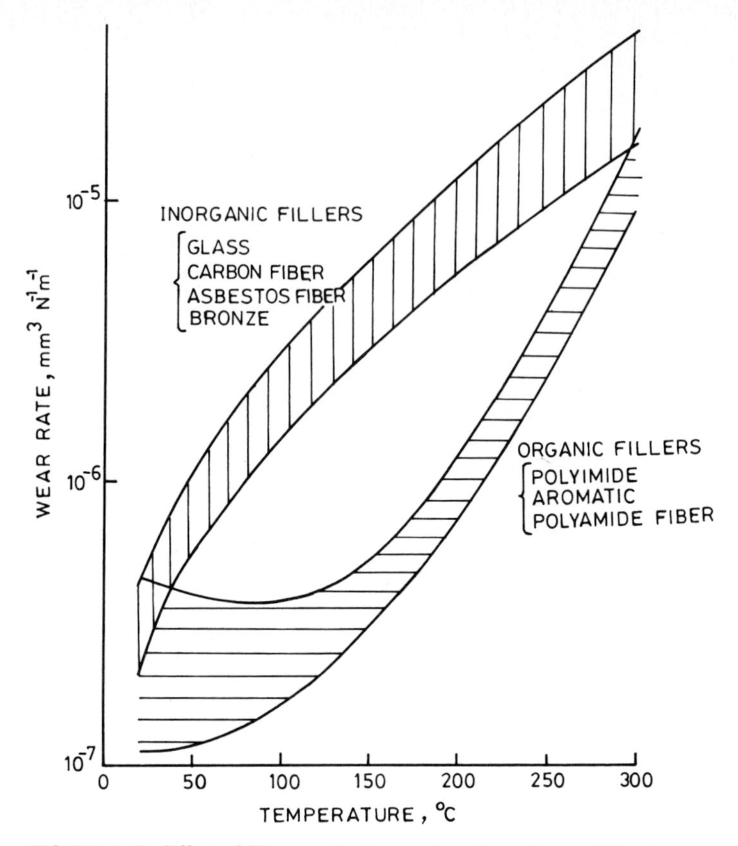

FIGURE 5.15 Effect of fillers on the temperature dependence of the wear rate of PTFE composites versus stainless steel. [*Adapted from Evans and Lancaster (1979).*]

then increases. This indifferent filler behavior is due to differences in nucleation and growth of the transfer film on the mating surfaces, inorganic fillers being relatively abrasive.

Istomin and Khruschov (1973) have studied the effect of graphite-fiber orientation in PTFE. They have reported that the wear rate of filled PTFE is highest when the basal planes of the graphite fibers are oriented parallel to the sliding surfaces and lowest when they are oriented normal to the surface. Figure 5.16 illustrates the effect of fiber orientation on carbon fiber–reinforced polyester sliding against a relatively smooth metal surface. Both the coefficient of friction and the wear rate are lower when the fibers are oriented normally to the sliding surface.

Certain fillers, such as glass and carbon are harder than the common construction metals, such as cast iron and mild steel, and thus may cause damage to the metal counterface. Cumulative damage caused by the composite can be attributed to the abrasiveness of the particular filler used and is an important factor to consider when selecting a material for use against a relatively soft metal such as aluminum. The degree of surface damage caused by a filler is mainly dictated by

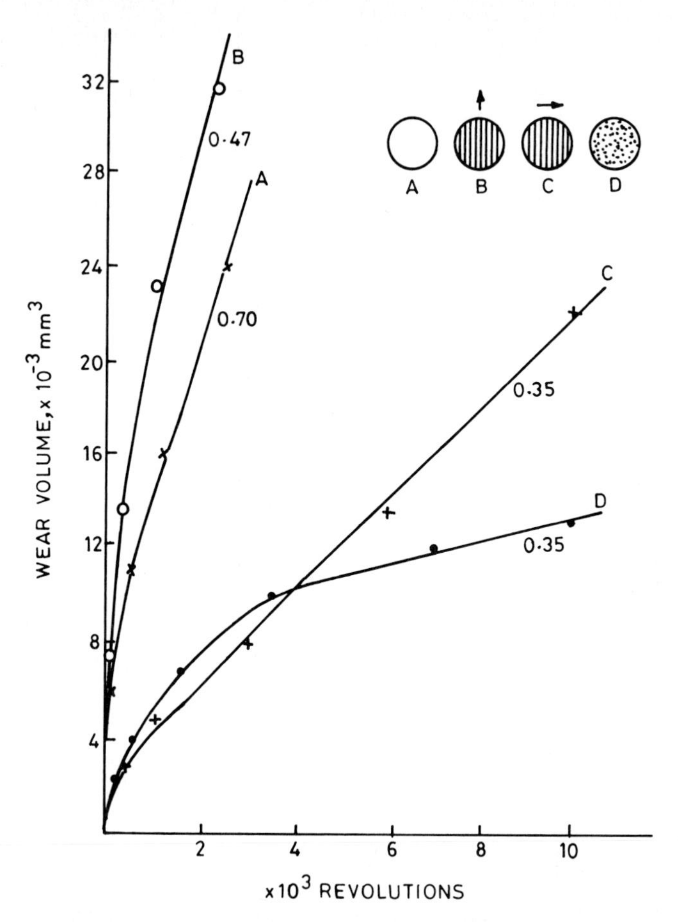

FIGURE 5.16 Effect of carbon fiber orientation in reinforced polyester resin (25 wt % fiber) slid against hardened tool steel. The coefficients of friction are given adjacent to each curve. [*Adapted from Lancaster (1968).*]

its hardness (Lancaster, 1972c) and its shape (Speerschneider and Li, 1962a, 1962b). Table 5.19 lists the relative abrasiveness of a variety of polymer composites as determined from wear measurements against brass balls. PTFE composites reinforced with glass and mineral fibers exhibit more abrasive characteristics than the carbon-, graphite-, or bronze-filled composites. Of the two carbon fiber fillers (high modulus, type I; and high strength with low modulus, type II), the former, which has a more graphitic structure, is comparatively less abrasive than the latter. The hardness of the counterface metal (ball) plays a key role in determining the abrasiveness of a particular filler, as shown in Figure 5.17 (Anderson and Davies, 1982). This helps designers select suitable polymer composite and counterface combinations for the least wear and friction.

Bhushan and Wilcock (1981, 1982) have carried out a systematic study of the effect of various fillers, such as PTFE powder, MoS_2 powder, graphite fiber and powder, Kevlar-aramid fiber, glass fiber, asbestos fiber mixed with several poly-

TABLE 5.19 Relative Abrasiveness of Polymer Composites Sliding Against a Brass Ball (200 HV)

Class	No.	Composite type	Wear coefficient vs. brass ball, $\times 10^{-15}$ $m^3 (N \cdot m)^{-1}$
	1	PTFE/mineral fibers	29
	2	PTFE/mineral fibers plus other filler	4.2
	3	PTFE/40 wt % glass fibers	16
	4	PTFE/25 wt % glass fibers	8.3
	5	PTFE/15 wt % glass fibers	6
Filled PTFE	6	PTFE/20 wt % glass fibers plus 5 wt % MoS_2	7
	7	PTFE/15 wt % glass fibers plus 10 wt % MoS_2	6
	8	PTFE/graphite-carbon	3
	9	PTFE/60 wt % bronze powder	0.2
	10	PTFE/25 wt % carbon	0.05
	11	PTFE/50 wt % graphite	0.02
Carbon fiber composites	12	Polyimide/type II (high strength) fibers	8.1
	13	Polyimide/type II (high strength) fibers	5
	14	PTFE/type I (high modulus) fibers	0.13
	15	Polyamide/type I (high modulus) fibers	0.1
Steel-backed thin-layer materials	16	Mineral-filled PTFE/bronze mesh	10
	17	Woven PTFE fiber/glass fiber	8
	18	PTFE/lead/bronze, sintered	0.75
	19	PTFE flock/Nomex fabric	0.17
Reinforced thermosets	20	Polyester resin/textile laminate	2.7
	21	Asbestos yarn/graphite	0.14
	22	Phenolic resin/cotton fabric/PTFE	0.04

Source: Adapted from Anderson and Davies (1982).

mers such as PTFE, polyimide, poly(amide-imide), PPS, linear aromatic polyester, and phenolic. Both commercially available and custom-made composites were used. Among the custom-made composites, PTFE was filled with either graphite or glass fibers and with CdO-graphite-Ag powder, which consisted of 65 percent by weight of 99.9 percent pure synthetic graphite with a particle size of about 2 μm, 25 percent by weight of 99.9 percent pure CdO with a particle size of about 95 to 200 μm, and 10 percent by weight of 99 percent pure Ag with a particle size of about 1 to 5 μm. The CdO-graphite-Ag powder composition was initially developed for solid-lubricant coatings (Bhushan, 1982). Fiber Materials, Inc. (FMI), New Biddford, Maine, produced a three-dimensional weave of graphite fibers and impregnated them with polyimide and phenolic resins. Graphite weaves impregnated with graphite and carbon and MoS_2 compacts (commercially available from Pure Carbon Company, St. Marys, Pa.) also were tested for comparisons.

Accelerated wear tests (to simulate engine seals) were conducted using a reciprocating tester operating at 1200 cycles per minute with a stroke of 40 mm in a nitrogen environment. A circular flat-ended polymer pin was slid against a flat, hard-nitrided steel plate in reciprocating motion at a normal stress of about 0.7 MPa and an average sliding speed of 1.6 m s^{-1}. Thermocouples about 0.13 mm in diameter were embedded in the pins at a distance of 0.64 mm from the rubbing

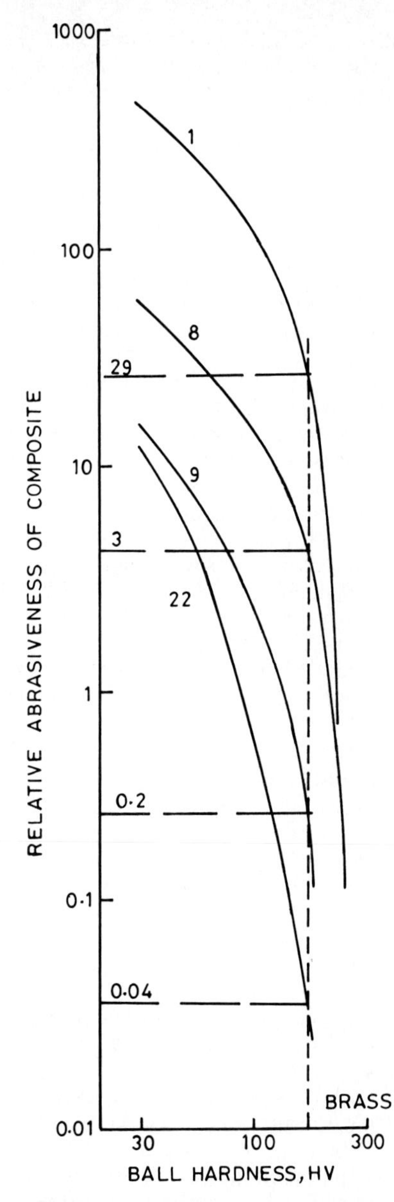

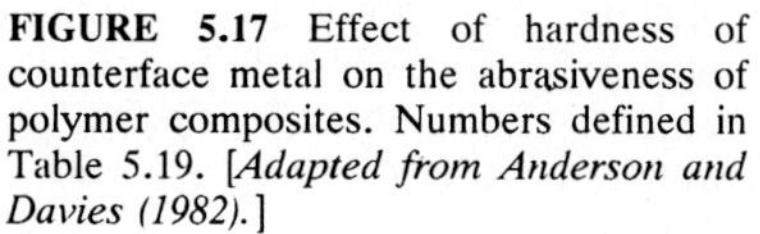

FIGURE 5.17 Effect of hardness of counterface metal on the abrasiveness of polymer composites. Numbers defined in Table 5.19. [*Adapted from Anderson and Davies (1982).*]

surface to measure temperature rise due to frictional heat. Test duration typically was 100 hours. The coefficient of friction was measured using a unidirectional sliding tester with a conforming pin sliding on a rotating cylinder at a normal load of about 0.7 N (or stress ~0.7 MPa) and a sliding velocity of 1.2 m s^{-1}. The friction and wear data for various plastics are summarized in Table 5.20.

Bhushan and Wilcock (1981, 1982) noted that the coefficient of friction of PTFE composites was essentially independent of *PV* limit in the range from 0.2 to 2.2 MPa · m s^{-1} (also see Tanaka, 1977). Temperature rise in the pin was expected to be related to the coefficient of friction of frictional energy input (top of Fig. 5.18) and thermal conductivity of the pin. Assuming a one-dimensional heat flow in the pin, the friction should be a linear function of thermal conductivity times the temperature rise. The data are plotted at the bottom of Figure 5.18 for the materials whose thermal conductivities were known. The correlation was generally good, except for M_8. Based on the correlations observed, the temperature rise can be used as a crude indicator of the coefficient of friction, provided the thermal conductivities of the materials tested are known.

The coefficient of friction and wear data reported in Table 5.18 are plotted in Fig. 5.19. It is not surprising to note that there is no definite correlation between friction and wear characteristics. However, PTFE composites with low friction also had low wear rates, with the exception of PTFE filled with carbon powder. In all other composites, those with high wear had a high coefficient of friction, with the exceptions of PPS filled with 15 percent PTFE and 30 percent carbon fiber. Polymers with the lowest friction and wear* were (1) PTFE filled with 10 percent glass fibers and 15 percent CdO-graphite-Ag (M_{12}), (2) PTFE filled with 10 percent graphite fibers and 10 percent CdO-graphite-Ag (M_{14}), and (3) PTFE filled with glass fibers and other fillers (M_4, Rulon E, made by Dixon Corporation, Bristol, R.I.). Bhushan and Wilcock also noted that polymer materials with low friction and wear formed a thin, uniform, and tenacious transfer film on the mating metal surface. It is well known that among polymers, PTFE forms a transfer film most readily (Craig, 1964; Lancaster, 1973; Steijn, 1966). The transfer film thickness for PTFE composites was estimated to be about 0.5 to 2 μm. The film was adherent and could not be scraped off easily.

In most cases, some abrasion of the metal surface occurred. Lancaster (1973) reported that most fillers and reinforcing fibers are abrasive when used against metals, including lamellar solid-lubricant fillers such as graphite and MoS_2. In the experiments of Bhushan and Wilcock, wear of the metal surface was readily observed in the obliteration of the random polishing scratches on the surfaces. Figure 5.20 shows the differences in scratch depth between the end and middle sections of the rub track, as well as between the areas outside and within the track.

Bhushan and Wilcock also reported that wear rates of PTFE composites at ambient temperatures of 52 to 78°C were two to four times greater than those at room temperature (~22°C). Briscoe and Steward (1977) have shown that oxygen and water vapor increase the wear of carbon fiber–filled PTFE composites.

The surface roughness and orientation of the grooves on the mating material in relation to the direction of sliding have a significant influence on wear rate. Metallurgical examinations of worn surfaces by Bhushan and Wilcock showed that the rougher mating metal surface had a thicker transfer film buildup, which may be responsible for the lower friction. The influence of surface roughness on wear rate is illustrated by Figure 5.21. Initial wear is high with excessive surface

*At *PV* limits higher than those used here (>1.1 MPa · m s^{-1}), PTFE composites lose mechanical integrity and higher-temperature polymer composites would outperform PTFE composites.

TABLE 5.20 Friction and Wear Data for Plastic Composites from Reciprocating Sliding Tests

Material no.	Pin material	Coefficient of friction*		Pin temperature rise during test, °C	Specific plastic wear rate,† $\times 10^{-6}$ mm^3 $(N \cdot m)^{-1}$	CLA surface roughness of plate after the test, μm		Surface examination after test
		Static	Kinetic			Longitudinal‡	Transverse§	
M_1	PTFE with glass fibers and other fillers, radial cut (Rulon LD from Dixon)	0.31	0.30	10	0.40	0.05	0.05	Transfer film on plate, smooth all over, pin quite smooth
M_2	PTFE with glass fibers and other fillers, axial cut (Rulon LD)	0.28	0.31	9	0.518	0.051	0.076	Thick coating with patchy appearance on plate
M_3	PTFE filled with polyimide powder (Rulon J)	—	0.23	9	0.16	0.08	0.25	Transfer film nonuniform, longitudinal grooves, corresponding grooves on pin
M_4	PTFE with glass fibers and other fillers (Rulon E)	0.29	0.29	7	0.13	0.03	0.05	Coating very smooth and shiny, pin smooth
M_5	PTFE filled with 15 wt % graphite powder (Dixon)	—	—	7	29.9	0.30	0.50	Center of plate covered with nonuniform, patchy coating, pin smooth
M_6	PTFE filled with carbon powder (Dixon 7035)	0.36	0.41	8	0.07	0.08	0.10	Ultrathin coating, plate essentially unchanged, pin smooth
M_7	PTFE filled with 15 wt % graphite fiber (Dixon)	0.23	0.33	12	0.81	0.025	0.051	Very thin and uniform coating on plate, pin deformed during test
M_8	PTFE filled with 55 wt % bronze powder and 5 wt % MoS_2 (Crane Packing)	0.24	0.30	8	0.17	0.025	0.051	Thin and uniform coating on plate, pin smooth, bronze particles clearly visible

TABLE 5.20 Friction and Wear Data for Plastic Composites from Reciprocating Sliding Tests (*Continued*)

Material no.	Pin material	Coefficient of friction*		Pin temperature rise during test, °C	Specific plastic wear rate,† $\times 10^{-6}$ $mm^3 (N \cdot m)^{-1}$	CLA surface roughness of plate after the test, μm		Surface examination after test
		Static	Kinetic			Longitudinal‡	Transverse§	
M_9	PTFE filled with 25 wt % glass fiber (Crane Packing)	0.29	0.29	9	0.35	0.051	0.051	Very thin coating on plate, pin smooth
M_{10}	PTFE filled with glass fiber (GL-15 from Polypenco)	0.34	0.35	7	0.30	0.025	0.076	Thick and uniform plastic coating on plate
M_{11}	PTFE with 25 wt % glass fibers and additions of complex salts (Fluon VX1 from ICI)	0.29	0.31	8	0.54	0.025	0.102	Thin coating and some scratches on plate
M_{12}	PTFE filled with 10 wt % glass fibers and 15 wt % CdO-graphite-Ag (Dixon)	0.25	0.28	7	0.13	0.04	0.05	Ultrathin coating on plate, slightly thicker in the center, pin smooth
M_{13}	PTFE filled with 10 wt % glass fibers and 15 wt % CdO-graphite-Ag (Tribol)	0.30	0.31	9	0.19	0.025	0.05	Ultrathin coating, no apparent polishing of plate, pin smooth
M_{14}	PTFE filled with 10 wt % graphite fibers and 10 wt % CdO-graphite-Ag (Dixon)	0.15	0.22	8	0.15	0.025	0.05	Heavy and quite uniform coating on plate, fine grooves on pin
M_{15}	PTFE filled with 10 wt % graphite fibers and 10 wt % CdO-graphite-Ag (Tribol)	0.15	0.19	7	0.18	0.025	0.025	Thin, nonuniform coating on plate, pin smooth
M_{16}	Polyimide filled with 15 wt % graphite powder (SP-21 from Du Pont)	0.47	Erratic	20	0.38	0.30	0.20	Heavy coating in patches, pin generally smooth
M_{17}	Thermoplastic polyimide with powder fillers (Rulon II)	—	0.33	9	0.83	0.10	0.30	Coating very spotty and nonuniform, longitudinal grooves on pin

M_{18}	Polyimide filled with 15 wt % graphite powder (Envex 1315 from Rogers)	0.49	0.53 (erratic)	17	105.61	—	—	
M_{19}	Polyimide with 50 wt % graphite powder (Filimide from TRW)	—	—	7	0.194	0.152	0.254	Thick coating with patchy appearance on plate, metal worn in spots, pin smooth
M_{20}	Polyimide filled with 15 wt % CdO-graphite-Ag (Dixon)	0.37	0.38	9	0.21	0.089	0.152	Heavy coating on longitudinal patches, plate topography unchanged, longitudinal grooves on pin
M_{21}	Polyimide (PI 500) filled with 15 wt % CdO-graphite-Ag (Tribol)	0.38	0.39	8	0.30	0.102	0.23	Heavy plastic transfer in patches on plate, longitudinal grooves on pin
M_{22}	Poly(amide-imide) filled with 3 wt % PTFE and 20 wt % graphite powder (Torlon 4275 from Amoco)	0.34	0.34	11	0.42	0.05	0.05	Ultrathin but uniform film, many grooves on pin
M_{23}	Poly(amide-imide) filled with 3 wt % PTFE and 12 wt % graphite powder (Torlon 4301)	0.39	0.39	12	25.6	0.10	0.30	Pieces of black powder stuck on the plate, no evidence of film, some grooves on pin
M_{24}	Polyphenylene sulfide filled with 15 wt % PTFE and 30 wt % carbon fiber (Thermocomp OCL-4036 from LNP)	0.39	Erratic	19	0.16	0.05	0.05	Ultrathin coating, fine grooves on pin
M_{25}	Phenolic filled with 15 wt % graphite powder (Cadillac Plastic)	0.69	0.61	>13	14.1	0.20	0.23	Heavy and spotty coating on plate, longitudinal grooves on pin
M_{26}	Linear aromatic polyester filled with 20 wt % graphite powder (Ekkcel CL-1340 from Carborundum)	0.54	Erratic	19	25.7	0.30	0.48	Heavy coating in patches, no apparent polishing of plates, longitudinal grooves on pin

TABLE 5.20 Friction and Wear Data for Plastic Composites from Reciprocating Sliding Tests (*Continued*)

Material no.	Pin material	Coefficient of friction*		Pin temperature rise during test, °C	Specific plastic wear rate,† $\times 10^{-6}$ $mm^3 (N \cdot m)^{-1}$	CLA surface roughness of plate after the test, μm		Surface examination after test
		Static	Kinetic			Longitudinal‡	Transverse§	
M_{27}	Polyimide with 50 vol % three-dimensional graphite weave (FMI)	0.54	0.49 (erratic)	14	0.49	0.051	0.051	Plate has some longitudinal scratches, no plastic transfer, weave in plastic pin visible
M_{28}	Phenolic with 50 vol % three-dimensional graphite weave (FMI)	0.46	0.39 (erratic)	—	0.75	0.025	0.203	Numerous longitudinal scratches on plate, no plastic transfer, visible weave in plastic pin
M_{29}	Graphite with 50 vol % three-dimensional graphite weave (FMI) (for comparisons)	0.21	0.22	16	0.33	0.102	0.102	Streaked black coating on plate, numerous cracks on pin
M_{30}	Impregnated carbon graphite (P5N) (for comparisons)	0.27	0.26	7	0.122	0.025	0.051	Plate unchanged, grainy appearance of pin
M_{31}	Ta-based alloy filled with MoS_2 (Molalloy PM 107) (for comparisons)	0.45	Erratic	—	115.21	0.051	0.457	Graphite patches on plate

*Unidirectional sliding tests: pin-on-a-cylinder; no water cooling; load, 0.69 N; sliding speed, 1.2 m s^{-1}.

†Reciprocating sliding test: pin-on-a-plate; average sliding speed, 1.6 m s^{-1} (1200 strokes min^{-1}); speed range, 0–2.5 m s^{-1}; stroke length, 40 mm; normal stress, 0.70 MPa; duration of test, 100 h; environment, nitrogen atmosphere at room temperature; plate material, nitrided water-cooled SAE 7140 steel (hardness, 62 HRC); cla roughness of plates before testing, 0.05 μm; for the test conditions used, 1×10^{-6} $mm^3 (N \cdot m)^{-1} \approx 4$ μm h^{-1}.

‡Traverse, 3.8 mm.

§Traverse, 1.8 mm.

Source: Adapted from Bhushan and Wilcock (1982).

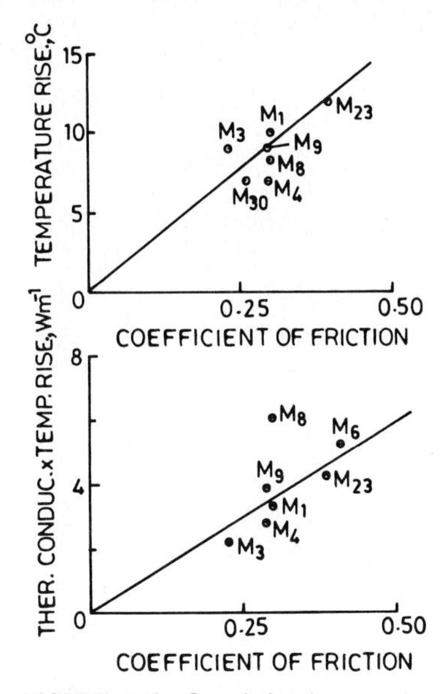

FIGURE 5.18 Correlation between interface temperature rise and coefficient of friction for plastic composites. [*Adapted from Bhushan and Wilcock (1981).*]

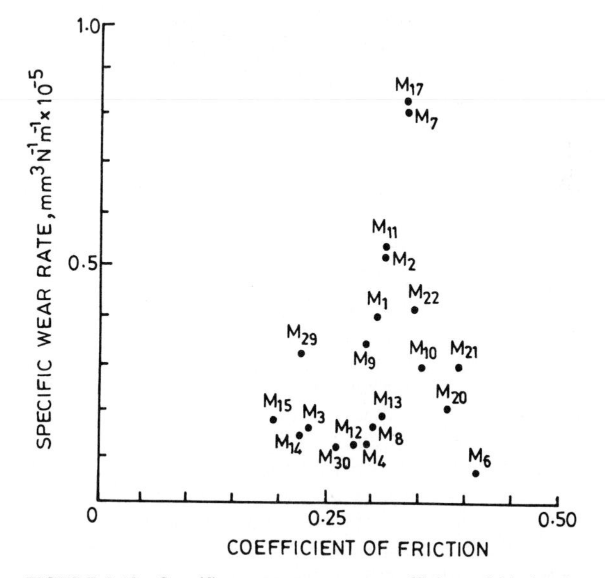

FIGURE 5.19 Specific wear rate versus coefficient of friction for the polymer composites. The numbers represent the material numbers in Table 5.20. [*Adapted from Bhushan and Wilcock (1982).*]

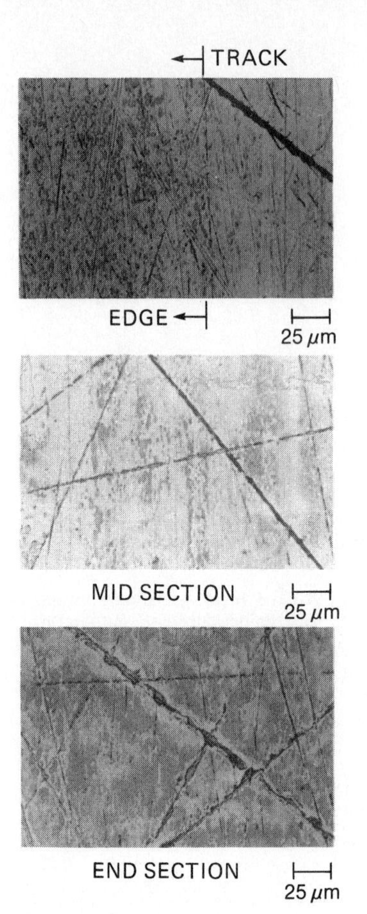

FIGURE 5.20 Optical micrographs of mating metal plate after test against PTFE, 10% glass fibers and 15 wt % CdO-graphite-Ag (M_{12}). [*Adapted from Bhushan and Wilcock (1981).*]

roughness because it takes more material to pack the grooves. Once the adherent transfer film is built up, the wear rate decreases and the subsequent wear rate of a rougher surface is lower than that of a smooth surface (Bhushan and Wilcock, 1982; Lancaster, 1973; Play and Godet, 1979). Transverse grooves have more initial and steady-state wear than longitudinal grooves owing to a more marked abrading (cutting) action. Materials with random roughness show high initial wear and longer time for buildup of the transfer film. Therefore, a mating surface with a relatively high surface roughness (~0.8 μm) for polymers at the conditions tested and grooves in the direction of sliding are recommended.

Fusaro (1988) tested various plastic composites both as solids and as coatings for space applications. A pin-on-disk test apparatus with the disk specimens made of polymers and the pin specimens made of AISI 440C stainless steel with

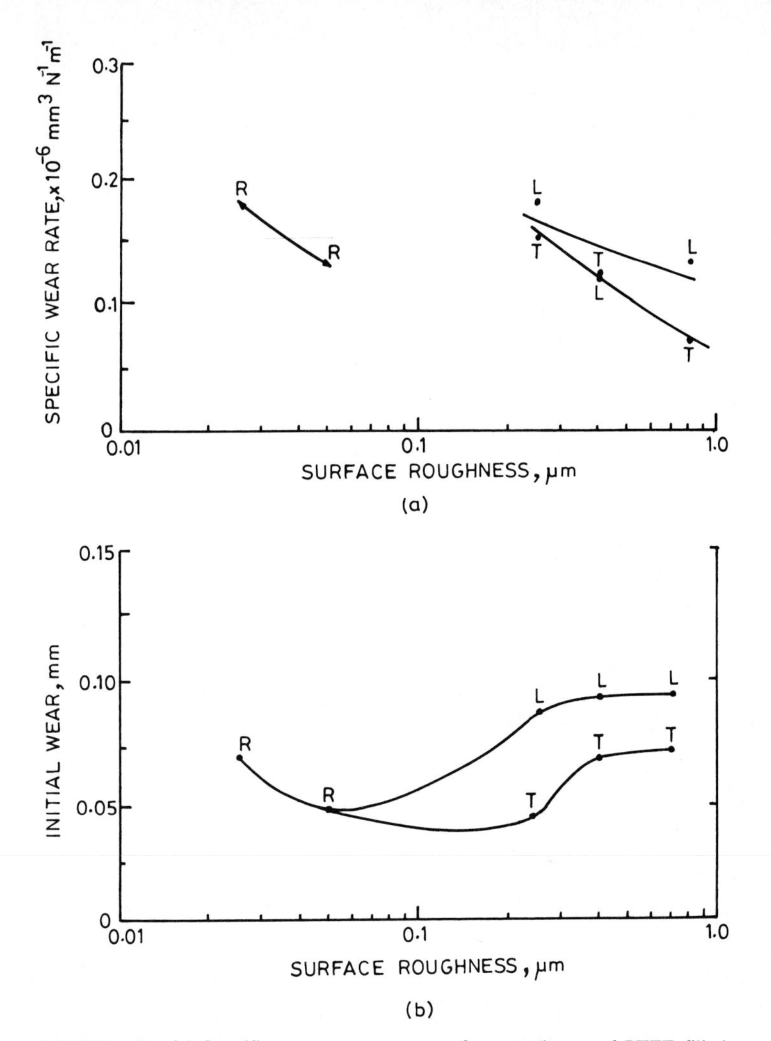

FIGURE 5.21 (*a*) Specific wear rate versus surface roughness of PTFE filled with 10% glass fiber plus 15 wt % CdO-graphite-Ag (M_{12}) slid against nitrided SAE 7140 steel in reciprocating mode in a nitrogen atmosphere. (*b*) Initial wear versus roughness of same combination in nitrogen atmosphere (*R*, random; *L*, longitudinal; and *T*, transverse). [*Adapted from Bhushan and Wilcock (1982).*]

a hardness of 58 HRC was used for this study. Figure 5.22 presents the friction and wear data for 10 polymers as solids and coatings tested in air (50 percent relative humidity) and vacuum (0.13 and 1.3×10^{-4} Pa or 10^{-3} and 10^{-6} torr) at room temperature. The results show that all polymer materials evaluated, except for 80 percent PMDA/20 percent BTDA polyimide, 100 percent PMDA polyimide, and polyimide filled with 50 percent graphite fibers, produced lower coefficients of friction in vacuum than in air. Polyimides without any solid-lubricant additives produced lower wear rates in vacuum than in air. Solid-

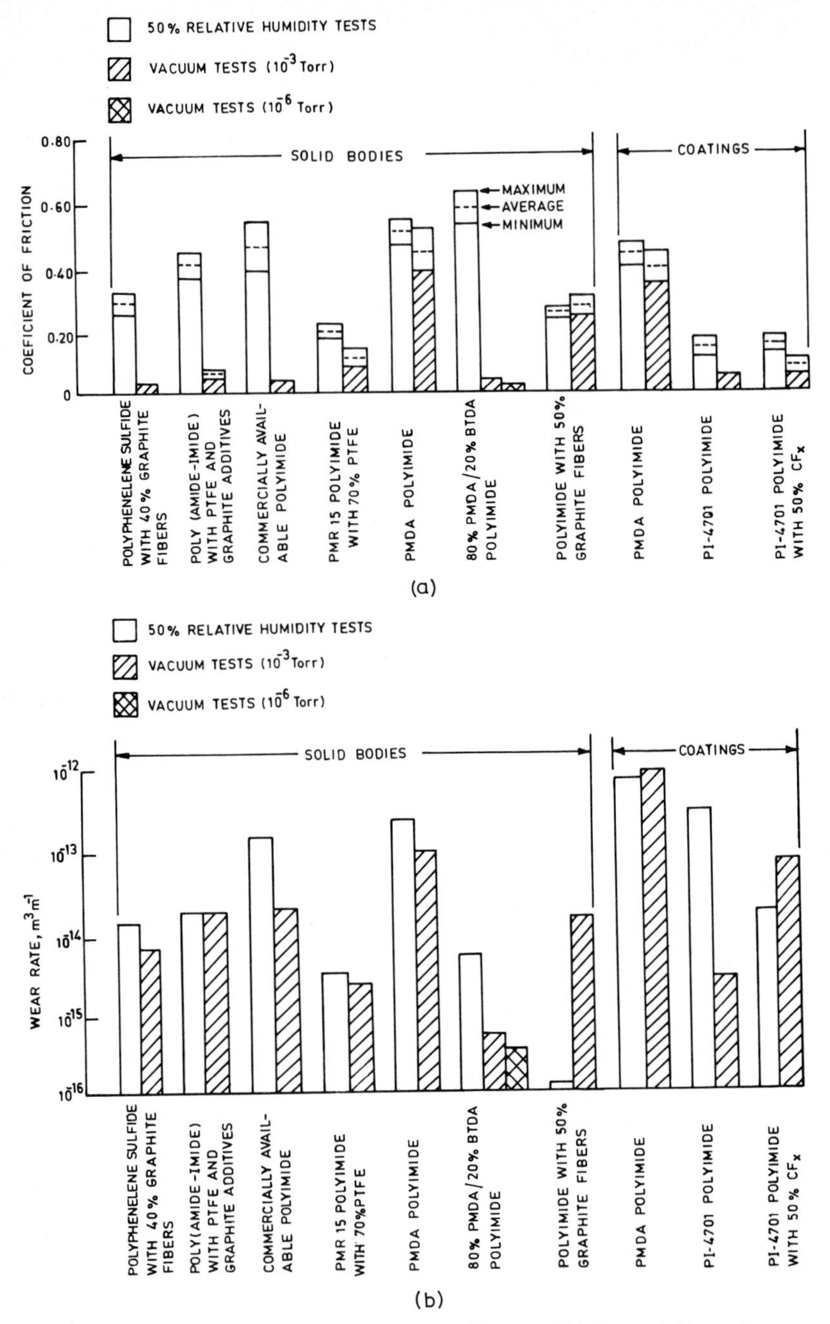

FIGURE 5.22 Comparisons of (*a*) steady-state coefficients of friction and (*b*) steady-state wear rates in air and in vacuum for different polymer materials sliding against an AISI 440C stainless steel (HRC 58) pin in a pin-on-disk test apparatus at a normal load of 9.8 N and a speed of 100 r/min. [*Adapted from Fusaro (1988).*]

lubricant additives in polymers tend to produce weak points, where cracks could be initiated and spallation could occur under sliding conditions. The friction and wear properties of most polyimides are good, and solid-lubricant additives are not necessary. Of the 10 polymer materials evaluated, the 80 percent PMDA/20 percent BTDA solid-body polyimide and the PI-4701 polyimide coating exhibited the best friction and wear performance in a vacuum (space) environment.

Gardos and McConnell (1981, 1982) developed a polyimide-based composite for the cage of a rolling-element bearing to be operated at high temperature (315°C) without an external lubricant. Type II carbon/graphite fibers were selected as the preferred reinforcement because of their better tribological properties and high strength at the elevated temperatures. A Ga-In-WSe_2 material that was developed as a high-temperature lubricant compact was found to be an excellent solid-lubricant additive. Dibasic ammonium phosphate [$(NH_4)_2HPO_4$] was demonstrated to be an excellent adjuvant for enhancing the properties not only of the fibers, but also of the composite itself. Three-dimensional, orthogonal, carbon-fiber-weave preforms were fabricated and then impregnated with the polyimide-based material containing powdered versions of the Ga-In-WSe_2 self-lubricating compact and the dibasic ammonium phosphate adjuvant. This cage material had a strength of about 210 MPa at 315°C and exhibited low friction and wear when tested against a nickel-based alloy (Rene 41) at a normal stress of about 175 MPa and a temperature of 315°C (Gardos and McConnell, 1982).

Bhushan and Winn (1979) and Bhushan and Dashnaw (1981) tested various plastic composites for oil-lubricated and seawater-lubricated (for the future) stern-tube bearings and face seals in commercial ships that operate at typical pressures ranging from 0.35 to 0.60 MPa and surface velocities ranging from 1.8 to 7.2 m s^{-1}. To date, tin-based babbitt or asbestos-reinforced, filament-wound phenolic-resin composites (made by Railko Company) are used as bearing materials, and mild steel (ABS grade 2) is used as journal material in the oil-lubricated environment. Face seals of the phenolic composite and mating rings usually made of bronze or Ni resist (a group of austenitic high-nickel or nickel-copper cast alloys) are used in the oil-lubricated environment. Bhushan and Winn selected thermosetting and high-temperature thermoplastics with high *PV* limits because these bearings and face seals operate at high *PV* limits. The plastics were reinforced with fibers and filled with solid-lubricant fillers for excellent high-temperature strength and lubrication properties (Craig, 1964; Hurren et al., 1963–64; Pinchbeck, 1962; Smith, 1967, 1973). Both commercial and custom-made composites were selected. A polyester-PTFE coating with the trade name Karon V (produced by Kamatics Corporation, Bloomfield, Conn.), nitrile elastomer, and tin-based babbitt also were included for comparison.

Selected plastics were subjected to soak (swell and accelerated aging) tests and accelerated friction and wear tests. The soak tests were designed to determine the geometric stability and compatibility of the polymers with the operating fluids. The specimens were soaked in oil and seawater at 90 and 120°C for oil and 40 and 90°C for seawater for 12 months. During the 12-month period, the Shore hardness, weight, thickness, and diameter of the specimens were measured periodically. The following materials were found to be poor in the soak tests: UHMW polyethylene (poor in seawater only); polyimide, 30 wt % PTFE; polyimide, 15 wt % graphite, 10 wt % PTFE (poor in seawater only); linear aromatic polyester, 20 wt % graphite (poor in seawater only); and acetal, 30 wt % PTFE. Another two polyimides, one with Kevlar-aramid fibers and the other with graphite powder, did well. It is believed that material degradation is due to high porosity in molding and/or wettability of the fillers at the interface. Some fillers absorb more water and do not bond well to the resin, resulting in high porosity and the absorption of more water.

Accelerated friction and wear tests were conducted in a unidirectional sliding mode at very high *PV* limits (a normal stress of 7.2 MPa and a sliding speed of 6.1 m s^{-1}). The tests were first conducted on plastics against mild steel (ABS grade 2) in the presence of petroleum-based turbo oil 78. Selected plastics were then tested against a corrosion-resistant Nitronic 50 steel in the presence of synthetic seawater. The friction and wear data are presented in Table 5.21. No wear of the mating metal surface was observed. The following three materials showed low

TABLE 5.21 Friction and Wear Data for Plastic Composites from Unidirectional Sliding[a] Tests in Oil and Seawater

Pin material	Static friction coefficient		Kinetic friction coefficient		Total linear wear, μm
	At start	At stop	Start of test	End of test	
Tested against mild steel (ABS grade 2)[b] in petroleum-based oil (Turbo oil 78)					
Phenolic, asbestos fiber filled[c] (Railko WA 80H)	0.021	0.15	0.020	0.14	536
Phenolic (with linen fabric), 15 wt % graphite powder filled (Cadillac Plastic)	0.025	0.047	0.011	0.018	5.1
Phenolic, 65 wt % Kevlar-aramid 49 fiber filled (Fiber Materials, Inc.)	0.07	0.09	0.037	0.026	223
Graphitized canvas-base phenolic laminate (Insurok T-768)	0.043	0.058	0.015	0.017	15.2
Polyimide, 15 wt % graphite powder filled (SP-21)	0.041	0.022	0.021	0.008	5.2
Polyimide, 30 wt % PTFE powder filled (Meldin PI-30x)	0.040	0.064	0.028	0.034	17.8
Polyimide, 15 wt % graphite and 10 wt % PTFE powder filled (PI 120)	0.040	0.057	0.023	0.014	5.1
Polyimide, 20 wt % Kevlar-aramid fiber filled (Tribol)	0.048	0.057	0.012	0.020	33.0
Poly(amide-imide), 3 wt % PTFE and 12 wt % graphite powder filled (Torlon 4301)	0.039	0.051	0.015	0.009	5.1
Polyphenylene sulfide, 25 wt % glass and 25% PTFE powder filled (Ryton NR 09)	0.026	0.052	0.024	0.021	17.8
Linear aromatic polyester, 20 wt % graphite powder filled (Ekkcel CL-1340)	0.042	0.048	0.007	0.008	7.6
UHMW Polyethylene (Cadco 1900)	0.017	0.030	0.028	0.021	45.7
Polyester with PTFE powder coating (Karon V)	0.011	0.085	0.041	0.063	17.8
Tin-based babbitt, ASTM grade 2[d] (for comparisons)	0.026	0.035	0.012	0.012	<2.5

TABLE 5.21 Friction and Wear Data for Plastic Composites from Unidirectional Sliding[a] Tests in Oil and Seawater (*Continued*)

Pin material	Static friction coefficient		Kinetic friction coefficient		Total linear wear, μm
	At start	At stop	Start of test	End of test	
Tested against Nitronic 50[e] in synthetic seawater					
Phenolic with linen fabric, 15 wt % graphite powder filled (Cadillac Plastic)	0.026	0.042	0.014	0.013	5.1
Polyimide, 15 wt % graphite powder filled (SP-21)	0.075	0.17	0.009	0.007	5.1
Polyimide, 15 wt % graphite and 10 wt % PTFE powder filled (PI 120)	0.09	0.25	0.012	0.012	15.2
Poly(amide-imide), 3 wt % PTFE and 12 wt % graphite powder filled (Torlon 4301)	0.05	0.13	0.007	0.009	2.5
Linear aromatic polyester, 20 wt % graphite powder filled (Ekkcel CL-1340)	0.056	0.11	0.006	0.007	10.2
Butadiene acrylonitrile Copolymer (Buna-N or Nitrile)[f] (for comparisons)	0.064	0.35	0.008	0.009	12.7

[a]Pin-on-cylinder test: sliding speed, 6.1 m s^{-1}; normal stress, 7.2 MPa; ambient temperature, 27°C; test duration, 24 h.
[b]Surface roughness, 0.25 μm cla.
[c]The test was repeated with fibers perpendicular to sliding direction as opposed to parallel to sliding direction in test 1, and the results were identical.
[d]Fe, 0.08%; Sn, 89%; Sb, 7.5%; Cu, 3.25%; Pb, 0.35%; Al, 0.005%; As, 0.1%; Bi, 0.08%.
[e]Surface roughness, 0.15 μm cla.
[f]Normal stress reduced to 3.76 MPa because specimen bent during test at 7.24 MPa.
Source: Adapted from Bhushan and Dashnaw (1981).

friction and wear both in oil and in seawater: polyimide filled with 15 wt % graphite powder, phenolic filled with 15 wt % graphite powder, and poly(amide-imide) filled with 3 wt % PTFE and 12 wt % graphite powder. It should be noted that the coefficients of friction in oil and seawater are comparable, but the wear rates in seawater are substantially higher than those in oil. Bhushan and Winn also reported friction and wear data from tests conducted against flame-sprayed Cr_2O_3 and Niresist type 2 (cast iron), and the data were comparable with those for mild steel.

Note also that polyimide, poly(amide-imide), and phenolic composites performed poorly in dry sliding (reported earlier), apparently because they do not form smooth and uniform transfer films, which appear to be necessary for dry lubrication.

5.4.1.2 ***Liquid-Lubricant Fillers.*** So far we have discussed the role of solid fillers. Liquid lubricants are also added to microporous polymers (referred to as *microporous polymer lubricants* or *MPLs*) to substantially reduce friction and wear. Proper lubrication with some oils produces a coefficient of friction as low as 0.02, which is much less than the lowest friction value for polymers (about

0.04), that for PTFE (Braithwaite, 1967). MPLs store and provide small quantities of lubricants to the interface at a controlled rate. MPLs provide the benefits and reliability of liquid-lubricant application, yet eliminate the storage, sealing, and pumping problems associated with liquid lubricants.

MPLs are spongelike, three-dimensional networks of plastics whose pores are filled with lubricating oils. The pores are so small (1 to 10 μm in diameter) that the oil can be retained within them by surface-tension forces. The ratio of oil to polymer can be adjusted from pure polymer to as much as 80 percent oil. The pores act as reservoirs to store oil adjacent to the interface (requiring lubrication) and dispense the lubricant as well (Jamison, 1985).

When MPLs are heated by frictional heat, the oil and polymer phases attempt to expand at rates dictated by their bulk thermal expansion. If the bulk thermal expansion of the polymer is less than that of the oil, the elastic polymer network will stretch and will exert a force that attempts to regain its unstretched dimension, thus creating an internal pressure in the oil. If the oil wets the polymer, the surface-tension forces will cause the oil to rise according to the wetting angle between the oil and the polymer. The effect of surface-tension forces in MPLs is to cause the oil to flow to the surface when the MPLs are performing their lubricant-dispensing functions and to reabsorb the excess lubricant during static periods.

The materials commonly used for MPLs are ultrahigh-molecular-weight polyethylene (thermoplastic) and synthetic hydrocarbon oils. Oil content in the range of 50 to 80 percent is commonly used. Polyethylene MPLs can be used at operating temperatures up to about 100°C. To manufacture MPLs, polyethylene and hydrocarbon oil are blended to form a gel. The polyethylene gel can then be formed by injection or compression molding into desired shapes. Alternatively, injection-molding techniques can be used to fill existing voids or specially manufactured chambers adjacent to bearing surfaces. For example, in a ball bearing, all the natural voids would be filled with an MPL instead of the usual grease pack. In a sleeve bearing, MPLs are employed as plugs that are inserted into holes drilled in the surface of the bearing. A commercial MPL based on polyethylene and synthetic hydrocarbon oil is sold by SKF Industries, King of Prussia, Pa., as Poly-Oil bearing a trademark of General Polymeric Corporation, Reading, Pa. Other MPLs sold commercially are Nyloil (Nylon and oil) and Nylasint (Nylon, MoS_2, and up to 25 percent hydrocarbon oil). Nylasint is a trademark of General Polymeric Corporation, Reading, Pa., and is produced by Fiberfil, Inc., Evansville, Ind.

Clauss (1972) has shown that oil-impregnated Nylasint can be used for higher PV conditions. It generally produces less wear, and a coefficient of friction as low as 0.02 can be achieved. The MPLs are in widespread use in bearings for steel mills and the mining industry, reducing maintenance costs.

5.4.2 Elastomer Composites

The friction and wear characteristics of elastomers also can be modified by adding fillers such as carbon black, graphite, MoS_2, PTFE, CdO-graphite-Ag, bronze, magnesium oxide, silica, and silicate powders (Bhushan, 1980b; Bhushan and Winn, 1979, 1981; Bhushan et al., 1982; Gray et al., 1985; Ksieski, 1973; Voet, 1980). Theberge and Arkles (1976) added glass fibers to thermoplastic elastomers such as polyurethane and olefin-based rubber to improve resistance to deformation. Fillers increase the tearing energy at low rates, and wear involving a tearing component is reduced; the effect tends to disappear at high rates. Fillers also improve the hardness, tensile strength, and elongation to break, and these, in turn, also will generally increase wear resistance. Solid-lubricant fillers reduce both friction and wear.

Zaprivoda and Tenenbaum (1965) proposed applying an antifriction coating to the working elastomeric seal surfaces. The antifriction PTFE coating is made in the form of a porous film that is pressed onto an elastomer base in the process of forming the sealing element, where adhesion of the coating with the base is accomplished by the rubber mass flowing into the pores of the film and by subsequent vulcanization. Nersasian (1980) added lubricating oil additives to the elastomers during compounding for a combination of excellent heat and fluid resistance and sealability that is used for many automotive applications. Bhushan et al. (1982) added migrating and lubricating oils to reduce friction and wear in elastomers.

Filled elastomers are used for numerous bearing and seal applications, operating dry or in the presence of water, oils, and other process fluids. For example, filled Buna-N per military specifications in MIL-B-17901 class 1 (32 to 33 wt % acrylonitrile thermoplastic and the balance butadiene rubber and fillers) is widely used for seawater-lubricated stern-tube bearings. Filled Buna-N and Viton 2282 are used for lip seals in merchant ships and submarines. (Buna-N bearings are produced by several companies, such as Waukesha Bearings, Waukesha, Wisc., and Johnson Rubber Company, Middlefield, Ohio). The most common liner materials for this application are 316 stainless steel, flame-sprayed Cr_2O_3, and plasma-sprayed Al_2O_3.

Bhushan and Winn (1979, 1981) conducted a screening study to compare the performance of various filled elastomers. Elastomers selected for this study were butadiene-acrylonitrile copolymer (Buna-N or Nitrile per MIL-B-17901 class 1, produced by Waukesha Bearing), vinylidene fluoride–hexafluoropropylene (Viton 2282, Du Pont), perfluoroelastomer (Kalrez 1050, Du Pont), phosphonitrilic fluoroelastomer (PNF R-211627, Firestone Rubber Company), and silicone rubber (MeViSi, SE 3808U with dicumyl peroxide curing agent, General Electric). Elastomers have a tendency to swell and age after extended exposure to seawater, moisture, and liquid lubricants (especially at elevated temperatures), which can modify their physical properties (such as geometry, hardness, tensile strength, and elongation). Therefore, the filled elastomers were subjected to a soak (swell and accelerated aging) test in addition to the friction and wear tests described previously in this section (Bhushan and Winn, 1979, 1981).

Bhushan and Winn found that all elastomers except silicone rubber survived 7-month soak tests. Silicone rubber became brittle and chipped off in seawater and gained weight in oil. PNF also lost some of its hardness in oil but was believed to be acceptable. Accelerated friction and wear tests of the various elastomers sliding against 316L stainless steel, flame-sprayed Cr_2O_3, and plasma-sprayed Al_2O_3 coated surfaces in petroleum-based oil and seawater also were conducted. A PTFE composite (Rulon LD) also was included for comparison. Representative data for various elastomers slid against a stainless steel liner in oil and seawater are presented in Table 5.22. From these data we note that the wear rates of Buna-N and Viton are comparable. PNF was excellent in oil but performed poorly in seawater. Kalrez performed well in both oil and seawater. The kinetic friction was lower in seawater than that in oil, whereas the static friction was higher.

Bhushan and Winn concluded that there were no commercially available elastomers that performed better than Nitrile and Viton. They felt that the friction and wear of Nitrile, which is used extensively and is inexpensive, can be

TABLE 5.22 Friction and Wear Data for Elastomer Composites from Unidirectional Sliding* Tests in Oil and Seawater

Pin material	Static friction coefficient		Kinetic friction coefficient		Total linear wear, μm
	At start	At stop	Start of test	End of test	
Tested against 316L stainless steel† in petroleum-based oil (Turbo oil 78)					
Butadiene-acrylonitrile copolymer with fillers (Buna-N, MIL-B-17901 class 1)	0.11	0.18	0.046	0.070	43
Fluoroelastomer with fillers (Viton 2282)	0.18	0.22	0.061	0.053	57.3
Perfluoroelastomer with fillers (Kalrez 1050)	0.12	0.15	0.062	0.07	25
Phosphonitrilic fluoroelastomer with fillers (PNF R-211627)	0.07	0.14	0.06	0.11	10
PTFE with glass and other fiber (Rulon LD) (for comparisons)	0.06	0.11	0.07	0.07	38
Tested against 316L stainless steel† in synthetic seawater					
Butadiene-acrylonitrile copolymer with fillers (Buna-N, MIL-B-17901 class 1)	0.33	0.28	0.022	0.021	31
Fluoroelastomer with fillers (Viton 2282)	0.26	0.36	0.027	0.027	18
Perfluoroelastomer with fillers (Kalrez 1050)	0.52	0.61	0.027	0.027	23
Phosphonitrilic fluoroelastomer with fillers (PNF R-211627)	0.13	0.28	0.013	0.013	46

*Pin-on-cylinder test: sliding speed, 6.1 m s^{-1}; normal stress, 2.76 MPa; ambient temperature, 27°C; test duration, 24 h.

†Surface roughness, 0.10 μm cla.

Source: Adapted from Bhushan and Winn (1981).

improved significantly by formulating it with solid-lubricant particles and liquid lubricants. The low friction of elastomers also may be useful in minimizing audible vibrations or the "bearing squeal" that is seen in water-lubricated bearings in marine propeller-shaft applications during operations at low shaft speeds (Bhushan, 1980c).

Bhushan (1980b), Bhushan et al. (1982), and Gray et al. (1985) selected Nitrile (or Buna-N) per military specifications MIL-B-17901 class 1* (32 to 33 wt % acrylonitrile thermoplastic and the balance butadiene) as the base elastomer. Based on their experience with solid-lubricant coatings and reinforced plastics, they selected MoS_2, graphite, and CdO-graphite-Ag fillers (Bhushan, 1982). Incompatible and lubricating oils also were added to the elastomers during com-

*Mil-B-17901 class 1 requirements: Shore A hardness = 85 ± 5, tensile strength = 10.35 MPa minimum, and elongation at break = 150 percent minimum.

pounding. As a result of aging, these oils migrate to the surface and provide a lubricant film. This concept is known as *blooming* (Bhushan et al., 1982). Filled elastomers were made by three companies: Astro Molding, Inc., Old Bridge, N.J., Minnesota Rubber Company, Minneapolis, Minn., and Johnson Rubber Company, Middlefield, Ohio.

The elastomer samples were subjected to soak tests and friction and wear tests. The soak tests were designed to determine the geometric stability and compatibility of the elastomers with the operating fluid. In these tests, samples were submerged in synthetic seawater at 0, 22, and 65°C for a period of 2 months, representative of ships and sea transport vehicles. Weight, volume, Shore A hardness, tensile strength, and elongation at break were measured before and after the test to determine any degradation in the physical and mechanical properties. Friction and wear tests were conducted using a unidirectional continuous sliding tester. These tests were conducted against a marine propellor shaft material that was 70 wt % Cu to 30 percent Ni (Monel 400) per Mil-C-15345 alloy 24 (casted) in the presence of seawater at a normal stress of 0.35 MPa and a sliding speed of 0.37 m s^{-1} for a test duration of 24 hours.

Soak tests showed that changes in the physical and mechanical properties seen in filled elastomers were comparable with those seen in unfilled elastomers. Therefore, lubricant fillers have no adverse effect on physical and mechanical properties during long-term seawater soak tests. Kinetic friction data are presented in Table 5.23. The measured total wear of different materials was very small, less than 25 μm, within the experimental error, and no discrimination could be made based on wear. Friction data shown in Table 5.23 indicate that the kinetic coefficient of friction of lubricant-filled elastomers varies from 0.015 to 0.06 lower than that of unfilled elastomers (0.07 to 0.09). CdO-graphite-Ag performed generally better than straight graphite and MoS_2. Elastomers with higher concentrations of lubricant fillers generally had slightly lower friction than those with lower concentrations of fillers. Bhushan et al. (1982) also noted that normal pressure and speed variations in the range tested (0.18 to 0.70 MPa at 0.37 m s^{-1} and 0.35 MPa at 0.18 and 0.55 m s^{-1}) had no influence on the friction.

Gray et al. (1985) further optimized the composition and content of solid-lubricant fillers for Nitrile rubber for seawater operation. They selected the following additives: several different graphites, both natural and synthetic, as well as a range of graphite particle sizes, various compositions of CdO + graphite + Ag, MoS_2 on its own and in various mixtures, and PTFE. Based on their study, the following three lubricant additive mixtures performed the best: CdO + graphite + Ag, CdO + MoS_2 + Ag, and MoS_2. It should be noted that the CdO-graphite-Ag mixture has performed very well as an additive to PTFE (see previous subsection), and it will be shown in Chapter 13 that this mixture is superior to graphite as a solid-lubricant coating.

5.4.3 Polymer Dispersions in Liquids

PTFE colloids dispersed in petroleum and synthetic lubricating oils and greases are used to minimize friction and wear in sliding and rolling contacts. A very serious difficulty encountered in the use of colloidal suspensions of PTFE has been settling and agglomeration. Stable PTFE colloids developed by Reick (1977, 1978, 1982) are pristine, the main source being the aqueous dispersions that result as PTFE is polymerized. With proper surface chemistry, they are extremely re-

TABLE 5.23 Friction Data for Elastomers from Unidirectional Sliding* Tests

Pin material	Kinetic friction coefficient	
	Start of test	End of test
Supplier no. 1		
Standard Nitrile rubber	0.21	0.09
Nitrile rubber with 5 wt % migrating lubricant	0.11	0.03
Nitrile rubber with 5 wt % no. 2 medium-flake graphite	0.15	0.05
Nitrile rubber with 10 wt % no. 2 medium-flake graphite	0.17	0.06
Nitrile rubber with 5 wt % CdO-graphite-Ag†	0.19	0.04
Nitrile rubber with 10 wt % CdO-graphite-Ag	0.17	0.04
Supplier no. 2		
Standard Nitrile rubber	0.22	0.07
Nitrile rubber with 20 phr‡ graphite (0.2 to 5 μm)	0.29	0.04
Nitrile rubber with 5 phr "slippery stuff"§	0.13	0.04
Nitrile rubber with 20 phr molybdenum disulfide (4 to 10 μm)	0.13	0.05
Supplier no. 3		
Standard Nitrile rubber	0.18	0.07
Nitrile rubber with 10 wt % graphite (0.2 to 0.5 μm)	0.17	0.03
Nitrile rubber with 20 wt % graphite (0.2 to 5 μm)	0.07	0.02
Nitrile rubber with 10 wt % CdO-graphite-Ag	0.06	0.03
Nitrile rubber with 20 wt % CdO-graphite-Ag	0.07	0.015
Nitrile rubber with 10 wt % MoS_2 (4 to 10 μm)	0.14	0.03
Nitrile rubber with 20 wt % MoS_2 (4 to 10 μm)	0.06	0.025

*Pin-on-cylinder test: lubricant, synthetic seawater; sliding speed, 0.37 m s^{-1}; normal stress, 0.35 MPa; ambient temperature, 22°C; test duration, 24 h; surface roughness of 70% Cu, 30% Ni (Monel) mating metal, 0.40 μm cla.

†20 wt % CdO (95 to 200 μm), 65% synthetic graphite (~2 μm), and 15 wt % Ag (1 to 5 μm).

‡Parts per hundred parts of Nitrile (base polymer).

§Proprietary material of Minnesota Rubber Company, Minneapolis, Minn.

Source: Adapted from Bhushan et al. (1982).

sistant to settling, showing the beginning of Brownian movement under the microscope. The dispersion of particles in the size range of 0.02 to 0.5 μm and smaller shows little tendency to agglomerate or to be trapped in oil filters.

The PTFE dispersions lubricate the metal surfaces in such a way that the very desirable low-friction characteristics of PTFE rubbing against PTFE can be obtained in situ. PTFE colloids reduce the metal-to-metal contact because of the presence of floating particles that roll and slide between the surfaces. A number of laboratory and engine tests have been conducted by several investigators such as Reick (1982) and Papay (1979), and these tests show that a normal engine oil (such as 10W40) containing PTFE colloids resulted in significantly longer wear life than an oil having no colloids. A dispersion of PTFE colloids for lubricating oils is commercially available under the trade name Tufoil from Fluoramics, Inc., Upper Saddle River, N.J.

Water-soluble polymer resins such as poly(ethylene oxide) and polyalkylene oxide are commercially available to reduce friction and wear in sliding applications. Polyox water-soluble resins are available from Union Carbide Corporation, Chemicals and Plastics Division, New York. Polyox resins are nonionic poly(ethylene oxide) homopolymers ranging in molecular weight from 100,000 to 5 million that are completely soluble in water up to its boiling point and in several organic liquids. Merril et al. (1966) and Hampson and Naylor (1975) have shown that turbulent fluid friction can be reduced by more than 50 percent by adding small amounts (on the order of 50 $\mu g\ g^{-1}$) of high-molecular-weight polymer (such as Polyox WSR 301) to a fluid (such as water and methyl alcohol). There is a gradual reduction in effectiveness as the polymer solution is repeatedly sheared. Degradation of the polymer is thus the limiting factor in the exploitation of this method of friction reduction.

5.5 REFERENCES

Acheson, E. G. (1907), "Deflocculated graphite," *Trans. Am. Electrochem. Soc.*, Vol. 12, pp. 29–37.

Anderson, J. C. (1982), "Wear of Commercially Available Plastic Materials," *Tribol. Int.*, Vol. 15, pp. 255–263.

———, (1986), "The Wear and Friction of Commercial Polymers and Composites," in *Friction and Wear of Polymer Composites* (K. Friedrich, ed.), pp. 329–362, Elsevier Science Publishers, Amsterdam.

———, and Davies, A. (1982), "Polymer Composite Abrasiveness in Relation to Counterface Wear," *Lub. Eng.*, Vol. 40, pp. 41–46.

Anonymous (1964), "Impregnation Improves Carbon-Graphite," *Mater. Design Eng.*, Vol. 60, Sept., pp. 93–95.

——— (1973a), *Design Ahead with Purebon Molalloy*, Pure Carbon Co., Inc., St. Marys, Pa.

——— (1973b), *Illustrated Mechanism of Molybdenum Disulfide Lubrication*, Dow-Corning Corporation, Midland, Mich.

——— (1973c), "Seal Reference Issue," *Machine Design*, Vol. 45, Sept. 13.

——— (1975), *Design Ahead with Purebon*, Pure Carbon Company, Inc., St. Marys, Pa.

——— (1976a), "Purebide Grades PE-6923 and PE-7130," Tech. Bull. No. 375, Pure Carbon Company, Inc., St. Marys, Pa.

——— (1976b), *DU Self-Lubricating Bearing*, Garlock Bearings, Inc., Thorofare, N.J.

——— (1976c), *Graphitar (Carbon-Graphite)*, Wickes Engineering Materials, Wickes Corporation, Saginaw, Mich.

——— (1978), *Fortified Polymers*, LNP Corporation, Malvern, Pa.

——— (1979), *Facts and Figures of the Plastic Industry*, The Society of the Plastics Industry, Inc., New York.

——— (1980), "Materials Reference Issue," *Machine Design*, Vol. 52, April 17.

——— (1983), *Cabot Carbon Blacks for Ink, Paints, Plastics, Paper*, Special Blacks Division, Cabot Corporation, Boston, Mass.

——— (1988), *Modern Plastics Encyclopedia*, McGraw-Hill, New York.

Arkles, B., Gerakaris, S., and Goodhue, R. (1974), "Wear Characteristics of Fluoropolymer Composites," in *Advances in Polymer Friction and Wear*, Polymer Science and Technology (L. H. Lee, ed.), Vol. 5B, pp. 663–688, Plenum, New York.

Badami, D. V., and Wiggs, P. K. C. (1970), "Friction and Wear," in *Modern Aspects of Graphite Technology* (L. C. F. Blackman, ed.), Chap. 6, Academic, New York.

Banerjee, J. K. (1981), "Some Effects of Molybdenum Disulfide (MoS_2)–Based Cutting Fluids on the Workpiece Surface Finish in Grinding and Conventional Machining Processes," *Proc. 3rd Int. Conf. on Wear of Materials*, ASME, New York.

Bartenev, G. M., and Lavrentev, V. V. (1961), "The Laws of Vulcanized Rubber Friction," *Wear*, Vol. 4, pp. 154–160.

———, and ——— (1981), *Friction and Wear of Polymers*, Elsevier, Amsterdam.

Bartz, W. J. (1971), "Solid Lubricant Additives: Effect of Concentration and Other Additives on Antiwear Performance," *Wear*, Vol. 17, pp. 421–432.

——— (1972), "Some Investigations on the Influence of Particle Size in the Lubricating Effectiveness of Molybdenum Disulfide," *ASLE Trans.*, Vol. 15, pp. 207–215.

——— (1980), "Lubricating Effectiveness of Oil-Soluble Additives and Molybdenum Disulfide Dispersed in Mineral Oil," *Lub. Eng.*, Vol. 36, pp. 579–585.

———, and Oppelt, J. (1978), "Lubricating Effectiveness of Oil-Soluble Additives and Graphite Dispersed in Mineral Oil," *Proc. 2d Int. Conf. on Solid Lubrication*, ASLE, Park Ridge, Ill.

Bates, J. J. (1973), "Carbon Fiber Brushes for Electrical Machines," in *Carbon Fibers in Engineering* (M. Langley, ed.), Chap. 6, McGraw-Hill, New York.

Bennington, T. E., Cole, D. D., Ghirla, P. J., and Smith, R. K. (1975), "Stable Colloid Additives for Engine Oils: Potential Improvement in Fuel Economy," SAE paper No. 750677, presented at SAE Fuels and Lubricants Meeting, Houston, Texas, June 3–5.

Bhushan, B. (1978a), "Wear Tests of Plastics Rubbing Against Aluminum in Transmission Oil at 250°F," Tech. Rep. MTI-78TR94, Mechanical Technology, Inc., Latham, N.Y.

——— (1978b), "Investigation of Material Combinations in Freon-22 and Refrigerant Oil Mixture," Tech. Rep. MTI-78TR87, Mechanical Technology, Inc., Latham, N.Y.

——— (1980a), "Wear Data for Plastic Bearings," *Machine Design*, Vol. 52, No. 5, pp. 261–264.

——— (1980b), "Development of Low-Friction Elastomers for Stern-Tube Bearings," Tech. Rep. MTI-80TR44, Mechanical Technology, Inc., Latham, N.Y.

——— (1980c), "Stick-Slip Induced Noise Generation in Water-Lubricated Compliant Rubber Bearings," *J. Lub. Technol.* (*Trans ASME*), Vol. 102, pp. 201–212.

——— (1981), "Solid Lubricant Fillers Are Replacing Liquids for Many Difficult Jobs," *Modern Plastics*, Vol. 58, No. 1, pp. 100B–100D.

——— (1982), "Development of CdO-Graphite-Ag Coatings for Gas Bearings to 427°C," *Wear*, Vol. 75, pp. 333–356.

——— (1987), "Overview of Coating Materials, Surface Treatments, and Screening Techniques for Tribological Applications Part I: Coating Materials and Surface Treatments," in *Testing of Metallic and Inorganic Coatings* (W. B. Harding and G. A. DiBari, eds.), Special Publication No. STP947, pp. 289–309, ASTM, Philadelphia.

———, and Dashnaw, F. (1981), "Material Study for Advanced Stern-Tube Bearings and Face Seals," *ASLE Trans.*, Vol. 24, pp. 398–409.

———, and Gray, S. (1978), "Material Study for High Pressure Seawater Hydraulic Tool Motors," Tech. Rep. No. CR-78-012, Civil Engineering Laboratory, Naval Construction Batallion Center, Port Hueneme, Calif.

———, and ——— (1979), "Investigation of Material Combinations under High Load and Speed in Synthetic Seawater," *Lub. Eng.*, Vol. 35, pp. 628–639.

———, and Wilcock, D. F. (1981), "Frictional Behavior of Polymeric Compositions in Dry Sliding," *Proc. Seventh Leeds-Lyon Symp. on Tribol.*, pp. 103–113, Westbury House, IPC Business Press, Guildford, England.

———, and ——— (1982), "Wear Behavior of Polymeric Composition in Dry Reciprocating Sliding," *Wear*, Vol. 75, pp. 41–70.

———, and Winn, L. W. (1979), "Material Study for Advanced Stern-tube Seals and Bearings for High Performance Merchant Ships," Tech. Rep. MA-920-79069, Office of Commercial Development, U.S. Maritime Administration, Washington, D.C.

———, and ——— (1981), "Material Study for Advanced Stern-tube Lip Seals," *ASLE Trans.*, Vol. 24, pp. 398–409.

———, Gray, S., and Graham, R. W., II (1982), "Development of Low-Friction Elastomers for Bearings and Seals," *Lub. Eng.*, Vol. 38, pp. 626–634.

Blow, C. M. (1971), "Determination of Seal/Fluid Compatibility Using the A.H.E.M.S.C.I.

Method," paper G1, presented at the 5th Int. Conf. on Fluid Sealing, Warwick County, England.

Boeke, P. J. (1974), "Polyphenylene Sulfide: A High Temperature Matrix for Self-Lubricating Composites," paper presented at SPE National Technical Conference on Plastics in Surface Transport, Detroit, Mich.

Boes, D. J. (1964), "New Solid Lubricants: Their Preparation, Properties, and Potential for Aerospace Applications," *IEEE Trans. Aerospace*, Vol. AS-2, pp. 457–466.

Bowden, F. P. (1950), "Frictional Properties of Porous Metals Containing Molybdenum Disulfide," *Research* (*Lond.*), Vol. 3, pp. 383–384.

———, and Young, J. E. (1951), "Friction of Diamond, Graphite, and Carbon and the Influence of Surface Films," *Proc. R. Soc. Lond.*, Vol. 208, pp. 444–455.

Bowers, R. C., Clinton, W. C., and Zisman, W. A. (1954), "Frictional Properties of Polymers," *Modern Plastics*, Feb., pp. 131–225.

Braithwaite, E. R. (1964), *Solid Lubricants and Solid Surfaces*, Pergamon, Oxford, England.

——— (1966), "Friction and Wear of Graphite and Molybdenum Disulfide," *Sci. Lub.* (*Lond.*), Vol. 18, pp. 17–21.

——— (1967), *Lubrication and Lubricants*, Elsevier, Amsterdam.

Briscoe, B. J.(1981), "Wear of Polymers: An Essay on Fundamental Aspects," *Tribol. Int.*, Vol. 14, pp. 231–243.

———, and Steward, M. D. (1977), "The Effect of Carbon-Filled Aspect Ratio and Surface Chemistry on the Wear of Carbon-Filled PTFE Composites," *Proc. Int. Conf. on Wear of Materials* (W. A. Glaeser, K. C. Ludema, and S. K. Rhee, eds.), pp. 538–540, ASME, New York.

Bryant, P. J., Gutshall, P. L., and Taylor, L. H. (1964), "A Study of Mechanisms of Graphite Friction and Wear," *Wear*, Vol. 7, pp. 118–126.

Brydson, J. A. (1982), *Plastic Materials*, Butterworth, London.

Buckley, D. H. (1971), "Friction, Wear and Lubrication in Vacuum," Special Publication No. SP-277, NASA, Washington, D.C.

———, and Brainard, W. A. (1974), "The Atomic Nature of Polymer-Atomic Interactions in Adhesion, Friction, and Wear," in *Advances in Polymer Friction and Wear*, Polymer Science and Technology Vol. 5A (L. H. Lee, ed.), pp. 315–328, Plenum, New York.

———, and Johnson, R. L. (1963), "Friction, Wear and Decomposition Mechanisms for Various Polymer Compositions in Vacuum to 10^{-9} Millimeter of Mercury," Tech. Rep. TN-D2073, NASA, Cleveland, Ohio.

———, and ——— (1964), "Mechanism of Lubrication for Solid Carbon Materials in Vacuum to 10^{-9} mm of Mercury," *ASLE Trans.*, Vol. 7, p. 97.

Bunn, C. W., and Howells, E. R. (1954), "Structure of Molecules and Crystals of Fluorocarbons," *Nature*, Vol. 174, pp. 549–551.

Campbell, M. E. (1972), "Solid Lubricants: A Survey," Special publication No. SP-5059(01), NASA, Washington, D.C.

———, Loser, J. B., and Sneegas, E. (1966), "Solid Lubricants," Special Publication, No. SP-5059, NASA, Washington, D.C.

Carson, R. W. (1965), "Selecting High Temperature Sleeve Bearings," *Product Eng.*, Vol. 36, Jan. 4, pp. 76–82.

Clauss, F. J. (1972), *Solid Lubricants and Self-Lubricating Solids*, Academic, New York.

Craig, W. D. (1964), "Operation of PTFE Bearings," *Lub. Eng.*, Vol. 20, pp. 456–462.

Devine, M. J., and Kroll, Q. E. (1964), "Aromatic Polyimide Compositions for Solid Lubrication," *Lub. Eng.*, Vol. 20, pp. 225–230.

Dillich, S., and Kuhlman-Wilsdorf, D. (1982), "Two Regimes of Current Conduction in Metal-Graphite Electrical Brushes and Resulting Instabilities," *Proc. IEEE 28th Holm Conf. on Electrical Contact Phenomena*, pp. 201–212, IEEE, New York.

Dumbleton, J. H., and Shen, C. (1976), "The Wear Behavior of Ultrahigh Molecular Weight Polyethylene," *Wear*, Vol. 37, pp. 279–289.

Economy, J., Nowak, B. E., and Cottis, S. G. (1970), "Ekonol: A High Temperature Aromatic Polyester," *SAMPLE J.*, Vol. 6, Aug./Sept., pp. 21–27.

Eiss, N. S., and Czichos, H. (1986), "Tribological Studies on Rubber-Modified Epoxies: Influence of Material Properties and Operating Conditions," *Wear*, Vol. 111, pp. 347–361.

Evans, D. C. (1976), "Polymer-Fluid Interactions in Relation to Wear," *Proc. 3d Leeds-Lyon Symp. Tribology* (D. Dowson, M. Godet, and C. M. Taylor, eds.), pp. 47–55, Mechanical Engineering Publishers, London.

———, and Lancaster, J. K. (1979), "The Wear of Polymers," in *Treatise on Material Science and Technology*, Vol. 13: *Wear* (D. Scott, ed.), pp. 85–139, Academic, New York.

Farr, J. P. G. (1975), "Molybdenum Disulphide in Lubrication: A Review," *Wear*, Vol. 35, pp. 1–22.

Feng, I. M. (1952), "Lubricating Properties of Molybdenum Disulfide," *Lub. Eng.*, Vol. 8, pp. 285–289.

Flom, D. G., and Haltner, A. J. (1968), "Lubricants for the Space Environment," Presented at the AIChE Materials Conference, Symposium on Materials for Reentry and Spacecraft Systems–Spacecraft Materials, Part 2, Philadelphia, March 31–April 4.

Frados, J. (1977), *The Story of the Plastics Industry*, The Society of the Plastics Industry, Inc., New York.

Fusaro, R. L. (1988), "Evaluation of Several Polymer Materials for Use as Solid Lubricants in Space," *Tribol. Trans.*, Vol. 31, pp. 174–181.

Gansheimer, J., and Holinski, R. (1973), "Molybdenum Disulfide in Oils and Greases under Boundary Conditions," *J. Lub. Technol.* (*ASME Trans.*), Vol. 95, pp. 242–248.

Gardos, M. N., and McConnell, B. D. (1981), *Development of High-Speed, High-Temperature Self-Lubricating Composites*, Part I: *Polymer Matrix Selection*; Part II: *Reinforcement Selection*; Part III: *Additives Selection*; Part IV: *Formulation and Performance of the Best Composition*, Special Publication No. SP-9, ASLE, Park Ridge, Ill.

——— and ——— (1982), "Self-Lubricating Composites for Extreme Environment Applications," *Tribol. Int.*, Vol. 15, pp. 273–283.

Giltrow, J. P. (1973), "Friction and Wear of Some Polyimides," *Tribol. Int.*, Vol. 6, pp. 253–257.

Grattan, P. A., and Lancaster, J. K. (1967), "Abrasion by Lamellar Solid Lubricants," *Wear*, Vol. 10, pp. 453–468.

Gray, S., and Wilson, D. S. (1961), "The Development of Lubricants for High Speed Rolling Contact Bearings Operating over the Range of Room Temperature to 1200°F," Progress Rep. No. 3 on Contract AF33 (616)-6589 by Fairchild Stratos Corporation, U.S. Air Force, WPAFB, Dayton, Ohio.

———, Cronin, M. J., and Graham, R. W., II (1985), "Low Friction Elastomer Selection of Stave-Type Bearings and Seals," *Lub. Eng.*, Vol. 41, pp. 659–669.

Hampson, L. G., and Naylor, H. (1975), "Friction Reduction in Journal Bearings by High-Molecular-Weight Polymers," *Proc. Inst. Mech. Eng.*, Vol. 189, pp. 279–286.

Harper, C. A. (ed.) (1970), *Handbook of Plastics and Elastomers*, McGraw-Hill, New York.

Hisakado, T., Tsukizoe, T., and Yoshikawa, H. (1983), "Lubrication Mechanism of Solid Lubricants in Oils," *J. Lub. Technol.* (*Trans. ASME*), Vol. 105, pp. 245–253.

Holinski, R., and Gansheimer, J. (1972), "A Study of the Lubricating Mechanism of Molybdenum Disulfide," *Wear*, Vol. 19, pp. 329–342.

Hopkins, V., and Campbell, M. (1977), "A Composite-Like Solid Film Lubricant," *Lub. Eng.*, Vol. 33, pp. 252–257.

Hurren, D. E., Midgley, J. W., and Pritchard, C. (1963–1964), "A Study of the Friction Characteristics of Resin-Bonded Bearing Materials Lubricated by Water," *Proc. Inst. Mech. Eng.*, Vol. 178, Part 3N, pp. 9–15.

Hutcheon, J. M. (1970), "Manufacturing Technology of Baked and Graphite Carbon Bodies," in *Modern Aspects of Graphite Technology* (L. C. F. Blackman, ed.), Academic, New York.

Ishikawa, T., and Shimada, T. (1969), "Application of Polycarbon Monofluoride," paper presented at the Fluorine Symposium, Moscow.

Istomin, N. P., and Khruschov, M. M. (1973), "The Effect of Cleavage Plane Orientation of Graphite on the Friction and Wear of Its Composition with PTFE," *J. Mater. Sci.*, Vol. 1, pp. 97–101.

Jain, V. K., and Bahadur, S. (1977), "Material Transfer in Polymer-Polymer Sliding," *Proc. Int. Conf. on Wear of Materials* (W. A. Glaeser, K. C. Ludema, and S. K. Rhee, eds.), pp. 487–493, ASME, New York.

Jamison, W. E. (1985), "New Developments in Microporous Polymer Lubricants," *Lub. Eng.*, Vol. 41, pp. 274–279.

Johnson, R. L., Swikert, M., and Bisson, E. (1950), "Friction and Wear of Hot-Pressed Bearing Materials Containing Molybdenum Disulfide," TN-2027, NASA, Washington, D.C.

Jost, H. P. (1956), "Pure Molybdenum Disulfide," *Sheet Metal Ind.*, Vol. 33, pp. 679–697.

Kanakia, M. D., and Peterson, M. B. (1987), "Literature Review of Solid Lubrication Mechanisms: Interim Report BFLRF No. 213," Southwest Research Institute, San Antonio, Tex.

Kay, E. (1965), "The Corrosion of Steel in Contact with Molybdenum Disulphide," Tech. Rep. No. 65219, Royal Aircraft Establishment, United Kingdom.

Kelly, B. T. (1981), *Physics of Graphite*, Applied Science Publishers, London.

Killefer, D. H., and Linz, A. (1952), *Molybdenum Compounds: Their Chemistry and Technology*, Interscience, New York.

King, R. F., and Tabor, D. (1953), "The Effect of Temperature on the Mechanical Properties and the Friction of Plastics," *Proc. Phys. Soc.*, Vol. B66, pp. 728–736.

Ksieski, K. J. (1973), "Sliding Friction of Elastomers," *Proc. 6th Int. Conf. on Fluid Sealing*, BHRA, Cranfield, United Kingdom.

Kyker, G. S., and Antkowiak, T. A. (1974), "Phosphonitrilic Fluoroelastomer: A New Class of Solvent-Resistant Polymers with Exceptional Flexibility at Low Temperatures," *Rubber Chem. Technol.*, Vol. 47, pp. 32–47.

Lancaster, J. K. (1968), "The Effect of Carbon Fiber Reinforcement on the Friction and Wear Behaviour of Polymers," *Br. J. Appl. Phys.*, Vol. 1, pp. 549–559.

——— (1972a), "Lubrication of Carbon Fiber-Reinforced Polymers: II. Organic Fluids," *Wear*, Vol. 20, pp. 335–351.

——— (1972b), "Polymer-Based Bearing Materials: The Role of Fillers and Fiber Reinforcement," *Tribol. Int.*, Vol. 5, pp. 249–255.

——— (1972c), "Friction and Wear," in *Polymer Science: A Material Science Handbook* (A. D. Jenkins, ed.), pp. 959–1046, North-Holland, New York.

——— (1973), "Dry Bearings: A Survey of Materials and Factors Affecting Their Performance," *Tribol. Int.*, Vol. 6, pp. 219–251.

——— (1979), "The Wear of Carbons and Graphites," in *Treatise on Material Science and Technology*, Vol. 13: *Wear* (D. Scott, ed.), pp. 141–175, Academic, New York.

———, Play, D., Godet, M., Verrall, A. P., Waghorne, R. (1980), "Third Body Formation and the Wear of PTFE Fiber-Based Dry Bearings," *J. Lub. Technol.* (*Trans. ASME*), Vol. 102, pp. 236–244.

Lansdown, A. R. (1974), *Molybdenum Disulphide Lubrication: A Survey of the Present State of the Art*, University College of Swansea, United Kingdom.

——— (1976), *Molybdenum Disulphide Lubrication: A Continuation Survey 1973–74*. (Prepared for European Space Agency.) University College of Swansea, United Kingdom.

——— (1979), *Molybdenum Disulphide Lubrication: A Continuation Survey 1979–80*. (Prepared for European Space Agency.) University College of Swansea, United Kingdom.

Lavik, M. T. (1959), "High Temperature Solid Dry Film Lubricants," WADC-TR-57-455, Midwest Research Institute, Kansas City, Mo.

Lewis, R. B. (1964), "Predicting Wear of Sliding Plastic Surfaces," *Mech. Eng.*, Vol. 86, pp. 32–35.

——— (1969), "Wear of Polyimide Resin," *Lub. Eng.*, Vol. 25, pp. 356–359.

Lipp, L. C. (1976), "Solid Lubricants: Their Advantages and Limitations," *Lub. Eng.*, Vol. 32, pp. 574–584.

Lomax, J. Y., and O'Rourke, J. T. (1966), "TFE-Lubricated Thermoplastics," *Machine Design*, Vol. 38, June 23, pp. 158–164.

Low, M. B. J. (1979), "The Effect of the Transfer Film on the Friction and Wear of Dry Bearing Materials for a Power Plant Application," *Wear*, Vol. 52, pp. 347–363.

Ludema, K. C., and Tabor, D. (1966), "The Friction and Viscoelastic Properties of Polymeric Solids," *Wear*, Vol. 9, pp. 329–348.

MacLaren, K. G., and Tabor, D. (1963), "Friction of Polymers at Engineering Speeds: Influence of Speed, Temperature, and Lubricants," *Proc. Inst. Mech. Eng. Wear Convention*, Bournenouth, p. 210.

McCabe, J. T. (1966), "How the Automotive Industry Uses Solid Lubricants," *Metal Prog.*, Vol. 109, May.

Merril, E., Smith, K. A., Slim, H., and Mickley, H. S. (1966), "Study of Turbulent Flows of Dilute Polymer Solutions in a Couette Viscometer," *Trans. Soc. Rheol.*, Vol. 10, p. 335.

Midgley, J. W. (1956), "The Frictional Properties of Molybdenum Disulfide," *J. Inst. Petrol. Lond.*, Vol. 43, pp. 316–321.

Mitchell, D. C., and Pratt, G. C. (1957), "Friction, Wear, and Physical Properties of Some Filled PTFE Bearing Materials," *Proc. Conf. on Lubrication and Wear*, pp. 416–423, Institute of Mechanical Engineers, London.

Moore, A. W. (1973), "Highly Oriented Pyrolytic Graphite," *Adv. Chemistry and Physics of Carbon* (P. L. Walker and P. A. Thrower, eds.), Vol. 11, pp. 69–187.

Moore, D. F. (1972), *The Friction and Lubrication of Elastomers*, Pergamon, Oxford, England.

Nersasian, A. (1980), "The Effect of Lubricating Oil Additives on the Properties of Fluorohydrocarbon Elastomers," *ASLE Trans.*, Vol. 23, pp. 343–352.

Noda, T., Inagaki, M., and Yamada, S. (1969), "Glass-Like Carbons," *J. Noncrystal Solids*, Vol. 1, pp. 285–302.

Ogintz, S. M. (1978), "Perfluoroelastomer O-Ring Seals for Severe Chemical and High Temperature Application," *Lub. Eng.*, Vol. 34, pp. 327–333.

Oleesky, S. S., and Mohr, J. G. (1964), *Handbook of Reinforced Plastics*, Reinhold, New York.

Papay, A. G. (1979), "Energy-Saving Gear Lubricants," *Lub. Eng.*, Vol. 35, pp. 613–617.

Paxton, R. R. (1967), "Carbon and Graphite Materials for Seals, Bearings, and Brushes," *Electrochem. Technol.*, Vol. 5, pp. 174–182.

——— (1979), *Manufactured Carbon: A Self-Lubricating Material for Mechanical Devices*, CRC Press, Boca Raton, Fla.

Peterson, M. B., and Johnson, R. L. (1953), "Friction and Wear Characteristics of Molybdenum Disulfide I. Effect of Moisture," Tech. Rep. TN-3055, NACA, Washington, D.C.

———, and ——— (1955), "Factors Influencing Friction and Wear with Solid Lubricants," *Lub. Eng.*, Vol. 11, pp. 325–330.

Pinchbeck, P. H. (1962), "A Review of Plastic Bearings," *Wear*, Vol. 5, pp. 85–113.

Play, D., and Godet, M. (1979), "Self-Protection of High Wear Materials," *ASLE Trans.*, Vol. 32, pp. 56–64.

Pooley, C. M., and Tabor, D. (1972), "Friction and Molecular Structure: The Behavior of Some Thermoplastics," *Proc. R. Soc. Lond.*, Vol. A329, pp. 251–274.

Reick, F. G. (1977), "Stable Polytetrafluoroethylene Colloids for Lubricating Oils Part I: Microscopy," *Am. Lab.*, Vol. 9, pp. 53–56.

——— (1978), "Stable Polytetrafluoroethylene Colloids for Lubricating Oils Part II: Laboratory Tests," *Am. Lab.*, Vol. 10, pp. 57–67.

——— (1982), "Energy Saving Lubricants Containing Colloidal PTFE," *Lub. Eng.*, Vol. 38, pp. 635–646.

Reynard, K. A., Vicic, J. C., Sicka, R. W., and Rose, S. H. (1974), "Elastomers for Service with Engine Lubricants and Hydraulic Fluids," Contract N00019-73-C-0406, Final Report NTIS-AD-781715, Horizons, Inc., Cleveland, Ohio.

Reynolds, W. N. (1973) "Structural and Physical Properties of Carbon Fibers," in *Adv. in Chemistry and Physics of Carbon* (P. L. Walker and P. A. Thrower, eds.), Vol. II, pp. 1–68, Marcel Dekker, New York.

Reznikovskii, M. M. (1960), "Relation Between the Abrasion Resistance and Other Mechanical Properties of Rubber," in *Abrasion of Rubber* (D. I. James, ed.), p. 119, McLaren & Sons, London.

Rhee, S. H., and Ludema, K. C. (1978), "Mechanisms of Formation of Polymeric Transfer Films," *Wear*, Vol. 46, pp. 231–240.

Richardson, H. M. (1957), "Heavy Duty Bearings of Phenolic Plastic," *Inst. Mech. Eng. Proc. General Discussion on Lubrication and Lubricants*, pp. 249–259, Institute of Mechanical Engineers, London.

Risdon, T. J. (1987), "Properties of Molybdenum Disulfide," Amax Mineral Sales, Amax Center, Greenwich, Conn.

Robertson, J. (1986), "Amorphous Carbon," *Adv. Phys.*, Vol. 35, pp. 317–374.

Rowe, G. W. (1960), "Some Observations on the Friction Behavior of Boron Nitride and of Graphite," *Wear*, Vol. 3, pp. 274–285.

Sauer, J. A., and Richardson, G. C. (1980), "Fatigue of Polymers," *Int. J. Fract.*, Vol. 16, pp. 499–532.

Savage, R. H. (1948), "Graphite," *Lubrication*, Vol. 19, pp. 1–10.

———, and Schaefer, D. L. (1956), "Vapor Lubrication of Graphite Sliding Contacts," *J. Appl. Phys.*, Vol. 27, pp. 136–138.

Schallamach, A. (1952), "Abrasion of Rubber by a Needle," *J. Polymer Sci.*, Vol. 9, pp. 385–404.

——— (1953), "The Velocity and Temperature Dependence of Rubber Friction," *Proc. Phys. Soc.*, Vol. B66, pp. 386–392.

——— (1958), "Friction and Abrasion of Rubber," *Wear*, Vol. 1, pp. 384–417.

——— (1963), "Abrasion and Tyre Wear," in *Chemistry and Physics of Rubber-Like Substances* (L. Bateman, ed.), McLaren & Sons, London.

Schlosser, A. (1963), "Development of Gas-Entrained Powder Lubricants for High-Speed and High-Temperature Operation of Spur Gears," Progress Rep. No. 3, Fairchild Stratos Corporation, Contract AF-33(657)-8625, U.S. Air Force, WPAFB, Dayton, Ohio.

Seeley, S. B. (1964), "Natural Graphite," in *Encyclopedia of Chemical Technology* (R. E. Kirk and D. F. Othmer, eds.), Vol. 4, 2d Ed., pp. 304–335, Wiley, New York.

Serafini, T. T., Delvigs, P., and Lightsey, G. R. (1972), "Thermally Stable Polyimides from Solutions of Monomeric Reactants," *J. Appl. Poly. Sci.*, Vol. 16, pp. 905–916.

Seymour, R. B. (1990), "Polymer Composites: New Concepts," in *Polymer Science*, C. R. H. I. de Jonge, U.S.P. Zeist, The Netherlands.

Sharples, A. (1972), "Crystallinity," in *Handbook of Polymer Science* (A. D. Jenkins, ed.), North Holland, Amsterdam.

Shooter, K. V., and Tabor, D. (1952), "The Friction Properties of Plastics," *Proc. Phys. Soc.*, Vol. B65, pp. 661–671.

Short, J. N., and Hill, H. W. (1972), "Polyphenylene Sulfide Coating and Molding Resins," *Chem. Technol.*, Vol. 2, pp. 481–485.

Sliney, H. E. (1979), "Some Load Limits and Self-Lubricating Properties of Plain Spherical Bearings with Molded Graphite Fiber-Reinforced Polyimide Liners to 320°C," *Lub. Eng.*, Vol. 35, pp. 497–502.

——— (1982), "Solid Lubricant Materials for High Temperatures: A Review," *Tribol. Int.*, Vol. 15, pp. 303–314.

Smith, W. V. (1967), "Wear Properties of Plastics as Ship Bearings," In *Testing of Polymers 3*, pp. 221–244, Interscience, New York.

——— (1973), "Material Selection Criteria for Water Lubrication," *Wear*, Vol. 25, pp. 139–153.

Speerschneider, C. J., and Li, C. H. (1962a), "The Role of Filler Geometrical Shape in Wear and Friction of Filled PTFE," *Wear*, Vol. 5, pp. 392–399.

———, and ——— (1962b), "Some Observations on the Structure of Polytetrafluoroethylene," *J. Appl. Phys.*, Vol. 33, pp. 1871–1875.

Steijn, R. P. (1966), "The Effect of Time, Temperature, and Environment on the Sliding Behavior of Polytetrafluoroethylene," *ASLE Trans.*, Vol. 9, pp. 149–159.

——— (1967), "Friction and Wear of Plastics," *Metals Eng. Q.*, Vol. 7, pp. 371–383.

Stock, A. J. (1966), "Evaluation of Solid Lubricant Dispersions on a Four-Ball Tester," *Lub. Eng.*, Vol. 22, pp. 146–152.

Sung, N. H., and Suh, N. P. (1979), "Effect of Fiber Orientation on Friction and Wear of Fiber-Reinforced Polymer Composites," *Wear*, Vol. 53, pp. 129–141.

Tanaka, K. (1977), "Friction and Wear of Glass and Carbon Fiber-Filled Thermoplastic Polymers," *J. Lub. Technol. (Trans. ASME)*, Vol. 99, pp. 408–414.

———, and Kawakami, S. (1982), "Effect of Various Fillers on the Friction and Wear of Polytetrafluoroethylene-Based Composites," *Wear*, Vol. 79, pp. 221–234.

———, and Uchiyama, Y. (1974), "Friction, Wear, and Surface Melting of Crystalline Polymers," in *Advances in Polymer Friction and Wear*, Vol. 5A (L. H. Lee, ed.), pp. 499–529, Plenum, New York.

———, and Yamada, Y. (1983), "Effect of Sliding Speed on Transfer and Wear of Semicrystalline Polymers Sliding Against Smooth Steel Surface," *Proc. Int. Conf. on Wear of Materials* (K. C. Ludema, ed.), pp. 617–624, ASME, New York.

———, Uchiyama, Y., and Toyooka, S. (1973), "The Mechanism of Wear of Polytetrafluoroethylene," *Wear*, Vol. 23, pp. 153–172.

Theberge, J. E., and Arkles, B. (1974), "Wear Characteristics of Carbon Fiber Reinforced Thermoplastics," *Lub. Eng.*, Vol. 30, pp. 585–589.

———, and ——— (1976), "Now, Fiber Reinforcement Provides Stronger Thermoplastic Elastomers," *Machine Design*, Vol. 48, Feb. 12, pp. 113–116.

Thorp, J. M. (1982), "Friction of Some Commercial Polymer-Based Bearing Materials Against Steel," *Tribol. Int.*, Vol. 15, pp. 69–74.

Tsuya, Y., Umeda, K., and Kitamura, M. (1976), "Optimum Concentration of Solid Lubricant in Compact," *Lub. Eng.*, Vol. 32, pp. 402–410.

Uchiyama, Y., and Tanaka, K. (1980), "Wear Laws for Polytetrafluoroethylene," *Wear*, Vol. 58, pp. 223–235.

Vaughan, R. W., Jones, R. J., O'Rell, M. K., and Zakrzewski, G. A. (1976), "Development of Autoclave Moldable Addition-Type Polyimides," TRW Rep. No. 26446-6015-RU-00, Tech. Rep. No. CR-134900, NASA, Cleveland, Ohio.

Voet, A. (1980), "Reinforcement of Elastomers by Fillers: Review of Period 1967–1976," *J. Poly. Sci. Macromol. Rev.*, Vol. 15, pp. 327–373.

White, H. S. (1956), "Small Oil-Free Bearings," *J. Res. Nat. Bur. Stand.*, Vol. 57, pp. 185–204.

Winer, W. O. (1965), "Molybdenum Disulfide as a Lubricant: A Review of the Fundamental Knowledge," *Wear*, Vol. 10, pp. 422–452.

Wolverton, M., and Theberge, J. E. (1981), "Wear Behavior of Plastics on Plastics," *Machine Design*, Vol. 53, Feb. 12.

———, ———, and McCadden, K. L. (1983), "How Plastic Composites Wear at High Temperatures," *Machine Design*, Vol. 55, Feb., pp. 111–115.

Wood, K. B. (1961), "Molybdenum Disulfide as a Lubricant," ASME Paper No. 61-MD-21, presented at the Design Engineering Conference, Detroit, Mich., May. 22–25.

Zaprivoda, A. I., and Tenenbaum, M. M. (1965), "Method of Increasing the Wear Resistance of Rubber Seals," USSR Patent No. 2,27,807.

APPENDIX 5A: WEAR BEHAVIOR CHARACTERIZATION OF SOLID LUBRICANTS AND SELF-LUBRICATING SOLIDS

There are two methods of characterizing the wear behavior of solid lubricants and self-lubricating solids (Lewis, 1964), i.e., wear coefficient K and PV limit. The wear coefficient K, defined by the wear equation presented in Chapter 2, is the volume of wear per unit normal load per unit sliding distance and has the units of inverse pressure. The PV limit is the product of normal pressure and the surface speed. The PV limit is used to define the onset of catastrophic failure of material resulting from mechanical degradation by frictional heating.

Two methods are generally accepted for establishing PV limit. Test method A maintains that limiting PV relates directly to friction and temperature at the material interface. In this test method, step loading is carried out at constant velocity while friction force and/or interface temperature are monitored (Fig. 5.A.1). At some advanced load increment, friction force and/or temperature will no longer stabilize. The highest load value where stabilization occurs is multiplied by the constant speed to establish the PV limit at the test velocity. This method minimizes specimen wear as a variable and is useful only in establishing the *maximum short-term PV rating*. Tests at several velocities establish points for plotting limiting PV curves as a function of velocity. Note that the PV limit *generally* (not always) decreases with an increase in sliding velocity.

Test method B is based on the theory that the wear coefficient is practically constant below the limiting PV. Thus a plot of wear coefficient versus load at constant velocity defines limiting PV by establishing the load at which the wear coefficient is no longer constant (see Figure 5.A.1). The value established by test method B is below the test method A value. Only data established by test method B should be used in designing *PV limits for sustained operation*.

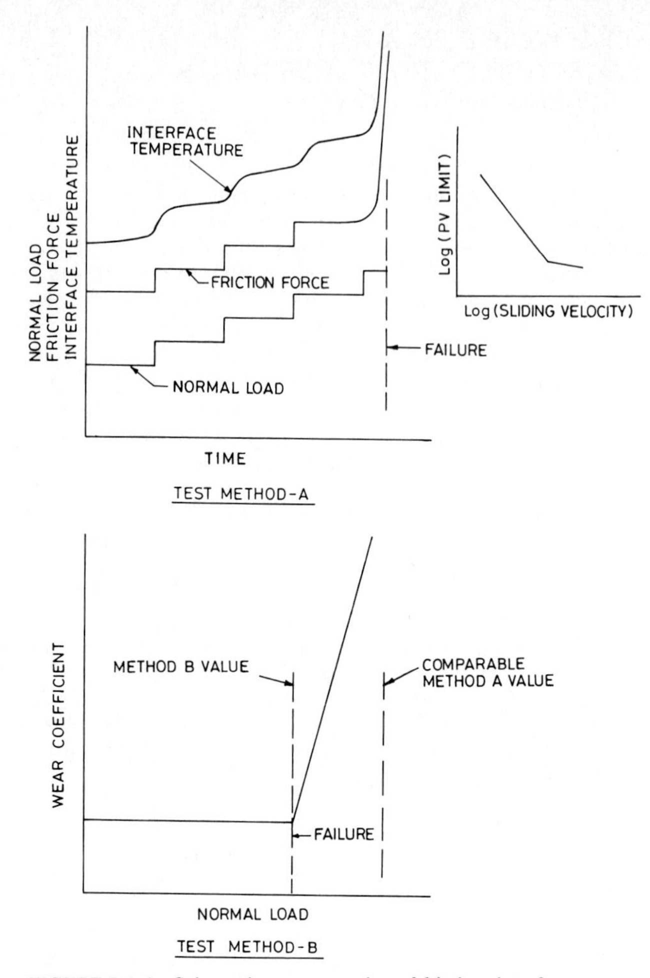

FIGURE 5.A.1 Schematic representation of friction, interface temperature, and wear coefficient changes during determination of *PV* limit for materials.

CHAPTER 6
COATING DEPOSITION AND SURFACE TREATMENT TECHNIQUES: CLASSIFICATION

Many coating deposition techniques and surface treatments are available for the modification of surface characteristics. If the material is added to the surface, the process is referred to as a *coating deposition process*; however, if the surface microstructure and/or chemistry is altered, then the process is referred to as a *surface treatment process*. The effectiveness (load-carrying capacity, wear resistance, and coefficient of friction) of coatings and surface treatments depends on the particular surface modification technique (Anonymous, 1981, 1982, 1983; Bhushan, 1980a, 1987; Bunshah et al., 1982; Chapman and Anderson, 1974; Clauss, 1972; Donovan and Sanders, 1972; Lang, 1983; Maissel and Glang, 1970; Smart, 1978; Vossen and Kern, 1978).

Selection of the coating deposition or surface treatment technique depends on the functional requirements, shape, size, and metallurgy of the substrate, availability of the coating material in the required form, adaptability of the coating material for the technique, level of adhesion desired, availability of coating equipment, and cost. The coating deposition and surface treatment techniques must be compatible with the substrate; e.g., surface modification at high temperature, say 1000°C, may warp thin substrate or modify the heat treatment of metallic substrates which can lead to poor mechanical characteristics.

Figure 6.1 shows commonly used coating deposition techniques and surface treatments (Bhushan, 1980a, 1987). The selection of a suitable coating deposition or surface treatment technique often becomes difficult, and it requires a thorough understanding of the merits and demerits of each technique. The various coating deposition and surface treatment techniques employed for tribological applications are briefly discussed along with their limitations in this chapter. Surface preparation techniques to achieve better adhesion of coating to the substrate and the details of various deposition and surface treatment techniques are presented in Chaps. 7 to 12.

6.1 COATING DEPOSITION TECHNIQUES

In recent years a number of new technologies have been developed for coating deposition that offer considerable flexibility and process economics in tailoring

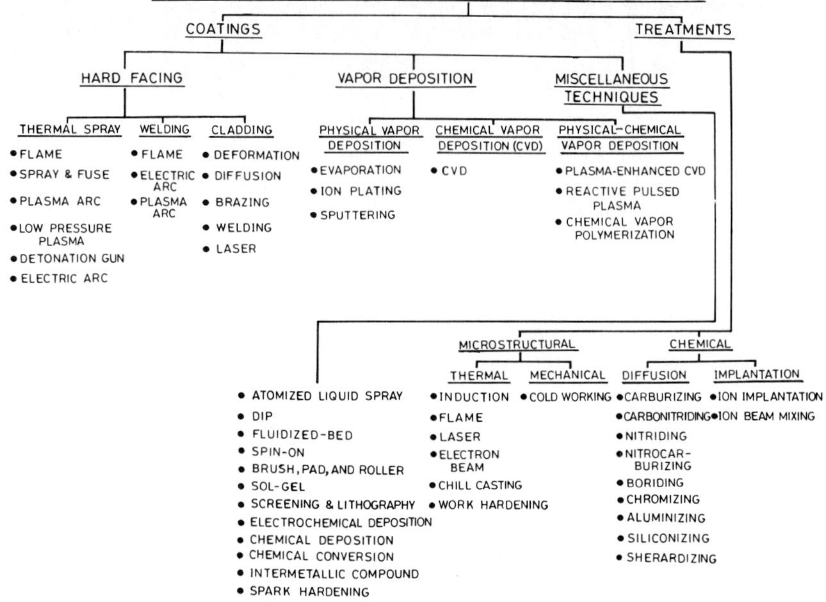

FIGURE 6.1 Classification of various coating deposition and surface treatment techniques.

surface characteristics for tribological applications (Anonymous, 1982, 1983; Appleton, 1984; Bhushan, 1980a, 1987; Booser, 1983, 1984; Bunshah et al., 1982; Chapman and Anderson, 1974; Chopra, 1979; Clauss, 1972; Elliott, 1978; Lang, 1983; Maissel and Glang, 1970; Peterson and Winer, 1980; Vossen and Kern, 1978; Waterhouse, 1965). Table 6.1 presents the comparison of selected process parameters of various coating deposition and surface treatment techniques.

Hard facing is used to deposit thick coatings (typically 50 μm or thicker) of hard wear-resistant materials by thermal spraying and welding. Welded coatings can be applied to substrates which can withstand high temperatures (typically >750°C). A lining of wrought material can be applied by cladding. Thermal sprayed coatings are used extensively in various applications requiring resistance to wear and corrosion.

Many soft and hard coatings are deposited by vapor deposition processes. Usually these techniques are used to deposit thin and reproducible coatings (a few micrometers or less) with significant flexibility and excellent adhesion. However, special equipment is required for deposition. The coating adhesion is superior in ion beam type of ion plating and ion beam type of sputtering and *chemical vapor deposition* (CVD) processes followed by plasma-enhanced CVD process then by glow-discharge type of ion plating and glow-discharge sputtering deposition processes. Evaporation probably results in the least adhesion compared to other vacuum deposition processes; however, it has high throughput. CVD can be used if the substrate can withstand high temperatures (typically >800°C); however, *plasma-enhanced CVD* (PECVD) can be used for deposition at or near room temperature. Ion beam type of process is enormously complex and expensive and is only used in the exceptional cases.

TABLE 6.1 Selected Process Parameters of Various Coating Deposition and Surface Treatment Techniques

(a) Coating deposition techniques								
Deposition technique	Type of coating material	Type of substrate	Deposition temp., °C	Thickness range, μm	Pretreatment	Posttreatment	Relative adhesion	Comments
Thermal spray	Almost any material	Almost any	100–150	50–500	Grit blast	Grind/lap	Fair to good	Relatively thick substrate (≥250 μm) to avoid warpage
Welding	Most metals and alloys	Most metals and alloys	900–1400	750–3000	Mechanically clean	Grind/lap	Good	Relatively thick substrate (≥500 μm) to avoid warpage
Cladding	Most metals and alloys	Most metals and alloys	50–1200	50–100,000	Mechanically clean	None	Excellent	To provide bulk material-like properties of a wear surface, can be used for very large continuous area
Evaporation	Almost any material	Any nongassing material	200–1600	0.1–1000	Mechanically and chemically clean	None	Fair to good	Normally used for thin coatings
Ion plating (glow-discharge/ion beam)	Almost any material	Any nongassing material	100–500	0.02–10	Mechanically and chemically clean	None	Good to excellent	Normally used for thin coatings; complex process
Sputtering (glow-discharge/ion beam)	Almost any material	Any nongassing material	100–500	0.02–10	Mechanically and chemically clean	None	Good to excellent	Normally used for thin coatings; complex process
Chemical vapor deposition	Almost any material	Any material that can withstand deposition temperature	150–2200	0.5–1000	Mechanically and chemically clean	None	Excellent	Normally used for thin coatings

TABLE 6.1 Selected Process Parameters of Various Coating Deposition and Surface Treatment Techniques (*Continued*)

(a) Coating deposition techniques

Deposition technique	Type of coating material	Type of substrate	Deposition temp., °C	Thickness range, μm	Pretreatment	Posttreatment	Relative adhesion	Comments
Plasma-enhanced chemical vapor deposition	Almost any material	Any nongassing material	100–600	0.5–10	Mechanically and chemically clean	None	Good to excellent	Normally used for thin coatings; complex process
Atomized liquid spray/dip/spin-on/brush, pad, and roller	Solid lubricants, metals, and metal compounds	Many materials	Typically 25–200	Typically 7.5–50	Grit blast/etch	Burnish/lap	Poor	Very economical, fast
Fluidized bed	Polymers	Almost any	90–300	20–500	Grit blast/etch	Burnish	Poor	Very economical, fast
Sol-gel	Ceramics	Metals and alloys	25–850	0.3–10	Grit blast/etch	Burnish	Poor	Very economical, fast
Electrochemical deposition	Metals, alloys, and composites	Conducting materials	25–100	0.1–5000	Chemically clean	None	Fair	Very economical, fast
Chemical deposition	Metals, alloys, and composites	Almost any	25–100	0.1–50	Chemically clean	None	Fair	Economical, can coat insulating surfaces
Chemical conversion	Phosphate, chromate, oxide, anodizing, sulfide conversion, and metalliding coating	Many metals	Typically 0–600	Typically 0.1–10	Chemically clean	None	Good	Limited to few substrates
Intermetallic compound	Sn, Cu-Sn, Cu-In-Zn, Sb-Sn-Cd	Iron, steel, brass, bronze, aluminum bronze, aluminum alloys	120–600	20–50	Chemically clean	None	Good	Proprietary process
Spark hardening	Carbide	High-speed steel, Ni, Ti	~1000	—	Chemically clean	Lap	Good	Requires postfinishing

(b) Surface treatment techniques

Treatment	Suitable substrates	Diffusion elements	Treatment temp., °C	Case depth, μm	Comments
Microstructural					
Thermal	Steel and iron	—	Typically 750–1300	Typically 250–3000	Fast, cheap, can be automated; high process temperature
Mechanical	Steel and iron	—	20	100–500	Only used for work hardening steels
Chemical					
Diffusion	Steel and iron	C, N, B, Zn, Cr, Al, Si	Typically 500–1300	Typically 5–500	Fast, cheap, higher hardness than with microstructural changes; high process temperature
Ion implantation	Almost any	Many	50–200	0.01–0.5	Large capital investment, processed at ambient temperature, line-of-sight process, ultra-thin case depths

There are many miscellaneous deposition processes. Atomized liquid spray, dip, spin-on, and brushing techniques are extensively used for application of (generally soft) organic and inorganic coatings from a liquid carrier. Spray pyrolysis is used to apply oxides, sulfides, and carbides of various metals. Spray fusion processes are used to apply specialty coatings for high-temperature applications. Hot dipping processes are used to apply metallic and organic coatings from a bath of molten metal and organic materials with solvent, respectively. Metallic coatings produced by chemical and electrochemical processes are commonly used for wear resistance. Chemical conversion coatings are extensively applied for protection against corrosion as well as lubrication and wear resistance. These processes do not require sophisticated equipment but provide relatively poor adhesion and in some cases little control.

6.1.1 Hard Facing

Hard facing is used to deposit thick coatings of hard wear-resistant materials by thermal spraying or welding on either a worn component or a new component that is to be subjected to wear in service (Anonymous, 1982, 1983; Bunshah et al., 1982; Chapman and Anderson, 1974; Lang, 1983). Hard facing by cladding is another widely used process to apply thick lining of wrought material (Anonymous, 1982). These processes are line-of-sight processes; therefore, it is difficult to coat complex-shaped parts.

As a rule, thermal spraying processes are preferred for applications requiring hard coatings applied with minimum thermal distortion of the workpiece and with good process control. Welding processes, however, are preferred for applications requiring dense, relatively thick coatings with high bond strength. Thermal spraying processes are used to deposit practically any material, but weld deposition and cladding processes are used to deposit mostly metals and alloys. Clad surfaces produced by cladding with wrought material exclude the problems of porosity, nonstoichiometry, and lower hardness of the coating materials as compared to bulk. Parts produced by thermal spraying and welding processes generally need to be postfinished to achieve acceptable roughness, but clad parts generally need not be postfinished.

6.1.1.1 Thermal Spraying. The thermal spraying processes are one of the most versatile methods available for deposition of coating materials. In thermal spraying, coating material is fed to a heating zone where it becomes molten and then is propelled to the pretreated base material. Although the temperature of the molten particles striking the base material is several hundred degrees Celsius, the base material remains below 150°C. The thermal energy necessary to melt the spraying material can be produced by a flame created by combustion gases, by detonation of the combustion gas by a spark plug, electric plasma produced by an electric discharge, or an electric arc. The coating materials are mostly in final powder form, but sometimes the materials can be in rod or wire form. Almost any material capable of being melted without decomposition, vaporization, or sublimation and then resolidified can be thermally sprayed. Most processes are also independent of the composition or condition of the substrate material, provided it can be prepared to offer a receptive surface. Substrates for thermal spraying need to be preroughened by mechanical or chemical methods in order to improve coating adhesion (Anonymous, 1982; Bunshah et al., 1982; Chapman and Anderson, 1974).

In the flame spraying process, fine powder or wire (usually of a metal) is carried in a gas stream and is passed through an intense combustion flame, where it becomes molten. The gas stream, expanding rapidly because of the heating, then sprays the molten powder onto the substrate, where it solidifies. Flame spraying equipment is transportable for in situ work and is widely used by industry for the reclamation of worn or out-of-tolerance parts. However, coatings produced by flame spraying processes are relatively porous.

The spray-and-fuse process is a modification of the powder flame method. The spray-and-fuse process is a two-step process in which the powdered material is deposited by a flame spray gun (or plasma spray gun) and subsequently fused by using an oxyacetylene torch, induction heating, or heat treatment in vacuum or reducing-atmosphere furnaces. The major advantages of the spray-and-fuse process are the virtually fully dense microstructure of the coatings and a metallurgical bond at the interface. In general, nickel- or cobalt-based alloys are sprayed by this process.

A limitation of flame spraying is that only those materials (usually metals) which will melt in the combustion flame can be successfully deposited. This limitation is not found in other thermal spraying processes. The plasma spraying process is similar to flame spraying, except that the powder is now passed through an electric plasma produced by low-voltage, high-current electric discharges. By this means, even refractory materials can be deposited. An inert atmosphere and low-pressure plasma spraying are a variation of conventional plasma spraying process, in which the plasma gun and substrates are enclosed in a vacuum tank. The coating spraying is carried out in an inert atmosphere at low pressure to maintain coating stoichiometry and to produce extremely high bond strength. Complex coating alloys (such as the MCrAlY type for high-temperature applications) are applied by this process.

Detonation-gun coating is somewhat similar to flame spraying; a measured amount of powder is injected into what is essentially a gun, along with a controlled mixture of oxygen and acetylene. The mixture is ignited, and the powder particles are heated and accelerated to extremely high velocities with which they impinge on the substrate, resulting in probably the best adhesion of the thermal spraying process. The process is repeated several times a second. With this process, very hard metals and nonmetal substrates cannot be coated because the high-velocity hot gas stream produces surface erosion.

Finally, in the electric arc spraying, an electric arc is struck between two converging wires close to their intersection point. The high-temperature arc melts the wire electrodes which are formed into high-velocity molten particles by an atomizing gas flow; the wires are continuously fed to balance the loss. The molten particles are then deposited onto a substrate, as with the other spraying processes. With this process, only relatively ductile and electrically conductive materials which can be produced in the wire form can be deposited.

6.1.1.2 ***Welding.*** In the welding technique, the coating is deposited by melting the coating material onto the substrate by gas flame, electric arc, or plasma arc welding processes (Anonymous, 1983; Gregory, 1976; Lang, 1983). The coating material is supplied in the form of powder, paste, rod, strip, or wire. Any material that can be melted and cast may be utilized in the form of welding rod. Arc processes require that the substrate and the coating material be electric conductors.

In contrast to thermal spraying processes, which do not penetrate the substrate metal, welding processes melt a portion of the surface, and this fusion zone becomes mixed or diluted in the weld metal. Mixing of a proportion of substrate metal with the weld, referred to as *dilution*, can affect the composition and the microstructure and hence the wear resistance. However, because of the melting of the substrate metal in the fusion zone, on solidification of the weld there is metallurgical continuity (or graded interface) across the fusion boundary which should ensure the best mechanical properties. Dilution is generally more of a problem in arc welding processes than in gas welding processes. Dilution in gas welding is less than 5 percent, and in arc welding it can range from 5 percent to as much as 30 to 60 percent. However, arc welding processes could be readily automated. The welding processes are capable of depositing coatings onto small components of intricate shape as well as large areas of flat or cylindrical shape.

In *oxyfuel gas welding* (OGW), the heat from a gas flame is used to melt coating materials onto the workpiece. The gas flame is created by the combustion of a hydrocarbon, typically oxyacetylene gas, and oxygen. In this process, the dilution is low, and it requires low-cost equipment which is portable. However, the process is labor-intensive. In the electric arc processes, the heat is supplied by an arc maintained between the coating material and the workpiece. Among electric arc welding processes, the *shielded metal arc welding* (SMAW) process is the most widely used, because this process requires the least capital expenditure. In this process, electrodes with welding flux are used to protect the molten weld metal from atmospheric contamination. In *submerged arc welding* (SAW), a layer of finely divided free-flowing granular flux is placed by gravity ahead of the continuously fed consumable strip or wire. The flux becomes molten, and the arc is submerged beneath the layer of the fusible flux material, thus protecting the molten weld metal from atmospheric contamination. The deposition rates of this process are very high (lower than only electroslag welding). Porosity-free deposits with excellent appearance can be produced. However, dilution of deposit by this process is extremely high—as much as 60 percent. In *electroslag welding* (ESW), an arc is initiated beneath a layer of granular flux as in SAW. As soon as a sufficient layer of flux is melted to form the slag, all arc action stops and current passes from the electrode to the base metal through the conductive slag. The heat generated from passing the electric current through the slag is used to melt the edges of the base metal and filler wire. The deposition rates of this process are the highest of the welding processes. However, the process is expensive and used only for specialized applications.

In *self-shielded metal arc welding* (SSMAW), the workpiece is shielded by a self-generated or self-contained gas or flux. The process is capable of extremely high deposition rates, is cheap, and is portable. However, since the arc is not shielded externally, some porosity can be expected with this process. In *gas metal arc welding* (GMAW), the weld pool is protected by a gaseous shield provided by a gas or mixture of gases fed through the welding gun. This process requires expensive equipment. However, the high cost is generally offset by the higher deposit quality.

In *gas tungsten arc welding* (GTAW), heat is produced by an arc between a nonconsumable tungsten electrode and the workpiece. The electrode, the weld pool, and the arc are protected from atmospheric contamination by a flow of shielding inert gas fed through the torch. The use of a separate filler rod or wire enables greater operator control of deposit shape. However, the deposits produced by this process result in high dilution compared to those by OGW. Its main application is in coating small intricate parts and other specialized applications. Plasma arc welding is similar to GTAW in that both processes employ a gas-shielded electric arc, called a *transferred arc*, between a nonconsumable tungsten electrode and the workpiece as the primary heat source for weld deposition. The process differs in that the PAW process maximizes the use of plasma as a secondary heat source. Another arc is employed between the tungsten electrode and a copper annulus surrounding it. This results in ionization of gas surrounding the electrode, thereby producing a high-temperature plasma. The stiffer arc intrinsic to PAW enables greater control of deposit profile and position. The dilution with this process is relatively low. High deposition rates are possible, and the large parts can be coated. The process has the ability to accommodate reactive substrates, as GTAW does. However, the equipment is very expensive and is used for specialized applications.

6.1.1.3 Cladding. In the cladding process, a metallic foil or sheet (ranging from a few tens of micrometers to a few millimeters or more in thickness) is metallurgically bonded to a metallic substrate to produce a composite structure; the bond is usually stronger than the weaker of the parent metals (Antony et al., 1983; Lang, 1983). The metallic powders or other fillers can also be clad to the metallic substrates. Most metals and alloys can be clad. Metals and alloys are clad by deformation cladding (which includes mechanical methods—roll cladding and coextrusion cladding, explosive cladding, and electromagnetic impact bonding), diffusion bonding, braze cladding, weld cladding, and laser cladding.

Clad surfaces produced by cladding with wrought material experience no problems of porosity, nonstoichiometry, or lower hardness of the coating materials as compared to bulk. The difference between cladding and coatings deposited by conventional deposition techniques is the thickness. Cladding usually denotes the application of a relatively high thickness (typically 1 mm or more) of clad metal whereas a coating is usually thinner. Limitations of this process are that cladding material in many cladding processes must be available in sheet form, and it is difficult to clad parts having complex shapes and extremely large sizes.

In deformation cladding, the metals are bonded by solid-phase methods which rely on a combination of deformation (gross plastic flow) by pressure or impact and heat to generate intimate contact and intermixing with minimum diffusion and to rapidly (within seconds) produce a true metallurgical bond. Mechanical methods are limited to relatively ductile metals and alloys. In diffusion bonding processes, coalescence of contacting surfaces is achieved with minimum macroscopic deformation by diffusion—controlled processes that are induced by applying heat and pressure for a finite time. Bonding by deformation is usually accomplished within seconds, whereas diffusion bonding typically requires contact durations ranging from minutes to hours. Cladding materials in the form of sheet or powder can be applied.

In braze cladding, the surfaces to be clad are sandwiched by a brazing material in the form of powder, paste, rod, wire, strip, or foil, and metallurgical bonding is obtained by heating the stack by furnace, torch, resistance, dip in molten salt bath, induction, or laser beam. The selection of the heating method depends on the coating material, substrate material, coating thickness and shape, and size of the substrate. An advantage of brazing cladding over other methods is that strong joints can be produced at relatively low heat input. However, the brazing material may not be as strong as the parent metals; thus other techniques may produce a bond to withstand higher loads and temperatures.

In the weld and laser cladding processes, the metal is clad by melting it and fusing it to the substrate. The clad metal in the form of cast rod, strip, wire, or powder is melted by electric arc or plasma arc in the case of weld cladding and by a laser beam in the case of laser cladding. In laser cladding, local heating of the parts avoids the distortion and other metallurgical changes. In addition, since only the substrate surface skin is heated, this allows the cladding of high-melting-point metals to low-melting-point substrates.

6.1.2 Vapor Deposition

Many soft and hard coatings can be applied from the vapor phase. There are three classes of vapor deposition techniques: *physical vapor deposition* (PVD), chemical vapor deposition (CVD), and *physical-chemical vapor deposition* (P-CVD) (or plasma-enhanced CVD). Each deposition technique has its advantages and disadvantages as well as a range of preferred applications. PVD processes are line-of-sight processes. The CVD process can be used if the substrate can withstand high temperatures (typically >800°C); however, plasma-enhanced CVD can be used for deposition at or near room temperature.

Vapor deposition processes are versatile in the composition of coating materials. They have the ability to produce rather thin (as low as a couple of nanometers thick) coatings with high purity, high adhesion, and unusual microstructure or self-supported shapes at high deposition rates. Most nongassing substrate materials which can withstand the deposition temperature can be coated. Vapor deposition processes reproduce surface topography and generally do not require postfinishing. A major disadvantage of vapor deposition processes is the high capital cost and processing costs associated with vacuum systems.

6.1.2.1 Physical Vapor Deposition. Physical vapor deposition is used to apply coatings by condensation of vapors in high vacuum (10^{-6} to 10 Pa or 10^{-8} to 10^{-1} torr) atomistically (atom by atom) at the substrate surface. These processes involve PVD: (thermal) evaporation, activated reactive evaporation, glow-discharge or ion beam ion plating, and glow-discharge or ion beam sputtering. Ion beam ion-plating processes give better coverage and higher deposition rates compared to sputtering processes. The high kinetic energy of the coating particles in the ion beam type of process results in superior adhesion compared to glow-discharge-type processes and evaporation without the need for high substrate temperatures (Bhushan, 1987; Bunshah et al., 1982; Chapman and Anderson, 1974; Chopra, 1979; Maissel and Glang, 1970; Powell et al., 1966; Sundgren and Hentzell, 1986; Vossen and Kern, 1978). Evaporation probably results in the least adhesion compared to other vacuum deposition processes; however, it has high throughput.

In the evaporation process, the coating material is generated in the vapor phase by heating the source material to a temperature (typically 1000 to 2000°C) in vacuum (10^{-6} to 1 Pa or 10^{-8} to 10^{-2} torr) such that the vapor pressure (>1 Pa or 10^{-2} torr) significantly exceeds the ambient pressure and produces sufficient vapor for condensation. The coating material is in an electrically neutral state and is expelled from the surface of the source at thermal energies typically 0.1 to 0.3 eV. The substrate must be preheated (typically 200 to 1600°C) to ensure coating adhesion and to obtain proper coating structure. Evaporation in the presence of reactive gases (reactive evaporation) is used to prepare alloy and compound coatings with proper stoichiometry. A plasma is sometimes included in the reactive evaporation in the pressure range of 10^{-2} to 1 Pa (10^{-4} to 10^{-2} torr), to enhance the reaction between the reactants and/or to cause ionization of both the coating metal and gas atoms in the vapor phase. This process is known as *activated reactive evaporation* (ARE). The concurrent ion impingement by an ion beam of inert or reactive gases is also used to improve the adhesion and mechanical properties of evaporated coatings.

The source material is normally in the form of powder, wire, or rod. Direct resistive heating and induction heating are used to apply metallic materials having a low melting point (<1200°C). Refractory materials having a high melting point or high rate of evaporation to deposit thick coatings require high-power density methods such as electron beam heating. Electron beam evaporation can be used to deposit materials with melting temperatures up to 3500°C. Vacuum arc and laser beam heating methods have been used on occasion. The evaporation process is simpler and cheaper compared to other vacuum deposition processes. Evaporation rates as high as 25 μm min^{-1} can be obtained. Coatings or self-supported structures up to 1 mm can be deposited. The adhesion of evaporated coatings may be somewhat poor, in view of the low kinetic energy of the evaporated species, inadequate for many tribological applications.

In the ion-plating process, partial ionization of the metal vapor is used to increase the adhesion of the coating to the substrate. During deposition, the substrate surface and/or the depositing coating is subjected to high-energy flux of ions and energetic neutrals sufficient to cause changes in the interfacial regions or coating properties compared to the nonbombarded deposition. Ion-plating processes can be classified in two broad categories: glow-discharge (plasma) ion plating performed in low vacuum (0.5 to 10 Pa or 5×10^{-3} to 10^{-1} torr) and ion beam ion plating performed in high vacuum (10^{-5} to 10^{-2} Pa or 10^{-7} to 10^{-4} torr). In the glow-discharge ion-plating processes, the material to be deposited is evaporated in a manner similar to ordinary evaporation but passes through a gaseous glow discharge on its way to the substrate, thus ionizing the evaporated atoms in the plasma. Condensation of the vapor takes place under the action of ions from either a carrier gas or the vapor itself on the substrate biased to a high negative potential (typically 2 to 5 kV). A *radio frequency* (rf) potential has to be used for insulating surfaces. In ion beam ion-plating processes, the ion bombardment source is an external ionization source (gun). These utilize single or cluster ion beams. Ion beams can be of inert gas ions or ionized species of coating materials. The substrates must withstand ion bombardment heating. Many metals, alloys, and compounds can be ion-plated.

Depending on the type of plating to be performed, a broad range of vapor sources are used: evaporation by electric resistance, rf induction, electron beam or cathodic arc, and metal-bearing gas or hydrocarbon gases (for carbon). A reactive gas or a beam of desired ionized species can be used to produce alloy or compound coatings. This process is referred to as *reactive ion plating*. Adhesion of ion beam ion-plated coatings is found to be excellent even at low substrate temperature, compared to other vapor deposition processes, because of the extremely high energy of arrival of the deposit ions (10 eV to 10 keV). However, ion beam deposition processes have poorer deposition rates compared to those of glow-discharge deposition processes.

Sputtering is probably the most widely used vapor deposition technique. In this process, the coating material is dislodged and ejected from the solid surface by energetic particles. The high-energy particles are usually positive ions (and energetic neutrals) of a heavy inert gas or reactive gases or species of the coating material. The sputtered material is ejected primarily in atomic form from the source of coating material, called a *target*, which is negatively charged in the case of glow-discharge sputtering. The substrate is positioned in front of the target so that it intercepts the flux of sputtered atoms. The processes used for producing ions are glow-discharge ionization at a pressure range of 0.5 to 10 Pa (5×10^{-3} to

10^{-1} torr) and ion beam sources which are used at pressures ranging from 10^{-5} to 10^{-2} Pa (10^{-7} to 10^{-4} torr). In the ion beam sputtering process, the ion bombardment source is an external ionization source. Ion beams can be of inert or reactive gas ions or ionized species of coating materials. Ion beam sputtering permits independent control over the kinetic energy and the current density and directional control of the bombarding ions.

The bombarding ion energies in glow-discharge sputtering are typically in the range of 100 to 1000 eV, and for external ion beams they range from 100 eV to 10 keV. The average energies of the sputtered atoms are in the range of 10 to 40 eV (up to 100 eV in some processes). Thus the sputtering process is a relatively high-energy process compared to evaporation. Since the coating material is passed into the vapor phase by a mechanical process rather than a chemical or thermal process, virtually any material is a candidate for coating.

6.1.2.2 Chemical Vapor Deposition. The chemical vapor deposition process utilizes a volatile component(s) of coating material which is thermally decomposed (pyrolysis) or chemically reacted with other gases or vapors to form a coating atomistically (atom by atom) on the hot substrate surface. The chemical reactions generally take place in the temperature range of 150 to 2200°C at a pressure ranging from about 65 Pa (0.5 torr) to atmospheric pressure (0.1 MPa or 760 torr). The substrate is generally heated by electric resistance, inductance, or infrared heating. The CVD process at low pressure allows the deposition of coatings with superior quality and uniformity over a large substrate area. The use of a laser as the heat source offers the advantage of heating a very small portion of the substrate, if necessary. The CVD reactions can take place in the presence of electron bombardment which facilitates activation of precursor gases, allowing deposition at lower substrate temperatures.

CVD is a versatile and flexible technique used in producing coatings at high deposition rates. Practically any gas or vapor can be used as a precursor material. CVD coatings usually exhibit excellent adhesion compared to those deposited by evaporation and glow-discharge deposition processes, but requirements of high substrate temperature limit their application to substrates which can withstand high processing temperatures (Bryant, 1977; Bunshah et al., 1982; Chapman and Anderson, 1974; Glaski, 1972; Hintermann, 1981, 1983; Vossen and Kern, 1978).

6.1.2.3 Physical-Chemical Vapor Deposition. Physical-chemical vapor deposition processes are the hybrid processes which use glow-discharge plasma to activate the CVD process. These are broadly referred to as plasma-enhanced CVD (PECVD) or *plasma-assisted CVD* (PACVD). Many phenomena of PECVD are similar to those conventional high-temperature CVD processes. Likewise, much of the physical description of glow-discharge plasmas for sputtering carries over to PECVD. PECVD processes allow coating application to take place at significantly lower substrate temperatures (100 to 600°C) because of the ability of high-energy electrons produced by glow-discharge plasma (at pressures ranging from 1 to 500 Pa or 0.01 to 5 torr) to break chemical bonds and thereby promote chemical reactions. Practically any gas or vapor including polymers can be used as precursor materials. However, the PECVD process is more complex than thermal CVD (Bunshah et al., 1982; Ojha, 1982; Věprek, 1985; Vossen and Kern, 1978; Wright, 1967).

Another method, known as the *reactive pulsed plasma* (RPP) deposition process, uses a high-energy pulsed plasma rather than continuous discharge and restricts the substrate to room temperature (Michalski et al., 1985). The discharge between two coaxial electrodes at a chamber pressure of 10 to 10^2 Pa (0.1 to 1 torr) is struck by discharging the capacitors. The high temperatures produced during the pulsed discharge evaporate and ablate the electrode material, which reacts with the pulse-injected gas in the plasma to form compounds that are deposited on the substrate.

Chemical reactions in the deposition of polymer coatings, called *chemical vapor polymerization*, can be activated by ultraviolet irradiation and electron beam methods as well as by glow-discharge plasma (Shen, 1976; Yasuda, 1978).

6.1.3 Miscellaneous Techniques

This section describes miscellaneous coating deposition techniques: atomized liquid sprayed coating; dip coating; fluidized-bed coating; spin-on coating; brush, pad, and roller coating; sol-gel coating; screening and lithography; electrochemical deposition; chemical deposition; chemical conversion coating; intermetallic compound coating; and spark hardening. These techniques have been widely used for applying polymer coatings, metallic and nonmetallic coatings, and composite coatings for wear and corrosion resistance.

The adhesion and wear resistance of coatings produced by most of the miscellaneous coating techniques described in this section are not as good as those of vapor deposited coatings, and coating adhesion is sometimes not adequate for severe tribological applications. However, they are cheaper deposition processes, and a variety of coatings are used under various situations such as soft, low-shear-strength solid-lubricant and metallic coatings for sliding-wear applications. Chemical conversion processes produce graded interface with good adhesion. These are extensively used for applications requiring corrosion resistance, mild wear resistance, and adhesion promoters for subsequent coating deposition.

6.1.3.1 Atomized Liquid Spray Coatings. Liquid spray processes are most economical and are widely used for paint application and application of organic and inorganic solid lubricants (Bhushan, 1987; Clauss, 1972; Lang, 1983). Several types of spray applications are available: compressed air, hot, airless, and electrostatic spray. All the spray techniques atomize the fluid into tiny droplets and propel them to the substrate. There is usually a significant waste of material due to overspraying. Electrostatic spraying is effective in reducing these losses. The electrostatic sprayer imparts a negative charge on each coating droplet. The part to be coated must be conductive and connected to the ground. The liquid spray processes are extensively used for applying a blend of solid-lubricant powders and solvent system. These mixtures are applied on a preroughened surface, cured at room temperature or elevated temperatures, and burnished.

Spray pyrolysis and spray fusion are the other two techniques in which substrate heating during or after spraying results in improved (chemical) adhesion. In the spray pyrolysis process, the coating suspension is sprayed onto heated substrate (typically 300 to 700°C) at which chemical compounds undergo pyrolytic (endothermic) decomposition and form the coating (Chamberlin and Skarman, 1966; Gupta and Agnihotri, 1977). This process is based on a thermally stimulated chemical reaction between a cluster of liquid or vapor atoms of different chemical species at the substrate surface. Spray pyrolysis is used to apply oxides, sulfides, and carbides of various metals.

In the spray fusion process, a coating is prepared by the fusion of the coating material to the metal substrate. The liquid suspension of the coating mixture is sprayed onto the metal surface, and the coated sample is fired at elevated temperatures (typically 900 to 1100°C) so that the coating melts and fusion results (Bhushan, 1980b). CaF_2-BaF_2 eutectic and PbO-SiO_2 based coatings have been produced by the spray fusion process for high-temperature applications.

6.1.3.2 Dip Coatings. As the name implies, in a dip coating or hot dip coating (also called *hot dipping*) process, the substrate is literally dipped into a liquid bath (Anonymous, 1982; Clauss, 1972; Paul, 1985). The bath consists of a mixture of pigments, resin, and solvents for application of solid-lubricant coatings and molten metals for metallic coatings. The solid-lubricant coatings are usually dried and baked after withdrawal from the bath. Most coatings which can be applied by the spray process can also be applied by the dipping process. The dipping process is desirable for achieving a continuous automated manufacturing operation, and it has the advantage over spray processes of ensuring coverage of parts with complex shapes and inaccessible recesses. The dip process is used in coating large parts.

6.1.3.3 Fluidized-Bed Coatings. A unique variation of dip coating is the fluidized-bed process. In this process, solid polymer powder is levitated in a chamber by air passing through a perforated plate in the bottom of the chamber. So levitated, the particles have much lower bulk density than the quiescent powder, and the "fluidized" powder actually appears to be gently boiling. Coating is accomplished by preheating a solid substrate to a temperature higher than the melting point of the polymer powder and immersing the substrate in the chamber. As the powder strikes the hot substrate, it melts and coalesces into a continuous coating on the surface. As the substrate cools, the polymer solidifies.

6.1.3.4 Spin-On Coatings. Spin-on coating is used to deposit coating on circular substrates. In this technique, a turntable is used that holds the substrate one at a time. The coating suspension/solution is metered onto the substrate near its center, and the turntable is accelerated very rapidly to several thousand revolutions per minute, which spreads the coating suspension/solution.

6.1.3.5 Brush, Pad, and Roller Coatings. Brush, pad, and roller (or roll) coating processes are the mechanical processes commonly used in many industries (Chapman and Anderson, 1974; Paul, 1985). Brushing is a slow method and is used primarily where the areas to be coated are small and somewhat uniquely located. Roller coatings are applicable to flat surfaces. Automated roller coating of both sides of an object is very fast and gives uniform coating. Most coatings which can be applied by spray, dip, and spin-on coating processes also can be applied by brush, pad, and roller coating processes.

6.1.3.6 Sol-Gel Method. The sol (colloidal dispersion) is applied by a suitable method such as dipping, spin coating, or electrophoresis; then dried (transformed to a gel or solid); and finally fired, typically at up to 850°C for a short period (on the order of 15 min). Colloidal dispersions of hydrous oxides or hydroxides have been used as precursors for the preparation of oxide ceramic coatings.

6.1.3.7 Screening and Lithography. In screening and lithography processes, the coating is deposited in selected areas. Screening is used to produce features larger than about 100 μm, whereas lithography is used to produce features as small as 1 μm or less (Thompson et al., 1983).

6.1.3.8 ***Electrochemical Deposition.*** Electrochemical deposition (also known as *electrolytic deposition, electrodeposition*, or *electroplating*) is the most convenient method of applying coatings of metals with high melting points. In this process, the coating is applied by making the substrate the cathode and the donor material the anode in an electrolytic bath. An electric potential to the cell results into electrochemical dissolution of the donor material which gets coated to the substrate (Anonymous, 1982; Bunshah et al., 1982; Chapman and Anderson, 1974). Coatings can be applied to any metal surface and, with suitable preparation, to plastics and many other nonconducting substrates at room temperature or slightly higher temperature (<100°C).

Codeposition of two or more metals is possible under suitable conditions of electric potential and polarization. Composites containing insoluble hard particles in a metal matrix can be deposited by codepositing particles from suspension in agitated electrolytic solutions.

Electrophoretic coating is closely related to an electrodeposition process except that it is primarily concerned with the deposition of colloidal particles rather than ions. The part to be coated is immersed in an aqueous dispersion which dissociates into negatively charged colloidal particles and positive cations. An electric field is applied with the substrate as the anode; the colloidal particles are transported to the anode, where they are discharged and form a coating. Electrophoresis is commonly used as an industrial paint application method.

6.1.3.9 ***Chemical Deposition.*** Chemical deposition or chemical plating differs from the electrochemical deposition process in that no electric current source is required. Metallic coatings by this process can be applied to nonconducting substrates. Coatings of many metals are produced by homogeneous chemical reduction or autocatalytic chemical reduction (electroless) of metal ions in aqueous solution by a reducing agent (Anonymous, 1982; Chapman and Anderson, 1974). The solutions containing metal ions and reducing agents are mixed and sprayed onto substrates which react together, and deposition occurs. Most metals except Pb, Cd, Sn, and Bi can be plated. Hard coatings of metals such as Ni-P and composite coatings of very fine hard particles dispersed in an electroless nickel matrix are commonly used for wear applications.

The application of electroless deposition is usually chosen because of one or more of the following advantages over electrodeposition: The deposits are very uniform, less porous, and more corrosion-resistant than electroplated deposits; by mixing other elements in the coating, the coating can be hard and more wear-resistant than electroplated deposits, and nonconductors can be plated with metallic coatings; complex shapes including holes and inside-diameter surfaces can be plated. The cost of reducing and other complexing agents used in the chemical plating bath makes it more expensive than electroplating processes. The deposition rates of chemical deposition processes are lower than those of electrodeposition processes.

6.1.3.10 ***Chemical Conversion Coatings.*** In chemical conversion processes, nonmetallic coatings are produced by the transformation of the outer atomic layers of a metal surface into new, nonmetallic forms with different sets of properties from the original surface. This is accomplished by means of a reaction induced with an artificial environment in situ wherein the metal itself forms one member of the compound with other members supplied from the environment (Waterhouse, 1965). These coatings have a graded interface, whereas electrodeposited or sprayed coatings merely lie on top of the substrate. These coatings can be produced by chemical or electrochemical deposition at low substrate temperatures (typically under 200°C). The most commonly applied coatings include phosphate coatings, chromate coatings, oxide coatings, anodizing, sulfidized coatings, and metalliding. The chemical conversion coatings are extensively applied for protection against corrosion for the underlying metal as well as for lubrication and wear resistance. Some of the coatings provide the preconditioning to improve adhesion between the underlying metal and the subsequently applied coating.

Phosphate coating or phosphating is the treatment of cast iron, steel, galvanized steel, or aluminum with a dilute solution of phosphoric acid and other chemicals (accelerators) at 80 to 100°C. The surface of the metal, reacting chemically with the phosphoric acid media, is converted to a dense, continuous, nonmetallic, insoluble layer of either crystalline or amorphous phosphate which is an integral part of the surface. Three principal types of coatings are in use: zinc, iron, and manganese phosphate.

Chromate conversion coatings are formed on metal surfaces as a result of chemical attack at room temperature which occurs when a metal is immersed in or sprayed with an aqueous solution of chromic acid, chromium salts such as sodium or potassium chromate or dichromate, hydrofluoric acid or hydrofluoric acid salts, phosphoric acid, or other mineral acids. The chemical attack causes the dissolution of some surface metal and the formation of a protective film containing complex chromium compounds. Generally, the chromate coating is formed by reaction between the metal and the hexavalent Cr; the latter oxidizes the metal and is itself reduced to a trivalent state.

Ferrous metals and aluminum and its alloys are commercially treated to produce oxide coatings by chemical conversion of their surfaces. The most commonly used method for blackening steels and stainless steels (black oxide coating) is the immersion of the metal in a hot (up to 350°C), highly alkaline solution containing oxidizing agent and treatment in an atmosphere of steam at 300–600° C, to convert the surface to black oxide (Fe_3O_4).

In the anodizing or anodic oxidation process, the surface of the material, such as aluminum, is converted to its oxide electrolytically. The liquid anodizing process consists of making the part to be treated the anode in an electrolytic solution through which controlled electric current passes to the cathode (metal container for the electrolyte), converting the surface of the anode to an insulating, tough, amorphous metal oxide. The liquid electrolyzing bath can be replaced by a low-pressure glow discharge. This is known as *gaseous anodization*. The metal substrate to be anodized is positioned in the most conductive region of the discharge and is positively polarized with respect to the anode.

The process of applying a sulfide conversion coating to iron and steel electrolytically is known as *sulfidizing* or *sulfurizing*. In a commercial Sulf BT (or Caubot) process, the iron surface is converted to iron sulfide by using a molten mixture of sodium and potassium thiocyanates as the electrolyte. The parts to be treated are made the anode, and the electrolyte is held at about 190°C.

In metalliding, the diffusion coatings of solid solutions and intermetallic compounds are produced, mostly in the solid state, of metals and metalloids ranging from beryllium to uranium. The diffusing metal, serving as an anode, and the receptor metal, serving as a cathode, are suspended in an electrolyte bath (solution) of molten fluoride salts at 500 to 1200°C. The most useful metalliding reactions appear to be berylliding, boriding, siliciding, aluminiding, titaniding, zirconiding, chrominiding, tantaliding, and yttriding.

6.1.3.11 Intermetallic Compound Coatings. Intermetallic compound coatings (with a graded interface) involve the prior electrodeposition of a nonferrous alloy (for example, Sn, Cu-Sn, Cu-In-Zn) onto ferrous and nonferrous alloy substrates followed by a diffusion treatment of the coated substrate at a predetermined temperature, which gives rise to the formation of intermetallic compounds. Engineering components made of iron, steel, brass, bronze, aluminum bronze, and light aluminum alloys using proprietary processes can be given considerably increased resistance to scuffing, wear, and sometimes corrosion.

6.1.3.12 Spark Hardening. In the spark hardening process, a thin and hard skin integral with substrate is generated by the action of electric sparks struck between the surface and a scanning electrode. A positively charged electrode of the hard material (such as carbide) is vibrated against a negatively charged substrate. Each time a contact is made, current discharges from a condenser, the substrate material is rapidly heated and quenched, and electrode material is transferred and diffused into the substrate, resulting in increased adhesion. Spark hardening has been used for a broad range of machine tools and components to improve their wear resistance (Welsh and Watts, 1962).

6.2 SURFACE TREATMENT TECHNIQUES

Surface treatment processes may be classified into two categories (Anonymous, 1981; Appleton, 1984; Bhushan, 1980a, 1987; Bunshah et al., 1982; Chapman and Anderson, 1974; Dearnaley and Hartley, 1978; Draper and Ewing, 1984; Elliott, 1978; Gabel and Donovan, 1980; Gregory, 1976; Kossowsky and Singhal, 1984; Lang, 1983; Thelning, 1984; Tsaur, 1980):

1. Those in which only the surface structure is changed by thermal or mechanical means: induction hardening, flame hardening, laser hardening, electron beam hardening, chill casting, and work hardening
2. Those in which both the surface composition and the structure are changed by diffusion or ion implantation: carburizing, carbonitriding, nitriding, nitrocarburizing, boriding, chromizing, aluminizing, siliconizing, sherardizing, ion implantation, and ion beam mixing.

Selected process parameters of various surface treatments are compared in Table 6.1.

Surface treatment processes do not produce coating-to-substrate interface as found in coatings. The interface produced in coatings is often subjected to mechanical weakness and interface corrosion. Many of the surface treatment techniques can be applied very economically to large and intricate components and are preferred over coatings for cost reasons. The surface treatments are widely used mostly for ferrous-based materials. Ion implantation can be used to incorporate virtually any kind of ion species into any material without developing a set of diffusion conditions or considering the control of chemical constraints. However, the depth of the transformed layer (~0.01 to 0.5 μm) is generally very small compared to that of other surface treatment processes. Ion implantation requires large capital investment and high running costs. A related technique, ion beam mixing, is used to improve coating performance.

Microstructural treatments are cheap and do not require special materials, unlike chemical diffusion treatments. Surface-heating processes have the ability to selectively harden surfaces, are especially suitable for treatment of large components, and are cheap. Among various surface-heating processes, laser hardening allows a very thin surface layer to be hardened with close control. We note that chemical diffusion processes generally result in higher hardness than surface-heating processes do. Boriding produces the highest hardness compared to any other chemical diffusion process. If the processing is desired at lower treatment temperatures (500 to 600°C), then nitrocarburizing and nitriding processes are preferred. Nitriding and nitrocarburizing do not require quenching and use lower temperatures; thus they can be applied to finished components.

The surface treatment processes are generally used to increase the surface hardness in order to improve the wear resistance. Some processes also improve fatigue resistance and corrosion resistance. The surface treatments are normally applicable to ferrous alloys. The practical thickness that can be applied varies with both the process and the material. The heavier the case, the higher the cost and the greater the dimensional changes. Case depth versus wear resistance is generally ranked as shown in Table 6.2 (Murray, 1984).

6.2.1 Microstructural Treatments

In most microstructural treatments, a thin surface shell of ferrous metal (steels and cast irons) is case-hardened by localized surface heating and quick quenching to produce a hard, wear-resistant martensitic structure with a tough, ductile core (Cottrell, 1964; Fleischer, 1964; Peckner, 1964). The fatigue strength is increased with this process because the treated surface is left with compressive residual stresses. Two heat treatment techniques are being widely used: induction heating and flame hardening (Anonymous, 1981; Thelning, 1984). In addition, laser and electron beam hardening are also becoming popular (Anonymous, 1981; DesForges, 1978; Draper and Ewing, 1984). Flame hardening does not lend itself to close control, but it is particularly suitable for large parts. The other three processes can be closely controlled by varying the energy input. Laser hardening allows a very thin surface layer to be hardened with close controls. A few old surface hardening techniques—chill casting and work hardening in addition to induction, flame, laser, and electron beam hardening—are also used.

The heat-treated parts are used in severe applications requiring wear resistance even at very high compressive stresses, such as bearing surfaces, rotating shafts, valves, gears, cams, and piston rings.

TABLE 6.2 Typical Ranking of Wear Resistance as a Function of Case Depths of Surface Treatments

Case depth		Functional performance
Light case	<50 μm	Good wear resistance at low stresses
Medium case	50–200 μm	Good wear resistance at higher stresses
Heavy case	200–500 μm	Sliding and abrasive wear resistance; resists crushing and fatigue
Extra heavy case	>500 μm	Wear and shock resistance at severe conditions

6.2.1.1 ***Induction Hardening.*** In induction hardening, the medium-carbon steels containing 0.3 to 0.5 percent carbon are generally treated. In this process, a high-frequency alternating current induces eddy currents in the surface of the steel components which consequently become heated. The depth of penetration of the heat is influenced by the power density and frequency of the alternating current applied in the conductor coil and the time for which the current is applied. The depth of penetration is inversely proportional to the square root of the frequency. Quenching, depending on the alloy content, can be achieved either by immersing the heated part in a tank or by spraying the component after heating has been completed. The time required for induction hardening is short (1 to 30 s), and components can be heat-treated with practically no scaling. However, the capital cost of induction hardening is high.

6.2.1.2 ***Flame Hardening.*** Flame hardening is similar to induction hardening in terms of materials processed and their application. But instead of using an induction current to rapidly heat the surface of a ferrous base metal, heating is achieved either by direct impingement of a high-temperature flame or by high-velocity combustion product gases. Flame hardening is particularly useful for components which would be too large for conventional induction surfaces. The surfaces can be selectively hardened. However, it lends to poor control and excessive scaling.

6.2.1.3 ***Laser Hardening.*** In laser hardening, a laser (typically 1 to 15 kW CO_2) beam (which can be scanned) irradiates the surface for a short time and, on absorption, melts a very thin layer (1 to 100 μm) adjacent to the surface. Since the melting occurs in a very short time (within a fraction of a second) and only at the surface, the bulk of the material remains cool, thus serving as an intimate heat sink. Quench rates as great as 10^{11}°C s^{-1} can be realized. Because of rapid heating and the restricted area of heating, there is minimum distortion of the substrate. The laser hardening process is normally applied to machine-finished or ground surfaces.

6.2.1.4 ***Electron Beam Hardening.*** In electron beam hardening, the surface is heated by direct electron bombardment of an accelerated stream of electrons in vacuum ($\sim 10^{-2}$ Pa or 10^{-4} torr). After the heat treatment cycle, ranging typically from 0.5 to 2.5 s, the electron flow is stopped abruptly to allow the component being processed to self-quench. Electron beam hardening is a very efficient process (90 percent) compared to laser hardening (10 percent). Like laser hardening, this process is normally adapted for finished or ground surfaces. The limitations of both laser and electron beam hardening processes are the restrictions on the size of the components, and the line-of-sight nature restricts the applications.

6.2.1.5 ***Chill Casting.*** Chill casting is also used to harden critical surfaces on cast-iron parts which contain about 3 percent carbon. Instead of allowing slow cooling with formation of graphite flakes, chills are used to cool the cast iron rapidly, and cementite (Fe_3C) is formed. Other carbide-forming elements such as chromium and vanadium are also added to promote surface hardness.

6.2.1.6 ***Work Hardening.*** Ferrous and nonferrous metals and alloys can also be strengthened at the surface considerably by plastic deformation through work hardening. Work hardening can be achieved by metalworking and shot peening.

Metalworking refers to the large plastic deformation of metal under conditions of temperature and strain rates that induce strain hardening. In the metalworking process, an initially simple part—a billet or a blanket sheet, e.g.—is plastically deformed between tools (or dies) to obtain the desired final configuration. Various metalworking processes include drawing, extrusion, forming, and forging (Anonymous, 1988).

In shot peening, the metallic parts are plastically deformed by the impingement of a stream of shot, directed at the metal surface at high velocity under controlled conditions. This process increases the strength of a ductile crystalline material by coalescence of dislocations; this interrupts their motion and thus decreases dislocation mobility. The result is increased hardness and reduced toughness. The interaction of dislocations, reduced grain size, develops residual stress (compressive in nature) which increases the yield strength and flow stress. The major purpose of shot peening is to increase fatigue strength.

6.2.2 Diffusion Treatments

A variety of commercial thermochemical diffusion treatments increases the hardness of metals, particularly steel and iron parts. At elevated temperatures, chemical species can be diffused into most iron and steel alloys at significant rates. This fact is used to advantage by exposing heated ferrous parts to an appropriate medium, which may be a solid, liquid, gas, or plasma. Atom-by-atom transfer occurs from the medium to the part, thus modifying the surface chemistry and creating a new alloy at the surface (Cottrell, 1964; Fleischer, 1964; Peckner, 1964). Surface treatments such as carburizing, carbonitriding, nitriding, nitrocarburizing, boriding, ion implantation, and ion beam mixing are the most prominent for tribological applications, and many modifications are available to achieve specified changes in surface chemistry and metallurgical structure. Other treatments which are also used include chromizing, aluminizing, siliconizing, and sherardizing. Extensive information is available about various processes in the literature (Anonymous, 1981; Child, 1983; Elliott, 1978; Gregory, 1976; Thelning, 1984) and from the suppliers.

*6.2.2.1 **Carburizing.*** During carburizing, the carbon content of a low-carbon (0.1 to 0.2 percent C) steel surface, which can be from free-cutting mild steels to high-strength nickel-chromium steels, is increased by heating the steel to about 925°C (so that it is in the austenitic condition) and holding it in the presence of a carbon-rich medium, which can be a solid (charcoal), molten salt, gas, or plasma. These parts are then directly quenched or cooled, reheated, and again quenched in oil or water, depending on the alloy content of the steel. During case hardening, it is usual to obtain a surface carbon composition of 0.65 to 0.8 percent, which produces a high surface hardness. Carburizing can produce very thick cases with a hardness of up to 850 HV, which is accompanied by improved wear resistance. However, in the hardening process, the structure transforms to martensite which results in some growth. In addition, high-temperature treatment causes some distortion.

*6.2.2.2 **Carbonitriding.*** Carbonitriding is a modified form of carburizing, rather than a form of nitriding. In the carbonitriding process, carbon and nitrogen are introduced simultaneously into the surface of the steel in the austenitic condition at temperatures between 800 and 870°C via either liquid or gaseous medium and then quenched. In carbonitriding, the diffused amount of carbon is greater than that of nitrogen. Nitrogen increases hardness more effectively than diffused carbon in the case. Carbonitriding can be applied to any of the steels suitable for carburizing. However, the case depth is limited up to 750 μm, whereas carburizing can produce very thick (up to 6-mm) cases.

6.2.2.3 *Nitriding.* Nitriding treatments are ferritic thermochemical treatments for low-alloy and tool steels, and they usually involve the introduction of atomic nitrogen into the ferritic phase in the temperature range of 500 to 590°C. The nitrogen is introduced into the surface of steel by reaction with a gas or solid phase or by a nitrogen-containing plasma. Nitriding is most effective when applied to only a limited range of alloy steels containing elements that form stable nitrides, for example, Al, Cr, Mo, V, and W.

The process times are long compared to carburizing; for example, 100 h at 520°C is necessary to obtain a case depth of 0.5 mm, and at the end of this time parts are cooled slowly in the furnace under a protective atmosphere. Quenching is not needed, unlike with carburizing; therefore, distortion is minimal, and treatment of finished surfaces is possible. A recent development is the ion nitriding process. Hardnesses as high as 1300 kg mm^{-2} can be obtained by this process. Ion nitriding also reduces the treatment time, even at lower temperatures, which reduces distortion of the substrate.

6.2.2.4 *Nitrocarburizing.* Nitrocarburizing is a modified form of nitriding rather than a form of carburizing. This process involves the introduction of nitrogen and carbon, occasionally with small amounts of sulfur, into the surface of steel in the ferritic condition at temperatures from 540 to 600°C via either liquid or gaseous medium or nitrogen- and carbon-containing plasma. In nitrocarburizing, the amount of diffused nitrogen is very high compared to that of carbon. Nitrocarburizing treatment is carried out on mild steel in the rolled condition or normalized conditions as well as on alloy and stainless steels. Nitrocarburized steel surfaces offer exceptional resistance to wear and fatigue.

6.2.2.5 *Boriding.* Boriding, also known as *boronizing*, applies mainly to plain carbon and low-alloy steels. But this process can also be applied to nonferrous metals such as nickel- and cobalt-based alloys and refractory metals. Boriding can be carried out in either pack, liquid, or gas media at a temperature generally between 900 and 1100°C to form stable borides such as Fe_2B, CrB_2, MoB, and NiB_2. The borided steels exhibit extremely high hardness, typically 1500 to 2300 HV.

6.2.2.6 *Chromizing.* Chromizing involves diffusion of chromium into high-carbon steel by heating at 950 to 1300°C in chromium-rich (solid, liquid, or gaseous) media. The treated surface provides oxidation and wear resistance. The typical hardness achieved for chromized cases ranges from 1200 to 1300 HV for high-carbon steel.

6.2.2.7 *Aluminizing.* In aluminizing, also known as *calorizing*, aluminum is diffused into the steel surface at 750 to 1000°C in the aluminum-rich (solid or liquid) media. The treated surfaces impart resistance to corrosion and oxidation. The typical hardness ranges from 200 to 500 HV.

6.2.2.8 *Siliconizing.* In the siliconizing process, silicon diffuses into the surface of steel or malleable cast iron, which acquires a high resistance to wear, scaling at elevated temperatures, and corrosion. In the process, steel is treated at 950 to 1200°C in the presence of solid or gaseous media. The siliconized layer exhibits high brittleness, moderate hardness of 200 to 300 HV, and high porosity.

6.2.2.9 ***Sherardizing.*** In sherardizing, zinc is diffused into the iron or steel surface at 350 to 450°C in zinc dust media. The treated surface provides corrosion protection.

6.2.3 Implantation Treatments

6.2.3.1 ***Ion Implantation.*** Ion implantation has been used successfully as a surface treatment (Appleton, 1984; Kossowsky and Singhal, 1984) to improve the wear (Dearnaley and Hartley, 1978; Hartley, 1979), fatigue (Lo Russo et al., 1979), and corrosion resistance (Ashworth et al., 1976) of materials. Ion implantation allows the controlled introduction of one or more species into the surface of a substrate by using high-energy ion beams of the species to be introduced in a vacuum chamber (10^{-3} to 10^{-4} Pa or 10^{-5} to 10^{-6} torr). Ion beam energies are in the range of 10 to 200 keV. The conversion layers of ion-implanted surfaces are very shallow (0.01 to 0.5 μm) compared to other surface treatments. Therefore, ion implantation is used only for lightly loaded tribological applications where thicker coatings and treatments are not feasible. However, since the altered layer is extremely thin and the implantation is normally carried out at low temperature, ion implantation can be carried out on delicate and finished substrates. Ion implantation is a line-of-sight process. Ion implantation requires high capital and running costs.

The implantation of nonmetallic (for example, B, N, C, P) and metallic (for example, Cr, Ti, Zr, Ni, Fe, Co, Sn, In, Sb, As) species into metals, cermets, ceramics, and polymers offers a whole new field of technological exploration since the injection of atomic species A into substrate B is not limited by thermodynamic criteria.

6.2.3.2 ***Ion Beam Mixing.*** The coating characteristics such as interfacial adhesion can be dramatically enhanced by irradiating or bombarding the coating surface with ion beams. In ion beam mixing, a 10- to 100-nm-thick coating is first applied by conventional vacuum deposition process and then bombarded (or concurrently bombarded during the total deposition period) vigorously with a beam of energetic ions (typically 100 to 500 keV) of easily generated reactive or nonreactive gaseous species such as Ar^+, Kr^+, O^+, and N^+ (Appleton, 1984). Under optimum conditions, either partial or complete intermixing takes place, which produces changes in microstructure and hardness and increased wear resistance (Chevallier et al., 1986; Köbs et al., 1986).

6.3 CRITERIA FOR SELECTING COATING MATERIAL/DEPOSITION AND SURFACE TREATMENT TECHNIQUES

The selection of a coating material/deposition or surface treatment technique for an engineering component can be made only when all engineering and environment requirements are known. In selecting a material for a particular engineering function, three requirements must be satisfied (Bhushan, 1987):

- The mechanical and tribological properties—such as hardness, ductility, fatigue strength, shear strength; resistance to wear, oxidation, and corrosion; thermal and impact shock—must be adequate to provide a satisfactory operating life,
- It must be feasible to fabricate the component economically in numbers likely to be needed,
- The material must be capable of withstanding the environmental conditions imposed upon it during its design lifetime.

It would be a rare service situation that exhibits a single environment feature needing consideration, and most engineering components are subject to a complex range of factors, each affecting coating material and treatment selection. For example, complex demands are placed on turbine blade and ring materials for jet engines, where resistance to creep, fatigue, high-temperature oxidation, erosion, fretting, thermal shock, thermal cycling, and thermal stability is required of a single component. Similarly complex demands are placed on many components in highly sophisticated modern engineering plants. In selecting materials for wear-resistant components, the first requirement normally is to identify the type of wear expected and to prescribe against it.

Coating thicknesses range from a fraction of micrometer to few millimeters. Generally, a thicker coating is desirable for longer life, but in some applications thin coatings (from submicrometers to few micrometers) are more desirable for the following reasons (Bhushan, 1987):

- Thin coatings accurately reproduce the substrate topography, removing the need for finishing (grinding).
- They tolerate thermal expansion mismatches better.
- They do not change the mechanical properties of the substrate (important in thin substrates).
- When the substrate is flexible, they can bend freely without cracking.

If due consideration is given to the environmental factors and the wear processes are considered at the design stage, then selection of coating deposition and surface treatment techniques will be relatively easy.

Table 6.3 categorizes coating deposition and surface treatment techniques used to solve wear problems (based on a table initially presented by Murray, 1984). One obvious conclusion is that in many applications two types of coatings with entirely different physical properties might provide equally satisfactory service. Table 6.4 presents typical examples of coatings and surface treatments for solving specific wear problems.

TABLE 6.3 Practical Applications for Coating Deposition and Surface Treatment Techniques

Coating or treatment technique	Adhesive	Abrasive	Fretting	Erosion	Impact sliding	Corrosive
Hard facing						
Thermal spray	A	A	B	A	A	C
Welding/cladding	B	A	B	A	A	A
Physical vapor deposition	A	—	C	—	—	C
Chemical vapor deposition	A	B	—	A	—	C
Miscellaneous coating deposition						
Atomized liquid spray	A	B	B	—	—	—
Dipping	A	B	B	—	—	C
Fusion	A	B	B	—	—	—
Electrochemical	C	—	C	—	—	A
Chemical	B	B	C	—	—	A
Chemical conversion	C	—	—	—	—	A
Thermal treatments	A	C	C	—	C	—
Mechanical treatments	A	—	—	—	—	—
Chemical treatments	A	A	B	—	C	C

A = major usage, B = frequent usage, C = occasional usage.

TABLE 6.4 Summary of Approaches to Selection of Coatings and Surface Treatments for Controlling Specific Type of Wear

Purpose of coating	Basic approach	Coating process	Limitations	Use on
Reduce adhesive wear		Electroplating or electroless composite	Must be lubricated	Most metals
	Change composition of surfaces (increased hardness generally beneficial)	Chemical conversion	Limited wear life	Cast iron and steel
		Solid-lubricant coatings, organic coatings	Liquid lubricants may soften coatings	Most metals
		Sprayed coatings	Quality control	Most metals
		Anodizing	Cracking because of substrate deformation	Al, Be, Mg
		Malcomizing (nitriding)	Very thin cases	Austenitic stainless steels
		Soft nitriding	Thin case	Most steels
		Sputtering	Cost, best for high-precision parts	Most metals
		Chemical vapor deposition	High-temperature process	Temperature-tolerant materials
Improve conformability, prevent scuffing	Soft, conformable coating	Electroplated soft metal (e.g., Sn or Pb alloy)	Lubricated applications	Electroplating—most metals
	Chemical conversion coating	Phosphate, oxide, or sulfurize	Fluid film lubrication should result	Chemical conversion—generally cast iron or steel
Reduce abrasive wear	Embeddable material or elastomer coating	Babbitt lining or rubber bonded to metal backing	Life determined by: hardness, shape and size of abrasive, and coating thickness	Steel, bronze, aluminum, etc.
Light abrasion	Increase surface hardness	Thermal or diffusion coatings		Generally steels and cast iron
		Chrome plate		Most metals
		Sprayed coatings of cermets or ceramics		
Moderate abrasion	Increase surface hardness	Heavy case carburizing	Same as above, high-temperature processes	Steel and cast iron
		Sprayed, fused coatings		Steel and other high-melting-point alloys
		Hard facings		

TABLE 6.4 Summary of Approaches to Selection of Coatings and Surface Treatments for Controlling Specific Type of Wear (*Continued*)

Purpose of coating	Basic approach	Coating process	Limitations	Use on
Heavy abrasion	Increase hardness	Hard facings, e.g., high-chrome steels	Same as above	Steel and other high-melting-point alloys
Abrasion plus oxidation or corrosion	Increase hardness and corrosion resistance	Cobalt- or nickel-based hard facings	Same as above	
Abrasion plus impact	Use work hardenable fatigue-resistant alloys	Facings of austenitic manganese steels	Same as above	
Reduce corrosive wear	Change surface composition	Electroplating	Wear life of coatings	Most metals
		Nitriding		Steels
		Chemical conversion		Cast iron, steel
		Ion implantation		Most metals
		Sprayed metals		Most metals
		Siliconizing		Steel
Reduce fretting wear	Increase hardness and reduce adhesion	Thermal or diffusion coatings	Wear life	Steel or cast iron
	Provide lubricant reservoirs	Spark hardening		Most metals
		Porous sprayed coatings		
	If dry, provide lubricant	Bonded solid lubricants	Wear life; may be affected by liquid lubricants	Most metals
Reduce solid-particle erosive wear	Use hard or elastic surface coatings*	Diffusion	For low-angle impact	Steel, cast iron
		Thermal spray and fuse	Same as above	Most metals
		Sprayed metals (annealed)	For high-angle impact	Most metals
		Elastomers and plastics	For high-angle impact	Most metals
Reduce cavitation-erosive wear	Sacrificial metal overlays	Sprayed, welded, or electroplated	Wear life of coatings and adhesion to substrate	Most metals
	Elastic surface coatings	Elastomers or plastics		Most metals

*Some very hard coatings are promising for high-angle impact.
Source: Adapted from Murray (1984).

6.4 REFERENCES

Anonymous (1981), *Metals Handbook*, Vol. 4: *Heat Treating*, 9th ed., American Society for Metals, Metals Park, Ohio.

——— (1982), *Metals Handbook*, Vol. 5: *Surface Cleaning, Finishing and Coating*, 9th ed., American Society for Metals, Metals Park, Ohio.

——— (1983), *Metals Handbook*, Vol. 6: *Welding, Brazing and Soldering*, 9th ed., American Society for Metals, Metals Park, Ohio.

——— (1988), *Metals Handbook*, Vol. 14: *Forming and Forging*, 9th ed., American Society for Metals, Metals Park, Ohio.

Antony, K. C., Bhansali, K. J., Messler, E. W., Miller, A. E., Price, M. O., and Tucker, R. C. (1983), "Hardfacing," in *Metals Handbook*, Vol. 6: *Welding, Brazing, and Soldering*, 9th ed., pp. 771–803, American Society for Metals, Metals Park, Ohio.

Appleton, B. R. (1984), "Ion Beam and Laser Mixing: Fundamentals and Applications," in *Ion Implantation and Beam Processing* (J. S. Williams and J. M. Poate, eds.), pp. 189–261, Academic, New York.

Ashworth, V., Grant, W. A., and Procter, R. P. M. (1976), "The Applications of Ion Beams to Corrosion Science," *Corrosion Sci.*, Vol. 16, pp. 661–675.

Bhushan, B. (1980a), "High Temperature Self-Lubricating Coatings and Treatments—A Review," *Metal Finish.*, Vol. 78, May, pp. 83–88, and June, pp. 71–75.

——— (1980b), "High Temperature Self-Lubricating Coating for Air Lubricated Foil Bearings for the Automotive Gas-Turbine Engine," Tech. Rep. No. CR-159848, NASA Lewis Research Center, Cleveland, Ohio.

——— (1987), "Overview of Coating Materials, Surface Treatments and Screening Techniques for Tribological Applications Part I: Coating Materials and Surface Treatments," *Testing of Metallic and Inorganic Coatings* (W. B. Harding and G. A. DiBari, eds.), Special Publication STP 947, pp. 289–309, ASTM, Philadelphia, Pa.

———, Ruscitto, D., and Gray, S. (1978), "Hydrodynamic Air-Lubricated Compliant Surface Bearings for an Automotive Gas Turbine Engine Part II: Materials and Coatings," Tech. Rep. No. CR-135402, NASA Lewis Research Center, Cleveland, Ohio.

Booser, E. R. (1983), *Handbook of Lubrication: Theory and Practice of Tribology,* Vol. I: *Application and Maintenance*, CRC Press, Boca Raton, Fla.

——— (1984), *Handbook of Lubrication: Theory and Practiceof Tribology,* Vol. II: *Theory and Design*, CRC Press, Boca Raton, Fla.

Bryant, W. A. (1977), "The Fundamentals of Chemical Vapor Deposition," *J. Mater. Sci.*, Vol. 12, pp. 1285–1306.

Bunshah, R. F., Blocher, J. M., Bonifield, T. D., Fish, J. G., Ghate, P. B., Jacobson, B. E., Mattox, D. M., McGuire, G. E., Schwartz, M., Thornton, J. A., and Tucker, R. C. (1982), *Deposition Technologies for Films and Coatings*, Noyes, Park Ridge, N.J.

Chamberlin, R. R., and Skarman, J. S. (1966), "Chemical Spray Deposition Process for Inorganic Films," *J. Electrochem. Soc.*, Vol. 113, pp. 86–89.

Chapman, B. N., and Anderson, J. C. (eds.) (1974), *Science and Technology of Surface Coatings*, Academic, New York.

Chevallier, J., Olesen, S., Sørensen, G., and Gupta, B. (1986), "Enhancement of Sliding Life of MoS_2 Films Deposited by Combining Sputtering and High Energy Ion-Implantation," *Appl. Phys. Lett.*, Vol. 48, pp. 876–877.

Child, T. H. C. (1983), "Surface Treatments for Tribology Problems," in *Coatings for High Temperature Applications* (E. Lang, ed.), pp. 395–433, Applied Science Publishers, New York.

Chopra, K. L. (1979), *Thin Film Phenomenon*, Robert E. Krieger Publishing Co., New York.

Clauss, F. J. (1972), *Solid Lubricants and Self-Lubricating Solids*, Academic, New York.

Cottrell, A. H. (1964), *Theory of Crystal Dislocations*, Gordon and Breach, New York.

Dearnaley, G., and Hartley, N. E. W. (1978), "Ion Implantation into Metals and Carbides," *Thin Solid Films*, Vol. 54, pp. 215–232.

DesForges, C. D. (1978), "Laser Heat Treatment," *Tribol. Int.*, Vol. 11, pp. 139–144.

Donovan, M., and Sanders, J. L. (1972), "Surface Coatings," *Tribol. Int.*, Vol. 5, pp. 205–224.

Draper, C. W., and Ewing, C. A. (1984), "Review Laser Surface Alloying; A Bibliography," *J. Mater. Sci.*, Vol. 19, pp. 3815–3825.

Elliott, T. L. (1978), "Surface Hardening," *Tribol. Int.*, Vol. 11, pp. 121–128.

Fleischer, R. L. (1964), *The Strengthening of Metals*, p. 93, Reinhold, New York.

Gabel, M. K., and Donovan, J. M. (1980), "Wear Resistant Coatings and Treatments," in *Wear Control Handbook* (M. B. Peterson and W. O. Winer, eds.), pp. 343–371, ASME, New York.

Glaski, F. (ed.) (1972), *Proc. 3rd Int. Conf. on Chemical Vapor Deposition*, American Nuclear Society, Hinsdale, Ill.

Gregory, E. N. (1976), "Hard Surfacing by Welding for Improved Wear Resistance," *Surfacing J.*, Vol. 7, pp. 5–10.

Gupta, B. K., and Agnihotri, O. P. (1977), "Effect of Cation-Anion Ratio on Crystallinity and Optical Properties of Cadmium Sulfide Films Prepared by Chemical Spray Deposition Process," *Solid State Commun.*, Vol. 23, pp. 295–300.

Hartley, N. E. W. (1976), "Tribological Effects in Ion-Implanted Metals," in *Application of Ion Beams to Materials* (G. Carter, J. S. Colligen, and W. A. Grant, eds.), Institute of Physics, London.

——— (1979), "Friction and Wear of Ion-Implanted Metals—A Review," *Thin Solid Films*, Vol. 64, pp. 177–190.

Hintermann, H. E. (1981), "Tribological and Protective Coatings by Chemical Vapor Deposition," *Thin Solid Films*, Vol. 84, pp. 215–243.

——— (1983), "Surface Treatments," in *Proc. Int. Conf. on Science of Hard Materials* (R. K. Viswanadham, D. J. Rowcliffe, and J. Gurland, eds.), pp. 357–391, Plenum, New York.

Köbs, K., Dimigen, H., Hubsch, H., Tolle, H. J., Leutenecker, R., and Ryssel, H. (1986), "Improved Tribological Properties of Sputtered MoS_x Films by Ion Beam Mixing," *Appl. Phys. Lett.*, Vol. 49, pp. 496–498.

Kossowsky, R., and Singhal, S. C. (1984), *Surface Engineering: Surface Modification of Materials*, Martinus Nijhoff Publishers, Dordrecht, The Netherlands.

Lang, E. (1983), *Coatings for High Temperature Applications*, Applied Science Publishers, New York.

Lo Russo, S., Mazzoldi, P., Scotoni, I., Tosello, C., and Tosto, S. (1979), "Effect of Nitrogen-Ion Implantation on the Lubricated Sliding Wear on Steel," *Appl. Phys. Lett.*, Vol. 34, pp. 627–629.

Maissel, L. I., and Glang, R. (1970), *Handbook of Thin Film Technology*, McGraw-Hill, New York.

Michalski, A., Sokolowska, A., and Legutko, S. (1985), "The Useful Properties of TiN_x-Ti Coatings Deposited onto Drills at 500 K Using the Reactive Pulse Plasma Method," *Thin Solid Films*, Vol. 129, pp. 249–254.

Murray, S. F. (1984), "Wear Resistant Coatings and Surface Treatments," in *Handbook of Lubrication: Theory and Practice of Tribology,* Vol. II: *Theory and Design* (E. R. Booser, ed.), pp. 623–644, CRC Press, Boca Raton, Fla.

Ojha, S. M. (1982), "Plasma-Enhanced Chemical Vapor Deposition of Thin Films," in *Physics of Thin Films* (G. Hass, M. H. Francombe, and J. L. Vossen, eds.), Vol. 12, pp. 237–296, Academic, New York.

Paul, S. (1985), *Surface Coatings: Science and Technology*, Wiley, New York.

Peckner, D. (ed.) (1964), *The Strengthening of Metals*, Reinhold, New York.

Peterson, M. B., and Winer, W. O. (1980), *Wear Control Handbook*, ASME, New York.

Powell, C. F., Oxley, J. H., and Blocher, J. M. (eds.) (1966), *Vapor Deposition*, Wiley, New York.

Shen, M. (ed.) (1976), *Plasma Chemistry of Polymers*, Marcel Dekker, New York.

Smart, R. F. (1978), "Selection of Surfacing Treatments," *Tribol. Int.*, Vol. 11, pp. 97–104.

Sundgren, J.-E., and Hentzell, H. T. G. (1986), "A Review of the Present State of the Art in Hard Coatings Grown from the Vapor Phase," *J. Vac. Sci. Technol.*, Vol. A4, pp. 2259–2279.

Thelning, K. E. (1984), *Steel and Its Heat Treatment: Bofors Handbook*, Butterworth, London.

Thompson, L. F., Willson, C. G., and Bowden, M. J. (eds.) (1983), *Introduction to Microlithography,* Symp. Series No. 19, American Chemical Society, Washington, D.C.

Tsaur, B. Y. (1980), "Ion-Beam-Induced Interface Mixing and Thin-Film Reactions," in *Proc. of Symp. on Thin-Film Interfaces and Interactions*, Vol. 80-2, pp. 205–231, The Electrochemical Society, Princeton, N.J.

Věprek, S. (1985), "Plasma-Induced and Plasma-Assisted Chemical Vapor Deposition," *Thin Solid Films*, Vol. 130, pp. 135–154.

Vossen, J. L., and Cuomo, J. J. (1978), "Glow Discharge Sputter Deposition," in *Thin Film Processes* (J. L. Vossen and W. Kern, eds.), pp. 12–75, Academic, New York.

———, and Kern, W. (eds.) (1978), *Thin Film Processes*, Academic, New York.

Waterhouse, R. B. (1965), "The Formation, Structure and Wear Properties of Certain Non-metallic Coatings on Metals," *Wear*, Vol. 8, pp. 421–447.

Welsh, N. C., and Watts, P. E. (1962), "The Wear Resistance of Spark-Hardened Surfaces," *Wear*, Vol. 15, pp. 289–307.

Wright, A. N. (1967), "Surface Photopolymerization of Vinyl and Diene Monomers," *Nature*, Vol. 215, pp. 953–955.

Yasuda, H. (1978), "Glow Discharge Polymerization," in *Thin Film Processes* (J. L. Vossen and W. Kern, eds.), pp. 361–400, Academic, New York.

CHAPTER 7
SURFACE PREPARATION FOR COATING DEPOSITION

Under usual deposition applications, the atoms of the coating material arrive at the substrate surface one at a time, nucleate, and gradually coalesce to form a film (Walton, 1966). The adhesion* of a deposited coating to a substrate depends on the type and structure of the interfacial region between the coating and substrate as well as the nature and extent of bonding across this region. An understanding of the relationship between the interface and adhesion enables specification of the causes of poor adhesion and potential solution of adhesion problems.

The types of coating-substrate interfacial regions formed may be classified into four categories according to typical characteristics: (1) A mechanical interface may be formed if the surface contains roughness or large pores. In this case, the coating material fills the valleys of the roughness or the porosity, and mechanical interlocking provides the adhesion. No chemical reaction between the materials is necessary, and the strength of the bond depends on the physical properties of the material and the roughness or porosity. (2) A monolayer-to-monolayer interface is characterized by an abrupt change from the coating material to the substrate material in a distance on the order of the separation between atoms (0.2 to 0.5 nm). This type of interface will be formed if there is no diffusion of the depositing atom into the substrate surface and little chemical reaction between the atoms of the coating material and the substrate. (3) A compound interface is characterized by a constant composition, many lattice parameters thick. The compound is formed by the chemical interaction of the coating and substrate material. (4) A diffusion interface is characterized by a gradual change in composition and lattice parameters across the interfacial region between the coating and the substrate. The formation of this type of bond requires that there be diffusion between the coating and substrate material, which means that there must be some degree of solubility and that energy must be available to overcome the activation energy for bulk diffusion (1 to 5 eV). Obviously, a combination of several types of interfacial regions is possible.

Several factors may change the mode or extent of nucleation and interface formation and may prevent strong bonding. Barrier layers in the form of gross contamination, physically adsorbed gases, chemisorbed gases, and compound layers prevent intimate contact across the interface. The substrate surface structure is

*The term *adhesion*, as commonly used, refers to the ability of two materials to remain in contact under a tensile force.

important in determining the type and extent of nucleation and interface formation. Residual stress in the interface region will decrease the bond strength. The structure of the interface region, such as defect level, also determines the ease of propagation of a fracture (Mattox, 1965).

Obviously the way to overcome adhesion problems is to change the conditions which cause poor adhesion. Control of surface contamination is extremely important for good adhesion (Bhushan, 1978, 1979a, 1979b; Bhushan et al., 1978; Jellison, 1979; Mattox, 1965, 1976, 1978; Mittal, 1979). Even a very minute level of undesirable material on the surface can lead to gross bond failures. Surface modification allows the depositing atom to find a different environment on which to deposit. An intermediate material may be used between the coating and substrate which is adherent to both coating and substrate. Enhanced adhesion may also be achieved by changing the extent and/or type of interfacial region. For example, substrate surface roughening has been shown to improve the mechanical interface and increase the wear life of some solid-lubricant coatings by as much as a factor of 10 (Bhushan, 1978, 1979a, 1979c; Stupp, 1959). A monolayer-to-monolayer type of interface where stresses and imperfections are confined to a plane may be converted to a diffusion type of interface if the substrate is heated and the materials are mutually soluble. Heating also increases chemical reaction between the coating and substrate material. It might also be desirable to create a highly defective surface, to increase the number of nucleation sites and enhance diffusion. If the materials are nonsoluble, it is impossible to have diffusion. In this case, in addition to having a clean and highly defective surface, it would be advantageous to create a pseudodiffusion type of interface by some technique. This would allow the formation of an interfacial region of appreciable extent.

In this chapter, we focus our attention on the surface preparation techniques: cleaning and roughening in order to achieve better adhesion of coating to the substrate and result in better endurance of a coated surface. Various surface cleaning processes utilized to obtain better-quality coatings are discussed in Sec. 7.1. The cleaning processes which also can be used for roughening and other roughening processes are discussed in Sec. 7.2.

7.1 CLEANING

Cleaning is defined as the removal, by physical and/or chemical means, of soils and contaminants and their replacement, if necessary, with an intermediate coating that will have negligible interference. *Surface contaminants* are defined as undesirable materials with chemical characteristics different from those of the surface of the substrate material. Surface contamination could be present either in the coating form or as particulates of organic and inorganic materials. A substrate surface which has been cleaned of gross contaminants, such as grease and dirt, will still not be clean on an atomic scale. The remaining contamination may be in the form of physically adsorbed gases which are bound to the surface with 0.1 to 0.5 eV, chemisorbed gases which are bound with 1 to 10 eV, or compound layers several lattice parameters in extent having bonding of 1 to 10 eV (Mattox, 1965). Clearly the atoms which are deposited having only thermal energies will be incapable of affecting these layers. Only techniques such as sputtering, where the kinetic energy of the depositing atoms is on the same order as the binding energy of the adsorbed gases to the surface, may appreciably affect the surface. Therefore, a low-energy atom impinging on a contaminated surface finds itself not on the substrate but in contact with a contaminated layer.

The contamination often gets introduced during processing of the materials such as shaping (e.g., machining and grinding), joining (e.g., soldering and brazing), heat treatment, and exposure to reactive atmospheres. Surface contaminants can also originate during handling and use, such as from adsorption of impurity from the environment. Typical contaminants are machining oils and greases, thin oxide layers from heat treatments or hot working, dust, lint, fingerprints, metallic contaminants, and the residues from reactions of outgassing products in the assembly of metals and nonmetals (plastics) (Anonymous, 1971). Contamination may also be accidental (rust, corrosion). In polymeric surfaces, an additional source of surface contamination is due to leaching of low-molecular-weight fractions or plasticizers. Table 7.1 lists typical contaminants and their possible sources.

A clean surface is one that contains only a small fraction of monolayers of foreign material. Indeed, no manufacturing material is without some degree of contamination. The level of cleanliness depends on the application (Anonymous, 1971, 1982; Beal, 1978; Fransworth, 1967, 1970; Goldfinger, 1970; Maissel and Glang, 1970; Mattox, 1978; Meserve et al., 1982; Snorgen, 1974). A particular surface may be considered a highly contaminated surface by someone interested in doing fundamental adsorption studies, whereas the same surface could be totally acceptable as a clean surface for other applications. Sometimes, however, deliberate addition of foreign matter may be helpful in improving the solid-solid bonding. Intermediate conversion coatings are sometimes added to metal surfaces for improved adhesion and surface passivation.

General cleaning technique often involves removing some of the surface material as well as contaminants, e.g., sputter cleaning, electropolishing, and chemical etching. Often a cleaning technique will involve a combination of procedures used in sequence. Unless the cleaning procedure is designed to take advantage of the strong points and weak points of each step, one cleaning step may defeat another. The following cleaning techniques are commonly used (Anonymous, 1971;

TABLE 7.1 Typical Contaminants and Their Sources

Contaminant	Possible sources
Fibers (nylon, cellulose, etc.)	Clothing, paper towels, tissues
Silicates	Rocks, soil, sand, fly ash
Oxides and scale	Oxidation products of substrate metals
Oils, greases	Oils from machining, fingerprints, body greases, lotions, ointments, human skin, vacuum pump exhausts
Silicones	Hair sprays, cosmetics, hand lotions
Metals	Metal powder from grinding, machining, fabrication of metal parts, particles from metal storage tanks
Ionic residues	Perspiration, fingerprints, residual from cleaning solutions containing ionic detergents, residual from various chemical steps such as etching and electrochemical deposition
Nonionic residues	Resin fluxes, organic processing materials
Solvent residues	Cleaning solvents and solutions

Mattox, 1978; Meserve et al., 1982): solvent cleaning (cold solvent cleaning and vapor degreasing), emulsion cleaning, alkaline cleaning, acid cleaning, pickling, salt bath descaling, ultrasonic cleaning, abrasive blasting (sand blasting and vapor honing), barrel finishing, mechanical polishing and buffing, chemical etching, electroetching, and plasma etching or ion etching. Several processes such as abrasive blasting, barrel finishing, mechanical polishing and buffing, chemical etching, and electroetching can also cause appreciable surface roughening in addition to cleaning by optimizing their operating variables; these are discussed in Sec. 7.2. The cleaning processes employed for various types of contaminants are listed in Table 7.2, where the contaminants have been broadly classified into six groups.

The design of the cleaning process for a surface will depend on (1) the type of contaminant and its level of adherence to the surface, (2) the extent to which the contaminants should be removed, i.e., the degree of cleanliness required and subsequent operation to be performed, and (3) the nature of the substrate surface and the importance of the condition of the surface or structure to the ultimate use of the component (Meserve et al., 1982). Cleaning processes may be categorized as specific or general. Specific cleaning processes are designed to remove specific types of contaminants without changing the surface of interest, e.g., solvent cleaning techniques which dissolve or emulsify contaminants without attacking the surface (Anonymous, 1982).

One of the factors in designing a cleaning procedure is identification of the type of contaminant and its level of adherence to the substrate. A water-soluble contaminant and some organic contaminants can be removed with a detergent, by scrubbing with an abrasive cleaner or by ultrasonic agitation. The common shop oils and greases can be effectively removed by several different cleaners and vapor degreasers (by employing hot vapors of a chlorinated solvent). Vapor degreasing is an effective and widely used method for removing a variety of oils and greases and is well adapted to clean oil-impregnated parts such as bearings. Small quantities of oil and grease can be removed by rubbing the surface with methyl alcohol or acetone.

Metal surfaces often have oxide layers which can be removed by mechanical scrubbing, blasting with abrasive particles, or etching. Surfaces can be prepared by polishing with a silica paper or sand blasting. Thin surfaces can be prepared by vapor honing, chemical etching, or electroetching. If subsequently the coating has to be applied by the vapor deposition process, the surface can be cleaned by ion bombardment (sputter etching or sputter cleaning). Sputter etching has the advantage that it may be done in situ in the deposition system just prior to the deposition process. The removal of the more tenacious surface barrier layers of chemisorbed gases, physisorbed gases, and compound layers and the creation of an atomically clean surface require chemical cleaning, sputter etching followed by *ultrahigh vacuum* (UHV) heating, and cleaning in UHV.

The cleaning of polymers poses some unique problems because of their complex nature. The mechanism of adhesion between metals and polymers is not well defined. In general with polymers the surface is composed of low-molecular-weight fractions which fractures easily. The contaminants such as dust from the polymer surfaces may be removed by a detergent wash or solvent cleaning. However, some solvent may be absorbed into the polymer structure, causing it to expand and possibly craze on drying, and may leach some of the low-molecular-weight fractions from the surface. For good adhesion the surface should be modified so that it does not represent a zone of weakness at the interface. Thus in polymers, cleaning more precisely means modifying the interfaces for better adhesion.

TABLE 7.2 Cleaning Processes Employed for Various Types of Contaminants

	Contaminants					
Cleaning process	Pigmented drawing compounds	Unpigmented oil and grease	Chips and cutting fluids	Polishing and buffing compounds	Rust and scale or oxide	Adsorbed impurity, e.g., hydrocarbons
Cold solvent cleaning	—	—	X	—	—	X
Vapor degreasing	X	X	X	X	—	X
Emulsion cleaning	X	X	X	X	—	—
Alkaline cleaning	X	X	X	X	X	—
Acid cleaning	X	X	X	X	X	—
Pickling	—	—	—	—	X	—
Salt bath descaling	—	—	—	—	X	—
Ultrasonic cleaning	—	—	—	X	X	X
Abrasive blasting	—	—	—	—	X	—
Barrel finishing	—	—	X	—	—	—
Mechanical polishing	X	—	—	—	X	—
Chemical etching and buffing	X	—	—	—	X	—
Electroetching	X	—	—	—	X	—
Ion etching	X	—	—	X	X	X

Another factor in designing a cleaning procedure is the degree of cleanliness required. Abrasive blasting produces the lowest degree of cleanliness. Solvent cleaning, solvent vapor degreasing, emulsion soaking, alkaline soaking, alkaline electrocleaning, alkaline plus acid cleaning, and finally ultrasonics each produce a progressively cleaner surface. In addition to these conventional methods, very exotic and highly technical procedures have been developed in the electronics and space industries to produce clean surfaces far above the normal requirements for industrial use.

In addition to designing a cleaning procedure to remove undesirable contaminants from the surface, one must be concerned with the effect of the cleaning procedure on the surface. The cleaning procedure may change the surface composition to something which may be undesirable. For example, a high-temperature, high-vacuum bake of stainless steel will volatilize the protective chromium oxide and allow the surface to rust on exposure to the atmosphere. Cleaning may also affect the surface morphology by selective etching, etc. This change in morphology may affect subsequent processing. The cleaning process may leave some undesirable residuals on the surface. For example, some electropolishing solutions leave inorganic compounds on the surface.

7.1.1 Solvent Cleaning

Solvent cleaning is a surface preparation process, which is commonly used to remove organic compounds such as grease or oil from the surface of a metal (Clement et al., 1982). Organic compounds are easily made soluble by solvents and then removed from the substrate. Solvent cleaning is ineffective where the contaminants are insoluble, such as metallic salts and oxides, forging, welding and heat-treated scales, carbonaceous deposits, and many other inorganic soldering, brazing, and welding fluxes. The efficiency of the solvent cleaning process is mainly dependent on (Jackson, 1976)

- Selection of the solvent
- Operating temperature of the solvent
- Nature of the contaminant to be removed

The solvent cleaning process is carried out at room-temperature baths, called *cold cleaning*, or by using hot vapors in vapor degreasing techniques. The typical organic solvents used for cold cleaning to remove oil, grease, loose metal chips, and other contaminants include oliphatic petroleums, chlorinated hydrocarbons, chlorofluorocarbons, and their blends. Sometimes ultrasonic agitation along with solvent cleaning helps in removing soils from deep recesses and complex shapes.

The vapor degreasing process uses hot vapors of chlorinated or fluorinated solvents to remove the contaminants, particularly oils, greases, and waxes, from a metallic surface. A vapor degreasing unit consists of an open steel tank with a heated solvent reservoir or sump at the bottom and a cooling zone at the top. The hot vapors of solvent generated at the bottom condense at the substrate and then dissolve the contaminants. This process is well suited for cleaning all industrial metal and multimetallic parts and is used for a wide variety of materials including glass, plastics, elastomers, and combinations of materials that are not affected by solvents. Halogenated solvents are used extensively for the vapor degreasing process because of their high solvency, low heat of vaporization, high vapor density, nonflammability, and noncorrosiveness. The most commonly used solvents are trichloroethylene; perchloroethylene-1,1,1-trichloroethylene; and methylene chloride and trichlorotrifluoroethane.

Figure 7.1 shows the principal systems employed for cleaning metallic surfaces by vapor degreasing. In the vapor phase, the substrate to be cleaned is lowered into the vapor zone [Fig. 7.1(*a*)], where the relative coolness of the substrate causes the vapor condensation. The solvent vapor can be sprayed into deep recesses or blind holes [Fig. 7.1(*b*)]. In small parts with thin sections which may quickly attain elevated temperature before condensation, the warm liquid–vapor system is recommended. The substrate is lowered into the warm liquid and then transferred into the vapor zone [Fig. 7.1(*c*)]. For substrates having very adherent contaminants, the boiling liquid–warm liquid system is recommended. The substrate is dipped first into boiling solvent, where violent boiling action scrubs off most of the adherent contaminants, and then into warm liquid, which removes the remaining dirty solvent and lowers the substrate temperature; finally the substrate is transferred to the vapor zone for vapor condensation. The efficiency of the vapor degreasing process can be significantly improved by introducing ultrasonic agitation.

Table 7.3 lists a wide range of applications in which vapor degreasing processes are employed. However, the typical limitations of this cleaning process

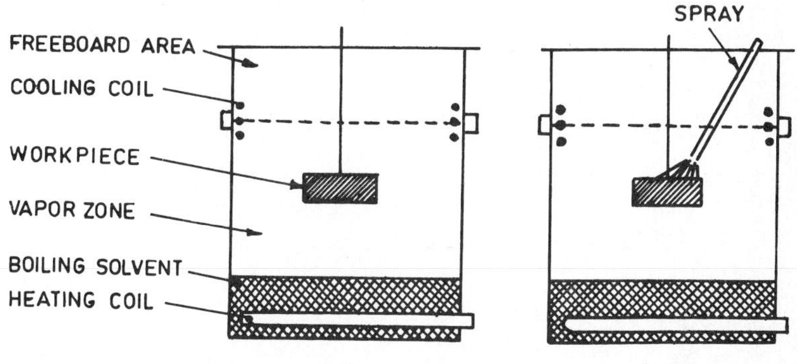

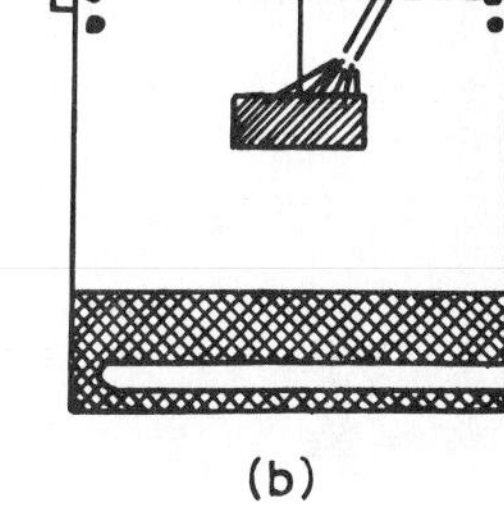

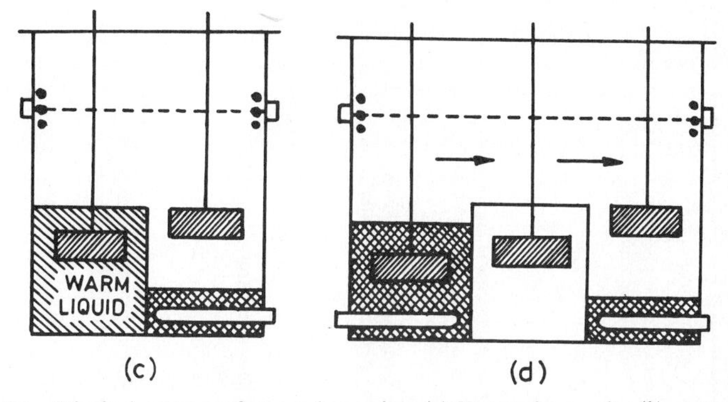

FIGURE 7.1 Principal systems of vapor degreasing. (*a*) Vapor phase only; (*b*) vapor-spray-vapor; (*c*) warm liquid–vapor; (*d*) boiling liquid–warm liquid–vapor.

TABLE 7.3 Various Applications of Vapor Degreasing System

Components	Metal	Production, kg h^{-1}	Contaminant removed	Subsequent operations	Notes on processing
Degreasing by vapor-spray–vapor system					
Valves (automotive)	Steel	540	Machining oil	Nitriding	Automatic conveyer
Valves (aircraft)	Steel	590	Machining oil	Aluminum coating	Automatic conveyer
Automatic transmission components	Steel	18,100	Machining oil, light chips, dirt	Assembly	Double monorail conveyer
Kitchen utensils	Aluminum	450	Buffing compounds	Inspection	Special fixture and conveyer
Degreasing by warm liquid–vapor system					
Speedometer shafts and gears	Steel, brass	340	Machining oil, chips	Inspection, assembly	Rotating baskets
Screws	Steel, brass	680	Machining oil, chips	Painting, finishing	Flat and rotating baskets
Electron-tube components	Steel	910	Light oils	Dry hydrogen fire	Conveyerized unit
Tractor gears and shafts	Steel	910	Machining oil, chips, quenching oil	Nitriding	Elevator-type conveyer handling work in heat treatment trays
Degreasing by boiling liquid–warm liquid vapor system					
Valves (automotive, aircraft)	Steel	450	Machining oil	Welding	Manual operations
Knife blades	Steel	820	Oil, emery	Buffing	Manual operations
Carbide-tip tool holders	Steel	910	Lubricant, chips	Recess milling	Conveyerized unit
Tubing, 60 cm long	Aluminum	910	Drawing lubricants, quench oils	Satin finishing	Conveyerized, tube handled vertically
Calculating machine components	Steel	1360	Stamping oil	Plating	Conveyerized unit
Hand-tool housings die-cast	Zinc-base	1360	Tapping oil, chips	Assembly	Automatic conveyer, racks
Screw machine products	Steel, brass	1360	Cutting lubricants, chips	Assembly	Flat and rotating baskets, conveyer
Cable fittings	Steel	1810	Light oils	Inspection	Conveyerized
Stampings	Steel	2270	Light oil, chips	Furnace brazing	Small stampings nested in baskets

Source: Adapted from Clement et al. (1982).

involve the material it can clean without any deterioration and the contaminants it can remove effectively. Several polymer materials are adversely affected by solvents generally used in this process.

7.1.2 Emulsion Cleaning

The emulsion cleaning process employs organic solvents dispersed in an aqueous medium aided by an emulsifying agent for removing heavy soils from the surfaces of metals and nonmetals (Roberts and Srinivasan, 1982). The medium is usually alkaline, having a pH of about 7.8 to 10.0, and cleaning is carried out at 10 to 80°C. The substrate is cleaned by dissolving and/or emulsifying the contaminants. An emulsion system contains two mutually insoluble or nearly insoluble phases, one of which is dispersed in the other in the form of globules. One phase is usually a hydrocarbon oil, and the other is water. The stability of emulsion cleaners depends on the properties of emulsifying agents that are capable of causing oil and water to mix and form a more stable system. The solvent used for emulsion cleaning is of petroleum origin such as heterocyclic (M-pyrol), naphthenic, aromatic, and hydrocarbon nature (kerosene). The emulsifiers include nonionic polyethers and high-molecular-weight sodium or amine soaps of hydrocarbon sulfonates, amine salts of alkyl aryl sulfonates (anionic), fatty acid esters of polyglycerides, glycerols, or polyalcohols. The two conventional techniques—spray cleaning and immersion cleaning—are adapted for emulsion cleaning, which depends on the shape and size of the substrates.

The emulsion cleaning process is effective for removing a wide variety of contaminants such as pigmented drawing lubricants, unpigmented oils and greases, cutting fluids, and residues due to buffing, polishing, and magnetic particle inspection. Emulsion cleaning is followed by alkaline cleaning in order to remove all traces of the organic solvents and other cleaning by-products, to avoid any contamination during the coating application. The emulsion cleaning process is not preferred for substrates having high porosity, deep recesses, and blind holes because it is difficult to remove the emulsifying agents from the substrate completely.

7.1.3 Alkaline Cleaning

Alkaline cleaning is a mainstay of industrial cleaning and is widely used to remove oils, grease, dust, metallic particles, carbon deposits, silica, etc. The alkaline cleaners are applied by soaking, spraying, and immersion, and the cleaning takes place through saponification, emulsification, dispersion, or their combinations at room temperature and elevated temperatures (60 to 100°C) (Murphy, 1982). The alkaline cleaners consist of builders such as alkali metal orthophosphate, condensed phosphates, hydroxides, carbonates, bicarbonates, silicates and borates, and surfactants such as sodium alkylbenzene sulfonates and ammonium chloride.

Electrolytic cleaning is a modification of alkaline cleaning in which an electric current is imposed on the part to produce vigorous gassing on the surface, to promote the release of soils. Electrocleaning can be either anodic or cathodic. An alkaline cleaner called a *deruster* is extensively used by electroplaters. This process is adapted to remove light oxides on ferrous and nonferrous metals such as steels, aluminum, zinc, etc.

7.1.4 Acid Cleaning

Acid cleaning is used more often in conjunction with other steps than by itself. Acids have the ability to dissolve oxides, which are usually insoluble in other solutions. Acid cleaning processes are used to remove oil, grease, and light metal oxides (Pilznienski, 1982).

Some acid cleaners are also used to produce inert phosphate coatings for corrosion protection of the underlying substrate (Anonymous, 1978, 1979; Haltner and Oliver, 1960; Yaniv et al., 1979). The phosphate coatings are covered in detail in Chapter 10. The distinction between acid cleaning and acid pickling (to be discussed later) is a matter of degree, and there is some overlap in the use of these terms. Acid pickling is a more severe treatment for the removal of scale from semifinished mill products, forgings, or castings, whereas acid cleaning generally refers to the use of acid solutions for final or near-final preparation of metal surfaces before plating, painting, or storage.

Acid cleaning is a process in which a solution of an inorganic (mineral) acid, organic acid, or acid salt, in combination with a wetting agent and detergent, is used with or without the application of heat (Blume, 1973; Pilznienski, 1982; Stiteler, 1982). Organic acids such as citric, tartaric, acetic, oxalic, and gluconic acid and acid salts such as sodium phosphate, ammonium persulfate, sodium acid sulfate, and bifluoride salts are used in various combinations. Solvents, such as ethylene glycol monobutyl ether and other glycol ethers, and wetting agents and detergents, such as alkyl aryl, polyether alcohols, antifoam agents, and inhibitors, may be included to enhance the removal of soil, oil, and grease. Many acid cleaners are available as proprietary compounds in the form of liquid concentrates or powders to be mixed with water. In this cleaning process, the contaminants are removed by etching, wetting, emulsification, or dissolution.

The typical methods used for acid cleaning are wipe on/wipe off, spray, immersion, and rotating-barrel and electrolytic cleaning methods. The wipe on/wipe off process is simply carried out by hand and does not involve any equipment. Spray cleaning is preferred when the substrate area is very large. Immersion cleaning is used extensively because of its versatility as its operation varies from hand dipping a single component to highly automated installations operating at elevated temperatures with controlled agitation. Barrel cleaning is often used for large quantities of small substrates, to achieve continuous high production. Perforated barrels of parts are immersed and rotated in tanks of cleaning solution. In electrolytic cleaning, the contaminants are removed by the evolution of gas and chemical reduction of surface oxide films when used anodically. Sulfuric acid baths are most commonly electrolyzed and are used as a final cleaner before plating.

The acid cleaning process is incapable of removing heavy adherent deposits of soil and grease and other metal oxides, and it attacks the substrate surface even when inhibitors are mixed. However, organic acids such as acetic acid, citric acid, and gluconic acid offer better cleaning efficiency and less damage to the metal surface. This process is employed for cleaning boilers, which involve the removal of iron oxide, copper oxide, etc.; stainless-steel tanks; trucks; railroad cars; nuclear power plant decontamination; etc.

7.1.5 Pickling

This process removes heavy scale (surface oxides) and dirt through chemical reaction of the substrate metal with an aqueous acid solution. This is an efficient process for removing scale, i.e., oxides from very large-area substrates available in the shape of bars, blooms, billets, sheet, strip, wire, and tubing (Hudson et al., 1982). The general mechanism of removal is that acid penetrates through cracks in the scale and the acid reacts with the metal, which generates hydrogen gas. As the hydrogen gas pressure increases, a point is reached where the scale is blown off the metal surface. The typical acids employed for pickling are sulfuric acid and hydrochloric acid, and sometimes aided by a wetting agent. The cleaning process is carried out at room temperature or slightly elevated temperature, say up to 100°C, depending on the material and the oxide layer to be removed. This process has been adapted to clean low-carbon steel, high-carbon steel, cast iron, stainless steel, etc. The substrates are precleaned through alkaline cleaning in order to remove oils, soaps, lubricants, carrier coating, and other contaminants which may not react with the acid. The level of cleanliness and efficiency of pickling affected by the acid concentration, temperature, and agitation; substrate temperature; and concentrations of ferrous sulfate and the inhibitor.

The pickling process gives rise to certain defects such as pitting, blistering, and hydrogen embrittlement. Sometimes overpickling causes porosity of the transverse surfaces and leads to roughening, discoloring, and dimension reduction of the substrate. Indeed this process requires careful monitoring of the cleaned surface and a controlled automatic system for various operations.

7.1.6 Salt Bath Descaling

In salt bath descaling, the cleaning is done in a salt bath through oxidizing, electrolytic, and reducing phenomena (Wood and McCardle, 1982). Complete removal of scale requires the use of an acid dip or acid pickling after salt bath descaling. The oxidizing process is more important industrially because of wider applications and its simplicity of operation. The electrolytic process usually provides more thorough scale removal than the reducing process, and the electrolytic process also has the capability of functioning in both an oxidizing and a reducing mode. Because the reducing process employs a bath that is operated at a lower temperature, it is sometimes advantageous for descaling metals that undergo a change in properties at high temperature.

Metal finishing operations involving stainless steels, superalloys, and titanium metals usually require an oxidizing salt conditioning treatment. The operating temperatures of these salts vary from 200 to 475°C to exhibit high chemical activity to remove complex oxides and scales developed in hot forming operations. The molten salt baths are not complete cleaning systems. They must be used in conjunction with acid pickling solutions.

Fused salts that are neither chemically oxidizing nor chemically reducing can be activated to produce either condition by the input of electric energy. This process is very effective to remove sand from iron castings. Another contaminant on cast-iron surfaces, which is most effectively removed by fused salts, is graphitic carbon or graphite smears developed in machining. The salts which are used in this process include sodium nitrate, sodium hydroxide, and sodium carbonate.

The reducing or sodium hydride descaling process is performed in a fused caustic bath in which sodium hydride (NaH) is generated in situ from metallic sodium and dissociated ammonia. The sodium hydride descaling process has limited application in modern fused salt descaling. It does not lend itself well to continuous lines because of the necessity of keeping the surface protected from contact with the atmosphere. It is quite effective in removing heavy scales from 400 series stainless steel, some carbon steels, and titanium.

Generally, the salt bath descaling process is adapted for cleaning materials which are not reactive to common acidic and alkaline solvents. Some general classes of materials that require molten salts for complete removal are as follows: oxides—M_xO_y (hot work and heat treatment scales); organics—free graphite, lubricants, binders; ceramics—SiO_2 (glass-type lubricants, mold, and core material); and combination coatings—MO + C (lubricant and protective coatings, core wash). The salt bath descaling process sees widespread use for the removal of complex coring from castings, flame sprayed and plasma sprayed coatings, and all types of inorganic and organic coatings.

7.1.7 Ultrasonic Cleaning

One of the most practical and most effective cleaning techniques, when properly used, is ultrasonic cleaning. It is quite effective in removing loosely adherent particulate matter and material soluble in the cleaning solvent. It does not work on attached burrs, nor will it effectively remove contaminants such as hardened epoxy or adhesives.

Ultrasonic cleaning uses sound waves passed at very high frequency through liquid cleaners, which could be alkaline, acid, or even organic solvents. The passage of ultrasonic waves through the liquid media creates tiny gas bubbles, which offer vigorous scrubbing action on the parts being cleaned. This action brings about the most efficient cleaning possible.

For the ultrasonic process, basically two elements are required: a generator to produce electric oscillations and a transducer to convert the electric oscillations to sonic waves. Ultrasonic energy can be used in conjunction with several types of cleaners, but it is most commonly applied to chlorinated hydrocarbon solvents, water, and water with surfactants. The efficiency of ultrasonic cleaning depends on the characteristics of the ultrasonic equipment and the choice of fluid.

Ultrasonic cleaning is used as a final cleaner only, after most of the soil or contaminant has been removed by another method. Ultrasonic cleaning is more expensive than traditional cleaning methods, because of the higher initial cost of equipment and higher maintenance cost, and consequently its use is largely restricted to applications in which other methods have proved inadequate. Part size is a limitation, and it is used for small parts.

7.1.8 Plasma Cleaning

Plasma cleaning (ion etching, or sputter cleaning) provides substrate cleaning in situ in the deposition system just prior to the vapor deposition. This is adapted in materials that are difficult to etch by chemical methods and in vacuum deposition processes. The ion bombardment results in removal of materials by a momentum transfer process called sputtering. The contaminant removal rate depends on atom bonding, the ion employed, and the bombardment energy (Anonymous, 1971; Bhushan, 1979b; Kominiak and Mattox, 1977; Mattox, 1978; O'Kane and Mittal, 1974). As the bombardment energy increases, the depth of penetration increases and the sputtering yield increases. However, very high bombardment energy ($\sim$200 keV) results in implantation of the bombarding gas ions. Generally two modes of ion etching are used for substrate cleaning:

- Physical sputtering with an ion beam under high vacuum
- Physical cathode sputtering by an inert- or reactive-gas discharge ion bombardment under vacuum

In the cathode sputtering and ion-plating process, plasma cleaning is an integral step in the deposition process.

In certain cases, gas incorporation in the sputter-etched surface may be released during subsequent processing, causing loss of adhesion of a coating applied on a sputter-cleaned surface. This can be avoided by heating the substrate during or after cleaning. Reactive plasma formed in a *radio frequency* (rf) discharge is effective in removing contaminant and barrier layers from a variety of metals. The advantages of reactive plasma etching over inert-gas sputter etching are the higher etching rates that can be obtained and the minimal substrate surface damage.

The plasma cleaning processes provide the cleanest surface as well as very precise control over the contaminant removal. The plasma cleaning technique creates a striking improvement in the adhesion of sputtered films compared with organic solvent cleaning. In the case of aluminum sputtered onto iron tensile bars, the substrates which had been cleaned by reactive plasma showed no loss of adhesion when pulled to the onset of substrate necking (extension about 35 percent) (Kominiak and Mattox, 1977).

7.2 ROUGHENING

If the primary mechanism of coating adhesion is mechanical, then preroughening of the surface after cleaning of foreign matter is beneficial. Surface roughening forms microgrooves, coating material fills the grooves, and mechanical interlocking provides the adhesion. No chemical reaction between the material is necessary, and the strength of the bond depends on the physical properties of the materials and the texture of surface roughness. Generally, resin-bonded coatings on comparatively rough surfaces exhibit better adhesion (Bhushan, 1978, 1979c).

Commonly, the surfaces are roughened by one of the following processes (Anonymous, 1982; Beadle, 1972; Bhushan, 1979a; Mattox, 1978; Von Fraunhofer, 1976): abrasive blasting (sand blasting, vapor honing), barrel finishing, mechanical polishing and buffing, chemical etching, and electrolytic etching. Sand blasting can roughen the surface quite rapidly. The abrasive grit size and blast air pressure can be tailored to obtain a specific roughness level. This technique is usually adapted for thick substrate. Vapor honing uses a finer grit size and lower air pressure and can be used to roughen very thin substrate (foils) very gently. Surfaces can also be roughened by mechanical hand polishing by choosing an abrasive of suitable material and a suitable particle size. Chemical etching and electroetching roughen the surfaces by attacking the grain boundaries and/or the grains. The roughness level in electroetching can be controlled very precisely.

The treatments adapted for better adhesion of coatings applied to polymer surface include roughening, activation, and cross-linking. The polymer surfaces are roughened either by abrasive or grit blasting or by etching with proper chemicals. For instance, the ABS (acrylonitrile-butadiene-styrene) polymer may be etched by using a mixture of chromic and sulfuric acids.

The activation of polymer surfaces is achieved by chemical treatments where organometallic complexes are formed at the interface (Berman and Silverstein, 1975). Typical activation treatments include iodine treatment of polyamides and sodium treatment of PTFE. Activation may also be accomplished by adding a small amount of material which reacts with both the substrate and the coating. These materials may be highly polar, low-molecular-weight polymers.

Cross-linking is used to increase the strength of the surface and to reduce the amount of undesirable low-molecular-weight components. The cross-linking process is encouraged by corona discharge (Henderson, 1965), exposure to plasma (Sowell et al., 1972), and exposure to ultraviolet radiation (Oster et al., 1959). These treatments, not much different from each other, cause bond scission and add carbonyl groups to the polymer surfaces. In ion-plating and sputtering techniques, cross-linking may be done in situ prior to the coating applications.

7.2.1 Abrasive Blasting

The abrasive blasting process, also called *abrasive blast cleaning*, uses abrasive grains (small, sharp particles), propelled by an airstream or water jet to impinge on the surface, removing contaminants and roughening by the resulting impact force (Leliaert et al., 1982). The abrasive blasting process is used for cleaning and/or roughening such as

- Removal of mold sand, heat treatment scale, rust, carbon deposits, and other solid contaminants
- Surface roughening of metals, such as stainless steel, for painting in order to produce a mechanical lock for adhesion
- Removal of surface irregularities to improve surface finish, e.g., by elimination of directional grinding lines, burns, etc.
- With fine abrasive (1250-mesh) suspended in a slurry, for honing multitooth hubs, broaches, and cutting tools to increase their life

The substrates which can be cleaned by abrasive blasting include ferrous and nonferrous castings, weldments, thermoplastic and thermosetting polymers, etc. The abrasive blasting is often preferred for removing heavy scale and paint, especially on large and otherwise inaccessible areas. Abrasive blasting is also frequently the only allowable cleaning method for steels sensitive to hydrogen embrittlement.

Abrasive blasting may be divided into dry and wet processes. The two methods used in dry blasting are (1) mechanical blasting, in which the abrasive is directed by centrifugal force to the substrate from a power-driven, rapidly rotating bladed wheel, and (2) air pressure blasting, which includes both the direct pressure and the suction (induction) types. Mechanical dry blasting is more popular because of the rapid removal of surface contaminants by increasing both the abrasive particle velocity and the quantity of sand impacts per unit of time. Air pressure blasting is used when low production requirements or intermittent operations are anticipated. Wet blasting normally consists of suspending fine abrasive particles in chemically treated water to form a slurry which is forced by compressed air to nozzles fixed against the substrate surface. Abrasive particles used in wet blasting are usually much finer (20- to 5000-mesh) than those in dry blasting. In further contrast to dry blasting, wet blasting is not intended for gross removal of heavy scale, coarse burrs, or soil, but is intended to remove trace surface contaminants and minute burrs on precision parts, to blend surface finish lines to produce satin or mat finishes, and to prepare the surface for coating (Leliaert et al., 1982).

The performance of the abrasive blasting process depends on the abrasive characteristics. Typical abrasives for dry blasting include steel shot, angular grit, slug product, aluminum oxide, silicon carbide, glass beads, and plastic abrasives. Typical abrasives for wet blasting include walnut shells and peach pits, novaculite (a soft type of silica), silica, quartz, garnet, aluminum oxide, glass beads, and other refractory abrasives. A range of 20 to 35 volume percent abrasive is satisfactory for most applications but may be modified because of particle size, surface tension, specific gravity, and the desired effect on the workpiece. The abrasive medium must be carefully selected for soft, ductile metals and their alloys, for example, Al, Mg, Cu, and Zn. The abrasive blasting may produce residual compressive stresses, warping in the case of thin substrate, and chipping in the case of hard materials. Further, the peening effect of abrasive particles may distort the flatness and surface finish.

Based on the grit size used, dry blasting is referred to either sand blasting or vapor honing. The sand blasting uses coarse grit that can roughen the surface quite rapidly. The final surface quality, i.e., roughness level, is dictated by the grit size, shape, and air velocity. The sand blasting process is usually adapted to clean or roughen thick substrates. With silica of 100-μm grit size at an air pressure of about 275 kPa, the surface can be roughened to 3.0 μm rms in about 10 min. Vapor honing is another similar process in which much finer grits (~10-μm grit size) are used. This process is very gentle to the treated surface and can be adapted for very thin substrates with a thickness in the range of 50 to 100 μm.

7.2.2 Barrel Finishing

Barrel finishing is a mass finishing process which is a simple, low-cost means of deburring and surface-conditioning components. It involves loading of substrates to be finished in a container or barrel along with abrasive media and water (Bernard et al., 1982). The barrel is normally loaded about 60 percent full with a mixture of parts, abrasive media, and water. As the barrel rotates, the substrate moves upward to a turnover point: then the force of gravity overcomes the tendency of the mass to stick together, and the top layer slides toward the lower area of barrel. The abrasive media rub against the surfaces, edges, and corners of the substrate during the rotating or tumbling of the container, i.e., barrel. This abrasive action may deburr, generate edge and corner radii, and clean the parts by removing rust and scale.

A wide variety of abrasives such as aluminum oxide, silica sand, limestone, and granite are used for barrel finishing. The selection of an abrasive depends on the size and shape of the substrate, hardness of substrate material, and purpose of finishing, e.g., to remove or finish recesses, angles, fillets, holes, and intricate contours. Soft metals such as aluminum, brass, and zinc, in which the surface edges are more easily affected, should be processed in smaller-size or lightweight plastic-bonded media. The common problems encountered in barrel finishing are excessive impingement on the substrate surface; rollover of edges, corners, and burrs; and persistent loading of abrasive media in holes, slots, or recessed areas.

7.2.3 Mechanical Polishing and Buffing

Mechanical polishing and buffing are abrading operations (Hignett et al., 1982). Polishing is the use of abrasives firmly attached to a flexible backing, such as a wheel or belt. Grinding is the use of abrasives firmly adhered to a rigid backing. Buffing or fine polishing is a process using abrasives or abrasive compounds that adhere loosely to a flexible backing, such as a wheel. Polishing operations usually follow grinding and precede buffing. In general, polishing permits more aggressive abrading action than buffing. It has greater capability to modify the shape of a component and has greater definition of scratches. Buffing achieves finer finishes, has greater flexibility, and follows the contours of components. Polishing and buffing processes are used in most metals and many nonmetals.

Mechanical finishing typically is a two-step process. First the rough polishing (rms ~1 to 10 μm) is done using an abrasive paper (50 to 600 mesh), and then fine polishing or buffing (rms ~0.1 to 1.0 μm) is done using an abrasive paste usually made of alumina particles of ~0.3 μm. For the purposes of surface pretreatment for coating application, desirable roughening may require one or both of the rough and fine polishing steps. The selection of an abrasive mainly depends on the substrate material, size and design of the substrate, initial surface finish, and final surface finish required.

The polishing is accomplished by bringing a substrate into direct contact with the abrasive. The abrasives and the backers for polishing are either abrasive (endless) belts or abrasive wheels. The abrasives are firmly attached to the backing material. In general, the backing for a belt is either paper or cloth. The backing for a wheel is made of abrasive impregnated, nonwoven nylon or other synthetic fibers. An adhesive, usually resin or glue, is used to bond the abrasives to the backing.

The natural abrasives used for polishing are emery and corundum. The emery is a naturally occurring mixture containing from 57 to 75 percent alumina with the remainder being iron oxide and impurities. Corundum is a naturally occurring mixture of fused aluminum oxide and silicon carbide. It is tougher and harder than emery. Aluminum oxide and silicon carbide are the most widely used synthetic abrasives. They are harder, more uniform, longer-lasting, and easier to control than natural abrasives. Because aluminum oxide grains are very angular and have excellent bonding properties, they are particularly useful for polishing tougher metals, such as alloy steels, high-speed steels, and malleable and wrought iron. Silicon carbide is harder than aluminum oxide and fractures easily, providing new cutting surfaces which extend the useful life of the abrasive. Silicon carbide is usually used in polishing low-strength metals such as aluminum and copper. It is also used in polishing hard, brittle materials such as carbide tools, high-speed steels, and chilled and gray cast irons. Usually, polishing is started with a coarse abrasive paper (~50 to 220 mesh) and followed by finer abrasives (~600 mesh) at a polishing speed of 10 to 30 m s^{-1}.

Lubrication of the cutting face of the polishing wheel with oil or grease is required in a number of cases to prevent heavy abrasion when a fine-polished surface is required. The most popular method of lubricating involves the use of a tallow grease mixture applied by friction from a bar to the rotating polishing wheel. Where polishing wheels can be lubricated automatically, sprayable liquid lubricants are used which are available in formulations that can be easily cleaned. In many cases, tap water is used in polishing.

The fine polishing (or buffing) process which follows rough polishing produces smooth, reflective surfaces. This is accomplished by bringing a substrate into direct contact with a revolving cloth or sisal buffing wheel (known as *buff*) which is charged with a suitable buffing compound. Compounds may be in liquid paste or bar form, and they contain bonding and lubricating agents. Liquid compounds are sprayed on the buff, and bar compounds are applied by hand or with an automatic applicator. The typical compounds used for buffing are tripoli, a cryptocrystalline silica, bobbing compound, coarse silica, green chromium oxide, alumina, emery paste, and diamond. Usually, buffing is started with a coarser grit size (say, 3-μm) compound and followed by finer grit size (say, 0.3-μm) compounds.

The limitations of mechanical polishing and buffing are associated with either the substrate shape and size or the ability of the process to achieve a specific surface finish, which depends on the abrasive grit sizes available.

7.2.4 Chemical Etching

Chemical etching, using either alkaline or acid solutions, produces a mat finish on metallic substrates (Elliott, 1959). The chemical etching process may be adapted for final finishing but is often used as intermediate treatment prior to the coating application. Chemical etching is also used extensively in conjunction with buffing. The chemical etching process removes oxide films and embedded surface contaminants and roughens the surface slightly. Chemical etchants are used either by immersing the substrate and gently agitating it or by swabbing the sample, usually at room temperature. The etching time is usually determined empirically.

Alkaline etching reduces or eliminates surface scratches, nicks, extrusion die lines, and other imperfections. It is also used to selectively roughen the substrate. Various alkaline etchants include sodium hydroxide, potassium hydroxide, trisodium phosphate, and sodium carbonates. This process is carried out at slightly elevated temperature, that is, 50 to 80°C. Sequestrants such as gluconic acid, sodium gluconate, the glucamines, and sorbital are added to alkaline solutions to increase the life of the bath by preventing the formation of scale and by reducing the accumulation of sludge in the tank.

Acid etching is commonly used for finishing and/or roughening substrates (Bhushan, 1978). Hydrochloric, nitric, phosphoric, chromic, and sulfuric acids are generally used in acid etching. Acids are mixed together and with salts to achieve cleaning for specific applications. For instance, sulfuric-chromic acid removes heat treatment stains with little etching of aluminum metal substrates; dilute sulfuric-nitric acid solutions produce bright, slightly mat textured surfaces; ferric chloride is used for deep etchings of aluminum metals; and nine parts hydrochloric acid and one part hydrogen peroxide are used for etching some nickel-based alloys (such as Inconel X-750). Generally acid etching is used alone, but under certain cases it is used in conjunction with alkaline etching either before or after.

7.2.5 Electropolishing or Electroetching

Electropolishing or electroetching is an electrochemical process, the reverse of electroplating, whereby the metal is removed by making the metal anodic or positive in a concentrated acid or alkaline solution (Winter, 1982). The most commonly used electropolishing solutions contain one or more of the concentrated inorganic acids such as sulfuric, phosphoric, and chromic acids and organic acids such as acetic, citric, tartric, and glyconic acids. The organic acids are said to assist in throwing power and permit electropolishing at lower current densities and temperatures. The smoothness and texture of the surface are influenced by the activity of etchants, current, and the time of etching. Conditions of industrial electropolishing operations using various acid electrolytes are described by Winter (1982).

The electropolishing process has been employed commercially for several decades. The typical substrates which are electropolished include aluminum alloys, brass, nickel, silver, copper, stainless steel, low-alloy steels, and zinc die-casting alloys. Surface conditions, regardless of the metal and alloy, determine to a large extent the degree of successful electropolishing. Various factors contributing to unsatisfactory results are large grain size, nonuniform structure, nonmetallic inclusions, directional roll marks, and scale contaminations.

The electropolishing process is also used to remove burrs from cut edges, because electrolytic current flow is greater at edges and protrusions. This process has been used successfully in the production of hydraulic pump cylinders and some fine machine parts. Electrodeburring is also used to manufacture various precise components such as aircraft and missile components, gyrohousing, drive gears, and drive shafts. For instance, for roughening of a thin (100-μm) Inconel X-750 foil (nickel-based alloy) preceding application of bonded lubricant coating, Bhushan (1978, 1979a) has used a solution of 25 g of sodium fluoride, 95 mL of sulfuric acid, and 1000 mL of water. The microstructures of Inconel X-750 foil at current densities ranging from 775 A to 4650 A m^{-2} are shown in Fig. 7.2. An optimum current density of 2325 A m^{-2} and a 2-min time provide more or less uniform grain boundary attack and virtually no attack at the grains [Fig. 7.2(*d*)].

7.3 MONITORING SURFACE CLEANING AND ROUGHENING

After a surface has been prepared, it should be checked for cleanliness and proper surface roughness. Techniques for monitoring the degree of surface cleanliness range from very simple to very sophisticated (Anonymous, 1971; Dwight and Wightman, 1979; Lindfors, 1979; Mattox, 1978; Miller, 1973; Mittal, 1979; Pudvin, 1973; Szkudlapski, 1973; White, 1977; Yang et al., 1975). These are the virtues of ideal tests of monitoring the degree of surface cleanliness:

- Objective
- Reproducible, reliable
- Quick, independent of operator experience
- Nondestructive to the sample
- Capable of identifying all types of contaminations (inorganic and organic)
- Quantitative, i.e., capable of yielding some number on contamination scale

The whole range of monitoring techniques may be classed as direct or indirect. In a direct method, one can directly identify the nature of the contaminant, whereas in an indirect method a certain property of the cleaned surface is used as an indicator of the presence of a contaminant. Some direct methods can provide information as to the state, chemical composition, physical nature, and quantity of the contaminants.

Direct monitoring techniques include various surface analytical techniques such as *Auger electron spectroscopy* (AES), *X-ray photoelectron spectroscopy* (XPS), *appearance potential spectroscopy* (APS), *ion-scattering spectroscopy*

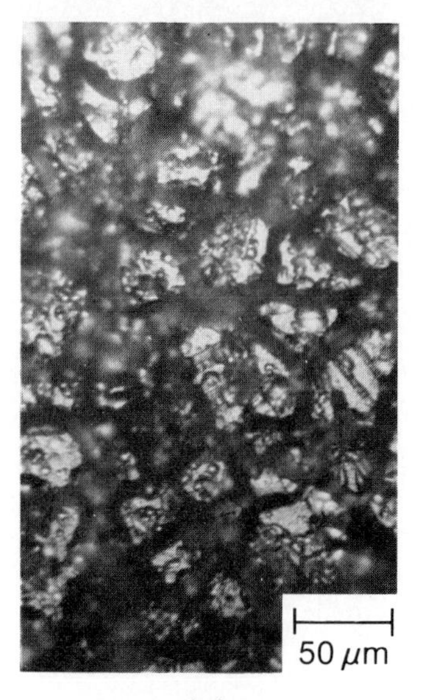

(a)

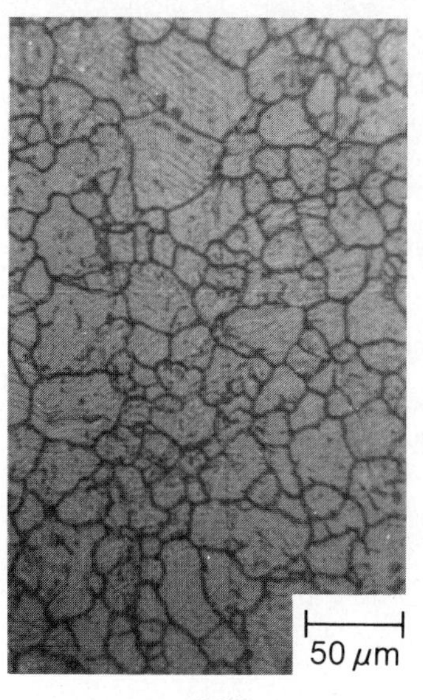

(b)

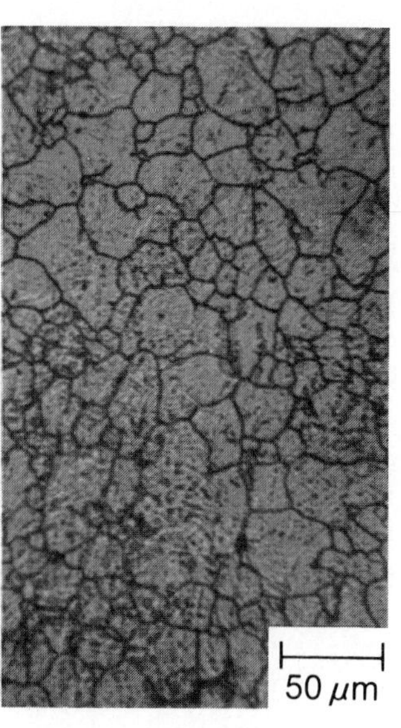

(c)

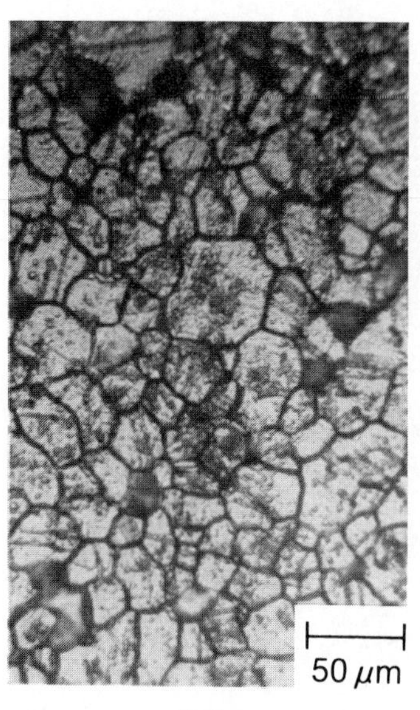

(d)

FIGURE 7.2 Microstructure of electroetched Inconel X-750 foil surfaces at (*a*) 2325 A m^{-2} for 6 min, rms roughness = 0.53 μm; (*b*) 775 A m^{-2} for 6 min, rms roughness = 0.20 μm; (*c*) 4650 A m^{-2} for 2 min, rms roughness = 0.25 μm; (*d*) 2325 A m^{-2} for 2 min, rms roughness = 0.28 μm. [*Adapted from Bhushan (1979a).*]

(ISS), *secondary ion mass spectrometry* (SIMS), and *Fourier transform infrared* (FTIR) analysis (Bryson et al., 1979; Dowsett et al., 1977; Dwight and Wightman, 1979; Miller and Czanderna, 1978; Ratner et al., 1979; Sparrow, 1977). Figure 7.3 demonstrates the effectiveness of SIMS, ISS, and AES techniques in analyzing the presence of contaminants. These surface analysis techniques have been very useful for unraveling the poor adhesion of a variety of substrates.

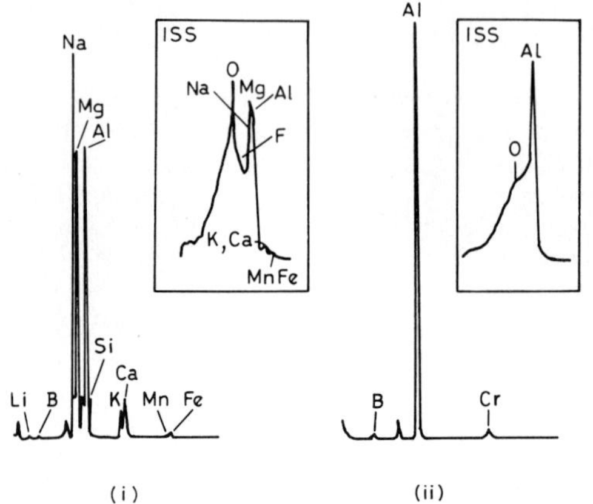

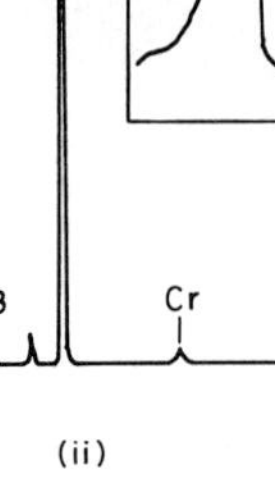

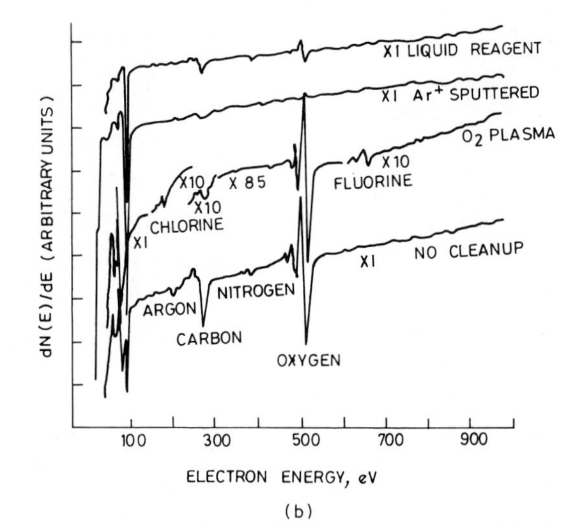

FIGURE 7.3 (*a*) Typical SIMS and ISS spectra of 1024 Al sheet (i) prior to and (ii) after cleaning procedures. (*b*) Typical Auger spectra prior to and after several cleaning steps. Liquid reagent cleanup sequence includes deionized water rinse, 10 percent HF acid dip, deionized water rinse, and isopropanol or Freon vapor dry. [*Adapted from Yang et al. (1975) and Sparrow (1977).*]

Indirect monitoring techniques utilize the behavior of the surface to imply the presence of contaminants, i.e., wetting or contact angle, adhesion, adsorption, microscopy, and ellipsometry (Bhushan, 1979a; Pudvin, 1973). The use of wettability or contact angle is the most common technique to monitor the cleanliness of a surface. The angle that the edge of a drop of relatively pure water makes with a surface, called the *contact angle*, is a very sensitive measure of the wettability of the surface. Clean metals have very low contact angle, and water spreads readily on the surfaces. The presence of even a partial monolayer of a hydrophobic film will cause a large increase in the contact angle (up to 180°) and will greatly inhibit spreading of the water on the surface. Several methods of applying the water are used to perform this type of test: atomizer spray and water-break (Anonymous, 1971). In the atomized spray, high-purity water is applied as an atomized mist from an artist's airbrush. On a clean surface, the droplets coalesce to form a thin uniform film, whereas a surface contaminated with hydrophobic material will take on a gray or diffuse appearance, caused by light scattering from the individual droplets. In the water-break test, the surface is immersed in pure water and then removed and allowed to drain momentarily. Hydrophobic contamination is evidenced by areas where the water film retreats and fails to wet. Although the contact-angle tests are qualitative in nature, the measurement of the contact angle gives numerical values. The values of contact angle θ are commonly used to follow the effectiveness of various surface cleaning techniques. This test will not detect hydrophilic contaminants.

The coefficient of friction or adhesion of a surface with cleaned or contaminated surface is another indirect but destructive method. Usually, with cleaned surfaces the coefficient of friction is high ($\mu \approx 1.0$), and high adhesion results. One can establish a correlation between the coefficient of friction and the contaminant level by doing some systematic experimentation.

The absorption of fluorescent dyes and radioactive trace material can also be used to detect the presence of contaminants that absorb these materials (Maissel et al., 1966; Slater and Donahue, 1959). Certain surface contaminants such as oils and greases can be easily detected by adding fluorescent dyes and then illuminating.

Light microscopy is an indirect, widely used, and simple method. An experienced microscopist can identify many microscopic objects just as all of us identify macroscopic objects, i.e., by shape, size, surface detail, color, luster, etc. The microscopy under reflection gives a direct hint about the presence of surface contaminants by a change in refractive index of the substrate surface (Cameron, 1961).

Ellipsometry, also known as *polarization spectroscopy*, is another indirect technique, which involves study of the state of polarization of light reflected at abnormal incidence from the cleaned or contaminated surface. The state of polarization is determined by the relative amplitude of the parallel (R_p) and perpendicular (R_s) components of radiation with respect to the plane of incidence (or the sample surface and phase difference between the two components $\Delta p - \Delta s$. The ratio of two amplitudes R_p/R_s and phase difference $\Delta p - \Delta s$ undergo changes which are influenced by the optical properties of the surface. Vedam (1975) has used this technique to characterize quantitatively the surface layer of contaminated surfaces. This technique has also been used to characterize the quality of the surface, i.e., smoothness and damage of the surface. Table 7.4 illustrates the effect of polishing materials, such as diamond and cerium oxide, and etching duration on the ellipsometric constants, that is, R_p/R_s and $\Delta p - \Delta s$, for vitreous silica samples. It is easy to establish that the silica sample polished by cerium oxide

TABLE 7.4 Ellipsometric Constants of Vitreous Silica

Surface preparation or treatment	Polished with diamond paste		Polished with cerium oxide	
	$\tan^{-1}$ (R_p/R_s) (in degrees)	$\Delta p - \Delta s$	$\tan^{-1}$ (R_p/R_s) (in degrees)	$\Delta p - \Delta s$
Ideal case	3.79	360.00	3.79	360.00
Mechanical polish	0.52	290.00	3.79	359.64
Etch in 10% HF:				
2 min	2.55	334.00	3.77	359.93
4 min	3.93	4.00	3.79	0.11
6 min	4.16	3.04	3.80	0.29

Source: Adapted from Vedam (1975).

caused less damage compared to that polished by diamond paste. Further, the ellipsometric constants for etched surfaces (in 10 percent HF) show better surfaces near the ideal case on moderate etching and a rough and damaged surface on continued etching.

Several other techniques, in addition to those described above, such as plasma chromatography (Carr, 1977), surface potential or contact potential difference (Guttenplan, 1978; Smith, 1979), and electrochemical technique (Srinivasan and Sawyer, 1970) have been used to examine the cleanliness level and to detect the presence of contamination.

The surface roughness can be quantitatively monitored by using a surface profiler or by looking at the surface in a scanning electron microscope. The visual inspection, i.e., by measuring the amount of reflected light, also gives a rough estimate of roughness level. Further details on various roughness measuring instruments can be found in Chap. 15.

7.4 REFERENCES

Anonymous (1971), "A Survey of Contamination (Its Nature, Detection and Control)," Report No. SC-M-70-303, Spectroscopy Division 5525, Sandia Laboratories, Albuquerque, N.Mex.

——— (1978), "Pre-paint Preparation of Metal with Oakite Cryscoat Processes," Oakite Products Inc., Berkeley Heights, N.J.

——— (1979), "Phosphatizing," Stauffer Chemical Company, Chemical Systems Division, Livonia, Mich.

——— (1982), *Metals Handbook*, Vol. 5: *Surface Cleaning, Finishing, and Coating*, 9th ed., American Society for Metals, Metals Park, Ohio.

Beadle, J. D. (ed.) (1972), *Product Treatment and Finishing*, Macmillan, Hampshire, United Kingdom.

Beal, S. E. (1978), "Cleaning and Surface Preparation," *Thin Solid Films*, Vol. 53, pp. 97–99.

Berman, S. M., and Silverstein, S. D. (eds.) (1975), "Energy Conservation and Window Systems," *AIP Conf. No. 25, Proc.—Efficient Use of Energy Pt. III*, pp. 245–299, American Institute of Physics, New York.

Bernard, J., Coffield, J., Johannesen, R. W., Kittredge, J. B., and Snell, G. A. (1982), "Mass Finishing," in *Metals Handbook*, Vol. 5: *Surface Cleaning, Finishing, and Coating*, 9th ed., pp. 129–137, American Society for Metals, Metals Park, Ohio.

Bhushan, B. (1978), "Surface Pretreatment of Thin Inconel X-750 Foils for Improved Coating Adherence," *Thin Solid Films*, Vol. 53, pp. 99–107.

——— (1979a), "Techniques for Removing Surface Contamination in Thin Film Deposition," in *Surface Contamination: Its Genesis, Detection and Control* (K. L. Mittal, ed.), pp. 393–405, Plenum, New York.

——— (1979b), "Development of RF-Sputtered Chrome Oxide Coatings for Wear Application," *Thin Solid Films*, Vol. 64, pp. 231–241.

——— (1979c), "Effect of Surface Roughness and Temperature on the Wear of CdO-Graphite and MoS_2 Films," *Proc. Int. Conf. on Wear of Materials* (K. C. Ludema, W. A. Glaeser, and S. K. Rhee, eds.) pp. 409–414, ASME, New York.

———, Ruscitto, D., and Gray, S. (1978), "Hydrodynamic Air Lubricated Compliant Surface Bearing for an Automotive Gas Turbine Engine, Part II; Materials and Coatings," Tech. Rep. No. CR-135402, NASA Lewis Research Center, Cleveland, Ohio.

Blume, W. J. (1973), "Role of Organic Acids in Cleaning Stainless Steels," Special Technical Publication No. STP 538, pp. 43–53, ASTM, Philadelphia.

Bryson, C. E., Scharpen, L. H., and Zajicek, P. L. (1979), "An ESCA Analysis of Several Surface Cleaning Techniques," in *Surface Contamination: Its Genesis, Detection and Control* (K. L. Mittal, ed.), Vol. 2, pp. 687–696, Plenum, New York.

Cameron, E. N. (1961), *Ore Microscopy*, Wiley, New York.

Carr, T. W. (1977), "Analysis of Surface Contaminants by Plasma Chromatography-Mass Spectroscopy," *Thin Solid Films*, Vol. 45, pp. 115–122.

Clement, R. W., Rains, J. H., Simmons, R., and Surprenant, K. (1982), "Solvent Cleaning," in *Metals Handbook*, Vol. 5: *Surface Cleaning, Finishing, and Coating*, 9th ed., pp. 40–57, American Society for Metals, Metals Park, Ohio.

Dowsett, M. G., King, R. M., and Parker, E. H. C. (1977), "SIMS Evaluation of Contamination on Ion Cleaned (100) InP Substrates," *Appl. Phys. Lett.*, Vol. 31, pp. 529–531.

Dwight, D. W., and Wightman, J. P. (1979), "Identification of Contaminants with Energetic Beam Techniques," in *Surface Contamination: Its Genesis, Detection and Control* (K. L. Mittal, ed.), Vol. 2, pp. 569–586, Plenum, New York.

Elliott, R. H. (1959), "Composition and Operating Conditions for an Alkaline Etching Solution," U.S. Patent 2,882,135.

Fransworth, H. E. (1967), "Atomically Clean Solid Surfaces—Preparation and Evaluation," in *The Solid-Gas Interface* (F. A. Flood, ed.), chap. 13, Marcel Dekker, New York.

——— (1970), "Characterization and Preparation of Atomically Clean Surfaces for Investigation in Ultrahigh Vacuum," in *Clean Surfaces—Their Preparations and Characterization of Interfacial Studies* (G. Goldfinger, ed.), pp. 77–95, Marcel Dekker, New York.

Goldfinger, G. (ed.) (1970), *Clean Surfaces—Their Preparation and Characterization of Interfacial Studies*, Marcel Dekker, New York.

Guttenplan, S. D. (1978), "SPD—A NDT Method for Detecting Surface Contamination," *Plating Surf. Finish.*, June, pp. 54–58.

Haltner, A. J., and Oliver, C. S. (1960), "Frictional Properties of Molybdenum Disulfide Films Containing Inorganic Sulfide Additives," *Nature*, Vol. 188, pp. 308–309.

Henderson, A. W. (1965), "Pre-treatment of Surfaces for Adhesive Bonding," in *Aspects of Adhesion* (D. J. Alner, ed.), Vol. 1, pp. 33–46, University of London Press, London.

Hignett, J. B., Andrew, T. C., Downing, W., Duwell, E. J., Belanger, J., and Tulinski, E. H. (1982), "Polishing and Buffing," in *Metals Handbook*, Vol. 5: *Surface Cleaning, Finishing, and Coating*, 9th ed., pp. 107–127, American Society for Metals, Metals Park, Ohio.

Hudson, R. M., Joniec, R. J., and Shatynski, S. R. (1982), "Pickling of Iron and Steel," in *Metals Handbook*, Vol. 5: *Surface Cleaning, Finishing, and Coating*, 9th ed., pp. 68–82, American Society for Metals, Metals Park, Ohio.

Jackson, L. C. (1976), "Solvent Cleaning Process Efficiency, Contaminant Removal in the Electronic Industry," *Adhesive Age*, July, pp. 31–34.

Jellison, J. L. (1979), "Effect of Surface Contamination on Solid Phase Welding—An Overview," in *Surface Contamination: Its Genesis, Detection and Control* (K. L. Mittal, ed.), Vol. 2, pp. 899–923, Plenum, New York.

Kominiak, G. J., and Mattox, D. M. (1977), "Reactive Plasma Cleaning of Metals," *Thin Solid Films*, Vol. 40, pp. 141–148.

Leliaert, R. M., Weightman, N., and Woelfel, M. M. (1982), "Abrasive Blast Cleaning," in *Metals Handbook*, Vol. 5: *Surface Cleaning, Finishing, and Coating*, 9th ed., pp. 83–96, American Society for Metals, Metals Park, Ohio.

Lindfors, P. A. (1979), "Application of Auger Electron Spectroscopy to Characterize Contamination," in *Surface Contamination: Its Genesis, Detection and Control* (K. L. Mittal, ed.), Vol. 2, pp. 587–594, Plenum, New York.

Maissel, L. I., and Glang, R. (1970), *Handbook of Thin Film Technology*, McGraw-Hill, New York.

———, Torgesson, D. R., and Wagner, H. M. (1966), "Surface Cleanliness Testing," *Electronic Packag. Prod.*, Vol. 6, p. 70.

Mattox, D. M. (1965), "Interface Formation and Adhesion of Thin Films," Sandia Corporation Monograph No. SC-R-65-852, Sandia Laboratories, Albuquerque, N.Mex.

——— (1976), "Thin-Film Adhesion and Adhesive Failure—A Perspective," in *Adhesion Measurements of Thin Films, Thick Films and Bulk Coatings* (K. L. Mittal, ed.), Special Technical Publication No. STP 640, pp. 54–62, ASTM, Philadelphia.

——— (1978), "Surface Cleaning in Thin Film Technology," *Thin Solid Films*, Vol. 53, pp. 81–96.

Meserve, H. O., Hicks, H., Lindblad, K., O'Keeffe, T. P., and Srinivasan, B. (1982), "Selection of Cleaning Process," in *Metals Handbook*, Vol. 5: *Surface Cleaning, Finishing and Coating*, 9th ed., pp. 3–21, American Society for Metals, Metals Park, Ohio.

Miller, A. C., and Czanderna, A. W. (1978), "Ion Scattering Analysis of Contaminated Copper Oxide Surfaces before and after Cleaning," *Vacuum*, Vol. 28, pp. 9–10.

Miller, R. N. (1973), "Rapid Method for Determining the Degree of Cleanliness of Metal Surfaces," *Mater. Prot. Performance*, May, pp. 31–36.

Mittal, K. L. (ed.) (1979), *Surface Contamination: Genesis, Detection and Control*, Vols. 1 and 2, Plenum, New York.

Murphy, D. P. (1982), "Alkaline Cleaning," in *Metals Handbook*, Vol. 5: *Surface Cleaning, Finishing and Coating*, 9th ed., pp. 22–32, American Society for Metals, Metals Park, Ohio.

O'Kane, D. F., and Mittal, K. L. (1974), "Plasma Cleaning of Metal Surfaces," *J. Vac. Sci. Technol.*, Vol. 11, pp. 567–569.

Oster, G., Oster, G. K., and Morson, H. (1959), "Ultraviolet Induced Crosslinking and Grafting of Solid High Polymers," *J. Polymer Sci.*, Vol. 34, pp. 671–684.

Pilznienski, J. (1982), "Acid Cleaning of Iron and Steel," in *Metals Handbook*, Vol. 5: *Surface Cleaning, Finishing, and Coating*, 9th ed., pp. 59–67, American Society for Metals, Metals Park, Ohio.

Pudvin, J. E. (1973), "Surface Cleaning Theory," in *Environmental Control in Electronic Packaging* (Western Elec. Series) (P. W. Morrison, ed.), pp. 146–178, Van Nostrand–Reinhold, New York.

Ratner, B. D., Rosen, J. J., Hoffman, A. S., and Scharpen, L. H. (1979), "An ESCA Study of Surface Contaminants on Glass Substrates for Cell Adhesion," in *Surface Contamination: Its Genesis, Detection and Control* (K. L. Mittal, ed.), pp. 669–686, Plenum, New York.

Roberts, W. J., and Srinivasan, B. (1982), "Emulsion Cleaning," in *Metals Handbook*, Vol. 5: *Surface Cleaning, Finishing and Coating*, 9th ed., pp. 33–39, American Society for Metals, Metals Park, Ohio.

Slater, M. N., and Donahue, D. J. (1959), "Radiotracers in Evaluating Part Cleaning," in *Cleaning of Electronic Device Components and Materials*, STP-246, pp. 146–154, ASTM, Philadelphia.

Smith, T. (1979), "Quantitative Techniques for Monitoring Surface Contamination," in *Surface Contamination: Its Genesis, Detection and Control* (K. L. Mittal, ed.), Vol. 2, pp. 697–712, Plenum, New York.

Snorgen, R. C. (1974), *Handbook of Surface Preparation*, Palmerton Pub. Co., Atlanta.

Sowell, R. R., DeLollis, N. J., Gregory, H. H., and Montoya, O. (1972), "Effect of Activated Gas Plasma on Surface Characteristics and Bondability of RTV Silicone and Polyethylene," *J. Adhesion*, Vol. 4, pp. 15–24.

Sparrow, G. R. (1977), "Characterization of Cleaned and Prepared Bonding Surfaces by ISS/SIMS," *Proc. 2nd Int. Conf. on Deburring and Surface Conditioning*, Society of Manufacturing Engineers, Chicago.

Srinivasan, S., and Sawyer, P. N. (1970), "Electrochemical Techniques for the Characterization of Clean Surfaces in Solution," in *Clean Surfaces: Their Preparation and Characterization for Interfacial Studies* (G. Goldfinger, ed.), pp. 195–218, Marcel Dekker, New York.

Stiteler, D. J. (1982), "A Chemical Cleaning Process to Remove Deposits from Nuclear Steam Generators," *Corrosion*, Vol. 82, March, pp. 22–62.

Stupp, B. C. (1959), "Effect of Surface Preparation on Wear Life of Solid Lubricant Films," presented at joint Air Force–Navy Industry Conf. on Lubricants, Dayton, Ohio, Feb. 17.

Szkudlapski, A. H. (1973), "Surface Cleaning Practice," in *Environmental Control in Electronic Packaging* (Western Elec. Series) (P. W. Morrison, ed.), pp. 180–206, Van Nostrand–Reinhold, New York.

Vedam, K. (1975), "Characterization of Surfaces," in *The Characterization of Materials in Research, Ceramic and Polymers* (J. J. Burke and V. Weiss, eds.), pp. 503–537, Syracuse University Press, Syracuse, N.Y.

Von Fraunhofer, J. A. (1976), *Basic Metal Finishing*, Elek Science, London.

Walton, A. (1966), "Nucleation," *Int. Sci. Tech.*, Vol. 3, December, pp. 28–39.

White, M. L. (1977), "Clean Surface Technology," in *Proc. 27th Annual Frequency Control Symp.*, pp. 79–88, Electronic Industries Association, Cherry Hill Road, N.J.

Winter, L. S. (1982), "Electropolishing," in *Metals Handbook*, Vol. 5: *Surface Cleaning, Finishing and Coating*, 9th ed., pp. 303–309, American Society for Metals, Metals Park, Ohio.

Wood, W. G., and McCardle, T. F. (1982), "Salt Bath Descaling," in *Metals Handbook*, Vol. 5: *Surface Cleaning, Finishing and Coating*, 9th ed., pp. 97–103, American Society for Metals, Metals Park, Ohio.

Yang, M. G., Koliwad, K. M., and McGuire, G. E. (1975), "Auger Electron Spectroscopy for Cleanup of Related Contamination of Silicon Surface," *J. Electrochem. Soc.*, Vol. 122, pp. 675–678.

Yaniv, A. E., Rona, M., and Sharon, J. (1979), "The Influence of Surface Preparation on the Behavior of Organic Coatings," *Metal Finish.*, Vol. 77, November, pp. 55–61; December, pp. 25–30.

CHAPTER 8
COATING DEPOSITION BY HARD FACING

Hard facing is used to deposit a wear-resistant material on either a worn component or a new component which is to be subjected to wear in service. These coatings are also used to provide protection from corrosion by oxidation or seawater. *Hard facing* is a generic term for a group of commonly used processes (thermal spraying, welding, and cladding) for applying thick (typically >50-μm) coatings of (usually hard) metallic and nonmetallic materials (Anderson, 1978; Antony et al., 1983; Bhushan, 1980a, 1987; Bucklow, 1983; Chapman and Anderson, 1974; Clare and Crawmer, 1982; Foroulis, 1984; Gregory, 1978, 1980; Hocking et al., 1989; Ingham and Shepard, 1965; Smart and Catherall, 1972; Steffens, 1983; Tucker, 1982). Various coating deposition techniques used for hard facing are shown in Fig. 8.1. The properties of the hard facings are influenced by several factors such as deposition technique, processing parameters, coating material, substrate material, and pre- and posttreatments of the substrate. Table 8.1 pre-

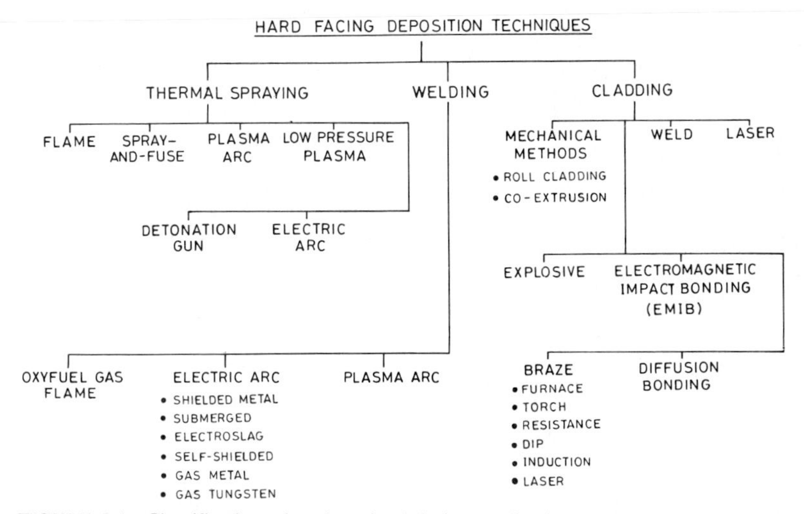

FIGURE 8.1 Classification of various hard facing application techniques. Wire flame spraying using metal in wire form and electric arc spraying processes are also called metallizing.

TABLE 8.1 Process Parameters of Various Coating Deposition Techniques Based on Hard Facing

Deposition techniques	Type of coating materials	Type of substrates	Deposition or treatment temperature, °C	Thickness range, μm	Rate of deposition, kg h^{-1}	Pretreatment	Posttreatment	Comments
Thermal spray	Most metals, alloys, ceramics, and cermets, e.g., Al, Mo, stainless steels, Ni- and Co-based alloys, carbides, nitrides, borides, silicides, and oxides	Almost any	100–150	50–500	1–10	Grit-blast	Grind/lap	Relatively thick substrate (≥250 μm) to avoid warpage
Welding	Most metals and alloys, e.g., martensitic, austenitic, and pearlitic steels; high-alloy irons; Ni- and Co-based alloys; carbides; and other specialized superalloys	Most metals and alloys	900–1400	750–3000	2–30	Mechanically clean	Grind/lap	Relatively thick substrate (≥500 μm) to avoid warpage
Cladding	Metals and alloys, e.g., irons, steels, Al, Cu, Ni, Co, Cr, and their alloys	Metals and alloys	50–1200	50–100,000	—	Mechanically clean	None (except in high-precision components)	—

sents process parameters of various hard facing deposition techniques, and Table 8.2 presents their advantages and disadvantages. These processes are line-of-sight processes; therefore, it is difficult to coat complex-shaped parts.

As a rule, thermal spraying processes are preferred for applications requiring thin, hard coatings applied with minimum thermal distortion of the workpiece and with good process control. Welding processes, on the other hand, are preferred for applications requiring dense, relatively thick coatings (due to extremely high deposition rates) with high bond strength. Welded coatings can be applied to substrates which can withstand high temperatures (typically >900°C). Thermal spraying processes most commonly use the coating material in the powder form, and almost any material capable of being melted without decomposition, vaporization, sublimation, or dissociation can be thermally sprayed. Welding processes, however, most commonly use the coating material in the rod or wire form. Thus materials that can be easily cast in rods or drawn into wires are commonly deposited. In arc welding, the substrate and the coating material must be electrically conductive. Welding processes are most commonly used to deposit primarily various metals and alloys on metallic substrates.

In contrast to thermal spraying processes, which do not penetrate the substrate metal, welding processes melt a portion of the surface, and this fusion zone becomes mixed or diluted in the weld metal. Mixing of a proportion of substrate metal with the weld can affect the composition, the microstructure, and hence the wear resistance. (Dilution generally reduces wear resistance of the coating alloy.) But because of the melting of the substrate metal in the fusion zone, on solidification of the weld, there is a metallurgical continuity (or graded interface) across the fusion boundary, which should ensure the best mechanical properties. Most thermal sprayed coatings (except spray-and-fuse) do not melt the substrate surface, and only a mechanical bond is produced from impingement of the coating particles with high kinetic energies onto a preroughened substrate. In general, thermal sprayed coatings do not have mechanical properties equivalent to those of welded coatings, which can tolerate higher impact stresses, such as may occur under shock loading.

Cladding processes are used to bond bulk material in foil, sheet, or plate form to the substrate to provide tribological properties. The cladding processes are used either where coatings by thermal spraying and welding cannot be applied or for applications which require surfaces with bulklike properties such as low porosity. Since relatively thick sheets can be readily clad to the substrates, increased wear protection may be possible compared to thermal spraying and welding. If the coating material is available in sheet form, then cladding may be a cheaper alternative to surface protection. It is difficult to clad parts having complex shapes and extremely large sizes. Metallic powders or other fillers can also be cladded to the metallic substrates. The difference between such cladding and coatings deposited by conventional techniques is the thickness; cladding usually denotes the application of a relatively thick layer (3 mm or larger).

Substrates for thermal spraying need to be preroughened by mechanical or chemical methods; therefore substrates must be relatively thick (≳250 μm), so that they do not warp. In welding, substrate needs to be even thicker so that it does not warp from its skin melting during the welding process. Fairly thin substrates can be processed by some cladding techniques. Parts produced by thermal spraying and welding techniques generally need to be postfinished to an acceptable roughness; however, clad parts generally do not need to be postfinished.

This chapter describes the principal processes used for hard facing and the alloys available. A vast range of properties are available from which the user may select those appropriate to his/her specific needs.

TABLE 8.2 Advantages and Disadvantages of Various Hard Facing Techniques

Deposition technique	Advantages	Disadvantages
Thermal Spraying		
Flame spraying	Transportable for in-situ work, low capital investment, high deposition rates and efficiencies, relative ease, and low cost of maintenance.	Lower bond strengths, higher porosity, a narrower working temperature range, and higher heat transmission to the substrate; cannot be used for materials with high melting point. Requires expensive combustible gases.
Spray-and-fuse	Fully dense microstructure of the coating and the metallurgical bond.	High temperature to which the substrate is exposed during the fusion step.
Plasma arc spraying	Wider range of high-melting-temperature materials, particularly ceramics and cermets, can be coated. High particle velocities result in higher bond strength and coating density.	Lower bond strength and density than detonation gun. Equipment not transportable.
Low-pressure plasma arc spraying	Prevents oxidation of some materials, higher bond strength of the coatings, excellent dimensional thickness control, higher deposition efficiency.	Higher operating and maintenance costs compared to flame and plasma spraying.
Detonation-gun spraying	Extremely high velocity results in higher density, greater internal strength, and superior bond strength. Process is fully automated.	Very hard substrates with poor ductility generally cannot be coated. Requires expensive combustible gases.
Electric arc spraying	Transportable, low capital investment, thick coatings at high deposition rates, lower substrate heating.	Limited to relatively ductile, electrically conductive coating metals available in wire form.

Welding		
Oxyfuel gas welding (OFW)	Versatile, very low dilution, simple, low cost of equipment, portable, very small areas or limited access can be coated.	High porosity, requires a high degree of welder skill, labor-intensive, low deposition rates, unsuitable for large areas, limited range of consumables available.
Shielded metal arc welding (SMAW)	Most versatile, low capital cost, portable, does not require operator skill, small and limited-access areas can be coated, does not require preheating.	Relatively high dilution, low deposition rates, substrate cracking may occur with hard coatings.
Submerged arc welding (SAW)	Fully automated process, very high deposition rates (only lower than ESW), excellent deposit appearance, minimum finishing required.	Very high dilution, expensive equipment, not easily portable, generally limited to simple round or flat components.
Self-shielded arc welding (SSAW)	Continuous process that can be semiautomated, fairly high deposition rates, simpler than GMAW.	Fairly high dilution, some porosity, expensive equipment, not easily portable.
Gas metal arc welding (GMAW)	Continuous process that can be semiautomated, very low porosity and contamination, used for high deposit quality with a high level of purity.	Fairly high dilution, expensive equipment, not easily portable, low deposition rates.
Electroslag welding (ESW)	Deposition rates higher than any other welding process.	Not versatile, only vertically positioned welds possible, expensive equipment, limited use.
Gas tungsten arc welding (GTAW)	Can be automated, fairly low dilution, high-quality deposit, used for very specialized applications (reactive materials).	Expensive equipment, not easily portable, low deposition rates and used for small areas, high dilution, limited range of consumables available.
Plasma arc welding (PAW)	Can be automated, fairly low dilution, high deposition rates, good process control, high-quality deposit can be used on reactive surfaces.	Expensive equipment, not easily portable, cannot be used for low-volume job shop.

TABLE 8.2 Advantages and Disadvantages of Various Hard Facing Techniques (*Continued*)

Deposition technique	Advantages	Disadvantages
Cladding		
Roll cladding	Cheapest process, simple.	Restricted to relatively ductile metals and alloys.
Coextrusion cladding	Cheap, simple.	Restricted to relatively ductile metals and alloys.
Explosive cladding	Largest range of material combinations can be processed, large areas can be bonded, simple, intrinsically very rapid, cheap.	Storage and handling of explosives require safety precautions, component must be thick to withstand pressures involved, requires postfinishing, can be used for only simple shapes, equipment bulky.
Electromagnetic impact bonding	Largest range of material combinations can be processed, large areas can be bonded, simple, intrinsically very rapid, cheap.	Components must be circumferentially continuous and of good electrical conductivity, component must be thick to withstand pressures involved, equipment bulky, and high capital costs.
Diffusion bonding	Awkward shapes can be clad, both flat surfaces and powders can be clad.	Slow and most expensive process, equipment bulky, not large capacity, postfinishing required, use confined to small components.
Braze cladding	Cheap, simple, requires relatively low heat input, does not require postfinishing.	Restricted to metals that can be brazed, braze bond may not be as strong as the bonds produced by parent metals.
Weld cladding	Large range of metals and alloys can be processed, powders can be clad.	Some processes are slow, generally a wrought surface cannot be obtained, requires postfinishing.
Laser cladding	Large range of metals and alloys can be processed, high-melting-point metals can be clad on low-melting substrates, powders can be clad, only substrate surface skin is molten.	A wrought surface cannot be obtained, needs laser source.

8.1 THERMAL SPRAYING

The thermal spraying processes are one of the most versatile techniques available for application of coating materials used for resistance to wear and corrosion from oxidation or seawater (Anderson, 1978; Antony et al., 1983; Clare and Crawmer, 1982; Foroulis, 1984; Gerdeman and Hecht, 1972; Gregory, 1978; Grisaffe, 1967; Henne et al., 1984; Ingham, 1969; Ingham and Shepard, 1965; Leckey, 1969; Li, 1980; Longo, 1972; Mansford, 1972; Marantz, 1974; Mock, 1974; Moss and Young, 1974; Sliney, 1979; Smart and Catherall, 1972; Smith, 1974a, 1974b; Steffens, 1983; Tucker, 1982; Wells, 1978; Wolfla and Tucker, 1978). Various thermal spraying processes are listed in Fig. 8.1.

The coatings can be applied by spraying from rod, wire, or powder; powder is the most common form of coating material used. In wire or rod form, material is fed into the heating zone axially from the rear, where it is melted in the few milliseconds available. The molten stock is then stripped from the end of the wire or rod and is atomized by a high-velocity stream of compressed air or other gas or by other means (detonation waves), propelling the material onto a water-cooled prepared substrate. In powder form, the material is metered by a powder feeder, into a compressed air or gas stream, which suspends and delivers the material to the heating zone, where it is heated to a molten or semimolten state and propelled to the water-cooled substrate. The distance from the workpiece surface to the gun tip generally falls in the range of 0.15 to 0.30 m. The molten particles accelerated toward the substrate are cooled to a semimolten or plasticlike condition. The particles splatter on the surface and are bonded instantly primarily by the mechanical interlocking. During coating deposition, a substantial amount of heat is transmitted to the substrate through carrier gas and the molten particles. Therefore, the substrate is water-cooled. Cooling air or CO_2 jets may be used as well. Under normal circumstances, the substrate temperature does not exceed about 150°C during coating.

The thermal energy needed to melt the spraying material can be produced in different ways. In flame spraying and spray-and-fuse processes, the material is melted by a flame produced from combustion gases; in detonation-gun spraying process, the combustion gas is detonated by a spark plug which produces heat. In plasma and low-pressure plasma spraying processes, the thermal energy of electric plasma produced by a low-voltage high-current electric discharge is utilized. In the electric arc spraying, an electric arc is struck between the two converging wires (coating feed material), and the high-temperature arc melts the wire.

Flame, plasma-arc, low-pressure plasma, and detonation-gun spraying are the most widely used processes. The flame spraying process is transportable and therefore is used for rework in situ. The limitation of the flame spraying process is that only those materials which will melt in the combustion flame (<3000°C) can be successfully deposited. This limitation is removed in plasma spraying. By this means, even refractory materials can be deposited. Recent developments in plasma spraying permit working under a controlled atmosphere at low pressure and permits the application of complex coating systems like the MCrAlY type (M = Co, Ni, Fe) with proper stoichiometry and structure, high coating adhesion, and low porosity. The detonation gun provides the molten particles hitting the substrate with a very high kinetic energy which results in higher hardness, higher density, and superior bond strength (see Chapter 14). With the detonation-gun process, metallic substrates with high hardness and nonmetallic substrates normally cannot be coated because the high-velocity hot gas stream produces surface erosion. Spray-and-fuse and electric arc spraying processes also find limited applications. With the spray-and-fuse process, the substrate must be able to withstand elevated temperature during the fusing step. In electric arc spraying, relatively ductile metals available only in wire form can be sprayed. Some important characteristics of the various deposition processes are compared in Table 8.3.

Almost any material that can be melted without decomposition, vaporization, sublimation, or dissociation can be thermally sprayed. The techniques are not limited to metals and are also used for the applications of nonmetallic materials, e.g., hard ceramics, cermets, and, to a limited extent, polymers. Typical coating thicknesses range from 50 to 500 μm, but in a few applications they may exceed 0.5 mm. Theoretically, there is no limit to the coating thickness which may be applied. However, in practical applications, internal (residual) stresses set up in coatings limit the maximum thickness for an adherent coating.

The thermal spraying processes, except detonation, are also independent of the composition or condition of the substrate material, provided it can be prepared to give a receptive surface. Thin, ductile materials may deform, or brittle materials may crack during surface pretreatment (grit blasting) or when impacted by very high-velocity spray particles (e.g., in detonation-gun spraying). Thermal sprayed coatings can be applied to most materials including glass and plastics. In general, bonding to metallic surfaces is superior to that of nonmetallics, although there are exceptions. The substrate is generally water-cooled and is seldom heated above 150°C. As a result, a part can be fabricated and fully heat-treated prior to coating. Usually there is no distortion, and the substrate microstructure and strength are not changed by the coating process.

Substrates are generally roughened by grit blasting or other techniques to a roughness of about 0.5 to 3 μm rms and are sometimes preheated to improve coating adhesion. It is advisable to warm the surface slightly (to about 125 to 150°C), usually with a pass of the spray torch without powder flowing, to remove adsorbed gases from the surface before the coating is applied. The interfacial bond of most thermal sprayed coatings (except spray-and-fuse coatings) is primarily mechanical.

The various thermal spray processes available for application of metallic and nonmetallic coatings are described in the following subsections.

8.1.1 Flame Spraying

This process utilizes combustion flame as a heat source to melt the coating material, available in wire, rod, or powder form (Anonymous, 1967; Ault, 1957; Ault and Wheildon, 1960; Ballard, 1963; Clare and Crawmer, 1982; Fisher, 1972; Grisaffe, 1967; Ingham and Shepard, 1965; Matty and Becker, 1956; Smith, 1974a). The term *metallizing* is also used to describe a flame spraying process which involves the use of metal in wire form (Anonymous, 1967; Gregory, 1978). Figure 8.2 shows schematically the wire and powder flame spraying processes. In the wire flame spraying process, shown in Fig. 8.2(*a*), the source comprises a nozzle, through the center of which (metallic) wire is fed at a controllable rate into the flame, which melts the wire tip. Compressed air is fed through the annulus around the outside of the nozzle, accelerating the molten or semimolten particles onto the substrate. The wire tip is continuously heated to its melting point and then is broken down into particles by the stream of compressed air. Any metal that can be drawn into wire form and will melt in the environment of com-

TABLE 8.3 Some Important Characteristics of Thermal Sprayed Techniques*

Deposition technique	Heat source	Propellent	Material feed	Typical temperature in spray gun, °C	Typical particle velocity, m s^{-1}	Average spray rate, kg h^{-1}		Application treatment temperature, °C	Type of coating materials	Coating porosity, % by volume	Relative bond strength
						Ceramics	Metals				
Flame spraying	Oxyacetylene/ oxyhydrogen	Air	Wire	3,300	240	—	2–30	100–135	Metallic	8–15	Good
			Powder		30–120	1.5–2.5	—	110–160	Metallic and ceramic	10–20	Fair
Spray-and-fuse	—	—	Powder	—	—	—	—	1000–1300	Fusible metals	<0.5	Excellent
Plasma spraying	Plasma arc	Inert gas	Powder	16,000	120–600	5.5–7	3–9	100–120	Metallic, ceramic, plastic, and compound	2–5	Very good to excellent
Low-pressure plasma spraying	Plasma arc	Inert gas	Powder	16,000	up to 900	—	—	—	Metallic, ceramic, plastic, and compound	<5	Excellent
Detonation-gun spraying	Oxygen/acetylene/ nitrogen gas detonation	Detonation shock waves	Powder	4,500	800	—	—	120–150	Metallic, ceramic, and compound	0.1–1	Excellent
Electric arc spraying	Arc between consumable electrodes	Air	Wire	6,000	240	—	5–55	50–120	Relatively ductile metals	8–15	Good

*Almost any substrate can be coated except for detonation gun in which the substrate should have some ductility.

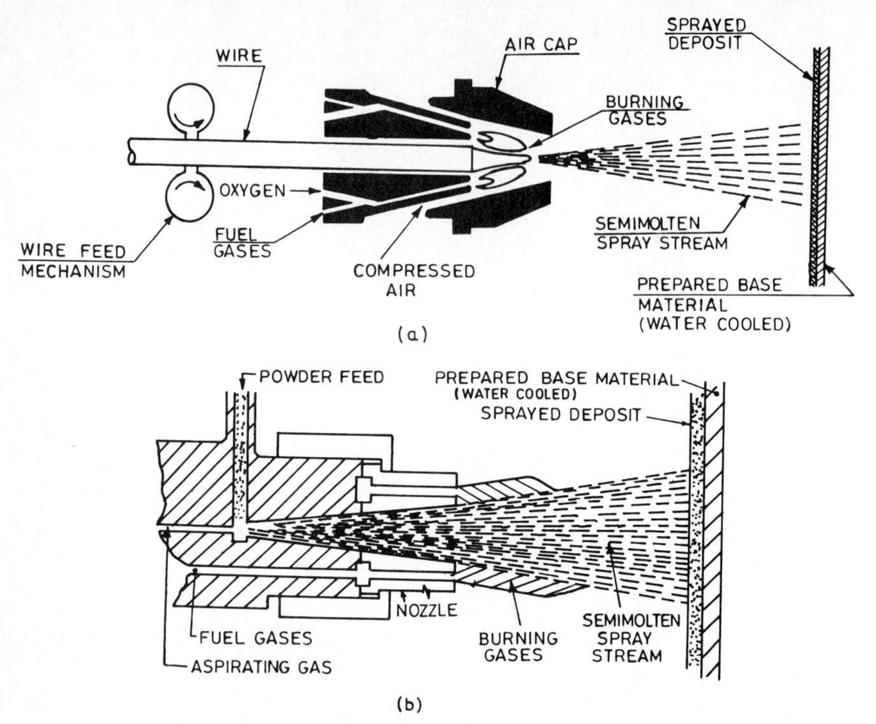

FIGURE 8.2 Schematics of the flame spraying processes: (*a*) wire form; (*b*) powder form.

bustion flame can be sprayed. Typical metals which are sprayed include aluminum, copper, nickel-zinc, bronze, steel, and molybdenum.

In the powder flame spraying process, shown in Fig. 8.2(b), the powdered material is held in a hopper atop the gun and is gravity-fed into the gun, where it is picked up by a carrier or aspirating gas and carried to the gun nozzle. Here the powder is melted and carried to the surface being sprayed by means of a siphon-jet arrangement at the gun nozzle. This process is also erroneously called *powder welding*. The powder flame spraying process extends the choice of materials to encompass those which cannot be produced in wire or rod form. Any material can be sprayed which will melt in the environment of an oxyfuel flame and does not decompose on melting. Oxidation-resistant metals and alloys and some ceramics are sprayed by the powder flame spraying process. Of materials commonly sprayed, molybdenum has the highest melting point, 2600°C. Materials which can be sprayed include Ni- and Co-based alloys, chromium carbide with nichrome, alumina, zirconia, and alumina-titania.

Sintered ceramic oxide rods may also be sprayed by the proprietary Rokide process (Norton Corp., Worcester, Massachusetts). Alumina, zirconia, chromia, zirconium silicate, magnesium aluminate, and various other ceramics are sprayed by this process.

The most commonly used combustible gases are oxyacetylene and oxyhydrogen. The flame temperature ranges from 3000 to 3350°C depending on the oxygen–fuel gas ratio. Mostly oxyacetylene flame is used for spraying of hard coatings, instead of oxyhydrogen, due to cost considerations. One advantage of oxyhydrogen spraying is that the flame is not as bright as an acetylene flame. Thus, details of the work can be observed more accurately during the spraying. Also oxyhydrogen spray coatings contain no oxide or slag, and the coating quality is superior.

The deposition rate and deposition thickness depend on the coating material. For example, certain oxide ceramic coatings applied by powder flame spraying would be deposited at 0.25 g s^{-1} whereas zinc coatings by wire flame spraying would be deposited at 7.5 to 8.8 g s^{-1}. There is a tendency for all coatings to shrink because of the rapid quenching of semimolten particles. These differences coupled with different bond strengths define what this thickness limit will be in practice. Certain materials are limited to a thickness of 500 to 700 μm, whereas others can be sprayed up to a thickness of about 1000 to 1500 μm. Typically in industry, flame sprayed coatings are used in thicknesses ranging from 125 to 1000 μm. The coating properties are influenced by the flame temperature, velocity of the carrier gas, etc. Flame spraying equipment is transportable for in situ work. The flame spraying process is characterized by very low capital investment, high deposition rates and efficiencies, and relative ease and cost of maintenance. The wire flame spraying (metallizing) apparatus is the least costly of all thermal spraying equipment. In general, flame sprayed coatings exhibit higher porosity, lower bond strength, a narrower working temperature range, and higher heat transmission to the substrate than other thermal spraying processes. Because the droplets from a wire or rod must be entirely molten before they pass through the spray gun, the coatings produced are slightly less porous than those produced by powder spraying.

The flame spraying process is widely used by industry for the corrosion resistance and reclamation of worn or out-of-tolerance parts. Coatings of zinc or aluminum produced by wire flame spraying are applied on large steel structures for protection against atmospheric and seawater corrosion, and the spraying of tin is used to patch defects of food processing vats. The oxidation resistance of stainless steel and nichrome produced by powder flame spraying sees many applications. Powder and rod flame sprayed coatings of ceramics also offer protection against heat and erosion in rocket nozzles, extrusion dies, and space hardware.

8.1.2 Spray-and-Fuse Process

The spray-and-fuse process is a modification of the powder flame spraying method previously mentioned. The spray-and-fuse process is a two-step process in which powdered coating material is deposited by conventional thermal spraying, usually by a flame spray gun (a plasma spray gun can be used instead) and subsequently fused (immediately after spraying) by an oxyacetylene torch, induction heating, or heat treatment in a vacuum or reducing-atmosphere furnace. Reducing-atmosphere furnaces are normally used to ensure a clean, well-bonded coating (Antony et al., 1983). Coating ranging from 500 to 2000 μm or greater in thickness can be made by building up several layers at a rate of 125 to 750 μm per pass. Typical deposition rates are 4 to 6 kg h^{-1}.

The coatings are usually made of self-fluxing, fusible materials which require postspray heat treatment. In general, these are nickel- or cobalt-based alloys which use boron, phosphorous, or silicon, either singly or in combination, as melting-point depressants and fluxing agents. These alloys generally fuse between 1000 and 1300°C, depending on the composition and produce a hardness as high as 65 HRC. Some powder manufacturers offer these alloys blended with tungsten carbide or chromium carbide particles to increase resistance to wear from abrasion, fretting, or erosion. Boron, phosphorous, and silicon act as fluxing agents, which during fusion permit the coating to react with any oxide film on the workpiece or individual powder particle surfaces, allowing the coating to wet and interdiffuse with the substrate and result in virtually full densification of the hard facing. Most workpiece materials can be hard-faced by using the spray-and-fuse process without taking special precautions, but others require special preheating or cooling procedures to prevent cracking of the coating. Spray-and-fuse coating may be deposited directly over an existing deposit if the workpiece surface is clean and free of oxide and is preheated to about 800°C before the initial deposition.

The major advantages of the spray-and-fuse process are the virtually fully dense microstructure of the coating and the metallurgical bond created between the coating and the substrate. The major disadvantage is the high temperature to which the substrate is exposed during the fusion step. This frequently causes undesirable changes in the structure of steels and other substrates. Spray-and-fuse coatings are widely used in applications requiring wear resistance and have been successfully used in the oil industry for sucker rod and in agriculture for plowshares. In most applications, fusible alloys make possible the use of less expensive substrate material.

8.1.3 Plasma Arc Spraying

In plasma arc spraying, also referred to as *arc plasma* and *plasma spraying*, the thermal energy of an electric arc and its associated plasma jet is utilized in melting and projecting material as a high-velocity spray onto substrate, where it solidifies as a coating (Anonymous, 1975a, 1976; Bhat and Herman, 1982; Clare and Crawmer, 1982; Gage et al., 1962; Gregory, 1978, 1980; Leckey, 1969; McPherson, 1981; Meyer and Muehlberger, 1984; Moss and Young, 1974; Safai and Herman, 1978, 1981; Scott and Cross, 1980; Scott and Woodhead, 1982). Figure 8.3 shows a schematic section of the plasma spray gun head. In this process, a dc electric arc is struck from a high-frequency arc starter between a central (tungsten) electrode in the torch and a water-cooled (copper) nozzle, which forms an anode, while a stream of inert gases (either argon or nitrogen, sometimes with the addition of other gases such as hydrogen or helium) is passed through this arc. This results in dissociation and ionization of the gases, thereby producing a high-temperature plasma stream from the gun nozzle. The spraying material is generally in powder form and is fed into the plasma flame by a carrier gas, where it melts and gains high velocity due to high plasma enthalpies and is propelled to the substrate. Wires or rods are rarely used instead of powder; but if they are electrically conductive, they may constitute a second anode (this mode of operation improves the thermal efficiency).

In a further modification to the plasma spraying, an arc may be transferred to a metallic workpiece in order to melt it and thus facilitate fusion penetration or bonding. The preheating also cleans the substrate surface. The transferred arc is achieved by applying a potential between the spray nozzle and the substrate (Fig. 8.3). This modified process is referred to as *transferred plasma arc spraying* or *plasma arc welding process*. This mode of operation is for substrates which are electrically conductive and can withstand some deformations. Furthermore, since substrate heating is part of the process, some alteration of the substrate microstructure is inevitable. This is described in more detail in the next section on welding.

The high temperature of the plasma flame, which can be up to 16,000°C, will

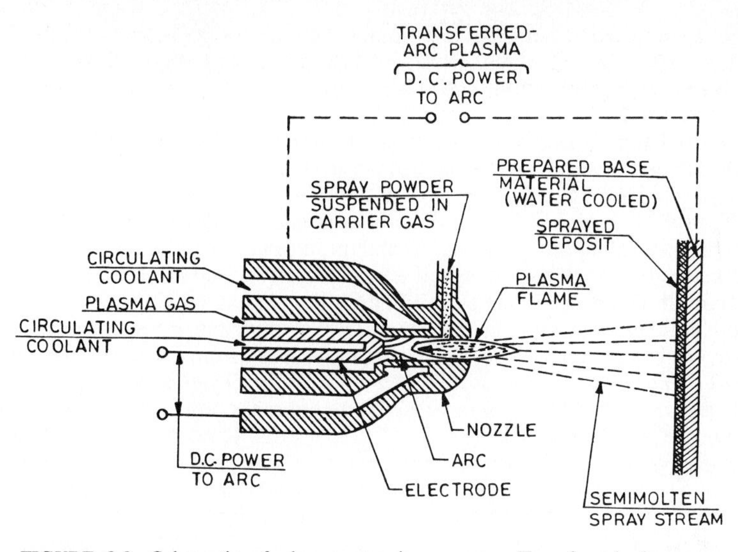

FIGURE 8.3 Schematic of plasma spraying process. Transferred plasma arc spraying (also called plasma arc welding) process is also shown.

melt any coating material whether a metal, an alloy, or ceramic. Therefore, this process has the widest range of materials of any spraying process. Standard plasma guns are rated at up to 40 kW (Anonymous, 1967) with powder particle velocities of about 120 to 400 m s^{-1}. The latest high-energy guns are rated at up to 80 kW (Anonymous, 1976; Muehlberger and Kremith, 1973) to produce high kinetic energies by the high enthalpies. The powder particle velocities average over 600 m s^{-1} (Mach 2) for the high-power guns. The high powder particle velocities in plasma spraying compared to flame spraying result in very high coating adhesion to substrate (bond strength) and high density. Because an inert gas is used as carrier gas in plasma spraying, high-purity deposits free of oxides are obtained. Tucker (1974) has reported that oxidation during deposition of stainless steel and aluminum in an air atmosphere resulted in a coating with lower density and lower tensile strength than the coating deposited in an argon atmosphere.

A broad range of materials, including materials with high melting points, can be plasma sprayed. The typical materials include metals such as nickel, chromium, and molybdenum; alloys such as nickel-, cobalt-, and chromium-based alloys; ceramics such as Al_2O_3, Al_2O_3-TiO_2, Cr_3C_2, ZrO_2-Y_2O_3; and cermets such as cobalt-bonded tungsten carbide (WC—12 wt % Co). A coating thickness of 0.5 to 5 mm and width of up to 30 mm can be achieved in a single pass at a powder feed rate of about 10 kg h^{-1}. Plasma spraying process is commonly used for applications requiring wear and corrosion resistance including bearings, seals, valve seats, aircraft engines, oil field components, mining machinery, and farm equipment.

8.1.4 Low-Pressure Plasma Spraying

Inert-atmosphere and *low-pressure plasma spraying* (LPPS), also referred to as *vacuum plasma spraying*, is a variation of the conventional plasma spraying process in which the plasma gun and workpiece are enclosed in a vacuum tank. Preheating of the workpiece (if any), spraying the coating, and subsequent heat treating (if any) are carried out in an inert atmosphere at low pressure (typically 7 kPa or 50 torr) (Clare and Crawmer, 1982; Eschnauer and Lugscheider, 1984; Gruner, 1984; Henne et al., 1984; Steffens et al., 1980; Steffens and Höhle, 1980; Taylor, 1975).

Inert atmosphere chamber spraying is used for two purposes: to confine hazardous materials and to restrict the formation of oxides that would occur in open spraying. Hazardous materials are grouped into two categories: toxic and pyrophoric. Toxic materials include beryllium and its alloys. Pyrophoric materials include magnesium, titanium, tantalum, lithium, sodium, and zirconium, which are highly reactive at elevated temperatures and tend to burn readily when in a finely divided form or when purified by the plasma process.

Inert atmosphere spraying in a low-pressure chamber offers several advantages over conventional plasma spraying in an inert atmosphere at atmospheric pressure. The spray maintains its chemistry and its original composition. The plasma jet size is considerably increased with the reduction in the spraying chamber pressure to 5 to 7 kPa (35 to 50 torr). In normal atmosphere, the plasma jet attains a length on the order of 45 mm whereas at low pressure it attains up to 0.4 to 0.5 m. Due to enlargement in plasma jet size, the heated surface area of the substrate is increased, which improves the coating quality. The deposition efficiency is increased because of increased particle dwell time in the longer heating zone of the plasma. On the other hand, higher losses of powder and a low-energy density of the plasma jet must be regarded as unavoidable in low-pressure plasma spraying. The plasma gas jet velocity can be increased as high as 3 times the speed of sound in the low-pressure spraying chamber. The increased velocity yields particles of high kinetic energy which provides better adhesion and lower porosity. To enhance the adhesion further, additional energy can be supplied by a transferred arc between the spray gun nozzle and the electrically conducting substrate.

Figure 8.4 shows a system arrangement of the low-pressure plasma spraying process. The water-cooled chamber is pumped down to about 55 Pa (0.4 torr) and then backfilled with inert gas to about 40 kPa (300 torr). Then the plasma spraying process is activated, and the chamber pressure is adjusted to the desired level for

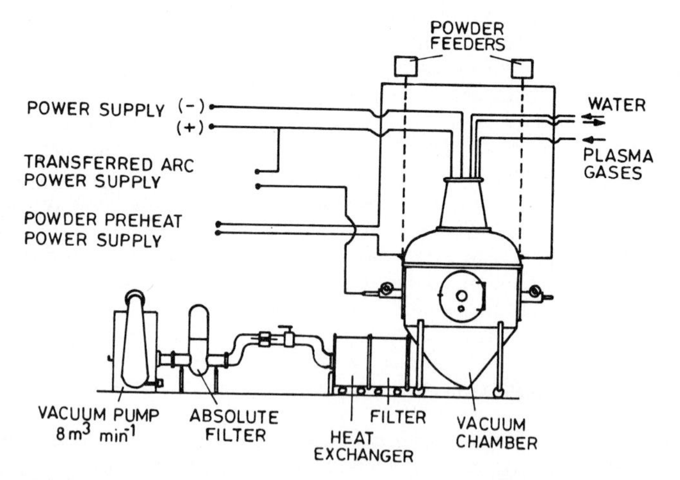

FIGURE 8.4 Schematic of low-pressure plasma spraying setup.

spraying. The entire spraying operation is accomplished in a soft vacuum of about 7 kPa (50 torr). The powder is transported by pressure to heating coils ahead of the plasma gun, where it is then injected into the nozzle of the gun. A filter and heat exchanger are used to clean and cool the gas, loaded with the spraying material.

The LPPS process opens up applications for a new set of materials which otherwise could be plasma sprayed. For example, this process has proved to be an effective means of applying chemically pure coatings of MCrAlY-type alloys on high-temperature aircraft engine components. Vinayo et al. (1985) reported that WC-Co coatings produced by open atmosphere Ar-H_2 plasma spraying showed a strong decomposition of the carbides. The decomposition was reduced, the coating microstructure improved, and the coating density increased by low-pressure Ar-He plasma spraying. Low-pressure plasma spraying at a chamber pressure of 8 kPa (60 torr) and at high power (50 kW) of WC + 20 percent cobalt gave a dense deposit consisting of carbide grains embedded in a cobalt matrix. Reardon et al. (1981) reported that the hardness of the Cr_3C_2 coating produced by low-pressure plasma spraying was higher than that produced by conventional plasma. Steffens and Höhle (1980) deposited titanium and tantalum coatings (used for corrosion protection in chemical plants) which can be sprayed only in an inert gas atmosphere. The deposited coatings by LPPS exhibited low porosity, good adhesion, and chemical purity. Typical applications of the LPPS process include coatings on turbine air foils, blade tips, and shroud segments and certain chemical environments that require impermeable coatings (zirconia or magnesium zirconate).

8.1.5 Detonation-Gun Spraying

The detonation-gun spraying process is somewhat similar to the oxyacetylene powder flame spraying process and was initially developed by Union Carbide Corporation (Poorman et al., 1955). In the detonation-gun spraying process, when measured quantities of oxygen, acetylene, and particles of coating material are metered into the firing chamber, a timed spark detonates the mixtures (Anonymous, 1975b; Bhushan et al., 1978; Cashon, 1975; Kharlamov, 1978; Land, 1974; Smith, 1974b; Tucker, 1982; Wolfla and Tucker, 1978). The resulting detonation consists of heat and pressure waves which create a hot, high-speed gas stream which instantly heats the particles to a molten state (about 4500°C) and hurls them at supersonic velocity (about 800 m s^{-1}) from the gun barrel to the substrate surface.

The detonation gun, shown schematically in Fig. 8.5, consists of a water-cooled barrel about 1 m long with an inside diameter of about 25 mm and associated gas and powder metering equipments. The noise level produced by the detonation is high, approximately 150 dB. Because of this, the detonation gun is housed in a double-walled soundproof room and is remotely controlled from outside by an operator. A mixture of oxygen and acetylene is fed via the poppet valve into the barrel along with a charge of powder in a carrier gas stream. The gas mixture is then ignited by an electric discharge or a spark plug. Detonation or shock waves, which follow within milliseconds after ignition, accelerate the powder to the substrate at about 800 m s^{-1} while heating it close to or above its melting point (Cashon, 1975; Smith, 1974b; Tucker, 1982). The gas mixture is then ignited by a spark plug. The gases burn and produce a detonation or shock wave, produced within a few milliseconds after ignition. As the detonation wave front

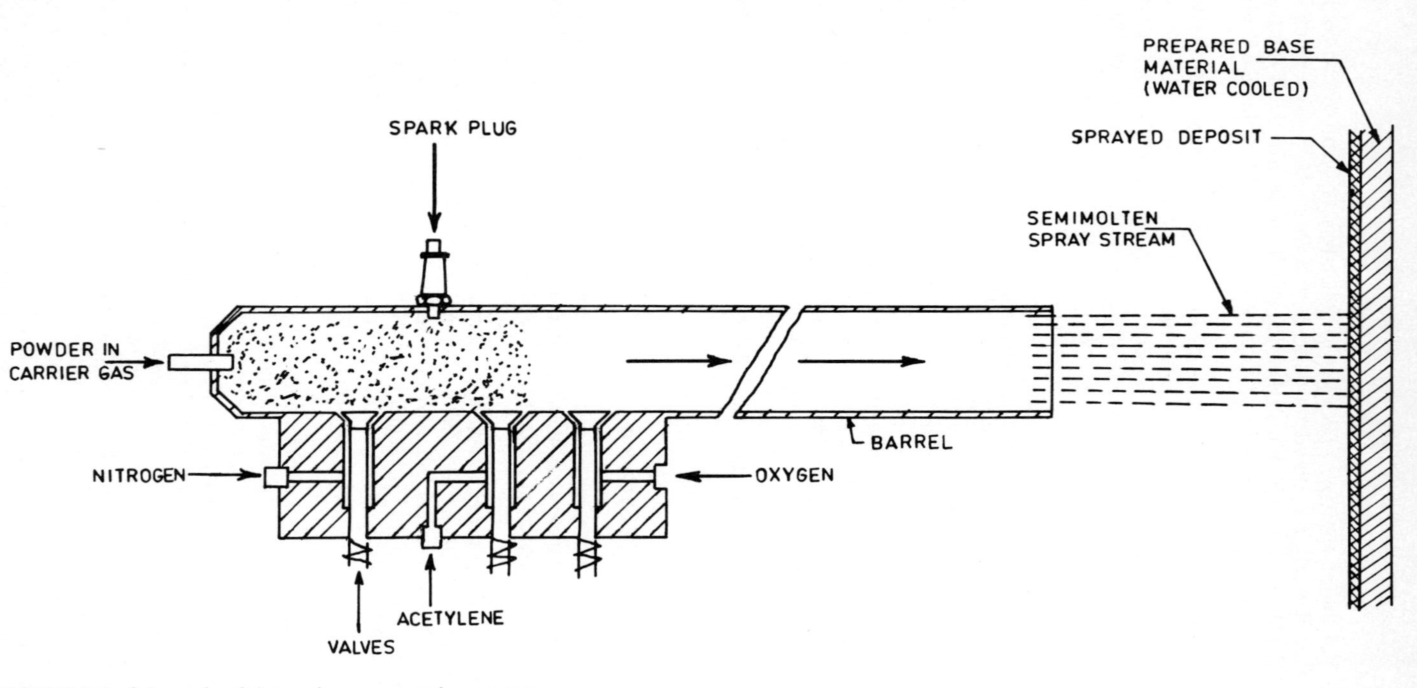

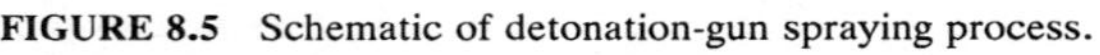

FIGURE 8.5 Schematic of detonation-gun spraying process.

passes through the suspended powder particles in the tube, they are heated, accelerated, and decelerated. The final high-temperature, high-pressure system set up in the reaction gas products in the tube on completion of the detonation picks up the molten particles and hurls them out of the open end of the tube at supersonic speeds.

The distance over which the powder is entrained in the high-velocity gas is much greater than that in a plasma or flame spraying process, which accounts for the very high particle velocity. After the powder has exited the barrel, a pulse of nitrogen gas purges the barrel. The cycle is repeated about 4 to 8 times per second. Because of the gases used in the detonation gun, the powder can be exposed to an oxidizing, carburizing, or essentially inert environment. The maximum free-burning temperature of the oxygen-acetylene mixture with 45 percent acetylene is about 3140°C, but under detonation conditions exceeds about 4500°C; so most, but not all, materials can be melted.

The uniform, closely packed, laminar structure of a detonation gun coating, produced by an extremely high-velocity hot gas stream, results in hardness, density, and bond strength higher than can be achieved with conventional plasma or flame spraying processes. The high kinetic energy of the particles is converted to additional heat on impact with the surface. This benefit of maintaining the molten state allows the particles to completely deform and wet the surface and closely follow the substrate contours, to produce extremely high bond strengths between coating and substrate, with tensile bond strengths in excess of 70 MPa. For example, the typical hardness for a cobalt–tungsten carbide detonation sprayed coating is 1300 HV, whereas the hardness of similar coatings applied by plasma spraying is on the order of 600 to 800 HV (Cashon, 1975; Tucker, 1974). The porosity of detonation-sprayed coatings is very fine, normally on the order of 10-μm diameter. The overall amount is usually between 0.25 and 1 percent by volume, and this is evenly distributed. However, these coatings contain a small percentage of interconnected porosity. In high-temperature applications, detonation coatings are often sufficiently dense as deposited that they seal themselves against internal oxidation and prevent attack of the substrate. For applications in corrosive environments, it is possible to seal the coating by the same methods used in plasma spray coatings.

All metallic materials with a surface hardness less than 58 to 60 HRC can be coated by the detonation-gun process. There is no deformation of these materials because the shock wave is dissipated before the coating powder contacts the substrate. Generally, nonmetallic substrates cannot be coated because the high-velocity hot gas stream produces surface erosion. A major disadvantage of the detonation-gun process is that it is not commercially available.

The types of coatings which can be deposited have been developed primarily for wear resistance and are capable of operating at elevated temperatures and under varying chemical environments. This has established coating technology around high-temperature ceramics, cermets, alloys, and, to a lesser degree, pure metals. The predominant materials include tungsten carbide with cobalt binder in varying percentages, mixed tungsten and chromium carbides, chromium carbide with nickel-chromium binder, aluminum oxide, chromium oxide with nickel-chromium binder, and mixtures of aluminum oxide–titanium oxide and aluminum oxide–chromium oxide.

These coatings have been successfully applied to critical areas of precision parts made from virtually all commercial alloys. They are particularly effective where close dimensional control must be maintained such as on valve components, compressor blade midspans, pump plungers, and compressor rods.

8.1.6 Electric Arc Spraying

The electric arc spraying process is used to spray metals in wire form. This spraying process along with the wire flame spraying process is also referred to as the *metallizing* process. This process differs from other thermal spraying processes in that there is no external heat source, such as a gas flame or electrically induced plasma (Clare and Crawmer, 1982; Marantz, 1974; Steffens, 1983; Steffens and Müller, 1974). Heating and melting occur when two oppositely charged wires, comprising the sprayed material, are fed together in such a manner that a controlled electric arc occurs between them at the intersection. Molten particles are atomized and accelerated onto the substrate by the compressed airstream.

Figure 8.6 shows a schematic of the electric arc spraying system. Two wires, which serve as consumable electrodes, are pushed through insulated flexible conduits to the electric arc gun. As the wires pass through the gun, they make intimate contact with the electrode tips which are energized by the dc power source. The gun is arranged so that the wires meet at a point with a small included angle between them. An atomizing gas jet or nozzle (usually air) is located directly in line with the intersecting wires. As the wires contact at a low contact pressure, an electric arc is drawn which continuously melts the wires as they are fed into the arc. The molten particles are atomized, accelerated, and propelled by the jet gas to the substrate. The use of an inert atomizing gas would inhibit oxidation and might improve the characteristics of some coatings. The open area behind the arc shield allows air to be pulled in by suction created by the flow of atomizing gas, which cools the electrode tips and forms an air sheath around the spray stream.

Due to the high current density in the arc plasma, extremely high temperatures are achieved. Temperatures of about 6000°C can be attained in an arc between iron electrodes with a current of 280 A (Marantz, 1974).

Electric arc spraying offers several advantages over other thermal spray processes. Differences exist in the resultant coatings, due to different heating processes involved. Substrate heating is lower than in other processes due primarily to the absence of a flame touching the substrate. Under comparable conditions, the adhesion of arc sprayed coatings is higher than that of flame sprayed coat-

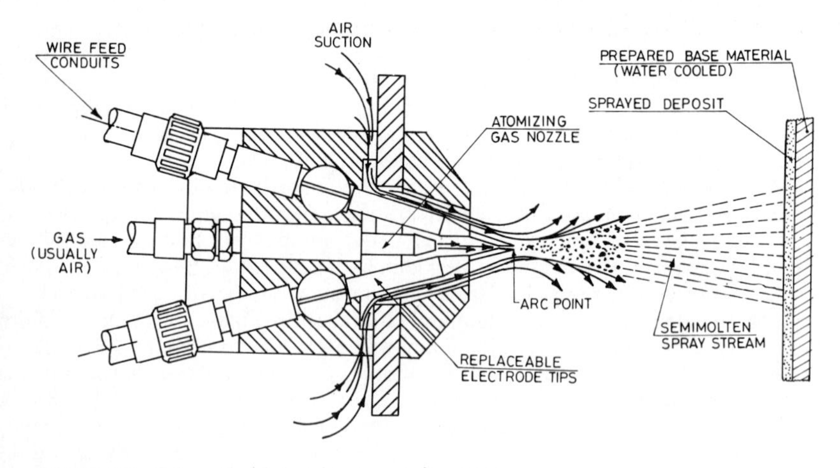

FIGURE 8.6 Schematic of electric arc spraying process.

ings; e.g., the bond strength of arc sprayed mild steel was 30 MPa compared to 20 MPa from flame spraying. The corresponding average particle velocities are 100 and 180 m s^{-1}, respectively (Steffens, 1983). This process provides high deposition rates, up to 55 kg h^{-1} for some nickel-based alloys. A less expensive wear surface can be deposited by using this process. The electric arc gun is lightweight and is easily held in the hand.

The electric arc spraying process can be used to produce coatings of mixed metals, often termed *pseudo-alloys*, by feeding dissimilar wires. For example, pseudo-alloys of copper and stainless steel may be produced, resulting in higher wear resistance due to steel and higher thermal conductivity due to copper. The electric arc spraying process is limited to relatively ductile, electrically conductive wire about 1.5 mm in diameter. Electric arc sprayed coatings of carbides, nitrides, and oxides are therefore not practical at present.

8.1.7 Coatings Deposited by Thermal Spraying

8.1.7.1 Coating Materials. In general, any material which does not decompose, vaporize, sublimate, or dissociate on heating can be thermal-sprayed. Consequently, a large class of metallic and nonmetallic materials (metals, alloys, ceramics, cermets, and polymers) can be deposited by thermal spraying processes. For wear resistance applications, coatings of hard metals, alloys, intermetallics, and nonmetals (including polymers) are commonly used. Metallic hard materials which are important for coating application are the binary compounds of the transition metals of groups IVa (Ti, Zr, and Hf), Va (V, Nb, and Ta), and VIa (Cr, Mo, and W) with carbon (carbides), nitrogen (nitrides), boron (borides), and silicon (silicides). The carbides are the most important hard materials. Nonmetallic hard materials are the oxides of various metals (for example, Al_2O_3, Cr_2O_3, TiO_2, and ZrO_2) and nonoxides of various metals (for example, B_4C, SiC, BN, Si_3N_4, and AlN). Cermets consist of two components, a metal or alloy and a ceramic.

Commonly used hard materials and alloys are listed in Table 8.4(*a*) (Antony et al., 1983; Bhat and Herman, 1982; Bhushan, 1980b; Bhushan et al., 1978; Bhushan and Gray, 1978, 1980; Bhushan and Heshmat, 1979; Chivers, 1985; Chuanxian et al., 1984; Eschnauer, 1980; Eschnauer and Lugscheider, 1984; Ferriss and Cameron, 1975; Foroulis, 1984; Furillo, 1980; Gray et al., 1981; Johnson and Farwick, 1978; Kieffer and Kalke, 1949; Lai, 1978; Mann et al., 1985; Moore and Ritter, 1974; Sangam and Nikitich, 1985; Sliney, 1979; Tucker, 1974, 1982; Van Bemst et al., 1981; Vinayo et al., 1985; Wilkins et al., 1976). Many of the hard materials are mixed with nichrome (80 percent Ni, 20 percent Cr)—typically 25 percent by weight—in order to improve the ductility of the coating (Eschnauer and Lugscheider, 1984; Foroulis, 1984; Longo, 1972; Sangam and Nikitich, 1985; Tucker, 1974, 1982). For low friction and wear, an equally large number of rather "soft" materials are thermal sprayed. Some of the commonly used not-so-hard materials are listed in Table 8.4(*b*). Properties and applications of some thermal sprayed coatings are presented in Table 8.5. Further details on the properties of thermal sprayed coatings can be found in Chapter 14.

For many applications, the use of multicomponent powders offer advantages. Multicomponent powders can be composed of coated particles, mixtures, and agglomerates of different chemical compounds or metallurgical phases. A number of methods are used to manufacture multicomponent powders, such as physical blending, agglomeration or micropelletizing, and coating.

TABLE 8.4(*a*) Some Examples of Hard Materials and Hard Alloys Used for Thermal-Sprayed Coatings

Hard materials		Hard alloys			
Metallic hard materials	Nonmetallic hard materials				
Carbides	*Oxides*	Matrix	Fe	Co	Ni
$WC-W_2C$					
Cr_3C_2 and $Cr_{23}C_6$	*Aluminum oxides*				
TiC	98% Al_2O_3-2% Cr_2O_3				
NbC	70% Al_2O_3-30% MgO				
TaC	97% Al_2O_3-3% TiO_2	Hard-material-forming elements	Cr	Cr	Cr
Carbide solid solutions	87% Al_2O_3-13% TiO_2		W	W	(Mo)
50% WC-50% TiC	60% Al_2O_3-40% TiO_2		Mo	Mo	
70% WC-30% TiC	*Chromium oxides*		V		
TaC-NbC	Cr_2O_3				
	50% Cr_2O_3-50% TiO_2				
	Titanium oxides				
	TiO_2				
Nitrides	*Zirconium oxides*				
TiN	95% ZrO_2-5% CaO	Metalloids	C	C	B
ZrN	90% ZrO_2-10% CaO		Si	Si	Si
TaN	70% ZrO_2-30% CaO		(B)	(B)	(C)
	97% ZrO_2-3% MgO				
	80% ZrO_2-20% MgO				
	(100-n)% ZrO_2-n% Y_2O_3 (n = 0–20%)				
Borides	65% ZrO_2-35% SiO_2				
TiB_2	*Others*	Other matrix elements	Mn	Ni	Cu
ZrB_2	$BaO-TiO_2$		Ni	Cu	Fe
LaB_6	$Cr_2O_3-Al_2O_3-TiO_2$				(Co)
Silicides	*Carbides*	Admixed hard materials		WC	
$MoSi_2$	B_4C			W_2C	
	SiC			Cr_3C_2	
				TiC	
	Nitrides				
	α-Si_3N_4				
	AlN				
	Borides				
	α-AlB_{12}				
	SiB_6				

Source: Adapted from Eschnauer (1980).

TABLE 8.4(*b*) Some Examples of Not-So-Hard Materials and Alloys Used for Thermal-Sprayed Coatings

Metals	Alloys	Intermetallic compounds	Polymers
W	Babbitt	Ni-Cr	Polyethylene
Ta	Bronze	Ni-Al	Polyamide (nylon)
Zr	Steel	M-Cr-Al-Y types	Polyvinyl chloride
Mo	High-alloy iron	(M-Ni, Co, Fe)	
Cr			
Ni			
Zn			
Al			
Cu			

8.1.7.2 ***Deposition Process Parameters.*** The process parameters that must be chosen for application of a coating of a given powder composition and size distribution include the type of gases and their flow rates, the torch design, the power level to be used, and, for some torches, the point of powder entry. All these factors vary with the specific torch model used. Most coating powders used for thermal spraying processes fall between 5 and 60 μm in size (Eschnauer, 1980; Tucker, 1974). To achieve uniform heating and acceleration of a single component powder, it is advisable to have as narrow a size distribution as possible. Generally speaking, fine powders produce denser and harder coating (see Table 8.6). They are accelerated and heated more rapidly in the plasma stream, but they tend to lose momentum more rapidly when spraying at longer distances (greater stand-offs). They generally yield a denser, but slightly more stressed coating (Tucker, 1974).

Oxide ceramic powders are available with particle sizes below 20 down to 5 μm (referred to as micron powders) and can be easily classified in very narrow ranges of particle size. Extremely dense coatings with very few pores have been produced by using these oxide-ceramic micrometer powders.

8.1.7.3 ***Coating Adhesion.*** Depending on the energy content (both thermal and kinetic) of the sprayed particles as well as the structural surface conditions and the energy state of the substrate material, the type of material combination, and the surrounding gas atmospheres, different mechanisms for adhesion of the sprayed layers may dominate. The most dominant mechanism for adhesion is due to mechanical interlocking of the coating with undercuts in the surface and also by flowing around roughness peaks and/or deposited particles. Another mechanism for the bonding of the coating to the base material could be van der Waals forces since the sprayed particles in some spraying processes have sufficiently high impact velocity. It is therefore supposed that contact between both materials approaches lattice dimensions. Wetting of the substrate occurs, and the surface layers of foreign material such as adsorbed gas or water are either adsorbed into the sprayed material or displaced by the impact of the particles, which facilitates van der Waals attraction.

A third mechanism results in a metallurgical bond which can originate from diffusion as well as reaction at the active centers of the surface due to thermal energy activation. In addition, diffusion may take place, carrying atoms of the base material into the deposit and atoms of the sprayed particle into the substrate

TABLE 8.5 Properties and Applications of Some Thermal Sprayed Coatings

Material	Useful temperature range, °C	Hardness (HV), kg mm^{-2}	Applications
Aluminum	300	60	Corrosion, shock resistance, electrical and thermal conductivity.
Molybdenum	425	450	Resistant to adhesive wear, undercoat for bonding metal and ceramic coatings for low-temperature use (350°C), also used for electrical contacts.
Aluminum bronze, babbitt	300	Soft	Soft bearing surfaces.
Steel and high-alloy iron	550–700	200–650	Good wear resistance, very low cost.
Nickel aluminide (Ni–20 wt % Al)	900	170	Mainly used as undercoat to steels, aluminum, cast iron, and titanium under final coatings of ceramics.
Nichrome (Ni–20 wt % Cr)	1000	200	Resistance to oxidation and corrosive gases in temperatures to 1000°C, also used as undercoat under final coatings of ceramics.
Nickel–15 wt % graphite	500	Soft	Abradable coatings up to 500°C for seals etc.
Ni-Co-Cr-Al-Y alloy	850	—	Superior oxidation resistance, also used as an undercoat for thermal barrier coatings.
Co-28 wt % Cr-3 wt % Ni-4 wt % W (Stellite 6)	650	400	Used for wear and high-velocity particle erosion.
Co-18 wt % Cr-28 wt % Mo-4 wt %Si (Triboloy 800)	1000	450	Superior friction and wear resistance, resistance to chemical corrosion and oxidation up to 1000°C.

Co-25 wt % Cr-10 wt % Ni-7.5 wt %W (cobalt alloy)	550–1000	350	Dense, strongly adherent coating, mixed with ceramics to produce cermets, used as high-temperature coating.
Tungsten carbide–12 wt % cobalt blend	425	800	Extremely hard coating for wear resistance, good wear and shock resistance.
Tungsten carbide–10 wt % nickel	425	550	Good wear resistance.
Chromium carbide–25 wt % nichrome	425–800	750	Good wear resistance.
Aluminum oxide (white)	1100	650	Good wear, heat, and corrosion resistance, stable in oxidizing or reducing atmosphere to 1600°C.
Aluminum oxide–titanium oxide	550	1000	Resistance to abrasion, fretting, and cavitation.
Chromium oxide–25 wt % nichrome	500–900	1200	Resistance to wear by abrasive grains, hard surfaces up to 900°C, good heat and corrosion resistance.
Zirconium oxide–8 wt % yttria	1700	450	Thermal barrier coating. Also used to resist high-temperature particle erosion above 900°C.
Magnesium zirconate	1100	—	Thermal barrier coatings up to 1100°C.
Molybdenum disilicide	1700	1200	Fire barrier coating.

TABLE 8.6 Effect of Powder Size on the Structure of Plasma Deposited Tungsten Carbide

Coating property	Powder size: Coarse, 10–105 μm	Medium, 10–74 μm	Fine, 10–44 μm
Apparent density, g cm^{-3}	10.5	13.0	14.2
Bulk density, g cm^{-3}	8.7	11.1	13.0
Percentage of theoretical density	60	77	89
Porosity, %	17.1	14.5	8.5
Hardness, HK_{500}	538	684	741

Source: Adapted from Tucker (1974).

surface. This diffusion process takes place at low temperatures due to the high concentration of lattice defects existing in both sprayed particles and the substrate surface. In addition to this diffusion, in some material combinations, exothermic reactions may occur, based on formation of intermediate phases at the same time as diffusion occurs. Adhesion strength of mechanical bonding is small compared to that of van der Waals and metallurgical bonding; the latter has the higher bond strength.

The adhesive strength of sprayed coatings is most frequently measured in a tensile test, per ASTM-C633, in which the coating is applied to the face of a 25-mm-diameter round bar and a mating bar is epoxied to it. The limit of the test is the strength of the epoxy, about 70 to 85 MPa. Bond strengths range from below 7 MPa for flame sprayed coatings to below 70 MPa for plasma sprayed coatings, but most detonation-gun coatings and a few plasma sprayed coatings have strengths that exceed the test limit, with the test serving only as a "proof" test. In a few instances where the mating bar is brazed to a detonation-gun coating, the tensile strengths exceeded 175 MPa (Tucker, 1974, 1982).

Residual stresses have a significant influence on the coating adhesion. Coatings are normally found to be in tension as a result of the residual stress. Residual stress occurs as a result of the cooling of individual powder particles or splats from above their melting point to the temperature of the workpiece. The magnitude of the residual stresses is a function of the spraying torch parameters, deposition rate, relative torch to workpiece surface speed, thermal properties of both the coating and the substrate, and amount of auxiliary cooling used.

8.1.7.4 Substrate Pretreatments. Since factors pertaining to the substrate influence adhesion of coating to the substrate, surface preparation is critical. Substrate cleanliness is extremely important in the adhesion of all coating systems. Thermal sprayed coatings rely mostly on mechanical bonding to the substrate. Therefore, the surface roughness of the substrate is an important factor. Most substrate surfaces are roughened by grit blasting or rough threading, to about 0.5 to 3 μm rms, but several nonmetallic substrates can be coated without much roughening. Grit blasting also results in the exposure of a very clean, highly active surface. This effect may be important in increasing bonding due to van der Waals forces. It is common practice to minimize the amount of time between grit blasting and coating.

Aluminum oxide and chilled iron are the most widely used abrasive grits for surface preparation by thermal spraying. Sand, crushed steel, and silicon carbide are also used as abrasive grit. Sand is commonly used for large exterior structures. Crushed steel grit is used on hardened steels. Alumina, sand, and especially silicon carbide may embed in softer metals such as aluminum, copper, and their alloys. For these metals, lower air pressures are recommended to minimize embedding. Chilled iron or crushed steel should be used in preparing surfaces to be flame sprayed and fused. Practical grit-size ranges are −10 to +30 mesh (coarse), −14 to +40 mesh (medium), and −30 to +80 mesh (fine). Because surface roughness is primarily the result of grit particle size, the grit size is determined by the required coating thickness. Coarse grit is used for coating thicknesses exceeding 250 μm and for best adherence. Medium grit size should be used for smoother finishes on coatings less than 250 μm and for fair adherence. Fine grit size should be used for the smoothest finish on coatings less than 250 μm to be used in as-sprayed condition. Grit blasting air pressures vary from 200 to 600 kPa with stand-off or working distances of 50 to 150 mm. Grit blasting nozzle openings are generally 6 to 10 mm in diameter. The blasting angle to the substrate should be about 90°. The substrate should be cleaned following grit blasting to remove residual dust. It is recommended that the surface be rinsed with Freon TF, alcohol, or a similar solvent.

Most coatings adhere better if the substrate is preheated. Preheating makes the substrate reactive and reduces the quench rate of the impacting particles. Interaction time is extended. Furthermore, substrate heating drives off adsorbed surface moisture which can inhibit bonding. Preheating also expands the substrate so that the differential contraction between it and the hot particles is minimized as the coated part cools. This reduces residual stresses. Reduction of residual stresses results in improved coating adhesion.

Some substrate materials may oxidize, and the coating may become less tenacious so that a protective underlayer may be useful. Certain materials will have an inherently high bond strength due to their ability to form an alloy (metallurgical) bond with the substrate. The coating produced from nichrome (80 percent Ni and 20 percent Cr) and that from nickel aluminide (80 to 95 percent Ni and 5 to 20 percent Al) are commonly used as transition layers to provide resistance to oxidation and corrosive gases, to provide a graded interface, and to improve the bond strength with subsequent coatings of other materials. Both these materials rapidly form adherent oxide films which prevent further oxidation. When these materials are sprayed in air, their oxides also form so that the final deposit contains about 70 percent metal and 30 percent oxide. This metal-oxide mixture produces a transition in properties between metal substrate and ceramic coating. Graded coatings are often helpful in ensuring a more gradual, controlled transition. When the nickel-aluminum material is thermal sprayed, its constituents react exothermically, thereby providing a higher particle energy level and a good bond. (During spraying, nickel and aluminum react to produce nickel aluminide and oxides of Ni and Al.) We note that nickel aluminide as an undercoating is less corrosion-resistant in water, but has better thermal shock resistance and higher bond strength than nichrome. Nickel aluminide can be used up to 750°C while nichrome can be used up to 1000°C (Bhushan, 1980a; Bhushan et al., 1978; Smith, 1974a). Molybdenum is another undercoating used for steel substrates because of its ability to form an alloy bond to steel. Molybdenum, however, is also subject to catastrophic oxidation above 500°C.

8.1.7.5 ***Coating Posttreatments.*** Thermal sprayed coatings usually exhibit two common features in the as-deposited condition: a sandpaperlike surface finish and a structure with inherent porosity. A typical surface roughness of a sprayed coating ranges from 5- to 13-μm arithmatic average (aa) which is too rough for most tribological applications. The porosity usually ranges from less than 1 to about 17 volume percent, depending on the process by which the coating is deposited and the material sprayed. The porosity is usually interconnected, making the coating permeable to gases and liquids.

Thermal sprayed coatings can be finished by using standard metal finishing techniques such as machining, grinding, lapping, or polishing to produce surfaces with surface roughness down to less than 50 nm rms. Considerable care must be taken in finishing a sprayed coating to avoid damage to the coating. Improper techniques during the finishing may result in excessive particle pullout, producing a severely pitted surface. It is essential that the sprayed particles be gently sheared and not pulled from the surface. The best surface finish that can be obtained is a function of not only the finishing technique, but also the coating composition and the deposition parameters. Machining can be used on some metallic coatings, but most coatings are ground with silicon carbide or diamond (diamond is usually preferred for detonation-gun coating). High-speed steel tools can be used to cut the softer sprayed materials such as low-carbon steel, copper-based alloys, and aluminum. Harder materials usually require the use of carbide tools or grinding. Grinding of coatings can be done either wet or dry: wet grinding is preferred when possible. In general, a wheel of low bond strength and relatively coarse structure is required. For many applications, it is desirable to treat the sprayed coating with a sealant prior to grinding. Sealing prevents the fine particles of grinding sludge from entering the pores of the coating. All sprayed coatings can be polished, but the softer materials (such as tin, zinc, and babbitt) are easily polished and buffed.

If the sprayed coatings are to be used for applications requiring corrosion resistance, they should be sealed. Sealing is the process by which the pores of a coating are filled to eliminate the possibility of infiltration by fluids or corrosive media that can contribute to premature failure. Sealing a coating also helps to reduce particle pullout from the surface during finishing. To ensure complete sealing for corrosion resistance, it is necessary to apply the seal material prior to surface finishing. Seal materials—organics (such as petroleum-based waxes, epoxy, phenolic varnish resin, dimethacrylate polyester resin, and vinyl) and inorganics (aluminum silicate and aluminum flake suspended in a vinyl lacquer)—are commonly used. Epoxy and phenolic sealers are usually more effective on coatings with higher porosity. One of the most effective methods of sealing coating porosity is vacuum impregnation. This method will usually fill all interconnected pores. To vacuum-impregnate, the coated part is immersed in a container of epoxy resin, and the container is placed in a vacuum chamber. A vacuum is drawn to pull air from the pores. When the vacuum is released, atmospheric pressure forces sealer into the evacuated pores.

8.1.7.6 ***Coating Microstructure.*** Thermal sprayed coatings consist of many layers of thin lenticular particles resulting from the impact of molten or semimolten powder particles. The impacting particles may split, with some small droplets branching out or separating from the central particle. Thus, the average splat volume may be smaller than the average starting powder size, and the total surface area may be much larger in the coating. Typically, a splat is about 5 μm thick and 10 to 50 μm in diameter.

The cooling rate of the impacting particles has been estimated to be 10^4 to 10^6 °C s^{-1} for oxides and 10^6 to 10^8 °C s^{-1} for metals (Wilms and Herman, 1976). Obviously the rates vary with the substrate material and coating thickness. As a

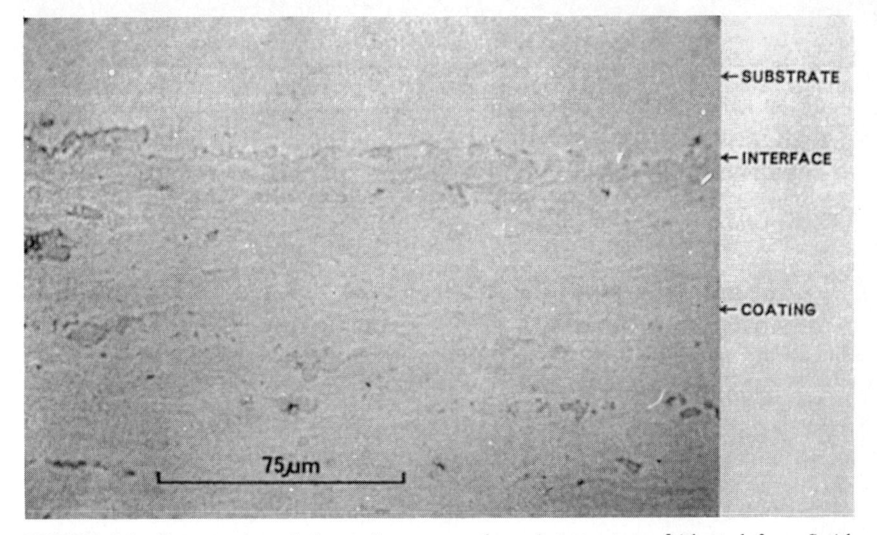

FIGURE 8.7 Cross-section of detonation-gun coating microstructure. [*Adapted from Smith (1974b).*]

result of the rapid cooling, some coatings have been found to have no crystallographic structure, and others have a thin amorphous layer next to the substrate followed by crystalline structure (Wilms and Herman, 1976). Many coatings form columnar grains within the splat in one or two layers perpendicular to the surface of the substrate.

In addition to a change in crystalline structure due to the rapid quench, some changes in composition may occur due to selective evaporation of one component in an alloy, decomposition of a gas, or reaction with the atmosphere. The reaction of the powder particles with their local environment in transit, particularly the extent of their oxidation, is very important to the properties of the coatings. The loss of carbon from tungsten carbide plasma coatings through oxidation of WC to form gaseous CO, W_2C, and free W has been reported.

Figure 8.7 shows a typical cross-section of a WC-Co coating produced by the detonation-gun process. At the substrate-coating interface, there is no separation or porosity. The amount of visible porosity is less than 0.5 percent. The general matrix of the coating, from which its hardness can be derived, is a mixture of various tungsten carbides and cobalt-rich binder phases. These are mainly W_2C and a range of tungsten carbides with varying percentages of carbon and cobalt. To give an idea of size, the tungsten monocarbide particles in Fig. 8.7 are approximately 6 μm in diameter.

8.2 WELDING

The application of a hard, wear-resistant surface layer of various metals and alloys on metallic substrates by welding, also called *fusion welding*, is one of the most versatile techniques (Anonymous, 1980a, 1980b, 1983; Antony et al., 1983; Avery, 1950, 1951, 1952a, 1952b, 1969; Bell, 1972; Bucklow, 1983; Gregory, 1978,

1980; Hazzard, 1972; Kretzschmar and Dollinger, 1979; Watson, 1973). The coating is laid down by melting the coating material, referred to as filler material (usually an alloy), onto the substrate and fusing the molten material to the substrate. The heat generated by a gas, electric arc, or plasma arc welding gun is used to melt the coating material, which is supplied in the form of powder, paste, rod, strip, or wire. Cast rods or wires are the most common form of filler material. The weld deposition rates are very high, ranging from 0.5 to 30 kg h^{-1} or even higher. Weld deposition processes are used in a variety of industrial applications requiring relatively thick alloy coatings ranging from 750 μm to a few millimeters. Most weld deposition processes are portable and relatively inexpensive. They are capable of depositing on small components of intricate shape as well as large areas of flat or cylindrical shape.

The less sensitive control of deposition rates or coating thickness, the deeper degree of penetration of the weld into the base metal, the inferior appearance of the deposited weld, and metallurgical limitations imposed by the different heat inputs are inherent characteristics of the weld deposition processes which have to be considered. Where heavy objects, components of irregular shape, and even fixed structures are to be given a heavy buildup, the economic benefits outweigh what may be considered as shortcomings of the weld deposition processes.

Welded deposits are, in effect, minicoatings characterized by variable composition (segregation) and solidification kinetics that influence deposit microstructure (Devletian and Wood, 1983). Composition variations also derive from base-metal dilution during welding, although carbon pickup in oxyfuel gas welding processes and solute volatilization during arc welding processes are additional factors. *Dilution* is the interalloying of the hard facing alloy and the base metal and is usually expressed as the percentage of base metal in the welded deposits. The wear resistance and other desirable properties of the hard facing alloys are generally thought to degrade as dilution increases. Welding should be selected so as to control dilution to process generally less than 20 percent. Dilution is acceptably low (1 to 5 percent) in the conventional oxyfuel gas welding process. Dilution is generally a greater problem in the arc welding process, ranging from about 5 percent in plasma arc to as much as 30 to 60 percent in submerged arc welding. High-dilution welding processes can be tolerated in applications requiring relatively thick coatings that can be applied in multiple layers. We note that dilution in thermal sprayed coatings is nil.

Base-metal composition, melting-temperature range, and thermal expansion and contraction characteristics have a significant effect on the selection of welding process, whereas thermal spraying processes do not put requirements on these properties of the base metal. Steels are generally suitable base metals for weld deposition. Age-hardenable base metals require solution and/or overaging heat treatments prior to weld deposition. Weld deposition requires that the base-metal melting-temperature range be greater than or at least equal to that of the coating alloy. Thermal expansion and contraction differences between the coating alloy and base metal must be minimal in weld deposition.

A comparison was made of major welding processes used for weld deposition by Antony et al. (1983). They reported that arc welding processes could be readily automated and hence are 3 to 5 times faster than gas welding. Equipment cost for automatic arc welding is much higher than for gas welding. Initial investment in equipment for plasma arc welding is about twice that for some electric arc welding processes (GTAW and GMAW, to be discussed later) and 10 times that for gas welding. However, labor cost is much higher for manual gas welding than for automatic arc welding.

The major basic processes utilized to produce welded coatings are described in the following subsections, and their process characteristics are compared in Table 8.7.

8.2.1 Oxyfuel Gas Welding

Oxyfuel gas welding (OFW) or flame welding is a process in which the heat from a gas flame is used to melt coating materials onto the workpiece. During deposition, the workpiece surface is superficially melted. The gas flame is created by the combustion of a hydrocarbon gas and oxygen. Acetylene is the most widely used fuel. In fact, OFW is commonly referred to as *oxyacetylene gas welding* (OAW) or just gas welding. In the OAW torch, oxygen and acetylene are fed to a blowpipe where they become mixed in the desired proportions and are ignited as they flow from the torch nozzle [Fig. 8.8(*a*)]. The coating materials are generally available in bare cast rod or tubular form and are fed into the oxyacetylene flame (Antony et al., 1983; Bell, 1972). The operation during this process is general (where the operator coordinates the feed rate of coating material into the flame). However, with some difficulty, it can be automated. Oxyfuel gas welding is a versatile process, readily adaptable to all coating product forms, including powder. OFW processes which use fillers in powder form are commonly referred to as *powder flame spraying processes*, described earlier.

The coatings applied by OFW are characterized by low dilution (typically 1 to 5 percent) since only the surface of the workpiece in the immediate area being coated is brought to the melting temperature. Porosity is probably the most common flaw which results from inadequate feeding or gas evolution during solidification. OFW equipment is simple, and capital costs are low. Furthermore, OFW equipment is portable and can be used in the field. Grooves and other recesses can be accurately coated, and very thin coatings can be applied. However, deposition by manual OFW requires a high degree of welder skill to obtain smooth, high-quality deposits, since both the welding rod and the torch flame have to be manipulated. The speed of welding and low deposition rates (typically 1 kg h^{-1}) often preclude the use of OFW for applications on large components that require large quantities of facing alloys, unless automation is warranted. A limited range of consumables are available for these processes.

Typical applications include seating surfaces on valves for combustion engines and compressors, rolls for precision rolling and forming of plastics and glass, power plant equipment parts such as steam valves and nozzles, and bearing sleeves for precision equipment operating in unusual environments of temperature and corrosion.

8.2.2 Electric Arc Welding

In deposition by *electric arc welding* (EAW), the heat from an electric arc is used to melt the filler material onto the workpiece surface. The electric arc is developed by impressing a voltage between a consumable electrode and the workpiece. The voltage required to sustain the arc varies with distance and the arc welding process. The filler material can derive directly from the electrode used to form the arc (referred to as *consumable* electrode processes) or can be externally introduced into the arc (referred to as *nonconsumable* electrode processes). As a rule, the latter processes are accomplished at lower power require-

TABLE 8.7 Some Important Characteristics of Various Welding Techniques*

Deposition technique	Heat source	Mode of operation	Forms of coating materials†	Minimum recommended thickness of deposit, mm	Dilution, %	Deposition rate, kg h^{-1}
Oxyfuel gas welding (OFW)	Oxyacetylene gas	Manual or automatic	Rod or powder	0.75	1–5	0.5–2
Shielded metal arc welding (SMAW)	Electric arc	Manual	Cast rod or tubular rod covered with flux	3	10–20	0.5–2
Submerged arc welding (SAW)	Flux-covered electric arc	Automatic	Bare solid or tubular wire	3	30–60	10–30
Electroslag welding (ESW)	Current flow through a layer of molten flux	—	Bare solid or tubular wire	3	—	150–200
Self-shielded arc welding (SSAW)	Electric arc	Semiautomatic	Tubular wire with flux or alloy core	3	15–30	5–15
Gas metal arc welding (GMAW)	Electric arc	Semiautomatic	Tubular wire fed into arc through a stream of inert gas	1.5	10–30	2–10
Gas tungsten arc welding (GTAW)	Inert-gas shielded electric arc	Manual or automatic	Bare cast rod or tubular rod	2.2	4–10	0.5–2
Plasma arc welding (PAW)	Inert-gas shielded plasma arc	Manual or automatic	Bare cast rod	2.2	4–10	4–20

*The substrate material (typically metals or alloys) must be an electrical conductor except in OFW.

†Any electrically conductive material (metals and alloys) which can be produced in a rod or wire form, can be weld-deposited except for OFW and PAW processes, where material does not have to be an electrical conductor and powder form can also be used.

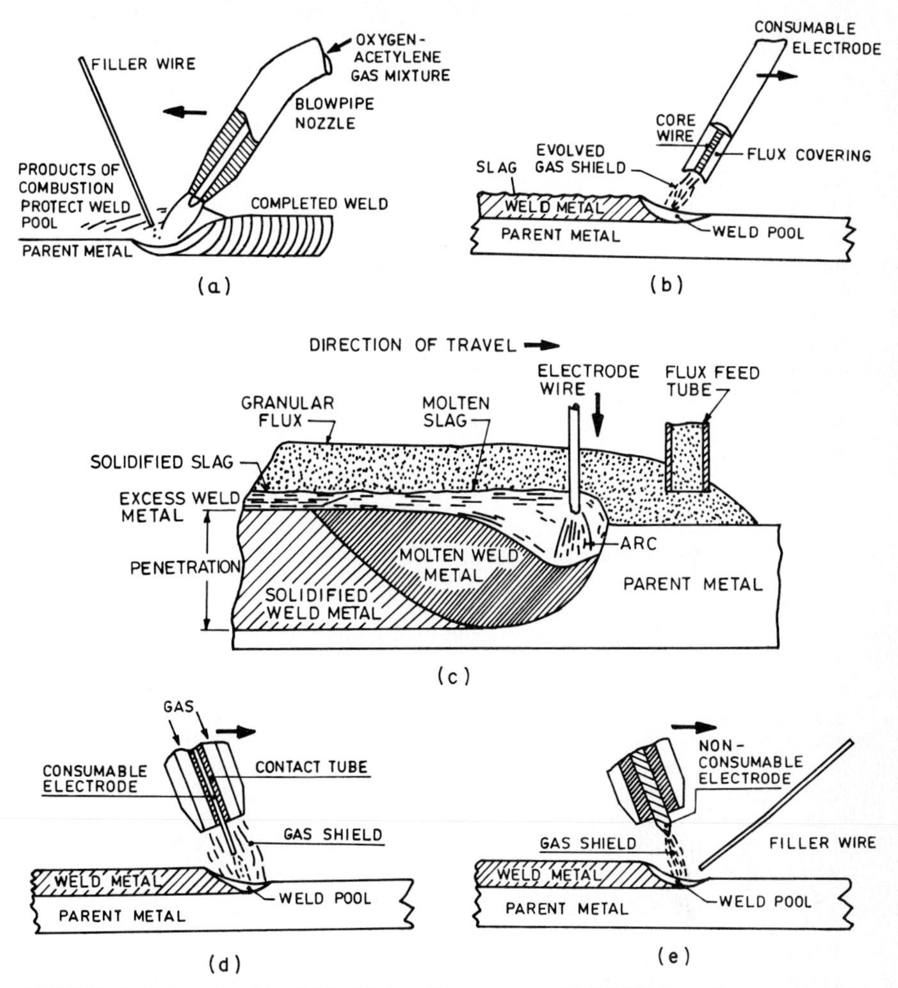

FIGURE 8.8 Schematic of the (*a*) OAW deposition process, (*b*) SMAW deposition process, (*c*) SAW deposition process, (*d*) GMAW deposition process, and (*e*) GTAW deposition process.

ments than the prior process, resulting in less dilution and lower deposition rates. The arc action in the nonconsumable electrode processes is generally smoother, quieter, and freer of weld spatter than in the consumable processes. The use of a separate filler rod or wire enables greater operator control of the deposit shape. Filler materials in consumable electrode processes are sometimes protected by fluxes, as in *shielded metal arc welding* (SMAW). A wide range of consumables (fillers) are available for electric arc welding.

8.2.2.1 Shielded Metal Arc Welding. The SMAW process is often called *manual metal arc welding, stick welding*, or *electric arc welding*. Shielded metal arc welding is the most widely used welding process, principally because, of all the processes, it requires the least capital expenditure. This process requires a power supply (an ac transformer and a generator or rectifier for direct current) which is inexpensive. In the SMAW process, an arc is struck between a hand-held consumable electrode covered with welding flux and the workpiece surface [Fig. 8.8(*b*)]. The high temperature resulting from the arc melts the electrode and workpiece surface, producing a weld deposit with a certain amount of substrate material diluted into the coating material and some alloying of the coating material into the substrate (Evans, 1962; Hazzard, 1972). Dilution is controlled by speed of travel, position of the electrode in relation to the weld pool, arc voltage, and amperage of the welding machine. Usually high current results in a faster deposition rate but also more dilution.

There are many advantages to the SMAW process. It is a most versatile process. The intense heat of the arc permits deposition on large parts without preheating. Small areas of parts finished to close tolerances can be coated without distorting the entire part. Difficult-to-reach areas are best accessed with coated electrodes. Another attractive feature of this process is the portability of the equipment. There are also some disadvantages to the SMAW process. Penetration tends to be high, and dilution rates of 20 percent are not uncommon. Because coated electrodes usually are deposited with no more than a small amount of preheating, cracking may occur, especially with hard overlays. The SMAW process is one of the least efficient processes in terms of deposition rates (typically 1.5 to 2 kg h^{-1}) compared to other arc deposition processes.

Typical applications include hard facing of steam shovel teeth, coke pusher shoes, catalyst valves in refinery applications, and crusher rolls.

8.2.2.2 ***Submerged Arc Welding.*** In *submerged arc welding* (SAW), the heat for welding is supplied by an arc maintained between continuously fed metal strips or tubular wires and the workpiece. A layer of finely divided, free-flowing granular flux (or slag) is placed by gravity ahead of the continuously fed metal strip or wire which acts as a consumable electrode for the arc. The flux becomes molten, and the arc is submerged beneath the layer of the fusible flux material, thus protecting the molten weld metal from atmospheric contamination [Fig. 8.8(*c*)]. The fluxes used for SAW are mineral compositions specially formulated to protect the molten filler metal without appreciable gas evolution.

A number of features make SAW particularly attractive for weld deposition (Matusek, 1959). SAW is a fully mechanized process. High deposition rates ranging from 10 to 30 kg h^{-1} are possible, which makes it attractive for coating of large areas. The process is capable of making porosity-free deposits with excellent appearance which are smooth and even on the surface and which either can be used as welded or require only a minimum of grinding to clean up. The major disadvantages of deposition by SAW are that it is generally limited to hard-facing simple cylindrical or flat workpieces and it is not easily adaptable to coating of small parts. Weld deposition in other than the flat position requires special flux retainers. Preheating the workpiece in excess of 350°C usually makes slag removal extremely difficult. Dilution is very high and can be as high as 30 to 60 percent; consequently, two or three layers of weld metal may be required for maximum wear resistance. The coating thickness is about 3 mm and upward. The SAW process is suitable for workshop use only in substantially fixed installation, although special portable equipment for in situ deposition is available. The equipment is very expensive and requires regular maintenance. Typical applications include the coatings of large valves and rolls and the lining of large chemical vessels for corrosion resistance.

8.2.2.3 ***Electroslag Welding.*** In the *electroslag welding* (ESW) process, heat is generated by passing an electric current through a pool of molten flux (slag) to melt the edges of the base metal and filler metal wire (electrode). The electrical resistivity of the molten flux is capable of continuously producing heat that is sufficient to continue the welding process. The slag pool and weld pool are contained in a cavity formed between the edges of the parts being joined and copper shoes (preferably water-cooled), used as dams. ESW may be performed with either dc or ac power (Thomas et al., 1983a).

Electroslag welding is begun by initiating an arc beneath a layer of granular flux, similar to SAW. As soon as a sufficient layer of flux is melted to form the slag, all arc action stops and current passes from the electrode to the base metal through the conductive molten slag. The process then becomes ESW—metal transfer is arcless. Welding progresses in a vertical direction, and retaining dams are used to mold the molten metal. The resulting weld solidification is highly directional, and nonmetallics are rejected upward into the slag pool.

In SAW, by comparison, the arc under granular flux creates the heat necessary for welding. The depth of flux is not sufficient to block arc action. In ESW, the slag pool is 30 to 50 mm deep, and because it offers a conductive path for the electric current between the filler wire and base metal, the molten slag suppresses the arc action without breaking the current flow.

The ESW process is used to weld low- and medium-carbon steels, low-alloy and structural steels, high-strength steels, stainless steels, and nickel-based alloys. Deposition rates are higher for ESW than for any other welding process—at least 25 to 50 percent higher than for SAW, which is usually the closest to ESW in deposition rate. The deposition rates for ESW range from 15 to 45 kg h^{-1}. Setup time for ESW is a limiting factor when compared to simpler, manually controlled processes, such as SSAW. The ESW process is used only for specialized applications. In eastern Europe it is used to rebuild large steel mill and tube-making rolls (Blaskovic, 1966).

8.2.2.4 ***Self-Shielded Arc Welding.*** The *self-shielded arc welding* (SSAW) process is also called *nonshielded arc welding, open arc welding*, or *flux-core arc welding*. SSAW is a semiautomatic process in which the arc between the electrode and workpiece is shielded by a self-generated or self-contained gas or flux. There is no externally applied shielding gas. The electrodes are continuous lengths or coils of filler material in tubular wire form containing a flux or alloying elements or both. The shielding of the arc and molten weld pool is by a gas evolved from the flux material contained in the core of the tubular electrode wire. High-alloy filler materials are generally shielded by a self-generated CO_2 gas derived from the reaction between carbonaceous materials contained in the core and the atmosphere.

Neither flux-handling nor gas-regulating equipment is required. Thus, the SSAW process is a continuous process that can be used semiautomatically with a hand-held gun or can be made fully automatic by traversing the gun or workpiece mechanically, requiring only a device to feed the continuous electrode wire. The process is simple and requires little training. This process is a high-energy process capable of much higher deposition rates than other processes. Most hard alloys can be deposited by the SSAW process without preheating. There are some disadvantages to the SSAW process. Because the arc is not shielded with inert gas or a blanket of granular flux, considerable spatter and some porosity in deposit can be expected. The amount of cracking is comparable to that of other arc welding processes.

8.2.2.5 Gas Metal Arc Welding. The *gas metal arc welding* (GMAW) process is also called *gas-shielded metal arc welding, metal inert-gas welding* (MIGW), or *CO_2 welding*. In this process, a consumable electrode in the form of bare tubular wire is fed continuously and mechanically to a hand-held welding gun [Figure 8.8(*d*)]. An electric arc is maintained between the consumable electrode wire and the workpiece. The arc and the weld pool are protected from atmospheric contamination by a gaseous shield provided by a gas or mixture of gases fed through the welding gun. Shielding gases generally consist of argon, carbon dioxide, or mixtures of the two. The deposit is visible at all times in MIGW, thereby enabling high-quality deposits. The GMAW process utilizes bare, small-diameter (generally 1.5-mm or less) filler wires. The dilutions range between 10 and 30 percent. The GMAW process can be used to a significantly higher arcing power than manual metal arc welding. This fact, combined with the higher current that can be employed, will give high deposition rates ranging from 2 to about 10 kg h^{-1}. The coating thicknesses are usually greater than 3 mm.

The GMAW process is suitable for semiautomatic deposition, and the versatility of the process lends it to the coating of complex shapes. However, an automated process cannot be easily used in either the overhead or vertical position. The use of auxiliary shielding gas and wire feed mechanisms adds to the cost of the equipment. However, the higher cost is generally offset by the higher deposit quality associated with GMAW. Typical applications include the overlay of seating surfaces on valve seats, the working areas on large piston rods, and pistons in which low friction and nongalling properties are required.

8.2.2.6 Gas Tungsten Arc Welding. This process, also known as *tungsten inert-gas welding* (TIGW) or *argon arc welding*, is a nonconsumable electrode arc process. In the *gas tungsten arc welding* (GTAW) process, heat is produced by an arc between a nonconsumable tungsten electrode and the workpiece. The electrode, the weld pool, the arc, and the adjacent heated areas of the substrate are protected from atmospheric contamination by a flow of shielding inert gas fed through the torch [Fig. 8.8(*e*)]. Thoriated tungsten is the preferred nonconsumable electrode material. Argon or helium is the commonly used inert gas. As with oxyacetylene welding, the filler material is applied separately in bare cast rod or wire form.

Weld deposition by manual GTAW is a useful alternative to deposition by OGW techniques, particularly when large components or reactive base metals are involved. GTAW is preferred for deposition on reactive base metals such as titanium-stabilized stainless steels or aluminum-bearing nickel-based alloys. GTAW also is preferred over OGW for applications in which carbon pickup is unacceptable or in applications involving filler materials that tend to boil under the influence of an oxyacetylene flame. The localized, intense heat of GTAW generally results in more base-metal dilution (4 to 10 percent) than that seen in OGW. Deposition rates are comparable in manual GTAW and OAW, despite the difference in base-metal dilution. Deposition by GTAW can be accomplished automatically by simply attaching the torch to an oscillating mechanism and using a mechanized device to feed the filler material into the arc region.

The arc action in GTAW is generally smoother, quieter, and free of weld spatter compared to other consumable electrode arc welding processes. The use of a separate filler rod or wire enables greater operator control of deposit shape. GTAW is more adaptable to coating small intricate parts and generally produces higher-quality deposits than other weld deposition processes. The intense heat of the arc permits the deposition on parts without preheating. It is slow and unsuitable for coating of large areas. The equipment is not portable and is generally restricted to a workshop. The GTAW equipment is expensive compared to that for some other weld deposition processes. GTAW process is used in some cases for depositing materials such as cobalt and high-nickel materials for specialized applications.

Typical applications include those for oxyacetylene welding, and the determining factor in choosing a process is usually the heat input into the base part (i.e., the part may be impractical to heat to the high temperature required for OAW and therefore demand the GTAW process with its higher temperature and more localized heat zone). The weld quality requirements may call for a high level of purity of deposits and thus rule out GTAW with its high dilution due to its heat concentration.

8.2.3 Plasma Arc Welding

Weld deposition by *plasma arc welding* (PAW) is similar to deposition by GTAW in that both processes employ a gas-shielded electric arc (referred to as *transferred arc*) between a nonconsumable tungsten electrode and the workpiece as the primary heat source for weld deposition. The processes differ in that the PAW process maximizes the use of plasma as a secondary heat source. Another arc is employed between the tungsten electrode and a water-cooled copper annulus surrounding it. This results in ionization of the gas flowing in a nozzle surrounding the electrode, thereby producing a high-temperature plasma stream from the nozzle. The electrode generally is recessed into the nozzle. The plasma gas emerges from a constricting-orifice arrangement.

The PAW process can use bare rod or wire as a hard facing consumable, but more often powder is used as the consumable. When powder is used, the process is often referred to as *plasma transferred arc welding*. A typical plasma transferred arc welding torch for powder consumables is illustrated in Fig. 8.9. Weld deposition by PAW using rod and/or wire filler materials is similar to deposition by GTAW. However, the stiffer arc intrinsic to PAW enables greater control of deposit profile and position than is possible with conventional GTAW. The choice of materials for surfacing is wide and includes a number of nickel-, cobalt-, and iron-based alloys either with or without additions of tungsten carbide.

Weld deposition by PAW is a relatively high-energy process. Deposition rates up to 4 kg h^{-1} are common. Still higher deposition rates ranging up to 20 kg h^{-1} are possible by using the plasma arc hot wire process, which utilizes wire filler consumables and two independently controlled systems operating in one weld pool to melt and fuse a coating alloy to a substrate. One system raises the filler wire to its melting point by resistance heating. The other system fuses the filler wire to the workpiece by using plasma arc energy. In a typical system shown in Fig. 8.9, two filler-metal wires are fed at a constant speed through individual hot wire nozzles to intersect in the weld pool beneath the plasma arc. The filler metals are energized by a power supply. Resistance heating occurs in each electrode between the wire contact tip and the weld pool (Antony et al., 1983).

Weld deposition by PAW has many advantages. The deposit quality is excellent. Deposits can range in size from 225 μm thick by 5 mm wide to approximately 6 mm thick by 40 mm wide by simply varying process parameters. It therefore can be used to coat large areas. The process is amenable to automation, which makes it well suited for high production. The dilution of less than 5 percent

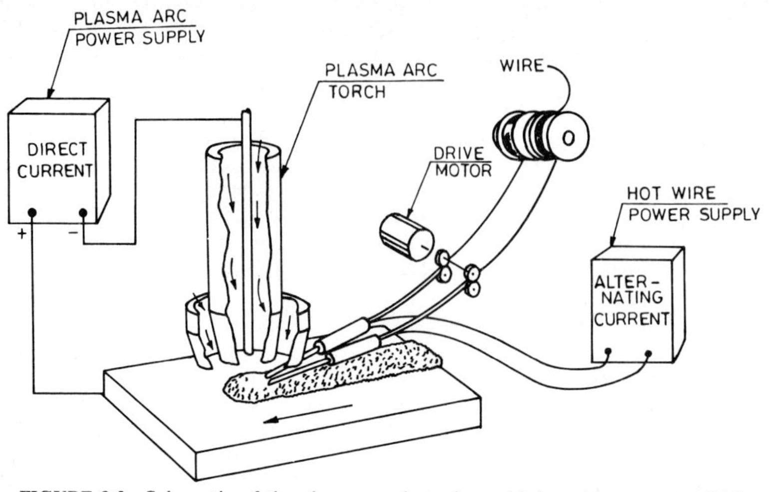

FIGURE 8.9 Schematic of the plasma arc hot wire weld deposition process. [*Adapted from Antony et al. (1983).*]

can be achieved with PAW process. PAW is comparable to GTAW with respect to its ability to accommodate reactive substrate materials. The PAW process has some disadvantages. The equipment is very expensive (compared to other processes) and is not portable. It is limited to straight-line or cylindrical parts except when special tooling is provided. Usually, several parts are needed for setup, which makes the process somewhat unattractive to the low-volume job shop.

Typical applications include edge on trim dies, flights on extrusion screws, process rolls, and components which are subjected to severe wear where composite alloys with high percentages of tungsten carbide are required.

8.2.4 Coatings Deposited by Welding

8.2.4.1 Coating Materials. Materials that can be melted and produced in (cast or otherwise) rods and wires can be weld-deposited. Furthermore, for weld deposition by arc, the substrate and the coating material must be electrically conductive. Thus, mostly metals and alloys are deposited on metallic substrates by weld deposition. Materials for weld deposition include a wide variety of alloys: martensitic, austenitic, and pearlitic steels; high-alloy irons; nickel-based alloys; cobalt-based alloys; special alloys; carbides; and borides. Microstructurally, the alloys generally consist of hard-phase precipitates such as carbides, borides, or intermetallics bound in softer iron-, nickel-, or cobalt-based alloy matrices. The copper-based alloys (such as brasses and bronzes) and tin-based alloys (babbitts) which are soft alloys are also used. Typical compositions of the consumables used for weld deposition are presented in Table 8.8 (Antony et al., 1983; Avery, 1952b; Bell, 1972; Bucklow, 1983; Foroulis, 1984; Gregory, 1980; Hazzard, 1972). The table also indicates the deposition method(s), typical hardness, and the temperature limitation of various alloys. Ceramics such as aluminum oxide can also be deposited as weld deposits in applications requiring extremely high wear and corrosion resistance.

Many of the compositions presented in Table 8.8 can have additions of fluxing agents such as silicon and boron at about 1 to 3 percent, which makes them suitable for the indirect weld deposition processes known as spray-and-fuse and powder welding. Silicon and boron act as fluxing agents and dissolve surface oxides on both substrate and coating materials, thus promoting wetting between the two and improving the coating adhesion and cohesion.

A broad comparison of properties of various alloys is made in Fig. 8.10. Low-cost iron-based alloys are most widely used generally for large equipment that undergoes severe wear such as crushing and grinding equipment and earth-moving equipment. Some typical applications of martensitic-alloy steels are listed in Table 8.9 (Gregory, 1978). Nickel-based alloys, cobalt-based alloys, and carbides are used for specialized applications. The nickel-based alloys are used where metal-to-metal wear or low stress abrasion is combined with heat and corrosion. Cobalt-based alloys are used in severe-wear applications requiring operation at high temperatures and corrosive environments (Anonymous, 1975b). They are used in diesel engines, steelworks, and chemical plants. Tungsten carbides are used for severe-wear applications where abrasion under high stress occurs, such as in agriculture, brick and clay manufacture, cement, coal mining, earth moving and road building, petroleum, and steel.

The nickel, nickel-copper, and nickel-iron alloys are soft and machinable (compared to other nickel-based alloys discussed here) and are used for weld deposition of cast irons and for buttering cast irons. The copper-based alloys (brasses and bronzes) are soft materials with modest abrasion resistance but good frictional properties. Copper-aluminum alloys are used for bearings, corrosion-resistant and wear-resistant surfaces. Tin bronzes are used for bearing surfaces and occasionally in corrosion-resistant applications. Silicon bronze and brasses are used for corrosion resistance only. The copper-based alloys are normally applied by gas tungsten arc welding.

*8.2.4.2 **Coating Filler Form.*** Coating alloys are available as powder, wire, bare cast rod, and covered electrodes. Virtually all coating alloys can be produced in powder form. Although many coating materials are sufficiently ductile to be

LOW ←	ABRASION RESISTANCE		→ HIGH	
14 wt % Mn steel (1.6) 14 wt % Mn–14 wt % Cr steel (1.7)	Low-alloy steel (1.2)	Arc deposited tungsten carbide (7)	Oxyacetylene tungsten carbide (7)	LOW ↑
	Martensitic Cr steels (1.3)			HEAT AND CORROSION RESISTANCE
Austenitic stainless steel (1.5)	High-speed steels (1.4)		Austenitic and martensitic irons (2.1, 2.2, 2.3, 2.4, 2.5)	↓
	Cobalt- and nickel-based alloys (4.1, 4.2, 4.3, 4.4, 3.4, 3.5, 3.6)			HIGH
HIGH ←	IMPACT RESISTANCE		→ LOW	

FIGURE 8.10 Comparison of properties of the major types of hard facing alloys. Figures in parentheses indicate group in Table 8.8 to which alloy belongs. [*Adapted from Gregory (1980).*]

TABLE 8.8 Classification and Typical Compositions of Consumables for Weld-Deposited Surfaces for Wear Protection

Classification		Typical composition, wt % (• indicates possible presence)						
Group	Type	Fe	C	Cr	Mn	Mo	V	W
1 Steels	1 Carbon steels	Balance	0.35					
	2 Low-alloy steels	Balance	0.1–0.5	•	•	•	•	•
	3 Martensitic Cr stainless steels	Balance	0.1–1.7	10–15	•	•	•	•
	4 High-speed steels	Balance	0.3–1.5	10 max	5	10	3	20
	5 Austenitic stainless steels	Balance	0.07–0.2	17–32	•	•		
	6 Austenitic Mn steels	Balance	0.5–1.0	•	11–16	•	•	
	7 Austenitic Cr-Mn steels	Balance	0.3–0.5	12–15	12–15	•	•	
2 Irons	1 Austenitic irons	Balance	4	12–20	•	•		
	2 Martensitic irons	Balance	1–4	1–10	•	•	•	•
	3 High Cr austenitic irons	Balance	3–6	20–40	•	•		
	4 High Cr martensitic irons	Balance	2–3	20–30	2 max			
	5 High Cr complex irons	Balance	2–5	20–40	•	•	•	•
3 Nickel alloys	1 Nickel	8	2		1			
	2 Nickel copper	3–6	0.35–0.55		2.5			
	3 Nickel iron	Balance	2		1			
	4 Ni-Mo-Cr-W	6	2.5	30 max		17		15
	5 Ni-Cr-B	•	•	5–25		•	•	
	6 Ni-Mo-Fe	5–20				20–30		
4 Cobalt alloys	1 Co-Cr-W low alloys		0.7–1.4	25–32				3–6
	2 Co-Cr-W medium alloys		1.0–1.7	25–32				7–10
	3 Co-Cr-W high alloys		1.7–3.0	25–35				11–20
	4 Co-Cr-W-Ni alloys		1.2–2.0	20–25				10–15
5. MCrAlY		M = Co, Ni, Ti; Cr = 25; Al = 6; Y_2O_3 = 0.1						
6. Copper alloys	1 Brasses	up to 40% Zn, balance Cu						
	2 Silicon bronzes	up to 4% Si, balance Cu						
	3 Aluminum bronzes	8–15% Al, balance Cu						
	4 Tin bronzes	4–12% Sn, balance Cu						
7. Carbides	1. Minimum 40% tungsten carbide granules in an iron base matrix, or can be in a copper, cobalt, or stainless-steel matrix							
	2. Chrome carbide/nichrome	Cr_3C_2 = 75–85; Ni/Cr (80/20) = 15–25						
8. Chromium boride paste	1			80–85				

(1) As deposited, (2) Work-hardened

OAW = oxyacetylene welding; SMAW = shielded metal arc welding; SAW = submerged arc welding; GMAW = gas metal arc welding; SSAW = self-shielded arc welding; GTAW = gas tungsten arc welding.

Source: Adapted from Gregory (1980).

drawn into wire and can be provided at relatively low cost, drawing usually becomes more difficult as the proportion of alloy increases and with carbon above certain limits. Carbon is an important element for conferring abrasion resistance, and the brittle, high-carbon alloys are generally cast. Most materials can be produced in rod form. With each welding rod an individual casting of such materials

Typical composition % (*Continued*)					Deposited	Deposition method •						Temperature limitation, °C
Co	Ni	B	Nb	Cu	hardness, HV	OAW	SMAW	SAW	GMAW	SSAW	GTAW	
					up to 250	•	•				•	
	•	•			250–650		•	•	•	•	•	
	•				350–650		•	•	•	•		550
12					600–700		•	•	•			600
	7–22				(1) 200							650
					(2) 500							
	•				(1) 200							
					(2) 600		•					600
	•				(1) 200		•	•		•		600
					(2) 600							
	•				300–600		•			•		500
	•	•			500–750		•			•		550
	•				500–750	•	•	•	•	•		550
					500–750	•	•	•	•	•		550
•	•	•			600–800		•	•	•	•		700
	85 min.				160		•					700
	60–70			25–30	130		•					500
	45–60				200		•			•		700
12	Balance		7		250–500	•	•		•		•	>800
	Balance	1–5			200–750	•	•				•	>800
	Balance				200–300		•					>800
Balance					350–400	•	•	•	•		•	>800
Balance					400–500	•	•	•	•		•	>800
Balance					500–650	•	•	•	•		•	>800
Balance	20–25				390–450	•	•		•		•	>800
						•					•	>800
					130				•		•	240
					80–100				•		•	240
					130–140		•		•		•	240
					40–110		•		•		•	240
					>1800 (granules)	•	•			•	•	600
											•	>800
		15–20			750–1100							>800

becomes expensive. To reduce manufacturing costs, less expensive tube rods, in which a steel tube or sheath is filled with alloying materials, are manufactured. Produced in coils, such tube electrodes may be useful for automatic welding. In some cases, such as rods containing tungsten carbide, this is the only feasible method of manufacture. The rods or tubes generally range in diameter from 3 to 10 mm and are up to 450 mm long. Some alloys can be cast in extra-long continuous lengths with diameter as low as 3 mm. The rods used for some welding processes are covered with a flux (slag) material (a specially formulated material) to

TABLE 8.9 General Properties and Applications of Martensitic Alloy Steel Hard Facing Alloys

Hardness, HV	Abrasion resistance	Impact resistance	Typical applications
250	Low	High	Metal-to-metal wear. Rails, shafts, axles, dog clutches. Machinable.
350	Low	High	Metal-to-metal wear coupled with abrasive particles. Track links, track rollers. Machinable.
450	Medium	Medium	Metal cutting and forming tools, e.g., shear blades for cropping steel bar. Machinable with special tools.
650	High	Low	Excavator buckets and teeth. Bulldozer blades, crusher jaws, and teeth. Agricultural implements. Cane and bean knives, ripper teeth. Unmachinable, must be ground.
800	High	Low	Brick-making equipment. Cane and bean knives, ploughshares, ballast tampers, mill hammers. Unmachinable, must be ground.

Source: Adapted from Gregory (1978).

protect the molten weld material and the substrate surface from atmospheric contamination (Antony et al., 1983).

8.2.4.3 ***Substrate Pretreatments.*** The surface to be built up should be thoroughly cleaned by grinding, machining, grit blasting, steam or chemical cleaning, or brush cleaning in order to get rid of dirt, scale, and other contaminants. If the surface cannot be properly cleaned, a good flux should be used when depositing the rod. To prevent cracks in the deposit or cracks in the alloy steel or cast-iron substrate, it is generally necessary to preheat the components to 300 to 450°C. Small parts may be preheated locally with the torch, but large parts should be preheated with gas burners in a refractory brick enclosure. Because of the preheating requirement, parts fabricated to close tolerances cannot be coated without distorting the entire part.

8.2.4.4 ***Coating Posttreatments.*** Cooling after weld deposition should be slow to minimize cracks and internal stresses, which may be set up due to variation in contraction from thermal expansion mismatch between the base metal and overlay, substantial variations in cross-sectional areas with changes in thermal masses of these, or too great a cooling of the substrate while the core remains at much higher temperature. As soon as deposition is completed, the entire part should be immersed in kieselguhr, powdered mica, vermiculite, or similar insulating material so that it will cool slowly.

Weld deposition generally produces a fairly rough surface with a peak-to-valley height of as much as 2 mm in shielded metal arc welding. Other processes, particularly the mechanized ones, can produce much smoother surfaces, but where a smooth surface is required (such as in valve seats), machining will be essential. If the deposited area is to be used for wear application, it is generally ground or machined to produce a smooth surface. The desired surface roughness depends on the deposit material, substrate, and wear application.

8.2.4.5 Coating Microstructure. The microstructure of the weld deposit depends on the selection of the deposition process and the process parameters. The microstructure and mechanical properties of weld deposits vary depending on solidification kinetics as well as dilution. Dilution is generally more of a problem in arc welding processes, ranging from approximately 5 percent in plasma arc welding to as much as 60 percent in submerged arc welding. Solidification generally is rapid in thermal spraying processes, with near amorphous structures being achieved with some thermal spraying processes. Solidification kinetics tend to be somewhat slower in conventional weld deposition processes. As a rule, deposits produced by OFW processes tend to solidify slower than deposits produced by arc welding processes. These differences in solidification rate produce widely different microstructures and widely different properties regardless of dilution (Antony et al., 1983; Avery, 1952b).

Figure 8.11 shows a micrograph of a cross section through a chromium-molybdenum martensitic iron gas weld deposit on SAE 1020 steel. Marked diffusion of carbon from above into the ferritic base metal occurs, despite the short welding time. The deposit is a mixture of hard carbide (gray) and martensite (white).

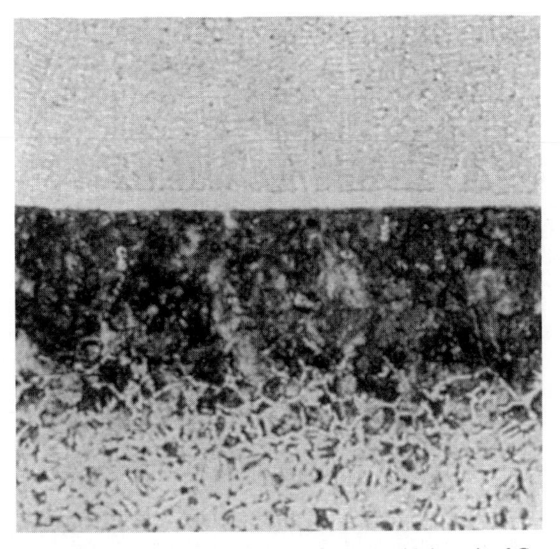

FIGURE 8.11 Photomicrograph of a gas weld deposit of Cr-Mo martensitic iron on SAE 1020 steel. The overlay is a mixture of hard carbide (gray) and martensite (white). Below the weld junction, a dark zone indicates that the steel has been carburized to a depth of about 0.5 mm. [*Adapted from Avery (1952b).*]

8.3 CLADDING

In the cladding process, a metallic foil or sheet (ranging from few tens of micrometers to few millimeters or more in thickness) is metallurgically bonded to a metallic substrate to produce a composite structure, and the bond is usually stronger than the weaker of the parent metals. Most metals and alloys can be metallurgically bonded by one of the cladding processes (Anonymous, 1985; Bucklow, 1983; Delagi, 1980; Edmonds, 1978; Jellison and Zanner, 1983; Thomas et al., 1983b). The metallic powders or other fillers can also be cladded to the metallic substrates by special and conventional deposition techniques; the difference between such cladding and coatings deposited by welding previously discussed in this chapter is that of the thickness. Cladding usually denotes the application of a relatively thick layer (typically 1 mm or more) of clad metal whereas a coating is usually thinner.

Some of the main features of cladding are as follows:

- The two materials can be bonded without disturbing the effect of cold work, dispersions, and precipitation hardening.
- This process has the ability to reduce significantly or to almost eliminate the formation of brittle intermetallic compounds at the bond zone, normally present in thermal spray and weld deposits.
- The clad surface produced by cladding with wrought material raises no concerns of porosity, stoichiometry, and hardness in the coatings.
- Bonding can be obtained quickly and cost-effectively over very large areas.

The typical limitations of this process are that the cladding material in many cladding processes must be available in sheet form and it is difficult to clad parts having complex shapes and extremely large sizes. We note that some cladding processes can use cladding material in powder form.

Metals are clad by deformation cladding [which includes mechanical methods—roll cladding and coextrusion cladding, explosive cladding, *electromagnetic impact bonding* (EMIB)], diffusion bonding, braze cladding, weld cladding, and laser cladding. In mechanical methods, explosive cladding, and EMIB, the metals are bonded by solid-phase methods, which rely on a combination of deformation (gross plastic flow) by pressure or impact and heat to generate intimate atomic contact and intermixing with minimum diffusion, and rapidly (within seconds) produce a true metallurgical bond. In diffusion bonding processes, coalescence of contacting surfaces is produced with minimum macroscopic deformation by diffusion-controlled processes induced by applying heat and pressure for a finite time interval. Bonding by deformation is usually accomplished within seconds, whereas diffusion bonding typically requires contact durations ranging from minutes to hours. Cladding materials in the form of sheet or powder can be applied.

In braze cladding, the surfaces to be clad are sandwiched by a brazing material in the form of powder, paste, rod, wire, strip, or foil, and metallurgical bonding is obtained by heating the stack, by furnace, torch, resistance, dip in molten bath, induction, and laser. In weld and laser cladding processes, the metal is clad by melting it and fusing it to the substrate.

Explosive cladding offers the largest range of material combinations. Mechanical methods are limited to relatively ductile metals and alloys. Coextrusion is very largely limited to tube and rods, with the accent on tube. If all processes are properly carried out, there is no difference between the resulting bond integrity. An advantage offered by braze cladding over other methods is that strong joints may be produced at relatively low heat input. Yet most production brazing processes (except laser brazing) require heating of the entire part to the brazing temperature. In addition, the brazing material may not be as strong as the parent metals, thus other techniques may produce a bond to withstand high loads and temperatures.

The diffusion cladding processes are very slow processes, and explosive cladding and EMIB are very fast processes. Mechanical methods, brazing, welding, and laser cladding processes, on the other hand, are medium-rate methods involving processing times on the order of 1 min. Roll cladding is probably the cheapest, and diffusion bonding is probably the most expensive with explosive cladding being relatively cheap. Because of the very high pressures involved in the explosive and EMIB processes, they can be used only to bond relatively thick layers (greater than about 0.3 mm) of clad metals. Weld cladding and laser cladding allow the use of metal to be clad in powder form.

Among other brazing techniques, in induction brazing and laser brazing, the parts are locally heated and thus avoid the distortion and other adverse metallurgical changes. In laser cladding, only the substrate surface skin is heated, and this allows the cladding of high-melting-point metals to low-melting-point substrates.

Some important characteristics of various cladding processes are presented in Table 8.10. Various cladding processes are discussed in this section.

8.3.1 Mechanical Methods

The mechanical methods for cladding are very old processes used to join two dissimilar metals. These medium-rate processes involve processing times on the order of 1 min. The mechanical methods employed for cladding are roll cladding and coextrusion cladding (Anonymous, 1985; Bucklow, 1983; Jellison and Zanner, 1983).

For metals and alloys to be clad by mechanical methods, they should have relatively high ductility and should be thermally stable at the processing temperature.

8.3.1.1 ***Roll Cladding.*** In roll cladding, also known as *roll welding* or *bonding*, two or more foils, sheets, or plates are stacked together and then passed through rolls until sufficient deformation has occurred to produce solid-state welds. The roll cladding can be performed either cold or hot (in order to reduce the processing pressure).

Two modes of roll cladding are common. In the first, the parts to be clad are merely stacked and passed through the rolls. Sometimes the parts are first tack-welded at several locations to ensure alignment during rolling. The second method, usually termed *pack rolling*, involves sealing the parts to be rolled in a pack or sheath and then roll-cladding the pack assembly. The first method is more generally employed in the cold rolling of ductile metals and alloys. The required deformation can be reduced by hot rolling, if the metals to be clad can tolerate preheating without excessive oxidation. Pack roll cladding is employed for hot rolling. In pack roll cladding, the plates to be clad are completely enclosed in a pack that is sealed, usually by fusion welding, and often evacuated to provide a vacuum atmosphere. All roll bonding is done hot, and many passes are usually

TABLE 8.10 Some Important Characteristics of Various Cladding Processes

Cladding technique	Substrate or cladding materials	Typical forms of cladding parts	Process parameters	Process times	Pretreatment	Posttreatment	Comments
Mechanical methods	Many metals and alloys with relatively high ductility	Foil, sheet, or plate (roll cladding); tubes and rods (coextrusion cladding)	Moderate to high pressure, ambient to high temperature	Medium (on the order of 1 minute)	Mechanically clean	Yes	Cheapest process
Explosive cladding	Almost any metals or alloys	Sheet or plate	Very high pressure (deformation), ambient temperature	Very fast (few seconds)	Mechanically clean	Yes	Relatively cheap process, can be used only for relatively thick layers ($\geq$0.3 mm) of cladded metals
Electromagnetic impact bonding (EMIB)	Almost any electrically conducting metals and alloys	Sheet or plate	Very high pressure, high temperature	Very fast (few seconds)	Mechanically clean	Yes	Can be used only for relatively thick layers of cladded metals
Diffusion bonding	Many metals and alloys	Foil, sheet, plate, or awkward shapes, powders	Moderate pressure, very high temperature	Very slow (minutes to hours)	Mechanically clean	Yes	Most expensive process. Awkward shapes can be cladded.

Braze cladding	Many metals and alloys	Foil, sheet, plate	Low pressure, moderate to high temperature	Medium	Chemically and mechanically clean	Generally none	Requires brazing material in between, bond may not be as strong as by other processes.
Weld cladding	Most metals and alloys	Rod, wire, powder	Ambient pressure, very high temperature (melting)	Medium	Mechanically clean	Yes	Can be used only for relatively thick substrates.
Laser cladding	Most metals and alloys	Rod, wire, powder	Ambient pressure, very high temperature (melting)	Medium	Mechanically clean	Maybe	High-melting-point metals can be clad to low-melting-point substrates, only substrate surface skin is molten.

required to achieve the necessary bond integrity. Generally hot rolling is performed above the recrystallization temperature. If the bonding is not adequate in rolling, it is possible to start with a thin sheet of backing-plate material explosively bonded to a much thicker cladding plate and then to roll-bond this composite to the main backing plate.

The most common metals that are roll-clad are low-carbon steels, aluminum, aluminum alloys, copper, copper alloys, nickel, tin, silver, and gold. Low-alloy steels, high-alloy steels, and nickel-based alloys have also been roll-clad. Some examples of easy-to-bond combinations are copper and steel, copper and nickel, copper and silver, copper and gold, aluminum and aluminum alloys, tin and copper, tin and nickel, and gold and nickel (Delagi, 1980). Bonded sheets are commercially available, typically 5 to 100 mm thick, up to 5 m wide, and in lengths up to 11 m. Cladding thickness ranges from 5 to 25 percent of the total.

Roll cladding is used for numerous applications, e.g., to produce corrosion-resistant stainless-steel-clad aluminum automobile trim, copper-clad stainless-steel cookware, copper-clad and stainless-steel-clad stainless-steel heat exchanger panels, copper-clad aluminum-lead frames for integrated circuits and wires, precision-metal-clad electronic connectors, and cupronickel-clad copper for coins (Anonymous, 1985).

8.3.1.2 ***Coextrusion Cladding.*** Coextrusion cladding, also known as *extrusion welding* or *bonding*, generally involves the coextrusion of two or more metal parts. This process is largely limited to tube and rod with the accent on tube. This process is typically conducted at elevated temperature not only to improve cladding but also to reduce extrusion pressures (Bucklow, 1983; Jellison and Zanner, 1983). Some cold extrusion cladding of aluminum and copper has been successfully performed. For extrusion of tubes, the tube is made by constructing a composite hollow billet from the two materials, one being machined to form a close fit inside the other. It is not often necessary to electroplate a barrier or bonding layer on one of the surfaces. The edges of the sleeves are then weld-sealed, the composite billet is heated to the extrusion temperature, and it is extruded in one pass over a mandrel in a tube. Further reduction is done by cold drawing. For reactive metals such as Zr, Ti, and Ta, the retort can be evacuated and sealed. A major advantage of extrusion cladding is that the high isostatic pressure associated with the process is favorable to the deformation cladding of low-ductility alloys.

The most common metals clad by the extrusion process include low-carbon steel, Al and Al alloys, Cu and Cu alloys, Ni and Ni alloys, Zr, Ti, and Ta. Some Ni-Cr-Fe alloy tubes clad with Pt or Zr have been produced for the chemical industry. The tubes, coated inside or outside, are commercially available, typically from 12-mm internal diameter up to 200-mm outside diameter in lengths up to 25 m according to diameter. Various wall thicknesses are available with cladding ratios from 5 to 20 percent. Coextrusion clad tubes are used in many industrial applications for corrosion and wear resistance.

8.3.2 Explosive Cladding

Explosive cladding, also known as *explosive welding* or *explosive bonding*, is a solid-phase welding process in which bonding is produced by an oblique, high-velocity collision between the two plates to be joined. The high-velocity collision is caused by the controlled energy of a detonating explosive (Bahrani, 1978;

Bahrani et al., 1967; Kaminsky, 1980; Pocalyko et al., 1983; Popoff, 1978; Prummer, 1977). This process does not require any external heat. Almost all combinations of metals and alloys can be bonded which otherwise cannot be bonded by any means. Great differences in melting point, ductility, and specific weight of the cladding and base materials are allowable. However, the process is restricted to relatively thick layers (greater than about 0.3 mm) of the cladding metals because during the cladding operation very high pressures, on the order of 3 GPa, acting on the metal sheets may rupture thin layers. The cladding of thickness as low as 1 μm can be bonded.

The cladding material, called the *flyer plate*, either is positioned at a small angle of incidence to the backing plate (touching at one end) or is held parallel to it at a distance greater than the thickness of the flyer plate. A protective buffer (such as rubber sheet) is placed on the top of the flyer plate, and the explosive, in the form of a sheet or slurry, is laid on the buffer with its detonator at one corner (parallel plate) or along the touching end (inclined plate). On firing, the progressive detonation wave deforms the flyer plate into contact at an angle to the backing plate surface [Fig. 8.12(*a*)], and if conditions are optimum, the rate of shear strain is so high that contacting surfaces of the flyer and backing plates behave as if they were liquid and are squirted out in a reentrant jet [Fig. 8.12(*b*)], which has a high velocity and a very efficient scouring action. Note that the contaminant surface films are plastically jetted off the substrate as a result of the high-pressure collision of the two metals, without melting, to leave two atomically clean surfaces that are forced into atomic contact to produce a strong metallurgical bond. Under optimum conditions, the metal flow around the collision point is unstable and oscillates, generating a wavy interface. Figure 8.13 shows an explosion clad interface between Incoloy 800 and carbon steel.

The quality of the explosion bond is directly related to jet formation. Typically jet formation is a function of several variables such as plate collision angle, collision point velocity, flyer plate velocity, pressure at collision point, and mechanical and physical properties of the flyer and backing plate. The detonation velocity and rate of energy release, provided by the explosive, must be high enough to accelerate the flyer plate to the appropriate velocity. It is very important that the right type of explosive be used and that the correct explosive–flyer plate weight ratio be employed. Too small a charge will not give a bond, while too large a charge or too high a detonation velocity will lead to spalling and interface melting. Usually a detonation velocity of 2000 m s^{-1} has been found to give satisfactory bonding, especially on thick plates. In general, the thickness of the backing plate should be twice that of the flyer plate and preferably not less than 10 mm. The surface of the plates is usually machined and ground and then kept clean, but no special preparation is required because the reentrant jet cleans the surface during bonding.

A very large number of combinations have been successfully joined, within the limits of the proviso that the metals have sufficient ductility to undergo a small amount of plastic deformation without spalling or shattering. The metals that are difficult or impossible to weld by conventional welding processes have been explosion-welded. Flat plates, solid rods, and outside and inside surfaces of the cylinder and pipes can be explosion-clad. Over 300 dissimilar metallic combinations have been clad, as well as numerous other combinations. The industrially useful combinations that are available in commercial sizes are shown in Fig. 8.14.

The explosive cladding process can also be used to have multilayer systems, i.e., a three-layer cladding to suit some special requirements, say of wear and

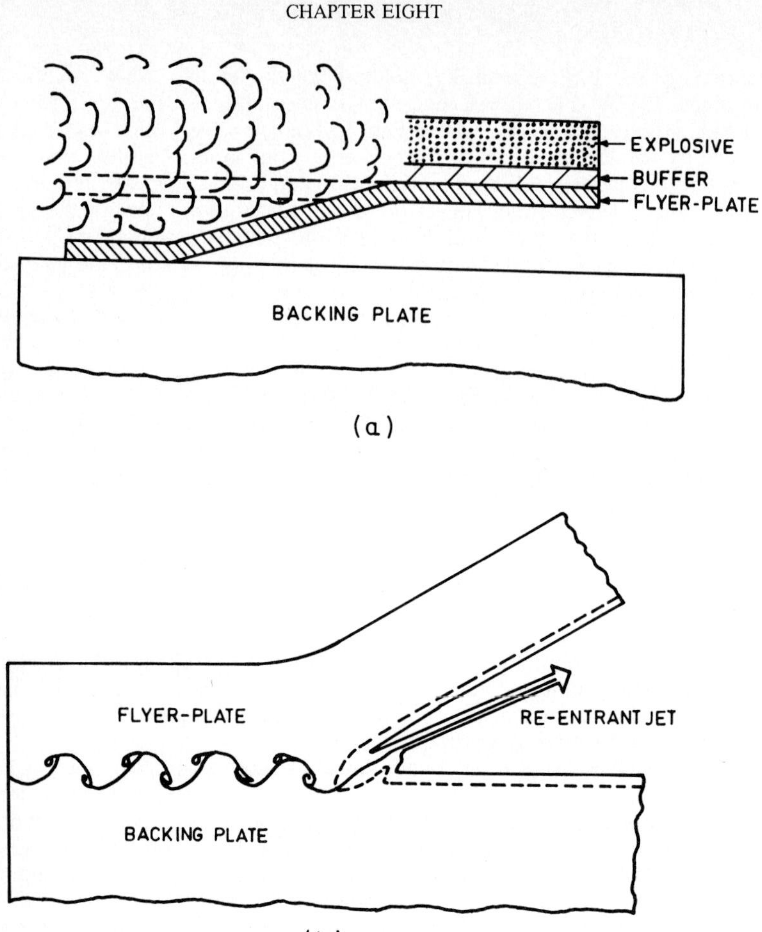

FIGURE 8.12 (*a*) Schematic of explosive cladding deformation of the (parallel) flyer plate. (*b*) Explosive cladding flow configuration in the region of impact.

corrosion resistance at elevated temperatures. For instance, vanadium or titanium or V-20/Ti has been bonded to Mo or Nb, which itself was bonded to copper for Tokomak fusion reactors (Kaminsky, 1980); similarly Pt or V or Mo has been bonded to Ni-based Cr/Fe or to Fe/22 Cr for use up to 1250°C.

Explosion-clad parts are used in the manufacture of process equipment for the marine, chemical, petrochemical, and petroleum industries, where corrosion resistance of an expensive metal is combined with the strength and cost of another. Typical applications include explosively clad titanium to Monel tube sheet and a tantalum–copper–carbon steel clad vessel to recover chlorine from waste hydrochloric acid. The explosion-clad metals are used in transition joints, electrical applications, and tube and pipeline welding.

The explosive bonding process has a number of characteristics that make it unique compared to other metal bonding processes. The inherent process flexibility provides the following advantages.

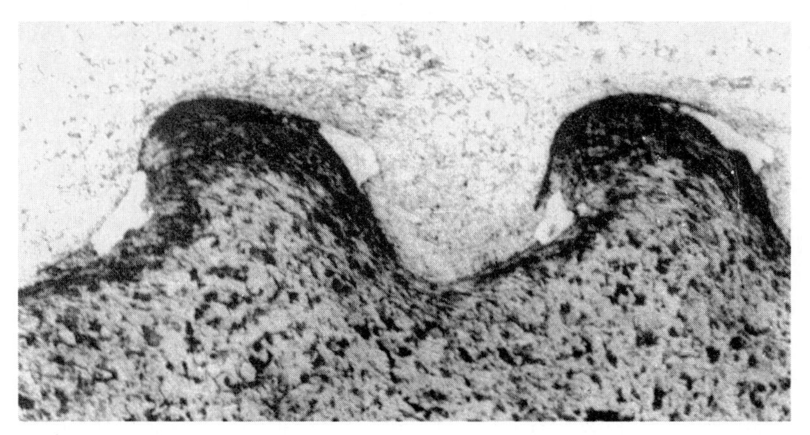

FIGURE 8.13 Explosive cladding: wave formation at an Incoloy 800 clad with SA-516-70 carbon steel. [*Adapted from Pocalyko et al. (1983).*]

- It is simple and intrinsically rapid.
- Very strong and continuous bonds can be produced.
- Almost all combinations of metals and alloys can be bonded. Bonds can be made between metals that cannot be bonded by any other means.
- Large areas are bonded; the bonding areas range from 12.5×10^{-6} to 35 m^2.
- Multilayered composite sheet and backing plate can be bonded in a single explosion cladding.
- Components are bonded without elaborate preparation.

The explosive bonding cladding process has the following limitations:

- Only simple shapes can be coated.
- The inherent hazards of storing and handling explosives entail safety regulations.
- It is difficult to obtain explosives with proper uniform energy, form, and detonation velocity.
- The undesirable noise and blast effects require that the process be performed in an isolated area.
- The process is restricted to relatively thick layers of cladding metals.
- Substrates often undergo some small deformation during bonding and must therefore be straightened.
- This process is not amenable to automated production technique.

8.3.3 Electromagnetic Impact Bonding

Electromagnetic impact bonding (EMIB) is another high-rate cladding process which has many characteristics similar to explosive cladding, but instead of chemical energy as the source of deformation, the necessary force is applied to

	ZIRCONIUM	MAGNESIUM	STELLITE 6B	PLATINUM	GOLD	SILVER	NIOBIUM	TANTALUM	HASTELLOY	TITANIUM	NICKEL ALLOYS	COPPER ALLOYS	ALUMINUM	STAINLESS STEELS	ALLOYS STEELS	CARBON STEELS
CARBON STEELS	●	●			●	●	●	●	●	●	●	●	●	●	●	●
ALLOYS STEELS	●	●	●					●	●	●	●	●	●	●	●	
STAINLESS STEELS			●		●	●	●	●		●	●	●	●	●		
ALUMINUM		●				●	●	●		●	●	●	●			
COPPER ALLOYS						●	●	●		●	●	●				
NICKEL ALLOYS		●		●	●			●	●	●	●					
TITANIUM	●	●				●	●	●		●						
HASTELLOY									●							
TANTALUM					●		●	●								
NIOBIUM				●			●									
SILVER						●										
GOLD																
PLATINUM				●												
STELLITE 6B																
MAGNESIUM		●														
ZIRCONIUM	●															

FIGURE 8.14 Combinations of commercially available explosion-clad materials. [*Adapted from Pocalyko et al. (1983).*]

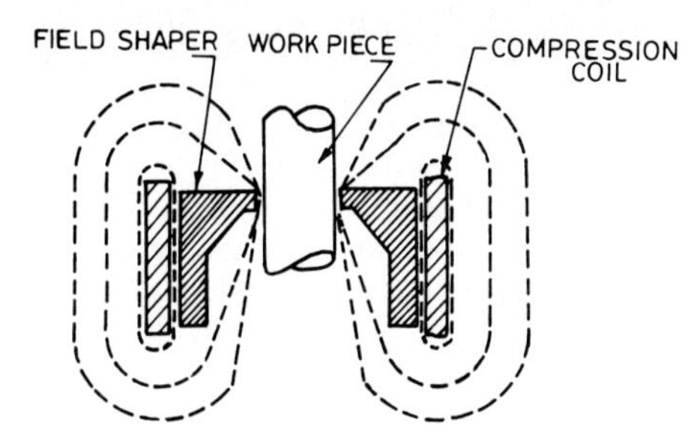

FIGURE 8.15 Schematic of electromagnetic impact bonding process.

the components by an intense magnetic field developed by a sudden surge of current through a coil (Westgate, 1978); see Fig. 8.15. In this process, the component must be circumferentially continuous and of good electrical conductivity, and the component being clad must be sufficiently thick to withstand the very high pressures involved in the electromagnetic impact. There are two types of processes: cold and hot EMIB.

In the cold EMIB process, the bonding force is developed by a surge discharge either through a permanent coil or through a fine-wire exploding coil. In the latter case, the electromagnetic force developed between the components is augmented by a shock wave as the coil vaporizes. This process is quite similar to explosive cladding.

In the hot EMIB process, a high-frequency one-turn induction coil is used to rapidly heat the component, at least to the recrystallization temperature. The coil is then switched to a capacitor bank which discharges in a few microseconds and thus develops a bonding force up to 350 MPa. This process is strictly a high-speed hot-pressure weld that involves some diffusion at the interface.

Surface preparation prior to bonding is no more than that required for explosive bonding, but the components must be in contact for the hot process or slightly inclined for the cold process. Equipment is bulky and capital costs are high, but the process has some of the advantages of explosive methods without the same safety restrictions. Component configurations are simple and are limited to rings, tubes, etc. The process is capable of high production rates on small, fairly simple shapes.

8.3.4 Diffusion Bonding

Diffusion bonding, also known as *diffusion welding, solid-state bonding, pressure bonding, isostatic bonding*, and *hot press bonding*, is accomplished by bringing the surfaces to be clad together under moderate pressure and elevated temperature in a controlled atmosphere, so that a coalescence of the interfaces can occur with minimum macroscopic deformation by diffusion-controlled processes induced by applying heat and pressure for a finite time (Anonymous, 1982; Bucklow, 1983; Jellison and Zanner, 1983; Nederveen et al., 1980).

Processing temperatures of about 0.50 to 0.75 of the melting point are used. The pressures used are sufficiently low to avoid appreciable bulk creep during the bonding period. When used as a welding technique, the applied pressure is uniaxial, but for cladding purposes a multidirectional application of pressure [such as *hot isostatic pressing* (HIP)] is needed. Pressures, times, and temperatures depend very much on the materials involved, but pressures up to 100 MPa at temperatures up to 1000°C for times up to several hours are not uncommon.

Isostatic pressure is usually applied via a gas to a deformable container within a heated pressure chamber. Within the container is a pressure-transmitting medium, usually a glass in the form of chips, and the substrate to be clad. The cladding material is either in the form of the sheet which may be sealed by welding to the substrate or in the form of a powder layer which has previously been applied, usually by spraying. Upon heating and pressurizing, pressure is uniformly transmitted to the cladding material which is forced into contact with the underlying substrate in the case of sheet or is consolidated in the case of sprayed powder or preplaced powder (Anonymous, 1982; Nederveen et al., 1980).

In general, the surface preparation is not very important in this process because surface contaminants such as oxides are often dissolved into the substrate matrix during heating as the externally applied pressure forces the two surfaces into intimate contact. Sometimes it is preferred to remove the oxide layer from the substrates. The process is very slow compared to explosive cladding and electromagnetic impact bonding processes. The diffusion bonding process involves bulk equipment with small capacity. Postfinishing operation is generally necessary for high-precision components. Its use is confined to relatively small components in the high-technology market.

Many materials that cannot be fusion-welded with ease (such as nickel-based alloys) can be diffusion-bonded. A summary of some material combinations that have been joined by diffusion bonding is given in Fig. 8.16. In this figure *direct* refers to contact between the faying surfaces of both materials with no interlayer materials, and *indirect* means that a separate interlayer material was used between the faying surfaces.

The most commonly used materials are nickel-based alloys, cobalt-based alloys, and the MCrAlY alloys, with some refractory metals, titanium and titanium alloys, and zirconium and zirconium alloys as well. The diffusion bonding process is used in many applications especially in aerospace and nuclear industries. Applications include Hastelloy X-finwall of TF30-P-100 jet-engine burner cans, rock drill bit cutters by joining a tungsten carbide cermet sandwiched with diamond to a cobalt matrix cermet stud, and niobium-alloy panels as a heat shield on a spacecraft.

8.3.5 Braze Cladding

In braze cladding, the surfaces to be clad are sandwiched by a brazing material in the form of powder, paste, rod, wire, strip, or foil. Metallurgical bonding is obtained by heating the stack by furnace, torch, resistance, dip in molten salt bath, induction, or laser beam. The selection of the heating method depends on the coating material, substrate material, coating thickness, and shape and size of the substrate.

Most metals and alloys and few nonmetals such as carbon and graphite can be brazed. Materials most commonly used as filler metals are silver alloys (BAg type), copper-zinc (RBCuZn type), and copper (BCu type). The compositions, product form, and brazing temperatures of various filler metals are presented in Table 8.11(*a*). Surface oxide films inhibit the wetting of the base metal by the filler metal and, therefore, the capillary flow of the filler metal in the joint. They also prevent the formation of a true metal-to-metal braze bond. Fluxes are used

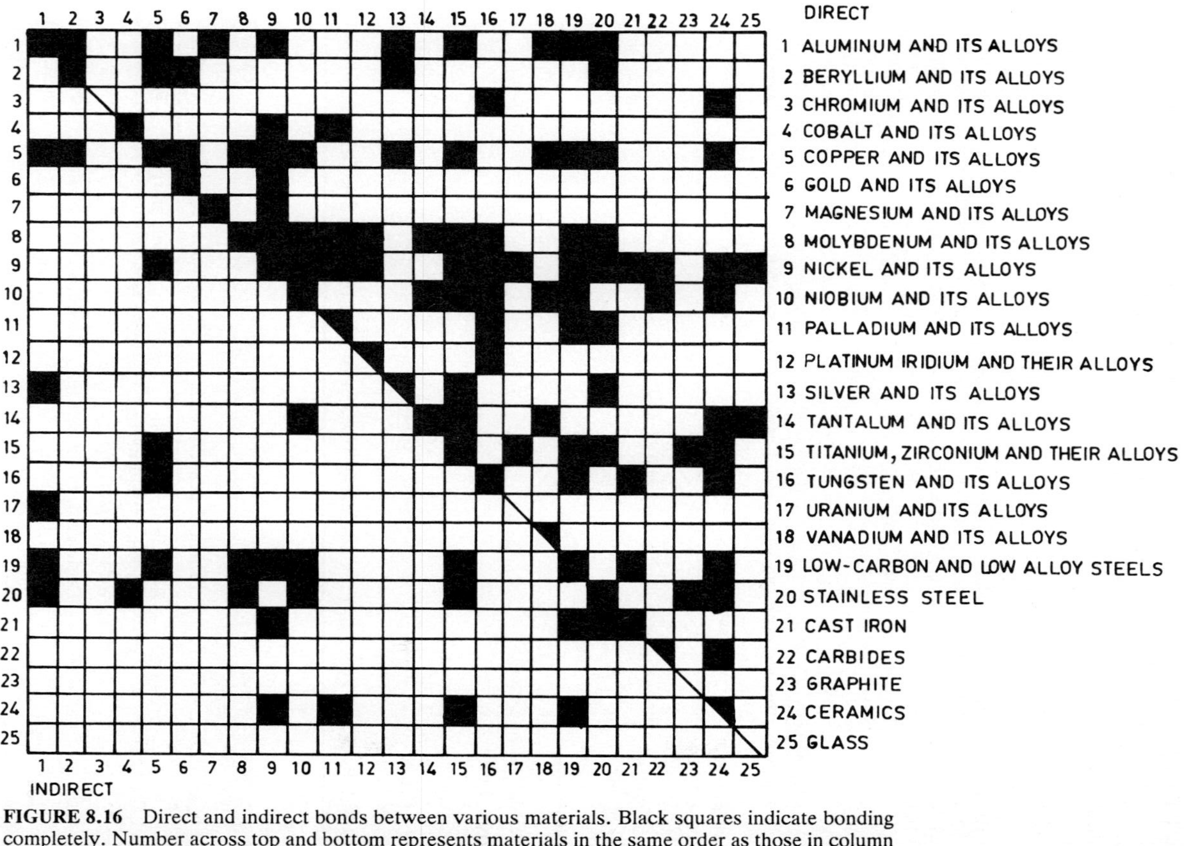

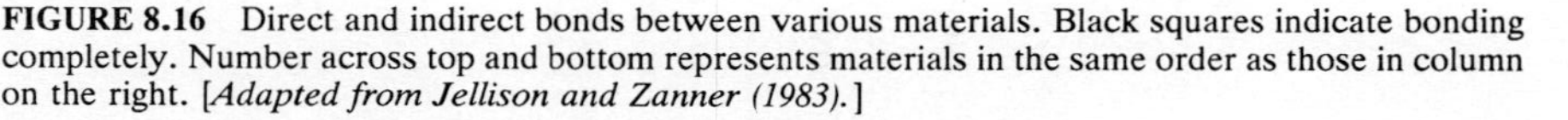

FIGURE 8.16 Direct and indirect bonds between various materials. Black squares indicate bonding completely. Number across top and bottom represents materials in the same order as those in column on the right. [*Adapted from Jellison and Zanner (1983).*]

TABLE 8.11(*a*) Typical Filler Metals for Brazing of Steels and Other Alloys

AWS classification	Product form	Nominal composition, % Ag	Cu	Zn	Cd	Ni	Sn	Si	P	Brazing temperature, °C
Silver alloys										
BAg-1	Strip, wire, powder	45	15	16	24	—	—	—	—	620–760
BAg-1a	Strip, wire, powder	50	15.5	16.5	18	—	—	—	—	635–760
BAg-2	Strip, wire, powder	35	26	21	18	—	—	—	—	700–760
BAg-2a	Strip, wire, powder	30	27	23	20	—	—	—	—	710–760
BAg-3	Strip, wire, powder	50	15.5	15.5	16	3.0	—	—	—	690–760
BAg-4	Strip, wire, powder	40	30	28	—	2.0	—	—	—	780–900
BAg-5	Strip, wire, powder	45	30	25	—	—	—	—	—	740–840
BAg-6	Strip, wire, powder	50	34	16	—	—	—	—	—	770–870
BAg-7	Strip, wire, powder	56	22	17	—	—	5.0	—	—	650–760
BAg-20	Strip, wire, powder	30	38	32	—	—	—	—	—	765–870
BAg-27	Strip, wire, powder	25	35	26.5	13.5	—	—	—	—	700–760
BAg-28	Strip, wire, powder	40	30	23	—	—	2	—	—	710–815
Copper-zinc alloys										
RBCuZn-A	Strip, rod, wire, powder	—	59	40	—	—	0.6	—	—	910–950
RBCuZn-D	Strip, rod, wire, powder	—	48	41	—	10.0	—	0.15	0.25	940–980
Copper										
BCu-1	Strip, rod, wire	—	99.0							1090–1150
BCu-1a	Powder	—	99.0							1090–1150
BCu-2	Paste (powder in a volatile vehicle)	—	86.5*							1090–1150

*And a maximum of 0.5 percent other metallics, 1.3 percent nonmetallic contaminants, and the remainder is cuprous oxide.

TABLE 8.11(*b*) Types of Flux Ordinarily Used in Brazing of Steels and Other Alloys

AWS type	Useful temperature range, °C	Principal constituents	Available form	Applicable filler metals
3A	565–870	Boric acid, borates, fluorides, fluoborates, wetting agent	Paste, liquid, slurry, powder	BAg
3C	565–980	Boric acid, borates, boron, fluorides, fluoborates, wetting agent	Paste, slurry, powder	BAg, RBCuZn
3D	760–1200	Boric acid, borates, borax, fluorides, fluoborates, wetting agent	Paste, slurry, powder	RBCuZn

in the brazing processes which must have sufficient chemical and physical activity to reduce or dissolve the base metal severely. A general list of the compounds used in brazing fluxes is presented in Table 8.11(*b*); of these, the boron and fluorine compounds are the active deoxidizing constituents.

Cleaning is almost always recommended before brazing because the presence of oil, grease, pigmented drawing lubricants, excessive amounts of oxide, and other surface contaminants in or near the brazed joint has a deleterious effect on the soundness and strength of the joint. Some contaminants interfere with the wetting action, so the normal flow of molten filler metal in the joint is prevented. Both chemical and mechanical cleaning methods are used, but chemical methods are more widely used.

Braze cladding is widely used in industries such as for masonry drills which are produced by brazing the sintered carbide tip to the low-alloy steel body.

8.3.5.1 ***Furnace Brazing.*** Furnace brazing is an inexpensive mass production process for cladding components of small assemblies by using a nonferrous filler metal as the bonding material (Gaines, 1983). In this process it should be possible to place and retain the filler material at the joint during heating in the furnace. This process also requires appropriate furnace atmosphere in order to protect the steel components from corrosion, oxidation, and decarburization during brazing and cooling. The proper brazing atmosphere facilitates proper wetting of the joint surfaces by the filler metal, usually without use of a brazing flux. Copper is a typical material used for brazing low-alloy steel, Ni-Cr, Cr-Mo, and Ni-Cr-Mo steels at about 800°C. Occasionally, silver-alloy filler metal is used which can be brazed at about 300°C. Copper is usually preferred because of its low cost and the high strength of the joints. The high brazing temperature necessary for copper is also advantageous when steel assemblies are to be heat-treated after brazing. The brazing atmosphere for steel mainly consists of exothermic- or endothermic-based atmospheres. The main content is usually nitrogen with minor additions of carbon dioxide, hydrogen, and methane. Vacuum is preferred because of the stability of surface oxides containing Cr, Mn, Ti, V, Al, and Si. High vacuum (about 10^{-1} Pa or 10^{-3} torr) is well suited for brazing base metal containing hard-to-dissociate oxides such as Ni-based superalloys. Partial vacuum (about 1 Pa or 10^{-2} torr) is used when the base or filler metal volatilizes at its brazing temperature under high vacuum.

8.3.5.2 ***Torch Brazing.*** Torch brazing is a brazing process in which the heat is obtained from a gas flame or flames impinging on or near the joint to be brazed. Torches used in this process may be of the hand-held type or may consist of fixed burners with one or more flames (Philp and Van Dyke, 1983). Acetylene, natural gas, and propane are the types of fuel gas most often used in torch brazing of steel. Torch brazing is used on stainless steels, cast irons, copper and copper alloys, and carbides. Fillers used for torch brazing of steel are the Ag alloys and the Cu-Zn alloys. Fluxes are used to inhibit surface oxidation. In some applications of torch brazing, flux can be directly applied through the gas flame. This is accomplished by bubbling the fuel gas through a small tank containing the flux in the form of a volatile liquid.

8.3.5.3 ***Resistance Brazing.*** In resistance brazing the surfaces to be clad, sandwiched by brazing filler material, are heated by resistance heating due to dc flow through the electrode and the substrate (Dixon, 1983). Parts of many different shapes can be resistance-brazed, provided that the surfaces to be joined are either flat or conform over a sufficient contact area. The joint area in most high-production resistance brazing is small, usually not more than 300 mm^2. The metals most frequently joined by resistance brazing are copper and copper alloys. Stainless steel, nickel alloys, and aluminum are resistance-brazed to a limited extent. In resistance brazing, compatible filler metal having the lowest brazing temperature is used because the maximum local temperature reached by the work must be kept as low as possible. Most common filler materials are silver alloys, aluminum-silicon alloys, and copper-phosphorus alloys.

8.3.5.4 ***Dip Brazing.*** In dip brazing, also known as *salt bath dip brazing* and *molten chemical bath dip brazing*, the sandwiched assembly to be clad is immersed in a bath of molten salt, which provides the heat and supplies the fluxing action for brazing as well (Mehrkam, 1983). The bath temperature is maintained above the liquidus temperature of the filler material and below the melting temperature of the base metal. The process is used to clad carbon and low-alloy steels with Ag alloy, Cu-Zn alloy, and Cu filler metals. Typical salt baths include $BaCl_2$, NaCl, KCl, etc. The primary advantages of the salt bath dip brazing are that the time for heating is about one-fourth that required in a controlled-atmosphere furnace and that more than one cladding operation can be brazed at the same time. However, parts need difficult cleaning after brazing. The filler metals most widely used for brazing of carbon and low-alloy steels include Ag alloys and Cu-Zn alloys.

8.3.5.5 ***Induction Brazing.*** In induction brazing, the surfaces to be clad are selectively heated to brazing temperature by electric energy transmitted to the substrate by induction (Capolongo, 1983). This process is quite similar to that discussed in Chap. 11 for induction hardening. The main advantage of induction brazing over other brazing processes discussed so far is its high-speed localized heating with minimum oxidation which subsequently reduces the cleaning operations. This process is capable of cladding most of the metals and alloys except Al and Mg by induction in air with relatively low temperatures (generally less than 850°C). Sometimes an inert atmosphere enclosure for brazing is required. Materials most commonly used as filler metals are silver alloys or BAg type with brazing temperatures ranging from 600 to 750°C. Flux or another means of oxygen exclusion is not required for induction brazing. Induction brazing is applied most conveniently to small- and medium-size assemblies. Assemblies of almost any shape can be heated for brazing by induction.

8.3.5.6 ***Laser Brazing.*** Laser brazing uses the thermal energy developed by laser beams to clad precision parts. This process has the ability to produce a brazing clad connection locally without heating the entire part at relatively low heat input (Witherell and Ramos, 1980). Another advantage is the high degree of control of the thermal energy of laser beams, including intensity, spot size, duration, and ability to be located or positioned precisely. The extremely rapid solidification and cooling rates (more than 10^{6}°C s^{-1}) characteristic of braze deposit produced by laser brazing result in an increase in deposit microhardness, with a nearly twofold increase for some compositions. The strength of laser brazed joints has, in some cases, exceeded that of fusion welds made in the same base materials. At the same time, the rapid rate of solidification characteristics of laser-melted deposits can lead to levels of brittleness that would be unsuitable in joints for many applications.

Most laser brazing is confined to line-contact types of joints where fillers are made to bridge the intersection of two surfaces or of a surface and edge. It is simply more convenient to fix the laser and move the part, which is usually very light and small. Both continuous-wave CO_2 and pulsed lasers can be used; but solid-state pulsed lasers, such as neodymium-doped yttrium-aluminum-garnet (YAG) pulsed lasers, appear to be more adaptable to the progressive overlapping-spot mode of brazing. Fillers such as silver alloys, nickel, and copper are commonly used for brazing of steel. Adequate atmospheric protection is required for all laser brazing, whether or not a flux is used.

Laser brazing generally should be reserved for precision jobs that require the strength of a brazed joint but cannot tolerate the heat, distortion, or other consequences of more conventional techniques. The best response in laser brazing is achieved with parts having thicknesses less than about 0.25 mm.

8.3.6 Weld Cladding

Weld cladding, also referred to as *weld overlays*, differs from the (fusion) welding described in Sec. 8.2 in terms of thickness. Weld cladding usually denotes the application of a relatively thick layer (3 mm or more, up to 100 mm) of weld metal whereas a weld-deposited coating produces a thinner coating. With very few exceptions, virtually any welding process can be used for weld cladding; however, certain processes are more applicable (Thomas et al., 1983b). Most commonly used processes are submerged arc welding (SAW), gas metal arc welding (GMAW), self-shielded arc welding (SSAW), and shielded-metal arc welding (SMAW). Less commonly used methods are gas tungsten arc welding (GTAW), plasma arc welding (PAW), and electroslag welding (ESW).

SAW with high deposition rates is used extensively to clad vessels with corrosion-resistant materials or to surface rolls with a wear-resistant overlay. GMAW and SSAW are particularly useful to produce final touch-up or repairs. SMAW can be used economically for small-area overlay, for one-of-a-kind or unusually shaped surfaces, or for repair overlay. In addition, field application of cladding is often done by SMAW. GTAW is used only in specialized applications for cladding. PAW is becoming more common. The addition of a powder spray to the plasma allows the deposition of a wide variety of cladding materials such as cobalt-tungsten carbide or ceramics which cannot be produced in wire form. ESW can be used to clad a wide variety of materials including those in powder form. The usual deposit thickness for GMAW is 10 to 20 mm, and for SAW and ESW it is 20 to 100 mm. Because of high surface temperatures, welding processes can be used only on relatively thick substrates. Based on the final application, the cladded surface may require postfinishing. For more details, see Section 8.2 on welding.

8.3.7 Laser Cladding

The laser cladding process differs from other surface alloying in that only enough substrate material is melted to ensure good bonding of the cladding material. The cladding materials, available in the form of cast rods, wires, or powders, can be melted under controlled conditions by using a laser beam, such as carbon dioxide lasers (with beam power up to 15 kW), to allow the molten alloy to freely spread and freeze over the substrate surface. Under optimum conditions, the laser beam melts a very thin surface layer of the substrate. This melted top layer mixes with the melted cladding material and subsequently freezes to form a metallurgical bond between the cladding surface and the substrate (Antony et al., 1983; Breinan et al., 1976). Laser cladding offers potential in applying clad materials with high melting points to low-melting-point substrates. The thickness of the clad layers typically ranges from about 1 to 6 mm or higher. Laser cladding is advantageous when thin claddings are desired or when access to the clad surface can be achieved more readily by a laser beam than an electrode or torch.

In the laser cladding process, prealloyed powders are generally applied to the substrate surface before laser processing with or without suitable bonding compounds. Self-fluxing alloy powders, such as nickel- or cobalt-based alloys with a high chromium concentration, along with a sufficient quantity of silicon and boron, provide a rapid method of hard facing when flame sprayed on the substrate and subsequently laser clad. Table 8.12 describes some of the experimental conditions involved in laser cladding.

This process is capable of cladding a very complicated shaped substrate, and the equipment is transportable.

8.4 REFERENCES

Anderson, J. C. (1978), *Proc. Int. Conf. Advances in Surface Coating Technology*, The Welding Institute, Cambridge, United Kingdom.

Anonymous (1967), "The Metco Flame Spraying Processes," Metco Inc., Westbury, N.Y.

——— (1975a), "Problem Solving Technology," Union Carbide Corp., Linde Division, Coatings Service Dept., 1500 Polco Street, Indianapolis, Ind.

——— (1975b), "Stellite Hard-facing Products," Stellite Division, Cabot Corporation, Kokomo, Ind.

——— (1976), "High Performance Coatings, Metco 7M Plasma Spray Process," Metco Inc., Westbury, N.Y.

——— (1980a), "Weld Surfacing and Hardfacing," The Welding Institute, Cambridge, United Kingdom.

——— (1980b), "Hardfacing Materials in Nuclear Power Plants," *Proc. Int. Colloq. Avignon, France*, September, Société Française d'Énergie Nucléaire, Paris.

——— (1982), "Trends in Powder Metallurgy Technology," *Met. Prog.*, Vol. 121, pp. 38–49.

——— (1983), *Metals Handbook*, Vol. 6: *Welding, Brazing, and Soldering*, 9th ed., American Society for Metals, Metals Park, Ohio.

——— (1985), "Clad Metals," Texas Instruments, Attleboro, Mass.

Antony, K. C., Bhansali, K. J., Messler, R. W., Miller, A. E., Price, M. O., and Tucker, R. C. (1983), "Hardfacing," in *Metals Handbook*, Vol. 6: *Welding, Brazing and Soldering*, 9th ed., pp. 771–803, American Society for Metals, Metals Park, Ohio.

Ault, N. N. (1957), "Characteristics of Refractory Oxide Coatings Produced by Flame Spraying," *J. Am. Ceramic Soc.*, Vol. 40, pp. 69–74.

———, and Wheildon, W. M. (1960), "Modern Flame-Sprayed Ceramic Coatings," in *Modern Materials*, Vol. 2 (H. H. Hausner, ed.), pp. 63–106, Academic, New York.

Avery, H. S. (1950), "Hot Hardness for Hard Facing Alloys," *Welding J.*, Vol. 29, pp. 552–578.

——— (1951), "Some Characteristics of Composite Tungsten Carbide Weld Deposits," *Welding J.*, Vol. 30, pp. 144–160.

TABLE 8.12 Some Experimental Parameters Involved in Laser Cladding

Cladding	Form of cladding materials	Base materials	Power depth, mm	Power width, mm	Powder application method	Preheat temp., °C	Laser beam size, mm × mm	Type of laser beam	Laser power, kW	Processing speed, mm s^{-1}	Shielding gas
Tribaloy T-800	Powder	ASTM A387	6	25	Slurry	20	14 × 14	Stationary	12.5	1.25	Helium
Haynes Stellite alloy No. 1	Cast rod, 3-mm dia.	AISI 4815	—	—	—	250	6.2 dia.	Stationary	3.5	4.25	Hydrogen and argon
Silicon	Powder, 44-μm size	AA390 Al-alloy	1	5	Slurry	20	5 dia.	Stationary	4.3	8.75	Hydrogen and argon
Tungsten carbide and iron	WC granules 0.05-mm size and iron powder 44-μm size	AISI 1018	1	19	Loose powder	20	12.5 × 12.5	Stationary	12.5	6.25	Helium
Alumina	Powder 0.3-μm size	2219 Al alloy	0.7	25	Loose powder	20	6.2 × 18.6	Oscillating (690 Hz)	12.5	8.7	Oxygen

Source: Adapted from Antony et al. (1983).

——— (1952a), "The Selection of Hardfacing Alloys," *Prod. Eng.*, Vol. 23, pp. 154–157.

——— (1952b), "Hardfacing for Impact," *Welding J.*, Vol. 31, pp. 116–143.

——— (1969), "Surfacing by Welding for Wear Resistance," in *Composite Engineering Laminates* (A. G. H. Dietz, ed.), pp. 255–303, M.I.T. Press, Cambridge, Mass.

Bahrani, A. S. (1978), "Explosive Cladding," *Surfacing J.*, Vol. 9, pp. 2–9.

———, Black, T. J., and Crossland, B. (1967), "The Mechanism of Wave Formation in Explosive Welding," *Proc. Roy. Soc. Lond.*, Vol. A296, pp. 123–136.

Ballard, W. E. (1963), *Metal Spraying and the Flame Deposition of Ceramics and Plastics*, Charles Griffin, London.

Bell, G. R. (1972), "Surface Coatings 3—Gas Welded Coatings," *Tribol. Int.*, Vol. 5, pp. 215–219.

Bhat, H., and Herman, H. (1982), "Plasma-Spray Quenched Martensitic Stainless Steel Coatings," *Thin Solid Films*, Vol. 95, pp. 227–235.

Bhushan, B. (1980a), "High Temperature Self-Lubricating Coatings and Treatments—A Review," *Met. Finish.*, Vol. 78, May, pp. 83–88; June, pp. 71–75.

——— (1980b), "High Temperature Self-Lubricating Coatings for Air-Lubricated Foil Bearings for the Automotive Gas-Turbine Engine," Tech. Rep. CR-159848, April, NASA Lewis Research Center, Cleveland, Ohio.

——— (1987), "Overview of Coating Materials, Surface Treatments, and Screening Techniques for Tribological Applications Part 1: Coating Materials and Surface Treatments," in *Testing of Metallic and Inorganic Coatings* (W. B. Harding and G. A. DiBari, eds.), Special Technical Publication No. STP 947, pp. 289–309, ASTM, Philadelphia.

———, and Gray, S. (1978), "Static Evaluation of Surface Coatings for Compliant Gas Bearings in an Oxidizing Atmosphere to 650°C," *Thin Solid Films*, Vol. 53, pp. 313–331.

———, and Heshmat, H. (1979), "Reliability Improvement Study for Compliant Foil Air Bearing in Chrysler Upgraded Automotive Gas Turbine Engine," Tech. Rep. MTI-79TR73, Mechanical Technology Inc., Latham, N.Y.

———, and Gray, S. (1980), "Development of Surface Coatings for Air Lubricated Compliant Journal Bearings to 650°C," *ASLE Trans.*, Vol. 23, pp. 185–196.

———, Ruscitto, D., and Gray, S. (1978), "Hydrodynamic Air-Lubricated Compliant Surface Bearings for an Automated Gas Turbine Engine Part II: Materials and Coatings," Tech. Rep. CR-135402, NASA Lewis Research Center, Cleveland, Ohio.

Blaskovic, P. (1966), "Electroslag Weld Surfacing of Pilger Work Rolls," *Zvaranie*, Vol. 15, No. 3, pp. 72–76.

Breinan, E. M., Kear, B. H., and Banas, C. M. (1976), "Processing Materials with Lasers," *Phys. Today*, Vol. 29, pp. 44–50.

Bucklow, I. A. (1983), "Bulk Coatings for Use at Elevated Temperatures," in *Coatings for High Temperature Applications* (E. Lang, ed.), pp. 139–167, Elsevier Applied Science Publishers, London.

Capolongo, P. (1983), "Induction Brazing of Steels," in *Metals Handbook*, Vol. 6: *Welding, Brazing, and Soldering*, 9th ed., pp. 965–975, American Society for Metals, Metals Park, Ohio.

Cashon, E. P. (1975), "Wear Resistant Coatings Applied by Detonation Gun," *Tribol. Int.*, Vol. 8, pp. 111–115.

Chapman, B. N., and Anderson, J. C. (eds.) (1974), *Science and Technology of Surface Coating*, Academic, New York.

Chivers, T. C. (1985), "Aspects of Fretting Wear of Sprayed Cermet Coatings," *Wear*, Vol. 106, pp. 63–76.

Chuanxian, D., Zatorski, R. A., Herman, H., and Ott, D. (1984), "Oxide Powders for Plasma Spraying: The Relationship between Powder Characteristics and Coating Properties," *Thin Solid Films*, Vol. 118, pp. 467–475.

Clare, J. H., and Crawmer, D. E. (1982), "Thermal Spray Coatings," in *Metals Handbook*,

Vol. 5: *Surface Cleaning, Finishing and Coating*, 9th ed., pp. 361–374, American Society for Metals, Metals Park, Ohio.

Delagi, R. G. (1980), "Designing with Clad Metals," *Machine Design*, Vol. 52, December, pp. 148–153.

Devletian, J. H., and Wood, W. E. (1983), "Principles of Joining Metallurgy," in *Metals Handbook*, Vol. 6: *Welding, Brazing, and Soldering*, 9th ed., pp. 21–49, American Society for Metals, Metals Park, Ohio.

Dixon, A. (1983), "Resistance Brazing," in *Metals Handbook*, Vol. 6: *Welding, Brazing, and Soldering*, 9th ed., pp. 976–988, American Society for Metals, Metals Park, Ohio.

Edmonds, D. P. (1978), "Cladding of Pressure Vessel Steels for Coal Conversion Applications: A Literature Review," Rep. No. ORNL/TM-6425, Oak Ridge National Laboratory, Oak Ridge, Tenn.

Eschnauer, H. (1980), "Hard Material Powders and Hard Alloy Powders for Plasma Surface Coating," *Thin Solid Films*, Vol. 73, pp. 1–17.

———, and Lugscheider, E. (1984), "Metallic and Ceramic Powders for Vacuum Plasma Spraying," *Thin Solid Films*, Vol. 118, pp. 421–435.

Evans, I. W. (1962), "Hardsurfacing of Mill Rolls and Mill Equipments," *Iron Steel Eng.*, Vol. 39, pp. 109–114.

Ferriss, D. P., and Cameron, C. B. (1975), "Wear Properties of Cast vs. Plasma-Sprayed Laves Intermetallic Alloys," *J. Vac. Sci. Technol.*, Vol. 12, pp. 795–799.

Fisher, I. A. (1972), "Variables Influencing the Characteristics of Plasma-Sprayed Coatings," *Int. Metall. Rev.*, Vol. 17, pp. 117–129.

Foroulis, Z. A. (1984), "Guidelines for the Selection of Hardfacing Alloys for Sliding Wear Resistant Applications," *Wear*, Vol. 96, pp. 203–218.

Furillo, F. T. (1980), "W-Mo Carbides for Hardfacing Applications," *Wear*, Vol. 60, pp. 183–204.

Gage, R. M., Nestor, O. H., and Yenni, D. M. (1962), "Collimated Electric Arc Powder Deposition Process," U.S. Patent 3,016,447, January 9.

Gaines, G. W. (1983), "Furnace Brazing of Steels," in *Metals Handbook*, Vol. 6: *Welding, Brazing, and Soldering*, 9th ed., pp. 929–949, American Society for Metals, Metals Park, Ohio.

Gerdeman, D. A., and Hecht, N. L. (1972), *Arc Plasma Technology in Materials Science*, Springer-Verlag, New York.

Gray, S., Heshmat, H., and Bhushan, B. (1981), "Technology Progress on Compliant Foil Air Bearing Systems for Commercial Applications," *Proc. Eighth Int. Gas Bearings Symp.*, pp. 69–97, *BHRA, Fluid Engineering*, Cranfield, Bradford, United Kingdom.

Gregory, E. N. (1978), "Hardfacing," *Tribol. Int.*, Vol. 11, pp. 129–134.

——— (1980), "Surfacing by Welding—Alloys Processes, Coatings and Material Selection," *Metal Constr.*, Vol. 12, pp. 685–690.

Grisaffe, S. J. (1967), "Simplified Guide to Thermal Spray Coatings," *Machine Design*, Vol. 39, No. 17, pp. 174–181.

Gruner, H. (1984), "Vacuum Plasma Spray Quality Control," *Thin Solid Films*, Vol. 118, pp. 409–420.

Hazzard, R. (1972), "Surface Coatings, 2—Arc Welding Coatings," *Tribol. Int.*, Vol. 5, pp. 207–214.

Henne, R., Schnurenberger, W., and Weber, W. (1984), "Low Pressure Plasma Spraying—Properties and Potential for Manufacturing Improved Electrolysers," *Thin Solid Films*, Vol. 119, pp. 141–152.

Hocking, M. G., Vasantasree, V., and Sidky, P. S. (1989), *Metallic and Ceramic Coatings: Production, High-Temperature Properties, and Applications*, Longman, London.

Ingham, H. S. (1969), "Flame Sprayed Coatings," in *Composite Engineering Laminates* (A. G. H. Dietz, ed.), pp. 304–322, M.I.T. Press, Cambridge, Mass.

———, and Shepard, A. P. (1965), *Metco Flame Spray Handbook*; Vol. I: *Wire Process*,

Vol. II: *Powder Process*, and Vol. III: *Plasma Flame Process*, Metco Inc., Westbury, N.Y.

Jellison, J. L., and Zanner, F. J. (1983), "Solid State Welding," in *Metals Handbook*, Vol. 6: *Welding, Brazing and Soldering*, 9th ed., pp. 672–691, American Society for Metals, Metals Park, Ohio.

Johnson, R. N., and Farwick, D. G. (1978), "Friction, Wear and Corrosion of Laves—Hardened Nickel Alloy Hardfacing in Sodium," *Thin Solid Films*, Vol. 53, pp. 365–373.

Kaminsky, M. (1980), "Clad Materials for Fusion Applications," *Thin Solid Films*, Vol. 73, pp. 117–132.

Kharlamov, Y. A. (1978), "Bonding of Detonation Sprayed Coatings," *Thin Solid Films*, Vol. 54, pp. 271–278.

Kieffer, R., and Kalke, F. (1949), "Tungsten Carbide-free Hard Metals," *Powder Metall. Bull.*, Vol. 4, pp. 4–17.

Kretzschmar, E., and Dollinger, A. (1979), "New Surfacing Methods," *Schweisstechnik* (in German), Vol. 29, No. 8, pp. 345–351.

Lai, G. Y. (1978), "Evaluation of Sprayed Chromium Carbide Coatings for Gas-Cooled Reactor Applications," *Thin Solid Films*, Vol. 53, pp. 343–351.

Land, J. E. (1974), "Application of Detonation Coatings," in *Science and Technology of Surface Coating* (B. N. Chapman and J. C. Anderson, eds.), pp. 281–286, Academic, New York.

Leckey, R. C. G. (1969), *Plasma Technology*, American Elsevier, New York.

Li, C. C. (1980), "Characterization of Thermally Sprayed Coatings for High Temperature Wear Protection Applications," *Thin Solid Films*, Vol. 73, pp. 59–77.

Longo, F. N. (1972), *Handbook of Coating Recommendations*, Metco Inc., Westbury, N.Y.

Mann, B. S., Krishnamurthy, P. R., and Vivekananda, P. (1985), "Cavitation Erosion Characteristics of Nickel-Based Alloy-Composite Coatings Obtained by Plasma Spraying," *Wear*, Vol. 103, pp. 43–55.

Mansford, R. (1972), "Surface Coatings, 4—Sprayed Coatings," *Tribol. Int.*, Vol. 5, pp. 220–224.

Marantz, D. R. (1974), "The Basic Principles of Electric-Arc Spraying," in *Science and Technology of Surface Coating* (B. N. Chapman and J. C. Anderson, eds.), pp. 308–321, Academic, New York.

Matty, A., and Becker, K. (1956), "An Experimental Investigation of the Metal Spraying Process," *Electroplat. Metal. Finish.*, April, Vol. 56, pp. 143–145.

Matusek, J. J. (1959), "Submerged Arc Welding Reclaims Blooming Mill Rolls," *Weld. Eng.*, Vol. 44, pp. 38–39.

McPherson, R. (1981), "The Relationship between the Mechanism of Formation, Microstructure and Properties of Plasma Sprayed Coatings," *Thin Solid Films*, Vol. 83, pp. 297–310.

Mehrkam, Q. D. (1983), "Dip Brazing of Steels in Molten Bath," in *Metals Handbook*, Vol. 6: *Welding, Brazing, and Soldering*, 9th ed., pp. 989–995, American Society for Metals, Metals Park, Ohio.

Meyer, P., and Muehlberger, S. (1984), "Historical Review and Update to the State of the Art of Automation for Plasma Coating Processes," *Thin Solid Films*, Vol. 118, pp. 445–456.

Mock, J. A. (1974), "Ceramic and Refractory Coatings," *Mater. Eng.*, Vol. 80, November, pp. 101–108.

Moore, G. D., and Ritter, J. E., Jr. (1974), "Friction and Wear of Plasma-Sprayed NiO Based Coatings," *J. Vac. Sci. Technol.*, Vol. 11, pp. 754–758.

Moss, A. R., and Young, W. J. (1974), "Arc Plasma Spraying," in *Science and Technology of Surface Coating* (B. N. Chapman and J. C. Anderson, eds.), pp. 287–299, Academic, New York.

Muehlberger, E., and Kremith, R. (1973), "New Sonic and Supersonic 80 kW Plasma Spray Systems," presented at Ninth Airlines Plating Forum, Montreal, Canada.

Nederveen, M. B., et al. (1980), *Proc. 9th Int. Thermal Spraying Conf.*, Institute Voor Lastechniek Den Haag, Netherlands.

Philp, C., and Van Dyke, C. (1983), "Torch Brazing of Steels," in *Metals Handbook*, Vol. 6: *Welding, Brazing, and Soldering*, 9th ed., pp. 950–964, American Society for Metals, Metals Park, Ohio.

Pocalyko, A., Kubota, A., and Linse, V. D. (1983), "Explosion Cladding," in *Metals Handbook*, Vol. 6: *Welding, Brazing and Soldering*, 9th ed., pp. 705–718, American Society for Metals, Metals Park, Ohio.

Poorman, R. M., Sargent, H. B., and Lamprey, H. (1955), "Method and Apparatus Utilizing Detonation Waves for Spraying and Other Purposes," U.S. Patent 2,714,553, August 2.

Popoff, R. A. (1978), "Explosion Welding," *Mech. Eng.*, Vol. 100, pp. 28–36.

Prummer, R. A. (1977), "Explosive Cladding of Thin Films," *Thin Solid Films*, Vol. 45, pp. 205–210.

Reardon, J. D., Mignogna, R., and Longo, F. N. (1981), "Plasma and Vacuum Plasma Sprayed Cr_3C_2 Composite Coatings," *Thin Solid Films*, Vol. 83, pp. 345–351.

Safai, S., and Herman, H. (1978), "Plasma Sprayed Coatings, Their Ultramicrostructure," in *Proc. Int. Conf. Advances in Surface Coating Technology* (J. C. Anderson, ed.), pp. 1–14, Welding Institute, Cambridge, United Kingdom.

———, and ——— (1981), "Plasma Sprayed Materials in Ultra-rapid Quenching of Liquid Alloys," *Treatise Mater. Sci. Technol.*, Vol. 20, pp. 183–214.

Sangam, M., and Nikitich, J. (1985), "Designing for Optimum Use of Plasma Coatings," *J. Metals*, Vol. 37, No. 9, pp. 55–60.

Scott, K. T., and Cross, G. (1980), "Fabrication of Ceramics by Plasma Spraying," in *Proc. 4th Int. Meeting on Modern Ceramic Technologies*, Elsevier Applied Science Publishers, New York.

———, and Woodhead, J. L. (1982), "Gel-Processed Powders for Plasma Spraying," *Thin Solid Films*, Vol. 95, pp. 219–225.

Sliney, H. E. (1979), "Wide Temperature Spectrum Self-Lubricating Coatings Prepared by Plasma Spraying," *Thin Solid Films*, Vol. 64, pp. 211–217.

Smart, R. F., and Catherall, J. A. (1972), *Plasma Spraying*, Mills and Boon Ltd., London.

Smith, C. W. (1974a), "The Basic Principles of Flame Spraying," in *Science and Technology of Surface Coating* (B. N. Chapman and J. C. Anderson, eds.), pp. 262–269, Academic, New York.

Smith, R. G. (1974b), "The Basic Principles of Detonation Coating," in *Science and Technology of Surface Coating* (B. N. Chapman and J. C. Anderson, eds.), pp. 271–279, Academic, New York.

Steffens, H. D. (1983), "Spray and Detonation Gun Technologies, Laser Assisted Technologies," in *Coatings for High Temperature Applications* (E. Lang, ed.), pp. 121–138, Applied Science Pub., New York.

———, and Höhle, H. M. (1980), "Further Development of the Thermal Spraying Technique by Low Pressure Plasma Spraying," *Proc. Ninth Int. Thermal Spraying Conf.*, May 19–23, 1980, The Hague, Netherlands.

———, ———, and Erturk, E. (1980), "Low Pressure Plasma Spraying of Reactive Materials," *Thin Solid Films*, Vol. 73, pp. 19–29.

———, and Müller, K. N. (1974), "Structure and Properties of Arc-Sprayed Titanium, Niobium and Molybdenum Coatings," *J. Vac. Sci. Technol.*, Vol. 11, pp. 735–740.

Taylor, T. A. (1975), "Phase Stability of Chrome-Carbide Ni-Cr Coatings in Low-Oxygen Environments," *J. Vac. Sci. Technol.*, Vol. 12, pp. 790–794.

Thomas, R. D., Amos, D. R., Edwards, G., Gix, G. A., Hannahs, J. R., Henderson, F. W., Leclair, G. A., Shutt, R. C., and Sims, J. E. (1983a), "Electroslag Welding," in *Metals*

Handbook, Vol. 6: *Welding, Brazing, and Soldering*, 9th ed., pp. 225–237, American Society for Metals, Metals Park, Ohio.

———, Barger, J. J., Farmer, H. N., Jones, J. E., Layo, W. E., Meyer, J. J., and Stark, L. E. (1983b), "Weld Overlays," in *Metals Handbook*, Vol. 6: *Welding, Brazing, and Soldering*, 9th ed., pp. 804–819, American Society for Metals, Metals Park, Ohio.

Tucker, R. C. (1974), "Structure Property Relationships in Deposits Produced by Plasma Spray and Detonation Gun Techniques," *J. Vac. Sci. Technol.*, Vol. 11, pp. 725–734.

——— (1982), "Plasma and Detonation Gun Deposition Techniques and Coating Properties," in *Deposition Technologies for Films and Coatings* (R. F. Bunshah et al., eds.), pp. 454–489, Noyes, Park Ridge, N.J.

Van Bemst, A., Niset, N., and Denirzere, R. (1981), "Structure and Phase Composition of Carbide Containing Cobalt, Chromium Surfacing Alloys," *Surfacing J.*, Vol. 12, No. 2, pp. 26–31.

Vinayo, M. E., Kassabji, F., Guyonnet, J., and Fauchais, P. (1985), "Plasma Sprayed WC-Co Coatings: Influence of Spray Conditions (Atmospheric and Low Pressure Plasma Spraying) on the Crystal Structure, Porosity, and Hardness," *J. Vac. Sci. Technol.*, Vol. A3, pp. 2483–2489.

Watson, H. N. (1973), "The Basic Principles of Welding as a Surface Coating Process," in *Science and Technology of Surface Coating* (B. N. Chapman and J. C. Anderson, eds.), pp. 215–221, Academic, New York.

Wells, J. C. (1978), "A Review of Surface Treatment and Coatings," *Surfacing J.*, Vol. 9, pp. 2–9.

Westgate, S. A. (1978), "Cladding by Electromagnetic Impact Bonding," *Surfacing J.*, Vol. 9, pp. 10–13.

Wilkins, C. R., Wallace, F. J., and Zajchowski, P. H. (1976), "Ceramic/Metallic Thermal Barrier Coatings for Gas Turbine Engine," in *8th Int. Thermal Spraying Conf.*, American Welding Society, Miami, Florida.

Wilms, V., and Herman, H. (1976), "Plasma Spraying of Al_2O_3 and Al_2O_3-Y_2O_3," *Thin Solid Films*, Vol. 39, pp. 251–262.

Witherell, C. E., and Ramos, T. J. (1980), "Laser Brazing," *Welding J.*, Vol. 59, pp. 267S–277S.

Wolfla, T. A., and Tucker, R. C. (1978), "High Temperature Wear Resistant Coatings," *Thin Solid Films*, Vol. 53, pp. 353–364.

CHAPTER 9
COATING DEPOSITION FROM VAPOR PHASE

The vapor deposition processes consist of *physical vapor deposition* (PVD) and *chemical vapor deposition* (CVD). The CVD processes, in addition, have been modified through glow-discharge plasma activation and are called *physical-chemical vapor deposition* (P-CVD). Each deposition technique has its advantages and disadvantages and its range of preferred applications. Figure 9.1 enumerates the variants of deposition techniques which are based on vapor deposition. Typical particle kinetic energy ranges for various vapor deposition processes are presented in Fig. 9.2. Particle energies are generally referred to in units of electronvolts. One electronvolt (eV) is the unit of energy that a particle with one unit of electron charge accumulates while passing through a potential difference of one volt. The kinetic energy of the vapor particles is known to have an important effect on the morphology, adhesion, and mechanical properties of the coatings. The kinds of materials that can be deposited, substrate materials, kinetic energy of vapors, deposition temperature, deposition pressure, coating thickness range, and deposition rate are summarized in Table 9.1. The main ad-

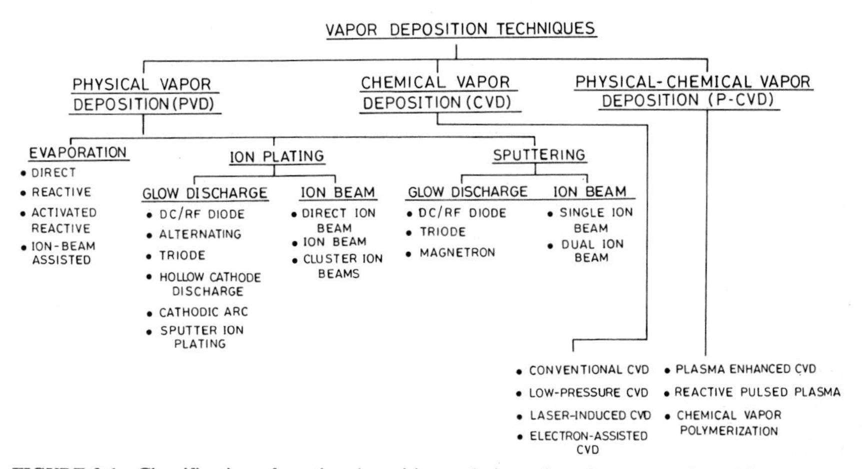

FIGURE 9.1 Classification of coating deposition techniques based on vapor deposition.

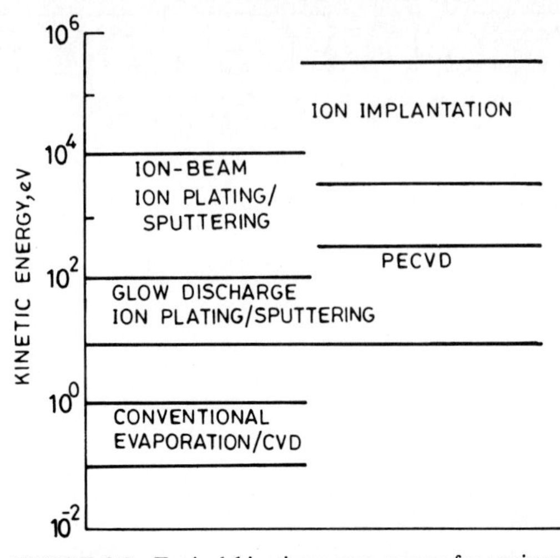

FIGURE 9.2 Typical kinetic energy ranges for various vapor deposition and surface modification processes. The CVD and PECVD processes involve chemical reactions at the interface.

vantages and disadvantages of various deposition processes are summarized in Table 9.2.

Physical vapor deposition is used to deposit coatings by condensation of vapors in high vacuum (10^{-6} to 10 Pa) 10^{-8} to 10^{-1} torr atomistically (atom by atom) at the substrate surface. PVD processes can also be used to deposit coatings on self-supported shapes such as sheet, foil, or tubing. PVD is extremely versatile and now covers a wide range of variants (Anonymous, 1972; Berry et al., 1968; Bhushan, 1980a, 1987; Bunshah, 1976; Bunshah et al., 1982; Chapman and Anderson, 1974; Chopra, 1979; Maissel and Glang, 1970; Powell et al., 1966; Reichelt and Jiang, 1990; Spalvins, 1985; Sundgren and Hentzell, 1986; Vossen and Kern, 1978). The basic processes involving PVD are thermal evaporation, activated reactive evaporation, glow-discharge or ion beam ion plating, and glow-discharge or ion beam sputtering. The evaporation and glow-discharge processes are extremely versatile in the composition of coatings and substrate materials. Virtually any metal, alloy, ceramic, intermetallic, and some polymeric-type materials and their mixtures can be easily deposited onto substrates of virtually any material (such as metals, ceramics, plastics, and paper) which are stable at operating temperatures in vacuum. The surfaces of substrates that evolve gas may be sealed with a lacquer before vapor coating. Ion beam deposition processes can be used to deposit many metals, alloys, and compounds. Substrates must withstand bombardment heating.

The evaporation process is relatively simple and cheap and is used to deposit thick coatings or produce self-supported structures up to 10 mm. Substrates generally must be rotated for uniform coating. Complex shapes can be coated by gas scattering technique. The high kinetic energy of the coating particles in the ion beam type of processes (ion-beam ion plating and dual ion-beam sputtering processes requiring direct impingement of ion beam on the coated surface) result in enhancement of diffusion and chemical reactions without the need for high

TABLE 9.1 Process Parameters of Various Coating Deposition Techniques Based on Vapor Deposition*

Deposition technique	Type of coating materials	Form of coating material source	Type of substrates	Kinetic energy, eV	Deposition pressure, Pa (torr)	Deposition temperature, °C	Coating thickness range, μm	Deposition rate, μm min^{-1}	Relative adhesion
Direct evaporation	Essentially any pure metal, many alloys, and compounds, e.g., Al, Au, Ag, Ni, Cr, Ti, Mo, W, stainless steel, Ni-Cr, Pb-Sn, MCrAlY, Al_2O_3, TiC, TiB_2	Solid	Any nongassing material	0.1–1	10^{-3}–10^{-6} (10^{-5}–10^{-8})	200–1600†	0.1–1000	1–25	Fair at elevated temperature
Activated reactive evaporation (ARE)	Almost any, e.g., Al_2O_3, SiO_2, Y_2O_3, TiO_2, SnO_2, In_2O_3, W_2C, TiC, Ni-TiC, ZrC, HfC, VC, NbC, TiC-CO, TiC-Ni, TiN, HfN, CBN, TiB_2, ZrB_2, TiC-TiB_2-Co	Solid	Any nongassing material	5–20	10^{-2}–1 (10^{-4}–10^{-2})	200–1600†	0.1–1000	1–75	Fair to good at elevated temperature
Glow-discharge ion plating	Almost any, e.g., Au, Ag, Pb, Sn, Pb-Sn, Al, Ni, Cr, Ni-Cr, CoCrAlY, SiO_2, Al_2O_3, Cr_2O_3, ZrO_2, TiO_2, WC, TiC, Cr_3C_2, B_4C, TiN, HfN, ZrN, Si_3N_4, and (TiAl)N	Solid	Substrate must withstand ion bombardment heating	10–100	5×10^{-1}–10 (5×10^{-3}–10^{-1})	100–300	0.02–10	0.02–5	Generally good
Ion beam ion plating	Almost any, e.g., Ni, Al, Au, i-C, Cr-C, Cr-N, TiN, Si_3N_4, AlN, and (TiAl)N	Solid	Substrate must withstand ion bombardment heating	100–10,000	10^{-5}–10^{-2} (10^{-7}–10^{-4})	100–500	0.02–10	0.02–2	Excellent

TABLE 9.1 Process Parameters of Various Coating Deposition Techniques Based on Vapor Deposition* (*Continued*)

Deposition technique	Type of coating materials	Form of coating material source	Type of substrates	Kinetic energy, eV	Deposition pressure, Pa (torr)	Deposition temperature, °C	Coating thickness range, μm	Deposition rate, μm min^{-1}	Relative adhesion
Glow-discharge sputtering	Essentially any, e.g., Au, Ag, Ni, Cr, Co, Mo, W, Al_2O_3, ZrO_2, Cr_2O_3, SiO_2, TiC, ZrC, HfC, B_4C, TiN, ZrN, HfN, Si_3N_4, CrB_2, Mo_2B_5, Cr_3Si_2, MoS_2, WS_2, PTFE, poly(amide-imide)	Solid	Any nongassing material	10–100	5×10^{-1}–10 (5×10^{-3}–10^{-1})	100–300	0.02–10	0.02–2	Generally good
Ion beam sputtering	Almost any, e.g., Ni, Cr, Mo, Ti, W, Zr, Ta, Si, Ni_3Al, Nb-Ti, i-C, Al_2O_3, SiO_2, Ta_2O_5, Cr-C, AlN, Si_3N_4, i-BN, PTFE	Solid	Any nongassing material	100–10,000	10^{-5}–10^{-2} (10^{-7}–10^{-4})	100–500	0.02–10	0.02–0.5	Good to excellent
Chemical vapor deposition (CVD)	Essentially any, e.g., Ni, Cr, Mo, Ti, W, Zr, Ta, Si, Cu, Fe, Pt, Al_2O_3, ZrO_2, Cr_2O_3, SiO_2, SnO_2, TiC, ZrC, HfC, B_4C, Cr_3C_2, WC, TiN, ZrN, HfN, Si_3N_4, BN, TiB_2, ZrB_2, MoB, FeB, $TiSi_2$, $MoSi_2$	Solid, liquid, gas/vapor	Any metal or nonmetal that can withstand deposition temperature and chemical attack by plating vapors	0.1–1	10–10^5 (0.1–760)	150–2200	0.5–1000	0.5–25	Excellent
Plasma-enhanced chemical vapor deposition (PECVD)	Essentially any, e.g., Si, Ni, C, diamond, TiO_2, TiC, B_4C, SiC, TiN, BN, Si_3N_4, polymers	Solid, liquid, gas/vapor	Any nongassing material that can withstand deposition temperature	10–500	1–500 (10^{-2}–5)	100–500	0.5–10	0.1–2	Good to excellent

*Substrate must be mechanically and chemically cleaned before deposition. The deposition faithfully reproduces the substrate topography, and no posttreatment is required.

†Substrate heated for good adhesion and dense microstructure.

TABLE 9.2 Advantages and Disadvantages of Various Vapor Deposition Techniques

Deposition technique	Advantages	Disadvantages
Direct/reactive evaporation	Simple, cheap, extremely high deposition rates, low pressure, no gas scattering, and incorporation of gas atoms	Low kinetic energy, inadequate adhesion, high substrate temperature for dense coatings, imprecise control of stoichiometry of compound coatings, substrate must generally be rotated for uniform coating
Activated reactive evaporation (ARE)	Extremely high deposition rates, variety of coating compositions, precise control of stoichiometry, better adhesion, denser microstructure than evaporation	High substrate temperature, addition of extra electrode make process slightly complicated as compared to evaporation, substrate must generally be rotated for uniform coating
Glow-discharge ion plating	Simple, excellent adhesion, graded interface, moderately high deposition rate, good coverage of complex shapes	Deposition is complicated due to substrate biasing as compared to sputtering and evaporation and ARE
Ion beam ion plating	Excellent adhesion, very high kinetic energy of ions, fine control of kinetic energy and incident angle of ions possible, low-pressure deposition, no gas atom incorporation	Needs separate ion source, poor throughput, low deposition rates, not suitable for industrial production
Glow-discharge sputtering	Simple, wide choice of coating materials, easy to obtain stoichiometric coatings through optimization of parameters, low substrate temperature, excellent uniformity over complex and large areas, but poor coverage and deep recesses	Relatively low kinetic energy
Ion beam sputtering	Independent control over kinetic energy and current density of ions, directionality control of ion beams, very high-kinetic-energy ion beam leads to excellent adhesion and high density, low deposition pressure, high deposition rates, no gas atom incorporation, no substrate heating, and unique coating properties	Needs separate ion source, cannot be used with very large substrate areas because of small ion bean size (~100-mm diameter)
Chemical vapor deposition	Wide choice of coating materials, extremely high deposition rates, near-theoretical density, controlled grain size, excellent adhesion, relatively uniform coating of complex shapes and over large areas possible	High substrate temperature, low kinetic energy, relatively high cost of starting materials, low utilization of starting materials
Plasma-enhanced chemical vapor deposition	Lower substrate temperatures as compared to CVD, high deposition rates, excellent adhesion due to high kinetic energy, and flexibility of substrate shape and size	More complex as compared to CVD because of inclusion of plasma, strong interaction of plasma with growing coating resulting in somewhat nonuniform deposition, deposition of pure materials very difficult

bulk temperatures. This results in superior adhesion as compared to evaporation and glow-discharge-type processes. The deposition rates of various processes range from 10^{-2} to 75 μm min^{-1}. The thickness of the coatings can vary from a couple of nanometers to a few millimeters. All PVD processes are line-of-sight processes.

The CVD process utilizes a volatile component of coating material which is thermally decomposed (pyrolysis) or chemically reacted with other gases or vapors to form a coating atomistically (atom by atom) on the hot substrate surface. The chemical reactions generally take place in the temperature range of 150 to 2200°C at a pressure of ranging from about 50 Pa (0.5 torr) to atmospheric pressure. CVD is a versatile and flexible technique used in producing a wide range of coatings at high deposition rates. CVD coatings usually exhibit excellent adhesion compared to the coatings deposited by evaporation and glow-discharge-type process, but requirements of high substrate temperature limit their application to substrates that can withstand high processing temperatures (Blocher, 1982; Bryant, 1977; Feist et al., 1969; Hocking et al., 1989; Powell et al., 1966; Vossen and Kern, 1978).

Physical-chemical vapor deposition processes are the hybrid processes which use glow-discharge plasma to activate CVD processes. These are broadly referred to as *plasma-enhanced chemical vapor deposition* (PECVD) or *plasma-assisted CVD* (PACVD) processes. Many of the phenomena of PECVD are similar to those of conventional high-temperature CVD. Likewise, much of the physical description of glow-discharge plasma for sputtering applies to plasma CVD as well. In PECVD processes, the coating can be applied at significantly lower substrate temperatures because of the ability of high-energy electrons produced by glow-discharge plasma to break chemical bonds and thereby promote chemical reactions. Practically any gas or vapor including polymers can be used as a precursor material (Bonifield, 1982; Hollahan and Rosler, 1978; Ojha, 1982; Věprek, 1985). Another method, known as the *reactive pulsed plasma* (RPP) deposition process, uses a high-energy pulsed plasma rather than continuous discharge and restricts the substrate to room temperature (Michalski, 1984; Michalski et al., 1985; Sokolowski et al., 1981). Chemical reactions in deposition of polymer coatings (called *chemical vapor polymerization*) can be activated by ultraviolet light irradiation and electron beam methods in addition to glow-discharge plasma (Holland, 1956; Shen, 1976; Yasuda, 1978). The application of vapor deposition processes ranges from decorative to utilitarian over significant segments of the engineering, chemical, nuclear, microelectronics, and related industries.

PVD and CVD processes have several advantages and disadvantages relative to other competitive processes.

Advantages

- There is great versatility in the composition of the coatings. Coatings can be produced with high purity, unusual microstructures, and new crystallographic modifications.
- Rather thin coatings (as low as a couple of nanometers thick) can be produced with extremely high adhesion or self-supported shapes at high deposition rates.
- Most nongassing substrate materials which can withstand deposition temperatures can be coated.

- Substrate temperature can be varied within very wide limits from subzero to high temperatures.
- Substrate surface topography is reproduced, and generally no postfinishing is required.

Disadvantages

- High capital cost and high processing costs are associated with vacuum systems.
- With certain exceptions, it is more difficult to deposit polymers.

A brief description of typical vacuum systems used for vapor deposition processes is presented in Appendix 9.B. Details of various vapor deposition processes follow.

9.1 EVAPORATION

Thermal evaporation is one of the oldest and most commonly used vacuum deposition techniques. In the evaporation process, coating material is generated in the vapor phase by heating the source material to a temperature (typically 1000 to 2000°C) in vacuum (10^{-6} to 1 Pa or 10^{-8} to 10^{-2} torr) such that the vapor pressure (>1 Pa or 10^{-2} torr) significantly exceeds the ambient chamber pressure and produces sufficient vapor for condensation. The coating material is in an electrically neutral state and is expelled from the surface of the source at thermal energies typically from 0.1 to 0.3 eV. The source-to-substrate distance ranges from 150 to 450 mm. Most materials (metals, alloys, intermetallic compounds, and refractory compounds) can be coated on almost any substrate material (including polymers) by using a suitable technique. The substrate may be floating, grounded or electrically biased. The substrate is normally preheated to elevated temperatures (200 to 1600°C) for dense equiaxed grain morphology of metal and refractory compound coatings. Coatings of many refractory compounds, deposited at higher substrate temperatures, exhibit higher adhesion, higher hardness, and higher wear resistance (Bunshah, 1976, 1982; Bunshah et al., 1978; Chopra, 1979; Holland, 1956; Maissel and Glang, 1970; Movchan and Demchishin, 1969a, 1969b; Paton and Movchan, 1978; Rigney, 1982; Schiller et al., 1982).

Most pure metals, many alloys, and compounds that do not undergo dissociation can be directly evaporated in vacuum. When many multicomponent alloys or compounds are thermally evaporated, the components may evaporate at different rates because of their different vapor pressures. Very few compounds evaporate without dissociation (Hoffman and Liebowitz, 1971, 1972). In the more general sense, when a compound is evaporated or sputtered, the material is not transformed to the vapor state as compound state but as fragments thereof. Subsequently, the fragments have to recombine, most probably on the substrate, to reconstitute the compound. Satisfactory methods of preparing alloys and compounds with proper stoichiometric coatings include reactive evaporation, multiple-source evaporation, and flash evaporation. The stoichiometry of the coating depends on several factors, including the deposition rate and the ratios of the various molecular fragments, the impingement of other gases present in the environment, and the reaction rate of the fragments on the substrate to reconstitute the compound. A plasma is sometimes included in the reactive evaporation to enhance the reaction between the reactants and to cause the generation of ions and energetic neutrals. This process is known as *activated reactive evaporation* (ARE). The ARE process is used to produce tribological coatings of a wide variety of materials with increased adhesion, superior mechanical properties, and high deposition rates. The concurrent ion impingement by an ion beam of inert or reactive gases is also used to improve the adhesion and mechanical properties of the evaporated coatings (Bunshah et al., 1982).

The source material is normally in the form of powder, wire, or rod. It is supported by various sources such as wire and metal foil sources and crucibles. The source material is heated by conduction, such as resistance, radiation, radiofrequency (rf) induction, electric arc, electron beams, and laser sources. Deposition by resistive heating is used to produce coatings of high-vapor-pressure materials or to produce thin coatings. Other conduction, radiation, and induction heating methods are used to deposit thicker coatings and coatings of intermediate-vapor-pressure materials. These are very simple and cheap processes. However, simple resistive heating suffers from the disadvantages of possible contamination from the support material and of the limitation of the input power which is incapable of evaporating high-melting-point (>1500°C) materials. Radiation heating has been used to deposit the materials which vaporize before melting. By using a sublimation source, the problem of finding nonreactive supports can be circumvented. Induction heating is more efficient but requires greater cost and increased space. More refractory substances and high-deposition rates require high-power-density methods such as electron beam and vacuum arc heating. Electron beam evaporation can be used to deposit materials with melting temperatures up to 3500°C. Laser beam heating methods have been used on occasion. In the electron beam or laser heating processes, portions of the evaporant next to the container can be kept at lower temperatures, thus minimizing interactions (contamination). Coatings produced by electron beam or laser beam are the purest. The laser beam heating has practical limitations and has not become popular.

Evaporation rates as high as 25 $\mu m\ min^{-1}$ are used, and rates up to 75 $\mu m\ min^{-1}$ have been achieved. A continuous feed system for the vapor source material is used for high deposition rates of special alloys and compounds. The deposition rate is mainly dependent on the heating temperature of the source and the evaporant area. Coatings up to a thickness of 1 mm can be deposited on self-supported structures. Typical coating thicknesses range from 0.1 to 100 μm. Most substrate materials can be coated. The primary requirement is that the material be stable at processing temperatures in vacuum. It must not evolve gas or vapor when exposed to the coating vapor. Substrates are normally preheated to elevated temperatures for dense, equiaxed grain morphology of metal and refractory compound coatings as well as for increased adhesion.

The major advantages of evaporation are that it is simpler and cheaper compared to other vacuum deposition processes and has high deposition rates. The evaporation process usually results in low coating adhesion because of the low kinetic energy of the vapor species; thus the evaporation is somewhat less desirable for tribological applications, compared to other vacuum deposition processes. However, the simplicity and low cost of the evaporation process and its high deposition rates makes it attractive for many applications. High substrate temperatures can warp thin substrates and modify the heat treatment of metallic substrates. The direct evaporation process is incapable of providing precise control of the stoichiometry of compound coatings. Many of the shortcomings of evaporation have been overcome by introducing the reactive gas during deposition, by ionizing the evaporant atoms, and by biasing the substrate.

The theory of vacuum evaporation, classification of evaporation processes, various vapor sources used to support the evaporant, various heating methods, continuous feed system, and coatings produced by evaporation are discussed in this section.

9.1.1 Theory of Vacuum Evaporation

The evaporant is in the electrically neutral state and is expelled from the surface of the source at thermal energies of 0.1 to 1 eV (typically 0.1 to 0.3 eV). The evaporant material atoms from the source surface undergo an essential collisionless line-of-sight transport in high vacuum prior to condensation on the substrate. The collisions may change the direction of the atoms and reduce their kinetic energy. To minimize the collisions of evaporant material atoms with ambient gas atoms, the source-to-substrate distance should be less than the mean free path of the gas atoms. The mean free path of residual gas molecules such as air molecules at 250°C at pressures of 10^{-2} to 10^{-4} Pa (10^{-4} to 10^{-6} torr) is about 0.45 to 4.5 m, respectively. Thus, pressures less than 10^{-4} torr are necessary to ensure a straight-line path for most of the emitted vapor atoms, for a substrate-to-source distance of about 450 mm in the vacuum evaporator.

The rate of free evaporation of vapor atoms from a clean surface per unit area in vacuum, m_e, is given by Langmuir's equation (Langmuir, 1913)

$$m_e = 5.85 \times 10^{-2} P\left(\frac{M}{T}\right)^{1/2} \quad \text{g cm}^{-2}\text{ s}^{-1} \tag{9.1}$$

where P = vapor pressure of evaporant under saturated vapor conditions at temperature T, torr or mm Hg

M = molecular weight of evaporant material

T = absolute temperature, K

Alternatively, we may write the evaporation rate as

$$N_e = 3.51 \times 10^{22} P\left(\frac{1}{MT}\right)^{1/2} \quad \text{molecules cm}^{-2}\text{ s}^{-1} \tag{9.2}$$

where N_e is the rate of evaporation of vapor molecules per unit area in vacuum.

Figure 9.3(*a*) shows the profile of vapor material condensed at different angles. The directionality of the evaporant molecules leaving the source is given by the well-known cosine law, as illustrated in Figure 9.3(*b*) for a small source surface area dA_1 emitting a vapor flux m. The material passing through a solid angle ψ at a direction of angle ϕ with the normal to the surface dA_1 per unit time is given by

$$dm = \frac{m}{\pi} \cos \phi \, d\psi \tag{9.3}$$

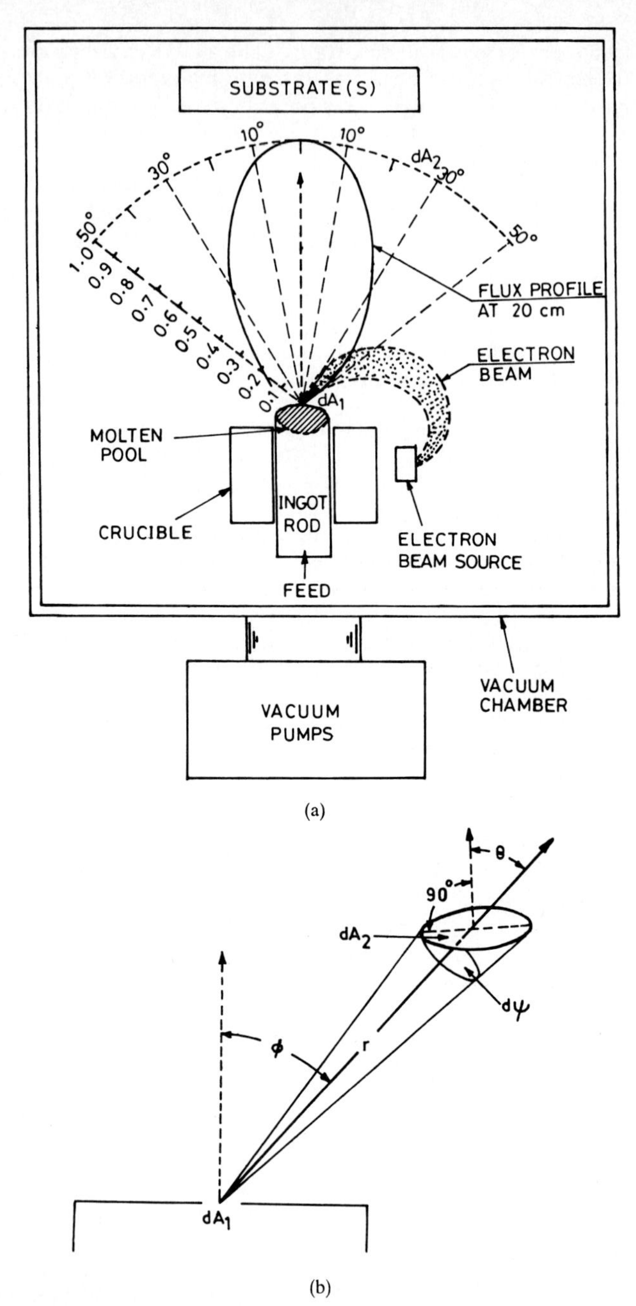

FIGURE 9.3 (*a*) Schematic of the direct evaporation process using electron beam heating. Directional dependence of coating thickness is shown. (*b*) Schematic of a surface element receiving coating from a small-area source.

Material arriving at a small substrate area dA_2 inclined at an angle θ to the vapor stream direction at a distance r from the source is given by

$$dm = \frac{m}{\pi r^2} \cos \phi \cos \theta \, dA_2 \tag{9.4a}$$

where m is the total mass evaporated. For a point source, Equation (9.4*a*) reduces to

$$\frac{dm}{dA_2} = \frac{m}{\pi r^2} \cos \phi \tag{9.4b}$$

For a uniform coating thickness, the point source must be located at the center of the spherical receiving surface such that r is a constant and $\cos \theta = 1$.

You may notice from Fig. 9.3 that the maximum amount of evaporant is condensed directly above the centerline of the source and decreases away from it (Graper, 1971). This problem is overcome by imparting a complex motion, e.g., planetary or rotary, or by introducing a gas at a pressure of about 15 to 200 mtorr into the chamber, so that the vapor stream undergoes multiple collisions during transport from the source to substrate, thus producing a coating on the substrate of reasonably uniform (±10 percent) thickness. The latter process is called *gas scattering* or *pressure plating* (Beale et al., 1973; Kennedy et al., 1971). The highest-quality coatings are made on surfaces nearly normal to the vapor flux. Coatings made on substrates at low angles to the vapor source result in fibrous, poorly bonded structures. Coatings resulting from excessive gas scattering are poorly adherent and amorphous.

9.1.2 Details of Evaporation Processes

9.1.2.1 Direct Evaporation. In the direct evaporation process, the metals, alloys, or refractory compounds are vaporized and deposited in the absence of any plasma intentionally generated. The process is carried out at pressures of less than 0.1 Pa (10^{-3} torr) and usually in vacuum levels of 10^{-2} to 10^{-4} Pa (10^{-4} to 10^{-6} torr). The substrate may be floating, grounded, or electrically biased. The substrate temperature ranges from about 200°C or less to as high as 1600°C (Bunshah et al., 1977b; Holland, 1970). Figure 9.3(*a*) is a schematic of a vacuum evaporation system illustrating electron beam heating. Figure 9.4 shows the details of an evaporation system using resistive-type heating filaments. Direct evaporation is commonly used to deposit pure metals, some alloys, and some compounds such as Al, Ag, Au, Ni, Cr, Ti, Mo, W, stainless steel, Ni-Cr, Pb-Sn, M-Cr-Al-Y, Al_2O_3, TiC, and TiB_2 (for typical melting conditions of various materials, see Table 9.3).

9.1.2.2 Reactive Evaporation. The process of direct evaporation of many compounds (such as oxides, carbides, and nitrides) results in a coating having a different composition from the source because of partial dissociation of the compound during the evaporation. For example, direct evaporation of Al_2O_3 results in an oxygen-deficient coating which can only be restored to stoichiometry by evaporation in the presence of a partial pressure of oxygen or possibly a postdeposition anneal in oxygen. TiC coatings prepared by direct evaporation are

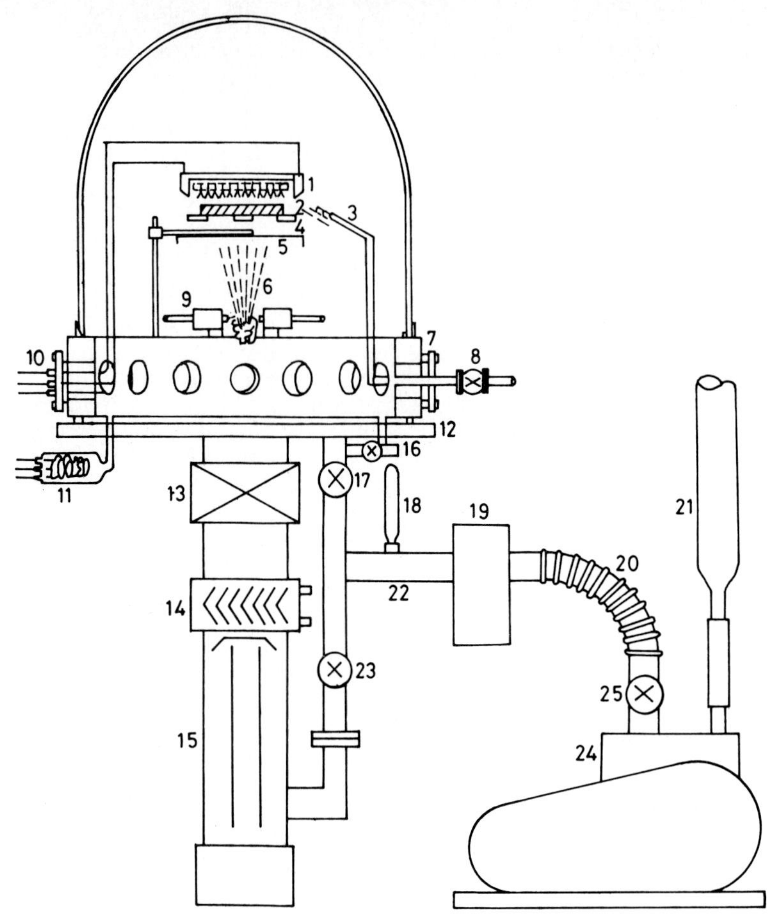

FIGURE 9.4 Overall schematic block diagram of a typical direct evaporation system using resistive heating of filaments and oil diffusion pump backed by mechanical rotary pump. 1 = Heater, 2 = substrate, 3 = quartz crystal for monitoring thickness, 4 = substrate mask, 5 = shutter, 6 = evaporant vapors, 7 = baseplate collar for feed-through, 8 = air admittance valve, 9 = electrodes with filament, 10 = multiple feed-through, 11 = Penning pressure gauge, 12 = baseplate, 13 = baffle valve, 14 = liquid-nitrogen trap, 15 = oil diffusion pump, 16 = Pirani pressure gauge, 17 = roughing valve, 18 = Pirani pressure gauge, 19 = fore line trap, 20 = rubber line, 21 = exhaust pipe, 22 = fore line, 23 = fore-line (backing) valve, 24 = mechanical rotary pump, 25 = fore line isolation valve.

found to be carbon-deficient, but this can be corrected by evaporation in the presence of a partial pressure of acetylene.

In reactive evaporation, metal atoms are evaporated from an evaporant source with a partial pressure of a reactive gas present in the chamber (Auwarter, 1960; Brinsmaid et al., 1957). The reaction between the metal atoms and the gas atoms takes place on the substrate in the vapor phase or on the evaporant source sur-

TABLE 9.3 Direct Evaporation of Various Materials

Material	Melting temperature, °C	Approximate temperature at which vapor pressure = 1.3 Pa (10^{-2} torr), °C
Al	660	1220
Ag	961	1030
Au	1063	1400
Fe	1535	1450
Ni	1450	1530
Cr	1900	1400
Ti	1670	1750
Mo	2610	2530
Ta	3000	3060
W	3380	3230
Ni-20Cr	1400	1620
ZrO_2	2700	1730
Al_2O_3	2030	1800
SiO_2	1730	1550
TiC	3065	1340
ZrC	3480	1700
TiB_2	2880	1550

face to form a compound (Hoffman and Liebowitz, 1972; Raghuram and Bunshah, 1972), e.g.,

$$2\text{Al(vapor)} + \frac{3}{2}\text{O}_2\text{(gas)} \rightleftharpoons \text{Al}_2\text{O}_3\text{(solid deposit)} \tag{9.5}$$

$$\Delta F = -429 \text{ kcal mol}^{-1} \text{ at 298 K}$$

or

$$\text{Ti (vapor)} + \frac{1}{2}\text{C}_2\text{H}_2\text{(gas)} \rightleftharpoons \text{TiC(solid deposit)} + \frac{1}{2}\text{H}_2\text{(gas)} \tag{9.6}$$

$$\Delta F = -76 \text{ kcal mol}^{-1} \text{ at 298 K}$$

The free energy of formation ΔF is more favorable for reaction (9.5) than for reaction (9.6). The rate and completeness of the reactions are dependent on the temperature of the substrate and the amount of reactive gas introduced. For high-melting-point compounds, such as Al_2O_3 and TiC, a high-power-density source has to be used to obtain appreciable evaporation rates, which result in a physical disintegration of many compound evaporant billets (prepared by pressing and sintering of powders), therefore, reactive evaporation is preferred to direct evaporation for compounds with high melting temperatures. Stoichiometric Al_2O_3, Cr_2O_3, Fe_2O_3, SiO_2, TiO_2, Ta_2O_3, ZrO_2, and Y_2O_3-ZrO_2 alloys have been formed by reactive oxidation evaporation. TiN and ZrN have been produced by using a nitrogen reactive gas.

9.1.2.3 Activated Reactive Evaporation. In activated reactive evaporation, the evaporation of a metal occurs in the presence of a reactive gas (with a partial pressure of about 10^{-2} to 1 Pa or 10^{-4} to 10^{-2} torr) and a plasma to deposit compounds with increased adhesion and increased deposition rates. The plasma is used to enhance the reaction between the reactants and/or to cause ionization of both the coating metal and gas atoms in the vapor phase. The plasma is generated by an interspace positively biased electrode when an electron beam evaporation source is used and by low-energy electrons injected into the vapor phase from a thermionically heated filament when a resistance-heated source is used (Bunshah and Raghuram, 1972; Bunshah et al., 1978; Nath and Bunshah, 1980).

As illustrated in Fig. 9.5, the metal is heated and melted by a high-acceleration-voltage (typically 5- to 20-kV) electron beam. The melt has a thin plasma sheath on top of it. The low-energy secondary electrons from the plasma sheath are pulled up into the reaction zone by an interspace electrode placed above the pool, biased to allow positive dc or ac potential (typically 20 to 100 V). The low-energy electrons have a high ionization cross section, thus ionizing or activating the metal and gas atoms. In addition, collisions between ions and neutral atoms result in charge exchange processes yielding energetic neutrals and positive ions. It is believed that these energetic neutrals, condensing on the substrate along with the ions and other neutral atoms, "activate" the reaction, thus increasing the reaction between the reactants.

The deposition of TiC with a carbon-to-metal ratio approaching unity has been achieved by this process for the reaction of titanium metal with C_2H_2 gas molecules (Bunshah, 1974a, 1974b; Raghuram and Bunshah, 1972). The hard refractory compounds such as Al_2O_3, SiO_2, W_2C, TiC, ZrC, HfC, VC, NbC, TiB_2, TiC-TiB_2, and TiC-TiB_2-Co have been deposited at substrate temperatures ranging from 500 to 1700°C (Bunshah and Raghuram, 1972; Bunshah et al., 1977b, 1978; Paton and Movchan, 1978).

A variation of the ARE process using a resistance-heated source instead of the electron-beam-heated source has been developed by Nath and Bunshah (1980) and is particularly useful for evaporation of low-melting-point metals (such as

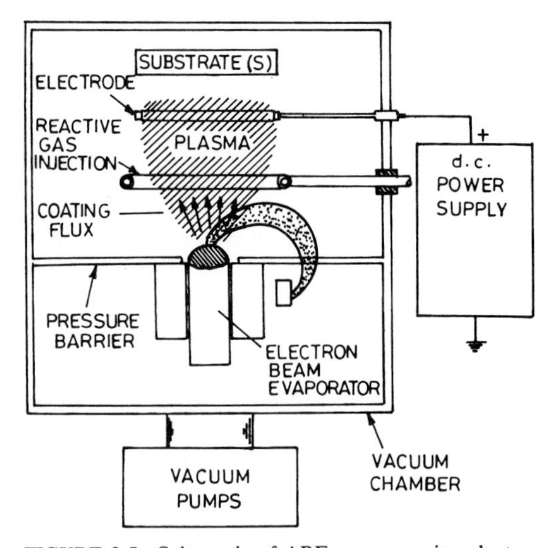

FIGURE 9.5 Schematic of ARE process using electron beam gun evaporation source. [*Adapted from Bunshah and Raghuram (1972).*]

indium and tin) where electron beam heating can cause splattering of the molten pool. In the experimental apparatus shown in Fig. 9.6, the metal vapors generated from the resistance-heated boat react with the reactive gas introduced to the chamber, the reaction being enhanced by a plasma generated by injecting low-energy electrons from a thermionic electron emitter (e.g., tungsten filament) toward a low-voltage anode assembly. A transverse magnetic field is applied to cause the electrons to travel in a spiral path, thus enhancing the ionization efficiency. This process has been used to deposit In_2O_3 and Sn-doped In_2O_3 transparent coatings for electronic applications.

The coating characteristics, such as structure, adhesion, and microhardness, have been improved by modifying the ARE process. The ARE process has substantial versatility since the substrate can be grounded, positively or negatively biased, or allowed to float electrically. Several investigators have modified the above-described ARE process, illustrated schematically in Fig. 9.7.

Low-Pressure Plasma Deposition Process. This is a modified ARE process, in which the substrate is positively biased (dc or ac potential) rather than using a positively biased interspace electrode (Nakamura et al., 1977). A wider deposition area can be obtained than by the ARE process. However, this version has a disadvantage, relative to the basic ARE process, of being unable to choose to ground the substrate, let it float, or bias it negatively (BARE process).

Enhanced ARE Process. Yoshihara and Mori (1979) have modified the ARE process by using electron beam heating with the addition of a thermionic electron emitter (e.g., a tungsten filament) for the deposition of refractory compounds at lower deposition rates than that obtained in the basic ARE process. The low-energy electrons from the emitter sustain the discharge, which would otherwise be extinguished since the primary electron beam (used to melt the metal) is so weak that it does not generate an adequate plasma sheath above the molten pool

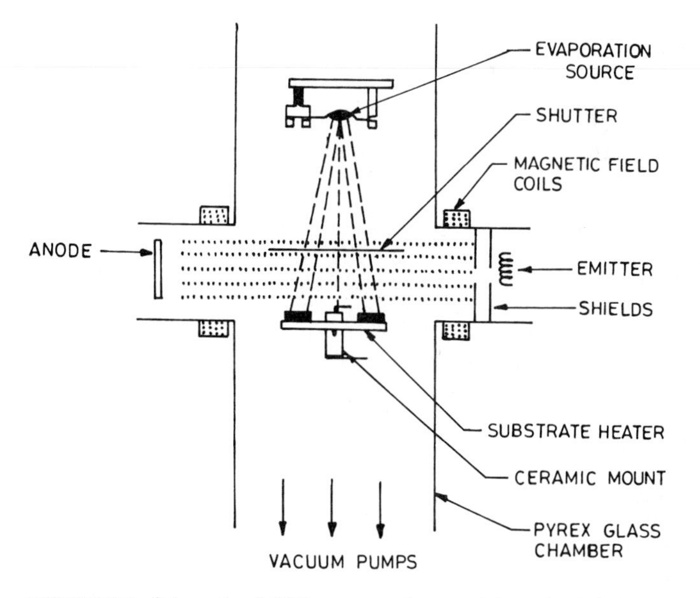

FIGURE 9.6 Schematic of ARE process using a resistance-heated evaporation source. [*Adapted from Nath and Bunshah (1980).*]

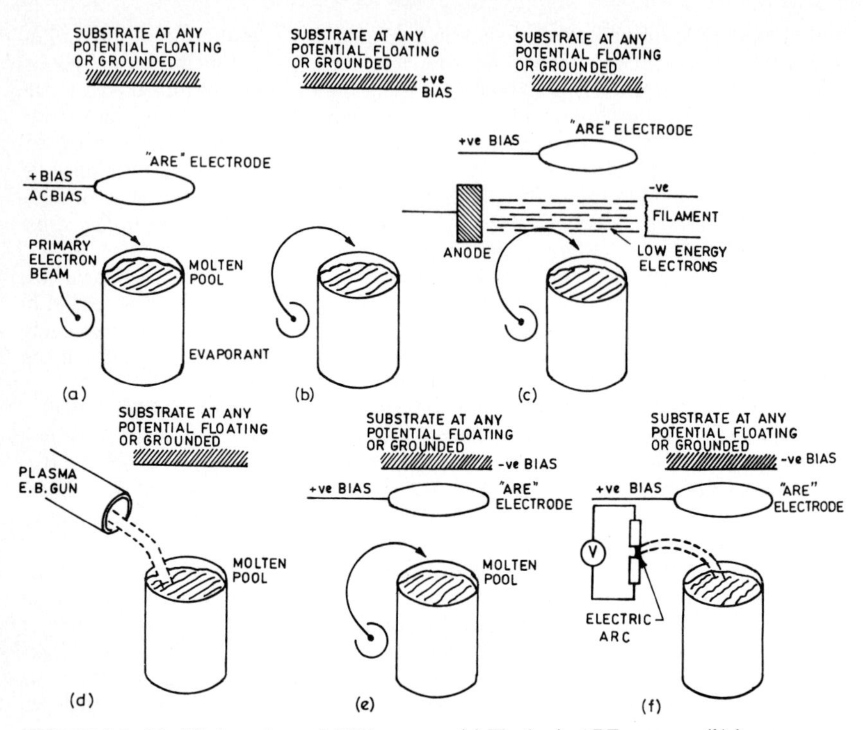

FIGURE 9.7 Modified versions of ARE process: (*a*) The basic ARE process, (*b*) low-pressure plasma deposition process, (*c*) enhanced ARE process, (*d*) ARE process with hot hollow cathode or cold cathode discharge electron beam gun, (*e*) biased ARE (BARE) process or reactive ion plating, (*f*) ARE process with an electric arc.

from which the low-energy secondary electrons can be extracted by a positively charged interspace electrode. The substrate may be biased, grounded, or floating.

Process Using Plasma Electron Beam Guns. The plasma electron beam gun can be used instead of the thermionic electron beam gun to carry out the ARE process. Komiya and Tsuruoka (1976) and Komiya et al. (1977a, 1978) have used a hot hollow cathode discharge gun (HCD gun) to deposit hard TiC, Cr-C, and Cr-N coatings. The HCD process is preferred to the cold cathode discharge gun because it requires only a low voltage to operate the HCD gun or to apply the substrate bias. Zega et al. (1977) have used a cold cathode discharge electron beam gun to deposit TiN coatings. In comparison with a hot cathode gun, the cold cathode gun is easy to make and needs only a high-voltage source, no filament transformer, and no differential pumping.

Biased ARE Process. In this version of the ARE process, the substrate is negatively biased to attract positive ions in the plasma (Beale et al., 1973). This process is also called *glow-discharge diode ion plating in the reactive mode* (or *reactive ion plating* or RIP), to be discussed later (Kobayashi and Doi, 1978).

ARE Process with an Arc Evaporation Source. Dorodnov (1978) and Snaper (1974) have applied nitride and carbide films by the evaporation of metals with a low-voltage arc in the presence of a plasma and a negatively biased substrate with nitrogen and hydrocarbon reactive gases. Since the arc naturally operates at low voltages, ionization processes are very efficient and a reaction plasma is very easily created.

9.1.2.4 Ion-Beam-Assisted Evaporation. This process combines the advantages of ion bombardment with the versatility and high rates of resistively heated or electron beam evaporation. The incorporation of concurrent impingement by an ion beam of inert or reactive gas improves the adhesion and mechanical properties of evaporated coatings dramatically. Figure 9.8 shows schematically an ion-beam-assisted evaporation. The details of ion sources and generation of ion beams are discussed in Section 9.2.

In this ion-beam-assisted growth of evaporated coating, each monolayer, or several-monolayer, region is subsequently modified by ion bombardment, followed by deposition of additional material. This process helps to achieve good adhesion of coatings and to tailor the composition of the compound coatings by using ion beams of reactive gases. For instance, stoichiometric coatings of SiO_2 (Dudonis and Pranevičius, 1976) and NbN (Cuomo et al., 1982) have been obtained by evaporating Si and Nb in the presence of oxygen and nitrogen ion beams, respectively. The ion bombardment is also found to alter the type of in-

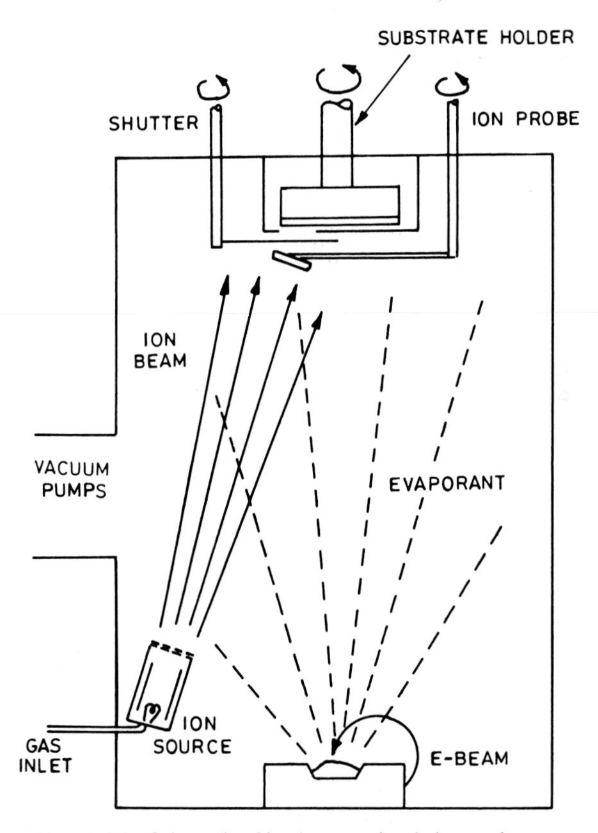

FIGURE 9.8 Schematic of ion-beam-assisted electron beam evaporation. [*Adapted from Cuomo et al. (1982).*]

ternal stress (say from tensile to compressive) of evaporated coatings. An ion dose of 11.5 keV Xe^+ (which corresponds to only 0.4 percent of the ion flux) was enough to change the intrinsic stress of Cr coatings from tensile to compressive (Hoffman and Gaerttner, 1980).

9.1.3 Vapor Sources to Support the Evaporant

The evaporation of materials in a vacuum system requires a vapor source or a container to support the evaporant and to supply the heat of vaporization while maintaining the charge at a temperature sufficiently high to produce the desired vapor pressure. To avoid contamination of the deposit, the support material itself must have negligible vapor and dissociation pressures at the operating temperature. Suitable materials are refractory metals and oxides. Further selection within these categories is made by considering the possibilities of alloying and chemical reactions between the support and evaporant materials. Additional factors influencing the choice of support materials are their availability in the desired shape (wire, sheet, crucible) and their amenability to different modes of heating. Comprehensive tables of some common evaporant materials along with containment materials commonly used are presented by Maissel and Glang (1970) and Rigney (1982).

The simplest vapor sources are resistance-heated wires and metal foils of various types (Fig. 9.9). These are commercially available in various shapes and sizes at low prices. Materials of construction are refractory metals (such as W, Mo, Ta, and Nb) which have high melting points and low vapor pressures. Pt, Fe,

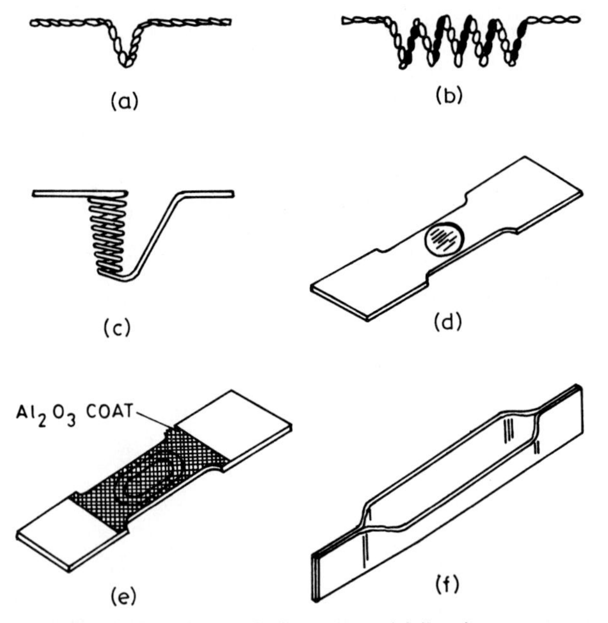

FIGURE 9.9 Schematics of wire and metal foil resistance sources for evaporation: (*a*) Hairpin wire, (*b*) wire helix, (*c*) wire basket, (*d*) dimpled foil, (*e*) alumina-coated dimpled foil, (*f*) canoe foil.

or Ni is sometimes employed for materials which evaporate below 1000°C. These sources serve the dual function of heating and holding the evaporant material. Sources—hairpin wire, wire helix, and wire basket shown in Fig. 9.9(*a*) to (*c*)—are used for materials which wet the source material on melting. The material used for the filament is determined by the wetting characteristics of the evaporant material and the evaporation temperature. Metal foils, shown in Fig. 9.9(*d*) to (*f*), have capacities of a few grams and are the most universal sources for small evaporant quantities. The dimpled sources have reduced width in the center to concentrate the heating at the center. Canoe or boat sources can be made in the laboratory by bending metal sheets into the desired shape. Wetting of the metal-foil surface by the molten evaporant is desirable for good thermal contact. However, molten metals may also lower the electrical resistance of the foil in the contact area, thereby causing the temperature to drop; this can be avoided by applying plasma sprayed coating of oxides (such as alumina) on metallic foil [Fig. 9.9(*e*)]. The power requirements for these sources are modest, usually between 1 and 3 kW. Nearly all elements except for many refractories (for example, Ag, Ni, Pd, Pt, Rh, Zn) can be evaporated in small quantities by using these sources.

Crucibles are the sources to support molten metals in quantities of few grams or more. Since the melt is in contact with the container for prolonged periods, the selection of thermally stable and noncontaminating support materials must be made very carefully. Crucible materials are usually refractory metals, oxides, nitrides, and carbon. Of the three refractory metals mentioned earlier, molybdenum is the least expensive and can be machined most easily. Mo crucibles are used for evaporation of Cu, Ag, and Au, for example. Refractory oxide crucibles are available in certain shapes and sizes. Refractory oxides which have been used for construction of crucibles include ThO_2, ZrO_2, BeO, MgO, Al_2O_3, TiO_2, and SiO_2. They can be used for operating temperatures ranging from 1600 to 2500°C (Maissel and Glang, 1970). Because of chemical interactions, Al and refractory metals W, Mo, and Ta cannot be evaporated from oxide supports. Examples of materials which are evaporated from oxide supports are As, Sb, Bi, Te, Co, Fe, Pd, Pt, and Rh. Boron nitride has excellent thermal stability (can be used up to 1600°C at 1 Pa or 10^{-2} torr), has good thermal conductivity, is relatively soft, and can be machined with ordinary cutting tools. Titanium diboride has a thermal conductivity similar to BN but is harder and melts at 2940°C. A composition consisting of 50 percent BN and 50 percent TiB_2 is a well-established crucible material. It is most commonly used for evaporation of aluminum. Another material frequently used for evaporation of metals is carbon.

The problem of nonreactive supports may be circumvented for materials that vaporize before melting. A sublimation source is presented in the next section. In electron beam heating, the evaporant is contained in a water-cooled copper hearth, thus eliminating the problem of crucible contamination.

9.1.4 Heating Methods

Thermal evaporation is achieved by a variety of physical methods.

9.1.4.1 Resistive Heating. This method consists of directly or indirectly heating the source material. The direct resistance heating method uses material with a resistively heated filament or boat. The indirect method includes the heating of crucibles.

The simplest sources are resistance wires and metal foils of various types, shown earlier in Fig. 9.9. The sources can be used with a wire, powder, or slug feed system for producing thick coatings on large-scale production. These vapor sources are cheap and are widely used for both experimental and production coatings. However, simple resistive heating suffers from the disadvantage of possible contamination from the support material, and the limitation of the input power makes it difficult to evaporate high-melting-point materials.

High-volume production of evaporated coatings of refractive metals and compounds is achieved by heating crucible sources which can contain a few grams of evaporant material. Heating is accomplished from a refractory heating element [Fig. 9.10(*a*) and (*b*)] by a combination of radiation and conduction.

9.1.4.2 Radiation Heating. If a material has a sufficiently high vapor pressure before melting occurs, it will sublime, and the condensed vapors form a coating. Several elements such as Cr, Mo, Pd, V, Fe, and Si can be evaporated directly from the solid phase. The evaporant material can be radiation-heated. A chro-

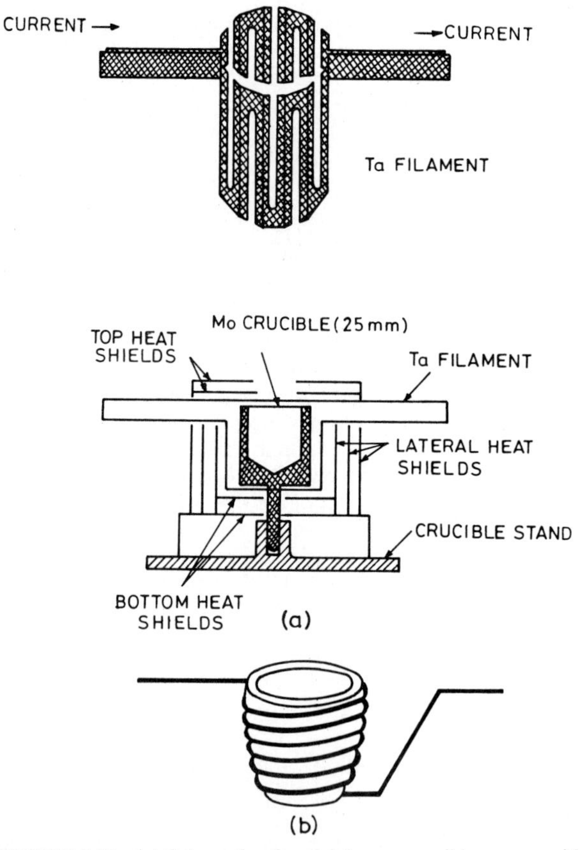

FIGURE 9.10 (*a*) Schematic of molybdenum crucible source with meandering tantalum sheet filament. (*b*) Schematic of oxide crucible with wire coil as heating element.

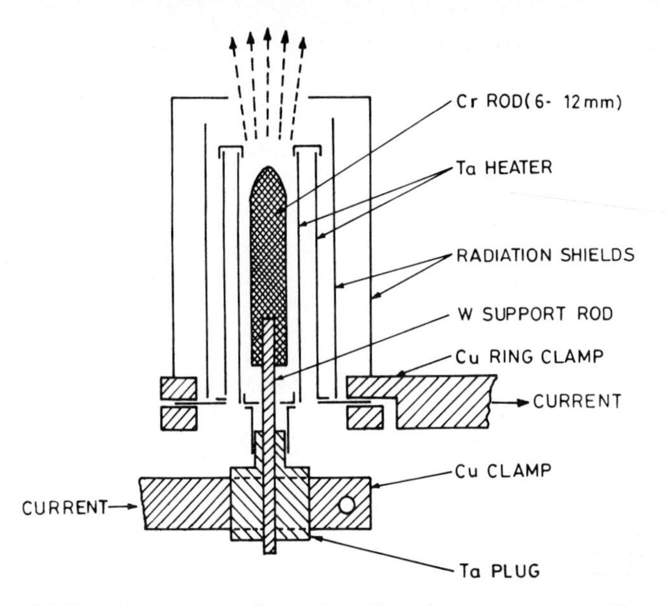

FIGURE 9.11 Schematic of chromium sublimation source evaporated by radiation. [*Adapted from Roberts and Via (1967).*]

mium sublimation source developed by Roberts and Via (1967) is shown in Fig. 9.11. A chromium rod is supported within a dual-wall cylinder which is made of tantalum sheet. The electric current flows through the tantalum cylinder, and sublimation occurs from the entire surface area of the rod, which is uniformly heated by radiation. The nonuniformity and variability of temperature experienced with a basket-type source [Fig. 9.9(*c*)] are largely eliminated in the sublimation source. These sublimation sources are successfully used to deposit thermally stable compounds available in the form of powder, loosely sintered chunks, etc.

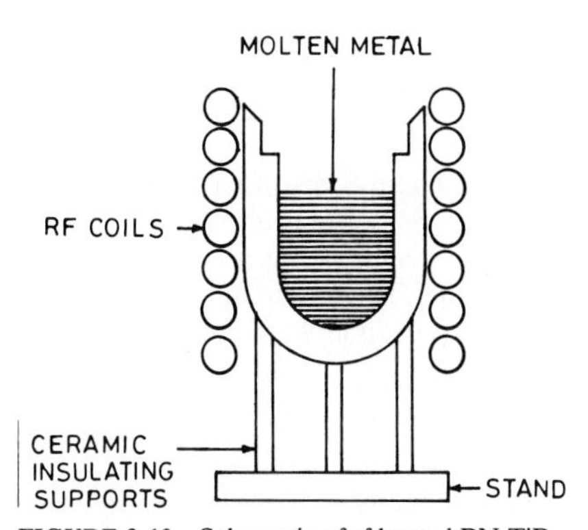

FIGURE 9.12 Schematic of rf heated BN-TiB_2 crucible. [*Adapted from Ames et al. (1966).*]

9.1.4.3 RF Induction Heating. The radio-frequency (rf) induction heating may be supplied to the source material in a crucible. The rf heating is capable of evaporating materials having melting points up to 2100°C. Pyrolytic boron nitride and sintered BN-TiB_2 crucibles have been extensively used to evaporate Al via rf heating (Fig. 9.12) (Ames et al., 1966).

RF induction heating has some advantages over conduction and radiation heating. Since at least part of the energy is coupled directly into the evaporant material, it is not necessary to produce filament or crucible temperatures in excess of the vaporization temperature for the purpose of maintaining a heat flow. By suitable arrangement of the rf coils, the induction-heated material can be levitated, thereby eliminating the possibility of contamination of the coating by the support material. Also the utilization of the energy supplied to the source is more efficient than in filament heating. The disadvantages are the greater cost and space requirements of the rf power supply.

9.1.4.4 Electric Arc Heating. By striking an arc between two electrodes of a conducting material, sufficiently high temperatures can be generated to evaporate refractive metals (such as Ti, Hf, Zr, Ta, and Nb) and refractory compounds. This method employs a standard dc arc welding generator connected to the two electrodes with a capacitor across the electrodes (Fig. 9.13). The arc thus initiated may last a fraction of a second. Deposition rates of approximately 5 nm s^{-1} or higher can be obtained for refractory metals (Kikuchi et al., 1965). This method is widely used for evaporation of carbon for electron microscope applications.

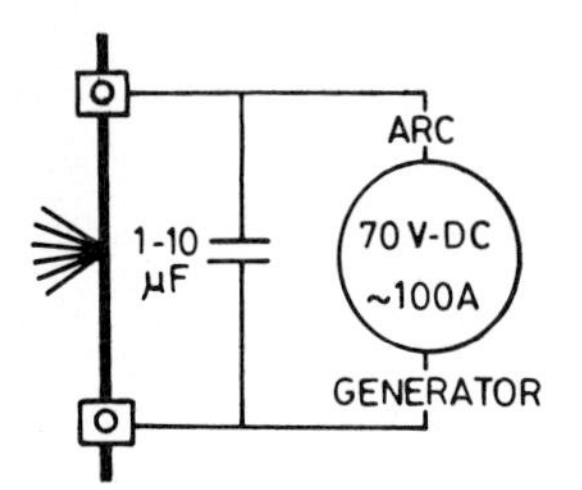

FIGURE 9.13 Schematic of arc evaporation.

9.1.4.5 Laser Heating. The high intensity of a laser is used to heat and vaporize materials having high melting points, by keeping the laser source outside the vacuum system and focusing the beam onto the surface of the source material (Fig. 9.14). Since the laser penetration depth is small (~10 nm), evaporation takes place at the surface only. Schwarz and Tourtellotte (1966) used a glass neodymium laser to deliver 80 to 150 J of energy per burst with a duration of 2 to 4 ms to evaporate $BaTiO_3$, $SrTiO_3$, ZnS, and Sb_2S_3. Deposition of several hundred nm per burst was obtained corresponding to a rate of $\sim 10^5$ nm s^{-1}. The temperature of the vapor source was estimated to be ~20,000 K (corresponding to an energy of 1 to 1.5 eV).

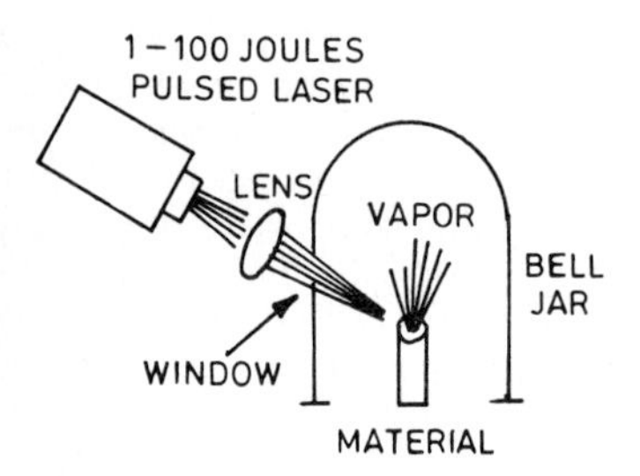

FIGURE 9.14 Schematic of laser evaporation.

There are practical limitations to the extended use of the laser beams (Bunshah, 1982):

- The laser would have to be placed inside the vacuum chamber or the beam transmitted through a window which is transparent for the laser wavelength for laser located outside the vacuum chamber. The former is not a practical probability and the latter suffers from the limited availability of suitable window material and the blockage of the window by an evaporated layer since the laser has to be in sight of the molten pool.
- The laser radiation wavelength must lie in the absorption band of the target material which is always not possible.
- The lasers have very low energy conversion efficiency.

All these problems preclude any serious consideration of the laser as a heat source for evaporation.

9.1.4.6 Electron Beam Heating In *electron beam* (EB) guns, a beam of electron is accelerated through a field of 5 to 20 kV and focused onto the evaporant material, where on impingement most of the kinetic energy of the electrons is converted to heat and temperatures as high as 4000°C can be achieved. Since the energy is imparted by charged particles, it can be concentrated on the evaporated surface while other portions of the evaporant are maintained at lower temperatures. Hence, interactions between evaporant and support materials are greatly reduced.

Electron beam heating sources have two major benefits. First, they have a very high power density, and hence extremely high temperatures and a wide range of control over evaporation rates, from very low to very high. Second, the evaporant is generally contained in a water-cooled copper hearth, thus eliminating the problem of crucible contamination. The coatings from electron beam heating with a water-cooled copper hearth are the purest. The EB guns are most commonly used for evaporation of materials having high melting points and for high deposition rates up to 25 μm min^{-1} (Pierce, 1954; Schiller et al., 1982).

Any EB gun must consist of at least two elements: a cathode and an anode. In addition it is necessary to contain these in a vacuum chamber ($<5 \times 10^{-2}$ Pa or 5×10^{-4} torr) in order to produce and control the flow of electrons, since they are easily scattered by gas molecules. A potential of several kilovolts is needed to accelerate the electrons emitted from the cathode. If the electric field is maintained between the cathode and the evaporant (the "work"), the structure is called *work-accelerated* or, much more usual in modern practice, is *self-accelerated* guns. These have a separate anode (an integral part of the gun) with an aperture through which the electron beam passes toward the work. Electron beam guns can be further subdivided into two types depending on the source of the electrons: thermionic (or hot-cathode) guns and plasma guns. The former are most commonly used for evaporation.

Thermionic (or Hot-Cathode) Guns. In the hot-cathode (or hot-filament) guns, the electrons are generated from a hot filament of a high-temperature alloy such as tungsten or tantalum. The principal self-accelerated hot-cathode guns in use are the linear-focus guns, which use electrostatic or electrostatic and magnetic focusing methods, and the bent-beam guns, which use a magnetic focusing method. The bent-beam EB guns are compact and have a large filament area which allows operation below 10 kV without sacrificing power, as experienced with the linear systems (Maissel and Glang, 1970).

Figure 9.15(*a*) shows two linear-focus guns. In (i), rough focusing is achieved electrostatically by means of a cylindrical shield and an anode disk. If the beam is to traverse longer distances, or if focusing onto a very small spot is desired, better control is obtained with the magnetic lens in (ii). In (ii), the electron beam is focused by a negatively biased filament shield, a conical anode, and a magnetic coil.

In the linear-focus guns, the path of the electron beam is a straight line. Therefore, either the gun or the substrates must be mounted off to the side. This restriction in the arrangement of electron source and substrate can be removed by bending the electron beam through a transverse magnetic field. In the example shown in Fig. 9.15(*b*), the transverse field is provided by an electromagnet which permits focusing during operation. Relatively large-area elongated cathodes are employed to increase the electron emission current.

Plasma Guns. In plasma EB guns, electrons are generated within a plasma. By proper application of electric potential, electrons can be extracted from the plasma to provide a useful energy beam similar to that obtained by hot-cathode

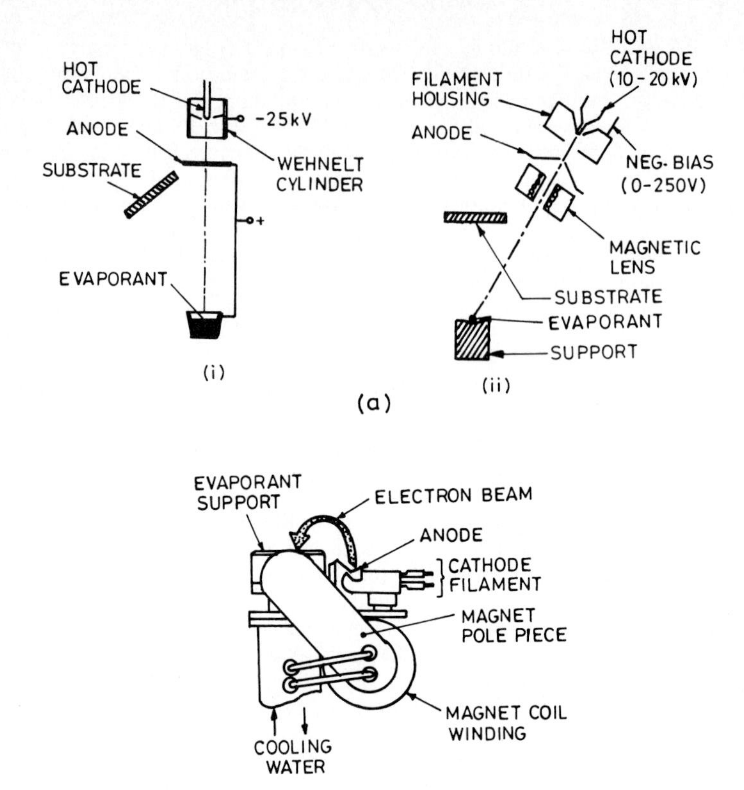

FIGURE 9.15 Schematics of self-accelerated hot-filament electron beam guns. (*a*) Linear-focus gun (i) electrostatically and (ii) electrostatically and magnetically focused. (*b*) Bent-beam electron gun with electromagnetic focusing and water-cooled evaporant support. [*Adapted from Reichelt and Mueller (1961) and Denton and Greene (1963).*]

guns. There are two types of plasma EB guns: cold-cathode guns and hot *hollow cathode discharge* (HCD) guns. The cold-cathode plasma EB gun has a cylindrical cavity made from a metal mesh or sheet (Fig. 9.16) containing the ionized plasma from which electrons are extracted through a small aperture in one end. The cathode is maintained at a negative potential relative to the workpiece and the remainder of the system, which are at ground potential. After evacuation of the system, a low pressure of ionizable gas in the range of 10^{-1} to 1 Pa (10^{-3} to 10^{-1} torr) is introduced. Depending on the high-voltage level, a long path discharge between the cathode and other parts of the system will occur in the gas at a particular pressure. Ionizing collisions in the gas then produce positive ions which are accelerated to the cathode, causing electrons to be released from the cathode surface.

In the HCD gun, the cathode must be constructed of a refractory metal since it operates at elevated temperature (Fig. 9.17). An ionizable gas, usually argon, is introduced to the system through the tube-shaped cathode. A pressure drop

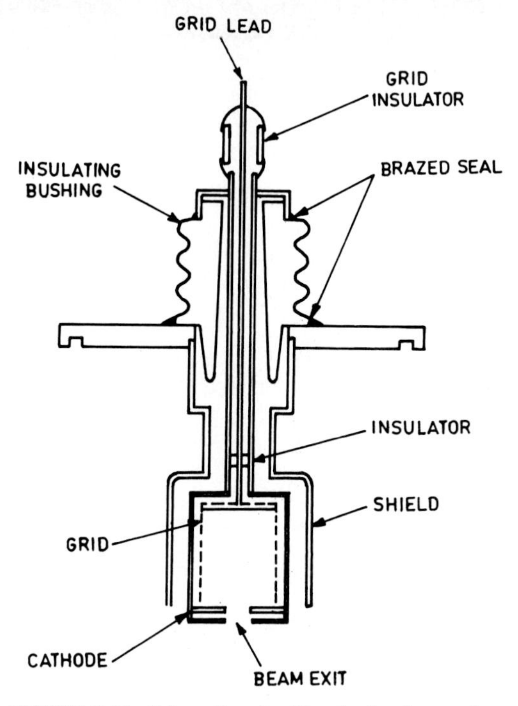

FIGURE 9.16 Schematic of cold-cathode plasma electron beam gun.

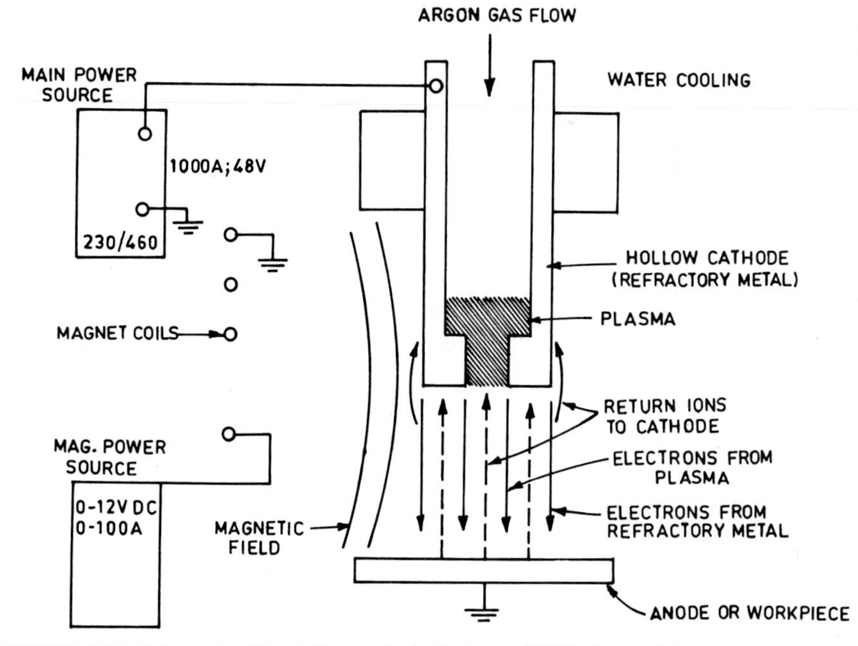

FIGURE 9.17 Schematic of hot hollow cathode discharge (HCD) electron beam gun.

across the orifice in the cathode provides a sufficient amount of gas inside the cathode to sustain the plasma, which generates the beam. A low-voltage, high-amperage dc power supply is used. The gun is biased to a negative potential to initiate a glow discharge (plasma). Continued ion bombardment of the cathode results in heating of the cathode and increased emission. The discharge appears as a low-power-density beam flowing from the cathode aperture and fanning out in conical shape into the chamber. A parallel axial magnetic field is imposed on the beam (Fig. 9.17) which then forms a high-power-density, well-collimated beam. The HCD guns can operate over the pressure range from 10^{-2} to 10 Pa (10^{-4} to 10^{-1} torr) (Bunshah, 1982).

We note that focusing of the beam spot is easier for the thermionic guns. The plasma guns have the advantage of being able to operate at higher pressures, which can be important for gas-scattering evaporation, reactive evaporation, and ion plating.

9.1.5 Evaporation of Multicomponent/Compound Coatings

9.1.5.1 Multiple-Source Evaporation. Two or more evaporant sources are simultaneously used to deposit multilayer and complex alloy coatings that cannot be easily deposited from a single source (Fig. 9.18). The material evaporated from each source can be a metal, alloy, or compound. Heating sources utilized are the same as those used for single-component evaporation. The process is complex because each source is independently controlled to produce the desired vapor flux. It is possible to evaporate each component sequentially, thus producing a multilayer deposit, which is then homogenized by annealing after deposition. Examples of binary alloys and compounds produced by dual-source evaporation are CdS, CdSe, PbSe, PbTe, Bi_2Te_3, AlSi, GeAs, Cr-SiO, Au-SiO, Nb_3Sn, and V_3Si.

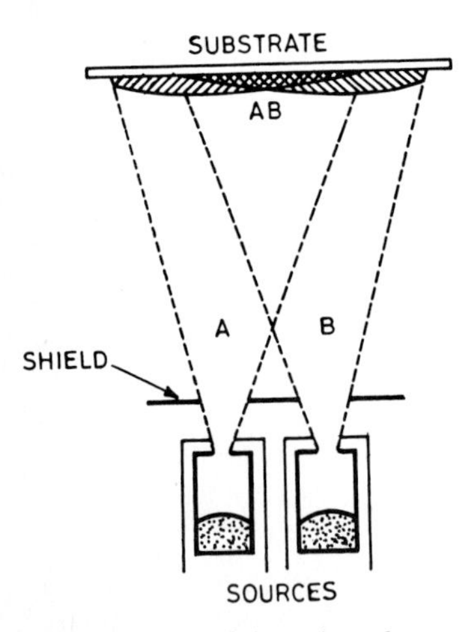

FIGURE 9.18 Schematic of two-source evaporation arrangement yielding variable coating composition. Two sources A and B are heated to temperatures T_1 and T_2.

9.1.5.2 Flash Evaporation. Flash evaporation is used to deposit small quantities of low-melting-point alloys with widely varying vapor pressures. Small quantities of materials are completely evaporated at one time (Bunshah, 1982). Although fractionation occurs, the coating exhibits compositional variations only within a few atomic layers. By providing a slow, constant feed of the evaporant, the inhomogeneity can be further minimized. To obtain maximum homogeneity, the evaporation temperature should be set to attain the vaporization of the least volatile constituent. Both powder and wire feed mechanisms have been used to deliver the alloy to the evaporator. Powder feeders are used most commonly and consist of a hopper to hold the powder from the hopper to a trough that delivers the powder to the heater at a constant rate. A typical feeder is shown in Fig. 9.19.

Intermetallic compounds such as GaAs, PbTe, and InSb have as their constituent el-

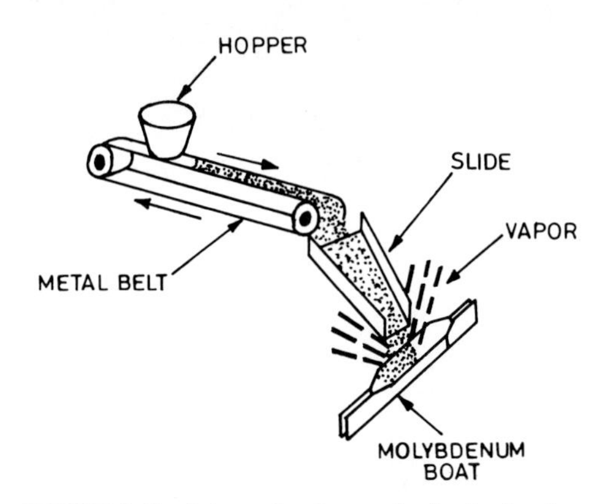

FIGURE 9.19 Schematic of a powder feeder for dispersing alloy powder for flash evaporation.

ements those with low melting points and high vapor pressures. These are deposited by flash evaporation or sputtering (to be discussed later).

9.1.5.3 Continuous Feed. When constant evaporation conditions have to be maintained for long periods, the crucibles of the evaporator must be fed during operation. This is important during operation at extremely high evaporation rates ($\sim 5\ \mu m\ min^{-1}$) or when great quantities have to be evaporated within one coating cycle. The evaporation of multiconstituent alloys calls for a feeding system to ensure a definite and constant vapor stream composition.

Continuous wire, rod, or bulk (powder) feeding replenishes the evaporant material (Fig. 9.20). At a given temperature, once a steady-state equilibrium has been attained, the vapor composition is equivalent to the feed composition. The molten pool in the crucible has a composition inversely proportional to the vapor pressure of each constituent. The more volatile the constituent, the less it is present in the molten pool. To maintain equilibrium, the volume, height, and temperature of the melt must be closely controlled. The method has been used to deposit Ni-20Cr, Ti-6Al-4V, Ag-10Cu, Ag-5Cu, Ni-20Cr-10Al-0.5Y, and a number of other four- and five-component alloys used for oxidation and hot corrosion protection coatings for gas-turbine superalloys (Boone et al., 1980; Smith et al., 1970).

9.1.6 Coatings Deposited by Evaporation

9.1.6.1 Coating Materials. A wide range of metals, alloys, intermetallic compounds, refractory compounds, and mixtures thereof can be applied by evaporation (Bunshah, 1981, 1982, 1983; Maissel and Glang, 1970; Rigney, 1982). These coatings may be broadly classified as decorative and functional. Decorative coatings are widely used in the automotive, home appliance, hardware, jewelry, and novelty fields. Functional coatings have numerous applications such as reflection, antireflection, and beam splitter coatings in optical instruments; current-carrying, dielectric, and semiconductor coatings on electronic components; hot corrosion-resistant coatings on aircraft and missile parts; oxidation-resistant coatings for gas-turbine blades and vanes; and wear-resistant coatings for cutting tools and bearings.

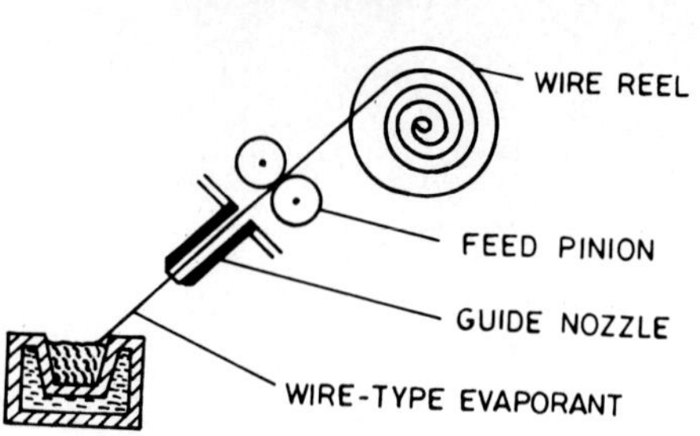

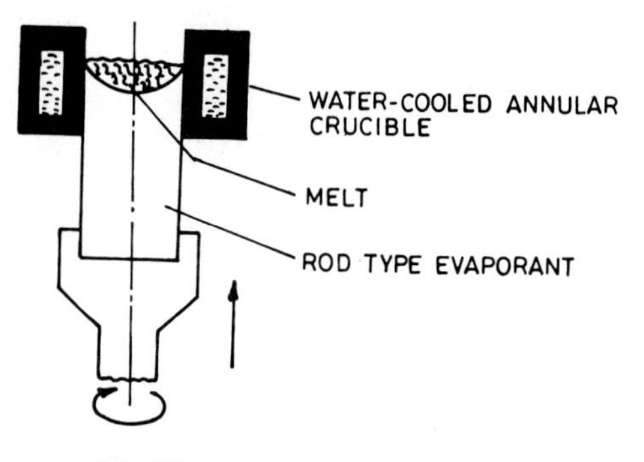

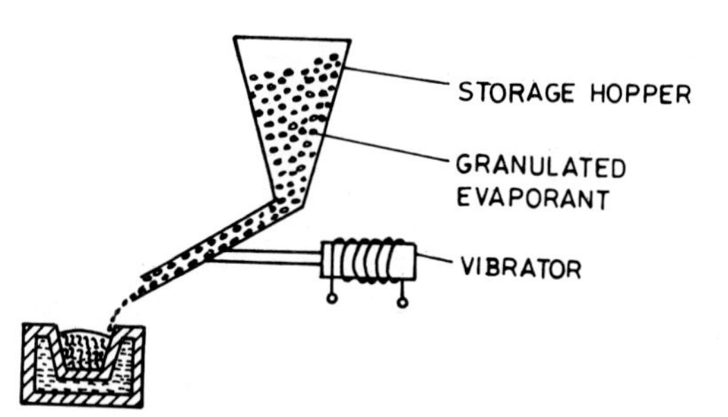

FIGURE 9.20 Schematics of continuous feed systems for evaporation: (*a*) wire feed, (*b*) rod feed, (*c*) bulk feed.

The coatings that are relevant to tribological applications include

Metals	Al, Cd, Cu, Ag, Au, Rh, Ni, Cr, W, Mo, Ti, Be, Fe
Alloys	Ag-Cu, Pb-Sn, Al-Zn, Ni-Cr, Ti-Al-V, M-Cr-Al-Y (M-Ni, Co, Fe or combination)
Refractory compounds	Al_2O_3, Cr_2O_3, SiO_2, ZrO_2, Y_2O_3-ZrO_2, TiO_2, SnO_2, In_2O_3, WC, W_2C, WC-Co, TiC, ZrC, HfC, NbC, Ta_2C, VC, TiC-ZrC, TiC-Co, TiC-Ni, TiN, Ti_2N, ZrN, HfN, MoN, TiB_2, ZrB_2, HfB_2, CrB_2, TiC-TiB_2, TiC-TiB_2-Co, ZrC-TiB_2

Coatings of single-metal components are generally applied by either resistive heating or electron beam heating. Refractory metals and compounds are generally applied by electron beam heating. Alloys and compounds with proper stoichiometry are generally deposited by using reactive evaporation and activated reactive evaporation. The coating thickness ranges from 0.01 to 10 μm or even greater (up to 300 μm) for oxidation-resistant applications. Aluminum coatings comprise about 90 percent of the volume of evaporation-deposited components. These coatings are widely used for decorative deposits, optically useful coatings, electrical and electronic coatings, and corrosion-resistant coatings. Cadmium coatings are used for corrosion resistance. Silicon monoxide coating is commonly used for wear-resistant applications. The coating is heated in air for conversion to silicon dioxide. SiO_2 coatings approximately 0.1 μm thick, when used as top coatings over aluminum, improve abrasion resistance by a factor of 10^4. For applications that are subject to severe atmospheric exposure and abrasion, rhodium, chromium, or chromium-nickel coatings are used. Four- and five-component corrosion- and oxidation-resistant MCrAlY alloys are deposited (ranging in thickness from 50 to 300 μm) via electron beam evaporation onto high-pressure turbine components of gas-turbine engines.

Coatings of various refractory compounds are deposited by activated reactive evaporation for metal cutting tools and other severe-wear applications. These have been deposited: oxides such as α-Al_2O_3, δ-Al_2O_3, TiO_2, Y_2O_3, In_2O_3, and Sn-In_2O_3 (Bunshah and Schramm, 1977; Colen and Bunshah, 1976; Grossklaus and Bunshah, 1975a; Nath and Bunshah, 1980); carbides such as W_2C, TiC, ZrC, HfC, NbC, Ta_2C, VC, and VC-TiN (Bunshah and Raghuram, 1972; Bunshah et al., 1977a; Grossklaus and Bunshah, 1975b); nitrides such as TiN, Ti_2N, HfN, and MoN (Bunshah and Raghuram, 1972); borides such as TiB_2, ZrB_2, HfB_2, TiB_2-ZrB_2, and TiC-TiB_2 (Bunshah et al., 1977b, 1978; Gebhardt and Cree, 1965); and cermets such as TiC-Ni, TiC-Co, and TiC-TiB_2-Co (Bunshah et al., 1977b, 1978; Nimmagadda and Bunshah, 1975, 1976; Sarin et al., 1977). For the deposition of the above coatings, the metal was evaporated from an EB gun heated source in the presence of an appropriate partial pressure of reactive gas. In the case of cermets like TiC-Ni and TiC-Co, the evaporants are Ti-Ni and Ti-Co rods, but only Ti reacts with C_2H_2 to form TiC while Ni and Co do not form the compound.

Vacuum metallization of plastics, glass, paper, textiles, and other substrates is common in producing finishes for both decorative and functional applications (Kut, 1974).

9.1.6.2 Microstructure and Stoichiometry of Evaporated Coatings. In the early stages of deposition, growth occurs at discrete nuclei, until the islands so formed link to form a continuous coating. Outward growth then proceeds to produce in many cases a "columnar" structure. The theories of formation (i.e., condensation, nucleation, and growth of thin coatings) are discussed by several authors (Bunshah et al., 1977b; Chopra, 1979; Maissel and Glang, 1970; Thornton, 1975). The structure of the coating is influenced by various parameters such as the nature and roughness of the surface, substrate temperature during deposition, angle of incidence/impingement of the vapor stream, and the gas pressure. The effects of substrate roughness and angle of incidence of the vapor stream have been studied by Dirks and Leamy (1977). Surface asperities or microprojections on the substrate are shadowed and cause localized rapid growth, which encourages the columnar defect and ultimately leads to a poorly bonded coating. Such defects are of major importance in overlay coating for gas-turbine blades (Boone et al., 1974). Clearly, high surface roughness will lead to many defects, and the effect is much more pronounced when the vapor stream reaches the substrate at a low angle to the substrate surface.

The adhesion of the coating can be significantly improved by increasing the mobility of vapor atoms during condensation, which will increase the diffusion.

The microstructure and morphology of thick single-phase coatings have been studied for a wide variety of metals, alloys, and refractory compounds. Movchan and Demchishin (1969a) have examined the influence of the substrate temperature T on the structure of thick (25- to 2000-μm) single-phase coatings of pure metals and oxides (Ni, Ti, W, Al_2O_3, and ZrO_2) deposited at high rates (1.2 to 1.8 μm min^{-1}) by EB gun evaporation. They divided the general features of the structure into three zones depending on T/T_m, where T_m is the evaporant melting point in kelvins. As shown in Fig. 9.21, the horizontal axis shows the scale of the melting point T_m(K). At 25 percent of the scale ($T_1 = 0.25T_m$) there is a boundary of the structure, and there is another at 45 percent ($T_2 = 0.45T_m$), dividing it into three zones.

At low temperatures (in zone 1), the coating atoms have limited mobility, and the structure is columnar, with tapered outgrowths. It is not a full-density structure and contains longitudinal porosity. It also contains high dislocation density and has a high level of residual stresses. As the substrate temperature is increased, the surface mobility increases. As the substrate temperature increases above about $0.3T_m$, even though the structure (zone 2) remains columnar, the columns tend to be finer, with parallel boundaries, normal to the substrate surface. These boundaries are stronger than in zone 1 structure and contain no porosity. Finally, at still higher temperatures, the structure shows an equiaxed grain morphology in zone 3. These structures have little residual stresses. Zone 2 structures have been associated with surface diffusion and zone 3 structures with bulk diffusion.

The texture of evaporated coatings is in general dependent on the deposition temperature. Most thick coatings exhibit a strong preferred orientation (fiber tex-

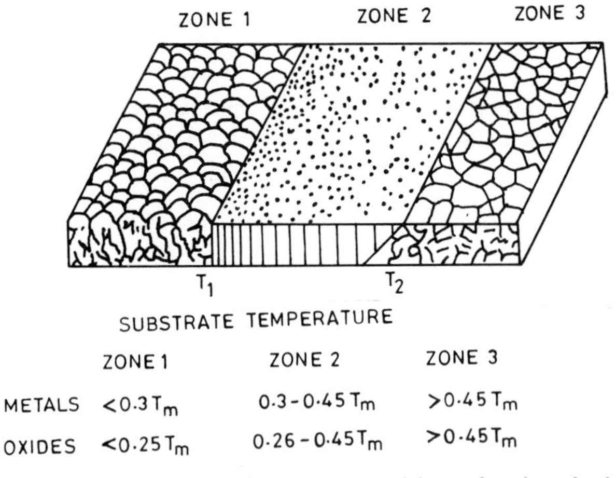

FIGURE 9.21 Thick coating structure model as a function of substrate temperature in electron beam evaporation of metals and oxides. T is the substrate temperature, and T_m is the melting point of the coating material in absolute degrees. [*Adapted from Movchan and Demchishin (1969b).*]

ture) at low deposition temperatures and tend toward a more random orientation with increasing deposition temperature (Bunshah, 1982). Based on Movchan and Demchishin's model, for dense equiaxed grain morphology of metal and ceramic coatings, the substrate should be preheated up to approximately half the melting point of the coating material. For instance, in the case of Al_2O_3 (melting point ≈ 2300°C) the substrate should be heated up to 1170°C for good adhesion and dense microstructure.

The microstructure of compound coatings such as TiC and (Ti-V)C can be modified through annealing at 1000°C (Jacobson et al., 1979a, 1979b). During the annealing process, recrystallization occurs with the result that the microstructure is transformed from fibrous grain bundles to equiaxed grains of about 10-nm diameter. The grain size was stabilized by an ultrafine network of cavities which were part of the original microstructure deposited at lower temperature (that is, 500°C).

Table 9.4 shows the effect of the supply of C_2H_2 gas on the deposition parameters, lattice parameter, carbon-to-metal ratio, and microhardness (Bunshah and Raghuram, 1972). This illustrates that the kinetic mechanism is no longer a rate-limiting step, and therefore the stoichiometry of the coating depends on the relative amount of the reactants. If the amount of C_2H_2 gas is further increased, the deposit now contains TiC and carbon. Similar trends have also been found for ZrC and (Hf-3%Zr)C coatings by Raghuram et al. (1974).

In systems containing more than one stable phase, as the presence of one of the reactants is increased while that of the other remains constant, the deposit composition varies in accordance with the equilibrium phases. For instance, in a Ti-N system, keeping the evaporation rate constant and increasing the partial pressure of N_2 gas yield a deposit consisting of the following successive layers:

α-Ti (with N_2 in solid solution)
α-Ti-Ti_2N
Ti_2N
Ti_2N-TiN
TiN

Further increase of N_2 partial pressure decreases the Ti-to-N_2 ratio in TiN.

TABLE 9.4 Effect of Partial Pressure of Reactive Gas on Deposition Parameters and Properties of TiC Coating Produced by ARE Process

$P(C_2H_2)$		Rate of evaporation, g min^{-1}	Rate of deposition, μm min^{-1}	Lattice parameter, nm	Carbon-to-metal ratio	Microhardness,* kg mm^{-2}
Pa	(torr)					
1.3×10^{-2}	(1×10^{-4})	0.67	4	—	Ti-TiC	—
4×10^{-2}	(3×10^{-4})	0.67	4	0.4308	0.5–0.6	2000
5.3×10^{-2}	(4×10^{-4})	0.67	4	0.4137	0.02–0.64	2550
6.6×10^{-2}	(5×10^{-4})	0.67	4	0.4322	0.64–0.70	2775
8–9 $\times 10^{-2}$	(7–8 $\times 10^{-4}$)	0.67	4	0.4329	0.73–0.96	2670

*Diamond pyramid hardness at a load of 50 g.
Source: Adapted from Bunshah and Raghuram (1972).

9.1.6.3 Properties of Evaporated Coatings. The properties of the evaporated coatings are primarily governed by the impurity content, microstructure, crystallinity, and imperfections which in turn are influenced by process parameters and deposition parameters including the substrate temperature (Bunshah, 1977, 1982; Bunshah et al., 1977a, 1977b; Raghuram and Bunshah, 1972).

The impurities in the coatings may be metallic or nonmetallic. Metallic impurities usually result from the reaction of the evaporant material with the source material. This can be avoided or reduced by using either ceramic-lined source or electron beam gun. The organic impurities may result from poor vacuum during deposition or adsorbed layers on the substrates. This can be avoided by sputter-etching the substrate surface prior to deposition.

Residual stresses in the evaporated metal coatings are tensile on the order of 1 GPa (Klokholm and Berry, 1968). The mechanical properties such as tensile strength, ductility, and hardness are influenced by the substrate temperature. Movchan and Demchishin (1969a) reported that the tensile strength of Ti, Ni, and W coatings decreases and ductility increases with increasing substrate temperature and larger grain size in zone 2. The strength and ductility values of specimens deposited in zone 3 were approximately the same as for recrystallized specimens produced from wrought material. The tapered crystallite morphology in zone 1 showed a high hardness, much greater (by a factor of 2 to 5) than that of annealed metal. The hardness decreased rapidly with increasing substrate temperature to a fairly constant value for zone 3 morphology, which corresponds to the hardness of recrystallized metals.

Bunshah (1982) reported the effect of deposition temperature on the grain size, tensile properties, and hardness of Ti, Ni, Nb, V, Mo, and Ni-20% Cr alloys for deposits in zones 2 and 3. They found that increasing deposition temperature produced large grain size, lower strength, higher ductility, and lower hardness.

Movchan and Demchishin (1969b) reported that the variation of microhardness with deposition temperature for two refractory compounds, Al_2O_3 and ZrO_2, is quite different in one respect from that of metals. The hardness falls when the structure changes from tapered crystallites (zone 1) to columnar grains (zone 2), as with metals. However, unlike metals, the hardness increases markedly as the deposition temperature rises from $0.3T_m$ to $0.5T_m$. Raghuram and Bunshah (1972) also show a marked increase in microhardness of TiC coatings on going from $0.15T_m$ (~500°C) to $0.5T_m$ (~1500°C). However, the rate of deposition decreases with an increase in substrate temperature (see Chapter 14). We believe that the decreased occurrence of defects at higher deposition temperatures may be responsible for higher hardness.

The microstructure of the coating influences its corrosion resistance. In general, the coatings with a columnar morphology are not as protective as those with an equiaxed morphology. We note that the real grain boundary area is much smaller in the columnar morphology than in the equiaxed morphology. Therefore, concentration of impurities per unit grain boundary area would be much higher in columnar morphology, which may account for the lower corrosion resistance. The occurrence of coating defects such as weak grain boundaries in the transition fibrous morphology would also reduce the corrosion resistance, as reported by Boone et al. (1974).

TiC/TiC and TiC/TiN coated couples (coated by the ARE process) showed very low coefficients of friction (0.1 to 0.2) and corresponding low wear rates. These hard coatings also show marked improvement in abrasive wear and impact erosive-wear conditions (Jamal et al., 1980). Hard coatings of carbides, nitrides, and oxides deposited by the ARE process on cemented carbide and high-speed steel cutting tools produce marked improvement by a factor of 2 to 20 in tool wear (Bunshah et al., 1977b; Kobayashi and Doi, 1978; Nakamura et al., 1977). For example, type M-10 high-speed steel drills coated with TiC and TiN by the ARE process have the tool life of 20 times the uncoated drills and 4.2 times the black oxide drills (Nimmagadda et al., 1981). Type M-50 high-speed steel used in tribological components such as bearings coated with Ti, Cr, Mo, Ni-20% Cr, TiC, and TiN by various evaporation processes eliminated the severe localized pitting corrosion of this steel when exposed to a hydrocarbon environment containing a few parts per million of chlorine ions (Bunshah, 1982).

9.2 ION PLATING

The ion-plating or ion vapor deposition technique using a glow discharge was first reported by Mattox (1964). In ion plating, partial ionization of the metal vapor is used to increase the adhesion of the coating to the substrate. During deposition, the substrate surface and/or the depositing coating is subjected to high-energy flux of ions and energetic neutrals sufficient to cause changes in the interfacial regions or coating properties compared to the nonbombarded deposition. Ion-plating processes can be classified into two broad categories: glow-discharge (plasma) ion plating performed in low vacuum (5×10^{-1} to 10 Pa or 5×10^{-3} to 10^{-1} torr) and ion beam ion plating performed in high vacuum (10^{-5} to 10^{-2} Pa or 10^{-7} to 10^{-4} torr).

In the glow-discharge ion-plating processes, the material to be deposited is evaporated similar to ordinary evaporation, but it passes through a gaseous glow discharge (at a chamber pressure of typically 10^{-1} to 10 Pa or 10^{-3} to 10^{-1} torr) on its way to the substrate, thus ionizing the evaporated atoms in the plasma. Condensation of the vapor takes place under the action of ions from either a carrier gas or the vapor itself on the substrate biased to a high negative potential (typically 2 to 5 kV) (Matsubara et al., 1974; Mattox, 1973, 1982; Schiller et al., 1977a, 1982). A rf potential has to be used for insulating substrate surfaces. Ion beam ion-plating processes are high-vacuum (10^{-5} to 10^{-2} Pa or 10^{-7} to 10^{-4} torr) deposition processes where the ion bombardment source is an external ionization source (gun). They utilize either single or cluster ion beams. Ion beams can be of mere gas ions or ionized species of coating materials (Mattox, 1982; Weissmantel et al., 1979a, 1979b, 1982). The substrates must withstand ion bombardment heating. Many metals, alloys, and compounds can be ion-plated. Depending on the type of ion plating to be performed, a broad range of vapor sources are used: evaporation by electric resistance heating, rf induction heating, electron beam, hollow cathode, cathodic arc, and metal-bearing gas or hydrocarbon gases (for carbon). The technique of using a gas as a vapor source (as compared to chemical vapor deposition) is often called *chemical ion plating* or *gas ion plating*. Resistively heated filaments are the most commonly used vapor sources for plating materials with a melting temperature of less than 1500°C. The electron beam evaporation provides the ability to deposit refractory materials or large quantities of materials easily; high-melting-point materials with melting temperatures up to 3500°C have been evaporated. A reactant gas or a beam of desired ionized species can be used to produce alloy or compound coatings. This process is referred to as *reactive ion plating*.

Prior to ion plating, the substrate is subjected to inert-gas ion bombardment for a time sufficient to remove surface contaminants and barrier layers (sputter cleaning) before coating deposition, as in sputter deposition. After the substrate surface is sputter-cleaned, the coating deposition is begun, without interrupting the ion bombardment. For a coating to form, it is necessary that the deposition rate exceed the sputtering rate. From the standpoint of adhesion, the principal benefits obtained from the ion-plating process are the ability to (1) sputter-clean the surface and keep it clean until the coating begins to form; (2) provide a high-energy flux to the substrate surface, giving a high surface temperature and thus enhancing diffusion and chemical reactions without necessitating bulk heating; and (3) alter the surface and interfacial structure by introducing high defect concentrations, physically mixing the coating and substrate material, and influencing the nucleation and growth of the deposited coating. In addition to modifying the substrate surface and interface formation, the ion bombardment of the growing coating may cause a modification of the morphology of the deposited material, changes in the internal stress of the deposited coating, or the modification of other physical and chemical properties.

Adhesion of ion beam ion-plated deposits is found to be excellent even at low substrate temperature, compared to other vapor deposition processes, because of the extremely high energy of arrival of the depositing ions (10 eV to 10 keV) and energetic neutrals. Some ion beam deposition processes use kinetic energy of the beams higher than any other deposition process except ion implantation. Cleaning of the substrate surface occurs by bombarding argon and coating-material ions before and during deposition. Another advantage of ion plating, in some situations, is the high "throwing power" and quite uniform coatings achieved without rotating the part. This high throwing power results from gas scattering, entrainment, and redeposition from the sputtered coating surface. This allows coatings to be formed in recesses and areas remote from the source-substrate line of sight, thus giving more complete surface coverage.

The glow-discharge techniques have the disadvantage in that the evaporation flux cannot be controlled in a defined manner with respect to the fraction of neutrals and ionized species, kinetic energy, and incident angle. The ion gun system has the distinct advantage that the ion bombardment parameters can be controlled independently of other deposition parameters. The operation of ion beam ion-plating processes in high vacuum reduces the likelihood of residual gas incorporation. The bombardment ions do not suffer collisions and, therefore, impinge on the substrate with high energy. Ion beam processes use beams of high kinetic energy—higher than that of the ions used in glow-discharge processes—resulting in excellent adhesion, but they have poor deposition rates compared to glow-discharge processes. Glow-discharge processes are simpler and more flexible compared to ion beam processes. The glow-discharge processes have been exploited for industrial production. Because of requirements of complex and expensive ion guns, ion beam ion-plating processes are often limited to situations where it is the only technique by which desired results can be obtained.

Tables 9.1 and 9.2 summarize the range of coating materials, kinetic energy of evaporant species, deposition (substrate) temperatures, deposition pressure, coating thickness, deposition rates, and advantages and disadvantages of various ion-plating processes. The details follow.

9.2.1 Glow-Discharge Ion Plating

In glow-discharge ion plating, the material to be deposited is evaporated similar to ordinary evaporation but passes through a gaseous glow discharge on its way to the substrate, thus ionizing the evaporated atoms in the plasma. Condensation of the vapor takes place under the action of ions from either a carrier gas or the vapor itself. The impingement of ions on the substrate surface or the deposited coating is combined with a transfer of energy and momentum. The glow discharge is produced by biasing the substrate (instead of the source target in the sputtering process) to a high negative potential (typically 2 to 5 kV) and admitting an inert gas, usually argon, at a pressure typically between 10^{-1} to 10 Pa (10^{-3} and 10^{-1} torr) into the chamber. The discharge potential is maintained as high as possible so that the ionized atoms are accelerated to the substrate (Mattox, 1973, 1982; Schiller et al., 1977a, 1982). The substrate potential may be dc or rf. Normally dc potential is used for conducting surfaces, and an rf potential may also be used with conducting surfaces, but is necessary for insulating surfaces.

By using various reactant gases in the ion-plating system in controlled amounts, alloys or compounds can be formed, known as *reactive ion plating*. The reactant gas, when introduced to the plasma, undergoes dissociation and ionization of the atoms of the reactant gas molecules. Evaporation or sputtering from multiple sources can also be used to form alloys or compounds (Murayama, 1975; Ohmae et al., 1974).

In glow-discharge ion plating, the substrate is bombarded by high-energy gas ions which sputter off the material present on the surface. This results in constant cleaning of the substrate (i.e., removal of gaseous and other impurities by sputtering) which is desirable for better adhesion and lower impurity content. However, this also means that some of the deposit itself will be sputtered away, thus decreasing the deposition rate, and the sputtering causes considerable heating of the substrate by the intense gas ion bombardment. The process parameters, therefore, have to be chosen correctly so that the deposition rate is higher than the sputtering rate.

9.2.1.1 Transport Kinetics Considerations. Figure 9.22 schematically depicts the processes that occur in the cathode region of the discharge (Mattox, 1963, 1973). The primary source of ionization of the discharge gas (G^0) is electron-atom collision

$$e^- + G^0 \rightarrow G^+ + 2e^- \tag{9.7}$$

The gas ions that are produced may then become part of plasma and ultimately be lost to the chamber walls, or they may be accelerated across the cathode dark space to the cathode. In traversing the cathode dark space, the accelerating ions may suffer charge exchange collisions with thermal neutrals:

$$G_1^+ + G_2^0 \rightarrow G_2^+ + G_1^0 \tag{9.8}$$

producing a spectrum of high-energy neutrals and high-energy ions. These high-energy atoms and ions that impinge on the cathode surface may generate secondary electrons, sputter cathode atoms (S) and surface contaminants, become neutralized and reflect from the cathode surface as high-energy neutrals or metastable atom (G*), or become incorporated into the cathode surface and change the surface properties.

In ion plating, the impinging particles may sputter coating atoms (M). The sputtered surface atoms can be scattered back to cathode, ionized by electron-atom collision or lost from the cathode region, or undergo metastable-atom collision

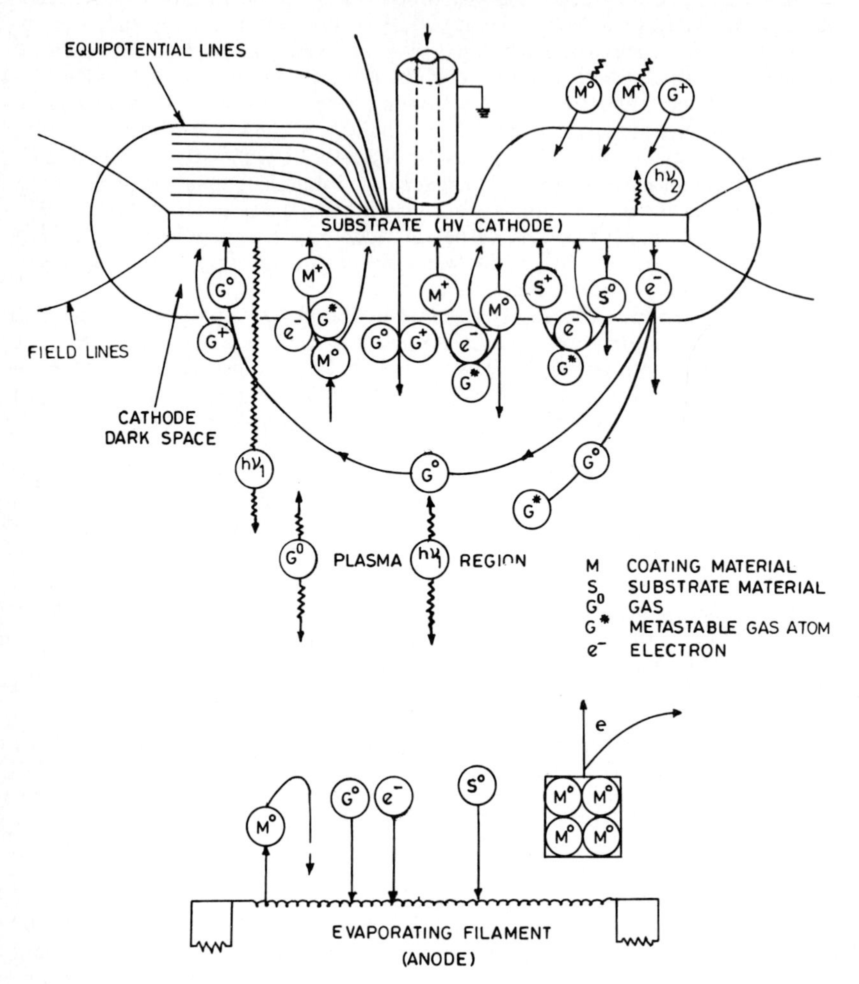

FIGURE 9.22 Schematic representation of glow discharge using an inert gas. Resistively heated filament is shown as heating source. [*Adapted from Mattox (1973).*]

$$S^0 + G^* \rightarrow S^+ + G^0 + e^- \tag{9.9}$$

and be accelerated back to the cathode and transported as high-energy neutrals to some other surface.

In ion plating, the coating atoms are injected into the glow discharge from an evaporant source such as a resistively heated filament (Fig. 9.22) and are deposited as neutrals on the substrate, ionized by electron-atom or metastable-atom collisions, and accelerated to the substrate; or they may collide with other coating atoms in the gas phase and nucleate into fine particles. The high-energy gas atoms that are trapped in the cathode surface may affect the coating nucleation and growth and its properties. The gas atoms that are incorporated in the surface during sputter cleaning may be released after coating formation and cause poor coating adhesion.

We note that in glow-discharge ion plating only a small percentage of the gas atoms (~1 percent) are ionized, so the cathode and other surfaces in the system are being continuously bombarded by thermal neutrals. The cathode surface wall becomes heated during sputtering since most of the energy of the bombarding ions is given up as heat and will radiate energy $h\nu_2$. The energy distribution of sputtered atoms is higher (1 to 100 eV) than that of thermally vaporized atoms (0.1 to 1 eV).

The energy and impulse transfer of the ions cause a change in the energy balance of the evaporant film during condensation. The energy transport equation of the vapor stream is

$$E_V = n_V e_V \tag{9.10}$$

where n_V = number of impinging particles per unit area per unit time
e_V = mean kinetic energy of the vapor particle = $(3/2)kT_V$
k = Boltzmann's constant
T_V = evaporation temperature, K

The energy transferred by the ions is given by

$$E_I = n_I e_I \tag{9.11}$$

where n_I = number of impinging ions per unit area per unit time
e_I = mean energy of an ion, which is $e\overline{U}_I$
$\overline{U}_I$ = mean accelerating voltage of impinging ions, V

The coefficient of energy, which is a measure of the energetic activation of the coating surface, is given by (Schiller et al., 1975)

$$\epsilon_I = \frac{E_I + E_V}{E_V} \tag{9.12a}$$

This coefficient gives the factor by which the condensation energy increases owing to ion action compared to condensation energy without such action. Since $E_V \ll E_I$, we get

$$\epsilon_I = (6 \times 10^3) \frac{\overline{U}_I}{T_V} \frac{n_I}{n_V} \tag{9.12b}$$

where $\overline{U}_I$ is measured in volts and T_V in kelvins.

Table 9.5 lists the values of coefficient of energy for various n_I/n_V ratios and mean accelerating voltage. A value of about 0.2 eV is typical for the mean energy e_V of a vapor particle obtained at an evaporation temperature of about 2000 K. Typical values of the mean energy e_I of the ions during ion plating are 50 eV to 5 keV. We note that the varied dependence of number of ions and ion energies, that are active during deposition. Ion-plating effects may differ even for a single value of ϵ_I. So in order to achieve high values of ϵ_I, an efficient ionization system is required which gives high values of n_I/n_V. However, the resputtering rate due to ions should not exceed the deposition rate, and the temperature rise of the substrate due to ion bombardment should be acceptable.

TABLE 9.5 Coefficient of Energy ϵ_I with Evaporation, Sputtering, and Ion Plating

Deposition technique	Coefficient of energy ϵ_I	Parameters		
Evaporation	1	e_V = 0.2 eV		
Sputtering (without bias)	5–10	e_S = 1–2 eV		
Ion plating	ϵ_I	n_I/n_V =	$\overline{U}_I$, V	$\overline{e}_{IP}$, eV
	1.2	10^{-3}	50	0.1–0.2
	3.5	10^{-2}	50	7
		10^{-4}	5000	
	25	10^{-1}	50	5
		10^{-3}	5000	
	250	10^{-1}	500	50
		10^{-2}	5000	
	2500	10^{-3}	5000	500

e_V = mean energy of vapor particle, e_S = mean energy of a sputtered particle, $\overline{e}_{IP}$ = mean energy of both vapor particles and ions, n_I/n_V = ratio of number of ions to the number of vapor particles, and $\overline{U}_I$ = mean ion acceleration voltage.

Source: Adapted from Schiller et al. (1975).

9.2.1.2 Details of Glow-Discharge Ion-Plating Processes. The glow-discharge (plasma) techniques can be classified based on the deposition system configuration, mode of production of vapor species, and method of enhancement of ionization of vapor species. The typical vapor sources include evaporation (by resistive heating, rf induction heating, electron beam evaporation, hollow cathode discharge, cathodic arc, etc.) and sputtering (Mattox, 1982; Schiller et al., 1977a, 1982). The vapor and ion actions can be separated by separating the ionization arrangement and vapor source and by a substrate motion relative to the action regions in alternating ion plating. The ionization efficiency and kinetic energy of vapor species can be controlled and enhanced by introducing additional sources of electron production and electric and magnetic fields (triode ion plating and sputter ion plating). A hollow cathode discharge (HCD) gun is used to enhance the ion yield at relatively low pressures (10^{-2} to 10^{-1} Pa or 10^{-4} to 10^{-3} torr). One of the main HCD gun features is that it uses a low-voltage high-current plasma electron beam. Very high plasma density can be achieved with the aid of powerful arc discharge. The arc discharge is sustained by a low voltage. Some important characteristics of various glow-discharge ion-plating processes are presented in Table 9.6. Details of various glow-discharge ion plating processes follow.

Conventional (Diode) Ion Plating. The dc diode configuration with a self-sustained discharge (cold cathode) is the most widely used ion-plating technique for coating of conducting substrates (Carpenter, 1974; Chambers et al., 1972; Fritz, 1972; Harker and Hill, 1972; Mattox, 1969, 1982; Teer, 1975). A typical system of this type is shown in Fig. 9.23. In a typical deposition, the system is evacuated up to 10^{-5} Pa (10^{-7} torr) and is backfilled with argon to about 5×10^{-1} to 10 Pa (5×10^{-3} to 10^{-1} torr; typically 5×10^{-1} to 1 Pa or 5×10^{-3} to 10^{-2} torr). A glow discharge is initiated by applying a negative dc potential of a few thousand volts (typically 2 to 5 kV) to the substrate (cathode). The coating material is evaporated in a manner similar to the evaporation process but passes

TABLE 9.6 Some Important Characteristics of Various Glow-Discharge Ion-Plating Techniques

Deposition technique	Typical vapor source(s)	Deposition pressure, Pa (torr)	Comments
DC/RF diode	Resistive, rf induction, electron beam heating	5×10^{-1}–10 (5×10^{-3}–10^{-1})	Simple, thickness uniformity over a large area, but relatively high deposition pressure
Alternating	Resistive, rf induction, electron beam heating	5×10^{-1}–10 (5×10^{-3}–10^{-1})	Vapor source and ionization arrangement can be optimized separately, but relative motion of substrate needed
Triode	Resistive, rf induction, electron beam heating	10^{-3}–10^{-1} (10^{-5}–10^{-3})	Low deposition pressure and relatively high deposition rates, but system more complex and does not produce uniform coating over large surface areas
Hollow cathode discharge	HCD electron-beam heating	10^{-2}–10^{-1} (10^{-4}–10^{-3})	Relatively low deposition pressure and high deposition rates, but system more complex
Cathodic arc	Cathodic arc heating	10^{-3}–10^{-1} (10^{-5}–10^{-3})	Low deposition pressure, very high plasma density can be achieved, very high deposition rates (up to 25 μm s^{-1}) but system more complex

through a gaseous glow discharge on its way to the substrate, thus ionizing the evaporated atoms in the plasma. Thus the substrate is simultaneously subjected to vapor and ions, with the ratio of impinging ions to impinging vapor particles generally on the order of 10^{-4} to 10^{-1}. In any of the dc-type discharges, an axial magnetic field of 50 to 200 Oe can be used over the substrate surface to increase the electron path length and thus increase the ionization probability, as in magnetron sputtering. Despite the rather low ion content in diode ion plating, however, uniform ion-plating effects can be obtained over the entire substrate surface. In the conventional mode of ion plating and for rather small substrates, it can be assumed that the densities of the vapor stream n_V and ions n_I are constant over the substrate surface and with respect to time. In the case of self-sustained gas discharges, the current densities are about 1 mA cm^{-2}. In a typical diode ion-plating operation, the power input may be several watts per square centimeter.

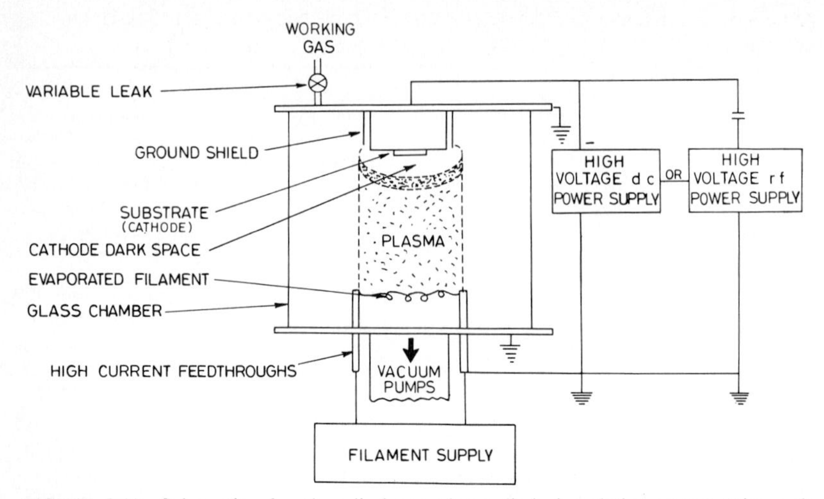

FIGURE 9.23 Schematic of a glow-discharge planar diode ion-plating system using a dc power supply or a capacitively coupled rf power supply. Resistively heated filament is shown as heating source.

If the substrate (cathode) is an insulator, it builds up a surface charge which prevents bombardment by positively charged particles. To permit ion bombardment of an insulator surface, a rf power supply is capacitively coupled to the substrate. This rf method develops a self-bias voltage that is negative relative to the plasma floating potential on any surface that is capacitively coupled to a glow discharge. In this case, a plasma of ions and electrons is generated by the rf discharge (see Appendix 9.A). If the rf supply has a high peak-to-peak voltage (typically >500 V), the surface bias will be sufficient to accelerate ions from the plasma to a high enough energy to cause sputtering. This technique is called *rf ion plating* and may be used to coat insulators or metals which tend to form dielectric layers on their surfaces or when depositing a dielectric coating (Harker and Hill, 1972; Mattox, 1982; Murayama, 1974, 1975). A dc potential with a superimposed rf can also be used to increase the sputtering from materials which form insulating surfaces.

Alternating Ion Plating. In the conventional mode of ion plating, the densities of the vapor stream and ions are constant with respect to time. With *alternating ion plating* (AIP), each point of the substrate is alternately subjected to vapors and ions. The densities of ions and, in general, the vapors are periodic functions of time for each point of a substrate, i.e.,

$$n_V = f_1(t) \qquad \text{and} \qquad n_I = f_2(t) \tag{9.13}$$

As shown in Fig. 9.24, this principle is readily put to use by local separation of vapor and ion actions and through a substrate motion relative to the action regions. The ionization arrangement and vapor source are separately mounted below the substrate. In this way, vapor and ions act on different points of the substrate at one and the same time. The relative motion must be carried out in such a way that, within each period, every spot on the substrate to be coated passes

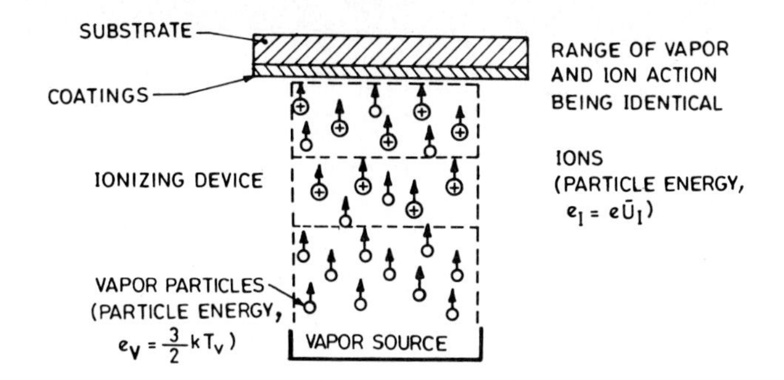

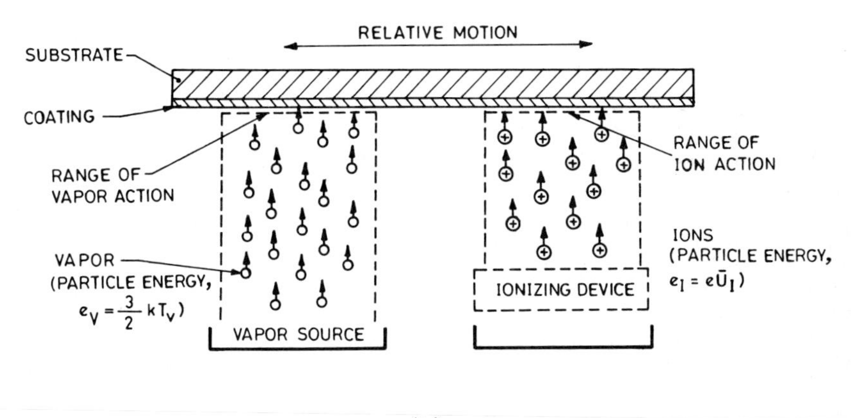

FIGURE 9.24 Principles of vapor stream and ion stream action in (*a*) conventional (diode) ion plating and (*b*) alternating ion plating.

once through a region of vapor action and once through a region of ion action (Schiller et al., 1975, 1982).

Figure 9.25 shows the schematic diagram of an AIP system that uses an electron beam (EB) evaporator to generate the vapor stream and a dc planar magnetron or plasmatron type of ionization device. The substrates are arranged on a rotary table and are thus alternately subjected to vapor and ion action. Ion current densities of more than 100 mA cm^{-2} can be achieved with the planar magnetron or plasmatron type of ionization devices. The combination of EB evaporator and energetic ion source leads to high deposition rates (~0.2 μm min^{-1} for Ni). Schiller et al. (1975) reported that ion action of AIP substantially reduces the porosity compared to conventional ion plating. They also reported that porosity decreases with an increase in the ion rate.

With AIP, it is possible to optimize the vapor source and ionization arrangement separately. By virtue of the alternating action of vapor and ions, no particular demands exist as far as the uniform local distribution of ionic current density

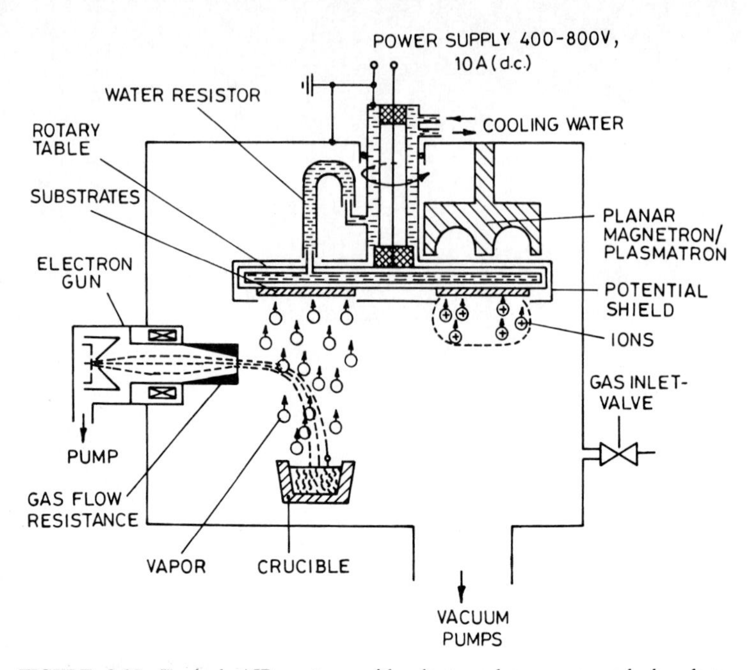

FIGURE 9.25 Typical AIP system with electron beam gun and dc planar magnetron/plasmatron type of ionizing device. [*Adapted from Schiller et al. (1982).*]

in the direction of the relative motion is concerned. However, a relative motion of the substrate is needed in the AIP system. But large-area coating generally calls for a relative motion of the substrate and coating device.

Triode Ion Plating. Triode ion plating uses a hot electron-emitting filament as the electron source (hot cathode) to enhance the ionization. In this technique, the plasma is generated between a hot electron-emitting filament and an anode. Ions are extracted from this plasma to bombard the negatively biased cathode. This type of discharge can be maintained at a lower gas pressure (10^{-3} to 10^{-1} Pa or 10^{-5} to 10^{-3} torr) than the diode configuration. The ions that bombard the cathode from such a low-pressure plasma probably have an energy spectrum in which most of the ions have energies close to the bias potential. A principal difficulty in triode ion plating is the maintaining of a uniform plasma density and composition. A schematic of a triode ion-plating system is shown in Fig. 9.26 (Kloos et al., 1981; Matthews and Teer, 1980, 1981; Teer and Delcea, 1978). Matthews and Teer (1980) used this technique under reactive mode to deposit TiN coatings at elevated temperatures (~600°C).

Ion Plating with Hollow Cathode Discharge. Coatings deposited in relatively low vacuum (10^{-1} to 10 Pa or 10^{-3} to 10^{-1} torr) by diode ion plating exhibit porosity due to gas scattering and inadequate adhesion due to the small percentage (~1 percent) of evaporant ions in the evaporant flux. The application of hollow cathode discharges with a sustaining voltage of about 100 V for the generation of vapors and ions is an approach to enhance the ion yield at relatively low pressures (10^{-2} to 10^{-1} Pa or 10^{-4} to 10^{-3} torr) (Chambers et al., 1972; Komiya and Tsuruoka, 1974, 1975, 1976; Komiya et al., 1978). One of the main HCD features

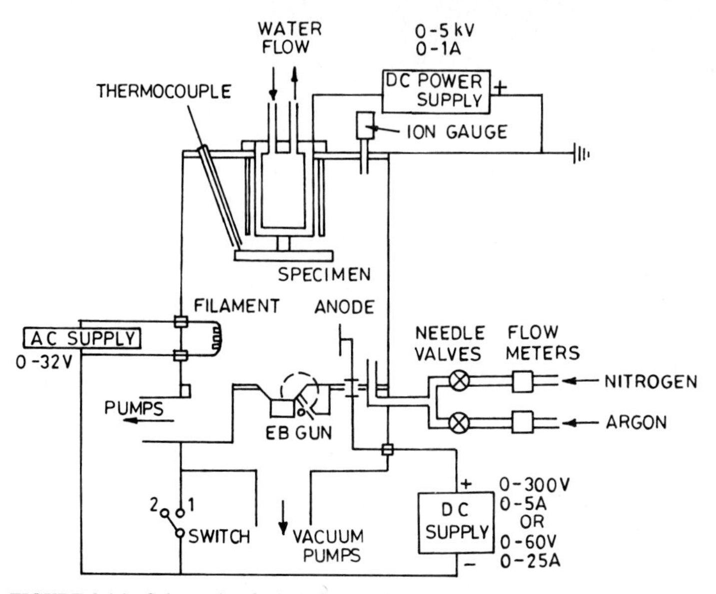

FIGURE 9.26 Schematic of triode ion-plating system.

is that it uses a low-voltage high-current plasma electron beam extracted from the HCD gun. If an active gas is introduced during deposition, gas molecules are very effectively activated and ionized in the space between the molten pool and the substrate. A compound coating can thus be deposited on the substrate at a low substrate bias voltage (Komiya and Tsuruoka, 1976).

Figure 9.27 shows a schematic of an ion-plating system with a (hot) HCD electron gun which produces a 90° bent-electron beam. The 90° bent-electron beam is used to provide a larger area for substrate free from the shadow of the gun assembly and to eliminate the effect of metal splash and vapor contamination at the hollow cathode tip.

Several authors have reported attempts to apply the HCD method for deposition of Cu and quartz (Morley and Smith, 1972), Ag (Komiya and Tsuruoka, 1975; Williams, 1974), and Cu and Cr (Komiya and Tsuruoka, 1975, 1976). The deposition rates range from 0.1 to 0.4 $\mu m\ min^{-1}$ in the case of Ag. Komiya et al. (1977a, 1977b, 1978) and Nakamura et al. (1977) have produced coatings of Cr-N, Cr-C, TiN, and TiC by using HCD in reactive mode.

Cathodic Arc Ion Plating. In the cathodic arc plasma deposition process, the evaporated species are produced by striking one or more electric arcs on the surface of the material to be evaporated, which is made the cathode of the arc discharge. Compared to electron beam evaporation with auxiliary discharge, much higher plasma density is achieved with the aid of powerful arc discharge. The arc discharge is sustained by a low-voltage range on the order of 15 to 150 V and typical arc current on the order of 20 to 200 A depending on the cathode material. The arc process is also utilized for the generation of plasma and therefore for the excitation and ionization of the vapor and the reaction gases. The degree of ionization is greater than 10 percent and, in extreme cases, approaches 100 percent as compared to about 1 percent in dc diode ion plating (Boxman et al., 1986; Burkhardt, 1939; Dorodnov and Petrosov, 1981; Freller and Haessler, 1987;

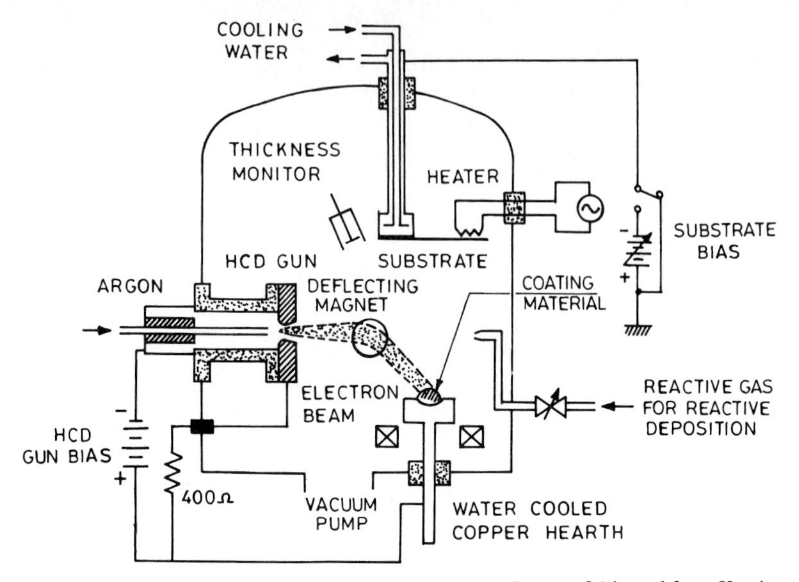

FIGURE 9.27 Schematic of ion-plating system using HCD gun. [*Adapted from Komiya and Tsuruoka (1976).*]

Hatto, 1985; Johnson, 1989; Nemchinski, 1979; Randhawa, 1987; Sanders and Pyle, 1987; Snaper, 1971).

The cathodic arc deposition process is inherently efficient, converting as much as one-third of the input power to evaporant, with high deposition rates up to 25 $\mu m\ s^{-1}$ (Boxman et al., 1986) in certain cases. However, the major disadvantage of the arc process is the release of microdroplets from the cathode which may result in rough surfaces and lead to failure due to cracking at boundaries. The evaporant species are removed from the cathode surface as a result of a rapid series of flash evaporation events as the arc spot scans randomly around it at speeds on the order of 100 $m\ s^{-1}$ (Daadler, 1983; Davis and Miller, 1969). The basic emission processes are schematically shown in Fig. 9.28. The arc source consists of a cathode (evaporant material), an anode, an arc igniter, and a means of confining the arc to the cathode surface. The arc spot of diameter 1 to 20 μm produces evaporant species, electrons, and micrometer-sized droplets (Martin et al., 1987; McClure, 1974). Electron-atom collisions result in the formation of positive ions in the region adjacent to the cathode surface. Some ions are accelerated toward the cathode and possibly create new emission points. Ions are also ejected from the positive ion cloud toward the anode. The high density of ions in the region adjacent to the cathode modifies the potential distribution which is sufficient to accelerate the ions to higher kinetic energies.

The trajectory of the arc spot is confined to and/or steered through magnetic fields. The application of magnetic field results in significant reduction in droplet content, acceleration of plasma, and enhancement in synthesis of compound formation in reactive mode. Sanders and Pyle (1987) reported that the magnetic and electrostatic field can be applied to direct the evaporant species to obtain controlled plasma, which results in an increase in the deposition rate by a factor of 4 and adhesion by a factor of 6.

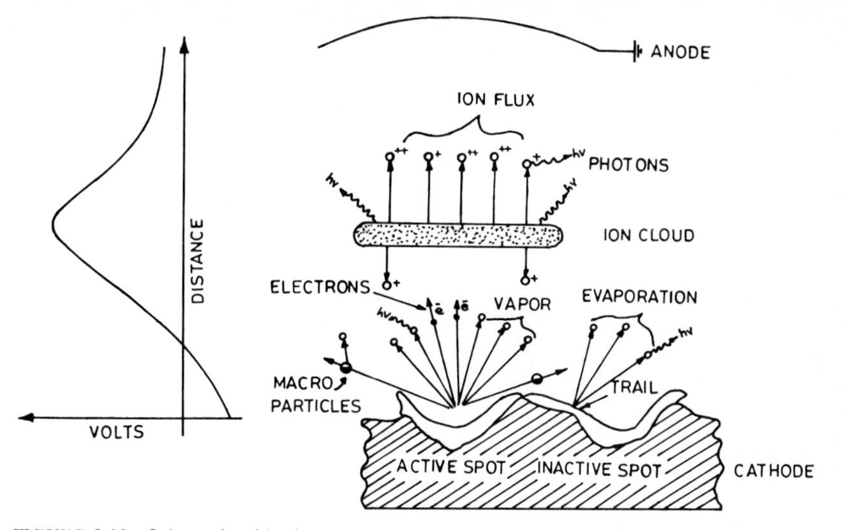

FIGURE 9.28 Schematic of basic arc evaporation process and potential distribution.

Figure 9.29 shows a schematic of an ion-plating setup using a cathodic arc evaporation system. The deposition takes place under high vacuum, 10^{-3} to 10^{-1} Pa or 10^{-5} to 10^{-3} torr, in a reactive or nonreactive gaseous environment.

Sputter Ion Plating. Sputter ion plating uses sputtering targets as the vapor source. Sputter ion plating is essentially the same as bias sputtering (typically magnetron or plasmatron type) except that the system is operated at relatively low pressures and higher bias voltages in order to produce intense flux composed of thermal neutrals, sputtered species, and energetic ions. The deposition conditions resemble those of more typical ion plating.

9.2.2 Ion Beam Ion-Plating Processes

In ion beam ion-plating processes, the ion bombardment source is an external ionization source (gun). These guns utilize various ion beams—single or cluster ion beams. Ion beams can be of inert-gas ions or ionized species of coating materials. Typical coating material vapor sources include evaporation, ionized coating material beams, and sputtering (Harper, 1978; Harper et al., 1982; Mattox, 1982; Weissmantel et al., 1979a, 1979b, 1982). Ion beam sputtering is discussed later in a sputtering section. By using a beam of desired ionized species, alloys or compounds can be formed, known as *reactive ion plating*. Ion beam plating is performed at high vacuum (10^{-5} to 10^{-2} Pa or 10^{-7} to 10^{-4} torr); therefore, this process is sometimes referred to as *vacuum ion plating*. To coat insulating surfaces, an ion gun system with a neutralizer filament is used.

When the ion energy is increased, say 10^3 to 10^4 eV, the ions penetrate a number of atomic layers into either the substrate or growing film, giving rise to a highly disturbed atomic region caused by collisions. The local activation simulates diffusion (Haff and Switkowski, 1977), phase transition, and chemical reactions in addition to defect creation and resputtering.

A description of various ion sources and various ion beam ion-plating processes follows.

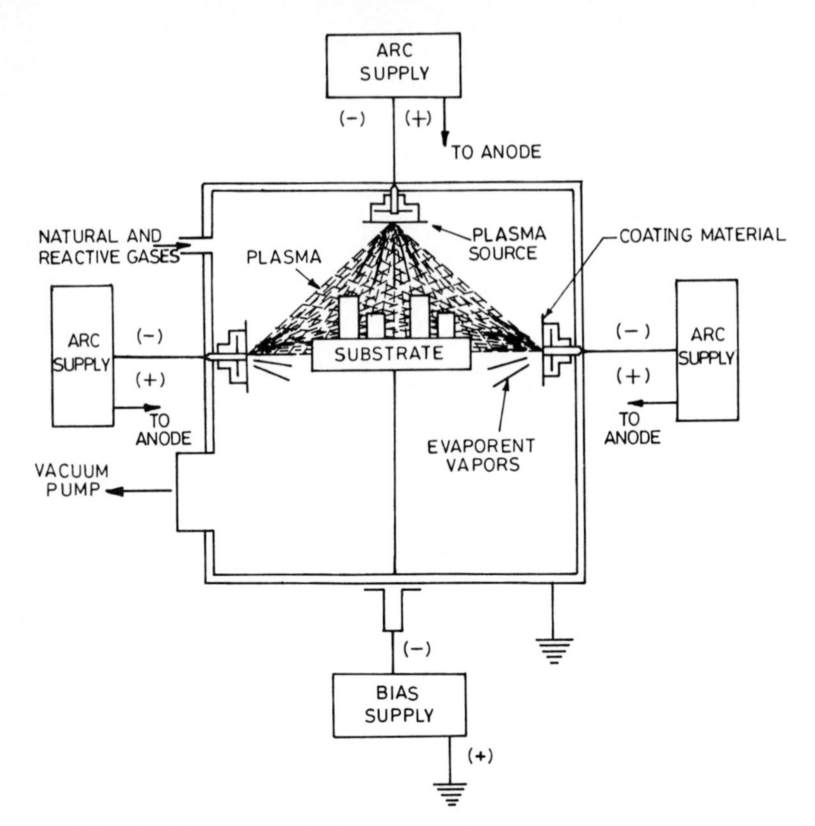

FIGURE 9.29 Schematic of cathodic arc ion-plating system.

9.2.2.1 Ion Sources. Various types of ion sources (guns) have been developed (Harper, 1984; Harper et al., 1982; Kaufman, 1978; Kaufman et al., 1982; Townsend et al., 1976), but Kaufman source and duoplasmatron are the most commonly used. The Kaufman source (Kaufman, 1978) is a broad-beam ion source whereby electrons from a heated filament (thermionic emission) are accelerated to the anode and ionize a noble gas such as argon while enroute (Fig. 9.30). A magnetic field inside the gun deflects the electrons in a cycloidal path for increased collision and ionization. Nearly monoenergetic ions are then ejected from the gun by an ion extraction focusing system, and a collimated accelerated ion beam is produced. When the ion beam is directed at an insulating target, a neutralizing element is added to the beam to compensate for the positive charge of the beam ions. Ion beams of most gases can be generated including inert gases such as Ar, Kr, and Xe, reactive gases such as O_2, N_2, and H_2, and more complex gases such as CH_4, CF_4, and $B_3N_3H_6$. The complex gases undergo fragmentation on discharge. The ionization efficiency is in the range of 50 to 75 percent. The extracted Ar ion current density from a 100-mm-diameter source is in the range of 1 to 2 mA cm^{-2} at a beam energy of 500 to 2000 eV (Reader and Kaufman, 1975). The ion energy is determined by an anode voltage and is varied from tens of electronvolts to several kiloelectronvolts. The ion beam intensity,

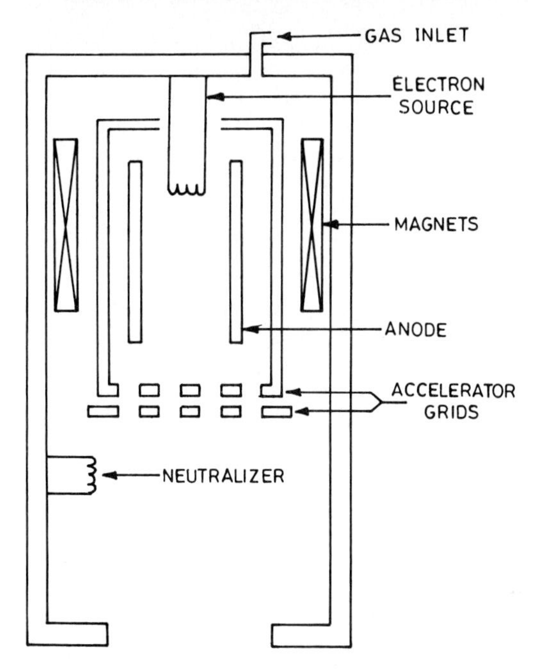

FIGURE 9.30 Schematic of a low-voltage Kaufman-type ion gun.

i.e., ion flux, is independently controlled by adjusting the heating current of the hot cathode.

The duoplasmatron is another widely used ion source which produces high-intensity narrow ion beams. The name refers to the dual manner of constricting the plasma in an arc discharge by both magnetic and electric fields to form a small region of high ion density adjacent to the extraction aperture (Von Ardenne, 1956). Argon ion currents of 2 mA and higher have been obtained at about 20-keV beam energy through an aperture of 0.3-mm diameter with an energy spread of 10 to 50 eV. The narrow beam allows deposition of thin coatings from milligrams of target material.

9.2.2.2 Details of Ion Beam Ion-Plating Processes. The ion beam ion-plating processes are classified according to whether they utilize single or cluster (or molecule) ion beams and their source of coating material. Among the various single-ion-beam processes, the ion beam may be used to post-ionize the evaporated flux (ion beam plating), or an ion beam of the coating material itself may be generated and used for deposition (direct ion beam deposition). Direct ion beam deposition utilizes an ion beam in the low energy range, 10 to 100 eV below or not too much above the thresholds of point defect creation and sputtering. In ion beam ion plating, the coating material vapors are ionized at ion acceleration voltages up to 10 kV. Cluster-ion-beam deposition involves multiatomic clusters of the coating material. The ionized macroparticles with ion energies up to 10 keV are utilized to obtain a very high deposition rate (from several nanometers per minute to several micrometers per minute. Details of various ion beam ion-plating processes follow.

Direct Ion Beam Deposition. In direct (or primary) ion beam deposition, also called *low-energy ion beam deposition*, ions of the coating material itself are generated in suitable sources and are condensed and neutralized on the substrate surface [see Fig. 9.31(*a*)]. The ion beams of the coating material in the low-energy range of about 10 to 100 eV are generated, i.e., below or not too much above the thresholds of point defect creation and sputtering. Of particular interest is the formation of alloys or compounds which possess metastable structures by this kind of low-energy implantation (Aisenberg and Chabot, 1971; Amano and Lawson, 1977; Gautherin and Weissmantel, 1978; Harper, 1978; Miyazawa et al., 1984; Spencer et al., 1976; Weissmantel, 1981; Weissmantel et al., 1979a; Yagi et al., 1977). Aisenberg and Chabot (1971), Spencer et al. (1976), Weissmantel (1981), Weissmantel et al. (1982), and Miyazawa et al. (1984) used C^+ ion beams to deposit diamondlike carbon coatings.

Ion Beam Plating. In ion beam plating, a fraction of the particles provided by a flux of evaporated neutrals is post-ionized and accelerated in the direction of the substrate. The film deposition occurs by the condensation of a mixture of ionized and neutral species of one or several components. The essential part is the ionization system which is comprised of a hot cathode, two negatively biased re-

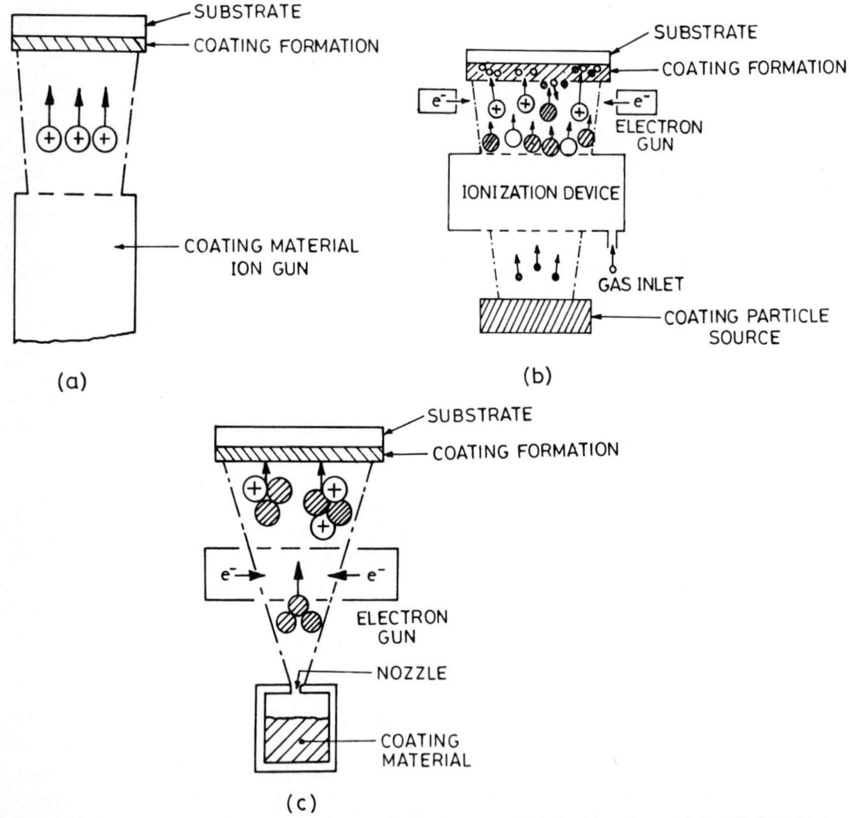

FIGURE 9.31 Schematics of ion beam configurations used for coating deposition: (*a*) direct ion beam deposition, (*b*) ion beam plating, (*c*) cluster (or molecule) ion beam deposition.

flecting electrodes, and a wire-mesh anode [see Fig. 9.31(*b*)]. An effective ionization frequency at rather low pressures of 10^{-3} to 10^{-2} Pa (10^{-5} to 10^{-4} torr) is accomplished (Weissmantel et al., 1979a). Weissmantel et al. (1979a) deposited Cr-C coatings from evaporated metal and benzene vapor and hard-carbon coatings from ionized hydrocarbon (benzene) vapors at ion acceleration voltages of about 1 keV.

Figure 9.32 shows an ion beam ion-plating system in which the ions are produced in a Kaufman ion source and then extracted at the desired energy (250 eV to 2 keV). After extraction through a multiaperture screen grid into the vacuum system, the ions impinge on the substrate. Deposition atoms are obtained by using a shielded evaporation source or by sputtering with a portion of the ion beam. The charge compensation filament minimizes space charge effects on the beam and charge buildup on insulator surfaces.

Cluster (or Molecule) Ion Beam Deposition. Multiatomic clusters (containing from 5×10^2 to 1×10^3 atoms) of the coating material are formed by adiabatic expansion of the vaporized material through the crucible nozzles into a high-vacuum region. The resulting cluster flux is ionized by passing through an electron bombardment system, and the cluster ions are accelerated in the direction of the substrate, where they condense [see Fig. 9.31(*c*)]. In addition to the

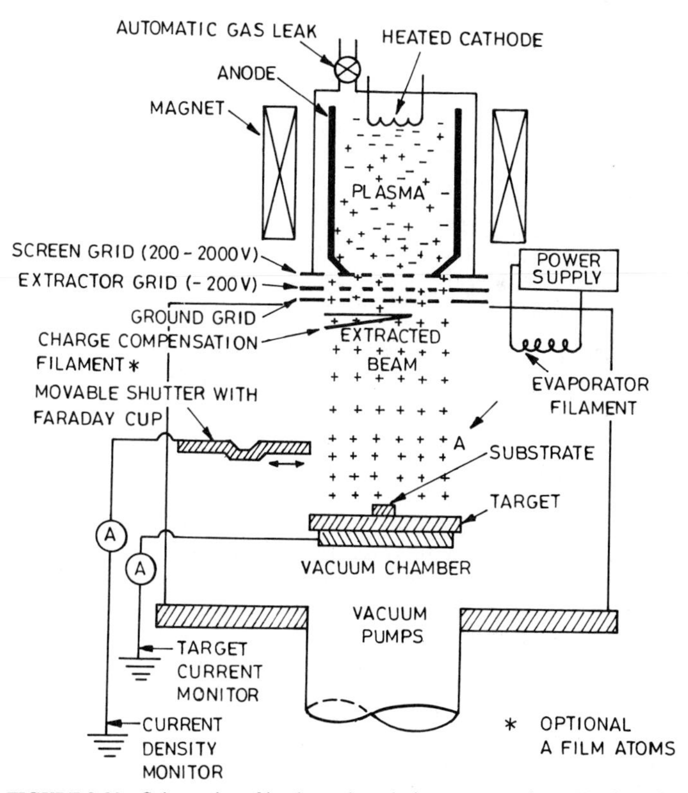

FIGURE 9.32 Schematics of ion beam ion-plating system using a Kaufman ion source and an evaporator filament source. [*Adapted from Mattox (1982).*]

ionized clusters, there is a flux of neutral aggregates. Since the kinetic energy is shared among many atoms, low energies per atom, say a few electronvolts, can be realized even at high bias voltages. The small charge-to-mass ratio reduces both space charge problems and the charging of an insulating surface during deposition of the layers. Another advantage is that the local activation caused by a cluster impact will extend over a larger region of the substrate or coating surface. Obviously the same effects can be obtained if stable multiatomic molecules are used instead of clusters. The ionized macroparticles are utilized to obtain a high deposition rate (from several nanometers to several micrometers per minute) and improved film quality (Takagi et al., 1976, 1979). Coatings of hard carbon have been deposited, at deposition rates up to 0.1 $\mu m\ min^{-1}$, by condensing ionized and accelerated molecules of benzene and condensed aromatic hydrocarbons. Diamondlike carbon coatings are obtained at ion energies from 300 to 1000 eV; a further increase promotes radiation damage (Weissmantel, 1981). Takagi et al. (1976) produced copper films on glass substrates with high adhesive strength.

9.2.3 Coatings Deposited by Ion Plating

9.2.3.1 Coating Materials. A wide variety of metallic and nonmetallic coatings have been applied onto metallic and nonmetallic (including polymer) substrates by ion-plating processes. Some of the common coatings applied by this process for tribological applications are as follows (Banks et al., 1985; Castellano et al., 1977; Chambers et al., 1972; Harker and Hill, 1972; Holland and Ojha, 1976, 1979; Johansen et al., 1987; Kashiwagi et al., 1986; Kobayashi and Doi, 1978; Mattox, 1982; Nakamura et al., 1977; Randhawa, 1987; Teer and Salama, 1977; Tsukizoe et al., 1977; Weissmantel et al., 1974, 1979a, 1980, 1982).

Metals	Ag, Au, Pb, Sn, Al, Cu, Ni, Cr, Co, Ti
Alloys	Cu-Pb, Cu-In, Ag-Pb, Ag-In, Cu-Ag-Pb-In, Pb-Sn, Al-Zn, Cu-Al-Fe, Co-Cr, Pt-W, Ti-6Al-4V
Nonmetals	Graphite, hard carbon, polytetrafluoroethylene
Oxides	Al_2O_3, Cr_2O_3, SiO_2, ZrO_2, TiO_2, Ta_2O_5, In_2O_3
Carbides	WC, TiC, Cr-C, SiC, B_4C, Si_3N_4, TiCN, (Ti-Ta-Nb)C, VC, NbC, Mo_2C
Nitrides	AlN, Cr-N, TiN, HfN, ZrN, NbN, TaN, WN, Si_3N_4, (Ti-Zr)N, (Ti, Al)N.

These coatings are applied by various ion-plating processes under various conditions of substrate biasing, evaporant species source, and degree of ionization.

Ion plating is much preferred to other coating processes because of the excellent uniformity of thickness, excellent adhesion, and good throwing power. Ion-plated coatings find use in optics, nuclear, aerospace, and electronics industries and metal cutting tools. Aluminum coating on uranium reactor parts was one of the first applications of ion plating. Aircraft and spacecraft steel and titanium fasteners are ion-plated with aluminum. A series of more unique applications include spacecraft telepoint rotors, missile launchers, high-strength-steel aircraft components, and aircraft door extension springs. Junction diodes have been made by the deposition of *n*-type silicon on a *p*-type substrate surface. Semiconductor materials, Si, GaAs, InSb, and PbSnTe have been ion-plated. Ion-plated metal cutting tools show superior wear performance.

9.2.3.2 ***Properties of Ion-Plated Coatings.*** The properties such as morphology, crystalline structure and grain size, stoichiometry, microhardness, and adhesion of the coating are strongly influenced by the process parameters such as ionization current density, ion energy, partial pressure of the reactive gases, deposition rate, substrate temperature, and substrate roughness. The morphology of a coating depends on how the evaporant atoms are incorporated into the existing structure. On a smooth surface, preferential growth of one area over another may result from varying surface mobilities of the evaporant atoms, usually as a result of grain orientation. Preferential growth leads to a dominant grain orientation and a surface roughening with thickness. As the surface gets rough, geometric shadowing will lead to preferential growth of the elevated regions, giving a columnar morphology to the coating. An elevated substrate temperature will affect the morphology by increasing surface mobility, enhancing bulk diffusion, and allowing recrystallization to occur (Movchan and Demchishin, 1969b). Structural order is generally controlled by the surface mobility of the atoms. Low mobility will lead to amorphous or small-grained material. High mobility will lead to large-grained material and more structural perfection.

The ion bombardment in ion plating usually leads to small grain size, high defect composition, low recrystallization temperature (metals), high intrinsic stresses, metastable crystallography and phase composition, and many other different features compared to sputtering and evaporation techniques. The ion-plating process also leads to the formation of a graded fused interface, regardless of the coating-substrate combination, which provides excellent coating adherence and an unusual ability to withstand interfacial stresses (Mattox, 1977; Spalvins, 1980a, 1980b; and Walls et al., 1978). The exact reaction mechanism which contributes to the formation of such an interface is not fully understood, but the controlling factors are the sputter-etched surface, diffusion, implantation, atomic mixing, and nucleation features. Some strengthening effects are observed which are believed to be due to a structural alteration of the crystal lattice and subsequent effects of dislocation interactions, where the coating acts as a barrier to the progress of dislocations near the surface, thus exhibiting pronounced mechanical strength effects such as increased yield, tensile, fatigue, and creep strengths. A morphological change from the normally columnar structure to a more dense, equiaxed grain structure is observed. This change gets more pronounced with an increase in ionization efficiency, which means that it is possible to produce dense, fine-grained, equiaxed coatings. Continuous coatings are obtained at lower nominal thicknesses than with other conventional deposition techniques.

Gabriel and Kloos (1982) have reported the influence of ionization current density, partial pressure of reactive gas, and deposition rates on the morphology of TiN coatings applied by triode ion-plating. The coating becomes fine-grained and smoother with increased current density from 0.7 to 1.5 mA cm^{-2} (Fig. 9.33).

The influence of substrate bias voltage and partial pressure of Ar gas during ion plating on microhardness is illustrated by Fig. 9.34 for copper coatings applied by a dc diode ion-plating system (Enomoto and Matsubara, 1975). The deposition rate ranged from 1.5 to 6 μm min^{-1}. Coatings deposited under enhanced ionization exhibit higher microhardness. The increase in hardness is attributed to the densification of the coating at high substrate bias and ionization levels.

9.3 SPUTTERING

Sputtering is a process whereby coating material is dislodged and ejected from the solid surface due to the momentum exchange associated with surface bom-

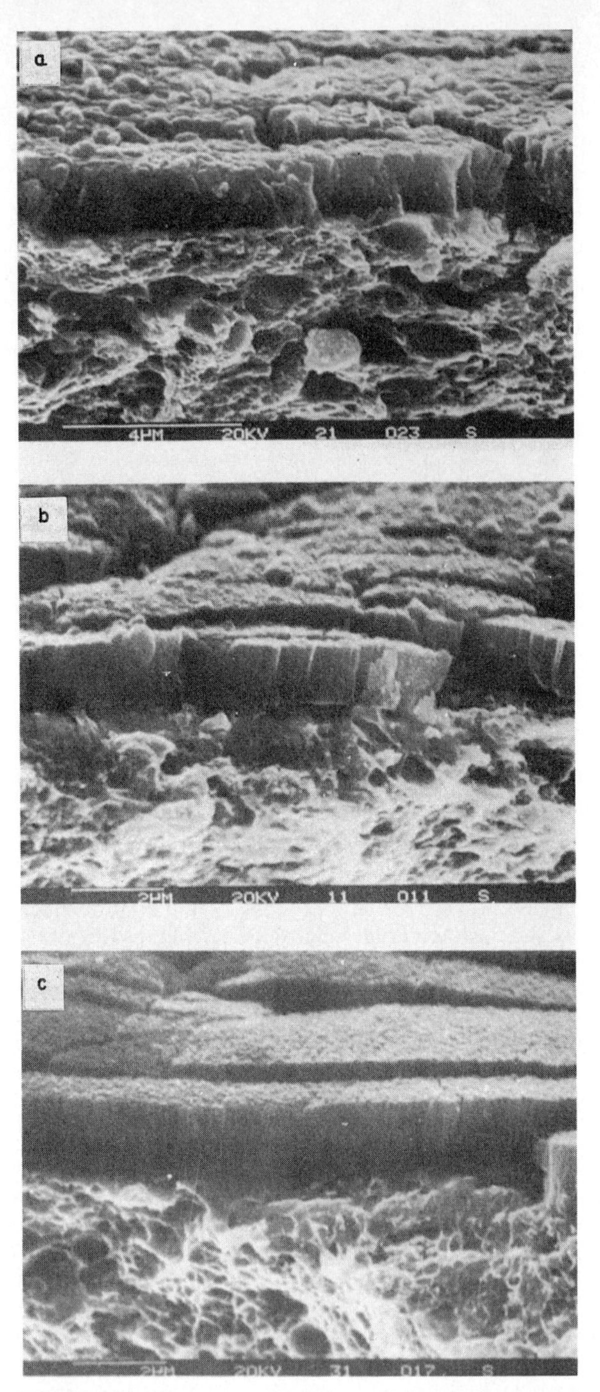

FIGURE 9.33 Fracture cross sections of glow-discharge triode ion-plated TiN coatings at substrate current densities of (*a*) 0.7 mA cm^{-2}, (*b*) 1.1 mA cm^{-2}, and (*c*) 1.5 mA cm^{-2}, under typical conditions: deposition rate of 10 nm s^{-1}, acceleration voltage of 1 kV, and deposition pressure of 0.56 Pa (4.2 mtorr). [*Adapted from Gabriel and Kloos (1984).*]

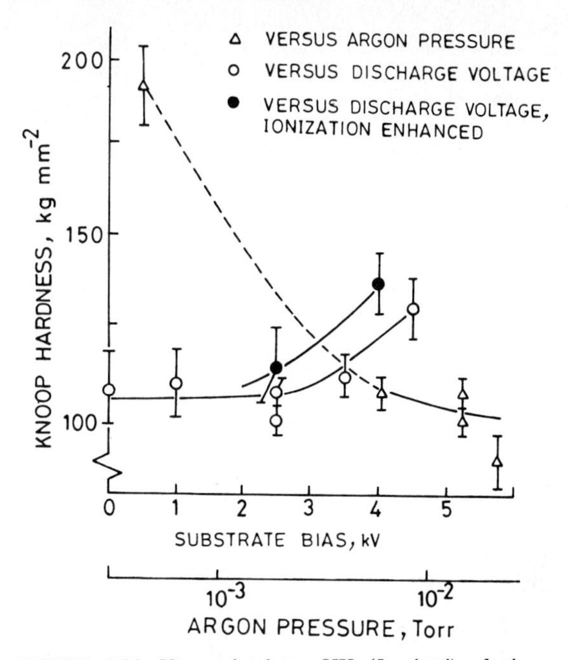

FIGURE 9.34 Knoop hardness HK (5-g load) of glow-discharge ion-plated copper coatings as a function of substrate bias V at 2 Pa (15 mtorr) and as a function of partial Ar pressure at −2.5-kV substrate bias. [*Adapted from Enomoto and Matsubara (1975).*]

bardment by energetic particles. The high-energy particles are usually positive ions (and energetic neutrals) of a heavy inert or reactive gas or species of coating material (argon is the most commonly used inert gas). The sputtered material is ejected primarily in atomic form from the source of the coating material, called the *target*. The substrate is positioned in front of the target so as to intercept the flux of sputtered atoms. The sputtering momentum exchange occurs primarily within a region extending only about 1 nm below the surface. The processes used for producing ions are glow-discharge ionization and ion beam sources (Berry et al., 1968; Carter and Colligen, 1969; Harper, 1978; Holland, 1974; Maissel, 1970; Takagi, 1983; Thornton, 1982; Townsend et al., 1976; Vossen and Cuomo, 1978; Weissmantel et al., 1979a, 1979b).

In the glow-discharge sputtering process, a source of coating material called a target is placed into a vacuum chamber which is evacuated to 10^{-5} to 10^{-3} Pa (10^{-7} to 10^{-5} torr) and then backfilled with a working gas, such as an inert gas, say, argon, to a pressure (5×10^{-1} to 10 Pa or 5×10^{-3} to 10^{-1} torr for conventional planar diode processes) adequate to sustain a plasma discharge. A negative bias (0.5 to 5 kV for dc device) is then applied to the target so that it is bombarded by positive ions from the plasma. Ions accelerated by this voltage impact on the target and cause ejection sputtering of the target material. The substrates are positioned so as to intercept the flux of sputtered atoms, where they condense to form a coating of the target material. In the ion beam sputtering process (or secondary ion beam deposition process), the ion bombardment source is an external ionization source used to sputter away the coating material from the target (at a pressure of 10^{-5} to 10^{-2} Pa or 10^{-7} to 10^{-4} torr). Ion beams can be of inert or reactive gas ions or ionized species of coating materials.

The bombarding ion energies in glow-discharge sputtering are typically in the range of 100 to 1000 eV,* and for external ion beams the range is typically from 100 eV to 10 keV. The atoms on the surface of a solid are typically bonded with energies (sublimation energies) of 2 to 10 eV. The average energies of sputtered atoms are in the range of 10 to 40 eV (even higher, up to 100 eV in the case of magnetron processes). This is in comparison to evaporation, where the evaporated atoms typically have energies of 0.1 to 0.3 eV. Thus the sputtering process is a relatively high-energy process compared to evaporation and many other coating technologies (Berry et al., 1968). Also target cooling is important, particularly at the high current densities used in some sputtering processes. The most striking characteristic of the sputtering process is its universality. The coating composition can be controlled with high precision. Since the coating material is passed into the vapor phase by a mechanical (momentum exchange) rather than a chemical or thermal process, virtually any material is a candidate for coating. The coating target can be formed by an almost unlimited collection of methods ranging from vacuum casting of metals and hot pressing of powders to forming composite targets by placing wires, strips, or disks of one material to cover a portion of the surface of another material.

Glow-discharge sputtering technology is limited, in the sense that the target current density and voltage cannot be independently controlled except by varying the working gas pressure. Ion beam sputtering permits independent control over the kinetic energy and the current density of the bombarding ions. Directionality controls of the beam permits a variable angle of incidence of the beam on the target and angle of deposition on substrate. Ions are generated in an external source which can be separated from the deposition chamber by pressure stages. Thus a low working pressure (10^{-5} to 10^{-2} Pa or 10^{-7} to 10^{-4} torr) can be maintained during the deposition process. Low pressure during deposition gives less gas incorporation, increased gas mean-free-path length, and controlled direction and higher mean energy of the ejected atoms striking the substrate owing to smaller gas collision losses (gas scattering). Since the target and substrates are not part of the power source circuit, substrate heating is minimized. However, because of the small ion beam sizes (~10-cm diameter), ion beam sputtering devices cannot compete as deposition sources with the very large substrate areas provided by magnetron-type glow-discharge sputtering devices; they are attractive for ion beam etching and for special deposition applications. Ion beam deposition with very high-energy ion beams leads to unique coating properties not obtained in conventional deposition methods. The relatively high energies of sputtered atoms in ion beam (and in some magnetron sputtering) processes may act to some degree to promote adhesion, for many of the sputtered atoms have sufficient energy to penetrate a few atomic layers into the substrate surface. Glow-discharge sputtering is the simplest and most commonly used technique for industrial applications.

Tables 9.1 and 9.2 summarize the range of coating materials, kinetic energy of evaporant species, application (substrate) temperatures, deposition pressures, coating thicknesses, deposition rates, and advantages and disadvantages of sputtering processes. In the following section, we describe the basic sputtering mechanisms and various sputtering configurations.

---

*The ionization energy for an argon atom is 15.76 eV.

9.3.1 Sputtering Mechanisms

The mechanism of ejection of target atoms into the vapor phase is largely a momentum transfer, involving the nuclei of the bombarding particles and the atoms in a relatively thin layer near the surface of the target (e.g., the penetration depth of a 1-keV argon ion incident on a copper target is only about three atomic layers) (Gillam, 1959; Holland, 1974; Moore, 1960). The distribution of atoms ejected from single-crystal or polycrystalline targets does not obey a cosine law. The noncosine behavior of the ejected material is illustrated in Fig. 9.35, which shows the angular distribution of molybdenum atoms ejected from a polycrystalline target by obliquely incident Hg^+ ions at 250 eV (Wehner, 1960). Wehner (1960) also reported that by bombarding single-crystal targets with low-energy ions (300 eV), the atoms are preferentially ejected in the directions of close packing. It is believed that the energetic ions cause chains of focused collisions similar to those which occur when one billiard ball strikes a tightly packed array of similar balls.

The sputtering process can be quantified in terms of *sputtering yield*, which is defined as the number of target atoms ejected per incident particle. The sputtering yield depends on target species, ion bombarding species, and its energy and angle of incidence. It has little dependence on the target temperature (Maissel and Glang, 1970). The sputtering process is also independent of whether the bombarding species is ionized or not, since ions have a high probability of being neutralized by a field-emitted electron prior to impact. Ions are generally used because of their ease of production and acceleration in a glow discharge. The yield tends to be greatest when the mass of the bombarding particles is of the same order of magnitude or larger than that of the target atoms. The use of inert gas avoids chemical reactions at the target and substrate. Accordingly, argon is gen-

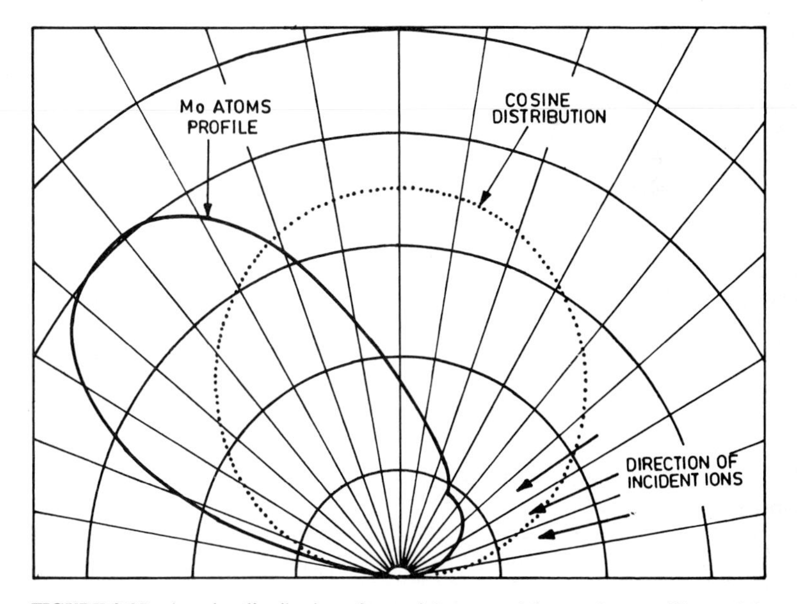

FIGURE 9.35 Angular distribution of material sputtered from polycrystalline molybdenum target under oblique incidence of Hg^+ ions. [*Adapted from Wehner (1960).*]

erally used because of its mass compatibility with materials of engineering interest and its low cost. However, krypton and xenon are sometimes used.

It is estimated that about 1 percent of the incidence energy of bombardment on the target surface produces ejected atoms, while about 75 percent causes target heating and the remainder is associated with emission of secondary electrons from the target (Maissel and Glang, 1970). Figure 9.36(*a*) shows yield versus argon ion energy data for several materials under normal ion incidence (Stuart and Wehner, 1962; Vossen and Cuomo, 1978). The yield dependence on the bombarding ion energy is seen to exhibit a threshold of about 10 to 30 eV, followed by a linear range which may extend to several hundred electronvolts. At higher energies the dependence is less than linear. Note that the yields of most metals are about unity and within an order of magnitude of one another. This is in contrast, for example, to evaporation where the rates for different materials at a given temperature can differ by several orders of magnitude. It is this universality that makes sputtering such an attractive process for many applications.

The general dependence of the yield on the ion angle of incidence is shown in Fig. 9.36(*b*) (Thornton, 1982). An ion which is incident on the target surface at an angle θ will, to first order, have its path length increased by a factor of Sec θ before it passes through the depth *d* where the primary sputtering momentum exchange occurs. At large angles of incidence, ion reflection dominates and the yield decreases. In glow-discharge sputtering devices, the ions generally approach the target in a direction normal to the target surface. The relationship shown in Fig. 9.36(*b*) is of particular significance when the target surface is highly irregular.

The sputtering momentum exchange process occurs primarily within a region extending only about 1 nm below the surface. To study the momentum exchange processes that occur during sputtering, we consider a particle of mass M_i and velocity V_i

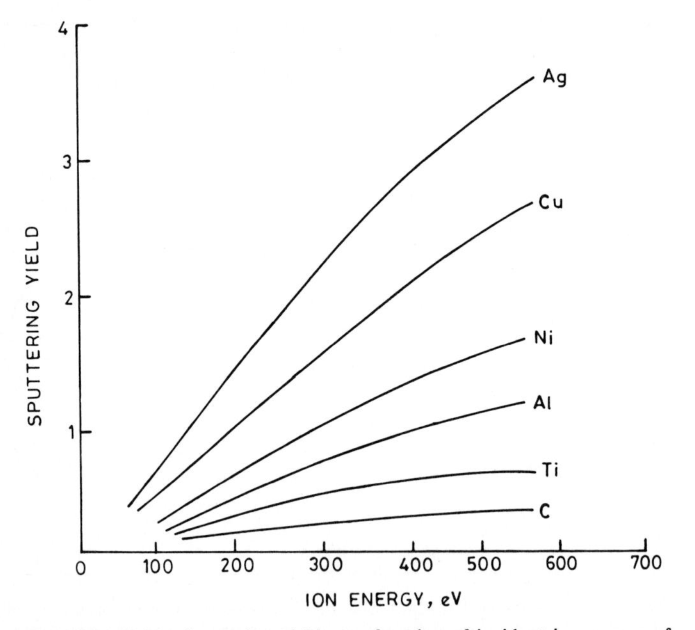

FIGURE 9.36(a) Sputtering yield as a function of incident ion energy of argon ions. [*Adapted from Stuart and Wehner (1962).*]

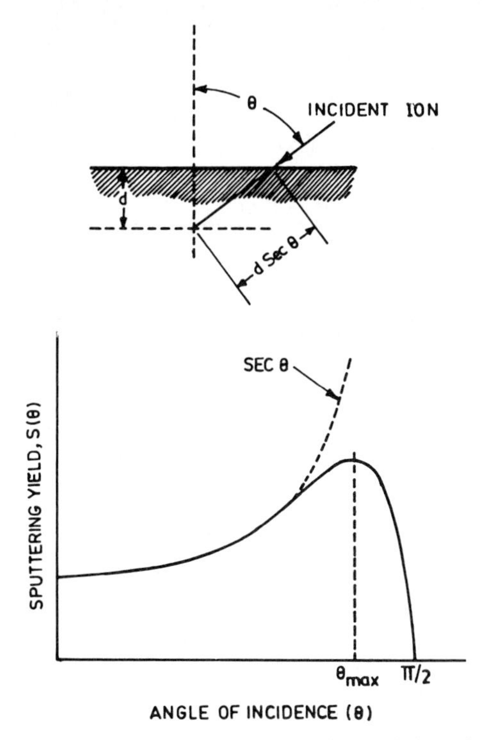

FIGURE 9.36(b) Schematic showing variation of sputtering yield with ion angle of incidence θ for constant ion energy; *d* is the depth of target in which momentum exchange occurs. [*Adapted from Thornton (1982).*]

which impacts on a line of centers with a target particle of mass M_t which is initially at rest ($V_t = 0$), as shown in Fig. 9.37(*a*). The momentum passed to the target particle drives it into the target. Thus, the ejection of a sputtered particle from a target requires a sequence of collisions so that a component of the initial momentum vector can be changed by more than 90°. The incident ion generally strikes two lattice atoms almost simultaneously. The low-energy knock-on receives a side component of momentum and initiates sputtering of one or more of its neighbors, as shown in Fig. 9.37(*b*). The primary knock-on is driven into the crystal, where it may be reflected and on occasion may return to the surface and produce sputtering by impacting on the rear of a surface atom, as indicated in the figure.

From a simple line-of-centers atomic collision a fraction (Moore, 1960)

$$\epsilon = \frac{4M_iM_t}{(M_i + M_t)^2} \tag{9.14}$$

of the kinetic energy of the incident particle which is transferred to the target particle. By assuming perpendicular ion incidence onto a target consisting of a random array of atoms with a planar surface, the sputtering yield S is given as (Thornton, 1982)

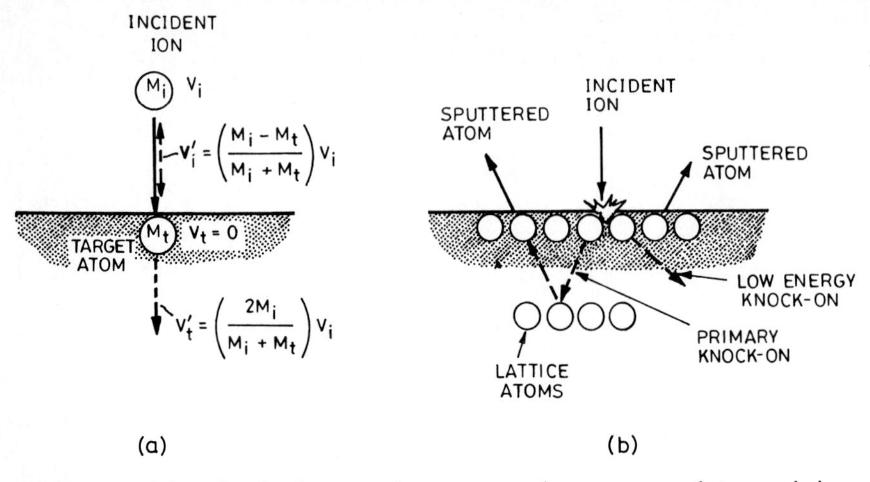

FIGURE 9.37 Schematics showing some of momentum exchange processes that occur during sputtering. V_i is ion velocity, V_t is target atom velocity, and prime denotes velocities after collision.

$$S = (\text{constant})\ \epsilon \frac{E}{U} \alpha \left(\frac{M_t}{M_i}\right) \tag{9.15}$$

The term $\alpha(M_t/M_i)$ is a near-linear function of M_t/M_i, E is the kinetic energy of the incident ion, and U is the heat of sublimation for the target material.

The sputtering (erosion) rate per unit area of target surface is given by (Thornton, 1973)

$$W = 6.23 \frac{JSM_A}{\rho} \quad \text{nm min}^{-1} \tag{9.16}$$

where J = ion current density, mA cm^{-2}
S = sputtering yield, atoms per ion
M_A = atomic weight of target material, g
ρ = density of target material, g cm^{-3}

Usually, the deposition rate is equal to W in the case of planar and long hollow cathodes. In long post cathodes, the deposition rate is equal to $(d_c/d_s)W$, where d_c and d_s are the diameters of the cathode and substrate mounting surfaces, respectively.

Another question of interest is the ultimate fate of the inert-gas ions that bombard the target. The probability of their being trapped in the target surface increases with ion energy above a threshold of about 100 eV. The amount of gas entrapped in the target can be large enough to influence the sputtering yield (Blank and Wittmaack, 1979).

9.3.2 Details of Sputtering Processes

The sputtering processes can be classified based on the means of producing high-energy ions: glow-discharge and ion beams from an ion source external to the sputtering system.

9.3.2.1 Glow-Discharge Sputtering. In the simplest self-sustained (planar diode) sputtering configuration, shown in Fig. 9.38, the target (a disk of coating material that is attached with solder or conducting epoxy to a backing plate which serves as a part of the cathode cooling channel) is connected to a negative voltage (dc) or rf power supply and faces the substrate to be coated. A gas is introduced to provide a medium in which a glow discharge can be initiated and maintained. In reactive sputtering, a reactive gas is also introduced to produce compound coatings. When the glow discharge is started, positive ions from the plasma strike the target with sufficient energy to dislodge the atoms by momentum transfer. The substrates are positioned so as to intercept the flux of sputtered atoms, which may collide repeatedly with the working gas atoms before reaching the substrate, where they condense to form a coating of the target material. The glow-discharge region in the sputtering deposition is of a form called *abnormal* glow discharge, to ensure the complete coverage of the cathode and thus a more uniform coverage of a substrate. A convenient rule of thumb is to keep the cathode-substrate distance about twice the thickness of the cathode dark space. Sputtering is an inefficient process from the viewpoint of energy. Most of the power input appears as target heating. Accordingly, the sputtering target is generally water-cooled to avoid interdiffusion of the target constituents.

The applied potential between the target and the substrate is typically from

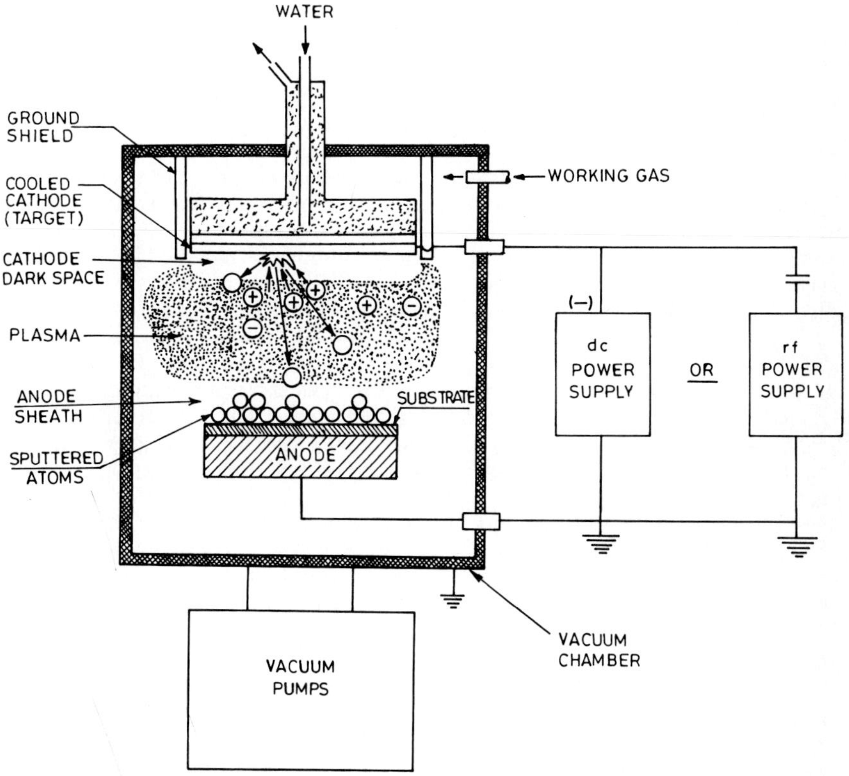

FIGURE 9.38 Schematic of a glow-discharge planar diode sputtering system using dc or capacitively coupled rf power supply.

500 V to 5 kV. If the target is made of an insulating material, it builds up the surface charge which prevents bombardment by positively charged particles. To permit ion bombardment of an insulating surface, a rf power supply is used which develops a self-bias voltage that is negative relative to the plasma floating potential on any surface which is capacitively coupled to a glow discharge (see Appendix 9.A). The discharge is operated at a frequency sufficiently high that significant ion charge accumulation does not occur during the cycle time when an electrode is serving as a cathode. Frequencies in the low megahertz range are required. Most apparatuses are operated at a frequency of 13.56 MHz. The rf sputtering system requires an impedance-matching network between the power supply and discharge chamber in order to set maximum power during glow discharge. At high frequencies, the matching network consists of *LC* networks with adjustable components, while low frequencies are matched through transformers (Jackson, 1970; Maissel and Glang, 1970; Vossen, 1971; Vossen and O'Neil, 1968).

The substrate is generally water-cooled for coating thermally sensitive substrates. By biasing the substrate to a negative potential (typically 100 to 1000 V) as an electrode prior to coating, conductive contamination is removed by sputtering, and coating nucleation sites are generated on the surface, minimizing the formation of interface voids. This process, known as *sputter cleaning* or *sputter etching*, is partially responsible for providing coatings with superior adhesion. An rf bias is required if the substrates are nonconducting. Typically sputter cleaning is performed for 5 to 20 min. Conditions are often selected to remove 20 to 500 nm of material from a surface. A suitable shield (e.g., shutter) should be used in sputter cleaning so that the contamination sputtered from the substrate does not pass onto the target.

Conventional glow-discharge diode-type sputtering essentially yields useful sputtering rates only in the 0.5 to 10 Pa (5- to 100-mtorr) pressure range since the density of ions required for sputtering falls rapidly with decreasing pressure. The lower concentration of trapped gas atoms, the increased gas mean-free-path length, and the controlled direction and higher mean energy of the ejected atoms striking the substrate owing to smaller gas collision losses (gas scattering) are some of the desirable features of sputtering at low pressures ($<10^{-1}$ Pa or 10^{-3} torr). Reasonable sputtering rates at low pressures may be obtained by increasing the supply of ionizing electrons and increasing the ionizing efficiency of the available electrons. Another alternative is ion beam sputtering.

Thermionically supported glow-discharge (triode) devices can be used to produce intense sputtering discharges at low pressures (10^{-3} to 10^{-1} Pa or 10^{-5} to 10^{-3} torr). In the triode system, ionization is increased by increasing the number of electrons available to produce ionizing collisions. Their deposition rates are higher than with planar diodes. A highly successful alternative means of increasing the ionization and thus the deposition rate is the magnetron. In the magnetron, the electrons are trapped by electric and magnetic fields, and they have extremely long paths along which they continue to cause ionization. The intense ionization is responsible for rapid sputtering, and the deposition rates (up to 2 μm min^{-1}) can be more than an order of magnitude larger than for planar diodes. Some magnetrons can be scaled to large sizes to provide uniform deposition over very large areas (few square meters). In addition, well-designed magnetrons virtually eliminate substrate heating due to electron and plasma bombardment. Thus magnetrons can yield the low substrate temperatures of 50 to 200°C that are necessary to deposit coatings on heat-sensitive substrates, such as plastics or porous casting materials.* Operating pressures in triode and

*Substrate heating is sometimes desirable to provide high substrate temperature for optimum coating structure and properties.

magnetron processes are sufficiently low (10^{-3} to 10^{-1} Pa or 10^{-5} to 10^{-3} torr) that the sputtered atoms may preserve much of their initial kinetic energy in passing to the substrates. Some important characteristics of various glow-discharge sputtering techniques are presented in Table 9.7. The descriptions of various sputtering configurations follow.

Various techniques are used to trap the impurities in coatings during deposition. Most commonly used techniques are bias sputtering and getter sputtering. Various techniques are used to produce multicomponent/compound coatings such as employing several targets simultaneously and reactive sputtering.

Diode Sputtering. The simplest self-sustained glow-discharge sputtering configuration is the planar diode, which consists of two planar electrodes, typically 2 to 30 cm in diameter, which are placed facing each other at a spacing of 5 to 10 cm, essentially as shown in Fig. 9.38. One of the electrodes contains the target

TABLE 9.7 Some Important Characteristics of Various Glow-Discharge Sputtering Techniques

Deposition technique	Deposition pressure, Pa (torr)	Deposition rate, μm min^{-1}	Deposition temperature,* °C	Comments
DC/RF diode	5×10^{-1}–10 (5×10^{-3}–10^{-1})	0.02–0.2	100–300	Simple, relative ease in fabrication and thickness uniformity over large area but relatively high deposition pressure and relatively high substrate temperature
Triode	10^{-3}–10^{-1} (10^{-5}–10^{-3})	0.05–0.5	100–150	Low deposition pressure, relatively high deposition rates and low substrate temperature. However, more complex, does not produce uniform coating over large surface areas, difficulty of scaling, vulnerability of thermionic emitter to reactive gases, and short filament life
Magnetron	10^{-3}–10^{-1} (10^{-5}–10^{-3})	0.2–2	100–150	Low deposition pressure, very high deposition rates, low substrate temperature, can be scaled up and commonly used for industrial production. However, more complex than planar diode systems

*With no external substrate heating.

(coating source material). It is usually water-cooled. The substrates are placed on the other electrode. The target mounting electrode is made the cathode when in operation with dc power. Applied potentials between the target and cathode are typically from 500 V to 5 keV. A rf power source capacitively coupled to the target can also be used. A rf power supply has to be used to produce coatings of insulating materials. It is difficult to sustain an intense plasma discharge in the planar diode electrode geometry. Thus working pressures are necessarily relatively high at 0.5 to 10 Pa (5 to 100 mtorr), and current densities are low (~1 mA cm^{-2}). The conductive substrate is biased for operation in bias sputtering mode. A reactive gas is used in the plasma for reactive sputtering to produce compound coatings with desirable stoichiometry.

A low-pressure glow discharge of a type known as abnormal glow discharge is maintained between the cathode and the adjacent anode. The current in such a discharge is carried in the vicinity of the negatively biased cathode, primarily by positive ions passing out of the plasma volume, and in the vicinity of the anode, by electrons passing from the plasma volume to the anode. Thus, a condition for sustaining the discharge is that the plasma volume be a suitable source of electrons and ions. At typical planar diode operating pressures, the motion of the ions across the cathode dark space is disrupted by collisions with gas atoms. In such collisions there is a high probability of charge exchange, particularly when noble-gas ions are passing through an atomic gas of their own species. A fast ion extracts an electron from a slow gas atom, as a result of which the fast ion becomes a "fast" neutral atom, while the "slow" atom becomes a slow positive ion, as indicated schematically in Fig. 9.39. Thus, instead of being bombarded by a current of ions having an energy equal to the potential drop across the cathode dark space, the target is bombarded with a much larger number of ions and atoms having an energy that is often less than 10 percent of the applied potential. This can cause a reduction in the rate of sputtering. The deposition rate in planar diodes is further reduced by gas scattering of the sputtered atoms.

In planar diode devices, the substrates are generally in contact with the plasma. Thus, even if the substrates are electrically isolated, they will be bombarded with plasma ions having kinetic energies of about 5 to 30 eV. The sub-

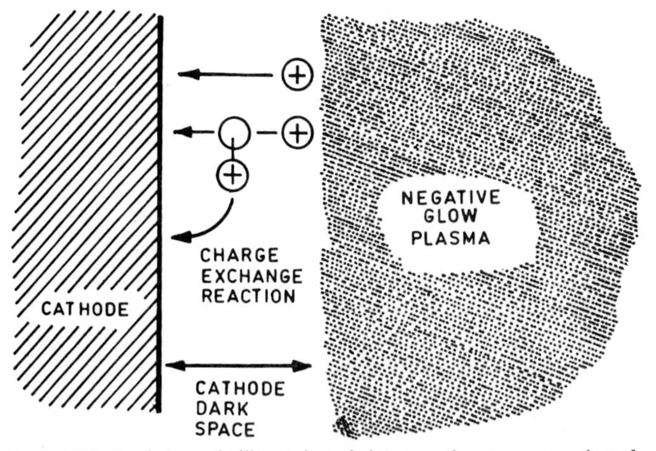

FIGURE 9.39 Schematic illustration of charge exchange process that affects ion transport across cathode dark space.

strates are also subjected to bombardment by energetic (100- to 1000-eV) primary electrons from the cathode (Thornton, 1982). These electrons are the primary source of substrate heating and cause damage in heat-sensitive substrates. In most cases, radiation is the primary cooling mechanism, even when the substrates are mechanically clamped to a cooled support. Thus even at moderate deposition rates (several tens of nanometers per minute) the substrate temperature can reach 300°C or higher in some cases.

Radio-frequency electromagnetic radiation can excite a discharge in a gas at low pressures ($<10^{-1}$ Pa or 10^{-3} torr) without the addition of any electrodes in the system. Gawehn (1962) has described a system using a high-frequency coil surrounding the sputtering chamber to induce the formation of a plasma. Using this system with a 500-V bias on the target and a 200-W input to the rf circuit, a deposition rate of 10 nm min^{-1} was reported. This technique offers the advantage over the hot-filament approach that there is no need for any additional electrodes in the system. However, the sputtering rates are very low.

Planar diodes are widely used despite the low deposition rates (typically 20 to 200 nm min^{-1}), substrate heating, and relatively small deposition surface areas. The reason is their simplicity and the relative ease with which planar targets can be fabricated from a wide range of materials to take advantage of the versatility of the sputtering process. With a planar diode configuration with large-area electrodes, thickness uniformity over a large area is relatively easy to attain. For example, with a circular target of 0.35-m diameter and a target-substrate distance of 75 mm, a thickness uniformity of better than 1 percent can be obtained.

Triode Sputtering. Coatings deposited by the diode sputtering process incorporate some impurities mainly because of high working discharge pressures. This pressure cannot be reduced because at low pressures, say 10^{-2} Pa (10^{-4} torr), the number of secondary electrons obtainable from the discharge is reduced to a great extent, and ultimately the glow discharge vanishes. It is possible to operate the glow discharge at low pressure by means of some other source of electrons other than those emitted through secondary emission of ions. This can be achieved in thermionically assisted (hot-cathode) glow-discharge triode mode by introducing a thermionic electron emitter, i.e., hot filament (hot cathode), and an anode held at about 50 to 100 V relative to the hot filament in the dc diode arrangement, as shown in Fig. 9.40. Electrons are emitted at the hot-filament surface by thermionic emission rather than ion bombardment. This relaxes the volume ionization requirement for sustaining the discharge. Consequently, hot-cathode triodes can be operated at low pressures (10^{-3} to 10^{-1} Pa or 10^{-5} to 10^{-3} torr). The ion currents may be several amperes. The density of ions in the plasma may be controlled by varying either the electron emission current or the voltage used to accelerate the electrons, while the energy of bombarding ions is controlled by controlling the target voltage (Berry et al., 1968; Maissel and Glang, 1970).

In triode devices, the substrates may or may not be in direct contact with the plasma. However, they will often be subjected to bombardment by energetic electrons originating from the target. Triodes operate at sufficiently low pressures that the sputtered atoms may preserve much of their kinetic energy in passing to the substrates. Furthermore, energetic neutral working gas atoms, which are generated at the cathode by the neutralization and reflection of ions, can also pass to the substrate with little loss of kinetic energy. However, these atoms can become trapped in the growing coating, particularly if the substrate temperature is low. Thus, the trapped argon content for coatings deposited with a triode has been found to be larger than that for similar coatings deposited with a planar diode

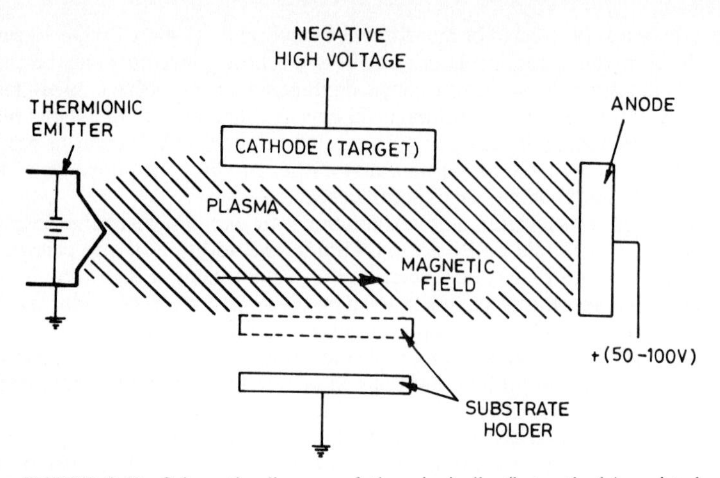

FIGURE 9.40 Schematic diagram of thermionically (hot-cathode) assisted glow-discharge triode arrangement.

(Lee and Oblas, 1975). Another disadvantage of triode sputtering is that this type of system does not produce a uniform coating from flat targets of large size, since the ion density (and thus the sputtering rate) is greatest along the axis of the electron beam and at the end nearest the hot filament.

Triodes permit high deposition rates (up to 0.5 μm min^{-1}) to be achieved even at low pressures (Tisone and Bindell, 1974; Tisone and Cruzan, 1975). Use of triodes has been limited by the complexity, difficulties of scaling, vulnerability (contamination or destruction) of the thermionic emitter to reactive gases in the case of reactive sputtering, and the short filament life. Consequently, magnetron sources (next subsection) are assuming primary importance as a high-rate sputtering process.

Magnetron Sputtering. The magnetically assisted (magnetron) glow-discharge sputtering process provides another method of increasing the ionization at low deposition pressures (10^{-3} to 10^{-1} Pa or 10^{-5} to 10^{-3} torr) and thus the deposition rate. Instead of increasing the number of electrons emitted at the cathode in the triode system, the efficiency of the available electrons is increased by a magnetic field, which allows sputtering at a lower pressure or provides greater current for a given applied voltage and pressure. One effect of the magnetic field is to cause electrons that are not moving parallel to it to describe helical paths around the lines of magnetic force. The radius of the helix decreases with increasing magnetic field. Thus, the electrons must travel over a longer path (along which they continue to cause ionization) to advance a given distance from the cathode. In this way, the magnetic field acts as though the gas pressure had been increased. Another effect of the magnetic field is the hindrance of radial diffusion of the electrons out of the glow, so that the loss of electrons which would otherwise be available for ionization is reduced (Vossen and Cuomo, 1978). The intense ionization is responsible for rapid sputtering and rates of deposition up to 2 μm min^{-1} are possible.

Magnetrons can be used for both dc and rf sputtering. Magnetrons operate at high current densities (~10 μA cm^{-2}), and deposition rates (up to 2 μm min^{-1}) can be more than an order of magnitude larger than those obtained with planar diodes. These sputtering rates are comparable to electron beam evaporation without the degree of radiation heating typical of thermal sources. Planar and cylindrical magnetrons can be scaled to large sizes to provide uniform deposition over very large areas (few square meters). In addition, the substrates are generally located beyond the magnetically confined plasma; therefore, well-designed magnetrons virtually eliminate substrate heating due to electron and plasma bombardment. Operating pressures are generally sufficiently low (compared to diode devices) that the sputtered atoms pass to the substrate with little loss of kinetic energy (Fischbein et al., 1972; Fraser, 1978; Holland, 1974; Münz and Hessberger, 1981; Münz et al., 1982; Schiller et al., 1977a; Thornton, 1978a, 1978b, 1981; Thornton and Penfold, 1978; Waits, 1978; Wasa and Hayakawa, 1967, 1969).

A variety of geometric arrangements give efficient magnetic effects such as the long rectangular planar magnetron (PM), cylindrical magnetron with post or hollow cathodes, and plasma ring sources, often referred to as *sputter gun* (trademark of Sloan Technology) and *S-gun* (trademark of Varian Associates). Figure 9.41 shows some of the more common types of magnetron sources (Fraser, 1978; Thornton and Penfold, 1978; Waits, 1978).

The most widely used is the planar magnetron, because of its simplicity of target construction [(Fig. 9.41(*a*)]. A magnetic field parallel to the cathode surface can restrain the primary electron motion and thereby increase the ionization efficiency (Schiller et al., 1977b; Waits, 1978). In essence, the planar magnetron is the classic dc or rf diode sputtering arrangement consisting of a planar (cathode) and its surrounding dark-space field with the essential addition of permanent magnets directly behind the cathode. Magnet arrangements can be varied substantially; all have in common a closed path or region where the magnetic field lines are parallel to the cathode surface. For instance, the paths may be single or multiple circles or ovals, concentric circles or ovals, etc. Permanent magnets, electromagnets, or a combination of both has been used. The maximum transverse component of the magnetic field in front of the target is typically in the range of 200 to 500 G, although the threshold flux density for the magnetron discharge may be as low as 80 to 90 G (McLeod, 1976). Planar magnetron sources usually are operating in argon at a pressure of 1 to 10 mtorr and at cathode potentials of 300 to 700 V. Under these conditions, current densities can vary from 4 to 60 mA cm^{-2} and power densities from 1 to 36 W cm^{-2}. Sputtering rates as high as 2 μm min^{-1} for Au and Cu can be achieved with planar magnetron sputtering.

The disadvantage of the planar magnetron is that the target erodes only in the transverse magnetic field region, which leads to a V-shaped erosion profile. This reduces the life of the target considerably. To overcome this, various means have been tried, such as targets of higher thickness, oscillating magnetic field, mechanical movement of magnets. Another disadvantage is coating thickness nonuniformity.

The distortion in plasma caused by $\vec{E}$ (electric flux density) × $\vec{B}$ (magnetic flux density) motion of the electrons in planar magnetrons can be avoided by using cylindrical cathodes which allow the $\vec{E} \times \vec{B}$ drift currents to close on themselves (Fraser, 1978; Thornton and Penfold, 1978; Tsukada et al., 1978). Electrons are again trapped within a plasma sheath surrounding the cylindrical cathode, approximately defined by the anode radius, and the cathode is eroded evenly. Two types—cylindrical post magnetron (CPM) and cylindrical hollow magnetron (CHM)—are used, but their performance is limited due to end losses. These losses are eliminated by proper shaping of the magnetic field or by using suitably placed electron reflecting surfaces, which result in the formation of an

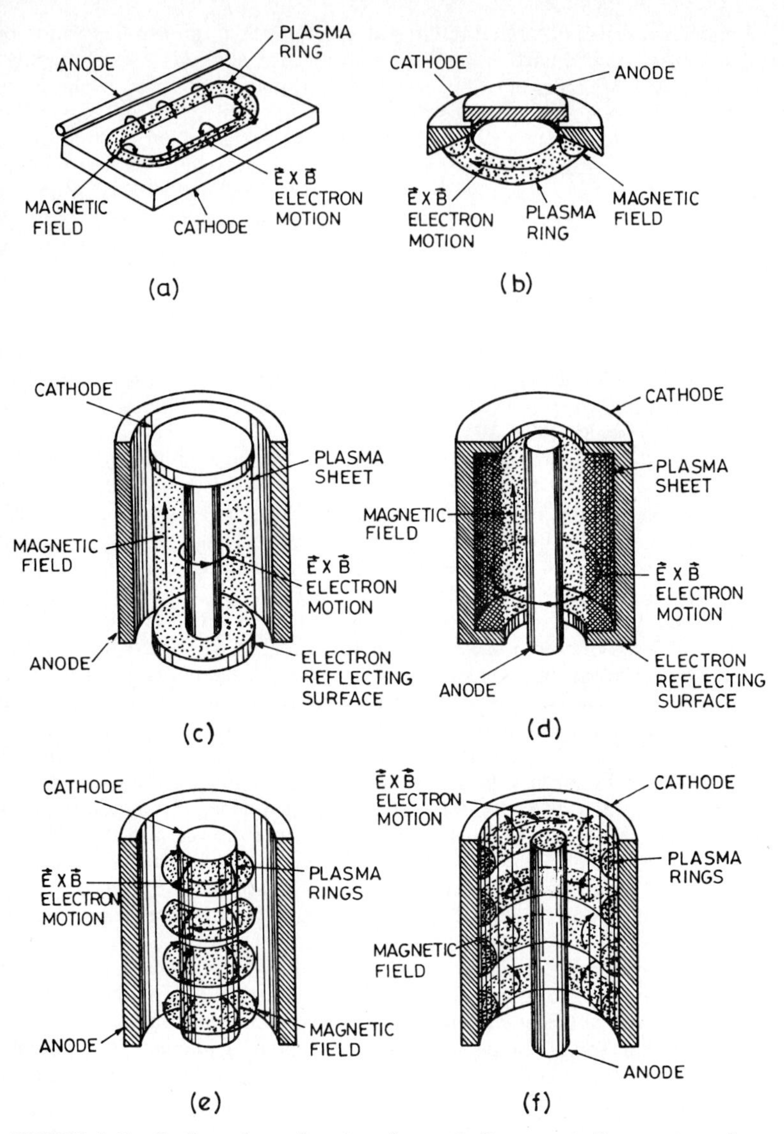

FIGURE 9.41 Configurations of various (magnetically supported) magnetron glow-discharge sputtering sources: (*a*) planar magnetron, (*b*) sputter gun type of magnetron, (*c*) cylindrical post magnetron with electrostatic end confinement (EEC), (*d* and *f*) cylindrical hollow (also called *inverted*) magnetron with EEC, (*e*) cylindrical magnetron with EEC. [*Adapted from Thornton and Penfold (1978).*]

intense uniform plasma sheet over the cathode, as shown in Fig. 9.41(*c*) and (*d*), or with curved magnetic fields which form a series of plasma rings over the cathode, as shown in Fig. 9.41(*e*) and (*f*). These are all cylindrical magnetrons, but the hollow-cathode types [Fig. 9.41(*d*) and (*f*)] are often known as *inverted magnetrons*. The term *cylindrical post magnetron* is used to describe type in Fig.

9.41(*c*), and *cylindrical hollow magnetron* describes the type in Fig. 9.41(*e*). The magnetic field can be applied externally by using electromagnets or permanent magnets or internally by putting a number of small permanent magnets inside the post cathode. When the magnetic mode is obtained from plasma rings sustained on short cylinders or within cylindrical surface cavities, as shown in Fig. 9.41(*b*), the devices are known as *sputter guns*.

Both planar and cylindrical magnetrons can be scaled up to large sizes. The cylindrical magnetron provides a convenient geometry for the batch processing of many substrates, while elongated planar magnetrons can be used for the continuous coating of large flat substrates, such as glass sheets carried over the plane target surface.

Trapping of Impurities in Coatings during Sputtering. The major sources of gaseous impurities in sputtering systems are residual gases (mainly H_2O) left after initial pump-out, wall desorption due to bombardment, the surface of the target (gases adsorbed when the system is vented), occluded gases in the sputtering target, backstreaming of pump fluids, leaks, and impurities in the gas supply itself. Many of these can be eliminated by a combination of good vacuum practice and extensive presputtering before coating deposition. Reactive impurities in the inert-gas sources can be reduced by passing the gas through hot Ti sponge before introducing it to the process chamber.

A significant improvement in the purity level in coatings during sputtering can be obtained by bias sputtering or getter sputtering. Note, however, that no methods are currently available to prevent nongaseous impurities present in the cathode from finding their way into the coating.

(a) BIAS SPUTTERING: The impurity content and structure of the coatings can be significantly controlled by applying a negative bias (typically up to 300 V) to the substrate at low ion energies (<200 eV) to cause surface ion bombardment during deposition, in order to remove loosely bonded contamination and create nucleation density which minimizes the formation of interface voids. This process is known as *bias sputtering*. The impurities are mainly the gaseous species such as Ar, O_2, N_2, H_2O, and hydrocarbons which result from the high working pressure (~1 Pa or 10^{-2} torr) during deposition. The basis for expecting bias sputtering to lead to coatings of high purity is that during resputtering most impurities (generally lighter atomic species) should be preferentially removed relative to the atoms of the main coating. The effectiveness of bias sputtering depends on the relative strength of the metal-to-impurity and metal-to-metal bonds. Several authors have reported improved coating density, electrical properties, magnetic properties, adhesion, and surface morphology (Christensen, 1970; Maissel and Glang, 1970; Westwood, 1976). In addition, the bias can influence the grain size, preferred orientation, structure, and adhesion of the coatings.

There are some disadvantages to the use of ion bombardment (bias sputtering). First, high ion flux is required. This means that the current density passed through the substrates must be of the same order of magnitude as that passed through the sputtering target. Second, the ion bombardment may change the chemical content of the coating by preferential sputtering. Finally, inert-gas ions with bombarding energies of more than 100 to 200 eV tend to become trapped in the growing coating. Inert-gas incorporation is generally greater in amorphous than in crystalline materials.

Closely related to bias sputtering is asymmetric ac sputtering. In this method, the coating is also resputtered during deposition, but this is accomplished by applying alternating current between the substrate (anode) and the target (cathode). By employing a suitable resistance network, a significantly larger current flows during the half cycle when the cathode is negative, so a net transfer of material to the substrate results. Asymmetric ac sputtering appears to be in all ways equivalent to bias sputtering except that, since the bias is applied for only half the total time, the concentration of trapped impurities will always be less in coatings deposited by bias sputtering than in ac sputtered coatings (Berry et al., 1968; Seeman, 1966).

(b) Getter Sputtering: This process reduces the amount of background contamination in the vicinity of the sputtering target and substrate by taking advantage of the fact that the depositing coating is an active getter for impurities. In a getter sputtering system, the sputtering gas is made to pass over an area of freshly deposited coating material before reaching the region where the coating which is to be retained is being deposited. In its simplest form, therefore, a getter sputtering system would consist of a very large cathode and an equally large substrate holder with the provision that the only coating material used be taken from what is deposited right near the center of this assembly. Any impurity atoms in the sputtering gas would have made a large number of collisions with depositing coating material before reaching the central region and would, therefore, have a very high probability of being removed. Such an arrangement thus cleanses the sputtering gas of its impurities, whether they were present in the original gas or generated through outgassing from the walls. However, the technique can do nothing about gaseous impurities that originate in the cathode target itself (Theuerer and Hauser, 1964; Vossen and Cuomo, 1978).

In a more practical getter sputtering arrangement shown in Fig. 9.42, the cathode is of finite size, but the walls of the sputtering chamber are arranged to be as close as possible to both the substrate and the cathode, subject only to not interfering with the glow discharge. The compact sputtering chamber (Fig. 9.42) is constructed so as to be moderately gastight and is placed inside the main vacuum chamber. This configuration allows the glow to emanate in all directions from the cathode. Argon is allowed to enter only at the top of the chamber so that most of the reactive gases entering with the argon reacted with the coating vapor and de-

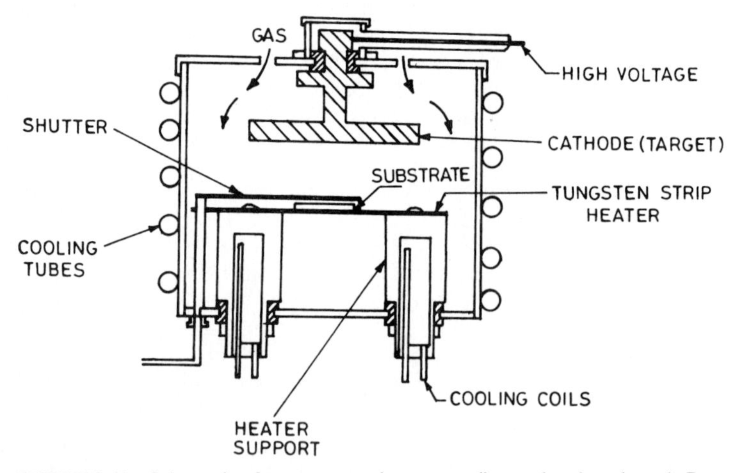

FIGURE 9.42 Schematic of getter sputtering system (inner chamber shown). Provisions were made to heat or cool the substrates. [*Adapted from Cook et al. (1966).*]

posited on the chilled walls of the chamber before reaching the deposition zone. Provision is also made as shown for heating and cooling the substrates. One of the most important features of the apparatus shown in Fig. 9.42 is the shutter used in sputtering before and after the deposition.

Sputtering of Multicomponent/Compound Coatings. Multicomponent/compound coatings can be formed by employing a target which is a mosaic of several materials, several targets simultaneously of different materials, a compound target, several targets sequentially to create a layered coating, or a reactive gas to introduce one or more of the coating materials into the chamber (known as reactive sputtering).

(a) Sputtering of Multicomponent Materials: An important feature of the sputtering process is that the chemical composition of a sputtered coating is often the same as that of the target from which it was sputtered. This is true even though the components of the system may differ considerably in their relative sputtering rates. The very first time the sputtering is performed from a multicomponent target, the component with the highest sputtering rate will come off faster, but an "altered region" soon forms at the surface of the target. This region becomes sufficiently deficient in the higher-sputtering-yield component to just compensate for its greater removal rate; so subsequent deposits have the composition of the parent material (Maissel and Glang, 1970; Vossen and Cuomo, 1978).

Compositional differences between sputtered coatings and their parent targets are found if a significant amount of resputtering occurs. In resputtering, material with the highest sputtering yield will be preferentially removed from the coating. Unlike the target, the surface of the growing coating is constantly being replenished with material of the target composition, so an altered layer cannot form. Another cause of compositional differences may be high target temperature. Not only can this cause diffusion into the altered layer, but also one or more of the components of the system may have a significant vapor pressure, which might lead to its evaporation from the target.

Multicomponent coatings can be synthesized by sputtering. Separate targets for each component can be used. All the sputtered materials may not be deposited directly beneath their point of origin on the target, but may end up some distance from this. This effect leads to a fair degree of intermingling of components sputtered from the interleaving targets. To obtain any particular alloy composition, the relative voltages to the targets (cathodes) are suitably adjusted.

(b) Reactive Sputtering: In reactive sputtering, the coating is synthesized into a compound by sputtering in a reactive gas atmosphere from a target of nominally pure metal, alloy, or mixture of species. This process allows the formation of complex compound coatings by using relatively easy-to-obtain metallic targets and without rf power supply. Even in sputtering the compound directly, it is usually necessary to add the reactive gas to compensate for the loss of the reactive component by dissociation. By changing the operating conditions during deposition, it is possible to form a graded interface at temperatures far below those required to cause significant thermal diffusion. A large number of reactive gases, as in activated reactive evaporation, are used to synthesize compound coatings from metallic targets.

A question of interest in reactive sputtering is where in the sputtering system the compound is formed. Does the reaction take place at the target surface, following which the reactive material is sputtered at the substrate surface, or does the reaction take place at the substrate itself? The bulk of available evidence suggests that the latter mechanism is more important. Reaction in the gas phase is unlikely because of problems of conserving momentum while also dissipating the heat of reaction. At low pressures of reactive gas and high target sputtering rate, virtually all the compound synthesis occurs at the substrate, and the stoichiometry of the coating depends on the relative rates of arrival of the target material and the gas atoms. Usually under such conditions, the rate of compound formation is much slower than the rate of sputtering of the target. However, as the reactive gas pressure is increased and/or the target sputtering rate is decreased, a threshold is reached at which the compound formation rate exceeds the sputtering rate. Thus the stoichiometry of sputtered coatings depends on the partial pressure of the reactive gas, which depends not only on glow-discharge conditions but also on the kinetics of compound formation (Maissel and Glang, 1970; Vossen and Cuomo, 1978).

A list of hard coatings applied by reactive sputtering and reactive ion-plating processes is presented in Table 9.8. In this table, those processes are included in which the starting material is a metallic target or metallic evaporant.

9.3.2.2 ***Ion Beam Sputtering.*** In ion beam sputtering (also called *secondary ion beam deposition*) processes, the ion bombardment source is an external ionization source (gun) used to sputter away the coating material from a target. The ion sources used to produce ions are the Kaufman and Duoplasmatron types. Ion beams can be of inert-gas ions (such as Ar^+ ion beams) or ionized species of coating materials. The energy of ion beams ranges typically from 100 eV to 10 keV. Ion beam sputtering processes are classified by whether they utilize a single beam or dual beams. In single-ion-beam sputtering, an ion beam is used for sputtering. Dual-ion-beam sputtering is a variation of ion beam sputtering in which the second gun bombards the substrate surface or the growing layer with relatively energetic ions (see Fig. 9.43). The resulting coatings are very hard with excellent adhesion (Castellano et al., 1977; Harper, 1978; Weissmantel, 1981; Weissmantel et al., 1979b). The coatings applied by ion beam sputtering at relatively low pressures in the range of 10^{-5} to 10^{-2} Pa (10^{-7} to 10^{-4} torr) are very pure with gas incorporation of less than 0.1 percent (Hinneberg et al., 1976).

Single-Ion-Beam Sputtering. In single-ion-beam sputtering, an ion beam of inert gas or reactive gas is used to sputter a target of coating material, which is condensed on a nearby substrate [see Fig. 9.43(*a*)]. An ion beam at energies ranging between 100 eV and 10 keV are used. The particles arriving at substrate comprise mainly sputtered atoms of some tens of electronvolts and usually a small fraction of sputtered ions and clusters. The deposition rate is on the order of several hundred nanometers per minute and is controlled by ion energy (Castellano et al., 1977; Gautherin and Weissmantel, 1978; Harper et al., 1982; Hinneberg et al., 1976; Mirtich, 1981; Weissmantel, 1981; Weissmantel et al., 1979a, 1982). Reactive ion beam plating can be used to deposit compounds of reactive species. In the first method, an inert-gas ion beam is used to sputter either the metal or the compound target, with reactive gas added to the deposition region. In the second method, an ion beam of the reactive gas itself or a mixture of inert and reactive gas is used. Usually, a mixture is preferable to the pure reactive gas in the beam to maintain a reasonable sputtering rate. Castellano et al. (1977) deposited Ni, Al, Ni_3Al, and Au coatings, and Weissmantel (1981) and Kaplan et al. (1985) deposited diamondlike carbon coatings. Weissmantel et al. (1979a) deposited Si_3N_4 coatings by sputtering a silicon target with a nitrogen ion beam.

Dual-Ion-Beam Sputtering. Dual-ion-beam sputtering is a variation of ion beam sputtering. In dual-beam deposition, two ion sources are used. The first gun is directed at the target at which a flux of sputtered species (neutrals) is generated

TABLE 9.8 Metallic Compound Coatings Deposited by Reactive Ion-Plating and Reactive Sputtering Techniques

Source	Gases	Source	Gases	Source	Gases
Ag	Air	Pb	Ar+O_2	Zn	Ar+H_2S
Ag	Ar+N_2	Pb	O_2	Zn	Ar+O_2
Ag	Ar+O_2	Pt	Air	Zr	Ar+N_2
Al	Ar+N_2	Pt	Ar+O_2	Zr	Ar+O_2
Al	Ar+O_2	Sb	Ar+O_2	Al-Si	Ar+O_2
Al	O_2	Si	Ar+C_2H_2	Au-Ta	Ar+O_2
Au	Air	Si	Ar+N_2	Au-W	Ar+O_2
Au	Ar+O_2	Si	Ar+NH_3	Ba-Ti	Ar+O_2
Bi	Ar+O_2	Si	Ar+O_2	Bi-Ta	Ar+O_2
C	N_2	Si	C_2H_2	Bi-Ti	Ar+O_2
Cd	Ar+H_2O	Si	N_2	Cd-Cu	Ar+H_2S
Cd	Ar+H_2S	Si	NH_3	Cd-Cu	Ar+O_2
Cd	Ar+O_2	Si	NH_3+SiH_4	Cd-In	Ar+H_2S
Cr	Ar+CH_4	Si	O_2	Cd-In	Ar+O_2
Co	Ar+O_2	Si	Ar+N_2 + O_2	Cd-Zn	Ar+H_2S
Cu	Ar+H_2S	Sn	Air	Cr-Mo	Ar+O_2
Cu	Ar+O_2	Sn	Ar+H_2S	Cu-Fe	Ar+O_2
Cu	Ne+O_2	Sn	Ar+N_2	Ga-Al	Ar+As
Fe	Ar+CH_4	Sn	Ar+O_2	Gd-Fe	Ar+O_2
Fe	Ar+O_2	Ta	Ar+CH_4	Hf-Ta	Ar+N_2
Ga	Ar+As	Ta	Ar+C_2H_2	In-Sn	Ar+O_2
Ge	Ar+N_2	Ta	Ar+CO	Li-Nb	Ar+O_2
Hf	Ar+C_2H_2	Ta	Ar+H_2	Nb-C	Ar+N_2
Hf	Ar+N_2	Ta	Ar+H_2O	Ni-Fe	Ar+O_2
Hf	Ar+O_2	Ta	Ar+N_2	Ni-Ti	Ar+C_2H_2
In	Ar+O_2	Ta	Ar+O_2	Pb-Te	Ar+O_2
Mg	Ar+O_2	Ta	Ar+O_2 + N_2	Pb-Ti	Ar+O_2
Mn	Ar+O_2	Ta	Ar+SiH_4	Pt-Ta	Ar+O_2
Mo	Ar+CH_4	Te	Ar+O_2	Pt-W	Ar+O_2
Mo	Ar+H_2S	Ti	Ar+CH_4	Si-Al	Ar+N_2
Mo	Ar+N_2	Ti	Ar+C_2H_2	Sn-In	Ar+O_2
Mo	Ar+O_2	Ti	Ar+H_2O	Sn-Sb	Ar+O_2
Nb	Ar+CH_4	Ti	Ar+N_2	Ta-Si	Ar+O_2
Nb	Ar+C_2H_2	Ti	Ar+NH_3	Ta-Ti	Ar+N_2
Nb	Ar+N_2	Ti	Ar+O_2	Ti-C	Ar+O_2
Nb	Ar+O_2	V	Ar+C_2H_2	Ti-Ni	C_2H_2
Nb	Ar+N_2 + CH_4	V	Ar+O_2	Zn-Cu	Ar+H_2S
Nb	Ar+N_2 + O_2	W	Ar+CH_4	Cd-Cu-In	Ar+H_2S
Ni	Ar+CH_4	W	Ar+O_2	Fe-Cr-Ni	Ar+CH_4
Ni	Ar+O_2	Y	Ar+O_2	Mg-Mn-Zn	Ar+O_2
Pb	Ar+H_2O	Zn	Ar+H_2O	Ti-Al-V	Ar+C_2H_2
Pb	Ar+H_2S			Pb-Nb-Zr-Fe-Bi-La	Ar+O_2
Pb	Ar+N_2				

Source: Adapted from Vossen and Cuomo (1978).

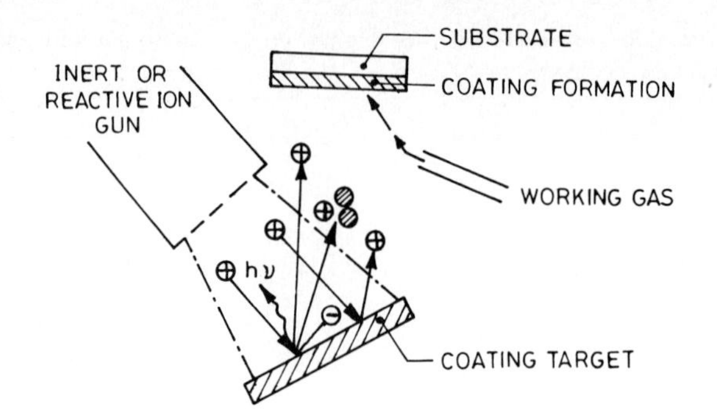

(a)

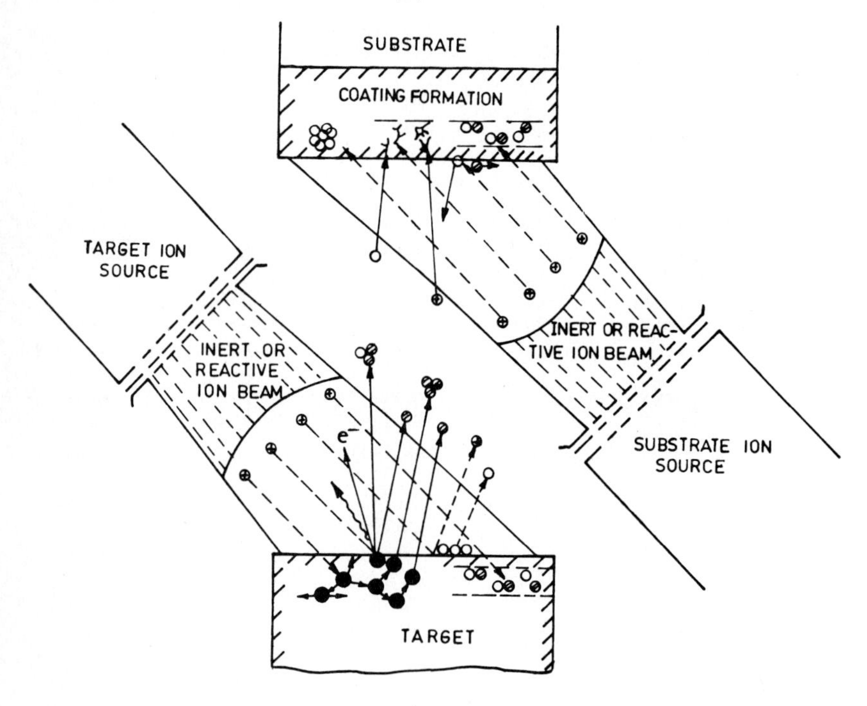

(b)

FIGURE 9.43 Schematics of (*a*) single-ion-beam and (*b*) dual-ion-beam sputtering configurations.

and directed onto the substrate, while the second gun generates an ion beam which bombards the substrate or the growing layer with relatively energetic ions [see Fig. 9.43(*b*)] (Gautherin and Weissmantel, 1978; Harper and Gambino, 1979; Harper et al., 1982; Marinov, 1977). If inert-gas ions (such as Ar^+ or Kr^+) are used, the impinging species will cause local activation by energy and momentum transfer. Alternatively, dual-beam compound synthesis is accomplished when ionized species of the second component are used. Direct ion bombardment of the substrate results in superior mechanical properties and adhesion similar to ion beam ion-plating processes. Weissmantel et al. (1979b, 1982) deposited coatings of metals, alloys, polymers, oxides, nitrides, and diamondlike carbon, using the dual-beam deposition method. For example, in the deposition of Si_3N_4 and AlN coatings, the first gun was operated with argon, thus producing a sputtered beam from the Si or Al target, whereas nitrogen was supplied to the growing film by its simultaneous bombardment with 0.68-keV N_2^+ obtained from the second ion source.

9.3.3 Coatings Deposited by Sputtering

9.3.3.1 Coating Materials. Sputtering is one of the most versatile techniques for applying coatings of metals, alloys, nonmetals, intermetallic compounds, refractory compounds, and their combinations. The typical coatings which are relevant to tribological applications are as follows (Anonymous, 1972; Bhushan, 1979, 1980b, 1981b; Castellano et al., 1977; Holland, 1974; Jackson, 1970; Johansson et al., 1984; Maissel and Glang, 1970; Morrison and Robertson, 1973; Münz et al., 1982; Savvides and Window, 1985; Shikama et al., 1983; Spalvins, 1980b; Sundgren, 1985; Thornton, 1982; Thornton and Cornog, 1977; Thornton and Ferriss, 1977; Vossen and Cuomo, 1978; Weissmantel, 1981; Weissmantel et al., 1979a, 1979b):

Metals	Al, Cu, Zn, Au, Ni, Cr, W, Mo, Ti, austenitic steel
Alloys	Ag-Cu, Pb-Sn, Al-Zn, Ni-Cr
Nonmetals	i-C, graphite, MoS_2, WS_2, PTFE, polyimide, poly(amide-imide)
Refractory oxides	Al_2O_3, Cr_2O_3, Al_2O_3-Cr_2O_3, SiO_2, ZrO_2, ZrO_2-Y_2O_3, TiO_2, Ta_2O_5, PbO, ZnO, SnO_2, In_2O_3
Refractory carbides	TiC, ZrC, HfC, NbC, Ta_2C, VC, Mo_2C, B_4C, SiC, WC, W_2C, Cr_3C_2, WC-Co, TiC-Ni, TiC-ZrC, ZrC-HfC
Refractory nitrides	TiN, Ti_2N, ZrN, HfN, MoN, Si_3N_4, i-BN, Cr-N, AlN, TiN-ZrN, TiN-AlN, TiN-AlN-ZrN
Refractory borides	TiB_2, ZrB_2, HfB_2, CrB_2, Mo_2B_5
Refractory silicides	$MoSi_2$, WSi_2, Cr_3Si_2

The coatings just mentioned have been applied by various sputtering techniques. Many of the refractory compound coatings are deposited by reactive sputtering.

The ability to control coating composition makes sputtering useful in the electronics industry for applications such as aluminum-alloy and refractory-metal microcircuit metallizing layers, oxide microcircuit insulating layers, piezoelectric transducers, thin-film resistors, thin-film capacitors, thin-film lasers, and microcircuit photolithographic mask blanks. Sputtered coatings are used for various metrological applications, e.g., thin-film strain gages and pressure transducers (Cheney and Samek, 1977). The combination of hardness and corrosion resistance of various sputtered coatings makes them suitable for many decorative and tribological applications. Gold-colored 0.5-μm-thick TiN coating is used instead of gold coating on fine jewelry such as watches (Münz and Hessberger, 1981). Titanium nitride coatings are also used as protective coatings on high-speed steel cutting tools. A 3-μm TiN coating with hardness of more than 2000 kg mm^{-2} on hobs shows a typical decrease in flank wear by a factor of 10, compared to the untreated tool (Münz and Hessberger, 1981; Sundgren, 1985). Sputtered Cr_2O_3 coatings in gas bearings have shown a significant improvement in wear life compared to that of other coated materials. A sputtered Cr_2O_3 coating on the bearing sliding against the detonation gun Cr_3C_2-coated journal has sustained more than 9000 start/stop cycles and 30,000 high-speed rubs (Bhushan, 1979, 1980b, 1981b). Sputtered coatings of solid lubricants such as MoS_2 exhibit good adhesion and low friction (down to 0.04) (Spalvins, 1980b). Even under hardest space conditions, these coatings offer more than 10^4 times the lifetime of conventional brushed-in MoS_2 coatings (Spalvins, 1983).

Magnetron sources have opened up more new applications because of their large-area capability, high sputtering rates, and reduced substrate heating. Large in-line systems with vacuum interlocks and roll coatings are used to coat 2- by 3.5-m architectural glass plates or thin plastic films for solar control applications. Magnetrons are also used on a production basis to deposit chromium decorative coatings on plastic automobile grilles and other exterior trims.

9.3.3.2 Microstructure of Glow-Discharge Sputtered Coatings. In the sputtering process, an atom undergoes a three-step process during the deposition. First, incident atoms transfer kinetic energy to the substrate lattice and become loosely bonded "adatoms." Second, the adatoms diffuse over the surface and exchange energy with the lattice and other atoms adsorbed on the surface until they either are desorbed by evaporation or back-sputtering or become trapped at low-energy sites. Finally, the incorporated atoms readjust their position within the lattice by bulk diffusion processes. The metallurgical structure of the coating that develops in a specific situation will be determined by the kinetics of the above processes, i.e., by the mobility of the adatom on the surface prior to finding a trapping site.

Movchan and Demchishin (1969b) represented evaporated coatings as a function of T/T_m in terms of three zones, each with its own characteristic structure and physical properties (see Fig. 9.21). Thornton (1974, 1977a) extended this model to sputtering in the absence of ion bombardment by adding an axis to account for the effect of the sputtering gas (see Fig. 9.44). The numbered zones correspond to those suggested by Movchan and Demchishin. In addition, a transition zone (zone T) was identified by Thornton between zones 1 and 2.

The zone 1 structure results when adatom diffusion is insufficient to overcome the effects of shadowing. It therefore forms at low T/T_m and is promoted by an elevated working gas pressure. It usually consists of tapered crystals with domed tops which are separated by voided boundaries. An elevated inert gas pressure at low T/T_m causes the zone 1 boundaries to become more open. Lower pressures give less gas incorporation (Carson, 1975). The crystal diameter increases with T/T_m. Shadowing introduces open boundaries because high points on the growing surface receive more coating flux than valleys, particularly when a significant oblique flux is present. The coating surface roughness can result from the shapes of initial growth nuclei, from preferential nucleation at substrate inhomogeneities, from substrate roughness, and from preferential growth. It can occur in amorphous as well as crystalline deposits. Therefore, it is difficult to achieve uniformly dense microstructures in coatings deposited over complex-shaped substrates at moderate T/T_m. The internal structure of the crystals is poorly defined, with a high dislocation density (Thornton, 1974).

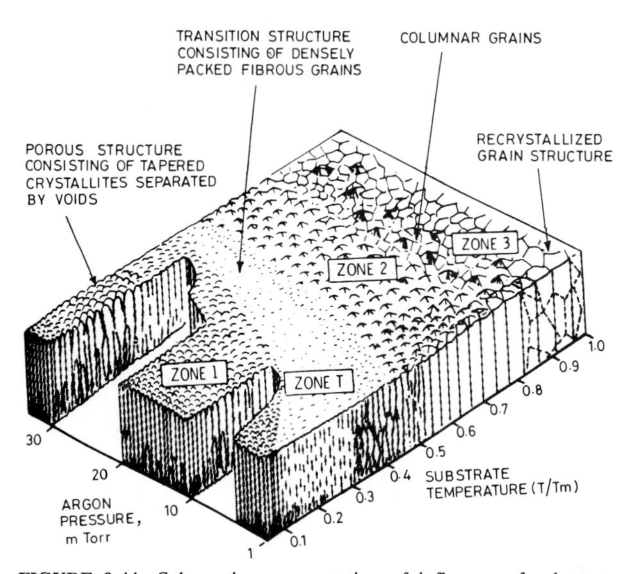

FIGURE 9.44 Schematic representation of influence of substrate temperature and argon deposition pressure on microstructure of metal coatings deposited by cylindrical magnetron sources in the absence of ion bombardment; T is substrate temperature, and T_m is melting point of the coating material in absolute degrees. [*Adapted from Thornton (1974, 1977a).*]

The zone T (transition) structure consists of a dense array of poorly defined fibrous grains without grossly voided boundaries. It has been defined as the limiting form of the zone 1 structure to zero T/T_m on infinitely smooth substrates (Thornton, 1977a). It is a columnar zone 1 structure with crystal sizes that are small and difficult to resolve and that appear fibrous. Coatings with structures approaching the zone 1 form grow under normal incidence on relatively smooth homogeneous substrates, at T/T_m values that permit the adatom diffusion to largely overcome the roughness introduced by the substrate and initial nucleation. Zone T is therefore viewed as a transition structure.

The zone 2 structure is defined as that range of T/T_m where the growth process is dominated by adatom surface diffusion (Thornton, 1977a). The structure consists of columnar grains separated by distinct, dense intercrystalline boundaries. Dislocations are primarily in the boundary regions. Grain sizes increase with T/T_m and may extend through the coating thickness at high T/T_m. Surfaces tend to be faceted.

The zone 3 structure is defined as that range of T/T_m where bulk diffusion has a dominant influence on the final structure of the coating (Thornton, 1977a). Movchan and Demchishin suggest a zone 3 structure consisting of equiaxed grains at T/T_m greater than about 0.5 for pure metals. However, the T/T_m value at which recrystallization occurs is dependent on the stored strain energy. It occurs typically at T/T_m above about 0.3 for bulk materials, but has been observed at room temperature in sputtered Cu coatings that were deposited with concurrent ion bombardment (Thornton, 1977a). It is believed that the high-temperature recrystallized structure need not be equiaxed. Sputtered coatings will generally be deposited with a columnar morphology. Recrystallization to equiaxed grains may be expected if points of high lattice strain energy are generated throughout the coating during deposition. Figure 9.45 shows examples of zone 1, T, 2, and 3 structures.

Role of Ion Bombardment and Bias Sputtering. Intense substrate ion bombardment during deposition can suppress the development of open zone 1 structures at low T/T_m. The deposits have a structure similar to the zone T type and appear to be free of nodular defects. The ion bombardment may suppress the zone 1 structure by creating nucleation sites for arriving coating atoms or by eroding surface peaks and redistributing material into valleys. Ion bombardment

CHROMIUM<110> T/T_m=0.04
Ar = 1 mTorr

(a)

ALUMINUM <100> T/T_m=0.08
Ar = 30 mTorr

(b)

COPPER <110> T/T_m=0.2
Ar = 30 mTorr

(c)

COPPER <111> T/T_m=0.2
Ar = 1 mTorr

(d)

(e)

COPPER <110> T/T_m= 0.5
Ar = 1 mTorr

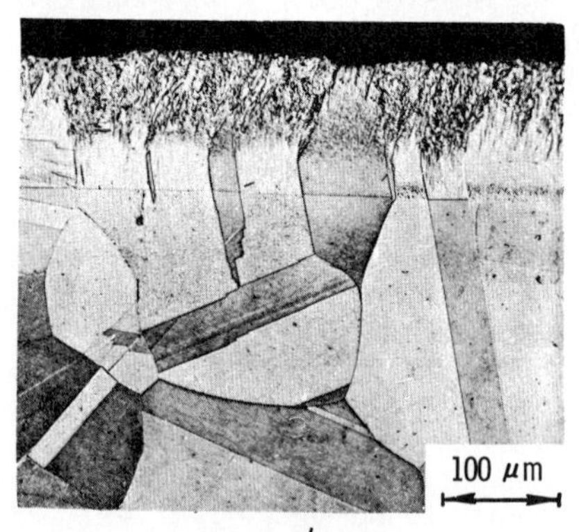

COPPER T/T_m= 0.5

(f)

FIGURE 9.45 Fracture cross sections of metal coatings deposited by using cylindrical hollow magnetrons at various substrate temperatures and argon pressures. (*a*), (*b*), (*c*) have zone 1 structure; (*d*) has zone T structure, (*e*) has zone 2 structure, and (*f*) has zone 3 structure. [*Adapted from Thornton (1974, 1977b).*]

has been found to be more effective at high working gas pressures in suppressing the influence of a rough substrate. Furthermore, the ion flux must be of such magnitude and energy that it is capable of backscattering a significant fraction of the arriving coating flux (30 to 60 percent) (Bland et al., 1974; Carson, 1975; Thornton, 1977b).

9.3.3.3 Residual Stresses. As-deposited coatings almost always contain residual stress. The total stress is composed of a thermal stress, due to the difference in the thermal expansion coefficients of the coating and substrate materials, and an intrinsic stress, due to the structural damage caused by the incorporation process (low-energy implantation) as well as atomic peening of the coating during growth by the impact of high-energy atoms or ions. For low-melting-point materials, the deposition conditions will generally involve sufficiently high values of T/T_m that the intrinsic stresses are significantly reduced by recovery during the coating growth. Thermal stresses are therefore of primary importance for such materials. Higher-melting-point materials are generally deposited at sufficiently low values of T/T_m (<0.25) that the intrinsic stresses dominate the thermal stresses.

A compressive state of stress is often desired, because many coating materials are stronger in compression than in tension. The interfacial bond must withstand the shear forces associated with the accumulated stresses throughout the coating. Since the intrinsic stress contribution increases with the coating thickness, it can be the cause of premature interface cracking.

Residual stresses in evaporated metal films are tensile on the order of 1 GPa (Klokholm and Berry, 1968). On the other hand, residual stresses in the sputtered metal films can be either tensile or compressive depending on the deposition parameters; lower deposition rates, lower gas pressure, and higher negative substrate bias result in higher values of compressive stress (Blachman, 1971; Thornton and Hoffman, 1981). Residual stresses in the refractory compounds (such as carbides, nitrides, and diamondlike carbon) are compressive and are on the order of 1 GPa, in some cases higher than the yield strengths of the coating material (Eizenberg and Murarka, 1983; Kuwahara et al., 1981; Pan and Greene, 1981).

9.3.3.4 Factors Influencing the Stoichiometry of Glow-Discharge Sputtered Compound Coatings. Process parameters have a significant influence on the stoichiometry of compound coatings (Carson, 1975; Dimigen et al., 1985; Sundgren et al., 1983; Thornton and Ferriss, 1977). The influence of partial pressure on the stoichiometry of rf sputtered MoS_2 is shown in Figure 9.46(*a*). The sulfur content decreases strongly with decreasing argon pressure, because some of the argon ions bombarding the target are neutralized and reflected to the target surface and thus travel to the substrate surface and resputter the more weakly bonded species of the growing layer (Gesang et al., 1976). Since these neutral atoms are repeatedly scattered by argon atoms on their way to the substrate, the residual energy at the substrate surface is mainly determined by the argon pressure. In the case of high argon pressure and therefore a short mean free path, the neutral atoms will lose all their energy before reaching the substrate surface, and no resputtering will occur.

Figure 9.46(*b*) illustrates the influence of sputtering power on the stoichiometry of TiN coatings applied by sputtering of the Ti target in a nitrogen atmosphere. A change in the net power from 600 to 1400 W is equivalent to a change in target voltage from 2.2 to 3.9 kV and in current density from 1.5 to 2.0 mA cm^{-2}, under typical conditions. At high partial pressure of 2×10^{-4} torr, no variation with power was noticed while at lower pressure ($<2 \times 10^{-4}$ torr) the ratio of N to Ti atoms decreases considerably (Sundgren et al., 1983).

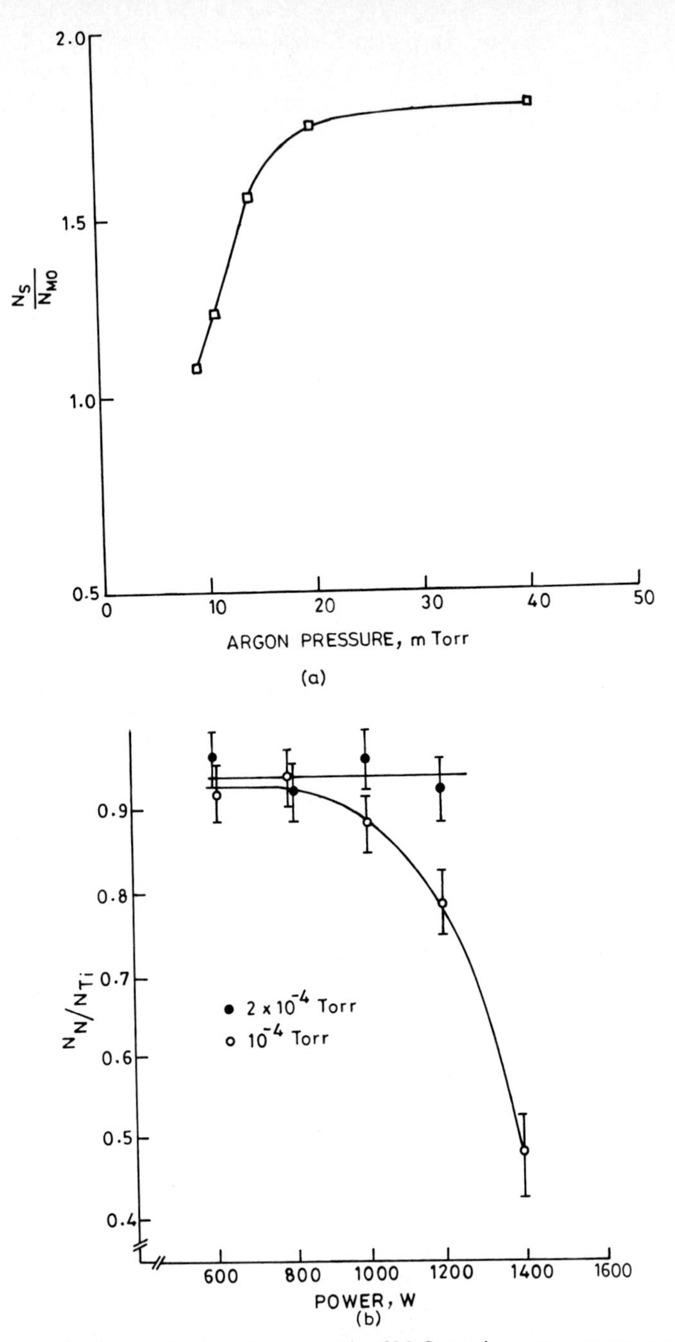

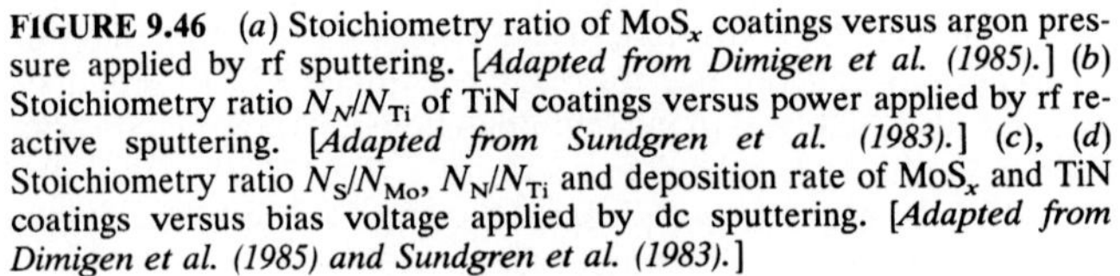

FIGURE 9.46 (*a*) Stoichiometry ratio of MoS_x coatings versus argon pressure applied by rf sputtering. [*Adapted from Dimigen et al. (1985).*] (*b*) Stoichiometry ratio N_N/N_{Ti} of TiN coatings versus power applied by rf reactive sputtering. [*Adapted from Sundgren et al. (1983).*] (*c*), (*d*) Stoichiometry ratio N_S/N_{Mo}, N_N/N_{Ti} and deposition rate of MoS_x and TiN coatings versus bias voltage applied by dc sputtering. [*Adapted from Dimigen et al. (1985) and Sundgren et al. (1983).*]

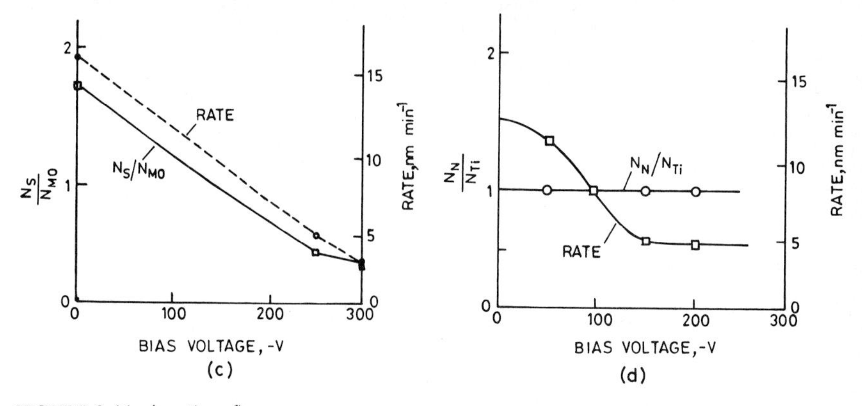

FIGURE 9.46 *(continued)*

The negative biasing of the substrate may resputter the lighter element of a compound film differently from the heavier element and therefore influence the stoichiometry considerably. This is illustrated by Figure 9.46(*c*) and (*d*) which shows the dependence of stoichiometry and the deposition rate of dc sputtered MoS_x and TiN_x coatings on biasing.

9.3.3.5 Factors Influencing the Deposition Rate in Glow-Discharge Sputtering. Deposition rate of most coatings increases relatively slowly with increasing ion energy. But the number of particles striking the cathode is proportional to the current density. Consequently, current is a significantly more important parameter for determining the deposition rate than is voltage. The power available is not without limit, and the only way to increase current without increasing power is to increase the gas pressure of the plasma. The deposition rate increases linearly with pressure. If the pressure is increased, however, there is a higher probability that sputtered atoms will return to the cathode by diffusion. As a matter of fact, at pressures in the range of 100 mtorr, only about 10 percent of the atoms sputtered from the cathode travel beyond the dark space. The net result of these observations is that the pressure chosen from sputtering rate (typically 7 to 8 Pa or 50 to 60 mtorr) considerations alone would be the highest pressure for which the yield is close to the maximum (Maissel and Glang, 1970; Sundgren et al., 1983). For the self-sustained dc glow discharge in argon, the best compromise for the sputtering pressure is 7 to 8 Pa or 50 to 60 mtorr.

Any of several impurity gases such as O_2, H_2, and He, when present in the argon during sputtering, may cause an appreciable reduction in the deposition rate. Gases such as carbon dioxide and water vapor which are decomposed in the glow discharge to produce oxygen had an effect similar to that of pure oxygen, while carbon monoxide showed a very slight tendency to increase the deposition rate, possibly because it removed traces of residual oxygen from the gas. Nitrogen has essentially no effect (Maissel and Glang, 1970). The presence of oxygen during the sputtering of metals results in the formation on the surface of the target of an oxide layer which will probably have a lower sputtering yield. The influence of hydrogen and helium on the deposition rate is due to a current-robbing effect. Because of oxygen's high mobility, the hydrogen and helium would be expected to carry a fraction of the current which is substantially larger than its concentration in the gas phase.

The deposition rate decreases with an increase in the target-to-substrate distance. The substrate should be as close as possible to the cathode, without disturbing the glow discharge, to collect a maximum of sputtered material. The coating becomes nonuniform at large spacing because the sputtered atoms are diverted and some of the kinetic energy is lost by interaction with ionized gaseous particles and other contaminants. For many materials, the deposition rate decreases with an increase in the substrate temperature; however, temperature dependence for materials deposited by reactive sputtering is reversed (Maissel and Glang, 1970). Thus, for maximum deposition rates we note that the process parameters such as sputtering power, discharge gas pressure, gaseous impurities, target-to-substrate distance, and substrate temperature must be optimized.

9.3.3.6 ***Properties of Glow-Discharge Sputtered Coatings.*** The properties of sputtered coatings are influenced considerably by parameters pertaining to starting materials, process parameters, and sputtering mode such as dc or rf diode, triode, magnetron, or ion beam (Bhushan, 1980b; Bhushan and Gray, 1980; Maissel and Glang, 1970; Spalvins, 1980b; Sundgren, 1985; Sundgren and Hentzell, 1986). The sputtering parameters are usually optimized for each coating system in order to obtain a smooth, dense coating with adequate adherence and desirable mechanical properties. The typical process parameters, which can be varied, include substrate temperature, partial pressure during deposition, gaseous impurities, target-to-substrate distance, power level, substrate biasing, substrate shape and topography, substrate hardness, and coating thickness. A discussion of the effect of various parameters and some examples of glow-discharge sputtered coatings follow.

Coatings deposited at low substrate temperatures (zone 1) are characterized by a high density of structural defects, while those deposited at elevated temperatures (zone 3) approach bulk properties (Thornton, 1977a). Simple metals with the fibrous zone T structure have high hardness but very little ductility (Thornton, 1974). Densely packed zone 1 structures have the same hardness (2 or 3 times that of annealed bulk material) but little lateral strength (Movchan and Demchishin, 1969b). Within the zone 2 and 3 regions, the sizes of the columnar and equiaxed grains increase with T/T_m, and the hardness and strength of simple metals decrease in accordance with the Hall-Petch relationship, approaching the values for annealed bulk materials (Bunshah, 1982). The ductility increases accordingly. In contrast to simple metals, ceramic compounds tend to have lower than bulk hardness when deposited at low T/T_m but approach bulk values at high T/T_m (Movchan and Demchishin, 1969b). Alloy deposits generally have smaller grain sizes and more highly dispersed phases than conventionally formed bulk materials. Columnar structures tend to promote surface irregularity and to be weak perpendicular to the growth directions. Thus, for maximum strength and smooth surface finish, deposition conditions should be chosen to give a fine-grained coating.

Bhushan (1979, 1980b, 1980c, 1981a, 1981b) has studied the effect of some process parameters on the morphology, adhesion, and wear performance of rf sputtered Cr_2O_3 coatings for foil gas bearing applications. Morphology was observed by scanning electron microscope (SEM) micrographs. Adhesion was measured by using a normal pull test and a 180° bend test. The wear was measured by using a start/stop wear test. Bhushan found that the coating is smoothest and has maximum adhesion when the pressure is lowest—5 mtorr (Fig. 9.47). When the pressure is high, there is increased diffusion of argon gas and other contaminants into the coating which makes the coating rough and less adherent (see increased crazing and minor flaking in Fig. 9.47(*c*). A lower target-to-substrate distance results in

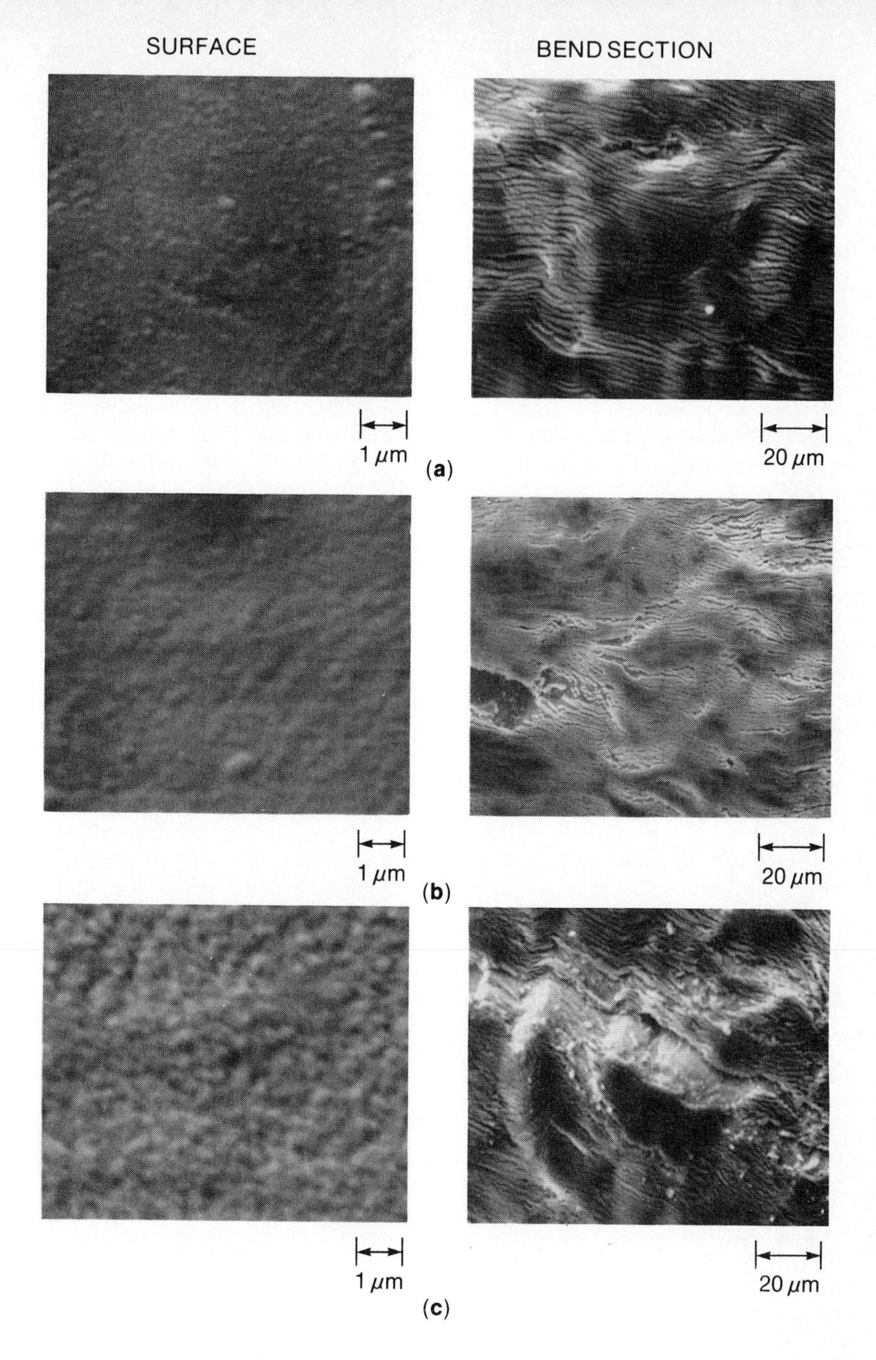

FIGURE 9.47 SEM micrographs of rf diode sputtered Cr_2O_3 deposited on heat-treated nickel-base alloy (Inconel X-750) at various argon gas pressure of (*a*) 0.65 Pa (5 mtorr), (*b*) 1.3 Pa (10 mtorr), (*c*) 4 Pa (30 mtorr). The micrographs on the left side are for as-coated specimens and on the right side for coatings after the bend test, carried out to measure adhesion. The other deposition conditions were: target-to-substrate distance = 41.3 mm; target size = 152-mm dia.; rf power = 390 W; sputtering time = 60 min; water-cooled substrate. [*Adapted from Bhushan (1979).*]

increased adhesion [see Fig. 9.48(*a*) and (*b*)]. At an increased target-to-substrate distance, the sputtered atoms are diverted and some of their kinetic energy is lost through interactions with ionized gas particles and other contaminants. The coatings sometimes become contaminated, resulting in poor adhesion and wear (Christy and Ludwig, 1979).

High target voltage generally results in increased density, bulklike structure, and improved interlayer adhesion (Bhushan, 1979) probably because of the higher energy of the bombarding ions. Berry et al. (1968) reported that tantalum coatings deposited at low voltages (<3 kV) can show a considerable porosity. In some cases, high power density results in soft coatings with poor wear resistance (e.g., Christy and Ludwig, 1979; Savvides and Window, 1985). If the power is increased, both target and substrate temperatures increase. High cathode (target) temperature degasses and desorbs the surface rapidly which may affect the coating purity. At high substrate temperatures, the coating sometimes outgasses, and there is a desorption of impurities during sputtering. Hence, this can provide cleaner deposition surface and thereby an adherent coating. In some cases, heating of the substrate may aid the formation of an intermetallic interface layer which facilitates adhesion (Greene et al., 1978; Hibbs et al., 1984). In addition, substrate heating may continuously anneal the growing coating and relieve the residual stresses developed during the deposition process; however, the annealing may also make the coating softer.

Substrate bias is generally found to give effective cleaning, brought about by surface bombardment which removes impurities in the coating resulting from chemisorption and condensation. Whether the impurities are removed depends on the relative strengths of the metal-to-impurity and metal-to-metal bonds. Substrate bias controls the flux of incidental ions or electrons bombarding the substrate, which may result in increased adhesion. Substrate bias may also affect the structure and stoichiometry of the coatings (Berry et al., 1968). Furthermore, substrate bias leads to problems when depositions are to be made on irregularly shaped substrate, with sharp edges and corners (Johansson et al., 1984). The rate of resputtering from the biased substrate is greater at areas closer to sharp corners. This edge effect can be so great in some cases that it causes complete resputtering of the deposited material.

The sputtered coatings usually replicate the surface roughness of the substrate. Surface defects (scratches, inclusions, impurities, etc.) on the substrate act as preferential nucleation sites, and as a result, isolated or complex nodular-type structures and various overgrowths are formed. Nodular growth within a coating can be minimized by decreasing the surface roughness (Spalvins and Brainard, 1974). A smoother coating will have improved wear life because high microprojections in a rough coating can be easily knocked off during sliding, resulting in catastrophic failure.

Substrate hardness has a significant effect on the coating adhesion. If the substrate is soft, the sputtering atoms penetrate farther into the surface and provide an extremely adherent coating [Fig. 9.48(*c*)]. If the substrate is very hard, the coating adhesion can be improved if the substrate is coated in the annealed condition and then heat-treated, provided that the coating can withstand the heat treatment cycle. Also the post heat treatment can promote diffusion at the interface which normally provides a stronger bond.

Finally, during a sliding contact, the wear life of coatings can be increased if the coatings are thick. However, if the sputtered coatings are too thick (more than a few micrometers), they spall.

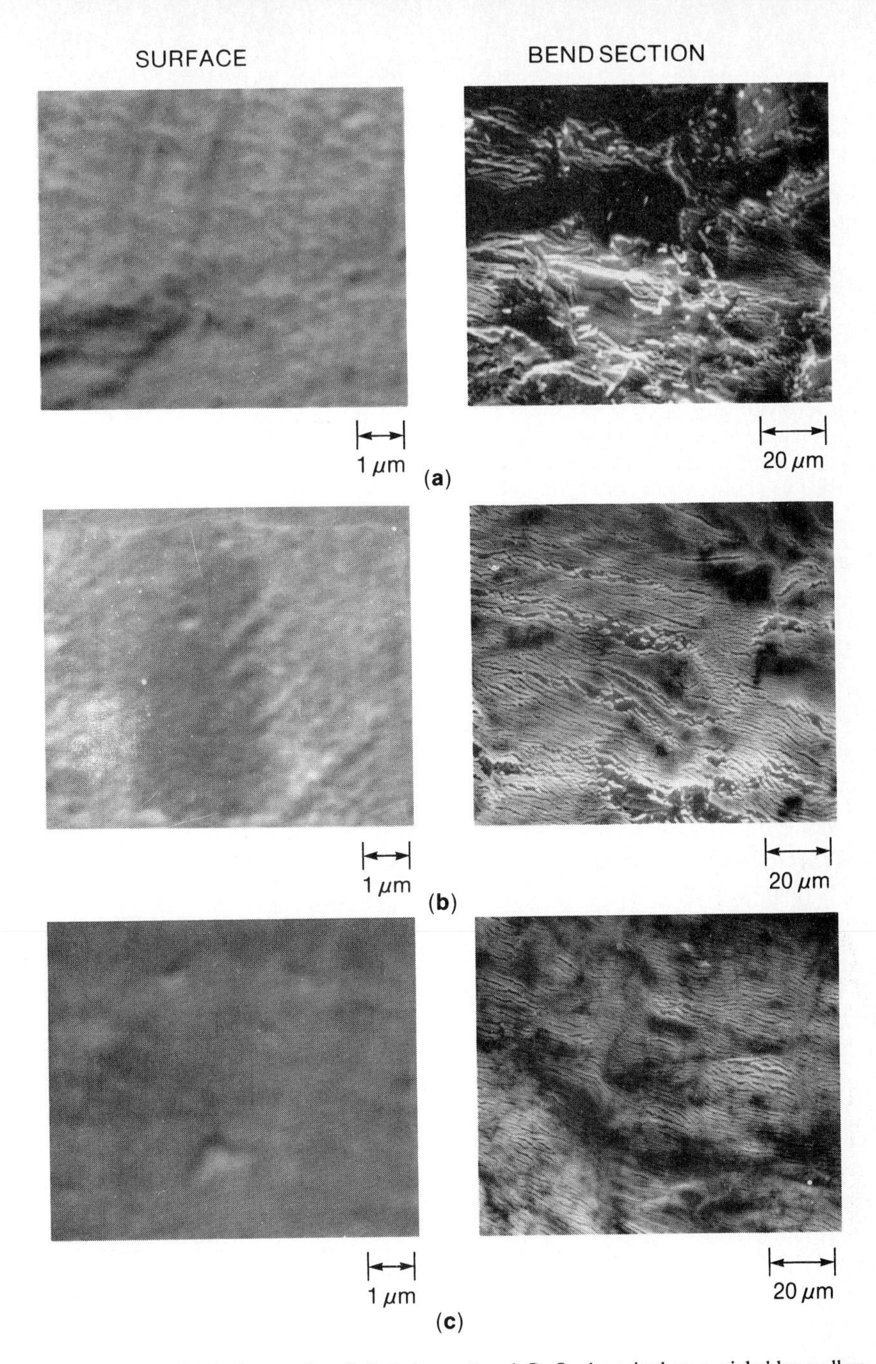

FIGURE 9.48 SEM micrographs of rf diode sputtered Cr_2O_3 deposited on a nickel-base alloy (Inconel X-750) at various target-to-substrate distances and substrate conditions: (*a*) target-to-substrate distance = 50.8 mm, heat-treated; (*b*) target-to-substrate distance = 41.3 mm, heat-treated; (*c*) target-to-substrate distance = 41.3 mm, annealed. The micrographs on the left side are for as-coated specimens and on the right side for coatings after bend test, carried out to measure adhesion. The other deposition conditions were: argon gas pressure = 1.3 Pa (10 mtorr); target size = 152-mm dia.; rf power = 390 W; sputtering time = 60 min; water-cooled substrate. [*Adapted from Bhushan (1979).*]

9.4 CHEMICAL VAPOR DEPOSITION

Chemical vapor deposition (CVD), also called *thermal CVD*, is another widely used coating deposition technique. In this technique, a volatile component of coating material is thermally decomposed (pyrolysis) or chemically reacts with other gases or vapors to form a coating atomistically (atom by atom) on the hot substrate surface (Fig. 9.49). No electric current or fields are required in the CVD process. The CVD reactions generally take place in the temperature range of 150 to 2200°C at a pressure of a 60 Pa (0.5 torr) to atmospheric pressure (0.1 MPa or 760 torr) of the reactants. Practically any gas or vapor can be used as a precursor material. The existence of diffusion zones can be observed at sufficiently high temperatures, resulting in coatings with excellent adhesion. The interdiffusion can produce alloys or compounds which necessarily contain substrate elements. The coating source material can be initially in solid, liquid, or vapor form and then converted to vapor phase by heating. The coating quality depends on the substrate cleanliness, compatibility of coating and substrate materials, thermodynamics, and kinetics of the reaction involved (Blocher, 1982; Blocher and Withers, 1970; Blocher et al., 1975; Bryant, 1977; Donaghey et al., 1977; Duret and Pichoir, 1983; Feist et al., 1969; Glaski, 1972; Hocking et al., 1989; Kern and Ban, 1978; Kern and Rosler, 1977; Powell et al., 1966; Schaffhauser, 1967; Wakefield and Blocher, 1973).

The gaseous phases of source material can be composed of metal halides, metal hydrides, metal carbonyls, organometallic compounds, and hydrocarbons. For a given coating, the choice of a source material will very much depend on the following practical considerations:

- The use of a volatile compound will greatly simplify the technology of the gas transport. Otherwise, the pressure must be held below atmospheric pressure.
- The source material must have sufficiently low vapor pressure to prevent its volatilization.
- It must not decompose when heated; otherwise an attempt must be made to evaporate at low temperature and low pressure (in the range of 10^{-2} to 1 Pa or 0.1 to 10 torr).
- It must decompose or react at a temperature generally not higher than about 1600°C. The lower the temperature, the more versatile is the method.

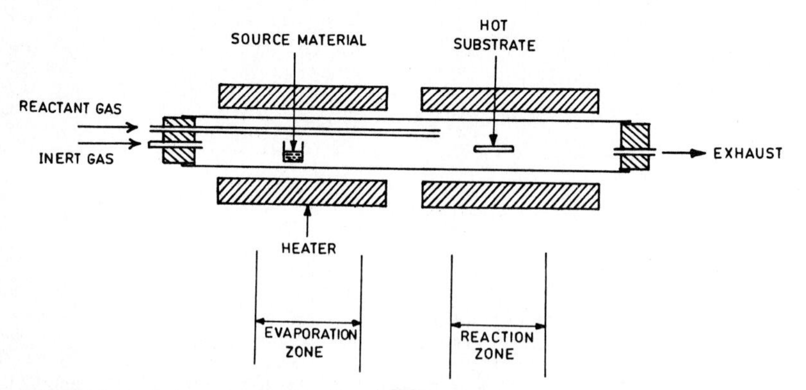

FIGURE 9.49 Schematic of conventional CVD process.

From the data available in the thermodynamic literature, a choice can be made at once of the metallic compound for the source material, the reactant gas, and the reaction temperature.

The CVD equipment includes a vapor source (a vaporizer or, if the reactant is a vapor at room temperature, a gas cylinder), a deposition chamber, and (assuming subatmospheric operation) a vacuum pump or an aspirator. Where the expense of an evacuation system is not warranted, deposition occurs at atmospheric pressure. Since the decreased reactant concentration is less likely to produce gas-phase nucleation and other unwanted coating structures, the partial pressures of reactants are kept low by operation at subambient pressures and/or by dilution of the stream with an inert gas such as argon, helium, nitrogen, or carbon dioxide. Hydrogen is frequently used as a carrier gas since it is a convenient reducant for many reactions. However, its density can be a problem in mixing higher-molecular-weight reactants.

One breakdown of reactor design can be made according to the wall temperature relative to the substrate temperature (Kern and Ban, 1978). In hot-wall reaction chambers, these temperatures are approximately equal while in cold-wall chambers the wall temperature is considerably lower and is often not much above ambient. In cold-wall reactors [Fig. 9.50(*a*)], the substrate is heated by direct electrical resistance (electrical resistance of a heater contained within the substrate), inductance or infrared heating. The foremost advantage of a cold-wall design is that deposition on the reactor walls is prevented or at least minimized. A further advantage is the reduced possibility of the occurrence of gas-phase powder formation since the stream temperature is usually much lower than that of the substrate. However, cold-wall methods of heating are not applicable to all systems because of the size, shape, and composition of the substrate.

In the hot-wall chamber shown in Fig. 9.50(*b*), the reactor tube is surrounded

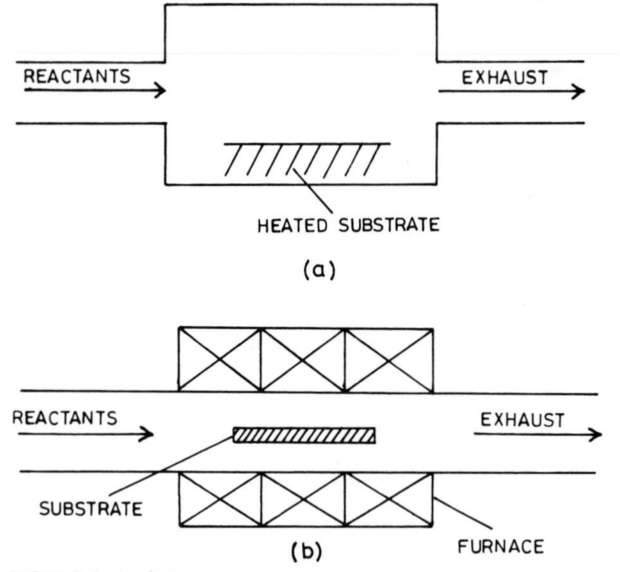

FIGURE 9.50 Schematic diagram of (*a*) a cold-wall reaction chamber and (*b*) a hot-wall reaction chamber.

by the tube furnace which heats the substrate by radiant heat. With this heating method, the chamber walls are also heated. Hot-wall reactors are frequently tubular; heating is most often accomplished by resistance elements surrounding the reactor tube. Hot-wall chambers offer certain advantages: Sharp temperature profiles can be achieved, and temperature control is relatively easy. However, unwanted deposition on the walls can prove to be uneconomical and increases the risk of contamination between the chamber wall and the vapor. The risk of contamination is low for coatings thinner than about 20 μm.

Many CVD processes are best carried out at reduced pressures where they are thermodynamically more favorable and provide dilution of reactant concentration. The CVD process at low pressure ranging from 10 to 10^3 Pa (0.1 to 10 torr), called *low-pressure CVD* (LPCVD), allows the deposition of coatings with superior quality and uniformity over a large substrate area. Using a laser as the heat source has the major advantage of heating a very small portion of the substrate, if necessary. The CVD reactions can take place in the presence of electron bombardment which facilitates activation of precursor gases, thus allowing deposition at lower substrate temperatures.

The substrate surface should be clean of oxides or other common contaminant surface films, and attention must be paid to the chemistry of the substrate surface with reference to the coating reaction; the surface preparation significantly affects the interlayer adhesion. Substrate surface preparation methods vary with the substrate surface and the coating material being used. The substrate must be mechanically and chemically cleaned by pickling, grit blasting, degreasing, etc. To remove gas adsorption, the substrate may be baked in situ for 1 to 2 h. For example, one gets rid of the oxygen adsorbed layer on the substrate surface by baking it in hydrogen flow.

Since a large variety of chemical reactions are available, CVD is a versatile and flexible technique used in producing a wide range of metallic and nonmetallic coatings on any nongassing substrate. The shape and size of the substrates are limited by the heating possibilities. It is possible to coat complex-shaped bulky pieces. Coatings can be easily deposited on holes and recessed areas. Refractory materials can be deposited at temperatures far below their melting points or sintering temperatures. The coating deposition rates are very high, and processing costs are generally relatively low. Deposition is possible at atmospheric pressure. CVD coatings usually exhibit near-theoretical or controlled density, controlled grain size, and excellent adhesion compared to evaporated or sputtered coatings; but the requirements of high substrate temperature limit their application to substrates which can withstand high processing temperatures without any adverse effects. The cost of typical starting materials may be high relative to the processing cost in cases where the chemical is in low demand. There is also low utilization of the coating material.

Tables 9.1 and 9.2 summarize the range of coating materials, deposition (substrate) temperatures, deposition pressure, coating thickness, deposition rates, and advantages and disadvantages of CVD processes. Their details follow.

9.4.1 Thermodynamic and Kinetic Considerations

The progress of CVD reactions is governed by thermodynamics which determines the driving force and kinetics which defines the available rate control mechanisms.

9.4.1.1 Thermodynamics. The thermodynamic variables which can be controlled experimentally are the substrate temperature, total pressure, and ratios of the chemical elements in the input vapor phase (or partial pressures of the reactants). The thermodynamic study of a CVD process allows the determination of the temperature and partial pressure ranges in which the reactions of coating formation take place (Blocher, 1982; Bryant, 1977).

In the simplest processes, the partial pressure under equilibrium conditions at a given temperature can be calculated by means of the reaction free energy. The reaction free energy is calculated from the free energy of formation of all the vapor and condensed constituents involved in the reaction. For instance, the free energy of the reaction

$$CrCl_2(g) + H_2(g) \rightarrow Cr(s) + 2HCl(g) \tag{9.17}$$

is given by

$$\Delta G = \Delta G^0 + RT \ln \frac{P_{HCl}^2}{P_{H_2}} \tag{9.18}$$

assuming that both $CrCl_2$ and Cr are solids. Here P is the partial pressure of the products and reactants, T is the reaction temperature, and R is the gas constant. Under equilibrium conditions at 800°C,

$$\Delta G = 0 \quad \text{and} \quad \Delta G^0 = -RT \frac{P_{HCl}^2}{P_{H_2}} = -14.8 \text{ kcal} \tag{9.19}$$

With a total pressure of 1 atm, $P_{HCl} = 0.031$ atm and $P_{H_2} = 0.069$ atm. Thus Cr deposition will take place as long as P_{HCl} is kept below 0.031 atm (Perry and Archer, 1980). Usually CVD deals with a multicomponent and a multiphase reaction. The ways to calculate the thermodynamics of multiphase systems are discussed by Kern and Ban (1978). Eriksson (1975) has developed methods for computing chemical equilibria of such complex reactions.

9.4.1.2 Kinetics. For successful coating deposition, the CVD reaction must be not only thermodynamically possible but also kinetically favorable. For instance, Al_2O_3 formation by the reaction between $AlCl_3$ and oxygen is thermodynamically possible, but the deposition rate of Al_2O_3 has been found to be extremely slow because of nonfavorable kinetics (Bryant, 1977).

The CVD reactions follow the following sequence of events to produce the coating: (1) formation of the gaseous phase, (2) gas transfer to the substrate, (3) adsorption, (4) decomposition of the adsorbed phase, (5) metal (or alloy or compound) deposition and desorption of the decomposition products, and (6) removal of these products. Stages 1, 2, and 6 generally include mass transport processes, and stages 2, 3, 4, and 5 are surface reaction processes. Mass transport rates depend mainly on the reactant concentration, diffusivity, and boundary-layer thickness, which is related to reactor configuration, flow velocity, distances from edges, etc. Surface reaction rates depend mainly on the reactant concentration and temperature.

In most systems, the deposition rate is controlled by both mass transport and surface reaction kinetics. The maximum deposition rate can be obtained only when mass transport is rate-controlling: however, the coating thickness may be nonuniform. Surface kinetics control can be used when extended areas, internal surfaces, etc., are to be coated (Blocher, 1982; Bryant, 1977).

9.4.2 Examples of CVD Reactions

The deposition reaction is generally either pyrolysis (or thermal decomposition) or a chemical reaction, in which case there is another reactant gas or vapor phase.

*9.4.2.1 **Pyrolysis (Thermal Decomposition).*** The thermal decomposition of volatile compounds to yield a deposit of stable residue is called *pyrolysis*. In pyrolysis, the substrate is normally heated above the gas decomposition temperature, and the reactant gases flow over the hot substrate, decomposing into coatings. Metal halides, metal hydrides, metal carbonyls, and organometallic compounds are particularly good starting materials for these reactions. Organometallic compounds decompose at low temperatures (<600°C), whereas metal halides decompose above about 600°C to yield metallic deposits.

The following are typical examples of thermal decomposition.

Nickel deposition (Mond's process):

$$\mathrm{Ni(CO_4)}(g) \xrightarrow{150^\circ\mathrm{C}} \mathrm{Ni}(s) + 4\mathrm{CO}(g) \tag{9.20}$$

Titanium deposition (Van Arkel's process):

$$\mathrm{TiI_4}(g) \xrightarrow{1200^\circ\mathrm{C}} \mathrm{Ti}(s) + 2\mathrm{I_2}(g) \tag{9.21}$$

Epitaxial Si deposition:

$$\mathrm{SiH_4}(g) \xrightarrow{1000\text{ to }1200^\circ\mathrm{C}} \mathrm{Si}(s) + 2\mathrm{H_2}(g) \tag{9.22}$$

Silane (SiH_4) at a pressure of less than 10^3 Pa (8 torr) decomposes completely above 770°C. Many of the organic silicates such as tetraethoxysilane vapors can be thermally decomposed in the 600 to 900°C range to give substantially pure SiO_2 coatings with dielectric properties close to those of silica. Deposition of alumina coatings has been obtained by pyrolysis of aluminum triethoxide in vacuo at 550°C and also of aluminum triisopropoxide. Si_3N_4 coatings have been deposited by pyrolysis of a silane-hydrazine mixture (SiH_4:N_2H_4) with mole ratio of 0.1 at temperatures between 550 and 1150°C. Deposition rates of 10 to 100 nm min^{-1} were obtained at a temperature above 700°C (Chopra, 1979; Perry and Archer, 1980).

*9.4.2.2 **Chemical Reactions.*** Chemical vapor deposition can also take place by chemical reactions of gaseous phases such as by reduction, oxidation, hydrolysis, and reactions with other gases—nitrogen, ammonia (NH_3), methane (CH_4), ethane (C_2H_4), and boric chloride (BCl_3). More complex reactions can be used to form simple alloy or compound coatings (Perry and Archer, 1980; Duret and Pichoir, 1983).

Reduction may be thought of as a pyrolysis reaction facilitated by the removal of one or more of the gaseous products of decomposition. The reaction temperature is lowered by several hundred degrees below that needed in the absence of a reducing agent. Hydrogen reduction of metallic halides or hydrides can form metallic coatings, such as

$$SiCl_4(g) + 2H_2(g) \rightarrow Si(s) + 4HCl(g) \quad (9.23)$$

$$TiCl_4(g) + 2H_2(g) \rightarrow Ti(s) + 4HCl(g) \quad (9.24)$$

$$CrCl_2(g) + H_2(g) \rightarrow Cr(s) + 2HCl(g) \quad (9.25)$$

$$WF_6(g) + 3H_2(g) \xrightarrow{350\text{–}1000^\circ C} W(s) + 6HF(g) \quad (9.26)$$

$$2TaCl_5(g) + 5H_2(g) \xrightarrow{700\text{–}1100^\circ C} 2Ta(s) + 10HCl(g) \quad (9.27)$$

More complex reactions can be used to form alloy coatings, such as codeposition of molybdenum and tungsten:

$$6xMoF_6(g) + 6(1 - x)WF_6(g) + 3H_2(g) \rightarrow 6Mo_xW_{1-x}(s) + 6HF(g) \quad (9.28)$$

Besides hydrogen, other reducing agents such as metallic vapors of Mg, Zn, Na, K, and Cd can be used with other vapors in CVD reactions, e.g.,

$$TiCl_4(g) + 2Mg(g) \xrightarrow{900^\circ C} Ti(s) + 2MgCl_2(g) \quad (9.29)$$

$$SiCl_4(g) + 2Zn(g) \rightarrow Si(s) + 2ZnCl_2(g) \quad (9.30)$$

Examples of reactions with metallic halides or hydrides vapors and with other gases are

1. O_2 to produce oxide coatings

$$SiH_4(g) + O_2(g) \rightarrow SiO_2(s) + 2H_2(g) \quad (9.31)$$

2. CO_2 and H_2 to produce oxide coatings by hydrolysis

$$Al_2Cl_6 + 3CO_2(g) + 3H_2(g) \xrightarrow{800\text{–}1400^\circ C} Al_2O_3(s) + 3CO(g) + 6HCl(g) \quad (9.32)$$

$$SiCl_4(g) + 2CO_2(g) + 2H_2(g) \xrightarrow{1200^\circ C} SiO_2(s) + 2CO(g) + 4HCl(g) \quad (9.33)$$

$$TiCl_4(g) + 2CO_2(g) + 2H_2(g) \xrightarrow{900^\circ C} TiO_2(s) + 4HCl(g) + 2CO(g) \quad (9.34)$$

3. CH_4 and H_2 or CH_4 or C_2H_4 to produce carbide coatings

$$WCl_6(g) + CH_4(g) + H_2(g) \rightarrow WC(s) + 6HCl(g) \quad (9.35)$$

$$TiCl_4(g) + CH_4(g) \xrightarrow{1200^\circ C} TiC(s) + 4HCl(g) \quad (9.36)$$

$$ZrCl_4(g) + CH_4(g) \xrightarrow{1200\text{–}1350^\circ C} ZrC(s) + 4HCl(g) \quad (9.37)$$

$$2SiH_4(g) + C_2H_4(g) \rightarrow 2SiC(s) + 6H_2(g) \quad (9.38)$$

4. Nitrogen and NH_3 to produce nitride coatings

$$2TiCl_4(g) + N_2(g) + 4H_2(g) \xrightarrow{550\text{–}950^\circ C} 2TiN(s) + 8HCl(g) \quad (9.39)$$

$$3SiCl_4(g) + 4NH_3(g) \xrightarrow{1200^\circ C} Si_3N_4(s) + 12HCl(g) \quad (9.40)$$

$$3SiH_4(g) + 4NH_3(g) \xrightarrow{500\text{–}1500^\circ C} Si_3N_4(s) + 12H_2(g) \quad (9.41)$$

$$B_2H_6(g) + 2NH_3(g) \rightarrow 2BN(s) + 6H_2(g) \quad (9.42)$$

$$BCl_3(g) + NH_3(g) \xrightarrow{500\text{–}1500°C} BN(s) + 3HCl(g) \tag{9.43}$$

5. BCl_3 and H_2 and B_2H_6 to produce boride coatings

$$TiCl_4(g) + 2BCl_3(g) + 5H_2(g) \rightarrow TiB_2(s) + 10HCl(g) \tag{9.44}$$

$$2TiCl_4(g) + 2BCl_3(g) + 7H_2(g) \rightarrow 2TiB(s) + 14HCl(g) \tag{9.45}$$

$$2TaCl_5(g) + 2B_2H_6(g) \rightarrow 2TaB_2(s) + 10HCl(g) + H_2(g) \tag{9.46}$$

6. With SiH_4 to produce silicide coatings

$$3WF_6(g) + 6SiH_4(g) \rightarrow 3WSi_2(s) + 18HF(g) + 3H_2(g) \tag{9.47}$$

$$2MoCl_5(g) + 4SiH_4(g) + H_2(g) \rightarrow 2MoSi_2(s) + 10HCl + 4H_2(g) \tag{9.48}$$

$$2TaCl_5(g) + 4SiH_4(g) \rightarrow 2TaSi_2(s) + 10HCl(g) + 3H_2(g) \tag{9.49}$$

7. And with various gases to form complex compounds

$$2TiCl_4(g) + 2CH_4(g) + N_2(g) \xrightarrow{700\text{–}900°C} 2Ti(C,N)(s) + 8HCl(g) \tag{9.50}$$

9.4.3 Details of CVD Processes

9.4.3.1 Conventional CVD Process. In the CVD process, the coating material, if not already in the vapor state, is formed by volatilization from either a liquid or a solid feed and is caused to flow by either a pressure differential or the action of carrier gas toward the substrate surface. In many cases, reactant gas or other material in vapor phase is added to produce either metal or compound coatings (Fig. 9.49). The coating is obtained by either thermal decomposition (pyrolysis) or chemical reaction of the gaseous phase(s) near the atmospheric pressure. Since the vapor will condense on any relatively cool surface that it contacts, all parts of the deposition system must be at least as hot as the vapor source. The reaction portion of the system is generally much hotter than the vapor source but considerably below the melting temperature of the coating. The CVD reactions generally take place in the temperature range of 150 to 2200°C and frequently between 500 and 1600°C at or just below atmospheric pressure of the reactants, in the case of conventional CVD.

The heat sources used for substrate heating play an important role in the CVD coating process (Blocher, 1982): Major means of substrate heating employed in conventional CVD processes are resistance, induction, and infrared heating. A resistively heated hot plate in direct contact with the substrate is shown in Fig. 9.51. A schematic of a single quartz tube heated with a tubular resistance furnace in a hot-walled reactor is shown in Fig. 9.52. The direct electrical resistance is limited to substrates of relatively small cross-sectional area and fairly high electrical resistivity. A pedestal heater using rf induction is shown in Fig. 9.52. RF induction heating is a very useful technique because it puts the heat where it is needed. It is most useful for axially symmetric forms which can be rotated in the induction field for uniformity of exposure. Substrates which are semiconductors or insulators cannot be inductively heated because of the high electrical resistivity. An infrared (ir) heating arrangement for direct heating of the substrate is shown in Fig. 9.53. Radiation can be passed through an ir-transmitting window and absorbed by the substrate or by the susceptor containing the substrates. The ir wavelength should be such that the reactant gas molecules are not absorbed for

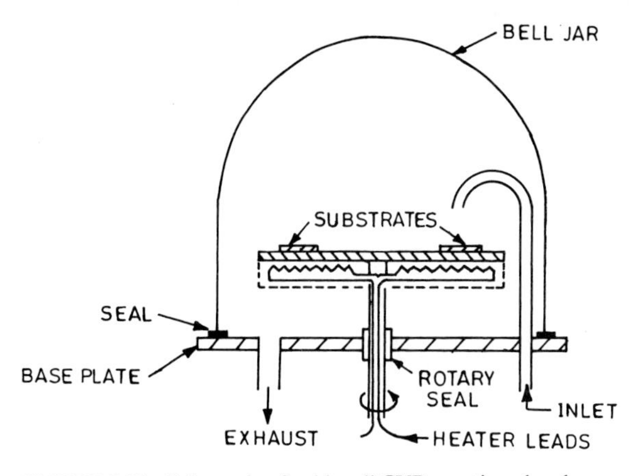

FIGURE 9.51 Schematic of cold-wall CVD reaction chamber using resistively heated plate.

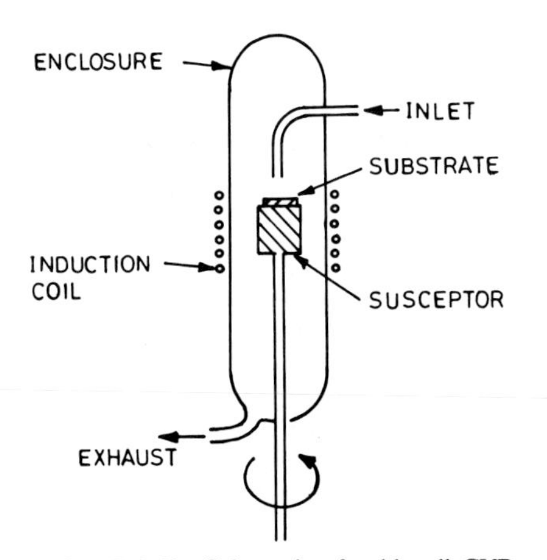

FIGURE 9.52 Schematic of cold-wall CVD reaction chamber using inductively heated pedestal.

premature reactions. The temperatures attainable by infrared heating are relatively low.

For uniformity of reactant composition across a substrate promoted by good mixing, we would like to operate a CVD reactor in the turbulent range (Reynolds number >2000). However, this advantage is outweighed in most cases by the following disadvantages: The shortened dwell time of the reactant gases in contact with substrate at the higher flow rates generally decreases the efficiency of reactant utilization per pass; very large volumes of gas would have to be handled to operate in the turbulent regime; and at the high rate of gas flow convective cooling of the substrate is increased. Accordingly, most of the CVD is done in the laminar-flow region, with considerable attention given to the gas flow patterns around the substrate being coated.

The coating deposition over a substrate of considerable surface area by the standard procedure of continually flowing the reactant gases over the substrate, may result in a nonuniform coating thickness. This nonuniformity of coating thickness is caused by the continual depletion of reactants and formation of product gas as the stream moves over the substrate. Reactant depletion affects the kinetics of deposition when the process is under mass transport control. Controlling deposition by surface kinetics, although helpful, does not provide the total solution, since at the lower pressures (in LPCVD processes, to be discussed later) which favor this control mode the stream reactant concentration can be quickly lowered to a level where mass transport becomes controlling. One alter-

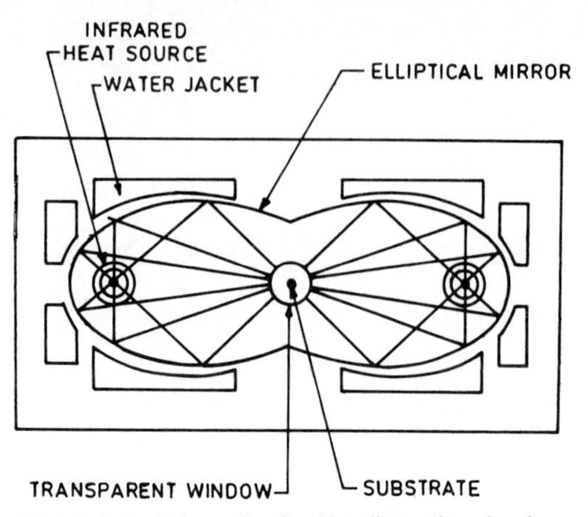

FIGURE 9.53 Schematic of cold-wall reaction chamber using dual elliptical infrared heating arrangement.

native approach involves altering the flow pattern of the gas, and there have been attempts such as periodically reversing the gas flow direction, rotating the substrate, combining gas flow reversal and substrate rotation, stirring the gas, periodically repositioning the substrate, imposing a temperature gradient over the length of the substrate, and tilting the substrate to increase its downstream projection into the boundary layer (Bryant, 1977).

*9.4.3.2 **Low-Pressure CVD.*** Many CVD processes are best carried out at reduced pressure where they are thermodynamically more favorable. Low-pressure CVD at pressures ranging from 10 to 10^3 Pa (0.1 to 10 torr) (typically 0.5 to 1 torr) produces coating deposition with superior quality and coating uniformity over a large substrate area. The fundamentals of depositing coatings at low pressure are basically the same as for those deposited near atmospheric pressure (conventional CVD). The major difference is the radical change of emphasis put on the rates of mass transfer of the gaseous reactant and by-product species relative to their surface reaction rate to form the coating. At normal atmospheric pressure, these rates are generally of the same order of magnitude. Therefore, to attain the objective of uniform coating thicknesses and properties over extended surfaces for large production-type loads, careful consideration must be given to both types of rate-determining steps. Mass transfer rates depend mainly on the reactant concentration, diffusivity, and boundary layer thickness, which is related to reactor configuration, flow velocity, distances from the edges, etc. Surface reaction rates depend on the reactant concentration and temperature.

The diffusivity of a gas is inversely related to pressure to the first power. Therefore, as pressure is lowered from its atmospheric value near 10^5 Pa to 50 to 10^3 Pa (760 torr to 0.5 to 10 torr), the diffusivity increases by a factor of 1000. This is only partially offset by the fact that the (laminar) boundary layer (distance across which the reactants must diffuse) increases at less than the square root of pressure. The net effect is more than an order-of-magnitude increase of the gas-phase transfer of reactants and by-products to the substrate surface. The rate-determining step in the sequence is the surface reaction. Therefore, in such a low-pressure operation, much less attention has to be paid to the mass transfer variables which are so critical at atmospheric pressure. The increased mean free path of the reactant gas molecules, at low-pressure operation, results in high throughput. The resulting coating exhibits improved uniformity, better step coverage and conformability, and superior structural integrity with fewer pinholes. The wide acceptance of LPCVD is due to its very favorable economics (low process cost) in addition to the high quality and uniformity of the coatings compared to the conventional CVD process.

A schematic diagram of LPCVD reactor with a tubular resistance-heated furnace is shown in Fig. 9.54 (Kern and Rosler, 1977). A round quartz tube, heated by standard wire-wound elements, forms the reaction chamber. This type of reactor is referred to as a *hot-walled reactor* in which the wall temperature and substrate temperature are nearly equal. The reactor is evacuated by mechanical vacuum pumps. Substrates, typically at 3- to 5-mm spacing, stand vertically facing the end of the tube. It is preferred to load and unload substrates at the gas inlet end of the tube to avoid contaminating them with hot unreacted gases and to avoid any inlet pressure change due to a different tube pressure drop caused by varying load size. The growth rate of LPCVD coatings as a function of axial substrate position in the reactor increases in the downstream direction as the pressure or the temperature is increased.

A LPCVD refinement in which the reactant gases are introduced in pulsating rather than in continuous flow manner has shown promise in further overcoming the problem of coating thickness uniformity (Bryant, 1976). In this technique, known quantities of reactant gases are injected into a previously evacuated reaction chamber, where they almost instantly blanket the substrate with a gas of uniform composition. Naturally, gas is confined to the reaction chamber, but the composition at any time is likely to be invariant with respect to position on the substrate. During evacuation of the spent gases, fresh reactant gases are being accumulated in reservoirs of known volume for the next injection. Successive repetitions of the evacuation-injection cycle are used to form a coating of uniform thickness.

In alloy deposition, the deposition kinetics of the separate elements are generally different to provide a compositional variation across the dimension of the coating paralleling the direction of gas flow, in addition to the thickness variability often found in the standard flow-through process. The difference in kinetics

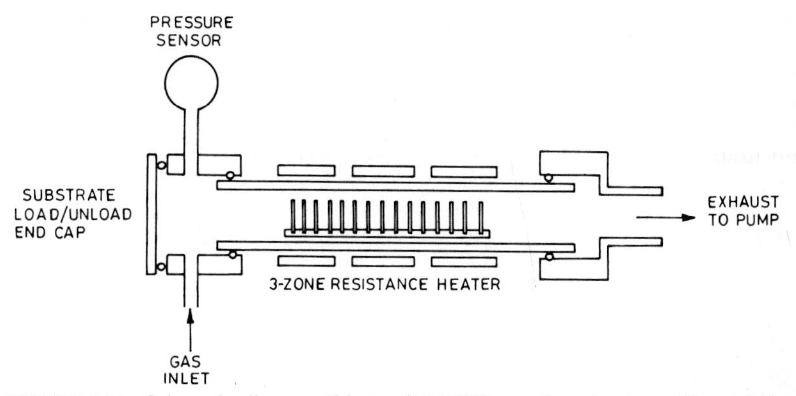

FIGURE 9.54 Schematic diagram of hot-wall LPCVD reaction chamber with a tubular resistance-heated furnace. [*Adapted from Kern and Rosler (1977).*]

stems from the difference in stabilities among the reactants. The pulsing variant of the process has been successfully used by Bryant (1976) to prepare homogeneous alloy coatings. In this technique, the reactant gases can be injected either simultaneously or sequentially. For ternary and higher-order alloys, some combination of simultaneous and sequential pulsing may prove best.

The LPCVD process has been successfully used for the deposition of polycrystalline Si coatings from SiH_4 at substrate temperature ranging from 600 to 650°C, Si_3N_4 coatings from SiH_2Cl_2 and NH_3 at 750 to 920°C, and for vitreous SiO_2 coatings from SiH_2Cl_2 and N_2O at 910°C (Rosler, 1977). The rate of deposition was about 20 nm min^{-1}.

9.4.3.3 Laser-Induced CVD. In conventional or low-pressure CVD processes, deposition usually occurs over the entire substrate surface, and subsequent masking and etching operations are used for removal of unwanted materials. Since large areas of substrate and susceptor are at the growth temperature, significant heating of the reactant gases occurs unless it is otherwise inhibited. The use of a laser as the heat source for CVD offers several distinct advantages: (1) spatial resolution and control, (2) limiting of coating growth to a very small portion of the substrate if necessary, although deposition over relatively large areas is also possible, (3) limited distortion of the substrate, (4) possibility of cleaner coatings due to the small area heated, (5) availability of rapid, i.e., nonequilibrium, heating and cooling rates, and (6) ability to interfere easily with laser annealing and diffusion of various devices and laser processing of metals and alloys (Young and Wood, 1982). The laser beam is used as a heat source in CVD processes operating at either near atmospheric pressure or low pressure. There are two ways to produce côatings by *laser CVD* (LCVD) processes. In laser-induced heterogeneous nucleation and reaction process, laser is focused onto an absorbing substrate, creating a localized hot spot at which the localized CVD reactions take place (Allen, 1981; Christensen and Lakin, 1978), see Fig. 9.55(*a*). In laser-induced homogeneous nucleation and reaction process, the reactant gases are heated by absorption of the laser beam (parallel to substrate) which react and condense on a substrate (Haggerty, 1984); see Fig. 9.55(*b*).

The type of the laser beam chosen is based on the absorptivities of the reactants and the substrate. A CO_2 laser source in a *continuous-wave* (CW) mode has been used to heat SiO_2 substrate in applying Ni, TiO_2, and TiC coatings (Allen, 1981). Deposition rates of few millimeters per minute have been observed for LCVD Ni and of about 20 μm min^{-1} for LCVD TiO_2. For small laser spot diameters, lateral conduction enlarges the heated surface area after a characteristic time. Therefore, a pulsed laser is desirable for obtaining deposition over very small areas. Christensen and Lakin (1978) achieved a heated area of about 110 μm in fused quartz with a 1-ms CO_2 laser pulse and deposited silicon by pyrolysis of silane at a temperature in the range of 1000 to 1200°C. Kitahama et al. (1986) synthesized diamond coatings by ArF excimer laser source using C_2H_2 diluted with H_2 as a source gas in the pressure range of 10^3 to 10^4 Pa (8 to 75 torr) and in the temperature range of 40 to 800°C.

9.4.3.4 Electron-Assisted CVD. In the *electron-assisted CVD* (EACVD), also called the *heated-filament-assisted CVD* (HFCVD), process, the CVD reaction takes place in the presence of electron bombardment which facilitates activation of the precursor gases, thus allowing deposition at lower substrate

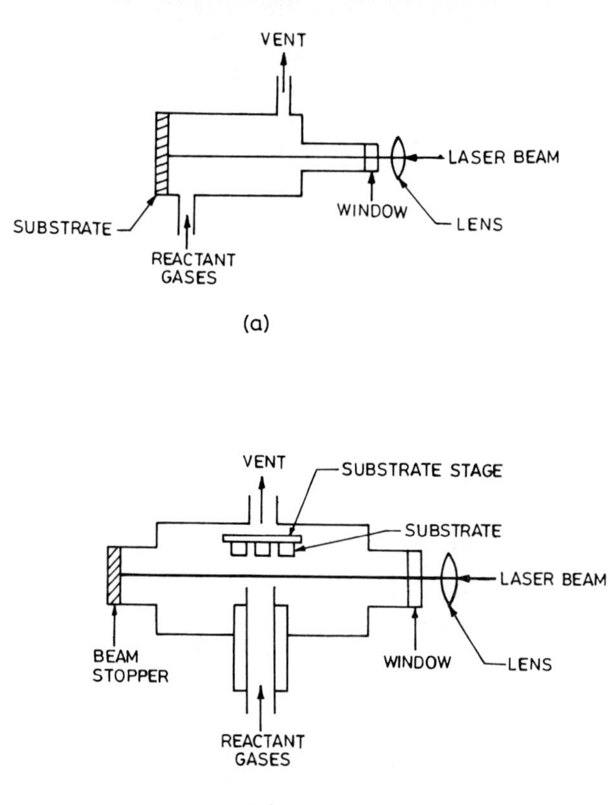

FIGURE 9.55 Schematic diagram of laser-induced CVD reactor using (*a*) heterogeneous nucleation and reaction and (*b*) homogeneous nucleation and reaction.

temperatures (Inuzuka and Ueda, 1968; Matsumoto et al., 1982; Sawabe and Inuzuka, 1986; Stirland, 1966). Figure 9.56 shows schematically the EACVD apparatus used for deposition of diamond coating with methane (CH_4) and H_2 mixed gases onto ceramic (SiC) with a deposition rate of 3 to 5 μm h^{-1} (Sawabe and Inuzuka, 1986).

In this apparatus, the electrons are produced thermionically (by the tungsten filament) and bombarded onto the substrate by applying a positive voltage (about 150 V) on the substrate holder. The decomposition of these gases occurs at the filament surface and at the substrate surface by electron bombardment. In comparison with conventional CVD, the decomposition of CH_4 and H_2 is accelerated by electron bombardment. During deposition, the concentration of CH_4 in the mixed gases was between 1 and 2 volume percent, the mixed gas pressure was 4 kPa (30 torr), the temperature of the filament was about 2000°C, and that of the substrate was about 850°C. The study of growth and nucleation of EACVD and CVD produced diamond coatings shows that the number of nucleation sites in EACVD coating is much higher than in CVD coating, which is also responsible for an improvement in the hardness (~9000 kg mm^{-2}).

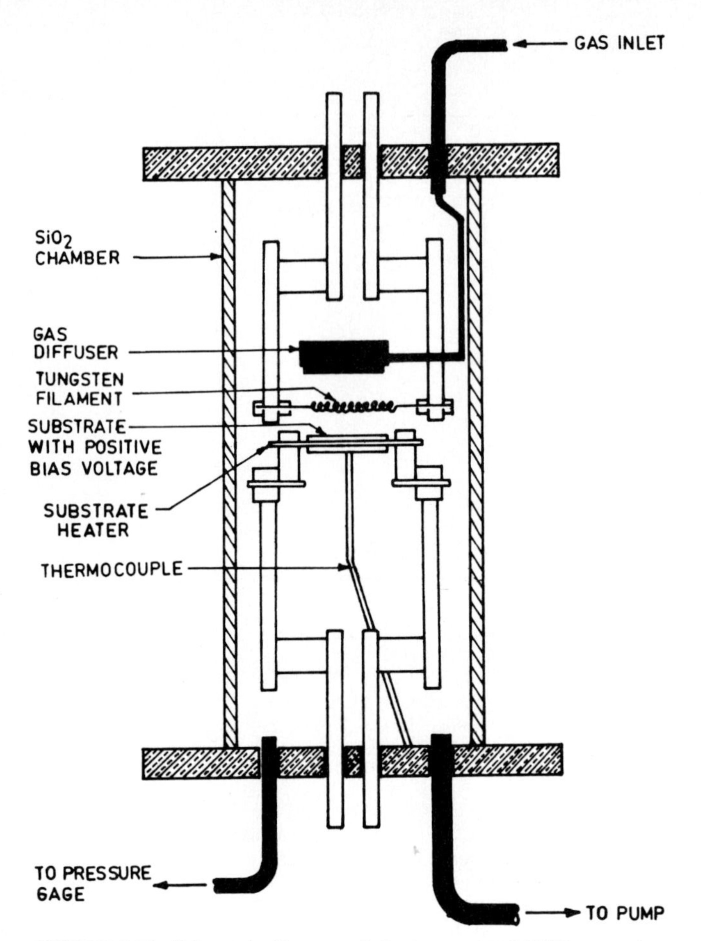

FIGURE 9.56 Schematic diagram of electron-assisted CVD apparatus. [*Adapted from Sawabe and Inuzuka (1986).*]

9.4.4 Coatings Deposited by CVD

9.4.4.1 Coating Materials. Chemical vapor deposition is one of the most versatile techniques for applying coatings of various materials. A partial list of metals and compounds which have been deposited by CVD is given in Tables 9.9 and 9.10 (Archer and Yee, 1978; Blocher et al., 1975; Brutsch, 1985; Bryant, 1977; Duret and Pichoir, 1983; Hintermann, 1980; Hocking et al., 1989; Jang and Chun, 1987; Johannesson and Lindstrom, 1975; Kern and Ban, 1978; Kern and Rosler, 1977; Kim and Chun, 1983; Perry and Archer, 1980).

The CVD coatings are widely used in semiconductor, structural, and tribological applications. Doped epitaxial layers of high-purity silicon on single-crystal substrates and polycrystalline silicon are grown from either the hydrogen reduction of trichlorosilane ($SiHCl_3$) or thermal decomposition of silane (SiH_4). Coatings of compound semiconductors such as CVD $GaAs_xP_{1-x}$ are used for

TABLE 9.9 Typical Metals Deposited by CVD

Material	CVD method	Temperature, °C
Aluminum	As metal. Thermal decomp. AlR_3, $AlHR_2$ (R = alkyl)	200–250
	As aluminide Al + AlX_3, (X = halide) in Al_2O_3 bed	700–1000
	$AlCl_3 + H_2$	900–1000
Antimony	Thermal decomp. $SbCl_3$	500–600
	Thermal decomp. SbH_3	~150
Arsenic	Thermal decomp. $AsCl_3$	300–500
	Thermal decomp. AsH_3	230–300
Boron	Thermal decomp. B_2H_6	400–500
	Thermal decomp. B Br_3	1000–1300
	$BCl_3 + H_2$	~1500
Bismuth	Thermal decomp. $BiCl_3$	~250
	Thermal decomp. BiH_3	230–300
Carbon	Thermal decomp. of hydrocarbons, e.g., CH_4 as carbon	1000–1500
	C_2H_6, C_2H_4, C_2H_2 as graphite	1500–2500
Chromium	As metal. Thermal decomp. of CrI_2	900–1000
	Thermal decomp. of $Cr(CO)_3$ and dicamene chromium	
	As chromium diffusion layer Cr + halide activated, e.g., NH_4Cl in Cr_2O_3 bed	400–1200
Cobalt	Thermal decomp. $Co(acac)_{2\ or\ 3}$	300–500
Copper	As metal. Thermal decomp. $Cu(acac)_2$	260–450
	As copper layer on steel $Cu_3Cl_3 + Fe \rightarrow Cu + FeCl_3$	500–1000
Germanium	Disproportion $GeI_2 \rightarrow Ge + GeI_4$	~450
	Thermal decomp. GeH_4	400–1100
Gold	Thermal decomp. $AuCl.PPh_3$	
	Disproportion $AuCl_2 \rightarrow Au + Au_2Cl_6$	
Hafnium	Thermal decomp. HfX_4(X = I, Br)	1100–1450
Lead	As powder from thermal decomp. of $PbEt_4$	300–500
Molybdenum	MoX_5 or $MoF_6 + H_2$	~650
	$Mo(CO)_5 + H_2$	300–600
Nickel	Thermal decomp. $Ni(CO)_4$	180–200
	Thermal decomp. $Ni(acac)_2$	350–450
Niobium	$NbCl_5 + H_2$	1000–1400
Osmium	$Os_3Cl_4 + H_2$	
	Thermal decomp. $Os_5(CO)_3Cl_2$	325–340
Palladium	Thermal decomp. $Pd(acac)_2$	350–450
Platinum	Thermal decomp. $Pt(CO)_2Cl_2$	~600
	Thermal decomp. $Pt(acac)_2$	350–450
Rhenium	$ReF_5 + H_2$	250–1100
	$ReCl_5 + H_2$	600–1200
	Thermal decomp. $ReOCl_4/ReCl_5$	1250–1500
Rhodium	Thermal decomp. $Rh(CO)_2Cl_2$	
Ruthenium	Thermal decomp. $Ru(CO)_5$	~200
	Thermal decomp. $Ru(CO)_2Cl_2$	
Silicon	$SiCl_4 + H_2$	900–1800
	$SiHCl_3 + H_2$	900–1800
	$SiHCl_3 + H_2$	950–1250
Tantalum	$TaCl_5 + H_2$	~1000
Tin	$SnCl_2 + H_2$	400–600
Titanium	$TiCl_4 + H_2$	1100–1400
	Thermal decomp. TiI_4	1200–1500
Tungsten	$WF_6 + H_2$	400–700
	$WCl_6 + H_2$	600–700
	$W(CO)_6$	350–600
Tungsten/Rhenium	W + Re fluoride + H_2	400–1000
Uranium	Thermal decomp. UI_6	1500
Vanadium	$VCl_2 + H_2$	800–1000

Source: Adapted from Perry and Archer (1980).

TABLE 9.10 Typical Compounds Deposited by CVD

Materials	CVD method	°C
Nitrides BN	$BCl_3 + NH_3$	1000–2000
	Thermal decomp. $B_3N_3H_3Cl_3$	1000–2000
HfN	$HfCl_x + N_2 + H_2$	950–1300
Si_3N_4	$SiH_4 + NH_3$	950–1050
	$SiCl_4 + NH_3$	1000–1500
TaN	$TaCl_5 + N_2 + H_2$	2100–2300
TiN	$TiCl_4 + N_2 + H_2$	650–1700
VN	$VCl_2 + N_2 + H_2$	1100–1300
ZrN	$ZrCl_4 + N_2 + H_2$	2000–2500
Oxides Al_2O_3	$AlCl_3 + CO_2 + H_2$	800–1300
SiO_2	$SiH_4 + O_2$	300–450
	Thermal decomp. $Si(OEt)_4$	800–1000
Silicon-oxynitride	$SiH_4 + H_2 + CO_2 + NH_3$	900–1000
SnO_2	$SnCl_4 + H_2O$	
TiO_2	$TiCl_4 + O_2$ + hydrocarbon (flame)	
Silicides V_3Si	$SiCl_4 + VCl_4 + H_2$	
MoSi	Mo (substrate) + $SiCl_2$	800–1100
Borides AIB_2	$AlCl_3 + BCl_3$	~1000
HfB_x	$HfCl_4 + BX_3$(X = Br, Cl)	1900–2700
SiB_x	$SiCl_4 + BCl_3$	1000–1300
TiB_2	$TiCl_4 + BX_3$(X = Br, Cl)	1000–1300
VB_2	$VCl_4 + BX_3$(X = Br, Cl)	1900–2300
ZrB_2	$ZrCl_4$ + B Br_3	1700–2500
Carbides B_4C	$BCl_3 + CO + H_2$	1200–1800
	$B_2H_6 + CH_4$	
	Thermal decomp. of Me_3B	~550
Cr_7C_3	$CrCl_2 + H_2$	~1000
Cr_3C_2	$Cr(CO)_5 + H_2$	300–650
HfC	$HfCl_4 + H_2 + C_7H_8$	2100–2500
	$HfCl_4 + H_2 + CH_4$	1000–1300
Mo_2C	$Mo(CO)_6$	350–475
	$Mo + C_3H_{12}$	1200–1800
SiC	$SiCl_4 + C_6H_5CH_3$	1500–1800
	$MeSiCl_3 + H_2$	~1000
TiC	$TiCl_4 + H_2 + CH_4$	980–1400
W_2C	Thermal decomp. $W(CO)_6$	300–500
	$WF_6 + C_6H_6 + H_2$	400–900
VC	$VCl_2 + H_2$	~1000

Source: Adapted from Perry and Archer (1980).

light-emitting diodes, and other CVD coatings are used for radiation detectors and piezoelectric transducers. Structural shapes of some materials are made by CVD simply because it is most economical. Thus vapor-deposited crucibles and the tungsten tubing are commercially available for use in fused-salt chemistry, etc. Because of superior throwing power, CVD-formed complex nickel molds are used in some pressure molding operations where the longer life of the molds justifies their higher cost. Significant advances in the area of composite materials have been possible through application of the CVD process in preparing the fibers of materials such as carbon, boron, and SiC, coating the fibers with a multitude of materials, and infiltrating open structures to provide a matrix.

For protection against wear, wet corrosion, hot corrosion, and molten metals, CVD coatings are used simply because no other method will give a dense, impermeable coating with excellent adhesion of a suitable material. Titanium nitride and alumina coatings on cemented carbide tool bits and various bearing applications provide increased wear resistance. Tungsten carbide coatings are used in journal gas bearings, air plug gauge, and extrusion dies for aluminum for increased wear resistance. Titanium carbide coatings are used in instrument bearings and rolling-element bearings in outer space. Silicon carbide coatings have a brilliant black appearance and are applied to watch cases and nozzles for increased wear resistance. Boron nitride and titanium diboride coatings on graphite are being used to form containers for the vacuum evaporation of liquid aluminum. The CVD coating of steel with tantalum is used in the chemical processing industry. A multiple refractory compound coating is used to improve the protection of niobium tubes in the core of a nuclear reactor from damage due to metallurgical corrosion by liquid steel at high temperature (1600°C).

9.4.4.2 Microstructure of CVD Coatings. As in condensation from the vapor phase without chemical reaction (PVD), the structure of the CVD deposited material depends on the substrate temperature and the supersaturation roughly as schematically shown in Fig. 9.57. The effective supersaturation is defined as the local effective concentration in the gas phase of the material to be deposited, relative to its equilibrium concentration. In a simple one-component system, the supersaturation is defined as the ratio of the actual pressure of the reactant in the vapor phase to the equilibrium vapor pressure (i.e., the pressure of the reactant in equilibrium with the condensed phase). The effective supersaturation depends not only on the concentration, but also on the substrate temperature, because the reaction is thermally activated. Since the effective supersaturation for thermally activated reactions increases with temperature, the opposing tendencies can lead in some cases to a reversal of the sequence of crystalline forms listed in Fig. 9.57, as the temperature is increased (Blocher, 1974). CVD coatings applied at high effective supersaturation and low temperature tend to be either amorphous or fine-grained whereas at low effective supersaturation and high temperature, the diffusion is significant and equiaxed grains result. The columnar structure promotes surface irregularity and tends to be weak perpendicular to growth direction. Therefore, the choice of a deposition reaction that can develop a high effective supersaturation at low temperature is desirable (Bryant, 1977).

Kim and Chun (1983) noted that the crystal size of the TiN coating is increased with an increase in the substrate temperature and the total flow rate of the reac-

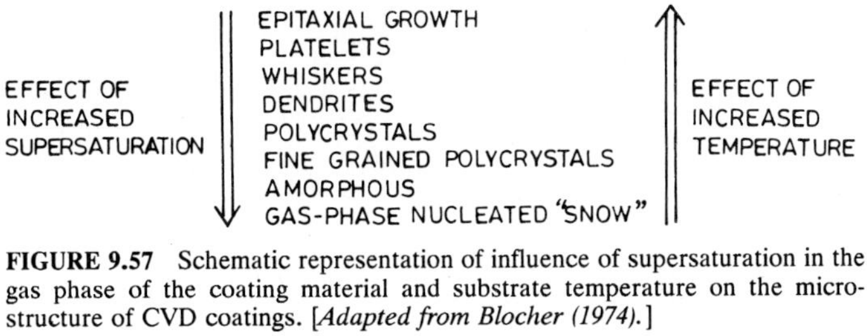

EFFECT OF INCREASED SUPERSATURATION ⇓

EPITAXIAL GROWTH
PLATELETS
WHISKERS
DENDRITES
POLYCRYSTALS
FINE GRAINED POLYCRYSTALS
AMORPHOUS
GAS-PHASE NUCLEATED "SNOW"

⇑ EFFECT OF INCREASED TEMPERATURE

FIGURE 9.57 Schematic representation of influence of supersaturation in the gas phase of the coating material and substrate temperature on the microstructure of CVD coatings. [*Adapted from Blocher (1974).*]

tant gases and is decreased with an increase in the partial pressure of $TiCl_4$ gas. Figures 9.58 and 9.59 show the morphologies of the TiN coating crystals at different substrate temperatures and partial pressure of $TiCl_4$ gas using a $TiCl_4$, H_2, and N_2 gas mixture.

The substrate material has a significant influence on the coating structure. Jang and Chun (1987) have shown that the morphology of TiC coating is quite different when applied on plain-carbon steel and on high-carbon steel.

The growth of columnar grains like those in Fig. 9.60 is characteristic of many materials in a certain range of conditions. This structure results from uninterrupted growth toward the source of supply. The presence of extrinsic "impurities" or dopants (those purposely added to the deposition system by the introduction of one or more additional reactant gases) can result in both the formation and the stabilization of an equiaxed structure (Fig. 9.60). The stabilizing microconstituent can be either a void (or bubble) or a second phase particle. For example, tungsten has been stabilized by the former means by the introduction of potassium, oxygen, and ammonia and by the latter means by the incorporation of silicon, carbon, and HfN (Bryant, 1977).

9.4.4.3 ***Factors Influencing the Deposition Rate.*** The substrate temperature and partial pressure of the reactants are closely interrelated and are of fundamental importance in controlling the deposition rate and coating properties (Kern and Rosler, 1977; Kim and Chun, 1983). With few exceptions, the deposition rate

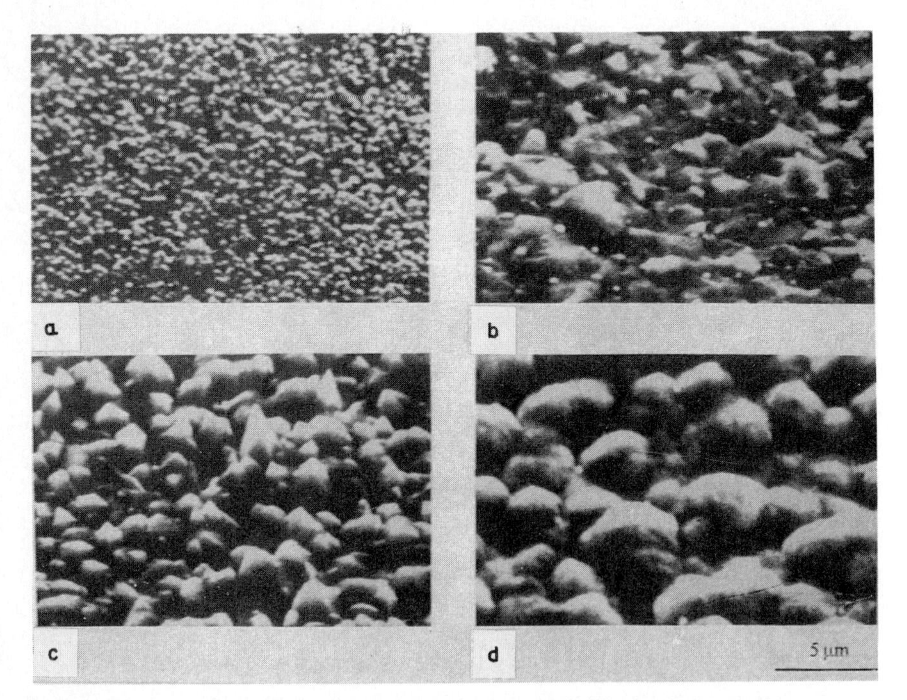

FIGURE 9.58 Scanning electron micrographs of the surface of TiN crystals deposited at various substrate temperatures applied at total flow rate of 600 cm^3 min^{-1}, P_{TiCl_4} = 0.03 atm, P_{H_2} = 0.485 atm: (*a*) 950°C, (*b*) 1050°C, (*c*) 1100°C, and (*d*) 1150°C. [*Adapted from Kim and Chun (1983).*]

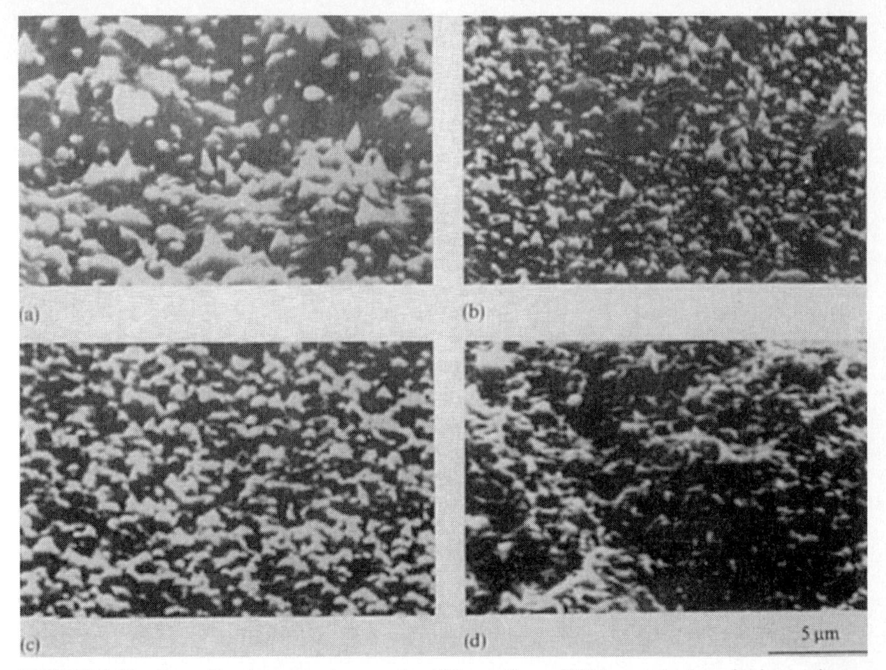

FIGURE 9.59 Scanning electron micrographs of the surface of TiN crystals deposited at various partial pressures of $TiCl_4$ applied at substrate temperature 1050°C, total flow rate of 500 cm^3 min^{-1}: (*a*) 0.015 atm, (*b*) 0.021 atm, (*c*) 0.027 atm, (*d*) 0.037 atm. [*Adapted from Kim and Chun (1983).*]

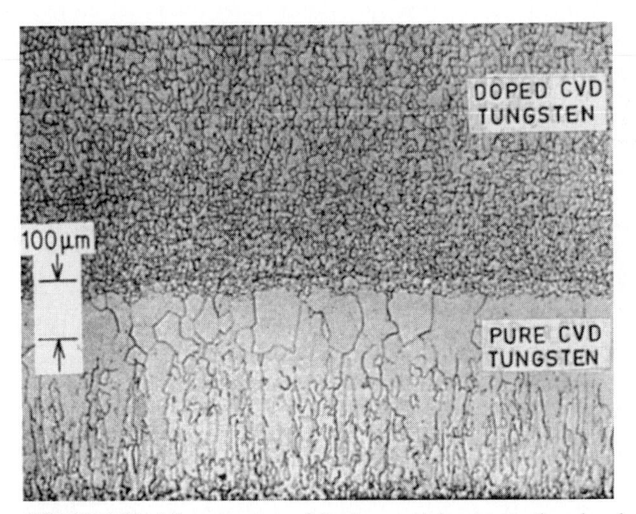

FIGURE 9.60 Microstructure of thorium acetylacetonate doped and undoped CVD tungsten coating followed annealing for 75 min at 2070°C. [*Adapted from Bryant (1977).*]

from CVD reactions increases with temperature and is limited by the rate of chemical reaction at the surface. As the temperature and deposition rate are increased, transport of material through the gas phase by diffusion adjacent to the substrate becomes a limiting factor, and the deposition rate becomes less dependent on temperature. The further increase in temperature may result in the decomposition of reactant gases, e.g., silane. For most CVD reactions, increasing the partial pressure of reactants tends to increase the amount of material transported per unit of atmospheric gas volume and thus the deposition rate. Because diffusivity in the reactant gas decreases with increased pressure, maximizing the deposition rate in this way also leads to decreased uniformity, as described previously. Thus, the compromise between increased uniformity and decreased deposition rate must be faced. Despite the convenience of operating at atmospheric pressure, use of a gas that is not one of the reactants should be avoided, for the sake of coating uniformity at a given partial pressure of reactants.

The quantity of reactant introduced into the reactor per unit time affects the deposition rate linearly up to some saturation level, which is dictated by the capacity of the reaction chamber. The diluent or carrier gas, if used, serves to reduce the concentration of each reactive gas so that their streams can be safely mixed without spontaneous reaction. The diluent gas also provides sufficient volume to the mixture to force it over the heated substrate which facilitates deposition. Too low or too high a flow rate depresses the deposition rate and causes poor coating uniformity. The type of diluent gas can affect the deposition rate. The exact shape and dimensions of the reaction chamber and the gas inlet-outlet configuration affect the gas flow dynamics and are therefore very critical with respect to both the deposition rate and the thickness uniformity of coating (Kim and Chun, 1983; Stull et al., 1969).

Chemical trace impurities on the substrate surface can act as growth catalysts or inhibitors. Finally, the substrate composition can affect the deposition rate. Jang and Chun (1987) have reported that the deposition rate of TiC coating on plain-carbon steel is approximately 4 times faster than that on the high-carbon steel under identical conditions. This is probably because of the interaction of carbon with other alloying elements in steel at high temperature. The carbon activity in steel depends on the ratio of the concentrations of carbide-forming elements (Cr, Mo, W, V) to the carbon content of the steel. In CVD of TiC on steel, carbon's activity is rate-controlling, and the deposition rate decreases as the carbon activity of the steel increases.

9.4.4.4 Properties of CVD Coatings. Properties of CVD coatings range widely depending on the material involved and the process conditions. Dense, fine-grained coatings have properties consistent with the wrought forms of that coating material. Porous or preferentially oriented coatings have properties of that material produced by other methods (Blocher, 1974; Bryant, 1977; Kern and Rosler, 1977; Kim and Chun, 1983).

The particulate contaminants in the gas stream can cause microbubbles, pinholes, and other localized structural defects in the coatings. Controlled incorporation of additives to the reactant is sometimes used to lower the residual stresses. The formation of brittle intermetallic compounds at the coating-substrate interface may result in poor coating adhesion. An additional source of poor adhesion is the chemical attack of the substrate by a gaseous product of the deposition reaction to produce a mechanically weak compound at the coating-substrate interface.

9.5 PHYSICAL-CHEMICAL VAPOR DEPOSITION

Physical-chemical vapor deposition (P-CVD) processes are the hybrid processes which use glow discharge to activate CVD processes. These are broadly referred to as *plasma-enhanced CVD* (PECVD) or *plasma-assisted CVD* (PACVD) processes. Many of the phenomena of PECVD are similar to those of conventional high-temperature CVD. Likewise, much of the physical description of glow-discharge plasmas for sputtering applies as well to plasma CVD. Instead of requiring thermal energy as in thermal CVD, the energetic electrons in the plasma (at pressures ranging from 1 to 5 × 10^2 Pa or 0.01 to 5 torr, typically less than 10 Pa or 0.1 torr) can activate almost any chemical reaction among the gases in the discharge at relatively low substrate temperatures, ranging from 100 to 600°C (typically less than 300°C). At the same time, the bulk of the gas and the substrates do not reach high temperatures because of the nonequilibrium nature of the plasma. The reactions proceed at high reaction rate in a system at low processing temperature where the desirable chemical reaction, although possible from the thermodynamic point of view, under the given conditions would proceed extremely slowly in the absence of plasma. A weak discharge at a discharge current density on the order of 1 mA cm^{-2} is sufficient to cause a significant enhancement of the overall reaction rate. Practically any gas or vapor including polymers can be used as a precursor material (Bell, 1980; Bonifield, 1982; Hollahan and Bell, 1974; Hollahan and Rosler, 1978; Ojha, 1982; Rosler and Engle, 1979; Sherman, 1984; Věprek, 1985; Vossen and Kern, 1978).

Another method, known as the *reactive pulsed plasma* (RPP) deposition process, uses a high-energy pulsed plasma rather than a continuous discharge and restricts the substrate to room temperature (Michalski, 1984; Michalski et al., 1985; Sokolowski, 1979; Sokolowski et al., 1977, 1979, 1981). The discharge between two coaxial electrodes at a chamber pressure of 10 to 100 Pa (0.1 to 1 torr) is struck by discharging the capacitors. The high temperatures produced during the pulsed discharge evaporate and ablate the electrode material, which reacts with the pulsed injected gas in the plasma to form compounds that are deposited on the substrate (not externally heated). The pulsed plasma is characterized by a high degree of ionization.

Chemical reactions in the deposition of polymer coatings (called *chemical vapor polymerization*) can be activated by ultraviolet light irradiation and electron beam methods as well as by glow discharge (Holland, 1956; Shen, 1976; Yasuda, 1978).

In contrast to thermal CVD process, the major advantage of PECVD is that the coating can be deposited at relatively low substrate temperatures and at sufficiently high deposition rates, enabling deposition on substrate that cannot withstand high temperatures. For example, most steels cannot be coated with the conventional thermal CVD techniques without suffering detrimental changes in properties. The coatings which are mismatched in the coefficient of thermal expansion with the substrate can be deposited with PECVD without severely stressing the coating upon cooling to room temperature. Some materials, such as some polymers, are unstable at the temperatures required to activate the deposition reaction. Another way of stating the temperature advantage is that at lower processing temperatures, PECVD often has much higher deposition rates than conventional thermal CVD. In addition to its low deposition temperature, the PECVD method can be used to coat large-area substrates. The kinetic energy of the ion bombarding the coating during growth is relatively high, ranging from 10 to 1000 eV (typically 50 to 300 eV), which facilitates coating adhesion. However, strong adhesion of thermal CVD coatings is facilitated by diffusion at very high substrate temperature.

The major limitation of PECVD process is the strong interaction of the plasma with the growing coating. Although, with proper understanding, the plasma interactions can be used to advantage, generally the deposition rates and some coating properties will depend on the spatial profile of the plasma. The result is often a nonuniform deposition across the reactor, so that the process and the equipment have to be adjusted for uniformity. The low substrate temperature is too low to give away the hydrogen involved in many CVD reactions, and it results in a hydrogen-containing coating. In addition, since the surface diffusion of adatoms is much retarded, the coating normally contains more defects and has a smaller density compared to a coating processed at high temperature. Another major limitation of PECVD is that it is more complex than thermal CVD but often more forgiving than other vacuum deposition techniques, such as evaporation or sputtering. Deposition of pure materials is very difficult with PECVD. Many of the gases used or made in PECVD are flammable (SiH_4, H_2, CH_4, etc.) or toxic. Scrubbers or "burn-offs" are required for safe handling of these gases in the exhaust of the vacuum pump.

Tables 9.1 and 9.2 summarize the range of coating materials, deposition (substrate) temperatures, deposition pressures, coating thicknesses, deposition rates, and advantages and disadvantages of PECVD processes.

9.5.1 Theory of PECVD

9.5.1.1 Glow-Discharge Plasmas. The glow discharges used in PECVD are basically the same type as used in glow-discharge sputtering. Many of the gases used in PECVD are polyatomic molecules (e.g., halides), and many have low ionization potentials. Gases used for sputtering tend to be just the opposite (e.g., argon). Also the gas pressures are usually higher in PECVD (typically 10 to 100 Pa or 0.1 to 1 torr) than in sputtering (typically 0.5 to 5 Pa or 0.005 to 0.05 torr), resulting in higher collision frequencies and shorter mean free paths for the electrons. Thus, electron energies are lower in PECVD (100 to 300 eV) and sensitive to the chemical nature of the gases and product gases in the discharge. Larger molecules generally result in lower electron energies.

The glow discharge produces a wide variety of chemical species, free radicals (neutrals), electrons, and ions and influences their thermodynamics and kinetics (Wagner and Věprek, 1982, 1983). In the low-pressure range (1 to 5×10^2 Pa or 0.01 to 5 torr), under typical conditions, for instance, electron and positive ion densities range from 10^9 to 10^{12} cm^{-3}. Positive ions are much less mobile than electrons and gain less energy. They also transfer more of their energy to the background gas during elastic collisions. The higher electron energy in the range of 2 to 10 eV is sufficient for ionization and dissociation of most types of gas molecules at high reaction rates. The high-energy electrons collide with neutral gas molecules and produce very reactive radicals, atoms, and ions by breaking the chemical bonds, thereby precluding the need for elevated temperatures to initiate reactions. The formation of a coating proceeds by adsorption of mostly free radical species to the substrate or coating surface and chemical bonding to other atoms on the surface.

The power for plasma generation is dc, high-frequency rf, or microwave (gigahertz) field. High-frequency rf discharges are most common, since the deposition of insulating coatings destroys the conduction path necessary for dc or very low-frequency rf discharges. The rf power supply can be (parallel-plate) capacitively coupled or inductively coupled to the powered electrode (Angus et al., 1986; Coburn and Kay, 1972). Capacitively coupled glow discharge is most commonly used for deposition. Microwave discharges are less common, because of the complication of using resonant cavity structures (Coburn and Kay, 1972; Vossen, 1979). At low frequencies (low kilohertz range), some degree of secondary electron emission from the electrodes is necessary to maintain the discharge. At higher frequencies (megahertz range), enough electrons gain the energy required to ionize gas molecules to maintain the discharge without secondary emission. It is shown in Appendix 9.A that if one of the electrodes of a diode arrangement in the presence of glow-discharge plasma is capacitively coupled to the rf power supply (known as powered electrode), it develops a dc self-bias potential that is negative relative to the plasma floating potential.

The coating can be deposited either on the powered electrode or on the grounded electrode. If coating is deposited on the powered electrode (with a negative dc self-bias), as in bias sputtering, the growing coating surfaces in glow discharge are concurrently bombarded by positive ions with energies as high as 500 eV (typically 100 to 300 eV) which improves the mechanical properties of coatings significantly. If the substrate is placed on the grounded electrode, ion bombardment can be achieved by external biasing the electrode with several tens of volts.

Several factors will determine the energy of the ions impinging on the coating during deposition. Besides factors such as discharge power, gas type, and pressure, which affect the plasma density, the rf voltage, rf frequency, and pressure determine what fractions of the plasma potential the ions can pick up as they cross the plasma sheath. At low pressures (less than 5 Pa or 0.05 torr) and low frequency ($\ll 1$ MHz), the ions can follow the rf field and are not seriously scattered in the sheath. They will then strike the surface with the full energy available to them. At high frequencies, the ions cannot respond to the rf fields and will drift through the sheath with less energy, seeing only an average plasma potential. At increased pressure, scattering in the sheath will reduce that energy further.

9.5.1.2 Transport Kinetics. The pressure and flow conditions used in PECVD are best described as slow, viscous, laminar flow. Except at the very low pressures used (less than 10 Pa or 0.1 torr), the mean free path of molecules in the gas, a small fraction of a millimeter, is much smaller than the dimensions of the flow channel of a reactor. Reynolds numbers are very small so that diffusion dominates over turbulent or convective mass transfer. Typical diffusion times are a few milliseconds, whereas the residence time of a reactor (pressure × channel volume ÷ total flow) is closer to a second, because of the low flows used (a few liters per minute). The transport kinetics of PECVD bear some resemblance to thermal, low-pressure CVD (LPCVD). The net flow velocity of the gas decreases to zero at the substrate surface, and mass transport of reactants to the surface relies on diffusion. The spatial distribution of free radical production in the discharge should be a dominant factor affecting the kinetics of coating growth (Bonifield, 1982).

9.5.2 Details of PECVD Processes

Block diagram of a PECVD reactor system is shown in Fig. 9.61. The most important subsystems are (1) the reactor or deposition chamber, (2) discharge

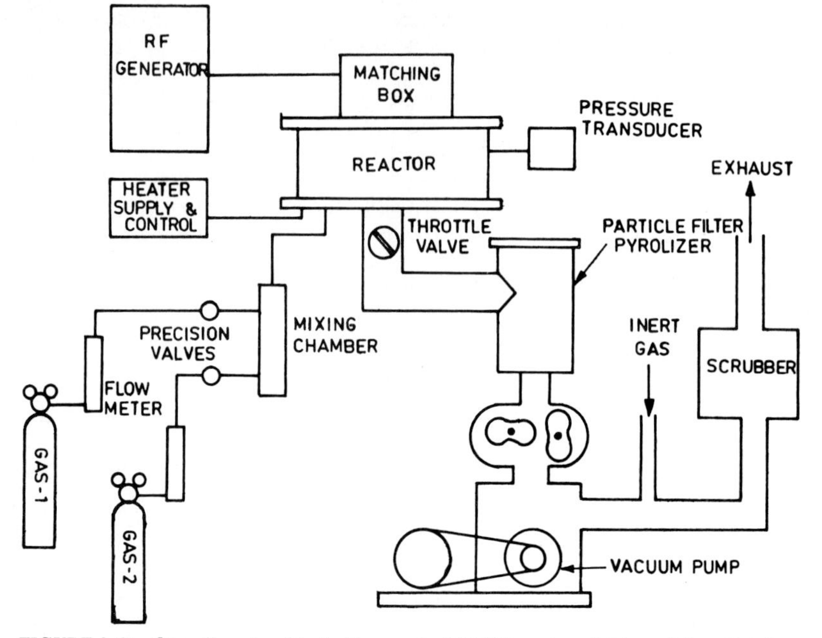

FIGURE 9.61 Overall system block diagram for PECVD system. [*Adapted from Reinberg (1979).*]

power supply, (3) vacuum system, and (4) gas supplies and their regulating and blending systems. The power for plasma generation is either dc, high-frequency rf, or microwave (gigahertz) field. A variety of reactor designs have been used (Angus et al., 1986; Hollahan and Rosler, 1978; Kern and Rosler, 1977; Ojha, 1982; Věprek, 1976).

Radio-frequency PECVD reactors can be divided broadly into two types: flat-bed (parallel-plate or planar diode) and tube-type (or tunnel) reactors (Vossen, 1979). In the parallel-plate reactors, the substrates usually lie flat (for heating purposes), while in the tube-type reactors substrates normally are in the stand-up position. Further division can be made depending on the way the electrical energy is coupled to the discharge—capacitively or inductively. In the tube-type plasma reactor, power for plasma generation is induced either inductively with an rf coil or capacitively with plates outside the reaction chamber [see Figs. 9.62(*b*) and 9.63]. The tubular designs lead to a significant radial variation in field density, and hence a large thickness variation across a substrate, if positioned vertically. In contrast, the field density in the parallel-plate reactor configuration [Fig. 9.62(*a*)], in which the substrate rests upon one plate of a two-plate capacitively coupled reaction zone, is reasonably uniform and controllable (Ojha, 1982). Its main advantage is the large substrate area available for deposition. Tube-type reactors, on the other hand, can take a higher throughput than parallel-plate reactors. The parallel-plate reactor configuration with capacitively coupled rf power supply is most commonly used for large-scale production-type systems for deposition.

Figure 9.62 shows schematics of two typical capacitively coupled rf PECVD reactors. The radial-flow circular parallel-plate system capacitively coupled with

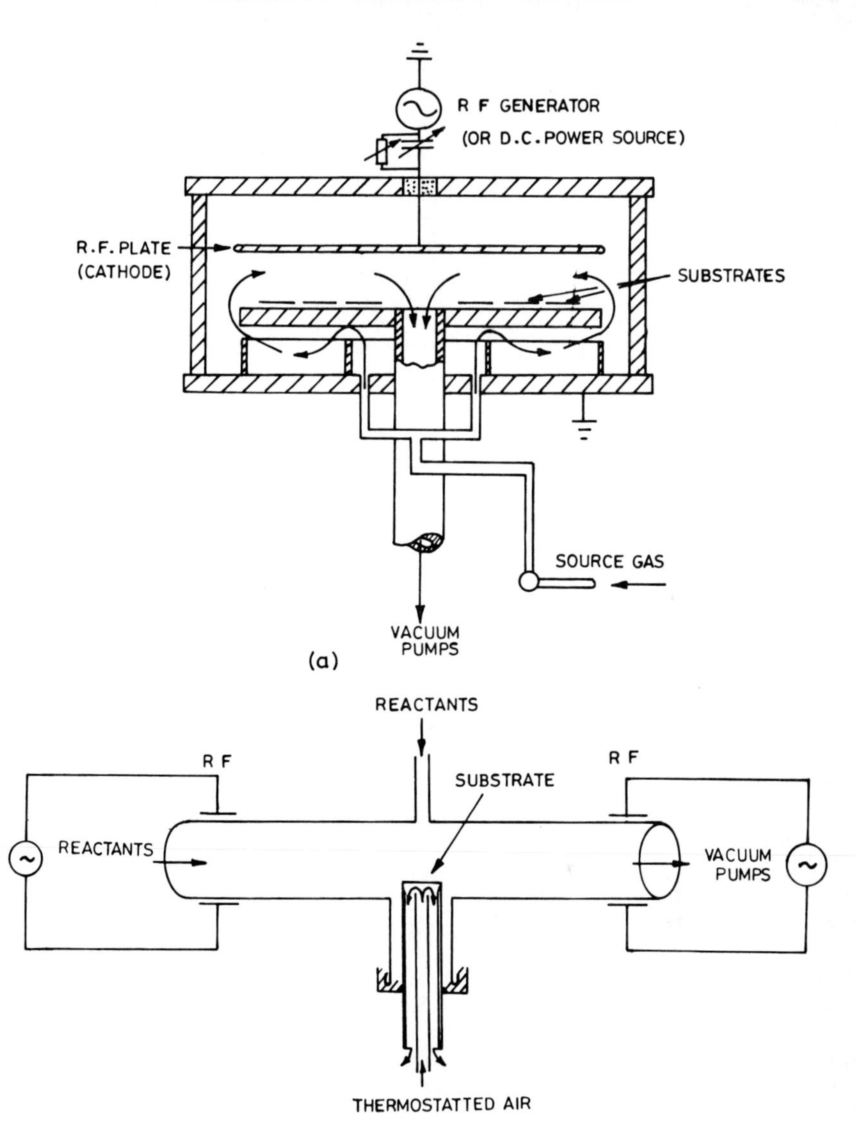

FIGURE 9.62 Schematic diagrams of capacitively coupled rf PECVD reactor systems: (*a*) radial-flow parallel-plate (flat-bed or planar diode) reactor, (*b*) tube-type (or tunnel) reactor with substrate located in plasma column. [*Adapted from Věprek (1985).*]

a top electrode and a heated lower grounded substrate plate is shown in Fig. 9.62(*a*). The plates are kept several centimeters apart, and the discharge is maintained at a total pressure from a few tens to a few hundred Pa (0.1 torr to a few torr) by means of an rf supply. The rf frequency is typically 13.56 MHz. The small biasing of the substrate and electrode with respect to plasma minimizes the

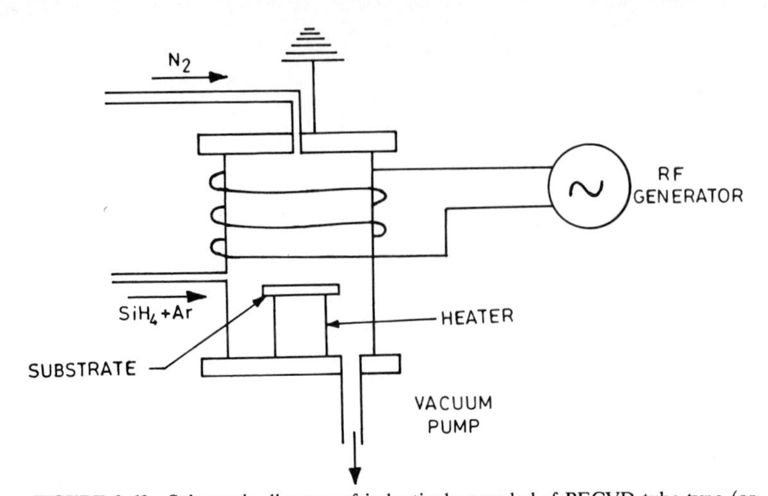

FIGURE 9.63 Schematic diagram of inductively coupled rf PECVD tube-type (or tunnel) reactor system used for deposition of Si_3N_4 coatings. [*Adapted from Helix et al. (1977).*]

bombardment of growing coatings by energetic ions, electrons, and soft X-rays (Věprek, 1976). The gas flows from the periphery of a circular, grounded substrate plate, across the substrates, and out the center of the reactor. As the initial gas is decomposed in the discharge, the direction of flow is such that the deposition occurs on decreasing areas of the substrate plate. Also the flow velocity increases toward the center of the reactor since the effective channel area of the flow is decreased. The higher flow velocity compensates for the reduced reactant concentration. These features, together with the cylindrical symmetry of the reactor, aid in maintaining deposition uniformity over large areas. This reactor configuration has been extensively used for deposition of Si_3N_4 coatings.

The discharge tube (Pyrex, silica glass, or alumina) shown in Fig. 9.62(*b*) with a small substrate which has a minimum capacitance with respect to the electrodes and to earth allows the bias to be minimized. Moreover, the bias can be perfectly controlled by using internal electrode and a dc discharge excitation, and it is even possible to obtain a positive bias of the substrate with respect to the plasma. The latter is not possible when a rf discharge is used (Coburn and Kay, 1972). Rosler and Engle (1979) developed a capacitively coupled rf PECVD tube-type system as an extension of the hot-wall LPCVD system shown in Fig. 9.54.

Figure 9.63 shows a schematic of a PECVD tube-type reactor that is powered inductively by a rf source. This setup has been used, for instance, to deposit silicon nitride coatings on a small heated substrate support pedestal (Helix et al., 1977).

Figure 9.64 shows a schematic of a *microwave plasma-enhanced CVD* (MPECVD) system. Microwaves generated by a magnetron are transmitted to a cavity out of the direct substrate region, through a set of wave guides, a power monitor, and a tuner. The typical frequency of microwaves used for deposition is 2.45 GHz. The MPECVD system has been used for deposition of diamond coatings (Kamo et al., 1983) and Si_3N_4 coatings (Shiloh et al., 1977).

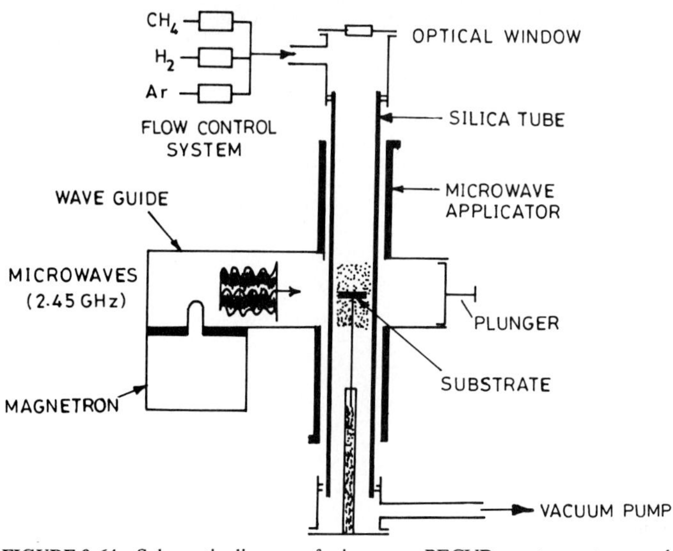

FIGURE 9.64 Schematic diagram of microwave PECVD reactor system used for deposition of diamond coatings. [*Adapted from Kamo et al. (1983).*]

9.5.2.1 Coatings Deposited by PECVD

Coating Materials. Because of the relatively low deposition temperature (<600°C), PECVD techniques are suitable for the coating deposition on a variety of substrates. Typical coatings applied by PECVD include amorphous and microcrystalline Si and Ge, W, Mo, Si_3N_4, SiO_2, SiO_xN_y, SiC, Al_2O_3, AlN, BN, TiC, TiN, TiC_xN_{1-x}, TiO_2, P_3N_5, WSi_2, i-carbon, and hydrocarbon and fluorocarbon polymers (Angus et al., 1986; Bonifield, 1982; Hess, 1984; Hollahan and Bell, 1974; Hollahan and Rosler, 1978; Kamo et al., 1983; Kay et al., 1980; Kern and Rosler, 1977; Ojha, 1982; Sundgren and Hentzell, 1986; Yasuda and Hsu, 1978b). Table 9.11 lists the coatings, deposition temperature, and deposition rate applied by various investigators in PECVD process (Věprek, 1985).

PECVD coatings are widely used in semiconductor electronics, nuclear, and metal-cutting industries. A number of dielectric films (SiO_2, Al_2O_3, Si_3N_4, etc.) and amorphous Si (a-Si) coatings have been applied in the fabrication of microelectronic devices, optical wave guides, and optical coatings. PECVD Si_3N_4 is extensively used as a final protective coating and passivation for integrated circuits because it is an excellent diffusion barrier against moisture and alkali ions. PECVD a-Si coatings are used in low-cost solar cells, thin-film field-effect transistors, liquid-crystal displays, etc. (Věprek, 1985). PECVD coatings are used for deposition of the first wall in devices for controlled nuclear fusion. PECVD coatings of hard ceramics (such as TiC, HfC, and TiN) on high-speed steel tools can significantly improve tool life. PECVD techniques are commonly used to produce diamondlike and diamond coatings from abundant hydrocarbons as the starting material.

Properties of PECVD Coatings. Coatings can be produced by PECVD process which are pinhole-free, are hard, and have excellent adhesion. The properties of PECVD coatings are affected by process variables—rf power and frequency, powered or grounded substrates, cathode and anode spacing, reactant

TABLE 9.11 Typical Temperature and Rate of Deposition for Various Materials Deposited by PECVD

Solid	Deposition temperature,* K	Deposition rate, nm s^{-1}	Reactants
a-Si	523–573*	0.1–1	SiH_4, SiF_4—H_2, Si(*s*)—H_2
μc-Si	523–673*	0.1–1	SiH_4—H_2, SiF_4—H_2, Si(*s*)—H_2
a-Ge	523–673*	0.1–1	GeH_4
μc-Ge	523–673*	0.1–1	GeH_4—H_2, Ge(*s*)—H_2
a-B	673*	0.1–1	B_2H_6, BCl_3—H_2, BBr_3
a-P, μc-P	293–473	≤100	P(*s*)—H_2
As	<373	≤10	AsH_3, As(*s*)—H_2
Se, Te, Sb, Bi	≤373	1–10	Me—H_2
Mo			$Mo(CO)_4$
Ni			$Ni(CO)_4$
i-C	≤523	0.1–100	C_nH_m
C (graphite)	1073–1273	≤100	C(*s*)—H_2, C(*s*)—N_2
CdS	373–573	≤10	Gd—H_2S
GaP	473–573	0.1	$Ga(CH_3)_3$—PH_3
Oxides			
SiO_2	≥523	0.1–10	$Si(OC_2H_5)_4$, SiH_4—O_2, N_2O
GeO_2	≥523	0.1–10	$Ge(OC_2H_5)_4$, GeH_4—O_2, N_2O
SiO_2/GeO_2	≈1273	≈3000	$SiCl_4$—$GeCl_4$—O_2
Al_2O_3	523–773	0.1–1	$AlCl_3$—O_2
TiO_2	473–673	0.1	$TiCl_4$—O_4, metallorganics
B_2O_3			$B(OC_2H_5)_3$—O_2
Nitrides			
Si_3N_4	573–773	0.1–1	SiH_4—N_2, NH_3
AlN	≤1273	≤10	$SiCl_4$—N_2
GaN	≤873	0.1–1	$GaCl_4$—N_2
TiN	523–1273	0.1–50	$TiCl_4$—H_2+N_2
BN	673–973		B_2H_6—NH_3
P_3N_5	633–673	≤50	P(s)—N_2, PH_3—N_2
Carbides			
SiC	473–773	0.1	SiH_4—C_nH_m
TiC	673–873	0.5–10	$TiCl_4$—$CH_4(C_2H_2)+H_2$
GeC	473–573	≈0.1	
B_xC	673*	0.1–1	B_2H_6—CH_4

*Optimum temperature.
Source: Adapted from Věprek (1985).

gas ratio, total pressure, and substrate temperature. Although the number of variables makes a process complicated, it allows a process to be adjusted for control of coating properties and for uniform deposition.

The influence of substrate temperature, rf power, reactant gas ratio, and total pressure on the coating composition, deposition rate, and refractive index for PECVD Si_3N_4 coatings is shown in Fig. 9.65. The complex deposition process is divided into three kinetic steps, i.e., radical generation, radical adsorption, and adatom rearrangement, to explain the observed change of the coating's chemical

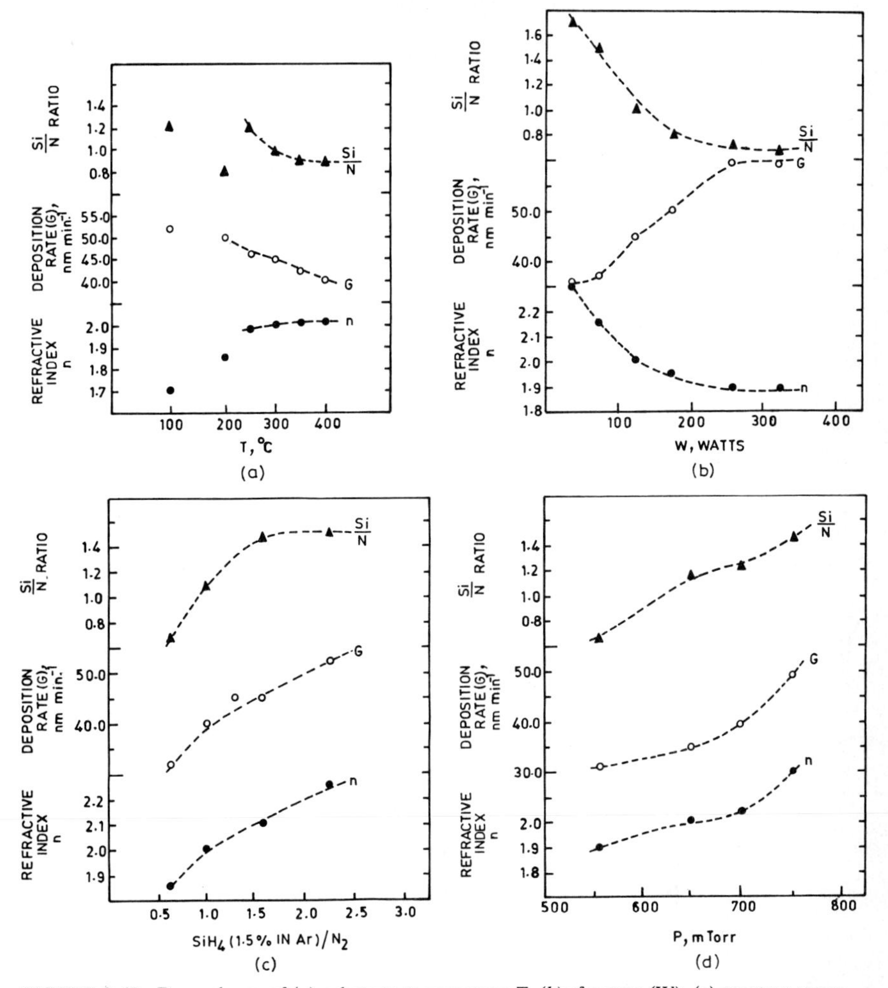

FIGURE 9.65 Dependence of (*a*) substrate temperature *T*, (*b*) rf power (W), (*c*) reactant gas ratio, and (*d*) partial pressure of gases P on refractive index *n*, deposition rate *G*, and stoichiometery Si/N of Si_3N_4 coatings prepared by PECVD with SiH_4 and N_2 as starting gases. [*Adapted from Dun et al. (1981).*]

and physical properties. The coating composition is determined not only by the reactant gas ratio, but also by a combined function of the rf power W and partial pressure of gases P in terms of W^x/P, with x a system-dependent factor. The dependence of coating composition on W^x/P can be related to the radical generation processes. The substrate temperature is found to affect the coating composition as well. Higher substrate temperature produces coatings with less hydrogen and more nitrogen and hence higher density (Dun et al., 1981).

In most cases, deposition rates are highest where the plasma density and reactant concentrations are the highest. These are dominant factors controlling the uniformity of the deposition. Often no deposition will occur outside the discharge plasma. Concurrent ion bombardment during coating growth has a pronounced effect on coating properties such as suppression of columnar growth, decrease in crystalline size, and increase in density, hardness, and coating-to-substrate adhesion. However, the high energy of ion bombardment can result in high compressive residual stresses and radiation damage due to ion implantation (Vandentop et al., 1990; Věprek, 1985). Increasing power and decreasing pressure increase the flux and energy of the ions. Because of the dc self-bias potential at the powered electrode, there is ion bombardment at the powered electrode. Ion bombardment on the powered or grounded electrode can be facilitated by a relatively small bias of several tens of volts.

9.5.3 Details of Reactive Pulsed Plasma Deposition Process

Sokolowski et al. (1979, 1981), Michalski (1984), and Michalski et al. (1985) developed the *reactive pulsed plasma* (RPP) process which restricts the substrate temperature to room temperature. It uses a high-energy pulsed plasma of a capacitor battery between two coaxial electrodes (accelerator) rather than a continuous discharge. In the RPP process, the metal vapors are generated by the electroerosion of hot electrodes, which react with pulsed injected gas in the plasma, and reaction products are deposited onto a substrate (not externally heated).

The schematic diagram of the RPP process is shown in Fig. 9.66. The mixture of reactant gases is pulse-injected into the chamber, previously evacuated to 10^{-5} Pa (10^{-7} torr). The electric discharge generating the plasma is appropriately synchronized with the operational gas injection in between the electrodes, each time the injection is performed prior to the start of the discharge. The electric discharge is generated by pulse-discharging a capacitor battery between the central electrode (made of coating source material) and the outer electrode (which con-

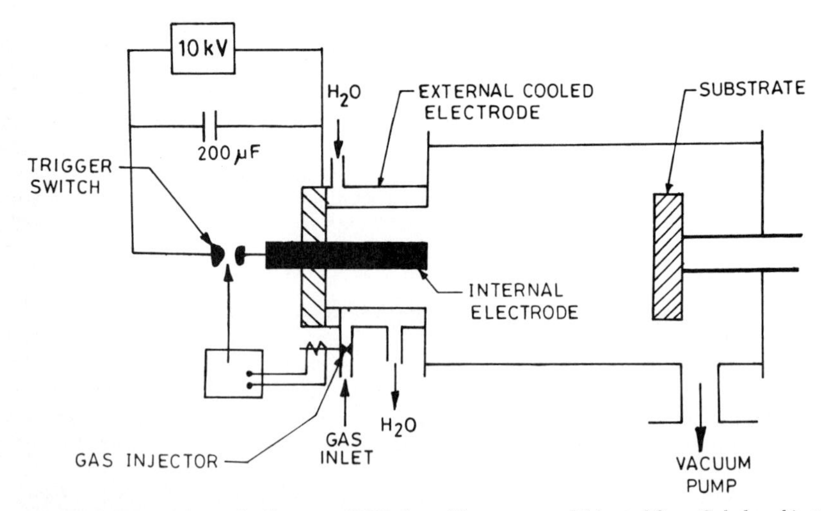

FIGURE 9.66 Schematic diagram of RPP deposition process. [*Adapted from Sokolowski et al. (1981).*]

sists of a water-cooled copper cylinder) at a chamber pressure of 10 to 10^2 Pa (0.1 to 1 torr). The plasma obtained with this method is not in equilibrium, and its temperature is on the order of 20,000 K. During the discharge (~0.1 ms), the central electrode undergoes electroerosion, being partially transformed to plasma with ions and excited atoms of the electrode material. Vapor products of the electroerosion of the central electrode react with the gas in the plasma to form compound materials which are deposited on the substrate. The vapors from the electrodes are ejected in the direction of the substrate at very high velocity, on the order of 10 km s^{-1}, which is responsible for excellent adhesion and very low porosity (Sokolowski, 1979). The substrate temperature increases to about 200°C during the deposition. The pulse duration typically used is on the order of 200 μs. The plasma pulses are generated on the order of every 20 s, and the energy released in a single pulse is on the order of 10 kJ. Extremely high deposition rates of 80 μm s^{-1} for TiN coatings have been reported by Michalski (1984).

Materials having high melting points, such as Al_2O_3, Ta_2O_5, TiN, AlN, and BN, are reported to be deposited by pulsing gases such as oxygen, nitrogen, or mixtures of nitrogen and hydrogen during the discharge (Sokolowski et al., 1981; Michalski et al., 1985). Alternatively, gases such as methane or a mixture of diborane and nitrogen are pulsed during the discharge, and dissociated products condense on substrates.

Ezumi et al. (1987) modified the RPP process and proposed a *shock wave deposition* (SWD) process. In this process, they used a T-shaped electromagnetic shock tube, in which vapors are produced from two sources—Joule's heating of pellet material placed between the two electrodes and the electroerosion of electrodes and pellets during the pulse discharge. The reaction between the metal vapors and carrier gases proceeds in the plasma produced by shock wave, and the reaction products coat the substrate. The SWD method, therefore, is capable of producing metal oxides, nitrides, and carbides containing more than one metal element.

Figure 9.67 shows the schematic diagram of the SWD process. The main part of the apparatus is a T-shaped electromagnetic shock tube which is made of Pyrex glass and two metallic electrodes. Metal pellets (source of coating material) are placed uniformly between the metal electrodes (generally of the same metal as the pellets), and reaction gases (source of coating material) under pressure of 1 to 10^2Pa (0.01 to 1 torr) are added to the chamber. To deposit coatings of NbN, Ezumi et al. (1987) used electrodes and pellets of Nb and N_2 or a mixture of N_2 and H_2 as the carrier gases. A discharge voltage of 3 to 5 kV was used. The half period of the discharge current measured was 80 μs, and the energy released in a single pulse varied from 1.8 to 5 kJ. The shock wave velocity ranged from 10 to 15 km s^{-1}.

9.5.4 Details of Chemical Vapor Polymerization Processes

In chemical vapor polymerization, deposition of polymeric coating takes place by means of a chemical reaction of monomers (molecular units) in vapor phase at the substrate surface. This process is used to apply thin (from about 10 nm to a few micrometers) continuous polymer coatings with superior properties, such as being pinhole-free, coherent, and hard and exhibiting excellent adhesion. The chemical reactions involved in polymers known as polymerization can be intensified (or activated) by ultraviolet (UV) light irradiation (called a *photolytic process*), electron beam, or plasma of a glow discharge (Holland, 1956; Shen, 1976).

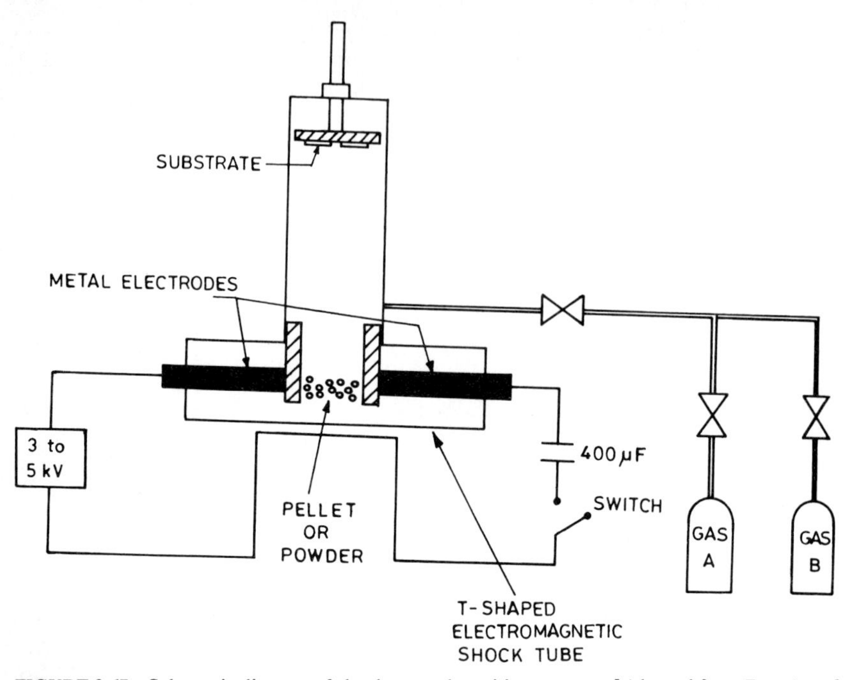

FIGURE 9.67 Schematic diagram of shock wave deposition process. [*Adapted from Ezumi et al. (1987).*]

This type of activation has a strong dependence on the kinds of monomers involved in the reaction. The glow-discharge (or plasma) polymerization process has been extensively used to apply polymer coatings for specialized applications.

In UV polymerization, the monomer gas is irradiated by UV irradiation ranging from 200 to 350 nm at a deposition pressure on the order of 500 Pa (5 torr). The monomers suitable for UV polymerization are those containing double bonds. Such bonds open up by a free-radical mechanism, allowing a head-to-tail coupling, with a formation of thin solid coating (Gregor and McGee, 1963; White, 1966, 1967; Wright, 1967). The coatings that can be applied by UV polymerization include vinyl monomers such as styrene, acrylonitrile, and methyl methacrylate and diene monomers such as 1,3-butadiene and 2,4-hexadiene (Wright, 1967).

In electron beam polymerization, the electron bombardment forms free radicals, ion radicals, and other activated species from liquid or gas source adsorbed on the substrate surface. These activated species are very short-lived and quickly combine into long-chain solid polymers. In this respect, this process is quite similar to UV polymerization. It is possible to apply thin polymer coatings on very fine lines and areas by employing an appropriate mask or by directing the electron beam on certain areas. The monomers that can be polymerized by electron bombardment are those containing double bonds (such as ethylenic groups). Thus monomeric tetrafluoroethylene with hexafluoropropylene, polyvinylidene fluoride, and other ethylenic polymers have been produced (Muehlberg et al., 1965).

In the glow-discharge polymerization or plasma polymerization process, an organic monomer vapor is injected into a glow discharge of an inert gas such as argon, or when a glow discharge of an organic vapor (without addition of inert gas) is created, polymer coating is deposited onto an exposed surface. Glow discharge is used to activate polymerization of the organic vapor phase and the polymer deposits (Baddour and Timmins, 1967; Gould, 1969; Hollahan and Bell, 1974; Kay and Dilks, 1981; McTaggart, 1967; Shen, 1976; Venugopalan, 1971; Yasuda and Hsu, 1978a). Glow-discharge polymerization is most frequently carried out in the pressure range of 1 to 500 Pa (0.01 to 5 torr, typically less than 10 Pa or 0.1 torr) using a dc or more commonly rf power supply in a plasma reactor. The substrate is generally not heated. The plasma reactor system for plasma polymerization is similar to that of PECVD (discussed earlier) except that the starting gas is an organic monomer.

In all these excitations, the reactive species, which consist of a mixture of ions and free radicals, are formed and adsorb onto the surface (Brochers and Konig, 1958). If the reactive species are deposited as active (excited) species, they undergo further reaction to form thin, adherent polymer coating; if they are deposited as inactive species, nonadherent powdery coating results (White, 1967). The characteristics of these polymer coatings depend on the condition of glow discharge (voltage, current, time, etc.) and on the nature of the surface and material of the substrate.

The major advantage of the glow-discharge method is the complete coverage of coatings onto substrates. The large spread in the angle of incidence with which the activated monomer gas molecules strike the surface allows the polymers to cover all surface imperfections and ultimately yields an integral coating. Other advantages include low substrate temperature (even at room temperature), comparatively low vacuum (~10 Pa or 0.1 torr), a wide choice of monomer gases as starting materials, a variety of physical and chemical properties of the polymer coatings, and excellent adhesion to the substrate. Polymer coatings can be deposited on virtually any substrate.

9.5.4.1 Mechanisms of Glow-Discharge Polymerization. Although the phenomenon of polymer formation in a glow discharge is referred to as glow-discharge polymerization, the term *polymerization* may not represent the actual process of forming a polymer, or the word *polymerization* may even be misleading. The conventional polymerization is molecular polymerization in which molecules of monomers (molecular units) are linked together by the polymerization process. Therefore, the resultant polymer is conventionally named by poly + the monomer. In contrast to conventional (or molecular) polymerization, polymer formation in glow discharge may be characterized primarily as elemental or atomic polymerization. That is, in plasma polymerization, the molecular structure of a monomer is not retained, and the original monomer molecules serve as the source of elements which will be used in the construction of large molecules (Yasuda, 1978).

The reactions involved in glow-discharge polymerization are quite complex and usually consist of two major types of polymerization mechanisms, i.e., plasma-induced polymerization and plasma polymerization (Fig. 9.68). Plasma-induced polymerization is essentially the conventional (molecular) polymerization initiated by reactive species that is created in an electric discharge, and so the starting monomer gas must contain polymerizable structure such as olefinic double bonds, triple bonds, or cyclic structure. The latter type, i.e., plasma polymerization (atomic), is a unique process which occurs only in a plasma state.

The degree of ionization in the plasma is on the order of one part in 10^5 to 10^6, and the numbers of free electrons and positive ions are almost equal. Since plasma polymerization is most frequently carried out in the pressure range of 10 to 10^3 Pa (0.1 to 10 torr), the plasma is maintained in a nonequilibrium state,

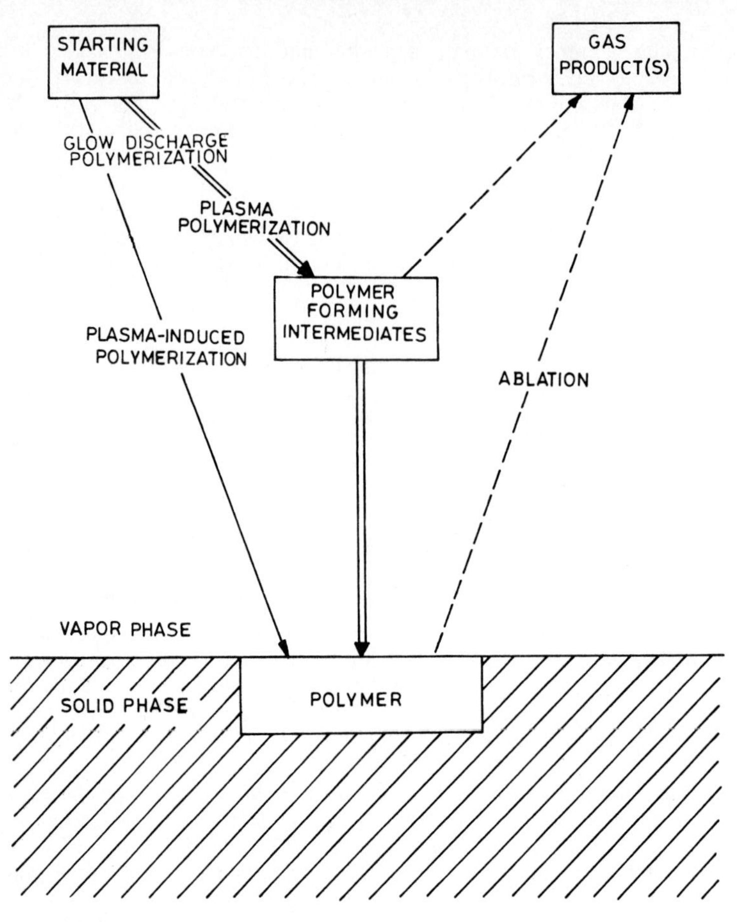

FIGURE 9.68 Mechanisms of glow-discharge polymerization. [*Adapted from Yasuda (1978).*]

which means that the electron temperature (on the order of 10^4 K) is 10 to 100 times greater than the ion temperature, which is nearly equivalent to the gas temperature (~300 K). When monomer gas enters the plasma, it immediately undergoes a series of reactions with the free electrons (Bell, 1976; Yasuda and Hsu, 1978b) and produces free radicals and positive ions. The polymerization is assumed to proceed in three consecutive steps: initiation, propagation, and termination. The reactive species are assumed to be free radicals. Initiation of monomer M is caused by its inelastic collisions with a high-energy free electron, while propagation and termination are analogous to conventional free-radical polymerization. This polymerization can be represented as follows:

Initiation or reinitiation:

$$M_i \rightarrow M_i^* \qquad M_k \rightarrow M_k^* \tag{9.51}$$

Propagation:

$$M_i^* + M_k^* \rightarrow M_i - M_k \quad (9.52)$$

Termination:

$$M_i^* + M_k \rightarrow M_i - M_k \quad (9.53)$$

Here i and k are the numbers of repeating units (that is, $i = k = 1$ for starting monomer) and M_i^* and M_k^* represent reactive species which can be an ion of either charge, free radical, or an excited molecule, produced from M but not necessarily retaining the molecular structure of the starting materials.

The first-generation free radicals and ions react further to produce species of increasing molecular weight. The principal mechanism of polymerization appears to occur via free radicals. A fraction of the free radicals and ions formed in the gas phase will diffuse to the substrate surface. In addition to the adsorption of radicals from the gas phase, surface free radicals can be formed through the release of energy on the surface by electron-ion recombination and bombardment by ions accelerated in the sheath which separates the bulk plasma from the surface.

Plasma-induced polymerization does not produce a gas phase by-product, since the polymerization proceeds via utilization of polymerizable structure. It may be schematically represented by a chain propagation mechanism as follows:

Propagation:

$$M^* + M \rightarrow MM^* \quad (9.54)$$

$$M_i^* + M \rightarrow M_i^* + 1 \quad (9.55)$$

Termination:

$$M_i^* + M_k^* \rightarrow M_i - M_k \quad (9.56)$$

The general polymerization in a glow-discharge consists of both plasma polymerization and plasma-induced polymerization. Which of the two polymerization mechanisms plays the dominant role in the polymer formation depends on not only the chemical structure of starting material but also the conditions of the discharge. The characteristics of gas-product plasma play the predominant factor in determining the extent of the ablation process. The polymer formation and properties of polymers formed by glow discharge are controlled by the balance among plasma-induced polymerization, plasma polymerization, and ablation, which depends on glow-discharge parameters such as modes of electric discharge, discharge current density, flow rate, system pressure, and geometric factors of the reactor. If the polymer deposition is allowed to occur on an appropriate substrate, the polymer coating is highly cross-linked and strongly bonded to the substrate.

9.5.4.2 Details of Glow-Discharge Polymerization Processes. The plasma reactor systems used for glow-discharge polymerization are similar to those of PECVD (discussed earlier) except that the starting gas is an organic monomer. The glow-discharge plasma is created in the pressure range of 1 to 500 Pa (0.01 to 5 torr, typically less than 10 Pa or 0.1 torr) using a dc or more commonly rf power supply inductively coupled or capacitively coupled to the monomer gas. Both tubular and flat-bed-type reactors are used (Connel and Gregor, 1965; Kay and Dilks, 1981; Shen, 1976; Yasuda, 1978). The rf capacitively coupled, parallel-plate (planar diode) reactors appear to be emerging among the most reliable and flexible to meet many technological needs. A schematic diagram of a capacitively coupled rf parallel-plate reactor is shown in Fig. 9.69. The substrate is set on the flat-bed electrode, and the chamber is evacuated to 10^{-2} Pa (10^{-4} torr) or less; then vaporized monomer is introduced in the chamber. The rf power supply for discharge is applied to deposit plasma polymerized coatings.

The glow-discharge polymerization process can also be used in the synthesis of metals containing polymer coatings by simultaneous plasma etching and polymerization in the same system. A typical reactor for this work is a rf capacitively coupled parallel-plate reactor arrangement. Substrate is placed on the grounded electrode. Plasmas are excited by means of high-frequency rf power applied to another electrode whose surface consists of the metal to be incorporated into the polymer coating. At high frequency, the excitation electrode attains a large self-biasing negative potential, relative to the plasma. Positive ions therefore arrive at its surface with sufficiently high kinetic energy so as to remove the metal from its surface, which later gets incorporated into the coating, polymerized simultaneously at the substrate placed on the grounded electrode (Kay and Dilks, 1981).

9.5.4.3 Coatings Deposited by Glow-Discharge Polymerization

Coating Materials. Nearly all organic compounds can be polymerized by glow-discharge polymerization. The typical families of materials include hydro-

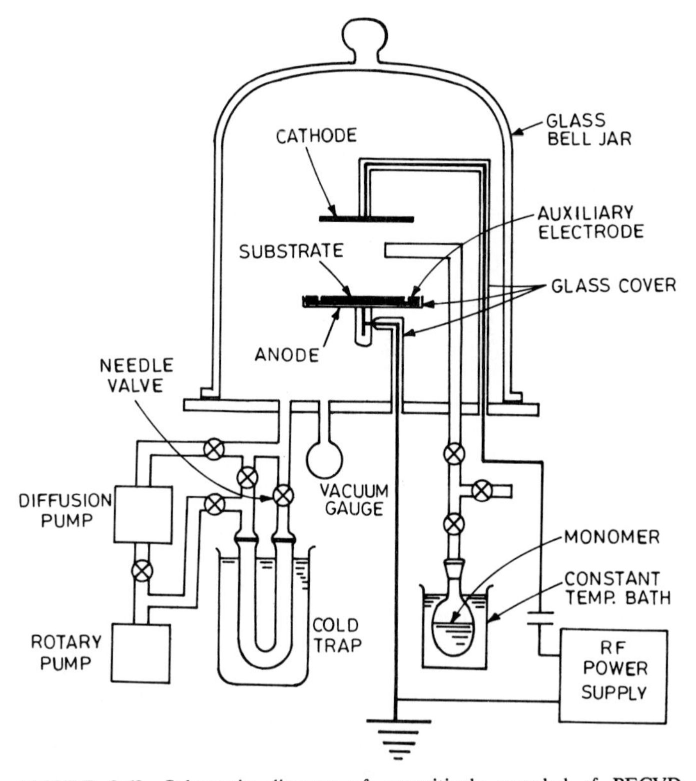

FIGURE 9.69 Schematic diagram of capacitively coupled rf PECVD parallel-plate reactor system used in deposition of plasma polymerized coatings. [*Adapted from Harada (1981).*]

carbons (triple-bond-containing and aromatic compounds, double-bond-containing and cyclic compounds), nitrogen-containing compounds, fluorine-containing compounds, oxygen-containing compounds, and silicon-containing compounds (Yasuda, 1978). In general, the elements which are reactive and exist in the gas phase in the normal temperature range favor the ablation process and do not contribute to polymerization. However, it is possible that many unusual elements which cannot be incorporated in conventional polymers could be incorporated into thin coatings produced by glow-discharge polymerization. The glow-discharge polymerization process can also be used in the synthesis of metal-containing polymer coatings. The range of metals which may be incorporated into glow-discharge polymer coatings seems to be essentially without limit.

Some of the coating materials which have been deposited by plasma polymerization include (Harada, 1981; Kay and Dilks, 1981; Shen, 1976; Washo, 1976; Yasuda, 1978) fluorocarbons such as polytetrafluoroethylene and hexafluoropropylene, toluene-disocyanate, pyrrolidine, polyethylene, and styrene. The plasma polymer coatings are used as protective coatings in semiconductor electronics and various tribological applications. Plasma polymer coatings are used for wear protection of thin-film rigid magnetic disks.

Properties of Glow-Discharge Polymerized Coatings. Characteristic polymer deposition by glow-discharge polymerization occurs on surfaces exposed to (directly contacting) the glow. The polymer deposition rates and their properties are strongly dependent on the modes of glow discharge, flow rate, deposition pressure, discharge power, and ratio of surface to volume of glow. Indeed all these operating parameters influence the glow-discharge polymerization in an interrelated manner.

(a) Modes of Glow Discharge: Glow-discharge polymerization can be obtained by the application of power sources having a wide range of frequencies, typically from zero (dc) to the gigahertz (microwave) range. A low-frequency power source requires internal electrodes. With high-frequency sources, external electrodes or a coil can be used. The use of internal electrodes has the advantage that any frequency can be used. In the low-frequency range, the glow discharge is more or less restricted between the internal electrodes. With a high-frequency range, the glow tends to stray away from the space between the electrodes; however, because of this tendency, polymer deposition onto a substrate placed in between the electrodes increases (Yasuda and Hsu, 1978b). The systems that employ external electrodes or a coil are suited for large-volume glow discharge. The relative location of the substrate with respect to electrodes in the case of internal electrodes and to the energy input region in the case of external electrodes or a coil affects the properties and uniformity of the polymer formed.

(b) Flow Rate and Deposition Pressure: The flow rate, i.e., feeding-in rate, of the starting monomer and pressure of the vacuum system strongly influence the deposition rate and uniformity of the coating. The rate of polymer deposition is generally proportional to the feed-in rate under ideal conditions. The apparent dependence of the polymer deposition rate on the flow rate is often characterized by a decrease of the polymer deposition rate after passing a maximum or a narrow plateau. Since the velocity of gas molecule is dependent on pressure, the equilibrium between flow rate and product gases and pumping rate is important in controlling the distribution of polymer deposition and the properties of polymers condensed in glow-discharge polymerization. The lower the deposition pressure, the larger the mean free path of gas molecules, and so the diffusion displacement becomes more efficient. Therefore, the polymer formation is not localized at either the region of excitation or the site of introduction of the starting material.

(c) GLOW-DISCHARGE POWER: The glow-discharge power plays a key role in the initiation of the polymerization process. For instance, a certain level of power (typically 60 W) in a given set of discharge conditions for one starting monomer (say ethylene) could not even initiate a glow discharge with another starting monomer (say *n*-hexane). In other words, a relative level of discharge power which varies according to the characteristics of the starting materials is needed to describe the discharge power for glow-discharge polymerization. Figure 9.70 shows the dependence of discharge power to obtain comparable levels of polymerization of various hydrocarbons as a function of the flow rate of the starting monomer. We note that the discharge power necessary for glow-discharge polymerization depends on both the molecular weight and the chemical structure of the compounds.

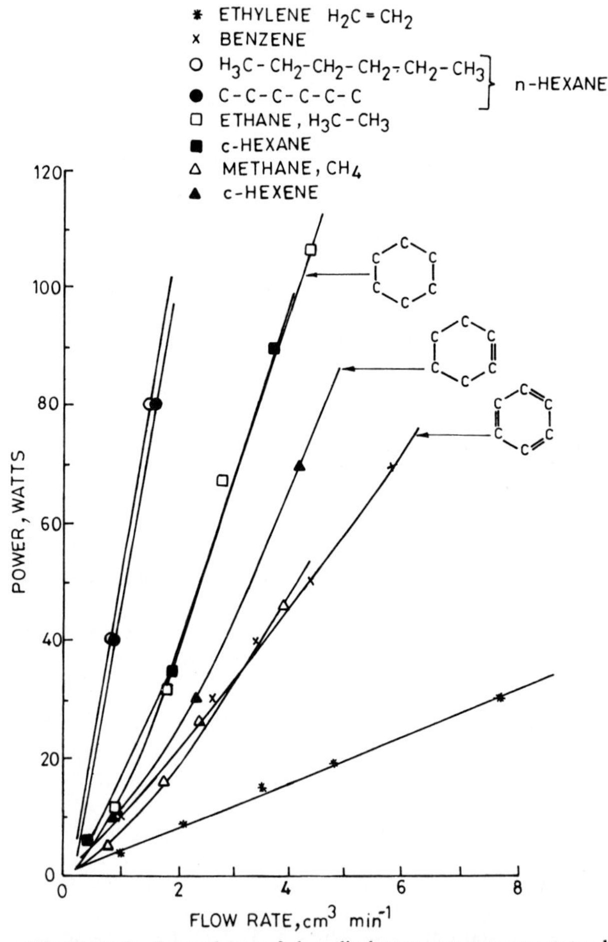

FIGURE 9.70 Dependence of glow-discharge power necessary to obtain comparable polymerization on the flow rate of starting monomers having different molecular weights and chemical structure. [*Adapted from Yasuda (1978).*]

The polymer deposition rate generally increases with discharge power in more or less linear fashion in a certain range of discharge power, and then it reaches a plateau. Further increase of discharge power often decreases the polymer deposition rate.

(d) GEOMETRIC FACTOR OF THE REACTOR: The location where the starting material is introduced is very important for polymer deposition. Not all starting materials fed into a glow-discharge polymerization reactor are utilized in the polymer formation. The *bypass ratio* represents the portion of flow which does not contribute to glow-discharge polymerization. Consequently, the higher the bypass ratio of a reactor, the lower the conversion of the starting material to the polymer. Clearly, this ratio depends on the ratio of the volume occupied by discharge to the total volume.

9.6 References

Aisenberg, S., and Chabot, R. (1971), "Ion-Beam Deposition of Thin Films of Diamondlike Carbon," *J. Appl. Phys.*, Vol. 42, pp. 2953–2958.

Allen, S. D. (1981), "Laser Chemical Vapor Deposition: A Technique for Selective Area Deposition," *J. Appl. Phys.*, Vol. 52, pp. 6501–6505.

Amano, J., and Lawson, R. P. W. (1977), "Thin Film Deposition Using Low Energy Ion Beams (3) Mg^+ Ion Beam Deposition and Analysis of Deposits," *J. Vac. Sci. Technol.*, Vol. 14, pp. 836–839.

Ames, I., Kaplan, L. H., and Roland, P. A. (1966), "Crucible Type Evaporation Source for Aluminum," *Rev. Sci. Instrum.*, Vol. 37, pp. 1737–1738.

Angus, J. C., Koidl, P., and Domitz, S. (1986), "Carbon Thin Films," in *Plasma Deposited Thin Films* (J. Mort and F. Jansen, eds.), pp. 89–127, CRC Press, Boca Raton, Fla.

Anonymous (1972), "Sputtering and Ion Plating," *Proc. of Technol. Utilization Conf.*, NASA, Washington, D.C.

Archer, N. J., and Yee, K. K. (1978), "Chemical Vapor Deposited Tungsten Carbide Wear-Resistant Coatings Formed at Low Temperatures," *Wear*, Vol. 48, pp. 237–250.

Auwarter, M. (1960), "Process for the Manufacture of Thin Films," U.S. Patent 2,920,002.

Baddour, R. F., and Timmins, R. S. (eds.) (1967), *The Application of Plasmas to Chemical Processing*, M.I.T. Press, Cambridge, Mass.

Banks, B. A., Mirtich, M. J., Rutledge, S. K., Swec, D. M., and Nahra, H. K. (1985), "Ion Beam Sputter-Deposited Thin Film Coatings for Protection of Spacecraft Polymers in Low Earth Orbit," Rep. NASA-TM 87051, NASA, Cleveland, Ohio, Presented at the 23d Aerospace Sci. Meeting, Reno, Nev., January 14–17, 1985.

Beale, H. A., Weiler, F., and Bunshah, R. F. (1973), "Evaporation Variables in Gas Scattering Plating Processes," *Proc. of 4th Int. Conf. on Vacuum Metallurgy*, pp. 238–241, Iron and Steel Institute of Japan, Tokyo.

Bell, A. T. (1976), "Fundamentals of Plasma Polymerization," in *Plasma Chemistry of Polymers* (M. Shen, ed.), pp. 1–13, Marcel Dekker, New York.

——— (1980), "The Mechanism and Kinetics of Plasma Polymerization," in *Topics in Current Chemistry; Plasma Chemistry III* (S. Věprek and M. Venugopalan, eds.), Vol. 94, pp. 43–68, Springer-Verlag, Berlin.

Berry, R. W., Hall, P. M., and Harris, M. T. (1968), *Thin Film Technology*, Van Nostrand, Princeton, N.J.

Bhushan, B. (1979), "Development of RF-Sputtered Chrome Oxide Coatings for Wear Application," *Thin Solid Films*, Vol. 64, pp. 231–241.

——— (1980a), "High Temperature Self-Lubricating Coatings and Treatments—A Review," *Metal Finish.*, Vol. 78, May, pp. 83–88; June, pp. 71–75.

——— (1980b), "High-Temperature Self-Lubricated Coatings for Air-Lubricated Foil Bearings for the Automotive Gas-Turbine Engine," Tech. Rep. CR-159848, NASA Lewis Research Center, Cleveland, Ohio.

——— (1980c), "Characterization of RF Sputter-Deposited Chrome Oxide Films," *Thin Solid Films*, Vol. 73, pp. 255–265.

——— (1981a), "Structural and Compositional Characterization of RF Sputter Deposited Cr_2O_3 + Ni-Cr Films," *J. Lub. Technol.* (*Trans. ASME*), Vol. 103, pp. 211–217.

——— (1981b), "Friction and Wear Results from Sputter-Deposited Chrome Oxide with and without Nichrome Metallic Binders and Interlayers," *J. Lub. Technol.* (*Trans. ASME*), Vol. 103, pp. 218–227.

——— (1987), "Overview of Coating Materials, Surface Treatments, Screening Techniques for Tribological Applications Part I: Coating Materials and Surface Treatments," in *Testing of Metallic and Inorganic Coatings* (W. B. Harding and G. A. DiBari, eds.), STP 947, pp. 289–309, ASTM, Philadelphia.

———, and Gray, S. (1980), "Development of Surface Coatings for Air-Lubricated Compliant Journal Bearings to 650°C," *ASLE Trans.*, Vol. 23, pp. 185–196.

Blachman, A. G. (1971), "Stress and Resistivity Control in Sputtered Molybdenum Films and Comparison with Sputtered Gold," *Metall. Trans.*, Vol. 2, pp. 699–709.

Bland, R. D., Kominiak, G. J., and Mattox, D. M. (1974), "Effect of Ion Bombardment During Deposition on Thick Metal and Ceramic Deposits," *J. Vac. Sci. Technol.*, Vol. 11, pp. 671–674.

Blank, P., and Wittmaack, K. (1979), "Energy and Fluence Dependence of the Sputtering Yield of Silicon Bombarded with Argon and Xenon," *J. Appl. Phys.*, Vol. 50, pp. 1519–1528.

Blocher, J. M. (1974), "Structure/Property/Process Relationships in Chemical Vapor Deposition," *J. Vac. Sci. Technol.*, Vol. 11, pp. 680–686.

——— (1982), "Chemical Vapor Deposition," in *Deposition Technologies for Films and Coatings* (R. F. Bunshah, J. M. Blocher, T. D. Bonifield, J. G. Fish, P. B. Ghate, B. E. Jacobson, D. M. Mattox, G. E. McGuire, M. Schwartz, J. A. Thornton, and R. C. Tucker, eds.), pp. 335–364, Noyes Publications, Park Ridge, N.J.

———, and Withers, J. C. (eds.) (1970), *Proc. of 2d Int. Conf. on Chemical Vapor Deposition*, American Electrochemical Society, Princeton, N.J.

———, Hintermann, H. E., and Hall, L. H. (eds.) (1975), *Proc. of 5th Int. Conf. on Chemical Vapor Deposition*, American Electrochemical Society, Princeton, N.J.

Bonifield, T. D. (1982), "Plasma Assisted Chemical Vapor Deposition," in *Deposition Technologies for Films and Coatings* (R. F. Bunshah, J. M. Blocher, T. D. Bonifield, J. G. Fish, P. B. Ghate, B. E. Jacobson, D. M. Mattox, G. E. McGuire, M. Schwartz, J. A. Thornton, and R. C. Tucker, eds.), pp. 365–384, Noyes Publications, Park Ridge, N.J.

Boone, D. H., Strangman, T. E., and Wilson, L. E. (1974), "Some Effects of Structure and Composition on the Properties of Electron Beam Vapor Deposited Coatings for Gas Turbine Superalloys," *J. Vac. Sci. Technol.*, Vol. 11, pp. 641–646.

———, Shen, S., and McKoon, R. (1980), "Electron Beam Evaporation of Low Pressure Elements in MCrAl Coating Composition," *Thin Solid Films*, Vol. 64, pp. 299–304.

Boxman, R. L., Goldsmith, S., Shalev, S., Yaloz, H., and Brosh, N. (1986), "Fast Deposition of Metallurgical Coatings and Production of Surface Alloys Using a Pulsed High Current Vacuum Arc," *Thin Solid Films*, Vol. 139, pp. 41–52.

Brinsmaid, S., Keenam, W. J., Koch, G. E., and Parsons, W. E. (1957), "Method of Producing Titanium Dioxide Coatings," U.S. Patent 2,784,115.

Brochers, A., and Konig, H. (1958), "Der Aufbau von glimmentladungs-polymerisaten verschiedener entstehungsbedingungen und seine veranderung durch Elektronenbestrahlung," *Z. Phys.*, Vol. 152, pp. 75–86.

Brown, S. C. (1966), *Introduction to Electrical Discharges in Gases*, Wiley, New York.

Brutsch, R. (1985), "Chemical Vapor Deposition of Silicon Carbide and Its Applications," *Thin Solid Films*, Vol. 126, pp. 313–318.

Bryant, W. A. (1976), "Producing Extended Area Deposits of Uniform Thickness by a New Chemical Vapor Deposition Technique," *J. Crystal Growth*, Vol. 35, pp. 257–261.

——— (1977), "The Fundamentals of Chemical Vapor Deposition," *J. Mater. Sci.*, Vol. 12, pp. 1285–1306.

Bunshah, R. F. (1974a), "High Rate Deposition of Carbides by Activated Reactive Evaporation," U.S. Patent 3,791,852.

——— (1974b), "Structure/Property Relationships in Evaporated Thick Films and Bulk Coatings," *J. Vac. Sci. Technol.*, Vol. 11, pp. 633–638.

——— (1976), "Physical Vapor Deposition of Metals, Alloys and Compounds," in *New Trends in Material Processing*, American Society for Metals, Metals Park, Ohio.

——— (1977), "Mechanical Properties of Thin Films," *Vacuum*, Vol. 27, pp. 353–362.

——— (1981), "The Activated Reactive Evaporation Process: Developments and Applications," *Thin Solid Films*, Vol. 80, pp. 255–261.

——— (1982), "Evaporation," in *Deposition Technologies for Films and Coatings* (R. F. Bunshah, J. M. Blocher, T. D. Bonifield, J. G. Fish, P. B. Ghate, B. E. Jacobson, D. M. Mattox, G. E. McGuire, M. Schwartz, J. A. Thornton, and R. C. Tucker, eds.), pp. 83–169, Noyes Publications, Park Ridge, N.J.

——— (1983), "Process of the Activated Reactive Evaporation Type and Their Tribological Applications," *Thin Solid Films*, Vol. 107, pp. 21–38.

———, and Raghuram, A. C. (1972), "Activated Reactive Evaporation Process for High Rate Deposition of Compounds," *J. Vac. Sci. Technol.*, Vol. 9, pp. 1385–1388.

———, and Schramm, R. J. (1977), "Alumina Deposition by Activated Reactive Evaporation," *Thin Solid Films*, Vol. 40, pp. 211–216.

———, Schramm, R. J., Nimmagadda, R., Movchan, B. A., and Borodin, V. P. (1977a), "Structure and Properties of Refractory Compounds Deposited by Direct Evaporation," *Thin Solid Films*, Vol. 40, pp. 169–182.

———, Shabaik, A. H., Nimmagadda, R., and Covey, J. (1977b), "Machining Studies on Coated High Speed Tools," *Thin Solid Films*, Vol. 45, pp. 453–462.

———, Nimmagadda, R., Dunford, W., Movchan, B. A., Demchishin, A. V., and Chursanov, N. A. (1978), "Structure and Properties of Refractory Compounds Deposited by Electron Beam Evaporation," *Thin Solid Films*, Vol. 54, pp. 85–106.

———, Blocher, J. M., Bonifield, T. D., Fish, J. G., Ghate, P. B., Jacobson, B. E., Mattox, D. M., McGuire, G. E., Schwartz, M., Thornton, J. A., and Tucker, R. C. (eds.) (1982), *Deposition Technologies for Films and Coatings*, Noyes Publications, Park Ridge, N.J.

Burkhardt, W. (1939), *Method of Coating Articles by Vaporized Coating Materials*, U.S. Patent 2,157,478.

Butler, H. S., and Kino, G. S. (1963), "Plasma Sheath Formation by Radio Frequency Fields," *Phys. Fluids*, Vol. 6, pp. 1346–1355.

Carpenter, R. (1974), "The Basic Principles of Ion-Plating," in *Science and Technologies of Surface Coatings* (B. N. Chapman and J. C. Anderson, eds.), pp. 393–403, Academic, New York.

Carson, W. W. (1975), "Sputter Gas Pressure and DC Substrate Bias Effects on Thick RF-Diode Sputtered Films on Ti Oxycarbides," *J. Vac. Sci. Technol.*, Vol. 12, pp. 845–849.

Carter, G., and Colligen, J. S. (1969), *Ion Bombardment of Solids*, Elsevier, Amsterdam.

Castellano, R. N., Notis, M. R., and Simmons, G. W. (1977), "Composition and Stress State of Thin Films Deposited by Ion Beam Sputtering," *Vacuum*, Vol. 27, pp. 109–117.

Chambers, D. L., Carmichael, D. C., and Wan, C. T. (1972), "Deposition of Metal and Alloy Coatings by Electron-Beam Ion Plating," in *Proc. of Technology Utilization Conf. on Sputtering and Ion Plating*, pp. 89–114, NASA, Washington, D.C.

Chapman, B. (1980), *Glow Discharge Processes*, Wiley, New York.

Chapman, B. N., and Anderson, J. C. (eds.) (1974), *Science and Technology of Surface Coating*, Academic, New York.

Cheney, R., and Samek, N. E. (1977), "Sputtered Thin Films for Pressure Transducers," *Res./Dev.*, Vol. 28, April, pp. 53–61.

Chopra, K. L. (1979), *Thin Film Phenomenon*, Robert E. Krieger Publishing Co., New York.

Christensen, O. (1970), "Characteristics and Applications of Bias Sputtering," *Solid State Technol.*, Vol. 13, No. 12, pp. 39–45.

Christensen, C. O., and Lakin, K. M. (1978), "Chemical Vapor Deposition of Silicon Using a CO_2 Laser," *Appl. Phys. Lett.*, Vol. 32, pp. 254–255.

Christy, R. I., and Ludwig, H. R. (1979), "R. F. Sputtered MoS_2 Parameter Effects on Wear Life," *Thin Solid Films*, Vol. 64, pp. 223–229.

Coburn, J. M., and Kay, E. (1972), "Positive-Ion Bombardment of Substrates in R. F. Diode Glow Discharge Sputtering," *J. Appl. Phys.*, Vol. 43, pp. 4965–4971.

Colen, M., and Bunshah, R. F. (1976), "Synthesis and Characterization of Yttrium Oxide (Y_2O_3)," *J. Vac. Sci. Technol.*, Vol. 13, pp. 536–539.

Connell, R. A., and Gregor, L. V. (1965), "Magnetically Shaped R. F. Discharge for Polymer Film Formation," *J. Electrochem. Soc.*, Vol. 112, pp. 1198–1200.

Cook, H. C., Covington, C. W., and Libsch, J. F. (1966), "The Preparation and Properties of Sputtered Aluminum Thin Films," *Trans. Met. Soc. AIME*, Vol. 236, pp. 314–320.

Cuomo, J. J., Harper, J. M. E., Guarnieri, C. R., Yee, D. S., Attanasio, L. J., Angilello, J., Wu, C. T., and Hammond, R. H. (1982), "Modification of Niobium Film Stress by Low Energy Ion Bombardment during Deposition," *J. Vac. Sci. Technol.*, Vol. 20, pp. 349–354.

Daadler, J. E. (1983), "Random Walk of Cathode Arc Spots in Vacuum," *J. Phys. D. (Appl. Phys.)*, Vol. 16, pp. 17–27.

Davis, W. D., and Miller, H. C. (1969), "Analysis of Electrode Products Emitted by DC Arcs in a Vacuum Ambient," *J. Appl. Phys.*, Vol. 40, pp. 2212–2221.

Denton, R. A., and Greene, A. D. (1963), "A Compact Electron Gun for Evaporation," *Proc. 5th Electron Beam Symp.* (J.R. Morley, ed.), Alloyd Electronics Corp., Cambridge, Mass., pp. 180–197.

Dimigen, H., Hubsch, H., Willich, P., and Reichelt, K. (1985), "Stoichiometry and Friction Properties of Sputtered MoS_x Layers," *Thin Solid Films*, Vol. 129, pp. 79–91.

Dirks, A. G., and Leamy, H. J. (1977), "Columnar Microstructure in Vapor Deposited Thin Films," *Thin Solid Films*, Vol. 47, pp. 219–233.

Donaghey, L. F., Rai-Chowdhury, P., and Tauber, R. N. (eds.) (1977), *Proc. of 6th Int. Conf. on Chemical Vapor Deposition*, American Electrochemical Society, Princeton, N.J.

Dorodnov, A. M. (1978), "Technical Application of Plasma Accelerators," *Sov. Phys. Tech. Phys.*, Vol. 23, pp. 1058–1065.

———, and Petrosov, B. A. (1981), "Physical Principles and Types of Technical Vacuum Plasma Devices," *Sov. Phys. Tech. Phys.*, Vol. 26, pp. 304–315.

Dudonis, J., and Pranevičius, L. (1976), "Influence of Ion Bombardment on the Properties of Vacuum Evaporated Thin Films," *Thin Solid Films*, Vol. 36, pp. 117–120.

Dugdale, R. A. (1971), *Glow Discharge Material Processing*, Mills and Boon, Ltd., London.

——— (1977), "DC Glow Discharge Techniques for Surface Treatment and Coatings," *Thin Solid Films*, Vol. 45, pp. 541–552.

Dun, H., Pan, P., White, F. R., and Douse, R. W. (1981), "Mechanisms of Plasma Enhanced Silicon Nitride Deposition Using SiH_4/N_2 Mixture," *J. Electrochem. Soc.*, Vol. 128, pp. 1555–1563.

Duret, C., and Pichoir, R. (1983), "Protective Coatings for High Temperature Materials: Chemical Vapor Deposition and Pack Cementation Process," in *Coatings for High Temperature Applications* (E. Lang, ed.), pp. 33–78, Applied Science Pub., London.

Eizenberg, M., and Murarka, S. P. (1983), "Reactively Sputtered Titanium Carbide Thin Films: Preparation and Properties," *J. Appl. Phys.*, Vol. 54, pp. 3190–3194.

Enomoto, Y., and Matsubara, K. (1975), "Structure and Mechanical Properties of Ion-Plated Thick Films," *J. Vac. Sci. Technol.*, Vol. 12, pp. 827–829.

Eriksson, G. (1975), "Thermodynamic Studies of High Temperature Equilibria," *Chem. Script.*, Vol. 8, pp. 100–103.

Ezumi, H., Endo, T., Yamada, H., Kuwahara, K., and Kino, T. (1987), "Formation of Thin Films by Shock Wave Deposition Method," *Thin Solid Films*, Vol. 163, pp. 261–265.

Feist, W. F., Steele, S. R., and Readey, D. W. (1969), "The Preparation of Films by Chemical Vapor Deposition," *Phys. Thin Films*, Vol. 5, pp. 237–322.

Fischbein, I. W., Alexander, B. H., and Sastri, A. (1972), "Sputtering Coating on Razor Blades with a Beta Tungsten Type Alloy," U.S. Patent 3,682,795.

Fraser, D. B. (1978), "The Sputter and S-gun Magnetrons," in *Thin Film Processes* (J. L. Vossen and W. Kern, eds.), pp. 115–130, Academic, New York.

Freller, H., and Haessler, H. (1987), "Ti_xAl_{1-x} Films Deposited by Ion Plating with an Arc Evaporator," *Thin Solid Films*, Vol. 153, pp. 67–74.

Fritz, L. (1972), "Ion-Plating; The Best of Sputtering and Evaporation," *Circ. Manuf.*, Vol. 12, p. 10.

Gabriel, H. M., and Kloos, K. H. (1982), "Morphology and Structure of Ion-Plated TiN, TiC and Ti(C, N) Coatings," *Thin Solid Films*, Vol. 118, pp. 243–254.

Gautherin, G., and Weissmantel, C. (1978), "Some Trends in Preparing Film Structures by Ion Beam Methods," *Thin Solid Films*, Vol. 50, pp. 135–144.

Gawehn, H. (1962), "On a New Cathode Sputtering Process," *Z. Angew. Physik.*, Vol. 14, pp. 458–462.

Gebhardt, J. J., and Cree, R. F. (1965), "Vapor-Deposited Borides of Group IVA Metals," *J. Am. Ceramic Soc.*, Vol. 48, No. 5, pp. 262–267.

Gesang, W. R., Oechsner, H., and Schoof, H. (1976), "Backscattering of Neutralized Noble Gas Ions from Polycrystalline Surfaces at Bombarding Energies below 1 keV," *Nucl. Instrum. Meth.*, Vol. 132, pp. 687–693.

Gillam, E. (1959), "The Penetration of Positive Ions of Low Energy into Alloys and Composition Changes Produced in Them by Sputtering," *J. Phys. Chem. Solids*, Vol. 11, pp. 55–67.

Glaski, F. A. (ed.) (1972), *Proc. on 3d Int. Conf. on Chemical Vapor Deposition*, American Nuclear Society, Hinsdale, Ill.

Gould, R. F. (ed.) (1969), "Chemical Reactions in Electrical Discharges, A Symposium," *Adv. Chem. Ser.*, No. 80, American Chemical Society, Washington, D.C.

Graper, E. P. (1971), "Evaporation Characteristics of Materials from an Electron Beam Gun," *J. Vac. Sci. Technol.*, Vol. 8, pp. 333–337.

Greene, J. E., Woodhouse, J., and Pestes, M. (1978), "A Technique for Detecting Critical Loads in the Scratch Test for Thin Film Adhesion," *Rev. Sci. Instrum.*, Vol. 45, pp. 747–749.

Gregor, L. V., and McGee, H. L. (1963), in *Proc. 5th Electrical Insulation Conference*, National Electrical Manufacturers Association and American Institute of Electrical Engineers.

Grossklaus, W., and Bunshah, R. F. (1975a), "Synthesis of Various Oxides in the Ti-O System by Reactive Evaporation and Activated Reactive Evaporation Techniques," *J. Vac. Sci. Technol.*, Vol. 12, pp. 593–597.

——— and ——— (1975b), "Synthesis and Morphology of Various Carbides in the Ta-C System," *J. Vac. Sci. Technol.*, Vol. 12, pp. 811–814.

Haff, P. K., and Switkowski, Z. E. (1977), "Ion Beam-Induced Atomic Mixing," *J. Appl. Phys.*, Vol. 48, pp. 3383–3386.

Haggerty, J. (1984), *Proc. Conf. on Emergent Process Methods for High-Technology Ceramics* (R. Davis, H. Palmour, III, and R. L. Porter, eds.), Vol. 17, p. 137, Science Research Series, Plenum, New York.

Harada, K. (1981), "Plasma Polymerized Protective Films for Plated Magnetic Disks," *J. Appl. Poly. Sci.*, Vol. 26, pp. 3707–3718.

Harker, H. R., and Hill, R. J. (1972), "The Deposition of Multicomponent Phases by Ion Plating," *J. Vac. Sci. Technol.*, Vol. 9, pp. 1395–1399.

Harper, J. M. E. (1978), "Ion Beam Deposition," in *Thin Film Processes* (J. L. Vossen and W. Kern, eds.), pp. 175–206, Academic, New York.

——— (1984), "Ion Beam Techniques for the Deposition of Ceramic Thin Films," in *Proc. Conf. on Emergent Process Methods for High Technology Ceramics* (R. F. Davis, H. Palmour, III, and R. L. Porter, eds.), Vol. 17, pp. 415–424, Materials Science Research Series, Plenum, New York.

——— and Gambino, R. J. (1979), "Combined Ion Beam Deposition and Etching for Thin Film Studies," *J. Vac. Sci. Technol.*, Vol. 17, pp. 1901–1905.

———, Cuomo, J. J., and Kaufman, H. R. (1982), "Technology and Applications of Broad-Beam Ion Sources Used in Sputtering Part II: Applications," *J. Vac. Sci. Technol.*, Vol. 21, pp. 737–756.

Hatto, P. (1985), "'Ion-Bond' TM and the Arc Evaporation Process," in *Proc. of Course on Nitride and Carbide Coatings Manufacturing Technology, Science and Applications* (H. T. G. Hentzell and J.-E. Sundgren, eds.), Neuchatel, Switzerland.

Helix, M. J., Vaidathan, K. V., and Streetman, B. G. (1977), *Extended Abstracts*, American Electrochemical Society, Princeton, N.J., pp. 77–82.

Hess, D. W. (1984), "Plasma Enhanced CVD: Oxides, Nitrides, Transition Metals, and Transition Metal Silicides," *J. Vac. Sci. Technol.*, Vol. A2, pp. 244–252.

Hibbs, M. K., Johansson, B. O., Sundgren, J.-E., and Helmersson, V. (1984), "Effect of Substrate Temperature and Substrate Material on the Structure of Reactively Sputtered TiN Films," *Thin Solid Films*, Vol. 122, pp. 115–129.

Hinneberg, H. J., Weidner, M., Hecht, G., and Weissmantel, C. (1976), "Electrical Properties of Ion Beam Sputtered Silicon Layers on Steel," *Thin Solid Films*, Vol. 33, pp. 29–34.

Hintermann, H. E. (1980), "Exploitation of Wear- and Corrosion-Resistant CVD Coatings," *Tribol. Int.* Vol. 13, pp. 267–277.

Hocking, M. G., Vasantasree, V., and Sidky, P. S. (1989) *Metallic and Ceramic Coatings: Production, High-Temperature Properties, and Applications*, Longman, London.

Hoffman, D. W., and Leibowitz, D. (1971), "Al_2O_3 Films Prepared by Electron Beam Evaporation of Hot-Pressed Al_2O_3 in Oxygen Ambient," *J. Vac. Sci. Technol.*, Vol. 8, pp. 107–111.

——— and ——— (1972), "Effect of Substrate Potential on Al_2O_3 Films Prepared by Electron Beam Evaporation," *J. Vac. Sci. Technol.*, Vol. 9, pp. 326–329.

——— and Gaerttner, M. R. (1980), "Modification of Evaporated Chromium by Concurrent Ion Bombardment," *J. Vac. Sci. Technol.*, Vol. 17, pp. 425–428.

Hollahan, J. H., and Bell, A. T. (eds.) (1974), *Techniques and Applications of Plasma Chemistry*, Wiley, New York.

———, and Rosler, R. S. (1978), "Plasma Deposition of Inorganic Thin Films," in *Thin Film Processes* J. L. Vossen and W. Kern, eds.), pp. 335–360, Academic, New York.

Holland, L. (ed.) (1956), *Thin Film Microelectronics*, Wiley, New York.

——— (1970), *Vacuum Deposition of Thin Films*, Wiley, New York.

——— (1974), "The Basic Principles of Sputter Deposition," in *Science and Technology of Surface Coating* (B. N. Chapman and J. C. Anderson, eds.), pp. 369–392, Academic, New York.

——— (1980), "A Review of Plasma Process Studies," *Surf. Technol.*, Vol. 11, pp. 145–169.

——— and Ojha, S. M. (1976), "Deposition of Hard and Insulating Carbonaceous Films on an RF Target in a Butane Plasma," *Thin Solid Films*, Vol. 38, pp. L17–L19.

——— and ——— (1979), "Growth of Carbon Films with Random Atomic Structure from Ion Impact Damage in a Hydrocarbon Plasma," *Thin Solid Films*, Vol. 58, pp. 107–116.

Inuzuka, T., and Ueda, R. (1968), "Nucleation of Gold Deposits on Alkali-Halide Crystals," *J. Phys. Soc. Jap.*, Vol. 25, pp. 1299–1307.

Jackson, G. N. (1970), "R. F. Sputtering," *Thin Solid Films*, Vol. 5, pp. 209–246.

Jacobson, B., Nimmagadda, R., and Bunshah, R. F. (1979a), "Microstructures of TiN and Ti_2N Deposits Prepared by Activated Reactive Evaporation," *Thin Solid Films*, Vol. 63, pp. 333–339.

———, Bunshah, R. F., and Nimmagadda, R. (1979b), "Annealing Studies of TiC and (Ti, V)C Structures Prepared by Activated Reactive Evaporation," *Thin Solid Films*, Vol. 63, pp. 357–362.

Jamal, T., Nimmagadda, R., and Bunshah, R. F. (1980), "Friction and Adhesive Wear of Titanium Carbide and Titanium Nitride Overlay Coatings," *Thin Solid Films*, Vol. 73, pp. 245–254.

Jang, D. H., and Chun, J. S. (1987), "The Effect of Substrate Steel on the Chemical Vapor Deposition of TiC onto Tool Steel," *Thin Solid Films*, Vol. 149, pp. 95–104.

Johannesson, T. R., and Lindström, J. N. (1975), "Factors Affecting the Initial Nucleation of Alumina on Cemented Carbide Substrates in the CVD Process," *J. Vac. Sci. Technol.*, Vol. 12, pp. 854–857.

Johansen, O. A., Dontje, J. H., and Zenner, R. L. D. (1987), "Reactive Arc Vapor Ion Deposition of TiN, ZrN and HfN," *Thin Solid Films*, Vol. 153, pp. 75–82.

Johansson, B. O., Sundgren, J.-E., Hentzell, H. T. G., and Karlsson, S.-E. (1984), "Influence of Substrate Shape on TiN Films Prepared by Reactive Sputtering," *Thin Solid Films*, Vol. 111, pp. 313–322.

Johnson, P. C. (1989), "The Cathodic Arc Plasma Deposition of Thin Films," in *Physics of Thin Films* (M. H. Francombe and J. L. Vossen, eds.), Vol. 14, pp. 130–200, Academic, New York.

Kamo, M., Sato, Y., Matsumoto, S., and Setaka, N. (1983), "Diamond Synthesis from Gas Phase in Microwave Plasma," *J. Crystal Growth*, Vol. 62, pp. 642–644.

Kaplan, S., Jansen, F., and Machonkin, M. (1985), "Characterization of Amorphous Carbon-Hydrogen Films by Solid-State Nuclear Magnetic Resonance," *Appl. Phys. Lett.*, Vol. 47, pp. 750–753.

Kashiwagi, K., Kobayashi, K., Murayama, A., and Murayama, Y. (1986), "Chromium Nitride Films Synthesized by Radio Frequency Reactive Ion Plating," *J. Vac. Sci. Technol.*, Vol. A4, pp. 210–214.

Kaufman, H. R. (1978), "Technology of Ion Beam Sources Used in Sputtering," *J. Vac. Sci. Technol.*, Vol. 15, pp. 272–276.

———, Cuomo, J. J., and Harper, J. M. E. (1982), "Technology and Applications of Broad-Beam Ion Sources Used in Sputtering, Part I. Ion Source Technology," *J. Vac. Sci. Technol.*, Vol. 21, pp. 725–736.

Kay, E., and Dilks, A. (1981), "Plasma Polymerization of Fluorocarbons in RF Capacitively Coupled Diode System," *J. Vac. Sci. Technol.*, Vol. 18, No. 1, pp. 1–11.

———, Coburn, J., and Dilks, A. (1980), "Plasma Chemistry of Fluorocarbons as Related to Plasma Etching and Plasma Polymerization," in *Topics in Current Chemistry: Plasma Chemistry III* (S. Veprek and M. Venugopalan, eds.), pp. 3–42, Springer-Verlag, Berlin.

Kennedy, K. D., Scheuermann, G. R., and Smith, H. R. (1971), "Gas Scattering and Ion-Plating Deposition Methods," *Res./Devel.*, November, Vol. 22, pp. 40–44.

Kern, W., and Ban, V. S. (1978), "Chemical Vapor Deposition of Inorganic Thin Films," in *Thin Film Processes* (J. L. Vossen and W. Kern, eds.), pp. 257–331, Academic, New York.

——— and Rosler, R. S. (1977), "Advances in Deposition Process for Passivation Film," *J. Vac. Sci. Technol.*, Vol. 14, pp. 1082–1099.

Kikuchi, M., Nagakura, S., Ohmura, H., and Oketani, S. (1965), "Structures of Metal Films Produced by Vacuum-Arc Evaporation Method," *Jap. J. Appl. Phys.*, Vol. 4, No. 11, p. 940.

Kim, M. S., and Chun, J. S. (1983), "Effects of the Experimental Conditions of Chemical Vapour Deposition on a TiC/TiN Double-Layer Coatings," *Thin Solid Films*, Vol. 107, pp. 129–139.

Kitahama, K., and Hirata, K., Nakamatsu, H., and Kawai, S. (1986), "Synthesis of Diamond by Laser-Induced Chemical Vapor Deposition," *Appl. Phys. Lett.*, Vol. 49, pp. 634–635.

Klokholm, E., and Berry, B. S. (1968), "Intrinsic Stress in Evaporated Metal Films," *J. Electrochem. Soc. Solid State Sci.*, Vol. 115, pp. 823–826.

Kloos, K. H., Broszeit, E., and Gaberial, H. M. (1981), "Tribological Properties of Soft Metallic Coatings Deposited by Conventional and Thermionically Assisted Triode Ion Plating," *Thin Solid Films*, Vol. 80, pp. 307–319.

Kobayashi, M., and Doi, Y. (1978), "TiN and TiC Coating on Cemented Carbides by Ion Plating," *Thin Solid Films*, Vol. 54, pp. 67–74.

Komiya, S., and Tsuruoka, K. (1974), "Production and Measurement of Dense Metals Ions for Physical Vapor Deposition by a Hollow Cathode Discharge," *Jap. J. Appl. Phys.*, Vol. 13, Suppl. 2, Pt. 1, pp. 415–418.

——— and ——— (1975), "Thermal Input to Substrate During Deposition by Hollow Cathode Discharge," *J. Vac. Sci. Technol.*, Vol. 12, pp. 589–592.

——— and ——— (1976), "Physical Vapor Deposition of Thick Cr and Its Carbide and Nitride Films by Hollow Cathode Discharge," *J. Vac. Sci. Technol.*, Vol. 13, pp. 520–524.

———, Yoshikawa, H., and Ono, S. (1977a), "Mass and Energy Analysis of Incident Ions to Substrate during Deposition by Hollow Cathode Discharge," *J. Vac. Sci. Technol.*, Vol. 14, pp. 1161–1164.

———, Ono, S., Umezu, N., and Narusawa, T. (1977b), "Characterization of Thick Chromium-Carbon and Chromium-Nitrogen Films Deposited by Hollow Cathode Discharge," *Thin Solid Films*, Vol. 45, pp. 433–445.

———, Umezu, N., and Narusawa, T. (1978), "Formation of Thick Titanium Carbide Films by the Hollow Cathode Discharge Reactive Deposition Process," *Thin Solid Films*, Vol. 54, pp. 51–60.

Kut, S. (1974), "Decoration by Vacuum Metallization of Plastics and Metals," in *Science and Technology of Surface Coating* (B. N. Chapman and J. C. Anderson, eds.). pp. 350–360, Academic, New York.

Kuwahara, K., Sumomogi, T., and Kondo, M. (1981), "Internal Stress in Aluminum Oxide, Titanium Carbide and Copper Films Obtained by Planar Magnetron Sputtering," *Thin Solid Films*, Vol. 78, pp. 41–47.

Langmuir, J. (1913), "The Vapor Pressure of Metallic Tungsten," *Phys. Rev.*, Vol. 2, pp. 329–342.

Lee, W. W. Y., and Oblas, D. (1975), "Argon Entrapment in Metal Films by DC Triode Sputtering," *J. Appl. Phys.*, Vol. 46, pp. 1728–1732.

Maissel, L. I. (1970), "Application of Sputtering to the Deposition of Films," in *Handbook of Thin Film Technology* (L. I. Maissel and R. Glang, eds.), pp. 4-1–4-44, McGraw-Hill, New York.

——— and Glang, R. (eds.) (1970), *Handbook of Thin Film Technology*, McGraw-Hill, New York.

Marinov, M. (1977), "Effect of Ion Bombardment on the Initial Stages of Thin Film Growth," *Thin Solid Films*, Vol. 46, pp. 267–274.

Martin, P. J., McKenzie, D. R., Netterfield, R. P., Swift, P., Filipczuk, S. W., Müller, K. H., Pacey, C. G., and James, B. (1987), "Characterization of Titanium Arc Evaporation Process," *Thin Solid Films*, Vol. 153, pp. 91–102.

Matsubara, K., Enomoto, Y., Yaguchi, G., Watanabe, M., and Yamazaki, R. (1974), "Study of Deposition Process and Film Property in Ion Plating," *Jap. J. Appl. Phys.*, Vol. 13, Suppl. 2, Pt. 1, pp. 455–458.

Matsumoto, S., Sato, Y., Tsutsumi, M., and Setaka, N. (1982), "Growth of Diamond Particles from Methane-Hydrogen Gas," *J. Mater. Sci.*, Vol. 17, pp. 3106–3112.

Matthews, A. (1985), "Developments in Ionization Assisted Processes," *J. Vac. Sci. Technol.*, Vol. A3, pp. 2354–2363.

——— and Teer, D. G. (1980), "Deposition of Ti-N Compounds by Thermionically Assisted Triode Reactive Ion-Plating," *Thin Solid Films*, Vol. 72, pp. 541–549.

——— and ——— (1981), "Characteristics of a Thermionically Assisted Triode Ion Plating System," *Thin Solid Films*, Vol. 80, pp. 41–48.

Mattox, D. M. (1963), "Interface Formation during Thin Film Deposition," *J. Appl. Phys.*, Vol. 34, pp. 2493–2394.

——— (1964), "Film Deposition Using Accelerated Ions," *Electrochem. Technol.*, Vol. 2, pp. 295–298.

——— (1969), "Interface Formation, Adhesion, and Ion Plating," *Trans. SAE*, Vol. 78, pp. 2175–2181.

——— (1973), "Fundamentals of Ion-Plating," *J. Vac. Sci. Technol.*, Vol. 10, pp. 547–552.

——— (1977), *Proc. 25th Meeting of Mechanical Failure Prevention Group, Engineering Design* (T. R. Shives and W. A. Willard, eds.), p. 311, SP-487, NBS, Gaithersburg, Md.

——— (1982), "Ion Plating Technology," in *Deposition Technologies for Films and Coatings* (R. F. Bunshah, J. M. Blocher, T. D. Bonifield, J. G. Fish, P. B. Ghate, B. E. Jacobson, D. M. Mattox, G. E. McGuire, M. Schwartz, J. A. Thornton, and R. C. Tucker, eds.), pp. 244–287, Noyes Publications, Park Ridge, N.J.

McClure, G. W. (1974), "Plasma Expansion as a Cause of Metal Displacement in Vacuum-Arc Cathode Spots," *J. Appl. Phys.*, Vol. 45, pp. 2078–2084.

McLeod, P. S. (1976), "Planar Magnetron Sputtering Method and Apparatus," U.S. Patent 3,956,093.

McTaggart, F. K. (1967), *Plasma Chemistry in Electrical Discharges*, Elsevier, Amsterdam.

Michalski, A. (1984), "Structure and Properties of Coatings Composed of TiN-Ti Obtained by the Reactive Pulse Plasma Method," *J. Mater. Sci. Lett.*, Vol. 3, pp. 505–508.

———, Sokolowska, A., and Legutko, S. (1985), "The Useful Properties of TiN_x-Ti Coatings Deposited onto Drills at 500 K Using the Reactive Pulse Plasma Method," *Thin Solid Films*, Vol. 129, pp. 249–254.

Mirtich, M. J. (1981), "Adherence of Ion Beam Sputter Deposited Metal Films on H-13 Steel," *J. Vac. Sci. Technol.*, Vol. 18, pp. 186–189.

Miyazawa, T., Misawa, S., Yoshida, S., and Gonda, S. (1984), "Preparation and Structure of Carbon Film Deposited by a Mass-Separated C^+ Ion Beam," *J. Appl. Phys.*, Vol. 55, pp. 188–193.

Moore, W. J. (1960), "The Ionic Bombardment of Solid Surfaces," *Am. Scientist*, Vol. 48, pp. 109–133.

Morley, J. R., and Smith, H. R. (1972), "High Rate Ion Production for Vacuum Deposition," *J. Vac. Sci. Technol.*, Vol. 9, pp. 1377–1378.

Morrison, D. T., and Robertson, T. (1973), "RF Sputtering of Plastics," *Thin Solid Films*, Vol. 15, pp. 87–101.

Movchan, B. A., and Demchishin, A. V. (1969a), "Study of Structure and Properties of Thick Vacuum Condensates of Nickel, Titanium, Tungsten, Aluminum Oxide and Zirconium Dioxide," *Phys. Metals Metallog.*, Vol. 28, pp. 83–90.

——— and ——— (1969b), "Issledovanie Struktury i Svojstv Toltych Vakuumnych Kondensatov Nikelja, Titana, Volframa, Okisi, Aljuminija i Dvuokisi Cirkonija," *Fiz. Met. Metalloved*, Vol. 28, pp. 653–660.

Muehlberg, P. E. et al. (1965), "Process for Making Tetrafluoroethylene Polymers," U.S. Patent 3,170,858.

Münz, W. D., and Hessberger, G; (1981), "Cathode Sputtering Coats Working Tools and Fine Jewelry," *Ind. Res. Devel.*, Vol. 23, No. 9, pp. 130–135.

———, Hofmann, D., and Hartig, K. (1982), "A High Rate Sputtering Process for the Formation of Hard Friction-Reducing TiN Coatings on Tools," *Thin Solid Films*, Vol. 96, pp. 79–86.

Murayama, A. (1974), "Structure of Gold Films Formed by Ion Plating," *Jap. J. Appl. Phys.*, Vol. 13, Suppl. 2, Pt. 1, pp. 459–462.

——— (1975), "Thin Film Formation of In_2O_3, TiN and TaN by RF Reactive Ion Plating," *J. Vac. Sci. Technol.*, Vol. 12, pp. 818–820.

Nakamura, K., Inagawa, K., Tsuruoka, K., and Komiya, S. (1977), "Applications of Wear-Resistant Thick Films Formed by Physical Vapor Deposition Processes," *Thin Solid Films*, Vol. 40, pp. 155–167.

Nath, P., and Bunshah, R. F., (1980), "Preparation of In_2O_3 and Ti Doped In_2O_3 Films by Novel Activated Reactive Evaporation Technique," *Thin Solid Films*, Vol. 69, pp. 63–68.

Nemchinski, U. A. (1979), "Motion of the Cathodic Spot of a Vacuum Arc," *Sov. Phys. Tech. Phys.*, Vol. 24, pp. 767–771.

Nimmagadda, R., and Bunshah, R. F. (1975), "Synthesis of Dispersion-Strengthened Alloys by the Activated Reactive Evaporation Process from a Single Rod-Fed Electron Beam Source," *J. Vac. Sci. Technol.*, Vol. 12, pp. 815–817.

——— and ——— (1976), "Synthesis of TiC-Ni Cermets by the Activated Reactive Evaporation," *J. Vac. Sci. Technol.*, Vol. 13, pp. 532–535.

———, Doerr, H. J., and Bunshah, R. F. (1981), "Improvement in Tool Life of Coated High Speed Steel Drills Using the Activated Reactive Evaporation Process," *Thin Solid Films*, Vol. 84, pp. 303–306.

Ohmae, N., Nakai, T., and Tsukizoe, T. (1974), "On the Application of Ion Plating Technique to Tribology," *Jap. J. Appl. Phys.*, Vol. 13, Suppl. 2, Pt. 1, pp. 451–454.

Ojha, S. M. (1982), "Plasma-Enhanced Chemical Vapor Deposition of Thin Films," in *Physics of Thin Films* (G. Hass, M. H. Francombe, and J. L. Vossen, eds.), Vol. 12, pp. 237–296, Academic, New York.

Pan, A., and Greene, J. E. (1981), "Residual Compressive Stress in Sputter-Deposited TiC Films on Steel Substrates," *Thin Solid Films*, Vol. 78, pp. 25–34.

Paton, B. E., and Movchan, B. A. (1978), "Production of Protective Coatings by Electron Beam Evaporation," *Thin Solid Films*, Vol. 54, pp. 1–8.

Penning, F. M. (1957), *Electrical Discharge in Gases*, Macmillan, New York.

Perry, A. J., and Archer, N. J., (1980), "Techniques for Chemical Vapor Deposition," *AGARD LS-106*, Materials Coating Techniques, Lisbon, March, 27–28.

Pierce, J. R. (1954), *Theory of Design of Electron Beams*, Van Nostrand, New York.

Powell, C. F., Oxley, J. H., and Blocher, J. M. (eds.) (1966), *Vapor Deposition*, Wiley, New York.

Raghuram, A. C., and Bunshah, R. F. (1972), "The Effect of Substrate Temperature on the Structure of Titanium Carbide Deposited by Activated Reactive Evaporation," *J. Vac. Sci. Technol.*, Vol. 9, pp. 1389–1394.

———, Nimmagadda, R., Bunshah, R. F., and Wagner, C. N. J. (1974), "Structure and Microhardness Relationships in Ti, Zr and Hf-3Zr Carbide Deposits Synthesized by Activated Reactive Evaporation," *Thin Solid Films*, Vol. 20, pp. 187–199.

Randhawa, H. (1987), "Cathodic Arc Plasma Deposition of TiC and TiC_xN_{1-x} Films," *Thin Solid Films*, Vol. 153, pp. 209–218.

Reader, P. D., and Kaufman, H. R. (1975), "Optimization of an Electron Bombardment Ion Source for Ion Machining Applications," *J. Vac. Sci. Technol.*, Vol. 12, pp. 1344–1347.

Reichelt, K., and Jiang, X. (1990), "The Preparation of Thin Films by Physical Vapor Deposition Methods," *Thin Solid Films*, Vol. 191, pp. 91–126.

Reichelt, W., and Mueller, G. F. P. (1961), *Trans. 4th AVS Symp.*, Pergamon, New York, p. 956.

Reinberg, A. R. (1979), "Plasma Deposition of Inorganic Thin Films," *Ann. Rev. Mater. Sci.*, Vol. 9, pp. 341–372.

Rigney, D. V. (1982), "Vacuum Coating," in *Metals Handbook*, Vol. 5: *Surface Cleaning, Finishing and Coating*, 9th ed., pp. 387–411, American Society for Metals, Metals Park, Ohio.

Roberts, G. C., and Via, G. G. (1967), "Monitored Evaporant Source," U.S. Patent 3,313,914.

Rosler, R. S. (1977), "Low Pressure CVD Production Process for Polynitride and Oxide," *Solid State Technol.*, Vol. 20, No. 4, pp. 63–70.

———, and Engle, G. M. (1979), "LPCVD-Type Plasma Enhanced Deposition System," *Solid State Technol.*, Vol. 22, No. 12, pp. 88–92.

Sanders, D. M., and Pyle, E. A. (1987), "Magnetic Enhancement of Cathodic Arc Deposition," *J. Vac. Sci. Technol.*, Vol. A5, pp. 2728–2731.

Sarin, V., Bunshah, R. F., and Nimmagadda, R. (1977), "Structure of TiC-Ni Coatings Synthesized by Activated Reactive Evaporation," *Thin Solid Films*, Vol. 40, pp. 183–188.

Savvides, N., and Window, B. (1985), "Diamondlike Amorphous Carbon Films Prepared by Magnetron Sputtering of Graphite," *J. Vac. Sci. Technol.*, Vol. A3, pp. 2386–2389.

Sawabe, A., and Inuzuka, T. (1986), "Growth of Diamond Thin Films by Electron-Assisted Chemical Vapor Deposition and Their Characterization," *Thin Solid Films*, Vol. 137,pp. 86–99.

Schaffhauser, A. C. (ed.) (1967), *Proc. of 1st Int. Conf. on Chemical Vapor Deposition of Refractory Metals, Alloys, and Compounds*, American Nuclear Society, Hinsdale, Ill.

Schiller, S., Heisig, U., and Goedicke, K. (1975), "Alternating Ion Plating—A Method of High-Rate Ion Vapor Deposition," *J. Vac. Sci. Technol.*, Vol. 12, pp. 858–864.

———, ———, and ——— (1977a), "Ion Deposition Techniques for Industrial Applications," *Proc. 7th Int. Vac. Congress and 3d Int. Conf. on Solid Surfaces*, Vol. 2, pp. 1545–1552, Vienna, Austria.

———, ———, and ——— (1977b), "Use of the Ring Gap Plasmatron for High Rate Sputtering," *Thin Solid Films*, Vol. 40, pp. 327–334.

———, ———, and Panzer, S. (1982), *Electron Beam Technology*, Wiley, New York.

Schwarz, H., and Tourtellotte, H. A. (1966), *13th National Vacuum Symp.*, p. 87, Macmillan, New York.

Seeman, J. M. (1966), "Bias Sputtering: Its Techniques and Applications," in *Proc. Symp. on the Deposition of Thin Films by Sputtering*; pp. 30–42, University of Rochester and Consolidated Vacuum Corp., Rochester, N.Y.

Shen, M. (ed.) (1976), *Plasma Chemistry of Polymers*, Marcel Dekker, New York.

Sherman, A. (1984), "Plasma-Assisted Chemical Vapor Deposition Process and Their Semiconductor Applications," *Thin Solid Films*, Vol. 113, pp. 135–149.

Shikama, T., Araki, H., Fujitsuka, M., Fukutomi, M., Shinno, H., and Okada, M. (1983), "Properties and Structure of Carbon Excess Ti_xC_{1-x} Deposited onto Molybdenum by Magnetron Sputtering," *Thin Solid Films*, Vol. 106, pp. 185–194.

Shiloh, M., Gayer, M., and Brinckman, F. E. (1977), "Preparation of Nitrides by Active Nitrogen, II: Si_3N_4," *J. Electrochem. Soc.*, Vol. 124, pp. 295–300.

Smith, H. R., Kennedy, K., and Boericke, F. S. (1970), "Metallurgical Characteristics of Titanium-Alloy Foil Prepared by Electron Beam Evaporation," *J. Vac. Sci. Technol.*, Vol. 7, pp. 548–551.

Snaper, A. A. (1971), "Arc Deposition Process and Apparatus," U.S. Patent 3,625,848.

——— (1974), "Arc Deposition Apparatus," U.S. Patent 3,836,451.

Sokolowski, M. (1979), "Deposition of Wurtzite Type Boron Nitride Layers by Reactive Pulse Plasma Crystallization," *J. Crystal Growth*, Vol. 46, pp. 136–138.

———, Sokolowska, A., Michalski, A., Gokieli, B., Romanowski, Z., and Rusek-Mazurek, A. (1977), "Crystallization from a Reactive Pulse Plasma," *J. Crystal Growth*, Vol. 42, pp. 507–511.

———, ———, ———, ———, ———, and ——— (1979), "Reactive Pulse Plasma Crystallization of Diamond and Diamond-like Carbon," *J. Crystal Growth*, Vol. 47, pp. 421–426.

———, ———, ———, Romanowski, Z., Rusek-Mazurek, A., and Wronikowski, M. (1981), "The Deposition of Thin Films of Materials with High Melting Points on Substrates at Room Temperature Using the Pulse Plasma Method," *Thin Solid Films*, Vol. 80, pp. 249–254.

Spalvins, T. (1980a), "Survey of Ion Plating Sources," *J. Vac. Sci. Technol.* Vol. 17, pp. 315–321.

——— (1980b), "Tribological Properties of Sputtered MoS_2 Films in Relation to Film Morphology," *Thin Solid Films*, Vol. 73, pp. 291–297.

——— (1985), "Status of Plasma Physics Techniques for the Deposition of Tribological Coatings," in *New Directions in Lubrication, Materials, Wear, and Surface Interactions: Tribology in 80's* (W. R. Loomis, ed.), pp. 691–712, Noyes, Park Ridge, N.J.

——— and Brainard, W. A. (1974), "Nodular Growth in Thick-Sputtered Metallic Coatings," *J. Vac. Sci. Technol.*, Vol. 11, pp. 1186–1192.

Spencer, E. G., Schmidt, P. H., Joy, D. C., and Sansalone, F. J. (1976), "Ion-Beam-Deposited Polycrystalline Diamond-like Films," *Appl. Phys. Lett.*, Vol. 29, pp. 118–120.

Stirland, D. J. (1966), "Electron-Bombardment-Induced Changes in the Growth and Epitaxy of Evaporated Gold Films," *Appl. Phys. Lett.*, Vol. 8, pp. 326–328.

Stuart, R. V., and Wehner, G. K. (1962), "Sputtering Yields at Very Low Bombarding Ion Energies," *J. Appl. Phys.*, Vol. 33, pp. 2345–2352.

Stull, D. R., Westrum, E. F., and Sinke, G. C. (1969), *The Chemical Thermo-dynamics of Organic Compounds*, Wiley, New York.

Sundgren, J.-E. (1985), "Structure and Properties of TiN Coatings," *Thin Solid Films*, Vol. 128, pp. 21–44.

——— and Hentzell, H. T. G. (1986), "A Review of the Present State of the Art in Hard Coatings Grown from the Vapor Phase," *J. Vac. Sci. Technol.*, Vol. A4, No. 5, pp. 2259–2279.

———, Johansson, B.-O., Karlsson, S.-E., and Hentzell, H. T. G. (1983), "Mechanism of Reactive Sputtering of Titanium Nitride and Titanium Carbide. I: Influence of Process Parameters on Film Composition. II: Morphology and Structure. III: Influence of Substrate Bias on Composition and Structure," *Thin Solid Films*, Vol. 105, pp. 353–393.

Takagi, T. (1983), "Role of Ions in Ion Engineering: New Approach to Studying the Ion-Assisted Technique," in *Proc. Int. Ion Engineering Congress, 7th Symp. on Ion Sources and Ion Assisted Technology, and 4th Int. Conf. on Ion and Plasma Assisted Techniques.* Institute of Electrical Engineers Japan, Tokyo, pp. 785–798.

———, Yamada, I., and Sasaki, A. (1976), "An Evaluation of Metal and Semiconductor Films Formed by Ionized-Cluster Beam Deposition," *Thin Solid Films*, Vol. 39, pp. 207–217.

———, Matsubara, K., Takaoka, H., and Yamada, I. (1979), "Ion Beam Epitaxial Technique and Applications," in *Proc. Conf. on Ion Plating and Allied Techniques*, pp. 174–185, CEP Consultants, Edinburgh.

Teer, D. G. (1975), "Ion-Plating," *Tribol. Int.*, Vol. 8, pp. 247–251.

——— and Delcea, B. L. (1978), "Grain Structure of Ion-Plated Coatings," *Thin Solid Films*, Vol. 54, pp. 295–301.

——— and Salama, N. (1977), "The Ion Plating of Carbon," *Thin Solid Films*, Vol. 45, pp. 553–561.

Theuerer, H. C., and Hauser, J. J. (1964), "Getter Sputtering for the Preparation of Thin Films of Superconducting Elements and Compounds," *J. Appl. Phys.*, Vol. 35, pp. 554–555.

Thornton, J. A. (1973), "Sputter Coating. Its Principles and Potential," *Trans. SAE*, Vol. 82, pp. 1787–1805.

——— (1974), "Influence of Apparatus Geometry and Deposition Conditions on the Structure and Topography of Thick Sputtered Coatings," *J. Vac. Sci. Technol.*, Vol. 11, pp. 666–670.

——— (1975), "Influence of Substrate Temperature and Deposition Rate on Structure of Thick Sputtered Cu Coatings," *J. Vac. Sci. Technol.*, Vol. 12, pp. 830–835.

——— (1977a), "High Rate Thick Film Growth," *Ann. Rev. Mater. Sci.*, Vol. 7, pp. 239–260.

——— (1977b), "The Influence of Bias Sputter Parameters on Thick Copper Coatings Deposited Using a Hollow Cathode," *Thin Solid Films*, Vol. 40, pp. 335–344.

——— (1978a), "Magnetron Sputtering—Basic Physics and Applications to Cylindrical Magnetrons," *J. Vac. Sci. Technol.*, Vol. 15, pp. 171–177.

——— (1978b), "Recent Developments in Sputtering—Magnetron Sputtering," *Proc. of Society of Vacuum Coaters*, Vol. 21, pp. 19–30, Society of Vacuum Coaters, Washington, D.C.

——— (1981), "High Rate Sputtering Techniques," *Thin Solid Films*, Vol. 80, pp. 1–11.

——— (1982), "Coating Deposition by Sputtering," in *Deposition Technologies for Films and Coatings* (R. F. Bunshah, J. M. Blocher, T. D. Bonifield, J. G. Fish, P. B. Ghate, B. E. Jacobson, D. M. Mattox, G. E. McGuire, M. Schwartz, J. A. Thornton, and R. C. Tucker, eds.), pp. 170–243, Noyes Publications, Park Ridge, N.J.

——— and Cornog, G. (1977), "Sputtered Austenitic Manganese Steel," *Thin Solid Films*, Vol. 45, pp. 397–406.

——— and Ferriss, D. P. (1977), "Structural and Wear Properties of Sputter-Deposited Laves Intermetallic Alloy," *Thin Solid Films*, Vol. 40, pp. 365–374.

——— and Hoffman, D. W. (1981), "Internal Stresses in Amorphous Silicon Films Deposited by Cylindrical Magnetron Sputtering Using Ne, Ar, Kr, Xe and Ar + H_2," *J. Vac. Sci. Technol.*, Vol. 18, pp. 203–207.

——— and Penfold, A. S. (1978), "Cylindrical Magnetron Sputtering," in *Thin Film Processes* (J. L. Vossen and W. Kern, eds.), pp. 75–129, Academic, New York.

Tisone, T. C., and Bindell, J. B. (1974), "Low-Voltage Triode Sputtering with a Confined Plasma, Part I. Geometric Aspects of Deposition," *J. Vac. Sci. Technol.*, Vol. 11, pp. 519–527.

——— and Cruzan, P. D. (1975), "Low Voltage Triode Sputtering with a Confined Plasma, Part V. Application to Backsputter Definition," *J. Vac. Sci. Technol.*, Vol. 12, pp. 677–688.

Townsend, P. D., Kelly, J. C., and Hartley, N. E. W. (1976), *Ion Implantation, Sputtering and Their Applications*, Academic, New York.

Tsukada, T., Hosokawa, N., and Kobayashi, H. (1978), "Increase of Substrate Temperature in High Rate Coaxial-Cylindrical Magnetron Sputtering," Jap. *J. Appl. Phys.*, Vol. 17, pp. 787–796.

Tsukizoe, T., Nakai, T., and Ohmae, N. (1977), "Ion-Beam Plating Using Mass-Analyzed Ions," *J. Appl. Phys.*, Vol. 48, pp. 4770–4776.

Vandentop, G. J., Kawasaki, M., Nix, R. M., Brown, I. G., Salmeron, M., and Somorjai, G. A. (1990), "The Formation of Hydrogenated Amorphous Carbon Films of Controlled Hardness from a Methane Plasma," *Phys. Rev.*, Vol. B41, No. 5, pp. 3200–3210.

Venugopalan, M. (1971), *Reactions under Plasma Conditions*, Wiley Interscience, New York.

Věprek, S. (1976), "Hetrogeneous Reactions in Non-isothermal Low Pressure Plasmas—Preparative Aspects and Applications," *Pure Appl. Chem.*, Vol. 48, pp. 163–178.

——— (1985), "Plasma-Induced and Plasma-Assisted Chemical Vapor Deposition," *Thin Solid Films*, Vol. 130, pp. 135–154.

———, Iqbal, Z., Oswald, H. R., and Webb, A. W. (1981), "Properties of Polycrystalline Silicon Prepared by Chemical Transport in Hydrogen Plasma at Temperatures between 80 and 400°C," *J. Phys. Chem.*, Vol. 14, pp. 295–308.

Von Ardenne, M. (1956), *Tabellen der Elektronenphysik Ionenphysik and Ubermicroskopie Dtsch*, Verlag, Wiss, Berlin, p. 554.

Von Engle, A. (1965), *Ionized Gases*, Clarendon Press, London.

Vossen, J. L. (1971), "Control of Film Properties by RF Sputtering Techniques," *J. Vac. Sci. Technol.*, Vol. 8, pp. S12–S30.

——— (1979), "Glow Discharge Phenomena in Plasma Deposition," *J. Electrochem. Soc.*, Vol. 126, pp. 319–324.

——— and Cuomo, J. J. (1978), "Glow Discharge Sputter Deposition," in *Thin Film Processes* (J. L. Vossen and W. Kern, eds.), pp. 12–75, Academic, New York.

——— and Kern, W. (eds.) (1978), *Thin Film Processes*, Academic, New York.

——— and O'Neill, J. J. (1968), "RF Sputtering Process," *RCA Rev.*, Vol. 29, pp. 149–179.

Wagner, J. J., and Věprek, S. (1982), "Kinetic Study of the Heterogeneous Si/H System under Low-Pressure Plasma Conditions by Means of Mass Spectroscopy," in *Plasma Chem.: Plasma Process.*, Vol. 2, pp. 95–107.

——— and ——— (1983), "Chemical Relaxation Study of the Heterogeneous Silicon-Hydrogen System under Plasma Conditions," in *Plasma Chem.: Plasma Process.*, Vol. 3, pp. 219–234.

Waits, R. K. (1978), "Planar Magnetron Sputtering," in *Thin Film Processes* (J. L. Vossen and W. Kern, eds.), pp. 131–173, Academic, New York.

Wakefield, G. F., and Blocher, J. M. (eds.) (1973), *Proc. of 4th Int. Conf. on Chemical Vapor Deposition*, American Electrochemical Society, Princeton, N.J.

Walls, J. M., Hall, D. D., Teer, D. G., and Delcea, B. L. (1978), "A Comparison of Vacuum-Evaporated and Ion-Plated Thin Films Using Auger Electron Spectroscopy," *Thin Solid Films*, Vol. 54, pp. 303–308.

Wasa, K., and Hayakawa, S. (1967), "Sputtering in a Crossed Electromagnetic Field," *IEEE Trans. on Parts, Mater. Packag.*, Vol. PMP-3, pp. 71–76.

——— and ——— (1969), "Low Pressure-Sputtering System of the Magnetron Type," *Rev. Sci. Instrum.*, Vol. 40, pp. 693–697.

Washo, B. D. (1976), "Surface Property Characterization of Plasma-Polymerized Tetrafluoroethylene Deposits," in *Plasma Chemistry of Polymers* (M. Shen, ed.), pp. 191–198, Marcel Dekker, New York.

Wehner, G. K. (1960), "Angular Distribution of Sputtered Material," *J. Appl. Phys.*, Vol. 31, pp. 177–179.

Weissmantel, C. (1981), "Ion Beam Deposition of Special Film Structures," *J. Vac. Sci. Technol.*, Vol. 18, pp. 179–185.

———, Rost, M., Fiedler, D., Erler, H.-J., Giegengack, H., and Horn, J. (1974), "Interaction of Ion Beams with Polymer Surfaces Leading to Etching and Sputtering," *Jap. J. Appl. Phys.*, Vol. 13, Suppl. 2, Pt. 1, pp. 439–442.

———, Reisse, G., Erler, H.-J., Henny, F., Bewilogua, K., Ebersbach, U., and Schurer, C. (1979a), "Preparation of Hard Coatings by Ion Beam Methods," *Thin Solid Films*, Vol. 63, pp. 315–325.

———, Erler, H.-J., and Reisse, G. (1979b), "Ion Beam Techniques for Thin and Thick Film Deposition," *Surf. Sci.*, Vol. 86, pp. 207–221.

———, Bewilogua, K., Breuer, K., Erler, H.-J., Hinnberg, H.-J., Klose, S., Nowick, W., and Reisse, G. (1980), "Ion-Activated Growth of Special Film Phases," *2nd Int. Conf. on Low-Energy Ion Beams*, Conf. Series No. 54, pp. 188–200, Institute of Physics, London.

———, ———, ———, Dietrich, D., Ebersbach, U., Erler, H. J., Rau, B., and Reisse, G. (1982), "Preparation and Properties of Hard i-C and i-BN Coatings," *Thin Solid Films*, Vol. 96, pp. 31–44.

Westwood, W. D. (1976), "Glow Discharge Sputtering," *Prog. Surf. Sci.*, Vol. 7, pp. 71–111.

White, P. (1966), "Preparation and Some Physical Properties of Very Thin Polymer Layers," *Electrochem. Technol.*, Vol. 4, Nos. 9–10, pp. 468–471.

——— (1967), "Preparation and Properties of Dielectric Layers Formed by Surface Irradiation Techniques," *Insulation*, May, pp. 52–58.

Williams, D. G. (1974), "Vacuum Coating with a Hollow Cathode Source," *J. Vac. Sci. Technol.*, Vol. 11, pp. 374–376.

Wright, A. N. (1967), "Surface Photopolymerization of Vinyl and Diene Monomers," *Nature*, Vol. 215, pp. 953–955.

Yagi, K., Tamura, S., and Tokuyama, T. (1977), "Germanium and Silicon Film Growth by Low-Energy Ion Beam Deposition," *Jap. J. Appl. Phys.*, Vol. 16, pp. 245–251.

Yasuda, H. (1978), "Glow Discharge Polymerization," in *Thin Film Processes* (J. L. Vossen and W. Kern, eds.), pp. 361–398, Academic, New York.

——— and Hsu, T. (1978a), "Plasma Polymerization Investigated by the Comparison of Hydrocarbons and Perfluorocarbons," *Surf. Sci.*, Vol. 76, pp. 232–241.

——— and ——— (1978b), "Distribution of Polymer Deposition in Plasma Polymerization. I: Acetyelene, Ethylene, and Effect of Carrier Gas. II: Effect of Carrier Gas. III: Effect of Discharge Power," *J. Polymer Sci.* (*Polymer Chem.*), Vol. 16, pp. 229–241, 313–326, 2587–2592.

Yoshihara, H., and Mori, H. (1979), "Enhanced ARE Apparatus and TiN Synthesis," *J. Vac. Sci. Technol.*, Vol. 16, pp. 1007–1012.

Young, R., and Wood, R. (1982), "Laser Processing of Semiconductor Materials," *Ann. Rev. Mater. Sci.*, Vol. 12, pp. 323–350.

Zega, B., Kornmann, M., and Amiguet, J. (1977), "Hard Decorative TiN Coatings by Ion Plating," *Thin Solid Films*, Vol. 45, pp. 577–582.

APPENDIX 9.A: GLOW-DISCHARGE PLASMAS

In the ion bombardment process by self-sustained glow discharge, the procedure is to backfill the evacuated chamber with a gas (typically an inert gas such as argon) to a pressure in the range of a few to several hundred millitorr and to apply a high potential between the cathode and anode, so that ionization of the working gas is produced. Such a low-pressure electric discharge is called a *glow discharge*, and the ionized gas is called *plasma* (Chapman, 1980; Dugdale, 1971; Penning, 1957).

9.A.1 Theory of DC Glow-Discharge Plasmas

The potential-current characteristics of a gas discharge using a dc power supply are shown in Fig. 9.A.1(*a*). Several specific regions can be distinguished, and the type which is observed between the two electrodes depends upon several factors: the gas pressure, applied voltage, and electrode configuration. If one applies a dc voltage between two electrodes in low-vacuum chamber, at low voltage little current is drawn and there is uniform potential gradient across the chamber. This region is known as the *Townsend region*. As the voltage is increased, the current increases slightly until the breakdown potential is reached, at which point the current increases rapidly and the potential drops to a lower value. At this voltage and gas pressure, one enters the normal glow-discharge region, where, with a constant voltage, the current is a function of gas pressure and the cathode current density is constant over the cathode spot. At higher pressures and higher currents, the cathode spot covers the whole cathode, and the current density becomes a function of the voltage at constant pressure. *Sputtering and ion-plating deposition are done in the abnormal glow-discharge region* to ensure complete coverage of the cathode and thus a more uniform coverage of a substrate. If the current is further increased, the voltage is further increased, and the voltage drops abruptly and the discharge becomes an arc (Vossen and Cuomo, 1978). To have a self-sustained dc diode gas discharge at several thousand volts in argon requires a minimum of about 0.5 Pa (5×10^{-3} torr) argon pressure. For neon, the pressure must be greater; but for a krypton discharge, the pressure can be less.

Figure 9.A.1(*b*) is a diagrammatic illustration of the appearance of glow discharge for a gas in the pressure region between a few to few hundred millitorr. Also shown are the plots of various parameters along its length. The plasma region is a low field region where there are equal numbers of ions and electrons, thus the net space charge is zero. This field strength is low, and the region is luminous (Berry et al., 1968; Brown, 1966; Dugdale, 1977; Holland, 1980; Matthews, 1985; Von Engle, 1965). Of primary interest are the cathode dark space (also called *Crookes' dark space*), where most of the potential drop is found, and the negative glow region, where most of the ionization occurs.

Since the mean free path of electrons is inversely proportional to the gas pressure (in actuality, gas density), it follows that the distance required for an electron to travel before it has produced adequate ionization to sustain the glow is also inversely proportional to the pressure. The thickness of the cathode dark space increases as the pressure is decreased. If the pressure is made sufficiently low, the cathode dark space will expand until the plane of the anode is reached, and the discharge will become extinguished. In the region of the normal glow, the product of the thickness of the dark space and the gas pressure is independent of

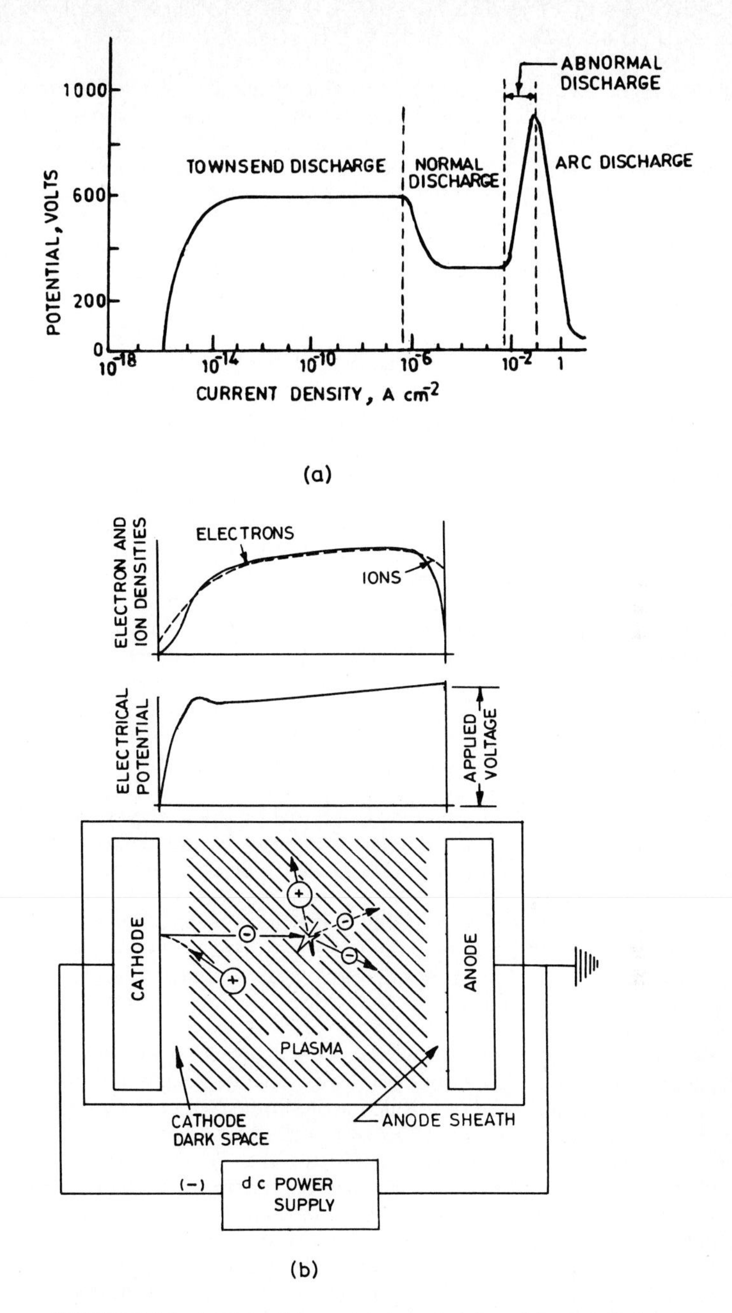

FIGURE 9.A.1 (*a*) Potential-current characteristics of dc glow discharge in neon gaseous atmosphere under low vacuum (100 Pa or 1 torr). (*b*) Schematic representation of electron density, ion density, and electric potential across a planar dc diode glow discharge for a gas at low pressure (~0.1 to 10 Pa or 1 to 100 mtorr).

current. For argon gas this product has a value of about 3 torr-millimeters. For example, at a pressure of 2 Pa (15 mtorr), the dark space would extend 200 mm from the cathode. The thickness of the dark space is a function of the voltage in the abnormal glow region. For example, at 5 kV (well into the abnormal glow) and a pressure of 2 Pa (15 mtorr), the dark space is found to extend only about 30 mm.

9.A.2 Potentials in RF Glow-Discharge Plasmas

If one of the electrodes of a diode arrangement in the presence of glow-discharge plasma is capacitively coupled* to the rf power supply, it develops a dc self-bias potential that is negative relative to the plasma floating potential, if the electrode area is smaller than the area of the grounded part of the system. The bias arises from the large difference in mobilities of the electrons and ions. When the rf field is applied to the capacitively coupled electrode, the electrode surface in contact with the plasma is alternatively polarized positive and negative. When negative, it attracts ions; when positive, it repels them. At high operating frequencies, the massive ions cannot follow the temporal variations in the applied potential. However, the electrons can. Thus the cloud of electrons that constitute the electron component of the negative glow plasma can be pictured as moving back and forth at the applied frequency in a sea of relatively stationary ions on the powered electrode.

The basis for this self-bias potential, which forms as a result of the differences in mobility between the electrons and ions, is shown in Fig. 9.A.2. The current-voltage characteristics for an electrode immersed in a plasma are shown in Fig. 9.A.2. Upon application of a rf voltage through the capacitor, a high initial electron current flows to the electrode. On the second half of the cycle, only a relatively small ion current can flow. Since no charge can be transferred through the capacitor, the voltage on the electrode surface must self-bias negatively until the net current (averaged over each rf cycle) is zero. This results in the pulsating negative potential shown in Fig. 9.A.2 (Butler and Kino, 1963; Chapman, 1980). The rf self-biasing works at frequencies in the megahertz range. Most apparatuses operate at a frequency of 13.56 MHz, since this is the frequency in the 10- to 20-MHz range that has been allocated by the Federal Communications Commission for industrial use.

Figure 9.A.3 illustrates the distribution of voltage in a rf glow discharge from a small capacitively coupled electrode to a large grounded electrode. The average dc self-bias V_B on the powered electrode (small) compared to the grounded electrode generally is slightly less than the zero-to-peak voltage V_0 of the rf signal applied. The most important discharge parameter, the sheath potential or wall potential V_S between the plasma and the powered electrode, is given by negative self-bias voltage, neglecting the very small contribution of plasma potential V_P. The reactor geometry, i.e., basically the ratio of the capacitively coupled electrode surface area (cathode, A_c) to the grounded part of the system (anode, A_a), is important for the potential distribution. The ratio of the sheath potential over the cathode and anode dark space ($V_{S,c}$ and $V_{S,a}$, respectively) depends on (Butler and Kino, 1963)

*If an insulating electrode is connected through a metal electrode, to the rf power supply, the insulating electrode provides the capacitive coupling.

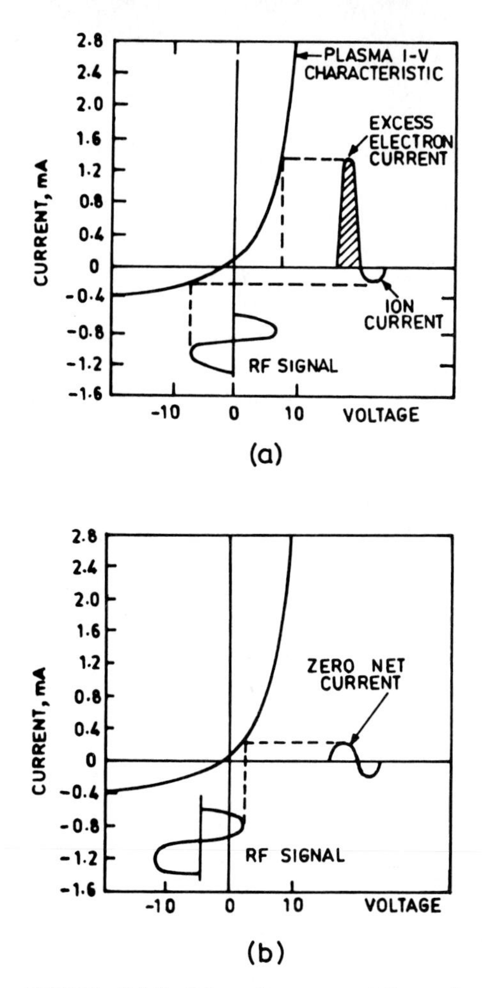

FIGURE 9.A.2 Schematic representation of mechanism by which a negative self-bias is developed on a capacitively coupled surface in a rf diode glow discharge. [*Adapted from Butler and Kino (1963).*]

$$\frac{V_{S,c}}{V_{S,a}} = \left(\frac{A_a}{A_c}\right)^m \tag{9.A.1}$$

where $1 \leq m \leq 4$. Clearly, to minimize bombardment of grounded fixtures, the area of all grounded parts should be very large in comparison to that of the target.

The rf discharges in planar diode apparatuses can be operated at considerably lower pressures (typically 0.5 to 2 Pa or 5 to 15 mtorr) than dc discharges can (typically 0.8 to 8 Pa or 10 to 60 mtorr). This is believed to be caused by a reduction in the loss of primary electrons and, at high frequencies, by an increase in the volume ionization caused by the oscillating electrons. Matching networks are needed to couple the rf power into the high-impedance discharge.

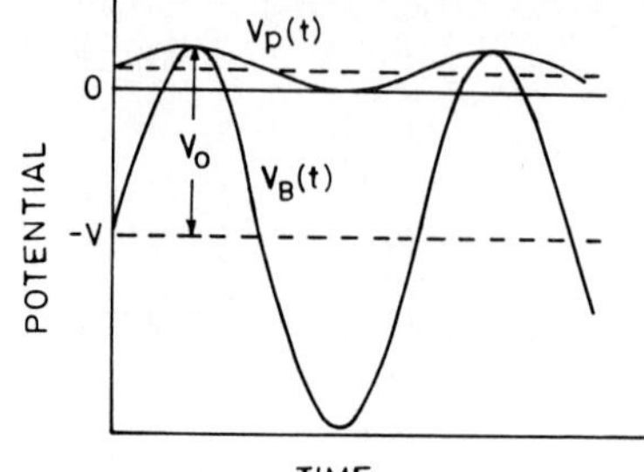

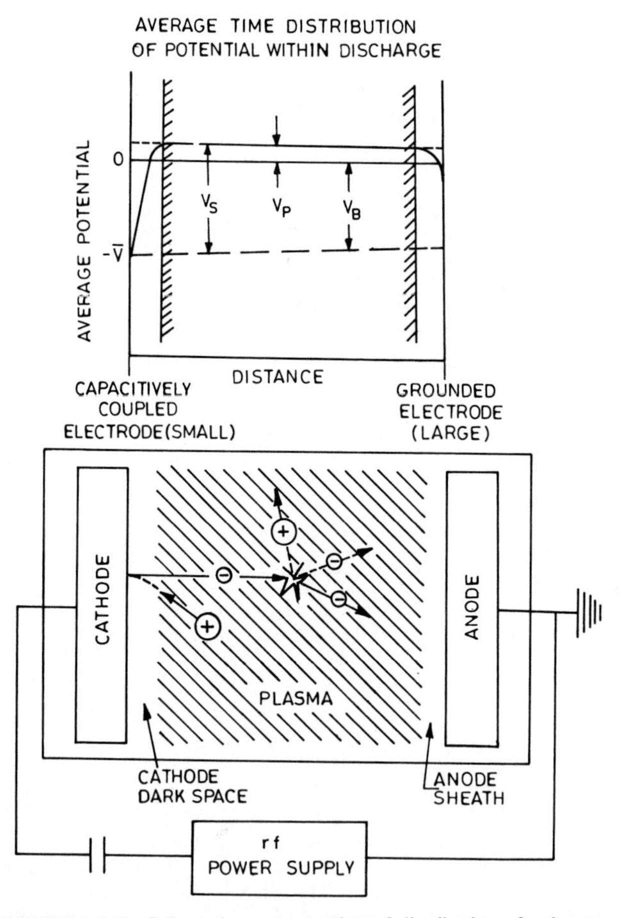

FIGURE 9.A.3 Schematic representation of distribution of voltages in rf diode glow discharge from small, capacitively coupled electrode to large, grounded electrode.

9.A.3 Enhanced Ionization Sources

Glow discharges are relatively inefficient ion sources. Only a few percent of the gas atoms in a glow discharge are ionized. Several techniques have been developed for increasing the ionization efficiency. Figure 9.A.4 shows some enhanced plasma sources that increase the plasma density by techniques other than ionization from secondary electrons arising from the ion bombardment, as in the glow-discharge technique. In Fig. 9.A.4(*a*), (*b*), and (*d*), the ionizing electrons are generated by a hot electron-emitting filament. In Fig. 9.A.4(*d*), the electron path length is increased by the use of a small magnetic field, and the shape is similar to that of triode sputtering. In Fig. 9.A.4(*c*), the secondary electrons emitted by the electron beam evaporation process are extracted toward the positively biased electrode and create ionization in the region above the evaporation hearth. The ionization in the gas phase can also be increased by the application of a magnetic field to increase the path length of electrons.

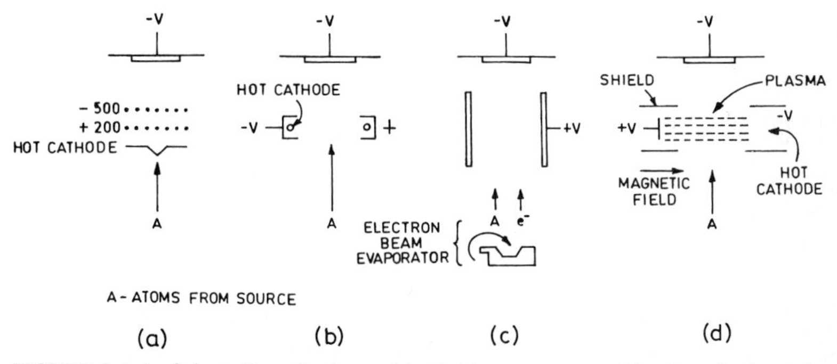

FIGURE 9.A.4 Schematics of enhanced ionization sources used in glow-discharge ion plating: (*a*) hot electron-emitting filament (hot cathode) with accelerating grid, (*b*) electron-emitting filament (hot cathode), (*c*) positive cylindrical electrode to attract secondary electrons from evaporation hearth, may be used with a magnetic field, (*d*) hot-cathode magnetically confined plasma. [*Adapted from Mattox (1982).*]

APPENDIX 9.B: VACUUM SYSTEMS USED FOR VAPOR DEPOSITION PROCESSES

The vapor deposition of thin coatings with controlled properties requires an operating environment which interfers as little as possible with the process of coating formation. Most vapor deposition takes place in high vacuum (typically 1 to 10^{-5} Pa or 10^{-2} to 10^{-7} torr) to minimize the interaction between residual gases and the surfaces of growing coatings. Normally, the deposition chamber is evacuated to high or ultra-high vacuum (up to 10^{-7} Pa or 10^{-9} torr) to remove contaminants and then is backfilled with high-purity gas to the desired deposition pressures.

Two different principles may be employed to reduce the pressure in a vacuum enclosure. The first involves physical removal of gases from the vessels and exhausting the gas load to the outside. Examples of this operating mode are the mechanical (oil-sealed rotary and roots) and the vapor stream pumps. These pumps are normally used to pump down to a low vacuum of about 10^{-1} Pa (10^{-3} torr). The other method of evacuation relies on condensation or trapping of gas molecules on some part of the inner surface of the enclosure without discharging the gas. Diffusion, cryogenic, cryosorption, sublimation, and getter-ion pumps belong in this category. These pumps can be used to pump down to ultrahigh vacuum up to 10^{-12} Pa (10^{-14} torr).

The most conventional combination of pumps used to evacuate up to about 10^{-5} to 10^{-7} Pa (10^{-7} to 10^{-9} torr) is the oil diffusion pump backed by the mechanical or rotary pump. The working principles of these are discussed by Maissel and Glang (1970). At the beginning of pump-down cycle or during the roughing cycle, the oil diffusion pump is isolated from the atmosphere by closing the baffle valve and backing valve. The chamber's pressure is reduced to 1 to 10^{-1} Pa (10^{-2} to 10^{-3} torr) by rotary pump through the roughing valve. After a vacuum of about 10^{-1} Pa (10^{-3} torr) is achieved, the diffusion pump is connected with the rotary pump and the chamber by closing the roughing valve and opening the backing valve and the baffle valve, and a high (or ultra-high) vacuum is achieved.

The diffusion pump usually incorporates a baffle which can be cooled by water or liquid nitrogen to prevent oil vapors from entering the vacuum chamber, due to backstreaming of the diffusion pump. Traps with molecular sieves cooled by liquid nitrogen are also essential between the rotary pump and the vacuum system. With adequate trapping, an oil diffusion pump can be used in the *ultra-high-vacuum* (UHV) region (that is, 10^{-7} Pa or 10^{-9} torr) which is preferable in order to obtain high purity and better adherence of the coatings. However, for the UHV system, common practice is to use an ion pump backed by a cryosorption pump. Recently turbomolecular and cryosorption pumping systems have been increasingly used for the contaminant-free coatings.

The pressure of the vacuum is monitored by various types of gauges such as ionization, Bayard-Alpert, Penning, Knudsen, Pirani, McLeod, capacitance manometer, and thermocouple gauges (Chopra, 1979). The type of gauge used depends on the pressure, but a typical system would use a Pirani, thermocouple, or capacitance manometer gauge for low vacuum (down to 10^{-1} Pa or 10^{-3} torr) and an ionization gauge for high and ultrahigh vacuum, that is, 10^{-3} to 10^{-12} Pa (10^{-5} to 10^{-14} torr). The mass spectrometer is used for measuring the partial pressure of contaminating gases and for leak detection of the vacuum system.

The residual gases in a system after pumpdown usually consist of H_2O, CO_2, N_2, H_2, and organic vapors of different molecular weights. Oil vapors reach a vacuum system from the rotary pump (the vapor pressure of oils used is $\sim 10^{-1}$ Pa or 10^{-3} torr at 40°C) during the roughing cycle and from the diffusion pump by backstreaming. With baffles and adequate trapping, contaminants can be minimized. We also note that a glass bell jar is not desirable for UHV and bakeable system; instead polished stainless-steel (type 304) and aluminum systems are used. At elevated temperatures, a considerable amount of water vapor diffuses from the interior of Pyrex glass to its surface. At temperatures above 350°C, glass decomposes in vacuo, liberating water vapors and alkali ions (for example, Na^+ and K^+) which are a source of contamination. Moreover, considerable permeation of Ne, H_2, and He through glass occurs to limit the upper limit of the vacuum attained.

CHAPTER 10
COATING DEPOSITION BY MISCELLANEOUS TECHNIQUES

Numerous coating deposition techniques based on particulate deposition and vapor deposition are reviewed in Chapters 8 and 9. This chapter includes several other important techniques: atomized liquid spray coating; dip coating; fluidized-bed coating; spin-on coating; brush, pad, and roller coating; sol-gel coating; screening and lithography; electrochemical deposition; chemical deposition; chemical conversion; intermetallic compound coatings; and spark hardening. Atomized liquid spray, dipping, spin-on coating, and brushing are also referred to as *wetting processes* in which the material is applied in liquid form and then becomes solid by solvent evaporation, drying, or cooling. Generally, any coating which can be deposited by spraying can also be applied by dipping, spin-on coating, and brushing. Solid polymer powders which can be levitated by the use of air can be deposited by fluidized-bed deposition technique. In the sol-gel technique, the sol (colloidal dispersion) is applied on a metal surface by a suitable method such as dipping, spin coating, or electrophoresis, then dried, and finally fired at elevated temperature. Screening and lithography processes are used to deposit coating in selected areas.

Metallic coatings can be applied by electrochemical deposition only on the metallic (electrically conducting) substrates. However, metallic coatings by chemical deposition can be applied even on nonconducting substrates. Chemical conversion type of coatings are the nonmetallic coatings produced by transformation of the outer atomic layers of a metal surface. Complex treatments are used to produce intermetallic compound coatings on ferrous and nonferrous metals. In the spark-hardening process, a thin and hard skin integral with the substrate is generated by the action of electric sparks struck between the surface and a scanning electrode.

Table 10.1 presents process parameters of various deposition techniques including the coating materials, substrates, deposition temperature, pretreatments, and posttreatments. Table 10.2 presents advantages and disadvantages of various deposition techniques.

10.1 ATOMIZED LIQUID SPRAY PROCESSES

Liquid spray processes are most economical and are widely used for paint application and application of organic and inorganic solid lubricants. Several types of

TABLE 10.1 Process Parameters of Miscellaneous Deposition Techniques

Deposition techniques	Type of coating materials	Type of substrates	Deposition/ treatment temp., °C	Thickness range, μm	Rate of deposition, kg h^{-1}	Pretreatment	Posttreatment
Compressed air/airless spray; dip; spin-on; brush; pad, and roller	Soft solid lubricants, e.g., PTFE, poly(amide-imide), PPS, MoS_2, graphite, CdO-graphite-Ag	Almost any	25–200	7.5–50	0.5–1.0	Grit blasting/ etch	Burnish
Electrostatic spray	Thermoplastics, thermosetting polymers	Conducting substrate	25–150	7.5–50		Grit blasting/ etch	Burnish
Spray pyrolysis	SnO_2, In_2O_3, CdS, PbS	Any material that can withstand deposition temperature	300–700	0.3–10		Chemically clean	None
Spray fusion	CaF_2-BaF_2-Ag, PbO-SiO_2-Ag	Metals and alloys	900–1000 (nonoxidizing atmos.)	7.5–50		Grit blasting/ etch	Lap
Hot dipping	Metals, e.g., Zn, Al, Al-Zn, Sn, Pb-Sn	Steels and cast irons	260–450	20–500		Grit blasting/ etch	Lap
Fluidized bed	Polymers, e.g., polyethylene, polyamide, and fluoropolymers	Almost any	90–300	20–500		Grit blasting/ etch	Burnish
Sol gel	Ceramics, e.g., ceria, silica, zirconia, titania, thoria, alumina, and gel glass	Metals and alloys	25–850	0.3–10		Grit blasting/ etch	Burnish
Electrochemical deposition	Metals, e.g., Au, Ag, Rh, Pb, Sn, In, Cu, Zn, Cd, Ni, Cr; alloys, e.g., Cu-Zn, Cu-Sn, Pb-Sn, Sn-Ni, Au-Mo, Fe-Ni, Cr-Mo; composites, e.g., Ni, Co, and Cr with Al_2O_3, SiC, WC	Conducting materials	25–100	0.1–5000	0.1–0.5	Chemically clean	None

Chemical deposition						
Chemical reduction	Ag, Au, Pb, Ni, Cu	Any material	Room temp. to 60°C	0.1–50	Chemically clean	None
Electroless	Ag, Au, Pd, Ni-P, Ni-B, Cu, Co-P, Ni-Co-P, Ni-P/SiC, Ni-P/diamond, Ni-P/PTFE	Materials catalytic to chemical reduction	Room temp. to 100°C	0.1–50	Chemically clean	None
Chemical conversion	Phosphate	Iron, steel, galvanized steel, Al	80–100	3–50	Chemically clean	None
	Chromate	Al, Mg, and Zn: electrodeposited coatings of Zn, Cd, Ag, Cu, Sn, and brass	25	5–200 (0.1–3.5 g m^{-2}):	Chemically clean	None
	Oxide	Ferrous metals and others	100–600	0.01–0.5	Chemically clean	None
	Anodizing	Al, Zr, Ta	0–80	5–100	Chemically clean	None
	Sulfide	Steels with up to 13% Cr	190	5–10	Chemically clean	None
	Metalliding (most metals)	Most metals	500–1200	5–1000	Chemically clean	None
Intermetallic compound	Sn, Cu-Sn, Sn-Cd, Sn-Sb-Cd, Cu-In-Zn	Iron, steel, brass, bronze, aluminum bronze, aluminum alloys	120–600	—	Chemically clean	None
Spark hardening	Carbide	High-speed steel, Ni, Ti	~1000	—	Chemically clean	Lap

TABLE 10.2 Advantages, Disadvantages, and Applications of Miscellaneous Deposition Techniques

Deposition technique	Advantages	Disadvantages
Atomized liquid spray		
Compressed air/airless spray	Fast, adaptable to varied shapes and sizes, simple and low-cost equipment.	Difficult to obtain complete coating on complex parts and to obtain uniform thickness and reproducible coverage.
Electrostatic spray	Highly efficient coverage and use of organic coating on complex parts.	High equipment cost, requires specially formulated coating.
Spray pyrolysis	Simple and inexpensive setup, easy control of deposition rate and doping level, composition and/or sequential deposition of different compounds yield gradient composition.	The coating material must be available in the form of soluble compounds; difficult to obtain uniform thickness and reproducible characteristics; substrate on heating may warp or deform.
Dip	Fast, simple, inexpensive, thorough coverage even on complex parts such as tubes.	Viscosity and pot life of dip must be monitored, speed of withdrawal must be regulated for consistent coating thickness.
Fluidized bed	Relative thick coating even in sharp edges, solventless operation with less cost.	Poor process control, limited to powders which can be levitated by air, requires preheating of part to above fusion temperature of coating.
Spin-on	Fast, simple, and inexpensive.	Can be used only to coat circular substrates, thickness nonuniformity can be a problem.
Brush	Fast, simple, and inexpensive setup.	Poor control on thickness, high labor cost.
Roller	High-speed, continuous process with excellent thickness control.	High equipment cost and setup time.

Sol gel	Simple and inexpensive, can be applied to complex parts.	Limited to few coating materials, poor adhesion.
Electrochemical deposition	Simple, very inexpensive, good control of thickness and uniformity, metals can be easily alloyed with other metals and ceramics.	Limited materials can be coated, availability of suitable chemicals, coating sometimes porous and nonadherent.
Chemical deposition		
Chemical reduction	Simple, inexpensive apparatus; metal can be deposited onto insulators.	Limited metals can be deposited.
Electroless	Simple, fast, inexpensive; metal can be deposited on insulators and alloyed or codeposited with ceramics easily.	Limited metals can be deposited, substrate must be catalytic to chemical reduction reactions.
Chemical conversion	Simple, inexpensive process, good throughput.	Limited metallic substrates can be coated; availability of suitable chemical compounds.
Intermetallic compound	Increased adhesion and wear resistance.	Proprietary process.
Spark hardening	Simple, inexpensive process.	Hardened surface is normally rough and requires postfinishing; less process control.

liquid spray applications are available: compressed air, hot, airless, and electrostatic spray (Anonymous, 1976; Bhushan, 1987; Campbell et al., 1966; Campbell, 1972; Clauss, 1972; Fish, 1982; Gross, 1970; Licari, 1970; Nylen and Sunderland, 1965; Paul, 1985). For completeness, we note that the thermal spray technique is also classified as a liquid spray technique because the coating material is molten when it hits the substrate. (See Chap. 8.) All the spray techniques discussed in this section atomize the fluid into tiny droplets and propel them to the substrate. There is usually a significant waste of material due to overspraying. Electrostatic spraying is effective in reducing these losses when metal is coated. The part to be coated must be conductive and connected to ground. The electrostatic sprayer imparts a positive charge on each coating droplet. The difference in charge and its mutual attraction causes about 90 percent of coating particles to flow to the substrate. After spraying the coating is air-cured or heat-cured and burnished.

Spray pyrolysis and spray fusion are the other two techniques in which substrate heating during or after spraying results in an improved (chemical) bond between the coating and the substrate. In the spray pyrolysis process, the coating suspension is sprayed onto heated substrate (typically 300 to 700°C) where chemical compounds undergo pyrolytic (endothermic) decomposition and form the polycrystalline coating. In the spray fusion process, the coating suspension is applied onto metal substrate. Then the coated sample is fired at elevated temperature, so that the coating is molten and fusion occurs. Spray pyrolysis is rarely used in tribological applications.

Various spray guns are manufactured by companies such as Binks Manufacturing Co., Chicago, Illinois; Ransburg Electro-Coating Corp., Indianapolis, Indiana; and Paasche Airbrush Co., Chicago, Illinois.

10.1.1 Air Spray

This process is based on the principle of atomizing the liquid containing coating powder particles (a few micrometers in size) into a fine mist, which is then directed to the substrate to be coated. Atomization is accomplished by the use of compressed air or pressurized volatile solvents. Figure 10.1 shows the basic mode of droplet formation in a typical air-atomizing nozzle. The liquid is directed into a free stream by striking it with a high-speed jet of compressed air. The speed, or energy content, of the airstream is partially transferred to the relatively slow-moving liquid. If this transfer of energy is sufficiently great, the liquid is accelerated to a speed above its critical point and atomization will result. There are two regions in the spray. Region A is in front of the nozzle, where the liquid is being lifted off the tip and accelerated into the main air cone turbulence–vortex region. In region B, the droplets are formed by the turbulence of the airstream. The vortex of the expanding helical shape of the rest of the aerosol envelope is seen at the confluence of regions A and B (Gross, 1970; Licari, 1970).

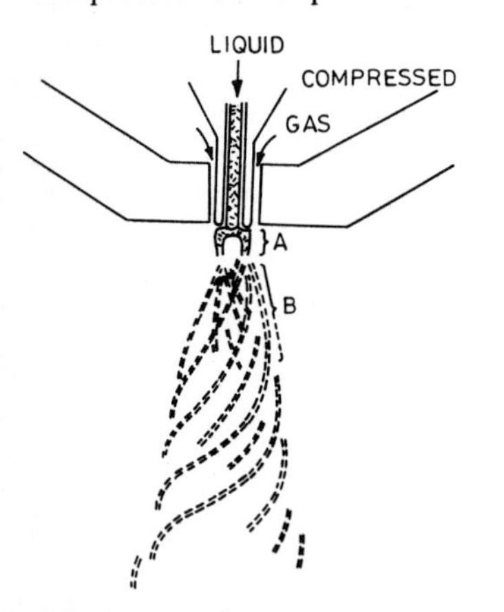

FIGURE 10.1 Basic mode of droplet formation in a typical air-atomizing nozzle.

In air-atomized spraying, only a small

amount of air at the nozzle is used for atomizing, and the remainder of the air pushes the coating material and controls the pattern and droplet size. Too high an atomizing air pressure will result in excessive vaporization in the spray cone; conversely, too low a pressure will result in a large droplet size with consequent runs or sags on the surface.

Aerosol spraying is an extension of compressed-air spraying where instead of a stream of air, suitable propellants are used. The aerosol containers hold coating powder and a liquefied propellant gas or gases, such as fluorinated hydrocarbons (arctons) or butane and butane-propane mixtures retarded with a proportion of fluorinated hydrocarbon. When these are sealed into the container, the system separates into a liquid and a vapor phase and soon reaches equilibrium. When the valve is released, the vapor pressure is sufficient to expel the liquid phase from the spray orifice into air at atmospheric pressure, which immediately vaporizes, leaving an aerosol of the coating.

Figure 10.2 shows a schematic of a conventional air spray gun. The mixing of air and liquid occurs outside the spray gun between the "horns" of the air nozzle. The needle valve (in conjunction with the fluid nozzle) acts as a stop/start valve for liquid flow through the spray gun. The air valve controls the airflow rate to

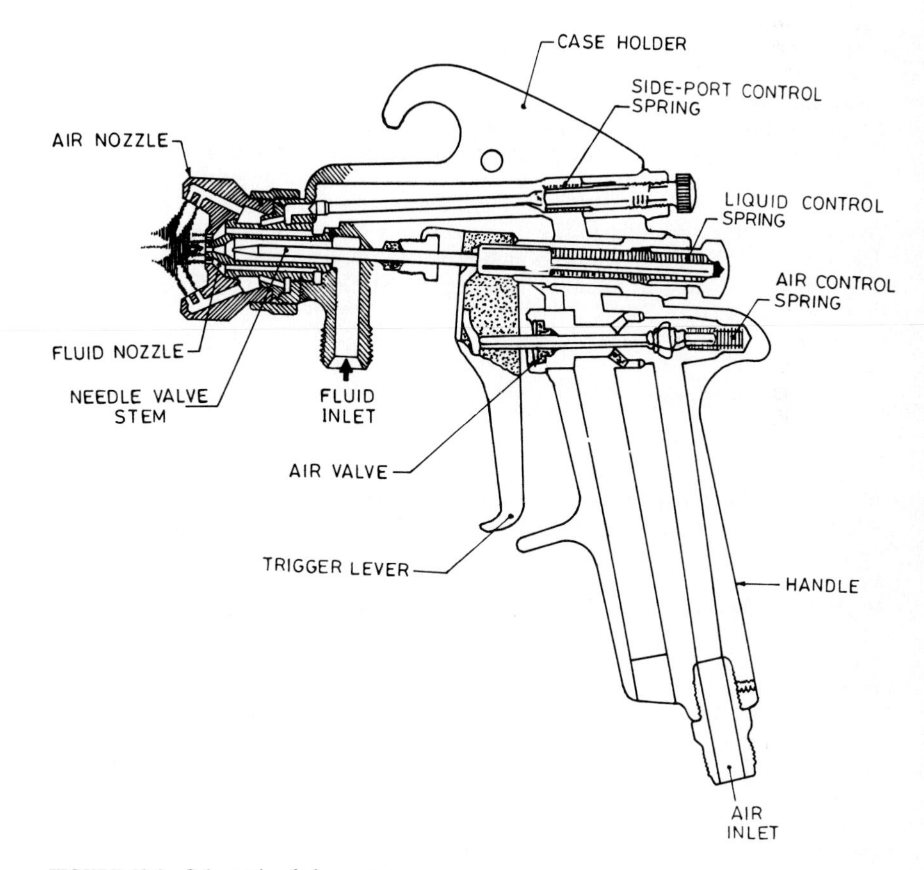

FIGURE 10.2 Schematic of air spray gun.

the air nozzle. The trigger lever operates and controls the side port control spring and liquid control spring. The side port control regulates the spray pattern width by controlling the air supply to the "horns" of the air nozzle. For maximum efficiency of atomization, the coating material volume should be controlled by adjusting the liquid pressure through the liquid control spring and changing the size of the liquid nozzle orifice.

Air spray is a very fast, simple, and inexpensive technique for applying solid-lubricant coatings (Bhushan, 1980, 1982, 1987; Bhushan et al., 1978; Clauss, 1972). The aerosol spray can is a very convenient way to package and dispense coatings, and almost every conventional varnish, enamel, paint, or one-component coating has been successfully packaged in an aerosol can. Aerosol is most economical for touch-up coatings and for coatings over small areas of components. The aerosol cans are commercially available for applying several solid-lubricant coatings such as PTFE, poly(amide-imide), polyphenylene sulfide (PPS), graphite, and MoS_2 and their combinations. These materials are sprayed, baked, and burnished to obtain a coating which can provide solid lubrication under moderate operating conditions.

10.1.2 Hot Spray

In this process (also called *low-pressure hot spray*) the air-atomized spray gun is used in conjunction with a heat exchanger to heat the coating material up to a predetermined temperature, which usually ranges from 60 to 80°C (Anonymous, 1976). This process utilizes heat to lower the viscosity of the coating material to the optimum range for spraying without solvents, which allows the material to be fully atomized at lower air pressures. Waste of coating due to rebound or overspray is thus minimized. The lower viscosity also allows application of coatings with higher solids contents which provides heavier coats in one application and thus saves time. The main disadvantage is that the heating unit increases capital investment.

10.1.3 Airless Spray

A main disadvantage of conventional air spray is the tendency for the blown cloud of atomized coating material to be swirled out of recesses, holes, and corners by the accompanying airstream, leaving these areas lightly coated to bare. Atomization by the airless method is achieved by forcing material at high pressure through an accurately designed small orifice in the spray cup. The combination of high velocity and the rapid expansion of the coating material, upon passing through the fine orifice into the atmosphere, causes the material to break up or atomize into a spray of fine particles. The pressures applied to the material are provided by a fluid pump with a pressure ratio of up to 30:1, thus giving final fluid pressures of up to 20 MPa. The pressurized coating material is fed to the spray gun via a spirally wound stainless-steel-reinforced PTFE or Nylon hose. The coating material, which is circulated continuously in the storage system, may be heated to reduce its viscosity and to volatilize solvents more efficiently (Gross, 1970; Muirhead, 1974).

The airless spray gun (Fig. 10.3) is similar in appearance to the conventional gun except that it has only a liquid inlet and no atomizing air inlet. The important components of the spray gun are the fluid tip and spray cap. These caps are spe-

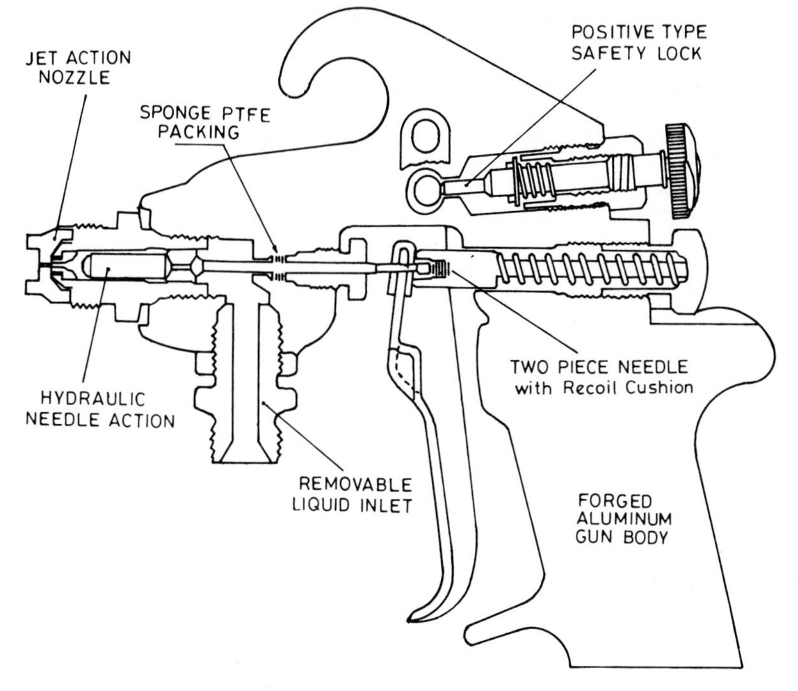

FIGURE 10.3 Schematic of airless spray gun.

cially shaped to give varying fan patterns. The fan pattern is predetermined by the angle at which the spray cap slots are machined in the respective caps.

Unlike the process of conventional spraying, there is no carrier or airstream to convey the atomized particles to the work surface after they leave the spray gun. The atomized particles of material have sufficient forward velocity, created by the high pressure exerted on the material, and once they are released from the tip of the spray cap, rapid expansion occurs. In addition to this hydraulic mechanism, atomization in airless spraying equipment is assisted by two other factors: the effect of the resistance of the atmospheric air surrounding the spray cap to the high-speed jet of material and the rapid evaporation of the volatile solvents present in the coating material.

The material is applied much wetter than in conventional spraying since the solvents are not subjected to the same degree of vaporization as in the conventional spray process, and therefore the coating material must be modified by using faster evaporating solvents than with conventional spraying. The airless spray process is highly efficient, especially for coating parts with many recesses, cavities, or projections. The main disadvantage of this process is that, unlike air spray guns, airless spray guns cannot be throttled.

10.1.4 Electrostatic Spray

In an electrostatic spray process, charged atomized particles (~30 to 120 μm) of organic material in a solvent are sprayed and attracted to the grounded substrate.

The coating material can be atomized by conventional air or airless spray techniques. The solvent used is evaporated to form a solid coating. Subsequently, the coating may be cured or heat-fused to the substrate (Gross, 1970; Licari, 1970; Spiller, 1974). The electrostatic spraying can also be used to deposit organic powders. In this case, no atomization stage is required; subsequent processing (heat-fused at 90 to 220°C for a few minutes) consolidates the powder into a homogeneous layer (Kut, 1974).

The electrostatic spraying of organic materials in liquid form is done by using one of two basic systems. In one, the electric charge is applied to coating particles (in solution) already atomized by either air spray or airless methods. In the other process, atomization occurs under the influence of an electrostatic field as the liquid flows from the edge of a spinning ball, forming a spray pattern of electrically charged particles (Spiller, 1974).

Figure 10.4 shows a schematic of an electrostatic air spray gun. Air is used to propel the coating powder from a reservoir to the gun. The powder particles are rendered electrostatic as they are propelled through a diffuser on the front of the spray gun and are then rapidly attracted to the substrate. The powder is emitted from the nozzle by a rotating diffuser driven by a small air motor located in the handle. Ionizing voltages range from 30 to 90 kV.

In all these systems, electric charges are conveyed by the coating particles to the substrate being sprayed. If the substrate is conducting (metals) and connected to the earth, the charge carried by the coating particles is dissipated to ground. However, if the sprayed substrate is nonconductive, the charge accumulates; and this charge accumulation on the substrate lowers the voltage drop between the atomizing device and the surface to be coated. This causes less effective deposition in the first method and less effective atomization and deposition in the second. Spiller (1974) has suggested several methods of physically modifying a nonconducting surface to obtain adequate conductivity for electrostatic spraying. These employ chemically absorbed gel layers, surface sulfonation, and photoconductive pigments.

This process is very amenable to thermoplastic polymer coatings, but some thermosetting types, such as epoxies, are also being applied successfully. The ad-

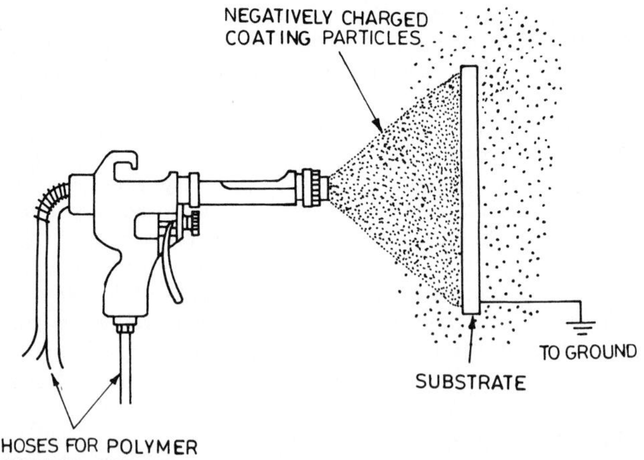

FIGURE 10.4 Schematic of electrostatic air spray gun.

vantages of the electrostatic spray process over other spray processes include high deposition efficiency, with very little material lost as overspray, and the high coating buildup that can be obtained in one application, reducing both time and cost. The main disadvantage of this process that the electrostatic attraction of the substrate for the coating material draws the material to the nearest edge or surface of the substrate, and it is difficult or impossible to get the coating material into deep recesses, corners, and shielded areas.

10.1.5 Spray Pyrolysis

The spray pyrolysis process, also known as *chemical spray*, is based on a thermally stimulated chemical reaction between a cluster of liquid or vapor atoms of different chemical species at the substrate surface (Chamberlin, 1966; Chamberlin and Skarman, 1966; Chopra et al., 1982; Feigelson et al., 1977). This process involves spraying a solution, usually aqueous, containing soluble salts of the constituent atoms of the required coating material onto a heated substrate (typically in the temperature range of 300 to 700°C). Every sprayed droplet reaching the hot substrate surface undergoes pyrolytic (endothermic) decomposition and forms a cluster of crystallites of the product. The other volatile by-products and the excess solvent escape in the vapor phase. The substrate provides the thermal energy for the thermal decomposition and subsequent recombination of the constituent species, followed by sintering and recrystallization of the clusters of crystallites. The result is a coherent coating chemically bonded to the substrate. The difference between spray pyrolysis and *chemical vapor deposition* (CVD) is that source material is not transported as a vapor, as in CVD, but as a solution in droplet form.

The chemical solution is atomized into a spray of fine droplets by a spray nozzle with the help of a carrier gas, which may (as in the case of SnO_x film) or may not (as in CdS film) play an active role in the pyrolytic reaction involved. The solvent liquid serves to carry the reactants and distribute them uniformly over the substrate area during the spray process. In most cases, the carrier liquid takes part in the pyrolytic process (Chopra et al., 1982).

For example, chemical reactions involved to obtain CdS coatings from a dilute aqueous solution of a water-soluble cadmium salt and a sulfoorganic salt (thiourea) are as follows:

$$CdCl_2 + (NH_2)_2CS + 2H_2O \rightarrow CdS\downarrow + 2NH_4Cl\uparrow + CO_2\uparrow \quad (10.1)$$

And to obtain optical-quality SnO_2 coatings from an aqueous chloride solution,

$$SnCl_4 + 2H_2O \rightarrow SnO_2\downarrow + 4HCl\uparrow \quad (10.2)$$

Figure 10.5 shows the schematic of spray pyrolitic systems using pneumatic (air) atomization and ultrasonic atomization. In pneumatic atomization, the carrier gas, which may be air or a reactive gas such as N_2, and the solution are fed into a spray nozzle at a constant pressure and flow rate [see Fig. 10.5(*a*)]. The droplet size and spray pattern mainly depend on the geometry of the spray nozzles. Large-area uniform coverage of the substrate may be obtained by employing several mechanical or electromechanical arrangements for scanning either the spray head or the substrate or both. The schematic for ultrasonic atomization is shown in Fig. 10.5(*b*), where the size of droplets is related to the solution con-

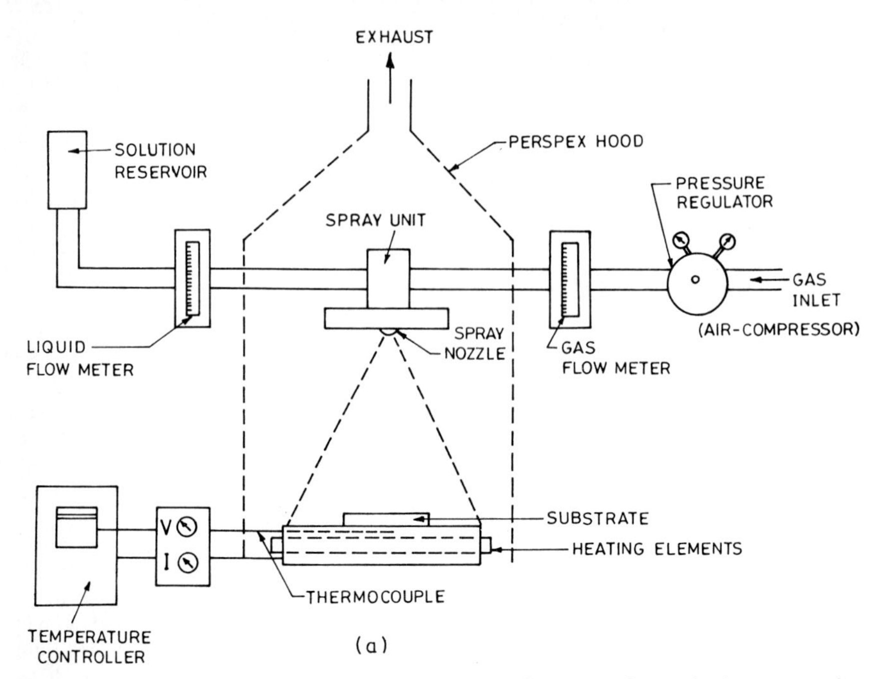

FIGURE 10.5 Schematic of spray pyrolytic system using (*a*) pneumatic atomization spray and (*b*) ultrasonic atomization spray.

centration and ultrasonic frequency. Usually ultrasonic atomization produces smaller droplets, and aerosol is independent of the gas flow.

The range of coatings that can be deposited by spray pyrolysis is smaller than that of CVD, because of the lower limit of substrate heating used in the spray pyrolysis. But this method offers several advantages over other processes:

- The source material or coating material is not heated.
- No early decomposition takes place until the aerosol approaches the substrate.
- Deposition of mixed compound and doped coatings can be obtained just by adjusting the solution composition.

10.1.6 Spray Fusion

In the spray fusion process, also known as the *fusion process*, inorganic coating is prepared by fusion of the coating powders to the metal substrate for high adhesion of the coating to the substrate. The fusion process is similar to that employed in applying porcelain enamel to metal surfaces. The liquid suspension of the coating mixture is sprayed on a metal surface. The coated sample is fired at elevated temperatures (typically 900 to 1000°C), so that the coating melts and fusion occurs. Oxides of the base metal which form during the firing process have an important effect on the bond strength and other properties of the coatings.

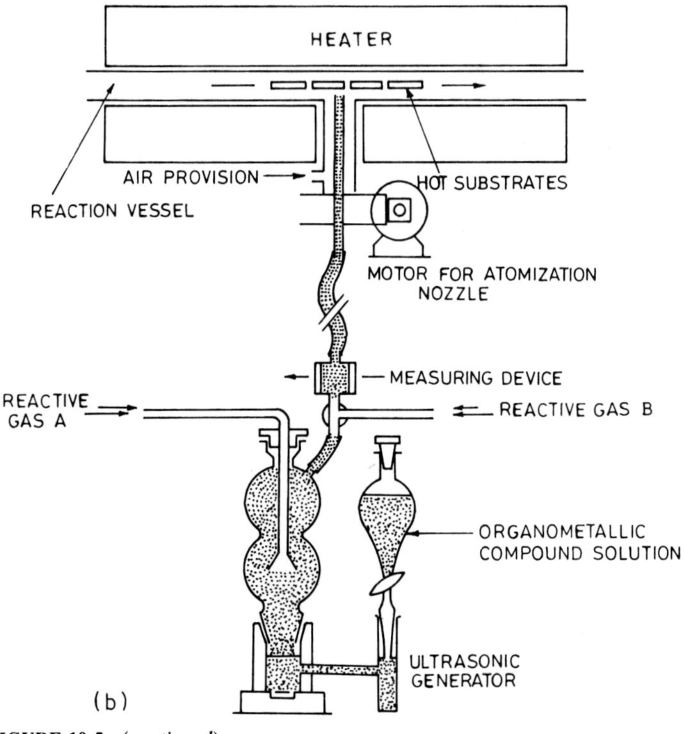

FIGURE 10.5 *(continued)*

10.1.7 Coatings Deposited by Spray Processes

Bonded coatings of many organic and inorganic solid lubricants are successfully applied by spray processes for friction and wear applications. The solid-lubricant and organic or inorganic binder particles are dissolved or dispersed by mechanical agitation in a solvent or water with a dispersant. The liquid suspension is sprayed onto a substrate, air- or heat-cured, and then burnished. The coating is usually sprayed on a preroughened surface to increase adhesion by mechanical interlocking. The coating thickness typically ranges from 7.5 to 50 μm. Several organic coatings [which are blends of primarily PTFE, poly(amide-imide), and/or polyphenylene sulfide] and several inorganic coatings (which are blends of primarily MoS_2 and graphite) with resin (organic) or inorganic binder and solvent can be purchased in an aerosol can. For more details, see Chap. 13.

Bhushan et al. (1978) and Bhushan (1980, 1982) developed air-sprayed CdO-graphite and CdO-graphite-Ag coatings with sodium silicate binder in a water carrier for foil journal bearing applications. High purity (99.9 percent pure) and fine powder (average particle size of about 2 μm) of synthetic graphite, commercially pure (99 percent pure) CdO with 95 percent of the particle sizes finer than 200-mesh (95–200), and high purity (99.99 percent pure) and fine powder (95–230) of Ag were mixed in distilled water and ball-milled for 4 h. Just before spraying, sodium silicate (also known as water glass—8.9 percent Na_2O, 28.7 percent SiO_2, and the balance water) as a binder and a drop of wetting agent (Absol 895—NPX

Industries) were added. The liquid suspension was air-sprayed to a thickness of about 25 μm onto a carefully prepared substrate. (The solid contents were 15 percent CdO, 45 percent graphite, 10 percent Ag, and 30 percent sodium silicate.) The substrate was preroughened for increased coating adhesion.

The coating was air-dried at room temperature for 30 min, then baked at 65°C for 2 h and next at 150°C for 8 h, and finally burnished to 8- to 10-μm thickness with a polishing paper or by rolling a dowel pin under pressure over the coated surface. The surface microstructure shows excellent dispersion of CdO, graphite, and Ag with the sodium silicate (see Figure 10.6). These coatings provided good bearing performance up to 430°C for a foil air bearing used in a gas-turbine engine.

Thermoplastic powders such as polyamide, polyethylene, cellulose, polyethers, polyvinyl chloride and polyacrylate, and thermosetting epoxy powders have been economically applied by electrostatic spray process. The coating typically ranges from 50 to 100 μm. These coatings are used for decorative, anticorrosive, and chemical-resistance purposes for a wide variety of industrial and consumer articles. Glass articles are coated with epoxy powders for decorative purposes. Steel furniture, in particular tubular articles and wireworks, such as chairs, lend themselves to electrostatic wraparound. Steel structures and cables used in aggressive atmospheres (marine, chemicals, processing plant) are coated by electrostatic spray.

The inorganic coatings that have been applied by spray pyrolysis of aqueous metal salt solutions and organic compounds are

- *Sulfides and selenides:* CdS, Cu_2S, PbS, CdSe, CdZnS, CuS, ZnSe, PbSe, Sb_2S_3
- *Oxides:* SnO_2, In_2O_3, ZnO, Fe_2O_3, Co_2O_3, Al_2O_3, MoO_3, TiO_2

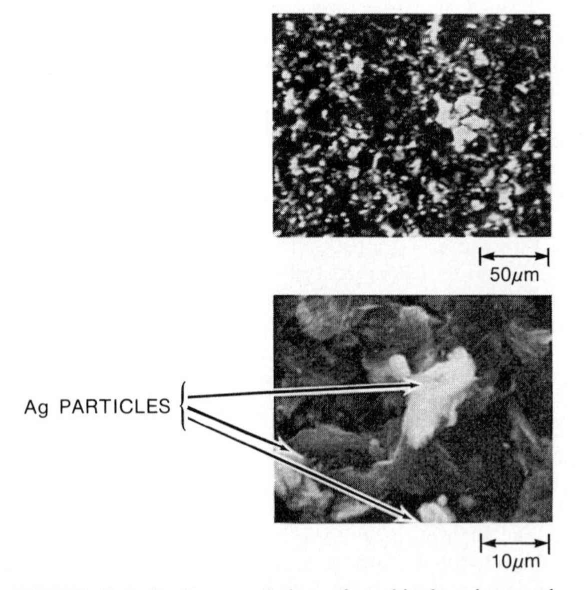

FIGURE 10.6 Surface morphology of graphite-based sprayed coatings of CdO-graphite-Ag. [*Adapted from Bhushan (1980).*]

Most of the above coatings have been prepared for solar cell applications (Feigelson et al., 1977; Gupta and Agnihotri, 1977; Gupta et al., 1978; Manifacier et al., 1979; Shanti et al., 1980a, 1980b), transparent heat mirror (Haacke, 1977; Hahn and Seraphin, 1978; Lampkin, 1979), and antireflection optical coatings (Hovel, 1978) etc. However, some coatings such as SnO_2 and In_2O_3 provide good resistance to scratches on window glasses and glassware and good resistance to erosion on aircraft glass windows.

Fused coatings based on CaF_2-BaF_2 eutectic and PbO-SiO_2 for substrates of nickel-based alloys and stainless steels have been developed by Bhushan et al. (1978), Bhushan (1980), Bhushan and Gray (1980), Sliney (1962, 1982), and Sliney and Johnson (1957) for low friction and wear at high environmental temperatures (260 to 650°C) and high speeds, which produce high surface temperatures. Ag powder was added to improve the room-temperature properties. Fluoride-based coating powder with a composition of 58.5 percent CaF_2-BaF_2 eutectic (31 percent CaF_2 and 69 percent BaF_2), 35 percent Ag, and 6.5 percent CaO-SiO_2 binder was mixed in a thin lacquer and was air-sprayed on stainless-steel (A-286) and nickel-based alloy (Inconel X-750) substrates with a coating thickness of about 10 μm. After baking in a vacuum oven overnight at 225°C to remove volatiles, the coatings were fused in an argon furnace at 995°C for 20 min. After 20 min of cooling in argon, the coatings were hand-burnished. Coating powder based on PbO-SiO_2 with a composition of 61.5 percent PbO, 3.5 percent SiO_2, 25 percent Ag, and 10 percent Fe_3O_4 was mixed with water, ball-milled for about 1 h, and air-sprayed after a dispersant was added to the liquid suspension. The sprayed coating was heated at 90°C to remove water and then fired at 900°C for 1 min after it melted.

10.2 DIP COATINGS

As the name implies, in a dip coating or hot dip coating (also called *hot dipping*) process the substrate is literally dipped (or immersed) into a liquid bath. After immersion, the substrate is withdrawn, and the excess coating material is removed by centrifugation or simply by gravity runoff (with occasional rotation to ensure removal from recesses). With some systems, *hot dipping* may be used to ensure greater wettability and to minimize pinholes. This variation of dip coating consists in preheating the part to an elevated temperature (about 75°C) and dipping it in a room-temperature bath. The thickness, uniformity, and adhesion of the dipped coating depend on the viscosity of the bath, rate of immersion and withdrawal, temperature of the bath, and number of dips. In the case of solvent-containing dip coatings, thicknesses of few micrometers to about 50 μm may be obtained. The thickness can, of course, be increased by multiple dipping (Gittings, 1982; Licari, 1970; Maykuth, 1982; Paul, 1985; Tonini et al., 1982).

Most coatings which can be applied by the spray process can also be applied by the dip process or by brushing, to be discussed later. The dip process is desirable for achieving a continuous automated manufacturing operation, and it has the advantage over spray processes of ensuring coverage of parts with complex shapes and inaccessible recesses. The dip process is also used in coating of large parts.

Solid-lubricant coatings applied by the dip process include PTFE, epoxy, silicone, and Nylon-based coatings. The bath consists of a mixture of pigments, resin, and solvents at ambient temperature for application of solid-lubricant coat-

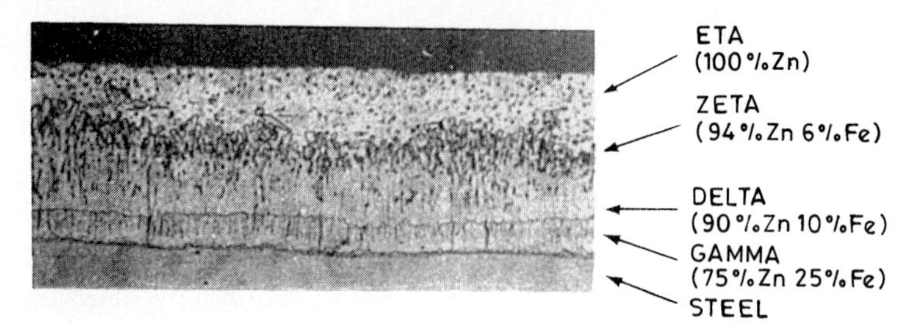

FIGURE 10.7 Cross-sectional view of hot dipped Zn coating on steel. [*Adapted from Tonini et al. (1982).*]

ings. The solid-lubricant coatings are usually dried and baked after withdrawal from the bath.

The metallic coatings applied by the hot dipping process include Zn, Al, Al-Zn, Sn, and Pb-Sn (babbitts) onto steel and cast-iron substrates (Gittings, 1982; Maykuth, 1982; Tonini et al., 1982). Because Pb alone does not alloy with iron, Sn, or other elements such as Sb (2 to 25 percent) must be added to form on the surface of the steel an Fe-Sn alloy to which the Pb will adhere. The chemical composition of irons and steels and even the form in which certain elements are present, such as carbon and silicon, determine the suitability of ferrous metals for hot dipped coatings. For example, steels that contain less than 0.25 percent C, 0.05 percent P, 1.3 percent Mn, and 0.05 percent Si individually or in combinations are generally suitable for hot dipped Zn coatings (known as *galvanizing*). Low-carbon steels containing less than 0.2 percent C and cast irons containing 3.5 percent C, 1.7 to 2.7 percent Si, 0.5 to 0.8 percent Mn, and up to 1.3 percent P are generally suitable for Sn coatings. These substrates must be properly cleaned/etched for better bonding. The hot dipped coatings are bonded to substrate through a metallurgical bond. Figure 10.7 illustrates how molten zinc is interlocked into the steel by the alloy reaction which forms a Zn-Fe alloy layer of varying compositions. Coatings closest to the basis metal are composed of Fe-Zn compounds; these, in turn, are covered by an outer layer consisting almost entirely of Zn (Tonini et al., 1982).

Hot dipped coatings of Zn (galvanizing), Al-Zn, and Al have been used to protect steel and iron parts against corrosion of the basis metal. Some major applications include structural steel for power generating plants and petrochemical facilities, automobile components, bridge structural members, highway guardrails, and marine pilings and rails. The hot dipped coating of tin is applied to iron and steel to provide protective coating for food handling, to facilitate the soldering of a variety of components used in electronic equipment, and to assist in bonding another metal to the basis metal. Hot dipped coatings of the Pb-Sn alloys, conventionally known as *babbitts* but also known as *terne* coatings, being soft, have excellent antifrictional properties and prevent galling and/or scoring of the shaft over a longer period of use. The Pb-Sn coatings also provide excellent solderability and good corrosion resistance in various environments and corrosive media.

10.3 FLUIDIZED-BED COATINGS

A unique variation of dip coating for polymer resins is the fluidized-bed process, sometimes called the *aerated-bed process*. In this method, solid polymer powder

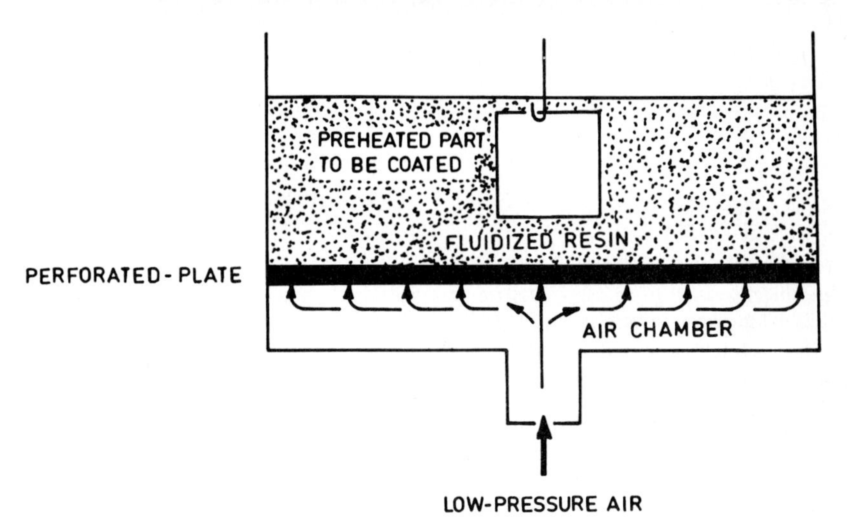

FIGURE 10.8 Schematic of fluidized-bed coating process.

is levitated in a chamber by the use of air passing through a perforated plate in the bottom of the chamber. So levitated, the particles have a much lower bulk density than the quiescent powder, and the "fluidized" powder actually appears to be gently boiling (see Figure 10.8). Coating is accomplished by preheating a solid substrate to a temperature higher than the melting point of the polymer powder and immersing it into the chamber. As the powder strikes the hot substrate, it melts and coalesces into a continuous coating on the surface. As the substrate cools, the polymer solidifies. If the polymer is reactive, such as an epoxy, it may need additional oven curing to complete the cross-linking. Clearly this method can be used only on materials that can be heated to a temperature typically greater than 90°C.

Several distinct advantages of fluidized-bed (or fluid) coating are that (1) relatively thick coatings can be applied, even over sharp edges, (2) it is a solventless operation that has cost and environmental advantages, and (3) polymers such as polyethylene, polyamides, and fluoropolymers that cannot be applied from solutions can be used for coatings.

10.4 SPIN-ON COATING

The spin-on coating process is used to deposit coating on circular substrates. In this technique, a turntable is used that holds one substrate at a time. The coating suspension/solution is metered onto the substrate near its center, and the turntable is accelerated very rapidly to several thousand revolutions per minute which spreads the coating suspension/solution by centrifugal action. The coating is then dried, baked, and burnished, if necessary. A coating ranging in thickness from 50 to 2000 nm can be formed in this way. The coating is deposited nonuniformly at high rotational speeds, and it is thicker near the outer radius of the substrate because of centrifugal effects. The spin-on coating processes are commonly used in semiconductor industry, e.g., to deposit photosensitive polymers to impart a pattern on to a circular silicon slice in a process called the photopolymer (2P) process, and in the magnetic media industry, the magnetic coating and protective overcoat are applied by a spin-on process (Bhushan, 1990).

10.5 BRUSH, PAD, AND ROLLER COATINGS

Brush, pad, and roller (or roll) coating processes are the mechanical processes commonly used in many industries such as in semiconductor, magnetic media, television, photographic, and paint industries (Bhushan, 1990; Fish, 1982; Paul, 1985). In some industrial applications, these may be the methods of choice. Brushing is one of the oldest methods used for the application of coatings such as paints [see Fig. 10.9(*a*)]. The quality of the coating depends on the quality of a brush which is controlled by the shape, length, and mechanical properties of the bristles and the nature of the bristle tops. Brushing is a slow method and is primarily used where the areas to be coated are small and somewhat uniquely located. Brushing is also used extensively for touch-up and repair work.

Roller coating, a highly mechanized and closely controlled method for coating flat stock, offers a means of applying a fixed amount of liquid coating to one or both surfaces of any sheet or strip requiring an even coating. In the pad and roller coating process, the liquid coating material is transferred from a reservoir pan onto a pad or transfer roll that ultimately contacts the flat substrate [see Fig. 10.9(*b*)]. Many variations in the process reflect the methods of coating the trans-

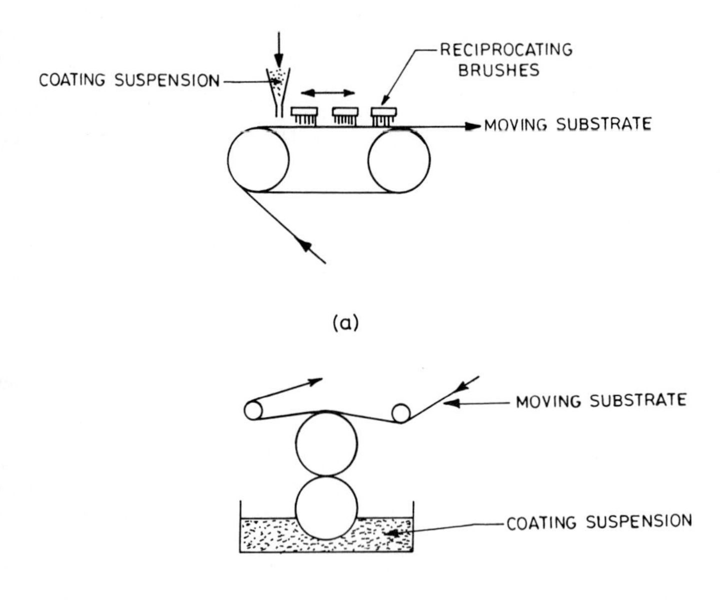

FIGURE 10.9 Schematic of (*a*) reciprocating brush coater and (*b*) reverse roll coater.

fer roll and ensuring that the coating is uniform and of the correct thickness. After the coating operation, the material passes through a zoned drying and curing oven, and the sheets are stacked or the strip is rerolled.

Most coating suspensions which can be applied by spray, dip, and spin-on coating processes can also be applied by the brush, pad, and roller coating processes. The detailed process of applying the coating depends on the type of coating material, substrate material to be coated, and size and configuration of the substrate.

Polymer and other solid-lubricant coatings are sometimes applied by brush, dip, and roller coating processes. For example, graphite coating is applied on the exterior and interior walls of the black-and-white television picture tube by brushing for electrical-contact purposes (DeGraaf, 1974). To apply the coating, the picture tube cone is fitted into a device which holds and rotates the tube. A specially designed, bendable brush is used. A small brush is fitted on a piece of spring steel which is fitted on a rod. By means of a lever and a wire, the spring can be bent, so that the brush will press on the inside of the cone. The brush is dipped into a flat basin containing the aqueous graphite dispersion. Then the brush is used to coat the rotating tube.

The roller coating process is used to apply magnetic coating on polyester substrate for magnetic media (Bhushan, 1990). It is used for applying photoresist coatings to copper-laminate stock on which etched circuitry is to be developed and for applying thin films on metal foil for use in capacitors. Roller coating with a lithographic plate is also a universally used method for decorating the metal stock for products such as cans, bottle caps, toys, and automobile instrument panels.

10.6 SOL-GEL COATING

Essentially, the sol-gel technique is based on the use of sols (colloidal dispersions) of a stable dispersion in a liquid (normally aqueous but organosols can be used) of colloidal units of hydrous oxide or hydroxides, ranging in size between 2 nm and 1 μm, for the production of oxide ceramic coating. On drying, i.e., removing the water between the colloidal units, the sol is transformed to a gel (solid). This process is normally reversible, and the gel can be reconverted to the sol by the mere addition of water. However, on firing at temperatures above 150°C the gel becomes consolidated as a ceramic, whose density is dependent on the degree of aggregation of the basic colloidal units in the sol. Nonaggregated colloidal units less than 10-nm diameter produce a ceramic layer of almost theoretical density even at modest temperatures ($\leq$850°C), used for wear-resistant application. In contrast, the same oxide derived from aggregated clusters consisting of 200 to 300 colloidal units has a considerably lower density (high porosity) and, as a result, a high effective surface area used for absorption and catalysis. The degree of aggregation is controlled by the aqueous sol (Brinker and Scherer, 1990; Nelson et al., 1981).

All the coatings are laid on a metal surface by a three-stage procedure. The sol is applied by a suitable method (such as dipping, spin coating, or electrophoresis), then dried, and finally fired, typically at up to 850°C, for a short period (on the order of 15 min). For extensive densification and wear resistance, high firing temperatures are used. The most striking advantage of the sol-gel technique is its inherent simplicity, and it can be readily applied even to substrates of complicated geometry.

Dense ceria, silica, zirconia, titania, thoria, alumina, and gel glass coatings have been produced. They are used for corrosion and wear protection. Sol-gel silica and ceria coatings have been evaluated for the fuel-cladding application. Sol-gel SiO_2 is applied by a sol gel method and used as a protective coating for magnetic thin-film rigid disks. In this method, tetrahydroxysilane dissolved in isopropanol alcohol solution (roughly 2% concentration) is spin-coated on the 130-mm-diameter disk surface at roughly 600 r/min and baked at temperatures ranging from 25 to 600°C for roughly 2 to 24 h. Higher temperature is used for increased hardness and wear protection (Bhushan, 1990).

10.7 SCREENING AND LITHOGRAPHY

In screening and lithography processes, the coating is deposited in selected areas. Screening is used to produce features larger than about 100 μm, whereas lithography is used to produce small features as small as 1 μm or less.

10.7.1 Screening

In the screening method, the coating is applied by squeezing the coating material through a silk, Nylon, or stainless-steel screen. To apply coating by screening, the surfaces must be very flat. The thickness of the coating may be varied from a few micrometers to more than a couple of hundred micrometers, depending on the screen mesh size and number of coats applied. To eliminate pinholes and other imperfections, as with other methods, two or three layers of coating are superimposed, with each layer partially or fully cured prior to application of the next one. A number of screenable epoxies, polyurethanes, and other coating types are available. Automated equipment for mass production is available. Screening process can be used to produce line thicknesses of about 100 μm with about 100-μm spacings. They are primarily used for printing applications such as color posters.

10.7.2 Lithography

Microelectronic circuit fabrication requires a method for forming accurate and precise patterns on the semiconductor (almost exclusively single-crystal silicon) substrates. These patterns delineate the areas for subsequent doping and/or internal interconnection. These areas are defined by lithographic processes in which the desired pattern is first defined in a resist layer (usually a polymeric coating which is spin-coated onto the substrate) from a mask and subsequently transferred (via techniques such as etching, ion implantation, and/or diffusion) to the underlying substrate. Lithography is also used to produce micropatterns for various other applications such as conductors, resistors, and capacitors in selected areas for fabrication of film circuits and film magnetic heads (Bhushan, 1990; Thompson et al., 1983).

The basic steps of the lithographic process are outlined in Fig. 10.10. The resist material is applied as a thin coating over some base and subsequently ex-

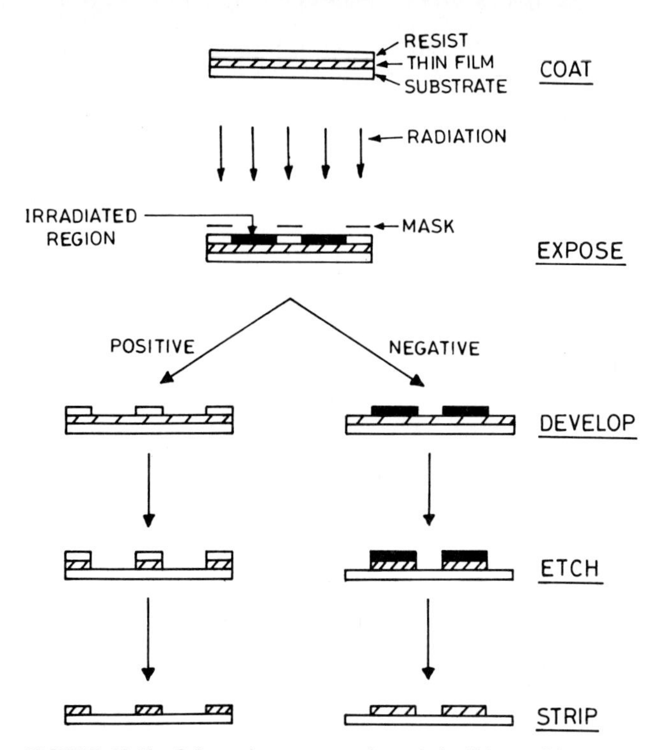

FIGURE 10.10 Schematic representation of the lithographic process sequence using positive and negative resists.

posed in an imagewise fashion (through a mask) such that radiation strikes selected areas of the resist material. The exposed resist is then subjected to a development step. Depending on the chemical nature of the resist material, the exposed areas may be rendered more soluble in some developing solvent than the unexposed areas, thereby producing a positive-tone image of the mask. Conversely, the exposed areas may be rendered less soluble, producing a negative-tone image of the mask. The areas that remain following the imaging and developing processes are used to mask the underlying substrate for subsequent etching or other image transfer steps. (The generic term *resist* evolved as a description of the ability of these materials to resist etchants.) Following the etching process, the resist is removed by stripping to produce a positive- or negative-tone relief image in the underlying substrate.

The method most commonly used is photolithography in which light rays are used with photoactive polymer known as *photoresist*. Typically, ultraviolet light in the wavelength region of 350 to 430 nm is used. Diffraction considerations limit the size of individual elements to about 0.5 to 1 μm, and various lithographic processes have been developed to achieve the dimensions and overlay accuracies required for future generations of devices (with features as small as 0.02 μm) such as short-wavelength photolithography, electron beam lithography, *X*-ray lithography, and ion beam lithography (Thompson et al., 1983).

The mask contains clear and opaque features that define the micropatterns. To produce a mask, a glass plate coated with a metal film (such as evaporated chromium) is spin-coated with a resist layer. The resist layer is exposed to a scanning radiation beam (generally an electron beam) whose motion is controlled by automated methods, to define the desired pattern. This process is followed by developing, etching, and stripping. Screens can be used to produce masks with very large features (several hundred micrometers in size).

A form of lithography is commonly used for offset printing. The lithographic plate (or mask) presents a flat metal surface on which the image appears. The metal areas of the plate are prepared to accept water, while the image areas accept ink and reject water. This demonstrates the principle that oil (which is contained in ink) does not mix with water; and nonimage areas are kept moist by the dampening system on the press while the ink rollers deposit ink on the image.

On the press, the image is transferred from the plate to the rubber blanket, then the blanket transfers to the paper. This indirect way of putting a print on paper is called "offset" printing (Sayre, 1974).

To create the image on the lithographic plate surface, the lithographic metal plate is coated with a resist layer. A film negative or positive is placed on the coated plate and brought into pressurized contact for exposure under ultraviolet lights. The coating is then removed from the nonprinting areas by washing with water or developing solutions, after which finishing solutions may be applied; then the plate is dried.

10.8 ELECTROCHEMICAL DEPOSITION

Electrochemical deposition (also known as *electrolytic deposition, electrodeposition*, or *electroplating*) is defined as the production of metallic coatings on solid surfaces by the action of an electric current. This process has been known for a long time, and there are many standard texts on the subject (Brenner, 1963; Duffy, 1962; Gaida, 1970; Henley, 1982; Kortum, 1965; Lowenheim, 1974; Ollard, 1969; Schwartz, 1982; Vagramyan, and Solev'eva, 1961). Electrodeposition is the most convenient method of applying coatings of metals with high melting points, such as chromium, nickel, copper, iron, silver, gold, and platinum. Electrodeposition of metals from aqueous solutions is mainly limited by the decomposition potentials of the metals to be deposited.

The application of electroplating for decorative purposes and corrosion resistance constitutes the greatest commercial use, finding wide applications in automotive hardware, household articles, furniture, jewelry, etc. Electroplated coatings also have wide applications in modifying the surface characteristics to improve wear resistance. A wide range of electrodeposited metals are used for wear resistance, such as Cu, Pb, In, Sn, and Zn in bearing field and Ag, Au, Rh, Cd, and Ni in electrical-contact applications, and hard chromium plating on piston rings, cylinder liners, cross-head pins, etc., in internal-combustion engines and compressors.

10.8.1 The Electrolysis Process

The electrodeposition or electroplating system consists essentially of an electrolytic bath, a dc power source, and two electrodes. The electrode connected to the positive pole is called the *anode*, and the one connected to the negative pole is the *cathode*. The electrolytic baths contain a conducting solution called the *electrolyte* whose principal component is a salt or other compound of the metals to be deposited. An electric potential is applied to the cell, and metal is deposited on the cathode by electrochemical dissolution from the donor metal (anode).

Figure 10.11 shows the simple electrodeposition system for the deposition of Ni from $NiSO_4$ electrolyte. When dissolved in water, $NiSO_4$ dissociates into positively charged nickel ions (cations) and negatively charged sulfate ions (anions):

$$NiSO_4 \rightleftharpoons Ni^{2+} + SO_4^{2-} \tag{10.3}$$

When the electrodes are connected through a dc power source, the cations, that is, Ni^{2+} ions migrate to the metal cathode, where they are discharged and deposited as metallic nickel

$$Ni^{2+} + 2e^- \rightarrow Ni\ (\text{metal}) \downarrow \tag{10.4}$$

Consequently nickel dissolves from the nickel anode to replace Ni ions in electrolyte, to maintain the electric neutrality

$$Ni \rightarrow Ni^{2+} + 2e^- \tag{10.5}$$

The overall process is known as *electrolysis*.

The electrolytic solution must contain ions for electrical conduction to occur. Usually salts are dissolved in water for the bath. The concentration of salt needed for a particular coating deposition is obtained from recipes tabulated in the handbooks. Generally, the electrolytes are developed by trial and error. A large variety of baths can be formulated for the specific elements to improve the adhesion, crystalline structure, current density, etc. A good electrolyte must deposit a coherent coating with good mechanical and chemical properties. Some electrolytes have sufficient metal-ion concentration such that it is not necessary to have an

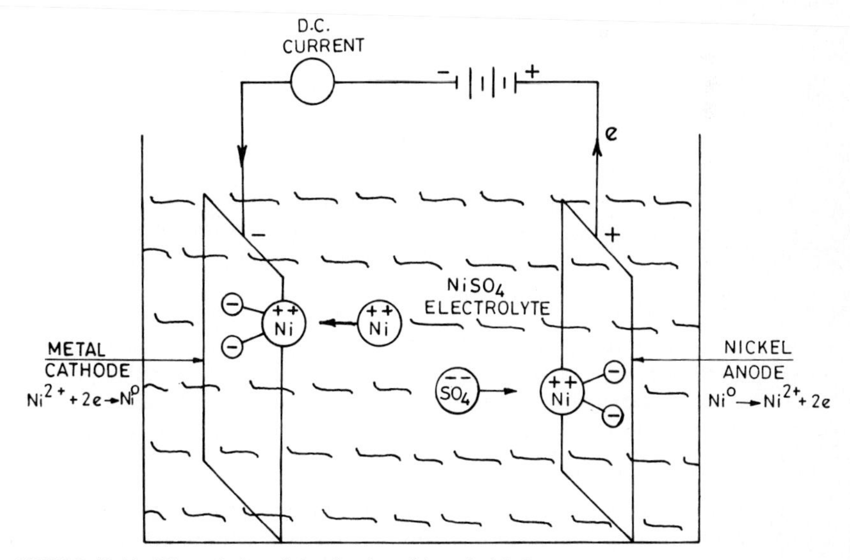

FIGURE 10.11 Electrolytic cell for the deposition of nickel.

anode of the material that is to be electrodeposited. An inert electrode such as carbon may be used for an anode in such cases.

10.8.2 Electrochemistry of Electroplating

The relationship between the weight of the material deposited under various conditions is given by *Faraday's law*: (1) The weight of the coating is proportional to the quantity of electricity passed, and (2) the weight of the metal deposited by the same amount of electricity is proportional to their chemical equivalent weights. Faraday's law may be expressed by

$$W = \frac{ItA}{FZ} \tag{10.6a}$$

where W = weight of deposit, g
I = current flow, amperes
t = deposition time, s
A = atomic weight of element deposited
Z = number of charges (electrons) involved (valence charge)
F = Faraday's constant (= 96,500 coulombs)

Here It is the quantity of electricity consumed (coulombs = ampere-seconds), and A/Z is the equivalent weight of the element. If the current is not constant, then it must be integrated:

$$\int_0^{t_1} I\,dt \tag{10.6b}$$

The current flowing through a conductor is driven by a potential difference or voltage, whose magnitude is determined by Ohm's law

$$E = IR \tag{10.7}$$

where E is the potential difference in volts and R is the resistance in ohms. The potentials within the electrolyte and, more importantly, the electrode-electrolyte interfaces are fundamental controlling factors and are not as straightforward as suggested by Ohm's law.

When a metal (M) is immersed in a solution containing its ions, an equilibrium condition is set up between the tendency for the metal to go into solution and the tendency for the metallic ions in solution to deposit on the metal

$$M^0 \rightleftarrows M^{n+} + ne^- \tag{10.8}$$

However, before this equilibrium is established, one of the reactions may be faster than the other, resulting in a "charge separation." If the reaction proceeding to the right is faster than that proceeding to the left, the metal surface will be negatively charged, and vice versa. This resulting potential between the metal and the solution (at unit activity where activity equals molar concentration times activity coefficient) is called the *standard electrode potential*. Since this is a half-cell reaction, a reference electrode, the saturated hydrogen electrode, is used to complete the circuit and is given the arbitrary value of zero potential. Potential measurements made in this manner result in a series known as the electromotive-force (EMF) series.

The magnitude of the potential difference between the metal and its ionic solution is given by the Nernst equation:

$$E = E_0 + \frac{RT}{ZF} \ln a_M{}^{Z+} \tag{10.9}$$

where E is the observed potential difference at the electrode surface (the electrode potential) in volts, E_0 is the standard emf, R is the gas constant, T is the absolute temperature in kelvins, Z is the valence of the metal ions M^{Z+} being deposited as metal, and a is the activity (apparent degree of dissociation). At unit activity of the depositing metal ions ($a_M{}^{Z+} = 1$), the electrode potential E from the above equation is equal to E_0.

The Nernst equation is derived from thermodynamic laws, from the considerations of the free-energy change for the reaction. This equation helps in determining electrode potentials for processes at reversible equilibrium, i.e., the conditions in which no net reaction takes place. Under practical conditions, nearly every electrode reaction is irreversible to some extent. This irreversibility causes the potential of the anode to become more noble, and the cathode potential less noble, than their respective static potentials calculated from Nernst equations. The overpotential is a measure of the degree of this irreversibility and is also called *activation overvoltage* or *polarization*.

The typical steps involved in the cathodic deposition of metals are migration of ions toward the cathode, electron transfer, adsorption of the cathode surface, and finally migration or diffusion to the growth point on the cathode surface (Lowenheim, 1974). The rate of deposition is determined by the slowest step. If migration is the rate-determining step, then the overpotential developed is referred to as *concentration polarization*. For proper electrodeposition, the polarization process must be understood well, to achieve the proper quality coating. For instance, if a metal is deposited at a high concentration polarization, defects in the appearance and crystal structure are usually observed because of the slow supply of the depositing species to the cathode surface.

10.8.3 Surface Preparation for Electroplating

The preparation of metal surfaces for plating involves the modification or replacement of interfering films to provide a surface upon which deposits can be produced with satisfactory adhesion. The substrate preparation cycles involved are designed to accomplish these objectives (Schwartz, 1982):

- Clean the surface.
- Pickle or "condition" the surface.
- Etch or activate the surface.
- "Stabilize" the surface. Strike.

We examine these steps separately. Cleaning involves the removal of bulk soils (oils, grease, dirt) and last "trace" residues. Removal of bulk soils may involve mechanical operations such as wet or dry blasting with abrasive media; brushing, scrubbing, or chemical cleaning with solvents (degreasing) or emulsions. Removal of the last trace residues usually involves chemical soaking (or spraying) and electrochemical cleaners. These cleaners may contain alkaline chemicals, surfactants, emulsifying or dispersing agents, and chelating agents.

Pickling or conditioning is an acid dip which neutralizes and solubilizes the residual alkaline films and microetches the surface. Common acid dips are either sulfuric acid (~5 to 15 percent vol %) or hydrochloric acid (5% to full-strength).

Etching or activating removes the undesirable metallurgical microconstitutents, e.g., silicides in aluminum alloys or nickel or chromium in stainless steels or superalloys. High nickel and/or chromium containing alloys usually have tenacious oxide or passive films which must be destroyed by strong acids or anodic etching in strong acids. Solutions containing 15 to 25 percent vol % or more sulfuric acid are usually employed at low current densities, 220 to 550 A m^{-2}.

Very active materials—alloys of aluminum, magnesium or titanium—tend to oxidize or adsorb gases readily, even during rinsing and transfer. These continue to interfere with adhesion of deposits. Therefore, a necessary step involving an immersion deposit of zinc or tin, electroless coating, or modified porous oxides is required to make the surface receptive to an adherent electrodeposit. Use of electrodeposits of thin coatings from special solutions called strikes can also be considered a stabilizing step, since it provides a new, homogeneous, virgin surface upon which subsequent deposits are plated. Most widely used strike is the cyanide copper strike. The next most common strikes are the Woods nickel strike, silver, and gold strikes. Electrodeposition of plastics for decorative and functional purposes has become common. Before plastics can be electroplated, they must be made electrically conducting by electroless plating. An organic solvent treatment is helpful in improving the adhesion of deposit on plastics. Next a conditioning process etches the plastic surfaces with chromic-sulfuric acid solutions followed by activation via immersion in a tin-palladium solution. The purpose of this conditioning process is to improve the wettability of the plastic surface and to give it a suitable structure for good adhesion of the plated metal.

The activation process deposits nuclei of tin and palladium which act as the starting materials for chemical deposition of materials such as nickel, cobalt, or copper by the reducing action of hypophosphite or other ions, such as borohydrides or boranes. The process after the deposition of the first layer of electroless metal is autocatalytic. A thickness of the electroless metal of 2 to 5 μm can be deposited in 15 to 30 min. The metal-coated plastic can then be electroplated with the required metal.

Good adhesion of the plated metal to the substrate is one of the most important requirements. For proper adhesion, the substrate must be thoroughly clean and free of any surface films. For optimum adhesion, the substrate and the deposited metal should interdiffuse with interlocking grains to give a continuous interfacial region. Alloy formation by the interdiffusion of the substrate and deposited metal is desirable; formation of an intermediate compound is not.

10.8.4 Electroplating of Metals

The structure and properties of the as-deposited metal depend on the type of solution used and the operating conditions. Apart from a few highly specialized applications in which organic solutions are used, electrolyte is usually from aqueous solution. The electrolytic solutions used can be categorized into acid, alkaline, and complex solutions. Figure 10.12 shows the electrolytic solutions for commonly used metals.

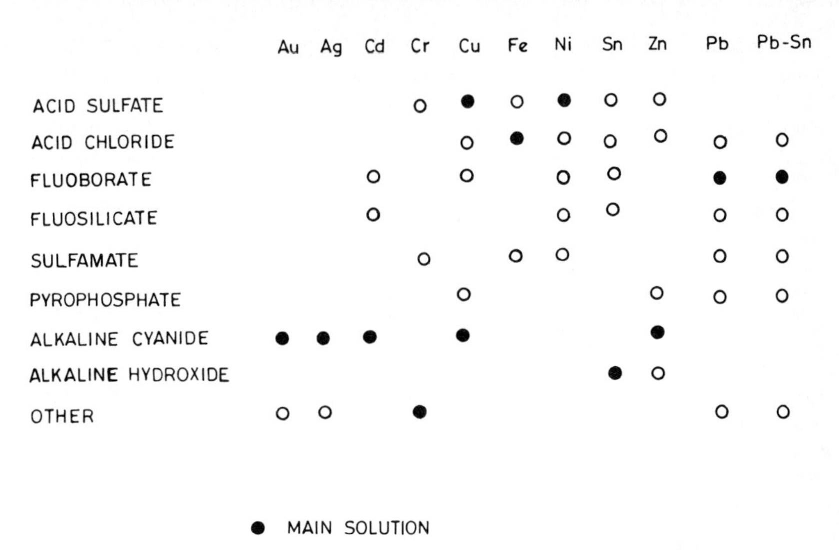

FIGURE 10.12 Electrolytic solutions employed for most common metals.

Acid solutions are used typically for applying Cu, Ni, Sn, and Zn based on the sulfate and/or chloride of the metal to be deposited. Chromium is also of this type, but it is based on the acid (chromic acid, CrO_3) rather than the salt. Such solutions are characterized by relatively simple metal ions, low deposition potential, and high current densities, but have generally poor throwing power.* Chromium solutions are ecologically bad.

Alkaline solutions include cyanide plating solutions for Cd, Cu, Au, Ag, Zn, and a number of alloy deposits. They are characterized by the formation of complex metal ions and relatively high deposition potentials. They are generally used at low current densities. These electrolytes exhibit better throwing power and coverage. Cyanides are the most dangerous chemicals used in electroplating. These are highly toxic.

Complex solutions typically include the sulfamate solutions for Ni and Pb, fluoborate and fluosilicate solution, Sn and its alloys, Pb and Ni. Commercially available alkaline pyrophosphate solutions for Cu also come under this category. These electrolytes exhibit much higher solubility of metal ions, allowing much higher current densities during deposition, and complex ion formation results in better covering and throwing power than for conventional acid electrolytes.

Plated metal coatings are used for decorative purposes, corrosion and wear resistance, and electrical applications. Coatings of Ag, Au, Rh, Ni, Cr, Cu-Ni-Cr, Cu-Zn (brass), and Cu-Sn (bronze) are used for decorative purposes such as household appliances and automotive trims. Coatings of Zn and Cd are used for corrosion-resistant applications such as fasteners, hardware fitting, and sacrificial coatings. Coatings of Ag, Au, Rh, and Ni are used for electrical and electronics applications, e.g., switch contacts and sliding conductors where wear-resistant

*Throwing power is defined as the ability of depositing coating on part of the substrate which is not facing (in line of sight) the source of coating material (i.e., anode in case of electroplating).

surfaces which maintain high conductivity are required. And Cu, Pb, In, Sn, Ag, Au, and Rh are used for bearings applications under moderate operational conditions. The use of precious metals is obviously restricted by their cost, and most applications are in areas where small surfaces are to be plated or where high costs are acceptable, such as in aerospace applications. Cr and Ni are used for extreme wear resistance.

Two most commonly used metal electrodeposits—Cr and Ni for wear resistance—are discussed in some detail.

10.8.4.1 ***Chromium.*** By far, chromium is the most widely used electroplated metal for wear resistance. The chromium is deposited from the chromic acid (CrO_3) and a catalytic anion via the hexavalent chromium ion (Cr^{6+}). Because chromium anodes would dissolve an unusable trivalent chromium, insoluble lead-tin or lead-antimony anodes are used. Therefore, additions of chromic acid must be made as required to keep the chromium-plated bath supplied with chromium metal ions (Cr^{6+}). We note that chromic acid solution does not deposit chromium unless a definite amount of catalytic anion is present. If there is either too much or too little catalyst, no chromium metal is deposited. The first acid anions used are soluble sulfates (SO_4^{2-}). Substitution of fluoride ions present in complex acid radicals for a portion of the sulfate marginally improves the chromium-plating operation. Reflecting this difference in catalyst, the principal types of baths are conventional sulfate and mixed catalyst (Chessin and Fernald, 1982).

Typical compositions for conventional sulfate baths are given in Table 10.3. Chromic acid (CrO_3) is obtained by dissolving chromic anhydride ($H_2Cr_2O_7$) in water. Soluble sulfates (SO_4^{2-}) are added as sulfuric acid or as soluble sulfate salt, e.g., sodium sulfate. The low concentration of the bath is widely used for hard chromium plating because it is capable of plating faster than the high-concentration bath. However, the high-concentration bath has the advantage of being less sensitive to concentration changes, and it is easier to control. The mixed-catalyst baths are similar to conventional sulfates in conductivity but produce harder deposits and have higher intrinsic current efficiency (consequently higher plating rate) than conventional baths. Bright coatings can be deposited on mat surfaces by using brightening additives. These are mostly organic compounds such as dextrose, saccharine, lactose, formaline, citrates, and tartarates.

Due to the low cathode efficiency, high current densities are required for chromium plating to achieve an economic plating rate. Typical plating rates are 65 $\mu m\ h^{-1}$ for sulfate-catalyzed (conventional) baths and 90 $\mu m\ h^{-1}$ for high-speed silicofluoride (mixed-catalyst) types (Chessin and Fernald, 1982). Chromium so-

TABLE 10.3 Typical Compositions and Operating Conditions of Conventional Sulfate Baths for Hard Chromium Plating

Process	Solution composition	Amount, $g\ L^{-1}$	Temperature, °C	Current density, $A\ mm^{-2}$
Low-concentration bath	Chromic acid, sulfate	250 2.5	52–63	0.31–0.62
High-concentration bath	Chromic acid, sulfate	400 4.0	43–63	0.16–0.54

Source: Adapted from Chessin and Fernald (1982).

lutions have poor covering and throwing power, tending to emphasize any surface defects. Workpiece preparation and electrode geometry must receive careful attention in order to achieve uniform current density on both macro- and microscale.

Hard chromium plating is differentiated from the decorative chromium plating in thickness and the type of undercoating used. The thickness of hard chromium coatings ranges from 2.5 to 500 μm, whereas the decorative chromium coatings are thin, ranging from 0.13 to 1.3 μm. Hard chromium coatings are applied directly on the basis metal, whereas the decorative chromium coatings are applied over undercoatings, such as nickel or copper and nickel, which impart a bright, semibright, or satin cosmetic appearance to the chromium (Chessin and Fernald, 1982; Gianelos, 1982).

The hardness of chromium plating is dependent on the current density, plating temperature, catalyst type and ratio, and impurities. Hardness increases with increasing deposition temperature up to a maximum, at 40 to 70°C depending on current density, and then decreases again at higher temperatures. The range of as-deposited hardness is about 450 to 1100 HV, reducing slightly at higher current densities (Brittain and Smith, 1960–1961). The high hardness of hard chromium plating is due to the trivalent chromium oxide inclusions. The hardness results from the lattice strains produced by the incorporation of impurities. Deposits with impurity incorporation are often fine-grained. In general, chromium plating in the bright range is optimally hard (900 to 1100 HV). We further note that the electrodeposited metal begins to decrease in hardness when it is exposed to a temperature above about 200°C. The hardness decreases progressively with an increase in temperature.

Chromium deposits have crystalline structure. They are normally cracked to a varying degree depending on the deposition conditions, and they possess varying amounts of internal stresses (Fry, 1955). A high-crack-count chromium plating usually has lower stress than a low-crack-count deposit. Crack-free deposits can be obtained, but they usually crack if heated or subjected to mechanical loading and give rise to potentially worse situations than a conventionally cracked or microcracked deposit. A high bath temperature of 50 to 55°C promotes crack-free deposit. The addition of selenium to the plating bath causes an increase in crack count and again gives a lower stress level in the coating.

The cracks (or porosity) are undesirable for resistance to corrosion and lower the fatigue resistance of the plated part. However, a porous structure can be advantageous in wear applications in which lubrication is required. Dense chromium plate has poor wettability by oil, and it does not retain the oil after initial lubrication. This problem can be overcome by producing porous chromium so that the crack structure is etched to produce an oil-retentive, scuff-resistant surface. The required amount and type of porosity are obtained by electrolytic or chemical etching of chromium after it is plated on a smooth surface or by preroughening the surface by mechanical methods (e.g., grit blasting, knurling), chemical methods, or electrochemical methods.

Despite the high cost of hard chromium plating relative to other electrodeposits, it finds wide use for extreme wear-resistant and low-friction applications. These coatings are used for piston rings, cylinder liners, cross-heads, etc., in internal-combustion engines and compressors, dies for injection and compression molding of plastic dies, high-temperature bearings, and tools.

10.8.4.2 Nickel. Nickel, with or without an underlying copper strike, is one of the oldest protective-decorative electrodeposits for steel, brass, and other basis metals. More recently, it has been applied to plastics. Nickel's as-plated hardness is relatively low (150 to 400 HV) compared with that of chromium, but it has good mechanical strength and ductility, is a good conductor of heat, and has good resistance to corrosion and high-temperature oxidation. The deposition rate is considerably higher (4 to 6 times) than for chromium up to 0.4 mm h^{-1} achieved from a sulfamate bath. Hard nickel deposits are used primarily for buildup or salvage purposes.

Table 10.4 lists typical composition and operating conditions of two hard nickel baths: sulfamate and sulfate for wear-resistant applications. This table also lists the mechanical properties of deposits. The hardness and tensile strength can be increased, with a corresponding decrease in ductility, by increasing the pH or decreasing the bath temperature. Sulfamate baths are useful for applications requiring low residual stress in the deposits, such as in electroforming, and for coating objects that are susceptible to fatigue cracking (Gianelos and McMullen, 1982). Table 10.4 also lists the Ni-P plating bath which produces deposits with high hardness. These solutions have found little commercial use because deposits of maximum hardness are brittle and highly stressed.

Pure nickel, although reasonably wear-resistant, is unsuitable in some applications because of its tendency to galling. The range of applications can be increased by the inclusion of a fine dispersion of wear-resistant particles, such as silicon carbide or alumina, by codeposition from a suspension in the plating solution (to be discussed later).

10.8.5 Electroplating of Alloys

Codeposition of two or more metals is possible under suitable conditions of potential and polarization. The necessary condition for the simultaneous deposition of two or more metals is that the cathode potential–current density curves (polarization curves) be similar and fairly close together. In the case of codeposition of metals with their standard potentials wide apart, the deposition potentials can be brought together by polarization of the more noble metal, by variation of solution concentration, and sometimes by employing complex ions. Occasionally interrupted current is used to plate alternate layers of metals (Brenner, 1963; Schwartz, 1982). Figure 10.13 summarizes the binary alloy compositions reported to June 1972 (Krohn and Bohn, 1971, 1973). Alloy deposits and composite deposits (to be discussed later) offer certain advantages over single-metal deposits: increased corrosion resistance due to greater density and finer grain structure, a combination of properties of the individual constituents, and "tailormade" properties by proper selection of the constituents. The disadvantages include the greater control required and the difficulty of reproducing the alloy composition.

Electrodeposits applied for corrosion resistance, e.g., tin, copper, cadmium, and zinc, are soft, and while they adequately resist corrosion, they may wear off due to handling and use. A number of alloy deposits are available which give improved wear resistance with good visual appeal and corrosion resistance, such as Fe-Zn, Sn-Zn, Zn-Ni, Cd-Ni, Pb-Sn, and Sn-Ni. And Cu-Zn (brasses) and Cu-Sn (bronzes) are used primarily for decorative purposes.

Several electrodeposited alloys such as Ag-Rh, Ag-Mo, Au-Rh, Au-Co, and Au-Ni have been electrodeposited for use in electronic contacts and wearing surfaces at elevated temperatures. Turns (1977) has produced Au-Mo alloy coatings which exhibit low coefficient of friction (0.1) at 600°C and low wear. Ni-Fe

TABLE 10.4 Typical Compositions and Operating Conditions of Hard Nickel and Ni-P Plating Solutions and Mechanical Properties of Electrodeposits

			Operating conditions			Mechanical properties of coatings			
Process	Bath composition	Amount, g L^{-1}	pH	Temperature, °C	Current density, A m^{-2}	Tensile strength, MPa	Microhardness, HV	Elongation, %	Residual stress, MPa
Sulfamate bath for hard Ni	Nickel chloride ($NiCl_2 \cdot 6H_2O$)	0–30	3–5	38–60	250–300	380–1070	150–400	3–30	3–110
	Nickel sulfamate [$Ni(SO_3NH_2)_2$]	260–450							
	Total nickel as metal	60–115							
	Boric acid (H_3BO_3)	30–45							
Sulfate bath for hard Ni	Nickel sulfate ($NiSO_4 \cdot 6H_2O$)	180	5.6–5.9	43–60	250–500	990–1100	350–450	4–9	—
	Total nickel as metal	40							
	Boric acid	30							
	Ammonium chloride (Nh_4Cl)	25							
Sulfate/chloride bath for Ni-P	Nickel sulfate	170 or 330	0.5–3	60–95	200–500	Brittle	300–900	—	—
	Nickel chloride	35–55							
	Total nickel as a metal	50–85							
	Boric acid	0–4							
	Phosphoric acid (H_3PO_4)	0–50							
	Phosphorous acid (H_3PO_3)	2–40							

Source: Adapted from Gianelos and McMullen (1982).

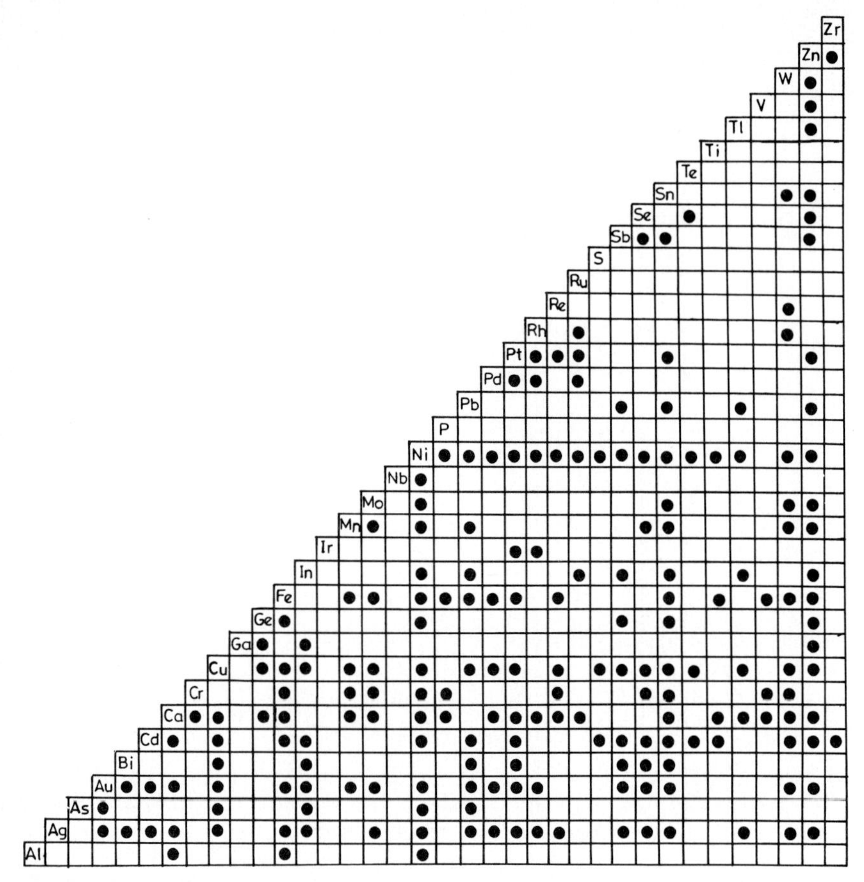

FIGURE 10.13 Binary alloy combinations which have been electrodeposited from aqueous solutions, reported up to June 1972. [*Adapted from Krohn and Bohn (1971, 1973).*]

(permalloy), Ni-P, Co-Ni, and Co-P have been electroplated and are used primarily for magnetic-recording applications (Bhushan, 1990).

Some alloy deposits are used specifically for wear resistance (Schwartz, 1982). Chromium with 1 to 2 percent Mo is used for some internal-combustion engine components to combat wear and corrosion (Kaichinger and Tauzin, 1974). Electrodeposited alloys of Ni with Co, Cr, or Fe are also used for wear and corrosion-resistant applications. Ni-Cr alloys have shown good corrosion resistance at elevated temperatures. Kalyuzhnaya and Pimenova (1962) produced ternery coatings of Fe-Ni-Cr for wear and corrosion-resistant applications. Cobalt-based alloys containing W or Mo have been used to improve the wear resistance of hot forging dies. Dennis and Still (1975) have shown that wear resistance decreases by increasing hardness, the hardness being proportional to the alloy content and increasing to about 900 HV at 18 percent Mo or 800 HV at 24 percent W.

Diffusion coating processes involve the deposition of coatings followed by a thermal diffusion treatment. Such techniques have been applied to improve the adhesion of deposits to difficult-to-plate substrates (diffusion bonding). They have not, however, been extensively applied to producing alloy coatings by deposition, possibly because of the temperature requirements and the formation of intermediate diffusion zones with undesirable properties such as brittleness (Schwartz, 1982). These coatings are discussed in some detail in Section 10.11.

10.8.6 Electroplated Composites

The incorporation of insoluble hard particles in a metal matrix allows further tailoring of the properties of electrodeposited coatings for wear-resistant applications. Strength, hardness, creep, and other properties of the deposit, including the retention of strength after thermal treatments, can be improved by composite coatings. Particles are codeposited from suspension in agitated electrolytic solutions (Foster and Cameron, 1976; Schwartz, 1982). This type of coating is referred to as *dispersion, inclusion, occlusion*, or *composite coating*. The requirements are simple: The particles must be insoluble; the particles must be compatible with solution, i.e., not produce any detrimental effects; the particle size is usually in the colloidal range (~0.005 to 0.2 μm) or slightly larger, usually less than 0.5 to 1.0 μm; and the particles must be dispersed either naturally (as colloidal size particles) or mechanically (stirring, agitation) in order to contact physically the surface being coated.

Successful composites have been achieved with Cr, Ni, and Co matrices and a wide range of ceramic powders, such as aluminum oxide, chromium carbide, silicon carbide, and tungsten carbide (Ghouse, 1984; Kedward and Kierham, 1967; Kedward et al., 1974, 1976; Sautter, 1963; Tomaszewski et al., 1969). Chromium-based composite coating containing zirconium diboride finds potential application for rocket motor and other similar applications.

Kilgore (1963) described various applications including (1) nongalling Ni deposits containing 1000-mesh silicon carbide for pistons and cylinder walls on internal-combustion engines, (2) inclusion of Cr in Ni deposits, producing nichrome by subsequent heat treatments, (3) 120-grit diamond dust in nickel to produce permanent abrasive grinding wheels. Greco and Baldauf (1968) found 2 to 15 percent Al_2O_3 to be the effective range for dispersion hardening of Ni deposits from a sulfamate bath. Schwartz (1982) has reported increased yield strength of 80 MPa for pure Ni deposits to 350 MPa for dispersed alloys containing 3.5 to 6.0 vol % Al_2O_3. Electroformed lead and lead alloys were strengthened only by additions of TiO_2.

10.8.7 Electrophoretic Deposits

Electrophoretic coating is closely related to an electrodeposition process except that it is primarily concerned with the deposition of colloidal particles rather than ions. In electrophoretic deposition, a composite coating is deposited on a conductive surface from a dispersion of colloidal or near colloidal size (that is, 0.2 to 40 μm) particles of one or more materials in a suspending medium (nonconducting or poorly conducting, such as water). The article to be coated is immersed in an aqueous dispersion which dissociates into negatively charged particles and positive cations. An electric field is applied with the article as the anode; the colloidal particles are transported to the anode, where they are discharged and form a coating. This coating process is very suitable for large production of uniform articles, where the whole of the surface is to be coated. The disadvantages include costly installation and the greater control required (Ortner, 1964; Shyne et al., 1955; With, 1974).

Electrophoresis is commonly used as an industrial paint application method. Articles of steel or other metals are immersed in an aqueous dispersion of resin and pigments, which are dissociated into negatively charged colloidal particles and cations consisting of ammonium or amines. The dispersion is stabilized at a pH of 8, say. When an electric field is applied, using the article as the anode and the tank as the cathode (or separate surrounding cathodes), the colloidal particles will migrate to the surface of the anode, where they will be discharged and coagulate, thus forming a coating. The paint coating is then rinsed with tapwater and deionized water and generally cured in an oven.

The electrophoretic process has been applied to the deposition of a variety of materials including metal powders, oxides, cermets, and other particles to metal substrates. Electrophoretic deposition has been applied to produce Ni, Ni-Cr, and Ni-Cr-Fe coatings to base metals as well as inclusion of such dispersoids as MoS_2 or SiC (Shyne et al., 1955). TaC-Fe-Ni coatings have been applied by electrophoretic deposition onto graphite sintered at 2300°C for the protection of rocket nozzle inserts and oxidation-resistant coatings for refractory alloys (Ortner, 1964).

10.9 CHEMICAL DEPOSITION

Chemical deposition or chemical plating differs from electroplating processes in that no electric current source is required. Metal coatings are produced by chemical reduction with the necessary electrons supplied by a reducing agent (RA) present in the solution (Campbell, 1970; Fields et al., 1982; Schlesinger, 1974; Schwartz, 1982)

$$M^{n+} + ne^{-}(\text{supplied by RA}) \rightarrow M^{0}\ (+\ \text{reaction products}) \qquad (10.10)$$

The chemical deposition techniques may be classified as

- Deposition by homogeneous chemical reduction, usually reduction of metal ions in solution by a reducing agent
- Autocatalytic (or electroless) deposition, which is similar to the above process except that the reduction takes place only on certain specific surfaces, called catalytic

In a homogeneous chemical reduction process, the reduction takes place throughout the solution, and deposition occurs over all surfaces in contact with the solution (e.g., silver mirroring). However, in autocatalytic (electroless) deposition, the deposition takes place only on the substrate surfaces which are catalytic to the reaction. The deposit itself continues to catalyze the reduction reaction so that the deposition process becomes self-sustaining or autocatalytic. These features permit the deposition of thick coatings. To prevent spontaneous reduction (decomposition), other chemicals are present; these are generally organic complexing agents and buffering agents. The electroless deposited coatings such as Ni find extensive use for wear-resistant applications.

Almost any metallic or nonmetallic, nonconducting surfaces including polymers, ceramics, and glasses can be plated. Those materials which are not catalytic to the reaction can be made catalytic by suitable sensitizing and nucleation treatments. The costs of the reducing and complexing agents used in chemical plating solution make them noncompetitive with electroplating processes. The applications of electroless plating is usually based on one or more of the following advantages over electroplating: Deposits are very uniform; deposits are usually less porous and more corrosion-resistant than electroplated deposits; by mixing other elements such as P and B in the coating, the coating can be hard and more wear-resistant than electroplated deposits; nonconductors can be plated with metallic coatings. The disadvantages compared to electroplating include solution instability, frequent replacement of tanks or liners, greater process control required, slower deposition rates, and more expensive.

10.9.1 Chemical Reduction

In chemical reduction process, the solutions containing metal ions and reducing agents are mixed and sprayed onto the surface, they react together, and deposition of the metallic coatings occurs by homogeneous chemical reduction of metal ions in solution by the reducing agent. Silver coatings are applied on glass and plastics by this process; that is, a mixture of silver salt (i.e., $AgNO_3$) as one solution and reducing agents (such as formaldehyde or hydrazine) as a second solution is sprayed on the surface. For nonmetallic surfaces, it may be necessary to sensitize the surface, usually by dipping in a solution of stannous chloride acidified with hydrochloric acid.

Copper and gold coatings can also be deposited onto glass and plastics in the similar way (Schwartz, 1982). The copper coatings have been extensively used in printed circuitry, and gold coatings are used for transparent heat mirrors. The adhesion of these coatings is usually inadequate for tribological applications.

10.9.2 Electroless Deposition

In this process, the coating is applied by an autocatalytic chemical reduction of metal ions on catalytic surfaces. The electroless deposition of some metals and their alloys on catalytic surfaces is well documented (Marton and Schlesinger, 1968; Pearlstein, 1974; West, 1965).

In this process a continuous coating of metal or alloy is deposited on immersion of the substrate in the electroless bath solution, which is a blend of different chemicals, each performing an important function (Fields et al., 1982):

- A metal salt as a source of coating metal
- A reducing agent to supply electrons for the reduction of metal ions produced by dissolving metal salts in water
- Energy (heat)
- Complexing agents (chelators) to control the free metal available to the reaction
- Buffering agents to resist the pH changes caused by the hydrogen released during deposition
- Accelerators (exultants) to help increase the speed of the reaction
- Inhibitors (stabilizers) to help control reduction

The composition of the deposit depends on the pH control of the bath during plating, its formulation (complexing and buffering agents), the plating temperature, and the ratio of metal to reducing agents. For example, in the case of Ni-P deposition, as the hypophosphite concentration with respect to Ni increases during the life of the bath, the phosphorus content of the deposit gradually increases. The deposition of the coating takes place all over the surface of the substrate, which has an access to the solution. This demonstrates the excellent throwing power of the process (Schlesinger, 1974).

The reduction of metal ions produced by dissolving metal salts in water occurs only on the surfaces which are catalytic. For this process to continue, the coating which is deposited must also be catalytic to the substrate surface. The surface of noncatalytic substrates can be made catalytic for electroless deposition by exposing it to activating solutions. For example, copper is more noble than nickel; it will not act as a catalyst to start the deposition of nickel. Copper is made catalytic by exposing it to $PdCl_2$ solution (typical composition: $PdCl_2$, 0.1 g L^{-1}; HCl, 0.1 mL L^{-1}) followed by a water rinse. The exposure to stannous chloride ($SnCl_2$) solution (typical composition: $SnCl_2$, 0.1 g L^{-1}; HCl, 0.1 mL L^{-1}) followed by a water rinse is often used for glass and many other nonconducting surfaces. The exposure to a solution of stannous chloride and palladium chloride in hydrochloric acid is commonly used for catalyzing of plastics. After rinsing, nuclei of metallic palladium, surrounded by hydrolyzed stannous hydroxide, are left attached to the surface.

The electroless deposition process has been used to apply coatings of several metals, for example, Ni, Co, Cu, Ag, Au, Pd, Pt, and a variety of alloys involving one or more of these metals plus P and B, metals embedded with hard particles (for example, Al_2O_3 and SiC), and metals embedded with solid-lubricant particles such as PTFE and graphite. Electroless plated coatings are used in printed circuits, magnetic media, and various other tribological applications (Baudrand, 1979; Bhushan, 1990; Feldstein, 1974; Shipley, 1984). For instance, electroless cobalt on polyester substrate may be prepared with magnetic characteristics applicable to digital-recording flexible media. Electroless nickel is an engineering coating normally used because of its excellent corrosion and wear resistance.

The preparation and properties of commonly used electroless deposited coatings are discussed briefly now.

10.9.2.1 Nickel-Phosphorous (Ni-P). Electroless Ni-P coating, commonly referred to as electroless Ni, and electroless Ni-B are the most widely used wear-resistant electroless deposits. There are many proprietary baths and processes available for depositing Ni-P coatings (Baldwin and Such, 1968; Fields et al., 1982; Gawrilov, 1979; Pearlstein, 1974; Shipley, 1984). Metal salts which have been used to deposit electroless Ni coatings are nickel chloride ($NiCl_2$) (alkaline bath) and nickel sulfate (acidic bath). A number of different reducing agents have been used in preparing electroless Ni baths including sodium hypophosphite (for Ni-P), aminoboranes (for Ni-B), sodium borohydride (for Ni-B), and hydrazine (for Ni). The majority of electroless Ni used commercially is deposited from acidic baths (nickel sulfate) reduced with sodium hypophosphite. The principal advantages of these solutions over those reduced with boron compounds or hydrazine include lower cost and better corrosion resistance of the coating. A typical composition of alkaline and acidic baths and operating conditions used for deposition of Ni-P coating are presented in Table 10.5 (Fields et al., 1982).

The most widely accepted mechanism for the chemical reactions which occur

TABLE 10.5 Typical Compositions and Operating Conditions of Electroless Ni-P and Ni-B Plating Solutions

Process	Bath composition	Amount, $g\ L^{-1}$	pH	Temperature, °C	Plating rate, $\mu m\ h^{-1}$
Hypophosphite-reduced					
Alkaline	Nickel chloride	30	8–10	90–95	8
	Sodium hypophosphite	10			
	Ammonium chloride	50			
	Ammonium citrate	65			
	Ammonium hydroxide	To pH			
Acid	Nickel sulfate	34	4.5–5.5	88–95	25
	Sodium hypophosphite	35			
	Malic acid	35			
	Succinic acid	10			
	Thiourea	1 ppm			
Aminoborane-reduced					
Bath A	Nickel chloride	30	5–7	65	7–12
	DEAB*	3			
	Isopropanol	50			
	Sodium citrate	10			
	Sodium succinate	20			
Bath B	Nickel chloride	24–28	5.5	70	7–12
	DMAB†	3–4.8			
	Potassium acetate	18–37			
Borohydride-reduced	Nickel sulfate	50	10	25	—
	DMAB†	3			
	Sodium pyrophosphate	100			

*DEAB = N-diethylamine borane.
†DMAB = N-dimethylamine borane.
Source: Adapted from Fields et al. (1982).

in sodium hypophosphite (i.e., sodium and H_2PO_2 ions in water) reduced electroless nickel-plating solutions is illustrated by the following equations:

$$(H_2PO_2)^- + H_2O \xrightarrow[\text{heat}]{\text{catalyst}} HPO_3^{2-} + 2H^+ + H^- \quad (10.11)$$

The hydride ion (H^-) on the catalytic surface can react with available nickel ions by electron transfer to produce a nickel deposit and hydrogen gas evolution:

$$2H^- + Ni^{2+} \rightarrow Ni^0\downarrow + H_2\uparrow \quad (10.12)$$

The overall reaction can be represented by

$$2(H_2PO_2)^- + 2H_2O + Ni^{2+} \rightarrow N_i^0\downarrow + H_2\uparrow + 4H^+ + 2HPO_3^{2-} \quad (10.13)$$

The efficiency of hydrophosphite usage may be decreased by reaction of hydrogen ion (H^+) with hydride ion (H^-) to produce hydrogen:

$$H^+ + H^- \rightarrow H_2 \quad (10.14)$$

The Ni coatings from the hypophosphite bath contain 3 to 15 percent P depending on bath composition and operating conditions. The surfaces that are catalytic to this process are Ni, Co, Rh, Pd, and steel.

The properties of Ni-P coatings applied by electroless deposition are quite different from those of pure Ni coatings applied by electrodeposition. These differences are primarily due to the codeposited P content in the Ni metallic matrix. Commercial Ni-P coatings typically contain 6 to 12 percent P dissolved in Ni and as much as 0.25 percent of other elements. Coatings containing more than 10 percent P and less than 0.05 percent impurities are typically continuous. Coatings with lower P content are often porous. Ni-P coatings are amorphous and exhibit high hardness of about 500 to 600 HV and significantly better wear resistance compared to electrodeposited Ni (crystalline structure with 99 percent Ni and hardness of 150 to 400 HV) (Randin and Hintermann, 1967). The hardness of Ni-P coatings considerably increases (1100 HV) on heating at 400°C for 1 h or at 280°C for 16 h. These Ni-P coatings have found use as an alternative to electrodeposited hard chromium for several wear-resistant applications. For more details, refer to Chap. 14.

10.9.2.2 Nickel-Boron (Ni-B). When aminoborane compounds, typically *N*-dimethylamine borane (DMAB) or *N*-diethylamine borane (DEAB), are used as reducing agents in nickel baths, the coatings contain boron typically between 0.4 and 5 percent, depending on the pH and/or the bath composition. Aminoborane reduced electroless baths have been formulated over a wide range of pH, although they are usually operated between pH 6 and 9. Operating temperatures for these baths range from 50 to 80°C, but they can be used as low as 30°C. Accordingly, aminoborane baths are very useful for plating plastics and nonmetals, which are their primary applications. Typical compositions of the plating baths are given in Table 10.5. The rate of deposition varies with pH and temperature, but is usually 7 to 12 $\mu m\ h^{-1}$ (Narcus, 1967).

Sodium borohydride is the most powerful reducing agent available for electroless nickel coating. The baths that utilize sodium borohydride as the reducing agent are operated at pH 12 to 14, because the hydrolysis of borohydride is very rapid at lower pH. Complexing agents are required to prevent precipitation of nickel hydroxide (Narcus, 1967). Alloys deposited from borohydride bath typically contain from 3 to 8 percent B. Typical compositions and operating conditions for baths used for deposition of Ni-B coatings with aminoborane borohydride reducing agents are presented in Table 10.5.

Ni-B coatings prepared from aminoborane and borohydride baths are similar to Ni-P coatings with few exceptions. As-deposited hardness of Ni-B coatings is about 650 to 750 HV higher than that of Ni-P. Coatings containing 3 percent B are somewhat harder (1200 HV) after heat treatment at 350 to 400°C than heat-treated Ni-P coatings. The wear resistance of these coatings is exceptional and after heat treatments exceeds that of electrodeposited hard Cr coatings. These coatings, however, are not completely amorphous (crystalline Ni mixed with Ni_2B glass) and have reduced corrosion resistance as well as being much more costly than Ni-P coatings. Ni-B coatings exhibit good lubricity, and their coefficient of friction vs. steel is 0.12 to 0.13 under lubricated conditions and 0.43 to 0.44 under dry conditions (Gawrilov, 1979). For more details, refer to Chap. 14.

10.9.2.3 ***Pure Ni.*** Hydrazine has also been used to produce nickel-rich Ni-B coatings. These baths operate at 90 to 95°C and 10 to 11 pH with a typical rate of deposition of 12 $\mu m\ h^{-1}$. Although the deposit from hydrazine reduced solutions is 97 to 99 percent Ni, it does not have a metallic appearance. The coatings produced by hydrazine bath are brittle and highly stressed and exhibit poor corrosion resistance. Unlike hypophosphite and boron reduced nickels, the hardness from a hydrazine reduced solution is not increased by heat treatment. Hydrazine reduced electroless nickel has very little commercial use (Gawrilov, 1979).

10.9.2.4 ***Other Metals.*** Electroless copper has been widely used for the production of printed-circuit boards (Feldstein, 1974). Many proprietary formulations which are extremely stable and permit steady-state operation have been offered. Most of the formulations involve formaldehyde as the reducing agent. A typical bath for electroless copper deposition consists of copper salt to provide copper in the bivalent form (usually $CuSO_4$), formaldehyde (HCHO), alkali (NaOH), a complexing agent, stabilizers, and other additives. For each gram atomic weight of copper at least 2 mol of formaldehyde and 4 mol of hydroxide is consumed, while at least 1 mol of hydrogen is evolved:

$$Cu^{2+} + 2HCHO + 4OH^- \xrightarrow[\text{surface}]{\text{catalyst}} Cu^0\downarrow + H_2\uparrow + 2HCOO^- + 2H_2O \qquad (10.15)$$

Thick electroless copper has been used for many years in making printed-circuit boards. Thin electroless copper is used in the functional plating of plastics, such as computer packages and other devices, for the purpose of controlling electromagnetic interference. Electroless copper is also now used for metallization of plastics for cosmetic appearance, such as automobiles and electronic instruments. The electroless copper is applied on plastics to provide a conducting surface for subsequent electrolytic buildup, in order to provide a durable, bright deposit. In general, a sequence of Cu, Ni, and Cr is carried out (Feldstein, 1974; Shipley, 1984). Cu finds limited use for tribological applications.

Cobalt is another metal that can be deposited by the electroless process. Reducing agents used for the deposition of electroless Co-P and Co-B coatings are the same as those used in the deposition of Ni-P and Ni-B, respectively. Electroless Co-P coating with high coercivity is used on magnetic media for magnetic-recording applications (Bhushan, 1990). Coatings with low coercivity are used for computer storage elements of high-speed switching devices (Shipley, 1984). Variations in bath constituents and operating conditions provide Co coatings ranging from high coercivity (for high-density recording) to low coercivity (for high-speed switching devices).

Palladium-phosphorous, gold, silver, and tin are other metals that can also be produced by electroless deposition. These coatings find applications in the electronics industry (Feldstein, 1974).

10.9.2.5 ***Polyalloys.*** A number of metals that cannot be deposited independently can be deposited in alloy form containing P or B. Electroless Ni-Cu-P alloy coatings containing only about 1 percent copper exhibit several unique properties, such as being microsmooth, bright, and more ductile and having better corrosion resistance than Ni-P coatings.

Ni-Co-P alloys in all proportions of Co and Ni can be produced by alkaline baths. Polyalloys of Ni, B, and Mo or W are suitable for ultrasonic bonding and solderability. These coatings have exhibited unique ductility and superior corrosion resistance. One of the Ni-Mo-B solutions will plate directly on copper without any previous activation steps. Ni-Sn-B and Ni-Sn-W-B alloys show outstanding chemical resistance while having favorable soldering characteristics (Baudrand, 1979).

10.9.2.6 ***Composites.*** Composite coatings can be deposited electrolessly by physically suspending finely divided particles in an electroless Ni-P bath during deposition (Feldstein, 1983; Parker, 1974). The electroless baths for composite coatings are similar to that as described above except that the bath is tailored with particles held in suspension. Optimum coating compositions are usually obtained by placing sufficient particles in the bath to result in 20 to 35 percent volume of entrapped particles in the coating. The particles are carefully sized and are normally 1 to 3 μm in diameter (Kenton, 1983; Sale, 1979; Vest and Bazzare, 1967). Figure 10.14 shows a microstructure of typical Ni-P, Ni-P/SiC, Ni-P/Al_2O_3, and Ni-P/CaF_2 coatings (Kenton, 1983). The hardness of the composite coatings increases after heat treatment; e.g., the hardness of Ni-P/SiC coating increases from 600 to 1000 HV after heat treatment at 425°C. For more details, refer to Chapter 14.

10.10 CHEMICAL CONVERSION COATINGS

Chemical conversion types of coatings are the nonmetallic coatings produced by "transformation of the outer atomic layers of a metal surface into new, nonmetallic forms with different sets of properties from the original surface, by means of a reaction induced with an artificial environment in situ wherein the metal itself forms one member of the compound with other members supplied from the environment" (Waterhouse, 1965). These coatings have graded interface, whereas electrodeposited or sprayed coatings merely lie on top of the surface.

These coatings can be produced by chemical or electrochemical deposition. Most commonly produced coatings by chemical deposition are phosphate, chromate, and oxide coatings, and those by electrochemical deposition (chemical reaction requiring the passage of current) are anodizing (or anodic oxidation), sulfide coatings, and metalliding. The chemical conversion coatings are extensively applied as a protection against corrosion for the underlying metal. Some of the coatings provide the preconditioning to improve adhesion between the underlying

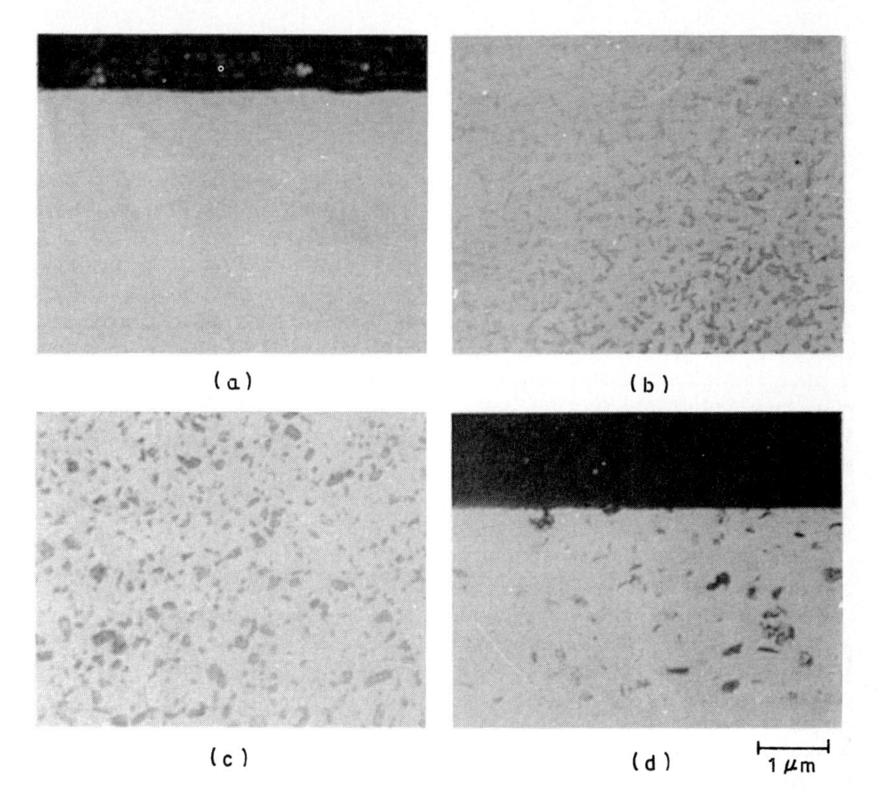

FIGURE 10.14 Cross-sectional view of typical (*a*) Ni-P, (*b*) Ni-P/SiC, (*c*) Ni-P/Al_2O_3, (*d*) Ni-P/CaF_2 composite coatings applied by electroless deposition. [*Adapted from Kenton (1983).*]

metal and the subsequently applied paint coat. Another application has been in metal-forming processes where very heavy deformation is produced; e.g., in wire drawing, one of the problems of such application is providing effective lubrication to minimize excessive wear of the tools and dies. These coatings, by virtue of the porosity which they usually contain, can act as a vehicle for a lubricant when applied to the material. The lubricant provides protection against abrasion, scuffing wear, and fretting corrosion.

10.10.1 Phosphate Coatings

Phosphate coatings, phosphating, or phosphatizing is the treatment of cast iron, steel, galvanized steel, or aluminum with a dilute solution of phosphoric acid and other chemicals (accelerators) such that the surface of the metal, reacting chemically with the phosphoric acid media, is converted to a dense, continuous, nonmetallic insoluble layer of either crystalline or amorphous phosphate which is an integral part of the surface. Phosphate coatings provide (1) temporary or short-time resistance to mild corrosion, (2) a base for subsequent painting or coating,

(3) a surface that facilitates cold forming, and (4) lubricity and resistance to wear, galling, or scoring of parts moving in contact, with or without oil.

Three principal types of coatings are in use: zinc, iron, and manganese phosphate. The weight and crystalline structure of the coating and the extent of penetration of the coating into the basis metal can be controlled by the method of cleaning before treatment; use of activating rinses; the method of applying the solution (simple immersion or spraying); temperature, concentration, and duration of treatments (Davis et al., 1982; Khaleghi et al., 1979; Perry and Eyre, 1977; Waterhouse, 1965).

Zinc phosphate coatings encompass a wide range of crystal characteristics ranging from heavy films with coarse crystals to ultrathin microcrystalline deposits. Iron phosphate coatings are formed from alkali-metal phosphate solutions operating at pH 4 to 5, which produce exceedingly fine crystals. The solutions produce an amorphous coating consisting primarily of iron oxides. The manganese phosphate coatings are tight and fine-grained. Zinc phosphate and iron phosphate coatings can be applied from 25 to 65°C by spraying or immersion or both. Manganese phosphate coatings are usually formed from high-temperature baths from 85 to 95°C only by immersion. Uniform coatings are normally produced on irregularly shaped articles, in recessed areas, and on threaded and flat surfaces, because of the chemical nature of the coating process. The coating thickness ranges typically from 3 to 50 μm.

All phosphate coatings are produced by the same type of chemical reaction; the acid bath containing the coating chemicals (such as alkali phosphate–phosphoric acid, ferrous phosphate, zinc phosphate, manganese phosphate) reacts with the metal to be coated, and at the interface a thin film of solution is neutralized because of its attack on the metal. In the neutralized solution, the solubility of the metal phosphates is reduced, and they precipitate from the solution as crystals. Crystals are then attracted to the surface of the metal by the normal electrostatic potential with the metal and deposit on the cathodic sites.

10.10.1.1 Crystalline-Type Phosphate Coating. When a metal is immersed in a weak acid solution of a primary phosphate of a heavy metal, deposition of the less soluble secondary phosphate or insoluble ternary phosphate occurs on the metal surface via the following reactions:

$$M(H_2PO_4)_2 \rightleftharpoons M^{2+} + 2H_2PO_4^- \tag{10.16}$$

$$H_2PO_4^- \rightleftharpoons HPO_4^{2-} + H^+ \rightleftharpoons PO_4^{3-} + H^+ \tag{10.17}$$

$$3M^{2+} + 2PO_4^{2-} \rightarrow M_3(PO_4)_2 \downarrow \tag{10.18}$$

The pH of such solutions, in practice, is about 2. The heavy-metal phosphates are usually those of zinc or manganese. The following are examples of the types of reaction which occur:

$$2H^+ + Fe \rightarrow Fe^{2+} + H_2 \uparrow \tag{10.19}$$

$$3Fe^{2+} + 3Zn^{2+} + 4PO_4^{3-} \rightarrow Zn_3(PO_4)_2 \downarrow + Fe_3(PO_4)_2 \downarrow \tag{10.20a}$$

or

$$Fe^{2+} + 2H_2PO_4^- \rightarrow Fe(H_2PO_4) \downarrow \tag{10.20b}$$

Such a coating thus always contains ferrous phosphate. If the bath is operated at near boiling point (80 to 100°C), the development of a satisfactory coating would require about 0.5 h. For manganese phosphate baths, the time would be even longer. For this reason, accelerated baths are more common. The most common addition to promote a faster reaction is an oxidizing agent, which acts by stimulating the cathode reaction through cathode depolarization and may produce passivation of the anodic areas as well. Common accelerators are nitrates, nitrides, chlorates, and hydrogen peroxide.

10.10.1.2 Amorphous-Type Phosphate Coating. Amorphous phosphate coatings are produced on steel by immersion in a solution of an alkali or ammonium phosphate in the presence of an oxidizing agent. The pH of such baths is nearly neutral. The following reactions indicate theories of their formation:

$$Fe + 2H_2PO_4^- + O \rightarrow FeHPO_4\downarrow + HPO_4^{2-} + H_2O \qquad (10.21)$$

$$2Fe + 3O \rightarrow Fe_2O_3 \qquad (10.22)$$

A number of phosphating solutions are commercially available for various applications (Anonymous, 1978, 1979). There are a variety of phosphating processes available commercially, under such trade names as Granodine (Imperial Chemical Industries Ltd., London), Parco Lubrite, and Bonderite.

10.10.1.3 Properties of Phosphate Coatings. Conversion of a metal surface to an insoluble phosphate coating provides a metal with a physical barrier against moisture. The affinity of heavy phosphate coatings for oil or wax is used to increase the corrosion resistance of these coatings. Phosphate coatings also improve coating adhesion in two ways: by preventing the undercoating corrosion resulting from a nick or scratch and by providing a surface more receptive to coating. Zinc and iron phosphates are commonly used for paint adhesion. Yaniv et al. (1979) have presented several examples in which the adhesion of organic and inorganic coatings (such as PTFE and resin-bonded MoS_2) to various steel and aluminum-alloy surfaces is improved by zinc phosphate or iron phosphate coating.

Zinc phosphate and iron phosphate coatings are used on cast iron, steel, galvanized steel, or aluminum surfaces to protect the metal against corrosion and to improve adhesion between the metal and the coating. Zinc phosphate is superior in corrosion resistance to iron phosphate. Zinc phosphate coatings of light to medium weight act as an aid in forming steel. The phosphated surface is coated with a lubricant (such as soap, oil, drawing compound) before the forming operation. The power used in deep drawing operations sets up a great amount of friction between the steel surface and the die. The phosphate coating of steel as a metal-forming lubricant, before it is drawn, reduces friction, increases the speed of drawing, reduces the consumption of power, and increases the life of tools and dies. When phosphate-coated steel is used in drawing stainless-steel tubing, the resulting decrease in friction is so pronounced that greater reduction of tube size per pass is possible. This reduction may be as great as one-half (Davis et al., 1982).

Manganese phosphating is a widely used method of reducing wear on machine elements. A phosphate coating permits new parts to be broken in rapidly by permitting retention of an adequate film of oil on surfaces at that critical time. In addition, the phosphate coating itself functions as a lubricant during the high-stress break-in. The coating is smeared between the parts and acts as a buffer to prevent galling on heavily loaded gears (Davis et al., 1982). Waterhouse (1965) has reported that the antifretting properties of mild steel under fatigue conditions are improved by manganese phosphate treatment in both dry and lubricated environments. Heavy manganese phosphate coatings (10 to 45 g m^{-2}) supplemented with proper lubrication (e.g., oiled, waxed) are used for wear resistance. Note that phosphate coatings are not suitable for long-term protection against wear, but that they are normally used on moving parts to help prevent scuffing during running-in. Some examples of parts coated with manganese phosphate for wear resistance include threaded fasteners made of cast iron or steel, bearing races made of high-alloy steel forgings, and piston rings and gears made of cast-iron and steel forgings; these phosphate components are oiled or waxed. Manganese phosphate coatings also provide corrosion protection.

10.10.2 Chromate Conversion Coatings

Chromate conversion coatings are formed on metal surfaces as a result of chemical attack that occurs when a metal is immersed in or sprayed with an aqueous solution of chromic acid, chromium salts such as sodium or potassium chromate or dichromate, hydrofluoric acid or hydrofluoric acid salts, phosphoric acid, or other mineral acids. The chemical attack causes the dissolution of some surface metal and the formation of a protective film containing complex chromium compounds. Generally, the chromate coating is formed by reaction between the metal and the hexavalent Cr; the latter oxidizes the metal and is itself reduced to trivalent state (Eppensteiner and Jenkins, 1977; Lasser et al., 1982).

Chromate treatments are of two types: those that are complete in themselves and deposit substantial chromate films on the substrate metal and those that are used to seal or supplement oxide, phosphate, or other types of nonmetallic protective coatings. Two chromate treatments for aluminum that are complete in themselves are chromium chromate and chromium phosphate. To form the chromium phosphate coating, phosphoric acid or phosphoric acid salts are required in addition to chromic and hydrofluoric acids or their corresponding salts.

For a chromate film to be deposited, the passivity that develops on a metal in a solution of strictly chromate anions must be broken down in solution in a controlled way. This is achieved by adding other anions (such as sulfate, nitrate, chloride, or fluoride) as activators to attack the metal. When attack occurs, some metal is dissolved, the resulting hydrogen reduces some of the chromate ion, and a slightly soluble hydrated chromium, chromate, is formed. This compound is deposited on the metal surface. Table 10.6 lists most common baths for chromate conversion coatings for Al substrates. Chromate coatings on aluminum range in weight from 0.1 to 3.5 g m^{-2} and may be stable up to 300 to 320°C.

A variety of metals and electrodeposited metal coatings can be chromate-conversion-coated. The chromate coatings are applied to electrodeposited coatings of Zn, Cd, Ag, Cu, Sn, and brass; Al, Mg, and Zn die castings; and the hot dipped galvanized parts (Eppensteiner and Jenkins, 1977).

Most conversion coatings dissolve very slowly in water and provide limited protection against corrosion in this medium. Chromate coatings are beneficial under marine atmospheric conditions or in humidity. These coatings can retard the formation of white corrosion products on zinc or cadmium when these metals are subjected to prolonged outdoor exposure, particularly in rural and seacoast environments. These chromate conversion coatings are selected when maximum resistance to corrosion is desired.

TABLE 10.6 Typical Compositions and Operating Conditions of Baths for Chromate Conversion Coating of Aluminum Surfaces

Process	Bath composition	Amount, g L^{-1}	pH	Temperature, °C	Treatment time
Process A*	CrO_3	6†	1.2–2.2	16–55	5 s–8 min
	NH_4HF_2	3			
	$SnCl_4$	4			
Process B‡	$Na_2Cr_2O_7$	7†	1.2–2.2	16–55	5 s–8 min
	× $2H_2O$	1			
	NaF	5			
	$K_3Fe(CN)_6$	§			
	HNO_3 (48°Be)				
Process C¶	H_3PO_4	64	1.2–2.2	40–80	1–10 min
	NaF	5			
	CrO_3	10			

*U.S. Patents 2,507,956 (1950) and 2,851,385 (1958).
†Desired range of hexavalent chromium ion, 1 to 7 g L^{-1}.
‡U.S. Patent 2,796,370 (1957).
§3 mL.
¶Process for Alodine, Alochrome, and Bonderite.
Source: Adapted from Lasser et al. (1982).

Chromate coatings on aluminum substrates exhibit low electrical resistivity (0.3 to 3.0 $\mu\Omega$ mm^{-2}). Their resistivity is low enough that a chromate-coated part can be used as an electric ground. The conductivity of the coatings at radio frequencies is extremely high. This permits the use of a chromate coating on electrical shields and waveguides. Thus, chromate conversion coating is widely used for treatment of aluminum articles for the electronics industry.

10.10.3 Oxide Coatings

The presence of an oxide layer on a metal surface, whether by accident or design, considerably reduces the adhesion with other metal surfaces, i.e., reduces adhesive wear. Indeed, the absence of such coatings can be disastrous. Oxide films can be formed on metal surfaces by heating them in oxidizing environments. Ferrous metals and aluminum and its alloys are commercially treated to produce oxide coatings (0.01- to 0.5-μm thick) by chemical conversion of their surfaces. The object of the oxide coatings is to improve the surface appearance, corrosion resistance, and wear resistance. As with phosphate coatings, when oiled or waxed, oxide coatings have a low coefficient of friction and offer resistance to scuffing. Next to phosphate coatings, oxide coatings are probably the most widely used conversion processes for iron and steel.

The most commonly used method for blackening steels and stainless steels (black oxide coating) is immersion of the metal in a hot, highly alkaline solution containing an oxidizing agent, to convert the surface to the black oxide (Fe_3O_4). Solutions are usually used at the boiling point. Another process, called Blackodizing, consists of immersion in two successive baths containing aqueous solutions of salts, probably caustic/nitrate solutions, operating at 140 to 160°C, with a total treatment time of 20 min. The parts which are oxidized are piston rings, needle roller bearings, and gears. Other methods of producing oxide coatings involve tempering in molten nitrate/nitride baths at about 350°C and treatment in an atmosphere of steam at 300 to 600°C. The later process is known as *steam tempering* and sold under various trade names, mainly for high-speed steel drills and tools. Ferrox is a similar process prescribed for piston rings.

The modified Bauer-Vogel (MBV), Erftwerk, (EW), and Alrok processes are the principal methods for oxide conversion coatings. Nominal compositions of the solutions used and typical operating solutions are given in Table 10.7. The MBV process is used on pure aluminum, as well as on aluminum-magnesium, aluminum-manganese, and aluminum-silicon alloys. The coating produced varies from a lustrous light gray to a dark gray-black. The EW process is used for aluminum alloys containing copper. The coating produced is usually very light gray. The Alrok process sees general-purpose use with all alloys and is often the final treatment for aluminum products. Coatings vary in color from gray to green and are sealed in a hot dichromate solution (Lasser et al., 1982).

10.10.4 Anodizing

Most of the metals form a coating if heated in a specific environment; e.g., heating in oxygen will give layers of oxide; in nitrogen, layers of nitride; and in carbon monoxide, layers of carbide. Usually, the coating thickness obtained in such cases is limited, up to 4 nm, and further growth of layers decreases exponentially with time because of its low ionic and electrical conductivity. However, by anodizing or anodic oxidation (an electrochemical process), the oxide film can be thickened (Waterhouse, 1965).

Anodizing closely resembles electrolytic deposition. It refers, however, to a process which occurs at the anode (hence its name) for a few specific metals limited by electrical and thermal considerations. Basically, the liquid anodizing or

TABLE 10.7 Typical Compositions and Operating Conditions of Baths for Oxide Conversion Coatings of Aluminum and Its Alloys

Process	Bath composition	Amount, g L^{-1}	Temperature, °C	Treatment time, min.	Uses
MBV I	Sodium chromate	15	96	5–10 or 20–30	Corrosion protection or foundation for varnishes or lacquers
	Sodium carbonate	50			
MBV II	Sodium chromate	15	65	15–30	In situ treatments of large objects with paint brush or spray
	Sodium carbonate	50			
	Sodium hydroxide	4			
EW	Sodium chromate	19	88–100	8–10	For copper-containing aluminum alloys
	Sodium carbonate	56			
	Sodium silicate	0.75–4.5			
Alrok	Sodium carbonate	20	88–100	20	Final treatment for aluminum products
	Potassium dichromate	5			

Source: Adapted from Lasser et al. (1982).

anodic-oxidation process consists of making the metal part to be treated the anode in an electrolytic solution through which controlled electric current passes to a cathode (metal container for the electrolyte). Sufficiently high voltage is applied to establish the desired polarization to deposit oxygen at the surface (O_2 overvoltage). The surface of the anode reacts with the oxygen, converting its surface to an insulating, tough, amorphous metal oxide (Campbell, 1967, 1970, 1974; Dell'Oca et al., 1971; Lasser et al., 1982; Schwartz, 1982; Young, 1961).

The reactions that occur at the anode and cathode can be represented by the following equations:

$$M + nH_2O \rightarrow MO_n + 2nH^+ + 2ne^- \quad \text{anode} \tag{10.23}$$

$$2ne^- + 2nH_2O \rightarrow nH_2\uparrow + 2nOH^- \quad \text{cathode} \tag{10.24}$$

where M stands for metal atoms and n is the valence of metallic atoms, for example, $n = 3$ for Al_2O_3. The above equations express the facts that metal is converted to metal oxide at the anode surface and evolution of hydrogen takes place at cathode. The equations imply the presence of water, and anodization usually occurs in aqueous electrolytes (e.g., a solution of phosphoric acid). However, only minute traces of water appear to be necessary, so that it is possible to use other electrolytic media such as certain pure alcohols or molten inorganic salts. A gaseous oxygen plasma and even solid ionic conductors may be substituted for the solution. The final oxide thickness will depend on the voltage applied. Oxide thicknesses of 5 to more than 100 μm in the case are readily available.

At the anode, the reaction with the larger electronegative potential will occur; with some materials this may not be the anodization reaction—the metal of the anode may go into solution, or oxygen may be evolved. The pH of the solution determines which reaction will occur. There is thus a limit to the number of metals that can be anodized. Even those which will anodize may not form useful layers because the oxide is nonadherent or too porous. The metals that can be anodized to give useful coatings are listed in Table 10.8. Anodizing is, in practice, almost entirely used on aluminum and its alloys.

Anodizing processes are used for the following applications (Lasser et al., 1982; Pritchard and Robinson, 1969):

TABLE 10.8 Typical Metals that Can Be Anodized to Give Nonporous Adhesive Oxides

Metal	Thickness-voltage ratio, nm V^{-1}	Maximum nonporous thickness attainable, μm
Al*	0.35	1.5
Ta	1.6	1.1
Nb	4.3	
Ti	1.5 (using aqueous electrolyte) 5.6 (using fused NaCl electrolyte)	
Zr	1.2–3.0	>1.0
Si	0.35	0.12

*The pH of the electrolyte must be correct, or parts of the oxide will redissolve. There is no thickness limit for porous coatings.

Source: Adapted from Gregor (1966).

- To increase corrosion resistance: Amorphous oxide is corrosion-resistant and impervious to atmospheric and saltwater attack.
- To increase paint adhesion: The tightly adhering anodic coating offers a chemically active surface for most paint systems.
- To permit subsequent plating: The inherent porosity of certain anodic coatings enhances electroplating.
- To improve adhesive bonding: A thin phosphoric acid coating improves bond strength and durability.
- To increase abrasion and scuffing resistances: The hard anodized coatings, with the inherent hardness of (aluminum) oxide, are thick enough for use in applications requiring abrasive and scuffing resistances.
- To improve the decorative appearance: All anodic coatings are lustrous and are used as the final finishing treatment.
- To provide unique, decorative colors: Colored anodic coatings are produced by dyeing, and integral color anodizing, depending on the alloy composition, is used to produce a range of stable earth-tone colors suitable for architectural applications.

10.10.4.1 Liquid Anodizing. The three principal types of anodizing processes used for aluminum and its alloys are chromic, in which the active agent is chromic acid; sulfuric, in which the active agent is sulfuric acid; and hard processes using sulfuric acid alone or with additives. Other processes, used less frequently or for special purposes, use sulfuric-oxalic, phosphoric, oxalic, boric, sulfosalicylic, or sulfophthalic acid solutions. Table 10.9 lists some conventional anodizing processes and their operating parameters. The chromic acid anodizing solutions contain from 3 to 10 weight percent CrO_3. The sulfuric acid anodizing solutions contain 10 to 20 weight percent sulfuric acid. The primary differences between the sulfuric acid and hard anodizing processes are the operating temperatures and the current density at which anodizing is accomplished. Hard anodizing produces a considerably heavier coating than conventional anodizing in a given time (Lasser et al., 1982).

The hard anodizing process uses a sulfuric acid bath containing 10 to 15 weight percent acid, with or without additives. The operating temperature of the bath ranges from 0 to 10°C, and the current density is between 200 and 360 A m^{-2}. High temperatures cause the formation of soft and more porous outer layers of the anodic coating, which has poor wear resistance. Several proprietary processes are commonly used: Martin hard coat, Alumilite 225 and 226, Alcanodox, Hardas, Kalcolor, and Lasser. Table 10.10 lists several proprietary baths and their operating parameters for hard anodizing process.

The anodized coating on aluminum consists of two layers (Fig. 10.15). The layer in contact with the metal is nonporous and is called the *barrier layer*. Its thickness does not exceed about 1 μm. The outer layer is the porous layer and is composed of hexagonal prisms, each with a pore along its axis. The number of pores and width of the prisms depend on the applied voltage. The distance between the pores is of the same order as the thickness of the barrier layer. Waterhouse (1965) has found that anodized coating consists of γ-Al_2O_3 of very small grain size so that it may be regarded as amorphous.

10.10.4.2 Gaseous Anodizing. The liquid electrolytic bath can be replaced by a low-pressure glow discharge (~1.3 Pa or 10 mtorr). This is known as *gaseous*

TABLE 10.9 Conventional Baths for Anodizing Processes of Aluminum and Its Alloys

Bath	Amount, wt %	Temperature, °C	Duration, min	Voltage, V	Current density, A m^{-2}	Film thickness, μm	Appearance properties	Remarks
Sulfuric acid bath								
Sulfuric acid	10	18	15–30	14–18	100–200	5–17	Colorless, transparent films	Hard, unsuitable for coloring, tensile strength design 2450–3630 N mm^{-1}
Water	90							
Alumilite								
Sulfuric acid	15	21	10–60	12–16	130	4–23	Colorless, transparent films	Good protection against corrosion
Water	85							
Oxydal								
Sulfuric acid	20	18	30	12–16	100–200	15–20	Colorless, transparent films	Good protection against corrosion, suitable for variegated and golden coloring
Water	80							
Anodal and anoxal								
Sulfuric acid	20	18	50	12–16	100–200	20–30	Colorless, transparent films	For coloring to dark tones, bronze, and black
Water	80							
Bengough-Stuart (original process)								
Chromic acid	3	40	60	0–50	30	5	Colorless to dark brown	Good chemical resistance, poor abrasion resistance; suitable for parts with narrow cavities, as residual electrolyte is not detrimental
Water	97							

TABLE 10.9 Conventional Baths for Anodizing Processes of Aluminum and Its Alloys (*Continued*)

Bath	Amount, wt %	Temperature, °C	Duration, min	Voltage, V	Current density, A m^{-2}	Film thickness, μm	Appearance properties	Remarks
Commercial chromic acid process								
Chromic acid	5–10	40	30–60	0 to increasing limit controlled by amperage	50–100	4–7	Gray to iridescent	Good chemical resistance, poor abrasion resistance; suitable for parts with narrow cavities, as residual electrolyte is not detrimental
Water	95–90							
Eloxal GX								
Oxalic acid	2–10	20–80	30–80	20–30	50–3000	5–60	Colorless to dark brown	Hard films, abrasion-resistant, some self-coloring dependent on alloy, 4410–4710 N mm^{-1} for tensile design
Water	98–90							
Oxal								
Oxalic acid	2–10	20–22	10–240	60	150	10–20 for 30 min., 30–40 for 50 min	Colorless to dark brown	Hard films, abrasion-resistant, some self-coloring dependent on alloy
Water	98–90							
Ematal								
Oxalic acid	1.2	50–70	30–40	120	300	12–17	Not transparent gray, opaque enamel-like	Hard and dense-type film possessing, extreme abrasion resistance
Titanium salt ($TiOC_2O_4K_2 \cdot H_2O$)	40							
Citric acid	1 g							
Boric acid	8 g							
Water	4 L							

Source: Adapted from Lasser et al. (1982).

TABLE 10.10 Proprietary Baths for Hard Anodizing Processes of Aluminum and Its Alloys

Process	Bath	Amount, wt %	Temperature, °C	Duration, min	Voltage, V	Current density, A m^{-2}	Film thickness, μm	Appearance properties	Remarks
Martin hard coat (MHC)	Sulfuric acid Water	15 85	−4–0	80	20–75	270	50	Light to dark gray or bronze	Very hard, wear-resistant
Alumilite 225 and 226	Sulfuric acid Oxalic acid Water	12 1 —	10	20, 40	10–75	2800	25, 50	Light to dark gray or bronze	Very hard, wear-resistant, allows a higher operating temperature over MHC
Alcanodox	Oxalic acid in water	—	2–20	—	—	—	20–35	Golden to bronze	—
Hardas	Oxalic acid Water	6 94	4	—	60 dc plus ac override	200	—	Light yellow to brown	—
Sanford	Sulfuric acid with organic additive	—	0–15	—	15–150 dc	120–150	—	Light to dark gray or bronze	—
Kalcolor	Sulfosalicylic acid Sulfuric acid Water	7–15 0.3–4 rem	18–24	—		150–400	15–35	Light yellow to brown to black	A self-coloring process, colors are dependent on alloy chosen, the colors produced are light, fast
Lasser	Oxalic acid Water	0.75 99.25	1–7	to 20	From 50–500 rising ramp	Voltage controlled	700	Colorless	Hard, thick coatings produced with special cooling processes

Source: Adapted from Lasser et al. (1982).

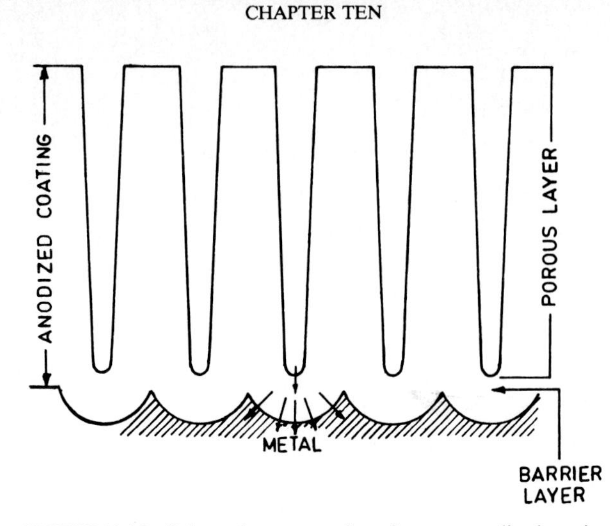

FIGURE 10.15 Schematic cross section of porous anodized coating (arrows indicate downward movement of metal surface and base of pore).

anodization (Campbell, 1970; Dell'Oca et al., 1971; Jackson, 1967). Figure 10.16 shows the schematic arrangement for gaseous anodization. The metal substrate to be anodized is positioned in the most conductive region of the discharge and is positively polarized with respect to the anode. The metals, anodized by this process, typically include Al, Ta, Li-Ti, etc. The thickness-voltage ratio is almost double (2.6 nm V^{-1} for Ta) that of conventional wet anodization, because of the high temperature of the anode in the gaseous discharge. This process has very low current efficiency and throwing power, which limits its applicability.

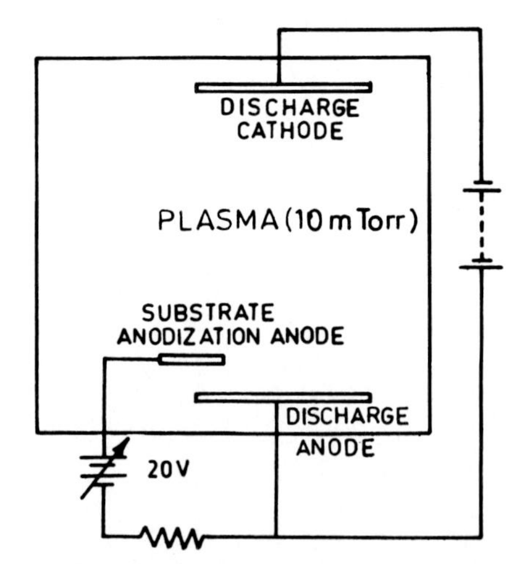

FIGURE 10.16 Schematic arrangement for gaseous anodization.

10.10.4.3 Properties of Anodized Aluminum. Anodized aluminum is produced primarily by the liquid process. These coatings exhibit significant protection against corrosion. The anodized coatings adhere very strongly to the underlying metal, but they have little ductility. Thus, an impact loading of anodized layer creates cracks (Waterhouse, 1965). The coating also has a low thermal conductivity; therefore frictional heat is dissipated only slowly, and a temperature rise can result which promotes disintegration of the coating. The hardness of anodized coatings is influenced by porosity and ranges from 400 to 1000 HV, but usually the barrier layer is found to be much harder than the outer porous layer. It offers some protection against fretting when run dry, but in certain cases (such as riveted joints) this protection can be further increased by impregnating the coating with suitable lubricants.

A commercial process combining anodizing with infusion of PTFE by an electrochemical process for aluminum substrates is called *Tuframed aluminum* (General Magnaplate Corporation, Linden, New Jersey). A commercial process combining anodizing with infusion of PTFE, MoS_2, or graphite for titanium substrates is called *Canadizing* (Anonymous, 1972). Bhushan (1978) reported that Tuframed 6061-T6 aluminum sliding against Meehanite cast iron (GA-40) in a thrust washer wear tests at 0.35 MPa and 5.5 m s^{-1} and in the Freon-22 vapor environment containing about 3 weight percent petroleum oil resulted in low friction and wear while being superior to conventional anodized aluminum sliding against cast iron.

10.10.5 Sulfide Coatings

The process of applying a sulfide conversion coating to iron and steel electrolytically is known as *sulfidizing, sulfiding*, or *sulfurizing*. This coating is integral with the ferrous substrate. Sulfides, by virtue of their crystal structures, and sulfur, because of its low melting point, have long been known to have good lubricating properties. There are many processes available to introduce sulfur or sulfides electrolytically onto or into the surfaces of metal parts (similar to anodizing process), these being used almost exclusively for application to ferrous metals containing up to 13 percent chromium (Gregory, 1974; Waterhouse, 1965). These coatings are applied to provide resistance to scuffing, sliding wear, fretting wear, and corrosion. Haltner and Oliver (1960) have reported that bonded MoS_2 coatings applied on presulfided steel surfaces exhibited improved adhesion.

There are two commercial processes (Hydromecanique et Frottment, Andrezieux Boutheon, 42160 France) specifically designed to produce sulfides at the surface of ferrous metal parts, these depending on different chemical reactions but both involving electrolysis. In the Sulf BT process (also known in the United States as the *Caubot process*), the iron surface is converted to iron sulfide by using a molten mixture of sodium and potassium thiocyanates as electrolyte (Amsallem, 1970). The parts to be treated are made the anode, and the metal container for the electrolyte is the cathode, the current density being about 320 A m^{-2} at a low voltage and the electrolyte being held at about 190°C. At the operating temperature, the thiocyanates are completely ionized, and the negative CNS ions containing sulfur migrate toward the anode, i.e., toward the parts being treated. Here the sulfur forms iron sulfide FeS at the surfaces of the parts being treated. The treatment time is normally 10 min which gives an FeS layer about 7.5 μm thick. The low treatment temperature is useful because hardened parts suffer no reduction in hardness and because parts such as cylinder liners can be treated with minimum distortion. The low treatment temperature is an advantage over nitriding and carburizing which require high treatment temperature (>500°C).

The structure of the sulfide coating is shown in Fig. 10.17, which is a photomicrograph of a field on a transverse section through a Sulf BT treated, case-hardened, and ground mild steel sample. The coating is clearly visible as a dark etching zone, and its complete integration with the hard martensitic substrate is also shown. There is some degree of porosity which can be an advantage in that it assists the surface to carry oil when used in lubricated contacts. Examples of parts treated by sulfidizing are cylinder liners, gears, bearings, heavy-duty rear-axle spiders, slitting machine saws, and textile machine parts.

In another commercial process Molynuz, molybdenum disulfide is synthesized in situ directly onto the surfaces of engineering components. This is done electrolytically from aqueous solutions of molybdenum and sulfur compounds and is said to increase the average life of the treated part more than 3 times. This process is used for many nonferrous metals as well as iron and steel.

In other processes such as sulfonitriding (Sulfinuz and Sursulf), sulfides are formed along with nitrides; they are discussed in a section on nitriding in Chapter 11.

Waterhouse (1965) has reported data of several friction and wear tests involving various treated and untreated samples. Based on dry and oil-lubricated tests, he concluded that a sulfide-coated mild steel disk sliding against itself produced lower wear than an untreated disk pair, but wear of sulfide-coated steel disks was greater than that of nitrided steel disks. Waterhouse further reported that the fatigue limit of a normalized mild steel is raised by 20 percent by sulfidizing. Where fretting is present, e.g., in a press-fitted hub-axle assembly, Waterhouse reported that sulfidizing produced an increase of 135 percent in the fatigue strength. The

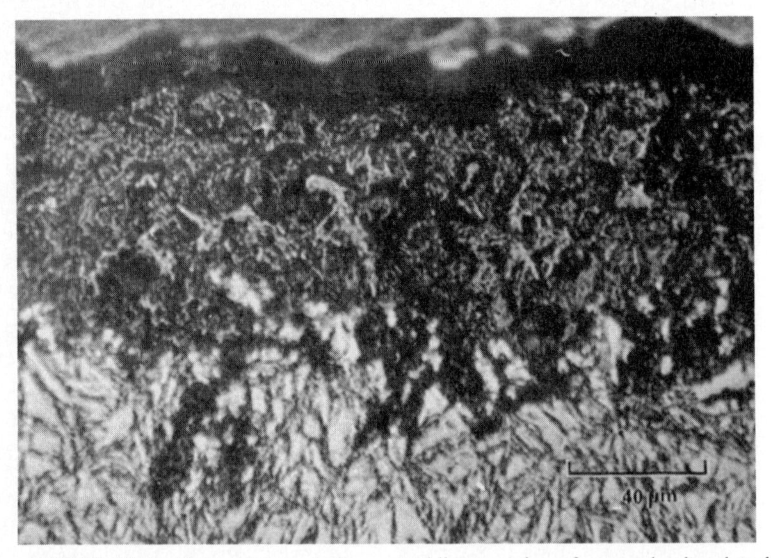

FIGURE 10.17 Photomicrograph from an oblique section of a case-hardened and ground mild steel sample subsequently treated by the Sulf BT process. The dark-etched surface layer is iron sulfide (FeS). Note its integration with the martensitic (light-etched) substrate. Etched in 2% Nital. [*Adapted from Gregory (1974).*]

improvement in fatigue strength was believed to be due to the presence of compressive stress in the coated surface as in nitrided and other diffusion treated parts.

10.10.6 Metalliding

Metalliding is a process to electrolytically produce diffusion coatings of solid solutions and intermetallic compounds from the interaction, mostly in the solid state, of metals and metalloids ranging from beryllium to uranium. The diffusing metal, serving as an anode, and the receptor metal, serving as a cathode, are suspended in electrolytic bath (solution) of molten fluoride salts. When direct current passes from the anode to the cathode, the anode material dissolves and is transported to the cathode. There the anode material is deposited and diffused deep into the cathode at high temperatures of the electrolytic bath, giving rise to an alloyed surface on metals from a few micrometers to tens of micrometers thick. The process is operated between 500 and 1200°C in metal vessels under inert atmospheres. Most of the metalliding reactions function as batteries and will proceed by themselves if the anode and cathode are connected. Generally, an external voltage is supplied to secure uniform, and higher, current densities than those provided by the battery action. Most of the coatings are formed to a thickness of 25 to 75 μm in 2 to 3 h. Almost without exception, increasing the process temperature speeds up the coating process (Cook, 1967, 1969; Cook et al., 1968). The graded interface in diffusion coatings minimizes the interlayer adhesion failures common to deposited coatings.

More than 25 different metals as alloying agents (diffusing metals) and more than 40 different metals as substrates (receptor metals), on which the alloy surfaces are formed, have been used. Hence metalliding appears to have the broadest range of all the surface-alloying methods. However, requirements of high process temperature limit their application to primarily severe-wear and corrosion applications.

Superficially, the metalliding process resembles electroplating. In a plating process, however, the electrolyte solution is generally aqueous, and it is at or slightly above room temperature. Plating works best in depositing the nobler, or less reactive, metals such as Au, Ag, Cu, and Ni; plating cannot be done with the more reactive metals such as Al, Ti, and Si because hydrogen ions always present in aqueous solutions are more readily plated out than the ions of the metals.

In the metalliding process, the molten fluorides of alkali metals and alkaline earth metals take the place of water as the solvent. The fluorides have melting points ranging from about 400 to 1350°C. These salts are stabler chemically than any of the fluorides dissolved in them for metalliding. They are ideally suited both as a solvent and as a medium from which metal ions can be deposited on metal surfaces. Moreover, the fluxing action of the fluorides dissolves from the surface of the cathode metal the oxide films that form in air on all metals except Au and Pt. The clean surface created by the fluoride solvents enables the atoms being electrolytically deposited to make direct contact with the atoms of the cathode's surface and allows diffusion to proceed at the maximum rate. Since most of the structural metals of the periodic table dissolve in other metals or react with them, and since molten fluoride solvents can be maintained at temperatures high enough to permit diffusion and reaction, surface alloying takes place.

The most useful fluoride solvents are those of lithium, sodium, potassium, calcium, and barium. These fluorides are not corrosive to most nickel-based and iron-based alloys (vessel materials) under inert atmospheres. The idling ion, formed from the anode metal in the electrolytic process, is usually present in the salt, at 0.1 to 0.5 mole percent concentration. Since metalliding reagents generally are more reactive electrochemically than the metals being treated, each reagent has a specific group of metals and alloys on which diffusion coatings can be formed, provided that it is soluble in those metals or forms compounds with them. The most active metals offer a wide range of metalliding possibilities. The most useful metalliding reactions appear to be berylliding, boriding, siliciding, aluminiding, titaniding, zirconiding, chrominiding, tantaliding, and yttriding.

The coatings produced by metalliding are comprised of solid solutions, intermetallic compounds, and mixtures of solutions and compounds. Multiple phases of solid solutions or different compounds can be formed in successive layers. The property changes that result are numerous and often very beneficial. They include improved hardness and erosion resistance, lower coefficient of friction, less galling and seizure tendency, and improved resistance to chemical and high-temperature corrosion.

Few examples will indicate the usefulness of the metalliding process. Titanium coated with beryllide is as hard as many of the Carboloy materials (about 1500 HK) and has a much lower coefficient of friction against steel than titanium alone. Beryllided titanium shows no sign of corrosion at 1000°C in air for 8 h. The metals that can be borided and silicided include most of the familiar structure metals such as V, Cr, Mn, Fe, Co, Ni, Cu, Nb, Mo, Rh, Pd, Ag, Ta, W, Pt, and Au. The salt mixture most commonly used for solution is a ternary composition of lithium fluoride, sodium fluoride, and potassium fluoride. Boride coatings are extremely hard; on steel they usually fall between 1500 and 3000 HK. The boride coatings have poor resistance to corrosion except on stainless steels. On molybdenum, boride coatings have measured above 4000 HK. These coatings have proved to be exceptionally good bearing surfaces for pumps handling liquid sodium and potassium at 650°C.

Silicide coatings are generally more brittle and less hard (500 to 1500 HK) than borides, but they strongly resist oxidation. Silicided molybdenum successfully resists burning in air for more than 750 h at 1500°C. Titanided coatings on mild steel, nickel-based alloys, and copper are highly resistant to corrosion, particularly by acids. Aluminided metals exhibit improved resistance to oxidation. Chromium diffuses rapidly at 1130°C into low-carbon steels, forming coatings 100 μm thick in less than an hour. Chrominided low-carbon steels are resistant to corrosion.

10.11 INTERMETALLIC COMPOUND COATINGS

Hydromécanique et Frottment (Andrezieux-Boutheon, 42610 France) and others found that under suitable conditions intermetallic compounds with useful properties could be produced with a graded interface, and based on these findings a series of commercial processes have been developed for treatment of ferrous and some nonferrous metals. All processes involve the prior electrodeposition of a nonferrous alloy (that is, Sn, Cu-Sn, Cu-In-Zn) onto ferrous and nonferrous alloy substrates followed by a diffusion treatment of the coated substrate at a predetermined temperature, which gives rise to the formation of intermetallic compounds. Engineering components made from iron and steel, brass, bronze, aluminum bronze, and light-aluminum alloys can be given considerably increased resistance to scuffing, wear, and sometimes corrosion (Caubet and Amsallem, 1974; Gregory, 1978). Table 10.11 summarizes the operating conditions, hardness, and typical uses of various coatings.

10.11.1 Coatings on Ferrous Metal Substrates

Typical intermetallic compound coatings based on the above principles are produced by Stanal and Forez processes, which confer different combinations of wear and corrosion resistance to iron and steel components.

The Stanal process involves prior deposition of an alloy based on Sn, followed by diffusion at 400 to 600°C, depending on the base metal, in the presence of nitrogen atmosphere [see Figure 10.18(*a*)]. This treatment results into an increase in surface hardness, and resistance to wear and corrosion. The increase in hardness is attributed to the formation of intermetallic compounds such as Fe_3Sn and Fe_3SnC. The increase in hardness is considerably influenced by the carbon content of base metal; e.g., a treated 0.85 percent carbon steel has a surface hardness of about 900 HV. During diffusion, other compounds such as $FeSn_2$ and FeSn are also formed, and originally deposited Sn alloy becomes integrated with the base metal. This process is preferred when the component is subjected to wear and corrosion attack simultaneously. For example, this process is used to combat scuffing, corrosion, and fretting corrosion of 1 percent C, 1.5 percent Cr steel pins used in a steam turbine at 100 to 300°C.

Forez is another similar process which involves the diffusion of super bronze coating into ferrous metal [see Figure 10.18(*b*)]. This process provides a surface which is capable of plastic deformation during running-in and allows replacement of many expensive brass and bronze components by treated ferrous metals. In this process, the initial electroplating consists of the deposition of a copper-tin alloy followed by the diffusion of coating in a nitrogen atmosphere at 580 to 600°C. The resultant surface layer consists of the α and δ phases, and their relative proportions can be optimized through process parameters. The soft α phase (170 HV) assist during running-in and the hard δ phase (550 HV) confer resistance to the ploughing action. The frictional characteristics are comparable to those of soft nitrided steels.

A typical successful application of the Forez process is in the treatment of steel pins (1 percent C, 1.5 percent Cr) used to fasten the martensitic stainless-steel components of the rotors of power station turbines. The untreated pin fails due to scuffing and seizure during dismantling for service. The other applications of Forez treated steels include worn wheels for reduction gears, bearings for presses, spline shafts, and wear rings of boring machines.

10.11.2 Coatings on Nonferrous Metal Substrates

Typical coatings developed for nonferrous metals and alloys are the Delsun and Zinal processes. The Delsun process is used for the treatment of brass, bronze, and aluminum bronze; the Zinal process is used for aluminum alloys to improve friction and wear characteristics.

The Delsun process involves the prior electrochemical deposition of Sn and Sn-Cd and then heating up to 400°C in the case of brass and bronze and 450°C in the case of aluminum bronze in air [see Fig. 10.18(*c*)]. During the treatment

TABLE 10.11 Process Parameters, Properties, and Applications of Complex Surface Treatments

Treatment	Suitable metals	Coating materials	Treatment temp., °C	Hardness, HV	Purpose	Applications
Stanal	Steels	Sn	400–600	600–900	Increased wear and corrosion resistance	Worm gears, pumps, bearings, pins
Forez	Steels	Cu-Sn	580–600	170–550	Increased wear and atmospheric corrosion resistance	Worm gears, splined shafts, marine engine connecting rods
Delsun	Cu alloys, Al bronze	Sn-Cd, Sn-Sb-Cd	400–450	450–600	Increased wear and corrosion resistance	Bearing cages, pistons, massive slideways (for turbochargers)
Zinal	Al alloys	Cu-In-Zn	150	200–500	Increased wear resistance	Diesel and petrol engine pistons, rockers arms

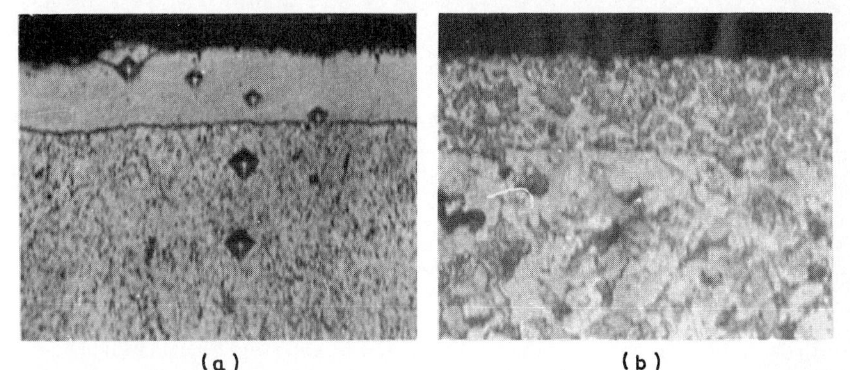

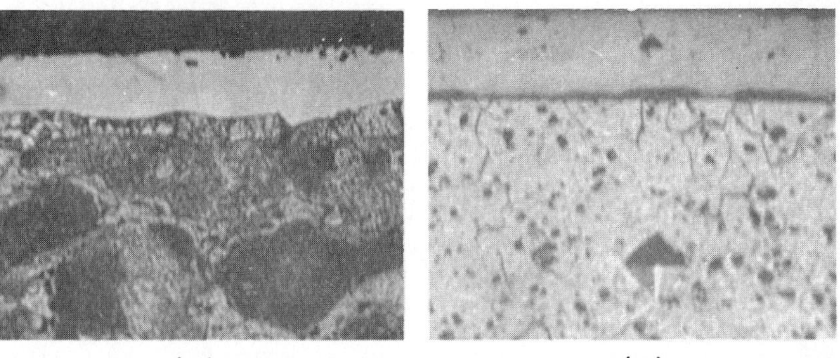

FIGURE 10.18 (*a*) Micrograph of Stanal treated hardened and tempered low-CrMo steel (transverse section ×415). (*b*) Micrograph of Forez treated 0.2 percent C normalized steel (transverse section ×330). (*c*) Micrograph of Delsun treated 60 percent Cu, 40 percent Zn brass (transverse section ×400). (*d*) Micrograph of Zinal treated Al-Cu alloy (transverse section ×330). [*Adapted from Gregory (1978).*]

intermetallic compounds are formed which makes the surface harder, while the surface layer remains ductile. The Sn-rich surface provides improved resistance to adhesive wear and scuffing, and the presence of hard intermetallic compound improves abrasion resistance. Salt spray tests have shown an improvement in corrosion resistance of treated brass and bronze. The Delsun treated surfaces also exhibit better performance under inadequate lubrication and load-carrying capacity. The typical applications of Delsun treated components include threaded parts of taps and valves, worm wheels for reduction gears, rolling mill components, compressor valves, electrical components, etc.

In the Zinal process the electrochemically deposited Cu-In-Zn is applied to aluminum alloys and heated at 120 to 150°C in air [see Figure 10.18(*d*)]. This process provides improved resistance to scuffing, seizure, and abrasion and does not affect corrosion behavior. The friction and wear characteristics of Zinal treated aluminum surfaces are as good as those of hard anodized aluminum surfaces. These surfaces are suitable for use with conventional lubricants and work well where lubrication is unreliable and can be used under vacuum conditions. The typical applications of the Zinal process include (Gregory, 1978)

- Treatment of Al-Si alloy automobile synchronizer rings to replace brass and bronze rings
- Treatment of moped cylinders made of an aluminum alloy
- High-tension cable supports made from an Al-Si alloy suffer scuffing and wear due to sliding of wires in contact with them as a result of stretching and thermal expansion and contraction in use. Zinal treatment of these supports has proved most effective.

The other applications include diesel and petrol engine pistons, rocker arms, clutch disk and brake shoes, and pulleys for V-belt drives.

A proprietary alloy of Ni-Zn called *Corronizing* deposited by electroplating was used commercially as an improved corrosion-resistant diffusion coating (Black, 1948). The substitution of Cd for Zn by Moeller and Snell (1946) produced a corrosion-preventive coating for jet-engine parts, enabling the use of low-alloy steels operating up to 550°C. The coating consisted of 5- to 10-μm Ni and 2.5- to 5-μm Cd diffused at 330°C (melting point of Cd is 320°C).

10.12 SPARK HARDENING

In the spark-hardening process, also called *carbide impregnation*, a thin and hard skin integral with the substrate is generated by the action of electric sparks struck between the surface and a scanning electrode. A positively charged electrode of the hard material (such as carbide) is vibrated against a negatively charged substrate. Each time contact is made, current discharges from a capacitor, and the substrate material is rapidly heated and quenched as well as the electrode material is transferred and diffused into the substrate, resulting in increased adhesion. This process was introduced by a Russian author, Mirkin (1955), and is used for a broad range of machine tools and components such as lathe tools, hacksaw blades, rock drills, and turbine blades. It claimed an improvement in wear life by a factor of 2 to 10. The main causes of improvement in hardness and wear resistance were identified as (Stein, 1974; Welsh, 1959; Welsh and Watts, 1962)

- Intense heating and rapid chilling of microvolumes of surface materials during sparks
- Surface alloying with matter derived from the atmospheres or from the electrode during sparks

In ferrous metals the thermal cycling during spark invokes the martensite change, but surface alloying confers much greater hardness and is the sole means by which most nonferrous metals can be hardened. The source of alloying of the greatest general importance is matter-transferred to the surface from the sparking electrode. In practice, the electrode is usually made of hard metal carbide, although many other materials have been tested.

The basic circuit diagram for spark hardening is illustrated in Fig. 10.19. The electrode (anode) and the workpiece (cathode) are connected to the poles of a condenser, charged from a rectifier current source through a resistor. For the treatment, the electrode is vibrated electromagnetically repetitively, making and breaking contact with the workpiece and discharging the condenser to produce the sparks. The time of discharge, which is on the order of milliseconds, is determined by the electrical parameters of the apparatus and the conditions of the

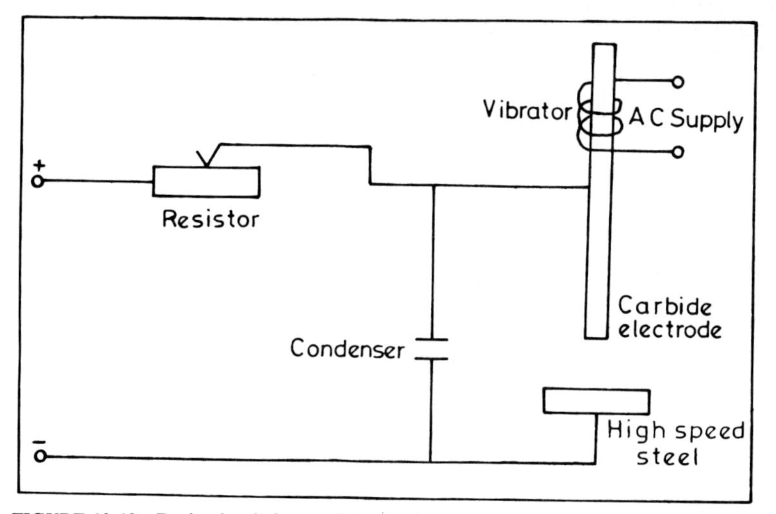

FIGURE 10.19 Basic circuit for spark hardening.

atmosphere. It is believed that at the time of discharge, the workpiece is heated to about 1000°C in the contact zone and is then quenched in an extremely short time. At the same time, material of the electrode melts and is transferred to the workpiece. The process causes significant physicochemical change in the structure of the surface layer and affects the hardness of the treated spots. The affected layer is well bonded to the base material. In this process, there is a net gain of weight of the component and a net loss of electrode weight. This depends on the sparking time and the power of spark. Figure 10.20 illustrates the dependence

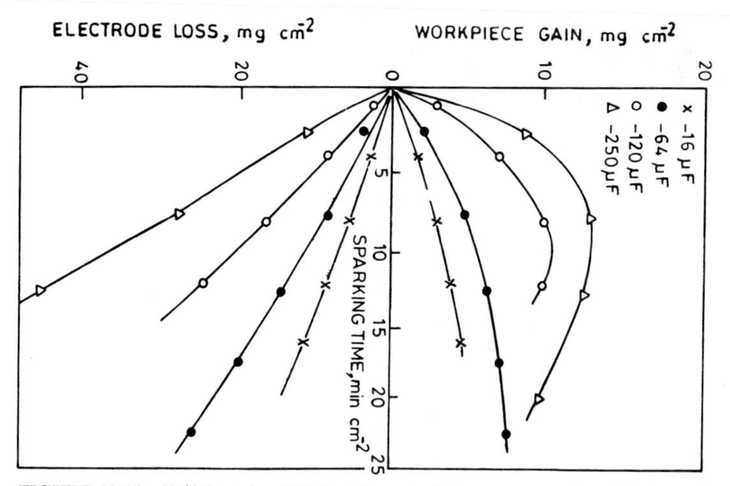

FIGURE 10.20 Weight gain of high-carbon-steel component and weight loss of sintered carbide electrode as a function of spark time at different powers. Capacitance values listed in the figure correspond to the spark power levels. [*Adapted from Welsh and Watts (1962).*]

of high-carbon-steel component weight gain and sintered carbide electrode weight loss as a function of sparking time and power. The figure reveals a general increase in transfer rate with increased spark power. Welsh (1959) has proposed three probable causes for the weight gain from the electrode to components:

- Recondensation of the vaporized material
- Ionic transfer under the influence of an applied electric field
- Asymmetric rupture of liquid or solid junctions when the electrodes separate

The spark-hardened layer was found to be consisting mainly of electrode material, usually carbide, followed by a diffusion layer.

Figure 10.21 illustrates a microsection of the high-speed steel sample treated with tungsten carbide electrode where the hardened layer has a thickness of about 25 μm and appears to be white. Because of this it is called the *white layer*. Below the white layer, a dark-appearing heat-affected zone can be recognized, and beneath this zone is the unaffected base material (Stein, 1974). The white layer is composed of nearly 80 percent electrode material and 20 percent steel base material. The structure of the tungsten carbide in the layer is different from that in the electrode. Because of the high temperatures during the transfer process and because the material is quenched in an extremely short time, the structure of the tungsten carbide changes. The tungsten carbide is disintegrated in the welding process and forms a complex combination with the base material. The average hardness of the hardened layer ranges from 1500 to 2000 kg mm^{-2}. The hardness increases with the spark energy.

The hardened layer has some pores and cracks. These faults are unavoidable. In some applications, the pores may act as lubricant reservoirs. The hardened surface is normally rough; a layer of about 30 μm has a roughness of about 20 μm centerline average (cla). Proprietary processes are available for better finishes.

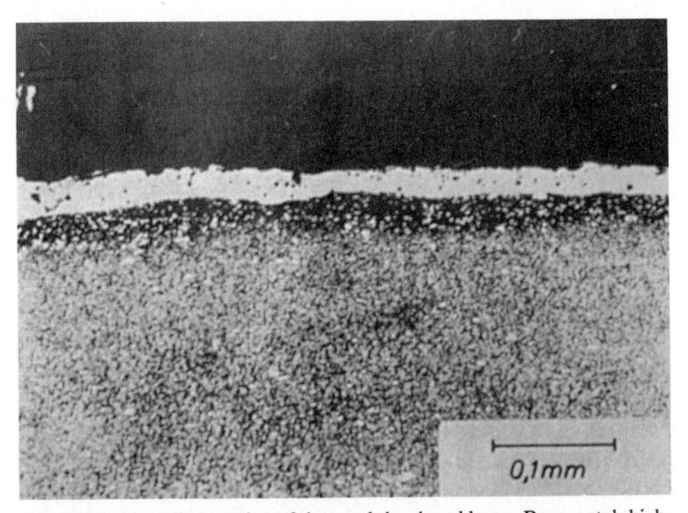

FIGURE 10.21 Microsection of the spark-hardened layer. Base metal: high-speed steel S18-1-2-5; electrode: tungsten carbide K10. [*Adapted from Stein (1974).*]

High-speed steel tools treated with tungsten electrode in spark hardening showed an improvement in the tool life up to a factor of 3 in rotation and milling tests (Stein, 1974). In a standard pin-on-ring test, various treated pin surfaces were tested against a ring made of tungsten carbide. In these tests, spark hardening resulted in an improvement of wear life of high-speed steel by 3 orders of magnitude. Nonferrous metals such as nickel and titanium treated with tungsten carbide electrode also showed significant improvement. Wolfe (1963) found that sparked silver coatings exhibited low wear. From a practical point of view, the spark-hardening process is extremely simple, involving a modest capital outlay and little technical skill.

10.13 REFERENCES

Amsallem, C. (1970), "Influence d'un traitement de surface par electrolyse en bain de sels fondus sur la resistance au grippage et a l'usure d'un acier cemente-trempe," *Revue de metallurg.*, Vol. 67, pp. 945–957.

Anonymous (1972), *A Guide to Hard Anodizing*, Acorn Hardas Ltd., Boreham Wood, Herts, England.

——— (1976), *Finishing Handbook and Directory*, 26th issue, chap. 4, Sawell Pub. Ltd., London.

——— (1978), "Pre-paint Preparation of Metal with Oakite Cryscoat Processes," Oakite Products Inc., Berkeley Heights, N.J.

——— (1979), "Phosphating," Stauffer Chemical Company, Chemical Systems Division, Livonia, Mich.

Baldwin, C., and Such, T. E. (1968), "The Plating Rates and Physical Properties of Electroless Nickel/Phosphorus Alloy Deposits," *Trans. Inst. Met. Finish.*, Vol. 46, pp. 73–80.

Baudrand, D. W. (1979), "Autocatalytic (Electroless) Nickel—Many and Varied," *Plat. Surf. Finish.*, Vol. 66, pp. 18–20, November.

Bhushan, B. (1978), "Investigation of Material Combinations in Freon-22 and Refrigerant Oil Mixture," Tech. Rep. MTI 78TR 87, Mechanical Technology Inc., Latham, N.Y.

——— (1980), "High Temperature Self-Lubricating Coatings for Air-Lubricated Foil Bearings for the Automotive Gas Turbine Engine," Tech. Rep. CR-159848, NASA Lewis Research Center, Cleveland, Ohio.

——— (1982), "Development of CdO-Graphite-Ag Coatings for Gas Bearings to 427°C," *Wear*, Vol. 75, pp. 333–356.

——— (1987), "Overview of Coating Materials, Surface Treatments, and Screening Techniques for Tribological Applications. Part I: Coating Materials and Surface Treatments," in *Testing of Metallic and Inorganic Coatings* (W. B. Harding and G. A. DiBari, eds.), Special Publication STP 947, pp. 289–309, ASTM, Philadelphia.

——— (1990), *Tribology and Mechanics of Magnetic Storage Devices*, Springer-Verlag, New York.

———, and Gray, S. (1978), "Static Evaluation of Surface Coatings for Compliant Gas Bearings in an Oxidizing Atmosphere to 650°C," *Thin Solid Film*, Vol. 53, pp. 313–331.

———, and ——— (1980), "Development of Surface Coatings for Air-Lubricated Compliant Journal Bearings to 650°C," *ASLE Trans.*, Vol. 23, pp. 185–196.

———, Ruscitto, D., and Gray, S. (1978), "Hydrodynamic Air-Lubricated Compliant Surface Bearings for an Automotive Gas Turbine Engine Part II: Materials and Coatings," Tech. Rep. CR-135402, NASA Lewis Research Center, Cleveland, Ohio.

Black, G. (1948), *Metal Finish.*, Vol. 44, p. 207; also U.S. Patent 2,315,740.

Brenner, A. (1963), *Electrodeposition of Alloys*, Vols. 1 and 2, Academic, New York.

Brinker, C. J., and Scherer, G. W. (1990). *Sol-Gel Science*, Academic, San Diego, Calif.

Brittain, C. P., and Smith, G. C. (1960–61), "The Structure and Hardness of Electrodeposited Chromium," *J. Inst. Metals*, Vol. 89, pp. 407–413.

Campbell, D. S. (1970), "The Deposition of Thin Films by Chemical Methods," in *Handbook of Thin Film Technology* (L. I. Maissel and R. Glang, eds.), pp. 5.1–5.25, McGraw-Hill, New York.

———(1974), "The Basic Principles of Anodization," in *Science and Technology of Surface Coating* (B. N. Chapman and J. C. Anderson, eds.), pp. 87–99, Academic, New York.

———(1972), "Solid Lubricants—A Survey," Special Publication SP-5059(01), NASA, Washington, D.C.

Campbell, M. E., Loser, J. B., and Sneegas, E. (1966), "Solid Lubricants," Special Publication SP-5059, NASA, Washington, D.C.

Campbell, W. J. (1967), "Anodic Finishes for Wear Resistance," *Conf. on Anodizing Aluminum, Nottingham*, Aluminum Development Association, Paper No. 11.

Caubet, J. J., and Amsallem, C. (1974), "Surface Treatments and Possibilities for Their Industrial Use," in *Science and Technology of Surface Coating* (B. N. Chapman and J. C. Anderson, eds.), pp. 121–128, Academic, New York.

Chamberlin, R. R. (1966), "Effect of Substrate on Films of Chemical Spray Deposited CdS," *Am. Ceram. Soc. Bull.*, Vol. 45, pp. 698–701.

———, and Skarman, J. S. (1966), "Chemical Spray Deposition Process for Inorganic Films," *J. Electrochem. Soc.*, Vol. 113, pp. 86–89.

Chessin, H., and Fernald, E. H. (1982), "Hard Chromium Plating," in *Metals Handbook*, Vol. 5: *Surface Cleaning, Finishing, and Coating*, 9th ed., pp. 170–187, American Society for Metals, Metals Park, Ohio.

Chopra, K. L., Kainthla, R. C., Pandya, D. A., and Thakoor, A. P. (1982), "Chemical Solution Deposition of Inorganic Films," in *Physics of Thin Films* (G. Hass, M. H. Francombe, and J. L. Vossen, eds.), Vol. 12, pp. 167–236, Academic, New York.

Clauss, F. J. (1972), *Solid Lubricants and Self-Lubricating Solids*, Academic, New York.

Cook, N. C. (1967), "Diffusion-Coating in Molten Fluoride Baths," *Proc. Int. Conf. on Protection Against Corrosion by Metal Finishing*, pp. 151–155, Forster, Zurich, Switzerland.

——— (1969), "Metalliding," *Sci. Am.*, Vol. 221, No. 2, pp. 38–46.

———, Fosnocht, B. A., and Evans, J. D. (1968), "Alloying of Metal Surfaces in Molten Fluorides," *Proc. 24th Int. Corrosion Conf. Series*, pp. 520–524, National Association of Corrosion Engineers, Houston, Tex.

Davis, J. W., Maurer, J. I., Steinbrecher, L. E., Varga, C. M., and White, J. K. (1982), "Phosphate Coating," in *Metals Handbook*, Vol. 5: *Surface Cleaning, Finishing, and Coating*, 9th ed., pp. 434–456, American Society for Metals, Metals Park, Ohio.

DeGraaf, D. (1974), "Use of Brush Applied Coatings in Black and White Picture Tubes," in *Science and Technology of Surface Coating* (B. N. Chapman and J. C. Anderson, eds.), pp. 209–214, Academic, New York.

Dell'Oca, C. J., Pulfrey, D. J., and Young, L. (1971), "Anodic Oxide Films," in *Physics of Thin Films* (M. H. Francombe and R. W. Hoffman, eds.), Vol. 6, pp. 1–79, Academic, New York.

Dennis, J. K., and Still, F. A. (1975), "The Use of Electrodeposited Cobalt Alloy Coatings to Enhance the Wear Resistance of Hot Forging Dies," *Cobalt*, Vol. 1, pp. 17–28.

Duffy, J. I. (1962), *Electrodeposition, Process, Equipment and Composition*, Noyes Publications, Park Ridge, N.J.

Eppensteiner, F. W., and Jenkins, M. R. (1977), in *Metal Finishing Guidebook Directory* (N. Hall, ed.), pp. 540–556, Metal Plast. Pub., Hackensack, N.J.

Feigelson, R. S., N'Diaye, A., Yin, S-Y., and Bube, R. H. (1977), "II–VI Solid Solution by Spray Pyrolysis," *J. Appl. Phys.*, Vol. 48, pp. 3162–3164.

Feldstein, N. (1974), "Electroless Plating in the Electronics Industry," *Plating*, Vol. 61, pp. 146–153.

——— (1983), "Selective O brushe Plating: A Put-on Tool, Part XIII: Mining Equipment," *Metal Finish.*, Vol. 81, pp. 35–37.

Fields, W. D., Duncan, R. N., Zickgraf, J. R., Baudrand, D. W., and Henry, R. A. (1982), "Electroless Nickel Plating," in *Metals Handbook*, Vol. 5: *Surface Cleaning Finishing, and Coating*, 9th ed., pp. 219–243, American Society for Metals, Metals Park, Ohio.

Fish, J. G. (1982), "Organic Polymer Coatings," in *Deposition Technologies for Films and Coatings* (R. F. Bunshah, J. M. Blocher, T. D. Bonifield, J. G. Fish, P. B. Ghate, B. E. Jacobson, D. M. Mattox, G. E. McGuire, M. Schwartz, J. A. Thornton, and R. C. Tucker, eds.), pp. 490–513, Noyes Publications, Park Ridge, N.J.

Foster, J., and Cameron, B. (1976), "The Effect of Current Density and Agitation on the Formation of Electrodeposited Composite Coatings," *Trans. Inst. Met. Finish.*, Vol. 54, pp. 178–183.

Fry, H. A. (1955), "A Study of Cracking in Chromium Deposits," *Trans. Inst. Met. Finish.*, Vol. 32, pp. 107–127.

Gaida, B. (1970), *Electroplating Science*, Robert Draper Ltd., Teddington, United Kingdom.

Gawrilov, G. G. (1979), *Chemical (Electroless) Nickel Plating*, Portucullis Press, Redhill, United Kingdom.

Ghouse, M. (1984), "Influence of Heat Treatment on the Bond Strength of Codeposited Ni-SiC Composite Coatings," *Surf. Technol.*, Vol. 21, pp. 193–200.

Gianelos, L. (1982), "Decorative Chromium Plating," in *Metals Handbook*, Vol. 5: *Surface Cleaning, Finishing, and Coating*, 9th ed., pp. 188–198, American Society for Metals, Metals Park, Ohio.

———, and McMullen, W. H. (1982), "Nickel Plating," in *Metals Handbook*, Vol. 5: *Surface Cleaning, Finishing, and Coating*, 9th ed., pp. 199–218, American Society for Metals, Metals Park, Ohio.

Gittings, D. O. (1982), "Hot Dip Lead Alloy Coating of Steel," in *Metals Handbook*, Vol. 5: *Surface Cleaning, Finishing, and Coating*, 9th ed., pp. 358–360, American Society for Metals, Metals Park, Ohio.

Greco, V. P., and Baldauf, W. (1968), "Electrodeposition of Ni-Al_2O_3, Ni-TiO_2, and Cr-TiO_2," *Plating*, Vol. 55, No. 3, pp. 250–257.

Gregor, L. V. (1966), "Gas Phase Deposition of Insulating Films," in *Physics of Thin Films* (G. Hass and R. Thun, eds.), Vol. 3, pp. 131–134, Academic, New York.

Gregory, J. C. (1974), "Conversion and Conversion/Diffusion Coating," in *Science and Technology of Surface Coating* (B. N. Chapman and J. C. Anderson, eds.), pp. 136–148, Academic, New York.

——— (1978), "Chemical Conversion Coatings to Metals to Resist Scuffing and Wear," *Tribol. Int.*, Vol. 11, pp. 105–112.

Gross, W. F. (1970), *Application Manual for Paint and Protective Coatings*, McGraw-Hill, New York.

Gupta, B. K., and Agnihotri, O. P. (1977), "Effect of Cation-Anion Ratio on Crystallinity and Optical Properties of Cadmium Sulfide Films Prepared by Chemical Spray Deposition Process," *Solid State Commun.*, Vol. 23, pp. 295–300.

———, ———, and Raza, A. (1978), "The Electrical and Photoconducting Properties of Chemically Sprayed Cadmium Sulfide Films," *Thin Solid Films*, Vol. 48, pp. 153–162.

Haacke, G. (1977), "Transparent Conducting Coatings," *Ann. Rev. Mater. Sci.*, Vol. 7, pp. 73–93.

Hahn, R. E., and Seraphin, B. O. (1978), "Spectrally Selective Surfaces for Photothermal Solar Energy Conversion," in *Physics of Thin Films* (G. Hass and M. H. Francombe, eds.), Vol. 10, pp. 1–69, Academic, New York.

Haltner, A. J., and Oliver, C. S. (1960), "Frictional Properties of Molybdenum Disulfide Films Containing Inorganic Sulfide Additives," *Nature*, Vol. 188, pp. 308–309.

Henley, V. F. (1982), *Anodic Oxidation of Aluminium and Its Alloys*, Pergamon, Oxford, England.

Hovel, H. J. (1978), "TiO_2 Antireflection Coatings by Low Temperature Spray Processes," *J. Electrochem. Soc.*, Vol. 125, pp. 983–985.

Jackson, N. F. (1967), "The Gaseous Anodization of Tantalum," *J. Mater. Sci.*, Vol. 2, pp. 12–17.

Kaichinger, A., and Tauzin, C. (1974), Contribution à L'amélioration de le Tenue des Chemises de Moteurs diesel Lents Hautement Suralimentes Assn Technique Maritime et Aeronautique; pp. 1–17.

Kalyuzhnaya, P. F., and Pimenova, K. N. (1962), "Electrolytic Coating of Metals with Iron-Nickel-Chromium Alloys," *J. Appl. Chem. USSR*, Vol. 35, p. 1020.

Kedward, E. C., and Kierhan, B. (1967), "Electrodeposited Composite Coatings for Wear Resistance," *Metal Finish.*, Vol. 65, p. 52.

———, Wright, K. W., and Tennent, A. A. B. (1974), "The Development of Electrodeposited Composites for Use as Wear Control Coatings on Aero Engines," *Tribol. Int.*, Vol. 7, pp. 221–227.

———, Addison, C. A., and Tennent, A. A. B. (1976), "Development of a Wear Resistant Electrodeposited Composite Coating for Use on Aero Engines," *Trans. Inst. Met. Finish.*, Vol. 54, pp. 8–16.

Kenton, D. J. (1983), "Particle Enhanced Electroless Coatings," *Int. J. Powder Metall. and Powder Technol.*, Vol. 19, pp. 185–195.

Khaleghi, M., Gabe, D. R., and Richardson, M. O. W. (1979), "Characteristics of Manganese Phosphate Coatings for Wear-Resistance Applications," *Wear*, Vol. 55, pp. 277–287.

Kilgore, C. R. (1963), *Prod. Finish*, Vol. 34.

Kortum, G. (1965), *Treatise on Electrochemistry*, Elsevier, Amsterdam.

Krohn, A., and Bohn, C. W. (1971), "Electrodeposition of Alloys—1965–1970," *Plating*, Vol. 58, No. 3, pp. 237–241.

———, and ——— (1973), "Electrodeposition of Alloys—Present State of the Art," *Electrodep. Surf. Treatment*, Vol. 1, No. 3, pp. 199–211.

Kut, S. (1974), "Product Finishing with Electrostatically Sprayed Powder Coatings," in *Science and Technology of Surface Coating* (B. N. Chapman and J. C. Anderson, eds.), pp. 43–51, Academic, New York.

Lampkin, C. M. (1979), "Aerodynamics of Nozzles Used in Spray Pyrolysis," *Prog. Crystal Growth Charact.*, Vol. 1, pp. 405–416.

Lasser, H. G., Schardein, D. J., Morris, A. W., Srinivasan, B., and Zelley, W. G. (1982), "Cleaning and Finishing of Aluminum and Aluminum Alloys," in *Metals Handbook*, Vol. 5: *Surface Cleaning, Finishing, and Coating*, 9th ed., pp. 571–610, American Society for Metals, Metals Park, Ohio.

Licari, J. J. (1970), *Plastic Coatings for Electronics*, McGraw-Hill, New York.

Lowenheim, F. A. (1974), *Modern Electroplating*, 3d ed., Wiley, New York.

Manifacier, J. C., Szpessy, L., Bresse, J. F., Perotin, M., and Stuck, R. (1979), "In_2O_3: (Sn) and SnO_2 (F) Films Application to Solar Energy Conversion Part I: Preparation and Characterization," *Mater. Res. Bull.*, Vol. 14, pp. 109–119.

Marton, J. P., and Schlesinger, M. (1968), "The Nucleation, Growth, and Structure of Thin Ni-P Films," *J. Electrochem. Soc.*, Vol. 115, pp. 16–21.

Maykuth, D. J. (1982), "Hot Dip Tin Coating of Steel and Cast Iron," in *Metals Handbook*, Vol. 5: *Surface Cleaning, Finishing, and Coating*, 9th ed., pp. 351–355, American Society for Metals, Metals Park, Ohio.

Mirkin, L. A. (1955), *New Methods of Electrical Working of Materials*, Mashgiz, Moscow.

Moeller, R. W., and Snell, W. A. (1946), "Diffused Nickel Cadmium as a Corrosion Preventive Plate for Jet Engine Parts," *Proc. Am. Electroplat. Soc.*, Vol. 42, pp. 189–192.

Muirhead, J. (1974), "The Basic Principles of Air and Airless Spraying," in *Science and Technology of Surface Coating* (B. N. Chapman and J. C. Anderson, eds.), pp. 248–256, Academic, New York.

Narcus, H. (1967), "An Introduction to Chemical Nickel Plating with Sodium Borohydride or Aminoboranes as Reducing Agents," *Plating*, Vol. 54, pp. 380–381.

Nelson, R. L., Ramsay, J. D. F., Woodhead, J. L., Cairns, J. A., and Crossley, J. A. A. (1981), "The Coating of Metals with Ceramic Oxides via Colloidal Intermediates," *Thin Solid Films*, Vol. 81, pp. 329–337.

Nylen, P., and Sunderland, E. (1965), *Modern Surface Coatings*, chap. 16, Inter-Science Pub., London.

Ollard, E. A. (1969), *Introductory Electroplating*, Robert Draper Ltd., Teddington, United Kingdom.

Ortner, M. (1964), "Recent Developments in Electrophoretic Coatings," *Plating*, Vol. 51, No. 9, pp. 885–889.

Parker, K. (1974), "Hardness and Wear Resistance Tests of Electroless Nickel Deposits," *Plating*, Vol. 61, pp. 834–841.

Paul, S. (1985), *Surface Coatings: Science and Technology*, Wiley, New York.

Pearlstein, F. (1974), "Electroless Plating," in *Modern Electroplating* (F. A. Lowenheim, ed.), 3d ed., Wiley, New York.

Perry, J., and Eyre, T. S. (1977), "The Effect of Phosphating on the Friction and Wear Properties of Grey Cast Iron," *Wear*, Vol. 43, pp. 185–197.

Pritchard, C., and Robinson, P. R. (1969), "Abrasion Resistance of Hard Aluminum," *Wear*, Vol. 13, pp. 361–368.

Randin, R. M., and Hintermann, H. E. (1967), "Electroless Nickel Deposited at Controlled pH, Mechanical Properties as a Function of Phosphorus Content," *Plating*, Vol. 54, pp. 523–532.

Sale, J. M. (1979), "Wear Resistance of Silicon Carbide Composite Coatings," *Metal Prog.*, Vol. 115, April, pp. 44–55.

Sautter, F. K. (1963), "Electrodeposition of Dispersion-Hardened Nickel-Al_2O_3 Alloys," *J. Electrochem. Soc.*, Vol. 110, pp. 557–560.

Sayre, I. H. (1974), *Photography and Platemaking for Photo-lithography*, Lithographic Textbook Publishing Co., Chicago.

Schlesinger, M. (1974), "The Basic Principles of Electroless Deposition," in *Science and Technology of Surface Coating* (B. N. Chapman and J. C. Anderson, eds.), Academic, New York, pp. 176–182.

Schwartz, M. (1982), "Deposition from Aqueous Solutions: An Overview," in *Deposition Technologies for Films and Coating* (R. F. Bunshah, J. M. Blocher, T. D. Bonifield, J. G. Fish, P. B. Ghate, B. E. Jacobson, D. M. Mattox, G. E. McGuire, M. Schwartz, J. A. Thornton, and R. C. Tucker, eds.), pp. 385–453, Noyes Publications, Park Ridge, N.J.

Shanti, E., Dutta, V., Banerjee, A., and Chopra, K. L. (1980a), "Annealing Characteristics of Tin Oxide Films Prepared by Spray Pyrolysis," *Thin Solid Films*, Vol. 71, pp. 237–244.

———, ———, ———, and ——— (1980b), "Electrical and Optical Properties of Undoped and Antimony Doped Tin Oxide Films," *J. Appl. Phys.*, Vol. 51, pp. 6243–6251.

Shipley, C. R. (1984), "Historical Highlights of Electroless Plating," *Plat. Surf. Finish.*, Vol. 71, June, pp. 92–99.

Shyne, J. J., Barr, H. N., Fletcher, W. D., and Scheible, H. G. (1955), "Electrophoretic Deposition of Metallic and Composite Coatings," *Plating*, Vol. 42, pp. 1255–1258.

Sliney, H. E. (1962), "Lubricating Properties of Ceramic-Bonded Calcium Fluoride Coatings on Nickel-Base Alloys from 750 to 1900°F," Tech. Note TND-1190, NASA, Washington, D.C.

——— (1982), "Solid Lubricant Materials for High Temperatures—A Review," *Tribol. Int.*, Vol. 15, pp. 303–314.

———, and Johnson, R. L. (1957), "Bonded Lead Monoxide Films as Solid Lubricants for Temperatures up to 1250°F," NACA RM E57B15, May, NACA, Washington, D.C.

Spiller, L. L. (1974), "The Basic Principle of Electrostatic Coating," in *Science and Technology of Surface Coating* (B. N. Chapman and J. C. Anderson, eds.), pp. 28–42, Academic, New York.

Stein, H. U. (1974), "Spark Hardening—A Technique for Producing Wear-Resistant Surfaces," in *Science and Technology of Surface Coating* (B. N. Chapman and J. C. Anderson, eds.), pp. 129–135, Academic, New York.

Thompson, L. F., Willson, C. G., and Bowden, M. J. (eds.) (1983), "Introduction to Microlithography," *Symp. Series*, Vol. 219, American Chemical Society, Washington, D.C.

Tomaszewski, T. W., Tomaszewski, L. C., and Brown, H. (1969), "Codeposition of Finely Dispersed Particles with Metals," *Plating*, Vol. 56, pp. 1234–1238.

Tonini, D. E., Belisle, S., Graff, H. F., and Pearce, D. C. (1982), "Hot Dip Galvanized Coatings," in *Metals Handbook*, Vol. 5: *Surface Cleaning, Finishing, and Coating*, 9th ed., pp. 323–332, American Society for Metals, Metals Park, Ohio.

Turns, E. W. (1977), "Electroplated Au-Mo Alloy as a Solid Film Lubricant," *Plat. Surf. Finish.*, Vol. 64, May, pp. 46–48, 52–54.

Vagramyan, A. T., and Solev'eva, Z. A. (1961), *Technology of Electrodeposition*, Robert Draper Ltd., Teddington, United Kingdom.

Vest, E. C., and Bazzare, F. (1967), "Codeposited Ni-MoS_2-Ag Self-Replenishing Solid Film Lubricant," *Metal Finish.*, Vol. 65, p. 52.

Waterhouse, R. B. (1965), "The Formation, Structure, and Wear Properties of Certain Nonmetallic Coating on Metals," *Wear*, Vol. 8, pp. 312–348.

Welsh, N. C. (1959), "Spark Hardening of Metals," *J. Inst. Met.*, Vol. 88, p. 103.

———, and Watts, P. E. (1962), "The Wear Resistance of Spark-Hardened Surfaces," *Wear*, Vol. 5, pp. 289–307.

West, J. M. (1965), *Electrodeposition and Corrosion Processes*, Van Nostrand, Princeton, N. J.

With, A. (1974), "The Basic Principles of Electrophoretic Coating—A Study," in *Science and Technology of Surface Coating* (B. N. Chapman and J. C. Anderson, eds.), pp. 60–68, Academic, New York.

Wolfe, G. F. (1963), "Effect of Surface Coatings on the Load-Carrying Capacity of Steel," *Lub. Eng.*, Vol. 19, p. 28.

Yaniv, A. E., Rona, M., and Sharon, J. (1979), "Preparation on the Behavior of Organic Coatings," *Metal Finish.*, November, pp. 55–61; December, pp. 25–30.

Young, L. (1961), *Anodic Oxide Films*, Academic, New York.

CHAPTER 11

SURFACE TREATMENTS BY THERMAL AND CHEMICAL PROCESSES

Surface treatment processes that are in daily use for improving hardness and wear resistance may be classified into three categories: (1) microstructural, (2) chemical diffusion, and (3) ion implantation. The treatments in the first two categories are used mostly for ferrous based materials (Bhushan, 1980, 1987; Dawes and Cooksey, 1965; Draper and Ewing, 1984; Elliott, 1978; Gabel and Donovan, 1980; Gregory, 1978; Hick, 1979; James et al., 1975; Thelning, 1984). The first category consists of selective surface-heating processes where surface microstructure is changed and core remains in the preheated conditions. The second category consists of chemical diffusion treatments that modify both the microstructure and the composition of the surface by incorporation of carbon, nitrogen, boron, and occasionally other elements through diffusion. The third category consists of ion implantation processes which modify the top layer surface (up to 0.1 μm) by implanting ionic species which form an alloy with the substrate material. This process is discussed in Chap. 12. There are many variations, both major and minor, of these basic approaches. Figure 11.1 enumerates various processes commonly used to impart high hardness on the surface of

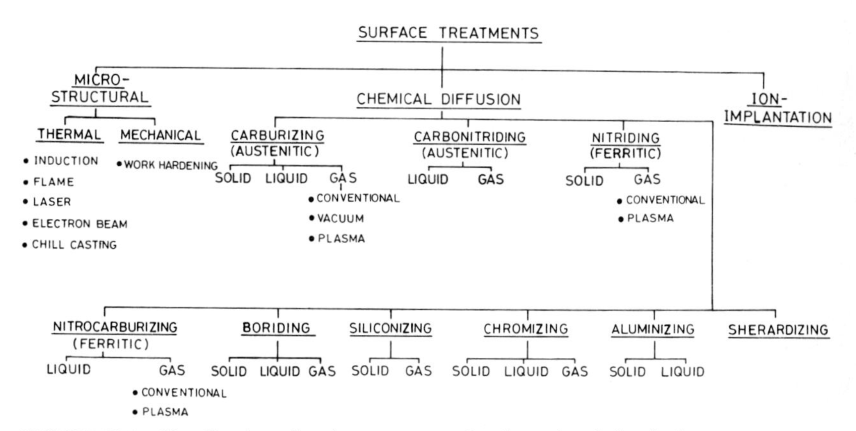

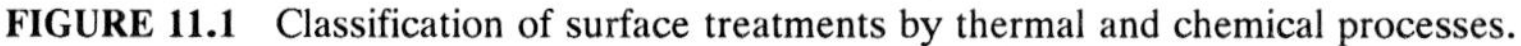

FIGURE 11.1 Classification of surface treatments by thermal and chemical processes.

ferrous metals. Various surface heat treatments of steels are compared in Fig. 11.2 in terms of process temperatures and case depths. Microstructural treatments are cheap and do not require special materials, unlike chemical diffusion treatments. However, the latter processes generally can produce higher hardness with better control of case depth and lower distortion.

Surface treatment processes do not build a coating over a surface; rather, such processes modify the surface itself. Thus surface treatments do not produce coating-to-substrate interface as found in coatings. The interface produced in coatings is often subject to mechanical weakness and interface corrosion. Typical properties which the surface treatments impart to a surface include high hardness, high wear resistance, high impact strength, high fatigue resistance, high corrosion and chemical resistance, and sometimes lower friction. The thickness of the transformed layer needs to be optimized. A thicker layer generally produces a higher level of compressive stresses which are desirable for increased wear resistance. In addition, a thicker layer lasts longer for a given wear rate. However, excessive stresses may result in component warpage and early catastrophic failures.

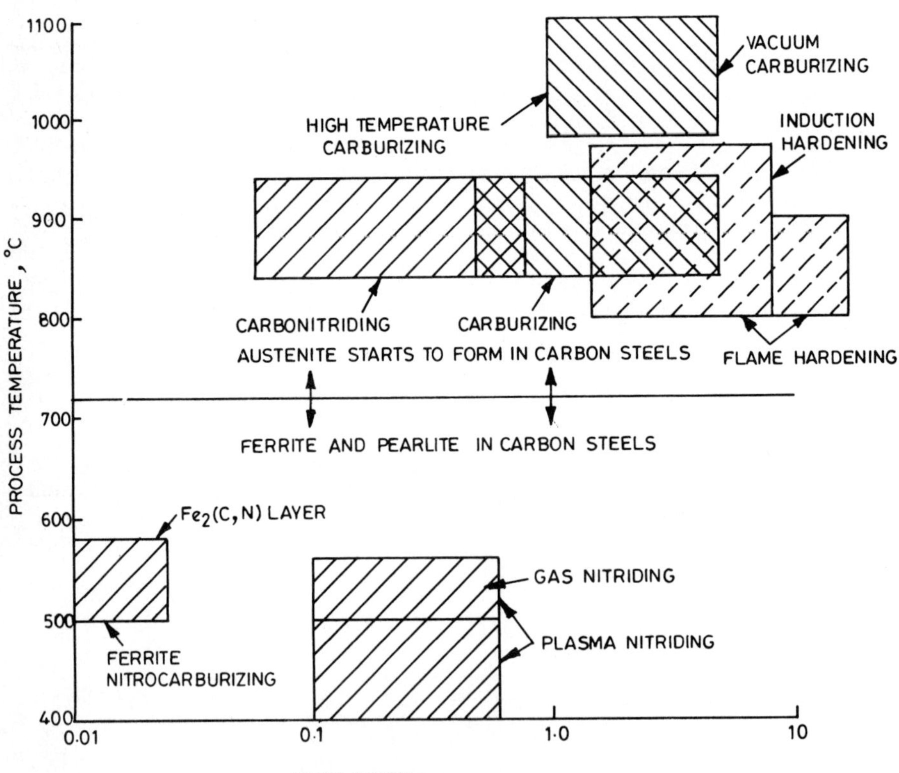

FIGURE 11.2 Comparison of treatment temperatures and case depth for various treatments of steels. [*Adapted from Child (1980a).*]

11.1 MICROSTRUCTURAL TREATMENTS

In most microstructural treatments, a thin surface shell of ferrous metal (e.g., steel) is rapidly heated to a temperature above the critical point (>720°C) so that steel changes its microstructure from a bcc lattice, ferrite, to an fcc lattice, austenitic at 720°C. The quick quenching of steel transforms austenite to body-centered tetragonal structure, martensite, while leaving the core of the component in its original form. In contrast, slow cooling causes transformation, as the temperature passes through the corresponding ranges, to pearlite, bainite, and martensite, with the ferrite structure being a combination of the three, resulting in a soft and ductile steel case (Thelning, 1984). Thus to obtain hard cases, the cooling must be very fast and effective so that it bypasses the first two transformation phases and directly transforms austenitic to martensite (a phase with high hardness). The workpieces thus treated are usually reheated to some intermediate temperature (150 to 200°C) for improvement of ductility with a compromise with high hardness and wear resistance. Quenching a steel not only increases the surface hardness but also produces a favorable residual stress distribution of compressive stresses in the surface. It is these stresses, together with the increased strength of the case, that greatly increase the fatigue strength of treated steel.

The transformation of the microstructure of steels to martensite phase can be achieved through several approaches and processes. The heat treatment process can be closely controlled by varying energy inputs and time in order to obtain a suitable case depth and hard structure. Two heat treatment processes are being widely used, induction hardening and flame hardening (Anonymous, 1981; Child, 1980a, 1980b, 1983; Stevens, 1981; Thelning, 1984). In addition, laser and electron beam hardening heating processes are also becoming popular (DesForges, 1978; Fiorletta, 1981; Sandven, 1981). Flame hardening does not lend itself to close control, but it is particularly suitable for large parts. Laser hardening allows a very thin surface layer to be hardened with close controls. Because the build-up of energy in both laser and electron beam hardening is rapid and well controlled, post heat treatment operations are not required in most cases.

With all forms of surface hardening, the case depth d depends on heat conduction and can be deduced approximately for steels from

$$d \sim k\sqrt{t} \tag{11.1}$$

and

$$k \sim 0.2$$

Where k is a constant that represents the penetration in the first second at a temperature, d is case depth in millimeters, and t is the time of heating in seconds.

During heat treatment, one must consider that the core of the material should remain cold and thus a faster cooling system is required. The reasonably faster cooling rate can be achieved by

- Immersion of the component in the quenchant during heating
- Spray quenching in heating chamber
- Spray quenching during progressive hardening just after heating
- Quenching in a cooling station

Usually, mild quenchants such as polymers and oils are recommended for high-carbon steels.

The processes for microstructural treatments are adapted for components where selective hardening is preferable, such as gear teeths where geometrical factors favor a local surface heat treatment. Long shafts which may distort with carburizing are ideal geometrically for local surface hardening. These processes can be readily used for large components such as lathe beds.

A few old surface hardening techniques—chill casting and cold hardening in addition to induction, flame, laser, and electron beam techniques—are described in the following subsection. Table 11.1 compares the characteristics of various hardening processes based on structural changes.

11.1.1 Induction Hardening

Induction hardening is the most extensively used process for surface hardening. This process exploits the principle of electromagnetic induction, i.e., the passage of an electric current through a conductor creating an electromagnetic field around itself, which in turn induces a current in another conductor placed within that field.

As high-frequency (500 to 15 $\times$ 10^6 Hz) alternating current flows through the inductor or work coil, a highly concentrated, rapidly alternating magnetic field is established within the coil. The magnetic field thus established induces an electric potential in the part to be heated, and because the part represents a closed circuit, the induced voltage causes the flow of current. The resistance of the part to the flow of the induced current causes heating (Guthrie and Archer, 1975; Morgan, 1971; Reinke and Gowan, 1978; Stevens, 1981; Thelning, 1984).

The depth of penetration of heat is influenced by the power density (10 to 20 W mm^{-2}) of the heated surface, frequency (10 to 500 kHz) of alternating current applied in the conductor coil (the higher the frequency, the shallower the depth), the time the current is applied, and the resistivity and permeability of the steel component which vary with temperature. The relationships between depth of the induced current penetration of the heat and the frequency can be calculated for steel up to the hardening temperature (Thelning, 1984) as

$$\text{Depth of penetration at cold state (20°C)}, d_c = \frac{20}{\sqrt{f}} \tag{11.2}$$

$$\text{Depth of penetration at hot state (800°C)}, d_h = \frac{500}{\sqrt{f}} \tag{11.3}$$

where d is the depth of penetration in millimeters and f is the frequency of power supply in hertz. Thus a high power density, with a high frequency and short heating time is used for shallow case depths, and a lower power density with a lower frequency and longer heating time is preferred for deep case depths. Table 11.2 lists the typical values of case depths at various frequencies and power input levels.

Owing to heat conduction in the material during heating, the overall depth of penetration is larger. It is possible to calculate the additional penetration due to heat conduction from Eq. (11.1). The total depth of penetration is obviously the depths given by Eqs. (11.1) and (11.2) or (11.3).

The induction hardening plant consists of three basic elements, i.e., a high-frequency generator, an inductor or coil, and a means of handling the components while they are being heated and quenched. For high or radio frequency (450 KHz), industrial thermionic valves are used, and low frequency (3 to 10 kHz) is

TABLE 11.1 Process Parameters, Properties, and Applications of Structural Heat Treatments

Treatments	Suitable substrate	Treatment temp., °C	Case depth, μm	Maximum hardness HV (kg mm^{-2})	Advantages/ disadvantages	Applications
Induction	Med. C, C-Mn, Cr-Mo, Ni-Cr-Mo steels, gray cast iron, Pearlitic/malleable cast iron, nodular cast iron	850–1100	250–5000	500–750	Fast, large case depth. However, high capital cost and cannot treat certain shapes.	Cams, rotating shafts, machine ways, swivel pins, steering joints, gears, valves and bearing surfaces
Flame	Med. C, C-Mn, Cr-Mo, Ni-Cr-Mo steels, gray cast iron, Pearlitic/malleable cast iron, nodular cast iron	750–1000	250–5000	400–650	Simple, cheap, flexible. However, lack of process control.	Large sections, e.g., rolls, machine beds, dies, track rails, and runner steels
Laser	Steels, cast irons	1000–1300	1–100	600–750	Local heating allows shallow case depth for less component distortion; can treat complex shapes.	Cylinder liners, steering box housings, gears, crankshafts
Electron beam	Plain C, med. C steels	1200–1300	1000–3000	600–750	Fast, large case depths. However, expensive.	Automobile transmission clutch cams
Chill casting	Cast irons	—	—	500	Simple, cheap. However, can be used for limited cast irons.	Cast camshafts
Work hardening						
Metal working	C-Mn steel (e.g., 1% C, 1.2% Cr, 13% Mn)	20°C or lower	—	600	A supplemental hardening process. Expensive and time-consuming.	Wear plates, earth-moving equipment, chute liners and piston bars
Shot peening	Plain C, low alloy steel, aluminum, Al-alloy	100°C or lower	100–500	—	A supplemental hardening process. Time-consuming.	Large ring gear, shaft, structural components of aircraft

TABLE 11.2 Typical Values of Case Depths at Various Frequency and Power Input Levels During Induction Hardening

Power input, W mm^{-2}	Frequency, kHz	Depth of hardening, mm
15–19	450	0.5–1.1
8–12	450	1.1–2.3
15–25	10	1.5–2.3
15–23	10	2.3–3.0
15–22	10	3.9–4.0
23–26	3	2.3–3.0
22–25	3	3.0–4.0
15–22	3	4.0–5.0

Source: Adapted from Guthrie and Archer (1975).

produced by either a motor or solid-state invertor. The inductor coil is usually made of low electrical resistive material, i.e., copper tubing which may be water cooled. Figure 11.3 illustrates various types of inductor coil employed for induction hardening. Depending on the area to be treated, the component may be held stationary, even though it may be spun to obtain uniformity of heat penetration, or alternatively, for such components as long shafts or machine beds, the induc-

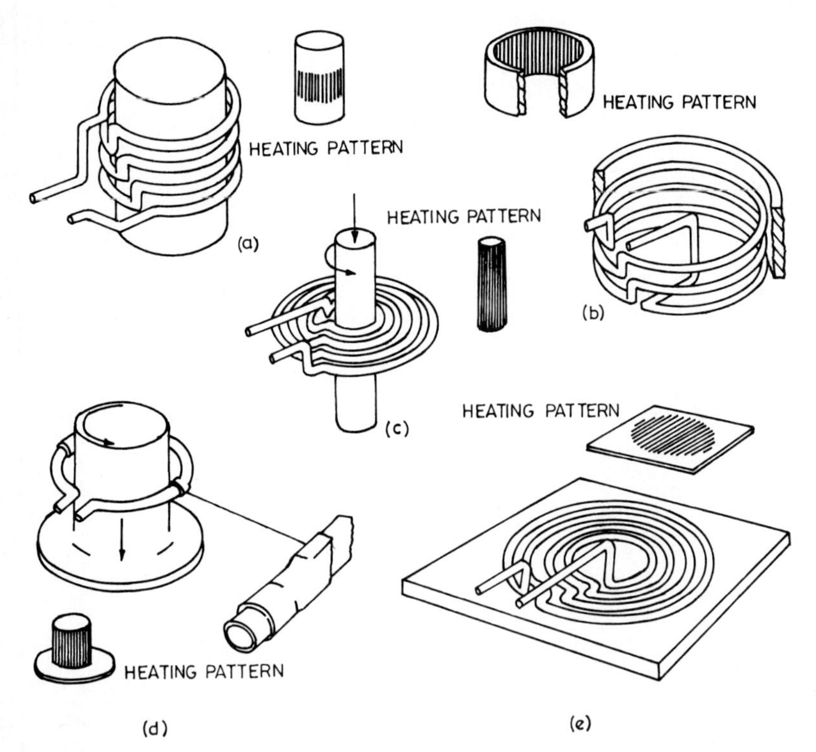

FIGURE 11.3 Some typical inductor coils and resulting heating patterns.

tor may be slowly swept along the component. Quenching, depending on the alloy content, can be achieved either by immersing the heated part in a tank or by spraying the component when heating has been completed. Steels to be induction hardened are of necessity those which austenitize readily because of the extremely short heating times. A fine uniform dispersion of carbide will dissolve more readily than coarse carbide, i.e., tempered martensite will austenitize readily while a coarse spheroidized structure represents a poor microstructure for induction hardening. Alloy steels, in which stable carbides form, must be heated to higher quenching temperatures for austenitization than carbon steels.

Medium-carbon steels containing from 0.3 to 0.5 percent C are most frequently induction hardened. With higher carbon contents, the risk of cracking is considerably increased, but with careful control, high-carbon steels can also be successfully hardened. Due to a high heating rate in induction hardening, on the order of several hundred degrees per second, the transformation of pearlite into austenite is shifted toward higher temperatures. Thus temperatures of transformation of bulk materials provide only a guideline. The steels are normally quenched in water or oil. In certain cases, the alloy steels can be cooled by means of an oil emulsion. The steels prior to induction hardening, should be in the hardened and tempered conditions, because their structure allows solution to carbon more easily than to an annealed structure. Fully annealed steel is not so suitable, because normally the time needed for dissolution of the carbides is longer than the normal heating-up time. Table 11.3 lists hardening temperatures and hardness values of steels and cast irons.

After being induction hardened, the steel is usually tempered at 150 to 200°C or higher if a lower hardness is desired. Parts that are to be ground should be tempered at least at 160°C. If the parts are tempered at lower temperatures there is considerable risk of cracking during grinding.

TABLE 11.3 Induction Hardening Temperatures and Expected Hardness for Various Ferrous Metals

Metals	Hardening temp., °C	Quench	Hardness, HRC (HV)
	Carbon and alloy steels		
0.30% C	900–925	Water	50 (513)
0.35% C	900	Water	52 (544)
0.40% C	870–900	Water	55 (595)
0.45% C	870–900	Water	58 (653)
0.50% C	870	Water	60 (697)
0.60% C	845–870	Water	64 (800)
		Oil	62 (746)
	Cast irons		
Gray iron	870–925	Water	45 (446)
Pearlitic/malleable iron	870–925	Water	48 (484)
Nodular iron	900–925	Water	50 (513)
	Stainless steels		
Type 420	1095–1150	Oil or air	50 (513)

Source: Adapted from Stevens (1981).

The induction hardening has numerous advantages over conventional methods of hardening such as

- Local hardening considerably reduces the power or energy requirements as compared to other heat treatment processes.
- Times required for induction hardening are short (1 to 30 s) and thereby increase the productivity.
- Components may be heat treated with practically no scaling so that the metal allowance for subsequent machining may be reduced.
- The process is capable of complete automation.

The few limitations of induction hardening are high capital cost of equipment and applicability of the process to certain shapes. Applications include cams, rotating shafts, machine ways, swivel pins, steering joints, gears, valves, and bearing surfaces.

11.1.2 Flame Hardening

Flame hardening is similar to induction hardening in terms of materials processed and their applications, but rather than using induction current to heat the ferrous metal, flame is used to heat the surface. Flame heating employs direct impingement of a high-temperature flame or of high-velocity combustion product gases. Gases used for heating are a mixture of fuel gas (natural gas, propane, or acetylene) with air or oxygen. The acetylene-oxygen mixture gives the most rapid heating rate and is used for shallow case depths or applications requiring a minimum transition zone. After heating, quenching is achieved by either spraying water or water-soluble oils onto the surface or by immersing the component in a tank (Anonymous, 1981; Elliott, 1978; Thelning, 1984).

The flame heat treatment is very simple—it can be as unsophisticated as the manual use of a welding torch to heat the required area to red heat, followed by a water quench. It can be applied by spinning the component (e.g., gear wheel) in the rotating chuck and heating the periphery. Progressive hardening is also possible by using a burner with a cooling spray system (Fig. 11.4). The rate of hardening is relatively low 50 to 200 mm min^{-1} and depends on the size of burner, and the required case depth of hardening is influenced by burner distance, and the duration of heating, usually short burner distance, results in deep hardening.

Steels commonly flame hardened are of the medium, 0.3 to 0.6 percent C range with alloy suitable for application. Flame hardening is particularly useful for components which would be too large for conventional induction furnaces, e.g., rolls, machine beds, and dies (Thelning, 1984). The main advantage of flame hardening is its simplicity. The few disadvantages of this process include overheating, spotty and uneven hardening, shallow depth of hardening, cracking or distortion of large components, and excessive scaling.

11.1.3 Laser Hardening

Industrial lasers (typically 1 to 15 kW CO_2) are used for laser hardening (Bennett, 1978; Carley, 1977a; Dawes and Cooksey, 1965; DesForges, 1977, 1978; Draper and Ewing, 1984; Gnanamuthu, 1979; Sandven, 1981; Seaman and Gnanamuthu,

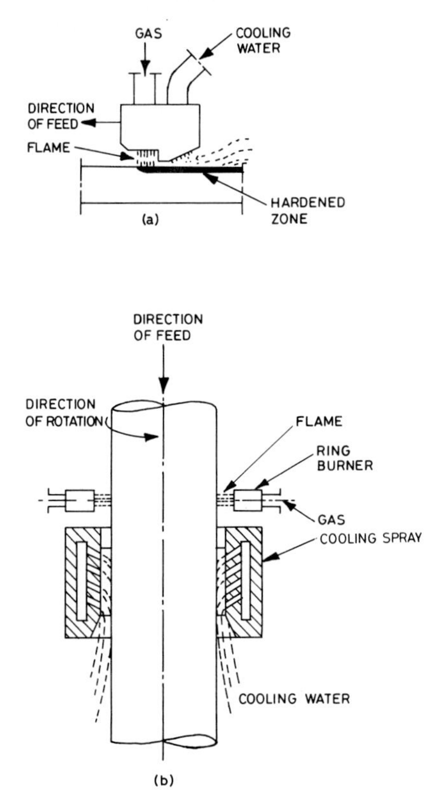

FIGURE 11.4 Illustration of: (*a*) Progressive flame hardening. (*b*) Progressive spin hardening. [*Adapted from Thelning (1984).*]

1975; Steen and Powell, 1981; Wick, 1976). In laser hardening, a laser beam (which can be scanned) irradiates, parts of its energy is absorbed as heat at the surface. If power density is sufficiently high, heat will be generated at the surface at a rate higher than heat conduction to the interior can remove it, and the temperature in the surface layer will increase rapidly. In a very short time, a thin surface layer will have reached austenitizing temperature, whereas the interior of the workpiece is still cool. Self-quenching occurs when the cold interior of the workpiece constitutes a sufficiently large heat sink to quench, with rates as great as 10^{11} °C s^{-1}, the hot surface by heat conduction to the interior at a rate high enough to allow martensite to form at the surface. By moving the laser beam over the workpiece surface, a point on the surface within the path of the beam is rapidly heated as the beam passes. This area is subsequently cooled rapidly by heat conduction to the interior after the beam has passed. Thus large areas can be hardened. With laser hardening, a wide variety of chemical and microstructural states can be retained because of rapid quench from the liquid phase.

A *laser* (an acronym for light amplification by stimulated emission of radiation) is a device for producing a beam of coherent, monochromatic radiations with wavelengths in the optical and infrared portions of the electromagnetic spectrum. In other words, a laser is a source of finely controlled energy, which, when in contact with material, produces a large amount of heat. Since the beam diameter is small, the power or energy density obtained from a laser is high (up to 10^9 W cm^{-2}). The ultimate effect of this energy depends upon the absorption by material and the time during which it interacts. The depth of hardening is controlled by the duration and power level of the laser beam. The power density is typically in the 1 to 2 kW cm^{-2} range. Hardening takes shorter times (fraction of a second) with lasers than with induction and flame hardening, and there is less heat effect on the surroundings and less distortion. Figure 11.5 shows some typical profiles of heat patterns produced by various beam optics.

Figure 11.6 shows the structure of a ductile cast iron after laser processing. The extent of carbon diffusion that occurred during the processing is clearly revealed by the region of martensite surrounding the graphite nodules.

The laser hardening process normally is applied to finish-machined or ground surfaces. Because the buildup of energy is rapid and well controlled, post heat treatment operations such as straightening or grinding are not required in most

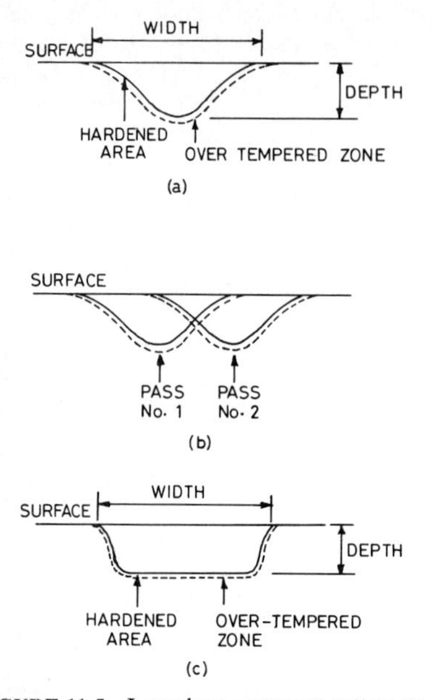

FIGURE 11.5 Laser heat treatment pattern results due to various beam optics: (*a*) Defocused beam, single pass. (*b*) Defocused beam, overlapping passes. (*c*) Oscillating beam, single pass. [*Adapted from Gilbert (1978).*]

instances. Further advantages include the ability to treat components with features which are not amenable to the induction hardening process, e.g., small bores and complex shapes. As a general rule, any component requiring localized surface regions of high hardness is a candidate for laser hardening; thus it competes with other thermal and chemical processes. Some of the disadvantages are that the capital cost is high and that the laser should have a line-of-sight access to the work.

This process has been adapted for several applications such as wear-resistant tracks on internal bores of cylinders (e.g., power steering housings of ferritic iron, malleable iron, and alloy cast iron, two-stroke railway diesel cylinder) in which distortion is a fundamental problem. The other components which can be laser treated are gear teeth, crankshafts, journals, cylinder bores, and pistons.

11.1.4 Electron Beam Hardening

In this process, the component surface is heated rapidly by direct bombardment or impingement of an accelerated stream of electrons. At the heat treatment cycle of 0.5 to 2.5 s, the electron flow stops abruptly to allow the component being processed to self-quench and to form a martensite structure with a compressive stress on the surface of the hardened area. Like laser hardening, this process is normally adapted for finished and/or ground surfaces because the buildup of energy is rapid and well controlled, and this negates the need of post heat treatments such as grinding and straightening (Carley, 1977b; Fiorletta, 1981; Gilbert, 1978; Schiller et al., 1982).

The electron beam hardening is a very efficient (90 percent) process as compared to laser hardening (10 percent). In addition, the operating cost of electron beam hardening is half as compared to laser hardening because of lower cost of capital equipment and consumables. However, depth of penetration of an electron beam is significantly higher than that of a laser beam. The electron beam hardening process is carried out in vacuum (0.5×10^{-2} Pa or $\sim 10^{-4}$ torr). The electron beam is produced by an electron beam gun (Schiller et al., 1982). The electron beam traverses a series of points (raster pattern) to harden a selective area. Figure 11.7 illustrates the various workpiece configurations and heating patterns for electron beam hardening. This figure shows that electron beam must have a line-of-sight access to the area to be treated. Electron beam guns having fine control over dwell time at any given point and beam energy (i.e., voltage and current), and focusing and scanning are commercially available. The power of the electron beam can be controlled through a program so that electron beam energy is high when the component is cold and energy reduces as components get heated. A typical relationship among case depth, surface temperature, and power

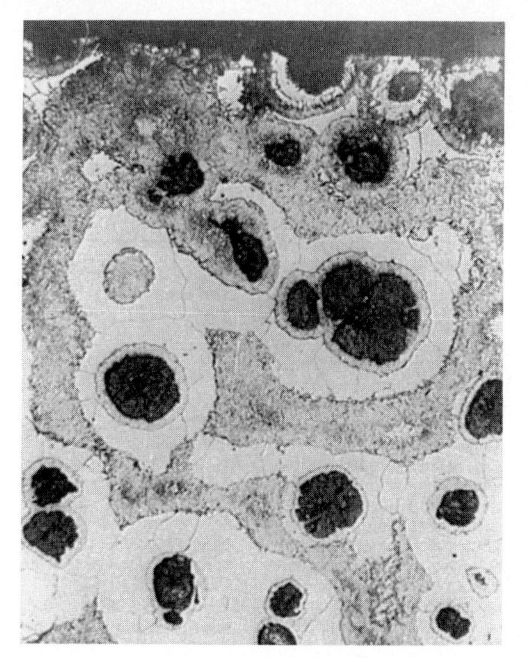

FIGURE 11.6 Microstructure of laser hardened ductile cast iron. [*Adapted from Sandven (1981).*]

inputs as a function of treatment time are shown in Fig. 11.8(*a*). The temperatures are sustained below the melting points and adjusted to produce a suitable or required case depth [Fig. 11.8(*b*)].

The electron beam hardening process offers several advantages such as

- Rapid flow of energy and restricted area of heating and considerable savings in energy as compared to induction and flame hardening.
- Minimum distortion due to rapid heating and fine control of electron beam.
- Due to self-quenching process, ductile austenite forms hard martensite from the subsurface to the surface.
- More efficient than laser hardening and generally deeper case depth than laser hardening.

The main disadvantages or limitations of this process are that there is a restriction on the size of the components and that the line-of-sight nature of this process restricts the process for simple-shape components only. The electron beam hardening process is used commercially for some precision components such as automatic transmission clutch cams which cannot be induction hardened because of distortion.

11.1.5 Chill Casting

The chill-casting process is used to obtain a peculiar change in the structure of cast irons. Usually the microstructure of conventional gray cast irons (approximately 3 to 3.5 percent C, 2 to 2.5 percent Si, 0.5 to 0.8 percent Mn, 0.1 percent

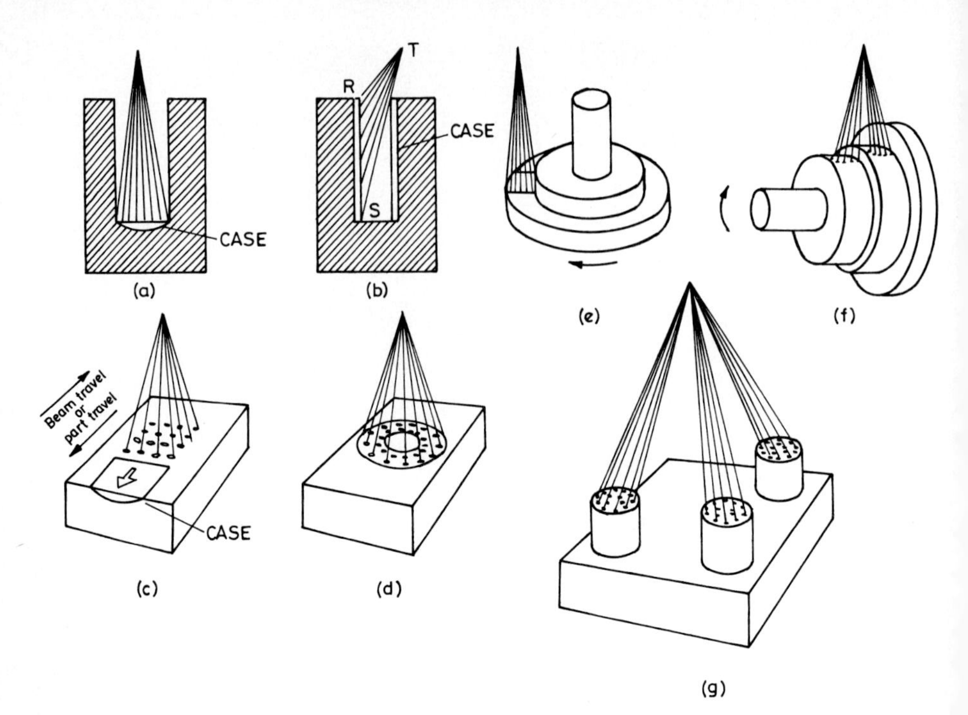

FIGURE 11.7 Workpiece configurations and heating patterns for electron beam heating. (*a*) Display static pattern within cavity in workpiece. (*b*) Maintain angle of workpiece rotation, RST, at 25° minimum. (*c*) Display static pattern and move the pattern or the workpiece to heat treat large areas. (*d*) Display static pattern. This annular pattern has well-defined inside and outside diameters. (*e*) Display static pattern and rotate workpiece. (*f*) Display more than one pattern and rotate workpiece. (*g*) Display multiple patterns on one workpiece or on a small group of workpieces for simultaneous hardening. Patterns may be similar or dissimilar in geometric shape. [*Adapted from Fiorletta (1981).*]

S, and less than 0.15 percent P), when cooled under normal conditions in a sand mold, consists of a soft structure (200/260 HV) with flakes of graphite and pearlite with some amount of ferrite. If the same casting of gray cast iron is cooled by chilling the mold, the resulting microstructure consists of hard brittle constituent Fe_3C (iron carbide) or cementite rather than soft graphite flakes. The depth of the chill during casting is very important for carrying out posttreatments such as grinding and lapping.

The microstructure of chill-casted gray cast iron is shown in Fig. 11.9 (Elliott, 1978). The microstructure of chill-casted cast iron is a mixture of plates of cementite and pearlite, which increases its hardness to 500 HV. The hardness is proportional to the amount of cementite present or, in other words, the level of carbon content. Additions of carbide-forming alloying elements such as Cr and V also increase the possibility of forming hard compound during chill casting and in turn increasing the hardness. In contrast, silicon has been found to promote graphite flake formation which decreases the hardness (see Chap. 4).

This chill-casting process is extensively followed for the production of cam shafts and cam followers for automobiles and rolls for rolling steel plates.

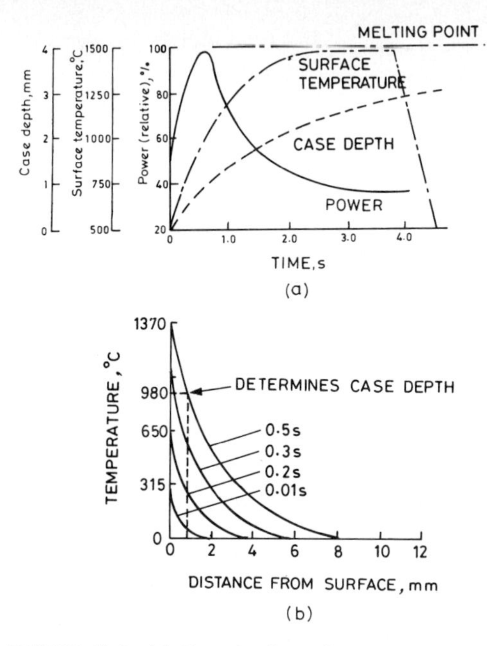

FIGURE 11.8 (*a*) Case depth, surface temperature, and power input as a function of treatment time of electron beam hardening. (*b*) Typical thermal gradients achieved. [*Adapted from Gilbert (1978).*]

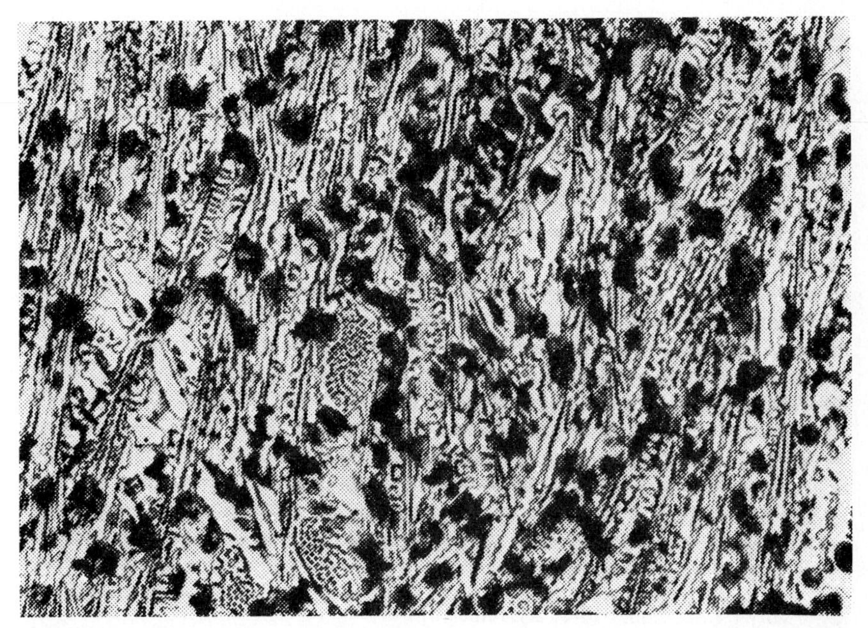

FIGURE 11.9 Microstructure of chilled-casted gray cast iron showing needles of hard carbides. [*Adapted from Elliot (1978).*]

11.1.6 Cold Working

When a ductile metal is deformed plastically below recrystallization temperature, such as by cold drawing, dislocations are produced in numbers proportionate to the degree of deformation. A complicated network of interlocking dislocations is thereby created which impedes the dislocation movement, and the metal is strengthened or work hardened (Cottrell, 1964; Peckner, 1964).

A high rate of work hardening implies mutual obstruction of dislocations gliding on intersecting systems. This can come about

- Through intersection of the stress fields of dislocations
- Through intersections which produce sessile locks
- Through the interpenetration of one slip system by another which results in the formation of dislocation jogs

The finer the dislocation network mesh becomes, the harder the material will be. Most evident is the increase in yield strength, hardness, and the simultaneous reduction in both toughness as measured by impact test and ductility as measured by the elongation and reduction of area. By heating cold-worked steel in the temperature range 100°C to 300°C, carbides and nitrides are precipitated on the dislocations which are thus impeded still more. Thereby an increased hardness and reduced toughness result. This process is called *work hardening, strain-age hardening*, or *cold working*.

Work hardening can be an advantage if the service requirements of a part allow its use in the as-formed condition. This is an important process for metals that do not respond well to heat treatments. Generally, the rate of work hardening is lower for hexagonal close packed (hcp) metals than for cubic metals. A significant reduction in the coefficient of friction (from 1.0 to 0.6 for Ag) and improvement in wear resistance (factor of 1.5 to 2.0 for Ag) of pure fcc metals such as A1, Cu, Au, and Ag by work hardening have been reported when sliding on abrasive surfaces such as steel files (Lin, 1969). Work hardening is used to improve the wear resistance in such materials as tool steels, high-carbon martensitic stainless steels, and carburized alloy steels. The typical applications include wear plates, earth-moving equipment, and chute liners.

We present here two methods of work hardening—metal working and shot peening.

11.1.6.1 ***Metal Working.*** Metal working refers to the large plastic deformation of metal under conditions of temperature and strain rates that induce strain hardening. In the metal-working process, an initially simple part—a billet or a blanket sheet, for example—is plastically deformed between tools (or dies) to obtain the desired final configuration. In the cold metal working, the workpiece enters at room temperature. Any subsequent increase in temperature, which may amount to several hundred degrees, is caused by the conversion of deformation work into heat. Metal working at high temperature is referred to as *hot working*. Various metal-working processes include drawing, extrusion, forming, and forging (Anonymous, 1988). An example of cold drawing–bar drawing is shown in Fig. 11.10. In the cold drawing of shapes, the basic contour of the incoming shape is established by cold rolling passes that are usually preceded by annealing. Annealing may occasionally be necessary after a number of drawing passes before the drawing operation is continued.

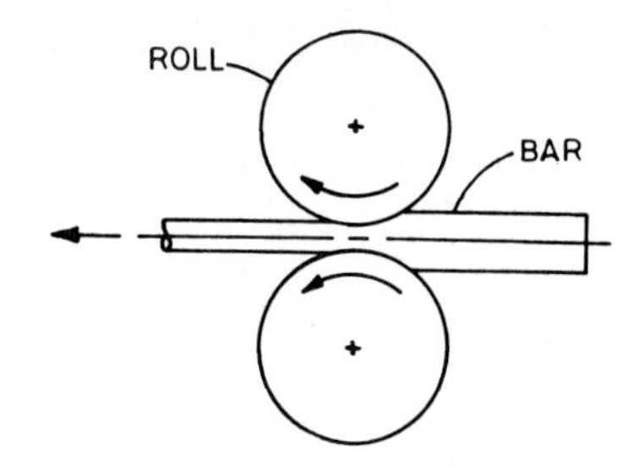

FIGURE 11.10 Cold drawing of a bar through undriven rolls.

Aluminum and its alloys, copper and its alloys, low-carbon and medium-carbon (up to 0.45 percent C) steels and stainless steels are the metals that are most commonly cold worked. The preceding listing is on the order of decreasing formability. Ductility of nickel-based alloys in the annealed condition makes them adaptable to virtually all methods of cold working. Some refractory metals (e.g., Mo, W, Nb, and Ta) and some platinum metals (e.g., Pt and Pd) can also be cold worked.

11.1.6.2 Shot Peening. Shot peening is a method of cold-working process in which compressive stresses are induced in the exposed metallic surfaces due to the impingement of a high-velocity stream of shots. When individual particles of shot in a high-velocity stream contact metal surface, they produce slight rounded depressions in the surface, thus stretching it radially and causing plastic flow of surface metal at the instant of contact. The effect usually extends to about 0.12 to 0.25 mm (up to 0.5 mm in some cases) below the surface. The metal beneath this layer is not plastically deformed. In the stress distribution that results, the surface metal has induced residual compressive stress parallel to the surface, while the metal beneath has reaction-induced tension. The surface compressive stress is usually several times greater than the tensile stress; this compressive stress offsets an imposed tensile stress such as that encountered in bending and considerably improves fatigue strength of the metal component. Peening action improves the distribution of stresses in surfaces that have been disturbed by grinding, machining, or heat treating (Mehelich et al., 1982).

The depth of the compressed layer depends upon the peening intensity, which is influenced by the velocity of stream, hardness, size, and weight of the shots and angle of impingement, and adequate exposure of the component surface. The quality and effectiveness of peening depend on the control of each of these independent variables. The equipment used in shot peening is essentially the same as that used in blast cleaning (see Chap. 7).

Shot peening itself is a finishing treatment, and usually no further processing of peened work is required, except for the application of a rust preventive for low-alloy steels. The as-peened surfaces of these steels are clean and chemically active and are highly susceptible to corrosion from fingerprints and other contaminants. Such surfaces are highly receptive to oils for rust prevention and lubrication and provide an excellent base for organic or inorganic coatings that do not require thermal treatment other than low-temperature baking. Temperatures high enough to relieve the beneficial compressive stresses imposed by peening must be avoided.

The major applications of shot peening are to increase fatigue strength, to strengthen the metals, and to relieve tensile stress that may contribute to stress corrosion cracking. The fatigue strengths of several carbon and low-alloy steels and aluminum alloys have been improved by the shot-peening process up to 30 to 60 percent (up to 87 percent in the case of 4340 cast steel) and 20 to 30 percent, respectively. However, the level of improvement of fatigue strength is considerably influenced by microstructure and alloying elements of the substrate and the velocity of stream of shots. The major limitations of shot peening are subsequent processes; for example, exposure to elevated temperatures and machining can nullify its beneficial results.

The typical applications of shot peening are improving fatigue strength of structural components of aircraft, precession collect, and salvaging parts such as large ring gears (restoring perfect roundness), shafts (restoring straightness).

11.2 Chemical Diffusion Treatments

Chemical diffusion treatments are defined as heat treatment processes in which the chemical composition of the surface layer of a component is deliberately modified by diffusion through the surface (Gregory, 1970, 1978; Gaucher et al., 1976; Stickels et al., 1981; Child, 1983; Thelning, 1984; Wensing et al., 1981). High hardness is obtained through the formation of compounds, and quenching is not required. By virtue of the core remaining comparatively soft and tough, the component as a whole shows high impact strength and development of compressive stresses in the surface layers during the chemical diffusion treatment, the fatigue strength of the component is also increased.

Chemical diffusion processes are applied almost exclusively to various irons and steels. The different approaches to the chemical diffusion processes are broadly classified according to the media used such as

- Pack process (e.g., carburizing in activated carbonaceous solid material)
- Salt-bath process (e.g., salt-bath carbonitriding)
- Gaseous process (e.g., gas carburizing and gas nitriding)
- Vacuum and plasma processes (e.g., vacuum carburizing and ion or plasma nitriding).

Gas and vacuum processes produce better-quality cases with precise control of diffused-element profile and minimize the possibility of oxidation which otherwise may occur during the transfer from furnace in pack (solid) and salt-bath (liquid) processes. Later processes, however, are cheaper and simpler to use. The plasma process produces yet better-quality cases in considerably reduced treatment time or temperature. The surface hardness of plasma-treated components is generally higher than that of other processes. Some processes used for tribological applications may require slight polishing to remove the top layer of a couple of microns as its relative roughness could abrade a softer bearing surface with which it is in contact, and the top layer (such as in nitriding) may be brittle.

The typical elements diffused in the surface include nonmetallic elements, e.g., C and N, metalloids, e.g., B and Si, and metals, e.g., Cr and Al. On an atomic scale, the diffusion process may be by an interstitial or a substitutional mechanism, and initially the diffusion element or elements go into solution in the metal surface. However, as the concentration of the diffused elements increases, their solubility limits are exceeded and the second phase may be precipitated. Initially these precipitations of second phases are very finely divided and well dispersed, but eventually the concentration of the second phases may build up to such an extent that it forms a continuous layer on the outer-exposed surface. Usually these hard layers are very hard and brittle. So, this practice leads to the formation of hard alloys only on the outer surface up to a maximum of a few millimeters of depth.

The most widely used diffusion processes for tribological applications of irons and steels are the carburizing, carbonitriding, nitriding, nitrocarburizing, and boriding processes in which carbon, nitrogen, or boron are diffused into the surface (to a depth of typically 25 to 75 μm) of the alloys of iron. Boriding provides the highest hardness compared to any other diffusion process. If the processing is desired at lower treatment temperatures (500 to 600°C), then nitrocarburizing and nitriding processes are preferred. Table 11.4 summarizes the main properties and functional characteristics of various chemical diffusion treatments. For comparisons, properties of ion implantation, WC/Co plasma sprayed, and electrochemically deposited chromium are also included for comparison in Table 11.4.

11.2.1 Austenitic Carburizing

The carburizing treatments (sometimes referred to as *case hardening*) are austenitic thermochemical treatments for steels in that they involve the diffusional addition of the interstitial alloying elements into the austenitic phase and rely on subsequent transformation of the austenite to martensite to produce a high surface hardness. The steel changes its structure from a bcc lattice, ferrite, to an fcc lattice, austenite, at about 720°C. Austenite is capable of dissolving carbon, and the longer the time or the higher the temperature, the deeper the carbon diffuses. If the steel is allowed to cool slowly from the carburizing temperature, the dissolved carbon will come out of solution to form a mixture of ferrite and iron carbide and pearlite. Fast cooling or quenching prevents this, that is, the carbon's being kept in solution, and the steel adopts a body-centered tetragonal structure, martensite. The hardness of the martensite will depend on its carbon content (Thelning, 1984).

In the carburizing process, carbon is diffused into the top surface layer of low-carbon steel by heating it to temperatures normally between 825 and 925°C in the presence of a carbon-rich medium, which can be a solid (charcoal), molten salt, or gaseous. Diffusion of the carbon from the surface toward the interior produces a case that can be hardened, usually by fast quenching from the bath. In double quenching, parts are also subsequently reheated to a lower temperature (780 to 820°C) and then requenched in oil or water. Double quenching is applied to steels that have not been fine-grain treated. During carburizing, it is usual to produce a surface with carbon composition of 0.65 to 0.8 percent, which produces a completely martensitic case with a high surface hardness (Elliott, 1978). Quenching of a carburized steel not only increases the surface hardness but also produces a favorable residual stress distribution of compressive stresses in the surface. Compressive stresses greatly increase the fatigue strength of case-hardened parts.

The carburizing process is carried out on low-carbon steels containing less than 0.2 percent C (typically 0.15 to 0.18 percent), which can range from free cutting mild steel to the high-strength nickel chromium alloys, so that the core is quite tough even in the as-quenched conditions without tempering. Cr, Mn, Mo, and Ni are the principal alloying elements used in carburizing steels and provide hardenability during the quenching operation in thick components. This process produces a surface hardness of up to about 850 HV, which is accompanied by improved wear resistance. It is normal to relieve stress at 180°C after quenching to reduce the inherent brittleness of the quenched martensite and its tendency to crack during grinding.

The rate of uptake and penetration of carbon depends mainly on the carburizing atmosphere, the time and temperature of operation, and the composition of steel. Diffusion rate increases considerably with temperature so diffu-

TABLE 11.4 Process Parameters, Properties, and Applications of Chemical Diffusion Treatments

Treatments	Suitable substrates	Diffusion elements	Media	Treatment temp., °C	Case depth, μm	Hardness, HV (kg mm^{-2})	Maximum service temp., °C	Advantages/ disadvantages	Applications
Carburizing	0.2%C, C-Mn, Cr-Mo, Ni-Mo, Ni-Cr-Mo steels	C	Solid, liquid, gas, vacuum, plasma	800–950	50–6000	600–850	200	Cheap, highest case depth possible. However, treatment causes component distortion.	Cams and camshafts, rockers and rocker shafts, bushes, and transmission gears
Carbonitriding	Mild, plain C, low-alloy steels, Cr-Mo, Ni-Cr-Mo steels	C, N	Liquid, gas	800–870	25–500	600–850	200	Carbonitriding causes less distortion than carburizing. Wear resistance and corrosion resistance better than by carburizing. Resistance to scuffing due to nitrogen. However, slower rate of case formation.	Bearing cages, gears, air-compressor, cylinders, camshafts
Nitriding	Steels containing Al, Cr, Mo, V, W, med.-C alloy steel, austenitic stainless steel	N	Solid, gas, plasma	500–590	5–250	900–1150, 1300 (plasma nitrided)	500	High hardness of case retained at high temperatures. Distortion reduced due to low process temperature and lack of quenching (distortion is almost zero). However, long process times required.	Crank and camshafts, transmission parts, bushes, rotating, and reciprocating parts, hydraulic and pump components, high-speed steel cutters
Nitro-carburizing	Mild steel, low-alloy steel, steel containing Cr, Ni-Cr	N, C	Liquid, gas, plasma	570	5–250	600–900	500	Lower hardness but better wear resistance.	Aluminum extrusion and die casting dies, gears, pumps, cylinder liners, crankshafts, cylinder heads, and plastic molding tools

Boriding	Steels containing Mo, Cr, and Ti, cast irons, Ni-, Co-based alloys, tool & die steels, refractory metals	B	Solid, liquid, gas	900–1100	5–250	900–2300	800	Highest hardness of any surface treatment process.	Designed for extreme abrasive wear and for tribooxidation. Used for tools such as circular cutter, four slid king pins, draw punch taps, bushings, molding dies, forming dies, wear plates.
Chromizing	High-C steels	Cr	Solid, liquid, gas	950–1300	100–200	1200–1300	600	Not brittle, good oxidation resistance, steels can be later heat treated.	Steam power equipments, water and steam fixtures, valves, connections, pipe.
Aluminizing	Low-C, medium-C steels	Al	Solid, liquid	750–1000	200–500	—	700	Resistance to corrosion and oxidation at high temperature.	Aircraft gas turbine blades, pyrometer protection tubes, flue stacks, boiler tubing.
Siliconizing	Steels, malleable iron	Si	Solid, gas	950–1200	500–1250	200–300	—	Improves wear resistance, but main purpose is corrosion and oxidation resistance. Steels can be heat treated afterward.	Pump shafts, pipeline, fitting nuts, bolts, rollers for bottle washing machine.
Sherardizing	Iron, steel	Zn	Solid	350–450	—	—	400	Low-temperature process, impart resistance to wear and corrosion.	Threaded parts, e.g., nuts, bolts, washers, scaffold clips.
Ion implantation	Any	Any	—	25–200	0.01–0.5	*	—	Produces ultrathin case depths. However, large capital costs and running expense.	Improve wear and fatigue resistance of steel and cemented tungsten carbide. Shown improvements in steel press tools, mill rolls, ring cutters, gear cutters, injection molding screws.
WC/Co plasma sprayed (for reference)	—	—	Solid	250	500	1600	500	—	—
Electrochemical deposited Cr (for reference)	—	—	Liquid	50	10–250	1000–1200	500	—	—

*Increase in hardness by a factor of about 1.5 to 4 in the case of steels.

sion times giving the same thickness of diffused layer decrease as diffusion treatment temperature increases. For normal carburizing, the case depth (CD) and time are related approximately as

$$\mathrm{CD} = k\sqrt{t} \tag{11.4}$$

and

$$k = 660 \exp(-8287/T)$$

where t is time in hours and T is temperature in Kelvins.

Quenching of the carburized component is normally carried out in agitated oil. Water quenching may cause cracking in the case of small components of alloy steels, but it is allowed for carbon steels and heavy components of alloy steels. Quenching is best performed without exposing the components to air, which may result in oxidation and decarburization. Quenching is most readily carried out in a gas-carburizing process.

Dimensional changes during carburizing are significant, and higher than those occurring during nitriding. The main sources for distortion may be as follows (Child, 1980a):

- Growth occurring during the hardening process as the normalized or annealed ferrite-carbide structure is transformed to martensite (4 percent volume increase)
- Growth occurring in the top surface layer because of the absorption of carbon in the lattice
- Distortion caused by creep or sagging because of ineffective jigging, and by the effect of thermal stress
- Concentration caused by precipitation of carbides during tempering

There is considerable variation in volume change depending on the steel composition, geometry, and heat treatment parameters. This distortion can be minimized by (Thelning, 1984)

- Carburizing at comparatively lower temperature and hence for longer time to obtain same case depth
- Normalizing or prehardening and tempering of the component before carburizing
- Hardening by reheating after carburizing for plug or press quenching in order to avoid distortion
- Use of quenchants better than water, such as hot oil

Carburizing can be carried out using particulate solid mixtures (pack cementation), liquid, or gaseous media. Gas carburizing is by far the most important chemical diffusion process for volume mass production. The more recent developments in carburizing processes include vacuum processing (Westeren, 1972), plasma-activated gas carburizing (Grube, 1980), and low-toxicity (noncyaniding) salt-bath treatments (Astley, 1975).

11.2.1.1 Pack Carburizing. Nowadays, this process is confined to components which are too large for gas-carburizing furnaces or where very heavy cases (>2.5 mm) are required. In this process, the component is heated in a sealed container packed in a granular solid mixture of hardwood charcoal or coke (grain size from 3 to 10 mm in diameter) and activator (typically 10 to 20 wt % barium or calcium or sodium carbonate) bound to charcoal or coke by oil, tar, or molasses. These energizers facilitate the reduction of carbon dioxide with carbon to form carbon monoxide. Oxygen present in the container reacts with the charcoal, and during the heating period a CO_2-rich mixture is produced which then continues to react with the charcoal as follows (Stickels et al., 1981):

$$CO_2 + C \rightarrow 2CO \tag{11.5}$$

As the temperature rises, the equilibrium is displaced to the right, i.e., the gas mixture becomes progressively richer in CO. At the steel surface, the CO breaks down as follows:

$$2CO \rightarrow CO_2 + C \tag{11.6}$$

The atomic or nascent carbon thus liberated is readily dissolved by the austenite phase of the steel and diffuses into the body of the steel, even to the extent of forming carbides (cementite). Since the amount of atmospheric oxygen in a charcoal packing can vary and may be insufficient to produce the carburizing gas, the charcoal is mixed with an energizer, usually barium carbonate, which reacts during the heating-up period as follows:

$$BaCO_3 \rightarrow BaO + CO_2 \tag{11.7a}$$

$$CO_2 + C \rightarrow 2CO \tag{11.7b}$$

The CO_2 originally formed then reacts with carbon in the charcoal, producing the active CO. If the temperature is increased but the pressure kept constant, the reaction proceeds from left to right, i.e., more CO is produced in Eq. (11.7*b*).

The post-pack-carburizing treatment required after this process depends on the carbon level developed at the surface and grain size produced in the steel. Fine-grain steel is best hardened from 780 to 820°C, at which excess carbide is not dissolved in the austenite, and difficulties associated with excessive carbon content, such as retained austenite, are avoided. Double hardening has been used in the past for coarse-grained steels. The component is first heated to 870 to 900°C, soaked and oil quenched, and then reheated to 780 to 820°C, soaked and oil quenched. The first hardening process austenitizes the core and refines the structure and, as a result, develops enough toughness, and the second hardening process ensures a satisfactory case structure.

The pack-carburizing process is incapable in those situations requiring

- Precise control of carbon profile
- Production of lower case depths
- Direct quenching

11.2.1.2 ***Liquid Carburizing.*** Liquid carburizing, also known as *salt-bath carburizing*, is a process in which steel parts are held above the austenizing temperature in a molten salt that will introduce carbon and nitrogen or carbon alone into the metal. It is a widely used process because of the low capital cost of the plant and because it has several advantages over gas and pack carburizing, such as

- Short heating times, because of the higher heat transfer coefficients in liquid salt
- Uniform bath temperature with convective circulation, ensuring uniform carburizing

- Exclusion of atmospheres
- Lower distortion and sagging because of buoyancy effect of salts
- Controlled quenching such as martempering

However, the cleaning of salt-bath carburized components is difficult especially when blind holes, recesses, and internal threads are present.

Most liquid-carburizing baths contain cyanide, which introduces both carbon and nitrogen into the case. One type of liquid-carburizing bath, however, uses a special grade of carbon, rather than cyanide, as the source of carbon. This bath produces a case that contains only carbon as the hardening agent. Liquid carburizing may be distinguished from cyaniding (which is performed in a bath containing a higher percentage of cyanide) by the character and composition of the case produced. Cases produced by liquid carburizing are lower in nitrogen and higher in carbon than cases produced by cyaniding. Cyanide cases are seldom applied to depths greater than 250 μm; liquid carburizing can produce cases as deep as 6 mm.

Cyanide-containing liquid-carburizing baths are used at low temperatures (from 845 to 900°C) for shallower depths (up to 250 μm) and high temperatures (900 to 955°C) for deeper case depths (above 500 μm). The active carburizing agent in the salt bath is sodium cyanide (NaCN) or potassium cyanide (KCN). Low-temperature baths are generally of the accelerated cyanogen type containing other constituents (10 to 23 percent NaCN, 20 to 40 percent NaCl, 30 percent max Na_2CO_3, 0.5 to 1 percent NaNCO, and KCl, other salts of alkaline earth metals and accelerators). In a low-temperature bath, several reactions occur simultaneously, depending on bath composition, to produce various end products and intermediates. These reaction products include: C, Na_2CO_3, N_2 or 2N, CO, CO_2, Na_2CN_2 (cyanamide), and NaCNO. Two of the major reactions believed to occur in the operating bath are "cyanamide shift" and the formation of cyanate (Wensing et al., 1981):

$$2NaCN \rightarrow Na_2CN_2 + C \tag{11.8}$$

and either

$$2NaCN + O_2 \rightarrow 2NaNCO \tag{11.9}$$

or

$$NaCN + CO_2 \rightarrow NaNCO + CO \tag{11.10}$$

Reactions that influence cyanate content proceed as follows:

$$NaNCO + C \rightarrow NaCN + CO \tag{11.11}$$

and either

$$NaNCO + 2O_2 \rightarrow 2Na_2CO_3 + 2CO + 4N \tag{11.12}$$

or

$$4NaNCO + 4CO_2 \rightarrow 2Na_2CO_3 + 6CO + 4N \tag{11.13}$$

Reactions that produce either CO or C, which are beneficial in producing the desired carburized case, are

$$Fe + 2CO \rightarrow Fe[C] + CO_2 \tag{11.14}$$

and

$$Fe + C \rightarrow Fe[C] \tag{11.15}$$

High-temperature baths consist of cyanide and a major proportion of barium chloride with or without supplemental acceleration from other salts of alkaline earth metals (6 to 16 percent NaCN, 30 to 55 percent $BaCl_2$, 30 percent max. Na_2CO_3, 0.5 percent max. NaNCO, and salts of other alkaline earth metals, KCl, NaCl, and accelerators). Although the reactions shown for low-temperature salts apply in some degree, the principal reaction is the so-called cyanamide shift:

$$Ba(CN)_2 \rightarrow BaCN_2 + C \tag{11.16}$$

In the presence of iron, the reaction is

$$Ba(CN)_2 + Fe \rightarrow BaCN_2 + Fe[C] \tag{11.17}$$

Liquid carburizing can be accommodated in a noncyanide bath containing a special grade of carbon. In this bath, carbon particles are dispersed in the molten salt by mechanical agitation. Operating temperatures for this type of bath (900 to 955°C) are generally higher than those for cyanide-type baths. The chemical reaction is thought to involve absorption of carbon monoxide on carbon particles. The carbon monoxide is generated by reaction between the carbon and carbonates, which are major ingredients of the molten salt. The absorbed carbon monoxide is presumed to react with steel surfaces much as in pack or gas carburizing. Parts that are slowly cooled following noncyanide carburization are more easily machined than parts slowly cooled following cyanide carburization.

11.2.1.3 Conventional Gas Carburizing. The conventional gas-carburizing process, often referred to as *case carburizing*, is considerably simpler and more widely used than the pack- and liquid-carburizing processes. It is accomplished by heating the component in a gaseous medium containing carbon. Usually, gas carburizing is carried out using an atmosphere of endothermic carrier gas (partially burnt natural gas or propane) enriched with methane or propane. The main constituents of the atmosphere are CO, N_2, H_2, CO_2, H_2O, and CH_4. Of these constituents, N_2 is inert, acting only as a dilutent. In such cases carburizing occurs principally as a result of the decomposition of carbon monoxide (Stickels et al., 1981) as

$$2CO \rightarrow (C)_{Fe} + CO_2 \tag{11.18}$$

and

$$CO + H_2 \rightarrow (C)_{Fe} + H_2O \tag{11.19}$$

A carburization process based solely on the decomposition of CO would require large flow rates of atmosphere gas. Methane or propane enrichment of endothermic gas provides the primary source of carbon for carburizing by slow reactions as

$$CH_4 + CO_2 \rightarrow 2CO + 2H_2 \tag{11.20}$$

and

$$CH_4 + H_2O \rightarrow CO + 3H_2 \tag{11.21}$$

which reduces the concentrations of CO_2 and H_2O, respectively. These reactions regenerate CO and H_2. The sum of reactions in Eqs. (11.18) and (11.20) and in (11.19) and (11.21) is reduced to

$$CH_4 \rightarrow (C)_{Fe} + 2H_2 \tag{11.22}$$

The carbon potential control of the furnace atmosphere is achieved by varying the flow rates of the hydrogen enrichment gas and maintaining a steady flow of endothermic carrier gas. With the increasing cost of hydrocarbons, from whatever source, synthetic carburizing gases are now being actively promoted.

With gas carburizing in sealed quench and continuous furnaces, it is possible to quench the charge in oil with no possibility of oxidation or decarburization which may occur during the transfer from the furnace in other processes.

11.2.1.4 Vacuum Carburizing. Vacuum carburizing is a high-temperature gas-carburizing process that is carried out at pressure below atmospheric pressure and at temperatures typically ranging from 980 to 1095°C. This process involves the use of conventional electrically heated cold-wall vacuum furnaces using graphite (or SiC) resistance-heating elements (Casper, 1977; Child, 1976). The component is heated to the carburizing temperature under a moderate vacuum of 7 to 50 kPa (50 to 375 torr). The furnace atmosphere usually consists solely of an enriching gas such as natural gas, pure methane, or propane. Nitrogen is sometimes used as a carrier gas.

Vacuum carburizing proceeds by the dissociation of hydrocarbon gas at the steel surface and by direct absorption of carbon; hydrogen gas is liberated (Stickels et al., 1981). For instance, the reaction of methane is

$$CH_4 \rightleftharpoons (C)_{Fe} + 2H_2 \uparrow \tag{11.23}$$

At normal carburizing temperatures, there is a relatively large driving force for the reaction in the above equation to proceed from left to right. The reaction favors rapid dissociation of the gas and rapid absorption of carbon by the hot steel surfaces. Because this reaction does not normally reach equilibrium, equilibrium carbon potential control normally is not used in vacuum carburizing.

Vacuum carburizing employs the so-called boost-diffuse method, a two-step process. In the first step, carburizing is carried out at saturation and then followed by a diffusion step (at 50 kPa or 400 torr) in which carbon diffuses at low pressure (10 kPa or 75 torr) causing the surface carbon to decrease and case depth to increase. The carburizing gas is used to maintain the pressure. The diffusion step is carried out at normal vacuum pressures of 10 kPa or 75 torr. Figure 11.11 shows the variations of temperature and pressure with time in vacuum carburizing (Westeren, 1972).

The advantages of vacuum carburizing over other conventional carburizing processes are as follows:

- Use of natural gas directly eliminates the need for carrier gas.
- Process times are short because of high carburizing temperature.
- High thermal efficiency of vacuum furnaces make the process economically viable.
- Precise control of process variables results in exceptional uniformity and repeatability.

11.2.1.5 Plasma Carburizing. Chemical diffusion treatment in plasma or dc glow discharge has been explored for carburizing (Grube, 1980; Grube and Gay, 1978; Yoneda and Takami, 1977). In plasma or ion carburizing, like vacuum carburizing, methane gas is ionized by applying dc voltage between electrodes

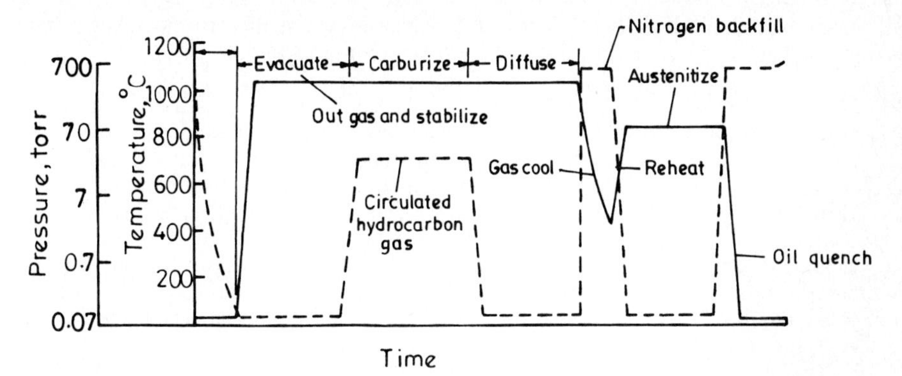

FIGURE 11.11 Variation of temperature and pressure in vacuum carburizing. [*Adapted from Westeren (1972).*]

and keeping the substrate as cathode. The additional electric field causes a net flux of charged particles and energetic neutral molecules toward the substrate. This increases the number of ions and/or molecules incident on the component surface, which in turn results in higher carburizing rates. Glow discharge provides some heat to the component. Usually plasma carburizing is carried out at 850 to 950°C. Figure 11.12 gives variations of temperature and pressure during plasma carburizing (Booth et al., 1983).

Plasma carburizing provides better control of diffusion and uniformity of

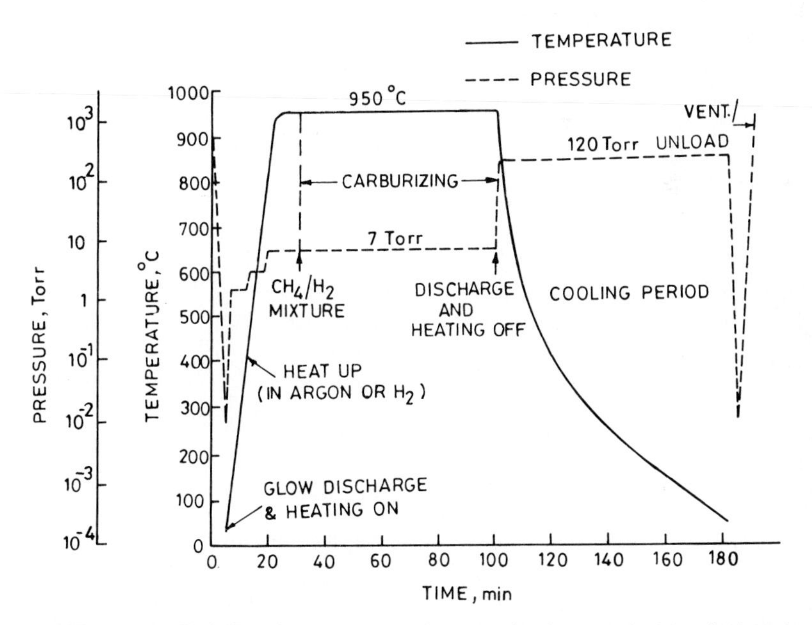

FIGURE 11.12 Variation of temperature and pressure in plasma carburizing. [*Adapted from Booth et al. (1983).*]

carburization all over the surface and low heat losses as the process takes place at comparatively lower pressures (100 to 500 Pa or 0.8 to 3.8 torr).

11.2.1.6 ***Properties of Carburized Steels.*** Carburizing produces a surface hardness of up to about 850 HV, and hardness is accompanied by improved wear resistance and fatigue strength. Some treated steels may also provide corrosion resistance. The surface hardness depends on the surface carbon content which varies slightly for different grades of alloy steels (Fig. 11.13). The carbon content decreases within the subsurface as distance increases. This depends upon the treatment temperature and the medium. Figure 11.14 demonstrates that an increased process rate is obtained at similar carburizing temperatures when using plasma carburizing. This figure also shows that much higher surface carbon concentrations are rapidly established in plasma carburizing.

The gas-carburizing process is more commonly used compared to other carburizing processes. Typical applications include cams and camshafts, rockers and rocker shafts, bushes, bearings, and transmission gears.

11.2.2 Austenitic Carbonitriding

Carbonitriding is a modified form of carburizing, rather than a form of nitriding. In the carbonitriding process, carbon as well as nitrogen are introduced simultaneously into the surface of the steel in the austenitic condition at temperatures between 800 and 870°C using either liquid or gaseous media and then quenched. In carbonitriding, the diffused amount of carbon is more than that of nitrogen; for instance, in a typical case of mild steel (0.08 percent C, 0.19 percent Si, and 0.4 percent Mn), the amount of C and N is found on the order of 0.80 percent and 0.26 percent, respectively. Nitrogen increases hardness more effectively than

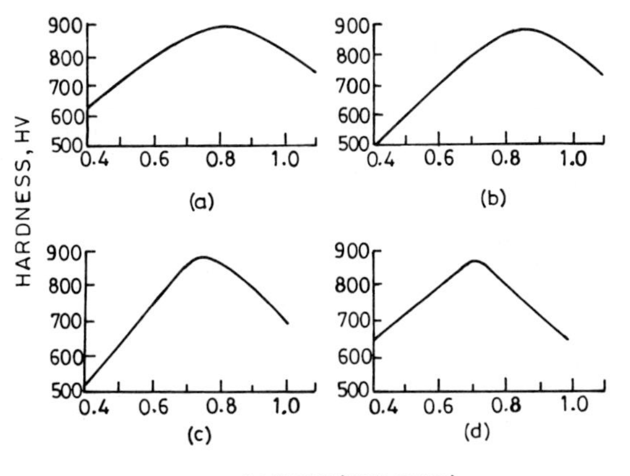

FIGURE 11.13 Influence of carbon content on surface hardness for carburized steels quenched from 925°C: (*a*) Nonalloyed steel. (*b*) C-Mn steel (214 A15). (*c*) Ni-Cr steel (635 M15). (*d*) Ni-Cr steel (637 M17). [*Adapted from Dawes and Tranter (1974).*]

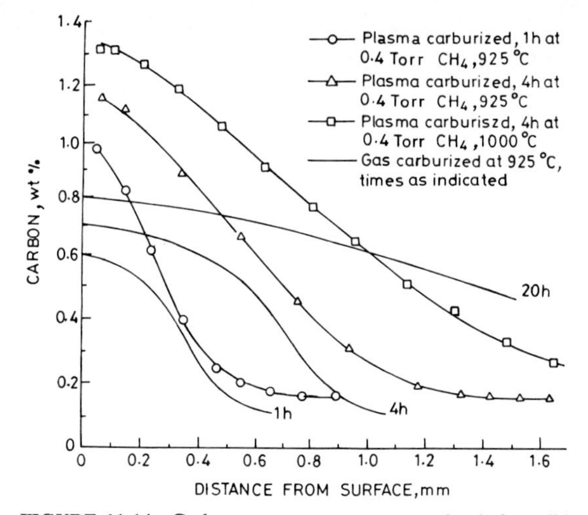

FIGURE 11.14 Carbon contents versus case depth for mild steel carburized in gas and plasma. [*Adapted from Dawes and Tranter (1974).*]

carbon diffusion in the case, and nitrogen is also an austenite stabilizer. Thus, carbonitriding is of use primarily to impart a hard wear-resistant case. Under certain cases, this process is used to correct skin hardenability after conventional carburizing that has resulted in internal oxidation and loss of alloying elements that impart hardenability to the austenite. Carbonitriding can be applied to any of the steels suitable for carburizing, for instance, mild, plain carbon or very low alloy steels. The total surface interstitial content for the treatment will be a function of both the carburizing and the nitriding potential of the surrounding media.

Although carbonitriding is a modified carburizing process, carbonitriding is largely limited to a case depth of 750 μm or less, while no such limitation is applied to carburizing. Two reasons for this are: (1) Carbonitriding is generally done at slightly lower temperatures (typically 870°C and below), whereas, because of the time factor involved, deeper cases are produced by processing at higher temperatures, and (2) the nitrogen addition is less readily controlled than is the carbon addition, a condition that can lead to an excess of nitrogen and, consequently, to high levels of retained austenite and case porosity when processing times are long. Carbonitriding causes less distortion than carburizing.

11.2.2.1 Liquid Carbonitriding. Liquid carbonitriding, also called *salt-bath carbonitriding* or *cyaniding*, is a process in which the steel is heated above austenizing temperature in a bath containing alkali cyanides and cyanates. The surface of the steel absorbs both carbon and nitrogen from the molten bath. When quenched in mineral oil, paraffin-base oil, water, or brine, the steel develops a hard surface layer, or case, that contains less carbon and more nitrogen than the case developed in activated liquid-carburizing baths.

The active carbonitriding agent in the salt bath is sodium cyanide (NaCN) with a concentration of 30 to 97 percent. The inert salts-sodium chloride and sodium carbonate are added to cyanide to provide fluidity and to control the melting points of all mixtures. A sodium cyanide mixture grade 30, containing 30 percent NaCN, 40 percent Na_2CO_3, and 30 percent NaCl, is generally used for cyaniding on a production basis. The cyaniding treatment results in the presence of carbon and nitrogen in the case (Wensing et al., 1981; Thelning, 1984). Carbon monoxide and nascent nitrogen are active hardening agents of the cyaniding bath, which are derived from sodium cyanate (NaNCO) which is formed from the molten cyanide (NaCN) in the presence of the air at the surface of the bath which in turn liberates carbon and nitrogen at the surface of the steel (Gregory, 1966; Nishida and Tanaka, 1968; Wensing et al., 1981)

$$4NaCN + 2O_2 \rightarrow 4NaNCO \tag{11.24}$$

$$4NaNCO \rightarrow Na_2CO_3 + 2NaCN + CO + 2(N)_{Fe} \tag{11.25}$$

$$2CO \rightarrow CO_2 + (C)_{Fe} \tag{11.26}$$

$$NaCN + CO_2 \rightarrow NaNCO + CO \tag{11.27}$$

The first and the last reactions take place at the interface between the salt bath and the atmosphere; the second and third take place at the interface between the salt bath and the steel. The nascent carbon and nitrogen diffuse into the steel, which produces a case of carbide and nitride.

The rate at which cyanate is formed and decomposes, liberating carbon and nitrogen at the surface of the steel, determines the carbonitriding activity of the bath. The amount of carbon and nitrogen picked up by the steel depends mainly on the cyanide content of the bath and its temperature. The carbon concentration increases whereas the nitrogen decreases as the cyanide content increases. With rising temperature, at a given concentration of NaCN, the concentration of carbon increases and that of nitrogen decreases (Thelning, 1984).

11.2.2.2 Gas Carbonitriding. This is a modified gas-carburizing process in which ammonia is introduced into the carburizing gas atmospheres (Stickels et al., 1981). The usual practice is to add specific amounts of ammonia by controlling gas inlet rates in the 3 to 8 percent range. The nitrogen content of the steel is a function of both temperature and ammonia content, 850°C is a typical temperature as less ammonia is required, and distortion is at minimum. The normal value for the C and N content of the surface layer are, for example, 0.8 percent and 0.4 percent, respectively, for a mild steel. After treatment, the steel is hardened in the same way as after gas carburizing.

11.2.2.3 Properties of Carbonitrided Steels. The properties achieved in carbonitrided steel are almost similar to carburized steel, but there are a few essential differences, as follows:

- The surface hardenability is improved, so that in some instances oil, rather than water, quenching can be used and quench cracking minimized.
- Higher wear and corrosion resistance for most service conditions is achieved than with carburizing.
- Resistance to tempering is higher than carburized cases (Bell, 1969).

The carbonitrided steels exhibit better resistance to adhesive wear than carburized steels. The mixed layer of carbides and nitrides of steel alloys also have greater stability and resistance to thermal softening at elevated temperatures. Case formation in carbonitriding occurs at a slower rate than carburizing. Thus, carbonitriding is best for small components and not used if heavy cases are desired. Typical applications include bearing cages, gears, air-compressor cylinders, and camshafts.

11.2.3 Ferritic Nitriding

Nitriding treatments are ferritic thermochemical treatments for low-alloy and tool steels and usually involve the introduction of atomic nitrogen into the ferrite phase in the temperature range of 500 to 590°C. Consequently no phase transformation occurs on cooling to room temperatures. The nitrogen is introduced into the steel surface by reaction with a solid (powder) or gas phase or more recently, by a nitrogen-containing plasma. The chemical reaction of nitrogen with steel produces hard, wear-resistant iron-alloy nitrogen compounds; therefore, the steels do not require quenching to achieve full hardness (Child, 1983; Stickels et al., 1981; Thelning, 1984).

At the nitriding temperatures customarily used, the nitrogen will dissolve in iron, but only up to a concentration of 0.1 percent. When the nitrogen content exceeds this value, γ'-nitride (Fe_4N) is formed. If the nitrogen concentration exceeds about 6 percent, the γ'-nitride starts to change into ϵ-nitride (FeN). Below 500°C, ζ-nitride (Fe_2N) begins to form. The nitrogen content of this phase is about 11 percent. When making observations in the optical microscope, the γ'- and ϵ-nitrides can be seen as a white surface layer, called the *white nitride layer*, Fig. 11.15. The nitrogen-rich white layer is very hard but brittle. Simultaneously with the increase in thickness of the white layer during nitriding, the nitrogen further diffuses into the steel. When the solubility limit is exceeded, nitrides are precipitated at the grain boundaries and along certain crystallographic planes.

FIGURE 11.15 White layer and diffusion layer with precipitated acicular iron nitride in steel containing 0.15% C after gas nitriding at 500°C for 10 h. [*Adapted from Thelning (1984).*]

Nitriding is applied almost exclusively to steel, although nitriding of titanium has been operated on a commercial basis under the trade name Tiduran (Kolene Corp., Detroit, Michigan). Unalloyed carbon steels are not well suited to nitriding, because they form an extremely brittle case that spalls readily, and the hardness increase in the diffusion zone is small. Nitriding is most effective when applied to only a limited range of alloy steels containing elements that form stable nitrides, e.g., Al, Cr, Mo, V, and W. The commercially available steels, suitable for nitriding, are nitralloy (aluminum containing low alloy) steels, medium-carbon, chromium containing low-alloy steels of the 4100, 4300, 5100, 6100, 8600, 8700, 9300, and 9800 series; hot-work die steels containing 5 percent chromium such as H11, H12, and H13; ferritic and martensitic stainless steels of the 400 series; austenitic stainless steels of the 300 series; and precipitation hardening stainless steels such as 17-4PH, 17-7PH, and A-286. All hardenable steels must be hardened and tempered at a high temperature, usually 650°C, to produce a fully stabilized tempered martensitic structure, prior to nitriding.

In practice, the nitriding temperature is restricted to a maximum of 590°C in order to avoid tempering during nitriding and thus reduce the core strength. This lower temperature, as compared to carburizing temperature, limits the diffusivity, and even at very long treatment times (100 h) case depth greater than 0.5 mm cannot be achieved.

The effect of nitriding time t on the case depth CD of liquid nitriding can, for low-alloy steels, be derived from the simple formula for diffusion, Eq. (11.4) (Thelning, 1984). Nitriding can be carried out in the presence of a nitrogen source in the solid or gaseous forms or in the plasma. Gas nitriding and liquid nitrocarburizing are most commonly used. Gas nitriding may be preferred in applications where heavier case depths and dependable stopoffs are required. Plasma nitriding permits lower times or temperatures (minimum about 380°C) due to the nature of the process.

*11.2.3.1 **Pack Nitriding.*** In pack or powder (solid) nitriding, the parts to be treated are packed into boxes in roughly the same way as in pack carburizing. About 15 percent by weight of an energizer is first placed on the bottom of the box, after which is laid a layer of nitriding powder on which the part is placed. More nitriding powder is then added, one layer of powder alternating with one layer of parts. The boxes are then closed as tightly as possible and are placed in a muffle furnace at a temperature of 520 to 570°C for less than 12 h (Birk, 1970).

Nitriding tests on steels BS 905 M39 (En 41B), BS 817 M40 (En 24), and H13 gave roughly the same values for hardness and case depth as those obtained during gas nitriding for corresponding times and temperatures. After powder nitriding, the steel surfaces can exhibit some pitting and spalling.

*11.2.3.2 **Conventional Gas Nitriding.*** Conventional gas nitriding is the most widely used production process of nitriding (Stickels et al., 1981). Gas nitriding is carried out in ammonia gas at approximately 530°C. The gas dissociates catalytically at the surface of the hot steel.

$$2NH_3 \rightarrow 2(N)_{Fe} + 3H_2\uparrow \tag{11.28}$$

At the instant of dissociation, nitrogen occurs in the atomic form, and as such, it diffuses into the steel surface and reacts with the surface to form various nitrides. Treatment times vary from 12 to 120 h. Higher temperatures result in increased dissociation. The low ammonia content gives a low nitrogen activity and thus results in a tougher case.

The nitrogen potential is measured by residuals of ammonia at the exit. The process is controlled by adjusting the flow rate of ammonia until the outlet gas is between 15 and 30 percent dissociated as measured with an absorption dissociation pipette. If the ammonia at the exit is too high, the flow rate is reduced. Under steady condition buildup, the residual ammonia level is reduced to about 50 percent (50 percent dissociation). Under these conditions, a white layer is formed during nitriding, which can be reduced by increasing the degree of dissociation of ammonia, by precracking or using a synthetic gas mixture containing 20 percent ammonia and 80 percent nitrogen. This will further lower the nitrogen activity and eliminate the white layer formation.

The differences between gas nitriding and liquid (or salt-bath) nitrocarburizing are as follows:

- Gas nitriding requires 12 to 120 h nitriding time whereas salt-bath nitrocarburizing requires 1 to 4 h.
- Gas nitriding makes the steel very hard and brittle, but this is not as much the case with salt-bath nitrocarburizing.

Gas nitriding is used for parts that require a case depth between 200 and 700 μm. Although gas-nitriding components have a good resistance to adhesive wear, the process is used more particularly to resist abrasion and to improve fatigue resistance. Examples are large crankshafts (to resist fatigue), high-quality cylinder liners, ejector pins for die casting, steering parts, valve stems, and grinding machine slides.

11.2.3.3 ***Plasma Nitriding.*** Plasma nitriding, also known as *ion nitriding* or *glow-discharge nitriding*, is an extension of the conventional gas-nitriding process using glow discharge (Bewley, 1977; Dixon et al., 1983; Ebersbach et al., 1984; Edenhofer, 1974; Edenhofer and Bewley, 1976; Hudis, 1973; Jindal, 1978; Jones et al., 1975, 1977; Korhonen and Sirvio, 1982; Robino and Inal, 1983; Spalvins, 1983; Sundaraman et al., 1983). The gas is generally a dilute ammonia gas or a mixture of nitrogen and hydrogen. The gas pressure is usually between 1 and 1.3 kPa (7 and 10 torr) during plasma nitriding. The component is made cathode with dc voltage (1 to 5 kV) applied between the component and the container walls. Gas becomes ionized, and the ionized nitrogen atoms and energetic neutrals bombard the surfaces of the component, which become heated and allow diffusion of nascent nitrogen inward to form nitrides. The use of plasma nitriding permits better-quality cases in considerably lower times or temperatures (380°C). This process has several advantages over conventional gas nitriding and liquid nitrocarburizing (Child, 1983):

- Transfer of nitrogen ions to the component occurs uniformly over the surface in contrast to preferred absorption at grain boundaries in the gas process.
- High nitrogen potential species, such as FeN, are formed in the gas phase from Fe sputtered from the metal surface and deposited on the surface.
- Can depassivate Cr-containing steels and remove contaminants prior to nitriding.
- There is a low distortion of complex components.

- Use of lower temperatures minimizes tempering.
- Improves properties, partly due to low temperature and the consequent development of maximum residual stresses in the surface.
- Selective nitriding can be accomplished easily using sheet metal shields (1 to 2 mm from the surface) to exclude the plasma.

11.2.3.4 Properties of Nitrided Steels. The case depth of nitrided surfaces is less than that for carburizing processes; however, the nitrided surfaces are harder. The case of nitrided steel is subjected to several significant changes due to alloy formation, and typical changes in properties include:

- High hardness surface layer (case) up to 1300 HV for ion-nitrided steel has high resistance to wear even under inadequate lubrication.
- Fatigue resistance is increased mainly by development of surface compressive stress.
- Low notch sensitivity is obtained due to residual stresses, and notched fatigue strength is much higher than that of unnitrided steels.
- Hot hardness and resistance to tempering are improved.
- In some steels, corrosion resistance is improved.

Because of the low process temperatures, the absence of a martensite transformation and lack of quenching distortion is virtually zero. Nitriding can be used on finished components. The slight growth takes place because of expansion due to the pickup of nitrogen and formation of alloy compounds. Typical growth per surface is 0.005 mm for a case of 0.25 mm and 0.01 mm for a case of 0.5 mm.

During all nitriding processes, a very thin (up to 10 μm thick) white layer develops which can adversely affect the mechanical properties. The white layer exhibits high brittleness and contains radial and concentric cracks which initiates fatigue of the alloy nitride layer. This may be removed through mechanical polishing or chemical solutions. The underlying diffusion zone containing dissolved nitrogen and iron (or alloy) nitrides increases the fatigue endurance limits. High hardness of nitrides increases the wear and corrosion resistance.

The hardness achieved at the surface is influenced by the carbon content and other surface alloying elements in the steel. The relationship between carbon content, tempering temperature before gas nitriding, and hardness after gas nitriding is illustrated in Fig. 11.16 (Hodgson and Baron, 1956). The relationship between temper resistance and the effectiveness of nitriding is important. Coincidentally, the alloying elements that form stable nitrides also promote temper resistance, so all nitriding steels exhibit this property to a reasonable degree. When austenitic heat-resisting steels are nitrided, temper resistance is not involved and higher nitriding temperatures are used. Figure 11.17 shows the level of hardness as a function of depth of case for tool and heat-resisting steels. Increased temperature results in increased case depth at constant time, but maximum hardness is obtained at lowest nitriding temperature (520°C).

11.2.4 Ferritic Nitrocarburizing

Nitrocarburizing is a modified form of nitriding rather than a form of carburizing (Bell, 1975; Smith, 1977). This process involves the introduction of nitrogen and carbon, occassionally with small amounts of sulfur, into the surface of steel in the

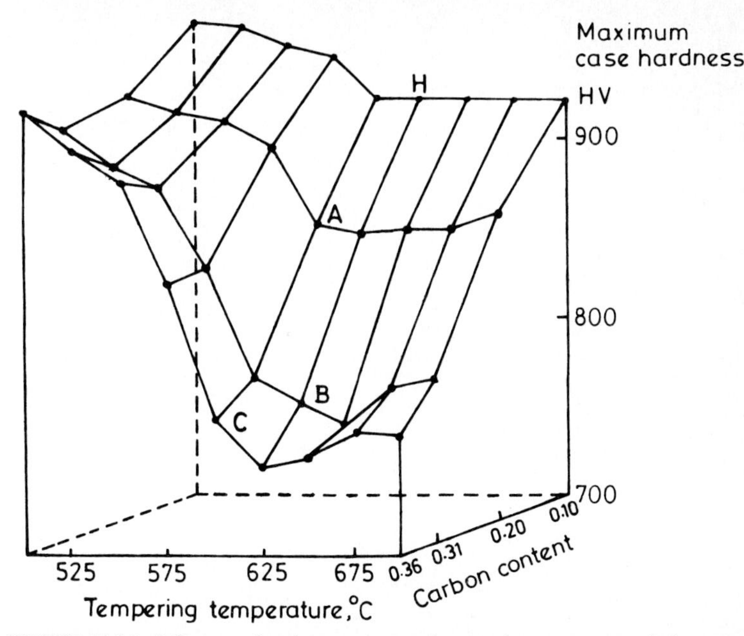

FIGURE 11.16 Influence of carbon content and tempering temperature before gas nitriding on the hardness after gas nitriding for En 240-type steels. Tempering time 100 h. Steel compositions: H—0.1% C, 2.9% Cr, 0.6% Mo, 0.14% Si, 0.53% Mn, bal. Fe. A—0.2% C, 2.8% Cr, 0.5% Mo, 0.24% Si, 0.56% Mn, bal. Fe. B—0.31% C, 2.7% Cr, 0.5% Mo, 0.3% Si, 0.55% Mn, bal. Fe. C—0.36% C, 2.9% Cr, 0.48% Mo, 0.29% Si, 0.65% Mn, bal. Fe. [*Adapted from Hodgson and Baron (1956).*]

ferritic condition at temperatures in the range 540 to 600°C by reaction with a liquid phase or gas and more recently plasma. In nitrocarburizing, the amount of diffused nitrogen is very high as compared to that of carbon. Nitrocarburizing treatment is carried out on mild steel in rolled or normalized conditions and as well as alloy and stainless steels. At the temperature (570°C) at which the nitrocarburizing treatments are carried out, which is below the eutectoid temperature, nitrogen is considerably more soluble in steel (i.e., ferrite) than carbon and therefore mainly diffuses into materials, typically, 1 mm in low-carbon materials in 90-min treatments.

The surface compound zone or white layer (approximately 10 to 25 μm thick) consists mainly of ductile epsilon iron carbonitride (1.25 percent C, 8 percent N). It is this zone which is responsible for the increased wear resistance. Below the compound zone is a nitrogen-rich layer (diffusion zone) which, if quenched immediately after processing, considerably increases the fatigue strength of the steel. This occurs as a result of the nitrogen not being allowed to precipitate from the ferrite lattice which is subsequently strained and strengthened. As with nitriding, distortion is virtually zero, but because the compound zone is so thin, parts should be finished to size prior to treatment.

The main advantages of nitrocarburizing over carburizing and/or carbonitriding are better wear resistance and antiscuffing characteristics resulting from the formation of a "white layer" (Harry, 1978).

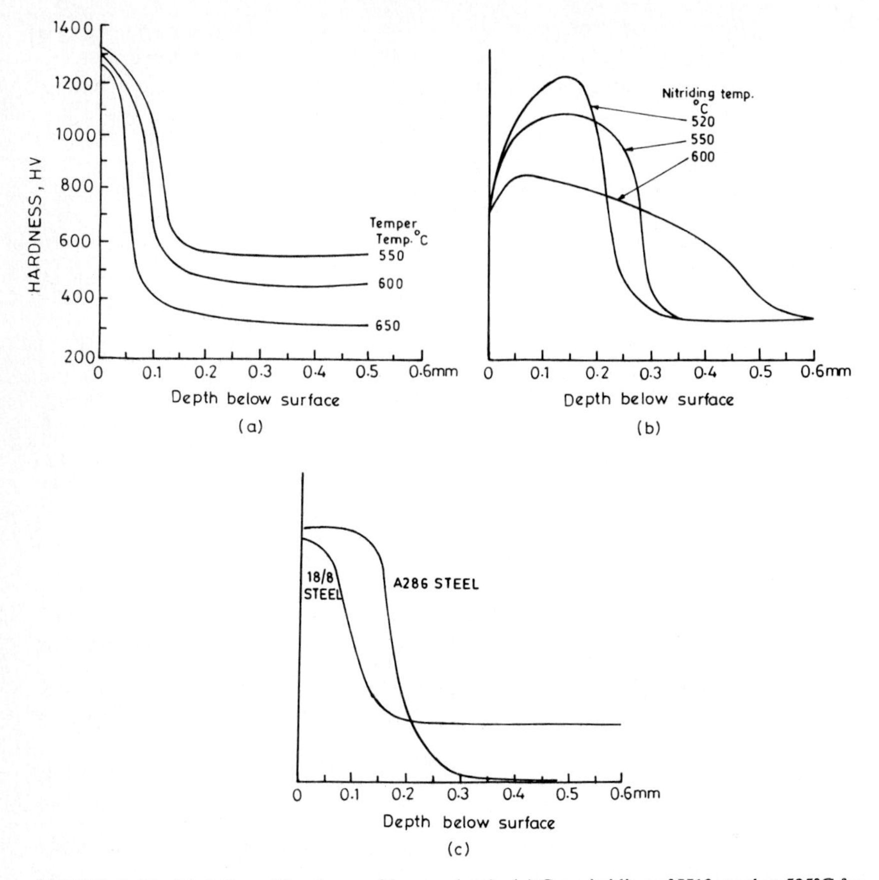

FIGURE 11.17 Variation of hardness with case depth: (*a*) Gas nitriding of H13 steel at 525°C for 10 h after hardening followed by tempering at different temperatures. (*b*) Gas nitriding of 13% Cr steel for 60 h at 520 to 600°C. (*c*) Gas nitriding of 18/8 steel and steel A 286 for 60 h at 600°C. [*Adapted from Thelning (1984).*]

11.2.4.1 Liquid Nitrocarburizing. Liquid nitrocarburizing, also called *salt-bath nitrocarburizing* (also called incorrectly *salt-bath nitriding*), is almost as old as the gas-nitriding method. The liquid nitrocarburizing is used for small lots of parts or for shallow cases (10 to 25 μm). The case hardening medium is a molten nitrogen-bearing, fused salt bath containing both cyanides and cyanates. A proprietary salt bath has the following composition by weight: 60 to 61 percent NaCN, 15 to 15.5 percent K_2CO_3, and 23 to 24 percent KCl. In addition, there are a few percent (~2.5%) of carbonates (Na_2CO_3) and about 0.5 percent cyanates (NaNCO). By aging the bath at 575°C for 12 h [Eq. (11.24)], the cyanate can be raised to the desired level, about 45 percent which produces nascent nitrogen and carbon at the steel surface [Eqs. (11.25) to (11.27)], which convert iron and steel surfaces to iron nitride as well as iron carbide. Steel parts are not immersed in the bath during the aging stage. The normal working temperature for treatment is 550 to 570°C, and the holding time is 1 to 4 h, which is much faster than gas nitriding (Thelning, 1984; Wensing et al., 1981). By injecting air into the bath, a better control of the cyanate control is obtained.

In some proprietary baths, sulfides (Na_2S) are added to the baths (such as "Sulfinuz" and "Sursulf" baths), and these processes are referred to as *sulfonitriding*. Sulphur and some sulphides by virtue of their crystal structures have good lubricating properties. In the sulfur-compound-containing baths, sulfur compounds impregnate the nitride-carbide coating with iron sulfide, which itself has antiscuffing properties. The addition of sulphur in the salt bath is found not only to increase the antiscuffing properties of treated parts but also to accelerate the nitriding action. With this process, ordinary steels with no special alloying elements can be nitrided.

Liquid nitrocarburizing processes employ proprietary solid additions, intended to serve several purposes such as accelerating the chemical activity of the bath, increasing the number of steels that can be processed, and improving the properties obtained as a result of nitriding. A variety of proprietary treatments are available with or without sulfur under trade names, for example, Tufftride (Tenifer) process (developed by Degussa, Frankfurt, Germany), and Sulfinuz. Nontoxic (especially cyanide free) salt baths based on thiocyanate (e.g., New Tufftride or Tufftride TF 1 and Sursulf) also produce similar results (Gregory, 1974, 1975).

Sursulf is based on a nontoxic bath made up entirely from cyanide-free salts, the nitriding action depending on the release of nascent nitrogen from alkaline cyanates which are present. The cyanates become oxidized to carbonates with the generation of nitrogen and carbon as carbon monoxide, as is the case in cyanide-containing baths, but the rate at which the oxidation occurs is reduced by 50 percent by the presence of lithium salts in the Sursulf bath. Sulphur is also present in the bath and produces sulphur compounds in the surface layers of treated parts.

After Tufftride or Sulfinuz treatment, the best surface is obtained if the steel parts are quenched in warm water. This rapid cooling results in a supersaturated solid solution of nitrogen in α-iron, which adds an extra contribution to the increase in fatigue strength, which is a characteristic feature of nitriding. However, the high internal stresses produced at the same time reduce toughness.

The liquid nitrocarburizing treatments are sometimes called *soft nitriding* on account of the nonbrittle (soft) epsilon carbonitride compound layer including epsilon FeN and iron carbide (Fe_3C), formed on the surface of the component. Its hardness is, however, in the range of 550 to 700 HV. Compared to plain-carbon or low-alloy steels, the hardness of the compound zone in medium-carbon, alloy and die steels is much greater, because of the nature of the more complex nitrides that are formed. Inspite of somewhat lower hardness, (liquid) nitrocarburized surfaces exhibit high wear and fatigue resistance. Tufftrided steels, for example, are produced by Lindberg Steel Treating Company, Chicago, Illinois. It is claimed that the Tufftride process can be used to improve friction and wear properties up to 600°C.

High-speed steel tools, stainless-steel parts, and special alloy components are commonly nitrocarburized to increase their surface hardness. Typical applications for which liquid nitrocarburizing proved superior to other case hardening processes include pumps, cylinder heads, pistons, crankshafts, valves, gears, bolts, nuts, and seat brackets (Wensing et al., 1981).

11.2.4.2 Conventional Gas Nitrocarburizing. In gas nitrocarburizing, hydrocarbons (up to about 50 percent) are added to nitriding gas atmosphere (ammonia or a mixture of hydrogen and nitrogen), usually in the form of pure propane or endothermic gas, produced from propane and air (Bell, 1975; Bell et al., 1981; Sachs and Clayton, 1979). This gas mixture is the same as that used for carbonitriding except the treatment temperatures are different in the two processes. The gas nitrocarburizing is performed at a temperature of roughly 570°C. The carbon diffuses into the steel at the same time as the nitrogen, and together they form a carbonitride mainly of the epsilon type.

There are many variations on this principle, but one of the best-known commercial processes is the Nitemper (Ipsen Industries, Rockford, Illinois) process. This uses a mixture of 50 percent ammonia and 50 percent endothermic gas in a furnace at 570°C, and components are usually treated for 1 to 3 h at 570°C followed by quenching into oil or cooling in a protective gas atmosphere. Other commercial processes are Nicarb and Trinitride.

Typical applications include rocker-arm spacers, textile machinery gears, pump cylinder blocks and jet nozzles, which are treated for wear resistance. Components such as crankshafts and driveshafts are treated to improve fatigue properties.

*11.2.4.3 **Plasma Nitrocarburizing.*** This is quite similar to plasma nitriding. The introduction of methane to ammonia or a mixture of nitrogen and hydrogen in plasma nitriding processes enables (epsilon) carbonitride compound layers to be produced under controlled and reproducible conditions. The process can be controlled in such a way that the surface layer is decarburized which reduces the brittleness of the layer.

*11.2.4.4 **Properties of Nitrocarburized Steels.*** The nitrocarburizing produces a thin layer of iron nitrocarbide and nitrides, the "white layer" or compound zone, with an underlying diffusion zone containing dissolved nitrogen and iron (or alloy) nitrides. The white layer enhances surface resistance to galling, corrosion, and wear. The diffusion zone increases the fatigue endurance limit significantly, especially in carbon and low-alloy steel.

The white layer is composed primarily of the epsilon carbonitride phase, with other nitrides and carbides. The exact composition is a function of the nitride-forming elements in the material and the composition of the atmosphere. The white layer has a reduced tendency to spall as compared to the white layer formed during conventional gas nitriding with ammonia. This has been attributed by most investigators to an essentially single-phase white layer of the less brittle epsilon phase, but it is also due in part to the fact that white layers are generally thinner (typically 10 to 15 μm thick) in nitrocarburizing. The diffusion zone beneath may extend to 250 μm or more.

The dimensional increase in nitrocarburizing is very small and is dependent on the depth of the compound and diffusion layer produced. The typical values of growth are 0.005 to 0.012 mm for Tufftride and even less for Sulfinuz. The hardness of the compound layer varies considerably and has been reported to be 600 to 900 HV for a wide range of steels and the type of processes. Figure 11.18 summarizes the results of increase in endurance limit of normalized, normalized and nitrocarburized, and Tufftrided steels having varying contents of carbon and chromium (Bell, 1975).

11.2.5 Boriding

Boriding, also known as *boronizing*, developed mainly within the aerospace industry and applied mainly to plain-carbon and low-alloy steels (Biddulph, 1974;

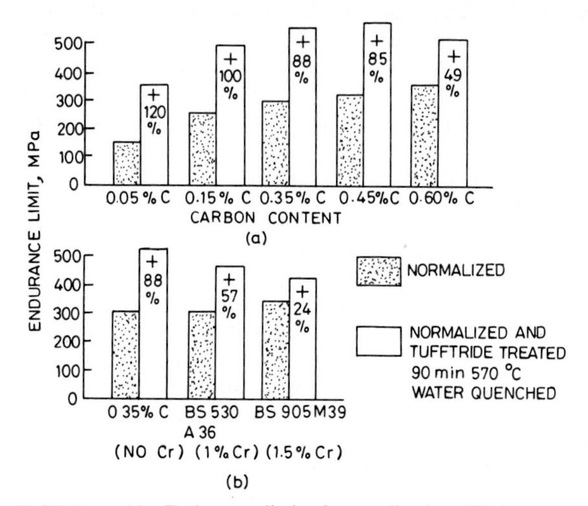

FIGURE 11.18 Endurance limit of normalized and Tufftrided unnotched pieces: (*a*) Effect of carbon content. (*b*) Effect of chromium contents. [*Adapted from Bell (1975).*]

Carbucicchio et al., 1980a, 1980b; Epik, 1977; Eyre, 1975; Eyre and Al Salim, 1977; Fichtl, 1982; Liliental et al., 1983; Linial and Hintermann, 1979; Rus et al., 1985; Singhal, 1977; Subrahmanyam and Gopinath, 1984; Tsipas and Perez-Perez, 1982). But this process can also be applied to austenitic alloy steels; nonferrous (nickel and cobalt-based) alloys; refractory metals such as Ti and Ta; and metal-bonded carbides, such as WC-Co, to improve their wear resistance (Chatterjee-Fischer and Schaaber, 1976; Sakai, 1985; Singhal, 1978). Extremely hard surfaces are formed provided they form borides. Boriding results in low wear if abrasion and tribooxidation are the main wear mechanisms.

Boriding can be carried out in either solid, liquid, or gas media at temperatures generally between 900°C to 1100°C for 1 to 6 h, depending upon the process and substrate. The most widely used method is pack boriding, in which the substrates are packed in compounds of either boron or boron carbides along with activators. The pack-boriding processes are technologically simple to carry out and particularly suitable for small-scale production. Liquid boriding and gas boriding have attracted less interest over the past few years, mainly because of high toxicity of chemicals used in the treatments. The case of borided steel can be as thick as 100 μm. It consists of iron borides, i.e., FeB (16.25 percent B) and Fe_2B (8.84 percent B), with the former concentrated at the surface, in a biphase layer. The FeB is harder on medium-carbon steel and shows a value of 1800 to 2000 HV while Fe_2B is softer, ductile, and more homogeneous in hardness between 1500 to 1600 HV. The FeB phase is brittle and in some applications, it is considered undesirable. The relative proportions of the two phases depends upon the availability of boron—the less boron, the greater the amount of Fe_2B is likely to be formed. In high-alloy steels, the occurrence of FeB in the outer layer cannot always be avoided.

When high-boron-availability conditions are applied, the coating usually consists of two clearly defined boride layers, an outer FeB layer and an inner layer of Fe_2B. The biphase layer can be converted to monophase Fe_2B layers by diffusion anneal of, for example, 2 h at 1000°C in argon.

Below the boride layer, boron is present mainly in solid solution but also as excess boron precipitates (borides and/or borocarbides) (Bozkurt et al., 1983). The boronized layer can either have a jagged, saw-toothed interface with the substrate or a relatively straight interface; the former occurs on plain-carbon steels and the latter on low-alloy steels (Fig. 11.19). The geometry of the interface has a critical effect on the adhesion of the layer. Experience shows that the straight interface obtained on alloyed steels often leads to spalling, particularly at sharp corners and under impact conditions (Eyre, 1975). A toothed structure appears beneficial for wear resistance because it increases substrate rigidity and enhances adhesion between boride layer and substrate (Whittle and Scott, 1984).

The growth and relative proportions of duplex metal borides are strongly influenced by the affinity of alloying elements such as Ni and Cr to boron (Goeuriot et al., 1981, 1982). Chromium reacts faster than iron with boron, and in grain-boundary regions of the metal, preferential growth of chromium-rich boride occurs as a result of faster diffusion of boron atoms along with metal grain boundaries than through the metal matrix. The ranking order of increasing affinity for boron, nickel < iron < chromium, is the same as that observed for other interstitial alloying elements such as nitrogen and carbon (Whittle and Scott, 1984). A toothed interface has been shown to be created by the preferential growth of boride at grain boundaries in steel alloys having Ni as alloying element (e.g., 20-2 austenitic steel, 20% Ni), in which inward movement of the boride-metal interface leaves behind isolated regions of Ni-rich metal. The toothed interface between boride and metal core has also been observed in case of ferritic steel. The depth of tooth has been shown to be affected by metal composition (Von Matuschka, 1980). For ferritic alloys steels with Cr > 6 percent in addition to reducing the boride layer thickness, inhibits the development of the toothed morphology.

The rate of boriding has been found to obey parabolic kinetics, i.e., case depth is linearly proportional to the square root of treatment time, Eq. (11.4) (Biddulph, 1974). The thickness of the boride layer is strongly influenced by substrate composition, i.e., alloying elements, composition of boriding powder-liquid mixtures, temperature, and duration of treatment. Boriding is used for extreme wear and

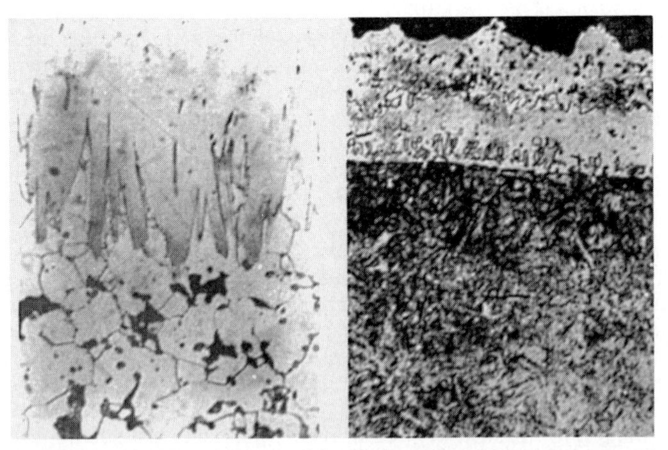

FIGURE 11.19 Microstructure of borided steels: (*a*) Plain carbon steel (EN1A), normalized and borided. (*b*) Low-alloy steel (20MoCr4), hardened and tempered and borided. [*Adapted from Eyre (1975).*]

tribooxidation applications. Typical applications include tools such as circular cutters, four slid king pins and draw punch taps, bushings, molding dies, forming dies, and wear plates.

11.2.5.1 Pack Boriding. In pack or solid boriding, the powder mixtures used mainly consist of a boron source such as boron carbide, ferroboron, amorphous boron, activators such as halogenides and fluoborates, and inert filler materials (Carbucicchio et al., 1980a; Carbucicchio and Palombarini, 1984; Goeuriot et al., 1982; Liliental et al., 1983; Palombarini and Carbucicchio, 1984; St. John and Sammells, 1981; Tsipas et al., 1988). The powder mixtures employed are roughly classified as low-boron or high-boron media, with reference to the well-known possibility of obtaining monophase Fe_2B or biphase layers of Fe_2B and FeB as reaction products. Borudif, a proprietary process, consists of three agents: a boriding agent (finely powdered boron carbide diluted with SiC), an activator (BF_3 gas), and an oxide moderator (silica: acid-washed sand). Different types of pack mixtures have been employed for different types of metals by various authors (Habig, 1980; Rus et al., 1985; Tsipas and Perez-Perez, 1982; Tsipas et al., 1988).

The composition of the pack mixture used by St. John and Sammells (1981), was 20 mol % B_4C (Cerac, −325 mesh, a boron compound), 5 mol % KBF_4 (Alpha product; an activator); and 75 mol % graphite (Fischer Acheson no. 38, a filler). Substrates of AISI 1008 steel were heated 1 to 5 h at 960 ± 10°C in an alumina or 304 stainless-steel containers. Rus et al. (1985) borided 304 stainless steel by using $B/NaF/Al_2O_3$ and $B_4C/SiC/KBF_4$ pack mixtures at 815 and 915°C respectively, for periods varying between 6 to 12 h. Tsipas and Perez-Perez (1982) borided low carbon (0.17 percent C) steel with $Li_2B_4O_7/B_4C$ in a different way. First, a solution of $Li_2B_4O_7$ in methyl alcohol was sprayed over the specimen at 100°C, which resulted in a uniform deposit of $Li_2B_4O_7$ (the amount of $Li_2B_4O_7$ was typically on the order of 1 percent of the substrate). Subsequently substrates were packed in a mixture of B_4C along with some activator such as NH_4Cl and BaF_2 and heated at 875°C in argon for periods varying 2 to 48 h. The following reaction leads to the formation of boride layers

$$Li_2B_4O_7 + B_4C \rightleftharpoons 8(B)_{Fe} + 6CO + Li_2O \tag{11.29}$$

Duplex boride layers consisting of FeB and Fe_2B have been found in all the pack mixtures. However, their proportions, thicknesses, porosity, etc., are considerably influenced by the activity of the boron compounds, activator, and composition of the substrate metal.

Similar pack mixtures have also been used to boride nonferrous metals such as titanium and tantalum and nickel alloys (Chatterjee-Fischer and Schaaber, 1976). Titanium and titanium alloys (Ti6Al4V) have been borided in either vacuum or argon atmospheres at 1000°C. The duplex boride layers consisted of the more brightly etched 15 to 20-μm-thick closed layer of TiB_2 and the darker-etched serrations of TiB. The layer was strongly interlocked with the base metal. The hardness of the boride layer for titanium and titanium alloy (Ti6Al4V) were on the order of 2500 and 3000 HV, respectively. Similar treatment of tantalum resulted in a smooth single-phase tantalum boride layer (45 μm thick) having a hardness of 3200 HV. The boriding of nickel-based alloy (IN 100 with 15 percent CO 10 percent Cr, 5.5 percent Al, 4.7 percent Ti, 3 percent Mo, 1 percent V, and 0.18 percent C, balance Ni) in amorphous boron at 940°C resulted in a 60-μm-thick layer of 1700 HV hardness.

11.2.5.2 ***Liquid Boriding.*** In liquid boriding, the metallic substrates are immersed in molten salt baths of boron compounds at 900 to 1000°C for 4 to 12 h. Qing'en and Zaizhi (1984) borided pure iron, plain carbon steel, austenitic stainless steel, and forged low-carbon alloy steel in the molten bath which consists of 80 to 85 percent borax and 15 to 20 percent silicon carbide at 950°C for 4 to 8 h. The top layer of borided substrates consisted mainly of Fe_2B but with different morphological growth patterns, which is perhaps due to the fact that some alloying elements such as Ni diminish or inhibit the boron diffusion.

Arai et al. (1975) used borax baths containing powders of metal alloys, carbides, oxides, and nitrides of such elements as Ca, Si, Al, Mn, rare earths, Mg, Zr, Ti, V, Nb, Cr, Zn, Ni, Co, and Cu to boride steel surfaces. They have observed that the tendency to form boride layers was greater in baths containing metal alloy powder with a high free-energy change ($-\Delta G$) of the oxide formation than the boron oxidation. Typical value of free-energy change for Ca-Si, Ca-Al-Si, and Ca-Al-Mn-Si is −88 kcal/mol, and for Fe-Ti, −36 kcal/mol.

Hosokawa (1976) and Hosokawa et al. (1972) have borided steel substrates at 1000°C for 1 to 5 h in a molten bath consisting typically of 24% B_4C, 16% BaO, 30% KCl, and 30% NaCl. The addition of BaO significantly improves the diffusion of boron, and a 0.3-mm-thick boride layer was obtained on treating for 5 h. The boride layer consisted of FeB and Fe_2B. Molten baths consisting of 50 percent ferroboron and 50 to 60 percent Na_2CO_3 were also tried, and the formation of Fe_2B without any pores was obtained for pure iron (0.005 percent C, 0.005 percent S, 0.005 percent P, 0.004 percent Mn, 0.005 percent Cu, bal Fe) substrates.

Fiedler and Sieraski (1971) borided tool steels by the electrodeposition of boron atoms on the metal from a bath of molten salts containing fluorides of lithium, sodium, potassium, and boron. The substrate surface to be borided was made cathode, and pieces of elemental boron in a copper basket constituted the anode of the electrolytic bath. The electrolytic bath was covered by argon and hydrogen to minimize contamination. The boron deposition and growth was controlled by current density and cell voltage. The current densities ranged from 50 to 250 A m^{-2}. The substrates were borided at temperatures ranging from 600 to 900°C for periods varying between 15 min and 5 h. The typical thickness of boride layers was on the order of 100 to 250 μm. Current density was found to have a significant influence on the microstructure and smoothness of the duplex boride layers.

11.2.5.3 ***Gas Boriding.*** In gaseous boriding, the boron-containing vapors are generated through the thermal decomposition of boron hydrides $Me(BH_4)_n$, and the process is carried out at elevated temperatures (Von Matuschka, 1980). The other alternative gaseous-boriding process involves hydrogen reduction of boron halides (e.g., BCl_3 and BBr_3). This process is particularly advantageous when case depth has to be controlled in respect to thickness and homogeneity, such as for precision components having complicated shapes, e.g., long narrow tubes and jet nozzles (Hintermann et al., 1973; Bonetti et al., 1975). A variety of proprietary gas-boriding treatments are available such as Boroloy (Lindberg Heat Treating Co., Melrose Park, Illinois) and Borofuse (Materials Development Corporation, Medford, Massachusetts).

11.2.5.4 ***Properties of Borided Metals.*** The borided steels exhibit high hardness (typically 1500 to 2300 HV) and reduced wear (up to 1/100) compared with untreated steel substrates (Linial and Lavella, 1973, 1974; Sridhar and Iyer, 1980). The duplex boride phases (FeB and Fe_2B) exhibit high hot hardness, for example, 1500 HV at 1000°C.

Habig and Chatterjee-Fischer (1981) have carried out wear studies extensively on borided steels alloyed with Cr, Mo, and V. These were: X20Cr13 (0.2% C, 0.4% Si, 0.3% Mn, 0.025% P or S, 13% Cr), Mo-Steel (0.37% C, 0.004% P, 0.006% S, 1.99% Mo, 0.08% Al), and 145 V 33 (1.45% C and 3.3% V). The boriding conditions, phase compositions, thicknesses, and roughnesses of boride layers selected for wear studies are summarized in Table 11.5. Data of some hardened, carburized, and nitrided steels are also included for comparisons. The adhesive wear studies were carried out on a pin-on-disk machine (load = 10 N; speed = 0.1 m s^{-1}; sliding distance = 1000 m) in a vacuum (10^{-3} to 10^{-4} Pa or 7×10^{-6} to 7×10^{-7} torr) at ambient temperatures. The abrasive (two-body abrasion) wear tests were performed on grinding apparatus with SiC paper (load = 17 N; speed = 125 r/min, 2 r/min; area = 900 mm^2; and duration = 6 min). The tribooxidative wear tests were carried out under the same conditions as that for adhesive wear tests, but at 50 percent relative humidity. An Amsler machine was used for surface fatigue wear tests, the two disks covered with oil rotated with 10 percent slip (load = 2000 N; speed = 4000 r/min; 362 r/min; number of cycles $\sim 10^4$ and in SAE 10 oil medium). Finally, wear tests under mixed lubrication, where all wear mechanisms superimpose (load = 400 N; speed = 0.1 m s^{-1}; sliding distance 1000 m; SAE 10 lubrication), were carried out on a pin-on-disk machine. The results of adhesive, abrasive, tribooxidative, surface fatigue, and mixed-lubricated sliding wear tests performed on borided steels, heat-treated carburized steel, and nitrided steel are compared in Fig. 11.20(*a*) to (*e*). We note from Fig. 11.20 that borided couples have low wear if abrasion and tribooxidation are the main wear mechanisms. Nitrided steel couples have the lowest wear in adhesion. Nitrided and carburized couples performed the best in surface fatigue wear mode. High fatigue wear of borided surfaces is connected with the formation of loose wear particles. The many holes present in the borided layers act as notches and promote crack formation. The difference in wear resistance between the variously treated steels, when tested under mixed lubrication, is not so large. The borided steels have a lower wear than carburized and nitrided steels.

Alloying elements of steels dissolved in boride layers have a marked influence on the wear characteristics. The influence is not uniform for different types of wear mechanisms. Adhesive wear resistance could be improved by adding Cr and Mo in the boride layer. Abrasive wear is reduced by solution of Cr, Mo, V, and, above all, vanadium carbide. The low resistance against surface fatigue is increased by Mo and V, while Cr seems to further reduce the wear resistance. Tribooxidative wear has least influence of alloying elements.

In another wear study of borided austenitic steels (321, 316, 255, and 20-2), a substantial improvement in wear resistance was achieved by Whittle and Scott (1984). However, wear rates on borided 316 and 255 steels, having 17 percent and 27 percent Cr, respectively, showed an abrupt increase when the load in a pin-on-disk machine exceeded 200 N, while it remained almost constant for 20-2 borided steels, having 20 percent Ni. This difference in wear behavior may be due to the formation of more deep-toothed morphology of the borided layer in 20-2 steel, which enhances substrate rigidity and reduces the tendency of the top surface to deform under high stresses. Eyre (1975) compared the wear life in a pin-on-disk wear test, of EN 1A steel and EN 8 steel and gray cast iron treated with various chemical diffusion processes including carburizing, nitrocarburizing (Tufftriding), and boriding. He reported that boriding was superior in adhesive wear to other diffusion treatments.

Boriding of nonferrous metals such as Ti, Ti6A14V, Ta, and Ni considerably improves their abrasive wear characteristics (Chatterjee-Fischer and Schaaber,

TABLE 11.5 Various Features such as Phase Composition, Hardness, Thickness of Case t, Roughness of Treated Surface of Borided, Carburized, and Nitrided Steels Selected for Wear Studies Having Dominance of Single Wear Mechanism Such as Adhesion, Abrasion, Triboxidative, Surface Fatigue, and Lubricated Wear at One Time

				Surface layer		Roughness R_t, μm			
Sign	Steel	Heat treatment	Phases	Hardness, HV (kg mm^{-2})	Thickness (t), μm	Pin	Disk ϕ42 mm	Disk ϕ70 mm	Base material, HV (kg mm^{-2})
				(*a*) Borided					
G	42CrMo4	Borided in E II 900°C, 2 h, slow cooling.	Fe_2B (FeB)	1500	56 ± 9	3.5	1.0	1.0	291
X	C 45	Borided in E II 900°C, 1 h, slow cooling.	Fe_2B	—	36 ± 9	2.6	0.9	1.1	188
Y	C 45	Borided in E II 900°C, 2 h, slow cooling.	Fe_2B	—	64 ± 14	1.9	1.0	1.3	195
BA	42CrMo4	Borided in E II 900°C, 4 h, slow cooling.	FeB, Fe_2B	1560	101 ± 5	1.5	—	1.0	274
BR	42CrMo4	Borided in E III 900°C, 6 h, slow cooling.	Fe_2B	1370	102 ± 27	0.9	0.7	—	301
BT	X38CrMoV 51	Borided in E II 900°C, 6 to 7.5 h.	FeB, Fe_2B	1920	55 ± 4	0.7 0.5	 0.8	0.7	515
ED	42CrMo4	Borided in E III 900°C, 6 h hardened, 850°C-oil 50°C, annealed 570°C, 2 h in argon.	Fe_2B	1600	101 ± 7	—	Amsler: 1.6 (R_z)	—	339
EX	X20Cr13	Borided in E IV 900°C, 3 h, slow cooling.	FeB, Fe_2B	1700	25 to 40	0.5 0.7	1.2 Amsler: 2.9 (R_z)	 1.2	200

EY	X 20Cr13	Borided in E V 900°C, 5 h, slow cooling.	Fe_2B (FeB)	1600	35 to 40	0.5 0.7	1.1 Amsler: 2.6 (R_z)	1.4	227
FB	145 V 33	Borided in E VI 900°C, 3.5 h, slow cooling.	Fe_2B, VC (FeB)	1600 to 1700	40 to 50	0.5 0.4	1.5 Amsler: 2.5 (R_z)	1.1	280
FL	Mo-steel (2% Mo, 0.37% C)	Borided in E VI 900°C, 2 to 3 h, slow cooling.	Fe_2B (FeB)	1620	40	0.7 0.6	1.6 Amsler: 2.1 (R_z)	1.6	188
			(*b*) Carburized and nitrided						
AN	C 45	Carburized 900°C, 3 h[1], hardened 780°C-10 percent NaOH, annealed 220°C, 30 min.	Martensite Fe_3C (Retained austenite)	832	—	0.5	0.8	0.6	188
AO	C 45	Hardened 830°C, 10 percent NaOH, annealed 220°C, 30 min.	Martensite (Retained austenite)	677	—	1.5	—	0.6	188
EE	16 MnCr 5	Carburized 930°C, 5 h to 0.67% C, quenched in oil, annealed 165°C, 1 h.	Martensite Fe_3C (Retained austenite)	798	—	—	Amsler: 1.6 (R_z)	—	—

TABLE 11.5 Various Features such as Phase Composition, Hardness, Thickness of Case *t*, Roughness of Treated Surface of Borided, Carburized, and Nitrided Steels Selected for Wear Studies Having Dominance of Single Wear Mechanism Such as Adhesion, Abrasion, Tribooxidative, Surface Fatigue, and Lubricated Wear at One Time (*Continued*)

				Surface layer		Roughness R_t, μm			
Sign	Steel	Heat treatment	Phases	Hardness, HV ($kg\ mm^{-2}$)	Thickness (t), μm	Pin	Disk ϕ42 mm	Disk ϕ70 mm	Base material, HV ($kg\ mm^{-2}$)
AR	42CrMo4	Hardened 830°C, oil 65°C, annealed 600°C, 1 h, salt-bath nitrided 570°C, 2 h.	ϵ-Fe_xN	—	10	0.6	1.8	—	—
BC	42CrMo4	Hardened 830°C, oil 50°C, annealed 600°C, 1 h, salt-bath nitrided 570°C, 2 h.	ϵ-Fe_yN	722	13	1.0	—	0.6	—
DU	34CrAlMo5	Hardened, annealed, gas nitrided 510°C, 90 h.	ϵ-Fe_xN γ'-Fe_4N	960	30	—	3	—	—
EG	42CrMo4	Hardened 850°C, oil 65°C, annealed 600°C, 1 h, salt-bath nitrided 580°C, 2 h[2]	ϵ-Fe_xN	731	16	—	Amsler: 1.5(R_z)	—	—

E II: Ekabor 1.
E III: 92.5% SiC; 5% KBF_4, 2.5% B_4C.
E IV: Ekabor 2.
E V: Ekabor 2 + 15% Cr powder.
E VI: Ekabor 2 + 5% Cr powder.
1. Proprietary carburizing powder Effge 453 from Gorig, Mannheim, Germany.
2. Proprietary Tufftride TF-1-nitriding bath from Degussa, Hanau, Germany.
Source: Adapted from Habig and Chatterjee-Fischer, 1981.

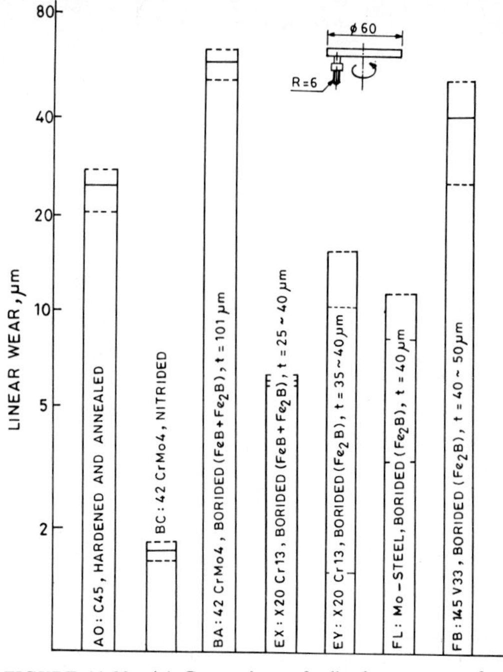

FIGURE 11.20 (*a*) Comparison of adhesive wear, performed on a pin-on-disk tester under dry condition, of borided steel with heat-treated, carburized, and nitrided steels (test conditions: load = 10 N; speed = 0.1 m s^{-1}; sliding distance = 1000 m; environment = vacuum (10^{-3} to 10^{-4} Pa or 7×10^{-6} to 7×10^{-7} torr); temperature = 23°C). [*Adapted from Habig and Chatterjee-Fischer (1981).*]

1976). The results of abrasive wear tests performed with a Faville tester (a rotating, loaded borided specimen against a SiC-coated disk and rotating in the opposite direction) are compared in Fig. 11.21(*a*) and (*b*). We note a significant improvement in the wear resistance of Ti and Ta on boriding.

11.2.6 Chromizing

Chromizing involves diffusion of chromium into high-carbon steel by heating at 950 to 1300°C in chromium-rich media (Seigle, 1984; Ullmann, 1969). The treated surface provides oxidation and wear resistance. In pack chromizing, the component is embedded in a mixture of source metal containing the element(s) to be transported (fine powders of ferrochromium), an activator (generally ammonium halides) and inert fillers (such as alumina, silica, and kaolin) and heated at 1100 to 1300°C for 12 to 15 h in an inert or reducing atmosphere. Usually this is done in a retort, but fluidized-bed versions of the process are also known. In a typical pack process, halides of the source element are formed by reaction of the activator (vaporizable ammonium halides are usually used) with powder in the pack containing the source element at a high activity level. Upon reaching the sample,

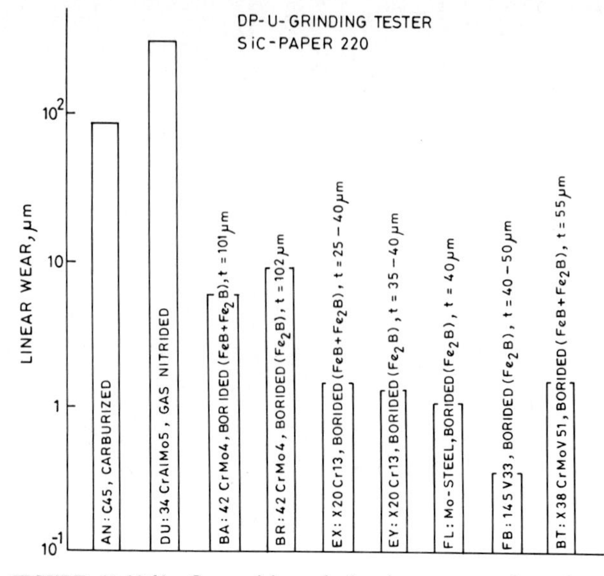

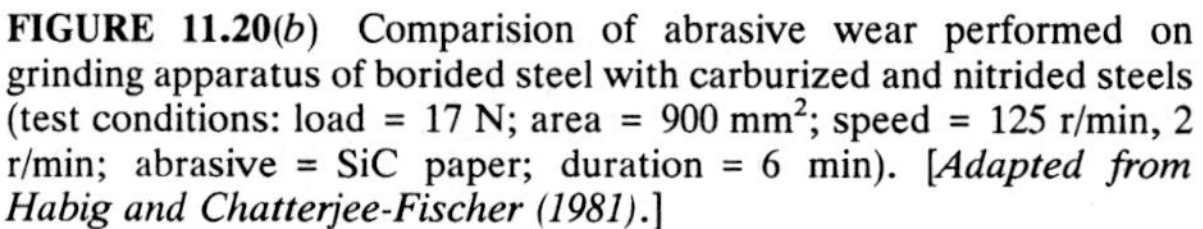

FIGURE 11.20(*b*) Comparision of abrasive wear performed on grinding apparatus of borided steel with carburized and nitrided steels (test conditions: load = 17 N; area = 900 mm^2; speed = 125 r/min, 2 r/min; abrasive = SiC paper; duration = 6 min). [*Adapted from Habig and Chatterjee-Fischer (1981).*]

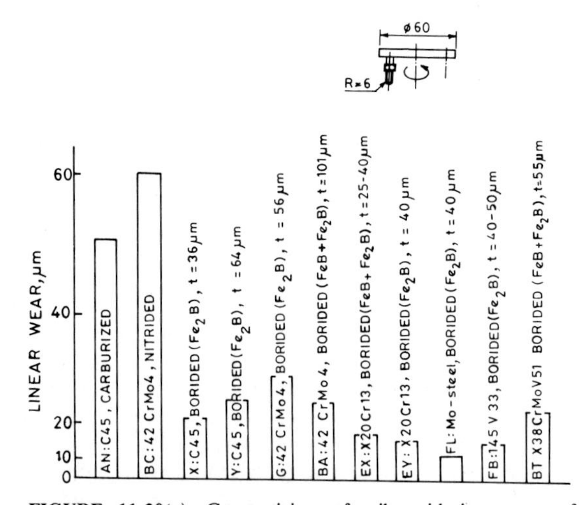

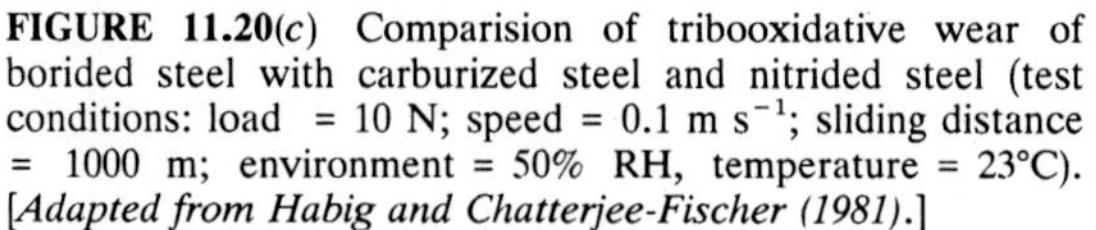

FIGURE 11.20(*c*) Comparision of tribooxidative wear of borided steel with carburized steel and nitrided steel (test conditions: load = 10 N; speed = 0.1 m s^{-1}; sliding distance = 1000 m; environment = 50% RH, temperature = 23°C). [*Adapted from Habig and Chatterjee-Fischer (1981).*]

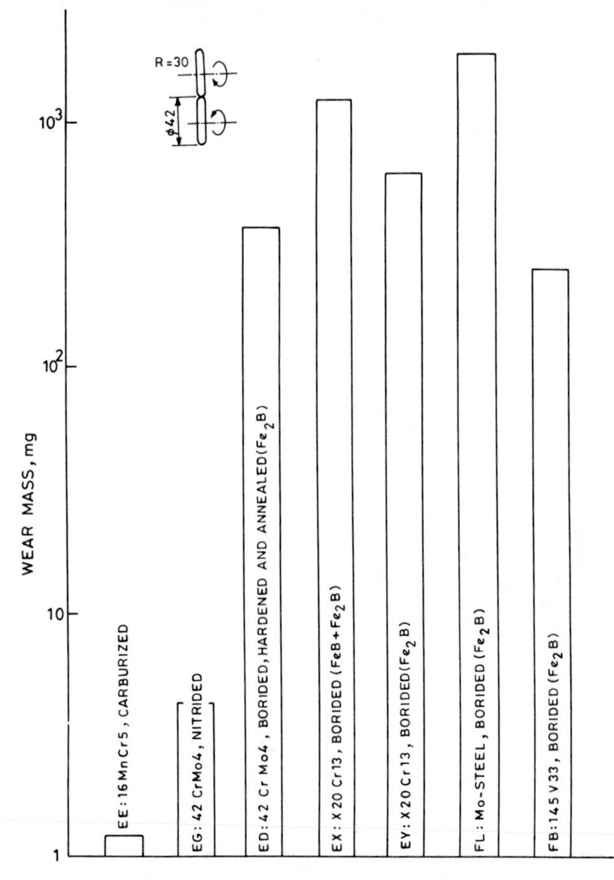

FIGURE 11.20(*d*) Comparision of surface fatigue wear, performed on Amsler machine, of borided steel with carburized steel and nitrided steel (test conditions: load = 2000 N; speed = 400 r/min; 362 r/min; number of cycles = 10^4 cycles (EE, EG = 10^5); medium SAE 10 oil; temperature = 23°C). [*Adapted from Habig and Chatterjee-Fisher (1981)*.]

these halides react at the surface to deposit source metal, which diffuses into the alloy to form the coating.

For example the chromium powder of the pack reacts with the vaporized ammonium iodide at high temperature:

$$Cr(solid) + 2NH_4I(gas) \rightarrow CrI_2(gas) + 2NH_3(gas) + H_2(gas) \quad (11.30)$$

The chromous iodide then reacts with the iron specimen in one or more ways as given below and diffuses into it.

Exchange reaction:

$$Fe + CrI_2(gas) \rightarrow FeI_2(gas) + (Cr)_{Fe} \quad (11.31a)$$

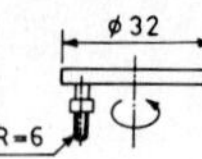

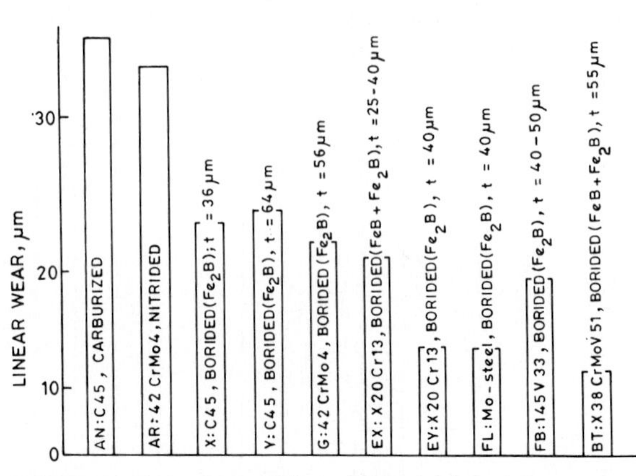

FIGURE 11.20(*e*) Comparision of mixed lubricated wear performed on a pin-on-disk tester, of borided steel with carburized steel and nitrided steel (test conditions: load = 400 N; speed = 0.01 m s^{-1}; sliding distance = 1000 m; medium SAE 10 oil; temperature = 23°C). [*Adapted from Habig and Chatterjee-Fischer (1981).*]

Reduction reaction:

$$H_2(\text{gas}) + CrI_2(\text{gas}) \rightarrow 2HI(\text{gas}) + (Cr)_{Fe} \tag{11.31b}$$

Decomposition reaction:

$$CrI_2(\text{gas}) \rightarrow 2I(\text{gas}) + (Cr)_{Fe} \tag{11.31c}$$

Depending upon whether iodides, bromides, chlorides, or fluorides are chosen halides and depending upon the treatment temperature any one of the three reactions may predominate. A temperature of 1300°C and 3 h of treatment result in a case depth of approximately 0.18 mm.

In liquid chromizing, the component is heated at 900 to 1000°C in the molten bath of $CrCl_2$ or ferrochromium (a source of Cr) and BaCl or $MgCl_2$ (which acts as an activitar to enhance reaction and improve diffusion). Gas chromizing is accomplished by heating the steel component in the presence of vapors of $CrCl_2$ and $CrCl_3$ and reducing gas (H_2) at 950 to 1050°C:

$$CrCl_2(\text{gas}) + H_2(\text{gas}) \rightarrow 2HCl(\text{gas}) + (Cr)_{Fe} \tag{11.32}$$

In high-carbon steels the chromized layer consists of complex chromium carbide $(FeCr)_7C_3$. The formation of a continuous carbide layer takes place due to diffusion of carbon from the inner layer to meet the chromium. The typical hardness achieved for chromized case in high-carbon steel ranges from 1200 to 1300 HV. Carbon in steel impedes the Cr diffusion to the iron. The addition of W, Mo, and Si to steel accelerates the Cr diffusion while Cr, Ni, and Mn reduce its rate (Fig. 11.22). Chromized surfaces exhibit excellent resistance to oxidation wear.

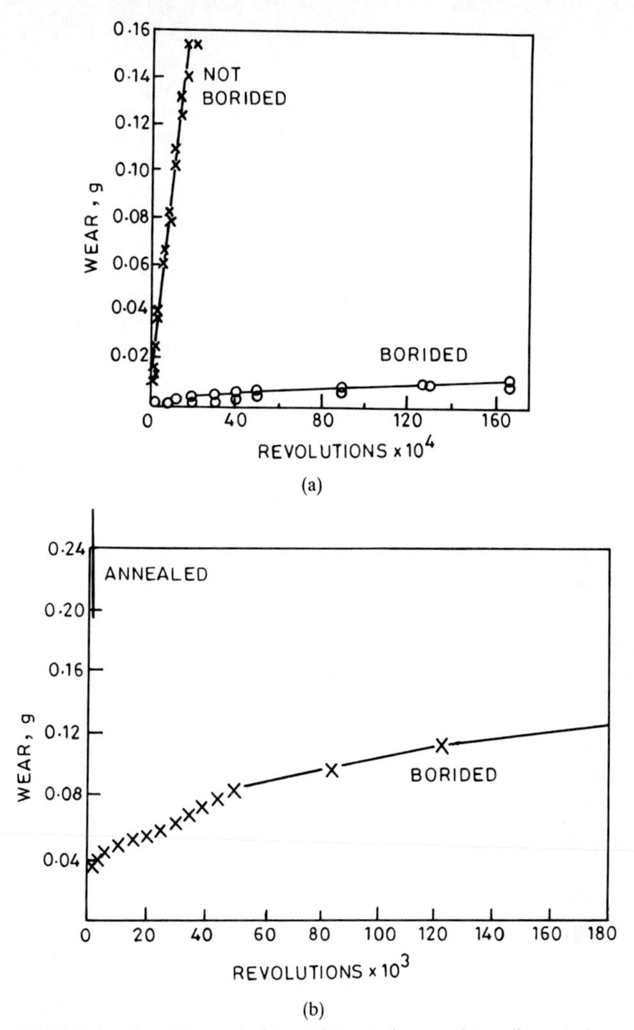

FIGURE 11.21 Influence of boriding of: (*a*) titanium, (*b*) tantalum metals on abrasive wear performed on Faville tester. [*Adapted from Chatterjee-Fischer and Schaaber (1976).*]

Chromized surfaces on low-carbon (0.1 to 0.2 percent C) steels are ductile and can be rolled, pressed, or drawn without damaging the chromized surface which flows evenly with the parent metal. Typical applications include turbine buckets, steel components to resist attack of nitric acid, and containers used for copper brazing.

11.2.7 Aluminizing

In aluminizing, also known as *calorizing*, aluminum is diffused into a low- and medium-carbon steel surface that imparts resistance to corrosion and oxidation at

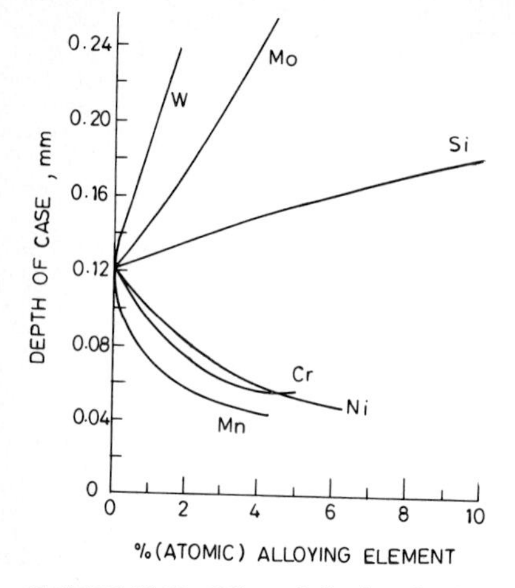

FIGURE 11.22 Effect of alloying elements in steel (0.45% C) on case depth during chromizing. [*Adapted from Lakhtin (1977).*]

high temperature. Pack aluminizing is obtained by heating the component in the container having the fine powders of ferroaluminum, ammonium chloride (activator), and inert fillers at 750 to 1000°C for 3 to 6 h (Seigle, 1984). The typical case depth in cases of low-carbon steel range from 0.2 to 0.5 mm. Pack aluminizing is followed by annealing at 800 to 980°C for 12 to 48 h, which reduces the brittleness of the aluminized layer and causes further penetration of aluminum by diffusion to depths ranging from 0.6 to 1 mm.

Liquid aluminizing is accomplished by immersion of a steel component in a bath of molten aluminum saturated with iron (6 to 8 percent) at 750 to 800°C. The iron is added to aluminum bath to prevent the aluminum from corroding (dissolving) the steel substrate. The heating for 1 to 1.5 h produces a case of 0.2 to 0.35 mm depth.

The aluminized layer is a solid solution of aluminium and iron. An increase in carbon content and other alloying elements in steel impedes the diffusion of aluminum (Fig. 11.23). This process is preferred when the component is subjected to elevated temperature operation. The aluminized layer gives excellent resistance to heat and corrosion by sulfurous gases. The usual content of 25 percent aluminium is enough to ensure adequate resistance to heat and corrosion. As a result of lower aluminum content, the alloy layer exhibits toughness and ductility. The typical application of aluminizing processes are aircraft gas turbine blades, burner pipes, pyrometer protection tubes, flue stacks, boiler tubing, high-temperature fasteners, chemical reactor tubing.

11.2.8 Siliconizing

In the siliconizing process, silicon diffuses into the surface of steel or malleable iron which acquires a high resistance to wear, scaling at elevated temperature, and corrosion (Kanter, 1964; Seigle, 1984). Pack siliconizing is accomplished by heating the component in a powder of ferrosilicon (50 to 95% Si) activated with ammonium chloride (2 to 3 percent) at 1100 to 1200°C for 2 to 24 h. The typical case depth ranges from 0.2 to 0.8 mm. This process is not commonly used because of the very high treatment temperatures.

Gas siliconizing is accomplished in retort furnaces in which steel is heated at 950 to 1050°C for 2 to 5 h in the presence of silicon carbide (40 to 70 mesh) and $SiCl_4$ vapors. $SiCl_4$ vapors react with steel to form a silicon-enriched case. The chlorine produced in the course of siliconizing action reacts with SiC:

$$2Cl_2 + SiC \rightarrow SiCl_4 + C \tag{11.33}$$

This cycle of reaction has the effect of reducing the concentration of free chlorine to minimize pitting of the steel workpiece and formation of iron chloride. The typical case depth obtained ranged from 0.5 to 1.25 mm. The siliconized case is a

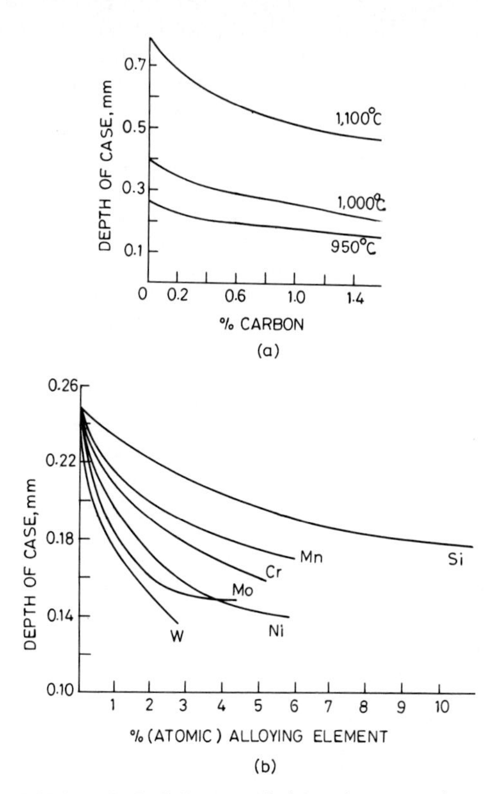

FIGURE 11.23 Influence of: (*a*) carbon at various processing temperatures and (*b*) alloying elements on case depth of steels during aluminizing. [*Adapted from Lakhtin (1977).*]

solid solution of Si in α-Fe (containing about 14 percent Si). Pearlite segregations are frequently observed beneath the diffusion layer. This is due to the expelling of carbon from the diffused layer since it is poorly soluble in siliceous ferrite. The siliconized layer exhibits high brittleness and moderate hardness, i.e., 200 to 300 HV and high porosity. Because parts are processed in the austenite range, some distortion is experienced.

The siliconized steel components reduce wear under poorly lubricated conditions and exhibit excellent corrosion and oxidation resistance. An example of an outstanding successful application of siliconizing is for rollers for bottle washing machines. The siliconized rollers have been unique in their ability to withstand the combination of wear and corrosion to which they are subjected as glass bottles are moved along in washing operations. Other applications include pump shafts, pipelines, and fitting nuts and bolts.

11.2.9 Sherardizing

Sherardizing is another diffusion process whereby the zinc metal is diffused into an iron or steel surface by heating the specimen in zinc dust media at temperatures in the range of 350 to 450°C. The specimens to be treated by the sherardizing process are enclosed in a sealed mild-steel container having a minimum of 95 percent metallic zinc dust (Evans, 1984). The sealed container is placed in a gas-fired furnace. Temperature uniformity is an important factor for producing the consistent alloying all over the surface. The particle size of the zinc powder is also important: a fine powder results in a smoother, less porous surface than a coarse powder. The case depth is controlled by processing time at the controlled treatment temperature.

The various phases of the iron-zinc alloys are formed by the interdiffusion of iron and zinc atoms. The zinc atoms penetrate the iron to form a solid solution of zinc and iron. The microstructure of the iron-zinc alloy produced at a constant temperature does not vary with the time of processing; only the proportions of the phases change. The microstructure of the iron-zinc alloy is markedly influenced by the temperature. At about 400°C treatment temperature, the alloy consists of a gamma layer at the steel interface containing 20 to 27 percent iron, and a delta compact layer, which makes the major portion of the alloy, containing 6 to 12 percent iron.

The mechanical properties of steel usually are not affected, and the temperatures employed are low enough to avoid many problems of heat distortions. This process is ideally suited for the surface treatment of high-tensile steels which are susceptible to hydrogen embrittlement. The iron-zinc alloy top layer provides a uniform and strongly adherent surface finish and protects iron and steel components against corrosion.

This process is used in the electrical domestic appliances, telecommunications, as well as the automotive industries. Typical applications include scaffold clips, hinges, nuts, bolts, chains, washers, screws, and nails used in the building construction and engineering industries.

11.3 REFERENCES

Anonymous (1981), "Flame Hardening," in *Metals Handbook*, Vol. 4: *Heat Treating*, 9th ed., pp. 484–506, American Society for Metals, Metals Park, Ohio.

——— (1988). *Metals Handbook*, Vol. 14, *Forming and Forging*, 9th ed., American Society for Metals, Metals Park, Ohio.

Arai, T., Mizutani, M., and Komatsu, N. (1975), "Diffusion Layers of Steel Surfaces Immersed in Fused Borax Baths Containing Various Kinds of Additives, *J. Jap. Inst. Metals* (in Japanese), Vol. 39 (no. 3), pp. 247–255.

Astley, P. (1975), "Tufftride—A New Development Reduces Treatment Costs and Process Toxicity." *Heat Treat. Metals*, Vol. 2 (no. 2), pp. 51–54.

Bell, T. (1975), "Ferritic Nitrocarburising," *Heat Treat. Metals*, Vol. 2, pp. 39–49.

———, Gladden, K. D., and Guy, D. N. (1981), "Gaseous Ferritic Nitrocarburizing" in *Metals Handbook*, Vol. 4, *Heat Treating*, 9th ed., pp. 264–269, American Society for Metals, Metals Park, Ohio.

Bennett, M. G. (1978), "Principles of Laser Heat Treatment," *Heat Treat. Metals*, Vol. 5 (no. 3), pp. 76–77.

Bewley, T. J. (1977), "Ion Nitriding," *Heat Treat. Metals*, Vol. 4 (no. 1), pp. 3–4.

Bhushan, B. (1980), "High Temperature Self-Lubricating Coatings and Treatments—A Review," *Metal Finish.*, Vol. 78, May, pp. 83–88; June, pp. 71–75.

——— (1987), "Overview of Coating Materials, Surface Treatments and Screening Techniques for Tribological Applications Part I: Coating Materials and Surface Treatments," in *Testing of Metallic and Inorganic Coatings* (W. B. Harding and G. A. DiBari, eds.), Special Publication STP 947, pp. 289–309, ASTM, Philadelphia.

Biddulph, R. H. (1974), "Boronising," *Heat Treat. Metals*, Vol. 1 (no. 3), pp. 95–97.

Birk, P. (1970), "Powder Nitriding, Theory and Practice," *Microtechnic* (in German), Vol. 25, p. 1.

Bonetti, R., Comte, D., and Hintermann, H. E. (1975), "Boronizing and the Deposition of Layers Containing Boron by the CVD Process," *Proc. of 5th Int. Conf. on Chemical Vapor Deposition* (J. M. Blocher, H. E. Hintermann, and L. H. Hall, eds.), pp. 495–508, The Electrochemical Society, Princeton, N.J.

Booth, M., Farwell, T., and Johnson, R. H. (1983), "The Theory and Practice of Plasma Carburising." *Heat Treat. Metals*, Vol. 2, pp. 45–52.

Bozkurt, N., Geckinli, A. E., and Geckinli, M. (1983), "Autoradiographic Study of Boronized Steel," *Mater. Sci. Eng.* Vol. 57, pp. 181–186.

Carbucicchio, M., and Palombarini, G. (1984), "Iron-Boron Reaction Products Depending on the Base Alloy Combination," *J. Mater. Sci. Lett.*, Vol. 3 (no. 12), pp. 1046–1048.

———, Bardani, L., and ——— (1980a), Mossbauer and Metallographic Analysis of Borided Surface Layers on Armco Iron," *J. Mater. Sci.*, Vol. 15, pp. 711–719.

———, ———, and Sambogna, G. (1980b), "On the Early Stages of High Purity Iron Boriding with Crystalline Boron Powder" *J. Mater. Sci.*, Vol. 15, pp. 1483–1490.

Carley, L. W. (1977a), "Laser Heat Treating." *Heat Treat.*, Vol. 9 (no. 2), pp. 16–21.

——— (1977b), "Electron Beam Heat Treating," *Heat Treat.*, Vol. 9 (no. 4), pp. 18–24.

Casper, S. E. (1977), "Vacuum Carburising," *Heat Treat. Metals*, Vol. 4 (no. 1), pp. 7–8.

Chatterjee-Fischer, R., and Schaaber, O. (1976), "Boriding of Steel and Non-ferrous Metals," *Proc. of Conf. on Heat Treatment*, pp. 27–30, The Metal Society, London.

Child, T. H. C. (1976), "Vacuum Carburising," *Heat Treat. Metals*, Vol. 3 (no. 3), pp. 60–65.

——— (1980a), "Surface Hardening of Steel," *Design Guide No. 37*, Design Council, The British Standards Institution, and Council of Engineering Institution, Oxford University Press, Oxford, England.

——— (1980b), "The Sliding Wear Mechanisms of Metals, Mainly Steels," *Tribol. Int.*, Vol. 13, pp. 285–293.

——— (1983), "Surface Treatments for Tribology Problems," in *Coatings for High Temperature Applications* (E. Lang, ed.), pp. 395–433, Applied Science Publishers, London.

Cottrell, A. H. (1964), *Theory of Crystal Dislocations*, Gordon and Beach, New York.

Dawes, C., and Cooksey, R. J. (1965), "Surface Treatment of Engineering Components," *Proc. Inaugural Meeting of Heat Treatment Joint Committee*, Iron and Steel Institute and Institute of Metals, London.

———, and Tranter, D. F. (1974), "Production Gas Carburising Control," *Heat Treat. Metals*, Vol. 1, pp. 121–130.

DesForges, C. D. (1977), "Laser Applications to Materials Processing Technology," *Engineering*, October, pp. I–VIII.

——— (1978), "Laser Heat Treatment," *Tribol. Int.*, Vol. 11, pp. 139–144.

Dixon, G. J., Plumb, S. A., and Child, T. H. C. (1983), "Processing Aspects of Plasma Nitriding," *Proc. of Conf. on Heat Treatment*, 1981, pp. 137–146.

Draper, C. W., and Ewing, C. A. (1984), "Review Laser Surface Alloying; A Bibliography," *J. Mater. Sci.*, Vol. 19, pp. 3815–3825.

Ebersbach, U., Henny, F., Winckler, U., Reisse, G., and Weissmantel, C. (1984), "Ion Beam Nitriding of High Purity Iron Surfaces," *Thin Solid Films*, Vol. 112, pp. 29–40.

Edenhofer, B. (1974), "Physical and Metallurgical Aspects of Ion Nitriding," *Heat Treat. Metals*, Vol. 1, pp. 23–28, and Vol. 2, pp. 59–67.

———, and Bewley, T. J. (1976), "Low Temperature Ion Nitriding; Nitriding at Temperatures below 500°C for Tools and Precision Machine Parts," *Proc. of Conf. on Heat Treatment*, pp. 7–13, The Metal Society, London.

Elliott, T. L. (1978), "Surface Hardening," *Tribol. Int.*, Vol. 11, pp. 121–128.

Epik, A. P. (1977), "Boride Coatings," *Boron and Refractory Borides*, pp. 597–612.

Evans, D. R. (1984), "The Sherardising Process," in *Coatings and Surface Treatment for Corrosion and Wear Resistance* (K. N. Strafford, P. K. Datta, and C. G. Googan, eds.), pp. 94–102, Ellis Harwood Ltd., Chichester, England.

Eyre, T. S. (1975), "Effect of Boronising on Friction and Wear of Ferrous Metals," *Wear*, Vol. 34. pp. 383–397.

———, and AlSalim, H. (1977), "Effect of Boronising on Adhesion Wear of Titanium Alloys," *Tribol. Int.*, Vol. 10, pp. 281–285.

Fichtl, W. (1982), "Boronizing and Its Applications," *Mater. Eng. (Surrey)*, Vol. 2(no. 6), pp. 276–286.

Fiedler, H. C., and Sieraski, R. J. (1971), "Boriding of Steels for Wear Resistance," *Metal Prog.*, Vol. 99, pp. 101–107.

Fiorletta, C. (1981), "Electron-Beam Heat Treating," in *Metals Handbook*, Vol. 4, *Heat Treating*, 9th ed., pp. 518–521, American Society for Metals, Metals Parks, Ohio.

Gabel, M. K., and Donovan, J. M. (1980), "Wear Resistant Coatings and Treatments," in *Wear Control Handbook* (M. B. Peterson and W. O. Winer, eds.), ASME, New York, pp. 343–371.

Gaucher, A., Guilhot, G., and Amsallem, C. (1976), "Surface Treatments for Iron and Steel," *Tribol. Int.*, Vol. 9, pp. 131–137.

Gilbert, C. L. (1978), "Computerized Control of Electron-Beam for Precision Surface Hardening," *Ind. Heat.*, Vol. 45 (no. 1), pp. 16–18.

Gnanamuthu, D. S. (1979), "Laser Surface Treatment," *Proc. of Conf. Applications of Lasers in Materials Processing* (E. A. Metzbower, ed.), pp. 177–221, American Society for Metals, Metals Park, Ohio.

Goeuriot, P., Thevenot, F., and Driver, J. H. (1981), "Surface Treatment of Steels: Borudif, a New Boriding Process," *Thin Solid Films*, Vol. 78, pp. 67–76.

———, Fillit, R., ———, ———, and Bruyas, H. (1982), "The Influence of the Alloying Element Additions on the Boriding of Steels," *Mater. Sci. Eng.*, Vol. 55, pp. 9–19.

Gregory, J. C. (1966), "A Salt Bath Treatment to Improve the Resistance to Ferrous Metals to Scuffing, Wear, Fretting and Fatigue," *Wear*, Vol. 9, pp. 249–281.

——— (1970), "Thermal and Chemico-Thermal Treatments of Ferrous Materials to Reduce Wear," *Tribol. Int.*, Vol. 3, pp. 73–83.

——— (1974), "Conversion and Conversion/Diffusion Coatings," in *Science and Technology of Surface Coating* (B. N. Chapman and J. C. Anderson, eds.), pp. 136–147, Academic, New York.

——— (1975), "Sursulf—Improving the Resistance of Ferrous Materials to Scuffing, Wear, Fretting and Fatigue by Treatment in a Non-toxic Salt Bath," *Heat Treat. Metals*, Vol. 2 (no. 2), pp. 55–64.

——— (1978), "Chemical Conversion Coatings of Metals to Resist Scuffing and Wear," *Tribol. Int.*, Vol. 11, pp. 105–112.

Grube, W. L. (1980), "Progress in Plasma Carburising," *J. Heat Treat.*, Vol. 1 (no.3), pp. 40–49.

———, and Gay, J. G. (1978), "High Rate Carburising in a Glow Discharge Methane Plasma," *Metall. Trans.*, Vol. 9A, pp. 1421–1429.

Guthrie, A. M., and Archer, K. C. (1975), "Induction Surface Hardening," *Heat Treat. Metals*, Vol. 1, pp. 15–21.

Habig, K. H. (1980), "Wear Protection of Steels by Boriding, Vanadizing, Nitriding, Carburizing and Hardening," *Mater. Eng.* (*Surrey*), Vol. 11, pp. 83–92.

———, and Chatterjee-Fischer, R. (1981), "Wear Behaviour of Boride Layers on Alloy Steels," *Tribol. Int.*, Vol. 14, pp. 209–215.

Harry, A. J. (1978), "Tufftride Treatment Improves Performance of Camera Components," *Tribol. Int.*, Vol. 11, pp. 127–128.

Hick, A. J. (1979), "Heat Treatment—The Present Position and Future Outlook," *Metallurgia*, Vol. 46 (no. 3), pp. 147–172.

Hintermann, H. E., Bonetti, R., and Breiter, H. (1973), "CVD of Boron," *Proc. of 4th Int. Conf. on Chemical Vapor Deposition* (G. F. Wakefield and J. M. Blocher, eds.), pp. 536–545, The American Electrochemical Society, Princeton, N.J.

Hodgson, C. C., and Baron, H. G. (1956), "The Tempering and Nitriding of Some 3% Chromium Steels," *J. Iron Steel Inst. (London)*, Vol. 182, pp. 256–265.

Hosokawa, K. (1976), "Boro-Carburising of Iron and Steel Immersed in Hot Bath Containing B_4C and BaO," *J. Jap. Inst. Metals* (in Japanese), Vol. 40 (no. 1), pp. 57–61.

———, Nishimura, K., Ueda, M., and Saki, F. (1972), "Boro-Carburization of Iron by Ferroboron and Na_2CO_3," *J. Jap. Inst. Metals* (in Japanese), Vol. 36 (no. 5), pp. 418–424.

Hudis, M. (1973), "Study of Ion-Nitriding," *J. Vac. Sci. Technol.*, Vol. 10, pp. 1489–1496.

James, D. H., Smart, R. F., and Reynolds, J. A. (1975), "Surface Treatments in Engine Component Technology," *Wear*, Vol. 34, pp. 373–382.

Jindal, P. C. (1978), "Ion-Nitriding of Steels," *J. Vac. Sci. Technol.*, Vol. 15, pp. 313–317.

Jones, C. K., Martin, S. W., Sturges, D. J., and Hudis, M. (1975), "Ion Nitriding," *Proc. Heat Treatment*, 1973, pp. 163–167, The Metal Society, London.

———, Sturges, D. J., and Martin, S. W. (1977), "Glow Discharge Nitriding in Production," *Source Book on Nitriding*, pp. 186–187, American Society for Metals, Metals Park, Ohio.

Kanter, J. J. (1964), "Siliconizing of Steel," in *Metals Handbook*, Vol. 2: *Heat Treating, Cleaning, and Finishing*, 8th ed., pp. 529–530, American Society for Metals, Metals Park, Ohio.

Korhonen, A. S., and Sirvio, E. H. (1982), "A New Low Pressure Plasma Nitriding Method," *Thin Solid Films*, Vol. 96, pp. 103–108.

Lakhtin, Y. (1977), *Engineering Physical Metallurgy*, Mir Publishers, Moscow.

Liliental, W., Tacikowski, J., and Senatorski, J. (1983), "Effect of Microstructure and Loading Pressure on Tribological Properties of Boride Layers on Steel," *Proc. of Conf. on Heat Treatment*, 1981, pp. 193–197, The Metal Society, London.

Lin, D. S. (1969), "The Effect of the Degree of Work Hardening on the Friction and Wear of Metals during Abrasion," *Wear*, Vol. 13, pp. 91–97.

Linial, A. V., and Lavella, J. (1973), "New Process for Obtaining Increased Metal Hardness and Reduced Friction Properties by Boronizing," *Ind. Heat.*, Vol. 40, No. 12, pp. 9–14.

———, and ——— (1974), "New Process for Obtaining Increased Metal Hardness and Reduced Friction Properties by Boronizing, Part II," *Ind. Heat.*, Vol. 41 (no. 1), pp. 28–31.

———, and Hintermann, H. E. (1979), "Boronizing Processes—A Tool, for Decreasing Wear," *Proc. of Conf. on Wear of Materials* (K. C. Ludema, W. A. Glaeser, and S. K. Rhee, eds.), pp. 403–408, ASME, New York.

Mehelich, C. S., Van Kuikon, L., and Woelfel, M. M. (1981), "Shot Peening," in *Metals Handbook*, Vol. 5: *Surface Cleaning, Finishing, and Coating*, 9th ed., pp. 138–149, American Society for Metals, Metals Park, Ohio.

Morgan, H. L. (1971), "Surface Hardening of Cast Iron by Induction and Flame Hardening Processes," BC IRA Report No. 1047.

Nishida, K., and Tanaka, Y. (1968), "Decomposition Reaction of Fused NaCN Salt and Its Relation to the Nitriding of Steel," *J. Jap. Inst. Metals*, Vol. 32, pp. 435–440.

Palombarini, G., and Carbucicchio, M. (1984), "On the Morphology of Thermochemically Produced Fe_2B/Fe Interfaces," *J. Mater. Sci. Lett.*, Vol. 3 (no. 9), pp. 791–794.

Peckner, D. (ed.) (1964), *The Strengthening of Metals*, Reinhold, New York.

Qing'en, M., and Zaizhi, C. (1984), "Boronization Process in Steel: Theory of Diffusion Channel Model of Boron Atoms," *Proc. Heat Treatment Shanghai 1983*, pp. 3.8–3.12, The Metal Society, London.

Reinke, F. H., and Gowan, W. H. (1978), "Recent Developments in Induction Hardening," *Heat Treat. Metals*, Vol. 5 (no. 2), pp. 39–45.

Robino, C. V., and Inal, O. T. (1983), "Ion Nitriding Behaviour of Several Low Alloy Steels," *Mater. Sci. Eng.*, Vol. 59, pp. 79–90.

Rus, J., DeLeal, C. L., and Tsipas, D. N. (1985), "Boronizing 304," *J. Mater. Sci. Lett.*, Vol. 4 (no. 5), pp. 558–560.

Sachs, K., and Clayton, D. B. (1979), "Nitriding and Nitrocarburising in Gaseous Atmospheres," *Heat Treat. Metals*, Vol. 6 (no. 2), pp. 29–34.

Sakai, T. (1985), "Diffusion Coating of Cemented Carbide WC-Co Systems I. Boronizing and Siliconizing," *J. Jap. Soc. Powder and Powder Metall.*, Vol. 32 (no. 2), pp. 48–54.

Sandven, O. (1981), "Laser Surface Transformation Hardening," in *Metals Handbook*, Vol. 4: *Heat Treating*, 9th ed., pp. 507–517, American Society for Metals, Metals Park, Ohio.

Schiller, S., Heisig, U., and Panzer, S. (1982), *Electron Beam Technology*, Wiley, New York.

Seaman, F. D., and Gnanamuthu, D. S. (1975), "Using the Industrial Laser to Surface Harden and Alloy," *Metal Prog.*, Vol. 108, pp. 67–74.

Seigle, L. L. (1984), "Thermodynamics and Kinetics of Pack Cementation Process," in *Surface Engineering; Surface Modification of Materials* (R. Kossowsky and S. C. Singhal, eds.), pp. 345–369, Martinus Nijhoff, Dordrecht, The Netherlands.

Singhal, S. C. (1977), "A Hard Diffusion Boride Coating for Ferrous Materials," *Thin Solid Films*, Vol. 45, pp. 321–329.

——— (1978), "An Erosion-Resistant Coating for Titanium and Its Alloys," *Thin Solid Films*, Vol. 53, pp. 475–481.

Smith, C. G. (1977), "Nitrocarburising," *Heat Treat. Metals*, Vol. 4 (no. 1), pp. 9–10.

Spalvins, T. (1983), "Tribological and Microstructural Characteristics of Ion-Nitrided Steels," *Thin Solid Films*, Vol. 108, pp. 157–163.

Sridhar, D., and Iyer, K. J. L. (1980), "Wear Studies of Boronized Steel," *Tool Alloy Steels*, Vol. 14 (no. 12), pp. 397–401.

Steen, W. M., and Powell, J. (1981), "Laser Surface Treatment," *Mater. Eng.*, Vol. 2 (no. 3), pp. 157–162.

Stevens, N. (1981), "Induction Hardening and Tempering," in *Metals Handbook*, Vol. 4: *Heat Treating*, 9th ed., pp. 451–483, American Society for Metals, Metals Park, Ohio.

Stickels, C. A., Byrnes, L. E., Dossett, J. L., Hughes, R. L., Gladden, K. D., Poteet, D. A., Riopelle, J. A., and Rogers, R. D. (1981), "Case hardening of steel-gas carburizing, carbonitriding, gas nitriding, pack carburizing, and vacuum carburizing, in *Metals Handbook*, Vol. 4: *Heat Treating*; 9th ed., pp. 133–226, 270–274, American Society for Metals, Metals Park, Ohio.

St. John, M. R., and Sammells, A. F. (1981), "Characterization of an Iron Boride Coating Produced by Pack Boronization of Low-Carbon Steel," *J. Mater. Sci.*, Vol. 16, pp. 2327–2329.

Subrahmanyam, J., and Gopinath, K. (1984), "Wear Studies on Boronized Mild Steel," *Wear*, Vol. 95, pp. 287–292.

Sundararaman, D., Kuppusami, P., and Raghunathan, U. S. (1983), "Some Observations on the Ion Nitriding Behaviour of a Type 316 Stainless Steel," *Surf. Technol.*, Vol. 18, pp. 341–347.

Thelning, K. E. (1984), *Steel and Its Heat Treatment*, *Bofors Handbook*, Butterworth, London.

Tsipas, D. N., and Perez-Perez, C. (1982), "A Boronizing Treatment for Low Carbon Steels," *J. Mater. Sci. Lett.*, Vol. 1, pp. 298–299.

———, Rus, J., and Noguerra, H. (1988), "Boronized Steel and Their Properties," *Proc. Heat Treatment 1987*, pp. 203–210, The Metals Society, London.

Ullmann, D. K. O. (1969), "Application of Chromising," *Eng. Mater. Design*, Vol. 12, pp. 897–950.

Von Matuschka, A. G. (1980), *Boronizing*. Hanser and Heyden, Munchen, Germany.

Wensing, D. R., Barber, D. R., Foreman, R. W., Greene, G., Kopietz, K. H., Laird, W. J., Mehrkam, Q. D., Payne, B. R., Swift, J. A., Wood, W. G., and Wilk, T. B. (1981), "Case Hardening of Steel-Liquid Carburizing and Cyaniding, and Liquid Nitriding," in *Metals Handbook*, Vol. 4: *Heat Treating*, 9th ed., pp. 227–263, American Society for Metals, Metals Park, Ohio.

Westeren, H. W. (1972), "Vacuum Furnace Carburizes Faster, More Efficiently," *Metal Prog.*, Vol. 102, pp. 101–103.

Whittle, R. D. T., and Scott, V. D. (1984), "Sliding Wear Evaluation of Boronized Austenitic Alloys," *Metals Technol.* Vol. 11 (no. 12), pp. 522–529.

Wick, C. (1976), "Laser Hardening," *Eng. Dig.*, Vol. 37 (no. 8), p. 13.

Yoneda, Y., and Takami, S. (1977), "Ion Carburising Process for Industrial Applications," *Proc. of Vacuum Metallurgy*, pp. 135–156, Science Press, Princeton, N.J.

CHAPTER 12
SURFACE TREATMENTS BY ION BEAMS

Ion implantation is the process of modifying the physical or chemical properties of the near surface of a solid (0.01 to 0.5 μm in depth; typically 0.1 μm) by embedding appropriate atoms into it from a beam of energetic ions (2 to 200 keV) in a high vacuum (10^{-3} to 10^{-4} Pa or 10^{-5} to 10^{-6} torr). The properties to be modified may be electrical, optical, or mechanical. The solid may be crystalline, polycrystalline, or amorphous and need not be homogeneous (Baumvol, 1984; Biasse et al., 1983; Carter et al., 1976; Chernow et al., 1977; Crowder, 1973; Dearnaley, 1978; Dearnaley et al., 1973; Eisen and Chaddertan, 1971; Hubler et al., 1984; Mayer et al., 1970; Morehead and Crowder, 1973; Pavlov, 1983; Picraux, 1984a, 1984b; Picraux et al., 1974; Poate, 1985; Townsend et al., 1976; Ziegler, 1984a).

Ion implantation allows the controlled introduction of one or more species into the surface of a substrate by using high-energy ion beams of the species to be introduced. When alternatives such as coatings and thermochemical treatments are impractical, ion implantation offers a straightforward and reproducible method for attaining a desired result. Like other surface treatment processes, the ion implantation does not produce an interface, unlike coatings which are desirable for increased wear performance. The thickness of the altered layer is extremely thin, and the implantation is normally carried out at low temperatures (typically less than 200°C); therefore, ion implantation can be carried out on delicate and finished surfaces without distortion.

Because ion implantation is a nonequilibrium process, solubility limits may be exceeded with or without subsequent precipitation. This makes it possible to incorporate any kind of ion without developing a set of diffusion conditions or considering the control of chemical constraints. Ion implantation was initially used in the fabrication of semiconductor components. For this application, ions were introduced at a relatively low concentration into the silicon substrate. However, it has been discovered that by introducing large quantities of implanted species, a broad range of chemical and mechanical properties of substrate material may be modified.

The implantation of nonmetallic (e.g., B, C, N, P, and S) and metallic (e.g., Cr, Ti, Zr, Ni, Fe, Co, Sn, In, Sb, and As) species into metals, cermets, ceramics, and polymers offers a whole new field of technological exploitation since the injection of atomic species A into substrate B is not limited by thermodynamic criteria (e.g., diffusivities and solubilities).

Typical properties which the ion implantation can impart to a component surface include improved hardness and density, improved wear resistance, improved fatigue resistance, and improved corrosion and chemical resistance, and sometimes reduction of sliding friction and/or adhesion (Baumvol, 1981a, 1981b, 1984; Bolster and Singer, 1980, 1981; Dearnaley and Hartley, 1978; Deutchman and Partyka, 1984; Dillich et al., 1984; Fayeulle et al., 1986; Hartley, 1975a, 1979, 1980, 1982; Herman, 1981; Hirvonen, 1978; Jones and Ferrante, 1983; Lempert, 1988; Singer et al., 1980; Sioshansi, 1989; Straede, 1989).

The ion implantation process has some superficial similarities with the ion-plating process in which the coatings species are also ionized prior to the condensation on the substrate. The implantation of a material results in a graded composition of the alloy of host material and implanted species which possesses no demarkable interface with respect to substrate, while in ion plating, deposition of a thick layer (typically 1 to 5 μm) takes place onto the substrate. The ionization energy involved in ion plating is much lower (100 eV) than implantation and with a poor degree of ionization (1 to 10 percent) of coating species. Therefore, these two processes yield entirely different effects on the properties of treated surfaces.

In ion beam mixing or ion mixing, which has grown out of the success of ion implantation, a thin coating (typically 0.01 to 0.1 μm) is first applied by conventional vapor deposition process and then bombarded (or concurrently bombarded during the total deposition period) vigorously with a beam of energetic ion species (typically 100 to 500 keV) of easily generated reactive or nonreactive gaseous species such as Ar^+, Kr^+, O^+, or N^+ (Appleton, 1984; Chevallier et al., 1986; Gaillaird, 1984; Mayer et al., 1981; McHargue et al., 1984; Nicolet et al., 1984; Shreter et al., 1984; Tsaur, 1981; Tsaur et al., 1979b). Under optimum conditions, either partial or complete intermixing takes place, and the end result resembles the effect of a high dose of implantations of coating atoms which improves the adhesion of the coating to substrate and coating mechanical properties. Ion beam mixing has also been found to increase the wear life and performance of several soft coatings such as MoS_2, WS_2, C, and Ag and hard coatings such as TiN, HfN, TiB_2 (Dearnaley et al., 1986; Hirano and Miyake, 1985; Hirvonen et al., 1987; Köbs et al., 1987; Mayer et al., 1980, 1981; Matteson et al., 1979; Mikkelsen et al., 1988; Padmanabhan and Sørensen, 1981; Padmanabhan et al., 1983).

This chapter will cover some relevant aspects of the physical processes, implanted alloy states, which accompany ion implantation, examples of their uses to modify friction and wear characteristics, and some of the details of equipment and advantages and disadvantages of the process. The effects of ion beam mixing on the friction and wear characteristics and endurance life of solid lubricant and hard coatings are also discussed.

12.1 ION IMPLANTATION

Ion implantation involves the production of ionized species (mostly singly charged) from gas, vapor, or a sputtered surface and acceleration of the ions by an electric field toward the component surface to be treated. Energies lie usually between about 10 keV and 200 keV, and since the ions mostly have a single electronic charge, these values correspond to the voltages required. A limit is set to the lowest energy which can be used by the fact that the ions must penetrate the thin oxide films normally present on metallic surfaces. The process is carried out in a moderately good vacuum of 10^{-3} to 10^{-4} Pa (10^{-5} to 10^{-6} torr), so that undue scattering or neutralization of the charged particles should not occur. The process clearly does not demand a high temperature, as is the case in the various thermochemical treatments for diffusing mobile species such as carbon, nitrogen, or boron.

By virtue of their kinetic energy, the ions penetrate the bombarded material and come to a stop at depths of fraction of a micron (0.01 to 1.0 μm, typically 0.1 μm) in the host material as a result of energy losses due to random collisions with host atoms. Because of statistical fluctuations in the energy loss process, the ions come to rest over a range of depths with a peak in their distribution at the *mean projected range* R_p. Usually, at low ion fluences (or doses, i.e., the number of ions per unit area), the ions are distributed in a more or less gaussion distribution while higher fluences may result in sputtering or inward ion beam-induced migration of atoms into the host matrix. Figure 12.1 represents the ion implantation process and typical gaussion distribution of implanted ions. For 100 keV nitrogen ions in steel, for example, the mean projected range is about 100 nm.

This is a very shallow depth, but in the case of relatively mobile atoms, such as carbon and nitrogen in steel, secondary effects may occur which carry the implanted material to greater depths. There will normally be a temperature rise on the order of 150°C during the implantation process, due to the electric power in the ion beam. With appropriate control of this temperature, the ensuing thermal diffusion of implanted atoms can spread out the distribution throughout a depth of about 500 nm. Excessive diffusion will carry many atoms to the surface, from which they may escape. Alternatively, radiation-enhanced diffusion brought about by the high concentration of vacancies created by intense ion bombardment may give rise to a redistribution of implanted material, to a depth of perhaps 1000 nm (Dearnaley et al., 1973).

The depths to which the composition is modified are still very small compared with those normally encountered in surface coatings or thermochemical treatments. We shall see, however, that this surface layer can have a strong effect on the friction, wear, and corrosion behavior of a material. The fact that renders ion implantation interesting as a practicable method of metal finishing is the existence of a set of mechanisms capable of transporting implanted atoms into the material to considerable depths as wear or corrosion proceeds (Dearnaley, 1978).

From the point of view of ion implantation, the range of a particle in a solid is more important than its specific energy loss, and the latter is generally used as an intermediate parameter. First, let us define what is meant by the *range*: When the energy of an ion falls to about 20 eV, it ceases to move through the solid, becoming trapped by the cohesive forces of the material, and the total distance traveled from the surface to this point is called the *total range* R_{total}. Since the ion will generally follow a devious route with many changes of direction, it is also convenient to define, besides a total range, a projected range R_p (measured parallel to the incident ion direction). The range distribution analysis has been presented by Lindhard et al. (1963), referred to as *LSS theory* in the remaining chapter (Dearnaley et al., 1973; Hirovonen, 1980; Mayer et al., 1970; Townsend et al., 1976; Ziegler, 1984b). The LSS theory provides an approximate relationship between the total and projected ranges and an expression for the total range:

$$R_{total} \approx \left(1 + \frac{M_2}{3M_1}\right) R_p \qquad (12.1a)$$

and

$$R_{total} = 0.6 \times 10^{-6} \frac{(Z_1^{2/3} + Z_2^{2/3})^{1/2} (M_1 + M_2) M_2 E}{Z_1 Z_2 M_1 \rho} \qquad (12.1b)$$

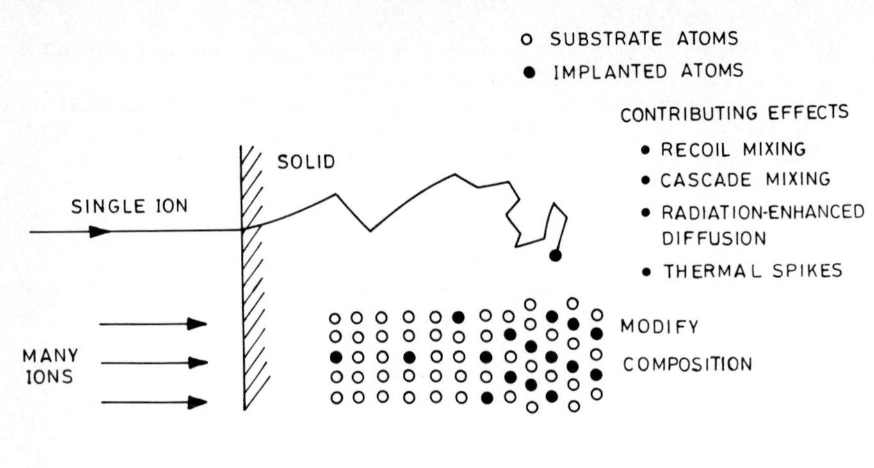

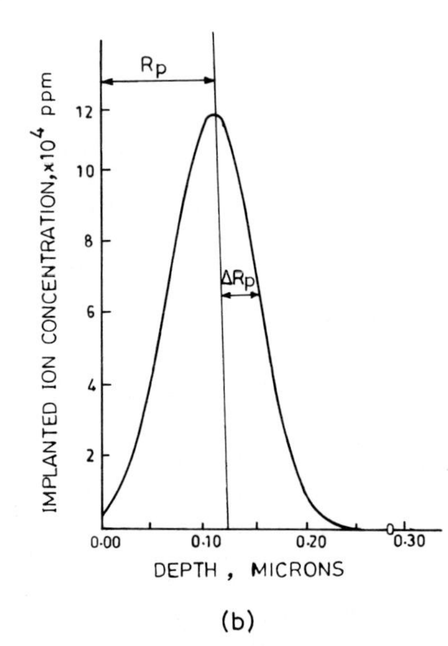

FIGURE 12.1 (*a*) Schematic of the ion-implantation process. (*b*) The implanted ion distribution for steel bombarded with 100-keV nitrogen ions to a fluence of 10^{17} ion cm^{-2}.

where M_1 and M_2 = atomic masses of incident ions and target atoms, respectively

Z_1 and Z_2 = atomic numbers of incident ions and target atoms, respectively

E = acceleration voltage, keV

ρ = target substrate density, g cm^{-3}

The constant 0.6×10^{-6} in Eq. (12.1*b*) is chosen so that E is measured in kiloelectronvolts, and R_{total} is given in centimeters.

Another important quantity is range straggling or mean square fluctuation in range or spread in the range (ΔR_p). The concentration at the peak of the profile corresponding to $x = R_p$ is given as

$$N_{\max} \sim \frac{N_s}{2.5\ \Delta R_p} \tag{12.2}$$

where N_s is the ion fluence (i.e., total number of ions impinging per unit area for the whole implantation). The ion concentration per unit area at depth x from the surface can be related to the maximum concentration as

$$N(x) \approx N_{\max} \exp - \left[\frac{(x - R_p)^2}{2\ \Delta R_p^{\ 2}}\right] \tag{12.3}$$

Projected range and range straggling tables for several implanted species versus host material combinations are available based on the LSS theory in many textbooks (Dearnaley et al., 1973; Hirvonen, 1980).

The depth profile of implanted ions can be evaluated by several surface analytical techniques such as Rutherford back scattering, Auger analysis, XPS, and nuclear reaction analysis.

12.1.1 Implantation Metallurgy

The alloy formation by ion implantation has several features which distinguish it from conventional thermochemical treatments such as nitriding. The typical features are as follows:

- High-energy ion (>10 keV) implantation leads to an intimate atomic mixture within the near-surface region whose composition is not dictated by thermodynamic constraints.
- Concentration versus depth profiles can be controlled through optimization of ion fluence and kinetic energy.
- Atomic displacement by impinging ion beam results in extensive lattice damage which may induce significant atomic transport.
- Ion implantation is an athermal process, which allows an independent control of thermally activated phenomena via the host temperature.

At the energies we are concerned with, ions lose energy in solids by elastic collisions with stationary atoms and by electronic excitation and ionization. Elastic collisions between implanted ions and host atoms frequently result in a recoil of the particle that has been struck. If the recoil energy exceeds about 20 eV, the substrate atom is likely to be displaced from its original site, creating a vacancy and corresponding interstitial: These are termed *point defects* (Brinkman, 1954). Impurity atom relocations as a result of direct interactions with the incident ions are called *primary recoil mixing*. The recoil energy is often enough to cause many secondary matrix atom recoils (displacements), and a so-called secondary recoil or collision cascade is produced which results in significant lattice damage [Fig. 12.2(*a*)]. A single heavy ion can lead to the displacement of many hundreds of

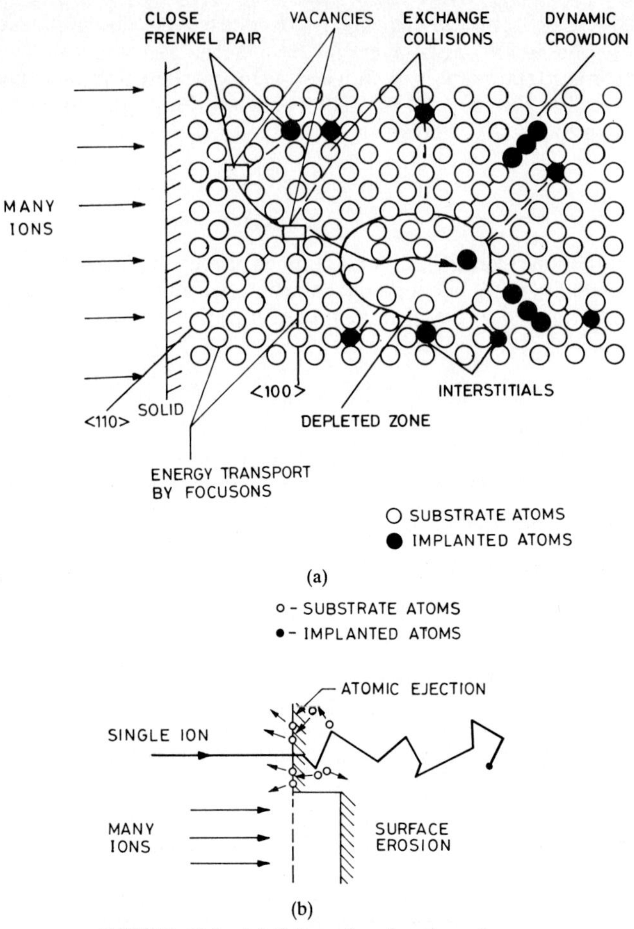

FIGURE 12.2 (*a*) Schematic of various defects and collision processes which may occur during collision cascade. [*Adapted from Dearnaley and Hartley (1978).*] (*b*) Schematic of the sputtering process which may occur during energetic ion beams.

lattice atoms within a volume surrounding the ion trajectory. The vacancy-interstitial pair (point defects) generation rates are usually high, as shown in Fig. 12.1, for the case of 10^{17} nitrogen ion cm^{-2} in iron, and it is clear that the surface is heavily disturbed.

In metals, the vacancy-rich zones tend to collapse, and the more mobile interstitials tend to aggregate, in each case forming platelets of defects, although implanted impurities interfere with these processes. Other defects migrate to sinks such as dislocations, grain bounderies, or the surface itself. Annealing of the damage results in a dense dislocation network like that induced by cold working, which produces resistance to dislocation movement (dislocation pinning) and increased hardness (Cottrell, 1964; Fleischer, 1964; Peckner, 1964).

The range distributions just described are appropriate as long as the bombardment is not too intense. Under some conditions, however, other mechanisms may alter the distribution as implantation proceeds. One of these is sputtering, introduced earlier as a process of erosion of the workpiece, and eventually this will bring about a situation of dynamic equilibrium in which the maximum concentration of impurity lies at the surface [Fig. 12.2(*b*)].

At high-dose rates (ion fluxes), the concentration of vacancies generated by irradiation may accelerate the normal thermally activated diffusion and may serve to transport material away from the implanted zone. This phenomenon is called *radiation-enhanced* or *defect-enhanced diffusion* (Dearnaley and Hartley, 1978). This is significant only at high temperatures (above *RT*) where thermally activated diffusion is significant.

The energy deposited in collision cascades is dissipated through atomic motions in times roughly 10^{-13} s (Seitz and Koehler, 1956). Therefore, following ion impact, the region of the collision cascade is hot for a very short time during which the compound grows by normal thermal reaction, and as the region cools by classical heat conduction, the process is halted. The rapid quenching effects are referred to as *thermal spike* model (Seitz and Koehler, 1956). If an energy E is deposited into a cascade volume V in a target of atomic density N and of thermal conductivity K and if it is imagined that the atoms in motion within the cascade can be treated as a hot gas in equilibrium, a spike temperature T_s can be defined as (Dienes and Vineyard, 1957; Nelson, 1973)

$$T_s \sim \left(\frac{2}{3}\right)\left(\frac{E}{KNV}\right) \tag{12.4}$$

It is usually assumed that the thermal energy is dissipated by lattice thermal conductivity, and typically a cascade volume can be expected to contain ≥1000 atoms.

Nelson (1973) has reported that based on the examination of thermal spike behavior in a number of solids, within 1 to 5×10^{-12} s the temperature of the spike (dense cascade) volume was a few hundred kelvins (up to 10^3 K) higher than the surrounding lattice. The high local temperature should cause some homogenization of the vacancy core-interstitial shell. The implanted atoms move isotropically by thermal diffusion, and their depth distribution can broaden indefinitely with peak of the distribution to shifting inward. The quenching rate of a spike cascade is very rapid and is estimated to be on the order of 10^{13} K s^{-1}. Many metastable and amorphous materials have been produced by the more conventional techniques of vapor quenching or splat cooling (liquid quenching) which have quenching rates of $\sim 10^{10}$ K s^{-1} and 10^6 K s^{-1}, respectively. Thus, a variety of new material properties can be expected from the extremely rapid quenching. The ultimate observable alteration, however, will depend on the defect interaction and annealing properties of the solid, as well as on the details of the cascade.

The injection of defects and of foreign atoms leads to a tendency for the bombarded zone to swell, and since this is restrained by the substrate, an intense lateral compressive stress (on the order of 5 to 50 GPa) may be generated (Hartley, 1975c). This is analogous to the effect of shot peening, and ion bombardment may thus be regarded as shot peening on an atomic scale. We shall see later that the resulting compressive stress may be beneficial for the fatigue life of components and may play a part during the initial stage of wear.

Depending upon the composition and temperature of the host material, lattice damage and the recovery process (spontaneous or thermally activated) crystallinity of the host material may be maintained, or alternatively the surface layer may become amorphous, which is always metastable, and recrystallization may occur above certain characteristic temperatures (Nelson, 1973). The majority of metallic systems remain crystalline on implantation at room temperature. Discussion on various alloy formation processes follows.

12.1.1.1 ***Physical Configuration of Implanted Species.*** An ion-implanted solid solution will be atomically dispersed prior to diffusion and will not contain the aggregates of intermetallic compounds, which often occur in conventionally formed alloys. However, annealing will induce the formation of precipitates where species have a limited solubility, and there will also be interactions between implanted atoms and defects. The latter may well influence the solubility. The lattice locations are well defined and unique, particularly when the implanted concentration is not more than several at. % and may be either substitutional or interstitial. The implantation at high temperature where all the constituents are mobile, the system is governed by thermodynamic equilibrium (Myers, 1978). The precipitation occurs in conformance to the phase diagram: the lattice damage anneals out.

12.1.1.2 ***Supersaturated Crystalline Solutions.*** Under negligible thermally activated atomic transport, the initially dispersed distribution of implanted atoms may persist for concentrations several orders of magnitude above the equilibrium solubility (Borders and Poate, 1976; DeWaard and Feldman, 1974; Kaufman et al., 1977, 1982; Liau and Mayer, 1978; Poate, 1978; Poate and Cullis, 1980; Sood and Dearnaley, 1976). Several highly saturated crystalline solutions have been obtained at room temperature for various metals such as Be, Al, Fe, Ni, and Cu (Potate and Cullis, 1980; Sood and Dearnaley, 1976). For example, implanted Ag^+ in Cu was found to be nearly 100 percent substitutional at 17 at. % although copper's room-temperature solubility is only <0.1 at. %.

12.1.1.3 ***Metastable Crystalline Structures.*** Under certain conditions, ion implantation leads to a metastable crystalline structure (Picraux and Choyke, 1982; Poate, 1978; Poate et al., 1977; Tsaur et al., 1979a). Such phases may range from randomized solutions with a wide composition range to stoichiometric compounds. Phase transformations have been observed in the metallic substrates of Ni, Fe, Mo, and Ti implanted with reactive species C, O, N, and with inert gas species He and Ar (Bykov et al., 1975). Structural transformations from bcc to fcc, fcc to hcp, and hcp to fcc were reported and interpreted as due to accumulation of radiation defects in the coatings.

Depending upon the nature of the system, the formation of metastable phases is attributed to

- High degree of lattice damage
- Mobility of only certain constituents of the material such as light interstitials
- Athermal production of composition

Some typical examples of formation of metastable phase include:

- P^+ implantation in fcc Ni apparently produced small hcp inclusions (Grant, 1978).

- Entailed implanting C^+ into Fe at room temperature followed by annealing at 330°C (Vogel, 1975). Martensite, a metastable, distorted bcc Fe lattice with interstitial C is reported at room temperature. Subsequent annealing produced precipitates of metastable Fe_3C. This feature corresponds to bulk carbon steels quenched and then tempered.

12.1.1.4 Amorphous Phases. The high atomic displacement rates and composition changes produced by ion implantation drive certain materials into amorphous phases apparently similar to those produced by extremely rapid quenching (Appleton, 1984; Grant, 1978). This method of amorphization is exceptionally effective (Table 12.1) as compared to other conventional techniques such as vapor quenching, splat cooling (liquid quenching), and laser glazing. Another important feature of this process is that both stoichiometry and degree of disorder can be varied continuously. Appleton (1984) ion implanted an Si (100) single crystal with 1×10^{16} As^+ cm^{-2} with 100 keV at room temperature. He reported that the sample surface became amorphous because (1) each incident ion created a dense cascade (thermal spike), which became amorphous as a result of the rapid quenching, and (2) Si is a solid with strong covalent bonding, which can stabilize in an amorphous structure at room temperature.

The amorphous metallic systems can be considered to fall into several broad categories such as: metalloid (B or P) implanted into transition metals (Ni, Co, or Fe), P into stainless steel; rare earth (Dy, Bi) implanted into transition metals (Ni, Fe) and Ta or W into Cu and Au into Pt; metalloid (Si) implanted into noble metal (Au); and pure metals.

A variety of interesting properties such as mechanical strength and unique combination of hardness and toughness are exhibited by amorphous alloys. The formation of amorphous phases on ion implantation may be due to the following reasons:

- *Disorder stabilization*: The atomic displacements created by the implanted atoms results in lattice disorder, which builds up; implanted species may help to stabilize this feature.
- *Thermal spike*: The high density of energy dissipated within a collision cascade

TABLE 12.1 Selected Methods of Amorphous-Phase Formation

Method	Process	Form	Quench rate, $K\ s^{-1}$
Condensation of vapor phase	Coevaporation or sputtering onto cold surface	Film	10^{15}
Quench of liquid phase	Stream of molten metal onto cold rotating drum	Ribbon	10^6–10^9
Splat cooling	Molten puddle onto cold surface	Slug	10^9
Laser glazing	High-power heat pulse onto surface	Surface layer	10^9
Ion implantation	Athermal atom introduction with "energy spike"	Surface layer	10^{14}

Source: Adapted from Picraux (1980b).

provides for local rearrangements, and rapid quenching freezes the disordered structure.

- *Phase transformation*: When an appropriate atomic concentration is reached, the amorphous phase may be the preferred structure in the absence of sufficient temperature to allow phase separation and thereby result in a diffusionless transformation.
- *Extremely high stress levels*, present in the implanted layer, may be a driving force for amorphization.

12.1.1.5 Equilibrium-Phase Alloys. The implantation at low temperatures provides diffusionless phase transition when the appropriate concentrations are reached unless appreciable enhanced migration due to implanted ion beam occurs (Picraux, 1980a, 1980b). Homogenous mixture of the implanted species and host matrix atoms is formed at temperature where implanted species's mobility is nil. If the resulting composition lies outside the single-phase diagram, then precipitation may result from heating. When implanted metals are heated so that the diffusion lengths [Dt (D = diffusion coefficient (or rate), t = time at high temperature] for all constituents are large compared to the lattice spacing, a stage close to thermodynamic equilibrium is usually achieved within the implanted region. Irradiation damage anneals out, and additional phases as per phase diagrams are observed to precipitate (Myers, 1978, 1980; Kant et al., 1979a). The high density and small size of precipitates is a unique feature of second-phase systems, formed by implantation, which results in surface hardening in most situations.

The formation and evolution of precipitates are illustrated here by considering implantation of Sb ions into single-crystal aluminum (Kant et al., 1979b). The Sb ion mobility is nil at room temperature and becomes significant at 120°C. The binary-phase diagram indicates only the fcc phase Al; hcp phase Sb and diamond lattice compound AlSb are allowed under equilibrium. The solid solubility of Sb in Al is very low ($\ll$0.1 at. %). Thus for a typical Sb concentration of 1 to 20 at. % Sb, one would expect to be in a two-phase region composed of Al and AlSb precipitates once the mobility of Sb is adequate. Figure 12.3 shows dark field micrographs and corresponding size distribution of AlSb precipitates formed by implanting Sb^+ ions into single-crystal Al at different fluxes and at 50 keV and 300°C (Kant et al., 1979b). At the higher flux, more and smaller precipitates are obtained. Kant et al. also reported that the precipitate mean size becomes larger and number density decreases with increasing fluence and increasing implantation temperature. Moreover, the flux and fluence dependences are stronger at higher temperatures. Similar conformance to the phase diagram following annealing has been found for various implantations in hosts Be, Fe, and Cu (Myers, 1978).

The implanted species diffuse into the underlying bulk of the host on annealing. But such continuing solute transport does not disturb the equilibrium in the implanted layer because diffusion-limited communication between closely spaced precipitates is rapid compared to long-range migration into the substrate (rates of these two processes typically differ by a factor of 10^4).

In the equilibrium regime, ion implantation provides a fine control over precipitation size (range typically from a few nanometers to a fraction of a micron), volume fraction, and depth distribution. Very high particle densities and volume fractions have been achieved, for example, TiC precipitates have been produced into Fe to a density of 10^{20} particle cm^{-3} and a volume fraction of ~30 percent.

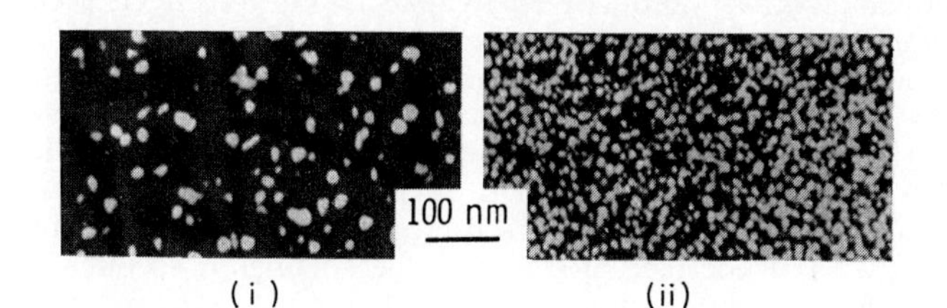

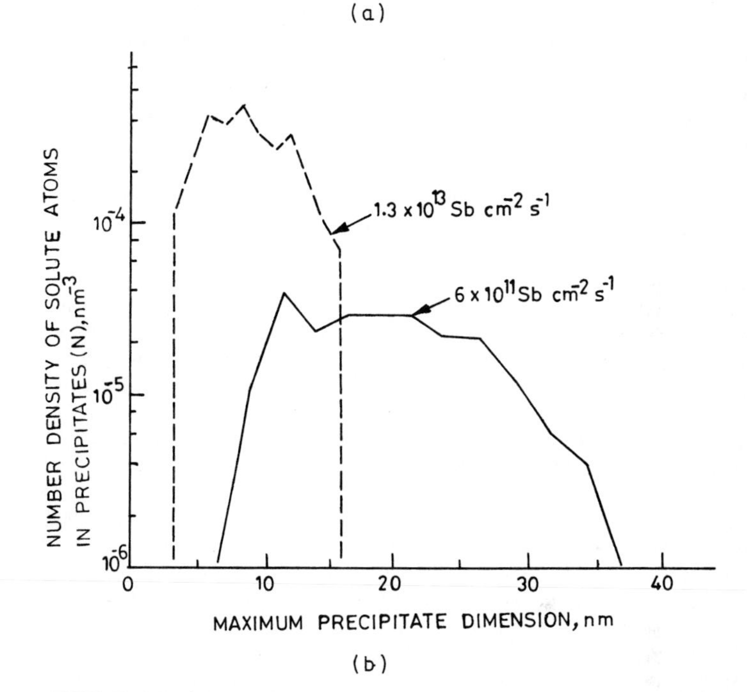

FIGURE 12.3 (*a*) Dark-field micrographs of AlSb precipitates formed by implanting Sb ions into single-crystal aluminum (5×10^{15} Sb cm^{-2} fluence at 50 keV) with (i) 6×10^{11} Sb cm^{-2} s^{-1} and (ii) 1.3×10^{13} Sb cm^{-2} s^{-1} ion fluxes at 300°C. (*b*) AlSb precipitate size distribution, i.e., N equals the number of precipitates per unit size interval per unit projected substrate area versus maximum projected precipitate dimension as a function of Sb ion flux at 300°C. [*Adapted from Kant et al. (1979b).*]

12.1.2 Advantages and Disadvantages of Ion Implantation

Ion implantation allows the controlled incorporation of any one or more atoms into any substrate material and does not depend upon the implanted species being able to diffuse or even dissolve in the host material. Novel nonequilibrium structures and metallurgical phases with properties that cannot be duplicated in bulk materials can be produced at the surface. In certain cases, amorphous or glassy phases can be formed. If the implanted atoms are mobile, inclusions and precipitates can be formed. Cooperative effects of two implanted species can occur; for example, implantations of both molybdenum and sulfur into steels seem to have an effect similar to lubrication with molybdenum disulfide.

Ion implantation is not a coating technique. The atoms implanted are introduced below the surface of the original material and exert their influence within the matrix. There is no change in the dimensions of the component. The implantation results into a graded composition of the alloy of host material and implanted species which possess no demarkable interface with respect to the substrate, Fig. 12.1. The interface produced in the coatings is so often subject to mechanical weakness and interface corrosion. Ion implantation is applied to finished components because it is a low-temperature process (carried out as a rule below 200°C), and the thickness of transformed layer is very small; therefore, there is no risk of distortion. Posttreatment is unnecessary and undesirable. In contrast, conventional thermochemical treatments usually involve temperatures of at least 500°C to in excess of 900°C in order to introduce a limited variety of atomic species (e.g., nitrogen, carbon, boron, and sulfur) by diffusion. We normally experience distortion from the high temperatures involved.

The energies of atoms during thermally activated diffusion processes are very low, and rarely more than 0.1 eV. Diffusion is frequently impeded by the presence of thin surface oxide films or other forms of contamination, and diffusion rates are affected by built-in strains and dislocations in a metal. For these reasons, diffusion is often poorly controlled. In contrast, the energies of atoms in ion implantation are typically a million times greater than those in thermal diffusion. The penetration results from the kinetic energy of these particles and is not dependent upon dislocations, surface films, or grain boundary effects (Dearnaley, 1982).

The ion implantation process is easily controlled through the electric signals applied to the accelerator. The process creates no problems of disposal of waste products, as does electroplating.

These advantages must be weighed against the known disadvantages of ion implantation. The depth of penetration of species is very shallow (0.01 to 0.5 μm) compared to that of thermochemical treatments (20 to 3000 μm) or many surface coatings (typically greater than a micron) which may limit the applications to light load and/or lubricated situations. It requires that substrate be manipulated in vacuo to implant surfaces of varying design. It is a line-of-sight process, lacking the throwing power. High capital cost limits the applicability to high-cost parts where a specific surface is extremely important, or to applications in fail-safe systems where high reliability is necessary. Such reliability can be achieved because of the improved corrosion and fatigue resistance, as well as the much reduced friction and/or wear which are characteristic of this process.

Major advantages and disadvantages of ion implantation technology in comparison to conventional surface treatments and coatings are summarized in Table 12.2.

12.1.3 Ion Implantation Equipment

The main parts of an ion implantation equipment are

- The ion sources which ionize solids, liquids, or gases
- The electrostatic extraction system which extracts ions from ion sources
- The acceleration systems, which impart high kinetic energy to ions

TABLE 12.2 Advantages and Disadvantages of the Ion Implantation Process

Advantages	Disadvantages
Versatility, as regards ion species and substrates	Shallow treatment
No interfacial adhesion problem of coatings	Line-of-sight process
Low-temperature process	Requires in vacuo manipulation
No buildup	High capital cost
Applied to finished components	
Monitored electrically throughout	
Controllability	
No toxicity	

- The analyzing system (mass separator) where the ions are separated according to their masses
- The scanning system which distributes the ions uniformly over the surface to be treated

Numerous excellent equipment surveys exist, e.g., those by Dearnaley (1978, 1983b); Dearnaley et al. (1973); Glawischnig and Naock (1984); Ryding et al. (1976); Ryssel and Glawischnig (1982, 1983); Townsend et al. (1976); and Ziegler and Brown (1984). This section gives a brief introduction to the description and operation of ion implantation equipment. Different applications impose different demands with respect to dose range, energy ion species, size and shape of component to be treated, as well as automation and control.

Figure 12.4 shows a schematic diagram of a typical research-type ion implanter used for chromium ion implantation. It consists of an ion source capable of producing low-intensity ion beams of almost any element, high-potential electrodes which extract ions from an ion source and accelerate through an evacuated acceleration column, ion analysis magnets which separate various ions, and electrostatic scanning plates which focus and steer ion beams toward the specimen chamber. The various constituents of the ion implanter are discussed in the following section.

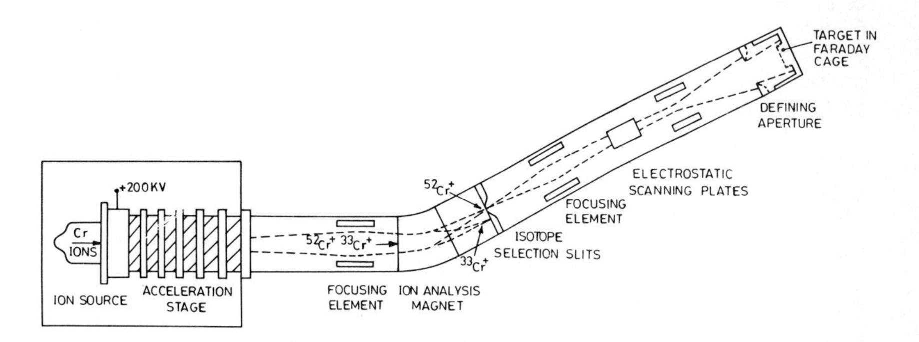

FIGURE 12.4 Schematic block diagram of an ion implanter employed for implanting Cr ions.

12.1.3.1 Ion Sources. The choice of ion source depends on the type of ion that is to be generated, the required ion current, the gas efficiency, the current fluctuations, the beam shape, and adaptability to automatic control, etc. Table 12.3 summarizes the details of several commonly used ion sources (Harper et al., 1982; Kaufman et al., 1982; Keller and Robinson, 1987; Stephens, 1984; Townsend et al., 1976). In addition, cold cathode and hot hollow cathode ion sources are used.

The typical ion sources used for low- and medium-current implanters are cold cathode penning sources (Bennett, 1969); small-sized hot hollow cathode sources (Sidenius, 1978); and a scaled-down version of the Freeman type (Freeman, 1963). The ion sources for high-current implanters are Freeman, multipole, solenoid (Penning Ionization Gauge or PIG), Duopigatrons, etc.

12.1.3.2 Ion Beam Extractor System. In ion sources, ions are produced at low energy, and they collect in quiescent plasma near the apertures of an ion extraction system. From this stage, the ions are extracted and accelerated to the required energy by a series of electrodes or biased aperture plates. The extraction voltage is typically fixed at 25 kV, and in some cases is adjustable up to 50 kV in order to obtain higher current and energies. The deceleration voltage is usually kept at −2 kV. In a typical triode extractor, ions are collected through holes in the first electrodes that border the plasma. The second electrode is biased negatively to attract ions which pass through it. The third electrode is biased positively relative to the second, mainly to prevent neutralizing electrons beyond the extraction system from streaming back into the plasma. This system is called an *accel-decel system*.

12.1.3.3 Accelerators and Analyzing Systems. Accelerators are used to produce ion beams with high kinetic energies. There are three principal implantation configurations. Each layout has advantages and disadvantages for implantation. The simplest and most conventional form of accelerator is a machine with the ion source at high voltage and the beam analysis and target chamber at earth potential. The advantages of such a facility are that the source is electrically isolated from the rest of the system and that there is free access to the target. The main limitations are the probable loss of beam quality when it is used at low energies, and in the case of the higher-energy machines, the need for large and powerful deflecting magnets to analyze the beam. In the second accelerator configuration with the analyzer at high voltage, the mass separation is carried out at low energy, typically tens of keV, and a small analyzer can be used. There is no deterioration of performance over a wide energy range. There is, however, the complication that the analyzer with its supplies must be at high voltage. The third configuration with the target chamber at high voltage may in its simplest form consist of a small conventional accelerator with a bias target stage to extend the energy range. It has the same advantage as the previous configurations, that a small analyzer can be used, but the immediate drawback to the use of high voltages on the target is that it restricts access and limits the experimental flexibility. It has high resolution, and intense beam currents are readily obtained.

The primary role of the analyzing systems (mass separators) of an ion accelerator is to separate the required dopant beams (implanting species) through mass analysis of the extracted ions which is a mixture of a different fraction of molecules and atoms of the source material (Hanley, 1983; Keller, 1981). Focusing is inherent in mass analysis with a bending magnet. The mass analysis is based on the following principle. When a charged particle passes through the field of the

TABLE 12.3 Types of Ion Sources and Their Characteristics

Source	Electron confinement	Ion extraction	Special features
Calutron	Magnetically confined arc	Side (cross-field)	Separate filament chamber
Bernas	Magnetically confined arc	Side	Single arc chamber
Freeman	Magnetron with end containment	Side	Simple structure; good gas efficiency
Multipole	Cusp magnetic fields	Field-free multiple aperture	Quiet with a high gas efficiency
Solenoid (Penning Ionization gauge)	Magnetic with electrostatic end confinement	End (parallel to field)	Filament as large as extraction area
Duopigatron	Solenoid or cusp	Parallel to field or field free	Strong electromagnetic field layer separating anode and cathode chambers
Microwave	Strong magnetic field	End	No filament; long operating life

Source: Adapted from Keller and Robinson (1987).

analyzing magnet, it is deflected into a circular trajectory, whose radius R is given by

$$R = \frac{144}{H}\sqrt{\frac{MU}{n}} \tag{12.5}$$

where M is ion mass (amu), U is the acceleration voltage (volts), H is magnetic field (gauss), and n is the charged state of the ion. Thus particles of the same charge state and energy but of differing mass originating from a common source will follow paths of different radius and thus can be separated.

In a typical system, sufficient resolution to separate two mass neighbors over the entire range with the highest resolvable mass of 75 As^+ or 121 Sb^+ can be achieved.

12.1.3.4 Ion Beam Scanning. Electrostatic scanning is used for low- and medium-current ion beams (~5 mA) and maximum energy up to 100 to 500 keV implanter. The electrostatic ion-scanning system usually consists of two or three pairs of deflection plates, including a neutral trap, causing a periodical sweep of the ion beam over the target. In case of high beam currents (~10 mA), the electrostatic scanning of the beam becomes difficult due to accumulation of space charge and unwanted target heatup. Therefore, for high-current ion beams (15 mA) and energy range up to 200 keV, implanter-, mechanical-, or hybrid-scanning systems are preferred in which the target changes in a batch and face the beam periodically.

In mechanical-scanning systems, the beam remains stationary, and the specimen makes three movements. The different types of commonly used mechanical-scanning systems are rotating disk, rotating chain, rotating carousel, and the Ferris wheel. In hybrid-scanning systems, the beam is deflected electrically across a carousel, and the other scan direction is mechanically achieved by rapid rotation of the carousel.

The use of ion implantation equipment for semiconductor and microelectronics applications are commercially available from several manufacturers. But at present only prototype ion implantation equipment is available to handle machine components for tribological applications. The requirements for tribological applications are significantly distinct as compared to semiconductor applications. For instance, in tribological applications,

- Higher ion fluences and intense ion beams (1 mA) are preferred.
- The components vary greatly in size and shape and are often complex in geometry.
- Usually metal ions Cr^+, Mo^+, and Ti^+ and nonmetal ions C^+, N^+, and B^+ are required for imparting wear resistance in metals and ceramics.

Several prototype ion implanters with a large working chamber size (8 m^3) have been developed at Atomic Energy Research Establishment, Harwell, U.K., which can provide ion beams of gaseous species over the range of 10 to 100 keV with high current up to 3 mA (Dearnaley, 1982).

12.1.4 Materials Treated by Ion Implantation

The ionic species of nonmetals such as C^+, N^+, B^+, P^+, S^+, O^+, Ar^+, and Xe^+, metals such as Ti^+, Cr^+, Ta^+, Mo^+, Ni^+, Fe^+, Al^+, Cu^+, and Sn^+, and rare earth metals such as Y^+ and Ce^+ have been implanted under varying fluences (typically 10^{15} to 10^{21} cm^{-2}) at different energies (50 keV to 3 MeV) in the following commonly used materials in order to increase hardness, fracture toughness, flexural strength, and to improve friction and wear characteristics.

- Metals: Fe, Cu, Al, Ti, Be, Mo, steels, phosphor bronze, co-based alloys, Ti, Ti6Al4V, Ni-Cr alloy
- Nonmetals: B
- Cermets: Co-WC
- Ceramics: Al_2O_3, sapphire (α-Al_2O_3), Al_2O_3-Cr_2O_3, ZrO_2, MgO, Si_3N_4, WC, SiC, TiB_2
- Polymers: polyimide, PTFE

The properties of the above materials can be tailored to specific needs by optimizing various ion implantation parameters.

12.1.5 Characteristics of Ion Implanted Materials

The microhardness, fracture toughness, flexural strength, friction, wear, fatigue, and corrosion characteristics of metals, cermets, ceramics, and polymers are greatly modified in the near surface region by implantation of foreign ionic species due to formation of compounds such as carbides, nitrides, borides, supersaturated solid solutions, metastable phases, and amorphous phases, depending upon the host material and implanted species (Antill et al., 1976; Dearnaley and Hartley, 1978; Dearnaley et al., 1985, 1986; Fayeulle et al., 1986; Goode et al., 1983; Hayashi et al., 1980; Hioki et al., 1986; Iwaki, 1983, 1987; Kant et al., 1979b; Kustas and Misra, 1984; Madakson, 1987; Madakson and Smith, 1983; Myers, 1980; Singer, 1984; Tsaur, 1980; Watanabe et al., 1989; Yust and McHargue, 1984).

The extent of improvement is dictated by

- The selection of proper composition (i.e., alloying elements) of host material and ionic species (i.e., lighter Ar^+ or heavier Xe^+)
- Microstructural phase, i.e., crystalline, polycrystalline, or amorphous of host material
- Implantation parameters, i.e., ion fluences and kinetic energy, implantation time, and temperature
- Post implantation treatments such as annealing at an elevated temperature (300 to 500°C in the case of steels)

In this section, we summarize selected published data which demonstrate the improvements by ion implantation (Tables 12.4 to 12.8).

12.1.5.1 Microhardness. Microhardness of implanted layer is increased at sufficiently high doses due to formation of harder compounds as a result of implantation, increase in dislocation density, and large residual compressive stresses produced due to the difference in molar volume of the intermetallic compounds. Hartley (1975c) has reported that the compressive stresses produced in N^+ implanted nitriding alloy (En4OB) steel at 10^{16} and 10^{17} ion cm^{-2} are about 1 and 50 GPa, respectively. The surface is observed to begin blistering at doses approaching 10^{17} ion cm^{-2} and eventually bursts to leave an implanted surface heavily pock-marked with small craters at about 10^{18} ion cm^{-2}.

The microhardness obtained by various researchers for treated surfaces are summarized in Table 12.4. There is no universal explanation which can explain the change in microhardness for a particular combination of implanted species and host materials. The significant difference in the improvement in microhardness of implanted materials is probably due to different implantation conditions and, to some extent, to the method and load of microhardness measurements. As the thickness of implanted layer is very thin (maximum 1 μm), the indentation must be very shallow, and thus very light loads (0.1 to 1 mN) should be used for microhardness measurements.

The influences of ion fluences/dose of N^+ ions into bearing steels (18 W, 4Cr, 1V, Fe) is illustrated in Fig. 12.5. There is very little change in microhardness at low doses (i.e., 10^{13} to 10^{15} N^+ cm^{-2}), but the increase is fourfold at higher doses (10^{17}). The large scatter in hardness value may be probably associated with large carbide precipitates on the implanted surface at higher doses. The negligible effect of postimplantation annealing treatment at 200°C for 1 h shows that an equilibrium phase is reached during implantation.

The hardness of implanted ceramics increased in most of the cases provided that the substrate remained crystalline. Figure 12.6 shows the influence of ion fluence on the relative hardness for single crystal Al_2O_3 (*c*-axis orientation) implanted with 280 keV Cr^+ ions at room temperature (McHargue, 1987). The amount of Cr^+ in the substitutional sublattice site (as viewed along the *c* axis) and residual disorder in Al sublattice are responsible for the increase in hardness. The

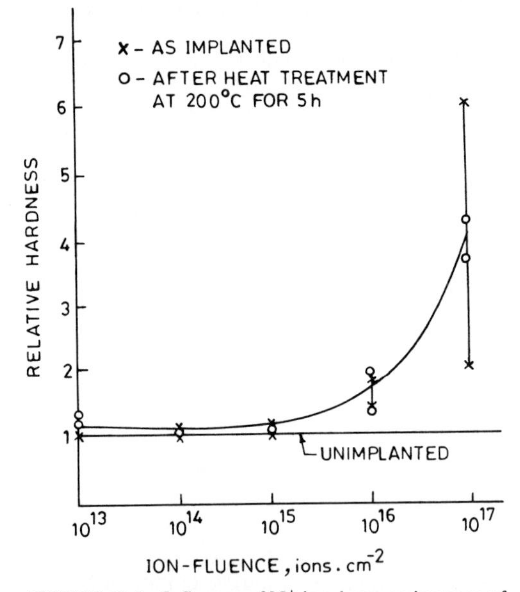

FIGURE 12.5 Influence of N^+ ion doses on increase of relative microhardness of (i.e. ratio of microhardness of implanted material to that of unimplanted material) 18 W-4Cr-1V, bearing steel implanted at 400 keV and room temperature. [*Adapted from Madakson (1987).*]

TABLE 12.4 Microhardness of Ion Implanted Materials Reported by Various Researchers

Materials	Implanted ions	Acceleration voltage, keV	Ion fluences, ion cm^{-2}	Load, N	Hardness of unimplanted material (HV)	Hardness of implanted material (HV)	References
Ferrous metals							
Pure iron (ferritic)	N^+	90	3.5×10^{21}	—	—	1.8*	Oliver et al. (1984), Pethica et al. (1983)
Low-carbon steel	N^+	150	10^{17}	0.02	310	650	Iwaki (1987)
	Cr^+	150	10^{17}	0.02	310	500	Iwaki (1987)
Cr-Mo steel	N^+	24	10^{17}	10	300	400	Kanaya et al. (1972)
Cr-Mo steel (42Cr 4Mo)/AISI 4140	N^+	100	10^{17}	0.02	575 ± 50	810 ± 80	Dimigen et al. (1985)
304 stainless steel (austenitic)	N^+	50	10^{17}	0.15	296	630	Suri et al. (1979)
	B^+	40	10^{17}	0.15	296	410	Suri et al. (1979)
316 stainless steel	N^+	150	5×10^{17}	0.15	164	238	Fu-Zhai et al. (1983)
416 stainless steel	N^+	100	10^{17}	0.02	380 ± 20	1225 ± 100	Dimigen et al. (1985)
52100 bearing steel (martensitic)	N^+	100	10^{17}	0.02	960 ± 60	1040 ± 50	Dimigen et al. (1985)
18W 4Cr 1V bearing steel	N^+	400	10^{17}	0.1	—	4*	Madakson (1987)
440B steel	N^+	100	10^{17}	0.02	330 ± 40	1130 ± 270	Dimigen et al. (1985)
H 13 tool steel	N^+	75	5×10^{17}	0.05	—	1.4*	Iwaki (1987)
	B^+	75	5×10^{17}	0.05	—	1.2*	Iwaki (1987)
Nonferrous metals							
Aluminum	N^+	50	10^{17}	0.15	100	128	Suri et al. (1979)
	B^+	40	10^{17}	0.15	100	131	Suri et al. (1979)
Beryllium	B^+	90–250	1.5×10^{17}	0.02	570 HK	1520 HK	Hirvonen (1978)
	B^+	90–250	10^{17}	0.05	—	4*	Kant et al. (1979a)
Titanium	N^+	50	10^{17}	0.15	230	300	Suri et al. (1979)
	B^+	40	10^{17}	0.15	230	330	Suri et al. (1979)
Ti6A14V alloy	N^+	90	3.5×10^{21}	—	—	2*	Oliver et al. (1984)
	N^+	40	2×10^{17}	0.1	370	440	Saritas et al. (1987)
	B^+	40	2×10^{17}	0.1	370	420	Saritas et al. (1987)
	C^+	40	5×10^{17}	0.1	370	575	Saritas et al. (1987)
	O^+	40	2×10^{17}	0.1	370	500	Saritas et al. (1987)
Hard electroplated chromium	N^+	90	3.5×10^{21}	—	—	1.3*	Oliver et al. (1984)

TABLE 12.4 Microhardness of Ion Implanted Materials Reported by Various Researchers (*Continued*)

Materials	Implanted ions	Acceleration voltage, keV	Ion fluences, ion cm^{-2}	Load, N	Hardness of unimplanted material (HV)	Hardness of implanted material (HV)	References
			Ceramics				
Alumina (Al_2O_3), single crystal	Cr^+	280	10^{16}–10^{17}	0.15	—	1.5*	McHargue and Yust (1984)
	Ti^+	300	7.7×10^{16}	0.25	—	1.75*	Burnett and Page (1987)
	Zr^+	300	1.7×10^{16}	0.25	—	1.4*	Burnett and Page (1987)
	Ni^+	300	10^{16}–10^{17}	0.25	—	1.3*	Hioki et al. (1989)
Magnesium oxide (MgO) (100) single crystal	Ti^+	300	2×10^{16}	0.10	—	2.3*	Burnett and Page (1984c)
	Cr^+	300	6×10^{16}	0.10	—	2.0*	Burnett and Page (1984c)
Y-stabilized zirconia	Al^+	190	10^{16}	0.5	—	1.28*	Legg et al. (1985)
	Ti^+	400	3×10^{16}	0.1	—	1.6*	Burnett and Page (1986a)
Silicon carbide (SiC) (*c* axis)	Cr^+	280	10^{15}	0.15	—	1.2*	McHargue (1987)
Silicon carbide (reaction bonded)	N^+	90	10^{15}	0.25	—	1.1*	Burnett and Page (1986b)
Sialon	Ti^+	300	10^{15}	0.25	—	1.05*	Burnett and Page (1986b)
Co-WC	N^+	100	10^{17}	0.25	—	1.1*	Burnett and Page (1986b)
	Ti^+	400	5×10^{16}	0.25	—	1.1*	Burnett and Page (1986b)
	N^+	—	10^{17}	—	—	4.5*	Fayeulle et al. (1986)
Titanium diboride (TiB_2) hot pressed	Ni^+	1000	10^{17}	0.15	—	1.7–2.1*	McHargue et al. (1986b)
			Polymers				
Polyimide	N^+	92	10^{17}	—	270	325	Watanabe et al. (1989)
Kapton film	N^+,C^+,B^+	500	3×10^{15}	—	—	3*	Lee et al. (1990)

*Relative hardness, i.e., ratio of hardness of implanted materials to that of unimplanted material.

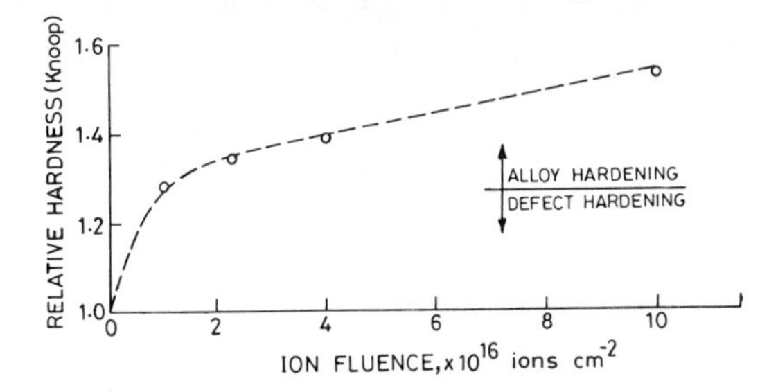

FIGURE 12.6 Influence of ion fluence on the increase in relative microhardness of single-crystal Al_2O_3 implanted with Cr^+ ions at 280 keV and room temperature. The Knoop indentations were made on the (0001) face with 0.15 N load. [*Adapted from McHargue (1987).*]

amount of disorder was almost constant for fluences from 10^{16} to 10^{17} ion cm^{-2} whereas the substitutional fraction of Cr increased to 0.55 at the maximum ion fluence. Therefore, it was estimated that radiation damage accounted for essentially all the increase at fluences up to about 2×10^{16} Cr^+ ion cm^{-2} and approximately half of it at 10^{17} Cr^+ ion cm^{-2} (McHargue, 1987). Implantations at higher temperatures (~350°C) result in smaller increases in hardness because of a lower amount of disorder or radiation damage.

Implantation-induced amorphization shows that transformation from crystalline to amorphous results in the softening of the ceramics, i.e., reduction in hardness. The data for Al_2O_3 (Hioki et al., 1986; McHargue et al., 1985, 1986a); MgO (Burnett and Page, 1984c); ZrO_2 (Burnett and Page, 1986a; Legg et al., 1985); and SiC, (McHargue et al., 1983; Roberts and Page, 1986; Spitznagel et al., 1986) indicate that the hardness of amorphous phase is about 50 to 60 percent that of the crystalline phase.

Ion implantation causes a significant improvement in the hardness of several polymers such as polytetrafluoroethylene (PTFE), polyimide (PA), polyetheretherketone (PEEK). Lee et al. (1990) have reported a dramatic increase (≈30 times) in the hardness of polyimide Kapton films, implanted with B^+, C^+ and N^+, near 100 nm depth. They have observed, by Rutherford Backscattering Spectroscopy, two-thirds depletion of the oxygen up to a depth of at least 200 nm and gradual depletion of carbon up to 100 nm. The improvement in hardness may result from the formation of a disordered carbon layer related to the depletion of hydrogen atoms and other heteroatoms such as N and O.

*12.1.5.2 **Fracture Toughness.*** The fracture toughness K_{IC} is a measure of resistance to propagation of a crack under stress once it has been initiated. The fracture toughness is a very important factor for ceramic materials and is one which makes them unfit for several tribological applications. The implantations of metallic ions such as Ni^+ into ceramics such as α-Al_2O_3, SiC, ZrO_2, MgO, and diamond improve the fracture toughness several times (Burnett and Page, 1984a, 1984b; Cochran et al., 1984; Farlow et al., 1984; Hasegawa et al., 1985; Hioki et al., 1984, 1985, 1986; Legg et al., 1985; McHargue, 1988; McHargue and Yust, 1984; McHargue et al., 1983, 1984, 1986a; Noda et al., 1987; Roberts and Page, 1982; Yust and McHargue, 1984). Table 12.5 summarizes the fracture toughness of some common implanted ceramics.

TABLE 12.5 Fracture Toughness of Ion Implanted Ceramics Reported by Various Researchers

Material	Implanted ion	Acceleration voltage, keV	Ion fluences, ion cm^{-2}	Relative fracture toughness, K^*_{IC}	References
Aluminum oxide (single crystal)	Ni^+	300 at 100 K	10^{17}	2.1	Hioki et al. (1986)
	Ni^+	300 at 520 K	10^{17}	1.8	Hioki et al. (1986)
Y-stabilized zirconia	Al^+	190	4×10^{17}	1.1	Legg et al. (1985)
Mg-stabilized zirconia	Ti^+	400	10^{16}	0.82	Burnett and Page (1986b)
Reaction-bonded SiC (polycrystalline)	N^+	180	10^{15}	1.25	Burnett and Page (1986b)
Sialon	Ti^+	300	10^{15}	0.7	Burnett and Page (1986b)
Titanium diboride TiB_2, hot pressed	Ni^+	1000	10^{17}	1.8	McHargue et al. (1986b)

*Relative fracture toughness is the ratio of fracture toughness of implanted ceramics to that of unimplanted ceramics under identical conditions.

Significant improvements (typically 15 to 100 percent) in indentation fracture toughness of ion implanted Al_2O_3 have been reported by several authors (Burnett and Page, 1985; Hioki et al., 1986; McHargue et al., 1983). Figure 12.7 illustrates the improvement in fracture toughness K_{IC} of α-Al_2O_3 (sapphire) implanted at 300 keV Ni^+ ions at 100, 300, and 523 K temperatures. This increase can be attributed to the following reasons:

- Accumulation of large residual compressive stress (~5 to 50 GPa) during implantation. This is high at low temperatures 100 K (Hartley, 1975*c*).
- Lattice damages due to implantations at higher influences results in amorphization which reduces hardness.

The fracture toughness of single-crystal SiC was increased (up to 30 percent) by metal ion implantation, both for damaged but crystalline layers and for amorphous layers (Burnett and Page, 1985). The incidence of lateral cracking was markedly reduced but there was little change in radial cracking. A large increase (80 percent) in indentation fracture toughness of the implanted single-crystal MgO was attributed to both the surface compressive stress and the change in mechanism of crack nucleation (Burnett and Page, 1984c).

The fracture toughness of yttria stabilized zirconia (fully) increased on Ti ion (Burnett and Page, 1986a) and Al ion (Legg et al., 1985) implantations at fluences below that necessary to induce amorphization. Lateral cracks were again suppressed by implantation. The improvement of implantation was found different for fully and partially stabilized zirconia. In contrast, fracture toughness of partially stabilized ZrO_2 reduced with increasing ion fluence of 400-keV Ti ions until

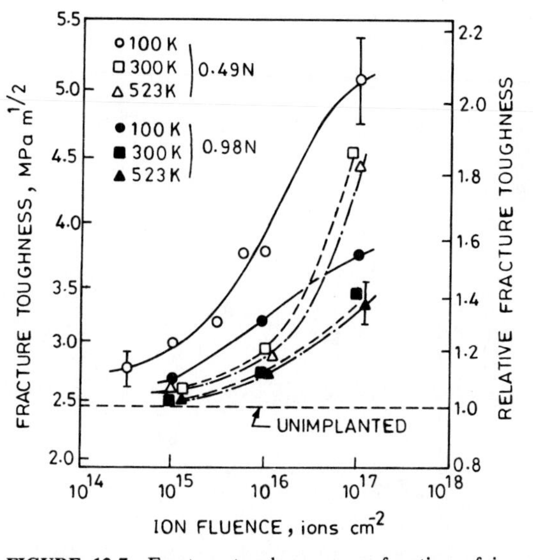

FIGURE 12.7 Fracture toughness as a function of ion fluence for the α-Al_2O_3 implanted with 300-keV Ni ions at 100, 300, and 523 K. The K_{IC} values evaluated from the Vickers indentations at loads of 0.49 and 0.98 N. [*Adapted from Hioki et al. (1986).*]

amorphous state was reached and then remained constant (Burnett and Page, 1986a).

12.1.5.3 ***Flexural Strength.*** Flexural strength or transverse rupture strength is another important property of ceramics which gives an indirect measure of tensile stress adequate for the mechanical failure. Generally, failure is initiated at the surface, which can be modified by conventional techniques such as chemical, thermal, and mechanical treatments and ion implantation which yield surface compression stress. Ion implantation modifies only a thin top layer, but significant improvements are observed on the strength of ceramics.

Hioki et al. (1986, 1989) reported a marked improvement in flexural strength of single-crystal and polycrystalline Al_2O_3 at N^+ and Ar^+ ion implantations at room temperature. Figure 12.8 shows a comparison of effect of ion fluences on flexural strengths of single-crystal and polycrystalline α-Al_2O_3 implanted with 800 keV Ar^+ and 400 keV N^+ ions. We note that an increase up to about 60 percent can be obtained for single-crystal Al_2O_3 and up to about 15 percent for the polycrystalline Al_2O_3. Hioki et al. reported increases of about 10 percent at 10^{17} ions cm^{-2} for the 300 and 523 K Ni^+ implantation into Al_2O_3. The increase was about 30 percent in the substrate implanted at 100 K and was almost same for ion fluences of 10^{15} to 10^{17} ion cm^{-2}. The implantation at 100 K produced an amorphous surface.

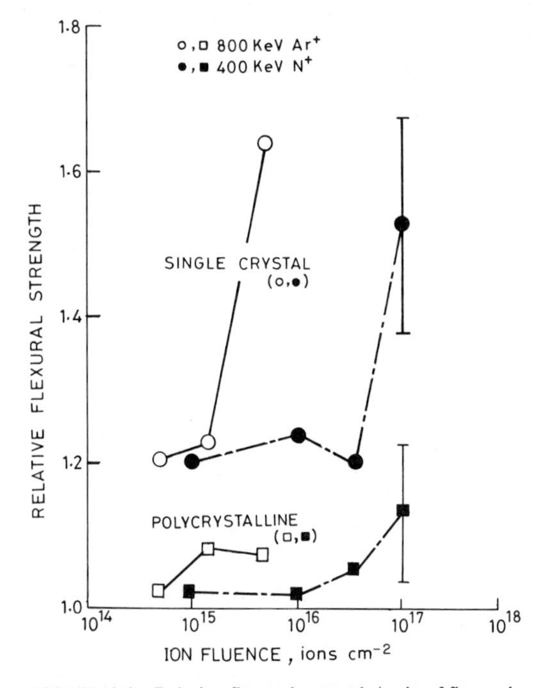

FIGURE 12.8 Relative flexural strength (ratio of flexural strength of implanted to unimplanted material) as a function of ion fluences for single-crystal and polycrystalline α-Al_2O_3 implanted with 800-keV Ar^+ and 400-keV N^+ ions. [*Adapted from Hioki et al. (1989).*]

The implantation of 400-keV nitrogen in the fluence range where the substrate remained crystalline increases the strength of single-crystal Y-ZrO_2 (by 57 percent at 1×10^{16} ion cm^{-2}). Similarly implantations that do not amorphize SiC and Si_3N_4 also increased the flexural strength (each by about 25 percent at 5×10^{15} Ar^+ cm^{-2}, 800 keV). However, in contrast to the behavior of Ni^+-implanted sapphire where amorphization produced an increase in strength, it caused decreases in the strength of SiC and Si_3N_4 (by about 20 percent at 5×10^{16} Ar^+ cm^{-2}, 800 keV).

12.1.5.4 Friction. Implantation of a variety of nonmetallic and metallic species into numerous metals, ceramics, and polymers has been observed to change the coefficient of friction in most cases. (Ion implantation, in general, is rarely used alone to reduce the friction.) The friction characteristics are dictated by the materials properties on the surface such as the resistance to the plastic flow of the weaker of the mating materials in shear and in compression (Chap. 2). The other important properties are microhardness, which limits the plowing by mating surface asperities, and surface energy, which is a material property guided by saturated bonds on the surface and which limits the adhesion of implanted surface to the mating surface. As discussed earlier, ion implantation results in increased hardness and formation of compounds of refractory nature which have low surface energy, and thus both factors favor the reduction in friction.

Hartley et al. (1973) first reported the macroscopic changes in friction of steel on implantation of various ions, i.e., Sn^+, In^+, Ag^+, Pb^+, Mo^+, and S^+ at fluences in excess of 10^{16} ion cm^{-2} and energies around 120 keV. The majority of ions reduce the coefficient of friction, which is probably due to dominant behavior of an ion-implanted species to weaken the shear strength of implanted layers by promoting the formation of oxide films. Implantation of molybdenum and sulphur (in the ratio of 1:2) simultaneously reduces the coefficient of friction to a greater extent than either species alone, which indicates the formation of MoS_2 on the surface as a result of ion implantation of Mo and S together. Iwaki (1983, 1987) has reported an increase in coefficient of friction by Ni, Cu ion implantations in low-carbon steels at 10^{16}, 10^{17} ions cm^{-2} fluences at 100 to 200 keV, and a decrease by ion implantation of Cr (Fig. 12.9). The reduction in coefficient of friction by Cr^+ implantation is attributed to the formation of Cr-based oxides which make the surface harder. Fischer et al. (1983) have reported a decrease in the friction by Ti^+ implantation at 4×10^{17} ion cm^{-2} fluences in ferritic AISI E52100 steel. They further reported that implantation of N^+ and Fe^+ increased the friction.

Dillich and Singer (1983) implanted 5×10^{17} Ti^+ ion cm^{-2} at 190 keV in a centrifugally cast cobalt-based alloy (Stoody 3). They conducted friction tests using a pin-on disk test with implanted stoody disk sliding against a pin of hard bearing steels such as 52100, 440C stainless steel and M50, as well as 302 stainless steel and low-carbon mild steel, and carbon-graphite. They found that titanium-implanted surfaces resulted in lower friction up to a factor of 2.

Ion implantation of single crystal Al_2O_3 (sapphire) to ion fluences less than those adequate for amorphization has resulted in an increase in the coefficient of friction for metal pins under lubricated sliding (Burnett and Page, 1987) and metal, sapphire, and diamond pins under dry sliding situations (Bull and Page, 1988a, 1988b). The onset of amorphization is accompanied by a decrease in coefficient of friction (e.g., from 0.24 to 0.04 for diamond pins at a normal force of 0.5 N (McHargue, 1987). This change in coefficient of friction is probably due to ploughing in the harder or softer implanted surfaces. Eskildsen et al. (1989) reported significant reductions in coefficient of friction from 0.45 to 0.26 to 0.38 (at

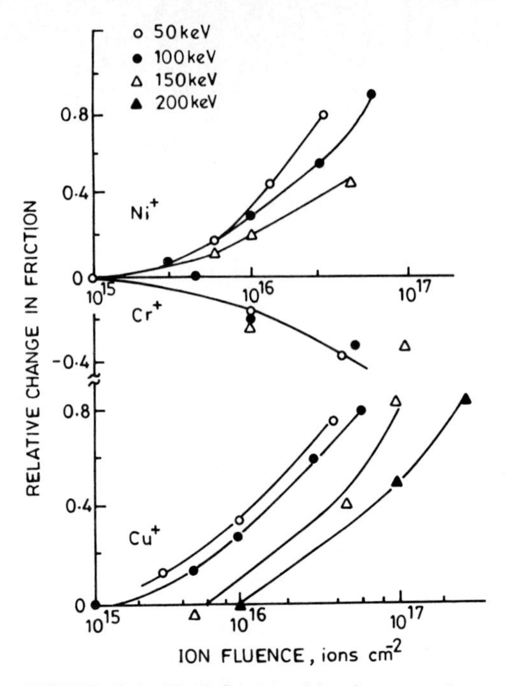

FIGURE 12.9 The influence of ion fluence and energy on the coefficient of friction of low-carbon steels implanted with copper, nickel, and chromium ions. Implanted low-carbon steel flat was slid against AISI 1025 pin in a Bowden Leben type of test apparatus. [*Adapted from Iwaki (1983, 1987).*]

a load of 10 N) between Al_2O_3 ball and polycrystalline sintered alumina disks implanted at 180 keV to 5×10^{16} Ti^+ and Ni^+ ion fluences.

Nastasi et al. (1988) also reported reduction in coefficient of friction between chromium steel balls and N^+ ion-implanted (at 195 keV and 77 K to 3×10^{17} ion cm^{-2} ion fluences) hot-pressed polycrystalline Al_2O_3, SiC, TiB_2, and B_4C. An initial low friction with similar values from both implanted and unimplanted substrates suggests the formation of a postimplantation surface layer that is chemically independent of nitrogen content. Similar results, i.e., reduction in the coefficient of friction have been found for Al_2O_3, CrC, TiC, WC (Shimura and Tsuya, 1977) and Si_3N_4 (Fisher and Tomizawa, 1987). Nitrogen-implanted-induced reductions in friction seem to be correlated with the thermodynamic tendency of the substrates to form nitridelike bonds (Nastasi et al., 1988).

Hioki et al. (1989) reported a reduction in the coefficient of friction from 0.6 to 0.15 between steel and polycrystalline silicon carbide implanted at 800 keV and room temperature to 1×10^{16} ion cm^{-2} fluences of argon ions. Figure 12.10 shows the coefficient of friction as a function of sliding cycle for SiC unimplanted and implanted with Ar ions under dry conditions at a normal load of 9 N and an ambient atmosphere of relative humidity of 50 to 70 percent. We note that the coefficient of friction for an unimplanted disk is almost constant (~0.6) with sliding cycle, while for an implanted disk, the coefficient of friction is high at the

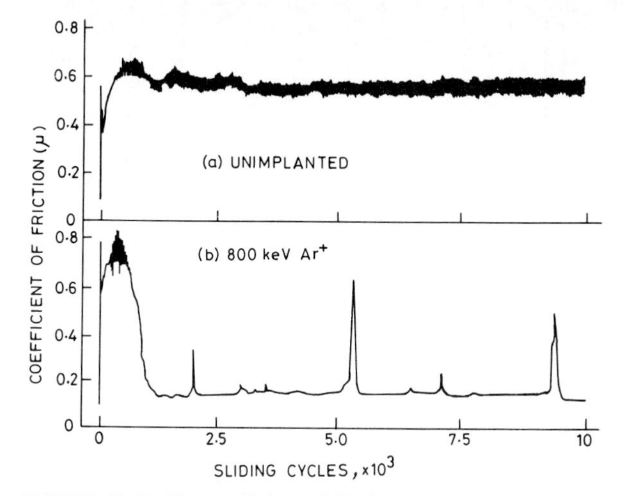

FIGURE 12.10 The coefficient of friction as a function of sliding cycle for SiC disk in sliding contact with steel pin under a normal load of 9 N dry condition at a speed of 0.05 m s^{-1} and ambient atmosphere of relative humidity of 50 to 70 percent: (*a*) Unimplanted SiC disk. (*b*) SiC disk implanted at 800 keV Ar^+ ions to 1×10^{16} ion cm^{-2} at room temperature. [*Adapted from Hioki et al. (1989).*]

initial stage of the test run, but it decreases gradually with cycling and approach to a value as low as 0.15. This remarkable reduction in coefficient of friction persists up to more than 10^5 cycles. The compositions analysis by Raman spectroscopy of implanted surface indicates amorphous phases of carbon and SiC, while of unimplanted surface indicates crystalline phase of SiC (Hioki et al., 1989).

The nitrogen ion implantations in Co-cemented tungsten carbides (with 6% Co) at 4×10^{17} ions cm^{-2} fluences reduce the coefficient of friction from 0.22 to 0.10 sliding against steel in the presence of white spirit (a paint thinner) (Dearnaley, 1983a).

Watanabe et al. (1989) reported a change in the friction behavior of polymers and polymer composites on ion implantation. The coefficient of friction of polyimide versus steel (SUJ2) was reduced from 0.5 to 0.35 with N^+ implantations at 92 keV and 10^{17} ion cm^{-2}. However, the coefficient of friction increased in PTFE, and graphite (15 wt %) filled polyimide on N^+ implantations. The change in friction characteristics of PTFE perhaps results due to depletion of fluorine which results in the disruption of molecular structure and carbon-rich or partially carbonized surface layer.

The friction characteristics of numerous combinations of ionic species and host materials at different fluences and energies are summarized in Table 12.6.

12.1.5.5 Wear. Various investigators have shown that ion implantation has a significant effect on the wear behavior of metal and ceramics (Table 12.7). Hartley (1975b) and Lo Russo et al. (1980) reported that the ion implantation provides a long-lasting wear resistance despite the very shallow initial penetration. Wear rates in which the track formed in the disk (in a pin-on-disk test) reached a depth of 12 μm showed no sign of a breakdown in wear resistance, and nitrogen was detectable in the base of the track at a level of 30 to 40 percent of the initial

TABLE 12.6 Coefficient of Friction of Ion Implanted Materials Reported by Various Researchers

Materials	Implanted ion	Acceleration voltage, keV	Ion fluence, ion cm^{-2}	Test conditions: Load, N	Test conditions: Environment	Test conditions: Lubrication	Mating surface material	Testing machine	Coefficient of friction of unimplanted material	Coefficient of friction of implanted material	References
Ferrous metals											
Pure iron (ferritic)	N^+	90	3.5×10^{21}	1–2	Air	Dry	Ruby ball	Ball on disk (sliding)	0.20	0.26	Oliver et al. (1984)
Low-carbon steel	Cr^+	150	10^{17}	0.5–2.0	Air	Dry	AISI 1025 pin	Bowden Leben type	0.5–0.6	0.42	Iwaki (1987)
	Cu^+	150	10^{17}	0.5–2.0	Air	Dry	AISI 1025 Pin	Bowden Leben type	0.5–0.6	1.1	Iwaki (1987)
En 352 (0.2% C, 0.35% Si, 0.5% Mn, 0.85–1.25% Ni, 0.6–1.0% Cr, 0.1% Mo)	Sn^+	380	3×10^{16}	20	Air	Dry	WC ball	Bowden Leben type	0.25	0.10	Hartley (1975a)
	Mo^+	400	3×10^{16}	20	Air	Dry	WC ball	Bowden Leben type	0.25	0.24	Hartley et al. (1973)
	S^+	400	6×10^{16}	20	Air	Dry	WC ball	Bowden Leben type	0.25	0.20	Hartley et al. (1973)
	$Mo^+ + S^+$	400	6×10^{16}	20	Air	Dry	WC ball	Bowden Leben type	0.25	0.19	Hartley et al. (1973)
304 stainless steel (austenitic)	N^+	50	10^{17}	4	Air	Dry	440 C steel ball	Pin on disk	0.42	0.17	Suri et al. (1979)
	B^+	40	10^{17}	4	Air	Dry	440 C steel ball	Pin on disk	0.42	0.39	Suri et al. (1979)
	$Ti^+ + C^+$	90–180	2×10^{17}	0.5	Air	Dry	440 C steel ball	Pin on disk	0.85	0.55	Yust et al. (1982, 1983)
52100 bearing steel (martensitic) (Fe-1.5% Cr, 1% C)	N^+	90	3.5×10^{21}	5	Air	Dry	Ruby ball	Pin on disk	0.19	0.19	Oliver et al. (1984)
	Ti^+	190	5×10^{17}	10	Air	Dry	AISI 52100 ball	Stick slip machine (ball on disk)	0.62	0.38	Singer et al. (1981)
18W4Cr1V bearing steel	N^+	400	10^{17}	10	Air	White spirit	--	Pin on disk (sliding)	--	0.8*	Madakson (1987)
440 C steel	$Ti^+ + C^+$	90–180	2×10^{17}	0.5	Air	Dry	440 C steel pin	Pin on disk	0.85	0.30	Follsaedt et al. (1983)
H 13 tool steel	N^+	75	5×10^{17}	0.5–2	Air	Dry	AISI 1025 pin	Bowden Leben type	0.55	0.5	Iwaki et al. (1987)
	B^+	75	5×10^{17}	0.5–2	Air	Dry	AISI 1025 pin	Bowden Leben type	0.55	0.17	Iwaki et al. (1987)

Nonferrous metals											
Aluminum	N^+	50	10^{17}	4	Air	Dry	440 C steel pin	Pin on disk.	1.0	1.17	Suri et al. (1979)
	B^+	40	10^{17}	4	Air	Dry	440 C steel pin	Pin on disk.	1.0	1.20	Suri et al. (1979)
Titanium	N^+	50	10^{17}	4	Air	Dry	440 Csteel pin	Pin on disk	0.70	0.87	Suri et al. (1979)
	B^+	40	10^{17}	4	Air	Kero-sene	440 C steel pin	Pin on disk	0.70	0.85	Suri et al. (1979)
Ti6Al4V alloy	N^+	90	3.5×10^{21}	2–3	Air	Dry	Ruby ball	Pin on disk	0.48	0.15	Oliver et al. (1984); Hutchings and Oliver (1983)
	N^+	40	2×10^{17}	—	Air	Dry	WC ball	Pin on disk	0.1	0.15	Saritas et al. (1987)
	B^+	40	2×10^{17}	—	Air	Dry	WC ball	Pin on disk	0.1	0.08	Saritas et al. (1987)
	C^+	40	2×10^{17}	—	Air	Dry	WC ball	Pin on disk	0.1	0.08	Saritas et al. (1987)
	O^+	40	2×10^{17}	—	Air	Dry	WC ball	Pin on disk	0.1	0.30	Saritas et al. (1987)
Co-based alloy Stoody 3 (31% Cr, 12.5% W, 2.2% C, bal. Co)	Ti^+	190	5×10^{17}	10	Air	Dry	52100 steel ball	Ball on disk	0.56	0.19	Dillich and Singer (1983)
	Ti^+	190	5×10^{17}	10	Air	Dry	Stellite 2 ball	Ball on disk	0.57	0.16	Dillich and Singer (1983)
	Ti^+	190	5×10^{17}	10	Air	Dry	WC ball	Ball on disk	0.62	0.27	Dillich and Singer (1983)
Ceramics											
Alumina (Al_2O_3) single crystal	Ti^+	300	7.7×10^{16}	5–25	Air	Alkanes	Pure iron pin	Pin on disk.	0.04–0.05	0.10	Burnett and Page (1987)
	Zr^+	300	1.7×10^{16}	5–25	Air	Alkanes	Pure iron pin	Pin on disk	0.04–0.05	0.09	Burnett and Page (1987)
Alumina (Al_2O_3) polycrystalline	Ti^+	180	5×10^{16}	1.0	Air	Dry	Alumina ball	Ball on disk	0.45	0.26–0.38	Eskildsen et al (1989)
	N^+	190	3×10^{17} at 77 K	0.17	Air	Dry	Cr-steel ball	Ball on disk (recip-rocating)	0.4–0.6	0.2	Nastasi et al. (1988)

TABLE 12.6 Coefficient of Friction of Ion Implanted Materials Reported by Various Researchers (*Continued*)

Materials	Implanted ion	Acceleration voltage, keV	Ion fluence, ion cm^{-2}	Test conditions: Load, N	Test conditions: Environment	Test conditions: Lubrication	Mating surface material	Testing machine	Coefficient of friction of unimplanted material	Coefficient of friction of implanted material	References
Ceramics (*cont.*)											
Silicon carbide (SiC)	Ar^+	800	1×10^{16}	9	Air	Dry	Steel (1% C) pin	Pin on disk	0.6	0.2	Hioki et al. (1989)
	N^+	190	3×10^{17} at 77 K	0.17	Air	Dry	Cr-steel ball	Ball on disk (reciprocating)	0.4	0.2	Nastasi et al. (1988)
Boron carbide (B_4C)	N^+	190	3×10^{17} at 77 K	0.17	Air	Dry	Cr-steel ball	Ball on disk (reciprocating)	0.75	0.3	Nastasi et al. (1988)
WC-Co	N^+	100	4×10^{17}	75	Air	White spirit	Steel pin	Pin on disk	0.22	0.10	Dearnaley (1983a)
Titanium diboride ($Ti B_2$)	N^+	190	3×10^{17} at 77 K	0.17	Air	Dry	Cr-steel ball	Ball on disk (reciprocating)	0.5	0.15	Nastasi et al. (1988)
Polymers											
PTFE	N^+	92	10^{17}	—	Air	Dry	SUJ2 steel	Ball on disk (reciprocating)	0.1	0.15	Watanabe et al. (1989)
Polyimide	N^+	92	10^{17}	—	Air	Dry	SUJ2 steel	Ball on disk (reciprocating)	0.5	0.35	Watanabe et al. (1989)

*Relative change in the coefficient of friction, i.e., ratio of coefficient of friction of implanted to that of unimplanted materials.

TABLE 12.7 Wear Characteristics of Ion Implanted Materials Reported by Various Researchers

Materials	Implanted ion	Acceleration voltage, keV	Ion fluence, ion cm^{-2}	Test conditions: Load, N	Test conditions: Environment	Test conditions: Lubrication	Mating surface material	Testing machine	Wear coefficient of unimplanted material	Wear coefficient of implanted material	References
Ferrous metals											
Pure iron (ferritic)	N^+	90	3.5×10^{21}	1–2	Air	Dry	Ruby ball	Ball on disk	—	1–2*	Oliver et al. (1984)
	N^+	1500	5×10^{17}	4.9	Air	n-hexadecane	M-50 steel disk	Pin on disk	—	1.4*	Jones and Ferrante (1983)
Mild steel	N^+	35	10^{18}	2.2-40	Air	White spirit	440 C steel pin	Pin on disk	1×10^{-5}	8×10^{-7}	Hartley (1975a)
	Mo^+	400	3×10^{16}	10–20	Air	White spirit	440 C steel pin	Pin on disk	7×10^{-5}	1×10^{-5} 5×10^{-6}	Hartley et al. (1974)
Carbon steel (1.2% C) (Ferritic)	N^+	30	3.2×10^{17}	15	Air	White spirit	—	Pin on disk	—	1–10*	Hartley et al. (1974)
Nitriding steel (En 40B)	N^+	30	2×10^{17}	15	Air	White Spirit	440 C steel pin	Pin on disk.	1×10^{-6}	2×10^{-7}	Hartley (1975)
	N^+	30	2.8×10^{17}	10	Air	White spirit	Stainless-steel pin	Pin on disk	3×10^{-7}	1×10^{-8}	Harding (1977)
	C^+	30	2×10^{17}	10	Air	White spirit	Stainless-steel pin	Pin on disk	3×10^{-7}	1×10^{-8}	Harding (1977)
	B^+	30	2×10^{17}	10	Air	White spirit	Stainless-steel pin	Pin on disk	3×10^{-7}	10^{-8}–10^{-9}	Harding (1977)
	Ne^+	30	2×10^{17}	10	Air	White spirit	Stainless-steel pin	Pin on disk	3×10^{-7}	3×10^{-8}	Harding (1977)
304 stainless steel (austenitic)	N^+	50	10^{17}	4	Air	Dry	440 C steel pin	Pin on disk	1.4×10^{-3}	1.3×10^{-3}	Suri et al (1979)
316 stainless-steel (austenitic) (17% Cr, 14% Ni, 2% Mo)	N^+	100	5×10^{17}	5–10	Air	Kerosene	GCr6 Steel pin	Falex and Timkin pin on disk	—	30–50*	Fu-Zhai et al. (1983)
416 stainless steel	N^+	40	10^{17}	20	Air	Boundary lubricant polyester	Unimplanted 416	Crossed cylinder	—	100–200*	Hirvonen (1978); Hirvonen et al. (1979)
	B^+, Ti^+, $Ti^+ + B^+$, $Ti^+ + C^+$	75	2×10^{17}	20	Air	Boundary lubricant polyester	Implanted 304	Crossed cylinder	—	100–300*	Hirvonen (1978); Hirvonen et al. (1979)
	N^+	100	3.5×10^{17}	2.2	Air	Dry	100Cr6 steel ball	Ball on disk	—	200*	Dimigen et al. (1985)

TABLE 12.7 Wear Characteristics of Ion Implanted Materials Reported by Various Researchers (*Continued*)

Materials	Implanted ion	Acceleration voltage, keV	Ion fluence, ion cm^{-2}	Test conditions: Load, N	Environment	Lubrication	Mating surface material	Testing machine	Wear coefficient of unimplanted material	Wear coefficient of implanted material	References
				Ferrous metals (*cont.*)							
4140 steel (42Cr, 4Mo)	N^+	100	3.5×10^{17}	2.2	Air	Dry	100Cr6 steel ball	Ball on disk	—	3*	Dimigen et al. (1985)
440 B steel (90Cr MoV18)	N^+	100	3.5×10^{17}	2.2	Air	Dry	100Cr6 steel ball	Ball on disk	—	1500*	Dimigen et al. (1985)
	B^+	100	3.5×10^{17}	0.45	Nitrogen	Dry	100Cr6 steel ball	Ball on disk	—	60*	Kluge et al. (1989)
Bearing 52100 steel (martensitic)	N^+	90	3.5×10^{21}	5	Air	Dry	Ruby ball	Ball on disk	—	<1*	Oliver et al. (1984)
	Ti^+	190	5×10^{17}	—	—	1–5 μm Diamond powder	—	Vibratory polisher for abrasion	—	5† at 100 nm depth 2† at 300 nm depth	Singer et al. (1980)
18W4Cr1V bearing steel	N^+	400	10^{17}	10	Air	White Spirit	—	Pin on disk	—	2.4*	Madakson (1987)
M 50 tool steel	N^+	1500	5×10^{17}	5	Air	Dry	Iron pin	Pin on disk	—	<1*	Jones and Ferrante (1983)
H 13 tool steel	N^+	75	5×10^{17}	21	Air	Dry	AISI 1025 pin	Pin on disk	—	60–70*	Iwaki (1987)
	B^+	75	5×10^{17}	21	Air	Dry	AISI 1025 pin	Pin on disk	—	~600*	Iwaki (1987)
M2 tool steel (4% Cr, 5% Mo, 6% W, 2% V, 0.85% C)	N^+	680	2×10^{17}	44	Air	Gear oil	M2 steel (implanted)	Ring and block		1.5*	Daniels and Wilbur (1987)
				Nonferrous metals							
Aluminium	N^+	50	10^{17}	4	Air	Dry	440 C steel pin	Pin on disk	2.5×10^{-5}	2×10^{-5}	Suri et al. (1979)
	B^+	40	10^{17}	4	Air	Dry	440 C Steel pin	Pin on disk	2.5×10^{-5}	7.0×10^{-6}	Suri et al. (1979)
Copper	B^+	40	10^{17}	2.2–15	Air	Dry	Graphite	Pin on disk	7×10^{-6}	$5–9 \times 10^{-7}$	Hartley (1979)
Titanium	N^+	50	10^{17}	4	Air	Dry	440 C steel pin	Pin on disk	3.5×10^{-4}	4.5×10^{-4}	Suri et al. (1979)
	B^+	40	10^{17}	4	Air	Dry	440 C steel pin	Pin on disk	3.5×10^{-4}	4×10^{-4}	Suri et al. (1979)

Phospher bronze	N^+	40	5×10^{17}–10^{17}	5–10	Air	Dry	—	Pin on disk	—	1–1.2*	Saritas et al. (1982)
	B^+	40	5×10^{17}	5–10	Air	Dry	—	Pin on disk	—	1.5*	Saritas et al. (1982)
	C^+	20	10^{17}	5–10	Air	Dry	—	Pin on disk	—	No change	Saritas et al. (1982)
	P^+	40	5×10^{17}	5–10	Air	Dry	—	Pin on disk	—	1.3*	Saritas et al. (1982)
Ti6A14V alloy	N^+	90	3.5×10^{21}	2.6	Air	Dry	Ruby ball	Pin on disk	—	300–500*	Hutchings and Oliver (1983)
	N^+	40	2×10^{17}	—	Air	Dry	WC ball	Pin on disk	—	1.2*	Saritas et al. (1987)
	B^+	40	2×10^{17}	—	Air	Dry	WC ball	Pin on disk	—	No change	Saritas et al. (1987)
	C^+	40	2×10^{17}	—	Air	Dry	WC ball	Pin on disk	—	1.2*	Saritas et al. (1987)
	O^+	40	2×10^{17}	—	Air	Dry	WC ball	Pin on disk	—	1.3*	Saritas et al. (1987)
Ceramics											
Alumina (Al_2O_3) polycrystalline	N^+	70	5×10^{16}	10	Air	Dry	52100 steel ball	Ball on disk	—	~5*	Eskildsen et al. (1989)
	N^+	70	5×10^{16}	10	Air	Dry	Alumina ball	Ball on disk	—	~10*	Eskildsen et al. (1989)
	Ti^+	180	5×10^{16}	10	Air	Dry	52100 steel ball	Ball on disk	—	~5*	Eskildsen et al. (1989)
	Ti^+	180	5×10^{16}	10	Air	Dry	Alumina ball	Ball on disk	—	~70*	Eskildsen et al. (1989)
Silicon carbide (SiC)	Ar^+	800	1×10^{16}	9.2	Air	Dry	Carbon steel (1% C) pin	Pin on disk	—	5*	Itoh et al. (1989)
WC-Co	N^+	100	4×10^{17}	100	Air	White Spirit	Steel pin	Pin on disk	—	~8–9*	Dearnaley (1983a)
WC-Co	N^+	40	10^{17}	200	Air	Water	Steel pin	Flat pin on cylinder	—	~1–5*	Fayeulle et al. (1986)
Polymers											
Polyimide	N^+	92	10^{17}	—	Air	Dry	SUJ2 steel disk	Disk on flat	—	4*	Watanabe et al. (1989)

*Reduction factor of wear volume, i.e., ratio of wear volume in unimplanted materials to that of implanted materials.

†Relative wear resistance, i.e., inverse of wear depth of implanted disks, normalized to that of carbon steel (400 HK hardness) unimplanted disk.

concentration. This effect has been attributed to the inward diffusion of the implanted species due to an increase in surface temperature during sliding. The mobile implanted species such as nitrogen, carbon, or boron atoms migrate interstitially to decorate subsurface dislocations generated as a consequence of local hertzian and shear stresses. The movement of dislocations is impeded by the impurities, and the metal is thereby hardened. As wear progresses, dislocations are drawn deeper and deeper, and this serves to carry the implanted atoms forward so as to re-create a hard surface of just the thickness which can be most effective.

Dearnaley and Hartley (1978) reported that ion-implanted metals result in small size of wear debris rather than platelike debris resulting from normal delamination and to the burnishing action which often occurs on ion-implanted metal. Such a hard surface would normally be far too brittle for practical use and would exhibit an ''ice-on-mud'' fracture when loaded, but the very thin film hardened by ion implantation retains the toughness of its substrate and shows no signs of brittle fracture.

Dimigen et al. (1985), Hartley (1975a, 1975b, 1976), Hirvonen (1978), and Hubler and Smidt (1985) have reported considerable wear reduction of steels implanted with N^+ at high fluences, above 10^{17} ion cm^{-2} which approximately corresponds to the introduction of about 20 at % of nitrogen. Jones and Ferrante (1983) reported a reduction in initial (~40 percent) and steady-state (~20 percent) wear rate of pure iron implanted with 1.5 MeV N^+ at high dose of 5×10^{17} ion cm^{-2}. Wear tests were conducted using the implanted iron pin sliding against an unimplanted M50 disk at a normal load of 4.9 N and sliding velocity of 0.07 m s^{-1} in the presence of *n*-hexadecane lubricant. Figure 12.11(*a*) shows dependence of relative decrease in total wear rate on N^+ fluences for a stainless steel pin sliding against an N^+ implanted nitriding alloy steel (En40B) disk. The reduction in wear rate is found due to formation of nitrides and carbonitrides (up to a depth of 75 nm) of iron which is confirmed by electron Mössbauer spectroscopy (Longworth and Hartley, 1977). The insignificant decrease in wear rate above 5×10^{17} ion cm^{-2} fluences may be due to complete saturation of surface microstructure with precipitates of nitrides and carbonitrides.

Harding (1977) compared the wear behavior of the three light interstitial species-C^+, N^+, and B^+-in nitriding alloy steel (En40B) in a pin-on-disk test, as shown in Fig. 12.11(*b*). It can be seen that they are all beneficial. He also showed that N^+ implantation reduced rates in mild steel (En8), tool steel (NSOH), nitriding alloy steel (En40B), and stainless steel (En58B), with the greatest improvement occurring when stainless steel is worn against itself. Ecer et al. (1983), Dearnaley and Hartley (1978), and Vardiman and Kant (1982) have reported that N^+ implantation into a previously nitrided steel can produce considerable increase in wear resistance. This is possible due to the fact that injection of many vacancies during ion bombardment together with a refinement of nitride precipitate can influence the mechanical properties.

In addition to N^+, C^+, and B^+, the other species such as P^+, Y^+, Ce^+, Ti^+, Ta^+, Cr^+, Mo^+, Ni^+, Fe^+, Sn^+, and Ar^+ have also found to modify the wear characteristics of steels (Baumvol, 1981b, 1984; Fischer et al., 1983; Hartley and Hirvonen, 1983; Shrivastava et al., 1987; Tosto, 1983). Dearnaley (1982) reported that rare earth species such as Y^+ and Ce^+ modify the growth mechanism of protective oxide formation of steel at the sliding interface and increase the adherence of the oxide to the metal resulting in improvement of wear resistance. Since rare earth ions and nitrogen act in quite different ways to increase wear resistance of steel surface, it is not surprising that a dual implantation of both species can provide a remarkably high degree of protection. Dual implantations of 5×10^{15} Y^+

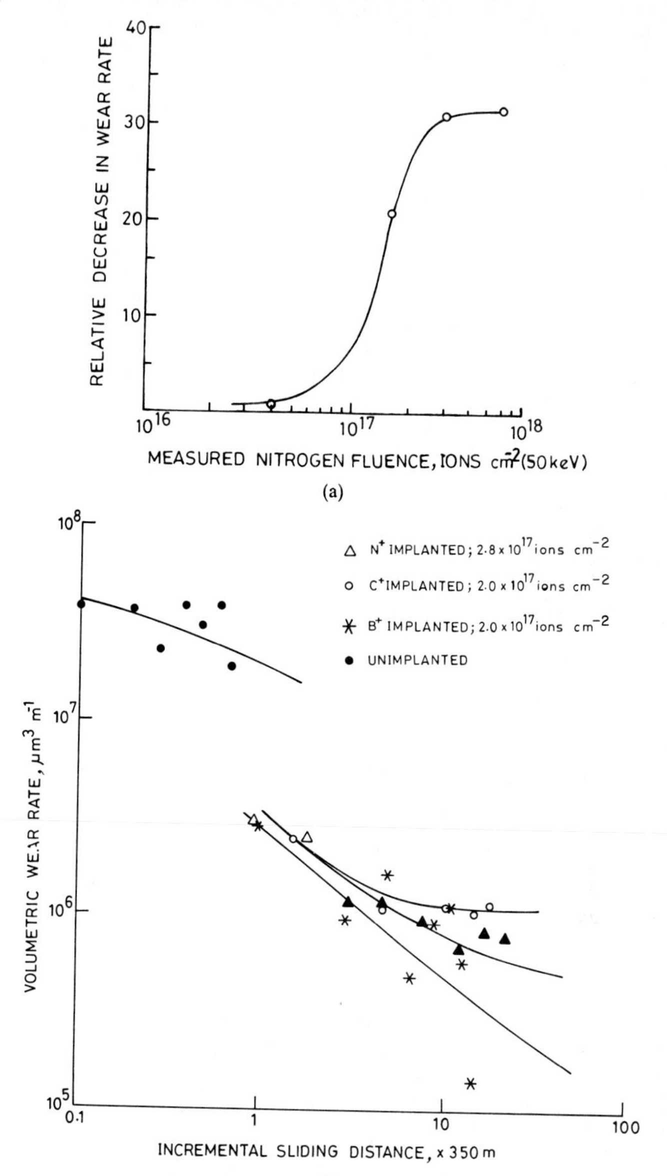

FIGURE 12.11 (*a*) Relative decrease in the wear rate as a function of ion fluence used in N^+ implantation in En4OB steel disk. A stainless-steel pin was slid against the implanted disk in a pin-on-disk test. [*Adapted from Dearnaley and Hartley (1978).*] (*b*) Volumetric wear rate as a function of sliding distance for nitriding alloy steel (En4OB) disk implanted with three ion species, C^+, N^+, and B^+, slid against a stainless-steel (En58B) pin in a pin-on-disk test (load = 10 N). [*Adapted from Harding (1977).*]

ion cm^{-2} and 2×10^{17} N^+ ion cm^{-2} in En58B stainless steel have been found to increase the wear resistance by a factor of 30 to 900 in a pin-on-disk test (Dearnaley, 1982). Ta^+ and Ti^+ are found to be very effective in increasing wear resistance, due to the possibility of formation of precipitates of TaC and TiC, respectively.

Improved wear characteristics of implanted single crystalline and polycrystalline ceramics such as Al_2O_3, ZrO_2 (fully stabilized), SiC, B_4C, Co-WC, and TiB_2 by nonmetallic (e.g., N^+, Ar^+) and metallic (e.g., Ti^+, Cr^+) ions have been reported by several investigators (Dearnaley, 1983a; Eskildsen et al., 1989; Hioki et al., 1989; McHargue, 1988; Nastasi et al., 1988). Generally, when a brittle ceramic material is scratched with a loaded diamond stylus, a number of lateral crackings and profuse radial crackings occur at and around the groove formed by scratching which lead to removal of material. Diamond scratch tests made on the implanted amorphous SiC (at 300 keV, 250°C, 4×10^{17} N^+ ion cm^{-2}) showed a reduction of wear volumes and a decrease in the level of lateral fracture as compared to unimplanted SiC (Nastasi et al., 1988). Similar results on Al_2O_3 were reported by Hioki et al. (1989), who observed a surface softening and formation of compressive surface stress following a 300-keV implantation of 2×10^{15} Ni^+ cm^{-2} at 100 K. Eskildsen et al. (1989) reported an improvement of an order of magnitude in wear volume of alumina balls sliding against Ti^+ ions implanted polycrystalline alumina disks (at 180 keV, up to 5×10^{17} ion cm^{-2}) under dry conditions at 10 N normal load and 0.1 m s^{-1} speed.

The improvement in wear resistance of N^+-implanted Co-cemented tungsten carbide (Co-WC) has been found by several investigators (Dearnaley, 1983a; Fayeulle et al., 1986; Greggi and Cossowsky, 1983; Oblas, 1984). Figure 12.12 illustrates the influence of N^+ ion fluences on wear rate of WC-Co composite slid against a high-speed steel cylinder in cylinder-flat pin geometry in water with normal loads of 200 and 500 N at a sliding speed of 0.36 m s^{-1}. No difference appeared between implanted and nonimplanted samples when the normal load was 500 N, probably because superficial layers are removed too quickly and no effects are seen. With a normal load of 200 N, wear was a function of the fluence. At the beginning of the test (under 4 h), all the fluences appeared to reduce the amount of wear. After 4 h, only the implantation of 10^{17} ion cm^{-2} improved the

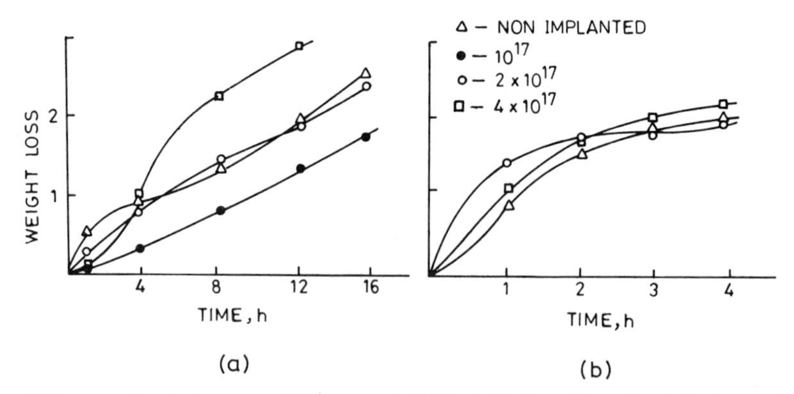

FIGURE 12.12 Weight loss of N^+ implanted WC-Co flat pin slid against a high-speed steel cylinder in water as a function of sliding time at normal loads of: (*a*) 200 N, (*b*) 500 N, and various ion fluences. [*Adapted from Fayeulle et al. (1986)*.]

amount of wear of carbide. Finally, after 8 h, the curves became parallel: implantation had no more effect. This appeared after a removal of 1 mg which corresponded for uniform wear to about 0.5 μm depth. Fayeulle et al. (1986) also reported that nitrogen remains in the samples beyond the initial implanted layer. Dearnaley (1983a) reported an improvement of disk wear by a factor of 8 to 9 on the WC-Co pin in a pin-on-disk test in the presence of white spirit.

The possible reasons for the wear resistance improvement may be as follows: martensitic transformation within the Co binder phase in N^+-implanted specimens and interstitial N^+ may distort the metal lattice and produce a hardening effect. We believe that the main effects of ion implantation occur in the cobalt binder phase and that reduced wear of this material lessens the undercutting and loss of carbide grains. The presence of dissolved nitrogen near the surface may also serve to lessen out-diffusion of cobalt into the worked metal.

A wide variety of tests have now been carried out on engineering components and tools under practical working conditions (Dearnaley, 1982). In one category of tools for cutting, punching, or shearing operations, good results were achieved by nitrogen implantation into chrome-carbon steels, tool steels, and chromium-plated coatings, the typical increase in life being by a factor of 4. In cobalt-cemented tungsten carbide knives used in an abrasive wear situation, a twelvefold increase in life resulted. Steel forming tools of 12 percent Cr–2 percent C steel for a heavy-duty application achieved a life which was 2.5 times greater than that with a chromium coating. Bearings of various kinds have also been shown to have an increased life. Dies of cemented carbide used for the drawing of copper or steel wire for extrusion, swaging, and metal-forming operations have been shown to last from two to six times longer as a result of ion implantation with carbon or nitrogen. In a particular application for drawing copper wire, dies implanted with C^+ achieved five times the normal throughput of wire and showed no abrupt failure despite the fact that a radial wear to the extent of 35 μm took place. This is about 10^3 times the depth of implantation (Dearnaley, 1982; Hartley, 1975b; Hirvonen, 1978).

Improved wear behavior of ion-implanted polymers and composites is reported by Lee et al. (1990) and Watanabe et al. (1989). The wear resistance of N^+ implanted polyimide, at 92 keV and 10^{17} ion cm^{-2}, with steel (SUJ2) is found two to three times more as compared to virgin polyimide. No wear track was observed by Lee et al. (1990) in N^+, C^+, and B^+ implanted polyimide Kapton films even after 10,000 reciprocating cycles of a Nylon ball with 1 N normal force. This noticable improvement in wear characteristics of implanted polymers could be due to formation of amorphous carbon layer and other hard compounds. For instance, very hard films have been obtained on implanting a silicon-containing polymer (dimethylsilene-co-methylphenylsilene) because of formation of silicon-carbide (Venkatesan, 1985). Thus the implantation of polymers and polymer composites with nonmetal and metal ion can produce novel materials for improved wear performance.

12.1.5.6 Fatigue. An increase in fatigue life by ion implantation has been experienced in a number of materials including carbon steel (Herman, 1981; Hu et al., 1978, 1980; Jata and Strake, 1983; Kao and Byrne, 1981), copper (Kujore et al., 1981), titanium alloys (Jata et al., 1983; Vardiman and Kant, 1982), and Ni-Cr alloy (Vardiman and Cox, 1985) (Table 12.8).

Generally fatigue failures are characterized by two stages, i.e., a crack nucleation stage which is often surface sensitive and a propagation-to-failure stage which is mainly dictated by properties of the bulk material and thus ion implan-

TABLE 12.8 High Cycle Fatigue Life of Ion Implanted Metals Reported by Various Researchers

Materials	Implanted ion	Acceleration voltage, keV	Ion fluence, ion cm^{-2}	Fatigue life measuring technique	Stress amplitude, MPa	Fatigue cycles of unimplanted materials	Fatigue cycles of implanted materials	References
				Ferrous metals				
AISI 1018 steel (0.18% C)	N^+	150	2×10^{17}	Rotating beam	345	10^6	2.5×10^6	Hu et al. (1978)
							1×10^8*	Herman (1981)
4140 steel (0.38% C)	N^+	100	2×10^{21}		400	8×10^3	8×10^4	Jata and Starke (1983)
304 stainless steel	N^+	3000	3×10^{17}	Rotating beam	—	—	No change	Bakhru et al. (1981).
(En 58A) stainless steel	Ti^+	200	2×10^{17}	Rotating beam	—	—	8–10‡	Dearnaley and Hartley (1978)
				Nonferrous metals				
Pure copper	Ne+	3000	5×10^{17}	—	—	6.1×10^6	9.6×10^6	Bakhru et al. (1981)
Copper (polycrystalline)	O^+	120	5×10^{16}	Servohydraulic push-pull type†	120	1.7–3.6×10^6	4.3×10^6	Mendez et al. (1982)
	Al^+	100	5×10^{19}	Servohydraulic close loop	100	4×10^6	1×10^7	Kujore et al. (1981)
	Cr^+	100	5×10^{19}	Servohydraulic close loop	100	4×10^6	8×10^6	Kujore et al. (1981)
Ti 6Al 4V alloy	N^+	75	2×10^{17}	Rotating beam	620	8×10^5	1.2×10^6	Vardiman and Kant (1982); Hirvonen et al. (1979)
	C^+	75	2×10^{17}	Rotating beam	620	8×10^5	5×10^6	Vardiman and Kant (1982); Hirvonen et al. (1979)
Ni-20Cr alloy	C^+	125	2×10^{17}	Servohydraulic close loop	700	1.5×10^5	2×10^6	Vardiman and Cox (1985)

*Fatigue cycles of implanted material when aged at room temperature for 4 months and at 100°C for 6 h.
†The measurements were carried out in a vacuum (10^{-4} Pa or 10^{-6} torr).
‡Relative fatigue life, i.e., life of implanted specimen, normalized to that of unimplanted specimen.

tation influences the first stage. The implanted layer restricts the dislocation movement near the surface of the alloy, which prevent persistent slip bands from forming a surface topography favorable to crack nucleation. Thus various mechanisms such as development of compressive stresses, solid solution hardening, and second-phase precipitation as a result of ion implantation, responsible for impeding dislocation movement and in turn strengthening the materials should also contribute to the improvement of fatigue life in rolling contact.

Herman (1981) and Hu et al. (1978) have reported significant improvements in high-cycle stress-controlled fatigue life (from 2.5×10^6 to 10^8 cycles at 345 MPa stress level) in aged (few months at room temperatures) nitrogen ion implanted carbon steel AISI 1018 (0.18 percent C) (at 150 keV to 2×10^{17} ion cm^{-2} ion fluences). Freshly implanted substrates showed no consistent improvements in fatigue life. The as-implanted substrates appeared to contain a nitrogen martensite. A finely dispersed metastable nitride of $F_{16}N_2$ type was also seen in the as-implanted substrate. Aging gives rise to both the relaxation of high residual stresses and to nitrogen diffusion to dislocation and metastable nitrides. The as-implanted nitrogen is likely to be limited up to 0.1-μm depth while aging permitted diffusion of N^+ into deeper regions of the substrate. Jata and Starke (1983) reported considerable increases in fatigue life of 4140 steels (0.38 percent C) on N^+ implantations (at 100 keV to 2×10^{21} ion cm^{-2}) but with no significant additional increase on annealing. Dearnaley and Hartley (1978) conducted rotating bend fatigue tests with stainless steel (En58A), and the titanium implanted with 2×10^{17} N^+ ion cm^{-2} at 200-keV energy increased the cycles to failure by factors between 8 and 10.

Vardiman and Kant (1982) reported improvements in fatigue life of Ti6Al4V alloy on N^+ and C^+ implantations at 75 keV to 2×10^{17} ion cm^{-2} ion fluences, Fig. 12.13. The unimplanted endurance limit of 500 MPa was increased approximately 10 percent by nitrogen implantation and 20 percent by carbon implantation. The effect of nitrogen implantation nearly disappears at higher stresses (690 MPa) while carbon implantation maintains a factor of 4 to 5 increases in endurance life even at 760 MPa stress level. No significant difference in improvement

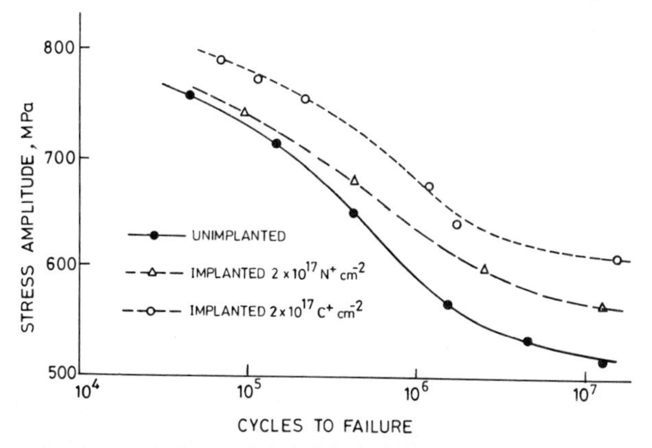

FIGURE 12.13 The cycles to failure versus maximum stress of Ti6Al4V alloy implanted by N^+ and C^+ at 75 keV to 2×10^{17} ion cm^{-2} ion fluences. The rotating-beam test was used to determine fatigue failure. [*Adapted from Vardiman and Kant (1982).*]

of fatigue life was observed after 4 h heat treatment of N^+ implanted alloy at 500°C and 1 h heat treatment of C^+ implanted alloy at 400°C. Very fine grained precipitates of TiC were identified by transmission electron micrographs (Vardiman and Kant, 1982) in as-implanted substrates which grew in size upon annealing. Only slight evidence for TiN precipitates was found for N^+ implanted substrates in as-implanted conditions and after annealing at 500°C. Because carbon and nitrogen are interstitial impurities, interstitial solid solutions and second phases may be the possible mechanisms for strengthening. Segregation to dislocations can also be expected.

White and Dearnaley (1980) reported the results of rolling-contact fatigue tests on En 31 standardized balls run in a shell four-ball tester. Tests were performed at room temperature after implantation of nitrogen to a fluence of 3×10^{17} N^+ ion cm^{-2}, with a temperature during implantation of 150°C for 6 h. Results are shown in Fig. 12.14, and a significant improvement in time to failure is observed. The 10 and 50 percent survival times are approximately doubled by implantation.

12.1.5.7 Corrosion. Small amounts of well-chosen species can have a profound effect upon the oxidation behavior of a metal, but it may be wasteful to introduce expensive additions (e.g., yttrium and rare earths) throughout the volume of the metal, and in some cases some bulk property (e.g., ductility, electrical conductivity, and nuclear properties) may be degraded as a result. The alternative would be to apply a coating, but then there exists an interface which is subjected to enhanced corrosion once the coating is pierced.

By ion implantation, chosen species can be introduced into the surface layers of material in order to combat oxidation without detriment to the bulk properties (Antill et al., 1976; Clayton, 1981, 1987; Dearnaley, 1980; Wang et al., 1979). Yttrium ion implanted into the surface of the chromium-rich (austenitic) stainless steel provided protection in carbon dioxide at 800°C for 8 months (which simulates a CO_2-cooled nuclear reactor system) (Fig. 12.15). The yttrium was found to

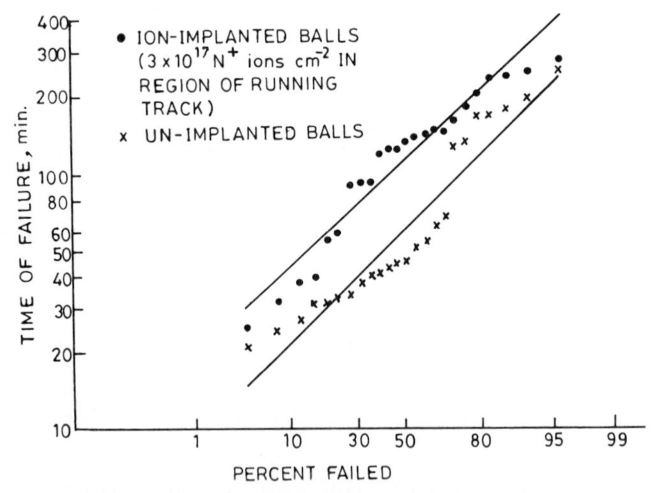

FIGURE 12.14 Times to failure against percentage failed of En31 standarized steel balls run in a rolling four-ball tester, with and without nitrogen implantation. [*Adapted from White and Dearnaley (1980).*]

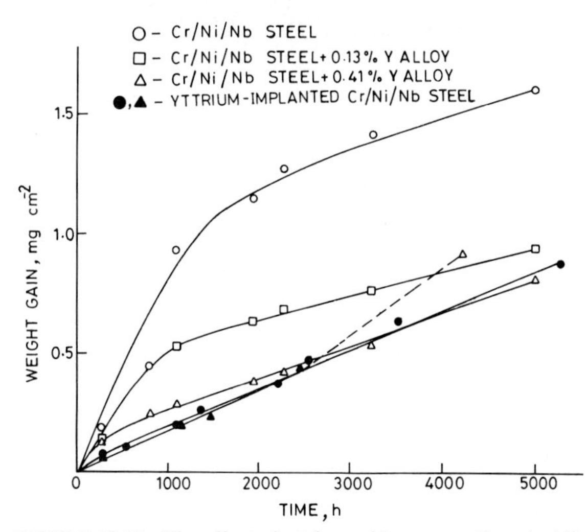

FIGURE 12.15 The effect of yttrium, either as an alloyed addition or ion implanted into the surface layers, upon the oxidation of 20% Cr-25% Ni-1% Nb stainless steel in carbon dioxide at 800°C. [*Adapted from Antill et al. (1976).*]

be carried forward during oxidation, remaining in the oxide close to the metal-oxide interface where it may act by formation of the $YCrO_3$ which is likely to serve as an impermeable barrier to chromium out-diffusion (Antill et al., 1976). Implanted cerium is found to be equally effective. The thermal oxidation of titanium has been reduced by implantation of alkaline earth and lanthanide elements (such as Sr^+, Eu^+, and Ba^+).

Turning now to the field of aqueous corrosion, a widely applied principle is to introduce species such as Cr, Al, Si, or rare earth atoms which have the property of forming highly stable and coherent oxides, by preferential surface oxidation, thus serving to protect the metal. Ashworth et al. (1975, 1980) have reported that implanted species such as chromium have beneficial effects in steel similar to that in the bulk materials, as measured by potentiokinetic methods. Tantalum, which is unsoluble in iron, has been shown to be even more effective than chromium. Molybdenum implanted into stainless steel was found to provide inhibitions against pitting corrosion, just as it does in molybdenum-containing alloys.

Valori et al. (1983) implanted single-element Cr or dual elements (Cr, and P or Mo) into M50 tool steel (commonly used in ball bearings). Based on simulated corrosion tests (two test pieces placed in contact and immersed in chloride-contaminated oil for 2 h) and potentiokinetic tests (in 1N-H_2SO_4 and 0.01 M or 0.1 M NaCl solution), they found that implantation substantially improved resistance to both general and pitting corrosion in M50 steel. Bearings with all surfaces implanted by various species were tested in the presence of liquid lubricant, and neither the bearing performance nor fatigue endurance life were adversely affected under the conditions tested.

Dearnaley (1982) has reported that the ion bombardment of steel (and other) surfaces in the presence of residual oxygen will bring about an anomalous oxidation. In the case of steels, the film consists of FeO_3 in a nonmagnetic, very finely divided form that may even be amorphous. The surprising feature of these attractively colored oxide films, produced at low temperatures on steel (≤200°C) is that they protect the surface against atmospheric or aqueous corrosion. The structure of the film may be such that grain boundary diffusion is impossible. The process serves to protect ion-implanted steel tools from rusting during the period before they are put into use, and it may reduce the corrosion of tools exposed to aqueous solution, e.g., razor blades.

12.2 ION BEAM MIXING

Ion beam mixing or ion mixing provides a method to form surface alloys by ion-beam-induced intermixing between thin coating (typically 50 to 200 nm) and substrate atomic species at the interface under energetic ion bombardment (Appleton, 1984; Dearnaley et al., 1986; Hirano and Miyake, 1985; Hirvonen et al., 1987; Hubler, 1987; Huber et al., 1984; Kobs et al., 1987; Matteson et al., 1979; Mayer et al., 1980, 1981; Nicolet et al., 1984; Tsaur, 1980; Tsaur et al., 1979b, 1981). The (easily available) gas ions at high energy knock some atoms out of the coating into the substrate where, in their turn, they are implanted. The energy of the implanted ions (typically 100 to 500 keV) is chosen such that a fraction would penetrate the coating-substrate interface and reach the substrate. This results in atomic mixing of the coating and the substrate and produces a graded interface, Fig. 12.16. The ion beam mixing is able to produce compound phases at temperatures well below those required to form phases by normal heat treatment.

So far, we have described ion beam mixing of a single-layered structure. Similarly, multilayered structures can also be treated with ion beam mixing. The coated surface can be bombarded after the initial stage of growth or after total deposition (referred to as *static ion beam mixing*) or concurrently bombarded during the entire deposition period (*dynamic ion beam mixing*). Ion beam mixing can be divided into categories, usually called *reactive ion beam mixing*, where the ion bombardment is capable of producing new chemical compounds or decomposing the compound of the coating in addition to mixing of the interface

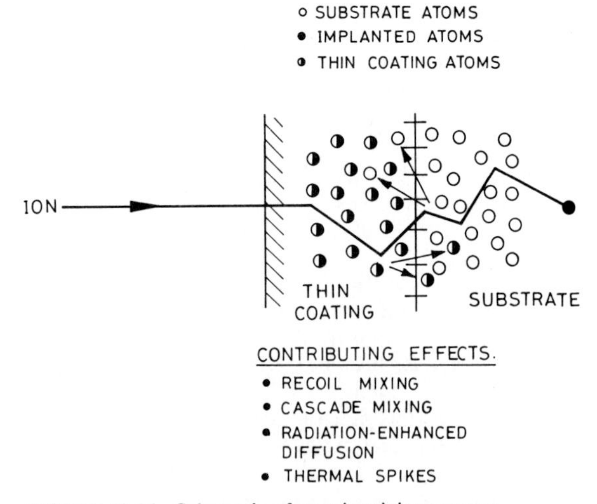

FIGURE 12.16 Schematic of atomic mixing process.

(Eskildsen and Sørensen, 1984; Guzman et al., 1984) and the category where this is not the case.

The special features of ion beam mixing process are

- A substrate can be implanted by ions, which are difficult or expensive to produce in the form of ion beams, by covering it with a suitable thin coating.
- Higher concentration can be obtained than by direct implantation of the alloying element because sputtering limits implanted concentrations.
- A required alloy concentration can be achieved at low ion fluences of easily available gaseous ion beams such as Ar^+, Kr^+, O^+, and N^+.
- The graded interface of the coatings with substrate increases its adhesion dramatically.
- The bombarded coatings have higher density and higher residual compressive stresses resulting in high hardness.

In ion beam mixing, the ions must penetrate the coating with appreciable energy, and the resulting mixing has usually been attributed to "cascade" or collisional mixing including recoil mixing. There is a close relationship between many high-dose implantation effects and ion beam mixing. In the bilayer systems (coating-substrate), the profile of long-range mixing is exponential. In the limiting case where the intermixed interfacial layer remains thin enough so as not to alter the interfacial characteristics significantly, the number of penetrating atoms increases as the square root of the fluence, and to be independent of the temperature at low temperatures (below *RT*) (Barcz et al., 1984; Nicolet et al., 1984).

Above a critical temperature (>300 K), perturbation produced in the bilayer is principally determined by the thermally activated diffusion of the point defects. In this regime, the details of the primary collisional process are largely obliterated by the subsequent thermally induced rearrangement of atoms. The outcome becomes sensitive not only to temperature but also to the specific chemical and structural properties of the irradiated material (Mayer et al., 1981). The thermal diffusion is enhanced by (irradiation-induced) defect interactions. These phenomena are called *radiation-enhanced* or *defect-enhanced diffusions*. The radiation-enhanced diffusion results in a much more efficient mixing (Nicolet et al., 1984).

Appleton (1984) and Shreter et al. (1984) suggest that the mixing mechanisms above 300 K are more complex than the two main mechanisms just mentioned (cascade mixing and radiation-enhanced diffusion). Radiation-enhanced diffusion would be expected to vary with the flux (rate of fluence) of the irradiation. Shreter et al. have reported that mixing of bilayers such as Ni/Si (above 300 K) is independent of the flux. They report that rapid dissipation energy from the dense cascade region (thermal spike) may be the most significant aspect of ion beam mixing. The quenching rate of a spike cascade can be estimated to be 10^{13} K s^{-1}. The thermal spike concept has been verified by Matteson et al. (1979). They showed that the thickness of the intermixed region increased in 3 nm Pt embedded in amorphous silicon after bombardment with As_1^+, As_2^+, and As_3^+ to a constant fluence (5×10^{15} As atoms cm^{-2}) and the same energy per atom (120 keV). By using As_2^+ and As_3^+, the energy density deposited in each collision cascade is, respectively, two and three times that deposited by As_1^+.

Figure 12.17 presents an example which shows the atomic mixing of Ni coated onto Si, where Ni atoms enter into Si and result in a concentration tail (shaded area in Ni scattering yield curve) and Si atoms enter into Ni coating. There is a formation of Ni_2Si compound at the interface, as evident from ion backscattering

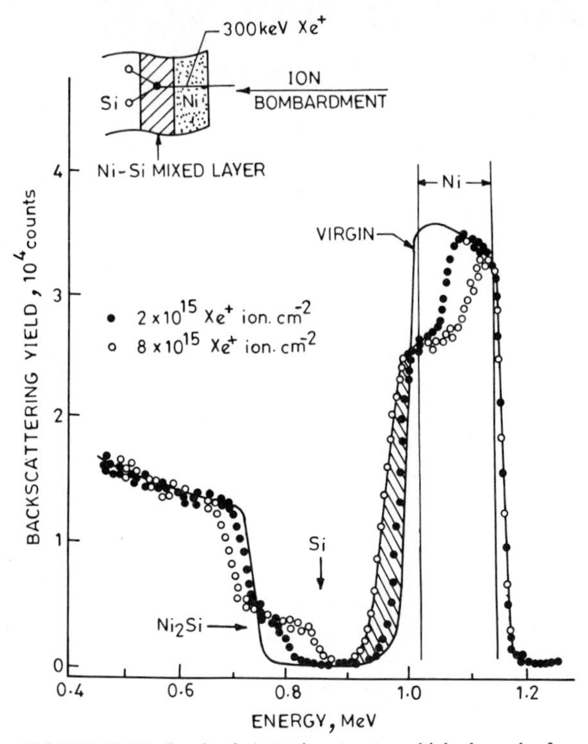

FIGURE 12.17 Ion backscattering spectra which show the formation of the Ni_2Si phase by implanting Xe^+ ions through a 50-nm-thick layer of Ni on top of an Si substrate at liquid nitrogen temperature. The ion-induced silicide phases are the same as those formed by thermal annealing of Ni layers on Si. [*Adapted from Tsaur (1980).*]

spectra (Tsaur, 1980). These results are quite similar to those observed in normal heat treatment where formation of the $NiSi_2$ phase is obtained at a temperature of about 200°C. The Ni concentration in the coating falls as determined by the Ni scattering level. This together with film thickness is used to determine the number of Si atoms cm^{-2} which have been mixed into the Ni coating, the mixing rate (Ni → Si).

Since ion beam mixing requires interdiffusion of atomic species during ion bombardment, one would expect that the substrate temperature during ion bombardment will have an influence on the atomic mobility (or diffusivity) and hence the amount of interface mixing (Matteson et al., 1979; Tsaur, 1980). As an example, Fig. 12.18 shows a plot of the amount of Si atoms intermixed with the Cr coating as a function of reciprocal temperature. The 30-nm-thick Cr coating on Si substrate was bombarded with 300 keV Xe^+ ions at various substrate temperatures (from liquid N_2 temperature to 270°C). As can be seen in the figure, the mixing is relatively insensitive to temperature at temperatures below room temperature (*RT*), but the reaction increases drastically above *RT*. This behavior can be interpreted as follows: at low temperatures (below *RT*), the mixing is domi-

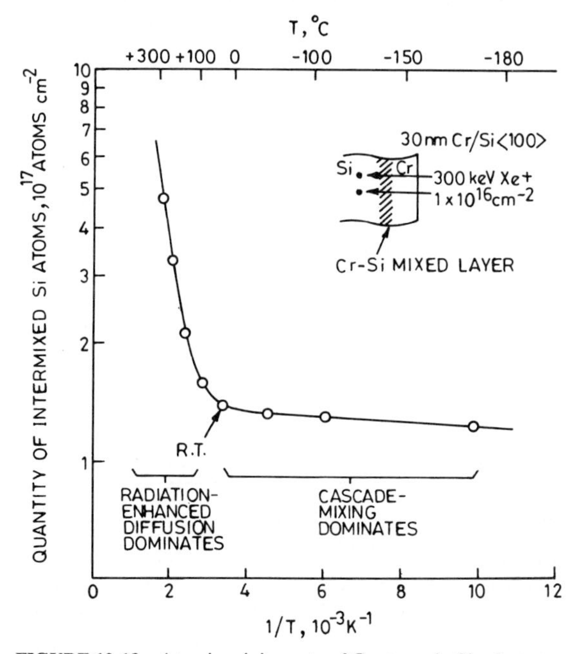

FIGURE 12.18 Atomic mixing rate of Cr atoms in Si substrate as a function of reciprocal implantation temperature. [*Adapted from Mayer et al. (1981).*]

nated by a dynamic collision process (cascade mixing including recoil mixing and radiation-enhanced diffusion) which depends mainly on the ion energy and the masses of ion and target atoms. This process is relatively insensitive to temperature. Tsaur (1980) also reported that phase transformation is unlikely at this temperature due to the lack of atomic mobility. In some cases, completely random (amorphous) structures are found in the mixed layers. In the high-temperature regime (above *RT*), chemical driving forces and radiation-enhanced diffusion become dominant, and the displaced atoms have enough mobility to migrate and form metallurgically distinct phases. Shreter et al. (1984) have argued that increased mixing at high temperature can be explained by the thermal spike model. The "onset temperature" of the transition at which drastic increase of mixing occurs and the slope of the curve in the higher temperature parts will vary from system to system.

Aside from dependence on intrinsic diffusion properties of target materials and implantation temperature, ion-induced interface reactions can also be affected by the projectile parameters such as ion energy, atomic mass of the implanted species, and fluence. In general, the phase obtained is independent of the energy and mass of the incident ions. However, the thickness of the intermixed layer was found to increase with ion energy and ion mass (Matteson et al., 1979; Tsaur, 1980). This can be expected because energetic heavy ions are more effective in initiating atomic collisions and generating defects in the material; therefore, the mixing process is more pronounced. As an example, Fig. 12.19 shows a plot of intermixed Si atoms cm^{-2} versus the square root of ion dose for 45-nm-thick Pt

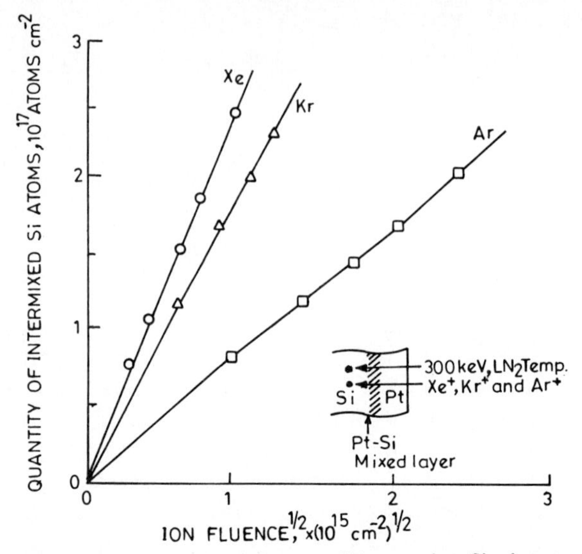

FIGURE 12.19 Atomic mixing rate of Pt atoms into Si substrate as a function of square roots of Ar^+, Kr^+, and Xe^+ fluence level at liquid nitrogen temperature. [*Adapted from Tsaur (1980).*]

coating on the Si substrate, bombarded with 300 keV Xe^+, Kr^+, or Ar^+ at liquid nitrogen temperature. The intermixed layer produced are nonuniform, which suggests the mixing is mostly contributed from the dynamic collision process. As shown in Fig. 12.19, a square root dependence of the amount of mixing on ion dose has been observed. The reaction is more pronounced for bombardment with heavy-mass ions. A square root of dose dependence suggests a diffusion-like process for the cascade mixing.

The ion beam mixing process has been used to produce several novel phases of metal alloys which were deposited in the form of multilayered stacks of different metals. For example, this process is used for the production of silicide films as contacts on semiconductor devices. Formation of the $NiSi_2$, $HfSi_2$, and V_3Si phases has been achieved this way at about 500°C, a temperature that is substantially lower than that (~800°C) required to form phases by thermal annealing of Ni or Hf layers on Si or V layers on SiO_2 (Tsaur, 1980). Several investigators (Liu et al., 1983; Mayer et al., 1981; Tsaur, 1980) have produced phase mixture of metals and observed the following:

- The mixing takes place and gives rise to a broadening of the domain of solubility.
- Compound formation can be thermally activated at temperatures well below the melting point (Pt/Si).
- Metastable phases can be obtained (Au/Ni, Au/Co).
- Amorphous alloys can always be formed when the constituents have a different crystalline structure.

The typical amorphous alloys which are formed by ion beam mixing at room temperature include (Gailliard, 1984)

- fcc/bcc : Ni/Mo, Al/Nb, Ni/Nb
- bcc/hcp : Mo/Co, Mo/Ru
- hcp/fcc : Ti/Au, Ti/Ni, Er/Ni

The effects of ion beam mixing on the mechanical properties and friction and wear behavior of soft and hard solid lubricant coatings and ceramic bulk materials are discussed in the following subsections.

12.2.1 Properties of Ion Beam Mixed Solid Lubricant And Hard Coatings on Metals

Ion beam mixing yields a distinct enhancement in microhardness, scratch adhesion, and wear resistance of several coatings such as MoS_2 (Chevallier et al., 1986; Kobs et al., 1987; Mikkelsen et al., 1988), WS_2 (Hirano and Miyake, 1985), Ag (Calliari et al., 1983), C (Padmanabhan et al., 1986), Ni and Sn (Guzman and Scotoni, 1984), Cu (Kant and Sartwell, 1984), TiN (Auner et al., 1983; Padmanabhan et al., 1983; Takano et al., 1987), HfN (Auner and Padmanabhan, 1985), TiC (Auner et al., 1983), WC (Dearnaley, 1982), TiB_2 (Auner et al., 1983; Padmanabhan and Sørensen, 1981). Guzman et al. (1984) produced a BN coating by N^+ implantation of evaporated B coating. The properties of ion beam mixed coatings are influenced by the ion fluences, temperature of ion implantation, atomic mass of the implanted species, and crystallinity and stoichiometry of the coating.

Figure 12.20 illustrates the influence of ion fluences and atomic mass of implanted ions on the adhesion and on endurance life of MoS_2 coatings deposited by radio frequency (rf) sputtering onto 304 stainless steel (Chevallier et al., 1986). Samples were first sputter-coated up to ~50 nm and were exposed to a 400-keV ion bombardment of Ar^+, Kr^+, or Xe^+ ions at various ion fluences. Then the sputter deposition process was continued to a total coating thickness of about 2 μm. Endurance life was improved primarily by the increased adhesion of the lubricant coating to the substrate because of the graded interface and modified crystallinity of the coating.

Similar improvements in adhesion and endurance life in vacuum of rf-sputtered MoS_2 coatings that were ion bombarded after deposition have also been obtained by Kobs et al. (1987). Figure 12.21 shows the depth profiles of S, Mo, and Fe for 50- and 200-keV nitrogen-implanted MoS_x coating (coating thickness, 0.47 μm) on AISI 52100 steel at fluence of 1×10^{16} ion cm^{-2} obtained by SIMS (Kobs et al., 1987). They observed no broadening of the interface for 50-keV ions compared with the as-sputtered coating. A markedly broadened interface of about 80 nm was detected after 200-keV implantation, revealing a strong correlation to the lifetime enhancement. The shift of the interface to the surface is due to the reduction in the coating thickness, to be described later.

Mikkelsen and Sørensen (1989a, 1989b) have reported that MoS_x ($x = 1.7$) coating is amorphized with the bombardment of 400 keV Ar^+ ions at a dose of 3×10^{15} to 1×10^{16} ion cm^{-2} (Fig. 12.22). They reported considerable reduction (20 to 50 percent) in coating thickness. As a result of ion bombardment, the coating density increased 100 percent to almost the bulk value for MoS_2. The endurance life of AISI 52100 steel ball sliding against amorphized MoS_x coated AISI 52100 steel disk in nitrogen with a load of 10 N was improved by a factor of 5 and coefficient of friction was found to be about 0.04 similar to that of the crystalline MoS_x coatings. These friction results for ion-beam mixed amorphous MoS_x coat-

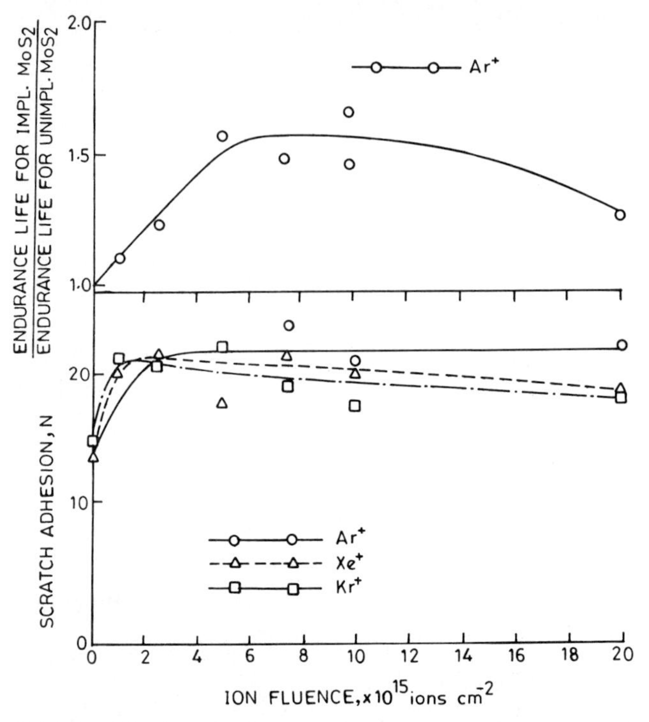

FIGURE 12.20 Scratch adhesion and relative change in endurance life (cycles required for friction failure) of rf-sputtered MoS_2 coatings on 304 stainless steel, mixed by ion beams at 400 keV. [*Adapted from Chevallier et al. (1986).*]

ings are contrary to amorphous MoS_2 coatings applied by rf sputtering by Spalvins (1978).

Padmanabhan et al. (1986) have reported improvements in microhardness, scratch adhesion, and wear resistance of diamondlike or hard carbon (a-C) coatings by the interaction of an ion beam with a concurrently deposited coating (dynamic ion beam mixing). The carbon coatings were produced on polished stainless-steel substrates by cracking of ethane in an rf glow discharge (plasma-enhanced chemical vapor deposition). The coating chamber was attached to the 600-keV heavy-ion accelerator. The carbon deposition rates were varied according to the rf power and the final thickness for all samples was kept constant at 1 μm. Figure 12.23 shows an increase in microhardness as a function of the Kr^+ implantation energy for different deposition rates (rf power). The considerable improvement at low carbon-deposition rates (33 nm min^{-1} at 10 W) may be due to the fact that the projected range of 100 keV ions is comparable (34 nm). The change in the relative microhardness as a function of the total Kr^+ ion flux implanted at 400 keV into the coating is presented in Fig. 12.24 where one sample was bombarded only at the initial stage of growth while the others were bombarded during the total deposition period. This clearly demonstrates that the concurrent ion bombardment has a beneficial effect on the coating microhardness and depends on the flux of the conconcurrent ion bombardment.

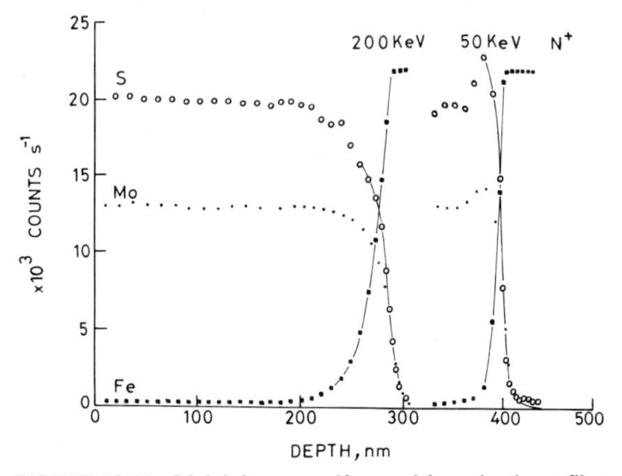

FIGURE 12.21 Molybdenum, sulfur, and iron depth profiles of 50 keV and 200 keV nitrogen-implanted MoS_x coating (thickness, 0.47 μm) on AISI 52100 steel obtained by SIMS; implanted fluence = 1×10^{16} ion cm^{-2}. [*Adapted from Köbs et al. (1987).*]

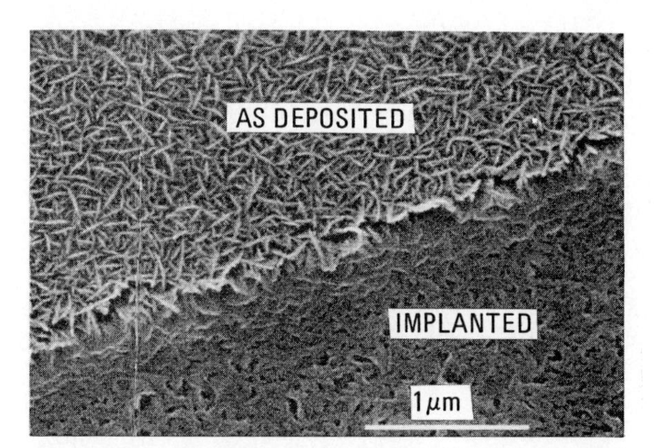

FIGURE 12.22 SEM picture of a 400-nm-thick MoS_x (x = 1.7) coating applied by rf sputtering on 52100 steel. One part of the coating is bombarded with 400 keV, 1×10^{16} Ar^+ cm^{-2}, and remaining part is as deposited. [*Adapted from Mikkelsen and Sørensen (1989a).*]

Padmanabhan et al. (1986) also reported that there was a marked improvement in the scratch adhesion of the carbon coatings as a result of ion bombardment. They reported that an improvement in adhesion can be obtained by concurrent ion bombardment during entire deposition or at the early stage of growth only. This suggests that mixing at the coating-substrate interface is responsible for the creation of a transition layer between the substrate and carbon coating.

The microhardness of implanted TiN coating produced by rf reactive sputtering onto stainless steel increases when bombarded with Kr^+ and Rb^+ ions of various fluences after coating deposition (Fig. 12.25). The projected range of Kr^+

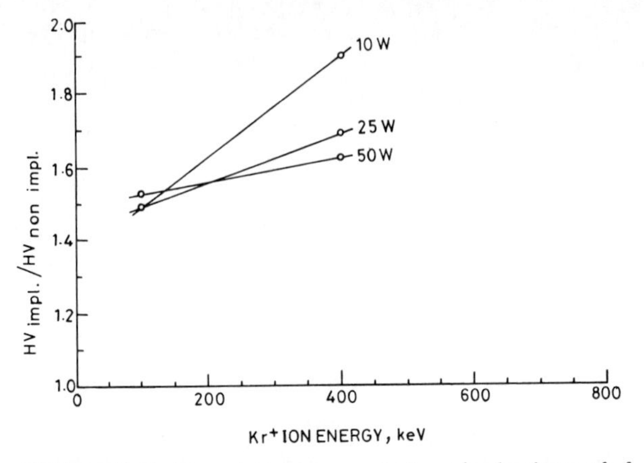

FIGURE 12.23 The relative increase in the microhardness of rf plasma carbon coatings as a function of Kr^+ energy and an ion flux of 400 $\mu C\ cm^{-2}\ min^{-1}$ for coatings applied at various deposition rates by controlling the rf power. The coatings were applied at 0.1 Pa (7×10^{-4} torr) on stainless-steel substrate. [*Adapted from Padmanabhan et al. (1986).*]

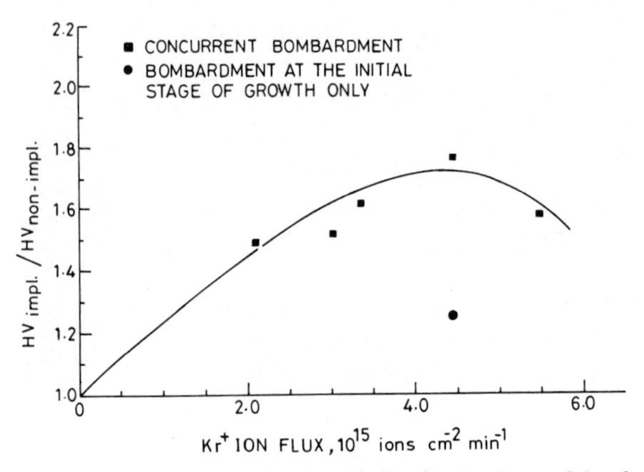

FIGURE 12.24 The relative increase in microhardness of the rf plasma carbon coatings bombarded with various Kr^+ ion fluxes at 400 keV. The coatings were deposited at 50 W rf power on stainless-steel substrates. [*Adapted from Padmanabhan et al. (1986).*]

ions with an energy of 450 keV in TiN was 106 nm. The hardness increased as high as 33 percent at ion fluence of 10^{16} ion cm^{-2} but further increase in ion fluence brings about the opposite effect (Auner et al., 1983; Padmanabhan et al., 1983). This is probably due to the fact that at higher fluences implanted inert gas ions, i.e., Ar^+ give rise to bubble formation yielding a more porous structure in the substrate below the interface and this may influence the hardness. This is further verified by implanting Rb^+ ions whose range distribution is similar to Kr^+. The increased microhardness of these samples has been found to depths which

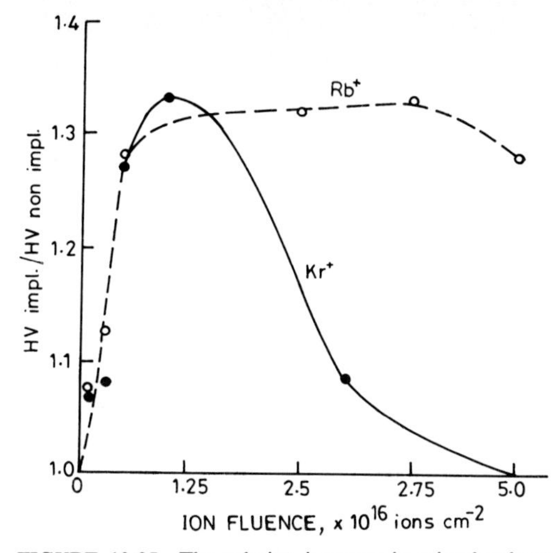

FIGURE 12.25 The relative increase in microhardness of TiN coatings (0.01 μm thick) on stainless steel (deposited by rf reactive sputtering from TiN target) as a function of ion fluences when bombarded with 500-keV Kr^+ and Rb^+ ions. [*Adapted from Padmanabhan et al. (1983).*]

are greater than the implanted ion range. This may be due to broadening of the interface and along with ion induced mixing, may result in a metastable nitride layer Fe-Ti-N.

Another striking effect is shown in Fig. 12.26 where the gain in microhardness and adhesion due to implantation as a function of the N_2 sputtering gas concentration for a reactively sputtered coating is presented. A sharp maximum is found for stoichiometric TiN.

Figure 12.27 shows the improvement in adhesion of a TiN-sputtered deposited with a thickness of 100 nm onto stainless steel implanted by Kr^+ at room temperature and elevated temperature of 200°C. The adhesion increase with ion fluence and further by a factor of 2 when ion beam mixing process occurs at 200°C. In both cases the increase in adhesion is obviously due to formation of graded interface of TiN and stainless steel. However, at 200°C, there is also an added contribution due to thermally assisted diffusion along with the ion-induced mixing.

Significant improvement in the hardness by ion implantation of reactively rf-sputtered TiC coatings has been reported by Auner et al. (1983), who have also reported an increase in hardness as a function of implantation temperature. For example, microhardness of sputtered TiB_2 coating increased by about 10 percent when implanted with Ni^+ ions at a temperature of 400°C compared to an increase of only about 2 percent at an implantation temperature of 200°C.

12.2.2 Properties of Ion Beam Mixed Hard Coatings on Ceramics

Recently Lankford et al. (1987), Wei and Lankford (1987), and Wei et al. (1987) have reported a dramatic reduction in coefficient of friction (from 0.25 to 0.06)

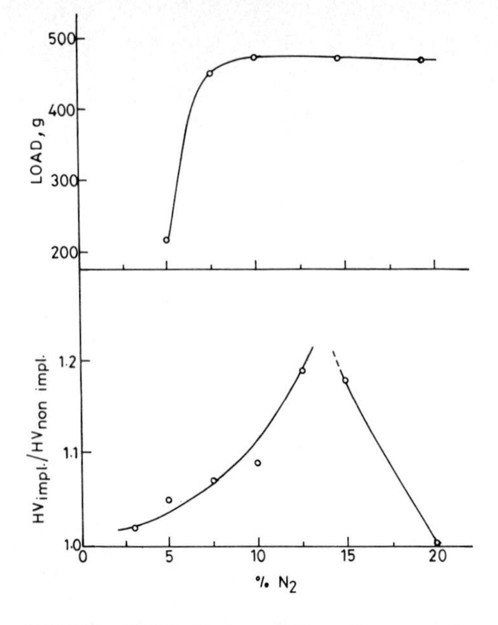

FIGURE 12.26 The relative increase in microhardness and adhesion of reactive sputtered TiN coatings (0.01 μm thick) ion implanted with 500-keV Kr^+ ions with a fluence of 5×10^{15} ion cm^{-2} as a function of the N_2 pressures in Ar/N_2 gas mixture during coating deposition. [*Adapted from Padmanabhan et al. (1983).*]

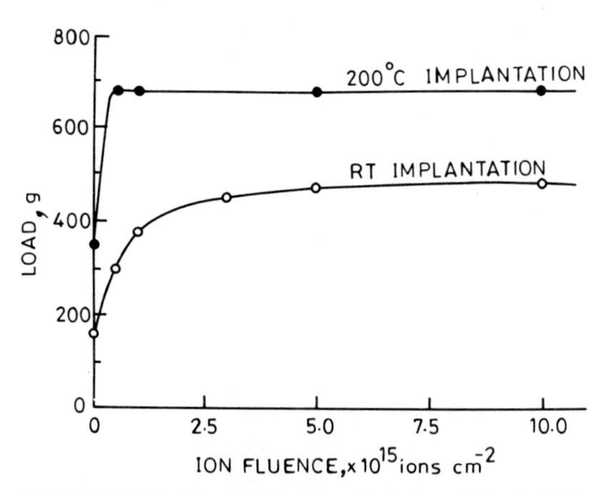

FIGURE 12.27 The effect of ion fluence and implantation temperature on the adhesion of sputtered TiN coating (0.01 μm thick) on stainless steel as measured by the scratch test. [*Adapted from Padmanabhan et al. (1983).*]

and wear between ceramic versus ceramic surfaces under dry conditions, due to ion beam mixing. In this case the most important ceramics, i.e., Si_3N_4 and partially stabilized ZrO_2 (from a tribological applications point of view) were vapor deposited by Ti-Ni or Co or Cr and then implanted with Ar^+ at 140 keV with a fluence of 10^{17} ion cm^{-2} and a flux of about 10^{12} ion cm^{-2} s^{-1}. The thicknesses of modified layers were about 400 nm or less. These metals were selected because they are known to form stable continuous metal oxides and provide effective lubrication at high temperatures.

Figure 12.28 shows the reduction of coefficient of friction at 800°C in a diesel environment (O_2, N_2, CO_2, H_2O) between TiC and TiC-5 Ni-5Mo (K 162 B) pins and Si_3N_4 and ZrO_2 disks modified by (Ti, Ni), Co, and Cr ion beam mixing. The least coefficient of friction as low as 0.06 comparable to conventional liquid-lubricated conditions occurs by the ion beam mixing of Ti-Ni or cobalt for certain pin-disk combinations (Ti-Ni/ZrO_2, Co/ZrO_2, and Ti-Ni/Si_3N_4).

Another major improvement resulting from ion beam modification of the ceramic disks was the reduction in wear of the pin-disk combinations in the diesel environment at 800°C to the extent that in many cases the amount of wear could not be measured either by weight change or by surface profiler. For ion beam mixed Ti-Ni on Si_3N_4 and ZrO_2 disks and Ni on Si_3N_4 disks, the wear track on the disks was barely discernible by a scanning electron microscope. The results of surface analyses indicated the formation of a lubricating oxide layer of Ti and Ni, which was responsible for the improvement in ceramic friction and wear at high temperatures.

The improvement in friction and wear characteristics of ceramic materials by ion beam mixing opens new ways to explore the possibility of using ceramics at high temperatures under unlubricated conditions. For instance, the typical applications are cylinder liners and piston rings for high-temperature adiabatic engines or high-temperature bearings in gas turbine engines.

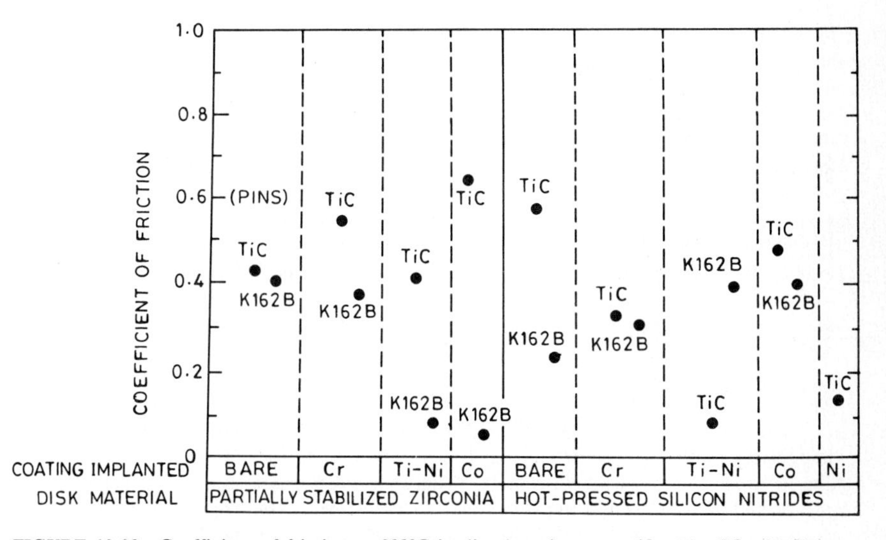

FIGURE 12.28 Coefficient of friction at 800°C in diesel environment (O_2, N_2, CO_2, H_2O) between TiC, Ni-Mo bonded TiC (K162 B) pins and ion beam mixed coatings on ZrO_2 and Si_3N_4 disks in a pin-on-disk test. [*Adapted from Wei et al. (1987).*]

12.3 REFERENCES

Antill, J. E., Bennett, M. J., Carney, R. F. A., Dearnaley, G., Fern, F. H., Goode, P. D., Myatt, B. L., Turner, J. F., and Warburton, J. B. (1976), "The Effect of Surface Implantation of Yttrium and Cerium upon the Oxidation Behavior of Stainless Steels and Aluminized Coatings at High Temperatures," *Corrosion Sci.*, Vol. 16, pp. 729–745.

Appleton, B. R. (1984), "Ion Beam and Laser Mixing: Fundamentals and Applications," in *Ion Implantation and Beam Processing* (J. S. Williams and J. M. Poate, eds.), pp. 189–259, Academic, New York.

Ashworth, V., Baxter, D., Grant, W. A., Procter, R. P. M., and Wellington, T. C. (1975), "The Effect of Ion Implantation on the Corrosion Behaviour of Fe," *Proc. Conf. Ion Implantation into Semiconductors* (S. Namba, ed.), pp. 367–374, Plenum, New York.

———, and Procter, R. P. M. (1980), "The Application of Ion Implantation to Aqueous Corrosion," in *Treatise on Materials Science and Technology, Vol. 18: Ion Implantation* (J. K. Hirvonen, ed.), pp. 175–256, Academic, New York.

Auner, G., and Padmanabhan, K. R. (1985), "Effect of Ion Bombardment on Thin Hafnium Nitride Films," *Thin Solid Films*, Vol. 123, pp. 315–323.

———, Hsieh, Y. F., and ——— (1983), "Effect of Ion Implantation on Thin Hard Coatings," *Thin Solid Films*, Vol. 107, pp. 191–199.

Bakhru, H., Gibson, W., Burr, C., Kumnick, A. J., and Welsch, G. E. (1981), "Modification of Fatigue Behaviour of Copper and Stainless Steel by Ion Implantation," *Nucl. Instrum. Meth.*, Vol. 182/183, pp. 959–964.

Barcz, A. J., Paine, B. M., and Nicolet, M.-A. (1984), "Ion Mixing of Markers in SiO_2 and Si," *Appl. Phys. Lett.*, Vol. 44, pp. 45–47.

Baumvol, I. J. R. (1981a), "Radiation Enhanced Diffusion as a Coating Method for Protection against Corrosion and Wear," *Phys. Stat. Sol.*, Vol. A67, pp. 287–298.

——— (1981b), "The Influence of Tin Implantation on the Oxidation of Iron," *J. Appl. Phys.*, Vol. 52, pp. 4583–4588.

——— (1984), "Ion Implantation Metallurgy," in *Ion-Implantation Science and Technology* (J. F. Ziegler, ed.), pp. 261–310, Academic, New York.

Bennett, J. F. J. (1969), "Ion Source for Multiply Charged Heavy Ions," *Proc. Int. Conf. on Ion Sources*, pp. 571–585.

Biasse, B., Destefanis, G., and Gailliard, J. P. (eds.) (1983), *Ion Beam Modification of Materials*, North Holland, Amsterdam.

Bolster, R. N., and Singer, I. L. (1980). "Surface Hardness and Abrasive Wear Resistance of Ion-Implanted Steels," *Appl. Phys. Lett.*, Vol. 36, pp. 208–209.

———, and ——— (1981), "Surface Hardness and Abrasive Wear Resistance of Ion-Implanted Steels," *ASLE Trans.*, Vol. 24, pp. 526–532.

Borders, J. A., and Poate, J. M. (1976), "Lattice-Site Location of Ion-Implanted Impurities in Copper and Other fcc Metals," *Phys. Rev.*, Vol. B 13, pp. 969–979.

Brinkman, J. A. (1954), "On the Nature of Radiation Damage in Metals," *J. Appl. Phys.*, Vol. 25, pp. 961–970.

Bull, S. J., and Page, T. F. (1988a), "The Friction and Wear of Ion Implanted Ceramics: The Role of Adhesion," *Nucl. Instrum. Meth. Phys. Res.*, Vol. B 32, pp. 91–95.

———, and ——— (1988b), "High Dose Ion Implantation of Ceramics: Benefits and Limitations for Tribology, *J. Mater. Sci.*, Vol. 23 (no. 12), pp. 4217–4230.

Burnett, P. J., and Page, T. F. (1984a), "Changing the Surface Mechanical Properties of Silicon and α-Al_2O_3 by Ion Implantation," *J. Mater. Sci.*, Vol. 19, pp. 3524–3545.

———, and ——— (1984b), "Surface Softening in Silicon by Ion Implantation," *J. Mater. Sci.*, Vol. 19, pp. 845–860.

———, and ——— (1984c), "Changing the Indentation Behaviour of MgO by Ion Implantation," in *Ion Implantation and Ion Beam Processing of Materials* (G. K. Hubler, O. W.

Holland, C. R. Clayton, and C. W. White, eds.), MRS Proc. Vol. 27, pp. 401–406, North-Holland, Amsterdam.

———, and ——— (1985), "An Investigation of Ion Implantation-Induced Near Surface Stresses and Their Effects in Sapphire and Glass," *J. Mater. Sci.*, Vol. 20, pp. 4624–4646.

———, and ——— (1986a), "Effects of Ion Implantation on the Mechanical Properties of Fully and Partially Stabilized Zirconias," *J. Am. Ceramic Soc.*, Vol. 65 (no. 10), pp. 1393–1398.

———, and ——— (1986b), "The Effect of Ion Implantation on the Surface Mechanical Properties of Some Engineering Hard Materials," *Proc. of the Int. Conf. on the Sciences of Hard Materials*, September 1984 (E. A. Almond, C. A. Brookes, and R. Warren, eds.), pp. 789–802. Institute of Physics Conf. Series No. 75, Adam Hilger Ltd., Bristol, England.

———, and ——— (1987), "The Friction and Hardness of Ion Implanted Sapphire," *Wear*, Vol. 114, pp. 85–96.

Bykov, V. N., Troyan, V. A., Zdorovtseva, J. R., and Khaimovich, V. S. (1975), "Phase Transformations at Bombardments of Thin Films with Ions," *Phys. Stat. Sol.*, Vol. A 32, pp. 53–61.

Calliari, L., Guzman, L., LoRusso, S., Tosello, C., Woey, G., and Zobele, G. (1983), "Effect of Ion Mixing on the Wear Behaviour of Silver," in *Ion Implantation Equipment and Techniques* (H. Glawischnig and H. Ryssel, eds.), pp. 357–363, Springer Verlag, Berlin, Germany.

Carter, G., Colligon, J. S., and Grant, W. A. (eds.) (1976), "Application of Ion Beams to Materials," *Conf. Series 28*, Institute of Physics, London.

Chernow, F., Borders, J. A., and Brice, D. K. (eds.) (1977), *Ion Implantation in Semiconductors*, Plenum, New York.

Chevallier, J., Olesen, S., Sørensen, G., and Gupta, B. (1986), "Enhancement of Sliding Life of MoS_2 Films Deposited by Combining Sputtering and High Energy Ion Implantation," *Appl. Phys. Lett.*, Vol. 48, pp. 876–877.

Clayton, C. R. (1981), "Modification of Metallic Corrosion by Ion Implantation," *Nucl. Instrum. Meth.*, Vol. 182/183, pp. 865–873.

——— (1987), "Chemical Effects of Ion Implantation on Oxidation, Corrosion, and Catalysis," in *Surface Alloying by Ion, Electron, and Laser Beams* (L. E. Rehn, S. T. Picraux, and H. Wiedersich, eds.), pp. 325–356, American Society for Metals, Metals Park, Ohio.

Cochran, J. K., Legg, K. O., and Baldau, G. R. (1984), "Microhardness of Nitrogen-Implanted Yttria Stabilized Zirconia," in *Emergent Process Methods for High Technology Ceramics* (R. Davis, H. Palmour, and R. Porter, eds.), pp. 549–557, Plenum, New York.

Cottrell, A. H. (1964), *Theory of Crystal Dislocations*, Gordon and Breach, New York.

Crowder, B. L. (ed.) (1973), *Ion Implantation in Semiconductors and Other Materials*, Plenum, New York.

Daniels, L. O., and Wilbur, P. J. (1987), "Effects of Nitrogen Ion Implantation on Load-Bearing Capacity and Adhesive Wear Behavior in Steels," *Nucl. Instrum. Meth. Phys. Res.*, Vol. B19/20, pp. 221–226.

Dearnaley, G. (1978), "Ion Implantation for Improved Resistance to Wear and Corrosion," *Mater. Eng. Appl.*, Vol. 1, September, pp. 28–41.

——— (1980), "Thermal Oxidation," in *Treatise on Materials Science and Technology*, Vol. 18: *Ion Implantation* (J. K. Hirvonen, ed.), pp. 257–320, Academic, New York.

——— (1982), "Practical Applications of Ion Implantation," *J. Metals*, Vol. 34, pp. 18–28.

——— (1983a), "Improvement of Wear Resistance in Cemented Tungsten Carbide by Ion Implantation," in *Proc. Int. Conf. on Science of Hard Materials* (R. K. Viswanadham, D. J. Rowcliffe, and J. Gurland, eds.), pp. 467–484, Plenum, New York.

——— (1983b), "Application of Ion Implantation in Metals," *Thin Solid Films*, Vol. 107, pp. 315–326.

———, Freeman, J. H., Nelson, R. S., and Stephen, J. (1973), *Ion Implantation*, North-Holland, Amsterdam.

———, and Hartley, N. E. W. (1978), "Ion Implantation into Metals and Carbides," *Thin Solid Films*, Vol. 54, pp. 215–232.

———, Goode, P. D., Minter, F. J., Peacock, A. J., and Waddell, C. N. (1985), "Ion Implantation and Ion-Assisted Coatings on Metals," *J. Vac. Sci. Technol.*, Vol. A3, pp. 2684–2690.

———, ———, ———, ———, Hughes, W., and Proctor, G. W. (1986), "Ion Implantation and Ion Assisted Coating of Metals," *Vacuum*, Vol. 36, pp. 807–811.

Deutchman, A. H., and Partyka, R. J. (1984), "Wear Resistance Performance of Ion Implanted Alloy Steels," in *Surface Engineering: Surface Modification of Materials* (R. Kossowsky and S. C. Singhal, eds.), pp. 185–196, Martinus Nijhoff Publishers, Dordrecht, The Netherlands.

DeWaard, H., and Feldman, L. C. (1974), "Lattice Location of Impurities Implanted into Metals," in *Application of Ion Beams to Metals* (S. T. Picraux, E. P. EerNisse, and F. L. Vook, eds.), pp. 317–351, Plenum, New York.

Dienes, C. J., and Vineyard, G. H. (eds.) (1957), *Radiation Effects in Solids*, Wiley, New York.

Dillich, S. A., and Singer, I. L. (1983), "Effect of Titanium Implantation on the Friction and Surface Chemistry of a Co-Cr-W-C Alloy," *Thin Solid Films*, Vol. 108, pp. 219–227.

———, Bolster, R. N., and ——— (1984), "Effects of Ion Implantation on the Tribological Behaviour of a Cobalt Based Alloy," in *Surface Engineering: Surface Modification of Materials* (R. Kossowsky and S. C. Singhal, eds.), pp. 197–208, Martinus Nijhoff Publishers, Dordrecht, The Netherlands.

Dimigen, H., Köbs, K., Leutenecker, R., Ryssel, H., and Eichinger, P. (1985), "Wear Resistance of Nitrogen-Implanted Steels," *Mater. Sci. Eng.*, Vol. 69, pp. 181–190.

Ecer, G. M., Wood, S. W., Boes, D., Schreurs, J. (1983), "Friction and Wear Properties of Nitrided and N^+ Implanted 17–4PH Stainless Steel," *Wear*, Vol. 89, pp. 201–214.

Eisen, F., and Chadderton, L. T. (eds.) (1971), *Ion Implantation*, Gordon and Breach, New York.

Eskildsen, S. S., and Sørensen, G. (1984), "On the Use of Reactive Ion-Beam Mixing for Surface Modifications," in *Surface Engineering: Surface Modification of Materials* (R. Kossowsky and S. C. Singhal, eds.), pp. 48–60, Martinus Nijhoff Publishers, Dordrecht, The Netherlands.

———, Vagø, F., Straede, C. A., and Sørensen, G. (1989), "Improved Mechanical and Tribological Properties of Ion Implanted Ceramics," *Proc. of the 5th Int. Congress on Tribology* (K. Holmberg and I. Nieminen, eds.), pp. 442–447, The Finnish Society for Tribology, Espoo, Finland.

Farlow, G. C., White, C. W., McHargue, C. J., and Appleton, B. R. (1984), "Behavior of Implanted α-Al_2O_3 in an Oxidizing Annealing Environment," *Proc. Mater. Res. Soc.*, Vol. 27, pp. 395–400.

Fayeulle, S., Treheux, D., Guiraldenq, P., Dubois, J., and Fantozzi, G. (1986), "Nitrogen Implantation in Tungsten Carbides," *J. Mater. Sci.* Vol. 21, pp. 1814–1818.

Fischer, T. E., and Tomizawa, H. (1987), "Interaction of Tribochemistry and Microfracture in the Friction and Wear of Silicon Nitride," *Wear*, Vol. 105, pp. 29–45.

———, Luton, M. J., Williams, J. M., White, C. W., and Appleton, B. R. (1983), "Tribological Properties of Ion-Implanted 52100 Steel," *ASLE Trans.*, Vol. 26, pp. 466–474.

Fleischer, R. L. (1964), *The Strengthening of Metals*, Reinhold, New York.

Follstaedt, D. M., Pope, L. E., Knapp, J. A., Picraux, S. T., and Yust, F. G. (1983), "The Microstructure of Type 304 Stainless Steel Implanted with Titanium and Carbon and Its Relation to Friction and Wear Tests," *Thin Solid Films*, Vol. 107, pp. 259–267.

Freeman, J. H. (1963), "A New Ion Source for Electromagnetic Isotope Separators," *Nucl. Instrum. Meth.*, Vol. 22, pp. 306–316.

Fu-Zhai, C., Li, H. D., Zhang, X. -Z. (1983), "Modification of Tribological Characteristics of Metal after Nitrogen Implantation," *Nucl. Instrum. Meth.*, Vol. 209/210, pp. 881–887.

Gailliard, J. P. (1984), "Recoil Implantation and Ion Mixing," in *Surface Engineering: Sur-*

face Modification of Materials (R. Kossowsky and S. C. Singhal, eds.), pp. 32–47, Martinus Nijhoff Publishers, Dordrecht, The Netherlands.

Glawischnig, H., and Noack, K. (1984), "Ion Implantation System Concepts," in *Ion Implantation Science and Technology* (J. F. Ziegler, ed.), pp. 313–373, Academic, New York.

Goode, P. D., Peacock, A. J., and Asher, J. (1983), "A Study of the Wear Behaviour of Ion-Implanted Pure Iron," *Nucl. Instrum. Meth.*, Vol. 209/210, pp. 925–931.

Grant, W. A. (1978), "Amorphous Metals and Ion Implantation," *J. Vac. Sci. Technol.*, Vol. 15, pp. 1644–1649.

Greggi, J., and Kossowsky, R. (1983), "A STEM Micro Analytical Investigation of Nitrogen Implanted Cemented WC-Co," in *Proc. 1st Int. Conf. on Science of Hard Materials* (R. K. Viswanadham, D. J. Rowcliffe, and J. Gurland, eds.), pp. 485–495, Plenum, New York.

Guzman, L., and Scotoni, I. (1984), "Ion Mixing in the Ag-Ni and Fe-Sn Systems," in *Surface Engineering: Surface Modification of Materials* (R. Kossowsky and S. C. Singhal, eds.), pp. 84–95, Martinus Nijhoff Publishers, Dordrecht, The Netherlands.

———, Marchetti, F., Calliari, L., Scotoni, I., and Ferrari, F. (1984), "Formation of BN by Nitrogen Ion Implantation of Boron Deposits," *Thin Solid Films*, Vol. 117, pp. L63–L66.

Hanley, P. R. (1983), "Physical Limitations of Ion Implantation Equipment," in *Proc. 4th Intl. Conf. on Ion Implantation: Equipment and Techniques* (H. Ryssel and H. Glawischnig, eds.), pp. 2–24, Springer Verlag, Berlin, Germany.

Harding, S. E. (1977), M.Sc. Thesis, Brighton Polytechnic, United Kingdom.

Harper, J. M. E., Cuomo, J. J., and Kaufman, H. R. (1982), "Technology and Applications of Broad-Beam Ion Sources Used in Sputtering Part II—Applications," *J. Vac. Sci. Technol.*, Vol. 21, pp. 737–756.

Hartley, N. E. W. (1975a), "The Tribology of Ion Implanted Metal Surfaces," *Tribol. Int.*, Vol. 8, pp. 65–72.

——— (1975b), "Ion Implantation and Surface Modification in Tribology, *Wear*, Vol. 34, pp. 427–438.

——— (1975c), Surface Stresses in Ion-Implanted Steel," *J. Vac. Sci. Technol.*, Vol. 12, pp. 485–489.

——— (1976), "Tribological Effects in Ion-Implanted Metals," in *Application of Ion Beams to Materials* (G. Carter, J. S. Colligen, and W. A. Grant, eds.), Conf. Series No. 28, pp. 210–223, Institute of Physics, London.

——— (1979), "Friction and Wear of Ion-Implanted Metals—A Review," *Thin Solid Films*, Vol. 64, pp. 177–190.

——— (1980), "Tribological and Mechanical Properties," in *Treatise on Materials Science and Technology Vol. 16: Ion Implantation* (J. K. Hirvonen, ed.), pp. 321–368, Academic, New York.

——— (1982), "Mechanical Property Improvements on Ion Implanted Diamond," in *Metastable Materials Formation by Ion Implantation* (S. T. Picraux and W. J. Choyke, eds.), pp. 295–302, Elsevier Science, New York.

———, Swindlehurst, W. E., Dearnaley, G., and Turner, J. F. (1973), "Friction Changes in the Ion Implanted Steels," *J. Mater. Sci.*, Vol. 8, pp. 900–904.

———, Dearnaley, G., Turner, J. F., and Saunders, J. (1974), "Friction and Wear of Ion Implanted Metals," in *Application of Ion Beams to Metals* (S. T. Picraux, E. P. EerNisse, and F. L. Vook, eds.), pp. 123–138, Plenum, New York.

———, and Hirvonen, J. K. (1983), "Wear Testing under High Load Conditions: The Effect of Anti-Scuff Additions to AISI 3135, 52100 and 9310 Steels Introduced by Ion Implantation and Ion Beam Mixing," *Nucl. Instrum. Meth.*, Vol. 209/210, pp. 933–940.

Hasegawa, H., Hioki, T., and Kamigaito, O. (1985), "Cubic-to-Rhombohedral Phase Transformation in Zirconia by Ion Implantation," *J. Mater. Sci. Lett.*, Vol. 4, pp. 1092–1094.

Hayashi, H., Iwaki, M., and Yoshida, K. (1980), "Ion Implantation into Metals," *Proc. of the 4th Int. Conf. on Prod. Eng., Tokyo*, Japan, pp. 1051–1056.

Herman, H. (1981), "Surface Mechanical Properties: Effects of Ion Implantation, *Nucl. Instrum. Meth.*, Vol. 182/183, pp. 887–898.

Hioki, T., Itoh, A., Noda, S., Doi, H., Kawamoto, J., Kamigaito, O. (1984), Strengthening of Al_2O_3 and ion-implantation, *J. Mater. Sci. Lett*, Vol. 3, pp. 1099–1101.

———, ———, ———, ———, ———, and ——— (1985), "Effect of Ion Implantation on Fracture Stress of Al_2O_3," *Nucl. Instrum. Meth. Phys. Res.*, Vol. B 7/8, pp. 521–525.

———, ———, ———, ———, ———, and ——— (1989), "Modification of the Mechanical Properties of Ceramics by Ion Implantation," *Nucl. Instrum. Meth. Phys. Res.*, Vol. B 39, pp. 657–664.

———, ———, Ohkubo, M., ———, ———, ———, and ——— (1986), "Mechanical Property Changes in Sapphire by Nickel Ion Implantation and Their Dependence on Implantation Temperature," *J. Mater. Sci.*, Vol. 21, pp. 1321–1328.

Hirano, M., and Miyake, S. (1985), "Sliding Life Enhancement of WS_2 Sputtered Film by Ion Beam Mixing," *Appl. Phys. Lett.*, Vol. 47, pp. 683–685.

Hirvonen, J. K. (1978), "Ion Implantation in Tribology and Corrosion Science," *J. Vac. Sci. Technol*, Vol. 15, pp. 1662–1668.

——— (ed.) (1980), *Ion Implantation: Treatise on Materials Science and Technology*, Vol. 18, Academic, New York.

———, Carosella, C. A., Kant, R. A., Singer, I., Vardiman, R. G., and Rath, B. B. (1979), "Improvement of Metal Properties by Ion Implantation," *Thin Solid Films*, Vol. 63, pp. 5–10.

———, Elvo, M. A., Mayer, J. W., and Johnson, H. (1987), "Ion Beam Mixing of Hydrogenated Multilayer Fe-Ti and Ni-Ti Films," *Mater. Sci. Eng.*, Vol. 90, pp. 13–19.

Hu, W. W., Clayton, C. R., Herman, H., and Hirvonen, J. K. (1978), "Fatigue Life Enhancement by Ion Implantation," *Scr. Metall.*, Vol. 12, pp. 679–698.

———, ———, ———, ———, and Kant, R. A. (1980), "Fatigue-Life Enhancement of Steel by Nitrogen Implantation," *Radiat. Effects*, Vol. 49, pp. 71–72.

Hubler, G. K. (1987), "Surface Alloying by Ion Beams: Mechanical Effects," in *Surface Alloying by Ion, Electron and Laser Beams* (L. E. Rehn, S. T. Picraux, H. Wiedersich, eds.), pp. 287–324, American Society for Metals, Metals Park, Ohio.

———, and Smidt, F. A. (1985), "Application of Ion Implantation to Wear Protection of Materials," *Nucl. Instrum. Meth. Phys. Res.*, Vol. B 7/8, pp. 151–157.

———, Holland, O. W., Clayton, C. R., and White, C. W. (eds.) (1984), "Ion Implantation and Ion Beam Processing," *Material Res. Soc. Symp. Proc.*, Vol. 27, Elsevier Science, New York.

Hutchings, R., and Oliver, W. C. (1983), "A Study of the Improved Wear Performance of Nitrogen-Implanted Ti6Al4V," *Wear*, Vol. 92, pp. 143–153.

Itoh, A., Hioki, T., and Kawamoto, J. (1989), "Tribological Properties of Ion Implanted SiC Ceramics," *Nucl. Instrum. Meth. Phys. Res.*, Vol. B 37/38, pp. 692–695.

Iwaki, M. (1983), "Surface Layer Characteristics of Ion Implanted Metals," *Thin Solid Films*, Vol. 101, pp. 223–231.

——— (1987), "Tribological Properties of Ion Implanted Steels," *Mater. Sci. Eng.*, Vol. 90, pp. 263–271.

Jata, K. V., Starke, E. A. Jr. (1983), "Surface Modification by Ion Implantation-Effects on Fatigue, *J. Metals.*, August, Vol. 35, pp. 23–27.

———, Han, J., Starke, E. A. Jr., Legg, K. O. (1983), "Ion Implantation Effect on Fatigue Crack Initiation in Ti-24V," *Scr. Metall.*, Vol. 17, pp. 479–483.

Jones, Jr., W. R., and Ferrante, J. (1983), "Tribological Characteristics of Nitrogen (N^+) Implanted Iron," *ASLE Trans.*, Vol. 26, pp. 351–359.

Kanaya, K., Koga, K., and Toki, K. (1972), "Extraction of High Current Ion Beams with Laminated Flow," *J. Phys. E. Instrum.*, Vol. 5, pp. 641–648.

Kant, R. A., Hirvonen, J. K., Kundson, A. R., and Wollam, J. S. (1979a), "Surface Hardening of Beryllium by Ion Implantation," *Thin Solid Films*, Vol. 63, pp. 27–30.

———, Myers, S. M., and Picraux, S. T. (1979b), "Experimental Study of Precipitation in an Ion-Implanted Metal: Sb in Al," *J. Appl. Phys.*, Vol. 50, pp. 214–222.

———, and Sartwell, B. D. (1984), "Surface Modification by Ion Beam Enhanced Deposition," in *Ion Implantation and Ion Beam Processing of Materials* (G. K. Hubler, O. W. Holland, C. R. Claytons, and C. W. White, eds.), Materials Res. Soc. Symp. Proc., Vol. 27, pp. 525–530, North Holland, New York.

Kao, P. W., and Byrne, J. G. (1980), "Ion Implantation Effects on Fatigue and Surface Hardness," *Fatigue Eng. Mater. Struct.*, Vol. 3, no. 3, pp. 271–276.

Kaufman, H. R., Cuomo, J. J., and Harper, J. M. E. (1982), "Technology and Application of Broad-Beam Ion Sources Used in Sputtering Part I: Ion Source Technology," *J. Vac. Sci. Technol.*, Vol. 21, pp. 725–736.

Kaufmann, E. N., Vianden, R., Chelikowsky, J. R., and Phillips, J. C. (1977), "Extension of Equilibrium Formation Criteria to Metastable Microalloys," *Phys. Rev. Lett.*, Vol. 39, pp. 1671–1676.

Keller, J. H. (1981), "Beam Optics Design for Ion Implantation," *Nucl. Instrum. Meth.*, Vol. 189, Part II, pp. 7–14.

———, and Robinson, J. W. (1987), "Ion Sources and Implantation Systems," *Mater. Sci. Eng.*, Vol. 90, pp. 423–432.

Kluge, A., Langguth, K., Ochsner, R., Kobs, K., and Ryssel, H. (1989), "A Comparison of Wear Behaviour of Ag, B, C, N, Pb and Sn Implanted Steels with 1.5% to 18% Chromium," *Nucl. Instrum. Meth. Phys. Res.*, Vol. B39, pp. 531–534.

Kobs, K., Dimigen, H., Hubsch, H., Tolle, H. J., Leutenecker, R., and Ryssel, H. (1987), "Enhanced Endurance Life of Sputtered MoS_x Films on Steel by Ion Beam Mixing," *Mater. Sci. Eng.*, Vol. 90, pp. 281–286.

Kujore, A., Chakraborty, S. B., Starke, E. A. J., Legg, K. O. (1981), "The Effect of Ion Implantation on the Fatigue Properties of Polycrystalline Copper," *Nucl. Instrum. Meth.*, Vol. 182/183, pp. 949–958.

Kustas, F. M., and Misra, M. S. (1984), "Application of Ion Implantation to Improve the Wear Resistance of 52100 Bearing Steel," *Thin Solid Films*, Vol. 122, pp. 279–286.

Lankford, J., Wei, W., and Kossowsky, R. (1987), "Friction and Wear Behaviour of Ion Beam Modified Ceramics," *J. Mater. Sci.*, Vol. 22, pp. 2069–2078.

Lee, E. H., Blau, P. J., Lewis, M. B., Allen, W. R., and Mansur, L. K. (1990), "Improved Surface Properties of Polymer Materials by Multiple Ion-Beam Treatments," *Proc. Surface and Near-Surface Structure of Polymer Interfaces*, Materials Research Society, Pittsburgh, Pa.

Legg, K. O., Cochran, J. K., Solnick-legg, H. F., and Mann, X. L. (1985), "Modification of Surface Properties of Yttria Stabilized Zirconia by Ion Implantation," *Nucl. Instrum. Meth. Phys. Res.*, Vol. B7/8, pp. 535–540.

Lempert, G. D. (1988), "Practical Application of Ion Implantation for Modifying Tribological Properties of Metals," *Surf. Coat. Technol.*, Vol. 34, no. 2, pp. 185–206.

Liau, Z. L., and Mayer, J. W. (1978), "Limit of Composition Achievable by Ion Implantation," *J. Vac. Sci. Technol.*, Vol. 15, pp. 1629–1635.

Lindhard, J., Scharff, J. M., and Schiott, H. E. (1963), *Kgl Danske Vid. Selsk. Matt. Fys. Medd.*, Vol. 33, no. 14.

Liu, B. X., Johnson, W. L., Nicolet, M.-A., and Lau, S. S. (1983), "Amorphous Film Formation by Ion Beam Mixing in Binary Metal Systems," *Nucl. Instrum. Meth.*, Vol. 209/210, pp. 229–234.

Longworth, G., and Hartley, N. E. W. (1977), "Mössbauer, Effect Study of Nitrogen-Implanted Iron Foils," *Thin Solid Films*, Vol. 48, pp. 95–104.

LoRusso, S., Mazzoldi, P., Scotoni, I., Tosello, C., and Tosto, S. (1980), "Fatigue Life Improvements by Nitrogen Ion Implantation on Steel: Dose Dependence," *Appl. Phys. Lett.*, Vol. 36, pp. 822–823.

Madakson, P. B. (1987), "Mechanical Properties of Nitrogen Implanted 18W-4Cr-1V Bearing Steel," *Mater. Sci. Eng.*, Vol. 90, pp. 287–290.

———, and Smith, A. A. (1983), "Friction and Wear of Ion Implanted Aluminum," *Nucl. Instrum. Meth.*, Vol. 209/210, pp. 983–988.

Matteson, S., Roth, J., and Nicolet, M.-A. (1979), "Ion-Induced Silicide Formation in Niobium Thin Films," *Radiat. Effects*, Vol. 42, pp. 217–225.

Mayer, J. W., Ericksson, L., and Davies, J. A. (1970), *Ion Implantation in Semiconductors: Silicon and Germanium*, Academic, New York.

———, Lau, S. S., Tsaur, B. Y., Poate, J. M., and Hirvonen, J. K. (1980), "High-Dose Implantation and Ion-Beam Mixing," in *Ion Implantation Metallurgy* (C. M. Preece and J. K. Hirvonen, eds.), pp. 37–46, Metallurgical Society of AIME, Warrendale, Pa.

———, Tsaur, B. Y., Lau, S. S., and Hung, L. S. (1981), "Ion-Beam Induced Reactions in Metal-Semiconductor and Metal-Metal Thin Film Structures," *Nucl. Instrum. Meth.*, Vol. 182/183, pp. 1–13.

McHargue, C. J. (1987), "Structure and Mechanical Properties of Ion Implanted Ceramics," *Nucl. Instrum. Meth. Phys. Res.*, Vol. B19/20, pp. 797–804.

——— (1988), "The Mechanical Properties of Ion Implanted Ceramics—A Review," *Defects Diffus. Forum*, Vol. 57/58, pp. 359–380.

———, and Yust, C. S. (1984), "Lattice Modification in Ion-Implanted Ceramics," *J. Am. Ceramic Soc.*, Vol. 67, pp. 117–123.

———, Lewis, M. B., Appleton, B. R., Naramoto, H., White, C. W., and Williams, J. M. (1983), "Alteration of Surface Properties by Ion Implantation," *Proc. 1st Int. Conf. Science of Hard Materials* (R. K. Viswanadham, D. J. Rowcliffe, and J. Gurland, eds.), pp. 451–465, Plenum, New York.

———, White, C. W., Appleton, B. R., Farlow, G. C., and Williams, J. M. (1984), "Ion Beam Modifications of Ceramics," in *Ion Implantation and Ion Beam Processing of Materials* (G. K. Hubler, O. W. Holland, C. R. Clayton, and C. W. White, eds.), *Materials Res. Soc. Symp. Proceeding*, Vol. 27, pp. 385–394, North Holland, New York.

———, Farlow, G. C., White, C. W., Williams, J. M., Appleton, B. R., and Naramoto, H. (1985), "The Amorphization of Ceramics by Ion Beams," *Mater. Sci. Eng.*, Vol. 69, pp. 123–127.

———, Farlow, G. C., White, C. W., Appleton, B. R., Williams, J. M., Sklad, P. S., Angelini, P., and Yust, C. S. (1986a), in *Application of Ion Plating and Ion Implantation to Materials* (R. Hochman, ed.), pp. 255–266, American Society for Metals, Metals Park, Ohio.

———, Yust, C. S., Angelini, P., Sklad, P. S., and Lewis, M. B. (1986b), "The Surface Mechanical Properties and Wear Behaviour of Ion Implanted TiB_2," in *Science of Hard Materials* (E. A. Almond, C. A. Brooks, and R. Warrens, eds.), pp. 803–812, Inst. Physics Conf. Series 75, Adam Hilger Ltd., Bristol, England.

Mendez, J., Violan, P., and Villain, J. P. (1982), "Modifications in Fatigue Surface Damage by Ion Implantation in Polycrystalline Copper," *Scr. Metall.*, Vol. 16, pp. 179–182.

Mikkelsen, N. J., and Sørensen, G. (1989a), "Modification of Molybdenum-Disulphide Films by Ion-Bombardment Techniques," *Proc. New Materials Approaches to Tribology: Theory and Applications* (L. E. Pope, L. Fehrenbacher, and W. O. Winer, eds.), MRS Symp. No. 140, pp. 265–269, Materials Research Society, Pittsburgh, Pa.

———, and ——— (1989b), "Ion-Beam Modification of MoS_x Films on Metals," *Mater. Sci. Eng.*, Vol. A115, pp. 343–347.

Mikkelsen, N. J., Chevallier, C., Sørensen, G., and Straede, C. A. (1988), "Friction and Wear Measurements of Sputtered MoS_x Films Amorphized by Ion Bombardment," *Appl. Phys. Lett.*, Vol. 52, pp. 1130–1132.

Morehead, F. F., Jr., and Crowder, B. L. (1973), "Ion Implantation," *Sci. Am.*, Vol. 228, pp. 65–71.

Myers, S. M. (1978), "Annealing Behaviour and Selected Applications of Ion-Implanted Alloys," *J. Vac. Sci. Technol.*, Vol. 15, pp. 1650–1665.

——— (1980), "Properties and Applications of Ion-Implanted Alloys," *J. Vac. Sci. Technol.*, Vol. 17, pp. 310–314.

Nastasi, M., Kossowsky, R., Hirvonen, J. P., Elliott, N. (1988), "Friction and Wear Studies in N-Implanted Al_2O_3, SiC, TiB_2 and B_4C Ceramics," *J. Mater. Res.*, Vol. 3, no. 6, pp. 1127–1133.

Nelson, R. S. (1973), "The Physical State of Ion Implanted Solids," in *Ion Implantation* (G. Dearnaley, J. H. Freeman, R. S. Nelson, and J. Stephan, eds.), pp. 154–254, North Holland, Amsterdam.

Nicolet, M.-A., Banwell, T. C., and Paine, B. M. (1984), "Ion-Mixing Processes," in *Ion-Implantation and Ion Beam Processing of Materials* (G. K. Hubler, O. W. Holland, C. R. Clayton, and C. W. White, eds.), *Materials Res. Soc. Symp. Proceeding*, Vol 27, pp. 3–12, North Holland, New York.

Noda, S., Doi, H., Hioki, T., Kawamoto, J-I., Kamigaito, O. (1987), "Sialon Formation by Si^+ and N_2^+ Ion Implantation into Sapphire," *J. Mater. Sci.*, Vol. 22, pp. 4267–4273.

Oblas, D. W. (1984), "The Characterization of Ion-Implanted WC/Co," in *Ion Implantation and Ion Beam Processing of Materials* (G. K. Hubler, O. W. Holland, C. R. Clayton, and C. W. White, eds.), *Materials Res. Soc. Symp. Proc.*, Vol. 27, pp. 631–636, North Holland, New York.

Oliver, W. C., Hutchings, R., Pethica, J. B. (1984), "The Wear Behaviour of Nitrogen-Implanted Metals," *Metall. Trans.*, Vol. 15A, pp. 2221–2229.

Padmanabhan, K. R., and Sørensen, G. (1981), "The Effect of Ion Implantation of rf Sputtered TiB_2 Films," *Thin Solid Films*, Vol. 81, pp. 13–19.

———, Hsieh, Y. F., Chevallier, J., and ——— (1983), "Modifications to the Microhardness, Adhesion, and Resistivity of Sputtered TiN Films by Ion Implantation," *J. Vac. Sci. Technol.*, Vol. A1, pp. 279–283.

———, Chevallier, J., and Sørensen, G. (1986), "The Influence of Ion Bombardment on Deposition of Carbon Films," *Nuc. Instrum. Meth. Phys. Res.*, Vol. B16, pp. 369–372.

Pavlov, P. V. (1983), "Implantation Phases in Solids," *Nucl. Instrum. Meth.*, Vol. 209/210, pp. 791–794.

Peckner, D. (ed.) (1964), *The Strengthening of Metals*, Reinhold, New York.

Pethica, J. B., Hutchings, R., and Oliver, W. C. (1983), "Composition and Hardness Profiles in Ion Implanted Metals," *Nucl. Instrum. Meth.*, Vol. 209/210, pp. 995–1000.

Picraux, S. T. (1980a), "Equilibrium Phase Formation by Ion Implantation," in *Site Characterization and Aggregation of Implanted Atoms in Materials* (A. Perez and R. Coussement, eds.), pp. 307–324, Plenum, New York.

——— (1980b), "Formation of Nonequilibrium System by Ion Implantation," in *Site Characterization and Aggregation of Implanted Atoms in Materials* (A. Perez and R. Coussement, eds.), pp. 325–337, Plenum, New York.

——— (1984a), "Ion Implantation in Metals," *Ann. Rev. Mater. Sci.*, Vol. 14, pp. 335–372.

——— (1984b), "Physics of Ion-Implantation," in *Surface Engineering: Surface Modification of Materials* (R. Kossowsky and S. C. Singhal, eds.), pp. 3–31, Martinus Nijhoff Publishers, Dordrecht, The Netherlands.

———, and Choyke, W. J. (eds.) (1982), *Metastable Materials Formation by Ion Implantation*, North Holland, New York.

———, EerNisse, E. P., and Vook, F. L. (eds.) (1974), *Application of Ion-Beams to Metals*, Plenum, New York.

Poate, J. M. (1978), "Metastable Alloy Formation," *J. Vac. Sci. Technol.*, Vol. 15, pp. 1636–1643.

——— (1985), "Ion Implantation," in *Treatise on Heavy Ion Sources*, Vol. 6 (D. A. Bromley, ed.), pp. 133–166, Plenum, New York.

———, and Cullis, A. G. (1980), "Implantation Metallurgy—Metastable Alloy Formation," in *Treatise in Materials Science and Technology* Vol. 18: *Ion Implantation* (J. K. Hirvonen, ed.), pp. 85–133, Academic, New York.

———, Borders, J. A., ———, and Hirvonen, J. K. (1977), "Ion Implantation as an Ultrafast Quenching Technique for Metastable Alloy Production: The Ag-Cu System," *Appl. Phys. Lett.*, Vol. 30, pp. 365–368.

Roberts, S. G., and Page, T. F. (1982), "The Effects of N_2^+ Ion Implantation on the Hardness and Wear Behaviour of Brittle Materials," in *Ion Implantation into Metals* (V. Asworth, W. A. Grant, and R. P. M. Procter, eds., pp. 135–146), Pergamon, Oxford.

———, and ——— (1986), "The Effects of N_2^+ and B^+ Ion Implantation on the Hardness Behaviour and Near Surface Structure of SiC," *J. Mater. Sci.*, Vol. 21, pp. 457–468.

Ryding, G., Wittkower, A. B., and Rose, P. H. (1976), "Features of a High Current Implanter and a Medium Current Implanter," *J. Vac. Sci. Technol.*, Vol. 13, pp. 1030–1036.

Ryssel, H., and Glawischnig, H. (eds.) (1982), "Ion Implantation Technique," *Springer Ser. in Electrophys., Vol. 10,* Springer Verlag, Berlin, Germany.

———, and ——— (eds.) (1983), "Ion Implantation: Equipment and Techniques," *Springer Series in Electrophys. Vol. 11*, Springer-Verlag, Berlin, Germany.

Saritas, S., Procter, R. P. M., Ashworth, V., and Grant, W. A. (1982), "The Effect of Ion Implantation on the Friction and Wear Behaviour of a Phosphor Bronze," *Wear*, Vol. 82, pp. 233–255.

———, Proctor, R. P. M., and Grant, W. A. (1987), "The Effect of Ion Implantation to Modify the Tribological Properties of Ti 6Al4V Alloy," *Mater. Sci. Eng.*, Vol. 90, pp. 297–306.

Seitz, F., and Koehler, J. S. (1956), "Displacement of Atoms during Irradiation" in *Solid State Physics—Advances in Research and Applications*, Vol. 2 (F. Seitz and D. Turnbull, eds.), pp. 305–448, Academic, New York.

Shimura, H., and Tsuya, Y. (1977), "Effects of Atmosphere on the Wear Rate of Some Ceramics and Cermets," in *Proc. of Int. Conf. on Wear of Materials* (W. A. Glaeser, K. C. Ludema, and S. K. Rhee, eds.), pp. 452–461, ASME, New York.

Shreter, U., So, F. C. T., Paine, B. M., and Nicolet, M.-A. (1984), "Investigation of a Thermal Spike Model for Ion Mixing of Metals with Si," in *Ion Implantation and Ion Beam Processing of Materials* (G. K. Hubler, O. W. Holland, C. R. Clayton, and C. W. White, eds.), *MRS Symp. Proc.*, Vol. 27, pp. 31–36, North Holland, New York.

Shrivastava, S., Jain, A., Sethuramiah, A., Vankar, V. D., Chopra, K. L. (1987), "Wear Behaviour of Nitrogen Implanted Stainless Steel," *Nucl. Instrum. Meth. Phys. Res.*, Vol. B21, pp. 591–594.

Sidenius, G. (1978), "Gas and Vapour Ion Sources for Low-Energy Accelerators," *Conf. Ser. No. 38*, pp. 1–11, Institute of Physics, London.

Singer, I. (1984), "Tribomechanical Properties of Ion Implanted Metals," *Ion Implantation and Ion Beam Processing of Materials* (G. K. Hubler, O. W. Holland, C. R. Clayton, and C. W. White, eds.), *MRS Symp. Proc.*, Vol. 27, pp. 585–595, North Holland, New York.

———, Bolster, R. N., and Carosella, C. A. (1980), "Abrasive Wear Resistance of Titanium and Nitrogen-Implanted 52100 Steel Surfaces," *Thin Solid Films*, Vol. 73, pp. 283–289.

———, Carosella, C. A., and Reed, J. R. (1981), "Friction Behaviour of 52100 Steel Modified by Ion Implanted Ti," *Nucl. Instrum. Meth.*, Vol. 182–183, pp. 923–932.

Sioshansi, P. (1989), "Surface Modification of Industrial Components by Ion Implantation," *Nucl. Instrum. Meth.*, Vol. B37/38, pp. 667–671.

Sood, D. K., and Dearnaley, G. (1976), "Ion-Implanted Surface Alloys in Copper and Aluminum," in *Applications of Ion Beams to Materials* (G. Carter, J. S. Colligen, and W. A. Grant, eds.), *Conf. Series No. 28*, pp. 196–203, The Institute of Physics, London.

Spalvins, T. (1978), "Coatings for Wear and Lubrication," *Thin Solid Films*, Vol. 53, pp. 285–300.

Spitznagel, J. A., Wood, S., Choyke, W. J., Doyle, N. J., Bradshaw, J., and Fishman, S. G. (1986), "Ion Beam Modification of 6H/15R SiC Crystals," *Nucl. Instrum. Meth. Phys. Res.*, Vol. B 16, pp. 237–243.

Stephens, K. G. (1984), "An Introduction to Ion Sources," in *Ion Implantation Science and Technology* (J. F. Ziegler, ed.), pp. 375–432, Academic, New York.

Straede, C. A. (1989), "Practical Applications of Ion Implantation for Tribological Modifications of Surfaces," *Wear*, Vol. 130, pp. 113–122.

Suri, A. K., Nimmagadda, R., and Bunshah, R. F. (1979), "Influence of Ion Implantation and Overlay Coatings on Various Physico-mechanical and Wear Properties of Stainless Steel, Titanium and Aluminum," *Thin Solid Films*, Vol. 64, pp. 191–203.

Takano, I., Isobe, S., and Takemoto, M. (1987), "Properties of TiN Thin Films on Several Substrate Plates with a High Current Ion Mixing Machine," *J. Vac. Sci. Technol.*, Vol. A5, pp. 2191–2193.

Tosto, S. (1983), "Wear Behaviour of 38NCD4 N-ion Implanted Steel," *J. Mater. Sci.*, Vol. 18, pp. 899–902.

Townsend, P. D., Kelly, J. C., and Hartley, N. E. W. (1976), *Ion Implantation, Sputtering, and Their Applications*, Academic, London.

Tsaur, B. Y. (1980), "Ion-Beam-Induced Interface Mixing and Thin-Film Reactions," in *Proc. Electrochem. Soc. Symp. on Thin Film Interfaces and Interactions*, Vol. 80–2, pp. 205–231, The Electrochemical Society, Princeton, N.J.

———, Liau, Z. L., and Mayer, J. W. (1979a), "Formation of Si-Enriched Metastable Compounds in the Pt-Si System Using Ion Bombardment and Post Annealing," *Phy. Lett.*, Vol. 71A, pp. 270–272.

———, Lau, S. S., ———, and ——— (1979b), "Ion-Beam-Induced Intermixing of Surface Layers," *Thin Solid Films*, Vol. 63, pp. 31–36.

———, ———, Hung, L. S., and Mayer, J. W. (1981), "Microalloying by Ion Beam Mixing," *Nucl. Instrum. Meth.*, Vol. 182/183, pp. 67–77.

Valori, R., Popgoshev, D., and Hubler, G. K. (1983), "Ion Implantation Bearing Surfaces for Corrosion Resistance," *J. Lub. Technol.* (*Trans. ASME*), Vol. 105, pp. 534–541.

Vardiman, R. G., and Kant, R. A. (1982), "The Improvement of Fatigue Life in Ti6Al4V by Ion Implantation," *J. Appl. Phys.*, Vol. 53, pp. 690–694.

———, and Kox, J. E. (1985), "A Study of the Mechanisms of Fatigue Life Improvement in an Ion Implanted Nickel-Chromium Alloy," *Acta. Metall.*, Vol. 33, no. 11, pp. 2033–2039.

Venkatesan, T. (1985), "High Energy Ion Beam Modification of Polymer Films," *Nucl. Instrum. Meth. Phys. Res.*, Vol. B 7/8, pp. 461–467.

Vogel, F. L., Jr. (1975), "Implantation of Carbon into Thin Iron Films," *Thin Solid Films*, Vol. 27, pp. 369–376.

Wang, Y. F., Clayton, C. R., Hubler, G. K., Lucke, H. W., and Hirvonen, J. K. (1979), "Application of Ion Implantation for the Improvement of Localized Corrosion Resistance of M50 Bearing Steel," *Thin Solid Films*, Vol. 63, pp. 11–18.

Watanabe, M., Shimura, H., and Enomoto, Y. (1989), "Effect of Nitrogen Ion Implantation on the Friction and Wear Properties of Some Plastics," in *Mechanics of Coatings* (D. Dowson, C. M. Taylor, and M. Godet, eds.), *Proc. 16th Leeds-Lyon Symp. on Tribology*, Elsevier Science Publishers, Amsterdam, The Netherlands.

Wei, W., and Lankford, J. (1987), "Characterization of Ion Beam Modified Ceramic Wear Surfaces Using Auger Electron Spectroscopy," *J. Mater. Sci.*, Vol. 22, pp. 2387–2396.

———, ———, and Kossowsky, R. (1987), "Friction and Wear of Ion-Beam Modified Ceramics for Use in High Temperature Adiabatic Engines," *Mater. Sci. Eng.*, Vol. 90, pp. 307–315.

White, G., and Dearnaley, G. (1980), "The Influence of N_2^+ Ion Implantation on Rolling Contact Fatigue Performance," *Wear*, Vol. 64, pp. 327–332.

Yust, C. S., and McHargue, C. J. (1984), "Microstructural and Mechanical Properties of Ion Implanted Ceramics," in *Emergent Process Methods for High Technology Ceramics* (R. F. Davis, M. Palmour, and R. L. Porter, eds.), pp. 533–547, Plenum, New York.

Yust, F. G., Pope, L. E., Follstaedt, D. M., Knapp, J. A., and Picraux, S. T. (1982), "Friction and Wear of Stainless Steel Implanted with Ti and C," in *Metastable Material Formation by Ion Implantation* (S. T. Picraux and W. J. Choyke, eds.), Elsevier, Amsterdam.

———, Picraux, S. T., Follstaedt, D. M., Pope, L. E., and Knapp, J. A. (1983), "The Effect of N^+ Implantation on the Wear and Friction of Type 304 and 15-5PH Stainless Steel," *Thin Solid Films*, Vol. 107, pp. 287–295.

Ziegler, J. F. (1984a), *Ion Implantation Science and Technology*, Academic, New York.

——— (1984b), "The Stoping and Range of Ions in Solids," in *Ion Implantation Science and Technology* (J. F. Ziegler, ed.), pp. 51–108, Academic, New York.

———, and Brown, R. L. (eds.) (1985), *Proc. 5th Int. Conf. on Ion Implantation: Equipment and Techniques, Nucl. Instrum. Meth. Phys. Res.*, Vol. B6, pp. 1–585.

CHAPTER 13
SOFT COATINGS

Soft lubricant coatings provide low friction and low wear and have been successfully used at relatively low loads and low speeds and typically from cryogenic temperatures to high temperatures in air, hydrogen, helium, and vacuum environments (Anonymous, 1971, 1972, 1978, 1984a; Benzing, 1964; Bhushan, 1978, 1980a, 1987a, 1987b; Bisson and Anderson, 1964; Braithwaite, 1964, 1967; Booser, 1983; Bowden and Tabor, 1950, 1964; Clauss, 1972; Campbell, 1972; Campbell et al., 1966; Kanakia and Peterson, 1987; Lansdown, 1974, 1976, 1979; McConnell, 1972; Lipp, 1976; Peterson and Winer, 1980; Sliney, 1982; Spalvins, 1987). In many instances overlays of soft lubricant materials (graphite, MoS_2, PTFE, Au, Ag, etc.) are applied onto hard coatings to provide protection during running-in and to prolong life (Bhushan, 1980a, 1980b, 1981a, 1981b). The mating part sliding against a part coated with soft coating should be smooth and free of burrs that might cut or abrade the coating on the mating part. The effectiveness of many soft coatings as a lubricant (reduction of its shear strength) is associated with the formation of adherent transfer films on the mating surfaces.

The soft lubricant coatings include: conventional layered lattice compounds such as graphite, graphite fluorides, and MoS_2; nonlayered lattice compounds such as $PbO\text{-}SiO_2$, CaF_2, BaF_2, and $CaF_2\text{-}BaF_2$ eutectics; polymers such as polytetrafluoroethylene (PTFE), polyphenylene sulfide (PPS), poly(amide-imide) and polyimide and metallic coatings such as Au, Ag, Cu, Pb, Sn, In, and Al and their alloys. Table 13.1 summarizes the operating temperature range of various soft lubricant coatings and commonly used techniques for their application. According to some estimates, MoS_2, PTFE, and graphite (MoS_2 and PTFE in particular) continue to represent over 90 percent of the total solid film lubricant usage at this time.

The maximum temperature for lubrication with graphite and MoS_2 in air atmosphere is limited to oxidation to about 430°C and 315°C, respectively. At low-humidity conditions including vacuums, the friction and wear properties of MoS_2 coatings are superior to that of graphite coatings. Bonded coatings of graphite and MoS_2 are extensively used to achieve longer wear lives and high load capacities (Bhushan, 1982; Clauss, 1972; Fusaro and Sliney, 1973; Willis, 1980). Sputtered MoS_2 is commonly used for space applications requiring sliding operation in a vacuum (Bhushan, 1980b; Buckley, 1971; Spalvins, 1971a, 1971b). Nonlayered lattice compounds such as CaF_2, BaF_2, and $CaF_2\text{-}BaF_2$ eutectic-based coatings have been used for high-temperature sliding applications (up to 900°C) under very heavy loads (Sliney, 1982). Fluoride-based coatings are generally deposited by spray fusion or plasma spraying processes (Sliney, 1979, 1982).

There are a number of polymer coatings such as PTFE, polyphenylene sulfide,

TABLE 13.1 Details of Commonly Used Soft Solid Lubricant Coatings

Material	Common deposition technique(s)	Operating temperature range, °C		Typical coefficient of friction		Remarks
		In air	In N_2 or vacuum	In air	In N_2 or vacuum	
Graphite	Air spray (resin bonded)	−240 to 430	Unstable in vacuum	0.1–0.2	0.1–0.5	Most effective in the presence of humidity (air); high friction and wear in vacuum; electrically conductive.
Graphite-CdO-Ag (HL-800-2)	Air spray (sodium silicate bonded)	−240 to 430	—	0.1–0.2	—	Lower friction and wear than graphite alone in operating temperature regime.
Graphite fluoride $(CF_x)_n$	Air spray (polyimide resin bonded)	−240 to 430	—	0.05–0.2	—	Lower friction and wear than graphite.
MoS_2 (Surf-kote 1284)	Air spray (resin bonded)	−240 to 315	−240 to 500	0.05–0.2	0.05–0.15	Very low friction and wear in vacuum and low humidity. Can promote metal corrosion at high humidity.
MoS_2	Sputtered	−240 to 315	−240 to 500	0.05–0.1	0.03–0.1	Lower friction and wear in vacuum than of resin bonded coating.
MoS_2-graphite-Au (MLF-5)	Air spray (sodium silicate bonded)	−240 to 430	−240 to 500	0.05–0.2	0.03–0.1	Low friction and wear at all humidities and vacuum.
MoS_2-graphite (Acheson)	Air spray (resin bonded)					
WS_2	Air spray (resin bonded)/ sputtered	−240 to 500	−240 to 700	0.1–0.2	0.05–0.15	Friction not as low as MoS_2, temperature capability in air and vacuum a little higher.
$WSe_2/NbSe_2$	Air spray (resin bonded)/ sputtered	−240 to 315	−240 to 1200	0.12–0.3	0.07–0.15	Temperature capability in vacuum higher than MoS_2.
$PbO-SiO_2-Ag$	Fusion	260 to 650	260 to 650	0.2–0.4	—	Very high load capacity, high-temperature lubricant, abrasive to metal surfaces.

CaF_2-BaF_2 eutectic-gold (AFSL-28)	Fusion (aluminum phosphate bonded)	260 to 650	260 to 650	0.2–0.3	0.2–0.3	Used for slow-speed applications, lower friction at high temperatures.
CaF_2-glass-Ag-nichrome (NASA PS 101)	Plasma spray (nichrome bonded)	20 to 900	20 to 900	0.2–0.3	0.2–0.3	Used for slow-speed applications.
CaF_2-Tribaloy 400-Ag (NASA PS 106T)	Plasma spray (Tribaloy 400 bonded)	20 to 500	20 to 500	0.2–0.4	0.2–0.4	Used for moderate sliding applications.
CaF_2-BaF_2 eutectic-Cr_3C_2-Ag-NiAl (NASA PS 212)	Plasma spray (NiAl bonded)	20 to 750	20 to 750	0.2–0.4	0.2–0.4	Good wear resistance, used for high-speed sliding applications.
PTFE	Air spray (resin bonded)/ sputtered	−240 to 170	− 240 to 170	0.03–0.1	0.03–0.1	Lowest friction of solid lubricants, sputtered PTFE used as dielectric layer in capacitors.
PTFE-PPS (Emralon 341)	Air spray (resin bonded)	−240 to 200	− 240 to 200	0.05–0.15	0.05–0.15	Moderate load capacity.
PTFE-poly(amide-imide) (Emralon 333)	Air spray (resin bonded)	−240 to 200	− 240 to 200	0.05–0.15	0.05–0.15	Moderate load capacity.
Polyimide (DuPont PI 4701)	Air spray	−240 to 300	− 240 to 300	0.1–0.2	0.1–0.2	Higher-temperature capability, moderate load capacity.
Au, Ag	Electrochemical/ sputtering/ion plating	−240 to 600	− 240 to 600	0.1–0.25	0.1–0.25	Low load capacity, commonly used in vacuum.
Pb, Pb-Sn-Cu	Electrochemical/ sputtering/ion plating	−240 to 200	− 240 to 200	0.1–0.25	0.1–0.25	Low load capacity, commonly used in vacuum.

poly(amide-imide), polyimide, and their blends in a resin binder and solvent system. Soft metallic coatings such as Au, Ag, Cu, Sn, Pb, In, and Al and their alloys are good lubricants at room and elevated temperatures (Anonymous, 1982; Buck, 1987; Gerkema, 1985; Inwood and Garwood, 1978; Sherbiney and Halling, 1977; Spalvins, 1978, 1987). These soft metallic coatings exhibit low coefficients of friction both in air and vacuum because of their intrinsic low shear strength. Metallic coatings are generally deposited by electrochemical deposition, sputtering, and ion-plating processes.

The friction and wear characteristics of various commonly used soft coatings are discussed in this chapter.

13.1 LAYERED LATTICE SOLID COATINGS

Thin coatings of layered lattice solids such as graphite, graphite fluoride, and dichalcogenides (for example, MoS_2 and WS_2) exhibit low coefficients of friction (0.05 to 0.1) because of their layered lattice. Most of these materials have a hexagonal layered structure (Chap. 5), and their shear properties are anisotropic with preferred planes for easy shear parallel to the basal planes of the crystallites. Common techniques used to deposit coatings are by burnishing solid lubricant particles on a substrate, atomized liquid spray, and sputtering. The choice of deposition technique is influenced by the requirements of the coating such as coating thickness, relative speed, applied load, temperature, and the environment during operation.

Layer lattice coatings of graphite and MoS_2, by itself or in combination with additives, are most commonly used in various applications requiring low friction and wear resistance. Graphite-based bonded coatings can be used up to 430°C; these coatings work best in the presence of some humidity and cannot be used in dry or vacuum environments. On the other hand, a low shear strength is intrinsic to the pure MoS_2 materials. MoS_2 performs satisfactorily regardless of the presence of humidity and/or contaminants including in vacuum. In air, MoS_2 oxidizes and can be used up to 315°C; however, in inert atmosphere (e.g., N_2 or vacuum) MoS_2 can be used up to a maximum temperature of about 500°C. Carbon fluoride $(CF_x)_n$ is another intercalation compound which shows a promise with friction and wear performance better than graphite up to 430°C. MoS_2 exhibits the lowest friction of the three lubricant compounds mentioned here; a relatively small concentration of graphite is added to MoS_2 to improve its low load performance.

13.1.1 Graphite

Graphite, having a hexagonal layered structure of carbon atoms, develops good lubricity in the presence of absorbed vapors such as water or hydrocarbons (for details see Chap. 5). The graphite coatings are usually applied by burnishing the graphite powder on a substrate or by atomized liquid spray. The effectiveness of the graphite for lubrication is associated with the formation of adherent transfer film on the mating surfaces. The presence of some oxides or salts improves the lubrication of graphite films over a wide temperature range (Peterson and Johnson, 1956; Sliney, 1966a). These salts also react chemically or alloy at their interface with the mating metal surface, which results in improved adherence.

Friction tests in the temperature range of room temperature to 540°C of powders of graphite and graphite mixed with several soft metallic salts and oxides (CdO, PbO, Na_2SO_4, and $CdSO_4$) were conducted by Peterson and Johnson (1956). They found that the graphite alone lubricated well at room temperature and above 425°C but not at the intermediate temperature [Fig. 13.1(*a*)]. This behavior is explained as follows:

- Adsorbed moisture at room temperature provides easy shear among graphite layers.
- At elevated temperatures, desorption of water takes place and makes the slip difficult.
- At temperatures higher than 430°C (high enough to promote oxidation of metal surfaces), the interaction of graphite with oxide film on the mating material probably leads to low friction.

Peterson and Johnson found that a mixture of graphite and cadmium oxide lubricated most effectively in the entire temperature range from room temperature to 540°C [Fig. 13.1(*b*)].

Bhushan (1980a, 1982) added Ag to CdO-graphite mixture to further improve its friction and wear performance in the entire temperature range from room temperature to 540°C and possibly to increase the maximum temperature limit. Silver has been proven to be a good lubricant in the temperature range from room tem-

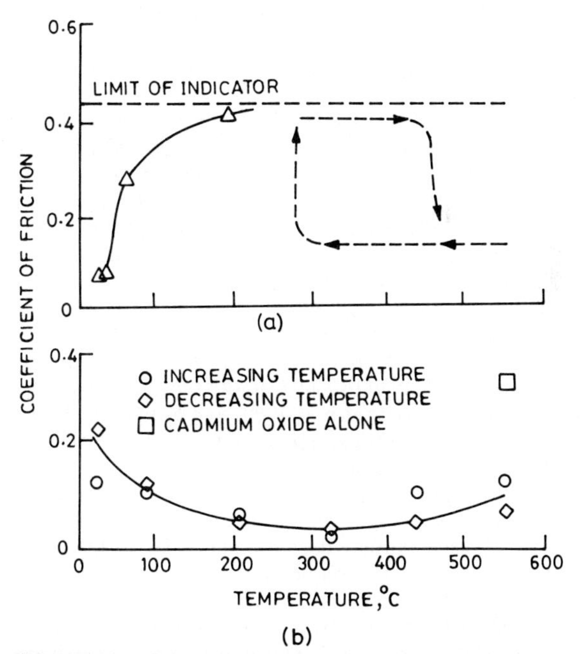

FIGURE 13.1 Effect of cadmium oxide adjuvants on the lubricity of graphite in a Inconel pin on Inconel disk sliding test at a sliding velocity of 29 mm s^{-1} and a normal load of 180 N: (*a*) Graphite alone. (*b*) Graphite-cadmium oxide (2:1) mixture. [*Adapted from Peterson and Johnson (1956).*]

perature to 260°C when added to fluoride coatings (Sliney, 1975). Silver has also been found to be a good lubricant at high temperatures in the range from 315 to 980°C by Peterson et al. (1959).

Graphite-based coatings are predominantly bonded coatings. Some attempts have been made to deposit them by physical vapor deposition processes (evaporation, ion plating).

13.1.1.1 ***Bonded Graphite Coatings.*** Bhushan (1980a) developed bonded coatings of CdO-graphite (HL-800) and CdO-graphite-Ag (HL-800-2) which were applied by air spraying techniques. Water glass or sodium silicate was used as the binder. The binder provides cohesive forces that hold the solid coating together and also provides adhesive forces that bond the coating to a substrate. Bhushan's experience showed that 3 parts graphite, 1 part CdO, and roughly 1 part Ag provide good lubricity. Higher contents of sodium silicate are not desirable as it is abrasive. Empirically, he found that about 30 percent by weight (water content not included) of sodium silicate gave adequate bonding for coating (Bhushan 1980c, 1981b, 1982; Bhushan and Gray, 1978, 1980; Bhushan et al., 1978). Composition of an optimum CdO-graphite-Ag coating (HL-800-2) is as follows: 15 percent by weight CdO (95 percent of the particle sizes finer than 200; 95 to 200), 45 percent by weight synthetic graphite (particle size of about 2 μm), 10 percent by weight of high purity (99.99 percent pure) and fine powder (95–230) of Ag, and 30 percent by weight sodium silicate (8.9 wt % Na_2O, 28.7 wt % SiO_2, and balance water) excluding water content. Water was used as the diluent, and a drop of wetting agent (Absol 895–NPX Industries) was added for good dispersion.

The coating solution is air sprayed and baked at 65°C for 2 h and then at 150°C for 8 h. The coating is then burnished using sandpapers to a roughness on the order of 25 to 100 nm rms. These coatings were applied onto foil (hydrodynamic) air bearing and journal for the automotive gas turbine engine and tested for several combinations of air sprayed CdO-graphite-Ag and Cr_3C_2 coating deposited by detonation gun (Bhushan, 1982). Bhushan reported that the CdO-graphite-Ag coating applied on a roughened Inconel X-750 foil surface (by electroetching) had better adherence than on a heat-treated foil with no further roughening (Chap. 7). Table 13.2 summarizes the results of start-stop tests carried out at 430°C and at a bearing normal stress of 14 kPa for several combinations of CdO-graphite, CdO-graphite-Ag, and detonation gun sprayed Cr_3C_2 on foil bearing and journal surfaces (Bhushan, 1979, 1982). Data for air sprayed MoS_2 are included for reference. The life of CdO-graphite coating can be improved by applying it on both sliding surfaces. The addition of Ag in CdO-graphite improves its wear performance and load-carrying capacity up to 430°C. Bhushan (1982) has reported that CdO-graphite-Ag can withstand high-speed (60 m s^{-1}) rub tests and has shown little wear up to an acceleration of 100 g.

Bhushan (1979) also start-stop tested foil surfaces coated with CdO-graphite and MoS_2 against plasma sprayed Cr_3C_2 and nitrided journal surfaces. He found that a smooth and polished journal surface does not scrape away the soft, solid lubricant coating from the mating surface; consequently the wear of the lubricant coatings rubbing against a smooth journal surface is generally better than a rough journal surface. However, an extremely smooth surface (<10 to 25 nm rms) may not allow formation of transfer film of the solid lubricant on the mating surface which may be undesirable for many soft solid lubricants (e.g., graphite, MoS_2, and PTFE).

Moisture could form in the bearing when operating under conditions of high humidity. In some applications, oil leaks could occur. These leaks could introduce oil droplets into the air and through the bearing. Solid lubricant coatings

TABLE 13.2 Start-Stop Test* Results for Air Sprayed Graphite-CdO Coating Combinations at a Maximum Temperature of 430°C

Test/foil journal coatings	Coatings thickness		Coefficient of kinetic friction		Surface roughness of journal (rms), μm		Bearing† normal load, kPa	Results and comments
	Foil	Journal	At start	At end	Before test	After test		
Air sprayed MoS_2 (Hohman, M1284)/detonation gun Cr_3C_2‡	7.5–10	63–88	0.13	0.21	0.48	0.16	14	Completed 9000 cycles. Rated successful.
Air sprayed CdO-graphite/detonation gun Cr_3C_2	7.5–10	63–88	0.22	0.59	0.15	0.13	14	Completed 9000 cycles, with loose CdO-graphite. Rated marginal.
Air sprayed CdO-graphite/air sprayed CdO-graphite with detonation gun Cr_3C_2 undercoat	7.5–10	CdO-graphite 7.5–10, Cr_3C_2, 63–88	0.18	0.41	Cr_3C_2, 0.2	—	14	After 27,000 cycles, several worn patches in the loaded zone of foil; journal had wear band on one edge. Rated successful.
Air sprayed CdO-graphite-Ag/detonation gun Cr_3C_2	7.5–10	63–88	0.16	0.26	0.13	0.13	14	After 27,000 cycles, several spots on foil with deep polishing and a few worn patches, journal had light foil coating transfer, coatings serviceable. Rated successful.
Air sprayed CdO-graphite-Ag/detonation gun Cr_3C_2	7.5–10	63–88	0.19	0.41	0.13	0.13	35	After 9000 cycles, several worn patches and deep polishing in the loaded zone of foil, journal had foil coating transfer. Rated successful.

*Start-stop test sequence included 500 start-stop cycles at room temperature, 500 cycles at maximum temperature, and 500 cycles at a controlled thermal gradient from room temperature to maximum temperature.

†Average bearing clearance, 71 μm.

‡Test conducted at 290°C.

Source: Adapted from Bhushan (1979, 1982).

should be able to withstand these contaminants during operation. Bhushan (1982) immersed CdO-graphite-Ag coated foils in water and Dextron II automatic transmission oil at room temperature for 30 days. The coating surfaces were rubbed before and after the soak tests with a polishing paper and examined for any change in coating cohesion. During the rub test the surface layer of the coating that had been soaked for 30 days came off rather easily, but the bulk of the coating was unchanged. The test shows that the CdO-graphite-Ag coating has a very high resistance to the two fluids.

Graphite blended with a sodium silicate binder (water based) is commercially available from Hohman Plating and Manufacturing, Inc., Dayton, Ohio, under a trade name of Surf-Kote LOB-1800-G. The coating is air sprayed and baked at 150°C for 1 h. Graphite and mixtures of graphite and other solid lubricants (such as MoS_2) dispersed in a resin solution are commercially available, for example, from Acheson Colloids Co., Port Huron, Michigan. The resins most commonly used are the thermosetting resins, such as alkyd, phenolic, and epoxy types, and are baked at approximately 150°C. The graphite solid content of the coating ranges from 40 to 60 percent by weight. The coating solution is sprayed on surfaces from the aerosol containers, baked, and burnished before use. General Magnaplate Corp. in Linden, New Jersey, markets a high-temperature coating (Hi-T-lube) with an unknown composition, good to 450°C.

Bonded coatings of graphite are used in many industrial applications, for example, business machine parts, automotive components, bearings, seals, and many sliding contacts (Bhushan, 1987a). Graphite-treated (gold, silver, or palladium plated) electrical connectors are used for superior wear resistance. In the graphite treatment, graphite particles in a nitrogen stream are sprayed at clean contact surfaces; the graphite adheres only a few molecules thick, enough to lubricate but not enough to raise contact resistance (Anonymous, 1977).

13.1.1.2 Ion-Plated Graphite Coatings. Various carbon and graphite coatings have been deposited using physical and chemical vapor deposition processes (also see Chap. 14). Teer and Salama (1977) produced graphite coating by ion plating. Two methods were used to produce carbon vapors. In the first method, contacting pointed carbon rods were used through which a high current was passed. In the second method, an electron beam gun was used to evaporate the carbon. The high surface temperature during deposition was believed to be necessary for the growth of highly graphitic carbon.

13.1.2 Graphite Fluoride

Graphite fluoride $(CF_x)_n$, also known as *carbon monofluoride* (when x = 1), is a layered lattice intercalation compound of graphite. It is produced by direct reaction of graphite and fluorine gas at controlled temperature and pressure (Bisson and Anderson, 1964; Fusaro, 1977a, 1981, 1982a; Fusaro and Sliney, 1970, 1973; McConnell et al., 1977; Peterson et al., 1959). It can also be produced from fluorination of coke. However, friction and wear performance of fluorinated graphite is superior to that of fluorinated coke. The structure of graphite fluoride consists of covalent bonds between the fluorine and carbon atoms and a tetrahedral arrangement of the carbon bonds with the fluorine atoms located between the disordered basal planes. The true hexagonal structure of the carbon atoms been distorted. Not only has the distance between the carbon layer planes has been expanded, but also the C–C distance has been stretched (Table 13.3). The distance

TABLE 13.3 Selected Physical Properties of Graphite Fluoride and Graphite

Powder	Density, kg m^{-3}	Carbon-to-carbon inlayer distance, nm	Distance between carbon layer planes, nm	Electrical resistivity, ohm cm
$(CF_{1.085})_n$	≥2390	0.142	0.82	$>3 \times 10^3$
$(CF_{0.988})_n$	2670	0.154	0.66	$>3 \times 10^3$
$(CF_{0.996})_n$	2780	0.149	0.60	$>3 \times 10^3$
Graphite	1400–1700	0.142	0.34	5×10^{-3}

between the basal planes is expanded from 0.34 nm in graphite to 0.75 ± 0.15 nm in $(CF_x)_n$ (Rudorff and Rudorff, 1947). A comparison of crystal lattice expansion and the distortion of the carbon basal planes in $(CF_x)_n$ are shown in Fig. 13.2. For $x \geq 1$, this compound is pure white, nonwettable by water (hydrophobic), and electrically nonconductive (Table 13.3). $(CF_x)_n$ is not known to oxidize in air, but it decomposes thermally above about 450°C to form carbon tetrafluoromethane, other low-molecular-weight fluorocarbon, and carbon.

The burnished coatings of graphite fluoride perform better or equivalent in air than burnished coatings of MoS_2 and graphite. The improved performance of $(CF_x)_n$ has been observed by mixing it with polyimide (PI) or silicate, or epoxy-phenolic binders (Gisser et al., 1972; Fusaro, 1978, 1981; and Ishikawa and Shimada, 1969). It is generally believed that fluorination of graphite weakens the material in shear plane; therefore, the load-carrying capacity of $(CF_x)_n$ may be somewhat poorer than of graphite.

The performance, i.e., the coefficient of friction and wear life, of graphite fluoride, is markedly influenced by the burnishing technique (hand burnished or machine burnished), the burnishing atmosphere, fluorine-to-carbon ratio, and temperature and moisture content of the operating environment. Figure 13.3

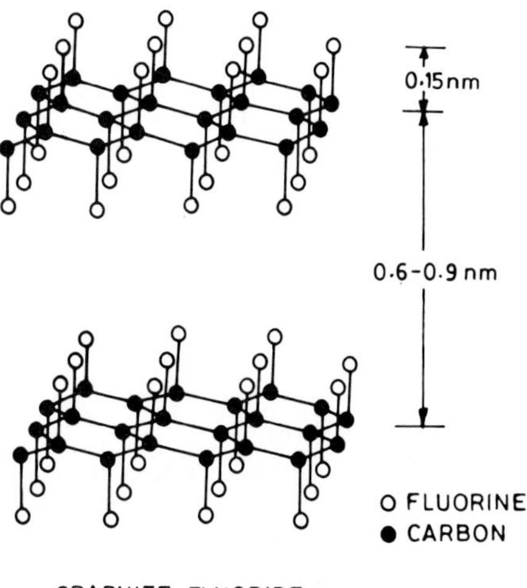

FIGURE 13.2 Crystal structure of graphite fluoride.

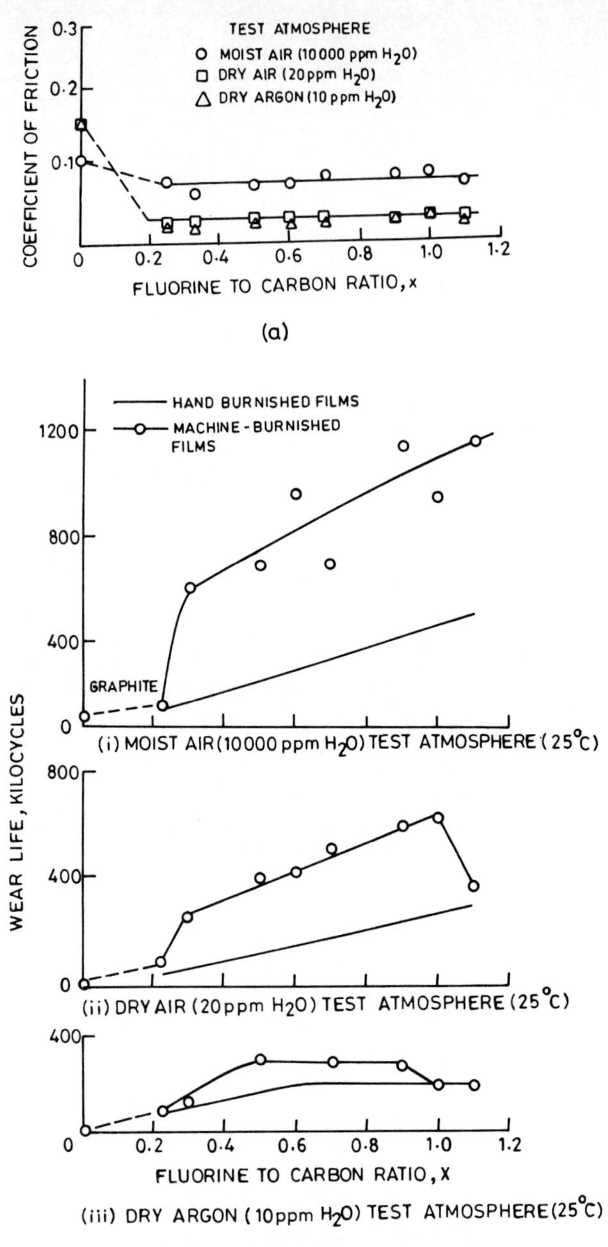

FIGURE 13.3 Effect of burnishing technique, fluorine content, and operating atmosphere (at 25°C) on: (*a*) Coefficient of friction (no difference between hand- and machine-burnished coatings). (*b*) Wear life of hand- and machine-burnished $(CF_x)_n$ coatings. The tests were carried out on a pin-on-disk machine with a pin made of 440C stainless steel, and a coated disk at a normal load of 9.8 N and a speed of 1000 r/min. [*Adapted from Fusaro (1977a).*]

illustrates the dependence of the coefficient of friction and wear (evaluated with a pin-on-disk machine) of burnished $(CF_x)_n$ coatings on fluorine content, water content in the operating environment, and burnishing technique (Fusaro, 1977a). The coatings were applied on a 440C stainless steel disk surface which had been roughened by sandblasting to an rms value of about 1 μm. The coatings were applied by rubbing the powder onto the disk surface with the back of an open-weave fabric. The following observations can be made:

- Wear life of $(CF_x)_n$ coatings increase with fluorine content; however, in dry Ar it increases up to $x = 0.6$ and then saturates.
- Longer wear life results in moist air, rather than dry air or dry Ar; however, the coefficient of friction is slightly higher.
- Better lubrication results when $(CF_x)_n$ coatings are burnished by machine rather than by hand.

Fusaro and Sliney (1969, 1970) reported that the $(CF_{1.12})_n$ burnished coatings exhibit a coefficient of friction (versus 440C stainless steel) of 0.06 up to about 430°C in dry air (20 ppm H_2O) (Fig. 13.4). Wear lives of both coatings seemed to decrease exponentially with increasing temperature. Wear life of $(CF_{1.12})_n$ at room temperature and at elevated temperatures was roughly an order of magnitude higher than that of burnished MoS_2.

Fusaro and Sliney (1973) studied friction and wear life of burnished and bonded $(CF_x)_n$ films. For the bonded films, Pyrolin 4701, polyimide (PI) varnish manufactured by the Dupont company, was used as a binder. The coefficient of friction and wear life of burnished $(CF_{1.1})_n$ and PI bonded $(CF_{1.1})_n$ [3 parts by weight of $(CF_{1.1})_n$ powder and 2 parts by weight of PI solids] coatings are compared with that of burnished MoS_2 and PI bonded MoS_2 (3 parts by weight of MoS_2 powder and 1 part by weight of PI solids) coatings in Fig. 13.5. The PI bonded $(CF_{1.1})_n$ coatings exhibit longer wear life as compared with burnished $(CF_{1.1})_n$ and burnished or PI bonded MoS_2 coatings. These PI bonded $(CF_{1.1})_n$ coatings are found remarkably durable in start-stop tests of compliant gas bearings at temperatures up to 350°C (Sliney, 1982).

$(CF_x)_n$ powders are available from various sources such as Allied Chemical, Inc., Morristown, New Jersey; Ozark-Mahoning Co., Tulsa, Oklahoma; and Nippon Carbon, Japan.

13.1.3 Molybdenum Disulfide

Molybdenum disulfide (MoS_2) is a hexagonal layered lattice compound which structurally quite closely resembles graphite (Chap. 5). In MoS_2, a low shear strength is intrinsic to the material; therefore, unlike graphite, MoS_2 performs satisfactorily without humidity and/or contaminants such as in vacuum. In fact, MoS_2 works better at lower relative humidity (such as dry air or vacuum conditions) than at high humidity. MoS_2 generally exhibits lower friction than graphite at ambient conditions, lower temperature-humidity conditions, and in vacuum. Many times, a small concentration of graphite is added to MoS_2 to improve its low load performance. Sometimes, small percentages of soft metals such as gold, silver, indium, and lead are also added. MoS_2 coatings can be applied: (1) by burnishing MoS_2 powder, (2) by air spraying, dipping, or brushing MoS_2 powder in a resin solution (bonded coating), or (3) by sputtering MoS_2. The sputtering technique provides a greater flexibility for tailoring the properties such as crystallin-

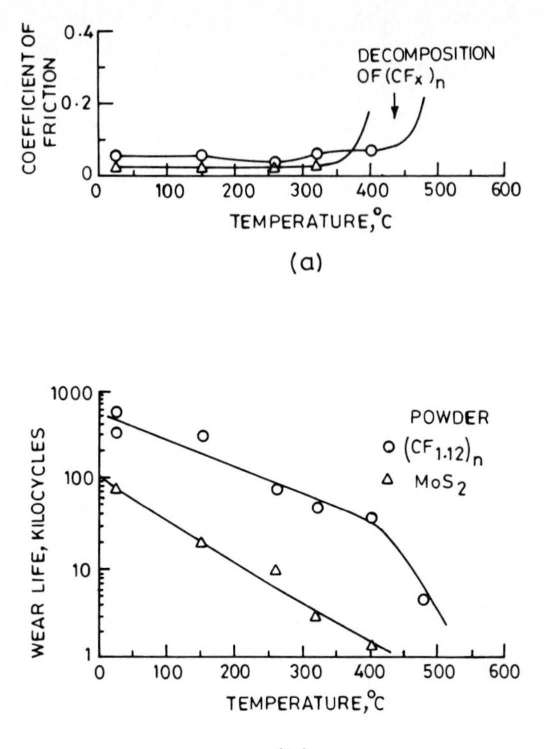

FIGURE 13.4 Effect of temperature on: (*a*) Coefficient of friction. (*b*) Wear life of hand-burnished $(CF_{1.12})_n$ and MoS_2 coatings. The tests were carried out on a pin-on-disk machine with a pin made of 440C stainless steel and a coated disk at a normal load of 4.9 N, speed of 600 r/min (1.6 m s^{-1}) and atmosphere of dry air (20 ppm H_2O). [*Adapted from Fusaro and Sliney (1970).*]

ity, stoichiometry, surface morphology, and adhesion of MoS_2 coatings to the substrate. Very thin coatings (as low as a few tens of nanometers) can be applied by the sputtering process. In addition, sputtered coatings are free of organic resins normally used to prepare bonded coatings, which are undesirable in high-vacuum applications.

13.1.3.1 Bonded MoS_2 Coatings. The bonded MoS_2 coatings consist of fine MoS_2 powder and a bonding agent. Bonded coatings of MoS_2 exhibit markedly superior wear life as compared to that obtained with unbonded MoS_2 powder. The binder provides cohesive forces that hold the solid coating together and also provides adhesive forces that bond the coating to a substrate. Bonded coatings of MoS_2 generally have better friction and wear characteristics than bonded films of other solid lubricants at room temperatures (Table 13.4). As a result of their excellent combination of wear life and friction, bonded coatings of MoS_2 are used in many industrial applications including automobiles, aircraft, missiles, satellites,

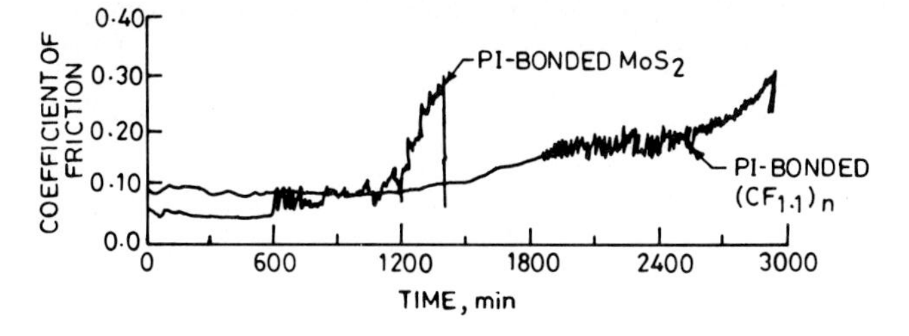

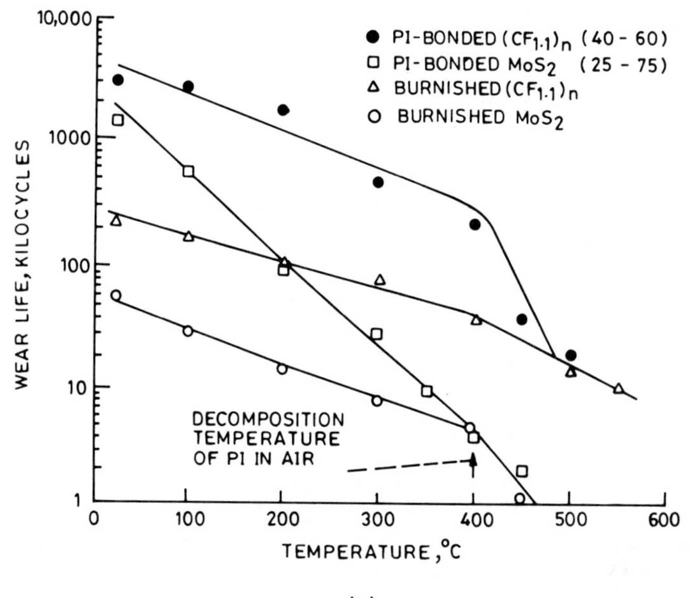

FIGURE 13.5 (*a*) Coefficient of friction at 25°C. (*b*) Wear life of PI bonded and burnished $(CF_{1.1})_n$ and MoS_2 coatings as a function of temperature. The tests were carried out on a pin-on-disk machine with a pin of 440C stainless steel and a coated disk at a normal load of 9.8 N and a speed of 1000 r/min. [*Adapted from Fusaro and Sliney (1973).*]

and nuclear power plants where extreme environments are encountered (Braithwaite, 1967; Clauss, 1972; Bhushan, 1987a). The resin-bonded MoS_2 and graphite-coated cutting tools have also shown improvement in wear life (Hartley and Wainwrights, 1981; Koury et al., 1979). In many of these cases, the coatings are the sole source of lubrication; in others, they are used in conjunction with oils or greases.

The performance of bonded MoS_2 coating is influenced by several factors such as (Bartz et al., 1986; Bartz and Xu, 1987; Campbell, 1972; Campbell et al., 1966; DiSapio, 1960, 1978; Fusaro, 1982b; Godfrey and Bisson, 1951; Johnson et al., 1948; Mitchell and Fulford, 1957):

TABLE 13.4 Friction and Wear Lives of Bonded Films of Solid Lubricants*

Solid lubricant†	Kinetic coefficient of friction	Wear life, cycles
Molybdenum disulfide	0.036	103,700
Tungsten disulfide	0.034	100,400
Graphite	0.080	8,700
Cadmium iodide	0.088	4,300
Boron nitride	0.148	400

*Test conditions: Solid lubricants were bonded with a phenolic resin to a Timken test cup made of manganese phosphated 4130 steel surface mounted on a rotating shaft and sliding against two flats of 4130 steel blocks; rotational speed = 72 r/min (0.15 m s^{-1}), test temperature = room temperature.

†Coating thickness varied from 7.5 to 13 μm.

Source: Adapted from Stupp (1958).

- Substrate material characteristics such as hardness and surface pretreatment
- Coating deposition parameters such as binder material, ratio of binder to MoS_2, curing conditions, and coating thickness
- Operating conditions, such as mating part, humidity, temperature, sliding velocity, load, partial pressure of gases, and composition of gases present in the atmosphere

The dependence of performance, i.e., coefficient of friction and wear life of MoS_2 coatings, on the various factors listed above is discussed in the following paragraphs. We note that the following discussion also applies to the bonded coatings of many other solid lubricants such as graphite and various polymers.

Substrate Material. The MoS_2 coatings can be applied onto ferrous, nonferrous, and molded polymer materials provided that their surfaces are properly prepared (roughened and cleaned) prior to coating deposition.

Hardness. The MoS_2 coatings applied onto harder surfaces exhibit a lower coefficient of friction and longer wear life. This is because hard substrate provides a nonbuckling surface and avoids any crack formation and defoilation at the coating-substrate interface during operation.

Surface Pretreatment. Surface pretreatment of the substrate is necessary to provide good physical-chemical bonding of the coating to the substrate (Chap. 7; Appendix 13.A). Chemical bonding provides a superior adhesion and wear life. The surface roughness of the substrate typically ranges from 25 to 250 nm rms for good adhesion of bonded MoS_2 coatings. Many metals must be chemically etched to provide a rougher surface or must have a chemical deposit on the surface that improves adhesion and increases the wear life (see Chap. 7 on surface pretreatments for details). The difference between a good and bad pretreatment can result in an order-of-magnitude difference in the wear life of coatings. Haltner and Oliver (1960) have shown that improved performance of resin-bonded MoS_2 coatings was obtained for coatings having metal sulfide additives or for coatings prepared on presulfided steel surfaces.

Binder Material. Although the binder material does not lend much to the lubricating ability of bonded MoS_2 coating, it is necessary for its application. The binder material should be thermally stable at operating temperatures. Two types of organic binders are commonly used—heat curable thermosetting material and air-dryable material. The binders used in most commercial coatings today are thermosetting-type organic resins such as the phenolics and epoxies (silicones and alkyds are also used) that require baking at temperatures up to 200°C. Air-drying films are also available; while these are sometimes more convenient to use, their load-carrying capacity or wear life is inferior to the heat-cured coatings. Table 13.5 illustrates the range of friction and wear values of bonded coatings made with different resins (Stupp, 1958). Inorganic binders include sodium silicate, aluminum-phosphate, and ceramic oxides. These binders are useful in more extreme environments that cannot be tolerated by organic binders such as very high temperatures (Anonymous, 1972; Lipp, 1976).

Ratio of Binder to MoS_2. The coefficient of friction and wear varies considerably with the amount of binder material. Usually, too small a ratio of binder to MoS_2 results in poor adhesion and too high a ratio deteriorates the lubricity. A binder-to-MoS_2 ratio typically ranging from 1 : 1 to 1 : 2 is used for several coatings. Mitchell and Fulford (1957) showed a decrease in friction for phenolic-bonded MoS_2 from 0.2 to 0.15 as MoS_2 content is increased from 5 to 33 percent. Crump (1956, 1957) also showed improved wear life with lower binder contents.

Other Additives. Many inorganic and organic additives have been used in the composition of MoS_2 coatings. Some examples are graphite, Ag, and Au. In some compositions, Sb_2O_3 is added to provide corrosion resistance and to improve wear life.

Coating Thickness. Thickness has been reported to be an important parameter affecting the performance of bonded MoS_2 coatings (Bahun and Jones, 1969; Bartz et al., 1986; DiSapio, 1963, 1978; Hopkins and Campbell, 1969; Murphy and Meade, 1966). Wear life is maximum for a specific coating thickness, and it usually ranges from 5 to 30 μm, the optimum being about 12.5 μm. Too thick a

TABLE 13.5 Friction and Wear Lives of MoS_2 Coatings Bonded with Various Types of Organic Resins*

Type of resin binder†	Kinetic coefficient of friction	Wear life, cycles
Phenolic	0.034	130,680
Phenolic-fluorocarbon	0.024	120,600
Phenolic-vinyl copolymer	0.040	102,660
Phenolic-vinyl acetate	0.040	96,120
Di-isocyanate-castor oil	0.060	96,750
Phenolic-acrylonitrile	0.045	86,400
Di-isocyanate-phenolic	0.050	86,400
Corn-syrup	0.031	85,080
Phenolic-rubber	0.064	69,120
Phenolic-neoprene	0.035	68,000
Phenolic-acrylic	0.060	50,400
Phenolic-epoxy	0.063	36,000
Phenolic-amide	0.074	23,760
Phenolic-vinyl butyl	0.065	21,600
Vinyl chloride	0.070	21,600
Silicone	0.054	15,120

*Test conditions: Coatings applied over a Timken test cup (manganese phosphated 4130 steel) mounted on a rotating shaft and sliding against two flats of 4130 steel blocks. Rotational speed = 72 r/min (0.15 m s^{-1}), test temperature = room temperature.

†Coating thickness varied from 7.5 to 13 μm.

Source: Adapted from Stupp (1958).

coating will cause the coating to peel or flake off with sliding, whereas too thin a coating may result in premature failure due to rupture of the coating. Figure 13.6 illustrates the effect of coating thickness on the wear life (number of cycles necessary to fail the coating) of bonded coating with MoS_2 and graphite in inorganic binder as measured on an LFW-1 tester.

Humidity. Wear life of bonded MoS_2 coatings can either increase or decrease with a change in humidity. For instance, the wear life of epoxy-bonded MoS_2 improves from 150,000 to 210,000 rev (50 percent increase) as relative humidity increases from 40 to 100 percent, while under identical conditions, wear life of MoS_2 + graphite + vinyl-butyl binder reduces by 60 percent (DiSapio, 1960). We have seen in Chap. 5 that in the case of an interface lubricated with powdered MoS_2, friction and wear rates increase with an increase in the relative humidity.

Temperature. The wear life usually remains unchanged with temperatures up to the curing temperature of the bonded coating and then decreases as temperature increases further.

Sliding Velocity. The coefficient of friction and wear life of bonded MoS_2 coatings usually decrease rapidly with increasing sliding velocity in the beginning and then decrease slowly with a further increase in sliding velocity. Figure 13.7*a* shows a variation of coefficient of friction with sliding velocity for MoS_2 coatings bonded with several organic binders (Johnson et al., 1948). Figure 13.7*b* shows variations of wear life with sliding velocity for MoS_2/graphite (47/53 percent) bonded with an organic binder (Bartz and Xu, 1987). Wear life decreases exponentially at higher sliding speeds due to generation of high temperatures at the contact which may decompose the binder material.

Load. The coefficient of friction and wear life usually decreases with an increase in normal load (Bartz et al., 1986; Bartz and Xu, 1987). Figure 13.7*b* shows the variation of wear life with load for MoS_2/graphite (47/53 percent) bonded with an organic binder. We note that wear life decreases exponentially with an increase in load.

Low Pressure (Vacuum). Bonded MoS_2 coatings are successfully used in vacuum. The necessary condition is that the binder material must have a low va-

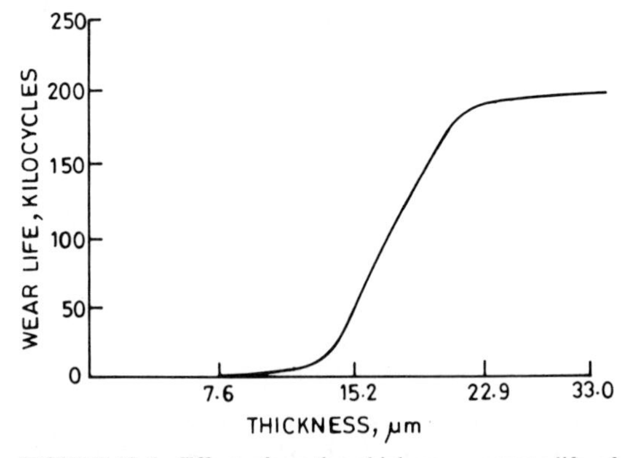

FIGURE 13.6 Effect of coating thickness on wear life of bonded MoS_2 + graphite coated steel ring rubbing against 60 HRC steel block in a LFW-1 tester (72 r/min, 2800 N load). [*Adapted from DiSapio (1978).*]

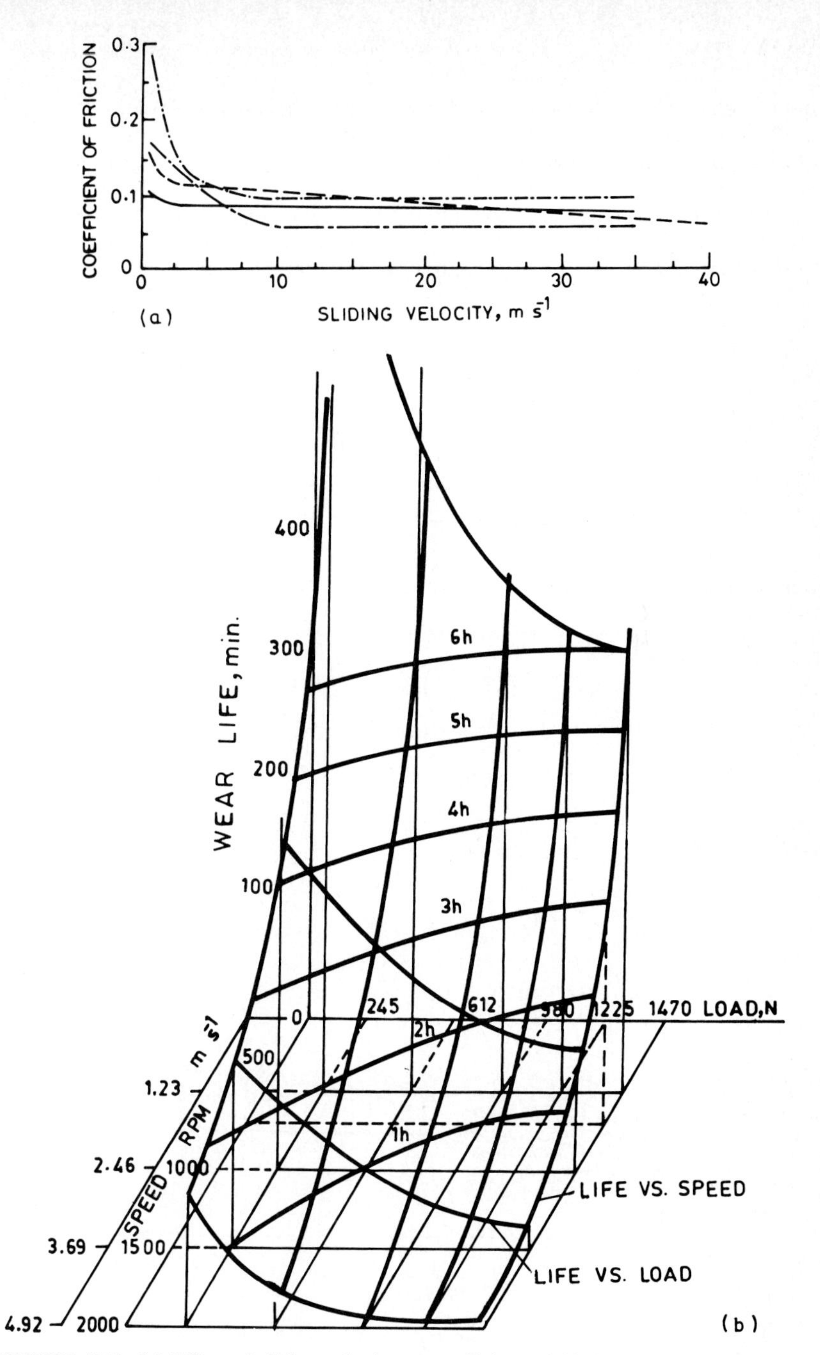

FIGURE 13.7 (*a*) Effect of sliding velocity on coefficient of friction of steel specimen coated with bonded MoS_2. Different curves correspond to different resin binders. [*Adapted from Johnson et al. (1948).*] (*b*) Effect of sliding velocity and normal load on wear life of bonded MoS_2 + graphite coating rubbing against 60 HRC steel block in a rotating 47-mm diameter ring/stationary block arrangement. [*Adapted from Bartz and Xu (1987).*]

por pressure (McKannan and Demorest, 1964). Matveevsky et al. (1978) tested MoS_2 coatings bonded with an inorganic and an organic (polyamide) binder, in vacuum at various temperatures. The tests were conducted using a pin-on-disk tester. The coefficient of friction as a function of the number of cycles at various temperatures in vacuum are shown in Fig. 13.8. We note that the bonded coatings of MoS_2 exhibit thermal stability in vacuum up to a maximum temperature of 500°C. This is consistent with the limiting temperature of about 500°C previously

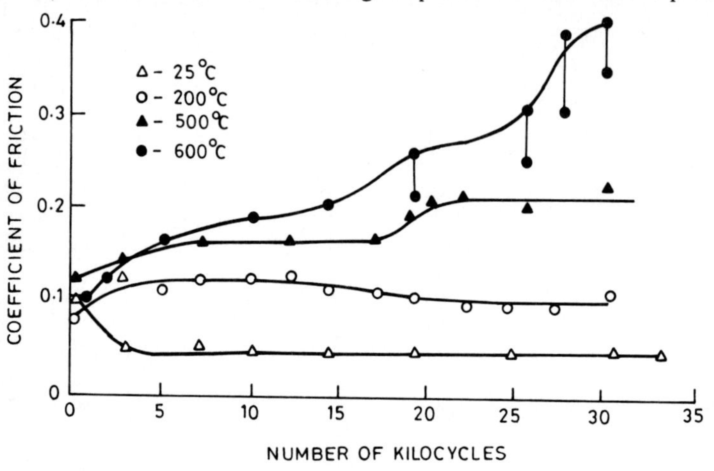

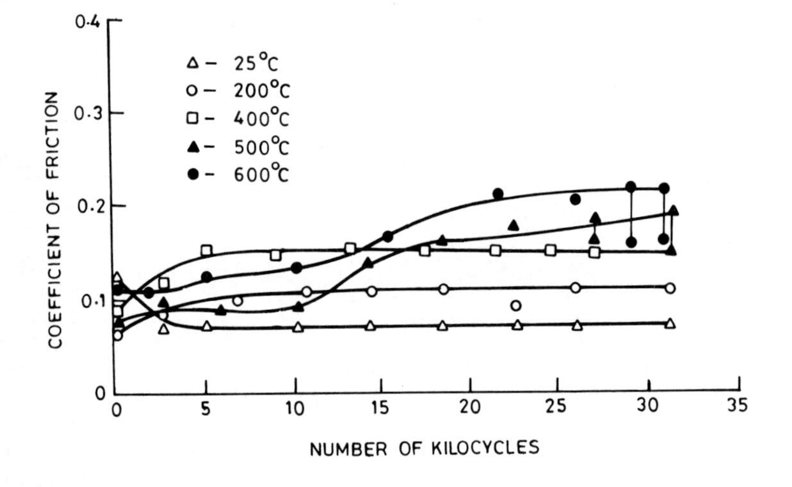

FIGURE 13.8 Coefficient of friction of MoS_2 coatings bonded with an (*a*) inorganic binder and (*b*) organic (polyamide) binder as a function of number of cycles at various test temperatures in vacuum. Tests were conducted using a pin-on-disk test. [*Adapted from Matveesky et al. (1978).*]

shown in Chap. 5 based on oxidation kinetics data of MoS_2 powders and friction data of burnished MoS_2 coatings.

Mixtures of the solid lubricant and binder suspended in a liquid carrier are commercially available, in aerosol containers ready to be sprayed. Dow Corning Corp., Midland, Michigan, and Acheson Colloids Co., Port Huron, Michigan, sell several products containing MoS_2 alone or MoS_2 with other solid lubricants. Hohman Plating & Manufacturing, Inc., of Dayton, Ohio, sells several products with various binder materials and other additives, for example, Surf-Kote A-2178A, an air-drying lubricant in a solution of organic resins; Surf-Kote M-2036, a lubricant in a solution of heat-curing resin; and Surf-Kote M-1284, an air-drying matrix-bonded lubricant. Alpha Molykote Corp., Stamford, Connecticut, sells several products under a trade name of Molykote such as Molykote X106. E/M Lubricants, Inc., of West Lafayette, Indiana, sells several products under a trade name of Perma-Slik with various resin binders, additives, and solvents. For example, Perma-Slik C and S combines MoS_2 and graphite and MoS_2 alone with an air-drying resin in a quick-drying solvent system. Tiodize Co., Inc., of Huntington Beach, California, sells products such as Tiolube 460 which consist of MoS_2 powder suspended in a thermosetting resin.

Several specialty lubricants were developed in the 1960s for aerospace applications (Anonymous, 1972; Campbell, 1972; Lipp, 1976). A coating mixture of 10 wt % graphite, 60 wt % MoS_2, and 30 wt % gold powder in a binder of sodium silicate (MLF-5) was developed by researchers at Midwest Research Institute, Kansas City, Missouri, for roller bearings, in high-temperature and high-load (aerospace) applications (Anonymous, 1972; McKannon and Demorest, 1964). MoS_2 and Sb_2O_3 with an epoxy-phenolic binder is designed to meet military specification MIL-L-46010. MoS_2 and Sb_2O_3 with a polyimide binder is designed for higher-temperature applications (proprietory 425). MoS_2, graphite, and Ag with a ceramic oxide binder is developed for very high temperature applications and is designated as Vitrolube 1220 manufactured by North American Aviation Company, Inc., Los Angeles (Hagon and Williams, 1964).

13.1.3.2 Sputtered MoS_2 Coatings. The sputtering process is ideally suited to applications which require close tolerances in mechanical devices because they are thin (a fraction of a micron to few microns thick) and they generally adhere well to most surfaces. Intermediate sputtered coatings of other materials can also be applied onto certain substrates to reduce corrosion and enhance adhesion. Spalvins (1969) reported that the sputtered coatings that were only 200 nm thick had good durability in a pin-on-disk tester (Fig. 13.9). The coefficient of friction was 0.05 to 0.10 in vacuum. We will see later that durability is much lower in air.

It is well established that the lubricating properties and wear life of rf (radio frequency) sputtered MoS_2 coatings are strongly influenced by the sputtering and substrate conditions (also see Chap. 9), which, in turn, influence the morphological growth (crystalline size and its orientation), stoichiometry, and adhesion of the coating to the substrate (Bhushan, 1980a; Buck, 1983; Christy and Ludwig, 1979; Christy, 1980; Dimigen et al., 1979, 1985; Fleischauer, 1984, 1987; Fleischauer and Bauer, 1987; Gardos, 1976; Lavik and Campbell, 1972; Nishimura et al., 1978; Spalvins, 1971, 1974, 1976, 1978, 1980, 1982; Stupp, 1981; Uemura et al., 1987).

Surface Morphology. The crystalline structure of MoS_2 coatings is highly dependent on the substrate temperature. Spalvins (1976, 1978) found that the coatings prepared at a cryogenic temperature (−195°C) are quench condensed and exhibit characteristic amorphous structure, and the coating morphology is

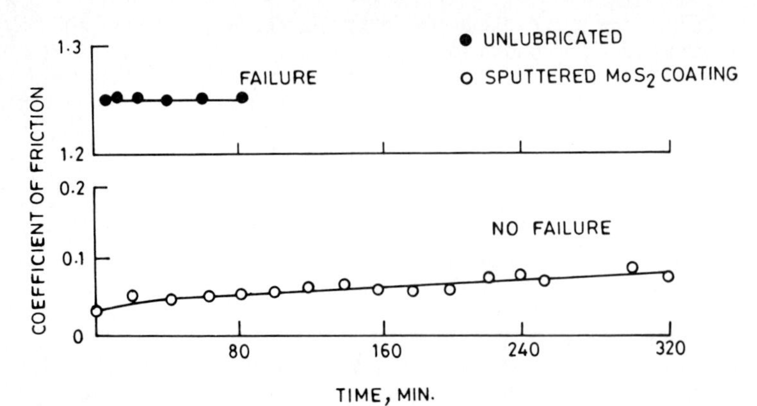

FIGURE 13.9 Coefficient of friction of niobium pin sliding on a nickel-chrome alloy disk specimen coated with 200-nm-thick sputtered MoS_2. Test conditions: Sliding velocity = 0.03 m s^{-1}; Normal load = 2.5 N; room temperature; vacuum = 1.3×10^{-9} Pa (10^{-11} torr). [*Adapted from Spalvins (1969).*]

continuous. At ambient or elevated temperatures (320°C), the coatings have crystalline structure with pronounced ridge formation (Fig. 13.10).

As the coating thickness increases to over about 0.08 μm, the ridge-type structure transforms into an equiaxed, dense transition zone, before it grows into a columnar fiberlike structural network, as shown schematically in Fig. 13.11. The equiaxed zone is basically pore free and has a densely packed structure between the equiaxed crystallites and normally does not exceed over 0.2 μm in thickness. The columnar fiber structure consists of vertical columns about 0.25 μm in diam-

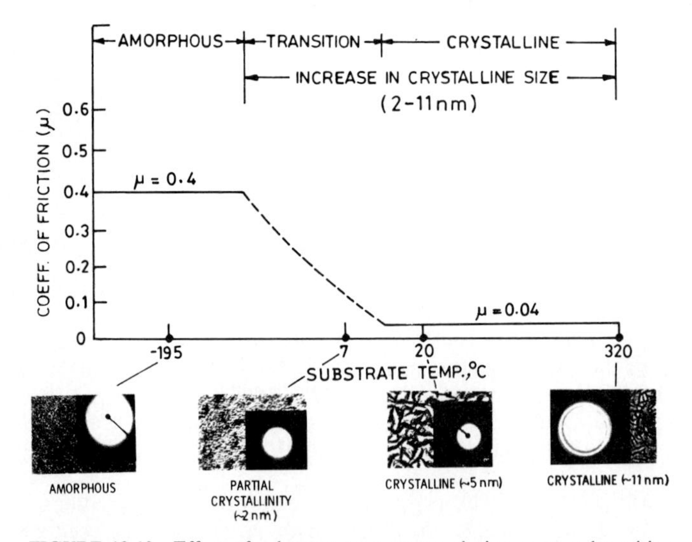

FIGURE 13.10 Effect of substrate temperature during sputter deposition on microstructure, crystallinity, and coefficient of friction of MoS_2 coatings. [*Adapted from Spalvins (1978).*]

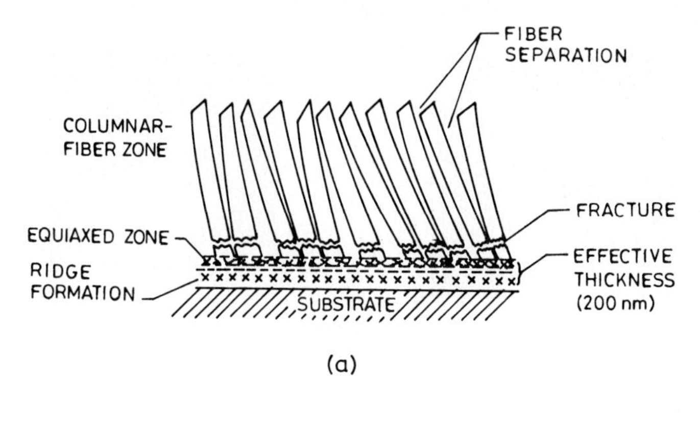

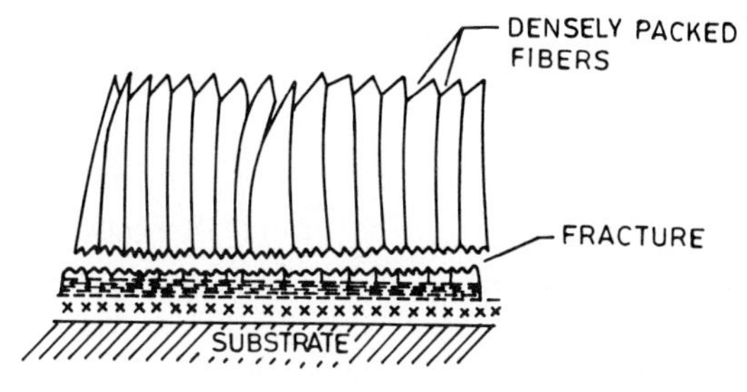

FIGURE 13.11 Schematic sketches of fracture and film fragmentation in the columnar zone of (a) an MoS_2 coating and (*b*) an Au-MoS_2 coating. [*Adapted from Spalvins (1984).*]

eter which are aligned perpendicular to the substrate and are separated by open voided boundaries a few tens of nanometers in width. The longitudinally low-density regions have a high dislocation density which exerts a high level of residual stresses.

The amorphous coatings (grain size ~1 nm) act like an abrasive and exhibit poor lubricity, i.e., high coefficients of friction (~0.4), whereas polycrystalline coatings (grain size 5 to 10 nm) with proper stoichiometry exhibit effective lubrication with a low coefficient of friction (~0.05) and long endurance life (Fig. 13.10) (Spalvins, 1976, 1978). The amorphous MoS_2 coatings have a highly dense structure and high hardness and brittleness whereas crystalline MoS_2 coatings are soft, greasy, and are found to have a featherly lamellar structure, i.e., the sheets that slide easily (Holinski and Gansheimer, 1972; Spalvins, 1980; Buck, 1983, 1987).

Coatings with different crystalline orientation with respect to the substrate plane [Fig. 13.12(*a*)] have drastically different chemical stability (oxidative reactivity at high humidities during storage) and significantly different lubrication properties. Fleischauer (1984, 1987) and Fleischauer and Bauer (1987) have shown that the coatings prepared initially in the coplanar orientation provide good lubrication and resistance to environmental attack [Fig. 13.12(*b*)]. Coatings

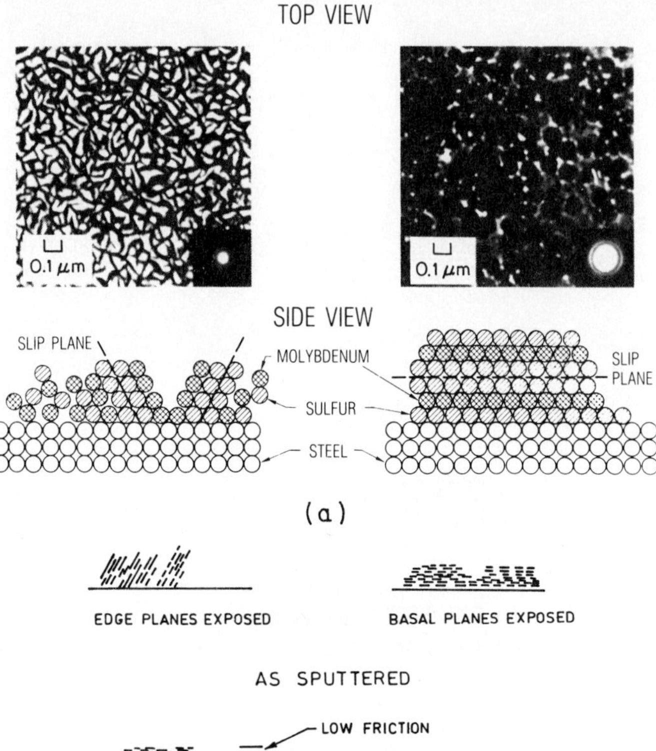

FIGURE 13.12 (*a*) Transmission electron micrographs and schematic of structure for two rf sputter-deposited MoS_2 coatings. (*b*) Schematic of structure for two rf sputter-deposited MoS_2 coatings. [*Adapted from Fleischauer (1987).*]

with random orientation allow the penetration of environmental elements and degradation of coatings (oxidation of MoS_2 into MoO_3). The plane representing the interface between crystallites of different orientations constitutes a region of failure during use. However, most randomly oriented coatings achieve some orientation but cannot be oriented completely on burnishing (rubbing); therefore, these coatings would not perform (low friction but inferior wear life) as well as the coatings initially prepared with orientation parallel to the surface. Fleischauer (1987) has shown that in a coating with their crystallites oriented perpendicular to the substrate surface, after substantial sliding (100,000 rev.), roughly 40 percent of the crystallites remain unoriented. We note that there would always be a zone of crystallites at the coating-substrate interface that would not become properly oriented because of the strength of the chemical bonding between the MoS_2 "edge" plane [for example, (100)] atoms and the substrate atoms.

Fleischauer (1984) further noted that the coatings with coplanar orientation providing good lubrication and resistance to environmental attack can be generally produced at selected sputtering conditions such as high substrate temperatures and slower sputtering rates, rates that approach thermodynamic conditions (Fleischauer, 1984; Fleischauer and Bauer, 1987; Stupp, 1981). Buck (1987) has suggested that the surface morphology (crystalline structure and orientation) is varied to a large extent by the variation of the amount of H_2O present in the plasma, which changes among other things with the sputtering parameters. Larger concentration of water vapor produces MoS_2 with a crystalline structure with coplanar orientation.

Recently, Mikkelsen et al. (1988) amorphized rf-sputtered MoS_2 coatings by ion bombardment with 400 keV argon ions. They found that amorphized coatings were dense and exhibited low friction (~0.04) comparable to that of crystalline coatings and with an improvement of wear life by roughly fivefold. These data are in contradiction to previously published results (Buck, 1987; Spalvins, 1987). Mikkelsen et al. argued that this large discrepancy may be explained by the process of amorphization used in other studies. They suggested that high friction coefficients in other studies could be attributed to water contamination rather than to the amorphous structure.

Stoichiometry. The coefficient of friction of sputtered MoS_2 coatings depend strongly on the stoichiometry, i.e., ratio of sulphur to molybdenum atoms (N_S/N_{Mo}) (Dimigen et al., 1979, 1985; Dimigen and Hubsch, 1983). Dimigen et al. (1985) found that a rather low argon pressure as well as a substrate bias potential cause a high sulphur deficiency in the sputtered coating compared with the target material; this can be compensated for by using an Ar-H_2S sputtering atmosphere. Figure 13.13 shows the dependence of the coefficient of friction (in dry nitrogen) of MoS_x coatings having x from 0.8 to 2.3. A low value of coefficient of friction is sustained even for a substantial Mo excess (N_S/N_{Mo} = 1), whereas a further decrease in S content leads to an abrupt increase to 0.5. Tests conducted at various partial pressures of oxygen show that stoichiometric MoS_x coatings exhibit friction independent of partial pressure of oxygen while nonstoichiometric coatings show dependence of friction on the actual partial pressure of oxygen [Fig. 13.14(*a*) and (*b*)]. This can be explained by the fact that the unsaturated bonds in nonstoichiometric coatings may cause adsorption of ambient gases which may strongly affect the coefficient of friction.

Coating Thickness. MoS_2 coating thickness has a very pronounced effect on the friction and wear. Essentially, the coefficient of friction and wear reach a minimum value at an effective or critical film thickness. Spalvins (1982) has made an attempt to determine the optimum film thickness on the basis of the morphological growth pattern during sliding conditions.

Figure 13.15 illustrates a typical wear track after a single-pass sliding on a 1.2-μm-thick coating on 440C stainless steel and glass substrate. Sputtered MoS_2 coatings as observed always have a tendency to break within the columnar-fiber-like region, as shown in Fig. 13.11(*a*). It should be pointed out that part of the columns remain on the substrate, and the roots of the columns are stronger than the cohesive force in the column. The strength of MoS_2 coatings certainly increases with an increase in the density as the structural zones change with the highest density in the equiaxed zone as illustrated in the wear tracks. The cohesive forces are weakest between the columns, above the equiaxed zone, as illus-

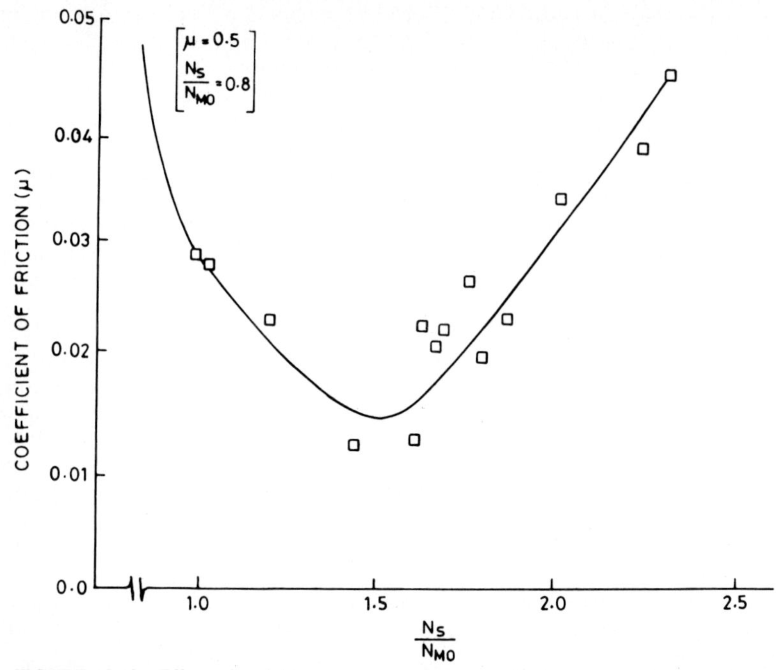

FIGURE 13.13 Effect of stoichiometry on coefficient of friction in dry nitrogen of sputtered MoS_x coated disk in a pin-on-disk (steel-steel) test. [*Adapted from Dimigen et al. (1985).*]

trated in the wear tracks. Actual lubrication is accomplished by the coating that remains on the substrate, which is roughly 0.2 μm thick. The ultimate lubricating coating thickness in the case of sputtered MoS_2 coating on the substrate is usually on the order of a fraction of a micron (typically 0.2 to 0.6 μm). The broken and separated columnar coating forms wear debris that can impair lubricating properties.

Effect of Environment. The performance of sputtered MoS_2 coatings deteriorates in the presence of oxygen and/or moisture. Usually, the coatings are said to decompose and to form oxidation products SO_2, H_2S, and H_2SO_4 (Kay, 1968; Lavik et al., 1965, 1968; Niederhauser et al., 1983).

The effect of environmental pressure on the coefficient of friction for sputtered MoS_2 coatings was evaluated using a pin-on-disk tester by Spalvins (1975). The changes in the coefficient of friction with changes in pressure from atmospheric pressure (0.1 MPa or 760 torr) at a relative humidity of 60 percent to high vacuum of 10^{-7} Pa (10^{-9} torr) are shown in Fig. 13.16(*a*). We note that a continuous change occurred in the coefficient of friction between 13 to 52 kPa (100 to 400 torr). Above 52 kPa (400 torr), the coefficient of friction was steady at ~0.15, and below 13 kPa (100 torr), it stabilized at 0.04. The friction tests of sputtered MoS_2 were also conducted in dry air, nitrogen, and argon. The coefficients of friction under these conditions were the same as in high vacuum (0.04).

Figure 13.16(*b*) shows the influence of relative humidity (RH) at room temperature (RT) on the wear life of sputtered MoS_2 coatings (Menoud et al., 1985). The

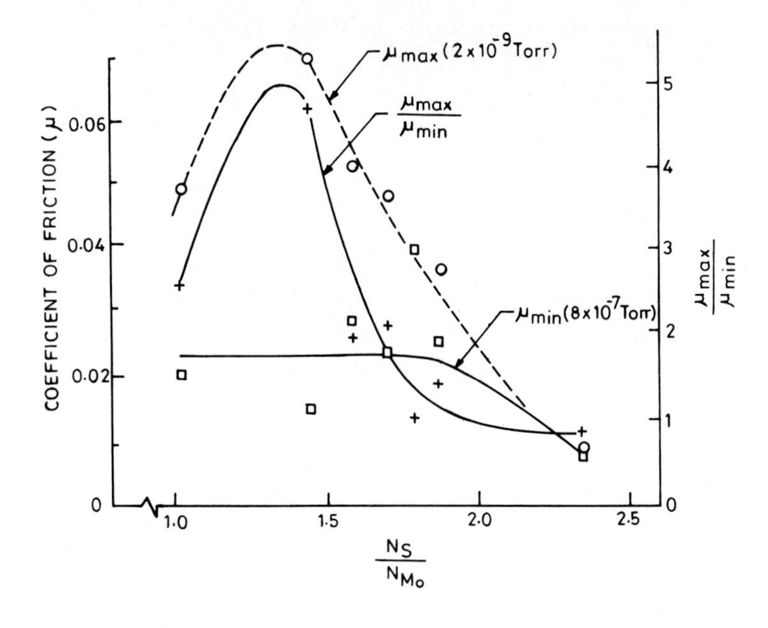

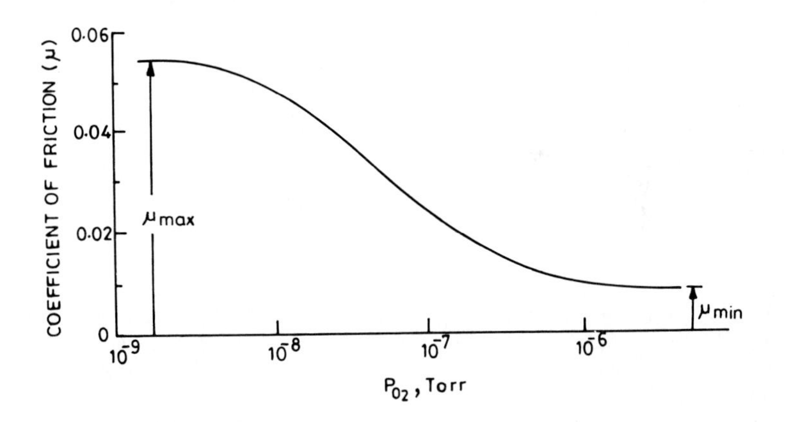

FIGURE 13.14 (*a*) Effect of stoichiometry on coefficient of friction (μ) and μ_{max}/μ_{min} of sputtered MoS_x coating measured at 3×10^{-7} Pa (2×10^{-9} torr) (μ_{max}) and 10^{-4} Pa (8×10^{-7} torr) (μ_{min}) in a pin-on-disk test. (*b*) Effect of actual oxygen partial pressure P_{O_2} on coefficient of friction of sputtered MoS_x ($x = 1.4$) coated disk in a pin-on-disk test. [*Adapted from Dimigen et al. (1985).*]

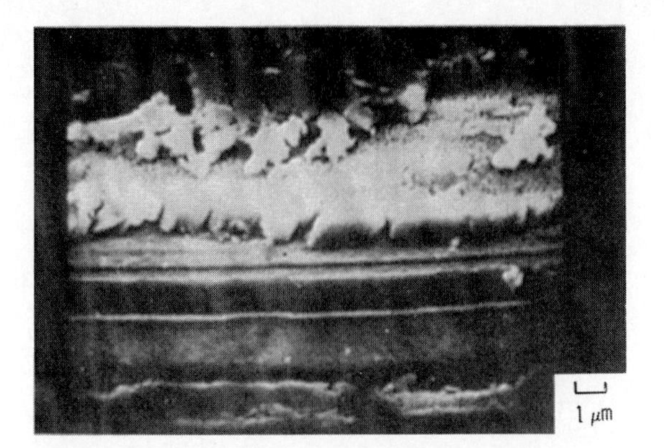

(a)

(b)

FIGURE 13.15 Sputtered MoS_2 coating after a single-pass sliding: (*a*) On a 440C stainless-steel substrate. (*b*) On a glass substrate. [*Adapted from Spalvins (1982).*]

wear life is measured by the number of revolutions, i.e., sliding distance in a pin-on-disk machine sufficient to cause an increase in the coefficient of friction abruptly. We note that wear life of the coating is at least an order of magnitude less than that of a low humidity or in vacuum.

Spalvins (1975) tested a ball bearing with sputtered MoS_2 coating on the cage, balls, and races. The MoS_2 coated bearing easily survived 1000 h of operation in vacuum at 1750 r/min and at 138 N radial load, but when air was admitted into the vacuum chamber, the bearing failed in less than an hour. In air, the thicker, bonded MoS_2 coatings are usually preferred, while the sputtered coatings are usually preferred for vacuum applications.

The performance of sputtered MoS_2 coatings in moist air can be dramatically improved by applying composite coatings of MoS_2 and water-repellent additives such as polymerized hydrocarbons or fluorohydrocarbons, e.g., PTFE. To increase the adhesion of such coatings and to prevent corrosive attack of the sub-

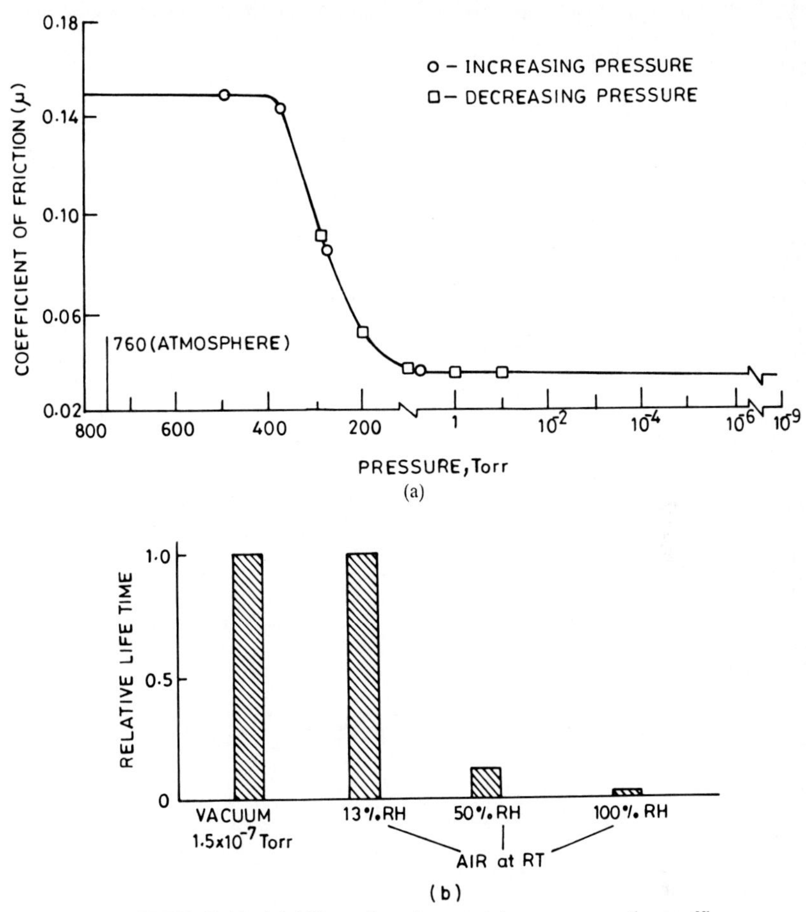

FIGURE 13.16 (*a*) Effect of environmental pressure on the coefficient of friction of sputtered MoS_2 coated disk in a pin-on-disk (Ni/Ni) test. Test conditions: normal load = 2.5 N; speed = 40 r/min; room temperature. [*Adapted from Spalvins (1975).*] (*b*) Effect of relative humidity (at room temperature) on wear life of sputtered MoS_2 coated disk in a pin-on-disk test. [*Adapted from Menoud et al. (1985).*]

strate, interlayers of corrosion-resistant materials are sometimes used. For example, Rh, Pd, Pd-Ni, Ru, W, Cr, Ni, Co, Au, Mo, TiN, TiC, or Cr_3C_2 on a steel substrate can be used. Niederhäuser et al. (1983) found that hydrocarbon (PTFE) additives improved the humidity resistance markedly. The friction coefficient in humid environments is reduced from 0.2 to 0.3 to around 0.1 by the addition of PTFE to MoS_2. The lifetime of these films in humid air is prolonged by a factor of about 10 (Table 13.6). When corrosion-resistant substrates are used or corrosion-resistant interlayers are applied on steel, the lifetime of both rubbing partners in the pin-on-disk tester is improved by a factor of about 1000 compared with that obtained when MoS_2-PTFE is deposited directly on steel (Table 13.6).

TABLE 13.6 Comparison of Wear Lives of Sputtered Multilayered MoS_2 Coatings in Moist Air*

Disk and lubrication system	Pin	Revolutions until $\mu > 0.3$	Mean time until failure,† min	Remarks
Pure MoS_2 on steel	Steel	510 ($\mu_b = 0.22$)	106	
MoS_2-PTFE on steel	Steel	4,800 ($\mu_b = 0.10$)	275	
MoS_2-PTFE on an Ni interlayer	Steel	8,100 ($\mu_b = 0.09$)	400	
Pure MoS_2 on steel	TiC-coated steel	350	9	Chemically vapor-deposited TiC on AISI 52100 steel
MoS_2-PTFE on an Ni interlayer	TiC-coated steel	63,000	1,880	
Pure MoS_2 on steel	Ruby	1,700	50	
MoS_2-PTFE on a Co interlayer	Ruby	39,800	1,360	
MoS_2-PTFE on a chemically activated interlayer of:				
Rh	Ruby	> 4,005,000	> 36,650	μ constant at 0.18
Cr	Ruby	13,000	300	
TiN	Ruby	6,100	290	Not activated
TiC	Ruby	2,800	45	Not activated
Au	Ruby	800	25	Not activated
MoS_2-PTFE on a chemically activated substrate of:				
Mo	Ruby	58,420	1,680	
Co	Ruby	> 40,800	> 1,390	
Ti	Ruby	32,000	1,180	
W	Ruby	19,400	540	
Ni	Ruby	8,400	220	
Cu	Ruby	5,500	450	—
Ag	Ruby	0	0	—

Note: μ = Coefficient of friction; μ_b = initial value of coefficient of friction.

*Steel for pins, AISI 440C unless otherwise stated; steel for disks; AISI D3, sliding velocity = 1 mm s^{-1}; normal load = 5 N; relative humidity = 52 percent.

†Defined as the time in which the friction noise increased to ±20 percent peak to peak of the average friction value, or the time at which the average friction increased to 1.5 times its value after run-in.

Source: Adapted from Niederhäuser et al. (1983).

Interfacial Chemistry. Strong chemical bonding between coating and substrate atoms is very desirable for good adhesion and long durability. The chemistry of the interface can be modified without affecting bulk properties of the coating. Similar to burnished MoS_2 coatings, improved wear performance (adhesion) of sputtered MoS_2 coatings is obtained for coatings having metal sulfide additives or for coatings made on presulfided steel surfaces (Niederhäuser et al., 1983). To further increase adhesion of MoS_2 coatings, Niederhäuser et al. chemically activated the substrate or the interlayer in a glow discharge with a plasma containing argon and H_2S. Lifetimes of such treated coatings were several times better than that of untreated coatings.

Fleischauer (1987) has suggested forming defects, specifically sulfur vacancies in MoS_2, within the basal plane or at step sites on the basal surface. The vacancies may act as anchor sites where substrate atoms could bond directly, or they might serve as sites where atoms are substituted to form bridge bonds between substrate and coating atoms. Sulfur vacancies can be formed by the bombardment of the basal planes of MoS_2 with ions in the energy range from 0.5 to 10 keV.

Influence of Metal Additives on Friction and Wear. Stupp (1981) and Spalvins (1984) introduced metallic dispersions into MoS_2 coatings to increase the densification and strengthening of the coatings in the columnar zone and also to increase the thickness of the remaining or effective coating. Stupp (1981) deposited Ni-MoS_2 and Ta-MoS_2 coatings by codeposition which was accomplished by simultaneously sputtering with two separate targets of Ni or Ta and MoS_2. The comparisons of performance of MoS_2 coatings with and without metal dispersions show that the MoS_2 coatings with metal dispersions have a coefficient of friction lower by about 35 percent, a higher degree of functional stability, and the wear life per unit thickness higher on the average by a factor of 2.

Spalvins (1984) prepared Au-MoS_2 sputtered coatings by using a compact Au-MoS_2 sputtering target. The cosputtered Au-MoS_2 coatings displayed a lower and more stable coefficient of friction and less wear debris generation compared with pure MoS_2 coatings. During sliding, the breakup of the coatings always occurs in the columnar zone for both Au-MoS_2 and MoS_2. Because of the higher columnar packing density in the Au-MoS_2 coatings, the separated Au-MoS_2 coatings after fracture do not disintegrate into a large volume of individually tapered crystallites as does the MoS_2 sputtered coating (Fig. 13.11). As a result, less wear debris is generated compared with pure MoS_2 coatings.

13.1.4 Other Layered Lattice Solid Coatings

The dichalcogenides, other than MoS_2 of interest as solid lubricant solids include the disulfides, diselenides, and ditellurides of Mo, W, Nb, and Ta (Boes, 1964; Bowden and Tabor, 1964; Flom et al., 1965; also see Chap. 5). These materials have crystallographic similarities with MoS_2 structure. Whereas MoS_2 occurs naturally, the other dichalcogenides are produced synthetically by direct combination of the elements at elevated temperatures. The composition is rarely stoichiometric, and the crystal structure is not wholly hexagonal. They are relatively expensive and specialized materials, and they can generally be considered as direct substitutes for MoS_2 where MoS_2 is limited by its relatively poor oxidation resistance above 315°C or by its relatively high electrical resistivity (Chap. 5). Some materials such as WS_2, TaS_2, WSe_2, and $TaSe_2$ exhibit better oxidation stability in air than MoS_2 (Chap. 5). WS_2 and $NbSe_2$ exhibit higher temperature capabilities (1200°C) in vacuum than MoS_2. WS_2 is an effective lubricant in vacuum; it behaves very similarly to MoS_2 (Lavik, 1959; Flom et al., 1965; Spalvins and Buckley, 1967).

Figure 13.17 summarizes data on the effects of speed and normal stress on the coefficients of friction of MoS_2, WS_2, WSe_2, and $NbSe_2$ coatings. These values were obtained with the solid lubricant coatings deposited by hand burnishing the powder onto metal surfaces (Anonymous, 1965).

Limited studies have been performed on WS_2 and $NbSe_2$ coatings deposited by sputtering (Spalvins and Buckley, 1967). WS_2 coatings exhibit a low coeffi-

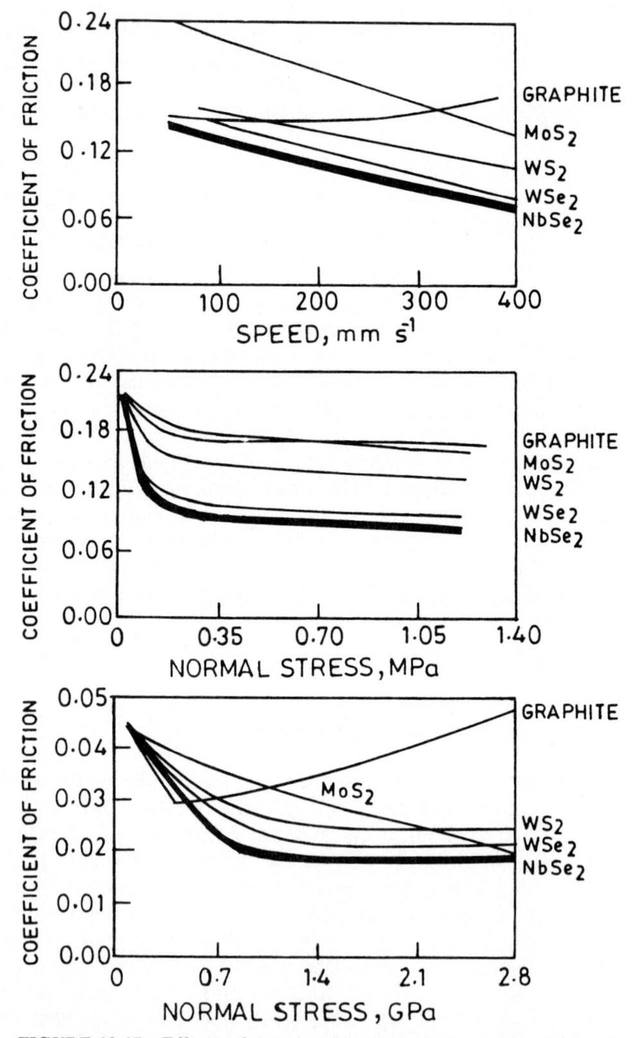

FIGURE 13.17 Effects of speed and normal stress on the coefficient of friction of hand-burnished coatings of graphite, MoS_2, WS_2, WSe_2, and $NbSe_2$ onto metal surfaces. [*Adapted from Anonymous (1965).*]

cient of friction (0.05) in vacuum and wear lives of over a million cycles with lubrication characteristics similar to MoS_2. But these coatings exhibit a high coefficient of friction (0.22) and a short wear life after only 1500 cycles when exposed to air. This failure is due to the formation of H_2SO_4 as a result of the reaction between sulfur in the coating with moisture and oxygen in atmosphere. Ramalingam et al. (1981) deposited WSe_2-In-Ga alloy by sputtering for heavily loaded rolling contact bearings. Comparative tests show that WSe_2-based coatings are as effective as sputtered MoS_2 coatings.

An air sprayed coating of WS_2 is available under a trade name Dicronite DL-5 from Rotary Components, Inc., Covina, California. The high cost and uncertainties about perfect layer lattice hexagonal structure and stoichiometry has precluded the widespread use of these coatings. However, WS_2 and WSe_2 have found limited applications in self-lubricating composites.

13.2 NONLAYERED LATTICE SOLID LUBRICANT COATINGS

Conventional solid lubricants such as graphite, molybdenum disulfide, and graphite fluoride oxidize or disassociate above about 500°C (Bhushan, 1987a; Lavik et al., 1968; Sliney, 1982). Numerous inorganic salts with low shear strength and film-forming ability have shown promise as solid lubricants for high-temperature applications (Amato and Martinengo, 1973; Bisson and Anderson, 1966; Bhushan and Gray, 1980; Johnson and Sliney, 1959; Sliney, 1974, 1979, 1982; Sliney et al., 1965). In fact, some of these solid lubricants exhibit high friction and wear at low temperatures.

For the high-temperature applications, the "quasi"-solid lubricants and the soft nonabrasive film-forming lubricants constitute the most important families. *Quasi-solid lubricants* are defined as lubricating materials that are solid at the bulk temperature of the sliding pair, but soft or molten from frictional heat at the point of sliding contact. Generally, the lubrication of this family of materials is obtained through soft ceramic coatings that readily glaze during sliding, and above its softening temperature shears by viscous flow. In this family, the lead monoxide-lead silicate, fusion-bonded coatings are the most promising compositions (Amato and Martinengo, 1973; Johnson and Sliney, 1959, Sliney, 1966a). Oxides are, of course, obvious candidates for consideration when oxidation-resistant materials are required. The hard oxides, typical of ceramic materials, such as alumina and silica have good wear resistance but generally high coefficients of friction. Further wear debris of hard oxides is abrasive.

Soft, nonabrasive film-forming lubricants are defined as lubricating materials with properties which can form a film with low friction at the points of sliding contact. In the family of these materials, fluorides of alkali metals and alkaline earth metals are the best choice because of their good chemical stability, high melting point, and nonabrasive property (Amato and Martinengo, 1973; Sliney, 1974, 1979, 1982; Sliney et al., 1965). Fluoride-based coatings have been deposited by either fusion or plasma spray processes. Of all these coatings, CaF_2 and CaF_2-BaF_2 eutectic-based plasma sprayed coatings are of most interest today for high-temperature applications. Some CaF_2-based coatings can be used from room temperature to 900°C. Discussion of the lead oxide and fluoride-based coatings follow.

13.2.1 PbO-Based Coatings

Soft oxides such as lead monoxide (PbO) are relatively nonabrasive and have relatively lower friction coefficients, especially at high temperatures where their shear strengths are reduced to the degree that deformation occurs by plastic flow rather than brittle fracture. Binary and ternary eutectic oxide systems are of interest because the melting-point suppression, which is the primary characteristic of the eutectic systems, tends to lower the shear strength relative to the individual oxides. If the second oxide is a vitrifying agent such as SiO_2, glaze (glass) formation is promoted at the sliding surface, and this tends to modify friction by introducing a viscous component of shear. This increases or decreases friction depending upon the viscosity of the glaze within the sliding contact. Therefore, the friction behavior of an oxide coating is controlled by a mechanism involving either or both crystalline shear and viscous drag (Sliney, 1982).

Increasing the surface temperature reduces both crystalline shear strength and glass viscosity within the sliding contact and, therefore, tends to reduce the coefficient of friction. This is illustrated in Figs. 13.18(*a*) and (*b*), which show the effect of ambient temperature and sliding velocity on the coefficient of friction and wear of a 440 C stainless-steel disk lubricated with a fusion-bonded coating of PbO-SiO_2 sliding against an Inconel alloy pin. At low sliding velocities, the PbO coating lubricated effectively over only a very small temperature range of about 500 to 650°C. With increasing sliding velocity, frictional heating rates increase, and low friction and wear are achieved at even lower ambient temperatures until at 6 m s^{-1}, coefficients of friction of 0.2 or lower are observed from room temperatures to 650°C. From frictional heating, the PbO-based coatings melt at the point of sliding contact to form a low shear strength film. This melting may be very localized and of very short duration during the time asperities are in sliding contact. For this reason, these lubricants are sometimes called *quasi-solid-lubricants*. Because of the narrow range of temperatures at which PbO lubricates effectively at low velocities, its use has been limited to high-speed, high-temperature applications such as lubrication of dies for high-speed wire drawing (Sliney, 1982).

In PbO-SiO_2 coating, about 5 percent of SiO_2 is primarily added to stabilize PbO against transformation to red lead (Pb_3O_4) which is not a good lubricant. PbO transformation to Pb_3O_4 occurs at lower temperatures (375 to 500°C) which reverts back to PbO above 550°C (Ott and McLaren, 1968). Below silica content of 6.7 percent, the phase present at equilibrium are PbO and tetralead silicate (4PbO-SiO_2). During firing of the coating process, the SiO_2 combines with molten PbO and forms lead tetrasilicate which, upon solidifying, acts as a binder for the PbO phase of the coating. It has also been noted that if the PbO-SiO_2 coating is applied on a substrate which does not form iron oxide (Fe_3O_4) during firing of the coating, the coating adherence is poor. It was observed that the iron oxide increases the fluidity of PbO which becomes homogeneous more readily than a highly viscous one, and this is reflected in improved uniformity after solidification resulting in a better bond. In addition, the presence of iron-oxide rich transition layer at the ceramic-metal interface is helpful in improved bond (Murray and Peterson, 1964). Johnson and Sliney (1959) added 10 percent by weight of Fe_3O_4 to the coating mixture when the coating was applied to the substrate which does not form iron oxide during firing, for example, 304 stainless steel and Inconel alloys.

Bhushan (1980a) and Bhushan and Gray (1980) added 10 percent of silver by weight to PbO-based coatings to further improve its low-temperature perfor-

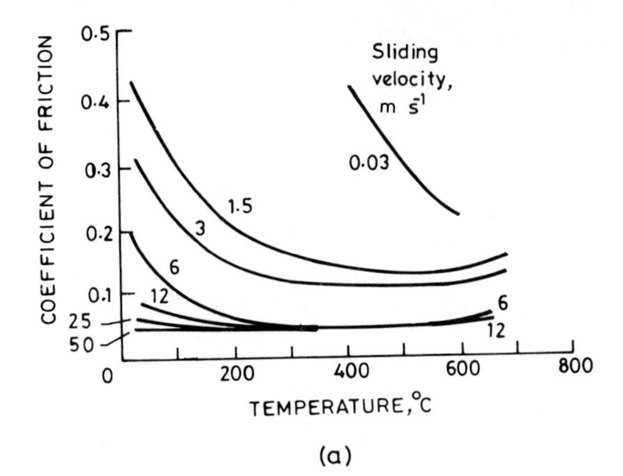

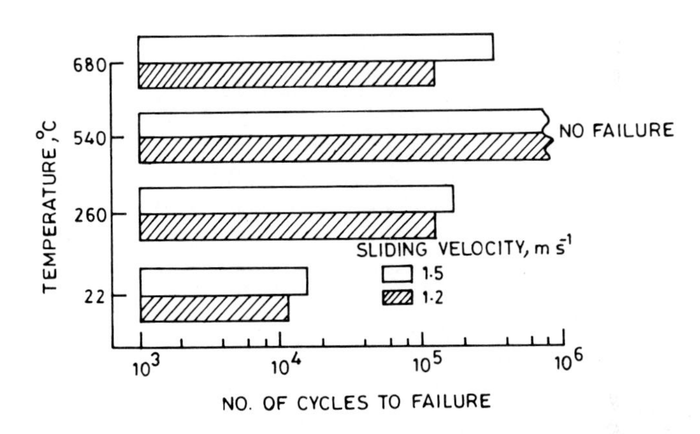

FIGURE 13.18 Effect of sliding velocity and temperature on: (*a*) Coefficient of friction. (*b*) Pin wear life of fusion bonded PbO-SiO_2 (95 wt %-5 wt %) coating of 30 μm thickness on a 440C steel disk substrate sliding against cast Inconel pin. Normal load = 9.8 N. [*Adapted from Johnson and Sliney (1959).*]

mance. Coatings were applied on foil bearings for air-lubricated automotive gas turbine engine. Coatings applied to A-286 journal did not contain Fe_3O_4 (85% PbO, 5% SiO_2, and 10% Ag), and the coatings applied to the Inconel X-750 foil substrate contained Fe_3O_4 (75% PbO, 5% SiO_2, 10% Ag, and 10% Fe_3O_4). The mixture was thoroughly dispersed in water, and the slurry was sprayed or brushed on the clean and preoxidized surfaces. The sprayed coating was first heated at 95°C to remove water and then fired at 800 to 900°C for about 1 min after melting with a total of 3 to 7 min followed by air cooling. Since the melting point of PbO-SiO_2 is 760°C, the mixture can be fired at a temperature slightly above 760°C, and the firing time depends on the substrate weight or the time it takes to heat the surface. Since silver melts at about 960°C, mixture with silver was fired at temperatures slightly below 960°C so that silver does not flow and make lumps. After baking, the coatings were burnished to a thickness of 15 to 25 μm. These PbO-based coatings have performed marginally for foil bearing applications (Bhushan, 1980a).

Amato and Martinengo (1973) studied the effect of coating hardness on the wear life of the $PbO-SiO_2$ coatings. In order to evaluate the influence of the hardness of $PbO-SiO_2$ coatings on their friction and wear properties, they mixed 2 percent by weight Fe, Al, or stainless steel (AISI 341 or 340) to the $PbO-SiO_2$ coatings. Figure 13.19(*a*) shows the hardness values as a function of temperature of the mixtures compared to the hardness of a mating surface (AISI 430 stainless steel). The coefficient of friction and pin wear (AISI 430) in a pin-on-disk test with disk coated with various PbO + SiO_2 + metallic elements as a function of temperature are compared in Figs. 13.19(*b*) and (*c*). The $PbO-SiO_2$-Fe coatings show lowest friction and minimum wear. These coatings exhibit good wear resistance and thermal stock resistance over a wide range of temperature, i.e., from 25 to 650°C. Based on the hardness, friction, and wear data, Amato and Martinengo speculate that the hardness of PbO-based coating should be slightly

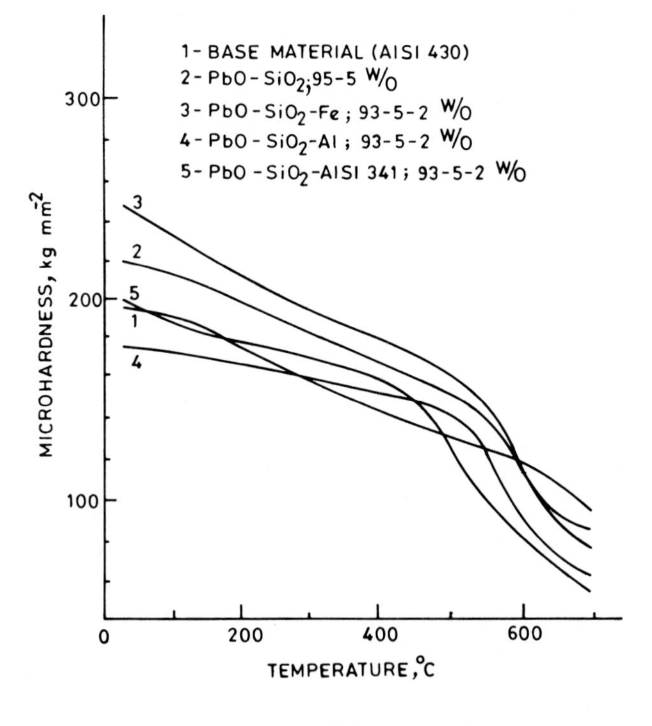

FIGURE 13.19 Effect of temperature on (*a*) hardness, (*b*) coefficient of friction, (*c*) pin wear of PbO-based coatings on annealed AISI-430 steel disk substrate sliding against a pin of AISI 430 steel at a normal load of 9.8 N, speed of 0.5 m s^{-1} for 1 h. [*Adapted from Amato and Martinengo (1973).*]

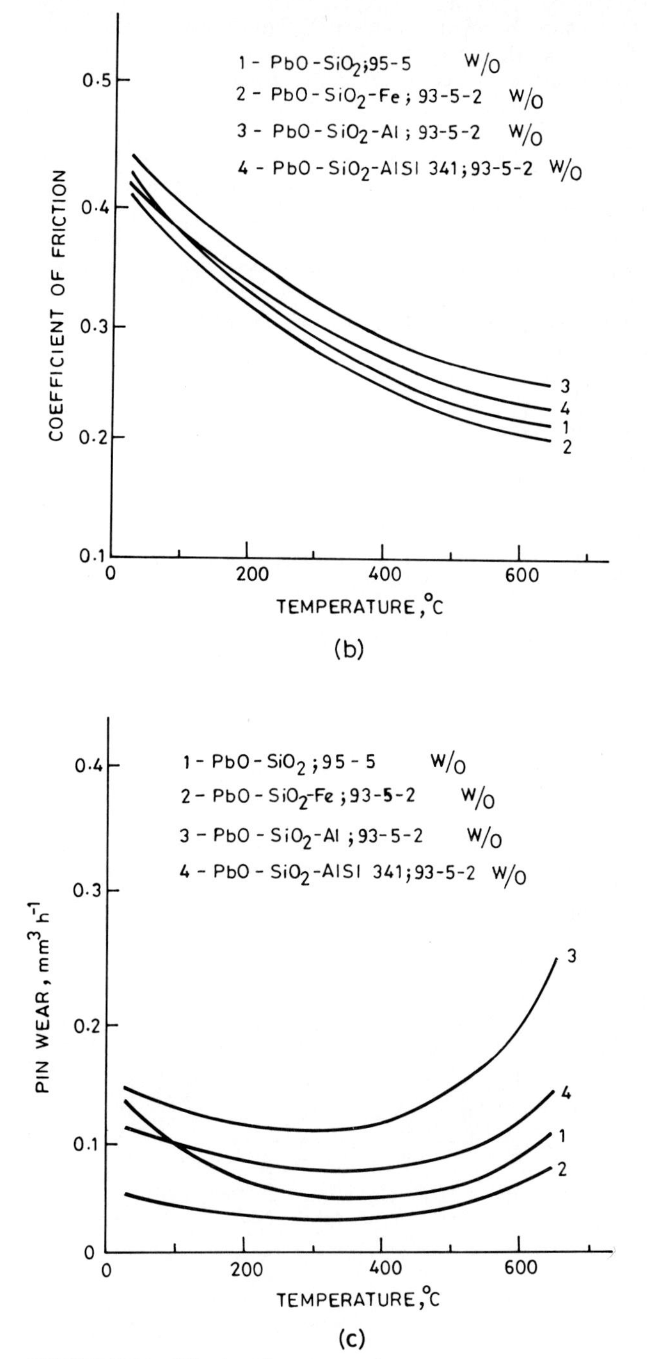

FIGURE 13.19 *(Continued)*

larger than that of the mating surface for good tribological performance. This speculation may not hold for other lubricants.

We note that other lead compounds PbS-based bonded coatings have been researched with very limited success (Lavik, 1960; McConnell, 1961). PbS coatings bonded with B_2O_3 (PbS-85%, B_2O_3-15%) provided a coefficient of friction of about 0.12 at 540°C but poor durability at high temperature and high friction at room temperature. To improve its room-temperature friction, MoS_2 and graphite were added to PbS. PbS-MoS_2-B_2O_3 (PbS-25%, MoS_2-50%, B_2O_3-25%), and PbS-graphite-B_2O_3 bonded coatings have also been tested by McConnell (1961). The wear life of these coatings was marginal especially at high temperatures, and these coatings are not used in any commercial applications. PbS and B_2O_3 are used as additives to solid lubricants. Sb_2O_3 is used as an additive to MoS_2 as a corrosion inhibitor.

13.2.2 Fluoride-Based Coatings

Chemically stable fluorides of some Group I and II metals, such as CaF_2, BaF_2, LiF, and their eutectics and mixtures lubricate effectively at extremely high temperatures ranging from 400 to 900°C (Amato and Martinengo, 1983; Bhushan, 1980a; Olsen and Sliney, 1967; Sliney, 1962, 1966a, 1966b, 1974, 1979; Sliney et al., 1965). Among the various fluorides, best properties have been obtained with CaF_2 and CaF_2-BaF_2 eutectics (Sliney, 1966a). CaF_2-BaF_2 eutectics have the lowest melting point of any other combination of the two fluorides, which gives greater plasticity and better lubrication ability at the desired temperatures (Fig. 13.20). However, for very high temperatures (above the melting point of eutectic)

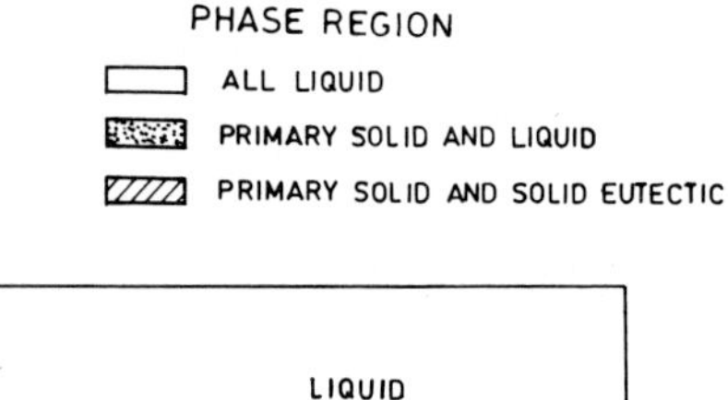

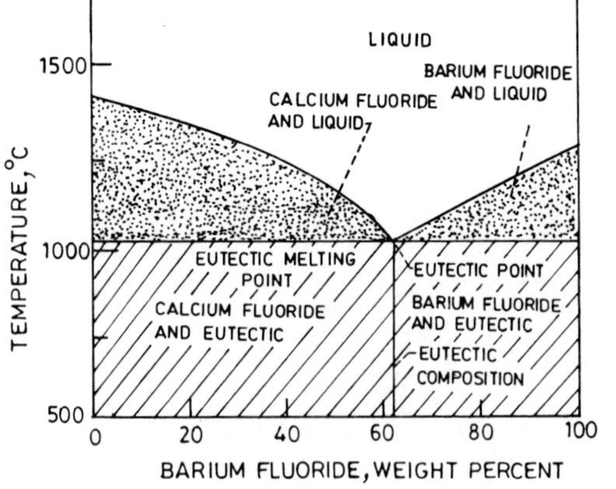

FIGURE 13.20 Phase diagram for calcium fluoride-barium fluoride.

applications, CaF_2 must be used. Fluorides have exceptionally good chemical stability in both strong oxidizing and reducing atmospheres and good matching of thermal expansion coefficients with the number of engineering alloys. Other attractive properties of these fluorides are their adequately high melting points (all above 800°C) and their nonabrasive property.

The coatings of the fluorides and various additives have been applied by ceramic bonding, fusion bonding, and plasma spray processes. The incorporation of finely dispersed Ag particles into the coating microstructure improved low-temperature friction and wear (Sliney, 1982). Silver gives good low-temperature friction since it can exhibit a high degree of plasticity even at ambient temperature. Metals, carbides, and oxides have been added in some coatings to improve their strength and ductility for improved wear life.

13.2.2.1 ***Ceramic-Bonded and Fusion-Bonded Coatings.*** Fluoride coatings have been bonded to metal surfaces in two ways: ceramic bonding and fused-fluoride bonding. In ceramic bonding, the coating slurry (water based or cellulose nitrate lacquer based) is applied by air spraying followed by firing a mixture of a fluoride with an oxide ceramic binder in an air atmosphere at a temperature above the melting point of the binder but below that of the fluoride. In fused-fluoride bonding, the coatings are applied by air spraying, followed by firing a mixture of fluorides at a temperature above the melting point of the fluoride. In many instances, fluoride coatings are fused in a hydrogen atmosphere to prevent oxidation of the base metal during firing (Amato and Martinengo, 1973; Bhushan, 1980a; Johnson and Sliney, 1959; Olson and Sliney, 1967; Sliney, 1962, 1966a).

CaF_2 coatings with a cobaltous oxide (CoO) binder (mp of binder 870°C, fired at 1090°C) with CaF_2 to binder ratio of 3 : 1 were tested by Sliney (1966a). Friction and wear data as a function of temperature, of the fluoride coatings on a disk of Inconel X sliding against a pin of cast Inconel are shown in Fig. 13.21. The pin wear and friction of uncoated specimens are included for comparison. The coating operated satisfactorily in the temperature ranging from 260 to 820°C. A fused all fluoride CaF_2-BaF_2 eutectic (38–62 wt %) (mp of eutectic 1025°C, fired at 1050°C) was tested by Sliney et al. (1965). Friction and wear data are also shown in Fig. 13.21. Over the entire temperature range the coatings reduced friction and wear relative to the uncoated metals. Similar trends have been reported by Sliney et al. (1965) for tests conducted in hydrogen-nitrogen and liquid sodium environments. Sliney et al. (1965) tested several other fused-fluoride coatings (LiF, 77 wt % CaF_2-23 wt % BaF_2, 80 wt % CaF_2-20 wt % BaF_2, 81.4 wt % CaF_2-18.6 wt % NaF). No one component showed an outstanding superiority in wear life; however, only the CaF_2-BaF_2 coatings were unaffected by sodium exposure.

Amato and Martinengo (1983), Bhushan (1980a), and Olson and Sliney (1967) added submicrometer-sized particles of silver to the fused-fluoride coating compositions to reduce friction and wear at low temperatures. In their study, Amato and Martinengo deposited fused-fluoride coatings on polished and sandblasted AISI-430 stainless-steel surfaces. Coefficient of friction and wear of CaF_2-BaF_2 eutectic (38-62 wt % with a melting point of 1025°C), CaF_2-BaF_2 (60-40 wt % with a liquidus temperature of 1230°C), CaF_2-LiF (77-23 wt % with a liquidus temperature of 1150°C) and CaF_2-BaF_2-Ag eutectic (45-30-25 wt %) coatings are shown in Fig. 13.22. We note that at low temperatures, the CaF_2-BaF_2 system have very high values of coefficient of friction, i.e., 0.6 and only at temperatures above 400°C it is less than 0.4, going at 700°C to 0.2. The CaF_2-LiF systems exhibit lower value of friction in the temperature range of 25 to 400°C, but higher in the temperature range of 400 to 700°C. The CaF_2-BaF_2-Ag system exhibits lower fric-

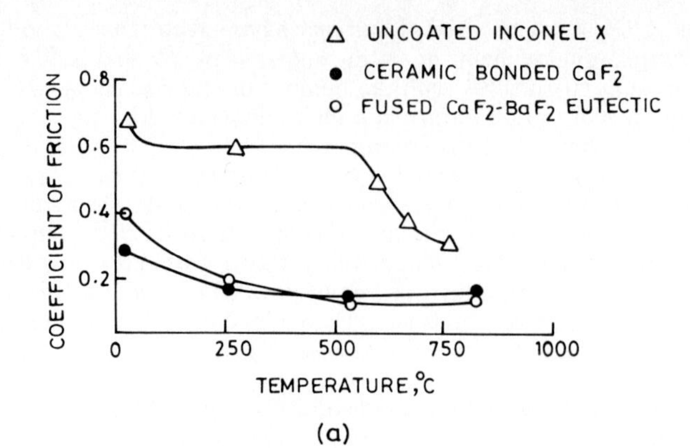

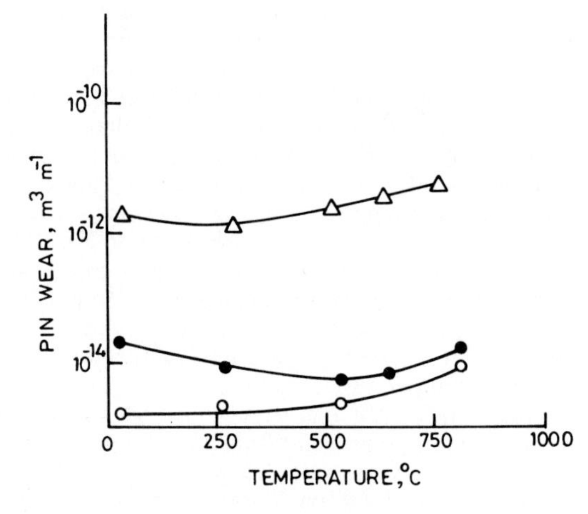

FIGURE 13.21 Effect of temperature on the (*a*) coefficient of friction and (*b*) pin wear rate of 50-μm-thick ceramic bonded CaF_2 coating and 30-μm-thick fused CaF_2-BaF_2 eutectic coating on Inconel disk substrate sliding against cast Inconel pin at a normal load of 4.9 N and sliding velocity of 2 m s^{-1}. [*Adapted from Johnson and Sliney (1959) and Sliney et al. (1965).*]

tion than others over the entire range of temperature. The wear behavior of CaF_2-based coatings is analogous to the friction coefficient obtained for these coatings. The wear increases with decreasing temperature, while for the CaF_2-BaF_2-Ag system it is almost constant and quite low over the entire temperature range. The friction and wear of the coating deposited on the polished surfaces were superior to that of sandblasted surfaces.

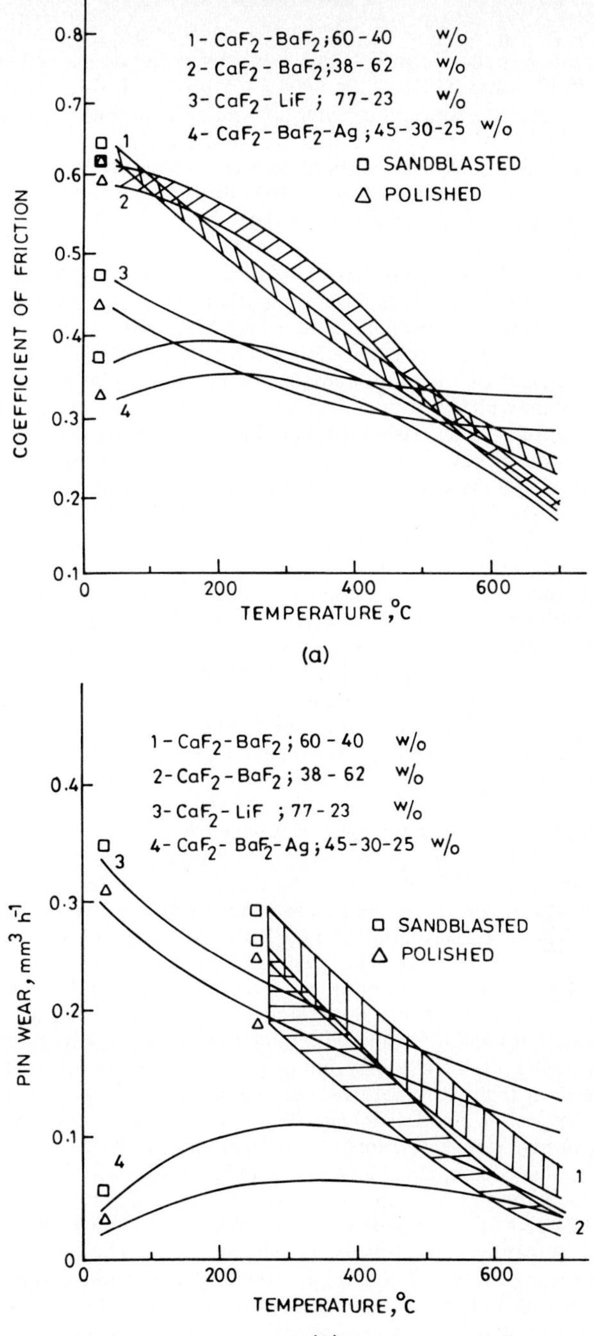

FIGURE 13.22 Effect of temperature on the (*a*) coefficient of friction and (*b*) pin wear rate of fused fluoride coatings on annealed AISI 430 steel disk substrate sliding against a pin of AISI 430 steel at a normal load of 9.8 N and speed of 0.5 m s^{-1} for 1 h. [*Adapted from Amato and Martinengo (1973).*]

Two ceramic-bonded fluoride-based coatings were developed in 1960s for aerospace applications: AFSL-28 and NASA OFS-6. AFSL-28 coating consists of CaF_2-BaF_2 eutectic, Au, and a water-soluble ceramic binder aluminum phosphate and is cured at 925°C for 1 min (Anonymous, 1972; Lavik and Moore, 1969). NASA OFS-6 coating consists of 58.5 wt % CaF_2-BaF_2 eutectic, 35 wt % Ag, and 6.5 wt % ceramic binder (CaO-SiO_2) mixed with a lacquer and is cured at 1000°C for 20 min in argon (Bhushan, 1980a).

13.2.2.2 ***Fluoride-Metal Composites.*** Composite bearing materials in which the solid lubricant is dispersed throughout the structure are advantageous when long lubricant life is required. In some cases, a thin, bonded solid lubricant coating is used as an overlay on the self-lubricating composite material. This assures the minimum coefficient of friction obtainable by enrichment of the composite surface with lubricant while providing long life due to the underlying, self-lubricating composite material. Sliney (1966b) prepared a self-lubricating composite by vacuum impregnation of porous (50 percent dense) nickel and nickel-chrome alloy (Inconel) matrix with 35 volume % molten CaF_2-BaF_2 eutectic at about 1000°C. This composite was then coated with a 25-μm-thick sintered film of CaF_2-BaF_2 eutectic by spraying the coating material from an aqueous slurry, then curing in argon atmosphere at about 950°C.

The friction and wear as a function of temperature of the impregnated Inconel composite with and without sintered overlay of eutectic and unlubricated composites are presented in Fig. 13.23. For wear life comparisons, wear life was arbitrarily taken as the time at which the coefficient of friction increased to 0.30. It is clear that the composites have a longer endurance life than the coatings. However, these composites are difficult and time-consuming to prepare. Plasma sprayed coatings are much easier to prepare and are preferred for commercial applications (Sliney, 1966b).

13.2.2.3 ***Fluoride-Metal Plasma Sprayed Composite Coatings.*** Composite coatings of CaF_2 with metal and glass have been produced by the plasma spraying process (Sliney, 1974, 1979, 1982). Typical composition of the two early coatings applied by plasma spray are designated as NASA PS100 (67 wt % nichrome, 16.5 wt % CaF_2, 16.5 wt % glass) and NASA PS101 (30 wt % nichrome, 30 wt % Ag, 25 wt % CaF_2, 15 wt % glass). The glass used for mixing in these coatings was sodium free with a composition of 58 wt % SiO_2, 21 wt % BaO, 8 wt % CaO, 13 wt % K_2O. Nichrome (NiCr) metal was added to increase strength, ductility, and thermal conductivity and to promote bonding to superalloy substrates. Glass was incorporated to harden the coatings and to act as a diffusion barrier to inhibit oxidation of NiCr at high temperatures. Ag was added to reduce friction at lower temperatures (<300°C). The PS-100 and PS-101 coatings (100 to 250 μm thick after finishing) are used in slow-speed applications operating in extreme environments, for example, the hot section of small jet engines and several space applications such as airframe bearings (of bay door) for use on the space shuttle orbiter.

Figure 13.24 gives the dependence of coefficient of friction on operating temperature for oscillating journal bearings (Sliney, 1979). The cylindrical bores of the bearings made of Rene 41 (a precipitation hardenable alloy) were coated with 250 μm of PS 100 and PS 101 compositions. All other bearing elements were made of Rene 41 with a hardness of HRC 32. A preoxidized, but otherwise uncoated, unlubricated bearing was included for comparison. The oxide film on the preoxidized bearing provided some protection against galling for the time being, but the bearing seized at 870°C. The addition of Ag in nichrome-CaF_2-glass

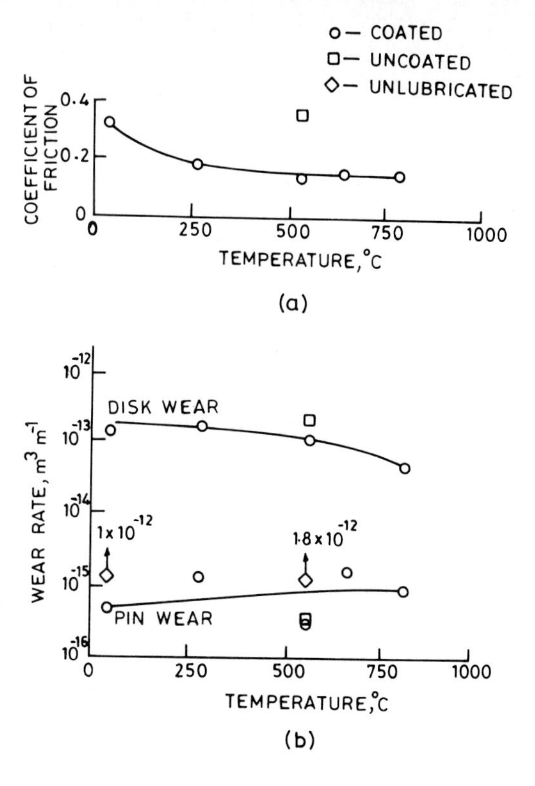

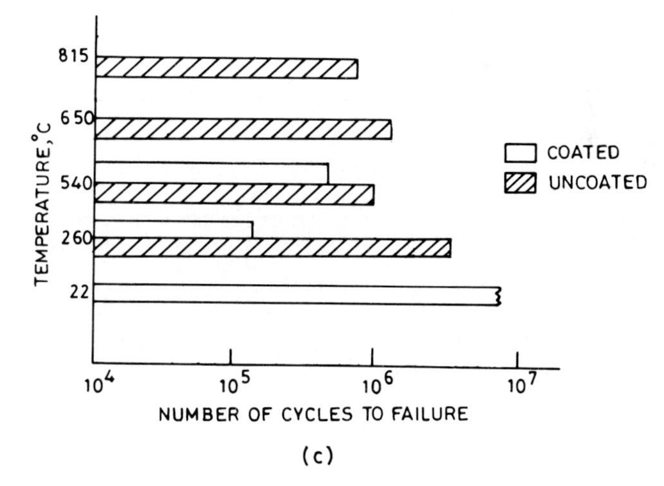

FIGURE 13.23 Effect of temperature on: (*a*) Coefficient of friction. (*b*) Disk and pin wear rates. (*c*) Wear life of fluoride-Inconel composites with and without coatings and unlubricated Inconel composites. Friction and wear tests were conducted using a pin-on-disk tester with a pin of cast Inconel at a normal load of 4.9 N and sliding velocity of 2 m s^{-1}. [*Adapted from Sliney (1966b).*]

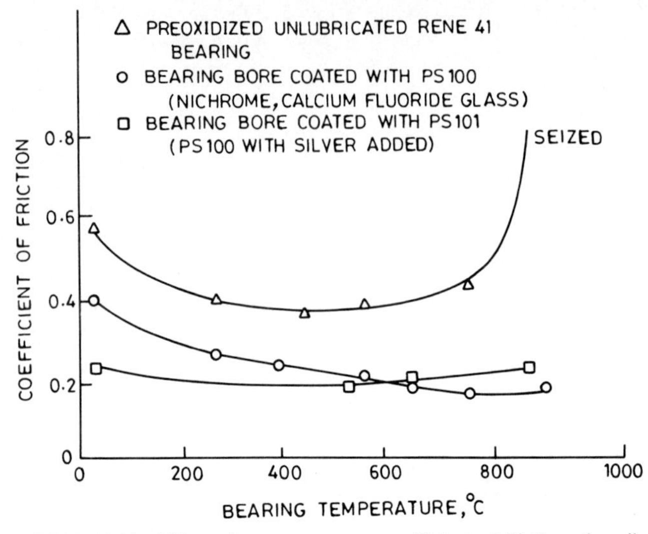

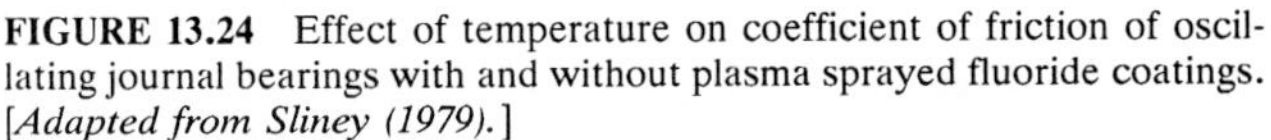
FIGURE 13.24 Effect of temperature on coefficient of friction of oscillating journal bearings with and without plasma sprayed fluoride coatings. [*Adapted from Sliney (1979).*]

showed lower values of friction coefficient and better lubrication in the temperature range of 500 to 900°C. The addition of Ag in nichrome-CaF_2-glass coatings resulted in lower friction (~0.25) in low temperature range (up to 600°C) without moderately reducing its high-temperature capability.

Table 13.7 presents the friction and wear data for the PS101 coatings which were tested in moderate vacuum, cold nitrogen gas, and in air. The lowest bearing friction (0.15) was observed in the 7 Pa (5×10^{-2} torr) vacuum. The wear rate remains almost constant over the entire temperature range. The wear rates are

TABLE 13.7 Wear Data for Oscillating Plain Slider Bearings* Self-Lubricated with a PS101† Plasma-Sprayed Coating in Various Atmospheres

Bearing temperature, °C	Ambient atmosphere	Typical friction coefficient	Increase in radial clearance, μm	
			After 100 cycles	After 5000 cycles
Room	Vacuum (5×10^{-2} torr)	0.15	13	45
−107	Nitrogen	0.22	3	38
Room	Air (760 torr)	0.24	5	70
540		0.19	5	60
650		0.21	3	25
870		0.23	3	25

*Normal pressure = 35 MPa, ±15° oscillation at 1 Hz.

†PS101 coating: 30 wt % NiCr, 30 wt % Ag, 25 wt % CaF_2, 15 wt % glass, coating thickness = 0.25 mm.

Source: Adapted from Sliney (1982).

low enough for slow-speed applications such as oscillating plain bearings for operation over a wide range of temperatures and atmospheric conditions.

For high-speed sliding applications, NiCr has been substituted with hard metal for longer lives. Sliney and Jacobson (1981) prepared several CaF_2 formulations which contained CaF_2, Ag, and two Laves phase cobalt-based alloys: Tribaloy 400, and Tribaloy 800 (see Table 13.8). NiCr-based fluoride, Tribaloy 400 and

TABLE 13.8 Chemical Compositions of Plasma-Sprayed Coatings* on Inconel 750 Disks

Coating code number PS-	Composition, wt %
T 400	62 Co-28 Mo-8 Cr-2 Si
T 800	52 Co-28 Mo-17 Cr-3Si
106	35 NiCr-35 Ag-30 CaF_2
106 T	35 T 400-35 Ag-30 CaF_2
120	60 T 400-20 Ag-20 CaF_2
121	60 T 800-20 Ag-20 CaF_2

*Undercoat: 80-μm-thick NiCr (80 wt % Ni, 20 wt % Cr); lubricant coating: 170-μm-thick composition indicated in the table.

Tribaloy 800 coatings were included for comparisons. The friction and wear tests were conducted using a tester with two flat rub shoes diametrically opposed and loaded against the outer rim of a rotating disk. The disks were Inconel 750 and coated, and the shoes were Rene 41. The disks were hardened to about HRC 35 before coating and the shoes to HRC 40.

Friction and wear data at room temperature, 400°C, and 650°C of various formulations are presented in Fig. 13.25. Tribaloy 400 and Tribaloy 800 coatings were poor both in friction and wear compared to CaF_2 coatings. For the most part, coefficient of friction of all CaF_2 coatings were in the range of 0.23 to 0.32 in all three temperatures. Wear of the cobalt-based fluoride coatings was consistently lower than with the nichrome-based fluoride coatings (PS 106). Both PS 106 and PS 106T showed a sharp increase in wear rate at 650°C. The increased wear at 650°C was attributed to the softening effect of the high content of silver (35 wt %) and CaF_2 (30 wt %) at this temperature. It is known that CaF_2 transforms from a brittle to a ductile state above about 500°C (Burn and Murray, 1962). In PS 120 and PS 121, silver and CaF_2 contents are reduced to 20 wt % each, and the wear did not exhibit an upward trend at 650°C. PS 120 gave low wear at all temperatures. PS 121 gave higher wear than PS 120 by a factor of 2 to 3 but lower wear than PS 106 at all three temperatures. All coatings were considerably oxidized after wear tests at 650°C although no evidence of spalling or delamination was apparent.

Therefore, Sliney and Jacobson conducted a brief oxidation study in which coatings of T400, PS 106, and PS 120 were held for 20 h at 650°C in air. The degree of oxidation was estimated by the weight increases of the specimens. The data are reported in Table 13.9. A thin film of oxide can be formed on the surface of silver at 200°C, but silver oxide decomposes into silver and oxygen by simple heating at temperatures above 300°C in air (Sneed et al., 1954). Therefore, silver did not oxidize during these experiments. It is also known that CaF_2 does not oxidize significantly at 650°C (Sliney et al., 1965). Therefore, all weight increases were attributed to oxidation of T400 or nichrome in the coating. Only about 2

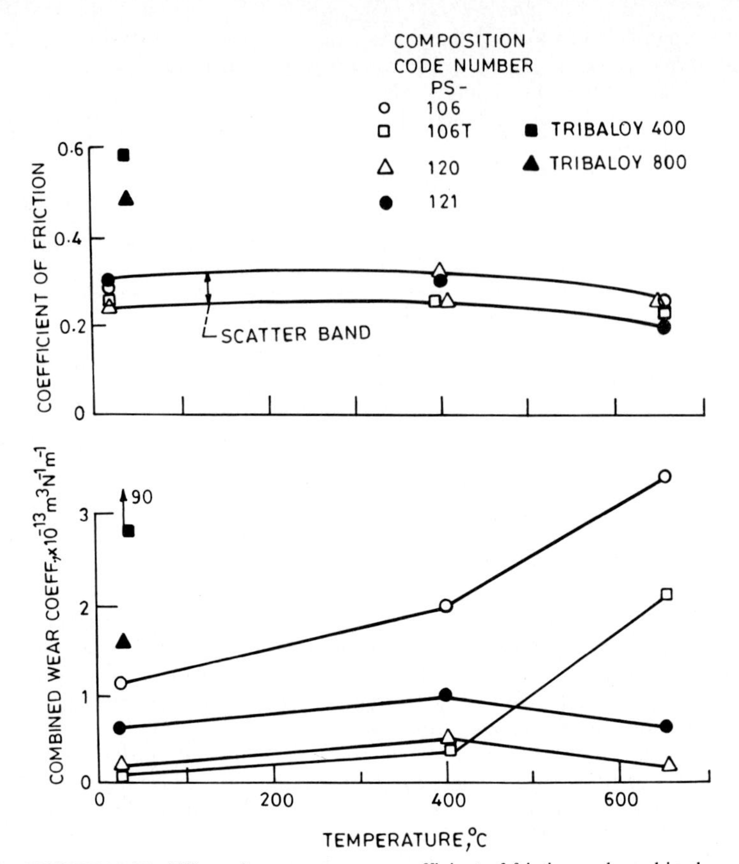

FIGURE 13.25 Effect of temperature on coefficient of friction and combined shoe and disk wear coefficient for two uncoated Rene 41 shoes diametrically opposed and loaded against the outer rim of a rotating (coated) disk. Sliding velocity = 0.27 m s^{-1}, normal load = 220 N per rub shoe. [*Adapted from Sliney and Jacobson (1981).*]

percent of the metal oxidized in the pure T400 coatings, but when the same alloy was formulated with Ag and CaF_2 in PS 120, about 70 percent of the alloy oxidized during the 20-h oxidation test. The accelerated oxidation appears to be either a catalytic effect of the silver content and/or a fluxing action of calcium fluoride on the oxidation of T400; the same is true for nichrome. Sliney and Jacobson estimated that the cobalt-based coatings could be used up to 500°C for long-duration and nichrome-based coating could be used up to 750°C for long duration.

The overall conclusions for the composite coatings with nichrome and cobalt alloys are that cobalt alloys are superior to nichrome as matrix metals in regard to wear resistance but are inferior in regard to oxidation resistance above 500°C.

13.2.2.4 Fluoride-Carbide Composite Plasma Sprayed Coatings. Ni-Cr bonded chrome-carbide coating is a thermally stable, wear-resistant coating material for

TABLE 13.9 Estimated Percentages of Cobalt or Nickel Alloy Content of Coatings that Oxidized in 20-h Furnace Exposure

Coating material	Weight increase, %	
	at 650°C	at 900°C
T 400	2	—
PS 106	40	78
PS 120	70	—

Source: Adapted from Sliney and Jacobson (1981).

temperatures to 650°C. As with most hard coatings, Cr_3C_2 is not effective in terms of low friction. Wagner and Sliney (1986) added two solid lubricants CaF_2-BaF_2 eutectic and Ag to the nickel-aluminide (Ni-Al) bonded Cr_3C_2 matrix. The bonded carbide matrix acts as a lubricant reservoir with a lower coefficient of thermal expansion than Ag-CaF_2/BaF_2 (Sliney, 1986). The composition of a coating, designated PS200, along with selected thermal properties, are presented in Table 13.10. The surface to be coated was first sandblasted, then plasma sprayed with a thin bond coat (75 μm) of Ni-Cr and then solid lubricant coating of PS200 (175 μm). This coating was applied on an Inconel journal and was tested against a preoxidized, Inconel X-750 foil in the start-stop mode, at a temperature sequence of 650°, 425°, 200°, and 25°C using a foil bearing tester. The foil was preoxidized in air to produce natural oxides-chromium oxide (Cr_2O_3) and di-iron nickel tetraoxide ($NiFe_2O_4$). A Ni-Al bonded Cr_3C_2 plasma sprayed coating was also tested for comparisons.

Figure 13.26(*a*) shows representative coefficients of starting friction as a function of start-stop cycles at various temperatures for Cr_3C_2 and PS200 coatings. Higher values of starting friction than the steady-state values are attributed to a "running-in" mechanism at the bearing-journal interface. There are several factors that explain the reduction of friction as the number of cycles accumulated: (1) As-sprayed coatings are porous and relatively rough, and sliding generates a

TABLE 13.10 Selected Thermal Properties of the Components and Composition of PS-200*

Component	Range of particle sizes, μm	Melting point, °C	Linear coefficient of thermal expansion at 650°C, $\times 10^{-6}$ °C^{-1}	Composition, wt %
CaF_2-BaF_2 (38-62) eutectic	5 to 20	1020	29.9	10
Ag	50 to 150	960	24.8	10
Cr_3C_2	10 to 70	1890	10.9	48
Ni-Al (95-5)	50 to 150	1640	15.8	32
Alloy PS 200			12.8†	

*Undercoat: 75-μm-thick NiCr; lubricant coating: 175-μm-thick PS-200.
†Weighted average.

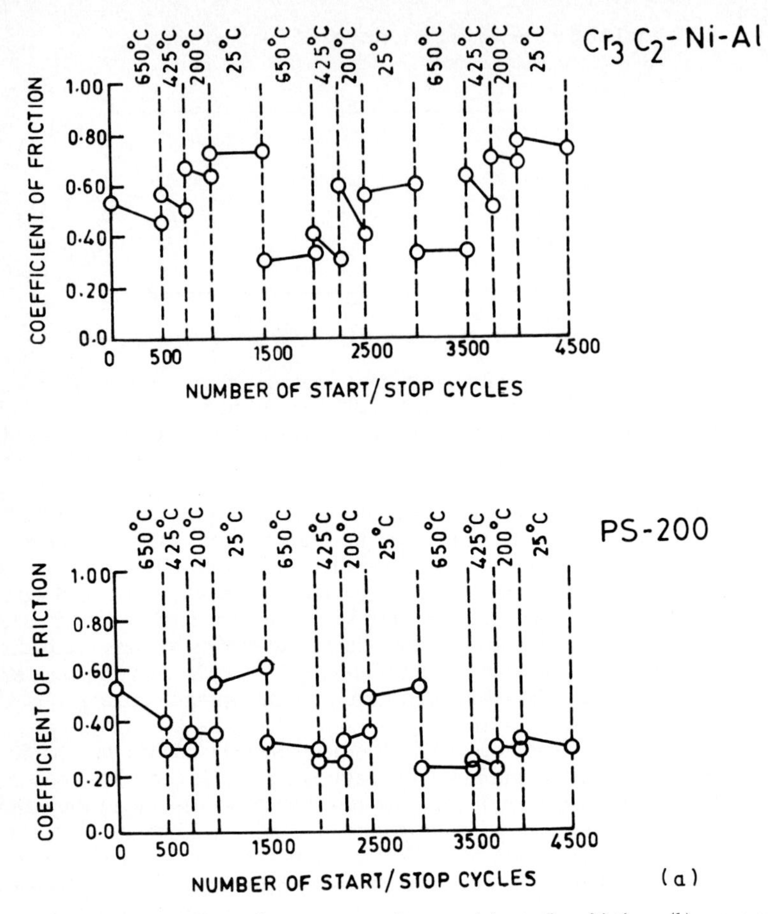

FIGURE 13.26 Effect of start-stop cycles on: (*a*) starting friction, (*b*) wear of preoxidized foil bearing against plasma sprayed Cr_3C_2-Ni-Al and PS-200 coated journals in a foil bearing tester at a test temperature range of 25 to 650°C, normal pressure of 14 kPa, and maximum surface velocity of 28 m s^{-1}. [*Adapted from Wagner and Sliney (1986).*]

smooth surface, which increases the effective radial clearance of the bearing, a factor which can be conducive to lower friction, and (2) in the case of PS200, the interspersed solid lubricants become exposed to the surface as sliding continues, decreasing the friction.

Coating endurance is defined as the number of start-stop cycles accumulated by a test bearing before a sharp rise in bearing friction was indicated or when foil wear exceeded 25 μm. Testing was terminated when friction increased to an extraordinarily high value (~0.8) at room temperature. PS200 successfully completed 9000 start-stop cycles, and Cr_3C_2-NiAl failed after 4500 cycles. Figure 13.26(*b*) shows the amount of foil bearing metal (initial thickness = 0.1 mm) removal as a function of the number of cycles. Again, bearing surface damage was excessive after 4500 cycles when tested against Cr_3C_2-NiAl. We note that there

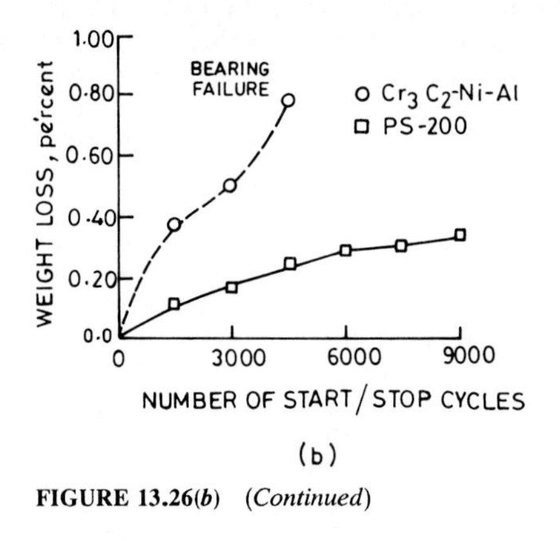

FIGURE 13.26(*b*) (*Continued*)

was no measurable wear of the coating material. Thus, PS200 gave much lower friction over the entire temperature range and lower wear than Cr_3C_2-NiAl coating without fluoride eutectic.

DellaCorte and Sliney (1987) conducted an empirical study to optimize composition of PS200 series for lowest friction and wear performance. Variations in PS200 series were tested in a pin-on-disk tester with pins made of four different materials: two stainless steels, Inconel, and Stellite 6B. Tests were conducted at a normal load of 4.9 N, sliding velocity of 2.7 m s^{-1}, and at three temperatures of 25°, 350°, and 760°C in hydrogen or helium environment. Tests of PS200 against four pin materials showed that Stellite 6B resulted in the lowest (at least an order of magnitude lower) pin and coating wear. Selected data of different coatings slid against Stellite 6B are presented in Table 13.11. The metal (Ni, Al, and Co) bonded Cr_3C_2 formulation with no added lubricants (PS218) provided adequate wear resistance for the counterface material, but the coating wear factors were higher than with

TABLE 13.11 Friction and Wear Data for Various Fluoride Eutectic Coatings Against Stellite 6B in Helium and Hydrogen*

Coating PS number	Atmosphere	Temperature, °C	Average coefficient of friction	Pin wear coefficient, $\times 10^{-16}$ m^3 (Nm)$^{-1}$	Coating wear coefficient, $\times 10^{-16}$ m^3 (Nm)$^{-1}$
PS 218	He	760	0.60	7	200
Cr_3C_2-Ni-Al-Co		350	0.50	0.2	40
(58-28-2-12)		25	0.55	0.1	140
PS 218	H_2	760	0.50	Not measur-	280
		350	0.48	able, coating	120
		25	0.35	transfer to	170
				pin surface	
PS 200	He	760	0.35	7	92
Metal bonded		350	0.25	2	8.2
Cr_3C_2-Ag-CaF_2-		25	0.38	5	1.7
BaF_2 eutectic (80-					
10-10)					
PS 212	He	760	0.26	40	50
Metal bonded		350	0.20	60	24
Cr_3C_2-Ag-CaF_2-		25	0.21	50	0.8
BaF_2 (70-15-15)					

*Pin-on-disk tester conditions: load = 4.9 N; sliding velocity = 2.7 m s^{-1}; chamber pressure = 40 kPa (300 torr).

Source: Adapted from DellaCorte and Sliney (1987).

the coatings with solid lubricants. Frictional coefficients with PS218 were very high and erratic, and there was a strong tendency to transfer coating material to the surface of the counterface material. The lowest friction and wear was achieved with PS212 coating (70 wt % metal-bonded Cr_3C_2, 15 wt % Ag, 15 wt % CaF_2-BaF_2 eutectic). Based on this study, DellaCorte and Sliney recommended PS212/Stellite 6B material combination for advanced heat engines such as adiabatic and Stirling engines.

DellaCorte et al. (1988) deposited thin silver films, 25 to 350 nm thick, onto PS200, to reduce run-in wear of the mating pin. They conducted friction and wear tests using a pin-on-disk tester with a pin made of Stellite 6B sliding against a coated disk. Tests were conducted at temperatures from 25 to 760°C in helium atmosphere with a sliding velocity of 2.7 m s^{-1} under a normal load of 4.9 N. Figure 13.27(*a*) shows pin wear volume versus sliding distance at three different temperatures for both PS200 with and without 100-nm-thick sputtered Ag. We note that the total pin wear after 3 h of sliding against PS200 coating with Ag overlay is about three times smaller than the PS200 coating. Similar trends were also shown for other thicknesses of silver. Thinner films did not seem to significantly reduce run-in wear, and plowing of the silver coating was observed when thicker coatings were tested. This plowing led to excessive silver transfer to the pin and higher friction especially at elevated temperatures. Pin-to-disk wear coefficient ratios as a function of Ag coating thickness at three temperatures are shown in Fig. 13.27(*b*). The larger the wear factor ratio, the worse the material couple is. We note that silver generally reduces the pin wear up to a film thickness of about 150 nm. Coatings thicker than 150 nm do not result in a further reduction of wear coefficient ratio.

Thus, we conclude that the application of 100- to 150-nm-thick Ag overlay

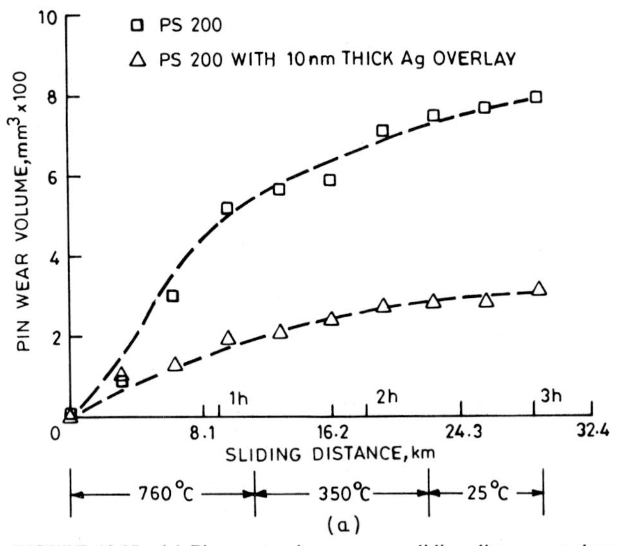

FIGURE 13.27 (*a*) Pin wear volume versus sliding distance at three test temperatures in a pin-on-disk wear test for a Stellite 6B pin sliding on a coated disk at a sliding velocity of 2.7 m s^{-1}, normal load of 4.9 N, and helium atmosphere. (*b*) Wear coefficient ratio of pin-to-disk coating versus silver coating overlay thickness at three test temperatures. [*Adapted from DellaCorte et al. (1988).*]

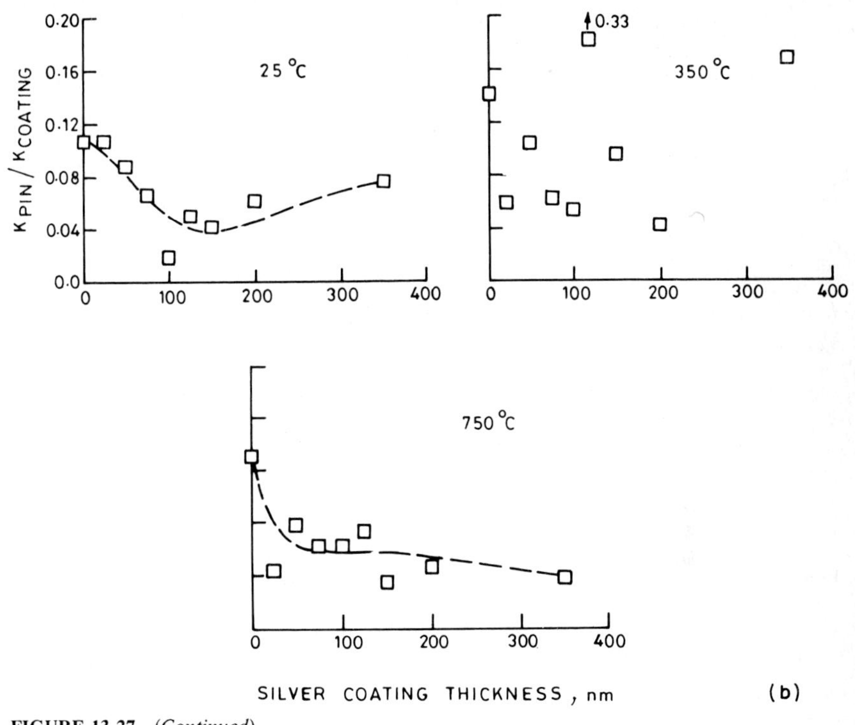

FIGURE 13.27 (*Continued*)

over PS200 reduced the run-in wear of the mating pin without causing plowing and excessive silver transfer to the pin surface. The silver film acted as a run-in lubricant which probably smoothed out the surface and reduced the initial abrasivity of the PS200 coating to the counterface material.

13.2.2.5 ***Fluoride-Oxide Composite Plasma Sprayed Coatings.*** Nonmetallic oxide coatings have drawn attention because of the limitations imposed by the oxidation of metal matrix in metal composite coatings for solid lubrication at high temperature. Moore and Ritter (1974) has reported good wear resistance at high temperatures for plasma sprayed coatings of NiO containing about 15 wt % CaF_2. This coating was deposited onto seal bars for regenerators that are used in automotive gas turbine engines to improve thermal efficiency. The porous regenerator core material used was lithium-aluminum silicate (LAS), magnesium-aluminum silicate (MAS), or aluminum silicate (AS).

The composite coatings of ZrO_2-CaF_2 (80-20) mixtures have been produced by plasma spray process (Sliney, 1982). Figure 13.28 gives the friction and wear data of plasma sprayed coatings. In these experiments, the coating was deposited on the cylindrical surface of a rotating disk and placed in sliding contact with two flat nickel-based superalloy (Inconel) rub blocks. The coatings have shown fairly high wear rates at room temperature and reasonably low wear at high temperature, i.e., 650°C. Wear of uncoated metal that slid against the coating was low in all cases indicating that the coatings were not particularly abrasive to the metal rub shoes. Friction coefficient was 0.4 at room temperature and 0.22 ± 0.04 at 650°C.

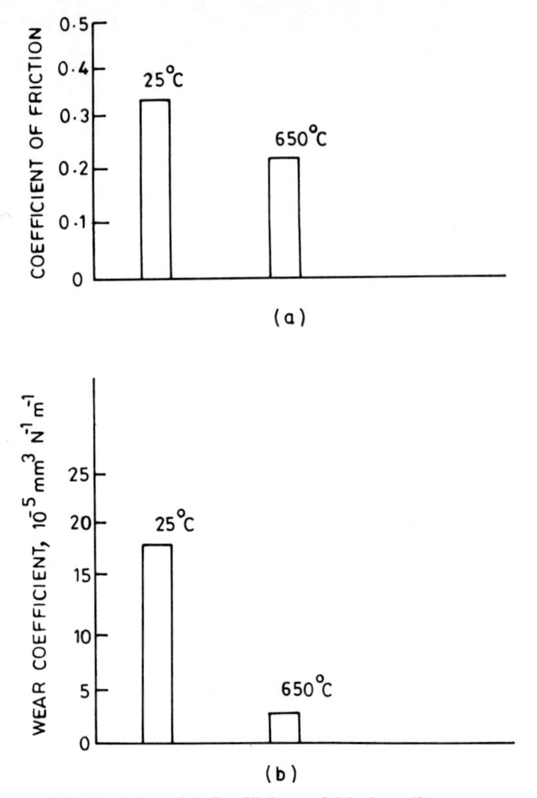

FIGURE 13.28 (*a*) Coefficient of friction, (*b*) wear coefficient of plasma sprayed coatings of ZrO_2-CaF_2 (80-20). Tests were conducted with double-rub shoe test with 215 N per Inconel shoe against coated disk at 0.3 m s^{-1} (150 r/min). [*Adapted from Sliney (1982).*]

13.3 POLYMER COATINGS

Polymer coatings are used to reduce friction and resist wear. Some polymer coatings can be subjected to operating conditions characterized by PV values greater than 2 MPa m s^{-1}. Polymer coatings can also resist abrasion and corrosion and provide thermal or electrical insulation. Typical examples of polymeric lubricants include polytetrafluoroethylene (PTFE), fluorinated ethylene propylene (FEP), polyphenylene sulfide (PPS), poly(amide-imides), polyamides (Nylons), and polyimides (Anonymous, 1983, 1984b; Behrens, 1981; Bhushan, 1987a; Clauss, 1972; Fish, 1982; Fitzsimmons and Zisman, 1958; Gaynes, 1967; Nylen and Sunderland, 1965; Paul, 1985; Willis, 1980). The bonded coatings have been applied onto metal and nonmetal surfaces by air spraying, electrostatic spraying (if available in powder form), brushing, and dipping. Polymer coatings are also applied by physical vapor deposition (evaporation, sputtering) and rf glow-discharge (plasma) polymerization for specialized applications. A bonded coating is compounded to meet specific requirements by varying solid lubricants, binders, and other additives. Coating hardness typically ranges from Shore D60 to D90.

Surface preparation (usually roughening and solvent cleaning and/or primer coat) is a major consideration for better adhesion of bonded polymer coatings (Willis, 1980; Yaniv et al., 1979). The substrate must be clean and free of oil and grease. Often a solvent wipe or vapor degreasing is adequate. However, for maximum coating adhesion, the substrate must be either grit-blasted, alkaline washed, rinsed with a mild acid, or coated with phosphate (Chap. 7; Appendix 13.A). The use of phosphates is reported to improve the wear life of PTFE coatings (Kay and Tingle, 1958).

Polymer coatings applied on harder metallic substrates usually exhibit lower coefficients of friction than bulk polymers. Metallic substrates conduct away frictional heat, and thus the performance of bulk polymers can be improved by applying in thin coating form on metals. The polymer coatings [such as PTFE, polyphenylene sulfide, poly(amide-imide), and polyimide] are extensively used in lightly loaded and low-speed bearing and other sliding applications up to an operating temperature of 200 to 300°C.

Elastomeric coatings effectively prevent erosive and abrasive wear under certain conditions. As long as the velocity of impinging solid particles is not too high, the elastomeric surface deforms and recovers elastically with no damage. Two abrasion-resistant polymer coatings of nonsolvented polyurethane and epoxy have provided unique durability on hulls of icebreakers (Calabrese et al., 1980). Polymer coatings are also used as barrier (water-repellant) coatings to protect metal parts and electronic components from corrosion. Most of these coatings belong to the large family of fluorocarbons which, as a group, are more inert to chemicals and solvents than are all other polymers. The most effective barrier coatings among fluorocarbons for a wide variety of corrosive conditions are fluorinated ethylene propylene (FEP) copolymer, perfluoroalkoxy (PFA), ethylene chlorotrifluoroethylene (ECTFE) copolymer, and polyvinylidene fluoride (PVDF). In addition to this group, PPS and Nylon 11 are also good barriers (Behrens, 1981). Many polymeric coatings are used in electrical and electronic applications that require good insulation resistance, dielectric strength, dielectric constant, and dissipation factor.

Typical applications of polymer coatings include bearings, seals, pumps, impellers, slurry tanks, punches, extrusion screws, fastening hardware, wear plates, air and hydraulic power tools, mold release media, automotive components, cookware, food processing machinery, aerospace and aircraft hardware, naval and undersea hardware, chemical process equipment, welding nozzles, welding equipments, and electronic components.

13.3.1 Resin-Bonded Polymer Coatings

The bonded coatings are a blend of solid lubricant particles in an organic resin binder, other additives, and a fast-drying solvent system. A coating is compounded to meet specific requirements by varying solid lubricants, binders, and other additives. Solid lubricants most commonly used are PTFE, polyphenylene sulfide, poly(amide-imide), polyamide, polyethylene, acetal, or a combination of these. They can be used generally up to about 200°C; softening occurs above this. Polyimide coatings are high-temperature materials and can be used up to 300°C. The solid lubricants are available as dry powders of fine particle size that can be dispersed in a binder. The binder provides cohesive forces that hold the solid coating together and also provide adhesive forces that bond the coating to a substrate. The ratio of lubricating solids to binder determines the lubricity of bonded coating. Commonly used binder materials are acrylic, polyester, phenolic, epoxy, polyimide, and PPS resins. Various additives can be included in the coating to control or improve specific coating properties. For example, wear properties can be altered by including calcium carbonate, glass beads, silica, or mica.

Many companies make resin-bonded coating solutions which are sprayed either by conventional air spray equipment or from aerosol cans, cured (by either air drying or baking at elevated temperatures), and then burnished. The polymer materials available in powder form are applied by electrostatic spray or fluidized-bed techniques. The finished thickness of a burnished coating, used to reduce friction and wear, typically ranges from 7.5 to 20 μm. A corrosion-resistant barrier coating is generally thick ranging from 15 to 250 μm; the thicker coatings are many times applied in two to three successive layers, with drying time between the layers being 5 to 10 min (Yaniv et al., 1979).

A two-step cure process for PTFE/FEP resins produces the "stratified" fluorocarbon-binder structure with superior friction and wear performance. In this two-step process, the thermosetting binder is first set at about 200°C. Then the thermoplastic fluoropolymer is stratified (floated) by a second cure at 315°C. Thus, the more durable, nonstick fluoropolymer resin forces its way to the surface, where it concentrates in the top layers to improve the hardness and low-friction properties; the lower layers retain a binder that strengthens adhesion to the substrates (Fig. 13.29). Coatings with a uniform dispersion of fluoropolymer and resin binder throughout are known as *suspended coatings*. When cured, the suspended coating produces a three-layer film with a fluorocarbon-rich surface a uniform fluorocarbon-binder matrix, and a molecular binder layer at the substrate surface (Fig. 13.29). Unlike stratified coatings, suspended coatings provide lubricity throughout the wear life of the coating. Teflon S finishes produced by DuPont for antistick cookware are stratified coatings. These are a series of water-based dispersions of PTFE and FEP resins.

Table 13.12 lists the types of resins and applications of some resin-bonded PTFE coatings (Stock, 1965). Phenolic-bonded PTFE coatings can be used up to 170 to 200°C. The coefficient of friction for the resin-bonded PTFE coatings range from 0.05 to 0.15. The PTFE coatings are free from pinholes and provide superior protection to corrosion in addition to wear. Resin-bonded PTFE-PPS and PTFE-polyamide-imide coatings have been successfully used for high-speed, low-load capacity applications (e.g., foil bearings) (Bhushan and Heshmat, 1979; Clauss, 1972; Fish, 1982; Stock, 1965).

Polyimide is another polymer which finds applications as solid lubricant coatings at elevated temperatures (300°C). It is available from several manufacturers, for example, polyimide varnish is available from DuPont Company, Wilmington, Delaware, under the trade name of PI 4701. It can be brushed on a workpiece or can be sprayed by diluting it with a thinner consisting of N-methylpyrrolidone and xylene. The friction and wear characteristics of polyimide and polyimide mixed with MoS_2 and $(CF_x)_n$ are shown in Fig. 13.30. These data were generated using a 440C steel pin on a coated 440C steel disk tester under dry Ar, dry air, and air containing 10^4 ppm water vapors at 9.8 N load and 3 m s^{-1} speed (Fusaro and Sliney, 1973; Fusaro, 1977b, 1978, 1987). The results indicate that at 25°C, moisture is beneficial in providing a longer wear life and a lower coefficient of friction. The polyimide coatings exhibit good friction and wear characteristics above 100°C in all the atmospheres. This transition in the tribological properties of polyimide has been attributed to second-order relaxation in the molecular

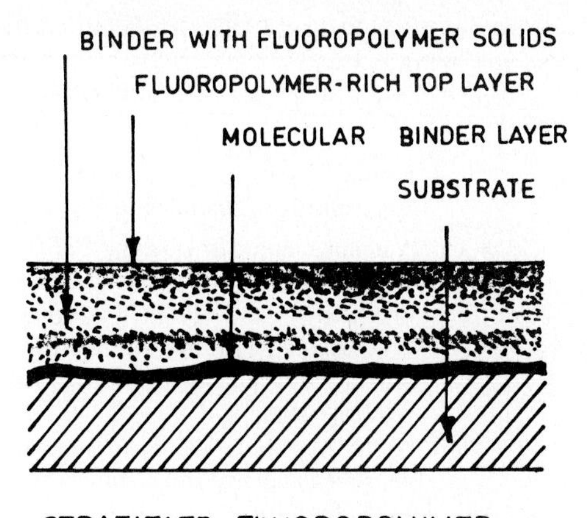

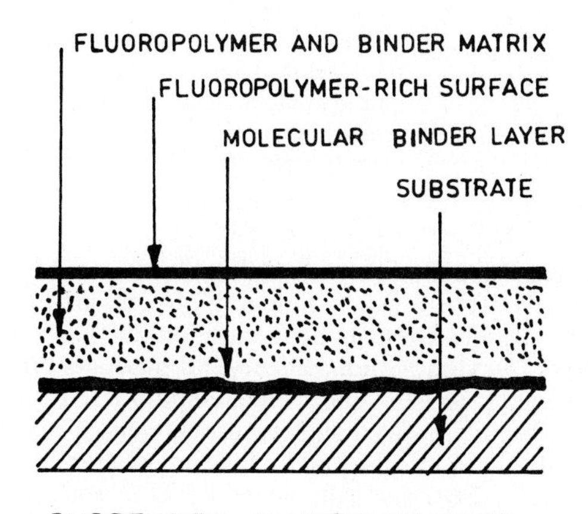

FIGURE 13.29 Schematics of fluoropolymer distribution in stratified- and suspended-type fluoropolymer coatings.

bonds of the polymer between 25 and 100°C (Fusaro, 1977b). The polyimide coatings exhibit good friction and wear characteristics above 100°C (Fusaro, 1977b). The additions of MoS_2 and $(CF_x)_n$ in polyimide improves the wear life and friction at room temperatures. In addition, this also avoids the undesirable transition in friction and wear characteristics at the transition temperature of 100°C. The fact that 75 wt % MoS_2 added to 25 wt % PI gave decreased life above 100°C indicates that this may not be an optimum binder-to-lubricant ratio.

The ready-to-spray solutions of various polymers are commercially available from a number of manufacturers. Acheson Colloids Company of Port Huron, Michigan, sells, for example, Emralon 341 which contains PTFE and PPS pig-

TABLE 13.12 Types and Uses of Resin-Bonded PTFE Dry-Film Lubricants

Type	Applications
Phenolic resin:	
Solution I	Low-baking PTFE dry-film lubricant designed for heat-sensitive substrates (baked at 150°C for 1 h).
Solution II	A nonpigmented low-baking PTFE dry-film lubricant developed specifically for food-handling and food-processing equipment (baked at 150°C for 1 h).
Acrylic resin latex	Low-baking PTFE dry-film lubricant developed primarily for flexible substrates such as O-rings and weather-stripping (baked at 150°C for 30 min).
Epoxy resin solution	PTFE dry-film lubricant for applications requiring improved stability and chemical resistance (baked at 175°C for 1 h).
Thermoplastic resin:	
Solution I	Air-hardening PTFE dry-film lubricant for heat-sensitive substrates (air dry 2 h at room temperature).
Solution II	A nonpigmented, air-hardening PTFE dry-film lubricant developed specifically for food-handling and food-processing equipment (air dry 2 h at room temperature).

Source: Adapted from Stock (1965).

ment in an organic resin binder and a water-dilutable solvent carrier and Emralon 333, which contains PTFE and poly(amide-imide) pigment in an organic resin binder and a solvent carrier. The coating is cured at 200°C for 30 min. Hohman Plating and Mfg., Inc., Dayton, Ohio, sells several products under a trade name of Surf-Kote, for example, 359 and 360, which contain PTFE pigments in a phenolic resin binder and modified alkyd binder, respectively. The sprayed coating is cured at 150°C for 30 min. Penn Dixon Company (Division of Dixon Industries Corporation) of Atlanta, Georgia, sells several PTFE and other fluorocarbon coatings under the trade names Penn Dixon T-5, T-7, T-10, TP-20, TS-35, and TS-40; PPS coatings under the trade name of Penn Dixon P-90; PTFE dispersed in a PPS-based binder under a trade name of Rulon R-95; and thermoplastic polyurethane (elastomeric) coatings under a trade name of Penn Dixon U-410. According to the manufacturer, Rulon R-95 is cured at 370°C and can be used up to 260°C in continuous service. Penn Dixon also sells several chemical-resistant coatings, for example, Penn Dixon H-610, coatings which contain ethylene-chloro-trifluoroethylene copolymer, and Penn Dixon K-710 coating, which contains polyvinylidene fluoride. Tiodize Co., Inc., of Huntington Beach, California, sells Tiolon A20 which contains PTFE pigments suspended in a fast-drying thermoplastic resin. Chemplast, Inc., of Wayne, New Jersey, sells Fluoroglide which contains PTFE pigments and organic resin in a fast-drying Freon propellant. Kamatics Corporation at Bloomfield, Connecticut sells a coating solution under the trade name of Karon V which consists of PTFE and polyester pigments.

General Magnaplate Corp. of Linden, New Jersey, sells synergistic coatings that combine the advantage of anodizing or plating with controlled infusion of

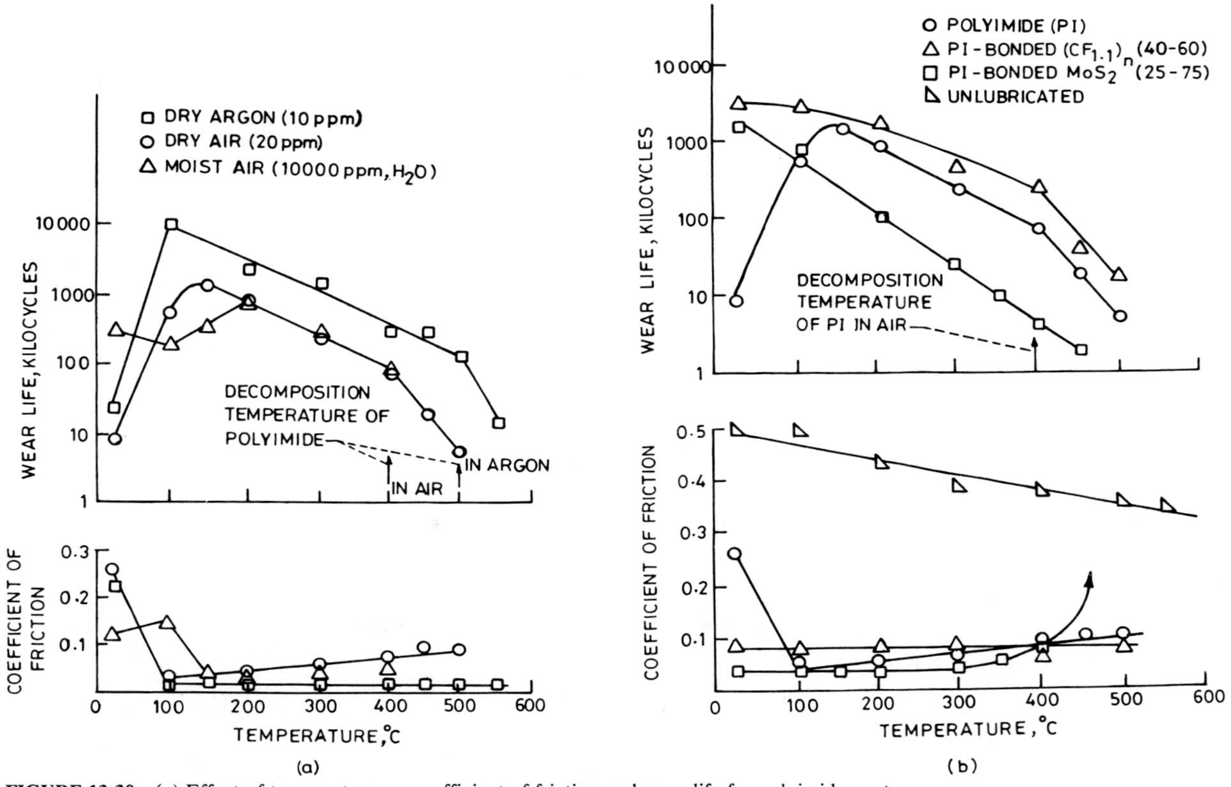

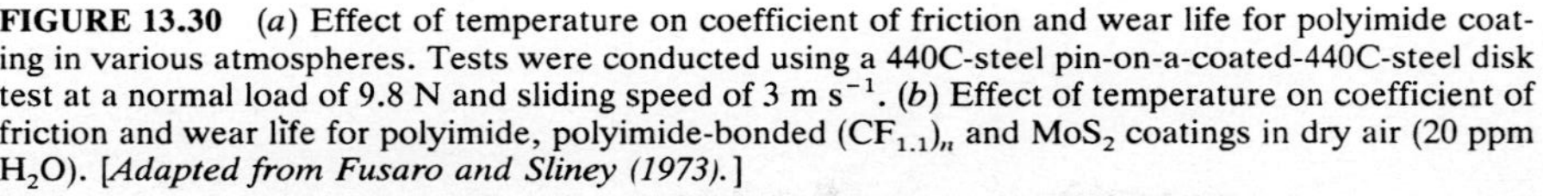

FIGURE 13.30 (*a*) Effect of temperature on coefficient of friction and wear life for polyimide coating in various atmospheres. Tests were conducted using a 440C-steel pin-on-a-coated-440C-steel disk test at a normal load of 9.8 N and sliding speed of 3 m s^{-1}. (*b*) Effect of temperature on coefficient of friction and wear life for polyimide, polyimide-bonded $(CF_{1.1})_n$ and MoS_2 coatings in dry air (20 ppm H_2O). [*Adapted from Fusaro and Sliney (1973).*]

PTFE. For example, they deposit Nedox coating on all ferrous and copper alloys which is controlled infusion of PTFE within hard chrome nickel alloy plating of electroless nickel. They also deposit Tufram coating on aluminum alloy which is controlled infusion of PTFE on an anodized aluminum surface.

13.3.2 Physical Vapor Deposited Polymer Coatings

Thin coatings of plastics with good electrical properties are used as dielectric layers for use in the capacitance industry. The polymer coatings applied by rf sputtering include PTFE and polyimide. PTFE has also been deposited by evaporation (Luff and White, 1968). The sputtered PTFE coatings exhibit excellent adherence and uniformity and pin-hole-free not only on metal substrate but also on glass, wood, and paper surfaces (Lee, 1971; Morrison and Robertson, 1973; Harrop and Harrop, 1969).

The sputtered PTFE coatings deposited by Harrop and Harrop (1969) were yellow in color and had an amorphous structure with a microhardness (>100-μm-thick coating) of about 12.5 kg mm^{-2}. We note that commercially available PTFE is white in color and has a crystalline structure with a microhardness of about 5 kg mm^{-2}. Darkening of color, amorphous structure, and increased hardness of sputtered coatings are believed to be due to a crosslinked polymer with appreciable molecular damage in glow discharge. Harrop and Harrop measured the static coefficient of friction of the sputtered PTFE coated block sliding against a flat mild steel plate at various coating thicknesses and normal loads. Results are shown in Fig. 13.31. We note that the coefficient of friction has a strong influence of coating thickness and load. The static coefficient of friction of these coatings is comparable with that of the original PTFE (~0.08) when the film is thicker than the asperities on the substrate (~1 μm). Between 2 and 12 μm, friction increases

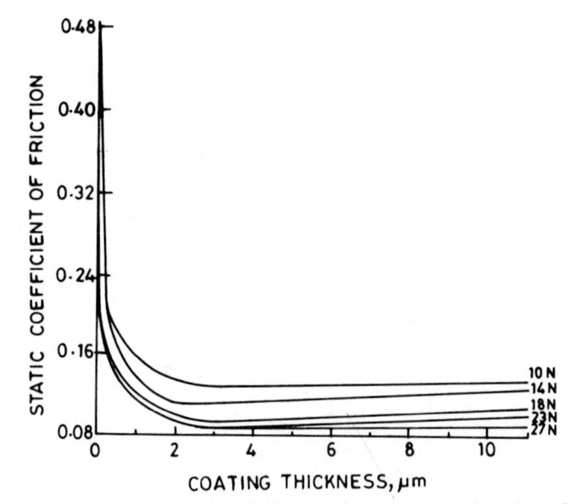

FIGURE 13.31 Static coefficient of friction as a function of thickness of sputtered PTFE coating at various normal loads in a coated block sliding against a flat mild steel plate. [*Adapted from Harrop and Harrop (1969).*]

slightly with thickness, probably due to an increased volume of polymer shearing. Below 0.5 μm, the coefficient of friction rises to that of the metal. At any given thickness, friction decreases markedly with an increase in load for most polymers (Harrop and Harrop, 1969). Harrop and Harrop heat-treated some coated samples to 450°C, cooled and tested for friction. Surface stylus measurement showed a 5 to 15 percent decrease in coating thickness. The static friction data as a function of load for treated and untreated coating are given in Fig. 13.32. Friction values have been corrected for the thickness change. We note that friction was lowered to below that expected for the new thickness. It is possible that this change in friction is due to increased crystallinity bringing the material nearer to commercial PTFE in frictional behavior.

13.3.3 Plasma Polymerized Coatings

Plasma (rf glow discharge) polymerization is a superior technique for forming thin polymer coatings from almost all organic vapors (Baddour and Timmins, 1967; Hollahan and Rosler, 1978; Shen and Bell, 1979; Yasuda and Hsu, 1978). Plasma polymerized polymer (PPP) coatings are, in general, highly crosslinked, insoluble, and chemically inert as well as thermally stable. Uniform coatings on flat metal substrates are expected from plasma polymerization because of a relatively small mean free path of monomer gas in the discharge.

Some of the coating materials which have been deposited by plasma polymerization include: fluorocarbons (such as PTFE, hexafluoropropylene), polyethylene, styrene, toluene-diisocyanate, and pyrrolidone. The plasma polymerized coatings are used as protective coatings in semiconductor electronics and various tribological applications (Bhushan, 1990; Kay and Dilks, 1981; Yasuda and Hsu, 1978). Plasma polymerization can also be used in synthesis of metal containing polymer coatings.

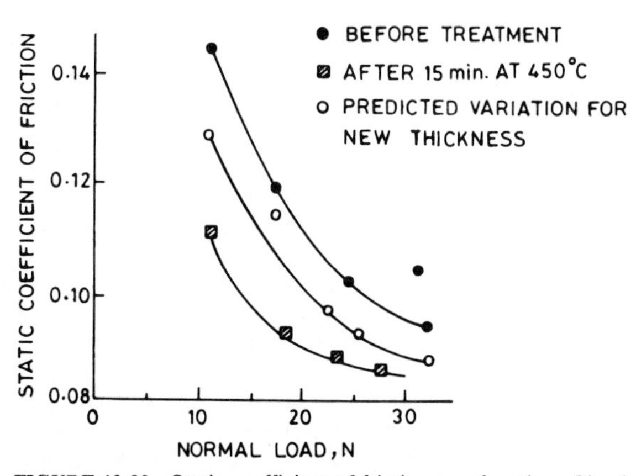

FIGURE 13.32 Static coefficient of friction as a function of load before and after heat treatment of sputtered PTFE coating at 450°C for 15 min compared with the predicted variation for new thickness (after heat treatment). Tests were conducted using a coated block sliding against a flat mild steel plate. [*Adapted from Harrop and Harrop (1969).*]

PPP coatings with a thickness ranging from 60 to 400 nm have been made from monomers-toluene-2, 4 diisocyanate (TCy), α-pyrrolidone (α-P_{yr}), tetrafluoroethylene (4 FE), and hexafluoropropylene (6 FP) by Harada (1981) for wear protection of magnetic disks. Harada tested friction and wear performance of thin protective coatings using a contact start-stop test with a magnetic slider sliding on a coated disk. He found that the coefficient of friction of PPP films ranged 0.3 to 0.4 and the minimum friction was achieved for the coating thickness equal to or larger than about 130 to 150 nm. Best durability was achieved with coatings with a thickness of about 100 nm; thicker coatings had poorer durability. Plasma polymerized toluene-2, 4-diisocyanate coatings showed good durability ($>$10,000 start-stops) because of their strong adhesion to metal substrates.

13.4 SOFT METALLIC COATINGS

Low shear strength (soft) metallic coatings such as Au, Ag, Cu, Pb, Sn, In, and Al and their alloys have received most attention as thin film solid lubricants at light loads in various environments including vacuum and elevated temperatures (Bowdon and Tabor, 1964; Buck, 1987; Buckley, 1971; Gerkema, 1985; Halling, 1983; Sherbiney and Halling, 1977; Spalvins and Buckley, 1967; Spalvins, 1978, 1987; Spalvins and Buzek, 1981; Takagi and Liu, 1967). Soft metals, when present as a thin coating between hard sliding surfaces, can function as a solid lubricant coating because of their very marked ductility. This extreme ductility allows coatings to plastically shear between the sliding surfaces and thereby provide a lubricating function. However, if the soft metals are too thick, they will deform excessively under the normal load resulting in a large frictional force. The soft metallic coatings are effective for wear reduction when adhesive wear dominates. The soft metallic coatings for tribological applications have been applied by electrochemical deposition and physical vapor deposition (evaporation, sputtering, and ion-plating) processes. They have also been applied by thermal spraying and weld deposition processes.

Soft metallic coatings have the inherent capability of operating at higher temperatures and more normal pressures than conventional organic coatings. They have been used successfully as lubricants both in equipment that has operated while exposed to the vacuum conditions of space and in equipment that has been evacuated and sealed off. Unlike many resin bonded solid lubricant coatings, soft metallic coatings appear to be compatible with liquid lubricants. Thus, these coatings could be used in vacuum for an enhanced degree of boundary lubrication with low-volatility synthetic oils.

Some soft metals can be used as lubricating solids to near their melting points in addition to low temperatures. It is possible to use silver and gold over the large temperature range not only because of their desirable shear properties but also because of their thermomechanical stability (particularly oxidation resistance) at high temperatures. Gold does not oxidize under normal conditions. Silver oxide is stable only at moderate temperatures. It has been found that silver oxide is very easily decomposed into silver and oxygen by simple heating, and the decomposition begins at temperatures as low as 100°C, proceeds to a substantial extent, and goes to completion at 300°C (Sneed et al., 1954).

Ag and Au coatings have been found successful in lubricants in ultrahigh vacuum in conditions of space capsules, ball bearings for rotating anode x-ray tubes that have been evacuated and sealed, high-performance jet engines, and high-speed machines operating under highly loaded conditions. Silver coatings have been used as screw thread lubricants up to a couple of hundred degrees centigrade (Evans and Flatley, 1963; Flatley, 1964; Kessler and Mahn, 1967; Voigtlander, 1963).

Electroplated silver has been used for many years on the mild steel retainers of oil-lubricated rolling-element bearings. Steel-backed electroplated silver bearings have been used in hydrodynamically lubricated crankshafts and main bearings in high-performance reciprocating engines for aircraft.

Electroplated and vacuum deposited lead and other soft metal coatings are used for space-borne bearings of satellite mechanisms such as solar array drives, despin assemblies, and gimbals, and such systems have successfully completed several years of running (Dalley and Warwck, 1978; Todd and Bentall, 1978). Indium-coated bearings are used in aircraft, diesel, and gasoline engines because of their excellent heavy-duty performance. A lead coating impregnated with Indium on steel-backed silver-lead bearings has been available commercially for several years.

Soft alloy coatings such as Sn-Pb, Au-Co, and Ag applied by electrochemical deposition, have been used on electrical contacts in order to control the increase in contact resistance due to oxidation and fretting corrosion which occur as a result of vibratory motion (Antler, 1982; Rabinowicz, 1983; Saka et al., 1984). Electroplated silver and gold coatings have also been applied in slip rings used to transfer power between rotary joints (rotating component in contact with a stationery component) both in air and vacuum, for example, satellite applications (Walton et al., 1975).

Table 13.13 summarizes shear strength, hardness, melting point, vapor pressure at melting point, and evaporation rates at 250°C of some soft metallic coatings. Their hardnesses lie between 1 and 3 on the Moh's scale (Haltner and Flom, 1983; Peterson et al., 1959).

13.4.1 Electrochemically Deposited Soft Metals

A wide range of electrochemically deposited metals and alloys are used for wear resistance. Microhardness ranges for various metals and alloys are presented in

TABLE 13.13 Physical Properties, Vapor Pressures, and Evaporation Rates of Soft Metals

Metal	Hardness, Moh's	Shear strength, kg mm^{-2}	Melting point, °C	Vapor pressure at melting point: Pa	Vapor pressure at melting point: Torr	Evaporation rate at 250°C, g cm^{-2} s^{-1}
In	1	0.75	155	—	—	10^{-8}
Tl	1.2	1.12	304	5.3×10^{-6}	4×10^{-8}	—
Pb	1.5	0.68	328	4×10^{-7}	3×10^{-9}	10^{-8}
Sn	1.8	0.77	232	—		$<10^{-10}$
Cd	2	1.9	321	13.3	10^{-1}	—
Au	2.5	4.5	1063	2.6×10^{-3}	2×10^{-5}	2×10^{-4}
Ag	2.5–3	4.7	961	4×10^{-1}	3×10^{-3}	10^{-8}
Pt	4.3	5.8	1755	2×10^{-2}	1.5×10^{-4}	—
Rh	4.5–5	—	1966	—	—	—

Fig. 13.33 (Safranek, 1974). The electrochemically deposited thin coatings, also known as overlays of lead-based alloys (e.g., Pb-Sn, Pb-Sn-Cu and Pb-In), have been used to improve the compatibility and corrosion resistance of leaded bronze bearings (Anonymous, 1982; Inwood and Garwood, 1978). These coatings also provide good dirt embeddability and conformability. Pb-Sn and Pb-Sn-Cu are codeposited directly whereas Pb and In are deposited consecutively and diffused together by a heat treatment. The hardness of these coatings are slightly higher (12 to 15 HV for 87 % Pb, 10 % Sn, 3 % Cu) than bulk hardness. The thickness of overlay used is a compromise between increased thickness for embeddability and tolerance to misalignment, and decreased thickness for high fatigue strength. Figure 13.34 shows the influence of thickness on the fatigue strength for Pb-Sn coatings. For a specific thickness, Pb-In coating gives slightly higher fatigue strength than Pb-Sn or Pb-Sn-Cu coatings.

The relative wear resistance for various coatings measured on dynamically loaded bearing test apparatus are shown in Fig. 13.35. Pb-Sn-Cu coatings have a higher wear resistance than Pb-Sn and Pb-In coatings (Schaefer, 1949). The relatively poor wear resistance of Pb-Sn coating negates to some extent its high fatigue rating, since for a given wear life, a thicker coating is required which leads to a lower fatigue strength.

Electrodeposits of cadmium and zinc are used extensively to protect steel and cast iron against corrosion. Because cadmium and zinc are anodic to iron, the underlying ferrous metal is protected at the expense of the cadmium plating even if the coating becomes scratched or nicked, exposing the substrate (Anonymous, 1982).

Copper is often used in engineering applications to combat fretting and embeddability. For example, a thin layer of copper may be used on cast-iron- or chromium-plated piston rings to reduce the tendency to scuff easily. Electroplated brasses (from pure zinc to pure copper) and bronzes (alloys of copper and tin) are used in many applications requiring low friction; they are also used in decorative applications.

The use of precious metals is obviously restricted by their high cost, and their use is confined mainly to the specialty bearing applications, particularly in the aerospace industry, and for contacts in electronics and electrical applications (Anonymous, 1982; Antler, 1979, 1982; Antler and Drozdowicz, 1979; Inwood and Garwood, 1978). Silver retains its bearing properties to higher temperatures (above 200°C) at fairly high load than lead-tin-type materials. The use of alloying additions such as Pb, Pd, Cu, and Ni can increase the maximum operating temperatures possible. It should be noted, however, that silver is unsuitable for use with some extreme pressure lubricant additives because the sulfur compounds present cause excessive corrosion of silver. Switch contacts and sliding connectors require wear-resistant surfaces which maintain high electrical conductivity; silver, gold, palladium, or rhodium platings are commonly used for these applications. Rhodium plating has been increasingly used in recent years because of its high hardness (HV800) and excellent wear resistance, with freedom from oxidation and tarnishing. It is used in switching applications for very low noise, or for operation in severe environments and at high temperatures.

Turns (1977) has developed a gold alloy Au-Mo coating with long durability at elevated temperatures. Tests made at 620°C showed that among various combinations, Au-Mo coating was very durable (Fig. 13.36). He also applied MoS_2, WS_2, and WSe_2 lubricants to the various gold and silver alloy coatings. He found that the solid lubricant coatings applied over metallic coatings had little utility above about 315°C.

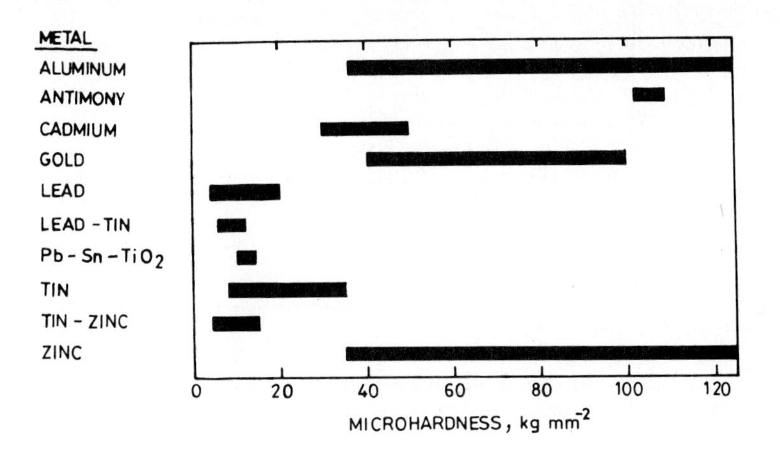

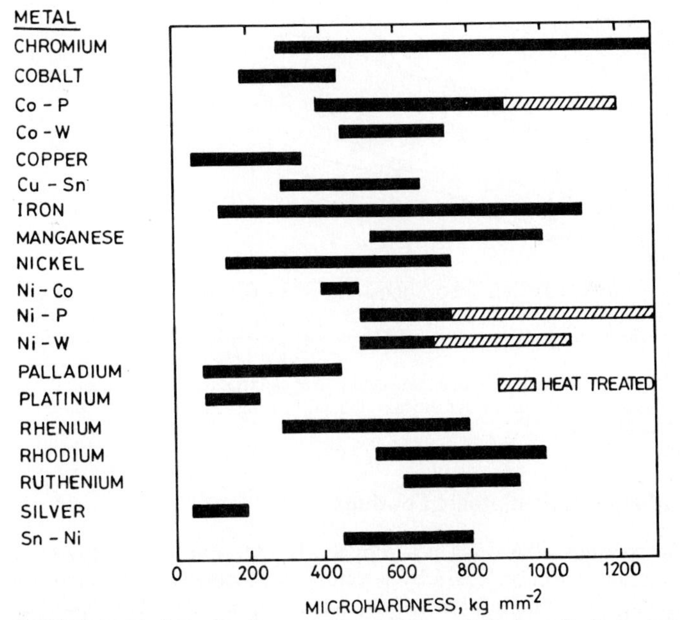

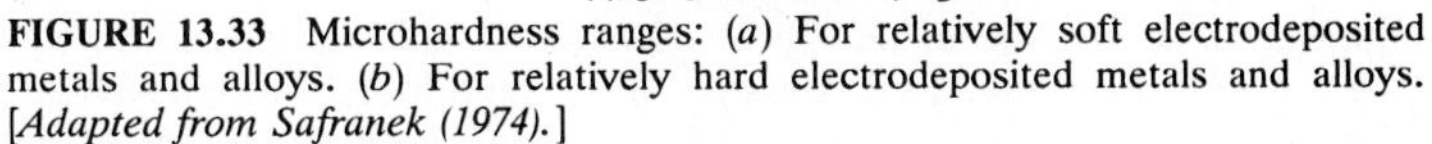

FIGURE 13.33 Microhardness ranges: (*a*) For relatively soft electrodeposited metals and alloys. (*b*) For relatively hard electrodeposited metals and alloys. [*Adapted from Safranek (1974).*]

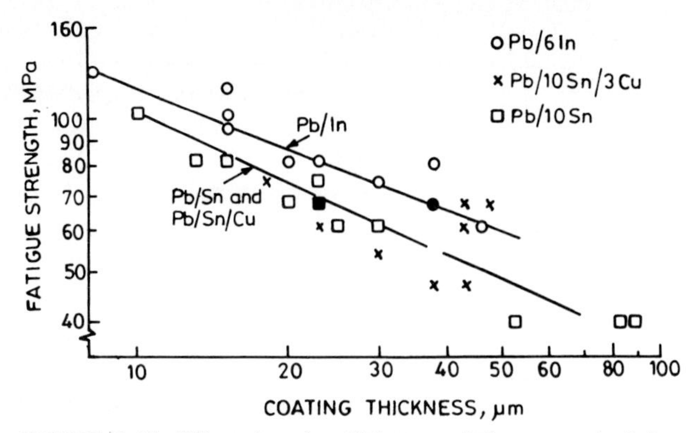

FIGURE 13.34 Effect of coating thickness on fatigue strength of electrochemically deposited metallic coatings. [*Adapted from Schaefer (1949).*]

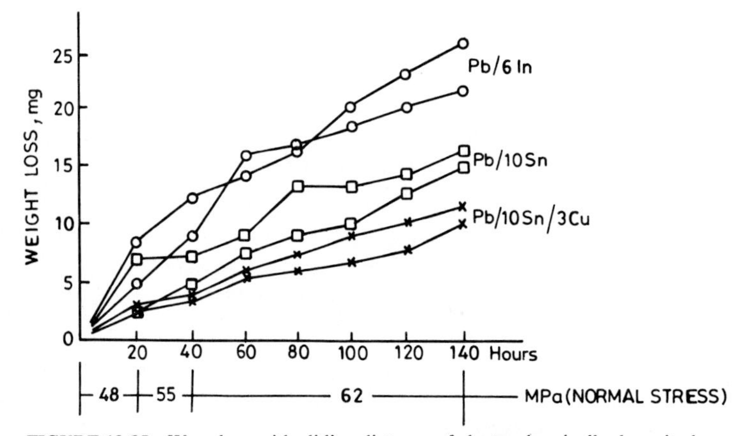

FIGURE 13.35 Wear loss with sliding distance of electrochemically deposited metallic coatings measured on dynamically loaded bearing test apparatus. [*Adapted from Schaefer (1949).*]

13.4.2 Ion-Plated Soft Metallic Coatings

Soft metallic coatings of Au, Ag, Cu, Pb, Sn, In, Al, and their alloys have been applied by the ion-plating process by several investigators (Arnell and Soliman, 1978; Gerkema, 1985; Gupta et al., 1987; Kloos et al., 1981; Kulshrestha et al., 1986; Ohmae, 1976; Ohmae et al., 1978; Sherbiney and Halling, 1977; Spalvins, 1978, 1981, 1987; Spalvins and Buzek, 1981; Sundquist and Myllyla, 1981; Sundquist et al., 1980). Evaporation and sputtering techniques have also been used to deposit soft metallic coatings (Christou and Day, 1973; Spalvins, 1987). The ion-plated soft metallic coatings such as Au and Pb have been shown to have longer wear life and lower friction than thermally evaporated coatings and electrochemically deposited coatings.

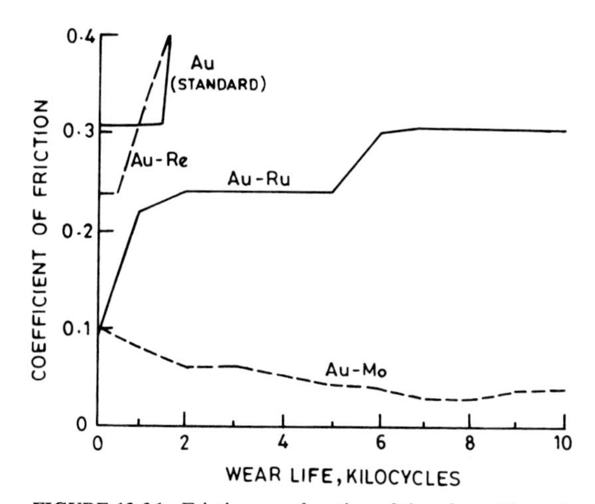

FIGURE 13.36 Friction as a function of time for gold matrix electrodeposits at 620°C in a Hohman A-6 tester with two steel rub shoes in contact with a plated disk. [*Adapted from Turns (1977).*]

Figure 13.37 compares the coefficient of friction and wear lives of evaporated and ion-plated gold coatings. We note that the wear life of ion-plated gold is nearly twice the life of evaporated gold. Further, the coefficient of friction of ion-plated gold coating increases, gradually giving adequate indication of impending failure while the vapor deposited coating failed abruptly without warning. The longer wear life of the ion-plated coating is attributed to its superior adherence to the substrate metal because kinetic energy of the particles impinging the substrate is higher than that in evaporation (Spalvins, 1978). The ion-plating process provides a graded interface which improves the coating adhesion. The use of ion-plated lead films for lubrication of ball bearings in vacuum has been reported by Todd and Bentall (1978). The ion-plated coating generated less debris and torque variation than vapor deposited lead, which was again attributed to the superior adherence of ion-plated films.

According to Sherbiney and Halling (1977) and El-Sherbiny and Salem (1979, 1986), the wear mechanism of soft metallic coatings appears to be an initial microcutting of the coating by the mating surface asperities followed by an abrasive wear mode promoted by the hard wear debris of the substrate, and finally by a much less severe form of fatigue wear due to asperity-substrate contacts. The performance of soft metallic coatings is strongly influenced by coating to substrate adhesive bond because essentially the shearing and fracture take place in the metal coating (Sherbiney and Halling, 1977; Jahanmir et al., 1976). An attempt has been made by introducing an interlayer of Cu between Pb coating and substrate and by mixing small amounts of Cu, Pt, or Mo into Pb coatings, to achieve lower wear and longer wear life (Gerkema, 1985). The other factors which influence the friction and wear characteristics are coating thickness, substrate roughness, sliding speed, and load.

13.4.2.1 Coating Thickness Effects. The coating thickness has a pronounced effect on the coefficient of friction. It is shown by Bowden and Tabor (1950, 1964)

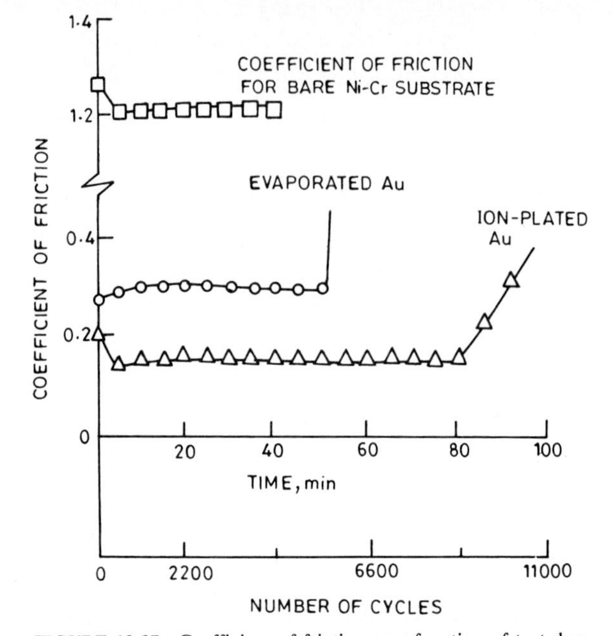

FIGURE 13.37 Coefficient of friction as a function of test duration of niobium pin sliding on Ni-Cr alloy disk with gold deposited by evaporation and ion plating about 200 nm thick at a sliding velocity of 0.03 m s^{-1}, normal load of 2.5 N, room temperature, and vacuum (1.3×10^{-9} Pa or 10^{-11} torr). [*Adapted from Spalvins (1978).*]

and Sherbiney and Halling (1977) that soft metal coatings having thicknesses in the range of 0.1 to 1.0 μm exhibit the least coefficient of friction.

Figure 13.38 schematically illustrates the influence of thickness on the friction of ion-plated soft metallic coating (Spalvins and Buzek, 1981). In ultrathin regions the high coefficient of friction is due to the breakthrough in the coating by mating asperities. In the higher-thickness region, the load-carrying capacity of the substrate decreases, and an increase in the apparent area of contact leads to a higher coefficient of friction.

13.4.2.2 Substrate Surface Roughness Effects. The dependence of the coefficient of friction on coating thickness is affected by the substrate surface roughness and coating deposition technique. Usually, with rough surfaces, the coefficient of friction increases with increasing thickness without any dip in the curve of friction versus coating thickness as shown in Fig. 13.39 (Arnell and Soliman, 1978; Sherbiney and Halling, 1977).

13.4.2.3 Effect of Sliding Speed. Figure 13.40 shows the influence of sliding speed on the coefficient of friction and wear life of ion-plated coatings of Ag, In, and Pb having different thicknesses. In almost all coatings, the coefficient of friction decreases with an increase in sliding velocity. This change in friction characteristic is attributed to thermal softening of coating material in the contact region. The thermal softening results in reduction of shear strength at the contact

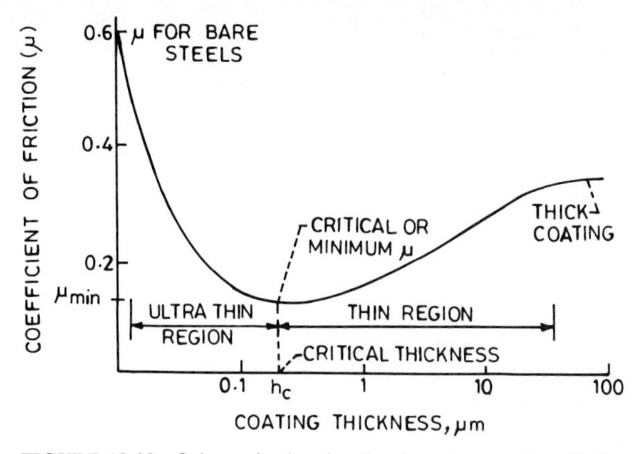

FIGURE 13.38 Schematic showing the dependence of coefficient of friction on coating thickness of ion-plated soft metallic coatings. [*Adapted from Spalvins and Buzek (1981).*]

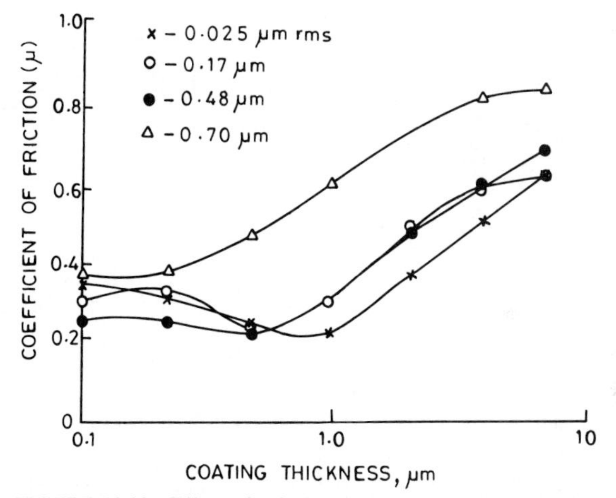

FIGURE 13.39 Effect of substrate surface roughness on coefficient of friction of ion-plated Pb coatings in a 52100 steel pin-on-a-coated-52100-steel-disk test. [*Adapted from Arnell and Soliman (1978).*]

with a significant reduction in load-carrying capacity, thereby reducing the coefficient of friction.

13.4.2.4 Effect of Load. Bowden and Tabor (1964) reported that the coefficient of friction of thin soft metallic coatings does not obey Amonton's law of friction. An elastic analysis of the Hertzian contact of thin film has shown that the relation between the load and the area of contact is markedly nonlinear (El-Sherbiney and Halling, 1976). Figure 13.41 demonstrates the dependence of the coefficient of friction and wear life on load for ion-plated Ag, Pb, and In coatings. The wear

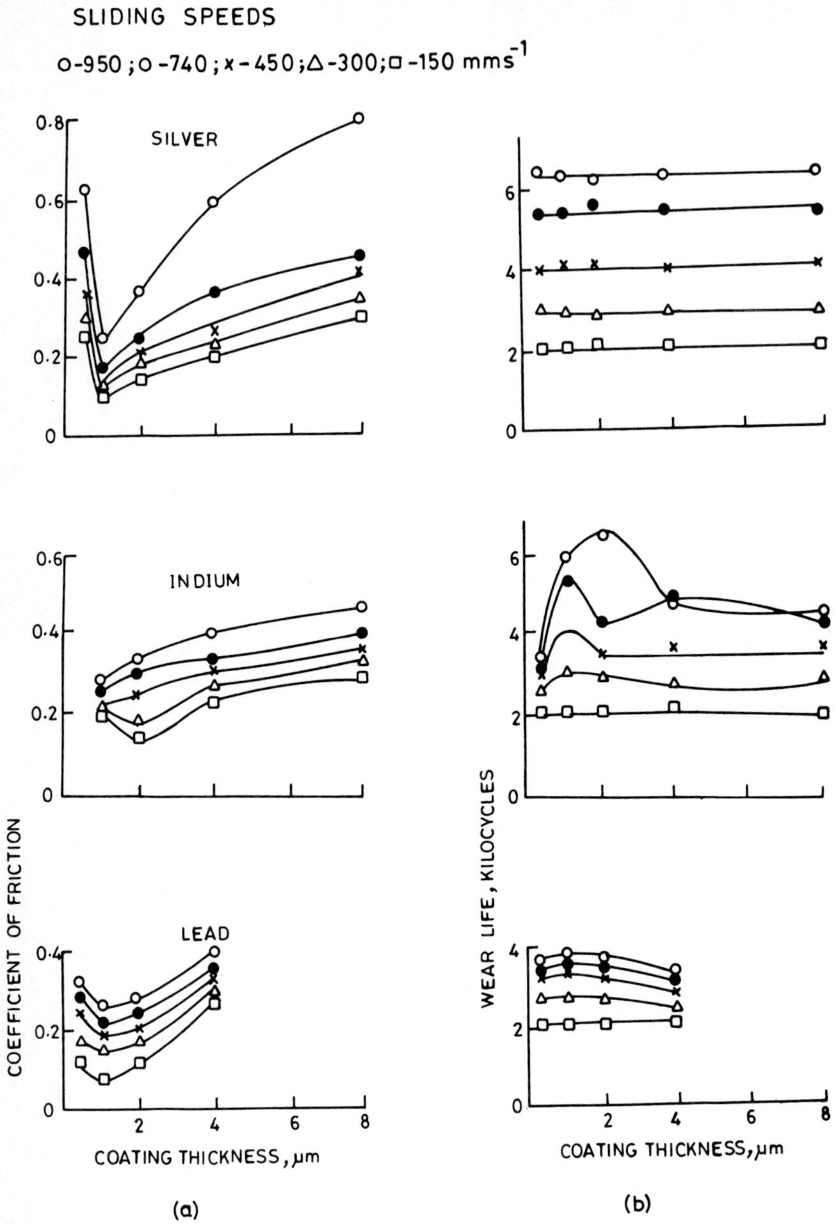

FIGURE 13.40 Effect of sliding speed on: (*a*) Coefficient of friction, (*b*) wear life of ion-plated Ag, In, and Pb coatings having different thicknesses. [*Adapted from Sherbiney and Halling (1977).*]

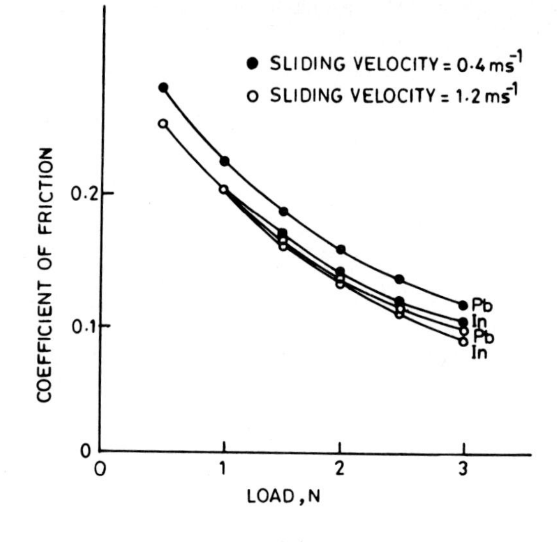

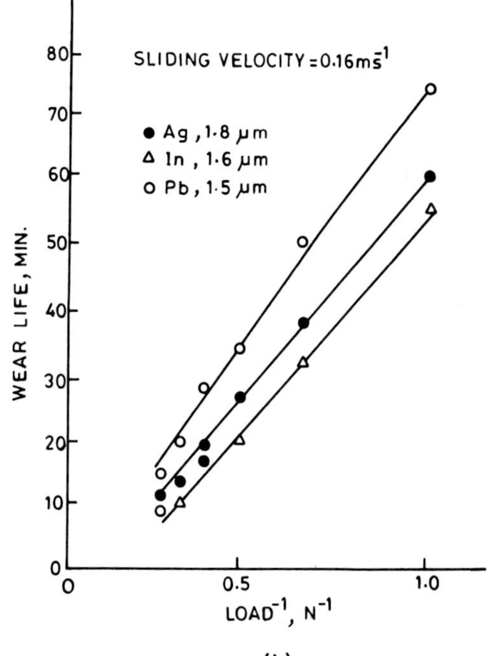

FIGURE 13.41 Effect of load on: (*a*) Coefficient of friction, (*b*) wear life of ion-plated Ag, In, and Pb coatings. [*Adapted from Sherbiney and Halling (1977).*]

life is inversely proportional to load. At higher loads the wear debris is found to contain substrate material particles.

13.4.2.5 ***Effect of Alloying Elements.*** Pb coatings mixed with Sn, Ag, Cu, Pt, or Mo have been applied by ion-plating and sputtering by Gerkema (1985) and Gupta et al. (1987). Cu and Mo differ from Ag and Pt in that Pb and the Cu and Mo metals are mutually insoluble whereas small amounts of Ag and Pt are soluble in Pb. Table 13.14 summarizes the coefficient of friction and wear life of a steel ball sliding on a steel disk, both sputter coated with Pb mixed with several metals and tested in high vacuum (6×10^{-5} Pa or 5×10^{-7} torr) at room temperature and 200°C. The additions of Ag in Pb do not improve the wear life, whereas additions of Cu and Pt substantially improve the wear life at room temperature as well as at 200°C. The Mo additions also improve wear life, although it has an extremely poor solubility in Pb. The additions of alloying elements improve the bonding between Pb-Pb atoms and adhesion and shear stress characteristics of the Pb alloyed coatings.

13.4.2.6 ***Effect of Interlayers.*** Duplex layers of Pb/M (M is a thin metallic layer between lead coating and steel substrate) were sputtered deposited on steel substrates by Gerkema (1985). Table 13.15 summarizes the coefficient of friction and wear life of a steel ball sliding on a steel disk, both sputter coated with Pb/M double layers (with M = Mo, Ta, W, Ag, and Cu) and at room temperature, 200°C, and 300°C. Variations in experimental temperatures (25°, 200°, and 300°C) show that the sliding distance decreases with increasing temperatures and become rather short at 300°C. From the friction and wear behavior point of view, we note that duplex layers Pb/Mo, Pb/Ta, Pb/W, and Pb/Ag are comparable with the friction and wear behavior of single Pb coating on steel. However, the wear life of Pb/Cu (~4500 m) is far beyond the wear life of a single Pb film, and the coefficient of friction still equals the value for single Pb films. When the load is doubled to 10 N, the wear life remains long. Gerkema also reported that the wear life of Pb/Cu coatings deposited by sputtering and electrodeposition were comparable.

TABLE 13.14 The Influence of Additives into Sputter Deposited Pb Coating on Coefficient of Friction and Wear Life in Sliding Distance*

Additive metal	Temperature, °C	Coefficient of friction	Sliding distance, $\times 10^2$ m
None	25	0.08	6.5
7.1 at. % Ag	25	0.09	1.5
0.46 at. % Cu	25	0.07	60
3.8 at. % Cu	25	0.07	23, >145
1.1 at. % Pt	25	0.07	135
0.63 at. % Mo	25	0.09	65
None	200	0.06	0.8
7.1 at. % Ag	200	—	0.08
0.46 at. % Cu	200	0.06	6
3.8 at. % Cu	200	0.06	1.5
1.1 at. % Pt	200	0.10	4.5
0.63 at. % Mo	200	0.30	67

*Steel ball and steel disk roughnesses are 50 nm cla, normal load = 5 N, sliding velocity = 25 mm s^{-1}, and test atmosphere = 6×10^{-5} Pa (5×10^{-7} torr).
Source: Adapted from Gerkema (1985).

TABLE 13.15 Coefficient of Friction and Wear Life in Sliding Distance of Sputtered Double Layers of Pb/M (M = Mo, Ta, W, Ag, Cu) at Various Temperatures*

Interlayer metal	Temperature, °C	Coefficient of friction	Sliding distance, $\times 10^2$ m
None	25	0.08	5.5
	200	0.06	0.8
	300	0.06	0.08
Mo	25	0.09	4.5
	200	0.05	0.06
	300	0.11	0.4
Ta	25	0.11	8.5
	200	0.05	3.6
	300	0.07	0.05
W	25	0.09	4.5
	200	0.06	1.9
	300	0.08	0.6
Ag	25	0.11	20
	200	0.05	0.9
Cu	25	0.05	>45, >48, >43
	25†	0.06	12, >25
	200	0.04	18, >20, >123

*Interlayer thickness is about 0.1 μm and steel ball and steel disk roughnesses are 50 nm cla; normal load = 5 N, sliding velocity = 25 mm s^{-1} and test atmosphere = 6×10^{-5} Pa (5×10^{-7} torr).

†Normal load = 10 N.

Source: Adapted from Gerkema (1985).

Gerkema noted that Mo, Ta, and W have no lubricating ability but Ag and Cu do. Apparently, the Cu interlayer with its lubricating capability decreases significantly the interaction of steel-steel contact, hence retarding the coating failure process. However, Ag shows a solid lubricant behavior similar to Cu, but Ag dissolves readily in Pb (whereas Cu is insoluble) and is not effective in reducing wear. From this, Gerkema concluded that a substantial improvement in wear life of Pb coating can be reached with interlayers if (1) the interlayers have a lubricating capability even without a Pb film and (2) lead and the underlying layer are mutually insoluble.

13.4.3 Thermal Sprayed and Weld Deposited Soft Metals

Copper-based alloys (leaded brasses and leaded-, tin-, silicon-, and aluminum-bronzes), and tin- and lead-based alloys (babbitts) are applied by flame (metallizing) and plasma spraying processes (Grisaffe, 1967; Gregory, 1978, 1980; Watson, 1973). Copper-based alloys are also deposited by shielded metal arc and gas tungsten arc welding processes (Hazzard, 1972; Bell, 1972). These alloys are deposited as overlays and inlays for corrosion-resistant and wear surfaces such as bearings. Various silicon-, aluminum-, and tin-bronzes are the alloys most frequently used for components such as gears, cams, sheaves, wear plates, and dies. In one bronze, a 16% Pb addition is made to produce a porous deposit and self-lubricating properties.

13.5 REFERENCES

Amato, I., and Martinengo, C. (1973), "Some Improvements in Solid Lubricant Coatings for High Temperature Operations," *ASLE Trans.*, Vol. 16, pp. 42–49.

Anonymous, *Recent World Research on the Lubricative Properties of Silver*, The Silver Institute, 10001 Connecticut Ave. NW, Washington, D.C.

——— (1965), *Technical Data Sheet 207*, Bemol Inc., Newton, Mass., June.

——— (1971), *Proc. 1st Int. Conf. on Solid Lubrication*, SP-3, ASLE, Park Ridge, Ill.

——— (1972), *Lubrication Handbook for Use in Space Industry; Part A—Solid Lubricants; Part B—Liquid Lubricants*, Contract No. NAS 8—27662; Midwest Research Institute, Kansas City, Mo.

——— (1977), "Advances in Connector Technology Deliver Longer Life, Better Performance," *Prod. Eng.*, July, pp. 27–28.

——— (1978), *Proc. 2nd Int. Conf. on Solid Lubrication*, SP-6, ASLE, Park Ridge, Ill.

——— (1982), *Metals Handbook*, Vol. 5, *Surface Cleaning, Finishing, and Coating*, 9th ed., American Society for Metals, Metals Park, Ohio.

——— (1983), Surface Coatings, Vol. 1: Raw Materials and Their Usage, 2d ed., Chapman and Hall, London.

——— (1984a), *Proc. of 3rd Int. Conf. on Solid Lubrication*, SP-14, ASLE, Park Ridge, Ill.

——— (1984b) *Surface Coatings*, Vol. 2: *Paints and Their Applications*, 2d ed., Chapman and Hall, London.

Antler, M. (1979), "Electrical Connectors," in *Encyclopedia of Chemical Technology* (R. E. Kirk and D. F. Othmer, eds.), Vol. 8, pp. 641–661, Wiley, New York.

——— (1982), "Fretting of Electric Contacts: An Investigation of Palladium Mated to Other Materials," *Wear*, Vol. 81, pp. 159–173.

———, and Drozdowicz, M. H. (1979), "Wear of Gold Electrodeposits: Effect of Substrate and of Nickel Underplate," *Bell Systems Tech. J.*, Vol. 58, no. 2, pp. 323–349.

Arnell, R. D., and Soliman, F. A. (1978), "The Effects of Speed, Film Thickness, and Substrate Surface Roughness on the Friction and Wear of Soft Metal Films in Ultrahigh Vacuum," *Thin Solid Films*, Vol. 53, pp. 333–341.

Baddour, R. F., and Timmins, R. S. (eds.) (1967), *The Application of Plasmas to Chemical Processing*, M.I.T. Press, Cambridge, Mass.

Bahun, C., and Jones, J. R. (1969), "Influence of Load, Speed, and Coating Thickness on the Wear Life of Bonded Solid Film Lubricants," *Lub. Eng.*, Vol. 25, pp. 351–355.

Bartz, W. J., Holinski, R., and Xu, J. (1986), "Wear Life and Frictional Behaviour of Bonded Solid Lubricants," *Lub. Eng.*, Vol. 42, pp. 762–769.

———, and Xu, J. (1987), "Wear Behavior and Failure Mechanism of Bonded Solid Lubricants," *Lub. Eng.*, Vol. 43, pp. 514–521.

Behrens, J. F. (1981), "Coatings that Take a Beating," *Machine Design*, Vol. 53, Aug. 20, pp. 71–75.

Bell, G. R. (1972), "Surface Coatings 3-Gas Welded Coatings," *Tribol. Int.*, Vol. 5, 215–219.

Benzing, R. J. (1964), "Solid Lubricants," in *Modern Materials* (B. W. Gonser and H. H. Hausner, eds.), Vol. 4, Academic, New York.

Bhushan, B. (1978), "High-Temperature Self-lubricating Coatings and Treatments—A Review," *Metal Finish.*, Vol. 78, May, pp. 83–88; June, pp. 71–75.

——— (1979), "Effect of Surface Roughness and Temperature on the Wear of CdO-Graphite and MoS_2 Films," *Proc. Int. Conf. on Wear of Material* (K. C. Ludema, W. A. Glaeser, and S. K. Rhee, eds.), pp. 409–414, Dearborn, Mich., ASME, New York.

——— (1980a), "High-Temperature Self-lubricating Coatings for Air Lubricated Foil Bearings for the Automotive Gas Turbine Engine," Tech. Report CR-159848, NASA Lewis Research Center, Cleveland, Ohio.

——— (1980b), "Coating Development and Friction Measurements for Solid-Lubricated Rolling-Element Bearings," Hughes Aircraft/DARPA Sponsored Tech. Rep. No. MTI-80 TR 30, Mechanical Technology Inc., Latham, N.Y.

——— (1980c), "High Temperature Low Friction Surface Coatings," U.S. Patent 4,227,756, October 14.

——— (1981a), "Friction and Wear Results from Sputter-Deposited Chrome Oxide with and without Nichrome Metallic Binders and Interlayers, *J. Lub. Technol.* (*Trans. ASME*), Vol. 103, pp. 218–227.

——— (1981b), "High Temperature Low Friction Surface Coatings and Method of Application," U.S. Patent 4,253,714, March 3.

——— (1982), "Development of CdO-Graphite-Ag Coatings for Gas Bearings to 427°C," *Wear*, 75, pp. 333–356.

——— (1987a), "Overview of Coating Materials, Surface Treatments, and Screening Techniques for Tribological Applications Part 1: Coating Materials and Surface Treatments," in *Testing of Metallic and Inorganic Coatings* (W. B. Harding and G. A. Di Bari, eds.), Special Publication STP 947, pp. 289–309, ASTM, Philadelphia.

——— (1987b), "Overview of Coating Materials, Surface Treatments, and Screening Techniques for Tribological Applications Part 2: Screening Techniques," in *Testing of Metallic and Inorganic Coatings* (W. B. Harding and G. A. Di Bari, eds.), Special Publication STP 947, pp. 310–319, ASTM, Philadelphia.

——— (1990), *Tribology and Mechanics of Magnetic Storage Devices*, Springer-Verlag, New York.

———, and Gray, S. (1978), "Static Evaluation of Surface Coatings for Compliant Gas Bearings in an Oxidizing Atmosphere to 650°C," *Thin Solid Films*, Vol. 53, pp. 313–331.

———, and ——— (1980), "Development of Surface Coatings for Air-Lubricated Compliant Journal Bearings to 650°C," *ASLE Trans.*, Vol. 23, pp. 185–196.

———, and Heshmat, H. (1979), "Reliability Improvement Study for Compliant Foil Air Bearing in Chrysler Upgraded Automotive Gas Turbine Engine," Chrysler Corp. Tech. Rep. MTI-79 TR 73, Mechanical Technology Inc., Latham, New York.

———, Ruscitto, D., and Gray, S. (1978), "Hydrodynamic Air-Lubricated Compliant Surface Bearing for an Automotive Gas Turbine Engine Part II—Materials and Coatings," Tech. Rep. CR-135402, NASA Lewis Research Center, Cleveland, Ohio.

Bisson, E. E., and Anderson, W. J. (1964), *Advanced Bearing Technology*, Special Publication SP-38, NASA, Washington, D.C.

Boes, D. J. (1964), "New Solid Lubricants: Their Preparation, Properties, and Potential for Aerospace Applications," *IEEE Trans. Aerospace*, Vol. AS-2, pp. 454–466.

Booser, E. R. (1983), *Handbook of Lubrication: Theory and Practice of Tribology*, Vol. I: *Application and Maintenance*, CRC Press, Boca Raton, Fla.

Bowden, F. P., and Tabor, D. (1950), *Friction and Lubrication of Solids*, Part I, Clarendon Press, Oxford, England.

———, and ——— (1964), *Friction and Lubrication of Solids*, Part II, Clarendon Press, Oxford, England.

Braithwaite, E. R. (1964), *Solid Lubricants and Surfaces*, Pergamon, Oxford, England.

——— (1967), *Lubrication and Lubricants*, Elsevier, Amsterdam.

Buck, V. (1983), "Morphological Properties of Sputtered MoS_2 Films," *Wear*, Vol. 91, pp. 281–288.

——— (1987), "Preparation and Properties of Different Types of Sputtered MoS_2 Films," *Wear*, Vol. 114, pp. 263–274.

Buckley, D. H. (1971), "Friction, Wear and Lubrication in Vacuum," Special Publication SP-277, NASA, Washington, D.C.

Burn, R., and Murray, G. T. (1962), "Plasticity and Dislocation Etch Pits in CaF_2," *J. Am. Ceramic Soc.*, Vol. 45, pp. 251–252.

Calabrese, S. J., Buxton, R., and Marsh, G. (1980), "Frictional Characteristics of Materials Sliding against Ice," *Lub. Eng.*, Vol. 36, pp. 283–289.

Campbell, M. E. (1972), "Solid Lubricants—A Survey," Special Publication SP-5059(01), NASA, Washington, D.C.

———, Loser, J. B., and Sneegas, E. (1966), "Solid Lubricants," Special Publication SP-5059, NASA, Washington, DC.

Christou, A., and Day, H. (1973), "Structure and Thermal Stability of Sputtered Au-Ta Films," *J. Appl. Phys.*, Vol. 44, pp. 3386–3393.

Christy, R. I. (1980), "Sputtered MoS_2 Lubricant Coating Improvements," *Thin Solid Films*, Vol. 73, pp. 299–307.

———, and Ludwig, H. R. (1979), "R.F. Sputtered MoS_2 Parameter Effects on Wear Life," *Thin Solid Films*, Vol. 64, pp. 223–229.

Clauss, F. J. (1972), *Solid Lubricants and Self-Lubricating Solids*, Academic, New York.

Crump, R. E. (1956), *Recent Developments in Solid Film Lubricants*, Air-force-Navy-Industry Lubricant Conf., Texas.

——— (1957), "Factors Influencing Wear and Friction of Solid Film Lubricants," *Prod. Eng.*, Vol. 28, pp. 24–27.

Dalley, K. R., and Warwck, M. G. (1978), "Research Programme into the Lubrication of Bearings and Slip Rings in Vacuum, Rep. No. T (E) S/161, ESA-CR (P)-1128, Marconi Space & Defence System, Inc., Camberly, England.

DellaCorte, C., and Sliney, H. E. (1987), "Composition Optimization of Self-Lubricating Chromium Carbide Based Composite Coatings for Use to 760°C, *ASLE Trans.*, Vol. 30, pp. 91–99.

———, ———, and Deadmore, D. L. (1988), "Sputtered Silver Films to Improve Chromium Carbide Based Solid Lubricant Coatings for Use to 900°C, *Tribol. Trans.*, Vol. 31, pp. 328–333.

Dimigen, H., and Hubsch, H. (1983), "Applying Low Friction Wear-Resistant Thin Solid Films by Physical Vapor Deposition," *Philips Tech. Rev.*, Vol. 41, pp. 186–197.

———, ———, Willich, P., and Reichelt, K. (1979), "Lubrication Properties of R.F. Sputtered MoS_2 Layers with Variable Stoichiometry," *Thin Solid Films*, Vol. 64, p. 221.

———, ———, ———, and ——— (1985), "Stoichiometry and Friction Properties of Sputtered MoS_x Layers," *Thin Solid Films*, Vol. 129, pp. 79–91.

DiSapio, A. (1960), "Bonded Coating Lubricate Metal Pair," *Prod. Eng.*, Vol. 31, pp. 48–53.

——— (1963), "Bonded Lubricant Coatings," *Machine Design*, Vol. 35, May, pp. 167–172.

——— (1978), "Thickness and Coverage of Bonded Solid Film Lubricants: Measurement and Effect," *Lub. Eng.*, Vol. 34, pp. 94–99.

El-Sherbiney, M. G., and Halling, J. (1976), "The Hertzian Contact of Surfaces Covered with Metallic Films," *Wear*, Vol. 40, pp. 325–337.

———, and Salem, F. B. (1979), "Initial Wear Rates of Soft Metallic Films," *Wear*, Vol. 54, pp. 391–400.

———, and ——— (1986), "Tribological Properties of PVD Silver Films," *ASLE Trans.*, Vol. 29, pp. 223–228.

Evans, H. E., and Flatley, T. W. (1963), "High Speed Vacuum Performance of Gold Plated Miniature Ball Bearings with Various Retainer Materials and Configurations," Tech. Note TND-2101, NASA, Washington, D.C.

Fish, J. G. (1982), "Organic Polymer Coatings," in *Deposition Technologies for Films and Coatings*, (J. M. Blocher, T. D. Bonifield, J. G. Fish, P. B. Ghate, B. E. Jacobson, D. M. Mattox, G. E. McGuire, M. Schwartz, J. A. Thornton, and R. C. Tucker, eds.), pp. 490–513, Noyes Publications, Park Ridge, N.J.

Fitzsimmons, V. G., and Zisman, W. A. (1958), "Thin Films of Polytetrafluoroethylene as Lubricants and Preservative Coatings for Metals," *Ind. Eng. Chem.*, Vol. 50, pp. 781–784.

Flatley, T. W. (1964), "High Speed Vacuum Performance of Miniature Ball Bearings Lubricated with Combinations of Barium, Gold, and Silver Films," Tech. Note TND-2304, NASA, Washington, D.C.

Fleischauer, P. D. (1984), "Effect of Crystalline Orientation on Environmental Stability and Lubrication Properties of Sputtered MoS_2 Thin Films," *ASLE Trans.*, Vol. 27, pp. 82–88.

——— (1987), "Fundamental Aspects of the Electronic Structure, Material Properties and Lubrication Performance of Sputtered MoS_2 Films," *Thin Solid Films*, Vol. 154, pp. 309–322.

———, and Bauer, R. (1987), "The Influence of Surface Chemistry on MoS_2 Transfer Film Formation," *ASLE Trans.*, Vol. 30, pp. 160–166.

Flom, D. G., Haltner, A. J., and Gaulin, C. A. (1965), "Friction and Cleavage of Lamellar Solids in Ultrahigh Vacuum," *ASLE Trans.*, Vol. 8, pp. 133–145.

Fusaro, R. L. (1977a), "Graphite Fluoride Lubrication: The Effect of Fluorine Content, Atmosphere, and Burnishing Technique," *ASLE Trans.*, Vol. 20, pp. 15–24.

——— (1977b), "Molecular Relaxation, Molecular Orientation, and the Friction Characteristics of Polyimide Films," *ASLE Trans.*, Vol. 20, pp. 1–14.

——— (1978), "Effect of Thermal Exposure on Lubricating Properties of Polyimide Film and Polyimide-Bonded Graphite Fluoride Films," Tech. Paper 1125, NASA, Washington, D.C.

——— (1981), "Mechanisms of Lubrication and Wear of a Bonded Solid-Lubricant Film," *ASLE Trans.*, Vol. 24, pp. 191–204.

——— (1982a), "Effect of Load, Area of Contact and Contact Stress on the Wear Mechanisms of a Bonded Solid Lubricant Film," *Wear*, Vol. 75, pp. 403–422.

——— (1982b), "Effect of Substrate Surface Finish on the Lubrication and Failure Mechanisms of Molybdenum Disulfide Films," *ASLE Trans.*, Vol. 25, pp. 141–156.

——— (1987), "How to Evaluate Solid-Lubricant Films Using a Pin-on-Disk Tribometer," *Lub. Eng.*, Vol. 43, pp. 330–338.

———, and Sliney, H. E. (1969), "Preliminary of Graphite Fluoride $(CF_x)_n$ as a Solid Lubricant," NASA TND-5097, NASA, Washington, D.C.

———, and ——— (1970), "Graphite Fluoride $(CF_x)_n$: A New Solid Lubricant," *ASLE Trans.*, Vol. 13, pp. 56–65.

———, and ——— (1973), "Lubricating Characteristics of Polyimide Bonded Graphite Fluoride and Polyimide Thin Films," *ASLE Trans.*, Vol. 16, pp. 189–196.

Gardos, M. N. (1976), "Quality Control of Sputtered MoS_2 Films," *Lub. Eng.*, Vol. 32, pp. 463–480.

Gaynes, N. I. (1967), *Formulation of Organic Coatings*, Van Nostrand, Princeton, N.J.

Gerkema, J. (1985), "Lead Thin Film Lubrication," *Wear*, Vol. 102, pp. 241–252.

Gisser, H., Petronio, M., Shapiro, A. (1972), "Graphite Fluoride as a Solid Lubricant," *Lub. Eng.*, Vol. 28, pp. 161–164.

Godfrey, D., and Bisson, E. E. (1951), "Bonding of Molybdenum Disulfide to Various Materials to Form a Solid Lubricating Film II—Friction and Endurance Characteristics of Films Bonded by Practical Methods," Tech. Note TN 2802, NACA, Washington, D.C.

Gregory, E. N. (1978), "Hard Facing," *Tribol. Int.*, Vol. 11, pp. 129–134.

——— (1980), "Surfacing by Welding-Alloy Processes, Coatings and Materials Selection," *Metal Constr.*, Vol. 12, pp. 685–690.

Grisaffe, S. J. (1967), "Simplified Guide to Thermal Spray Coatings," *Machine Design*, Vol. 39, No. 17, pp. 174–181.

Gupta, B. K., Kulshrestha, S., and Agarwal, A. K. (1987), "Friction and Wear Behaviour of Ion-Plated Lead-Tin Coatings," *J. Vac. Sci. Technol.*, Vol. A5, pp. 358–363.

Hagon, M. A., and Williams, F. J. (1964), "The Development and Testing of a New Ceramic-Bonded Dry Film Lubricant," *Proc. USAF-SWRI Aerospace Bearing Conf. 91*, March.

Halling, J. (1983), "The Tribology of Surface Films," *Thin Solid Films*, Vol. 108, pp. 103–115.

Haltner, A. J., and Oliver, C. S. (1960), "Frictional Properties of Molybdenum Disulfide Films Containing Inorganic Salt Additives," *Nature*, Vol. 188, pp. 308–309.

———, and Flom, D. G. (1983), "Lubrication in Space," in *Handbook of Lubrication: Theory and Practice of Tribology*, Vol. 1: *Application and Maintenance* (E. R. Booser, ed.), CRC Press, Boca Raton, Fla.

Harada, K. (1981), "Plasma Polymerized Protective Films for Plated Magnetic Disks," *J. Appl. Poly. Sci.*, Vol. 26, pp. 3707–3718.

Harrop, R., and Harrop, P. J. (1969), "Friction of Sputtered PTFE Films," *Thin Solid Films*, Vol. 3, pp. 109–117.

Hartley, D., and Wainwright, P. (1981), "Industrial Metal Cutting with Inorganically Bonded MoS_2 Coated Cutting Tools," *Proc. Third Int. Conf. on Wear of Materials* (S. K. Rhee, ed.), pp. 489–495, ASME, New York.

Hazzard, R. (1972), "Surface Coatings: 2, Arc Welding Coatings," *Tribol. Int.*, Vol. 5, pp. 207–214.

Holinski, R., and Gansheimer, J. (1972), "A Study of the Lubricating Mechanism of Molybdenum Disulfide," *Wear*, Vol. 19, pp. 329–342.

Hollahan, J. R., and Rosler, R. S. (1978), "Plasma Deposition of Inorganic Thin Films," in *Thin Film Processes* (J. L. Vossen and W. Kern, eds.), pp. 335–360, Academic, New York.

Hopkins, V., and Campbell, M. E. (1969), "Film-Thickness Effect on the Wear Life of a Bonded Solid Lubricant Film," *Lub. Eng.*, Vol. 25, pp. 15–22.

Inwood, B. C., and Garwood, A. E. (1978), "Electroplated Coatings for Wear Resistance," *Tribol. Int.*, Vol. 11, pp. 113–120.

Ishikawa, T., and Shimada, T. (1969), "Application of Polycarbon Monofluoride," *Proc. Fluorine Symposium*, Moscow, USSR.

Jahanmir, S., Abrahamson, II, E. P., and Suh, N. P. (1976), "Sliding Wear Resistance of Metallic Coated Surfaces," *Wear*, Vol. 40, pp. 75–84.

Johnson, R. L., Godfrey, D., and Bisson, E. E. (1948), "Friction of Solid Films on Steel at High Sliding Velocities," Tech. Note TN 1578, NACA, Washington, D.C.

———, and Sliney, H. E. (1959), "High Temperature Friction and Wear Properties of Bonded Solid Lubricant Films Containing Lead Monoxide," *Lub. Eng.*, Vol. 15, pp. 487–496.

Kanakia, M. D., and Peterson, M. B. (1987), "Literature Review of Solid Lubrication Mechanisms," *Interim Report BFLRF No. 213*, Southwest Research Institute, San Antonio, Tex.

Kay, E. (1968), "The Corrosion of Steel in Contact with Molybdenum Disulfide," *Wear*, Vol. 12, pp. 165–171.

———, and Tingle, E. D. (1958), "The Use of Polytetrafluoroethylene as a Lubricant," *Br. J. Appl. Phys.*, Vol. 9, pp. 17–25.

———, and Dilks, A. (1981), "Plasma Polymerization of Fluorocarbons in rf Capacitively Coupled Diode System," *J. Vac. Sci. Technol.*, Vol. 18, pp. 1–11.

Kessler, G. R., and Mahn, G. R. (1967), "Factors Affecting Design and Operation of Rolling Element Bearings in Rotating Anode X-ray Tube," *Lub. Eng.*, Vol. 23, pp. 372–379.

Kloos, K. H., Broszeit, E., and Gabriel, H. M. (1981), "Tribological Properties of Soft Metallic Coatings Deposited by Conventional and Thermionically Assisted Triode Ion-Plating," *Thin Solid Films*, Vol. 80, pp. 307–319.

Koury, A. J., Gabel, M. K., and Wijenayake, A. P. (1979), "Effect of Solid Film Lubricant on Tool Life," *Lub. Eng.*, Vol. 35, pp. 315–316, 329–338.

Kulshrestha, S., Agarwal, A. K., and Gupta, B. K. (1986), "Friction and Wear Characteristics of Ion-Plated Zinc-Aluminum Coatings," *Proc. National Conf. on Industrial Tribology*, Vol. 2, BHEL, Hyderabad, India.

Lansdown, A. R. (1974), *Molybdenum Disulphide Lubrication: A Survey of the Present State of the Art*, University College of Swansea, United Kingdom.

——— (1976), *Molybdenum Disulphide Lubrication: A Continuation Survey 1973–74*. (Prepared for European Space Agency.) University College of Swansea, United Kingdom.

——— (1979), *Molybdenum Disulphide Lubrication: A Continuation Survey 1979–80*. (Prepared for European Space Agency.) University College of Swansea, United Kingdom.

Lavik, M. T. (1959), "High Temperature Solid Dry Film Lubricants," WADC-TR-57-455, Midwest Res. Inst., Kansas City, Mo.

——— (1960), "Ceramic-Bonded Solid-Film Lubricants," WADC-TR-60-530, Wright Patterson Air Development Center, WPAFB, Ohio.

———, and Campbell, M. E. (1972), "Evidence of Crystal Structure in Some Sputtered MoS_2 Films," *ASLE Trans.*, Vol. 15, pp. 233–234.

———, Gross, G. E., and Vaughn, C. A. (1965), "Friction and Cleavage of Lamellar Solids in Ultrahigh Vacuum," *ASLE Trans.*, Vol. 8, pp. 133–145.

———, Medved, T. M., and Moore, G. D. (1968), "Oxidation Characteristics of MoS_2 and Other Solid Lubricants," *ASLE Trans.*, Vol. 11, pp. 44–55.

———, and Moore, G. D. (1969), "The Friction and Wear of Thin Fused Fluoride Films," *Proc. AFML-MRI Conf. on Solid Lubricants*, AFML-TR-70-127, September.

Lee, S. M. (1971), "Deposition of Barrier Coating Films on Semiconductor Devices by Glow Discharge Polymerization of Fluorinated Organic Gas Monomers," *Insul./Circ.*, Vol. 17, pp. 33–36.

Lipp, L. C. (1976), "Solid Lubricants—Their Advantages and Limitations," *Lub. Eng.*, Vol. 32, pp. 574–584.

Luff, P. D., and White, M. (1968), "Thermal Degradation of Polyethylene and Polytetrafluoroethylene during Vacuum Evaporation," *Vacuum*, Vol. 18, pp. 437–440.

Matveevsky, R. M., Lazovskaya, O. V., and Popov, S. A. (1978), "Temperature Stability of Molybdenum Disulfide Solid Lubricant Coatings in Vacuum," *Proc. 2nd Int. Conf. on Solid Lubrication*, Special Publication SP-6, pp. 41–44, ASLE, Park Ridge, Ill.

McConnell, B. D. (1961), "Solid-Film Lubricants for Extreme Environments," *Prod. Eng.*, Vol. 32, pp. 70–73.

——— (1972), *Assessment of Lubrication Technology*, ASME, New York.

———, Snyder, L. E., and Strang, J. R. (1977), "Analytical Evaluation of Graphite Fluoride and Its Lubrication Performance under Heavy Load, *Lub. Eng.*, Vol. 33, pp. 184–190.

McKannon, E. C., and Demorest, K. E. (1964), "Dry Film Lubrication of Highly Loaded Bearings," *Lub. Eng.*, Vol. 20, pp. 134–141.

Menoud, C., Zeming, G., Kocher, H., and Hintermann, H. E. (1985), "Morphology and Lifetime Investigation of Dry-Lubricating MoS_2 Films Deposited by rf Sputtering," *Proc. 5th Int. Conf. Ion and Plasma Assisted Techniques* (H. Oechsner, ed.), pp. 277–282, CEP Consultants, Edinburgh, United Kingdom.

Mikkelsen, N. J., Chevallier, J., and Sørensen, G. (1988), "Friction and Wear Measurements of Sputtered MoS_x Films Amorphized by Ion Bombardment," *Appl. Phys. Lett.*, Vol. 52, pp. 1130–1132.

Mitchell, D. C., and Fulford, B. B. (1957), "Wear of Selected Molybdenum Disulfide Lubricated Solids and Surface Films," *Proc. Conf. Lubrication & Wear*, Institute of Mechanical Engineers, London, pp. 376–383.

Moore, G. D., and Ritter, J. E., Jr. (1974), "The Friction and Wear Characteristics of Plasma-Sprayed NiO-CaF_2 in Rubbing Contact with a Ceramic Matrix," *Lub. Eng.*, Vol. 30, pp. 596–604.

Morrison, D. T., and Robertson, T. (1973), "RF Sputtering of Plastics," *Thin Solid Films*, Vol. 15, pp. 87–101.

Murphy, G., and Meade, F. S. (1966), "Improving the Endurance Life and Corrosion Protection Provided by Solid Film Lubricants," *Lub. Eng.*, Vol. 22, pp. 276–280.

Murray, S. F., and Peterson, M. B. (1964), "Solid Lubricant Coatings and Coating Compositions," U. S. Patent No. 3,158,495.

Niederhäuser, P., Hintermann, H. E., and Maillat, M. (1983), "Moisture-Resistant MoS_2-based Composite Lubricant Films," *Thin Solid Films*, Vol. 108, pp. 209–218.

Nishimura, M., Nosaka, M., Suzuki, M., and Miyakawa, Y. (1978), "The Friction and Wear of Sputtered Molybdenum Disulfide Films in Sliding Contact," *Proc. 2nd Int. Conf. on Solid Lubrication*, pp. 128–138, ASLE, Park Ridge, Ill.

Nylen, P., and Sunderland, E. (1965), *Modern Surface Coatings*, Interscience Publishers, London.

Ohmae, N. (1976), "Recent Work on Tribology of Ion-Plated Thin Films," *J. Vac. Sci. Technol.*, Vol. 13, pp. 82–87.

———, Tsukizoe, T., and Nakai, T. (1978), "Ion-Plated Thin Films for Antiwear Applications," *J. Lub. Tech.* (*Trans. ASME*), Vol. 100, pp. 129–135.

Olsen, K. M., and Sliney, H. E. (1967), "Addition to Fused Fluoride Lubricant Coatings for Reduction of Low Temperature Friction," Tech. Note TND-3793, NASA, Washington, D.C.

Ott, W. R., and McLaren, M. G. (1968), "DTA/TGA Investigation of Reactions in Sodium-Lead Silicate Glass Batch," *Proc. of the 2nd Int. Conf. on Thermal Analysis*, Worcester, Mass.

Paul, S. (1985), *Surface Coatings—Science and Technology*, Wiley, New York.

Peterson, M. B., and Johnson, R. L. (1956), "Friction Studies of Graphite and Mixtures of Graphite with Several Metallic Oxides and Salts at Temperatures to 1000°F," Tech. Report TN 3657, NACA, Washington, DC.

———, Murray, S. F., and Florek, J. J. (1959), "Consideration of Lubricants for Temperatures above 1000°F," *ASLE Trans.*, Vol. 2, pp. 225–234.

———, and Winer, W. O. (eds.) (1980), *Wear Control Handbook*, ASME, New York.

Rabinowicz, E. (1983), "The Importance of Electrical Contacts in Tribology," *Proc. Holm Semin. on Electric Contacts*, Illinois Institute of Technology, Chicago, Ill.

Ramalingam, S., Winer, W. O., and Bair, S. (1981), "Tribological Characteristics of Sputtered Thin Films in Rolling Hertzian Contacts," *Thin Solid Films*, Vol. 84, pp. 273–279.

Rudorff, W., and Rudorff, G. (1947), "Structure of Carbon Monofluoride," *Z. Anorg. Chem.*, Vol. 253, pp. 281–296.

Safranek, W. H. (1974), *The Properties of Electrodeposited Metals*, Elsevier, New York.

Saka, N., Liou, M. J., and Suh, N. P. (1984), "The Role of Tribology in Electrical Contact Phenomena," *Wear*, Vol. 100, pp. 77–105.

Schaefer, R. A. (1949), *Electroplated Bearings: Sleeve Bearing Materials* (R. W. Dayton, ed.), ASM, Cleveland, Ohio.

Shen, M., and Bell, A. T. (1979), "Plasma Polymerization," *ACS Symposium Series 108* (M. Shen and A. T. Bell, eds.), pp. 1–33, American Chemical Society, Washington, D.C.

Sherbiney, M. A., and Halling, J. (1977), "Friction and Wear of Ion-Plated Soft Metallic Films," *Wear*, Vol. 45, pp. 211–220.

Sliney, H. E. (1962), "Lubricating Properties of Ceramic-Bonded Calcium Fluoride Coatings on Nickel-Base Alloys from 75° to 1900°F," Tech. Note TND-1190, NASA, Washington, D.C.

——— (1966a), "Solid Lubricants for Extreme Temperatures," Tech. Memo. TMX-52214, NASA, Washington, D.C.

——— (1966b), "Self-lubricating Composites of Porous Nickel and Nickel Chromium Alloy Impregnated with Barium-Fluoride-Calcium Fluoride Eutectic," *ASLE Trans.*, Vol. 9, pp. 336–347.

——— (1974), "Plasma-Sprayed Metal-Glass and Metal-Glass Fluoride Coatings for Lubrication to 900°C," *ASLE Trans.*, Vol. 17, pp. 182–189.

——— (1979), "Wide Temperature Spectrum Self-Lubricating Coatings Prepared by Plasma Spraying," *Thin Solid Films*, Vol. 64, pp. 211–217.

——— (1982), "Solid Lubricant Materials for High Temperatures—A Review," *Tribol. Int.*, Vol. 15, pp. 303–314.

——— (1986), "The Use of Silver in Self-Lubricating Coatings for Extreme Environments," *ASLE Trans.*, Vol. 29, pp. 370–376.

———, and Jacobson, T. P. (1981), "Friction and Wear of Plasma Sprayed Coating Containing Cobalt Alloys from 25 to 650°C in Air," *ASLE Trans.*, Vol. 24, pp. 257–263.

———, Strom, T. N., and Allen, G. P. (1965), "Fluoride Solid Lubricants for Extreme Temperatures and Corrosive Environments," *ASLE Trans.*, Vol. 8, pp. 307–322.

Sneed, M. C., Maynard, J. L., and Brasted, R. C. (1954), *Comprehensive Inorganic Chemistry Vol. 2—Copper, Silver and Gold*, pp. 145–146, Van Nostrand, New York.

Spalvins, T. (1969), "Deposition of MoS_2 Films by Physical Sputtering and Their Lubrication Properties in Vacuum," *ASLE Trans.*, Vol. 12, pp. 36–43.

——— (1971), "Lubrication with Sputtered MoS_2 Films," *Proc. 1st Int. Conf. on Solid Lubrication*, pp. 360–370, ASLE, Park Ridge, Ill.

——— (1974), "Structure of Sputtered Molybdenum Disulfide Films at Various Substrate Temperatures," *ASLE Trans.*, Vol. 17, pp. 1–7.

——— (1975), "Bearing Endurance Tests in Vacuum for Sputtered Molybdenum Disulfide Films," Tech. Rep. TMX-3193, NASA, Washington, D.C.

——— (1976), "Morphological Growth of Sputtered MoS_2 Films," *ASLE Trans.*, Vol. 19, pp. 329–334.

——— (1978), "Coatings for Wear and Lubrication," *Thin Solid Films*, Vol. 53, pp. 285–300.

——— (1980), "Tribological Properties of Sputtered MoS_2 Films in Relation to Film Morphology," *Thin Solid Films*, Vol. 73, pp. 291–297.

——— (1981), "Frictional and Morphological Characteristics of Ion-Plated Soft Metallic Films," *Thin Solid Films*, Vol. 84, pp. 267–272.

——— (1982), "Morphological and Frictional Behavior of Sputtered MoS_2 Films," *Thin Solid Films*, Vol. 96, pp. 17–24.

——— (1984), "Frictional and Morphological Properties of Au-MoS_2 Films Sputtered from a Compact Target," *Thin Solid Films*, Vol. 118, pp. 375–384.

——— (1987), "A Review of Recent Advances in Solid Film Lubrication," *J. Vac. Sci. Technol.*, Vol. A5, pp. 212–219.

———, and Buckley, D. H. (1967), "Importance in Lubrication of Various Interface Types Formed during Vacuum Deposition of Thin Metallic Films," *Trans. 9th Vacuum Metallurgy Conf.*, p. 247, American Vacuum Society, New York.

———, and Buzek, B. (1981), "Frictional and Morphological Characteristics of Ion-Plated Soft Metallic Films," *Thin Solid Films*, Vol. 84, pp. 267–272.

Stock, A. J. (1965), "PTFE Dry Film Lubricants Reduce Friction and Wear," *Mater. Design Eng.*, Feb., pp. 97–99.

Stupp, B. C. (1958), "Molybdenum Disulfide and Related Solid Lubricants," *Lub. Eng.*, Vol. 14, pp. 159–163.

——— (1981), "Synergistic Effects of Metals Co-sputtered with MoS_2," *Thin Solid Films*, Vol. 84, pp. 257–266.

Sundquist, H. A., Matthews, A., and Teer, D. G. (1980), "Ion-Plated Aluminum Bronze Coatings for Sheet Metal Forming Dies," *Thin Solid Films*, Vol. 73, pp. 309–314.

———, and Myllyla, J. (1981), "Wear of Ion-Plated Aluminum Bronze Coatings under Arduous Metal-Forming Conditions," *Thin Solid Films*, Vol. 84, pp. 289–294.

Takagi, R., and Liu, T. (1967), "The Lubrication of Steel by Electroplated Gold," *ASLE Trans.*, Vol. 10, pp. 115–123.

Teer, D. G., and Salama, M. (1977), "The Ion Plating of Carbon," *Thin Solid Films*, Vol. 45, pp. 553–561.

Todd, M. J., and Bentall, R. H. (1978), "Lead Film Lubrication in Vacuum," *Proc. 2nd Int. Conf. on Solid Lubrication*, SP-6, pp. 148–157, ASLE, Park Ridge, Ill.

Turns, E. W. (1977), "Electroplated Au-Mo Alloy as a Solid Film Lubricant," *Plat. and Surf. Finish.*, Vol. 64, pp. 46–54.

Uemura, M., Okada, K., Mogami, A., and Okitsu, A. (1987), "Effect of Friction Mechanisms on Friction Coefficient of MoS_2 in an Ultrahigh Vacuum," *Lub. Eng.*, Vol. 43, pp. 937–942.

Voigtlander, W. (1963), "Soft Metal Film Lubricant Study," *Final Technical Progress Report LMSD-TR-61*, Lockheed Missiles and Space Co., Sunnyvale, Calif., March.

Wagner, R. C., and Sliney, H. E. (1986), "Effects of Silver and Group II Fluoride Solid Lubricant Additions to Plasma Sprayed Chromium Carbide Coatings for Foil Gas Bearings to 650°C," *ASLE Trans.*, Vol. 42, pp. 594–600.

Walton, A. J., Hintermann, H. E., and Mailatt, M. (1975), "Performance of Dry Lubricated Slip Rings for Satellite Applications," Presented at the European Space Tribology Symposium, Frascati, April.

Watson, H. N. (1973), "The Basic Principles of Welding as a Surface Coating Process," in *Science and Technology of Surface Coating* (B. N. Chapman and J. C. Anderson, eds.) pp. 215–221, Academic, New York.

Willis, D. P. (1980), "Fighting Friction with Bonded Coatings," *Machine Design*, Vol. 52, April, pp. 123–128.

Yaniv, A. E., Rona, M., and Sharon, J. (1979), "The Influence of Surface Preparation on the Behaviour of Organic Coatings," *Metal Finish.*, Vol. 79, Nov., pp. 55–61; Dec., pp. 25–30.

Yasuda, H., and Hsu, T. (1978), "Plasma Polymerization Investigated by the Comparison of Hydrocarbons and Perfluorocarbons," *Surf. Sci.*, Vol. 76, pp. 232–241.

APPENDIX 13.A SURFACE PRETREATMENTS AND BINDER MATERIALS FOR BONDED COATINGS

The bonded coating materials are a blend of solid lubricant particles (such as graphite, MoS_2, and polymers), a binder, other additives, carrier liquid, and dispersant. The fine particles of solid lubricant and the binder are dissolved or dispersed in a solvent or water in the presence of a dispersant. The particle size should be in the colloidal range (~0.005 to 0.2 μm) or slightly larger, usually less than 0.5 to 2.0 μm. The particles are dispersed either naturally (colloidal-size particles) or mechanically (stirring, agitation). Water-borne coatings require more mixing and do not remain in suspension as well as solvent-borne coatings do. The liquid suspension can be applied by atomized liquid spray, dip, or brush. Spraying appears most satisfactory. Coating is then cured at the ambient or elevated temperature and then burnished. The coating thickness ranges from 5 to 30 μm, the optimum being about 12.5 μm.

Before the bonded coating can be applied, the substrate must be pretreated for low friction and long durability. In the appendix, we describe surface pretreatments and binder materials used for various bonded coatings.

13.A.1 Surface Pretreatments

In the application of all types of bonded solid lubricants, especially the resin and silicate type of binders, pretreatment of the base substrate is very important for low friction and long durability. Methods used in the pretreating of metals include both chemical and mechanical processes. The type of pretreatment will depend on the base metal employed. A list of pretreatments for various metal surfaces is presented in Table 13.A.1. The most common pretreatments are mechanical processes, phosphating, and anodizing preceded by chemical cleaning.

TABLE 13.A.1 A List of Pretreatments for Various Metal Surfaces*

Type of metal surface	Pretreatment(s)
Steel (except stainless)	Vapor or sandblast or other mechanical process, phosphating, sulfiding
Stainless steel	Vapor or sandblast, chemical etch
Aluminum and aluminum alloys	Anodize, chemical film
Copper and copper alloys	Bright dip
Magnesium and magnesium alloys	Dichromate treatment
Titanium and titanium alloys	Vapor or sandblast
Chromium and nickel plating	Vapor or sandblast
Cadmium plating	Phosphating
Zinc plating	Phosphating

*Pretreatments listed here should be preceded by degreasing and removing corrosion films, as by immersion in liquids such as perchloroethylene or trichloroethylene.

13.A.2 Binder Materials

The binder provides cohesive forces that hold the coating together and also provides adhesive forces that bond the coating to a substrate. The binders can be organic and inorganic. Both organic and inorganic binders are used with inorganic solid lubricants (Bhushan, 1980a, 1987a; Campbell, 1972; Campbell et al., 1966). Organic binders are used for organic solid lubricants (Campbell, 1972; Campbell et al., 1966; Fish, 1982; Paul, 1985; Willis, 1980). Organic binders generally exhibit low friction but can be generally used only up to a temperature range of 200 to 300°C. Inorganic binders, on the other hand, can be used at high temperatures (in some cases, greater than 300°C) as well as in high vacuum.

13.A.3 Organic Binder Materials

Two types of organic (or resin) binders are used in most coatings—thermosetting and thermoplastic. These binders can be heat curable and/or air curable. The thermosetting-type binders include heat-curable polyester,* polyimide, phenolic, epoxy, and silicone and air-curable alkyds (a particular type of polyester with a lower amount of monomer) and polyester resins. Thermoplastic-type binders include air-curable acrylics (polymethyl methacrylate) and cellulosics, and heat-curable polyphenylene sulfide resins. A few organic resin binders are oxidatively stable and have glass transition temperatures above 300°C, such as polyquinoxilies, polybenzimidazoles, and polyimides. Of these, polyimides are by far the most commonly used.

Alkyd, cellulosic, and acrylic resins are relatively inexpensive and do not require heat cure and can therefore be used on substrates that cannot be baked. They produce hard film, but do not have good resistance to solvents. Phenolics have good surface adhesion and are harder than the alkyds but require a high-temperature (150 to 200°C) curing cycle. Epoxy resins have excellent solvent resistance and very good adhesion but are softer than phenolics. Modified epoxy-phenolics combine good properties of both materials. Silicones offer a high operating temperature but are softer and have only fair adhesion. Normally, they are used only for high-temperature service. Polyimides, on the other hand, offer a high operating temperature and high hardness.

Typical cure temperature and operating temperature ranges for various resin binders are shown in Fig. 13.42. Curing times vary from a few minutes to a couple of hours. The operating temperature range for the coating depends on the solid lubricant and the resin binder. The selection of a binder determines the relationship between cure temperature and cure time. The heat-cured materials are superior to the air-drying materials and should be used where high load-carrying ability or long life is required. Resins with high curing temperatures generally have higher service temperatures than those that are cured at low temperatures. Coatings can be cured over a broad temperature range; generally, high cure temperature means shorter cure times. Many coatings produced from aqueous solutions and cured at elevated temperatures (200 to 400°C) produce sintered coatings. The sintered coatings are generally more durable than the coatings cured at lower temperatures (20 to 200°C). Since the sintered coatings have to be cured at

*Some polyester resins can be air dried or force dried at temperatures slightly above room temperature.

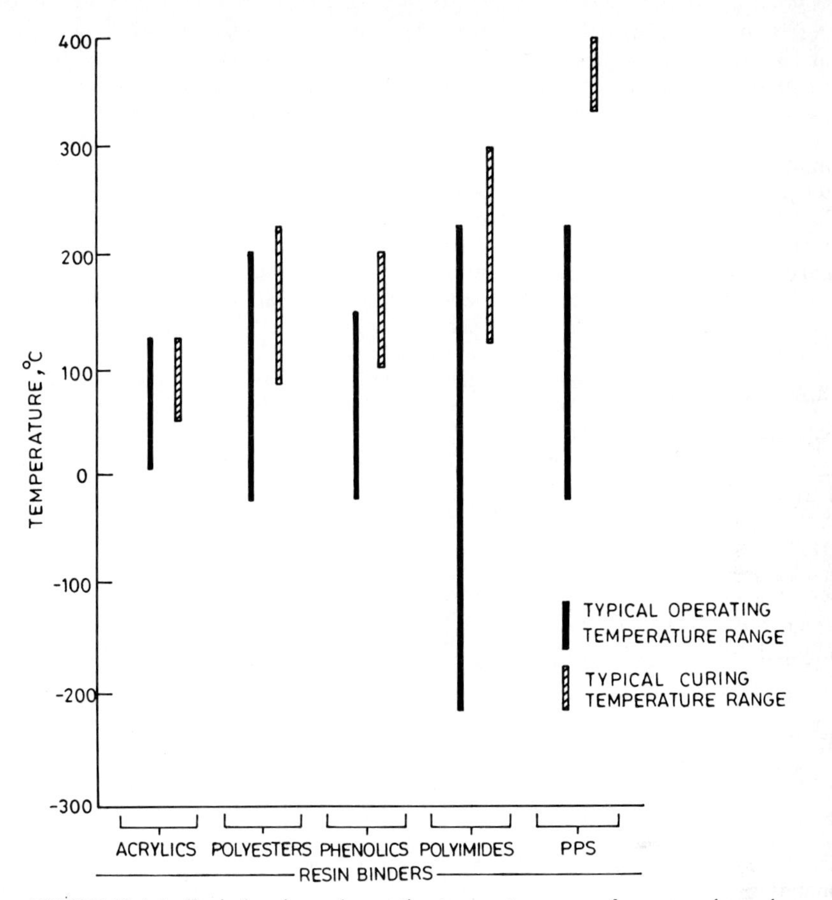

FIGURE 13.A.1 Typical curing and operating temperature ranges for commonly used organic resin binders.

elevated temperatures, they cannot be applied on temperature-sensitive substrate materials.

Solvents are volatile, and they evaporate as the coating dries. Water-borne coating solutions have not been extensive because few resin binders are compatible with water. Binders that have proved to be compatible include acrylic, polyimide, and PPS resins. Water-borne coatings require more mixing and do not remain in suspension as well as solvent-borne coatings do.

13.A.4 Inorganic Binder Materials

Inorganic binders used to produce bonded coatings of graphite, MoS_2, and other inorganic solid lubricants are useful at the highest service temperatures. They consist of salt-based and other ceramic oxide (glass) binders.

Salt-based binders include water glass (or sodium silicate with 8.9% Na_2O,

28.7% SiO_2, and balance water) and water-soluble aluminum phosphate $Al(PO_4)$ and borates. Baking temperature for silicates is generally in the same temperature range as the resin bonded materials but is often a two-step process, an example being 65°C for 2 h followed by a bake at 150°C for 8 h for sodium silicate binder. However, phosphate binder used with CaF_2-BaF_2 eutectic is cured at 925°C for 1 min. Salt-based binders produce a hard coating that tends to be somewhat brittle when cured. In general, they can be used at high loads (in excess of 700 MPa) and high temperatures (600°C or higher). However, in applications where movement is a prime design consideration, their wear life is not as good as resin bonded coatings. They have two advantages: They are not subject to outgassing in a high vacuum ($\sim 10^{-7}$ Pa or 10^{-9} torr), and they are compatible with liquid oxygen. Both these advantages have made them useful lubricants in the space programs. However, they do not afford corrosion protection.

The ceramic oxide binders are glasses and melt on heating. On cooling, they serve as a binding agent for the dispersed lubricant. They include silica (SiO_2), cobaltous oxide (CoO), and CaO-SiO_2. Coating slurry can be water based or solvent based. They are cured at a temperature above their melting point which is typically 870 to 1000°C. The curing time varies from about a minute to about 20 min. Their principal advantage is their good strength at elevated temperatures up to 900°C. Ceramic-oxide materials, as a class, do not perform as well as the resin bonded materials at lower (room temperature to about 200°C) temperatures; they generally exceed the resin film's capability by a considerable amount at higher temperature. There are exceptions, however, one being when the lubricants are run at high speeds which result in high temperatures being generated in the contact zone. In such cases, the ceramic oxide bonded coatings will generally outperform the resinous coatings. One problem in the use of ceramic oxide materials is the thermal expansion of the cured coating. This must be matched closely with the expansion of the base material.

CHAPTER 14
HARD COATINGS

Hard coatings provide low wear and are generally capable of sustaining heavy loads, high speeds, and high temperatures for extended periods without any deterioration in performance. The hard coatings necessarily do not have very low friction like the soft solid lubricant coatings discussed in Chap. 13. Hard coatings, used for protection against wear and corrosion at extreme operating conditions (such as elevated temperatures and corrosive environments) include a wide range of metals and ceramic materials. The materials most commonly used for hard coatings are listed in Table 14.1 (Alper, 1970a, 1970b, 1970c; Anonymous, 1976, 1979, 1981; Bucklow, 1983; Eschnauer and Lugscheider, 1984; Holleck, 1986; Hove and Riley, 1965; Lynch et al., 1966; Pope et al., 1989; Samsonov and Vinitskii, 1980; Storms, 1967; Toth, 1971). These are classified into three categories: ferrous (iron)-based alloys, nonferrous metals and alloys and (inorganic) nonmetals, i.e., ceramics (for more details, see Chap. 4). Some hard alloys consist of a metal matrix in which the hard material (e.g., carbide phase) is embedded for increased wear resistance. A wide variety of coating deposition techniques have been used to apply the hard coatings (Anonymous, 1982, 1987; Bhushan, 1980a, 1980b, 1980c, 1982, 1987; Bhushan et al., 1978; Bhushan and Gray, 1978, 1980; Bunshah, 1977; Gregory, 1978, 1980; Hess, 1984; Hintermann, 1981, 1983, 1984; Holleck, 1986; Mock, 1974; Murray, 1984; Peterson and Winer, 1980; Sundgren and Hentzell, 1986). In many instances, overlays of soft lubricant materials (MoS_2, Au, Ag, PTFE) are applied onto hard coatings to provide protection during running-in and to prolong life (Bhushan, 1980c, 1980d, 1981b, 1987; Bhushan et al., 1978).

Iron-based alloys (steels and cast irons) are used where heavy wear is encountered, under conditions that impose mechanical and thermal shock. They are used for corrosion and oxidation resistance generally up to 650°C. Although cobalt- and nickel-based alloys are superior in hardness (wear resistance) at elevated temperatures and corrosion and oxidation resistance, iron-based alloys are more widely used than Ni- and Co-based alloys because of the low cost of iron-based alloys. The iron-based alloy coatings are generally applied by weld deposition and thermal spraying processes.

Electrochemically deposited chromium coatings have been extensively used for decoration and protection against wear and corrosion in many engineering applications. These coatings exhibit high hardness (1000 to 1100 HV) up to a temperature of about 400°C. Chromium coatings can be made very hard (up to 3000 HV) by doping with oxygen, carbon, and nitrogen during the sputtering process.

Nickel is the most widely used wear-resistant coating after hard chromium. Compared with chromium, its as-coated hardness is relatively low, but it has

TABLE 14.1 Commonly Used Materials for Hard Coatings

1. *Ferrous alloys*
   High alloy cast irons
   Pearlitic steels
   Austenitic steels
   Martensitic steels

2. *Nonferrous metals and alloys*

Refractory Transition metals

IVa	Va	VIa	VIIIa
Ti	V	Cr	Ni
Zr	Nb	Mo	
Hf	Ta	W	

Intermetallic alloys

Co-based

Co-Cr-W-C (e.g., Haynes alloy 25, Stellites)
Co-Cr-Mo-Si (e.g., Tribaloy 800)

Ni-based

Ni-Cu (e.g., Monel)
Ni-Al
Ni-Cr
Ni-Mo-C (e.g., Hastelloy B)
Ni-Cr-Mo-C (e.g., Hastelloy X and C, Inconel 625)
Ni-Cr-Fe-C (e.g., Haynes alloy 711 and 716, Inconel X-750)
Ni-Cr-B-Si-C (e.g., Haynes alloy 40)
Ni-Mo-Cr-Si (e.g., Tribaloy 700)

Others

M-Cr-Al-Y (M = Ni, Co, Fe, or combinations thereof)

3. *Ceramics*

Metallic	*Refractory transition metals*			
Carbides	IVa	Ti	Zr	Hf
Nitrides	Va	V	Nb	Ta
Borides	VIa	Cr	Mo	W
Silicides				
Solid solutions				

Nonmetallic (ionic) oxides

Al_2O_3, Cr_2O_3, ZrO_2, HfO_2, TiO_2, SiO_2, BeO, MgO, ThO_2

Nonmetallic (covalent) nonoxides

B_4C, SiC, BN, Si_3N_4, AlN, diamond

Cermets

good mechanical strength and ductility, is a good conductor of heat, and has good resistance to corrosion and high-temperature oxidation. The deposition rate of plated Ni is four to six times higher than typical chromium-plating rates and therefore gives a more economical process. The electroless Ni coatings doped with P or B (typically 4 to 12 wt %) exhibit superior wear resistance and higher hardness (as coated up to 700 HV and hardened up to 1200 HV) as compared to electrodeposited Ni.

Molybdenum coatings applied by electrodeposition and plasma spraying processes find applications on piston rings and slideways. Magnetron sputtered Mo coatings doped with C and N exhibit very high hardness (typically three to six times higher than undoped) in the range of 2500 to 3000 HV.

Cobalt- and nickel-based alloys are used to protect against wear in corrosive conditions. Stellites, Tribaloys, Haynes alloys, Hastelloys, and MCrAlY (M = Ni, Co, or Fe) alloys exhibit high resistance to high-temperature oxidation, other corrosive environments, and wear. The composition of these multicomponent coatings strongly influences the hardness and ductility and provide flexibility of tailoring the properties as per specific needs.

Hard coatings of various ceramic materials such as oxides, carbides, nitrides, borides, and silicides have been extensively studied and used to protect against wear. Many ceramic coatings are blends or composites; cermets, for example, are made from a blend of ceramics and metals. These coatings have been generally applied by thermal spraying, physical vapor deposition (PVD), and chemical vapor deposition (CVD) processes. These processes provide a wide range of flexibility for tailoring performance and properties by optimizing composition and hardness, etc. The typical hard ceramic coatings are Al_2O_3, Cr_2O_3, ZrO_2, SiO_2, TiC, WC, Cr_3C_2, HfC, SiC, B_4C, TiN, Si_3N_4, HfN, CBN, TiB_2, ZrB_2, and $MoSi_2$. Certain nitride coatings such as TiN, HfN, and ZrN produce golden luster, and thus they find potential applications for decoration in addition to wear protection. The properties of these coatings widely vary with the composition, alloying element, and deposition process conditions.

The mechanical properties of the coatings such as hardness, ductility, and elastic modulus, porosity, and interlayer adhesion of the coating to the substrate are crucial for good friction and wear performance. Intrinsic (residual) stresses can cause coating rupture, loss of adhesion, substrate cracking, and changes in physical and chemical properties of the coatings. On the other hand, compressive residual stresses increase coating hardness significantly and provide resistance to tensile failure. Therefore, compressive stresses to a certain value can be beneficial. Deposition parameters need to be optimized for best coating performance.

The mechanical properties such as ductility, thermal expansion matching with the substrate, and adhesion of coating to substrate for better adhesion of hard coatings such as Cr_2O_3, Cr_3C_2, and WC applied by sputtering, thermal spraying, or other coating deposition techniques can be considerably improved by the addition of (~25 wt %) metallic binder such as nichrome (80 wt % Ni, and 20 wt % Cr) (Bhushan, 1980a, 1980c).

Some substrates may oxidize, and the coating may therefore become less tenacious. In these cases, a protective underlayer may be useful. In addition to oxidation resistance, some underlayers may provide a graded interface and provide high bond strength due to their ability to form an alloy (metallurgical bond) with the substrate. The coating of nichrome or nickel-aluminide (80 to 95 wt % Ni and 5 to 20 wt % Al) are commonly used as a transition layer to provide oxidation resistance and to act as a bond coat for subsequent coating of other materials (Bhushan et al., 1978). Both of these materials rapidly form adherent oxide films

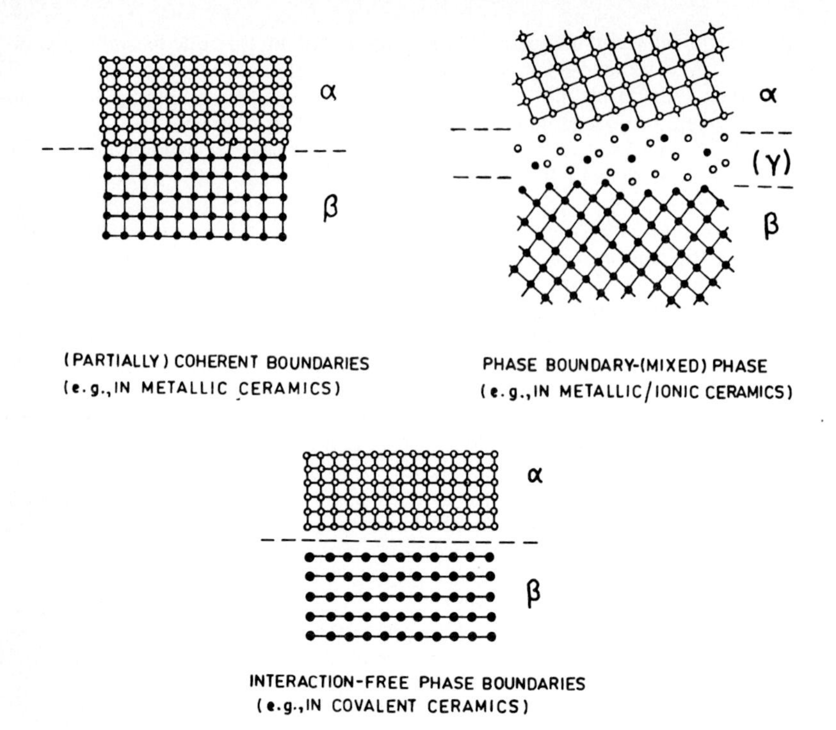

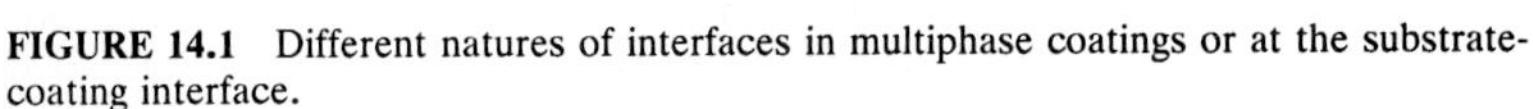

FIGURE 14.1 Different natures of interfaces in multiphase coatings or at the substrate-coating interface.

which prevent further oxidation. When these materials are, for example, thermal sprayed in air, their oxides form, so that the final deposit contains as little as 70 percent metal and 30 percent oxide. This metal-oxide mixture produces a transition in properties between metal substrate and ceramic coatings. Graded coatings are often helpful in ensuring a more gradual controlled transition. An undercoat of nickel-aluminide is less corrosion resistant in water but has better thermal shock resistance and higher bond strength* than nichrome. Temperature capabilities of nickel aluminide are up to 750°C while nichrome can be used up to 1000°C (Bhushan et al., 1978; Sangam and Nikitich, 1985).

Because of complex requirements such as hardness and toughness, weak adhesion at the surface, and at the same time, good adherence at the coating-substrate interface, multilayer or multiphase coatings sometimes seem to be the best compromise. Considering the special characteristics of the different hard materials, it is possible to construct coatings where the inner layer provides good adherence to the substrate, where one or more intermediate layers are responsible for hardness and strength, and where an outer layer reduces friction, adhesion, and reactivity. Figure 14.1 suggests three different possibilities for the constitution of interfaces (Holleck, 1986): coherent or partially coherent interfaces with a boundary phase and interaction free interfaces. Metallic ceramics are able to form coherent or semicoherent interfaces with metals or other metallic ceramics. As a result, low-energy interfaces

*In thermal spraying of the nickel and aluminum powder mixture, the constituents react exothermically, thereby providing a higher particle energy level and a very good bond.

with optimum adherence can be obtained (e.g., TiC/TiB_2). Interfaces between metallic and ionic ceramics often show intermediate regions of variable compositions (e.g., TiC/Al_2O_3). The behavior is strongly dependent on the constitution and structure of the boundary phase. Interfaces between covalent ceramics seem to be quasi-interaction-free with the consequence of bad adherence of the phases (e.g., B_4C/B_4C or B_4C/Al_2O_3) (Holleck, 1986).

The multiphase concept is already very common in the coating technology. Layers like TiC and TiN, Al_2O_3 and AlN or with coherent interfaces (TiC or TiN and TiB_2) to get sufficient bonding between the layers. Tool bits coated with multilayer coatings show much better performance than most of the one-layer-coated materials. The important role of interfaces for the wear behavior also become evident when one performs abrasive tests with one phase and two (or multiphase) bulk materials. Holleck (1986) has shown that high abrasion resistance is not only due to high hardness. For example, B_4C is the material with the highest hardness, whereas a B_4C containing composite material with lower hardness has a three times better abrasion resistance. This concept also works in the layered coatings.

Friction and wear characteristics of hard coatings applied by various deposition techniques are discussed in this chapter.

14.1 FERROUS-BASED ALLOY COATINGS

The composition, density, hardness, and wear data of various steel and high-alloy cast iron coatings applied by the weld deposition process are shown in Table 14.2 (Antony et al., 1983; Bell, 1972; Hazzard, 1972). Avery (1952) conducted wear tests on various ferrous and nonferrous bonded alloy coatings applied by the weld deposition process. Figure 14.2 presents data which show the erosion resistance as determined by dry quartz sand rubber wheel abrasion tests. Data for resistance to high stress or grinding abrasion are presented in Fig. 14.3. Figure 14.4 provides a simplified classification of various alloys, together with the notes about their salient properties. We note that abrasion resistance follows a trend opposite to that of impact resistance (in the sense of toughness). The first two alloys in Fig. 14.4, tungsten carbide and high chromium irons, are suitable for very light impacts, if gas welded. Neither is as appropriate for light impact, as the martensitic iron. Cobalt-based alloys should be selected primarily for hot hardness and oxidation resistance. They are expensive for impact and abrasion resistance. The nickel-based alloys are most appropriate for corrosion resistance.

The pearlitic steels are generally softer but not necessarily tougher than the martensitic steels. They are more appropriate for build-up welding than for surfacing. The austenitic stainless steels work harden, are nearly as tough, and can frequently substitute for austenitic manganese steel. Martensitic steels probably represent the best combination of relatively high hardness (abrasion resistance), relatively high toughness (impact resistance), and low cost. The martensitic steels are recommended for medium-impact applications.

14.2 NONFERROUS METALLIC COATINGS

14.2.1 Chromium Coatings

Chromium coatings are generally applied by electrochemical deposition and physical vapor deposition and less commonly by chemical vapor deposition.

TABLE 14.2 Composition, Hardness, and Wear Data of Some Ferrous-based Alloy Coatings Applied by Weld-Deposition

Alloy	Composition, %	Density, kg m^{-3}	Nominal hardness		Adhesive wear coefficients*	Abrasive wear coefficients†
			HV	HRC		
Pearlitic steels	Fe-2Cr-1Mn-0.2C	7850	318	32	6.6×10^{-5}	5.6×10^{-4}
	Fe-3.5Cr-2Mn-0.2C	7260	446	45	6.9×10^{-5}	5.8×10^{-4}
	Fe-1.7Cr-1.8Mn-0.1C	7600	372	38	9.9×10^{-5}	1.1×10^{-3}
Austenitic steels	Fe-14Mn-2Ni-2.5Cr-0.6C					
	As-deposited	7860	188	HRB 88	2.8×10^{-5}	5.1×10^{-4}
	Work-hardened		458	46		
	Fe-15Cr-15Mn-1.5Ni-0.2C					
	As-deposited	7840	230	18	2.5×10^{-5}	8.2×10^{-4}
	Work-hardened		485	48		
Martensitic steels	Fe-5.4Cr-3Mn-0.4C	7600	544	52	9.8×10^{-5}	9.3×10^{-4}
	Fe-12Cr-2Mn-0.3C	7690	577	54	6.7×10^{-5}	1.1×10^{-3}
High-alloy cast irons	Fe-16Cr-4C	7610	595	55	7.9×10^{-5}	2.5×10^{-4}
	Fe-26Cr-2.5C	7720	544	52	1.3×10^{-5}	8.8×10^{-4}
	Fe-26Cr-4.6C	7170	633	57	1×10^{-5}	2.4×10^{-4}
	Fe-29Cr-3C-3Ni	7560	697	60	1.8×10^{-5}	2.6×10^{-4}
	Fe-30Cr-4.6C	7330	560	53	5.3×10^{-5}	2.6×10^{-4}
	Fe-36Cr-5.7C	7690	633	57	2.4×10^{-5}	2.4×10^{-4}

*Unlubricated wear tests conducted against 4620 rings on Dow Corning LFW-1 tests at 80 r/min for 2000 rev.
†Dry sand rubber wheel abrasion tests using AFS test stand, 134 N load, and 1440 m sliding distance.
Source: Adapted from Antony et al. (1983).

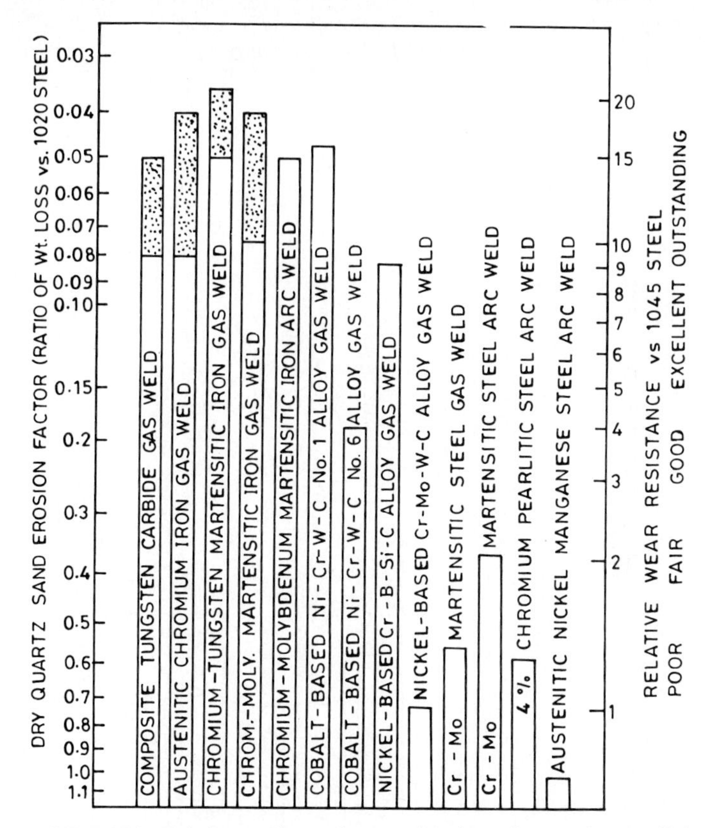

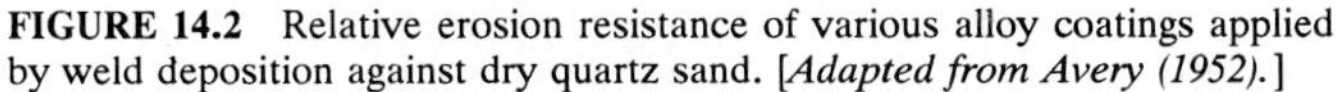

FIGURE 14.2 Relative erosion resistance of various alloy coatings applied by weld deposition against dry quartz sand. [*Adapted from Avery (1952).*]

Hard chromium coatings applied by electrochemical deposition are most commonly used for decorative purposes and for environmental corrosion and wear resistance (Chessin and Fernald, 1982; Gianelos, 1982; Inwood and Garwood, 1978). Decorative chromium coating is differentiated from hard chromium coating (for wear resistance) in thickness and the type of undercoating used. Decorative chromium coatings are very thin, usually not exceeding an average thickness of 1.25 μm. Decorative chromium is applied over undercoatings, such as nickel or copper and nickel, which impart a bright, semibright, or satin cosmetic appearance to the chromium. The choice of undercoatings, as well as the type of chromium applied, can also provide corrosion protection.

Some of the uses of hard chromium plating applied by electrochemical deposition for wear-resistance applications include automotive, aerospace, computers, mining, and metal-cutting tools. Its most extensive usage is on piston rings, cylinder liners, cross-head pins, etc., in internal combustion engines and compressors (Taylor and Eyre, 1979). Sputtered chromium coatings are used in various electronic applications and plastic automotive grilles.

14.2.1.1 Cr Coatings Applied by Electrochemical Deposition. Most hard chromium coatings are produced by electrodeposition from a solution containing

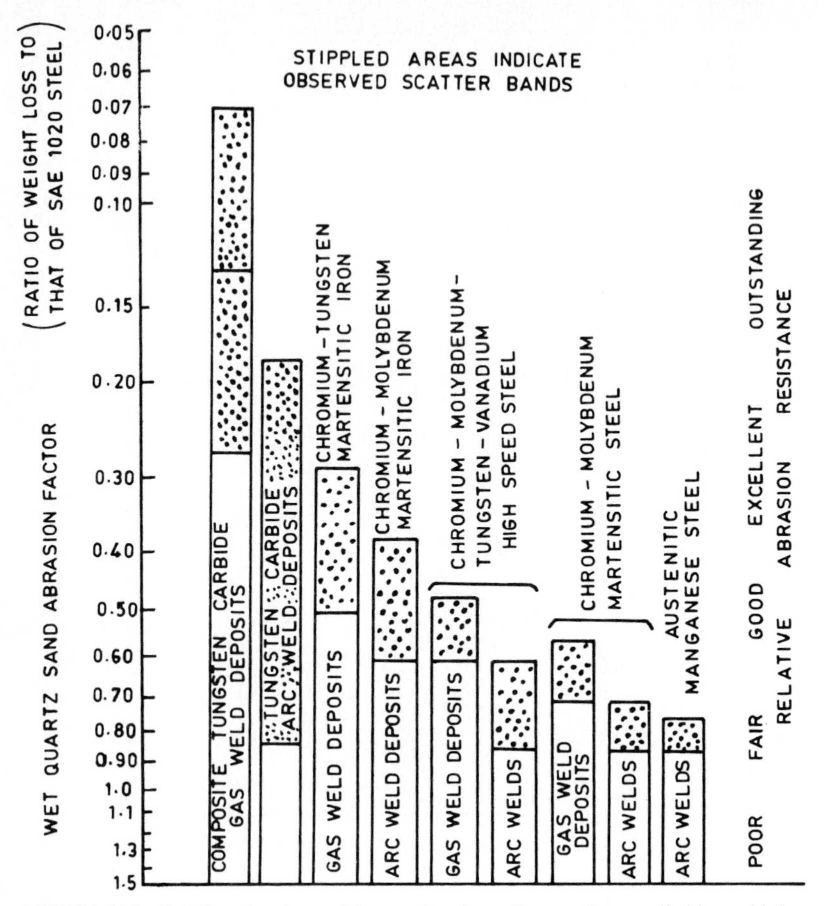

FIGURE 14.3 Relative abrasion resistance of various alloy coatings applied by weld deposition. [*Adapted from Avery (1952).*]

chromic acid (CrO_3) and a catalytic anion in proper proportion. Catalysts that have proved successful are acid anions, the first of which used was sulfate. Substitution of fluoride ions present in complex acid radicals for a portion of the sulfate marginally improves the chromium-plating operation. Reflecting this difference in catalyst, the principal types of baths are designated as conventional sulfate and mixed catalyst (Chessin and Fernald, 1982; Dennis and Such, 1972). The metal so produced is extremely hard and corrosion resistant. Various chromium alloy coatings are commercially available, e.g., from The Electrolyzing Company at Providence, Rhode Island.

Hardness of electrochemically deposited Cr is strongly influenced by deposition conditions and electrolytic bath. In general, Cr coatings deposited with a bright appearance are optimally hard. Bright Cr coatings deposited from conventional baths such as sulfate bath, have hardness in the range of 900 to 1000 HV, and those from mixed catalyst bath have hardnesses of 1000 to 1100 HV or

Major Alloy Coatings
Listed in Order of Decreasing Abrasion Resistance
and Increasing Toughness

(← Increasing toughness | Increasing abrasion resistance →)

Alloy coating	Properties sensitive to carbon content and structure
1. Tungsten carbide composites	Maximum abrasion resistance, worn surfaces become rough
2. High-chromium irons	Excellent erosion resistance, oxidation resistance
3. Martensitic irons	Excellent abrasion resistance, high compressive strength
4. Cobalt-based alloys	Oxidation resistance, corrosion resistance, hot strength and creep resistance
5. Nickel-based alloys	Corrosion resistance, may have oxidation and creep resistance
6. Martensitic steels	Good combinations of abrasion and impact resistance, good compressive strength
7. Pearlitic steels	Inexpensive, fair abrasion and impact resistance
8. Austenitic steels, stainless steels, manganese steels	Work hardening, corrosion resistance, Maximum toughness with fair abrasion resistance, good metal to metal wear resistance under impact

FIGURE 14.4 Comparison of properties of the major types of ferrous and nonferrous alloy coatings applied by weld deposition process. [*Adapted from Avery (1952).*]

higher. The hardness of hard Cr coatings remains almost constant up to 400°C and then decreases with an increase in temperature.

The wear life of a piston ring coated with hard Cr is improved by approximately five times. Piston rings for most engines have a chromium plating thickness of 100 to 175 μm on the wearing face, although thicknesses up to 250 μm are prescribed for heavy-duty engines.

Typical physical properties of hard electroplated chromium are presented in Table 14.3. The high hardness, good mechanical strength, and low coefficient of friction against many materials give chromium plating good resistance to wear by adhesion or abrasion, while its good thermal conductivity enables rapid dissipation of frictional heat (Gologan and Eyre, 1974). Under normal conditions, the chromium plat-

TABLE 14.3 Selected Physical Properties of Electroplated Nickel and Chromium Coatings and Electroless Nickel-Boron and Nickel-Phosphorus Coatings*

Property	Electroplated chromium	Electroless nickel-boron†	Electroless nickel-phosphorus‡	Electroplated nickel
Density, kg m^{-3}	6900–7200	8250	7750	8900
Melting point, °C	1890	1080	890	1455
Electrical resistivity, $\mu\ \Omega$ cm	15–50	89	90	7
Thermal conductivity, W $m^{-1} \cdot K^{-1}$	51	—	4	64
Coefficient of thermal expansion, (22–100°C) × 10^{-6} °C^{-1}	8.1	12.6	12	7.5
Magnetic properties	Nonmagnetic	Very weakly ferromagnetic	Nonmagnetic	Nonmagnetic
Internal stress, MPa	120–400	110	Nil	90–200
Tensile strength, MPa	100–350	110	700	350–800
Ductility, % elongation	—	0.2	1.0	5–30
Modulus of elasticity, GPa	100–200	120	200	210
As-deposited hardness, HV	900–1100	700	500	150–400
Heat-treated hardness, 400°C for 1 h, HV	No effect	1200	1100	No effect
Coefficient of friction vs. steel				
Unlubricated	—	0.43	0.45	Galling
Lubricated	—	0.12	0.13	0.2
Wear resistance, as-deposited	Excellent	Excellent	Excellent	Fair

*Properties are for coating in the as-deposited condition, unless noted.
†Borohydride reduced electroless nickel containing approximately 5% boron.
‡Hypophosphite reduced electroless nickel containing approximately 10.5% phosphorus.

ing has a dense, hard oxide layer on the surface, which is responsible for its low coefficient of friction and good corrosion resistance (Rabinowicz, 1967).

Use of electrochemically deposited hard Cr coatings is limited because of hydrogen embrittlement of high-strength steels, its low rate of deposition, and poor throwing power (Inwood and Garwood, 1978).

14.2.1.2 Cr Coatings Applied by PVD. Hard chromium coatings are applied by evaporation, ion plating, and magnetron sputtering. Sundgren and Hentzell (1986) have reported that as the substrate temperature is increased, the hardness of metal and alloy coatings applied by PVD decreases in contrast to the opposite behavior of refractory compound coatings to be discussed later. Thus, when depositing pure metals, low substrate temperatures are required to obtain hard coatings. Other process parameters such as deposition rate, energetic particle bombardment, and residual gas content influence the coating hardness (Bunshah, 1974).

Komiya et al. (1977a) electron-beam-evaporated Cr using a hollow cathode with substrates held at 190 and 520°C and obtained hardness values of 1040 and 950 kg mm^{-2}, respectively. Nilsson et al. (1977) reported that the hardness of evaporated Cr coatings increased with an increase in the deposition rate (Fig. 14.5). The hardness was about 775 kg mm^{-2} at a substrate temperature of 50°C and the deposition rate was 150 nm min^{-1}; it increased to 1250 kg mm^{-2} at a deposition rate of 600 nm min^{-1} (Nilsson et al., 1977). Differences in the microstructure of the evaporated chromium for the deposition rates of 200 nm min^{-1} and 400 nm min^{-1} are shown in Fig. 14.6. Similar trends of an increase in hardness with an increase in the deposition rate have been reported for rf sput-

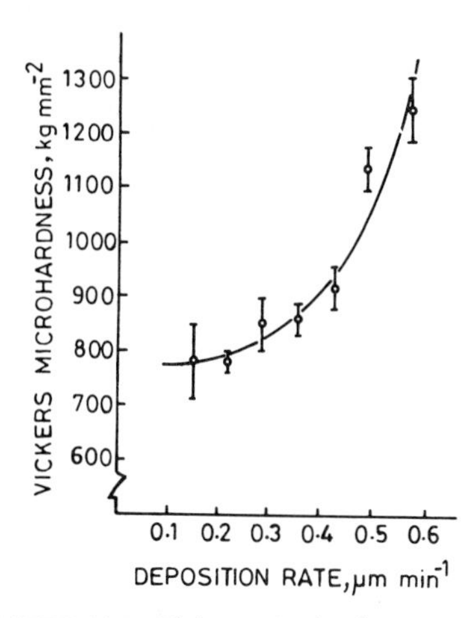

FIGURE 14.5 Vickers microhardness versus deposition rate for an evaporated chromium coating at a substrate temperature of 50°C. [*Adapted from Nilsson et al. (1977).*]

tered Cr coating by Sundgren and Hentzell (1986). Based on the materials analysis, Nilsson et al. (1977) explained the increase in hardness by an increasing defect density with increasing deposition rate. Curves of hardness versus deposition rate for higher temperatures show a behavior similar to Fig. 14.5. The hardness, however, is lower, and the increase in hardness begins at higher deposition rates (Nilsson et al., 1977).

Klokholm and Berry (1968) measured internal (residual) stresses in the evaporated Cr coatings. They reported tensile stresses ranging from 1.5 to 0.8 GPa for coating thicknesses of 20 and 160 nm, respectively (also see Hoffman and Gaerttner, 1980).

Hoffman and Thornton (1977) and Thornton and Hoffman (1981) measured internal stresses in the sputtered Cr coatings. They reported that Cr coatings sputtered onto glass at a deposition rate of 1 nm s^{-1} and 0.15 Pa (1.1 mtorr) of argon pressure incurred high compressive stresses (~1 GPa) (Fig. 14.7). Faster deposition rates at 0.15 Pa (1.1 mtorr) of argon decreased the magnitude of the integrated compressive stress in inverse proportion to the logarithm of the rate. Raising the pressure of argon from 0.15 to

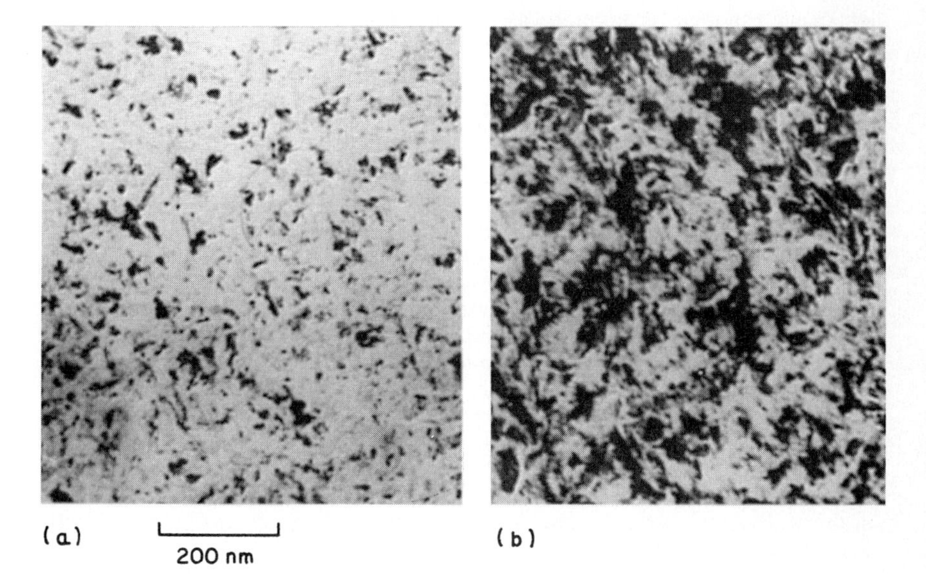

FIGURE 14.6 TEM micrographs of evaporated chromium coating at a substrate temperature of 50°C and deposition rates of: (*a*) 0.2 μm min^{-1}. (*b*) 0.8 μm min^{-1}. [*Adapted from Nilsson et al. (1977).*]

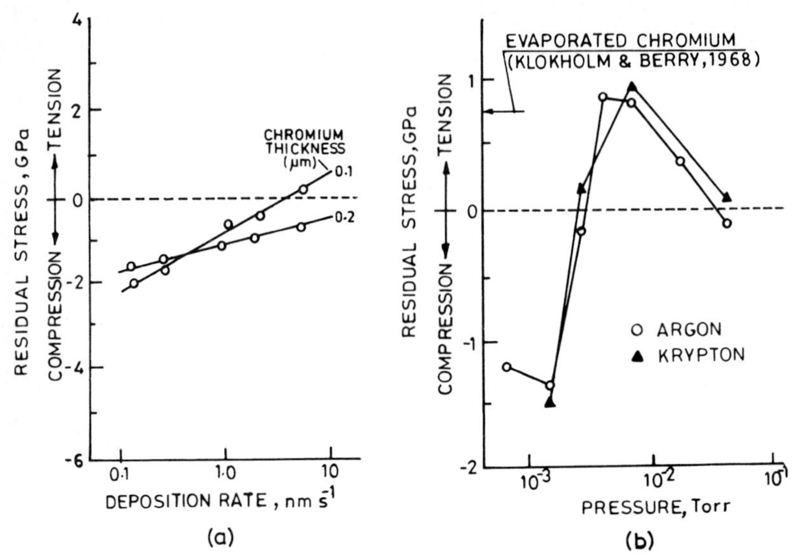

FIGURE 14.7 The residual stress as a function of: (*a*) Deposition rate at 1.1 mtorr of argon. (*b*) Working gas pressure at a deposition rate of 1.0 nm s^{-1} of sputtered chromium coating to a thickness of 0.25 µm in both argon and krypton gases. [*Adapted from Hoffman and Thornton (1977).*]

0.4 Pa (1.1 to 3 mtorr) reverses the stress from compressive to tensile, and further increase in pressure reduces the tensile stress more gradually to zero. Similar results were reported for krypton gas. Measured stress includes the thermal component of stress from substrate heating during deposition which was found to be negligible. Tensile residual stresses in rf sputtered Cu films (~0.1 GPa) were reported by Kuwahara et al. (1981). It is believed that the compressive stresses developed in the low pressure are caused by the location of the noble gas itself in the lattice and the structural damage caused by the incorporation process (low-energy implantation) as well as atomic peening of the coating during growth by the impact of high-energy atoms or ions.

The hardness of the sputtered coatings has been increased by doping Cr with C or N_2 (Aubert et al., 1983, 1985; Cholvy and Derep, 1985). Carbon or nitrogen was introduced either through the gas phase by reactive sputtering (in the presence of CH_4 or N_2) or by direct sputtering of sintered composite targets on a heated substrate (200 to 400°C). Chemical and structural analyses of coatings revealed the formation of supersaturated solid solutions for carbon or nitrogen concentrations up to several weight percent. The hardness of sputtered Cr coatings doped with C and N_2 reached about 2500 HV (coatings containing 4 at % C), and about 3000 HV (coatings consisting of 24 at % N). These hardness values are quite high as compared to electrochemically deposited Cr. X-ray diffraction analysis showed that the carbon and nitrogen are present as supersaturated solutions in the chromium matrix. Figure 14.8 shows the hardness of Cr-C coatings with thicknesses greater than 10 µm as a function of carbon content.

Table 14.4 summarizes the coefficient of friction of electrochemically deposited Cr and sputtered Cr coatings doped with C and N, sliding against M2 high-speed steel in water at 0.55 m s^{-1} speed and at different loads (P) in increments

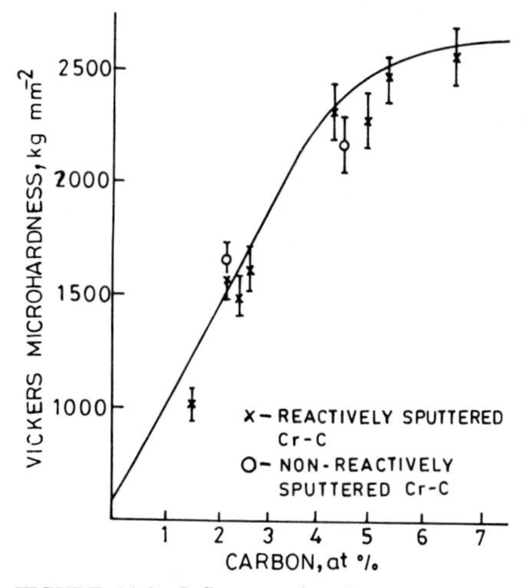

FIGURE 14.8 Influence of carbon content on microhardness of sputtered Cr-C coatings. [*Adapted from Aubert et al. (1985).*]

TABLE 14.4 Coefficient of Friction for Various Chromium Coatings (Applied by Electrodeposition and Sputtering Processes) Sliding against M2 High-Speed Steel Rotating Ring

Coating	Vickers micro-hardness (HV), kg mm^{-2}	Coefficient of friction					
		P = 100 N	P = 200 N	P = 300 N	P = 400 N	P = 500 N	P = 600 N
Electroplated hard Cr	1100	0.40	0.35	0.35-S			
Sputtered Cr	600	0.50	S				
Sputtered Cr-C	1500	0.23	0.22	0.21	0.21	0.22	S
Sputtered Cr-C (NR)	1600	0.24	0.22	0.22	0.22	0.22-S	
Sputtered Cr-C	2400	0.23	0.23	0.22	0.23	0.23-S	
Sputtered Cr-N	1300	0.23	0.22	0.22	0.22	0.23	S
Sputtered Cr-N	3000	0.33	0.29	0.28	0.30	0.30	0.35-S

Note: NR: nonreactive sputtering. S: seizure. Test conditions: Contact geometry-coated plane-rotating cylinder, medium: water, sliding speed: 0.55 m s^{-1}, load (P) increased by 100 N after every 10 min.

Source: Adapted from Aubert et al., 1985.

of 100 N every 10 min until seizure. The sputtered Cr coatings doped with C and N_2 exhibit a higher wear resistance and a lower coefficient of friction as compared to pure Cr coatings deposited by sputtering and electrochemical deposition.

14.2.1.3 Cr Coatings Applied by CVD. Hanni and Hintermann (1981) deposited Cr by CVD. They reported that CVD Cr coatings deposited onto carbon-containing steels transform into chromium carbide by carbon diffusion from thesubstrate. The hardness varies between 1500 and 2200 kg mm^{-2} depending on the carbon content in the substrate used.

14.2.1.4 Cr Composite Coatings Applied by Electrochemical Deposition. The incorporation of dispersed wear-resistant particles in the Cr electrodeposit allows further tailoring of its properties to achieve wear resistance. Particles are codeposited from the suspension in the agitated electroplating solution. Composites with a cobalt matrix and a wide range of ceramic powders such as alumina, boron nitride, chromium carbide, silicon nitride, and tungsten carbide, have been made (Kedward et al., 1976). The most successful combination so far is cobalt-chromium carbide composite used for high-temperature wear resistance of aircraft components.

14.2.2 Nickel Coatings

Nickel coating is one of the oldest electrochemically deposited metallic coating on various metallic and nonmetallic substrates and is used for decorative purposes and for environmental corrosion and wear resistance. Nickel coatings have been applied to metals and plastics (with an undercoat) by electrochemical and electroless deposition and less commonly by evaporation, and chemical vapor deposition. The coatings applied by electroless deposition are very hard and are most commonly used in many engineering components to combat wear and corrosion. They have found many applications, including those in petroleum, chemicals, plastics, optics, printing, mining, aerospace, nuclear, automotive, electronics, computers, textiles, and food machinery.

14.2.2.1 Ni Coatings Applied by Electrochemical Deposition. Electrochemically deposited nickel coatings possess a wide variety of properties, depending on the composition of electrolytic bath and operating conditions. For instance, the hardness of Ni coatings applied by Watts, sulfamate, and fluoborate baths range from 100 to 250 HV, 125 to 450 HV, and 125 to 300 HV, respectively. Most of these as-coated coatings are amorphous, and they have no crystal or phase structure (Gianelos and McMullen, 1982; Dennis and Such, 1972).

14.2.2.2 Ni Coatings Applied by Electroless Deposition. In the electroless deposition of nickel, the addition of phosphorus or boron into Ni plating increases hardness typically up to 700 HV. Electroless nickel plating is used to deposit nickel without the use of an electric current. The coating is deposited by an autocatalytic chemical reduction of nickel ions by hypophosphite compounds for Ni-P coatings and by aminoborane or borohydride compounds for Ni-B coatings. Hot acid hypophosphite-reduced baths are most frequently used to coat steel and other metals, whereas warm alkaline hypophosphite baths are used for coating plastics and nonmetals (Fields et al., 1982). The electroless bath incorporates P or B (typically 4 to 12 wt %) into nickel during deposition. Usually, electroless Ni-P and Ni-B coatings are uniform, hard and relatively brittle, low friction, low wear, easily solderable, and less porous (more corrosion resistant) compared to electroplated Ni.

The properties of electroless Ni-P coatings are similar to those of electroless Ni-B coatings with few exceptions. The hardness of Ni-B alloys is very high (650 to 700 HV) higher than that of Ni-P, and these alloys can be heat treated to levels greater than that of hard chromium. Ni-B coatings have outstanding resistance to wear and abrasion. These coatings, however, are not completely amorphous and have reduced resistance to corrosive environments, very weakly ferromagnetic, as well as being much more costly than Ni-P coatings. The physical and mechanical properties of Ni-P and Ni-B coatings are summarized in Table 14.3.

Electroless-nickel coatings are supersaturated solid solution or amorphous metastable alloys and can be precipitation hardened being converted to crystalline nickel and nickel phosphide (Ni_3P) or nickel boride (Ni_3B). As electroless Ni-P is heated to temperatures above 220 to 260°C, structural changes begin to occur. First, coherent and then distinct particles of nickel phosphite (Ni_3P) form within the alloy. Then at temperatures above 320°C, the deposits begin to crystallize and lose their amorphous structure. With continued heating, nickel phosphite particles conglomerate and a two-phase alloy forms. An increase of Ni_3P phase in hardened coating is responsible for increased hardness and wear resistance (Parker, 1974; Randin and Hintermann, 1967). Hardened coatings produce wear resistance equal to that of commercial hard chromium coatings.

Figure 14.9 shows the effect of 1-h treatments at various temperatures on the hardness of Ni-P coatings containing 10.5 percent P (Gawrilov, 1979). Maximum hardness of about 1100 HV can be obtained through thermal treatments at 400°C for 1 h. For some applications, high-temperature treatments cannot be tolerated because parts may warp, or the strength of the substrate may be reduced. For those applications, longer times and lower temperatures are sometimes used to obtain desired higher hardness. For instance, 1-h treatment at 350 to 400°C, Ni-B coating hardness (650 to 700 HV) increases hardness to 1200 HV; long-term treatments (30 to 40 weeks) at temperatures between 200 to 300°C can produce hardness values of 1700 to 2000 HV. This dramatic increase in hardness value is attributed to the finer dispersion of nickel boride into nickel coating.

Frictional properties of electroless nickel coating is excellent and similar to those of chromium. The coefficient of friction of electroless Ni-B coating sliding against steel is about 0.12 under lubricated conditions and about 0.43 under dry

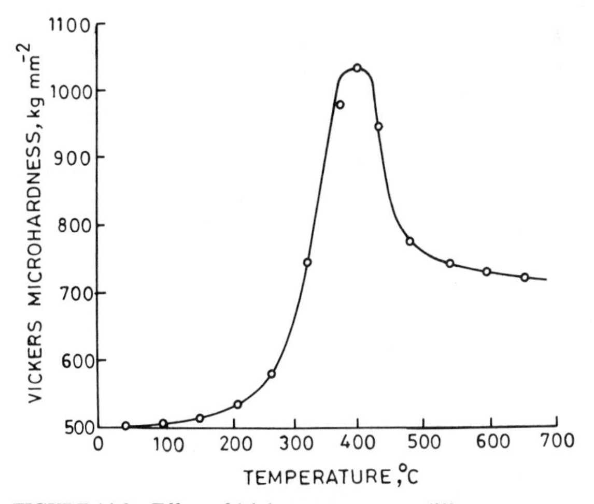

FIGURE 14.9 Effect of 1-h heat treatment at different temperatures on the microhardness of Ni-P (10.5% P) electroless coating. [*Adapted from Gawrilov (1979).*]

conditions (Gawrilov, 1979). Friction is practically insensitive to P or B content and heat treatment.

Because of their high hardness, electroless nickel coatings have excellent resistance to abrasive wear, both in the as-coated and hardened conditions. Table 14.5 summarizes the Taber abrasive index values for various electroless nickel coatings and electrodeposited Ni and Cr. We note that electroless Ni-B exhibits superior wear resistance compared to Ni-P, and their heat treatment further increases the wear resistance comparable to that of hard chromium. Figure 14.10 shows the wear behavior of Ni-P coatings with various concentrations of P and various heat treatments. The wear tests were conducted using a ball-disk machine. We note that the wear resistance of Ni-P coatings is found to be strongly influenced by P content and the heat treatment. The wear of the as-plated coating shows a maximum between 6 and 7 wt % P. The deposits heat treated at 250°C and of low P content are less worn than those with high P content. At 400°C, the phenomenon is the opposite: The wear is inversely proportional to the P content. At higher temperatures, the slope of the curve decreases for P content higher than 8 to 10 wt %. The wear resistance increases up to a temperature of about 400°C, then it increases only for coatings with a high P content. Wear resistance is believed to be proportional to the volume percent of the Ni_3P phase of the coating. It has also been reported that alloys containing more than 10% P and less than 0.05% impurities are typically continuous and consequently exhibit increased corrosion resistance than those with lower P contents (with high porosity) (Fields et al., 1982).

14.2.2.3 Ni Coatings Applied by CVD. Ni-B coatings having boron content from 0.05 to 37.4 at % have been applied by CVD by Skibo and Greulich (1984) and Mullendore and Pope (1987) for the hard surfacing of vapor-formed nickel dies. Up to 13.0 at % B, the Ni-B alloys were fine-grained, single-phase, supersaturated solid solutions. On adding 37.4 at % B, the alloy became completely amorphous (Mullendore and Pope, 1987). Vickers microhardnesses as a function of B content in the Ni-B coatings as deposited and after 350°C anneal are shown in Fig. 14.11*a*. Hardness increased with an increase in the boron content up to 13.0%,

TABLE 14.5 Comparison of Wear Resistance (Taber Abrasive Index) of Various Nickel and Chromium Coatings

Coating	Heat treatment for 1 h, °C	Taber abrasive wear index,* mg/1000 cycles
Electroplated nickel†	None	25
Electroless Ni-P‡	None	17
Electroless Ni-P‡	300	10
Electroless Ni-P‡	500	6
Electroless Ni-P‡	650	4
Electroless Ni-B§	None	9
Electroless Ni-B§	400	3
Electroplated hard chromium	None	2

*CS-10 abrasive wheels, 10 N normal load, determined as average weight loss per 1000 cycles for a total test of 6000 cycles.

†Produced in watts bath.

‡9% P reduced in hypophosphite bath.

§5% B reduced in borohydride bath.

Source: Adapted from Fields et al. (1982).

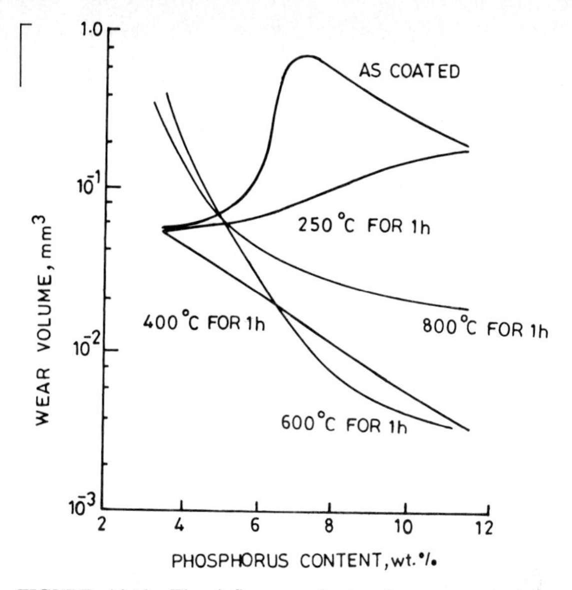

FIGURE 14.10 The influence of phosphorous content in electroless Ni-P coatings and heat treatment on wear resistance. These wear tests were carried out using a steel ball sliding against a coated disk in a ball-disk machine at 20 N load and 8 mm s^{-1}. [*Adapted from Randin and Hintermann (1967).*]

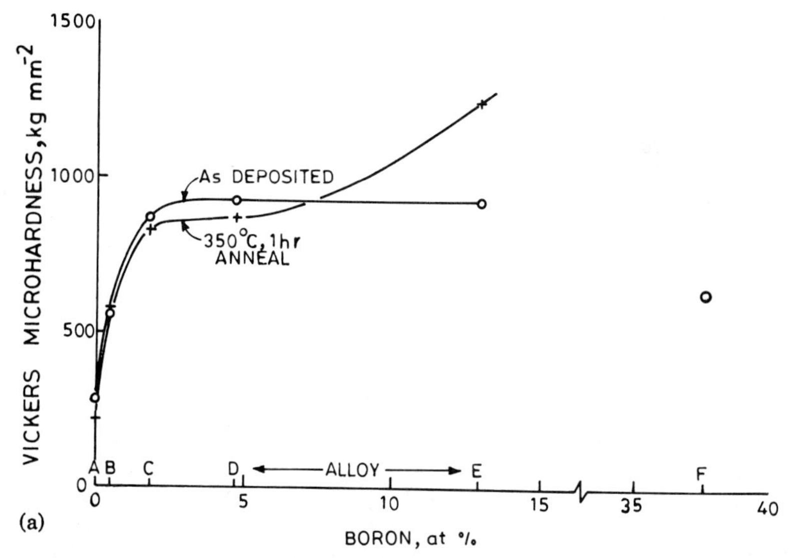

FIGURE 14.11 (*a*) Influence of boron content on microhardness for Ni-B coatings applied by chemical vapor deposition. (*b*) Coefficient of friction versus number of cycles for Ni-B coatings with varying amount of B. (*c*) Wear depth versus microhardness for Ni-B coatings having varying amount of B. [*Adapted from Mullendore and Pope (1987).*]

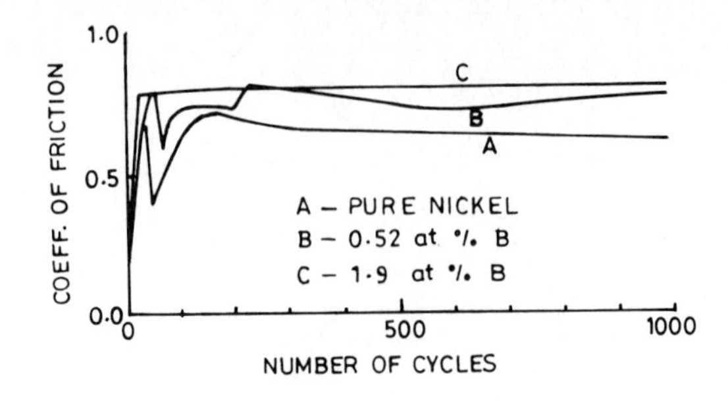

(b)

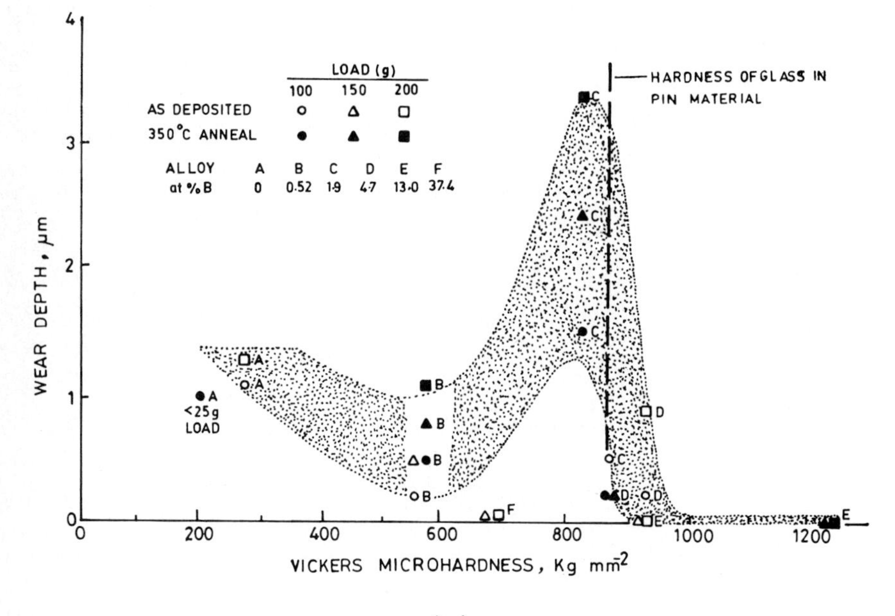

(c)

FIGURE 14.11 (*Continued*)

the 37.4 at % B amorphous alloy has a lower hardness (610 kg mm^{-2}). The hardness of the annealed 13.0 at % B alloy, which exhibited precipitation hardening, increased as a result of annealing. Figures 14.11*b* and *c* show friction and wear characteristics of Ni-B coatings having different amounts of boron and of coatings as deposited and after annealing. The friction and wear data were obtained with a pin-on-disk tester in which the pin was a glass-filled epoxy. The coefficient of friction as a function of number of cycles was found to increase from an initial value of 0.2 to higher steady-state value (0.5 to 0.9). The greatest wear depth was observed for the annealed alloy C (1.9 at % B) and appears to be associated with a hardness near but slightly lower than the hardness of the glass in the mating pin material. The wear depth for other alloys generally decreased with increasing hardness and increasing boron contents. The wear for alloys with 13.0 at % B and 37.4 at % B was negligible.

14.2.2.4 Ni Composite Coatings Applied by Electroless Deposition. Pure nickel, although reasonably wear resistant, is unsuitable in some applications because of its tendency to galling. The range of applications can be increased by the inclusion of a fine dispersion of hard, wear-resistant particles, such as silicon carbide, tungsten carbide, chromium carbide, titanium carbide, alumina, or diamond, by codeposition from a suspension in the plating solution. The hardness of the composite coatings increases after heat treatment; for example, the hardness of Ni-P/SiC coating increases from 600 to 1000 HV after heat treatment at 425°C. Composite coatings, having SiC, when heat treated at 430°C, exhibited improved lubricity, presumably due to formation of SiO_2 coating on the surface (Parker, 1974).

Commercial coatings containing hard particles are available, e.g., diamondizing and Nye-Carb from Electro-Coating, Inc., in Benton Harbor, Michigan (Dreger, 1977; Sale, 1979) and composite coatings from EC Technologies in Emeryville, California (Kenton, 1983). The Nye-Carb coatings contain 20 to 30 % silicon carbide particles by volume with particle sizes ranging from 10 to 15 μm, suspended in a electroless plated nickel alloy matrix. The matrix material is an alloy which is 90 to 93 % Ni and 7 to 10 % P. Diamondizing is a composite coating consisting of 20 to 30 % synthetic diamond particles by volume with particle sizes ranging from 2 to 4 μm, suspended in a nickel alloy matrix (Dreger, 1977). Coating thickness for most applications is 5 to 40 μm.

Solid lubricant particles, such as graphite, PTFE, and calcium fluoride, are also added to electroless nickel coatings to reduce friction and wear. Calcium fluoride and PTFE particles were codeposited by Kenton (1983) and the composite coating exhibited very low friction (from 0.3 to 0.6) and wear. General Magnaplate Corp. in Linden, New Jersey, offers a Nedox coating for all ferrous and copper alloy substrates. This coating consists of controlled infusion of PTFE within electroless nickel with a total thickness of about 5 to 75 nm. A heat treatment cycle assures thorough infusion of surface with PTFE and increases hardness of the matrix (~750 HV). The Nedox coating combines the advantages of high strength and toughness of electroless Ni with self-lubricating properties of PTFE. The coating also exhibits greatly improved corrosion resistance over conventional electroless nickel. The coating can be used from −200 to 250°C.

14.2.3 Molybdenum Coatings

Molybdenum coatings have been applied by electrochemical deposition, thermal spraying, evaporation, sputtering, CVD, and PECVD techniques. Molybdenum coatings are difficult to apply by electrochemical deposition because formation of natural oxides, during deposition, interfere with the adherence of the coatings. Chromium has been a favored strike coating because its coefficient of thermal expansion is similar to that of molybdenum. It has good as-plated adherence, and the two metals diffuse to form a solid solution at elevated temperatures. The plasma spraying process is preferred to apply Mo coatings. These coatings provide better performance on piston rings as compared to hard chromium coatings. The plasma sprayed Mo coating exhibits hardness typically in the range of 500 to 800 HV (Ingham, 1969).

The molybdenum coatings have also been applied by evaporation, dc sputtering, and magnetron sputtering (Aubert et al., 1985; Blachman, 1971; Bunshah, 1974; Glang et al., 1965). The doping of Mo coating with C and N improves the hardness as high as three to six times up to 2800 HV and subsequently wear resistance by several fold. Table 14.6 summarizes the coefficient of friction of doped Mo coatings sliding against M 2 high-speed steel in water at 0.55 m s^{-1} sliding speed and load applied in increments of 100 N every 10 min until seizure. The doped Mo coating exhibits higher hardness, lower coefficient of friction, and lower wear. The adhesion of sputtered Mo coating is much better as compared to electrochemically and plasma sprayed Mo coatings (Aubert et al., 1985).

Klokholm and Berry (1968) measured residual stresses in evaporated Mo coating. They found tensile stresses roughly 1 GPa for 100-nm-thick coating. Blachman (1971) and Glang et al. (1965) measured the effect of substrate bias on residual stresses in dc sputtered Mo. They found that a 0.1- and 0.3-μm-thick coating of Mo sputtered at a moderate rate (0.5 nm s^{-1}) and high pressures (5 Pa or 37 mtorr) are essentially unstressed, in agreement with the results presented for sputtered Cr. The application of an increasing negative voltage to the substrate causes first an increasing tensile stress and then a sudden reversal at −100 V from high tension to high compression (Fig. 14.12). This oscillation bears a remarkable resemblance to the stress variation with decreasing pressure for sputtered Cr of Fig. 14.7. This similarity suggests that the mechanism responsible for the stress reversal may be common to both increasing negative bias and decreasing pressure. Blachman (1971) also found that entrapment of Ar increased with an increase in the substrate bias. Therefore, Blachman (1971) suggested that the increasingly compressive stress in Mo coatings sputtered in Ar is associated with increasing argon content of the coatings and increasing impinging energy of the Ar ions at the substrate due to bias.

Molybdenum coatings have been deposited by CVD and PECVD primarily for semiconductor applications (Hess, 1984; Jaeger and Cohen, 1972). Stable, nonporous Mo coatings can be produced by plasma-enhanced reaction of hydrogen and molybdenum hexafluoride (MoF_6) with H_2 to MoF_6 ratio of at least 7:1 at a substrate temperature above 200°C. Mo coatings have also been produced with

TABLE 14.6 Coefficients of Friction for Various Molybdenum Coatings (Applied by Sputtering) Sliding against M2 High-Speed Steel Rotating Ring

Sputtered coating	Vickers microhardness (HV), kg mm^{-2}	Coefficient of friction					
		P = 100 N	P = 200 N	P = 300 N	P = 400 N	P = 500 N	P = 600 N
Mo	500	0.36	S				
Mo-C	1500	0.19	0.19	S			
Mo-C	2300	0.20	0.14	0.14	S		
Mo-C	2500	0.23	0.20	0.20	0.24	0.29	S
Mo-N	1000	0.16	0.17	0.17	S		
Mo-N	2800	0.20	0.18	0.17	0.20	S	

Notes: S: Seizure. Test conditions: Contact geometry-coated plane-rotating cylinder, medium: water, sliding speed: 0.55 m s^{-1}; load: increased by 100 N after every 10 min.

Source: Adapted from Aubert et al. (1985).

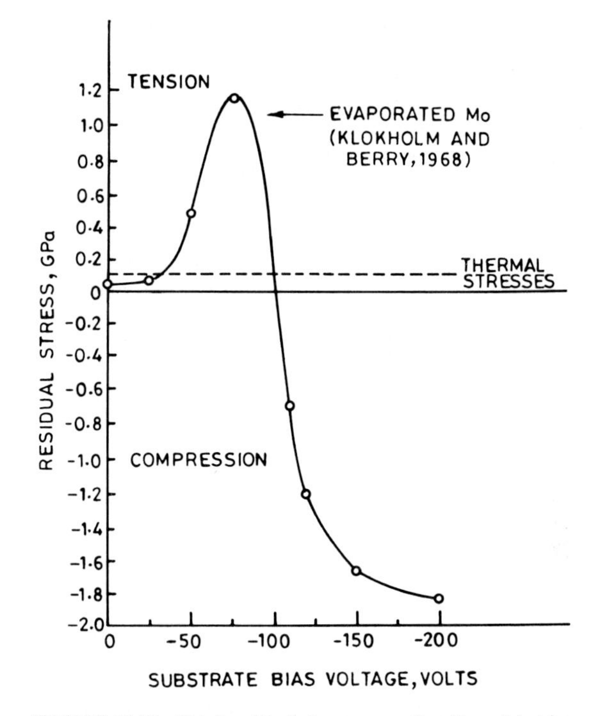

FIGURE 14.12 Total residual stresses as a function of dc bias for 10-nm-thick Mo coatings dc sputtered with the silicon substrate holder at 40°C. Thermal stress shows the value of thermal expansion mismatch stress. Residual stresses measured in a 100-nm-thick evaporated Mo coating by Klokholm and Berry (1968) are also shown. [*Adapted from Blachman (1971).*]

plasma enhanced reaction of hydrogen and molybdenum pentachloride ($MoCl_5$) at a substrate temperature between 170 and 430°C.

14.3 COBALT- AND NICKEL-BASED ALLOY COATINGS

In order to combat wear and corrosion, a large variety of alloys have become available over the years. The most important of these for extreme wear and/or high-temperature applications fall into two main classes: cobalt-based and nickel-based alloys. Coatings of these alloys are generally deposited by thermal spraying and welding (Antony et al., 1983; Bucklow, 1983; Foroulis, 1984) and less commonly by evaporation and sputtering processes (Krutenat et al., 1968; Pettit and Goward, 1983; Spalvins and Brainard, 1974; Thornton and Ferriss, 1977). Plasma sprayed and welded coatings are most common because of several advantages such as lower cost, good compositional control, and greater process flexibility.

The coating should be very hard and also ductile (i.e., not brittle) in order to have effective protection against wear. Usually, in most hard facing alloy coatings, the matrix provides the ductility and inclusions such as carbides provide the hard phase that increases the hardness. Thus it is a general practice to obtain a specific combination of hardness and ductility by optimizing the ratios of various inclusions in Ni or Co matrix. The addition of ceramic materials such as chromium oxide (Cr_2O_3) and aluminum oxide (Al_2O_3) (typically from 3 to 10 wt %) in cobalt-based alloys, significantly improves the wear resistance at elevated temperatures in the vicinity of about 1000°C (Wolfla and Tucker, 1978).

The hardening effect caused by alloying may be quite different in a thin coating compared to bulk material since the presence of alloying elements also will influence the growth and thus the microstructure of the coatings such as causing a finer-grain structure and formation of metastable phases. The hardness of alloy coatings may thus be considerably higher than bulk materials of the same alloy.

Table 14.7 lists some of the alloy compositions, which have been applied commonly by welding and plasma spraying.

14.3.1 Cobalt-Based Alloy Coatings

Two types of commercially available cobalt-based alloys for coating applications are: carbide-containing alloys and alloys containing Laves phase. The Stellite alloys ST-1, ST-6, and ST-12 (Anonymous, 1975) are cobalt-rich with different amounts of a dispersed hard second phase consisting of chromium, cobalt, and tungsten carbides. The amount of carbide phase varies from 29 wt % for alloy ST-1, to 12 wt % and 16 wt % for alloys ST-6 and ST-12, respectively (Silence, 1978). Silence (1978) has reported that abrasive wear resistance can be maximized through appropriate processing so as to maximize the particle size of the hardest constituents which usually decreases the hardness and the volume of the hard constituents. In metal-to-metal wear, the role of the hard phase (carbide) is less critical and the wear rate is less sensitive to structural features. He also reported that hardness is not a reliable indication of abrasive or metal-to-metal wear resistance.

Stellites can be applied by various welding and plasma spraying processes. Oxyacetylene hard facings have been found to be more wear resistant than most other types of deposits such as metal-arc deposits (Silence, 1978). Longer solidification times for oxyacetylene deposits, as compared to most other deposition methods, produce larger carbides, and result in increased wear resistance (Silence, 1978). Stellites are commonly used for components requiring resistance to metal-to-metal wear, and low-stress abrasion. They are used in hard facing of high-abrasion extruder screws, rock bits, bearing surfaces, slurry pumps, low-impact metal-forming dies, fluid valves, and diesel engine valves.

Wolfla and Tucker (1978) reported two families of Co-based alloys—coatings L-103 and LCO-17/LCO-19—developed at Union Carbide Corporation in Indianapolis for high-temperature wear and corrosion-resistance applications. The L-103 is a detonation gun coating with a cobalt-based matrix and a dispersion of Cr_2O_3 particles. The matrix of this coating contains Cr for corrosion resistance and W for both solid solution and carbide precipitation strengthening. LCO-17 and LCO-19 were developed specifically for wear resistance in a corrosive envi-

TABLE 14.7 Various Commonly Used Co- and Ni-Based Alloys for Wear and Corrosion Resistance

Type of coating	Nominal composition (wt %)	Preferred application process	Hardness	Maximum use temp., °C	References
		Co-based alloys			
Stellite alloy no. 1 (ST-1)	30% Cr, 12% W, 2.5% C, bal. Co	OAW/SMAW/GTAW/PS	54 HRC (575 HV)	800	Anonymous (1975)/Silence (1978)/Foroulis (1984)
Stellite alloy no. 6 (ST-6)	28% Cr, 4% W, 1% C, bal. Co	OAW/SMAW/GTAW/PS	42 HRC (410 HV)	800	Anonymous (1975)/Silence (1978)/Foroulis (1984)
Stellite alloy no. 12 (ST-12)	29% Cr, 8% W, 1.3% C, bal. Co	OAW/SMAW/GTAW/PS	44 HRC (435 HV)	800	Anonymous (1975)/Silence (1978)/Foroulis (1984)
L-103	28% Cr, 19% W, 5% Ni, 1% V, 3% C, 3% Cr_2O_3, bal. Co	DG	—	900	Wolfla and Tucker (1978)
LCO-17	25% Cr, 10% Ta, 7% Al, 0.5% Y, 2% C, 10% Al_2O_3, bal. Co	DG	—	900	Wolfla and Tucker (1978)
LCO-19	30% Cr, 10% Ta, 0.5% Y, 10% Al_2O_3, bal. Co	DG	—	1100	Wolfla and Tucker (1978)
Tribaloy 400 (T-400)	8% Cr, 2.5% Si, 28% Mo, 0.6% C, bal. Co	OAW/GTAW/PS/DG	52 HRC (535 HV)	800	Schmidt and Ferriss (1975)/Foroulis (1984)
Tribaloy 800 (T-800)	17% Cr, 3% Si, 28% Mo, 0.06% C, bal. Co	OAW/GTAW/PS/DG	58 HRC (630 HV)	800	Schmidt and Ferriss (1975)/Foroulis (1984)
		Ni-based alloys			
Nickel-copper (Monel)	25–30% Cu, 3–6% Fe, 3.5% Mn, 0.35–0.55% C, bal. Ni	SMAW	130 HV	500	Bucklow (1983)
Nickel aluminide	95% Ni, 5% Al	PS	170 HV	750	Sangam and Nikitich (1985)/Bucklow (1983)

TABLE 14.7 Various Commonly Used Co- and Ni-Based Alloys for Wear and Corrosion Resistance (*Continued*)

Type of coating	Nominal composition (wt %)	Preferred application process	Hardness	Maximum use temp., °C	References
		Ni-based alloys (*cont.*)			
Nichrome	80% Ni, 20% Cr	PS, SMAW	200 HV	1000	Sangam and Nikitich (1985)
Hastelloy C	17% Cr, 17% Mo, 0.12% C, bal. Ni	GTAW/PTAW/PS	95 HRB (200 HV)	—	Antony et al. (1983)
Haynes alloy No. 711 (H-711)	23% Fe, 27% Cr, 12% Co, 7% Mo, 3% W, 2.7% C, bal. Ni	OAW/GTAW	43 HRC (425 HV)	—	Bhansali et al. (1980)/ Foroulis (1984)
Haynes alloy No. 716 (H-716)	29% Fe, 26% Cr, 12% Co, 3% Mo, 3.5% W, 1.1% C, 0.4% B, bal. Ni	OAW/GTAW	35 HRC (345 HV)	—	Bhansali et al. (1980)/ Foroulis (1984)
Ni-Cr-B-Si-C alloys					
Haynes alloy No. 40 (H-40)	14% Cr, 4% Si, 3.4% B, 4% Fe, 0.75% C, bal. Ni	OAW/GTAW	56 HRC (615 HV)	650	Foroulis (1984)
	10% Cr, 2.25% Si, 2% B, 2.5% Fe, 0.45% C, bal. Ni	PW	35–40 HRC (345–390 HV)	—	Knotek et al. (1979)
	11.5% Cr, 3.75% Si, 2.5% B, 4.25% Fe, 0.65% C, bal. Ni	PW	45–50 HRC (445–515 HV)	—	Knotek et al. (1979)
	13.5% Cr, 4.25% Si, 3.0% B, 4.75% Fe, 0.75% C, bal. Ni	PW	56–61 HRC (615–720 HV)	—	Knotek et al. (1979)
	16% Cr, 4% Si, 4% B, 3% Cu, 3% Mo, 2.5% Fe, 0.5% C, bal. Ni	PS/sintering	820–975 HV	—	Mann et al. (1985)

WC+Ni-Cr-B-Si-C composite alloy	11% Cr, 2.5% Si, 2.5% B, 2.5% Fe, 0.5% C, 32% WC, bal. Ni	PS/sintering	740–1300 HV	—	Mann et al. (1985)
Tribaloy 700 (T-700)	15% Cr, 3.4% Si, 3% Fe+Co, 32% Mo, 0.06% C, bal. Ni	GTAW/PTAW/PS/DG	46 HRC (460 HV)	1000	Schmidt and Ferriss (1975)/Foroulis (1984)
Other alloys					
Ni-Co-Cr-Al-Y (PWA 276)	13.6% Al, 17.3% Cr, 20.3% Co, 0.5% Y_2O_3, 0.01% C, bal. Ni	PS	—	850	Hebsur and Miner (1987)
Ni-graphite (abradable coating)	15% graphite, bal. Ni	PS	Soft	475	Sangam and Nikitich (1985)

OAW: oxyacetylene gas welding.
PW: powder welding.
SMAW: shielded metal-arc welding.
GTAW: gas tungsten arc welding.
PTAW: plasma transferred arc welding.
PS: Plasma spray.
DG: Detonation gun.

ronment at higher temperatures than those for which L-103 is suitable. Ta was substituted for the W used in L-103 for the purposes of carbide and solid solution strengthening because of deleterious effects of W on hot corrosion resistance and because of its low oxidation resistance. Aluminum was added to LCO-17 in a sufficient amount to allow the coating to form a protective alumina scale in an oxidizing environment rather than to rely on a chromia scale as in L-103 and LCO-19. About 30 vol % Al_2O_3 dispersion was used in LCO-17 and LCO-19 instead of approximately 10 vol % Cr_2O_3 dispersion used to increase the wear resistance in L-103. To further improve the bond strength of these coatings for severe applications, coatings are heat-treated at 1080°C for 4 h in vacuum (designated by the postscript A such as L-103 A, etc.) in order to create a diffusion bond to the substrate. For L-103, the 780°C treatment for 1 h is done for stress relieving, after the diffusion heat treatments.

Tribaloys were developed by DuPont Corp., Wilmington, Delaware. They are free of carbide phase because of low carbon content. Their wear resistance is attributed to a hard Laves phase and plays a similar role as tungsten and chromium carbide phases play in the ST-1, ST-6, and ST-12 alloys. The Laves phase has a hexagonal structure similar to M_7C_3 carbides with a hardness of about 1200 HV (lower than that of carbides, consequently less resistant to abrasion compared to carbides) and solid solution matrix has a hardness in the range of 300 to 800 HV with a overall hardness in the range of 450 to 700 HV. At higher chromium content, the corrosion resistance is excellent in most environments (Schmidt and Ferriss, 1975). Tribaloys are used for wear resistance in such applications as: journal bearings, ball and roller bearings, seals, screw flights, pistons, piston rings, pump parts, vanes, tappet inserts, valves, and cams.

14.3.2 Ni-Based Alloy Coatings

Most commercially available nickel-based alloys for coating applications can be divided into four groups: nickel alloys, carbide-containing alloys, boride-containing alloys, and Laves phase-containing alloys. The 70% Ni/30% Cu alloys are widely used for their excellent corrosion resistance at temperatures below 450°C. The nickel-aluminide (Ni-Al) coatings are also used for oxidation and corrosion resistance up to about 750°C. The Ni-Cr-based alloys comprise possibly the widest range of materials available for operation above 500°C. Many of them exhibit a complex structure consisting of carbides or borides in a nickel-alloy matrix which gives good mechanical properties and oxidation resistance.

The carbide-containing Ni-Cr-Fe-C alloys-Haynes alloys 711 and 716 (developed by Cabot Corp., Kokomo, Indiana, and presently marketed by Stoody Deloro Stellite Inc., Industry, California under the trade name of Nistelle) with a small amount of cobalt were produced as an alternative to Stellites in order to reduce consumption of expensive cobalt. These alloys are applied by the weld deposition method. Metal-to-metal wear resistance of Haynes alloy No. 711 is better than ST-1. Haynes alloy No. 716 which contains 12% Co, compares to ST-6 in abrasive wear resistance and hot hardness. However, room-temperature hardness is lower. Another popular carbide-containing nickel-based alloy group is Ni-Mo-Cr-C systems such as Hastelloy C. Carbide-containing nickel-based alloys have poor oxyacetylene weldability.

The group of boride-containing nickel-based alloys is primarily composed of Ni-Cr-B-Si-C such as Haynes alloy No. 40 (developed by Cabot Corporation,

Kokomo, Indiana, and presently marketed by Stoody Deloro Stellite, Inc., Industry, California under the trade name of Deloro 60). Usually the boron content ranges from 1.5 to 3.5%, depending upon the chromium content, which varies from 0 to 15 percent. The higher chromium alloys generally contain a large amount of boron, which forms very hard chromium borides with hardness of approximately 1800 HV. The abrasion resistance of these alloys is a function of amount of hard borides present. The impact resistance decreases with an increase of borides. Ni-Cr-B-Si-C alloys have been deposited by welding and plasma spraying techniques for corrosion- and wear-resistance applications (Antony et al., 1983; Knotek et al., 1979; Mann et al., 1985). The coatings are sintered in vacuum furnaces at about 1050°C in order to produce a variation in the formation of hard phases and in the diffusion in the metal-coating interface. The formation of the various hard phases, the solubility of the nickel solid solution for silicon and boron, and the eutectic reactions in the Ni-B, Ni-Cr-B, and Ni-Cr-B-Si systems have been reported (Knotek et al., 1979). Mann et al. (1985) have reported that sintered coatings have significantly superior bond strength and cavitation erosion resistance compared to that of as-sprayed coatings.

Only one nickel-based alloy containing Laves phase is Ni-Cr-Mo-Si (Tribaloy 700). This alloy, like most nickel-based alloys, is difficult to weld by oxyacetylene welding but can be readily welded using GTAW, PTAW, or PS, DG. Most commonly, it is deposited by thermal spray techniques. Although Tribaloy 700 has excellent metal-to-metal wear resistance and modest abrasion resistance, it possesses poor impact resistance.

14.3.3 Other Alloy Coatings

The MCrAlY (M = Ni, Co, Fe, or combinations thereof) alloys and similar precipitation- or dispersion-hardened materials can be regarded as a separate group even though their major component is largely cobalt or nickel. Their excellent high-temperature properties are very attractive for extreme applications requiring oxidation and hot corrosion resistance up to 1200°C. As a group, the cobalt-based compositions exhibit much more hot corrosion resistance than that of nickel-based alloys. The iron-based coatings exhibit relatively poor hot corrosion resistance. Beneficial high-temperature oxidation properties of the MCrAlY coatings containing aluminum and chromium are based on the formation of alumina and chromia under oxidizing conditions.

The practical durability of these coatings is dependent on the adherence of the protective oxides. Oxygen-active element Y is added to promote oxide adherence nominally independent of substrate alloy composition. It is well known that SiO_2 has minimal solubility in highly acidic melts; therefore, silicon (typically 1.6 to 2.6 percent) is added to develop significant resistance to hot corrosion caused by acidic melts. An advanced NiCoCrAlY alloy is typified by a particular nominal combination of 18 wt % Cr, 23 wt % Co, 12 wt % Al, 0.5 wt % Y, balance Ni.

These alloys have been deposited by fusion of pastes or slurries (Packer and Clark, 1978), plasma spraying, electron beam evaporation, sputtering, and foil cladding processes. MCrAlY coatings deposited by electron beam evaporation or a plasma spraying process on blades and vanes of high-performance gas turbine engines are used to protect against oxidation and hot corrosion (Goward, 1979; Hebsur and Miner, 1987; Pettit and Goward, 1983). Also see Chap. 16 for further details of MCrAlY alloy coatings.

14.3.4 Sliding Wear and Corrosion Resistance of Various Hard Alloy Coatings

Foroulis (1984) conducted metal-to-metal sliding wear tests using a Falex ring and block machine on various cobalt-based and nickel-based alloys deposited by the inert gas tungsten arc deposition method. The wear tester uses a combination of a rotating test ring and a flat stationary block. The tests were carried out with a normal load of 935 N (hertzian stress = 360 kPa) and a surface velocity of 0.15 m s^{-1} for 2000 rev. A constant metal temperature of about 30°C was maintained by cooling the ring with water. Tables 14.8 and 14.9 summarize the coefficient of friction and wear coefficients for various metal-to-metal combinations. We note that coefficients of friction do not correlate directly with the degree of wear experienced. A plot of the wear data against the bulk hardness of the alloy coatings is shown in Fig. 14.13. This indicates that, in general, alloy coatings with a higher bulk hardness exhibited better wear resistance. However, this is not an exact correlation, and there is considerable scatter in the data plotted in Fig. 14.13.

From a close comparison of the data shown in Tables 14.8 and 14.9, we note the following: (1) Self-mated alloy T-400 shows superior wear resistance, (2) self-mated alloy ST-6 is ranked good but not as good as self-mated T-400, (3) alloys H-711 and H-716 exhibited generally poor performance either self-mated or in combination with other hard facing alloys, (4) nickel-based alloys such as H-40 show good performance when self-mated or in combination with T-400 or T-700, and (5) Ni-Mo alloys such as T-700 are subject to wear when self-mated and show good performance only when mated with alloy H-40 or alloy T-400.

As mentioned in Chap. 4, galling is an insidious form of adhesive wear. Seizure of mating alloy surfaces is frequently the result of severe galling when the contact stress is higher than the galling stress. We have seen in Chap. 4 that cobalt-based alloys such as T-400 and ST-6 have, in general, good galling resistance. However, several nickel-based alloys such as T-700 exhibit a very low threshold galling stress when self-mated or coupled to other similar alloys. Galling in general can be decreased by (1) decreasing the contact stress to below the threshold stress of the couple and (2) using alloy couples with an inherently better galling resistance (a higher threshold galling stress). Many cobalt-based alloys such as T-400 and ST-6 exhibit excellent galling resistance in addition to their wear resistance (Schumacher, 1981; Foroulis, 1984).

Foroulis (1984) also conducted corrosion tests to compare the relative corro-

TABLE 14.8 Coefficient of Friction for Similar and Dissimilar Alloy Coating Couples (Coatings Applied by Inert Gas Tungsten Arc Weld-Deposition Method)

Mating alloy	Coefficient of friction for the following couples							
	ST-1	ST-6	ST-12	T-400	T-700	H-40	H-711	H-716
ST-1	0.28	0.28	0.28	0.46	0.28	0.28	0.24	0.24
ST-6	0.28	0.48	0.28	0.28	0.28	0.28	0.21	0.24
ST-12	0.24	0.24	0.24	0.24	0.24	0.24	0.21	0.24
T-400	0.28	0.28	0.28	0.27	0.28	0.28	0.24	0.35
T-700	0.43	0.47	0.39	0.45	0.44	0.48	0.50	0.48
H-40	0.33	0.33	0.40	0.43	0.33	0.58	0.45	0.45
H-711	0.24	0.24	0.28	0.43	0.38	0.43	0.38	0.31
H-716	0.24	0.24	0.28	0.48	0.43	0.42	0.38	0.31

Source: Adapted from Foroulis (1984).

TABLE 14.9 Wear Coefficients for Several Similar and Dissimilar Alloy Coating Couples (Coatings Applied by Inert Gas Tungsten Arc Weld-Deposition Method)

Mating alloy	Wear coefficient k* ($\times 10^{-4}$) for the following couples							
	ST-1	ST-6	ST-12	T-400	T-700	H-40	H-711	H-716
ST-1	24.8	3.1	4.1	1.4	14.0	16.0	52.0	53.0
ST-6	3.1	1.5	3.7	1.6	5.2	6.6	31.0	38.0
ST-12	4.1	3.7	4.6	2.8	7.7	19.0	33.0	58.0
T-400	1.4	1.6	2.8	5.2	2.7	1.4	16.0	32.0
T-700	14.0	5.2	7.7	2.7	19.0	3.8	61.0	70.0
H-40	16.0	6.6	19.0	1.4	3.8	5.6	8.0	49.0
H-711	52.0	31.0	33.0	16.0	61.0	8.0	200	220
H-716	53.0	38.0	58.0	32.0	70.0	49.0	220	160

*Wear coefficient k is calculated from Archard's equation:

$$V = \frac{kWL}{3H}$$

where V is the volume of wear, W is the normal load, L is the sliding distance, and H is the indentation hardness of the softer of the two contacting surfaces.

Source: Adapted from Foroulis (1984).

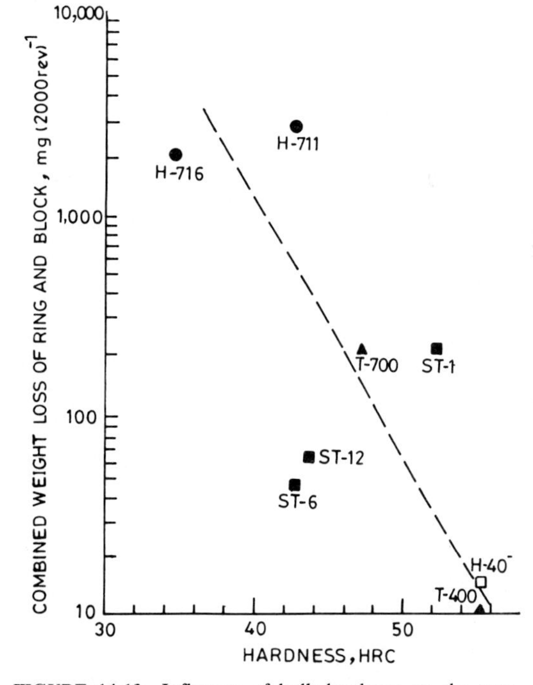

FIGURE 14.13 Influence of bulk hardness on the combined ring and block mass loss of self-mated alloy coating applied by inert gas tungsten arc weld deposition method. [*Adapted from Foroulis (1984).*]

sion resistance of the most effective coating alloys in a variety of corrosive environments which may be encountered in many process industrial applications. Such a comparison of the corrosion resistance of commonly used hard facing alloys is given in Table 14.10. Corrosion data are presented for two cobalt-based alloys, i.e., alloys T-400 and ST-6, and for three nickel-based alloys T-700, H-40, and H-711. In addition, for comparison, corrosion data are also given for several other commercial bulk alloys: several nickel-based alloys and two commonly used stainless steels. The data in Table 14.10 show that the cobalt-based coating alloys have, in general, very good corrosion resistance in a variety of environments. Their corrosion resistance is generally comparable with that of the best nickel-based alloys. Furthermore, the data also show that alloy T-400 has a better corrosion resistance than alloy ST-6 in most of the corrosive environments listed in Table 14.10 with the exception of the strong oxidizing acidic environments. The corrosion behavior of ST-1 and ST-12 alloys should be comparable with that of alloy ST-6 (Foroulis, 1984). Alloys H-711 and H-716 have a corrosion resistance comparable with that of Incoloy 800 and should thus be suitable, from a corrosion point of view, for many applications (Foroulis, 1984).

Thornton and Ferriss (1977) reported wear data for dc sputter-deposited Tribaloy 800 coatings. The coatings were deposited onto steel substrates at temperatures in the range 20 to 800°C, and the coating thickness ranged from 20 to 60 μm. Coatings deposited at a substrate temperature of about 800°C exhibited the Laves phase structure. Coatings deposited at lower temperatures exhibited fine-grained (amorphous-type) structure. Microhardnesses were about 900 kg mm^{-2}, independent of substrate temperature. In wear tests at room temperature with no lubricant, the wear properties of the steel rings (HRC 60) with the sputter-deposited coatings were superior to the uncoated ones when sliding against the tool steel, cobalt alloy, and cast Laves phase (Tribaloy 400) blocks.

Wolfla and Tucker (1978) conducted oxidation- and wear-resistance tests on the following cobalt-based alloy coatings at high temperature: L-103, LCO-17, and LCO-19 applied by detonation gun process. A nichrome bonded chromium carbide (LC-1B with 75% Cr_3C_2 and 25% NiCr), Tribaloy 400 (LDT-400), and Tribaloy 800 (LDT-800) were applied by detonation gun process. The coatings were applied on a nickel-based alloy (Hastelloy X) and on a cobalt-based alloy (Haynes alloy 188) substrate. Oxidation tests were conducted in hot air at 1000°C for 100 h. Fretting wear tests were conducted in a pin-on-disk configuration with some coating on both pin and disk, at 7 MPa and 775 cycles per minute at an amplitude of about 4 mm. During the tests, the sample was thermocycled with temperature changing between about 400 and 1065°C in 1 min. Oxidation and wear results are presented in Table 14.11 and Fig. 14.14, respectively. Based on these tests, we find that LCO-17A, an alumina, provides the best high-temperature performance. LCO-19A, a chromia, together with L-103A, should be considered at intermediate temperatures, and LC-1B, LDT-400, and LDT-800 at somewhat lower temperatures.

14.4 CERAMIC COATINGS

14.4.1 Oxide Coatings

Several oxide ceramic coatings such as Al_2O_3, TiO_2, Al_2O_3-TiO_2, Cr_2O_3, ZrO_2, and SiO_2 exhibit high hardness (750 to 2100 HV), high melting point, and extreme

TABLE 14.10 Corrosion Rates of Several Commercial Coating Alloys and Several Commonly Used Alloys in Acidic Environments

Alloy	Corrosion rates in the following corrosive environments ($\times 10^{-3}$ in $year^{-1}$)							
	Oxidizing		Nonoxidizing			Reducing		
	10% ferric chloride, room temperature	65% nitric acid, 66°C	85% phosphoric acid, 66°C	3.5% sodium chloride, 120°C	50% acetic acid, 100°C	5% hydrochloric acid, 66°C	45% formic acid, 100°C	10% sulfuric acid, 100°C
Coating alloys								
Alloy ST-6	80	8	0.3	0.3	<0.1	4100	20	8.5
Alloy H-40	930	2156	276	<1	70	2856	49	>7740
Alloy T-400	100	1350	0.3	0.3	<0.1	190	1	4.5
Alloy T-700	84	12	<0.1	<0.1	<0.1	30	—	2513
Alloy H-711	1116	17.7	32	<1	2.8	63	38	1646
Nickel-based alloys (for comparisons)								
Incoloy 800	45	<1	3.8	<1	60	800	87	185
Inconel 600	440	26	28	<1	14	123	37	66
Hastelloy C	1	22	0.3	<0.1	<0.1	18	3	<1
Hastelloy B	100	9860	0.5	0.6	<0.1	10	0.9	<1
Stainless steels (for comparisons)								
Type 304	>1000	8	1	<1	300	>1000	1700	1993
Type 316	>1000	11	0.5	<1	2	>500	500	148

Source: Adapted from Foroulis (1984).

TABLE 14.11 Oxidative Test Results*

Coating	Depth of oxidation, μm	Type of oxidation
LC-1B	25	Adherent scale
LCO-17 and LCO-17A	<5	Adherent scale
LCO-19 and LCO-19A	<5	Adherent scale
L-103 and L-103A	100	Adherent scale
LDT-400 and LDT-400A	225	Spalling
LDT-800 and LDT-800A	100	Internal oxide/spalling

*Test: 100 h at 1000°C in air: The samples were allowed to cool to room temperature each 24 h.
Source: Adapted from Wolfla and Tucker (1978).

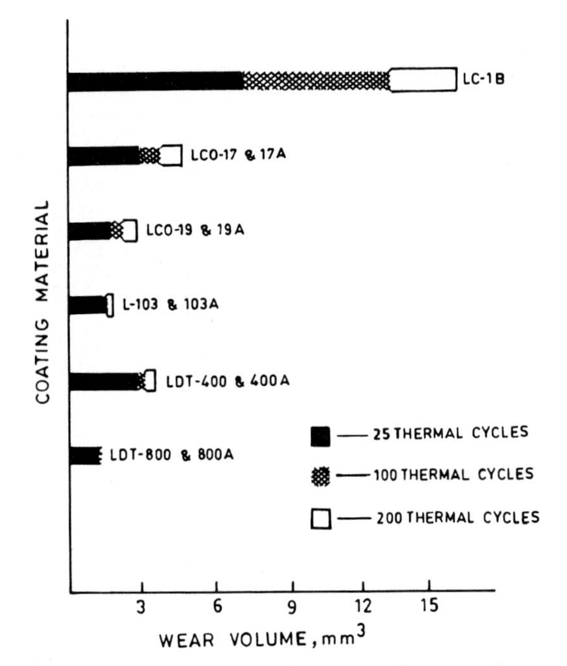

FIGURE 14.14 The comparative wear performance of several high-temperature coatings. [*Adapted from Wolfla and Tucker (1978).*]

chemical stability which make them a suitable candidate for corrosion- and wear-resistant applications. Coatings have been applied by thermal spray, PVD, CVD, and other proprietary processes. These coatings have been applied onto a variety of substrates such as metals, ceramics, and cermets (e.g., WC-Co).

14.4.1.1 Alumina and Alumina-Titania Coatings. Alumina coatings have been deposited by plasma spray, evaporation, sputtering, and CVD processes. Plasma sprayed alumina and alumina-titania coatings are commonly used for various wear applications at both ambient and high temperatures such as in a gas turbine engine shaft. Sputtered Al_2O_3 coatings are also used in various wear applications such as gas bearings (Bhushan, 1980c). These are also used for wear resistance and electrical insulation in the thin-film structure of magnetic heads for tape and disk drives. CVD Al_2O_3 is used in multilayer structures with TiN and TiC deposited on cemented carbide tool inserts.

Al_2O_3 and Al_2O_3-TiO_2 Coatings Applied by Plasma Spray. Chuanxian et al. (1984a, 1984b) conducted friction and wear tests on various oxide and carbide coatings applied by plasma spraying. The friction and wear tests were performed with an Amsler A135 abrasion testing machine. The coated surfaces were slid against stainless steel at 400 r/min under a load 250 N. Table 14.12 summarizes the properties of plasma sprayed oxide coatings and of selected plasma sprayed carbide and electrochemically deposited Cr for comparisons. (Comparison of plasma sprayed and detonation gun sprayed coatings will be presented later in Table 14.18.) Plasma sprayed coatings of several oxide ceramics Cr_2O_3, Al_2O_3, TiO_2, and Al_2O_3-TiO_2 exhibit coefficient of friction against stainless steel in the range of 0.1 to 0.2 and low wear. Figure 14.15 shows the wear characteristics of these coatings (Chuanxian et al., 1984b). The wear resistance of the coatings can be ranked in the following order: Cr_2O_3, WC-Co, Al_2O_3-TiO_2, Al_2O_3, TiO_2, and Cr_3C_2-NiCr. This ranking is based on the work by Chuanxian et al. (1984b) and should not be considered absolute. The wear resistance of different coating materials showed no obvious correlation with the microhardness or bending strength, e.g., Cr_3C_2-NiCr coating had a relatively high microhardness and high bending strength but had a poor wear resistance against steel.

The wear resistance of these coatings appears to be closely related to their thermal diffusivity characteristics (Fig. 14.16). The thermal diffusivity of Cr_2O_3 coating material is high as compared to other oxide materials. A coating with high thermal diffusivity (or thermal conductivity) rapidly dissipates the frictional heat and avoids local high temperature which would, otherwise, result in high local thermal stresses. The large local thermal stresses will induce spalling in these local regions; these act as the primary cause of abrasive wear during motion. Apart from thermal properties, the pore size, shape, and distribution as well as crack texture in the coating also affect significantly the wear resistance because of their influence on stress concentration.

Durmann and Longo (1969) reported that plasma sprayed mixtures of TiO_2 with other oxides such as Al_2O_3 and Cr_2O_3 have less porosity and better surface finish compared to that of Al_2O_3 and Cr_2O_3, desirable for oxidation and wear resistance. Since the melting point of TiO_2 is lower than that of Al_2O_3, it would be expected that the TiO_2-containing oxides would melt more easily and remain fluid for a longer period of time. The enhanced fluidity on the grit-blasted substrate gives more time for the solidifying oxide coating to become anchored into the substrate surface texture resulting in better bond strength.

Durmann and Longo described a composite powder (powder size −53 + 10 μm) for plasma spraying (sold by Metco, Inc., Westbury, New York) which was produced by cladding each Al_2O_3 particle with TiO_2 so that each composite particle has essentially the same chemistry. When this powder enters the plasma flame, the titania melts, flows, and readily wets alumina. The composite particles melt completely before impact and form a dense, tightly bonded coating. They showed that the coatings produced by composite powder exhibited higher bond strength (~30 MPa), higher density, higher hardness (65 to 70 HRC), high wear resistance and a higher surface finish (~25 to 50 nm rms) than is obtainable from coatings sprayed with blended powders (Table 14.13). Higher density is demonstrated by the dielectric strength, which is greater than shown by the other

TABLE 14.12 Properties of Plasma Sprayed Coatings of Oxides and Carbides

Coating material	Porosity, %	Bulk density, kg m^{-3}	Apparent density, kg m^{-3}	Vickers micro-hardness (HV), kg mm^{-2}	Bending strength, MPa	Thermal diffusivity, $m^2\ s^{-1}$	Coeffi-cient of friction	Wear resis-tance	Useful temp. range, °C
Cr_2O_3	4.7	4520	4720	900–1100	90–110	0.75–0.90 (700–1000°C)	0.14–0.15	Best	1000
WC-12 wt % Co	—	—	—	1230	115	—	0.11–0.13	Better	500
Al_2O_3-TiO_2	5.8	3320	3530	890–1060	87–91	0.73–0.65 (520–1020°C)	0.10–0.11	Better	550
Al_2O_3	5.7	3230	3430	940	77	0.61–0.48 (400–955°C)	0.13–0.20	Good	1000
TiO_2	6.0	4010	4260	910	84	0.50–0.57 (380–940°C)	0.10–0.15	Bad	—
Cr_3C_2-NiCr	5.7	6420	—	1075	128	—	0.13–0.14	Worst*	1100
Electroplated Cr	—	—	—	1100	—	—	0.30–0.38	Good	—

*Average wear mass loss of the Cr_3C_2-NiCr coating is about 18 mg cm^{-2} after a 250-min wear test.
Source: Adapted from Chuanxian et al. (1984b).

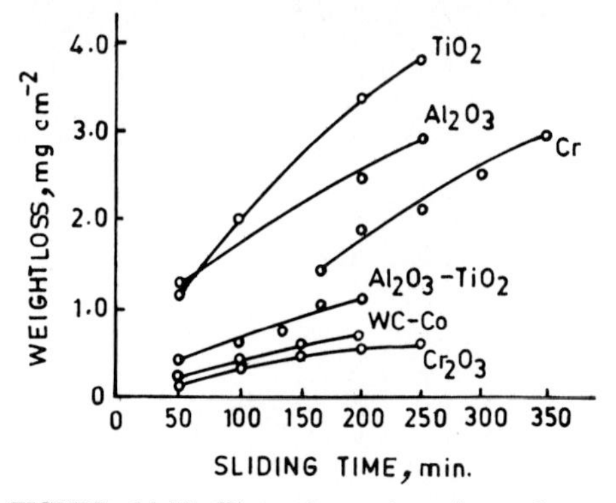

FIGURE 14.15 Wear data of various plasma sprayed coatings. Cr coating was applied by electrodeposition process. [*Adapted from Chuanxian et al. (1984b).*]

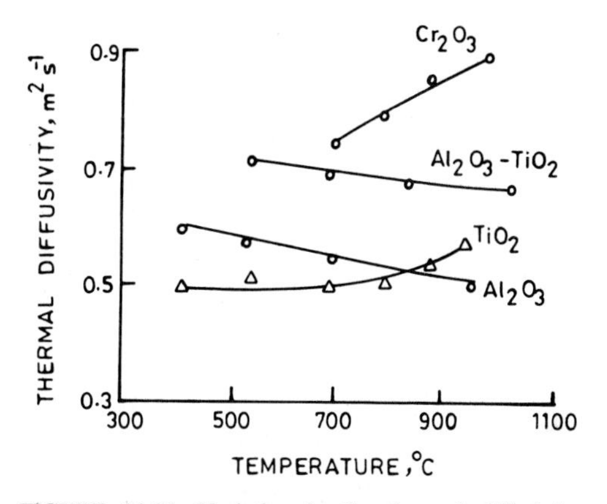

FIGURE 14.16 Variation in the thermal diffusivity with temperature for various plasma sprayed coatings. [*Adapted from Chuanxian et al. (1984b).*]

sprayed coatings, and by the better lapped finish. In the scribe test, close examination of the scratch path shows no particle pullout, but instead a wear track through particles; whereas, in the other coatings, the diamond point dislodged whole particles. These results demonstrate excellent interparticle bonding. The composite coatings exhibited excellent corrosion resistance to various chemicals.

Chuanxian et al. (1984a) have also shown that adhesion (bond) strength and microstructure of Al_2O_3-TiO_2 coatings depend on the processing technique for the powder and the composition of the powders used for plasma spraying. They examined two widely different powder types: agglomerated versus fused powders of mixed oxides based on Al_2O_3 and TiO_2. The agglomerated powders were produced by spray drying of the two constituent particles adhered by an organic binder, which burns off in the flame during thermal spraying. The fused powders were produced by melting the oxide components in given proportions and solid-

TABLE 14.13 Properties of Plasma Sprayed Mixed Oxide Coatings

Plasma sprayed coating	Powder size, μm	Hardness (HRC)	Surface finish lapped, nm	Tensile bond strength, MPa	Scribe test width, μm	Wear resistance dimensional loss, μm	Dielectric strength, V μm^{-1}
Al_2O_3-TiO_2 composite (13 wt % TiO_2)	−53 + 10	65–70	25–50	29	70	38	10
Al_2O_3	−53 + 10	55–60	200–225	10	125	25	8.6
Cr_2O_3-TiO_2 blend (55 wt % TiO_2)	−105 + 10	58–64	100–125	16	105	40	Conductive
Al_2O_3-TiO_2 blend (50 wt % TiO_2)	−53 + 10	40–45	250–4000	25	225	25	1.6–2

Source: Adapted from Durmann and Longo (1969)

ifying the melt at a relatively slow rate followed by mechanical reduction and sizing. The fused powders have high uniformity of chemistry and particle-to-particle melting point. This results in a broader melting zone within the flame and a coating with a larger fraction of major phase than would be obtained from a composite powder. The overall effect is a significantly higher adhesion strength, a more uniform cross section. The tensile bond strength of various sprayed Al_2O_3-TiO_2 coatings are plotted as a function of TiO_2 content in Fig. 14.17. For the coatings based on fused powders, the bond strength increases with the TiO_2 content. The behavior is more complex for coatings based on spray-dried powder; the bond strength for the Al_2O_3-13% TiO_2 mixed oxide is lower than that of pure Al_2O_3, but it is consistently lower than that of coatings with fused powders. In the coatings produced by fused powders, the β-Al_2TiO_5 phase was clearly visible with a considerable amount of γ phase.

Bond strength, density, microhardness, and modulus of rupture of detonation gun sprayed coatings are generally higher than that of plasma sprayed coatings (Tucker, 1974) data to be presented later. The kinetic energy of molten material in case of detonation gun spray is 3 to 5 times higher than that of plasma spray, responsible for improved properties.

Alumina Coatings Applied by PVD. Alumina coatings have been deposited by evaporation, activated reactive evaporation, and sputtering (Dragoo and Diamond, 1967; Bunshah and Schramm, 1977; Lee and Poppa, 1977; Movchan and Demchishin, 1969; Thornton and Chin, 1977). Alumina can exist in a number of metastable polymorphs in addition to the thermodynamically stable α-Al_2O_3 or corundum form as shown in Fig. 14.18 based on the work by Thornton and Chin (1977) on rf-sputtered Al_2O_3 coatings. The temperatures in Fig. 14.18 refer to substrate temperatures during deposition valid for both PVD and CVD processes. Amorphous Al_2O_3 coatings are soft and are mostly used as protection against chemical corrosion or as electrical insulators. Alpha Al_2O_3 coatings are hard and are primarily used for wear applications.

Movchan and Demchishin (1969) produced thick (25 to 2000 μm) coatings of

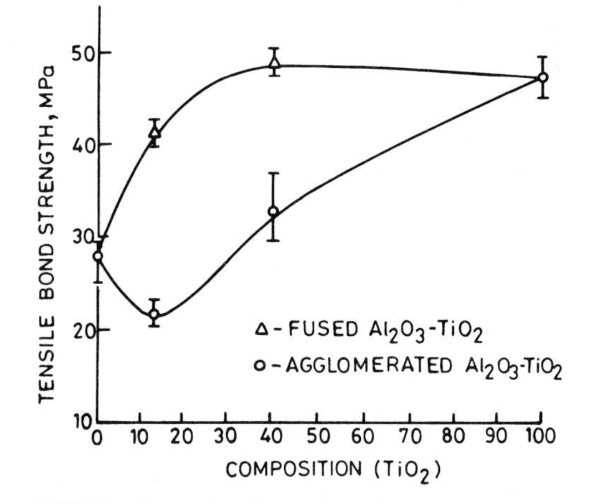

FIGURE 14.17 Tensile adhesion bond strengths of coatings of Al_2O_3, TiO_2, and Al_2O_3-TiO_2. [*Adapted from Chuanxian et al. (1984b).*]

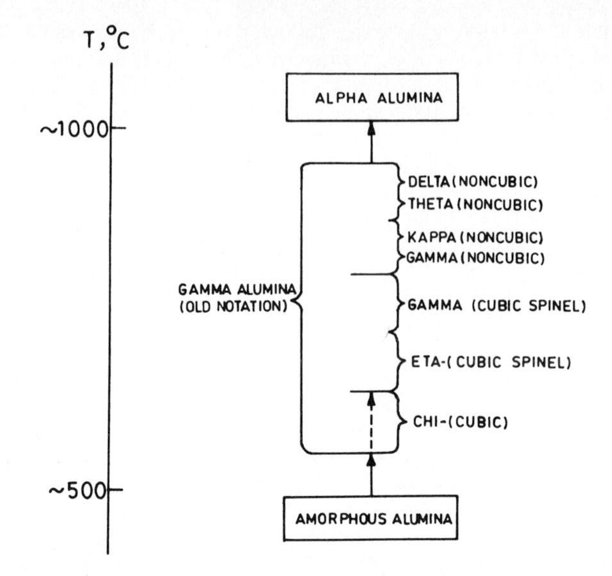

FIGURE 14.18 Schematic representation of aluminum oxide phases. Also shown is a temperature scale valid for deposition from the vapor phase. [*Adapted from Thornton and Chin (1977).*]

Al_2O_3, ZrO_2, Ni, Ti, and W deposited at high rates (1.2 to 1.8 μm/min) by electron beam evaporation. They found that the substrate temperature had substantial effect on the structure of the coatings. The general features of the resulting structures could be correlated into three zones depending on T/T_M, where T is the substrate temperature (K), and T_M is the coating material melting point (K) as shown in Fig. 14.19. Zone 1 is dominated by tapered macrograins with domed tops; zone 2 by columnar grains with denser boundaries; and zone 3, by equiaxed grains formed by recrystallization. For the case of evaporated alumina coatings, at substrate temperatures of 300 to 450°C, the coatings are clear deposits of gamma-Al_2O_3; from 450°C to 900°C, they are coarse flaky deposits of alpha-Al_2O_3; and above 900°C, they are dark deposits of alpha-Al_2O_3. These polymorphic forms along with those observed on rf-sputtered Al_2O_3 coatings by Thornton and Chin (1977) are shown in Fig. 14.19 for comparisons.

Figure 14.19 also shows the variation in microhardness with the deposition temperature. The hardness falls when the structure changes from tapered crystallites (zone 1) to columnar grains (zone 2). Hardness increases markedly as the deposition temperature rises from $0.3T_M$ to $0.5T_M$. This is attributed to a more "perfect" material produced at the higher deposition temperatures due to the "volume processes of sintering." Similar results were also reported by Movchan and Demchishin (1969) for evaporated ZrO_2 and Y_2O_3 deposits.

Thornton and Chin (1977) rf-sputter-deposited Al_2O_3 coatings from an Al_2O_3 target. The coatings of about 0.1 μm thickness were deposited on NaCl and Mo substrates at temperatures from 150 to 1200°C. The structure and annealing characteristics of the sputter-deposited alumina coatings are summarized in Fig. 14.20. The annealing characteristics of evaporated amorphous films obtained by the studies of Dragoo and Diamond (1967) are included for comparisons. We note that as-deposited structures were noncrystalline for substrate temperatures of

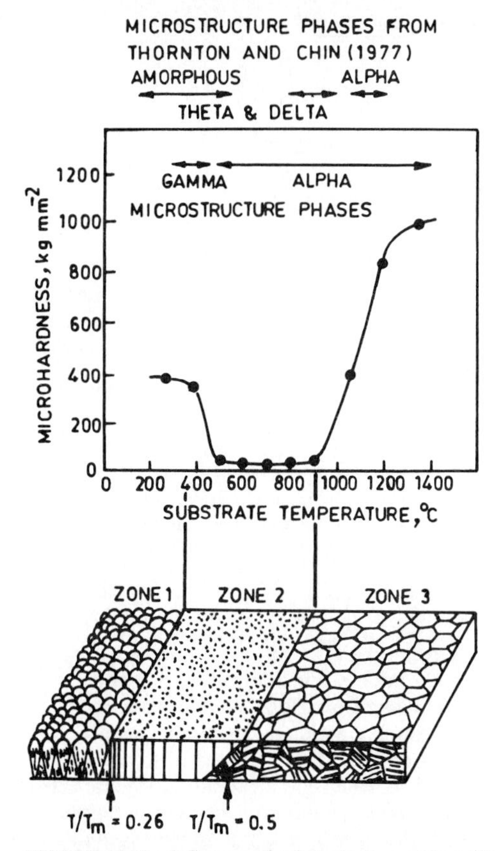

FIGURE 14.19 Influence of substrate temperature (T in K)/coating material melting point (T_M in K) on the microstructure phases and hardness of alumina deposited by electron beam evaporation. Polymorphic forms observed on rf-sputtered alumina by Thornton and Chin (1977) are also shown for comparisons. [*Adapted from Movchan and Demchishin (1969).*]

less than 500°C. Alpha-Al_2O_3 was directly deposited at a substrate temperature of about 1100°C. Intermediate temperatures yielded polymorphic phases. This observation is consistent with the studies of Al_2O_3 coatings applied by ARE process by Bunshah and Schramm (1977). Annealing tests indicate that the coatings deposited at low substrate temperatures in a noncrystalline form transform to alpha-Al_2O_3 more readily at 1200°C than do coatings deposited at higher substrate temperatures in crystalline metastable forms. Extended annealing at 1200°C would probably have yielded alpha-Al_2O_3 in all cases. The structures obtained at a given substrate temperature were generally consistent with those reported by Dragoo and Diamond (1967) after extended heat treatment of their amorphous alumina coatings at temperatures equivalent to substrate temperatures used in sputtering. The phases observed at various temperatures in this study are compared in Fig. 14.20 with that observed by Movchan and Demchishin (1969).

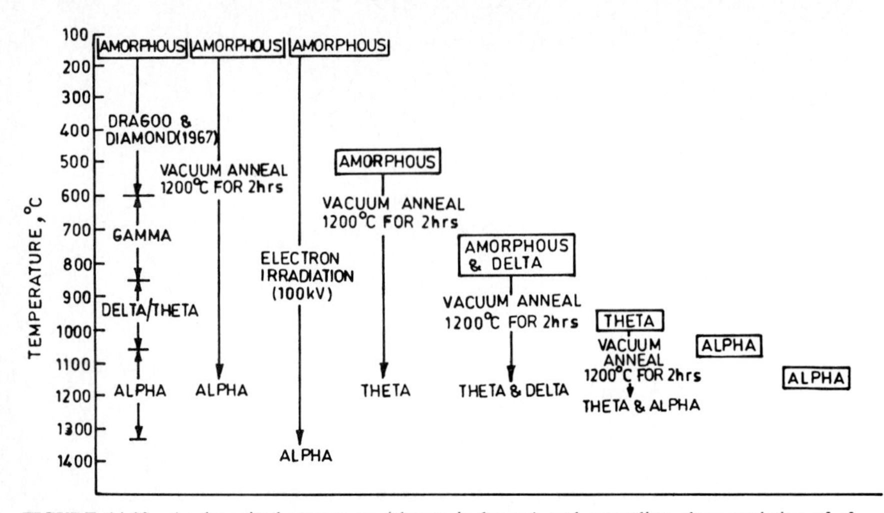

FIGURE 14.20 As-deposited structures (shown in boxes) and annealing characteristics of rf sputter-deposited aluminum oxide coatings compared with annealing characteristics of evaporated amorphous coatings (left side of figure) reported by Dragoo and Diamond (1967). [*Adapted from Thornton and Chin (1977).*]

The maximum hardness of PVD coatings of alpha-Al_2O_3 has been reported to be about 1000 kg mm^{-2} which is considerably lower than that of CVD coatings. This can probably be explained by different microstructures, with PVD coatings containing more defects such as voids and grain boundaries with a lower strength due to the higher growth rate used.

Kuwahara et al. (1981) measured the residual stresses in rf-sputtered Al_2O_3 coatings. They found that residual stresses were compressive; the average stresses decreased with increasing coating thickness and deposition rate. The compressive stress values were on the order of 0.5 GPa for a coating thickness ranging from 0.2 to 1 μm.

Al_2O_3 Coatings Applied by CVD. In order to obtain good mechanical properties of Al_2O_3 coatings, a gas mixture of $AlCl_3/H_2/CO_2$ is used in combination with a slow deposition rate and a low impurity level (Lux et al., 1986). The presence of trace impurities can considerably disturb the growth of Al_2O_3 and cause formation of whisker platelets or needlelike outgrowth as well as dendritically branched crystals. If the growth rate is slow (0.5 to 1 μm/h), the coating will consist of single-crystal grains. In the case of fast growth rates (>5 μm h^{-1}), formation of small-grained polycrystalline Al_2O_3 coatings containing many fine pores and defects occur.

The hardness of alpha-Al_2O_3 deposited by CVD is 2000 to 2100 kg mm^{-2} at 25°C, and upon heating to 1000°C, it rapidly decreases to 300 to 800 kg mm^{-2} (Oakes, 1983). Johannesson and Lindstrom (1975) and Lindstrom and Johannesson (1975) reported an increase in wear life of cemented-carbide cutting tool (WC + TiC, TaC, NbC + Co) by application of CVD Al_2O_3 coating. Oakes (1983) compared the lives of CVD coated cemented carbide tools (see section on HfN and ZrN coatings). They found that tool life of Al_2O_3 coated tool was better than that of TiC/TiN coated tool in continuous and interrupted turning of AISI 4340 steel at various cutting speeds.

An epitaxial layer of CVD Al_2O_3 on cold-pressed and sintered high-purity alumina has been shown to increase the wear life of ceramic (Al_2O_3) gas bearings (Rowe, 1968). CVD coatings of Al_2O_3 also have been used to provide an additional protective layer on plasma sprayed alumina coatings against the wear and corrosion (Mantyla et al., 1984). The CVD coated Al_2O_3 coatings on sprayed coatings completely cover the pores and partial cracks of thermal sprayed Al_2O_3 coating and avoid any failure due to thermal cycling and corrosion attack. Figure 14.21 illustrates the effect on surface microstructure of plasma sprayed alumina coating on CVD coated alumina coating (1 to 3 μm). The additional coating exhibits a significant improvement in the porosity of sprayed coating.

*14.4.1.2 **Chromia Coatings.*** Oxidation- and wear-resistant Cr_2O_3 coatings have been applied by plasma spray and rf sputtering. Plasma sprayed Cr_2O_3 coatings are used for various wear applications at ambient and elevated temperatures. Sputtered Cr_2O_3 coatings have been developed for a foil bearing which functions from room temperature to 650°C (Bhushan, 1980c). Details on plasma sprayed Cr_2O_3 were presented in the Al_2O_3 section. Details on sputtered Cr_2O_3 follow.

Rf-sputtered Cr_2O_3 coatings have been developed by Bhushan (1979, 1980a, 1980c, 1980d, 1981a, 1981b) for foil air bearing and rolling-element bearing applications. Sputtering parameters for Cr_2O_3 coatings were optimized by Bhushan (1979, 1980c). He found that low argon pressure (~0.6 Pa or 5 mtorr) high power (400 W over a 150-mm-diameter target), and a water-cooled substrate improves the coating adhesion and its surface smoothness. Coatings deposited with a substrate bias were not as adherent as with an unbiased substrate. One reason for poor adhesion in bias coating could be the presence of a native chromium-rich oxide layer present at the interface (which may be responsible for good adhesion) which gets removed during bias sputtering and another reason may be that bias coatings are less stoichiometric and highly stressed (Bhushan, 1979, 1980a). Substrate hardness was found to influence adhesion. Softer (annealed) substrates were found to give improved adhesion. "As-sputtered" coatings were amor-

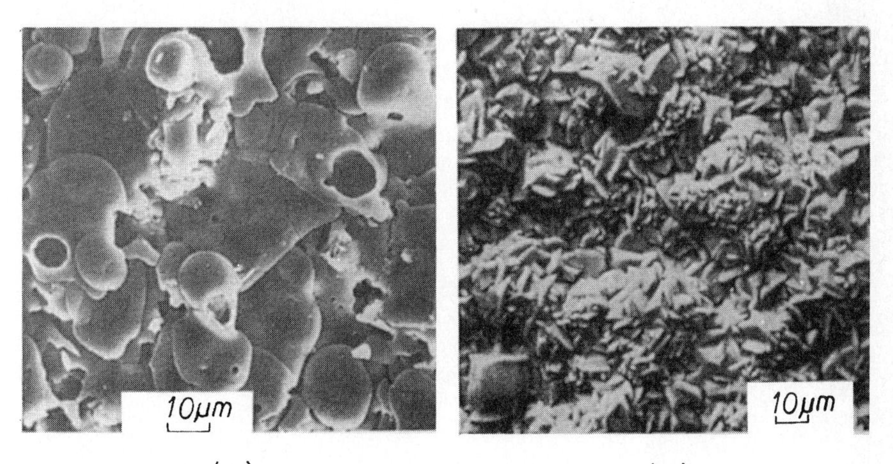

FIGURE 14.21 Microstructure of: (*a*) Plasma sprayed Al_2O_3 coating. (*b*) CVD Al_2O_3 over plasma sprayed Al_2O_3 coating. [*Adapted from Mantyla et al. (1984).*]

phous, and they crystallized after heat treatment (Bhushan, 1980a). A coating with a thickness of about 1 μm or less had good adhesion. Thicker coatings spalled (Bhushan, 1979).

In order to obtain good adhesion, good ductility, and proper matching of coefficient of thermal expansion of Cr_2O_3 coating on a metallic substrate, a multilayer coating approach was employed by Bhushan (1980c, 1981b) (Fig. 14.22). In this approach, the metallic substrate was first coated with a transition layer or interlayer of soft Ni-Cr (80-20 wt %) metal to improve adhesion by providing a graded interface and to act as an oxidation barrier to the substrate in the case of localized wear of the hard coating. The ductility and the thermal expansion of the Cr_2O_3 coating was improved by mixing 25 wt % Ni-Cr metallic binder in the coating material. In some cases, a sacrificial overlay coating of a soft lubricant (MoS_2, Ag, or Au) was applied over the hard coating to minimize wear of the hard coatings during run-in. Additionally, it was expected that the soft lubricant coating would transfer to the mating surface and prolong the wear life.

The performance of sputtered Cr_2O_3 and Ni-Cr mixed Cr_2O_3 was evaluated by start-stop wear tests by running Cr_2O_3 coated Inconel X-750 foil against Ni-Cr bonded Cr_3C_2 detonation gun spray coated journal (A-286, high Ni-Cr steel) (Bhushan, 1980c, 1981b; Bhushan and Gray, 1980). Table 14.14 summarizes the results of start-stop tests for various combinations of sputtered Cr_2O_3 (about 1 μm thick) with coated journal and compares with air sprayed MoS_2 coated foil bearings. The tests were conducted at ambient, intermediate, and high temperatures. The data on MoS_2 coating is intended for comparisons. The Cr_2O_3-Cr_3C_2 coating combination successfully survived for 9000 start-stop cycles at 14 kPa up to 650°C and for 3000 start-stop cycles at 35 kPa up to 650°C. In addition, the coating combination survived an impact of 100 g with the journal rotating at 30,000 r/min.

These oxide coatings were found to perform better at high temperature (650°C); for instance, the coefficient of friction at 650°C is half as compared to that at room temperature (Fig. 14.23). At 650°C, protective oxide films were formed, apparently by oxidation of the substrate alloys in the localized areas where the coating may have been worn. These oxides were found to reduce the coefficient of friction significantly.

Bhushan (1980d) also tested sputtered Cr_2O_3 for ball bearing applications. An overlay of MoS_2 of 200 to 300 nm thickness was deposited on sputtered Cr_2O_3 about 1 μm in thickness, to minimize wear during run-in period. In ball bearings,

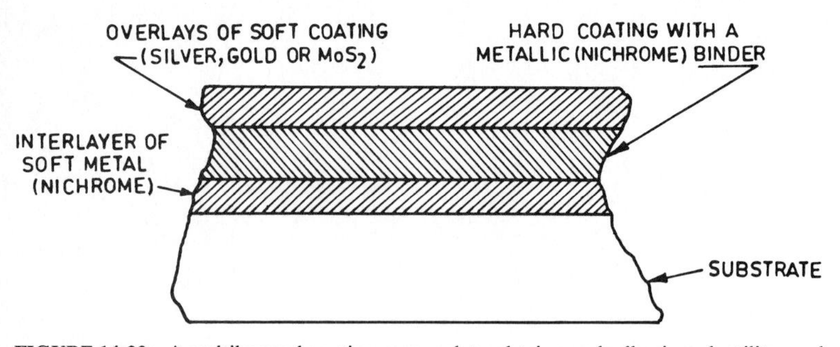

FIGURE 14.22 A multilayered coating approach to obtain good adhesion, ductility, and proper matching of coefficient of thermal expansion for hard ceramic coatings.

TABLE 14.14 Start-Stop Test Results of Coatings Applied on Foil Bearings*

Test no.	Foil and journal coatings	Maximum test temp., °C	Coefficient of friction: At start	After 499 cycles	After 999 cycles	After 1499 cycles	After 9000 cycles	rms surface roughness of journal, μm: Before	After	Comments
1.	Air sprayed MoS_2 versus det. gun sprayed Cr_3C_2 (for comparisons)	275	S[1] 0.13 K[2]	0.13	—	0.16 0.22	—	0.33	0.33	Completed 1500 cycles at RT, HT, and IT. Both surfaces serviceable after test. Rated successful.
2.	Sputtered Cr_2O_3 versus det. gun sprayed Cr_3C_2	650	S 0.19 K 0.26	— 0.22	— 0.45	— 0.56	0.28 0.40	0.05	0.56	500 cycles each at HT, IT, and RT polishing of foil coating over bumps in the loaded zone. Journal had fine scratches. After another 1500 cycles, journal looks the same, foil had more polishing spots. After 9000 cycles journal looks the same. Foil had several wear spots in the center and at outer edges, but the spots were reoxidized. Rated successful.
3.	Sputtered Ni-Cr-Cr_2O_3 versus det. gun sprayed Cr_3C_2	650	S 0.12 K 0.26	— 0.26	— —	0.37 0.40	—	0.046	0.61	500 cycles each at HT, IT, and RT; foil had three bands of deep polishing in loaded zone. Journal evenly polished. After 3000 cycles, journal looks the same. Foil had four wide and several small wear bands. Rated successful.

TABLE 14.14 Start-Stop Test Results of Coatings Applied on Foil Bearings* (*Continued*)

Test no.	Foil and journal coatings	Maximum test temp., °C	Coefficient of friction: At start	After 499 cycles	After 999 cycles	After 1499 cycles	After 9000 cycles	rms surface roughness of journal, μm: Before	After	Comments
4.	Sputtered Ni-Cr-Cr_2O_3 with Ni-Cr undercoat versus det. gun sprayed Cr_3C_2	650	S 0.08 K 0.41	— —	— —	— 0.81	—	0.05	0.31	500 cycles each at HT, IT, and RT. Journal unchanged. Foil worn some over bumps in loaded zone. After next 1500 cycles, journal OK. Foil worn in the loaded zone. Some scratches in worn area. Rated successful.
5.	Sputtered Cr_2O_3 versus det. gun sprayed Cr_3C_2 †	650	S 0.19 K 0.42	— 0.50	— —	— 0.95	— —	0.13	0.13	500 cycles each at HT, IT, and RT; coating on foil worn over bumps in loaded zone. After next 1500 cycles coating on foil worn significantly in the loaded zone, journal unchanged. Rated marginal.

*Load = 14 kPa based on bearing projected area.

†This test was conducted at a normal load of 35 kPa. Test cycle sequence: 500 HT (high temperature), 500 IT (intermediate temperature), and 500 RT (room temperature), repeated.

(1) Static breakaway at RT.

(2) Kinetic at test temperature.

Source: Adapted from Bhushan (1981b).

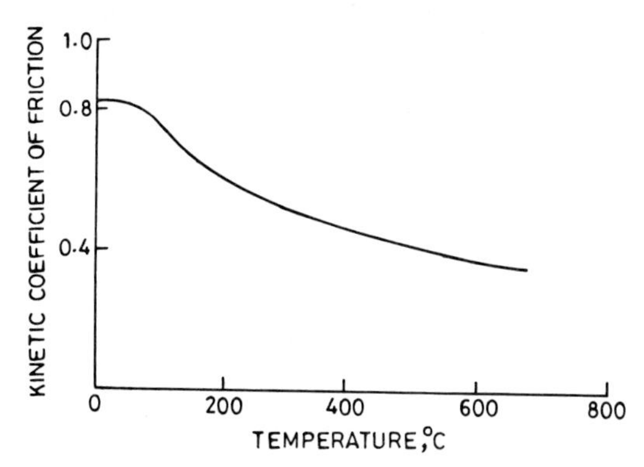

FIGURE 14.23 Kinetic coefficient of friction as a function of temperature for sputtered Cr_2O_3 coating on foil surface versus detonation gun sprayed Cr_3C_2 coating on journal surface in a foil bearing. [*Adapted from Bhushan (1981b).*]

most wear occurs during sliding component of the relative motion (commonly referred to as *slip*); therefore, accelerated sliding tests were conducted to simulate ball-race, ball-cage, and cage-race interactions. Table 14.15 summarizes the results for ball-race interactions. In these tests, a 6.764-mm-diameter ball was slid against the circumference of a 50.8-mm-diameter cylinder at various loads, speeds, and ambient temperatures. Results for a sputtered TiN (about 1 μm thick) with an overlay of sputtered MoS_2 are also listed in the table for comparisons. We note that the coefficient of friction was independent of speed but lower at higher temperatures and higher loads. TiN + MoS_2 and Cr_2O_3 + MoS_2 coated disks wore very little after a sliding test of 2 min; Cr_2O_3 + MoS_2 disk wore less than TiN + MoS_2 disk at 2070 MPa and speeds higher than 15 mm s^{-1}. We also note that the coated balls did not wear during the tests.

14.4.1.3 Zirconia Coatings. Zirconia coatings are generally applied by plasma spray and sputtering techniques. Plasma sprayed coatings of yttria-stabilized zirconia (such as Rokide Z from Norton's Abrasive Co., Worcester, Massachusetts), and ZrO_2-based coatings (calcium zirconate with 31 wt % CaO and ZrO_2 balance, and magnesium zirconate with 24 wt % MgO and ZrO_2 balance) are used as thermal barrier coatings in aircraft engine combustors, internal combustion engines, and as sensors in fuel cells (Fishman et al., 1985; Miller and Berndt, 1984; Pettit and Goward, 1983). Plasma sprayed coatings of partially stabilized zirconia (ZrO_2-Y_2O_3 and ZrO_2-MgO) have been used for tribological applications such as magnetic-tape heads (Bhushan, 1990) and gas turbine engine shafts.

Coatings of pure ZrO_2 by electron-beam evaporation were produced by Lee and Poppa (1977) and Movchan and Demchishin (1969), and those by rf sputtering were produced by Iwamoto et al. (1987), Pawlewicz and Hays (1982), and Rujkorakarn and Sites (1986). Pawlewicz and Hays (1982) reported that their coatings were monoclinic, whereas Lee and Poppa (1977) and Rujkorakarn and Sites (1986) reported that their coatings were amorphous, but crystallized into a mixed tetragonal and monoclinic phase at about 450°C. Iwamoto et al. (1987) have reported that rf-sputtered ZrO_2-Y_2O_3 coatings with 3 mol % Y_2O_3 were amorphous and remained amorphous even after annealing at 850°C for 15 min. Movchan and Demchishin (1969) showed that the hardness of evaporated ZrO_2

TABLE 14.15 Results of Friction and Wear Sliding Tests with Ball versus Circumference of a Cylinder*

Cylinder and ball coatings	Normal stress, MPa	Sliding speed, mm s^{-1}	Ambient temperature, °C	Kinetic coefficient of friction		Surface examination after test	General comments
				Start of test	End of test		
Sputtered $Cr_2O_3 + MoS_2$ vs. itself	520	15	22	0.30	0.27	No coating damage on cylinder and ball, slight polishing in the track.	—
		30		0.27	0.33		
		61		0.33	0.34		
		127		0.34	0.38		
Sputtered $Cr_2O_3+MoS_2$ vs. itself	2070	15	177	0.07	0.07	No change at all, could hardly see the track.	Friction went down with time.
		30		0.06	0.06		
		61		0.06	0.05		
		127		0.05	0.04		
Sputtered $Cr_2O_3+MoS_2$ vs. itself	2070	15	22	0.12	0.12	Debris buildup on the ball, coating spalled in the track at 127 mm s^{-1}.	Friction slightly increased during test. Coating was completely worn at 127 mm s^{-1}.
		30		0.12	0.15		
		61		0.15	0.18		
		127		0.18	0.18		
Sputtered $Cr_2O_3+MoS_2$ vs. itself	2070	15	177	0.07	0.07	Coating intact on ball and cylinder, very little buildup on the ball, light polishing in the track.	—
		30		0.07	0.07		
		61		0.07	0.07		
		127		0.07	0.08		
Sputtered $TiN+MoS_2$ vs. itself	520	15	22	0.32	0.30	Could not see any debris on ball, patches of coating deposited on cylinder, little coating worn.	Could see track at 61 mm s^{-1}.
		30		0.30	0.30		
		61		0.30	0.32		
		127		0.32	0.35		

Sputtered $TiN + MoS_2$ vs. itself	520	15	177	0.24	0.17	Hardly any change on ball, cylinder had three narrow streaks.	
		30		0.17	0.12		
		61		0.12	0.16		
		127		0.16	0.16		
Sputtered $TiN + MoS_2$ vs. itself	2070	15	22	0.19	0.41	The ball had wear debris pileup in the leading edge, most of the coating was worn away.	Cylinder coating was worn after 35 s at 30 mm s^{-1}, friction went down with time.
		30		0.37	0.15		
		61		0.15	0.15		
		127		0.15	0.15		
Sputtered $TiN + MoS_2$ vs. itself	2070	15	177	0.06	0.02	The ball had very little wear debris pileup, coating deeply polished, surface smooth.	Cylinder coating started to wear at 127 mm s^{-1}.
		30		0.02	0.02		
		61		0.02	0.01		
		127		0.01	0.01		

*Relative humidity = 41 ± 2 percent; Test duration at each speed = 30 s.

coatings increases from about 400 to 1100 kg mm^{-2} above 1000°C where the crystallographic structure is reported to change from monoclinic to the tetragonal structure.

Thin rf-sputtered ZrO_2-Y_2O_3 coatings ranging in composition from 3 to 15 mol % Y_2O_3 were produced by Knoll and Bradley (1984a, 1984b). The high-temperature cubic structure was quenched into the coatings during deposition, and neither yttria content nor substrate bias affected the structure (also see Pawlewicz and Hays, 1982; Yamashita et al., 1988). During annealing for moderate times at temperatures within the tetragonal + cubic field on the phase diagram (Chap. 4), tetragonal particles did not precipitate from the cubic phase and thus the sputtered zirconia is not "partially stabilized." Knoll and Bradley also reported that increased substrate bias reduced the coating's porosity but also introduced high compressive stress into the coatings. Pawlewicz and Hays (1982) reported that substrate bias allows manipulation of both the grain size and the preferred orientation of crystal structure. These trade-offs may be important in practical applications.

Thin films of zirconia show a high refractive index and low loss in the infrared region. These properties of ZrO_2 appear to render it suitable for fabricating high-reflectance mirrors and interference filters. In some types of zirconia, good mechanical and thermal properties have been obtained by controlling alloy components and deposition-process parameters; these offer good tribological properties. Sputtered ZrO_2 is used as wear- and corrosion-resistant coating in magnetic thin-film disks (Yamashita et al., 1988; Bhushan, 1990).

14.4.1.4 ***Other Oxide Coatings.*** Other coatings which have been studied are SiO_2, SiO_xN_y, TiO_2, and SnO_2. Sputtered and spin-coated SiO_2 coatings are used in semiconductor electronics applications and magnetic storage devices (Bhushan, 1990). Sputtered SiO_2 coatings are deposited by rf magnetron sputtering from an SiO_2 target. Spin-coated SiO_2 coating is produced by the sol-gel method. In this method, tetrahydroxysilane dissolved in isopropanol alcohol solution (roughly 2 percent concentration) is spin-coated and baked at temperatures ranging from 25 to 600°C for roughly 2 to 24 h, respectively. An increase in hardness and a reduction in porosity occurs with an increase in the baking temperature which results in better friction and wear properties. A maximum hardness of about 950 kg mm^{-2} has been reported (Bhushan, 1990). The sol-gel method has been used by Nelson et al. (1981) to deposit coatings of various ceramic oxides such as titania, ceria, and alumina for protective applications.

PECVD coatings of SiO_2, TiO_2, and SiO_xN_y have been reported by Hess (1984) and Reinberg (1979). PECVD SiO_2 coatings are deposited by a plasma-enhanced reaction of SiH_4 with nitrous oxide (N_2O) or CO_2 gases (Hess, 1984). PECVD TiO_2 coatings are produced by a plasma-enhanced reaction of O_2 with either titanium tetrachloride ($TiCl_4$) or with titanium isopropylate gases. PECVD SiO_xN_y are deposited by a plasma-enhanced reaction of SiH_4 and NH_3 source gases, with small amounts of NO, N_2O, O_2, or CO_2 gas added to serve as the oxygen source. PECVD SiO_2 is used as an intermediate dielectric coating and as a "scratch protection" layer on integrated circuits. PECVD TiO_2 coatings have an unusually high dielectric constant, excellent chemical stability and optical transmittance, and a high refractive index. Because of such properties, these coatings have been investigated for use as capacitor materials in microelectronics devices, as optical coatings, and as photoelectrochemical anodes for solar energy conversion. PECVD SiO_xN_y has found use in a variety of electronic device applications.

Transparent conductive layers of SnO_2 have been deposited by chemical vapor deposition which is carried out by the oxidation of stannous chloride at various substrate temperatures in the range 350 to 500°C (Sundaram and Bhagwat, 1981). SnO_2 coatings have also been deposited by plasma-assisted vapor reaction of CO_2 with an organic tin compound, such as tetramethyl tin. These coatings have been applied to fabricate CCD optical imager (Hynecek, 1979; Keenan and Harrison, 1985). These coatings are expected to find use in tribological applications.

*14.4.1.5 **Proprietary Oxide Coatings.*** Kaman Sciences Corporation in Colorado Springs, Colorado, produces a family of chemically bonded oxide coatings under trade names of K-ramic, K-ramicote, and K-aramite. These coatings are applied as a water-based slurry of the finely divided refractory materials. Slurry can be applied by means of an airbrush, dipping, or simply painting on with a fine brush. A moderately low-temperature cure cycle (450 to 650°C) follows application. During the cure cycle, strong oxide bonds are formed within the coating itself, as well as between the coating and substrate. In the case of metallic substrates or metal powders used in composite coatings, the oxide bond is made to the metallic oxide that forms on the metal surface. The coating thickness typically ranges from 25 μm to 0.75 mm. The hard dense coatings can be surface-finished to 75 nm rms. These coatings have been applied to a large variety of metals, ceramics, and glass substrates. These coatings can also be applied over a porous coating, such as a plasma sprayed coating, to reduce porosity and to increase wear resistance.

The exact chemical composition of these chemically bonded coatings may be varied by the use of different refractory materials in the slurry. The cured and bonded coatings may range from 100 percent refractory oxide compositions to those containing small to major percentages to metals, carbides, silicides, etc. Metal or ceramic fibers, whiskers, or other reinforcing materials can also be easily included in the formulation for improved flexural properties, impact strength, etc. Among the refractory oxides that have been used to produce hard, wear-resistant coatings are: alumina, chromia, beryllia, zirconia, titania, silica, tin oxide, and iron oxide. Other refractory materials that can be incorporated are carbides of tungsten, silicon, titanium, and boron, in addition to several of the silicides and beryllides. Well-bonded metal-ceramic composite coatings have been made using finely divided nickel, titanium, stainless steel, boron, chromium, iron, aluminum, and Inconel.

The coating formulations can be tailored to provide desired hardness, porosity, thermal shock, resistance to corrosive environments, and minimum as well as maximum operating-temperature limits. Most commonly used coatings are two basic materials: either alumina-chromia (CCA-1000) or silicon carbide-chromia (CCS-1000). The processing of K-ramic coatings does not require expensive hardware and is expected to be cheaper than other coating deposition techniques. Applications of these coatings include bearings, seals, pump-plungers, cylinder liners, valve seats, acid nozzles, ball valves, and others (Bhushan et al., 1978).

14.4.2 Carbide Coatings

Carbides generally have very high hardness with B_4C being the hardest carbide known. Several metal carbides such as B_4C, SiC, TiC, ZrC, HfC, Cr_3C_2, Mo_2C, WC, VC, NbC, and TaC have been applied onto steel, nonferrous metals, and cemented carbides to improve their life by several hundred percent. These coatings have been applied by a variety of processes such as plasma spray detonation gun, activated reactive evaporation, ion plating, sputtering, CVD, and plasma-enhanced CVD.

14.4.2.1 ***Titanium Carbide Coatings.*** Titanium carbide (TiC) is the most widely used ceramic carbide coating. TiC coatings have been applied onto tools to increase their life, onto balls of gyro-application ball bearings, and as first-wall materials in the tokamak-type nuclear fusion reactor. The properties of TiC coatings strongly depend on coating deposition technique and coating conditions (Fukutomi et al., 1982; Pan and Greene, 1981; Raghuram and Bunshah, 1972; Sundgren et al., 1983a, 1983b, 1983c; Torok, 1987). TiC coatings have been applied by plasma spraying, ARE, ion plating, sputtering, CVD, and PECVD techniques. CVD technique is used to those interfaces where the high temperatures during processing do not adversely affect the interface. The benefit of CVD coating is that it provides a graded interface with extremely high adhesion strength.

TiC Coatings Applied by ARE. TiC coatings by ARE have been developed for use on high-speed steel cutting tools and WC-Co cemented carbide tool tips (Raghuram and Bunshah, 1972; Nimmagadda and Bunshah, 1976; Sarin et al., 1977; Jamal et al., 1980). The addition of Ni (5 to 10 wt %) into TiC by evaporating Ti-Ni billet in partial pressure of C_2H_2, significantly improves the ductility and oxidation resistance of TiC. The TiC-Ni cermets have fine grain size and high hardness (HK-1800 to 2800 kg mm^{-2}). Bunshah and Shabaik (1975) coated HSS tools with TiC-Ni, and they reported an improvement of wear life by more than 300 percent.

The friction and wear characteristics of TiC and TiN coating applied by activated reactive evaporation onto 304 stainless steel, Ti, and Al disk and 440C stainless-steel pins are summarized in Table 14.16*a* for several combinations of coated and uncoated surface. These measurements were made on a pin-and-disk tester under dry and lubricated (kerosene oil) conditions. Microhardness of various coatings and substrates are presented in Table 14.16*b*. We note that the coefficient of friction and the wear were very much lower when the test couple consisted of a hard coating rubbing against a hard coating than when one or both of the components were uncoated metals, even in dry (unlubricated) conditions. In some instances of the hard-hard couple, a very low coefficient of friction (0.05 to 0.1) and extremely low wear in dry conditions were observed. From examination of the wear scar, the failure mode of the coating was observed to be microfragmentation resulting from the initiation and propagation of microcracks at the pin-disk interface.

Process parameters such as deposition rate, partial pressure, and substrate temperature can cause considerable changes in the microstructure, and especially high deposition rates and high pressures can cause open structures with weak grain boundaries, resulting in low coating hardness. An example of such a microstructure with very small grains and voids located in the grain boundaries is shown in Fig. 14.24 for (V, Ti) C coating produced by ARE on a substrate kept at 700°C (Jacobson et al., 1978). The hardness of this coating was only 1385 kg mm^{-2}. A decrease in the acetylene pressure by a factor of 2, and an increase in the substrate temperature to 1000°C resulted in coating with larger grains and void-free boundaries. These coatings exhibited a hardness of 2900 kg mm^{-2}.

The hardness of TiC coating produced by ARE as a function of substrate temperature is shown in Fig. 14.25 (Raghuram and Bunshah, 1972). The data for sputtered TiN and CVD TiB_2 coatings are also included for comparisons. The most commonly used argument to explain the hardness increase with substrate

TABLE 14.16*a* Coefficient of Friction and Wear of TiC and TiN Coatings Applied by ARE*

		Coefficient of friction			Pin wear†		Disk wear‡	
Pin	Disk	Static, dry	Kinetic, dry	Kinetic, lubricated	Dry	Lubricated	Dry	Lubricated
440C uncoated	304 uncoated	0.67	0.40	0.15	High	Medium	High	Low
440C uncoated	TiC/304	0.12	0.17	0.17	Medium	High	Low	Low
440C uncoated	TiN/304	0.50	0.75	0.37	High	High	Low	Low
440C uncoated	Ti uncoated	0.76	0.45	0.42	High	High	High	High
440C uncoated	TiC/Ti	0.17	0.27	0.27	Medium	High	Low	Low
440C uncoated	TiN + Ti_2N/Ti	0.20	0.35	0.27	High	High	Low	Low
440C uncoated	Al uncoated	0.82	0.40	0.17	High	High	High	Medium
440C uncoated	Ti + TiC/Al	0.50	0.76	0.40	High	High	Low	Low
440C uncoated	TiN + Ti_2N/Al	0.90	0.95	0.50	High	High	Low	Low
TiC/440C	304 uncoated	0.40	0.50	0.17	High	—	High	—
TiC/440C	TiC/304	0.22	0.20	0.10	Low	Very low	Low	Low
TiC/440C	TiN/304	0.25	0.20	0.25	Medium	High	Low	Low
TiC/440C	Ti uncoated	0.65	0.80	0.50	High	—	High	—
TiC/440C	TiC/Ti	0.27	0.17	0.17	Low	Low	Low	Low
TiC/440C	TiN + Ti_2N/Ti	0.27	0.20	0.10	Low	Low	Low	Low
TiC/440C	Al uncoated	0.50	0.85	0.07	—	—	High	—
TiC/440C	Ti + TiC/Al	0.81	0.30	0.12	High	Low	Low	Low
TiC/440C	TiN + Ti_2N/Al	0.42	0.35	0.17	High	Medium	Low	Low

TABLE 14.16a Coefficient of Friction and Wear of TiC and TiN Coatings Applied by ARE* (*Continued*)

		Coefficient of friction			Pin wear†		Disk wear‡	
Pin	Disk	Static, dry	Kinetic, dry	Kinetic, lubricated	Dry	Lubricated	Dry	Lubricated
TiN/440C	304 uncoated	0.29	0.41	0.17	High	Low	High	Low
TiN/440C	TiC/304	0.05	0.06	0.17	Low§	Very low	Low	Low
TiN/440C	TiN/304	0.65	0.45	0.15	High§	Low	Low	Low
TiN/440C	Ti uncoated	0.50	0.76	0.40	High	Medium	High	High
TiN/440C	TiC/Ti	0.05	0.02	0.15	Low§	Very low	Low	Low
TiN/440C	$TiN+Ti_2N/Ti$	0.10	0.05	0.12	Low	Very low	Low	Low
TiN/440C	Al uncoated	0.27	0.40	0.17	Medium	Low	High	Medium
TiN/440C	Ti+TiC/Al	0.20	0.17	0.17	Low	Low	Low	Low
TiN/440C	$TiN+Ti_2N/Al$	0.30	0.25	0.17	Medium	Medium	Low	Low

*Pin-on-disk test with the run distance of 500 m under a load of 4 N. The sliding velocity for the static test was $(1.2–1.5) \times 10^{-2}$ m s^{-1}; for the kinetic lubricated tests, it was 0.1–0.15 m s^{-1}.

†Pin wear: high is 5×10^{-3} mm^3 and higher; medium is $(1 \times 10^{-4})–(5 \times 10^{-3})$ mm^3; low is $(1 \times 10^{-5})–(1 \times 10^{-4})$ mm^3; and very low is less than 1×10^{-5} mm^3.

‡Disk wear: high is 2.5×10^{-3} mm and higher; medium is $0.25–2.5 \times 10^{-3}$ mm; low is less than 2.5×10^{-4} mm.

§Test run for 100 m under dry conditions.

Source: Adapted from Jamal et al. (1980).

TABLE 14.16*b* Microhardness of Various Substrates and Coatings

Material	Microhardness (HK), kg mm^{-2}
304 stainless steel	300
Ti	230
Al	100
440C stainless steel	315
TiC coating	2800–3000
$TiN+Ti_2N$ coating	2300–2800
Ti+TiC coating	2450–2650
TiN coating	1800–2000

Source: Adapted from Jamal et al. (1980).

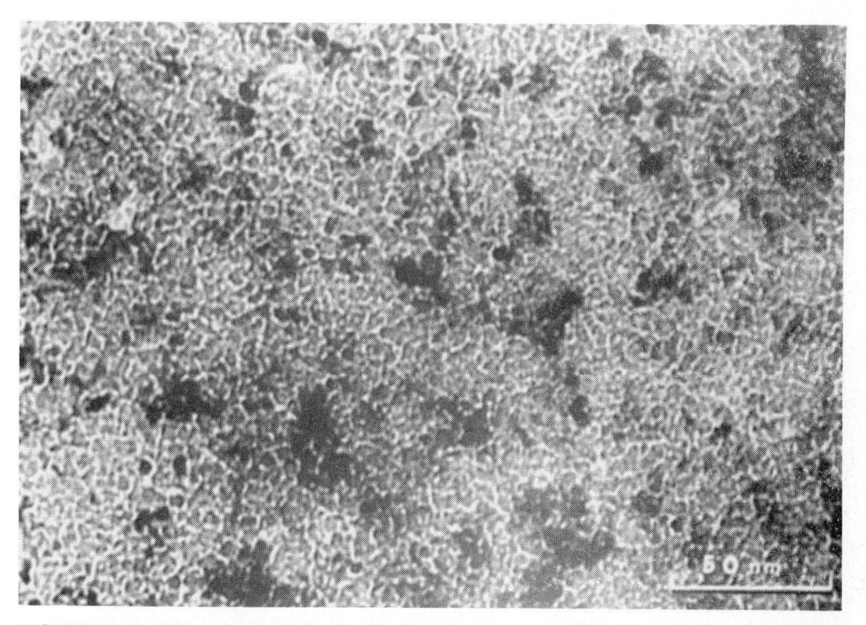

FIGURE 14.24 Microstructure of a (V, Ti) C coating grown by ARE. The micrograph is exposed in underfocused condition to reveal the fine cavity network along grain boundaries. [*Adapted from Jacobson et al. (1978).*]

temperature is the reduced defect content. Refractory compounds normally have a broad inhomogeneity range, and deviations from stoichiometry, and vacancies on either side of sublattices decrease the strength of these materials. However, for many of the cases quoted here, the hardness increases considerably above bulk values. This cannot be explained with a reduced defect content. Instead it seems more plausible that an increased strength of grain boundaries can explain the results. This would imply that as soon as the temperature is high enough for strong boundaries to develop, the hardness should reach a maximum value and then decrease toward the bulk value as the substrate temperature, and hence

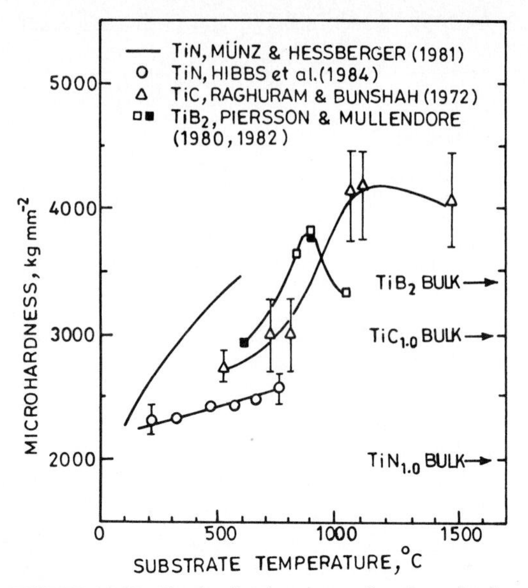

FIGURE 14.25 Coating hardness as a function of substrate temperature for ARE TiC, magnetron-sputtered TiN, and CVD TiB_2 coatings. Also shown are corresponding bulk values.

grain size, is increased further (Sundgren and Hentzell, 1986). This argument is valid for compounds where no phase transformation occurs as the temperature is raised. For example, Al_2O_3 and ZrO_2 go through several phase transformations in the temperature range of 500 to 900°C and their coatings have a minimum hardness in this temperature range (Thornton and Chin, 1977).

TiC Coatings Applied by Ion Plating. TiC coatings have been applied onto high-speed steel tools and stainless steel by using ion plating under a reactive mode (Kobayashi and Doi, 1978; Komiya, 1982). The TiC coating applied by ion plating using hollow cathode discharge, cathodic arc ion plating, and sputter ion plating exhibit almost similar values of microhardness and friction and wear characteristics. The hardness values and adhesion of TiC coatings are strongly dependent on the deposition conditions and substrate temperature. The microhardness of TiC coatings applied by cathodic-arc ion plating range from 2900 to 3300 kg mm^{-2} for coatings applied at a substrate temperature of 400 to 500°C, whereas coatings applied at 600°C exhibit high hardness, i.e., 3700 kg mm^{-2}.

TiC Coatings Applied by Sputtering. The sputtering process has been used to apply Ti_xC_{1-x} coatings on various substrates such as steel, stainless steel, Monel-400, copper, molybdenum, Ni-, Co-, and Ti-based alloys (Brainard and Wheeler, 1978, 1979a, 1979b; Caddoff et al., 1971; Carson, 1975; Dua et al., 1984; Obata et al., 1981; Shikama et al., 1983; Sundgren, 1985; Sundgren et al., 1983a, 1983b, 1983c). TiC can be sputtered from a target of TiC or by reactive sputtering of titanium in mixed Ar-CH_4 discharges.

The stoichiometry of TiC-sputtered coatings exhibits significant influence on the microstructure and physical properties (Caddoff et al., 1971; Eizenberg and Murarka, 1983; Shikama et al., 1983; Sundgren, 1985). Sundgren et al. (1983a,

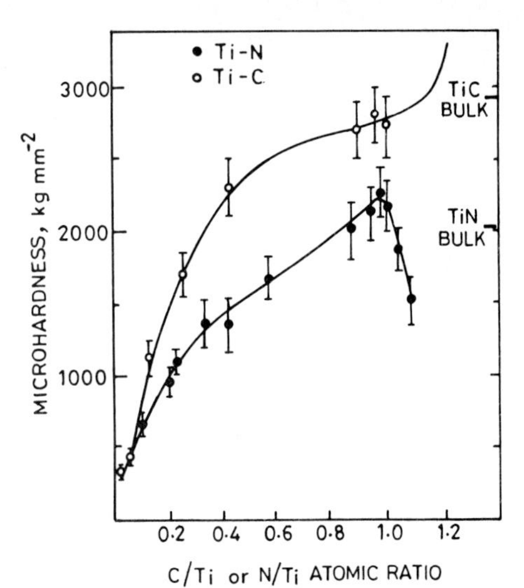

FIGURE 14.26 Coating microhardness for sputtered Ti-C and Ti-N coatings as a function of composition ratio. [*Adapted from Sundgren et al. (1983b).*]

1983b, 1983c) reactively rf-sputtered Ti in mixed Ar-CH_4 discharges on various substrates of polished steel, copper foil, fused silica, and alumina which were maintained at 500°C. They varied the composition by varying the partial pressure of CH_4 in the plasma. The microhardness as a function of composition ratio (C/Ti) is shown in Fig. 14.26. Data for Ti-N coatings are also included for comparisons. For Ti-C, the hardness increases with increasing C/Ti ratio, and for a ratio of 1 approaches 2800 kg mm^{-2} which is close to bulk values taken from Toth (1971). For high carbon concentrations, the hardness increases rapidly and approaches 4000 to 5000 kg mm^{-2}. Similar results are also reported by Caddoff et al. (1971).

Density and electrical resistivity also change with a change in the composition. Figure 14.27 shows the density of the coatings as a function of composition. The data for Ti-N are also included for comparison. For Ti-C, the density decreases gradually as the carbon content increases, and at a carbon concentration of 50 at. % (stoichiometric composition), a density of 3500 kg m^{-3} is obtained, and the bulk value for the TiC phase is 4900 kg m^{-3}. The difference in the densities is due to the open voided structure (porosity) of the coatings. The coatings were found to consist of TiC_x grains embedded in a porous matrix of carbon, where x is less than 1.

Eizenberg and Murarka (1983) measured the residual stresses in the TiC coatings produced by rf reactive sputtering process of Ti in the presence of Ar + CH_4 mixture. The results of the stress measurements show that all the Ti_xC coatings are in a state of compression. The maximum stress occurs around the stoichiometric composition to a value of about 8 GPa. This value was virtually independent of the nature of the substrate and the coating thickness. Residual stresses have been shown to be a function of deposition parameters. Pan and Greene (1981) have shown that residual stress in rf-sputtered TiC coatings are compressive on the order of 0.1 GPa (Fig. 14.28). Pan and Greene found that Ar

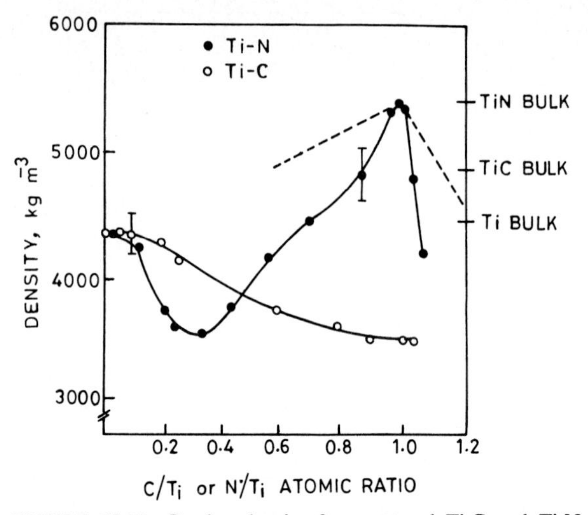

FIGURE 14.27 Coating density for sputtered Ti-C and Ti-N coatings as a function of composition ratio, showing also the bulk densities for Ti, TiC, and TiN. [*Adapted from Sundgren et al. (1983b).*]

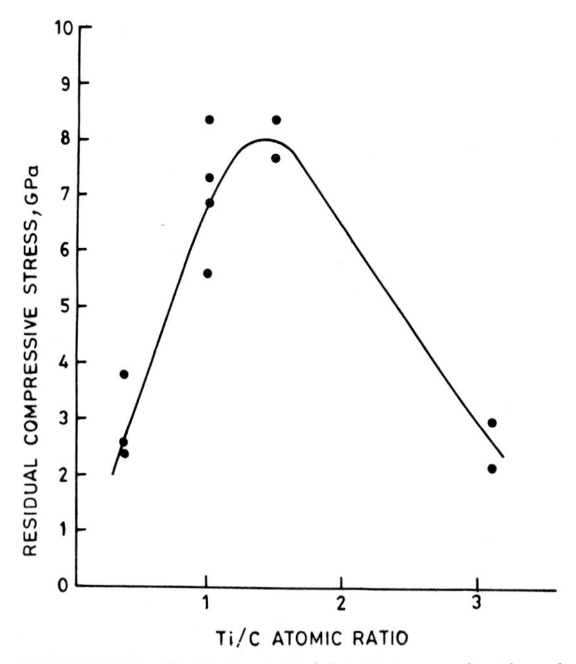

FIGURE 14.28 Residual compressive stress as a function of Ti/C atomic ratio in TiC coatings reactively sputtered at 1.3 Pa or 10 mtorr pressure and 500 W rf power. [*Adapted from Eizenberg and Murarka (1983).*]

incorporation up to 1.5 at. % is possible in TiC coatings depending on the residual gas pressure (Fig. 14.29). The Ar incorporation may cause compressive stresses in the coating. Kuwahara et al. (1981) also reported that residual stresses of magnetron-sputtered TiC and Al_2O_3 coatings were compressive on the order of 1 GPa. The values of compressive stresses were affected by the changes in the deposition rates and coating thicknesses (Fig. 14.30). Kuwahara et al. (1981) reported that though the residual stresses in sputtered coatings of TiC and Al_2O_3 were compressive, the stresses in sputtered Cu coatings were tensile (~0.1 GPa) consistent with other experiences on metal coatings (Cr and Mo).

It is believed that impurity content in the coating is partially responsible for the residual compressive stresses. It is both the location of Ar (or other noble gas atoms) itself in the lattice and the structural damage caused during the incorporation process (low-energy implantation) that cause these stresses (Pan and Greene, 1981). Thornton and Hoffman (1981) have suggested that the compressive stresses might be caused by the "atomic peening" action of energetic working gas particles (neutralized or reflected ions) which bombard the growing coating and produce an associative, not causative, entrapment of working gas. Very high compressive stresses (1 to 10 GPa) generally found in hard coatings could give rise to lattice distortion which would affect the dislocation propagation and thus an increase in the hardness of the coating. However, such high stresses may be detrimental to coating-to-substrate adhesion.

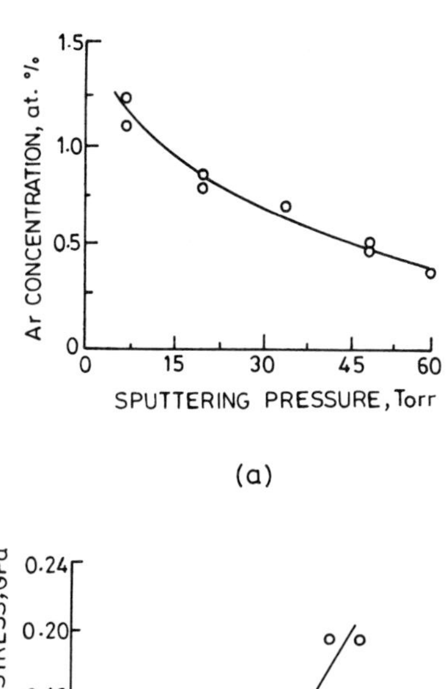

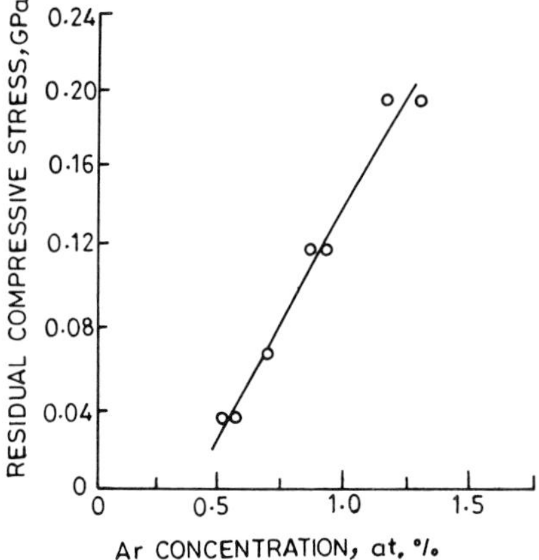

FIGURE 14.29 (*a*) Argon concentration as a function of sputtering pressure. (*b*) Residual compressive stresses as a function of argon concentration in TiC coatings (0.3 μm thick) produced by rf sputtering. [*Adapted from Pan and Greene (1981).*]

It has been reported that coatings adhere best when a mixed compound of coating and substrate metal is formed in the interfacial region (Blackburn et al., 1949; Katz, 1976; O'Brien and Chaklades, 1974). Brainard and Wheeler (1978) have shown that for certain refractory types of material, e.g., TiC, TiB_2, $MoSi_2$, and Mo_2C, significant improvement in adherence, and consequently wear resistance, could be obtained by promoting the formation of a "mixed-oxide" interface composed of iron oxide from the substrate and oxides of the constituents of the refractory compound being sputtered. Formation of a mixed-oxide interface can be promoted by the oxygen present in the sputtering system from

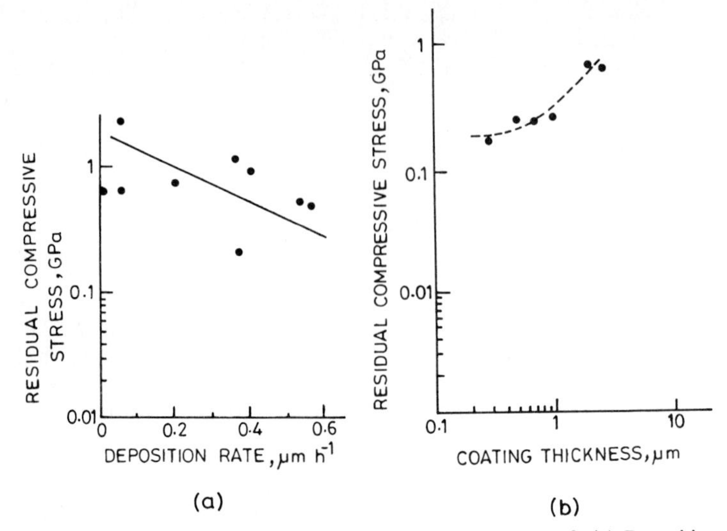

FIGURE 14.30 Residual compressive stress as a function of: (*a*) Deposition rate. (*b*) Coating thickness in TiC coatings produced by rf sputtering. [*Adapted from Kuwahara et al. (1981).*]

target degassing, etc., or by preoxidizing the surface (by furnace heating). Brainard and Wheeler (1978) and Wheeler and Brainard (1977, 1978) found that biasing, while generally improving bulk-film purity and stoichiometry (removal of oxide contaminants), and consequently wear performance, can adversely affect adherence by removing interfacial oxide layers. A coating prepared by applying a bias on an oxidized substrate or by applying the bias voltage after the first couple of minutes of deposition on a sputter-etched surface exhibited good adhesion and good wear performance (Fig. 14.31). Although some compounds were better bonded to oxidized disks, others (e.g., Mo_2B_5) showed little improvement. Figure 14.32 shows the XPS depth profile of Mo_2C rf-sputtered onto sputter-etched and oxidized substrates with and without bias. We note that oxidized substrate with bias and sputter-etched disk with no bias produced interfacial oxide region and a sputter-etched disk with bias did not.

Greene and Pestes (1976) have also shown improved adherence of sputtered TiC coatings on steel when oxides are formed at the interface. Their approach was to introduce a small percentage (2 percent) of oxygen into the argon plasma during initial deposition to promote the formation of a graded interface, which reportedly measured 125 nm in width. Brainard and Wheeler (1979b) introduced a small partial pressure of nitrogen (~0.5 percent) during the first couple of minutes of deposition, and this addition markedly improved the adherence, friction, and wear properties of TiC on 440C stainless steel substrate deposited by using a bias voltage, compared with coatings applied on sputter-etched, oxidized surfaces or in the presence of a small oxygen partial pressure during the first couple of minutes (Fig. 14.33). The improvements are related to the formation of an interface containing a mixture of the nitrides of titanium and of iron which are harder than their corresponding oxides. Brainard and Wheeler (1979a) found that adherence of TiC to the Ni-based alloy (Rene 41) improved using a partial pressure of acet-

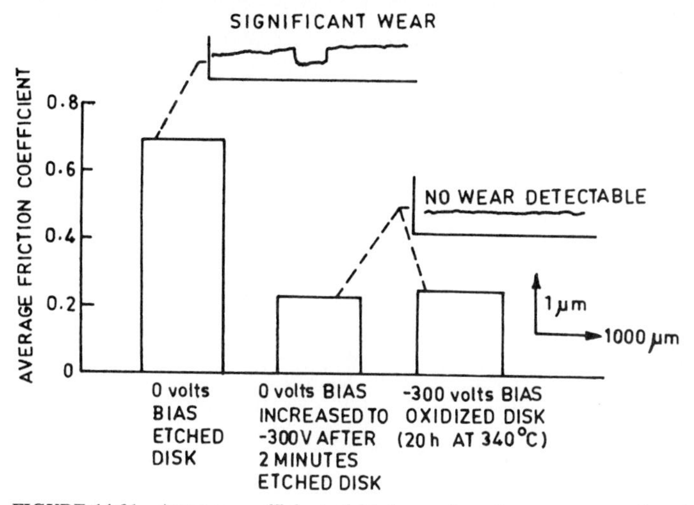

FIGURE 14.31 Average coefficient of friction and coating wear traces for rf-sputtered Mo_2C on 440C stainless steel disk sliding against 304 stainless-steel pin, normal load = 0.10 N, nitrogen atmosphere. [*Adapted from Brainard and Wheeler (1978).*]

ylene during the first couple of minutes of deposition. However, in the case of Ti-6Al-4V substrate, a titanium interlayer was effective in an entirely different way from that of the mixed compound layers in the other cases. In the case of Ti-6 Al-4V, there is already a passive Al_2O_3 layer, and Ti metal is able to form an adherent film on the Al_2O_3 passive layer and at the same time to provide a good substrate for the TiC coating.

Figure 14.34 compares the friction and wear data of sputtered coatings of various refractory carbide, silicide, and borides in a pin-on-disk test (Brainard and Wheeler, 1978). All coatings were applied on an oxidized 440C stainless steel disk, and the pin was made of 304 stainless steel with a radius of 4.76 mm. The tests were conducted in the dry nitrogen environment for a period of 30 min. All materials except $TiSi_2$ are significantly harder than 440C stainless steel (~700 kg mm^{-2}). The high friction and wear of the Mo_2B_5 and $TiSi_2$ coatings were due to coating failure (i.e., spalling) whereas the higher values for $MoSi_2$ and B_4C were a result of intrinsic coating properties. Based on these data, TiC and TiB_2 coatings on the steel surfaces, in the manner applied by Brainard and Wheeler, appear to have lowest friction and wear.

Carson (1975) produced coatings of titanium carbide (TiC), titanium oxycarbides ($TiO_{0.5}C_{0.5}$, $TiO_{0.25}C_{0.75}$), and titanium nitride (TiN) by rf sputtering from hot-pressed targets of TiC, TiO_2/TiC, and TiN materials. The coatings were deposited on cemented carbide cutting tools. Their cutting tests on heat-treated 4340 steel showed that the wear resistance of all four coated materials were within a factor of 2 of each other and very close to CVD TiC coated tools. All these materials exhibited more than an order of magnitude decrease in wear rate with respect to the uncoated tools.

TiC Coating Applied by CVD. Chemical vapor deposition (CVD) of TiC and TiN onto cemented carbide is a method widely used to improve the wear resistance of metal-working tools used in turning, i.e., in continuous cutting opera-

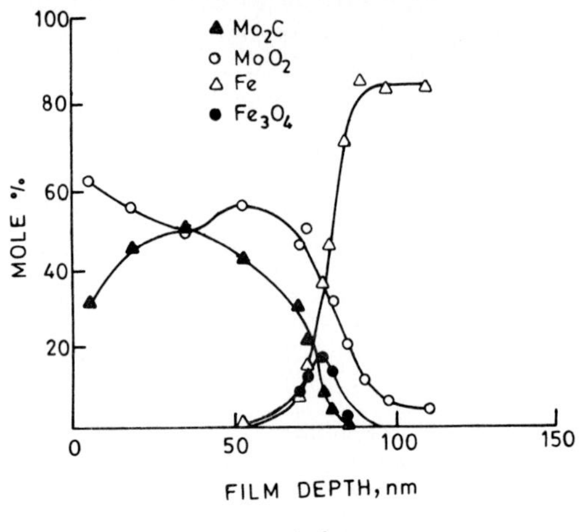

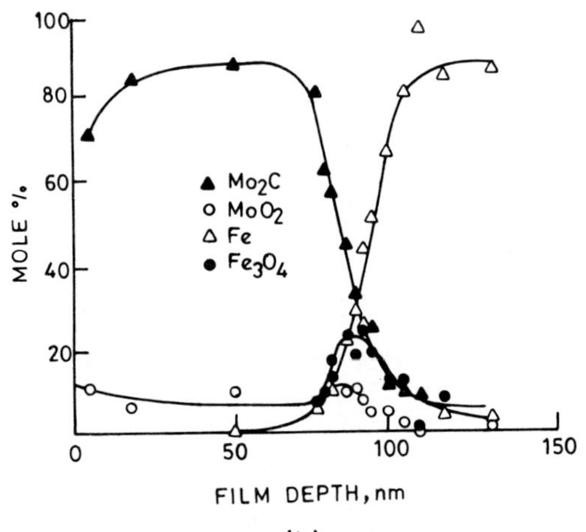

FIGURE 14.32 XPS depth profile of Mo_2C rf-sputtered onto: (*a*) Sputter-etched 440C stainless steel substrate, no bias. (*b*) Sputter-etched 440C stainless steel substrate with −300-V bias. (*c*) Oxidized 440C stainless steel substrate with −300-V bias. [*Adapted from Brainard and Wheeler (1978).*]

tions (Hintermann, 1980, 1981, 1984; Lee and Richman, 1974; Röser, 1981; Takahashi et al., 1967, 1970). TiC coatings are produced from the reactive gas mixtures of $TiCl_4$ and CH_4 at 900 to 1000°C. The reactive gas mixture is transported by a surplus of H_2 through heated tubes into the reactor. The CVD process is capable of applying adherent TiC coatings very economically on ceramics or cemented carbide cutting tools and with some care on steel, nickel- or cobalt-

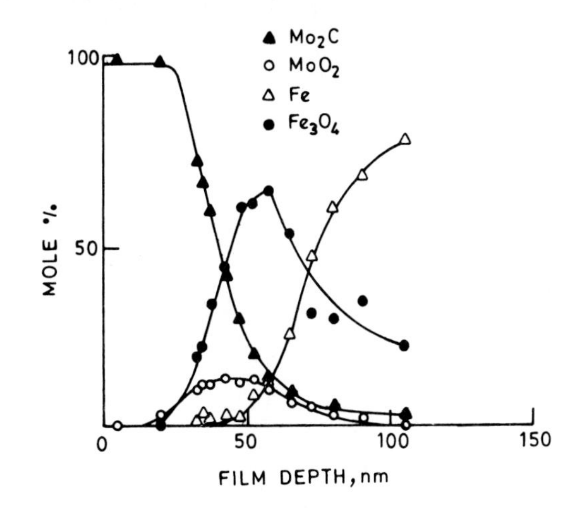

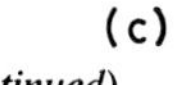
(c)

FIGURE 14.32 (*Continued*)

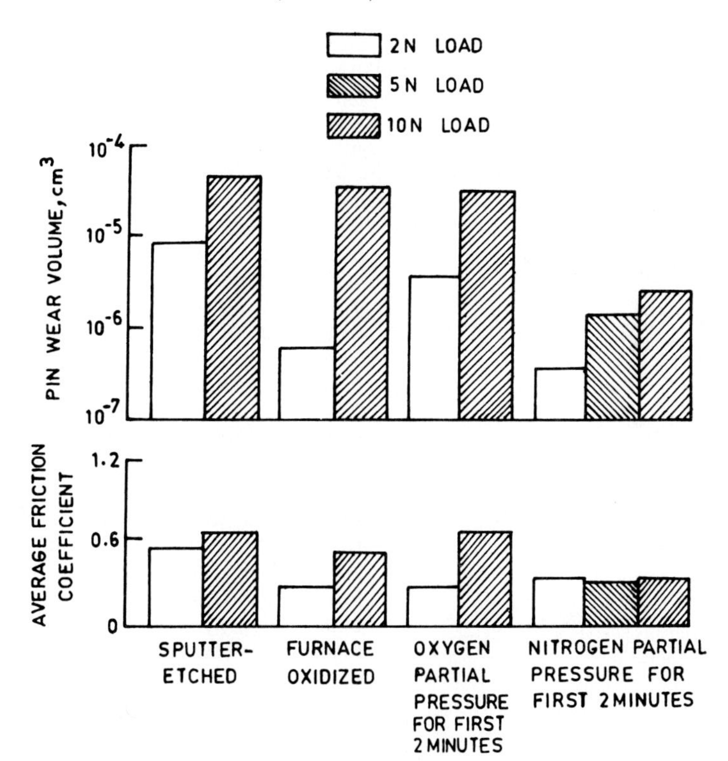

FIGURE 14.33 Coefficient of friction and pin wear volume for 440C stainless-steel disks sputtered with TiC, sliding speed = 0.25 m s^{-1}, 30-min run, N_2 atmosphere. [*Adapted from Brainard and Wheeler (1979b).*]

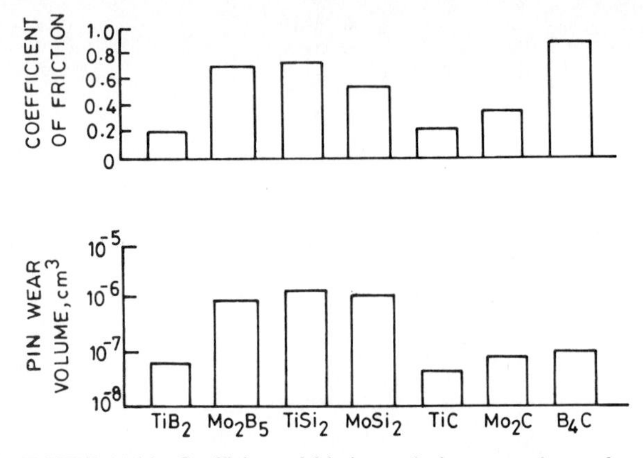

FIGURE 14.34 Coefficient of friction and pin wear volume after running on various sputtered coatings on oxidized 440C stainless-steel disk. Normal load = 0.25 N, speed = 0.25 $m\,s^{-1}$, 30-min run, nitrogen atmosphere. [*Adapted from Brainard and Wheeler (1978).*]

based alloys. The CVD TiC coatings have been applied to increase the life of tools and the load-carrying capacity of ball bearings, which are suitable for hostile environments such as vacuum, high-temperature, and corrosive atmospheres.

With CVD coatings, a very high bond strength to the substrate can generally be obtained. In almost every case the coating and substrate material form transition zones by interdiffusion facilitated by the high substrate temperature during processing. These can be a few micrometers thick in certain material combinations but also as thin as only a few atomic layers. Figure 14.35 shows the electron microprobe analysis of a TiC/TiN coated cemented carbide. The microprobe analysis reveals that an intermediate layer of a titanium carbonitride Ti(C, N) has formed; also tungsten from the substrate has diffused about 2.5 μm deep into the TiC layer, and nitrogen has diffused into the substrate. The strong Ti peak from the substrate is due to a mixed carbide (Ti, Ta, Nb, W) C grains in the cemented carbide. In order to improve adhesion further and in order to reduce the stresses, the functional coating is often deposited onto one or more intermediate layers. Peterson (1974) has reported that wear of cemented carbide tools (WC-TiC-TaC) is improved by at least an order of magnitude by CVD coating of TiC or TiN. TiC coated tools were slightly more wear resistant than TiN coated tools.

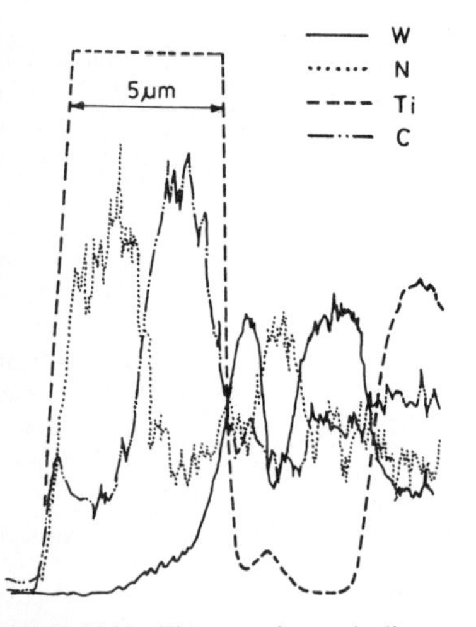

FIGURE 14.35 Electron microprobe line scan analysis of TiC/Ti (C, N)/TiN-coated cemented carbide. [*Adapted from Hintermann (1981).*]

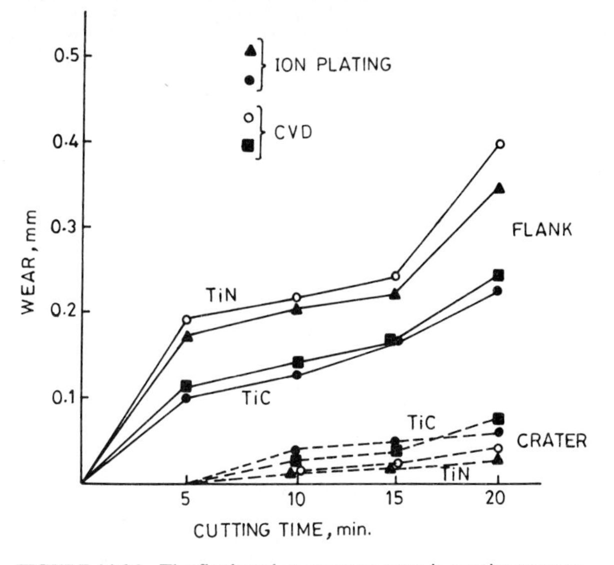

FIGURE 14.36 The flank and crater wear rates in continuous turning tests for cemented carbide inserts coated by ion plating and by the CVD processes. Work material: gray cast iron (BHN 190 to 240), speed: 150 m min^{-1}, feed = 0.305 mm r^{-1}, depth of cut: 2 mm. [*Adapted from Kobayashi and Doi (1978).*]

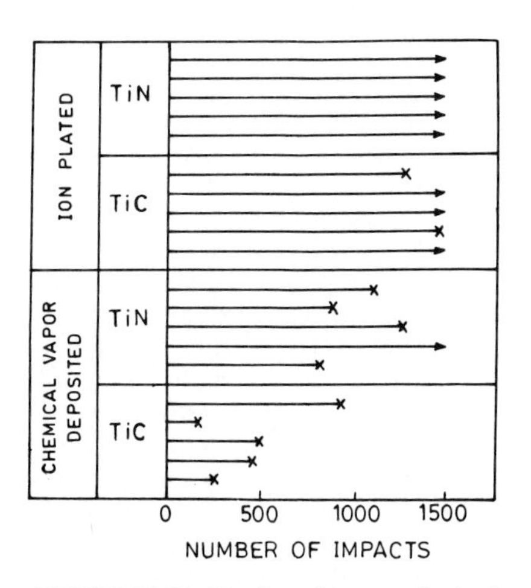

FIGURE 14.37 Number of impacts in the interrupted turning tests for cemented carbide inserts coated by ion plating and by the CVD processes. Work material: alloy steel (BHN 240 to 250), speed: 100 m min^{-1}, feed: 0.156 mm r^{-1}, depth of cut: 2 mm. [*Adapted from Kobayashi and Doi (1978).*]

Kobayashi and Doi (1978) conducted turning tests on cemented carbide (WC-2 % NbC-6 % Co, HV = 1570) tool inserts coated by the TiC and TiN using ion-plating and CVD processes (Fig. 14.36). We note that in continuous turning tests, the inserts coated by the ion-plating process with a 5 μm thick coating show a similar wear resistance to that of inserts coated by CVD. However, in interrupted turning tests, the inserts coated by the ion-plating process show a tool toughness superior to that of inserts coated by CVD (Fig. 14.37). In CVD process, the coating substance (e.g., $TiCl_4$) reacts not only with the reactive gas but also with the substrate (i.e., cemented carbide) so that a brittle decarburized layer of about 1 μm thick of Co_6W_6C (the η phase) tends to be formed between the coating (usually several micrometers thick) and the substrate. The brittle layer leads to instability of cutting performance which depends markedly on the toughness of the tool material. Sproul and Richman (1975) have proposed that the η carbide layer actually enhances the tool life by forming a diffusion barrier.

The CVD TiC coatings exhibit a low coefficient of friction against themselves and against ferrous materials and greatly reduce wear. The friction and wear characteristics of CVD TiC and other coatings measured on a pin-on-disk machine in dry or moist air are summarized in Table 14.17*a* (Hintermann, 1981). Material couples with low friction and low wear are summarized in Table 14.17*b*.

For use in the helium loops of nuclear reactors, the variation in the coefficient of friction for TiC against TiC with temperature in a quasi-inert atmosphere (helium) has been investigated. Figure 14.38 shows that the coefficient of friction of TiC against TiC in a pin-on-disk wear test remains remarkably low with increasing temperature up to 600°C. The increase of μ above 600°C is attributed due to softening of the substrate, i.e., cemented carbide. TiC coatings have also been tested for use in ball bearings for nuclear reactors. Ball bearings with TiC coatings either on the balls or races performed satisfactorily for 10^7 revolutions, as shown in Fig. 14.39.

In another application of navigation instrument bearings used for gyroscope motors, the angular-contact ball bearings were lubricated with oil and made with cemented carbide balls which were CVD coated with TiC and 440C stainless steel races. The outer and inner diameters of bearings tested were 19 and 6 mm, respectively, and the radial and axial loads were 2 and 10 N, respectively. The speed was 2400 r/min. Boving et al. (1983) reported that the bearing with coated carbide balls functioned smoothly for more than 20,000 h while stainless-steel lubricated ball bearings show wear tracks after 3000 h.

14.4.2.2 ***Tungsten Carbide Coatings.*** Since WC maintains its hardness at elevated temperatures and it is a major component in cemented carbides, it has also been studied in the form of thin coatings. In the tungsten carbide system, two phases exist: W_2C and WC. Both phases have a hexagonal structure. Tungsten carbide coatings have been applied to improve the life of extrusion dies, metrology parts, and bearings. These coatings have been applied primarily by thermal spray, sputtering, and CVD processes. The microhardness of WC/W_2C coatings range from 800 to 2100 HV. The properties such as microhardness and stoichiometry are strongly influenced by the deposition parameters.

Tungsten Carbide Applied by Thermal Spray. Plasma spray and detonation gun spray processes are used to apply thick WC coatings mixed with metallic binder such as CO, Ni, and Mo (Barnhart, 1968; Cashon, 1975; Eschnauer and Lugscheider, 1984; Ingham, 1967; Levy et al., 1968; Milewski, 1973; Mock, 1974; Most, 1969; Tucker, 1974; Vinayo et al., 1985). The microstructure of sprayed WC-CO coatings usually consists of a lamellar structure of thin lenticular particles or splats. The adhesion, hardness, and density of WC-Co coatings are strongly influenced by the microstructure of the coatings which in turn is dictated by the kinetic energy of the molten materials. The kinetic energy of molten material in case of detonation gun spray is 3 to 5 times higher than that of plasma spray (Tucker, 1974). Table 14.18 compares bond strength, density, hardness, and modulus of rupture of WC coatings with varying Co content for plasma and detonation gun sprayed coatings. Properties of alumina, alumina-titania, and chromium carbide are listed for comparisons.

The powder manufacturing process has a great influence on the chemical transformation experienced during atmospheric pressure spraying. Vinayo et al. (1985) studied three powder types of WC and W_2C: sintered and fused, agglom-

TABLE 14.17a Friction and Wear Results of CVD Hard Coatings at Various Relative Humidities*

Pin-disk	Coefficient of friction			Wear rate, $\times 10^{-6}$ mm^3 m^{-1}					
	RH† = 90%–95%	RH = 50%	RH = 0.5%–5%	RH = 90%–95%		RH = 50%		RH = 0.5%–5%	
				Pin	Disk	Pin	Disk	Pin	Disk
100Cr6-SiC		0.23		0.13	2.7			1.2	3.8
100Cr6-TiC		0.25				0.07	7.6		
100Cr6-TiN		0.49				95	0		
100Cr6-Fe_xB		0.76				58	0		
100Cr6-Cr_7C_3		0.79				1.1	76		
100Cr6-TiC‡		0.11				3.0	—		
TiC-TiN	0.16		0.18	2.5	<5			4.5	20
TiC-SiC	0.20		0.26	0.75	2			0.33	< 3
TiC-TiC	0.22		0.32	0	9			0.25	16
TiN-TiC	0.25		0.31	0	8			0	25
SiC-TiC	0.25		0.35	0	12			0	36
SiC-SiC	0.27		0.47	4.1	44			6	22
Al_2O_3 (polycrystalline)—100 Cr6‡		0.45				10	1500		
Al_2O_3 (polycrystalline)—TiC‡		0.19	0.37			0.1	10	0.7	20
Al_2O_3 (polycrystalline)—boronized cemented carbide‡		0.62				3.3	1.8		
Ruby-TiC‡ (for reference)		0.12				4.0	—		

*Pin-on-disk test with pin radius = 3 mm; velocity = 10 mm s^{-1}; load = 5 N; roughness = 0.1 μm arithmetic average; distance = 1 km; and temperature = 20°C.
†RH: relative humidity.
‡Measurement conditions same as for other tests except load = 20 N.
Source: Adapted from Hintermann (1981).

TABLE 14.17*b* Coating Couples Grouped According to Their Friction and Wear Behavior

Couples with lowest friction	Couples with lowest wear	Couples with lowest friction and wear
TiC-TiN	TiC-SiC	TiC-SiC
Al_2O_3-TiC	100Cr6-SiC	100Cr6-SiC
TiC-SiC	Al_2O_3-CC bor*	100Cr6-TiC
100Cr6-SiC	100Cr6-TiC	
100Cr6-TiC		

*Boronized cemented carbide.
Source: Adapted from Hintermann, 1981.

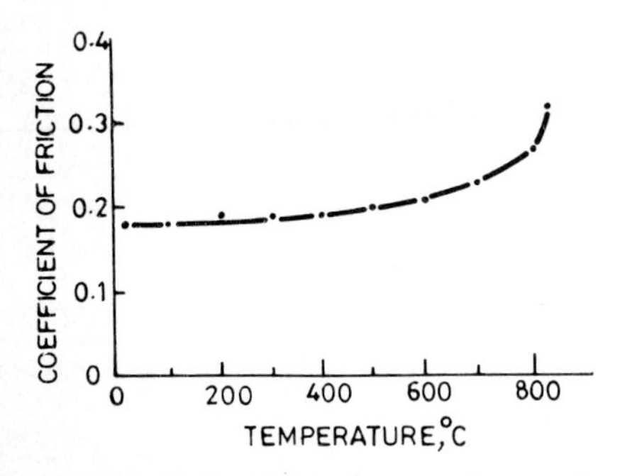

FIGURE 14.38 Effect of temperature on the coefficient of friction of CVD-coated TiC versus CVD-coated TiC in a pin-on-disk test at pressure = 90 MPa, sliding speed = 1 mm min^{-1}, and He atmosphere. [*Adapted from Hintermann (1981).*]

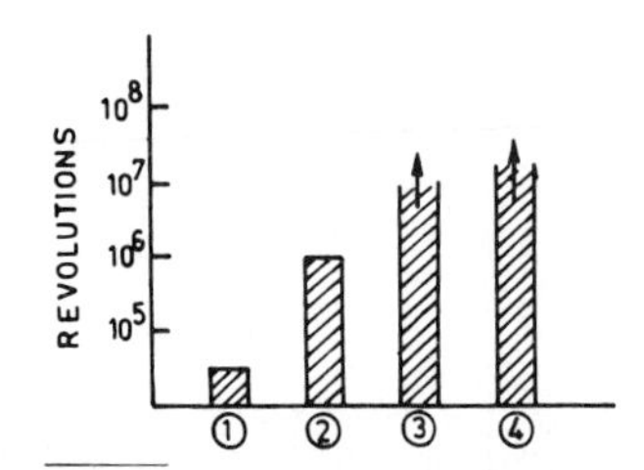

FIGURE 14.39 Lifetime of ball bearings (inner diameter of bearing = 25 mm) in helium at 100 r/min with a radial load of 500 N and ambient temperature of 300°C (nuclear reactor technology). (1) Standard steel ball-bearing-complete seizure. (2) Standard steel ball-bearing with MoS_2 spray-complete seizure. (3) Ball-bearing with CVD TiC-coated races-wear of balls but no seizure. (4) Ball-bearing with CVD TiC-coated balls-wear of races but no seizure. [*Adapted from Hintermann (1981).*]

erated, and composite. A sintered mixture of tungsten carbide-cobalt contained a small amount of complex Co-W-C phases. Agglomerated powders were a uniform distribution of WC and Co particles spherically agglomerated with an organic binder by a spray drying process. Composite (or coated) powders were prepared by applying an electrochemically deposited Co coating to the powder which was expected to protect the carbide core against decomposition and oxidation. Vinayo et al. (1985) found that composite and agglomerated particles experienced less decarburization during plasma spraying.

Tungsten Carbide Coatings Applied by Sputtering. Both coatings consisting of the monocarbide and of lower carbide W_2C have been grown (Archer and Yee, 1978; Srivastava et al., 1984). Also composite coatings of WC-Co have been produced (Eser et al., 1978a, 1978b). It has been claimed that the presence of Co is essential in WC coatings to reduce both friction and wear (Eser et al., 1978b; Taylor, 1977). WC + Co coatings on beryllium substrates were deposited by rf sputtering with selected values of dc bias by Eser et al. (1978a, 1978b). Compositional analysis showed variations in tungsten, cobalt, and carbon concentration as well as in Ar concentration in relation to the bias level. They found that the

TABLE 14.18 Properties of Selected D-Gun and Plasma Sprayed Coatings from Union Carbide Corp., Indianapolis, Indiana

Coating characteristics	LA-2	LA-6	LA-7	LW-1	LW-1N30	LW-11B	LC-1B
Composition (wt %)	>99 Al_2O_3	>99 Al_2O_3	$60Al_2O_3$ $40TiO_2$	91WC 9Co	87WC 13Co	88WC 12Co	$80Cr_3C_2$ 20NiCr (by volume)
Deposition method	D-gun	Plasma	D-gun	D-gun	D-gun	Plasma	D-gun
Vickers hardness (kg mm^{-2} 300 g load)	1100	825	950	1300	1150	750	700
Tensile bond strength, MPa*	70	—	62	175†	175†	70	125
Modulus of rupture, MPa	140	140	130	550	620	380	480
Modulus of elasticity, GPa	100	39	76	215	215	150	125
Density, kg m^{-3}	3400	3380	—	14,200	13,200	12,500	6,500
Metallographic porosity, volume %	2	3	1	0.5	0.5	2	0.5
Coeff. of thermal exp., $\times 10^{-6}$ °C^{-1}	6.8	6.8	—	8.1	8.1	8.1	11.5
Max. oper. temp. in oxidizing atmos., °C	1000	1000	700	540	540	540	1000

*Determined with ASTM method C633-69. Bond strength varies somewhat with the substrate composition, and the values shown are for either steel or aluminum.

†Bond strength of these coatings produced by the *detonation-gun process* is too high to be measured by ASTM method C633-69, which is used to measure bond strengths of plasma-applied coatings. Special laboratory methods for testing detonation coating bond strengths reveal values in excess of 175 MPa.

Source: Adapted from Tucker (1974) and Cashon (1975).

coating was amorphous at −112 V bias, and it became W_2C at bias levels of −250 V or higher. Eser et al. conducted friction and wear tests in a helium atmosphere. They found that the coefficient of friction decreased and the wear resistance increased with an increase in the substrate bias from −112 V up to a value of −300 V dc. These data suggest that crystalline structure of αW_2C is superior in tribological performance compared to the amorphous tungsten carbide.

Tungsten Carbide Coatings Applied by CVD. Uniform coatings of two tungsten carbides W_2C and W_3C have been applied by CVD up to a thickness of 50 μm, in the temperature range of 325 to 500°C with a hardness of 2000 to 2500 HV (Archer, 1975; Archer and Yee, 1978). The tungsten carbide coatings are used for high-precision steel components where there is a requirement for high surface hardness. Also, the low-temperature deposition makes it possible to coat a completely finished component, and the high coating thickness permits final finishing to the required size.

The formation of W_2C and W_3C phases by CVD are usually controlled by restricting the amount of carbon in the reaction mixture. The hardnesses of W_2C and W_3C are quite high (2000 to 2500 HV). These coatings have a greater degree of brittleness in addition to high hardness as compared to WC-Co coating or bulk. Generally, these coatings tend to fracture when subjected to repeatedly heavy loading, which is usually followed by flaking, because the coating-substrate bond fails at the same time. In a rolling four-ball wear test, the coating flaked rather rapidly (Archer and Yee, 1978). These coatings are most valuable in circumstances where hardness is more important than toughness.

An abrasive impact test was carried out by Archer and Yee (1978) to compare tungsten carbide with other materials—sintered tungsten carbide, detonation gun sprayed tungsten carbide, nitrided NK 7 steel, and hard chrome (Table 14.19*a*). In this test, cylinders of steel coated with a variety of hard materials were tumbled with abrasive particles. The CVD tungsten carbide resisted abrasion better than nitrided surfaces but was comparable with tungsten carbide applied with detonation gun process and poorer than others. In contrast to this, some abrasive wear tests, which involve countermoving surfaces with a ruby abrasive at low specific loads (Table 14.19*b*), show that W_2C has one of the lowest wear rates of all the materials tested. Archer and Yee (1978) also studied the adhesion of tungsten carbide with a variety of substrates. Very good adhesion was obtained with nickel and with nickel-plated materials.

CVD tungsten carbide coatings find use in applications which combine the requirements of high precision with high hardness in a low-load situation. These CVD tungsten carbide coatings have been used successfully on extrusion dies for aluminum. The life of the steel die can be considerably increased by the applica-

TABLE 14.19*a* Comparison of Abrasive Impact Wear of CVD W_2C Coating with Other Coatings and Treated Materials

Test material	Hardness (HV)	Mean depth of wear after 30 h, μm
Sintered tungsten carbide	1700	0.2
Hard chrome	840	0.5
Union carbide W1 tungsten carbide	1140	1.2
CVD W_2C	—	2
CVD TiC (Mutic process)	—	—
Nitrided NK 7 steel	1030	3.5

TABLE 14.19*b* Comparison of Abrasive Wear of CVD W_2C Coating with Other Coatings and Treated Materials

Test material	Coating/treatment	Hardness (HV)	Relative volumetric wear
	Chemical vapor deposition		
105WCr6	TiC (1050°C) 12 μm	3200	3.5
C100	W_2C (550°C) 30 μm	1900	4.8
100Cr6	Vanadized 30 μm	2400	7.5
C100	Vanadized 30 μm	2400	9
C100	Borided 100 μm	1600	12
100Cr6	Borided 100 μm	1600	14
	Electroplating		
Ck15EH	Hard chromium 50 μm	1050	23
CuZn40	Bright nickel 50 μm	580	97
	Anodizing		
AlMgSi	Ematel 20 μm	450	63
	Bath nitriding		
Ck 15	Tenifer 8–10 μm	640	73
	Spray coating		
NiCrBSi	Plasma, remelted	700–800	33
Mo	Flame sprayed	800	79
Cr_2O_3	Plasma	700–900	159
Al_2O_3	Plasma	850–950	127
WC/Co	Plasma	800–1000	106
Cr_3C_2/NiCr	Plasma	500	50
	Steel		
Ck 15 N	Annealed normally	180	165
100Cr6H	Hardened, tempered at 150°C	840	72

tion of tungsten carbide to the extrusion slot. The CVD process has the advantage that it is able to coat uniformly the very narrow slots which are sometimes involved in the production of thin-section aluminum extrusions.

14.4.2.3 Chromium Carbides. In the Cr-C system, three carbide phases exist: $Cr_{23}C_6$ which has a complex face-centered cubic structure, Cr_7C_3 which has a hexagonal phase, and Cr_3C_2 which is orthorhombic. Chromium carbide coatings are used in many industrial applications such as in resisting wear from hard surfaces (Cashon, 1975; Mock, 1974; Wedge and Eaves, 1980). Cr_3C_2 coatings are used for wear-resistance applications up to 800°C or higher where coatings such as WC and TiC tend to oxidize. The chromium carbide coatings have been applied by plasma spray, detonation gun spray, sputtering, and chemical vapor deposition processes. These coatings have been evaluated for use in helium- and sodium-cooled nuclear reactors (Lai, 1978, 1979; Reardon et al., 1981; Schwarz, 1980; Taylor, 1977; Wolfla and Johnson, 1975). Bhushan (1980c, 1981b) and

Bhushan et al. (1978) have used these coatings to reduce wear of journal of air lubricated foil bearings used in the gas turbine engine.

Chromium Carbide Coatings Applied by Thermal Spray. Plasma- and detonation-gun-sprayed chromium carbide-nichrome coatings have long been used for applications requiring superior wear resistance at temperatures up to 800°C or higher. Nichrome metallic binder is normally used to improve coating ductility and coating-to-substrate adhesion. These coatings are typically sprayed from mechanical blends of powder containing from 17 to 35 wt % nichrome (80 wt % Ni, 20 wt % Cr) (Tables 14.18 and 14.20). Usually, these coatings are susceptible to nonuniformity of microstructure because of segregation of the blended powders. Oxide formation such as Cr_2O_3 also occurs in both ambient atmosphere plasma and detonation-gun spray applied coating.

Reardon et al. (1981) developed a composite powder (size of −270 mesh + 5 μm) for application by the conventional plasma and vacuum plasma-spray processes. The developed material consisted of 50 wt % Cr_3C_2 clad with 50 wt % 80-20 Ni-Cr. Unlike powder blends, each Cr_3C_2 powder particle was clad with an essentially continuous layer of Ni-Cr. The coatings of the composite Cr_3C_2 powder were sprayed using the conventional arc plasma and the low-pressure, low-oxygen vacuum plasma-spray processes. These coatings were compared with coatings sprayed from a commercially available blend of 75 wt % Cr_3C_2-25 wt % Ni-Cr. The vacuum-plasma-sprayed composite coating combined superior wear resistance with low oxide (Table 14.20). Reardon et al. (1981) hypothesized that the nichrome cladding of the Cr_3C_2 particles places the metal and carbide in intimate contact, which in turn promotes solid solutioning of the carbide in the matrix.

The friction and wear characteristics of chromium carbide, i.e., Cr_3C_2 and $Cr_{23}C_6$ with Ni-Cr coatings sprayed by plasma and detonation gun spray, are shown in Fig. 14.40 and Table 14.21. Uncoated Hastelloy X sliding against itself is included for comparisons. The sliding wear tests were conducted on specimens in a button-on-plate arrangement (both mating surfaces, coated with identical coatings) at 815°C and 3.45 MPa in a helium environment which simulated HTGR (high-temperature gas-cooled reactor) (Li and Lai, 1979). We note that the coatings containing 75 or 80 wt % chromium carbide exhibited excellent wear resistance. As the chromium carbide content decreased from either 80 or 75 to 55 wt %, with a concurrent decrease in coating hardness, wear resistance deteriorated. The friction and wear behavior of the soft coating was similar to that of the bare

TABLE 14.20 Comparison of Cr_3C_2-Nichrome Coatings Applied by Plasma Spraying

	75 wt % Cr_3C_2-25 wt % nichrome blend (conventional plasma)	50 wt % Cr_3C_2-50 wt % nichrome composite (conventional plasma)	50 wt % Cr_3C_2-50 wt % nichrome composite (vacuum plasma)
Powder type	Blend	Composite	Composite
Microhardness HRC (HV)	50 (515)	49 (410)	60 (700)
Surface texture, μm aa*	4.5–6.3	5–6.3	5.3–6.5
Abrasive slurry wear resistance (X blend)	1	2.8	4.3

*aa: arithmetic average.
Source: Adapted from Reardon et al. (1981).

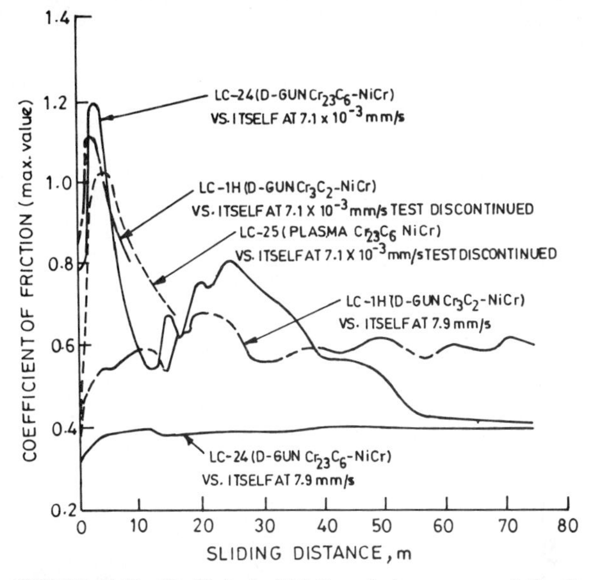

FIGURE 14.40 Coefficient of friction of plasma sprayed $Cr_{23}C_6$ and Cr_3C_2 coatings as a function of sliding distance at 815°C and 3.45 MPa in helium environment. [*Adapted from Li and Lai (1979).*]

metal, showing severe galling and a significant amount of wear debris. The friction characteristics of the hard coatings exhibited a strong velocity dependence with high friction coefficients in low sliding velocity tests and vice versa.

The performance of Cr_3C_2-NiCr sprayed coatings was evaluated by Bhushan (1980c, 1981b) and Gray et al. (1981) using a start-stop sliding test (Table 14.14). The Cr_3C_2 detonation-gun sprayed coating (on journal) shows adequate performance up to 650°C with Cr_2O_3 sputtered coating (on foil bearing). These coating combinations completed 9000 start-stop cycles (this represents useful life of an automotive gas turbine engine) at elevated temperatures.

The Cr_3C_2 coatings in some applications exhibit high wear under running-in conditions, which can be avoided by air spraying graphite (a soft lubricant) coating as an overlay (Bhushan, 1980c).

Chromium Carbide Coatings Applied by PVD. Chromium carbide coatings have been applied by physical vapor deposition (Cholvy and Derep, 1985; Komiya et al., 1977a; Spalvins, 1977). Komiya et al. (1977a) deposited ion-plated chromium carbide by evaporation of Cr using a hollow cathode discharge gun in the presence of acetylene. They obtained two-phase films consisting of Cr and Cr_7C_3 phases with hardness ranging from 1300 to 2100 kg mm^{-2}. Spalvins (1977) rf-sputtered Cr_3C_2 coatings on 440C stainless-steel substrates. He reported that sputtered Cr_3C_2 adhered strongly to the 440C stainless steel substrate. The coefficient of friction between a 440C stainless steel pin sliding against a coated 440C stainless steel disk increased linearly with an increase of the normal load. A coefficient of friction of less than 0.1 was achieved at lower loads (0.5 N).

Chromium Carbide Coatings Applied by CVD. CVD Cr_7C_3 coatings are applied for wear resistance to free wheels (e.g., bicycle free wheel) made of 52100

TABLE 14.21 Hardness and Wear Characteristics of Chromium Carbide-Nichrome Coatings Applied by Plasma and Detonation Gun Spray Techniques*

Type of coatings	Sliding velocity, $mm\ s^{-1}$	Sliding distance, m	Micro-hardness (HV)	Surface roughness, μm aa		Weight change, g	Max. wear depth, μm	Wear debris
				Pretest	Posttest			
LC-24† D-gun: $Cr_{23}C_6$-NiCr (80-20 wt %)	7.1×10^{-3}	75	680	2.44–4.06	0.91–2.74	+0.01151	15.0	No
LC-24 D-gun: $Cr_{23}C_6$-NiCr	7.9	75	680	3.05–4.06	1.32–2.03	+0.00018	10.0	No
LC-1H D-gun: Cr_3C_2-NiCr (80-20 wt %)	7.1×10^{-3}	8.3	860	2.39–4.57	1.02–2.03	+0.00370	20.0	No
LC-1H D-gun: Cr_3C_2-NiCr	7.9	75	860	3.25–4.06	1.93–2.24	+0.00040	5.0	No
T7A Plasma: $Cr_{23}C_6$-NiCr (55-45 wt %)	7.1×10^{-3}	26.4	390	2.03–5.08	5.08–10.16	−0.14678	216	Yes
T7A Plasma $Cr_{23}C_6$-NiCr	7.9	26.4	390	2.03–4.06	4.10–15.49	0.04430	203	Yes
Hastelloy X (0.06% C, 19.0% Fe, 22.3% Cr, 2.1% Co, 8.5% Mo, 0.61% Mn, 0.25% Si, 0.62% W, bal. Ni) uncoated	7.1×10^{-3}	23.7	520	0.15–0.20 (as-machined)	1.93–4.78	−0.09072	132	Yes

*The wear tests were conducted at 815°C and 3.45 MPa in a helium environment.
†Coating designations for Union Carbide Corp., New York.
Source: Adapted from Li and Lai (1979).

steel. Hintermann (1980) has reported an increase in lifetime by a factor of 20 by application of CVD Cr_3C_2 coatings.

14.4.2.4 Hafnium Carbide and Zirconium Carbide Coatings. HfC and ZrC coatings have been studied to a much lesser extent compared to TiC. These materials have been deposited by ARE and CVD processes.

HfC and ZrC Coatings Applied by ARE. The ARE coatings were produced by Raghuram and Bunshah (1972). They have shown that both HfC and ZrC could be obtained with hardness close to bulk values (1600 to 2800 kg mm^{-2}). In deposition of HfC by evaporation of hafnium in the presence of acetylene, the hardness increased as the ratio between the acetylene pressure and the deposition rate increased, thus as the amount of C in the coating increased. For HfC, a value as high as 3850 kg mm^{-2} was reported for coating deposited at the highest acetylene pressure. For ZrC coatings produced by a direct evaporation of ZrC billets, Bunshah et al. (1977, 1978) reported hardnesses which were lower than bulk values and depending on growth parameters, varied between 1000 and 2100 kg mm^{-2}.

HfC and ZrC Coatings Applied by CVD. HfC and ZrC coatings have been produced by CVD (Caputo, 1977; Ikawa, 1972; Kramer and Suh, 1980; Ogawa et al., 1979; Stolz et al., 1983). HfC coatings are produced by the reduction of hafnium halide ($HfCl_4$) in the presence of methane in the temperature range of 900 to 1100°C. HfC coatings have been deposited onto cemented carbide tools, and these coated tools along with tools coated with TiC, TiN, HfN, and Al_2O_3 have found some acceptance in the metal-cutting industry (Kramer and Suh, 1980).

Ikawa (1972) produced ZrC coatings by the reaction between methyl iodide vapor and zirconium sponge to form a methyl radical and zirconium iodides. The methyl radical serves as the carbon source. Caputo (1977) produced ZrC coatings by the reaction of zirconium tetrachloride and methane in the temperature range of 900 to 1400°C. Caputo (1977) found that ZrC deposited at 1050°C contained impurities up to 4% O and 1.5% Cl. As the deposition temperature increased, the impurity level decreased and the thermal properties of the coatings became more stable. ZrC coatings have been developed for cutting tools. These coatings are used on molybdenum grids for power tubes as a diffusion barrier (Stolz et al., 1983).

14.4.2.5 Silicon Carbide Coatings. Silicon carbide (SiC) is another promising coating because of some outstanding properties such as high hardness, good oxidation resistance at high temperatures, and inertness in contact with acids or other corrosive media. SiC coatings exist both in the cubic form (β-SiC) and a hexagonal form (α-SiC). SiC coatings are generally applied by ion plating and CVD. These coatings greatly reduce erosion of first wall components of fusion reactors.

SiC Coatings Applied by Ion Plating. SiC coatings up to 12 to 15 μm in thickness are applied by rf reactive ion plating in which Si is evaporated in C_2H_2 gas atmosphere (Fukutomi et al., 1984). The stoichiometry of SiC is controlled by varying the substrate temperature (~800°C) and evaporation rate of Si. The SiC coatings having excess carbon exhibit better thermal stability.

SiC Coatings Applied by CVD. SiC coatings up to a thickness of about 100 μm are applied on tool steel, hard metal (WC-Co cemented carbide), graphite, sapphire, and silicon nitride by CVD at 800 to 1200°C (Brutsch, 1985; Chin et al., 1977; Hintermann, 1980, 1981, 1984; Schlichting, 1980; Trester et al., 1983). Both tool steel and hard metal need an intermediate layer of a suitable hard material which acts as a diffusion barrier and avoids interaction between SiC and the substrate. The hard metals such as cemented carbide substrate suits very well to SiC coatings because of better matching of the thermal expansion coefficient of the two materials (5 to 6 × 10^{-6} °C^{-1}).

The carbon content of SiC coating mainly depends upon the C/Si ratio in gas phase, which can be controlled by selecting the appropriate silane or by adding a hydrocarbon to the reaction mixture. The carbon content in SiC coatings and the deposition conditions (temperature and pressure) strongly influence the microhardness. In one example, SiC coatings were prepared by a CVD process in a resistance-heated reactor in the temperature range of 800 to 1200°C at atmospheric pressure. The gas phase contained a volatile alkyl or arylchlorosilane and hydrogen as the reactants with nitrogen as the carrier gas. A hydrocarbon was added to increase the carbon contents of the gas phase. Maximum hardness was obtained for coatings having 30% C which corresponds to the stoichiometric composition and deposited at 800°C (Figure 14.41). The measured maximum microhardness is 6000 HV. Friction and wear behavior of CVD SiC coatings were included in Table 14.17. CVD SiC has a brilliant black appearance and can be used as a decorative scratch-proof coating for watch cases.

14.4.2.6 Other Carbide Coatings. Carbide coatings not discussed so far that have been studied for wear applications are TaC, VC, NbC, Mo_2C, and B_4C. These coatings have been produced primarily by ARE, rf sputtering, and CVD. Coatings of TaC have been produced by ARE and CVD (Caputo, 1977; Grossklaus and Bunshah, 1975; Takahashi and Sugiyama, 1974). Grossklaus and Bunshah (1975) have shown that both tantalum carbides Ta_2C and TaC can be grown using ARE. VC coatings have been produced onto steel substrates using a CVD process such as pack cementation by Horvath and Perry (1980). They have shown that dense equiaxed structures can be formed with hardnesses of 2600 kg

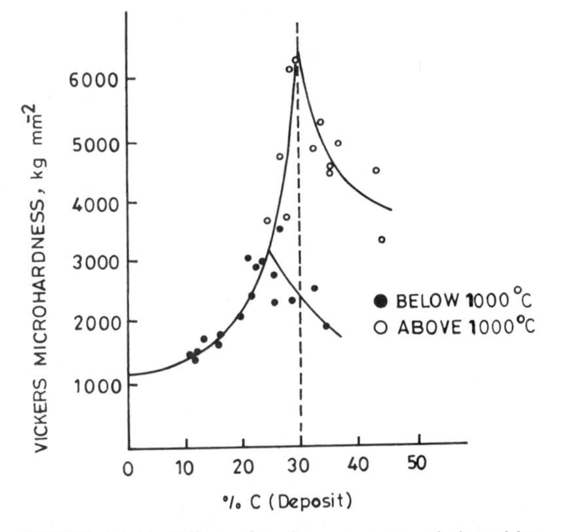

FIGURE 14.41 Effect of carbon content and deposition temperature on microhardness (HV) of SiC coatings applied by CVD on WC-Co hard cermet. [*Adapted from Brütsch (1975).*]

mm^{-2}. If the steel substrates contained more than about 1.4 wt % C, a cementite interlayer was formed. NbC coatings have also been produced because of their high superconducting transition temperatures.

Boron carbide coatings have been produced by CVD (Jansson and Carlsson, 1985). High substrate temperature in the range of 1000 to 1300°C are used. Several bulk phases exist depending on the purity, but under high-purity conditions, the rhombohedral $B_{13}C_2$ is most commonly reported. Brainard and Wheeler (1978) rf-sputtered B_4C and Mo_2C and compared their friction and wear performance with other carbides, borides, and silicides in a pin-on-disk tester (Fig. 14.34). They found that the wear of a 304 stainless-steel pin sliding against a B_4C and Mo_2C coated disk was relatively low, lower than that of Mo_2B_5, $TiSi_2$, and $MoSi_2$, but slightly higher than that of TiC and TiB_2. Mo_2C exhibited a coefficient of friction of about 0.4 while B_4C exhibited very high friction (~1).

14.4.3 Nitride Coatings

The extreme hardness of transition metal nitrides such as cubic boron nitride (CBN), Si_3N_4, TiN, ZrN, and HfN make them most popular wear-resistant coatings among nitrides. TiN is the most studied coating among nitrides. Because the hardness of HfN at elevated temperatures is much higher compared to that for TiN or TiC, this material is an interesting candidate for many wear-resistant applications such as cutting tools designed to operate at very high speeds and thus high working temperatures. CBN is, after diamond, the hardest material known. Si_3N_4 has good mechanical properties, e.g., a high hardness and good chemical resistance which makes it an interesting candidate for wear application. Si_3N_4 has also found applications as an electrical insulator in microelectronics due to its good dielectric properties. These materials are useful in increasing the wear life of steel parts such as cutting tools, punches, and bearing surfaces (Buhl et al., 1981; Jamal et al., 1980; Hatto, 1985; Hintermann, 1981; Knotek et al., 1987; Matthews, 1985; Matthews and Teer, 1980; Sundgren et al., 1983a, 1983b, 1983c).

The high-temperature strength along with good corrosion resistance also qualify these materials as high-temperature structural materials. The nitrides also have several interesting electronic and optical properties which make them suitable for integrated circuitry (Von Seefeld et al., 1980) and solar conversion devices (Karlsson et al., 1982) and superconducting devices (Spitz et al., 1974). The coatings of binary, ternary, and quaternary nitrides of Ti, Zr, and Al have been produced to tailor properties such as hardness, oxidation stability, and color according to specific needs.

14.4.3.1 Titanium Nitride Coatings. In the hard coatings category, which ranges from hard carbon coatings to transition metal carbides, nitrides, borides, and silicides, titanium nitride coatings are at present one of the most studied and used. TiN coatings are applied by a variety of PVD and CVD processes such as activated reactive evaporation, ion plating using HCD (hollow cathode discharge), cathodic arc ion plating and sputter ion-plating, rf/dc reactive-nonreactive sputtering and magnetron sputtering, and CVD. The different processes give rise to a large scatter in the microstructures and hence also in the properties of the coatings. Attempts have been made to produce titanium carbonitrides, titanium oxycarbonitrides, and titanium borocarbonitrides for maximum strength and hardness.

TiN is chemically extremely stable, is highly resistant to corrosion by strong acids, and exhibits low friction and wear. Therefore, TiN coatings are commonly used in cutting tools and various bearing and other sliding applications. TiN coating applied by CVD gained its early acceptance on carbide cutting tools because of high adhesion and wear resistance of the CVD coatings. High temperatures during CVD process promote diffusion and result in a graded interface partially responsible for excellent adhesion. The high temperature of the CVD process (700 to 1000°C) did not damage the carbide material, but CVD was not very successful on high-speed steels (HSS) because in this case it is necessary to reharden the tools after heat treatment. This is often not acceptable because heat treatment causes a certain amount of distortion. For example, twist drills and precision form cutters have not yet been coated successfully by CVD. Low-temperature PVD processes, namely, ion plating and sputtering, are commercially successful for applying TiN to HSS tooling (Buhl et al., 1981). The plasma-enhanced CVD can also be used to deposit coatings at much lower substrate temperatures (about 300 to 500°C) than that used in CVD.

Sputtered TiN coatings are also commonly used as a decorative coating material for watch cases, watch bands, eye glass frames etc., because of their golden color. Although the exact matching of hue and tone of TiN coatings with that of gold is difficult.

TiN Coatings Applied by ARE. Various phases of TiN have been deposited by the ARE process through the variation of Ti evaporation rate, nitrogen partial pressure, and substrate temperature (Jacobson et al., 1979, 1984). The hardnesses of TiN coatings having combinations of various phases are summarized in Table 14.22. We note that as the ratio of the evaporation rate of titanium to the partial pressure of N_2 decreases, the deposit changes from Ti to Ti_2N to TiN including two phase mixtures. The hardness of the deposit is influenced by the presence of the Ti_2N phase which produces the highest hardness levels. Jacobson et al. (1979) reported an increase in hardness with increasing substrate temperature. In these coatings, as substrate temperature increases, grain size increases significantly and the grain morphology changes from a faceted to a smooth topography. Coatings with small grains and weak grain boundaries with voids located along with them are produced at lower substrate temperatures which are responsible for lower hardness.

TiN Coatings Applied by Ion Plating. TiN coatings have been applied by ion-plating processes under reactive modes (Aoki et al., 1985; Buhl et al., 1981; Gabriel and Kloos, 1984; Gabriel et al., 1983; Hummer and Perry, 1983; Johansen et al., 1987; Kloos et al., 1982; Kobayashi and Doi, 1978; Komiya et al., 1979; Martin et al., 1987a, 1987b; Matthews and Teer, 1980; Matthews, 1985; Matthews and Lefkow, 1985; Zega et al., 1977). In the reactive ion plating, Ti is ion plated in the presence of N_2 atmosphere. The addition of carbon-containing gas, such as C_2H_2 into nitrogen, produces TiC_xN_y coatings with varying ratios of C and N. The properties such as hardness, density and adhesion and the stoichiometry of the TiN coatings are strongly influenced by the process parameters (partial pressure of the reactive gases, substrate temperature, degree of ionization of evaporant molecules) and substrate condition, i.e., smoothness and cleanliness prior to deposition. For example, in order to form a dense and a hard coating, it is necessary to enhance the ionization and increase the substrate current density. The microhardness of TiN coating applied by various ion-plating processes is precisely dictated by the phase of TiN coating and range from 2000 to 2600 kg mm^{-2}.

The friction and wear characteristics of TiN coatings applied by triode ion plating are shown in Fig. 14.42. These tests were carried out on a pin-and-disk machine, under dry conditions for various combinations of coated and uncoated

TABLE 14.22 Influence of Deposition Parameters on Microhardness of TiN Applied by ARE Process

Substrate	Evaporation rate,* g min^{-1}	Reactive gas pressure		Deposition temperature, °C	Deposition rate, μm min^{-1}	Thickness, μm	Hardness, HK (50 g)	Composition by x-ray diffraction
		mPa	(mtorr)					
Stainless steel	0.88	20	(0.15)	550	3.3	39	1155	$Ti + Ti_2N$
Stainless steel	0.43	23	(0.17)	550	1.0	15	2600	$Ti_2N + TiN$
Stainless steel	0.25	20	(0.15)	550	0.8	13	1245	TiN
Stainless steel	0.28	53	(0.40)	550	0.8	12	625	TiN
Tantalum	0.32	35	(0.26)	735	0.9	11	2800	$Ti_2N + TiN$
Tantalum	0.45	35	(0.26)	1000	0.9	19	2550	$TiN + Ti_2N$

*Evaporation rate of titanium.
Source: Adapted from Jacobsson et al. (1979).

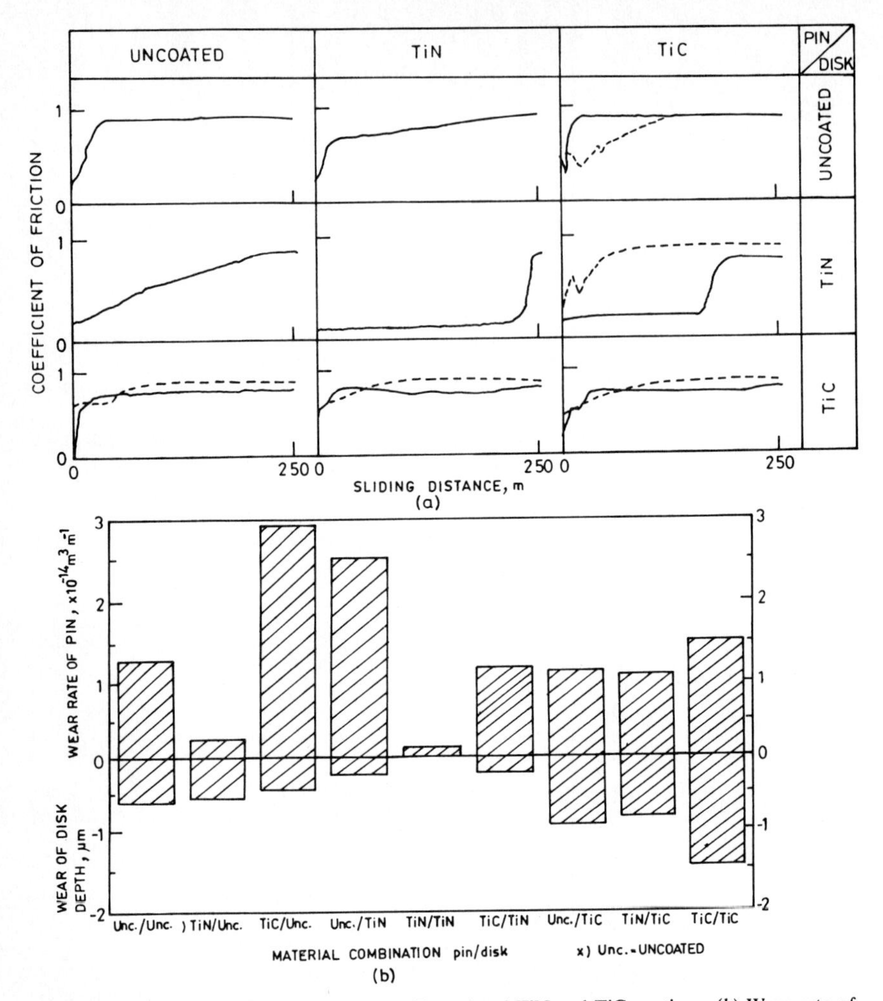

FIGURE 14.42 (*a*) Coefficient of friction of ion-plated TiN and TiC coatings. (*b*) Wear rate of ion-plated TiC and TiN coatings for various combinations of coated and uncoated sliding members in a pin-on-disk test. The upper part of Fig. 14.42(*b*) represents the wear of a pin ($\times\ 10^{-14}$ m^3 m^{-1}), and the lower part of the figure represents the wear of the disk which is expressed as the depth of track in micrometers. [*Adapted from Ulla and Juhani (1984).*]

pins and disks. The coefficient of friction (0.1) and wear rate are least when both the mating surfaces, i.e., pin and disk, are coated with TiN coatings. A comparison on the performance of TiC and TiN coatings on cemented carbide tools was presented in the TiC section.

TiN Coatings Applied by Sputtering. TiN coatings are applied by dc, rf, and magnetron sputtering and ion-beam sputtering under reactive and nonreactive modes by several authors (Bair et al., 1980; Bhushan, 1980d; Bucher et al., 1984; Clarke, 1977; Duckworth, 1981; Hibbs et al., 1983; Hochman et al., 1985; Je et al., 1986; Martin et al., 1982; Mumtaz and Class, 1982; Münz and Hessberger,

1981; Münz et al., 1982; Perry et al., 1987; Ramalingam and Winer, 1980; Poitevin et al., 1982; Rickerby et al., 1987; Sato et al., 1981; Sproul, 1985a, 1985b; Sproul and Rothstein, 1985; Sundgren, 1985; Sundgren et al., 1983a, 1983b, 1983c; Torok et al., 1987; Yamashina et al., 1983). There have been some attempts to produce titanium oxynitride (TiO_xN_y) by sputtering (Gabriel and Kloos, 1984), titanium oxycarbonitrides ($TiO_xC_yN_z$) by sputtering (Fark et al., 1983), and titanium borocarbonitrides ($TiB_xC_yN_z$) by sputtering for increased strength and hardness. Of all the PVD techniques, reactive sputtering is most commonly used for deposition of TiN coatings. In reactive sputtering of TiN deposition, titanium is generally sputtered in mixed Ar-N_2 discharges on the heated substrate (~500°C) (Sundgren et al., 1983a). The morphology, composition, and subsequently adhesion, hardness, and other properties are strongly influenced by process parameters. Münz and Hessberger (1981) measured an increase in hardness from approximately 1300 to 3500 kg mm^{-2} for magnetron-sputtered TiN as the substrate temperature increased from 100 to 600°C. Similar results have been reported by Hibbs et al. (1984) (Fig. 14.25). An increase in grain size and increased strength of grain boundaries are believed to be responsible for increased hardness. A maximum hardness of 4400 kg mm^{-2} was obtained for coatings with a composition of approximately $TiC_{0.34}O_{0.32}N_{0.24}$ (Fark et al., 1983).

The influence of composition ratio (N/Ti) on the microhardness and color tone of sputtered TiN obtained by several authors are compared in Figure 14.43 and Table 14.23. Table 14.23 includes hardness values of the coatings deposited by other physical vapor deposition processes. The large scatter in hardness values shown in Table 14.23 is attributed to different growth patterns and grain sizes that occur during different deposition conditions (Sundgren, 1985). The composition of the coating has significant influence on its hardness (Fig. 14.43). For understoichiometric coating (N/Ti <1), two trends can be distinguished: either the hardness increases up to a value of 3500 ± 500 kg mm^{-2} or up to a value of 1500 ± 500 kg mm^{-2}. For over stoichiometric coatings (N/Ti > 1), all results show decreasing hardness values as the ratio of N/Ti increases and values as low as 340 kg mm^{-2} are reported. The maximum hardness is achieved for a nearly stoichiometric coating with a composition ratio of N/Ti approximately equal to 1.0 (also see Fig. 14.26). The coatings with the highest hardnesses are grown at a rate considerably higher than those for other coatings with lower hardnesses.

Sundgren (1985) reported that the coatings with low hardness often show a columnar fibrous structure and they contain grain boundary voids. The generation of grain boundary voids, and thus weak grain boundaries, promotes low hardnesses. The high hardnesses of under stoichiometric coatings are manifested by the intrinsic compressive stresses in the coatings. Typical TEM micrographs of TiN coatings with a low hardness are shown in Fig. 14.44, and for comparisons there are also micrographs from stoichiometric TiN coatings with a hardness only slightly higher than the bulk values. TiN coatings produced by PVD methods are usually in a compressive stress state, and stresses as high as 10 GPa have been measured. These stresses are often verified by a decrease in the lattice parameter (Sundgren, 1985). Many mechanisms for the origin of residual stresses are proposed including high density of grain boundaries and defects such as dislocations, impurities located in the grain boundaries, incorporation of sputtering gas, and ion bombardment during growth (atomic peening) (Sundgren, 1985; Thornton and Hoffman, 1981).

The variation in coating density with composition is shown in Fig. 14.27. For Ti-N at low nitrogen concentrations, the density is slightly lower than that of bulk Ti. As the N concentration increases, the density decreases and reaches a minimum at 25 at % N. Above this concentration, the density increases again and

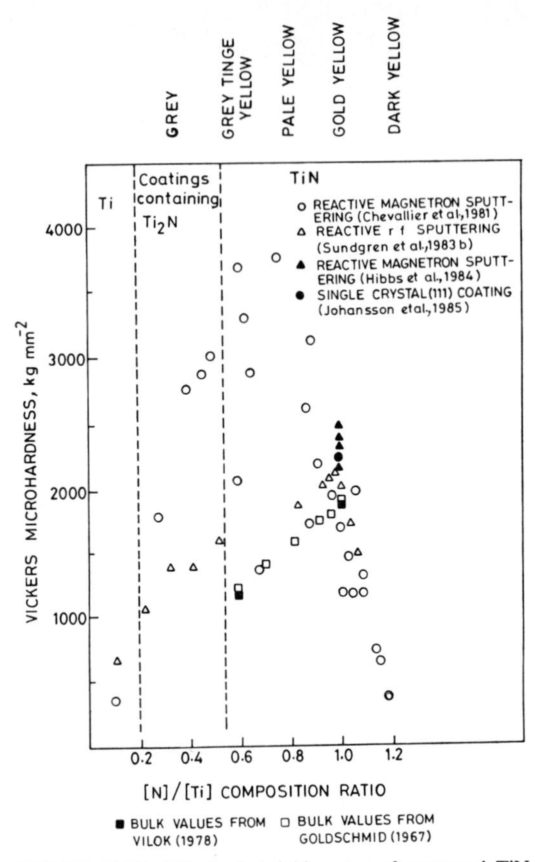

FIGURE 14.43 Effect of stoichiometry of sputtered TiN coatings on hardness and color. [*Adapted from Sundgren (1985).*]

approaches a maximum and equal bulk δTiN for the approximately stoichiometric coating (50 at % N). Above 50 at % N, a rapid decrease in density is noted. Sundgren et al. (1983b) reported that the grain size also decreases as the composition approaches stoichiometry. The grain size for the stoichiometry coatings ($TiN_{1.0}$) was about 35 nm; it is a factor of about 2 and 4 larger than that in the $TiC_{1.0}$ and $WC_{1.0}$ coatings, respectively.

The most successful application area for TiN is that of coatings on various carbide and high-speed steel tools. Sputtered TiN is especially attractive for high-speed steel (HSS) tools where CVD coatings cannot be easily used because of high substrate temperatures required during deposition. Sproul and Rothstein (1985) reported the data on accelerated cutting tests. They coated the M-7 HSS drills (65 HRC) using a magnetron-sputtering machine with Ti sputtered in the mixed Ar-N_2 mixture to a thickness of about 1 to 2 μm (hardness—2500 HV). They used coated and uncoated drills with or without a soluble oil cutting fluid to drill holes in 4130 steel. The results are reported in Table 14.24. TiN coated drills had

TABLE 14.23 Hardness of Ti-N Coatings Applied by Physical Vapor Deposition

Vickers hardness, kg mm^{-2}				
Ti + Ti_2N	Ti + Ti_2N + TiN or Ti + TiN	Ti_2N + TiN	TiN	References
1160–1270		2550–2800	465, 750	Jacobsson et al. (1979), Suri et al. (1980)
		1300–2800	340–1900	Nakamura et al. (1977)
1000		2000–2300	1700–2200	Sato et al. (1978)
600–700		2000–2800	1200–2000	Yoshihara and Mori (1979)
1100–1400		1700	1700–2200	Sundgren et al. (1983a)
	2800–3100		400–4000	Chevallier et al. (1981)
		2500	1800	Matthews and Teer (1980)
			1300–3500	Munz and Hessberger (1981)
			2400	Sproul (1985a)
			2500	Buhl et al. (1981)
			<2000	Zega et al. (1977)
			700–3100	Fleischer et al. (1979)
1400–2800		1800–2500	1400–1800	Matthews and Sundquist (1983)
		2400–3150	2100–2800	Gabriel (1983)
			1800–2600	Hibbs et al. (1984)
			2300	Johansson et al. (1985)

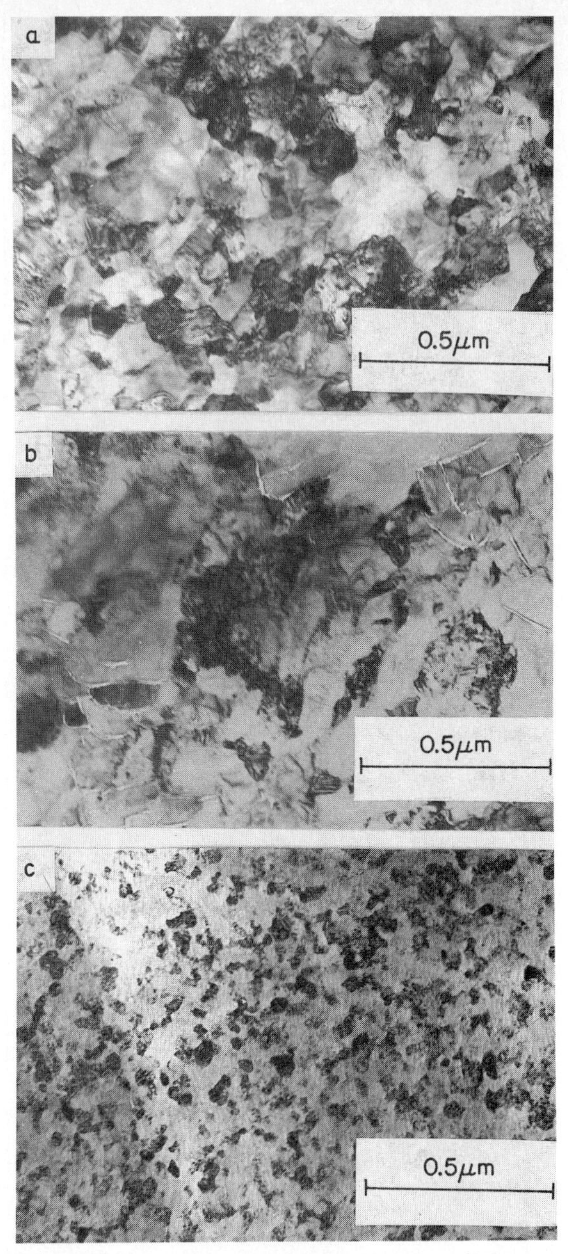

FIGURE 14.44 Transmission electron micrographs of TiN coatings produced by reactive sputtering. (*a*) Stoichiometric coating grown at a substrate temperature of 550°C on a stainless-steel substrate, hardness = 2400 kg mm^{-2}. (*b*) Stoichiometric coating grown at a substrate temperature of 650°C on a high-speed steel substrate, hardness = 1800 kg mm^{-2}. (*c*) Overstoichiometric coating (N/Ti ~ 1.1) produced at substrate temperature of 550°C on a low alloyed steel substrate, hardness = 1450 kg mm^{-2}. [*Adapted from Sundgren and Hentzell (1986).*]

TABLE 14.24 Tool Life of Uncoated and TiN-Sputter Coated M-7 HSS Drills

Conditions	Drill	Number of holes cut*	
		Average	Standard deviation
Dry	Uncoated	3.1	2.8
	TiN coated	150.0	82.8
Wet (soluble oil coolant)	Uncoated	6.0	0
	TiN coated	300.0	177.9

*Cutting conditions: 22.9 m min^{-1}; 0.21 mm rev^{-1}. Blind holes: 1.25 cm deep, 6.35 mm in diameter. Workpiece: 4130 steel (235 HB). Drill diameter: 6.35 mm.

Source: Adapted from Sproul and Rothstein (1985).

an improvement of 50 times over the uncoated drills both in the dry and lubricated conditions. Ramalingam and Winer (1980) conducted cutting tests on coated (reactive magnetron sputtered to 5 μm thickness) and uncoated cemented carbide (94 WC-6 Co) tools. These tools were used to machine a bar of AISI 1045 steel using a lathe. Flank and crater wear of the tools as a function of time are shown in Fig. 14.45. We note that coated tools experience significantly lower wear. These metal-cutting tools demonstrate that adherent thin films of TiN can be magnetron sputtered to withstand effectively severe sliding at elevated temperatures.

Bhushan (1980d) tested TiN coatings with a sputtered MoS_2 overlay for ball bearing operation in dry conditions. He ran sliding tests in a point-contact situation to simulate slip in a bearing in which most wear occurs. Results were presented in Table 14.15. The results showed that TiN + MoS_2 wore very little after low-speed sliding tests (15 to 127 mm/s) for a total of 2 min in the point-contact situation (a contact stress of about 2070 MPa). However, Cr_2O_3 + MoS_2 coating performed slightly better than TiN + MoS_2. Bair et al. (1980) and Ramalingam and Winer (1981) reported sliding wear data on coated and uncoated specimens both at low stress and very high hertzian stresses. They reported an improvement in wear volume by one to two orders of magnitude.

Hochman et al. (1985) and Ramalingam and Winer (1981) tested TiN coatings in rolling-contact-fatigue (RCF) tests. Hochman et al. deposited 0.2- and 0.8-μm-thick coatings by reactive sputtering in a dc magnetron-sputtering system on 440C stainless steel and AMS 5749 bearing steels. They found that the thin TiN coatings improved the rolling-contact-fatigue life by a factor of more than 10 at

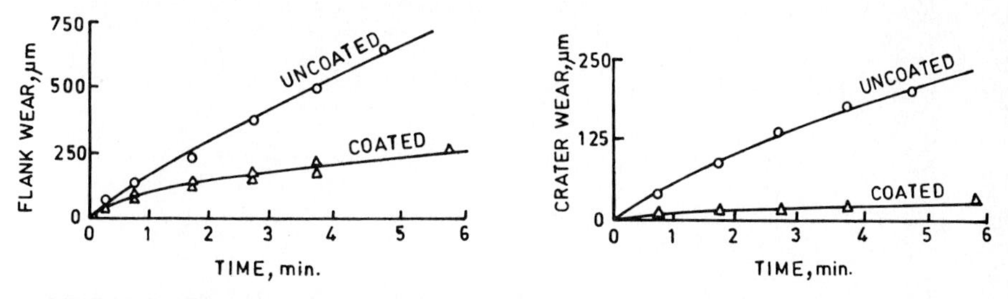

FIGURE 14.45 Wear of uncoated and rake face-coated cemented carbide (94% WC-6% Co) cutting tool insert when machining a 1045 steel bar at 117 m min^{-1} with a depth of cut of 1.5 mm and a feed of 0.264 mm r^{-1}. The reactive magnetron-sputtered TiN coating thickness was about 5 μm. [*Adapted from Ramalingam and Winer (1980).*]

the two stresses: 4 and 5.4 GPa tested. A parameter important in the RCF life is the maximum shear stress which is located at depths ranging from 50 to 150 μm for the two stresses tested, a lot deeper than the coating thickness. However, contrary to general belief, thin coating did improve RCF life. This suggests that RCF fracture in a composite (coated substrate) may be initiated at the surface and these coatings arrest or at least retard the initiation of fatigue microcracks.

TiN Coatings Applied by CVD. TiN is a commonly used CVD coating for cemented carbide tools and can also be applied on many types of steels (Blocher, 1974; Hintermann, 1981, 1983, 1984; Itoh et al., 1986; Laugier, 1983; Peterson, 1974; Stolz et al., 1983). There has been some work to produce titanium carbonitride Ti(CN) and TiC/TiN double layers (Pierson, 1977; Kim and Chun, 1983). TiN coatings are produced from the reactive gas mixtures of $TiCl_4$ (vapors from $TiCl_4$ liquid at 25 to 30°C), H_2 and N_2 at 600 to 800°C and Ti(CN) coatings are produced from the reactive gas mixtures of $TiCl_4$, N_2, H_2, and CH_4 at 700 to 900°C. However, if TiN is produced onto a hard metal substrate at temperatures on the order of 1000°C, an interfacial zone consisting of carbonitride is generally formed since carbon diffuses out of the substrate. If necessary, it can be avoided by increasing the pressure at which the coatings are produced.

Peterson (1974) found that low $TiCl_4$ partial pressure combines to produce columnar grains >10 μm in length and faster coating rates. Higher $TiCl_4$ partial pressures in the range of 0.05 to 0.10 atm produced randomly oriented grains of approximately 1 μm or less and slower cooling rates. $TiCl_4$ pressures greater than 0.30 atm at 1000°C nearly stopped the deposition of TiN. Figure 14.46 shows SEM micrographs of fracture surfaces of TiN coatings produced at 6.5×10^3 and 10^5 Pa (50 and 760 torr) (Sundgren and Hentzell, 1986).

The friction and wear characteristics of TiN coatings against themselves and other hard coatings in a ball-on-a-disk sliding test under dry and humid atmospheres are shown in Fig. 14.47. Figure 14.47(*b*) and (*c*) show the wear rate of ball and disk, respectively. The least coefficient of friction and wear was obtained in the case of TiN-coated pin (ball) versus TiN-coated disk. The wear data of

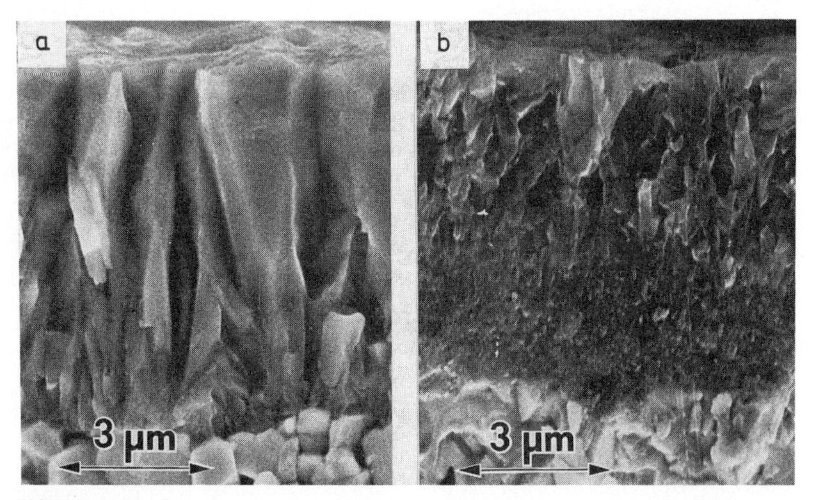

FIGURE 14.46 Scanning electron micrographs of TiN coatings produced by CVD on cemented carbide substrate at: (*a*) Pressure = 10^5 Pa (760 torr). (*b*) Pressure = 6.6×10^3 Pa (50 torr). [*Adapted from Sundgren and Hentzell (1986).*]

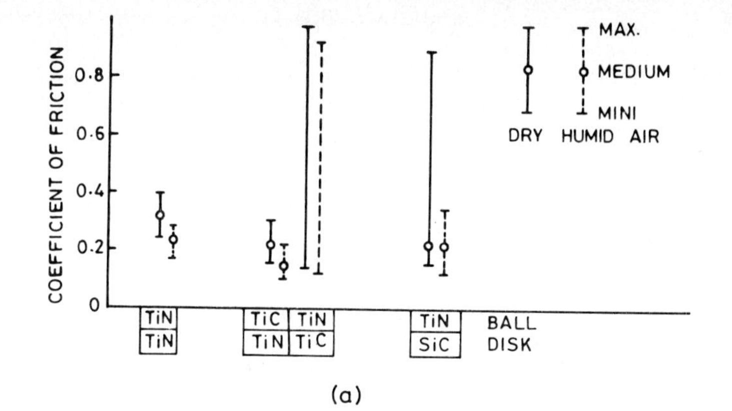

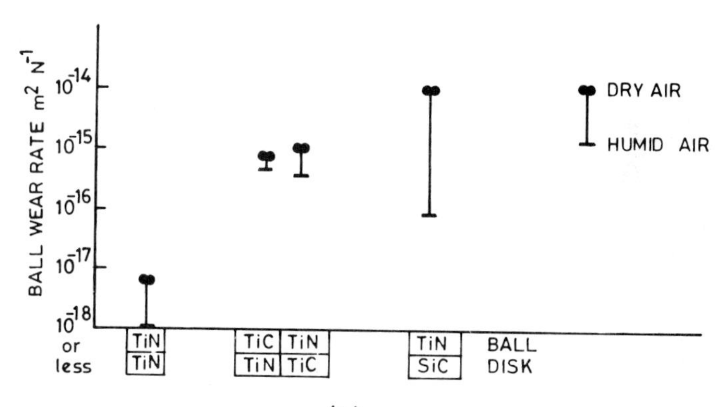

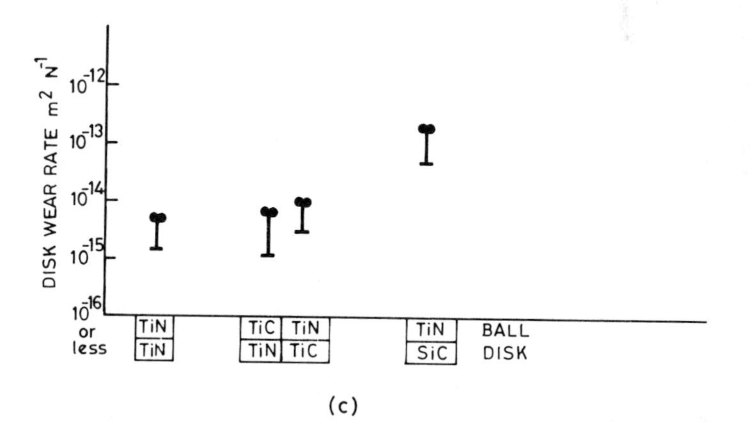

FIGURE 14.47 (*a*) Coefficient of friction of CVD TiN-coated ball versus disk under dry and humid conditions for various combinations. (*b*) Ball wear rate of various coating combinations. (*c*) Disk wear rate of various coating combinations. [*Adapted from Hintermann (1983).*]

CVD TiN coated and uncoated cemented carbide tools were presented earlier in the TiC section. TiN coatings exhibit good corrosion resistance in addition to good wear resistance (Hintermann, 1983). Itoh et al. (1986) subjected CVD TiN-coated steel to 6N HCl aqueous solution as a corrosion reagent for 17 h. Negligible corrosion (measured by weight loss) occurred after exposure, which suggests that the coating is well adhered to the steel substrate with no cracks and pin holes.

The plasma-enhanced CVD process is capable of producing TiN coating at much lower substrate temperatures (300–500°C) as compared to high temperatures (800 to 1000°C) involved in CVD. TiN coatings applied by PECVD at 400°C and TiC coatings at 500°C exhibit properties similar to those coatings obtained by conventional CVD (Archer, 1981).

*14.4.3.2 **Hafnium Nitride and Zirconium Nitride Coatings.*** HfN coatings have been applied by activated reactive evaporation, ion plating, reactive sputtering, and CVD (Aron and Grill, 1982; Chollet and Perry, 1985; Grill and Aron, 1983; Münz, 1985; Nimmagadda and Bunshah, 1979; Oakes, 1983; Perry, 1986; Perry and Chollet, 1986; Perry et al., 1983, 1984, 1985; Quinto et al., 1987; Randhawa et al., 1988; Sproul, 1984, 1985a, 1985b). The color of HfN coatings varies with nitrogen content and as the nitrogen level is increased the color changes from a light metallic gold to a yellow gold and then to a brownish gold. This feature closely resembles that of TiN and ZrN.

The hardness, adhesion, and other similar properties of HfN coatings have a strong influence on the nitrogen content and phase of HfN. Table 14.25 lists the various phases of HfN that can be obtained by the ARE process under various operating conditions. The hardness is found to be maximum (2420 HV) in the case of pure stoichiometric HfN coatings.

HfN coatings applied by CVD along with TiN, TiC, HfC, and Al_2O_3 have been extensively used commercially to coat cemented carbide tools to improve their performance and life. Some properties of CVD HfN, TiN, TiC, and Al_2O_3 are presented in Table 14.26. Of these compounds, HfN followed by Al_2O_3 have the highest hot hardness above 800°C which is the typical temperature of the cutting edge when steels or cast irons are machined at high speeds. Therefore, HfN or Al_2O_3 should be more effective resisting crater wear than either TiC or TiN (Kramer, 1983; Oakes, 1983; Quinto et al., 1987). TiC coatings ought to resist flank wear more effectively at lower speeds at which its hardness exceeds that of HfN and Al_2O_3. In addition to being hard, the coating should adhere well and provide thermal insulation to the substrate. These characteristics are related to mismatch in the coefficient of thermal expansion between the substrate and the coating and to the thermal conductivity of the coating. A smaller difference between the thermal expansion coefficients of the coating and the substrate allows thicker coatings to be applied with less risk of spalling caused by internal stresses developed during the coating application and during the machining process. A low thermal conductivity also reduces the service temperature of the substrate, making it more deformation resistant. HfN has the best combination of thermal conductivity and expansion coefficients of the four coating materials listed.

Quinto et al. (1987) have reported that the magnetron-sputtered coatings of TiN and HfN on cemented carbide tools exhibit significantly higher room-temperature microhardness than its CVD counterparts (Fig. 14.48). This microhardness advantage was much greater in HfN than in TiN. The microhardness advantage, however, decreases with temperatures and diminishes

TABLE 14.25 Hardness of HfN Coatings Applied by ARE Process

Substrate material	Evaporation rate, g min^{-1}	Reactive gas pressure		Deposition temperature, °C	Deposition rate, μm min^{-1}	Hardness (HK)	Phase present and lattice parameter of HfN, Å
		Pa	(mtorr)				
Tantalum	1.06	0.1	(0.75)	550	1.4	1870	Hf, Hf_3N_2, Hf_4N_3
Tantalum	0.76	0.1	(0.75)	550	1.0	1850	Hf, Hf_3N_2, Hf_4N_3, HfN
Tantalum	0.46	0.1	(0.75)	550	0.5	2340	Hf_3N_2, Hf_4N_3, Hf, HfN
Tantalum	0.31	0.1	(0.75)	1000	0.28	2145	Hf_4N_3, Hf_3N_2, Hf, HfN
Tantalum	0.20	0.1	(0.75)	700	0.17	2300	HfN, Hf_4N_3, Hf_3N_2, Hf
Tantalum	0.40	0.26	(2.0)	700	0.43	2350	HfN, Hf_4N_3, Hf_3N_2, Hf
Tantalum	0.32	0.26	(2.0)	1000	0.3	2365	HfN, Hf_4N_3, Hf_3N_2, Hf
Stainless steel	0.59	0.09	(0.7)	550	0.75	2135	HfN, traces of Hf_4N_3 4.541 ± 0.05
High-speed steel	0.44	0.09	(0.7)	550	0.5	2310	HfN
Stainless steel	0.27	0.09	(0.7)	550	0.25	2325	HfN 4.527 ± 0.001
Stainless steel	0.47	0.18	(1.4)	550	0.6	2325	HfN 4.528 ± 0.001
Stainless steel	0.20	0.13	(1.0)	550	0.2	2420	HfN 4.534 ± 0.001

Source: Adapted from Nimmagadda and Bunshah (1979).

TABLE 14.26 Some Physical Properties of CVD Coatings Applied on Cemented Carbide Tools

Coating material	Hardness HK at the following temperatures		Thermal conductivity, W $m^{-1} s^{-1}$	Coefficient of thermal expansion, $\times 10^{-6}$ °C^{-1}
	25°C	1000°C		
HfN	1700–2000	800–900	12.6	6.9
TiN	2000–2100	>200	29.3	9.5
TiC	2500–2800	200–300	33.5	7.6
Al_2O_3	2000–2100	300–800	4.2–16.8	8.3–9.0
Cemented carbide substrate	1600–2000	200–300	—	4.5–6.0

Source: Adapted from Oakes (1983).

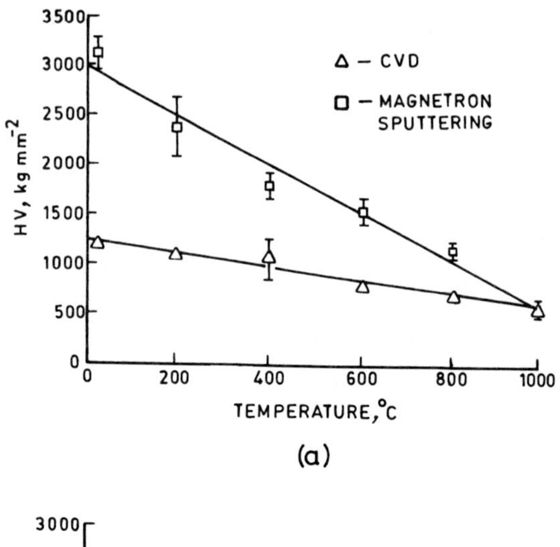

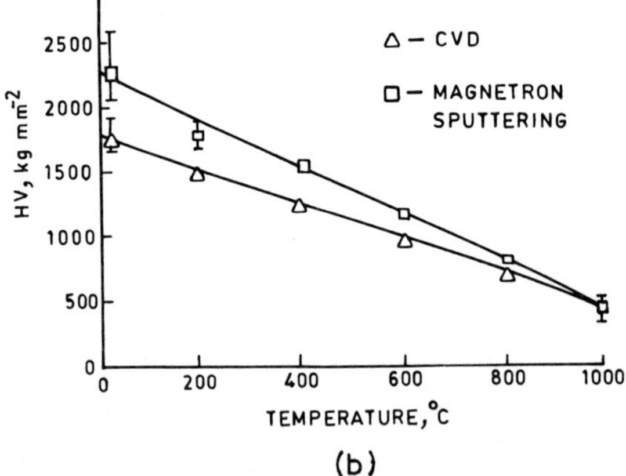

FIGURE 14.48 Dependence of microhardness on temperature for: (*a*) HfN coatings prepared by CVD and magnetron sputtering. (*b*) TiN coating prepared by CVD and magnetron sputtering. (*c*) Cemented carbide tool substrate. [*Adapted from Qunito et al. (1987).*]

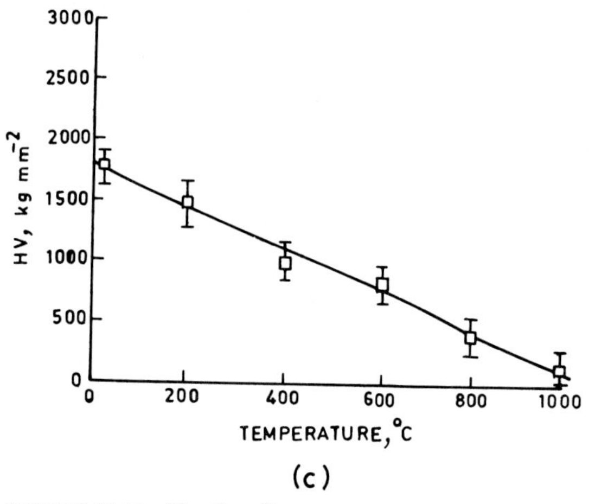

FIGURE 14.48 (*Continued*)

to approximately zero at 1000°C. Hot hardness (at 1000°C) of all these coatings is higher than that of cemented carbide tool substrate.

Kodama et al. (1978) coated cemented carbide tools with HfN and TiC by the activated reactive evaporation (ARE) process. They compared the tool life of the HfN- and TiC-ARE coated tools with the uncoated as well as ion-plated TiN/TiC double-coated tools (Fig. 14.49). They found that the tool life of HfN-ARE coated tools was about an order of magnitude better than that of uncoated tools. The crater tool life of the HfN-ARE coated tools was better than that of TiC-coated and ion-plated TiN/TiC double-coated tools. However, the flank wear rate of the TiC-coated tools was better than that of the HfN-coated tools.

Oakes (1983) compared the cemented carbide tool life coated with HfN, Al_2O_3, TiC/Al_2O_3, and TiC/TiN by CVD process (Fig. 14.50). HfN was found to be superior throughout the speed range from 0.3 to 0.5 m s^{-1} in the continuous turning of AISI-SAE 4340. HfN was comparable in performance with TiC/Al_2O_3 and TiC/TiN coatings in the continuous turning of gray cast iron at higher speeds. Furthermore, the interrupted tests indicate that HfN coating also protects the substrate from premature fracture more effectively than the TiC/Al_2O_3 or TiC/TiN coatings do.

This result could be explained by either the presence of a carbon-deficient brittle layer beneath the TiC or the improved adherence of the HfN to the substrate under repeated impacts. Adherence is facilitated by a better match of the thermal expansion coefficient between the coating and the substrate.

ZrN is another hard coating, which has shown very high potential as a substitute for TiN coating. ZrN coatings have been studied very little compared to TiN and HfN. They are applied by cathodic arc ion plating and reactive sputtering (Johansen et al., 1987; Johnson and Randhawa, 1987; Quinto et al., 1987; Sproul and Rothstein, 1983). Table 14.27 compares the adhesion and hardness of ZrN with TiN and HfN coatings applied under identical conditions of cathodic arc ion plating. This table shows the adhesion in terms of a load capable of setting up or initiating cracking and fracture. ZrN coatings show comparable adhesion to TiN and HfN. The ion-plated ZrN coatings exhibit higher hardness as compared to TiN coatings, but lower as compared to HfN. ZrN coatings are also of gold luster

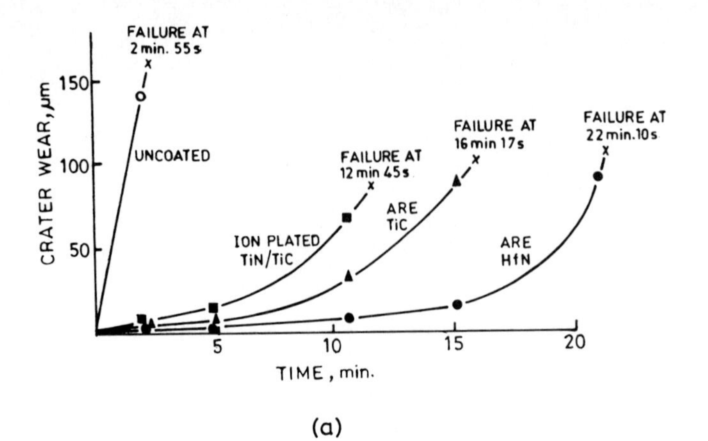

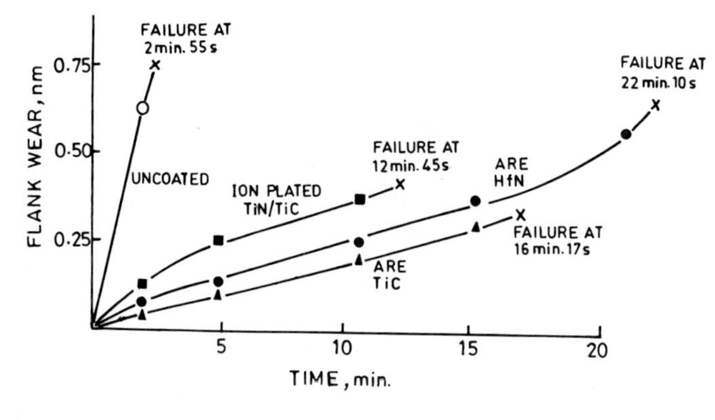

FIGURE 14.49 Wear depth versus cutting time for various cemented carbide-coated cutting tools when machining AISI 4340 steel HRC 36 at speed = 3.5 m s^{-1}, feed = 0.13 mm r^{-1}, depth = 1.25 mm. (*a*) Crater wear depth. (*b*) Flank wear depth. [*Adapted from Kodama et al. (1978).*]

as are TiN coatings and thus replace electrochemically deposited gold in a variety of applications.

Figure 14.51 compares the high-temperature microhardness of hard nitride compounds—TiN, HfN, and ZrN—which have been prepared using the same reactive, magnetron sputtering. It can be seen that incremental improvements over the microhardness of TiN were obtained, in increasing order, by ZrN and HfN. The microhardness converged at 1000°C in the case of all three coatings.

14.4.3.3 Silicon Nitride Coatings. The oxidation resistance of silicon nitride, principally due to formation of a protective silica layer on exposed surfaces combined with low thermal expansion coefficient, makes them a suitable candidate

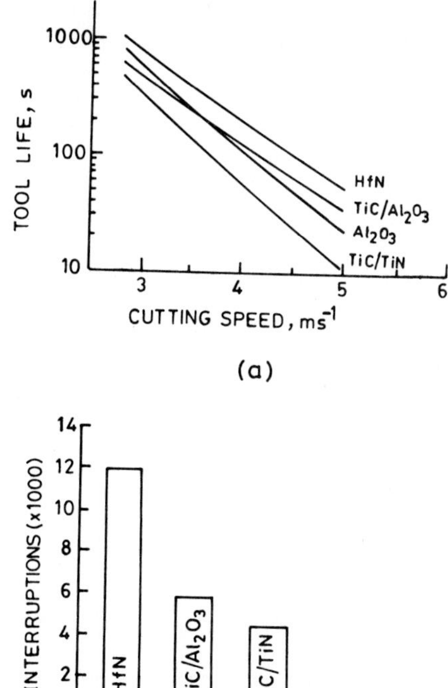

FIGURE 14.50 Tool life for CVD coated cemented carbide tools in: (*a*) Continuous turning of AISI 4340 steel (320 HB) at a feed rate 0.4 mm r^{-1} (depth of cut = 2.5 mm). (*b*) Interrupted turning of AISI 4340 steel (320 HB) at a feed rate = 0.25 mm r^{-1} (depth of cut = 2.5 mm). [*Adapted from Oakes (1983).*]

for high-temperature gas turbine components, dielectric layer in integrated circuits, and solar cells. Si_3N_4 has been deposited by sputtering, CVD, and PECVD.

Si_3N_4 Coatings Applied by Sputtering. Si_3N_4 coatings have been applied by sputtering from a Si_3N_4 target (Aron and Grill, 1982; Grill and Aron, 1983) or by reactive sputtering (Mogab and Lugujjo, 1976). The hardness reported for Si_3N_4 coatings varies over a large range. Aron and Grill (1982) and Grill and Aron (1983) have reported hardness values ranging from 500 to 3900 kg mm^{-2} for coatings sputtered from an Si_3N_4 target.

Si_3N_4 Coatings Applied by CVD and PECVD. Si_3N_4 coatings have been produced by CVD (Gebhardt et al., 1975; Hanni and Hintermann, 1981; Stiglich and Bhat, 1980) and PECVD (Bardos et al., 1983; Dun et al., 1981; Hess, 1984; Reinberg, 1979; Ron et al., 1983; Yokohama et al., 1980). The coatings by CVD and PECVD processes can be produced by reaction of silane with N_2 or NH_3 at

TABLE 14.27 Data for Adhesion (Critical Load) and Microhardness of Nitride Coatings Applied by Cathodic Arc Ion Plating

Deposition conditions†	Onset of cracking,* N	Lower critical failure, N	Knoop hardness (HK)
ZrN			
10 V, 0.066 Pa	9–10	29–30	3400–4400
100 V, 0.066 Pa	5–9	34–47	2800–3100
10 V, 3.3 Pa	7–8	41–42	2900–3900
100 V, 3.3 Pa	12–13	51–52	2600–3200
400 V, 3.3 Pa	10–11	60–61	2100–2300
TiN			
10 V, 0.066 Pa	15–18	47–49	3000–4000
100 V, 0.066 Pa‡	17–23	79–81	4000–5400
10 V, 3.3 Pa	4–5	38–43	2100–2500
100 V, 3.3 Pa	15–18	49–67	2100–3000
400 V, 3.3 Pa	4–7	61–67	2100–2400
HfN			
10 V, 0.066 Pa	8–12	10–15	4900–5600
100 V, 0.066 Pa	8–9	26–27	3700–4000
10 V, 3.3 Pa	4–5	22–23	3000–3200
100 V, 3.3 Pa	8–9	41–47	2600–3200
400 V, 3.3 Pa	10–11	52–63	2600–2800

*The values of minimum load are indicated which are capable of initiating cracking and fracture of the coating.

†Coating thickness was 2–3 μm.

‡Coating thickness on this sample was 16 μm.

Source: Adapted from Johansen et al. (1987).

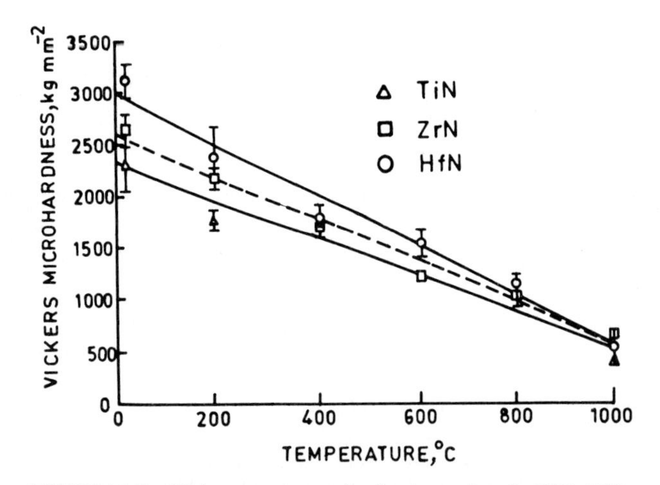

FIGURE 14.51 High-temperature microhardness plots for TiN, ZrN, and HfN coatings prepared by magnetron reactive sputtering. [*Adapted from Quinto et al. (1987).*]

1100 to 1500°C and 300 to 400°C substrate temperatures, respectively. Both crystalline and amorphous films have been produced. Amorphous films contain a considerable amount of hydrogen. For CVD coatings, Hanni and Hintermann (1981) have reported amorphous coatings up to temperatures as high as 1300°C. Ron et al. (1983) reported hardnesses in the range from less than 1000 to 2500 kg mm^{-2} for coatings produced by an rf plasma decomposition of $SiCl_4$. Ron et al. (1983) have pointed out the role of impurities. Because they used $SiCl_4$, the main impurity atom found was Cl, which was detected in concentrations as high as 15 at %. The maximum hardness was obtained at conditions resulting in a minimum in impurity content (~3 at % Cl).

Si_3N_4 coatings deposited for wear application onto steel or hard metals may, however, exhibit low adhesion especially for coatings produced at high temperatures. This is due to the low coefficient of thermal expansion of Si_3N_4 (Retajczyk and Sinha, 1980) and the high reactivity of Si with several transition metals resulting in silicide formation and an evolution of nitrogen of the interface. To avoid this, an intermediate layer of TiN should be used (Hanni and Hintermann, 1981).

The performance of CVD Si_3N_4 is compared with other ceramic material by monitoring erosive wear due to impingement of SiO_2 grit (50 μm size) at 22.5° and 90° (Fig. 14.52). In these cases Si_3N_4 coatings were deposited onto reaction bonded silicon nitride (RSBN) bulk material (Stiglich and Bhat, 1980). An improvement of six to eight times in erosion resistance was obtained by applying a CVD coating to reaction bonded Si_3N_4.

14.4.3.4 Cubic Boron Nitride (CBN) Coatings. It is well known that boron nitride forms several phases that are analogous to those of carbon: hexagonal α-BN resembling graphite and the sphalerite-type β-BN and Wurtzite-type γ-BN resembling cubic and hexagonal diamonds, respectively. The last two phases, like their carbon counterparts, are metastable under normal conditions. They form from the stable phase bulk material under only very high pressure and temperature (Kessler et al., 1987). Cubic boron nitride (CBN) coatings exhibit extremely high hardness (4000 HV or higher) and high thermal conductivity as diamond. These are highly insulating, chemically inert, and are resistant to oxidation at high temperature. CBN finds applications in semiconductor, optical, and tribological applications requiring operations at high temperatures in both ambient and space atmospheres. Hexagonal BN coatings are known to form a corrosion-resistant and insulating films and are used in corrosive environments.

BN coatings have been applied by ARE, rf sputtering, CVD, and PECVD processes (Satou and Fujimoto, 1983). The activated reactive evaporation (Chopra et al., 1985; Lin et al., 1987) and ion plating (Inagawa et al., 1987; Rother, 1985; Rother et al., 1986; Weissmantel et al., 1980) successfully synthesized cubic boron nitride coatings. BN coatings produced from γ-BN target in laser pulse vapor deposition (Kessler et al., 1987) were crystalline γ-BN (hexagonal). BN coatings produced by rf sputtering (Wiggins et al., 1984) from an α-BN target under N_2 or Ar/N_2 were found to be α-BN (graphite). BN coatings produced by reactive pulsed plasma by Sokolowski et al. (1981) were cubic, however, the coatings produced by Hyder and Yep (1976) and Miyamoto et al. (1983) were amorphous and involved a hexagonal-like short-range order. Coatings produced by CVD (Takahashi et al., 1979) were hexagonal in structure.

In the activated reactive evaporation process, CBN coatings are applied by evaporating boric acid (H_3BO_3) in the presence of NH_3 plasma at 450°C substrate temperature (Chopra et al., 1985; Lin et al., 1987). The ion-plating process in-

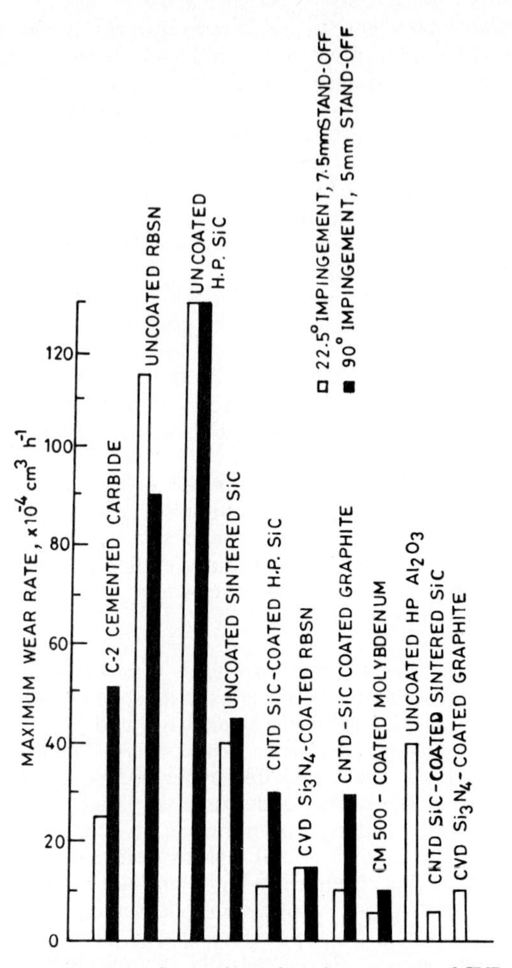

FIGURE 14.52 Comparison of erosive wear rate of CVD Si_3N_4 coating with several other ceramic coatings and bulk materials against 50-μm SiO_2 grit at 0.7 MPa. RBSN-reaction bonded silicon nitride, CNTD-controlled nucleation thermochemical deposition, (coated molybdenum) CM-500 (a W-C alloy). [*Adapted from Stiglich and Bhat (1980).*]

volves the deposition of CBN coating through the interaction of electron-beam-evaporated boron together with N_2, NH_3, or NH_3-Ar (Rother et al., 1986). This process is quite similar to the process suggested by Weissmental et al. (1980) for BN coatings in which ion beam synthesis of CBN takes place from borazine plasma. In rf reactive sputtering process, coatings are produced from a α-BN target in N_2 or Ar, N_2 discharges on a water-cooled substrate (Wiggins et al., 1984). The chemical vapor deposition of BN is done by using thermal decomposition of trichloroborazole (BCl_3) and reactions of NH_3 from the BCl_3-NH_3-H_2-Ar reactant system (Takahashi et al., 1979). In plasma-enhanced CVD, coatings are produced at a temperature below 300°C by the rf glow-discharge decomposition of a gas mixture of diborane (B_2H_6) and NH_3 diluted with H_2. Guzman et al. (1984) produced a BN layer by implanting evaporated boron coatings with 100 keV nitrogen ions at a dose of 6×10^{17} ion cm^{-2}. With the exception of sputter deposition and ion implantation, these processes require that the substrate be at high temperatures.

The microhardness of BN coatings has a strong dependence on the crystal structure. For instance, the cubic BN coatings exhibit very high hardness (4000 HV) as compared to hexagonal phase (~1000 HV). Figure 14.53 shows the relation between the hardness and the CBN content of the BN coatings prepared by electron beam evaporation (Inagawa et al., 1987). The microhardness values for CBN coatings reported by various authors are listed in Table 14.28. A large scatter in hardness values exists among different reports, which may be due to variations in the CBN coating nucleation and growth. The friction behavior of BN coatings synthesized using an ion beam from a borazine ($B_3N_3H_6$) plasma, a process similar to that used by Shanfield and Wolfson (1985), was studied by Miyoshi et al. (1985).

14.4.3.5 Other Nitride Coatings. In addition to the nitride coatings discussed so far, very limited studies have been reported on the following coatings: NbN, TaN, CrN, MoN, and WN. Singer et al. (1983) have reported on structure, hardness, and wear of NbN and Nb_2N coatings prepared by rf reactive sputtering. Depending on the deposition conditions, coatings with hardnesses ranging from 760 to 4000 kg mm^{-2} were obtained. They concluded that NbN coatings with distorted lattices but grown at high temperatures gave the highest hardnesses. Coatings consisting of a two-phase mixture of Ta_2N and TaN prepared through high-rate reactive sputtering was reported to have hardnesses as high as 4100 kg mm^{-2}

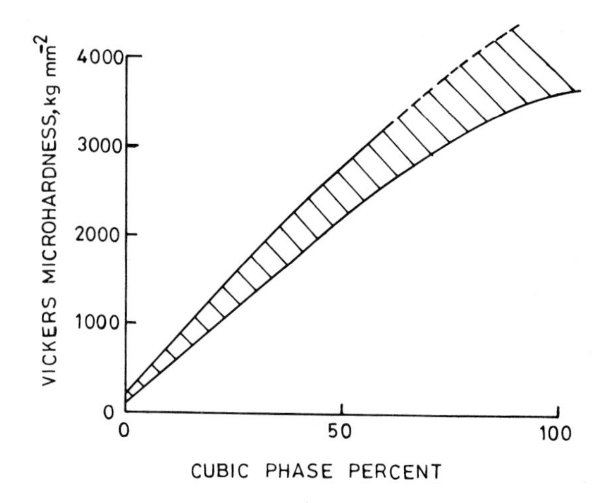

FIGURE 14.53 Relation between Vickers microhardness and percentage of cubic phase in BN coatings produced by electron beam evaporation. [*Adapted from Inagawa et al. (1987).*]

TABLE 14.28 Microhardness Values of CBN Coatings Reported by Various Authors

Deposition process	Substrate	Substrate temperature, °C	Structure of BN	Hardness, HV	Reference
Activated reactive evaporation	Stainless steel	450	Diamond cubic (zinc blende type)	2000–2200	Chopra et al. (1985)
Activated reactive evaporation	High-speed steel, quartz	450	Simple cubic NaCl type	3600–3800	Lin et al. (1987)
Ion beam ion plating (1 μm thick)	TiN-coated WC	—	Cubic	2200–3100	Shanfield and Wolfson (1983)
	Al_2O_3-coated WC	—	Cubic	1600–2200	Shanfield and Wolfson (1983)
	Ceramic	—	Cubic	2100–2800	Shanfield and Wolfson (1983)
Ion plating (HCD)	Single-crystal Si	400	Diamond cubic (zinc blende type)	4000 (Scratch)	Inagawa et al. (1987)
Reactive pulsed plasma	—	—	Diamond cubic	> 2000	Sokolowski et al. (1981)

(Sundgren and Hentzell, 1986). Since the bulk hardnesses of NbN and TaN are much less than the reported maximum hardness of bulk materials, the extreme hardnesses measured must be due to the highly distorted microstructure.

Among the group-VIa nitrides which have been studied for wear applications, CrN is most studied. In the Cr-N system, two nitride phases exist: Cr_2N (of hexagonal structure) and CrN (cubic structure). Cr-N coatings have been produced by ion plating and magnetron reactive sputtering (Fabis et al., 1990a, 1990b; Komiya et al., 1977a; Gobel, 1984). Komiya et al. (1977b) used a hollow cathode process to grow such two-phase coatings. The coatings had grain sizes increasing from 25 nm at a substrate temperature of 370°C to 70 nm at a substrate temperature of 760°C. In spite of this grain-size increase, the hardness remained constant at about 2200 kg mm^{-2}. After vacuum annealing, an increased hardness was observed, and a maximum value of 3540 kg mm^{-2} was measured. Gobel (1984) used reactive magnetron sputtering and produced both two-phase CrN + Cr_2N coatings and also single-phase CrN coatings. Both these types of coatings showed hardnesses in the range of 2000 to 2500 kg mm^{-2}.

Fabis et al. (1990a, 1990b) have deposited chromium nitride coatings on glass and carbon by reactive dc planar magnetron sputtering. Single-phase CrN coatings had homogeneous microstructures composed of columnar grains with a moderate defect density and compressive stresses up to 14 GPa. The dual phase CrN + Cr_2N coating has asymmetric grains, low structured homogeneity, a high defect density, and a stress range from tensile 2.15 GPa to compressive −1.14 GPa. The high stress level in these coating results is due to intercrystalline defects and can be optimized by changing sputtering parameters and substrate biasing.

Mo-N (Mo + Mo_2N) coatings have been reported by Sundgren and Hentzell (1986). These coatings were produced by sputtering on hardened steels and copper alloys for wear applications.

14.4.3.6 Polynitride Coatings. The binary, ternary, and quaternary nitrides of Ti, Zr, and Al have shown promising results to achieve significant improvement in mononitrides such as TiN. Several PVD processes such as cathodic arc evaporation, sputter ion plating, and reactive sputtering have been used to apply a wide range of materials which consist of more than one nitride such as (Ti, Al) N, (Ti, Zr) N, (Ti, Hf) N, (Ti, Al, V) N, and (Ti, Al, Zr) N (Fenske et al., 1987; Freller and Haessler, 1987; Knotek et al., 1987; Randhawa et al., 1987). Usually, these coatings are applied by producing metallic vapors either by evaporation or sputtering and allowing condensation to take place in the presence of reactive gas, i.e., N_2. The composition of such coatings varies from the original composition of the starting material, and is strongly influenced by deposition parameters such as substrate biasing. Figure 14.54 illustrates the change in aluminum content with substrate biasing for (Ti, Al) N coatings applied by cathodic arc ion plating. The microhardness of polynitride coatings are summarized in Table 14.29. A wide variation in colors of mixed coatings, such as of (Ti, Zr) N and (Ti, Al) N, has been observed. By suitable adjustment of composition of polynitride coatings, it is possible to obtain different color tones such as that of 10 K and 24 K gold.

The addition of Al in TiN (25 to 50 wt % Al) significantly improves the oxidation characteristics and wear characteristics of TiN coatings (Knotek et al., 1987). Figure 14.55 illustrates these characteristics for (Ti, Al) N and TiN coatings applied by sputter ion plating on 6-mm-diameter HSS drills. This figure shows a significant improvement in oxidation temperature and wear resistance by (Ti, Al) N over TiN coatings.

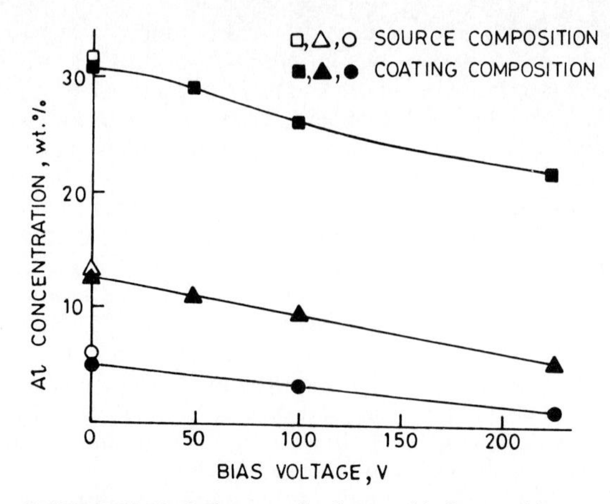

FIGURE 14.54 Influence of substrate biasing on Al content in (Ti, Al) N coatings applied by cathodic arc ion-plating. [*Adapted from Freller and Haessler (1987).*]

TABLE 14.29 Microhardness Values of Polynitride Coatings

Coating composition	Application process*	Microhardness, HV	References
(Ti, Al) N (50% Ti, 50% Al) N	Cathodic arc ion plating*	2450 ±200	Randhawa et al. (1988)
(Ti, Al) N (75% Ti, 25% Al) N	Cathodic arc ion plating	2650 ± 200	Randhawa et al. (1988)
(Ti6Al4V) N	Cathodic arc ion plating	2800 ± 200	Randhawa et al. (1988)
(Ti, Zr) N (50% Ti, 50% Zr) N	Cathodic arc ion plating	3000 ± 200	Randhawa et al. (1988)
(Ti, Zr) N (25% Ti, 75% Zr) N	Cathodic arc ion plating	3350 ± 200	Randhawa et al. (1988)
TiN	Cathodic arc ion plating	2500 ± 200	Randhawa et al. (1988)
ZrN	Cathodic arc ion plating	3200 ± 200	Randhawa et al. (1988)
(Ti, Al) N	Sputter ion plating	2200–2400	Knotek et al. (1987)
(Ti, Al, V) N	Sputter ion plating	2700	Knotek et al. (1987)
(Ti, Al, Zr) N	Sputter ion plating	3500	Knotek et al. (1987)
(Hf, Ti) N (50% Hf, 50% Ti) N	Reactive sputtering (dual-cathode configuration)	3760	Fenske et al. (1987)

*5–7-μm-thick coating.

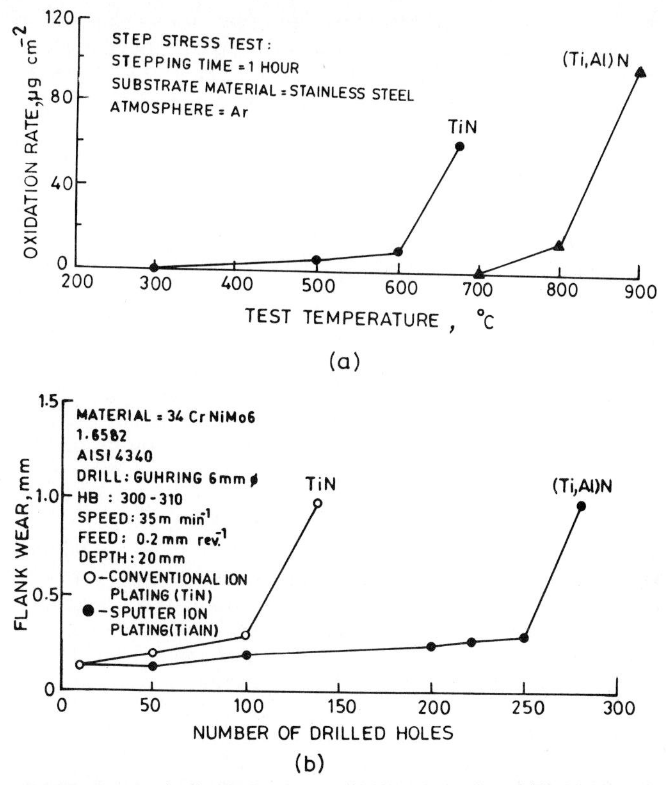

FIGURE 14.55 (*a*) Oxidational rate. (*b*) Wear behavior of (Ti-Al) N, and TiN coatings deposited by ion plating on high-speed steel drills. [*Adapted from Knotek et al. (1987).*]

The performance of these polynitride coatings can be further tailored to specific needs by applying mixed polynitride and polycarbide coatings such as (Ti, Al, V) C, N.

14.4.4 Boride Coatings

Borides of refractory metals such as TiB_2, ZrB_2, HfB_2, CrB_2, Mo_2B_5, and TaB_2 are important materials because of their very high hardness, high melting point, corrosion resistance, and abrasion resistance. Boride coatings have not been studied as much as oxides, carbides, and nitrides. These coatings have been applied by plasma spraying, activated reactive evaporation (ARE), rf sputtering, CVD, and electroplating. Despite the brittleness of borides, TiB_2 and ZrB_2 have been tried as wear-resistant coatings on carbide tools with limited success (Pierson and Randich, 1978).

Bunshah et al. (1977, 1978) synthesized the boride and carbide coatings by direct evaporation of TiB_2, ZrB_2, ZrC, 55% TiB_2-45% ZrB_2, and 60% ZrC-40% ZrB_2 from water-cooled crucibles using an electron beam as the heat source. The vapors were condensed on an Mo substrate. They found that the mechanical properties such as hardness are strongly influenced by stoichiometry and deposition conditions such as substrate temperature. Table 14.30 summarizes the phases and hardness of boride and carbide coatings applied at different substrate temperatures.

Brainard and Wheeler (1978) deposited TiB_2, Mo_2B_5, and various carbide and silicide coatings by rf sputtering on type 440C stainless steel from targets of respective materials. They found that the interfacial region (the transition between the bulk coating properties and bulk steel properties) was dependent on the coating material, substrate oxidation, and sputtering parameters. The formation of a mixed-oxide interface results in increased adhesion. The coefficient of friction and pin wear of sputter deposited TiB_2, Mo_2B_5, and other ceramic coatings were compared in Fig. 14.34. TiB_2 coatings exhibit low coefficients of friction (0.2) and pin wear as compared to Mo_2B_5 and other carbide and silicide coatings.

Gebhardt and Cree (1965) produced TiB_2, ZrB_2, and HfB_2 coatings by the CVD process. The coatings were produced by reaction of metal tetrahalide (MCl_4) and boron trichloride (BCl_3) at a substrate temperature of 1400°C. They reported that grain size, orientation, porosity, and stoichiometry are influenced by the process variations. CVD coatings of TiB_2 were produced at various substrate temperatures by Pierson and Mullendore (1980, 1982) and Pierson and Randich (1978). They produced very thick (more than 100 μm) and uniform coatings. TiB_2 grown by CVD on graphite is amorphous at substrate temperatures equal to or less than 500°C (Pierson and Mullendore, 1980). At these temperatures, the TiB_2 coatings are over stoichiometric. As the substrate temperatures were increased, the coating became crystallized at about 600°C but with a very small grain size. The grain size and hardness increased from 2900 kg mm^{-2} in hardness and 6 nm in grain size to 3700 kg mm^{-2} in hardness and 29 nm in grain size as the substrate temperature was raised from 600 to 900°C (Fig. 14.25). They also reported about the same magnitude of changes in the grain size as the concentrations of Ti and B or of H and Cl reactants were changed during the process.

Randich (1980) deposited TaB_2 coatings by CVD reaction of $TaCl_5$ and B_2H_6 gaseous mixture at about 600°C. Randich also found that stoichiometry and the hardness of the coating were a strong function of the deposition temperature (Fig. 14.56). We note that the B/Ta atomic ratio decreases carbon impurity content and that the hardness increases with an increase in the deposition temperature. A hardness of 2500 kg mm^{-2} or higher can be obtained at deposition temperatures of 600°C or higher. Randich (1980) reported that coating had an extremely small grain size and contained amorphous carbon when the deposition temperature was below 600°C but were substoichiometric in boron above this temperature. Carbon may be substituted for boron and thereby stabilizing the diboride structure at high deposition temperatures which may be responsible for high hardness.

Kirk et al. (1981) prepared electrodeposited coatings of TiB_2 for wear applications. The microhardness of these coatings ranged from 2200 to 5240 HV. They conducted wear tests of TiB_2 coated pins and Al_2O_3 single crystal (for comparison) sliding against SAE 415 steel of hardness 52 HRC in dry and lubricated conditions (SAE 30 oil). They found that TiB_2 coatings have a wear resistance comparable with Al_2O_3 single crystal under the high-speed (30 m s^{-1}) and unlubricated conditions. Under either the low-speed (less than 50 mm s^{-1}) or the high-speed lubricated conditions used, Al_2O_3 single crystal is the more wear-resistant material.

TABLE 14.30 The Influence of Phases of ARE Coated Boride and Carbide Coatings on Hardness

Deposition temperature, °C	ZrC		ZrB_2		TiB_2		55% TiB_2 + 45% ZrB_2 (solid solution)		60% ZrC + 40% ZrB_2	
	Phases	Hardness (HV)	Phases	Hardness (HV)	Phases	Hardness (HV)	Phases	Hardness (HV)	Phases	Hardness (HV)
700	ZrC	1250								
800	ZrC	1600					SS* + (TiB + ZrB)	3000		
900			ZrB_2+ (ZrB + Zr)	2800	TiB_2 + TiB	2900			ZrC+ZrB_2 + (ZrB + Zr)	1800
1000	ZrC + Zr	1100	ZrB_2+ (ZrB + Zr)	3000	TiB_2 + TiB	2600	SS+ (TiB + ZrB)	2800		
1100	ZrC + Zr	1150		3000					ZrC+ZrB_2 + (ZrB + Zr)	1600
1200			ZrB_2+ (ZrB + Zr)		TiB_2 + TiB	2400	SS+ (TiB + ZrB)	3100		
1300	ZrC + Zr	1250		2400					ZrC + ZrB_2 + (ZrB + Zr)	1710
1400					TiB_2 + TiB	2400	SS+ (TiB + ZrB)	2600	ZrC+ZrB_2 + (ZrB + Zr)	2030
1500	ZrC + Zr	1300	ZrB_2+ (ZrB + Zr)	2100	TiB_2 + TiB	2700				
1600	ZrC + Zr	1800					SS+ (TiB + ZrB)	3050	ZrC+ZrB_2 + (ZrB + Zr)	2450
Evaporant billet	ZrC		ZrB_2 + (ZrB)		TiB_2 + TiB					

*SS: solid solution.

Source: Adapted from Bunshah et al. (1977).

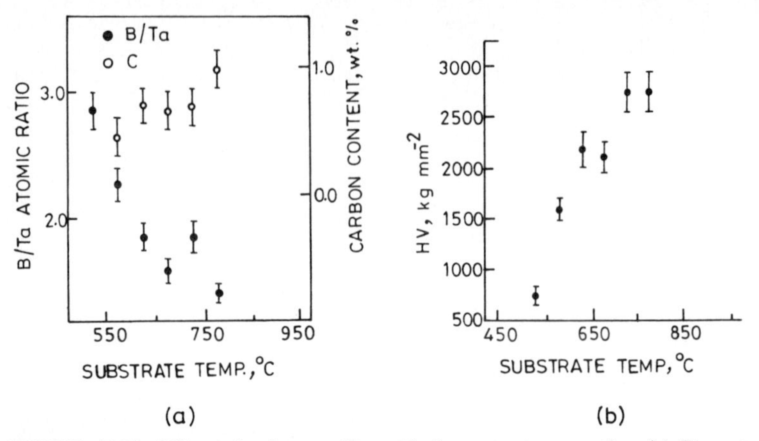

FIGURE 14.56 Effect of substrate (deposition) temperature on the: (*a*) Boron-to-tantalum atomic ratio and the carbon impurity content. (*b*) Microhardness of CVD TaB_2 coating. [*Adapted from Randich (1980).*]

14.4.5 Silicide Coatings

Silicides of refractory metals such as $MoSi_2$, $TiSi_2$, Cr_3Si_2, WSi_2, and $TaSi_2$ are important materials because of their very high corrosion (oxidation) resistance and good friction and wear properties. Refractory silicide coatings are used in very large scale integration (VLSI) technology because of the low resistivity of silicides desirable for VLSIs. Silicides have been the least studied compared to oxides, carbides, and nitrides. The silicide coatings have been primarily applied by plasma spray, evaporation, sputtering, CVD, and PECVD. Sangam and Nikitich (1985) produced $MoSi_2$ coatings by plasma spray primarily for fire barrier applications up to 1700°C. Thin coatings (30 to 150 nm thick) of $MoSi_2$ and WS_2 were produced by evaporation on Si (100) substrates for VLSI application by Torres et al. (1985). They studied silicide formation by two methods: (1) metal deposition onto substrates at room temperature followed by in situ annealing; (2) metal deposition onto heated substrates at temperatures ranging from 600 to 800°C, i.e., higher than the threshold temperature for silicide formation. They found that silicides produced by the second process grew expitaxially on Si (100) and the resistivity was related to the grain size and low impurity content. Retajczyk and Sinha (1980) prepared silicide coatings ($TiSi_2$, $MoSi_2$, WSi_2, and $TaSi_2$) with 250 nm thickness by cosputtering silicon and the appropriate metal in an S-gun magnetron using rotating planets and then annealing for 60 min in hydrogen at a temperature of 900°C for $TiSi_2$ and $TaSi_2$ and 1000°C for $MoSi_2$ and WSi_2. Brainard and Wheeler (1978) and Spalvins (1977) prepared silicide coatings ($MoSi_2$, $TiSi_2$, and Cr_3Si_2) by rf sputtering from respective compound targets for wear applications. Wieczorek (1985) produced $TaSi_2$ directly onto silicon and/or silica by CVD. In the CVD process, silane (SiH_4) and tantalum pentachloride ($TaCl_4$) were used with hydrogen as the carrier gas. Hess (1984) produced WSi_2 by the plasma-enhanced reaction between WF_6 and SiH_4 in a parallel plate reactor at 230°C for semiconductor applications.

Brainard and Wheeler (1978) measured friction and wear performance of sputtered $MoSi_2$, $TiSi_2$, and other carbide and boride coatings. Results were presented earlier in Fig. 14.34. Friction and wear of $MoSi_2$ and $TiSi_2$ were relatively high. Spalvins (1977) reported that Cr_3Si_2 sputtered coating adhered strongly to the 440C stainless-steel substrate. In a sliding test consisting of a 440C stainless steel pin sliding on a coated 440C stainless steel disk, he found that the sputtered Cr_3Si_2 coating exhibited a coefficient of friction two to three times lower than that of sputtered Cr_3C_2. He also coated 440C stainless steel ball bearing components—the bearing races and the cage. He found that bearings coated with 0.1 μm thick coating of Cr_3Si_2 and subsequently with 0.6 μm thick coating of MoS_2 had considerably longer (fivefold) endurance lives than bearings sputtered with only 0.6 μm thick coating of MoS_2.

14.4.6 Hard Carbon Coatings

Carbon occurs widely in its elemental form as crystalline and amorphous solids. Diamond and graphite are the two crystalline allotropes of carbon. The diamond crystalline structure is the modified face-centered cubic with interatomic distances of 0.154 nm. Each carbon atom is covalently bonded to its four carbon neighbors using (tetrahedrally directed) sp^3 atomic orbitals. The tetrahedral carbon-carbon bonds account for such properties as extreme hardness, chemical inertness, high electrical resistivity (a wide bandgap semiconductor), high dielectric strength, high thermal conductivity, high density, and optical transparency. Unusual combinations of these properties make them ideal candidates for optical and tribological coatings. The structure of graphite, a semimetal, is described as layers of carbon atoms with strong trigonal bonds (sp^2) with an interatomic distance of 0.1414 nm in the basal plane. The fourth electron in the outer shell forms a weak bond of the Van der Waals type between planes 0.34 nm apart and accounts for such properties as good electrical conductivity, lubricity, lower density, a graying-black appearance, and softness, which are in contrast to diamond. Carbon also exists in numerous amorphous forms, which can be characterized as degenerate or imperfect graphite structures.

Several forms of disordered carbons have been produced: glassy carbon formed by heating certain organic polymers (Jenkins and Kawamura, 1976; Noda et al., 1969); microcrystalline (μC) carbon, produced by irradiating graphite (Kelly, 1981); amorphous carbon (a-C) produced by evaporation in an arc, electron, or laser beam (Fink et al., 1984; Fujimori and Nagai, 1981; McLintock and Orr, 1973; Wada et al., 1980), ion plating (Weissmantel et al., 1982, 1983) or by dc or rf sputtering (Bewilogua et al., 1985; Cho et al., 1990b; Dimigen and Hübsch, 1983–84; Hauser, 1977; Hosokawa et al., 1981; Rubin et al., 1990; Savvides and Window, 1985; Tsai and Bogy, 1987; Tsai et al., 1988; Weissmantel et al., 1982); and hydrogeneted amorphous carbon (a-C:H) coatings produced by reactive sputtering (Craig and Harding, 1982; Hiraki et al., 1984; Jansen et al., 1985; McKenzie et al., 1982) and PECVD (Andersson and Berg, 1978; Angus et al., 1986; Berg and Andersson, 1979; Bubenzer et al., 1983; Enke et al., 1980; Fink et al., 1984; Holland and Ojha, 1976, 1979; Jones and Stewart, 1982; Kaplan et al., 1980; Moravec and Orent, 1981; Nyaiesh and Holland, 1984; Ojha, 1982; Robertson, 1986; Vandentop et al., 1990; Yamamoto et al., 1988).

Due to the unique properties of diamond, there have been numerous studies to produce diamond coatings. However, because diamond is in a metastable state at room temperature and atmospheric pressure, extreme condions in terms of pressure and temperature are required to form diamond. The diamond coatings have been attempted by CVD (Angus et al., 1986; Chauhan et al., 1976; Deryagin and Fedossov, 1975; Poferl et al., 1973), PECVD and hot filament CVD (Badzian et al., 1987; Deryagin and Fedoseev, 1975; DeVries, 1987; Fedoseev et al., 1984; Komatsu et al., 1987; Matsumoto, 1985; Matsumoto et al., 1982; Sawabe and Inuzuka, 1986; Spear, 1989; Spitsyn et al., 1981).

The amorphous carbon coatings with (a-C:H) or without hydrogen (a-C) produced by ion-beam assisted evaporation, sputtering, ion plating, and PECVD processes are very hard and are normally called *diamond-like carbon (DLC) coatings*. Because of the essential role of the ions in all preparation, i-C has been proposed as a generic term for these amorphous DLC coatings (Messier et al., 1987). The structure of DLC coatings is predominantly amorphous. However, small areas within the coatings show polycrystalline or even single-crystalline graphitic or diamond structure (Cho et al., 1989; Fink et al., 1983, 1984; Khan et al., 1988; Nadler et al., 1984*a*; Tsai et al., 1988). On the other hand, diamond coatings have a diamond crystalline morphology discernible by microscopy and other surface analytical techniques (Messier et al., 1987).

High-resolution electron microscopy, x-ray diffraction and transmission electron diffraction (TED), Auger electron spectroscopy (AES), x-ray photoelectron spectroscopy (XPS), electron energy loss spectroscopy (EELS), Raman scattering, infrared (IR) spectroscopy, and ellipsometry have been used to identify the bonding and structure of the amorphous network. Quantitative information on the relative concentration of sp^3 and sp^2 hybridized carbon has been obtained by transmission EELS, Raman scattering spectroscopy, IR spectroscopy, and nuclear magnetic resonance (NMR). IR spectroscopy provides information on the existence of hydrocarbon (CH, CH_2, or CH_3) groups and thus on reduction of carbon-carbon cross-linking by hydrogenation for the case of hydrogenated carbon coatings. Combustion analysis, proton NMR, and secondary ion mass spectroscopy (SIMS) are used to measure relative concentration of hydrogen in hydrogenated carbons (Angus et al., 1986; Robertson, 1986; Tsai and Bogy, 1987; Weissmantel et al., 1982).

Table 14.31 gives values of some key properties for diamond, graphite, and five forms of disordered carbons. Glassy carbons (and μC carbon) is essentially metallic. The a-C and a-C:H differ from glassy carbon in being truely amorphous and semiconducting. The presence of a semiconducting band gap is a crucial difference, and their structures are not so easily classified. Amorphous coatings are a random network of sp^3 and sp^2 bonded carbon with the relative fraction of each depending on the method of preparation and the process parameters. It is generally believed that glassy carbon contains approximately 100 % sp^2 sites, evaporated a-C 1–10 % sp^3 sites, a-C:H may comprise 30 to 60 at % hydrogen with perhaps 50 to 70 % of the carbon sites having an sp^3 configuration. Diamond coatings, on the other hand, have significant crystalline diamond structure (sp^3 sites). Hydrogenated carbon coatings have a density ($\sim$1800 kg m^{-3}) lower than that of graphite and diamond, high optical band gap $\sim$1.5–4 eV (less optical absorption), and high electrical resistivity ($\sim 10^{12}$ Ωcm) than that of unhydrogenated coatings.

DLC coatings (20 to 30 nm thickness) are used as wear- and corrosion-resistant coatings in thin-film magnetic rigid disks (Bhushan, 1990). DLC coatings have been studied for ball bearing applications. Deposition of diamond coatings requires very high substrate temperature ($\sim$1000°C), and the coatings as produced are rough; therefore, these coatings have so far found little application to tribology. The developments of diamond and diamondlike coatings offer the potential for exploiting their unique properties in applications ranging from wear-resistant coatings for bearings, abrasives, and tools to antireflection coatings for free-standing windows and lenses for visible and infrared (IR) transmission, to

TABLE 14.31 Selected Properties of Diamond, Graphite, and Five Forms of Disordered Carbon

Material/coating	Density, kg m^{-3}	Microhardness, kg mm^{-2}	Specific electrical resistivity, Ω cm	Refractive index at 0.6 μm wavelength	Optical band gap, eV
Diamond	3515*	8000–10,400	$10^7 - 10^{20}$	2.41	5.5
Graphite	2260*	Soft	5×10^{-5} (⊥ c) 4×10^{-3} (∥ c)	2.8 (⊥ c)	−0.04
Glassy carbon	1300–1550	800–1200	$10^{-2} - 10^{-3}$	—	10^{-2}
Evaporated a-C	1800–2200	20–50	$10^{-1} - 10$	—	0.4–0.7
Sputtered a-C	1800–2400	1000–2500†	$10^{-1} - 10^5$	2.0–2.7	0.4–2.0
PECVD/ion-plated a-C:H†	1500–2000	1200–3000 (PECVD) 4000–6000 (ion plated)	$10^7 - 10^{14}$	1.8–2.3	1.5–4.0
PECVD diamond	2800	6000–12,000 (thermal conductivity = 1100 W m^{-1} K^{-1})	$> 10^{13}$	—	—

*For reference, the density of dense pure carbon, dense hydrocarbon, and plasma polymer phases are 2800, 1500–2000, and 1200 kg m^{-3}, respectively.

†Typically contains 30 to 50 at % of hydrogen.

thin coatings for high-temperature, high-power semiconductor devices as a heat sink.

A summary of important characteristics of hard carbon coatings deposited by various deposition processes follows.

14.4.6.1 Amorphous Carbon Coatings. Disregarding the earlier attempts to deposit diamond coatings epitaxially on diamond, the first diamondlike carbon (DLC) coatings were deposited by a beam of carbon ions produced in an argon plasma on room-temperature substrates as reported by Aisenberg and Chabot (1971). Subsequent confirmation by Spencer et al. (1976) led to explosive growth of this field. Following the first work, several alternative techniques have been developed. The amorphous carbon coatings have been deposited using hydrocarbon gases and the solid carbon. Specifically, carbon coatings have been produced by evaporation, dc or rf sputtering, ion plating, and PECVD. Coatings with both graphitic and diamondlike properties have been produced. Evaporation and ion plating have been used to produce coatings with graphitic properties (low hardness, high electrical conductivity, very low friction, etc.). Ion-assisted evaporation, dc and rf sputtering, ion plating and PECVD have been used to produce coatings with diamondlike properties.

The prevailing atomic arrangement in these DLC coatings is amorphous or quasi-amorphous with small diamond (sp^3), graphite (sp^2) and other unidentifiable microcrystallites. DLC coatings produced by PECVD typically contain from a few to about 50 at % hydrogen. Sometimes hydrogen is deliberately incorporated in the sputtered and ion-plated DLC coatings to increase the optical band gap that is to reduce optical absorption. Dangling bonds, which cause energy levels in the band gap, are passified by hydrogen, thus removing these states from the gap. Optical band gap is controlled by the fraction of tetrahedral (sp^3) versus trigonal or graphitic (sp^2) bonding. Hydrogen helps to stabilize sp^3 sites (most of the carbon atoms attached to hydrogen have a tetrahedral structure); therefore, the sp^3/sp^2 ratio for hydrogenated carbon is higher* (Angus et al., 1986). The hydrogenated carbon, therefore, has a larger optical band gap and consequently has a high electrical resistivity (semiconductor) and has a lower optical absorption or higher optical transmission (lower refractive index). The hydrogenated coatings have a lower density probably because of reduction of cross-linking due to hydrogen incorporation. However, hardness decreases with an increase of the hydrogen even though the proportion of sp^3 sites increases (that is, as the local bonding environment becomes more diamondlike) (Kaplan et al., 1985). It is speculated that the high hydrogen content introduces frequent terminations in the otherwise strong three-dimensional network and hydrogen increases the soft polymeric component of the structure more than it enhances the cross-linking sp^3 fraction. Thus, structural trade-offs prevent the formation of amorphous diamond, i.e., a material with simultaneous extreme hardness and wide optical band gap (Jansen et al., 1985; Kaplan et al., 1985).

High-energy surface bombardment has been used to produce harder and denser coatings. With increasing ion energies, the extent of cracking of the hy-

---

*Optimal sp^3/sp^2 in random covalent network composed solely of sp^2 and sp^3 carbon sites (N_{sp^2} and N_{sp^3}) and hydrogen (Angus and Hayman, 1988):

$$\frac{N_{sp^3}}{N_{sp^2}} = \frac{6X_H - 1}{8 - 13X_H}$$

Where X_H is the atom fraction of hydrogen.

drocarbon species increases until finally all the weaker bonds (C-H) may be stripped off as a result of energetic impacts, since the C-C bond strength (607 kJ/mol) is greater than that of C-H bonds (337.2 kJ/mol) (Weast, 1983). Thus the more tightly bonded three- or fourfold coordinated carbon atom will survive over single-bonded hydrogen. Also, since double-bond hydrogenation is an exothermic process, graphitic bonding is favored over tetrahedral bonding in higher energy growth environments. Coatings thus formed should be less hydrogenated, mechanically harder, and should have a higher density than those formed in a lower-energy environment (Kaplan et al., 1985; Mori and Namba, 1983; Weissmantel et al., 1982; Yamamoto et al., 1988).

Table 14.32 compares the selected properties of amorphous carbon coatings prepared by various depositions techniques. We note that ion beam ion-plated coatings provide the highest hardness (up to 6000 kg mm^{-2}) followed by PECVD and sputtering. This can be partially explained by the impact ion energies with ion plating generally having the largest (100 to 1000 eV). The disadvantage of ion-plated deposition techniques is the increased process complexity and low rate of growth compared to that of PECVD. Sputtered carbon coatings are used because of simplicity and ease of control of the sputtering process. The details on coatings produced by various processes follow.

Evaporated Carbon Coatings. Evaporated carbon coatings are produced by evaporation in an arc, electron, or laser beam from powdered graphite (Fujimori and Nagai, 1981; McLintock and Orr, 1973; Wada et al., 1980). The arc evaporation method has been most widely used. Fujimori et al. (1982) have also produced carbon films by evaporating from powdered diamond source. Evaporated carbon coatings consist of relatively pure carbon in a highly disordered state in which there is short-range order, but no long-range order. In contrast, the purity of graphite is associated with an ordered state, and the disorder of chars is accompanied by the presence of large number of impurity atoms such as H and O. The evaporated coatings are dominated by sp^2 trigonal bonding, whereas the coatings formed at higher impact energies have a very significant amount of sp^3 tetragonal bonding. Evaporated carbon coatings are amorphous semiconductors and have been widely used as supports in electron microscopy and in switching and memory devices.

During coating deposition, simultaneous bombardment with Ar ions on a deposited substrate, coatings with higher light transmittance, resistivity, and hardness are obtained. This suggests that ion bombardment results in an increase of diamond bonds (sp^3 tetragonal bond) and the decrease of graphitic bonds in the random network structure of the coating (Fujimori and Nagai, 1981; Fujimori et al., 1982).

Sputtered Carbon Coatings. The deposition of DLC coatings by dc, rf, and magnetron sputtering have been reported by several authors (Bhushan, 1990; Banks and Rutledge, 1982; Bewilogua et al., 1985; Cho et al., 1990a, 1990b; Dimigen and Hubsch, 1983–84; Fujimori et al., 1982; Hauser, 1977; Hosokawa et al., 1981; Rubin et al., 1990; Savvides, 1985; Savvides and Window, 1985; Tsai et al., 1988; Weissmantel, 1977, 1981; Weissmantel et al., 1982). Several attempts have been made to produce hydrogenated carbon coatings by reactive sputtering in which both solid carbon and a hydrocarbon or hydrogen gas have been employed to produce dense coatings (Craig and Harding, 1982; Hiraki et al., 1984; McKenzie et al., 1982). Savvides (1985) and Savvides and Window (1985) deposited DLC by employing dc magnetron sputtering of a pyrolitic graphite (most-ordered form of synthetic graphite produced by heat and pressure treatments up to 3000°C) target in an ultrapure argon atmosphere at a pressure of 1 Pa (7.5 mtorr). In order to

TABLE 14.32 Selected Properties of Hard Diamond-like Carbon (a-C, a-C:H) Coatings Applied by Various Plasma Techniques

	Deposition conditions							
Deposition technique	Source of carbon	Ion energy	Density, kg m^{-3}	Electrical resistivity, Ω cm	Optical properties	Hardness (HV or HK)	Chemical inertness	References
Sputter deposition (a-C)	Carbon target in rf plasma	rf power 2.2 and 75 W	—	10^{-2}–10^{3}	$E_o \sim 0.8$ eV	—	—	Hauser (1977)
	dc magnetron sputtering of a graphite target	Power 10 W	2100–2200	1.5×10^{2}	$n = 2.3$ $E_o = 0.68$ eV	2400 HV	—	Savvides and Window (1985)
		50 W	1900	1.0	$n = 2.73$ $E_o = 0.5$	2095 HV	—	
		500 W	1600	0.2	$n = 2.95$ $E_o = 0.4$	740 HV	—	
	Carbon target in dc plasma (with simultaneous bombardment of substrate)	1–20 eV	2100–2200	$>10^{11}$	Reflectance = 0.2 Absorptance = 0.7	—	—	Banks and Rutledge (1982)
(a-C:H)	Carbon target in rf plasma of hydrogen	—	—	10^{12}	$E_o = 1.4$ to 2.8 eV	2000 HV	—	Hiraki et al. (1984)
Ion beam deposition (a-C:H)	Hydrocarbon in dc plasma and Ar ion beam	0.5-1.5-keV	~2000	10^{7}–10^{10}	—	4000–6000 HV	—	Weissmantel et al. (1982)
	Carbon target and Ar ion beam	1.2 keV	2250	—	$E_o = 0.7$ eV	>2000 HV	—	Kaplan et al. (1985)

PECVD deposition (a-C:H)	Hydrocarbon in rf plasma	40–100 eV	—	~10^{10}	n = 2.0	>glass	Resist HF for 40 h	Holland and Ojha (1979)
	Hydrocarbon in dc plasma	50–100 eV	—	—	n = 2.3 at λ = 5 μm	1850 HK	—	Moravec and Orent (1981)
	Hydrocarbon in rf plasma	200 eV	—	—	—	3000 HV	—	Yamamoto et al. (1988)
	Hydrocarbon in rf plasma	0.5–1.5 keV	1500–1800	10^{12}	n = 1.85 to 2.20 E_o = 1.0 to 1.6 eV	1250–1650	—	Bubenzer et al. (1983)
	Hydrocarbon in rf plasma	100 eV	—	—	—	2500 HK, modulus of elasticity – 120 MPa	—	Memming et al. (1986)

n = refractive index.
E_o = energy band gap.

prevent the contamination of coatings from atmospheric gases that are adsorbed by the graphite target, presputtering of the target at high power was carried out immediately prior to coating deposition. The substrates were placed at 30 mm below the target and were held at a temperature of about 25°C. Coatings were deposited at different power levels in the range 5 to 500 W at a rate of 1.7×10^{-4} $\mu m\ min^{-1}\ W^{-1}$ with an effective sputtering area of the target of 20 cm^2. The density of a-C coatings was determined using measurements of weight, surface area, and thickness. A four-probe geometry was employed to measure electrical conductivity. Microhardness measurements were made on 2- to 3-μm-thick coatings. Coatings for optical measurements were approximately 15 nm thick and were deposited on 0.5-mm-thick fused silica substrates. Coatings for electrical measurements were 0.5 to 1 μm thick and were deposited on glass substrates. The optical constants were determined in the wavelength range of 0.18 to 2.5 μm (photon energy range 0.5 to 7 eV) by analyzing reflectance and transmittance data (also see Cho et al., 1990b, and Rubin et al., 1990).

Table 14.33 summarizes the various properties as a function of power density. At the lowest sputtering power, the coatings have a density in the range of 2100 to 2200 kg m^{-3} which is comparable to that of single-crystal graphite reported to be 2260 kg m^{-3}. Pyrolytic graphite has a lower density, 1600 to 1950 kg m^{-3}, as a result of voids and open porosity. Electrical conductivity measurements show that the coatings prepared at lowest power possess insulating properties ($\sim 10^4$ Ωcm) which are similar to other extrinsic amorphous semiconductors such a-Si, a-Ge, or a-C:H, where the hydrogen content is <10 at %. Microhardness measurements show that the a-C coatings prepared at low power (≤ 20 W) are high, comparable to that of sapphire in their hardness. High hardness is desirable for low-friction and wear properties (Bhushan, 1990). The sputtering power has a significant influence on refractive index. The dependence of refractive index on power can be employed to "tune" the refractive index of the a-C coatings for IR and laser applications (lower refractive index for increased transmittence to light) requiring hard protective coatings. The optical band gap, a very useful parameter for characterizing amorphous semiconductors, is also found to be a strong function of the sputtering power.

Savvides et al. (1985) also derived information on the bonding of the carbon atoms from the dielectric constant data. They found that in a-C coatings prepared at low power, about three-fourths of the carbon atoms have tetrahedral coordination and one-fourth have trigonal coordination. As the sputtering power increases, the bonding configuration of the carbon matrix changes toward equal mixtures of tetrahedral and trigonal bonding.

Banks and Rutledge (1982) produced carbon coatings by dc sputtering with simultaneous bombardment with Ar ions on a deposited substrate. The density of simultaneously ion bombarded coatings was 2200 kg m^{-3} in comparison to 2100 kg m^{-3} for the sputter-deposited-only coating. The electrical resistivity of all coatings were at least 10^{11} Ωcm. The optical properties of both methods of deposition were very similar. Zelez (1983) produced DLC coatings with low internal stresses by a hybrid process involving rf bias sputtering with ultrapure carbon electrodes and plasma decomposition of normal butane. The low stress (<10 MPa) is believed to be related to the H_2 content of the coatings (<1 at %). The coatings had a high density (2800 kg m^{-3}), were good dielectrics, and were transparent in the visible spectrum and infrared (refractive index = 2.0 to 2.4).

Microstructural analyses of hard a-C (DLC) coatings show that the coatings are very smooth and featureless up to magnifications of 50,000X or higher. They show brittle fracture, and there is no evidence of columnar growth. Electron dif-

TABLE 14.33 Summary of Properties of a-C Coatings Prepared Using Various Sputtering Powers, and Properties of Diamond and Graphite

Power,* W	Density, kg m^{-3}	Microhardness, kg mm^{-2}	Specific electrical resistivity, Ω cm	Refractive index at λ = 0.6 μm	Optical band gap, eV	sp^3/sp^2 ratio	Average carbon coordination
5	—	—	2.5×10^3	2.20	0.74	76/24	3.76
10	2110	2340	1.7×10^2	2.23	0.68	75/25	3.75
20	2100	2160	11.0	2.35	0.62	70/30	3.70
50	1900	—	1.0	2.50	0.50	44/56	3.56
100	1850	1960	0.45	—	—	—	—
300	1730	1350	0.20	—	—	—	—
500	1600	—	0.20	2.72	0.40	44/56	3.44
Diamond	—	8000–10,400	10^7–10^{20}	2.41	5.5	1	4
Graphite	—	Very soft	5×10^{-5} (⊥C)	~2.8	~0	0	3

*Target area = 20 cm^2.
Source: Adapted from Savvides and Window (1985).

fraction analysis in a TEM confirm the amorphous nature of the DLC coatings (Savvides and Window, 1985; Tsai et al., 1988; Khan et al., 1988; Marchon et al., 1989). Raman scattering, electron-energy-loss spectroscopy, and ellipsometry suggest that the DLC coatings produced by dc magnetron sputtering are composed of small graphitic (sp^2) crystallites ($\leq$ 2 nm), randomly oriented, with a small percentage (0 to 5 percent) of tetrahedrally coordinated (sp^3) carbon bonds (Cho et al., 1990a, 1990b).

Khan et al. (1988) used scanning tunneling microscopy (STM) (Binning et al., 1982) to study the microstructure of DLC coatings. They made local barrier height (LBH) measurement to obtain information on the local value of the work function (Binning et al., 1982), which can provide chemical information in addition to the topographic map. In this mode, the tip to sample distance is modulated with an amplitude. The ac part of tunneling current leads to the local value of the work function. Work function values for graphite (001) basal plane and diamond (111) are 4.7 eV and 6.9 eV, respectively. A local increase of the LBH on the surface may, therefore, indicate a higher percentage of sp^3 type of carbon. Khan et al. (1988) studied two DLC coatings with different sputtering parameters resulting in different wear properties. Their analysis showed that the coating with better wear performance had a more uniform grain size distribution with a higher percentage (2 to 3 percent) of sp^3 bonded carbon atoms and have a more homogeneous work function distribution as compared to the coatings with poor wear performance.

McKenzie et al. (1982) and Craig and Harding (1982) prepared hydrogenated carbon (a-C:H) coatings by sputtering from a graphite target onto substrates at ambient temperature in a dc cylindrical-post cathode magnetron apparatus using mixtures of argon and hydrogen and argon-C_2H_2 as the sputtering gas, respectively. McKenzie et al. (1982) reported that coatings sputtered in the absence of hydrogen were graphitic in appearance and their resistivity was in the range of 8×10^{-3} to 4 Ωcm. The addition of hydrogen to the plasma (with a hydrogen partial pressure of about 0.6 Pa or 5 mtorr) had a profound effect on resistivity, increasing it by over nine orders of magnitude. Hiraki et al. (1984) and Miyasato et al. (1984) also reported that hydrogenated carbon, produced by rf sputtering a graphite target in a pure hydrogen atmosphere, had high resistivity ($\sim 10^{12}$ Ωcm). The H/C atomic ratio of these coatings was about 0.8 to 1.0 which decreased with increasing power or decreasing gas pressure for a given plasma gas mixture. Miyasato et al. (1984) reported that the hydrogenated coatings had a high ratio (more than 80 percent) of tetrahedrally coordinated (sp^3) carbon bonds based on Raman and IR spectroscopies. The optical band gap was 1.4 to 2.8 eV, which (and optical transmission) increased with the content of hydrogen in the coating. The hardness of the coating was about 2000 kg mm^{-2}.

Bhushan (1990) and Doan and MacKintosh (1988) reported friction and wear data of 20- to 30-nm-thick DLC coatings on magnetic rigid disks sliding against ceramic materials (Al_2O_3-TiC and Mn-Zn ferrite). Typically, the coefficient of friction at ambient conditions (<60 percent RH) ranged from 0.1 to 0.2. It increased rapidly above 40 to 60 percent RH (Fig. 14.57). The increase in friction was believed to be due to liquid-mediated adhesion (meniscus/viscous effects). Bhushan (1990) reported that hydrogenated carbons were generally more resistant to galvanic corrosion due to their high electrical resistivity. Dugger et al. (1990) measured the coefficient of kinetic friction as a function of sliding distance of a Mn-Zn ferrite pin sliding on a DLC-coated magnetic disk (Fig. 14.58). They conducted tests in three environment: vacuum (10^{-6} Pa or 10^{-8} torr), dry air, and 50% RH air. Wear life was defined by the period when the friction increased rap-

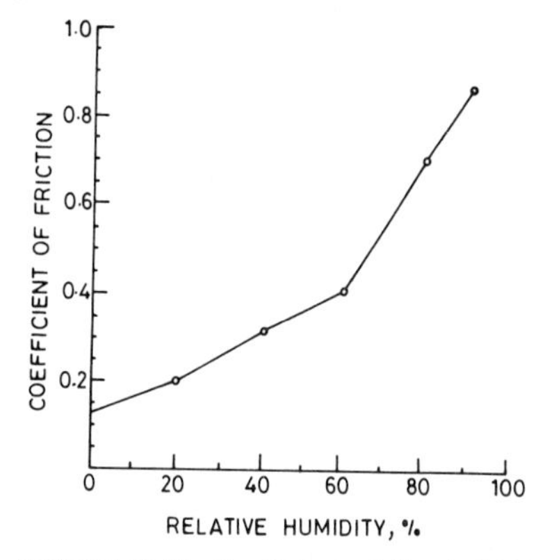

FIGURE 14.57 Coefficient of kinetic friction as a function of relative humidity of an Mn-Zn ferrite slider sliding on dc magnetron-sputtered (DLC) carbon-coated magnetic disk. Normal load = 100 mN; sliding speed = 1 m s^{-1}. [*Adapted from Doan and MacKintosh (1988).*]

idly. They found that the coefficient of friction was low in vacuum and high at 50 percent relative humidity consistent with the results reported by Dimigen and Hubsch (1983–84). However, trends of wear life as a function of the environment were reversed, that is, wear life was longest at 50 percent RH and shortest in vacuum. They reported that debris agglomerated in vacuum which resulted in higher wear rates.

Ion-Plated Carbon Coatings. Graphitic and hard (a-C) carbon coatings by ion plating have been produced by several investigators such as Bewilogua et al. (1979), Mirtich et al. (1985), Teer and Salama (1977), and Weissmantel et al. (1982). Graphitic carbon coatings were deposited using glow-discharge ion plating by Teer and Salama (1977). The carbon was produced in the vapor phase and deposited by two methods. In the first method, the carbon was produced by an electric arc between contacting pointed carbon rods. In the second method, a differentially pumped 15-kW electron beam gun was used to evaporate carbon. The coatings were deposited on copper and mild-steel substrates. These coatings were adherent, had low internal stresses, and were of high graphiticity.

Ion beam ion plating has been used to produce diamondlike carbon coatings. Aisenberg and Chabot (1971) produced hard carbon coatings on room-temperature substrates by direct ion beam deposition of carbon ions extracted from a carbon arc in an argon plasma and later confirmed by Spencer et al. (1976). Miyazawa et al. (1984) produced DLC coatings by using mass separated C^+ ion of 300 and 600 eV. The atomic density of the coating was close to that of diamond. The coatings had high resistivity and were transparent in the visible spectral region. Coatings were found to be amorphous by the transmission electron diffraction analyses. Based on electron microprobe and XPS analyses, they found that coatings consisted of tetrahedrally bonded carbon atoms. Since the sources of these coatings is the carbon ions, they are expected to have little hydrogen as an impurity. The hydrogenated and nonhydrogenated coatings produced by ion beam sputtering of a graphite target were reported by Jansen et al. (1985), Kaplan et al. (1985), and Weissmantel (1981). The data on these coatings will be reported in the next section and will be compared with the data of PECVD coatings.

Weissmantel et al. (1979a, 1979b, 1982) produced hard carbon (a-C:H) coatings from ionized benzene vapor using argon ion beams at acceleration voltage between 0.5 and 1.5 kV. Figure 14.59 illustrates the characteristic features of carbon coatings obtained by ion plating from benzene at various energies. The coatings condensed at low accleration voltages of some tens of volts were found to be polymerlike. The formation of a-C (DLC) requires the supply of considerable activation energy (several hundred electron volts) and momentum to the growing coating. Within that region, the hydrocarbon fragments become increasingly cracked. In contrast, coatings produced at acceleration voltages of over 1 kV

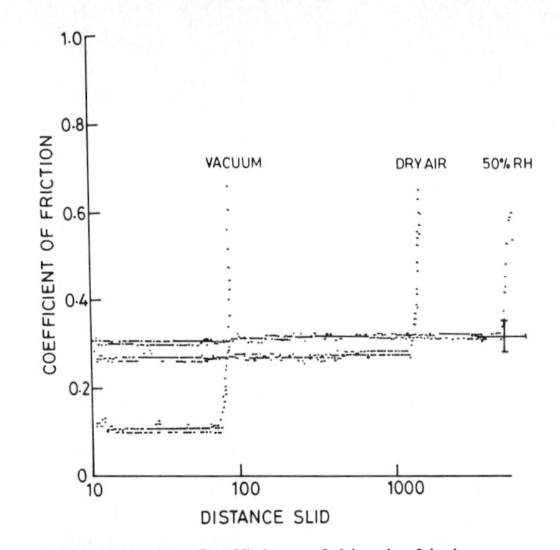

FIGURE 14.58 Coefficient of kinetic friction as a function of sliding distance of an Mn-Zn ferrite pin in various environments, sliding on a dc magnetron-sputtered (DLC) carbon-coated magnetic disk. Normal load = 150 mN; sliding speed = 200 r/min. [*Adapted from Dugger et al. (1990).*]

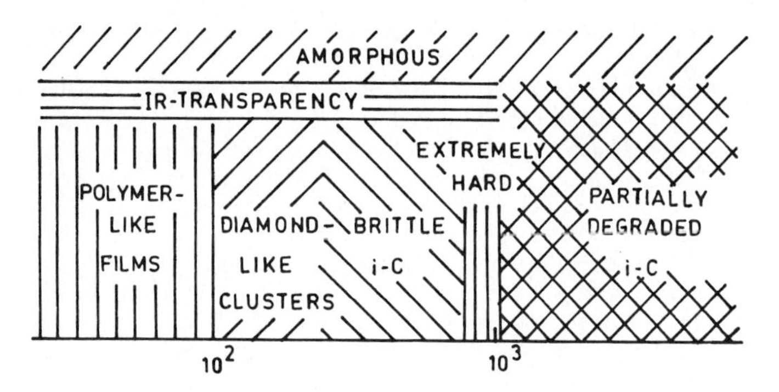

FIGURE 14.59 General properties of a-C coating deposited at various acceleration voltages by ion plating from benzene. [*Adapted from Weissmantel et al. (1982).*]

were found to be nontransparent, even in the IR region, and contained some graphite. Second radiation damage due to impinging energetic ions was believed to be responsible for these effects.

The majority of dense a-C:H (DLC) coatings were found to be amorphous. Sometimes crystalline phases in an amorphous matrix have been observed, especially in coatings of lower hydrogen content (Weissmantel et al., 1980, 1982). Graphite crystallites were identified by Mori and Namba (1983). In addition, other crystalline phases have been found, the structure of which have not been definitely established.

Figure 14.60 shows a plot of microhardness as a function of the acceleration voltage (Weissmantel et al., 1982). The distinct maximum indicates that extremely hard coatings (~6000 kg mm^{-2}) are obtained only within a rather limited range of ion energies (also see Mori and Namba, 1983). However, electrical resistivity decreased with increasing acceleration voltage. This was believed to be due to the decrease of hydrogen concentration. McKenzie et al. (1982) have reported that the ratio of fourfold to threefold coordinated carbon atoms increases when more hydrogen is present in the a-C coatings (also see Weissmantel et al., 1979a, 1979b).

Weissmantel et al. (1982) also measured the coefficient of friction of a-C coatings applied on steel in a cylinder-ring device. They found that these coatings had a low coefficient of friction (0.09 to 0.18). The a-C coatings on iron and steel substrates were found to be extremely inert against chemical attack by acids, alkalies, and organic solvents.

The ion beam deposition appears to produce DLC coatings with very high hardness due to very high energy surface bombardment during the deposition. One significant disadvantage of direct ion beam deposition is the low rate of growth compared to rf or dc PECVD methods. Typical growth rates are on the order of 0.1 nm s^{-1} compared to 2 nm s^{-1} or more for rf-discharge processes. Also, ion beam deposition takes place strictly in a line-of-sight from the source.

Plasma-Enhanced Chemical-Vapor Deposited (PECVD) Carbon Coatings. Dense carbon coatings have been produced by cracking of hydrocarbon gases using a variety of rf- and dc-discharge reactors (Andersson, 1977, 1981; Andersson et al., 1979; Andersson and Berg, 1978; Berg and Andersson, 1979; Bubenzer et al., 1983; Enke et al., 1980; Holland and Ojha, 1976; 1979; Jones and Stewart, 1982; Kaplan et al., 1980; Moravec and Orent, 1981; Nyaiesh and Holland, 1984; Ojha, 1982; Ojha and Holland, 1977*a*, 1977*b*, 1977*c*; Ojha et al., 1979). Carbon coatings applied by PECVD have several remarkable properties such as highly insulating (10^7 to 10^{14} Ωcm), negative temperature coefficient of conductivity

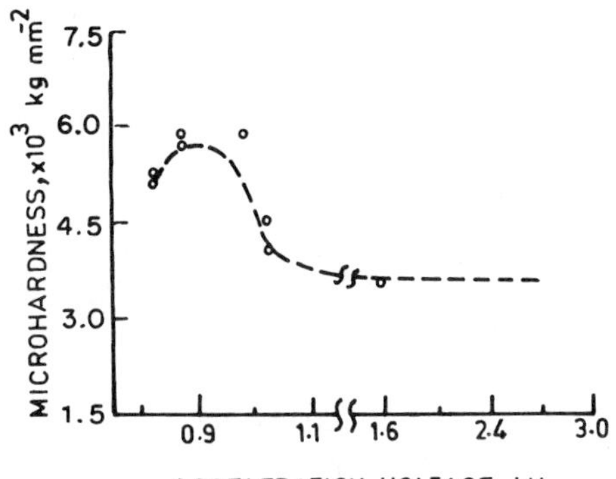

FIGURE 14.60 Microhardness as a function of the acceleration voltage of a-C coating deposited by ion plating from benzene. [*Adapted from Weissmantel et al. (1982).*]

(10^{-2} °C^{-1}), high dielectric strength (breakdown voltage is about 10^6 V cm^{-1}), chemical inertness toward acids and organic solvents, and extreme hardness (up to 3000 HV or higher).

The requirement to grow insulating coatings, sometimes on insulating substrates, strongly favors rf- over dc-discharge techniques. For this reason, dc biasing is ineffective since an insulating layer in contact with the plasma charges up to a floating potential which is close to the plasma potential. The capacitively coupled parallel-plate rf glow-discharge method is most commonly used (Andersson, 1981; Angus et al., 1986; Holland and Ojha, 1976).

Most experiments have been performed at 13.56 MHz. Hydrocarbon pressures in the range of ~1.5 to 15 kPa (10 to 100 mtorr), substrate temperatures in the range of 25 to 50°C, and rf powers of about 1 W cm^{-2} cathode area (resulting into self-bias voltages of typically 1000 to 2000 V) are used most often. Practically, any hydrocarbon gas or vapor can be used as a precursor material. The hydrocarbon gases most commonly used are methane, ethane, butane, propane, acetylene, ethylene, propylene, benzene, and others. The type of source gas influences the growth rate of the coatings for a given discharge power. Typical deposition rates range from about 0.5 μm h^{-1} (methane) to about 10 μm h^{-1} (benzene). There is only a minor influence of the type of source gas on the coating properties. For instance, a-C:H coatings prepared under similar conditions from benzene and *n*-hexane were found to have nearly identical properties, including the H/C ratio and the relative concentrations of sp^3, sp^2, and sp^1 hybridized carbon atoms (Angus et al., 1986; Wild et al., 1987). The hard, dense (density ~1800 kg m^{-3}) coatings consist of typically about two-thirds sp^3 bonded and one-third sp^2 bonded carbon, and contain typically 35 at % hydrogen.

The impact of energy of the impinging ions on the substrate has a major influence on the type of coating (Nyaiesh and Holland, 1984). Impact energy increases with an increase of the power density (or the self-bias voltage) and decreases with an increase of the hydrocarbon gas pressure (Bubenzer et al., 1983). Usually, at low power densities and rough vacuum pressures, the degree of hydrocarbon dissociation both in the gas phase and from ion impact at the receiver is insufficient to rupture all C-H bonds, and so accordingly a soft polymer film formation take place due to unsaturated hydrocarbon components (Nadler et al., 1984*a*, 1984*b*). This effect is illustrated in Fig. 14.61 for a carbon coating produced by PECVD method. As the rf power density increases and the gas pressure is reduced, one goes from plasma polymers to hard hydrocarbons to hard carbon. The C-H bonds decrease and C-C bonds (carbon content) increase as the rf power density in-

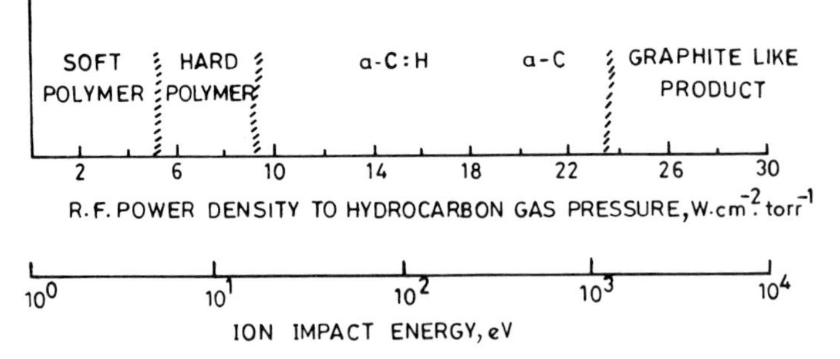

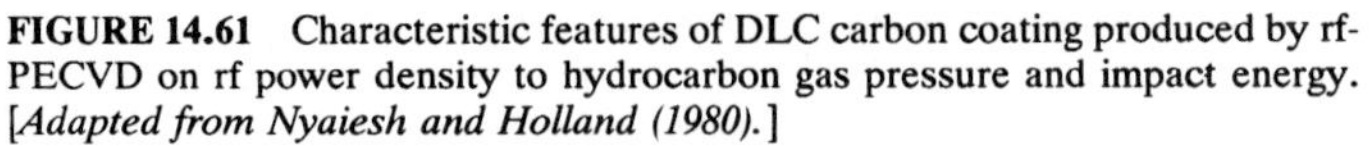

FIGURE 14.61 Characteristic features of DLC carbon coating produced by rf-PECVD on rf power density to hydrocarbon gas pressure and impact energy. [*Adapted from Nyaiesh and Holland (1980).*]

creases and gas pressure is reduced (Miyazawa et al., 1984; Zelez, 1983). The further increase in rf power density produces excessive heat which may lead to graphitization of carbon coatings. Angus et al. (1986) have reported that dense hydrogenated carbons are produced at ion impact energies of roughly 50 to 1000 eV. We note that, in general, the total applied potential is not equal to the ion impact energy.

It has been reported that the structure (sp^2/sp^3 bonding ratio) and coating properties, including hydrogen content, density, microhardness, refractive index, and optical band gap of DLC coatings can be varied by changing the average energy of ions impinging on the substrate which is proportional to bias voltage (or rf power)/(gas pressure)$^{1/2}$ (Bubenzer et al., 1983; Gambino and Thompson, 1980; Fink et al., 1983, 1984; Kurokawa et al., 1987; Nyaiesh and Holland, 1984; Pethica et al., 1985; Vandentop et al., 1990; Whitemell and Williamson, 1976; Yamamoto et al., 1988). Table 14.34 presents the properties of four hydrogenated coatings produced by the rf-PECVD technique using various processing conditions (Fink et al., 1984; Yamamoto et al., 1988). Yamamoto et al. (1988) deposited coatings at three substrate temperatures: − 100°C, room temperature, and 230°C. The substrate temperature had little effect on the coating hardness. The coating density in Table 14.34 ranges from 1490 to 1750 kg m^{-3} and thus is significantly lower than the crystalline carbon (diamond and graphite) but higher than in glassy carbon (~1500 kg m^{-3}) or hydrocarbon polymers (~1000 kg m^{-3}). Hydrogen concentration shows a strong dependence on the ion impact energy. A relative concentration of sp^2/sp^3 increases with a decrease of the hydrogen concentration.

The coatings produced at low self-bias voltage (lower rf power) have a large amount of hydrogen and have a low-density network structure; thus, these coatings are soft, though the proportion of sp^3 sites is very large. With increasing self-bias voltage, the coatings attain a high-density network structure, which is supported by decreasing hydrogen content, and so the hardness of the coatings increases.

The hardness and density decreases as the hydrogen content increases, though the proportion of sp^3 site increases, that is, as the local bonding environment becomes more diamondlike. This behavior may occur because hydrogen, which is monovalent, can serve only as the terminating atom on the carbon skeletal network. However, the average coordination number of carbon to other carbon atoms changes only slightly as the hydrogen content increases and remains close to the optimal value of 2.4 predicted by Angus and Jansen (1988). The large decrease in hardness and density is most likely caused by the development of regions of soft polymeric material and reduction in cross-linking due to increased hydrogen incorporation. When these soft regions have percolated through the entire structure, the hardness will decrease greatly.

The optical band gap of carbon with small concentration or no hydrogen remains relatively small (on the order of few tenths of an electron volt) and is only a small fraction of the band gap of sp^3 bonded carbon (5.5 eV). The optical band gap is controlled by the fraction of tetrahedral versus trigonal bonding. At higher concentrations of hydrogen, the sp^3/sp^2 ratio of carbon coating is higher and is responsible for higher band gap. Since unhydrogenated carbon coatings are optically highly absorbing and have a relatively high electrical conductivity. As the hydrogen content increases, the optical band gap increases, accompanied by an increase in optical transparency and resistivity (10^9 to 10^{14} Ω cm) and a decrease in the refractive index. The polymerlike coatings (with high hydrogen concentration) produced at low ion impact energy have a low refractive index ~1.8. We

TABLE 14.34 Properties for the Four a-C:H Coatings Produced by rf-PECVD*

Material	Negative self-bias voltage, V	Gas pressure, Pa	Bias volt./(gas press.)$^{1/2}$, V/Pa$^{1/2}$	Hydrogen conc., at %	sp^2/sp^3 ratio	Density, kg m^{-3}	Micro-hardness, kg mm^{-2}	Refractive index	Optical band gap, eV
Diamond†	—	—	—	0	0/100	3000–3500	8000–10,400	2.41	5.5
Coating	200	3.2	111	60	10/90	1490	—	1.78	2.05
Coating	400	5.0	180	50	13/87	1500	1200	1.82	1.60
Coating	800	3.6	420	40	20/80	1590	3000	1.93	1.20
Coating	1200	1.9	869	25	38/62	1750	2200	2.10	0.95
Graphite†	—	—	—	0	100/0	2300–2700	Soft	2.8 (⊥C)	0

*Yamamoto et al. (1988) reported that substrate temperature from −100 to 230°C had little effect on the properties.

†Literature values for bulk diamond and graphite are included for comparisons.

Source: Adapted from Fink et al. (1984), and Yamamoto et al. (1988).

note that hardness and optical band gap are diametrically opposed. It is not possible to deposit amorphous carbon coatings with the simultaneous properties of extreme hardness and high optical band gap.

Jones and Stewart (1982) and Meyerson and Smith (1980, 1981) reported that optical band gap decreased from 2.1 to 0.9 eV and resistivity decreased (11 orders of magnitude) with an increase in the deposition temperature from 25 to 375°C. Berg and Andersson (1979) reported that coatings produced at low deposition rates were highly insulating ($\sim 10^{14}$ Ωcm) whereas the high deposition rates were more conducting ($\sim 10^{8}$ Ωcm). Kaplan et al. (1985) measured the structure and properties of hydrogenated carbon coatings with various hydrogen concentrations, prepared by the ion beam sputtering and rf- and dc-PECVD techniques. The sp^2 and sp^3 bonding sites of hydrogenated carbon were measured by solid-state 13_C magic angle spinning nuclear magnetic resonance (NMR). Atomic percent hydrogen content was determined by combustion analysis and by proton NMR. Coating density measurements were performed using the float-sink technique. The hardness was measured by determining the load at which a stainless-steel tip of radius 50 μm scratched the coating surface. The optical absorption was measured as a function of wavelength using a spectrophotometer and the optical absorption data was used to calculate the optical band gap.

Table 14.35 summarizes the data. We note that the structure and properties of coatings depend critically on the deposition technique and the process parameters. We note that the hydrogen concentration changes by changing the process parameters. As seen previously, as the hydrogen incorporation increases, the sp^2/sp^3 ratio, density, and hardness decrease, and the optical band gap increases. The coating produced by ion beam sputtering (ion beam energy ~1.2 keV) produced a coating with high hardness and high density compared to that produced by PECVD. It has been suggested that during high-energy surface bombardment, weaker bonds are removed from the growing coating. Thus, the more tightly bonded three- or fourfold coordinated carbon atom will survive over singly bonded hydrogen. Also, since double-bond hydrogenation is an exothermic process, graphitic bonding is favored over tetrahedral bonding in higher energy growth environments. Coatings thus formed (e.g., ion beam sputtering) should be less hydrogenated, mechanically harder, and have a higher density than those formed in a lower energy environment (e.g., PECVD).

During carbon deposition by rf plasma, Padmanabhan et al. (1986) ion-bombarded the coated surface with a 100- and 400-keV krypton ions using a 600-keV heavy-ion accelerator. This increased coating adhesion and its hardness up to a factor of 2. Hardness increased with an increase in the Kr^+ energy.

The majority of dense (diamondlike) a-C:H coatings are found to be amorphous (Berg and Andersson, 1979; Fink et al., 1983, 1984; Kaplan et al., 1985; Khan et al., 1983; Memming, 1986; Memming et al., 1986; Mori and Namba, 1983; Nadler et al., 1984a). Sometimes crystalline inclusions in an amorphous matrix have been observed, especially in coatings with lower hydrogen content. In addition, other crystalline phases have been found, the structures of which have not been definitely established. With respect to the amorphous material, different structural models have been proposed (Angus et al., 1986).

To study the structure and bonding of the DLC coatings, Fink et al. (1983) used EELS. As the backscattering amplitude of carbon is much higher than that of hydrogen, only the carbon structure is reflected in the EELS. Carbon K-edge spectra of a DLC coating as deposited and after heat treatment at 600 and 1000°C for 1 h are shown in Fig. 14.62(*a*). Spectra for single-crystal diamond, single-crystal graphite, and amorphous evaporated carbon (AC) coating (produced by

TABLE 14.35 Properties of a-C:H Coatings Produced by PECVD and Ion Beam Sputtering

Material	H, at. %	sp^2/sp^3	Density, kg m^{-3}	Hardness, g to scratch	Optical band gap, eV
Diamond*	0	0/100	3000–3500	Extremely hard	5.5
rf anode PECVD	61	14/86	1020	160	4.1
rf cathode	58	20/80	1170	250	4.0
dc anode	47	50/50	1460	700	1.7
dc cathode	31	60/40	1700	1250	—
Ion beam sputtering (88% H_2 in Ar)	35	62/38	1530	>2000	1.2
Ion beam (0% H_2 in Ar)	0	—	2250	>2000	0.7
Graphite*	0	100/0	2300–2700	Soft	0

*Literature values for bulk diamond and graphite are included for comparisons.
Source: Adapted from Kaplan et al. (1985).

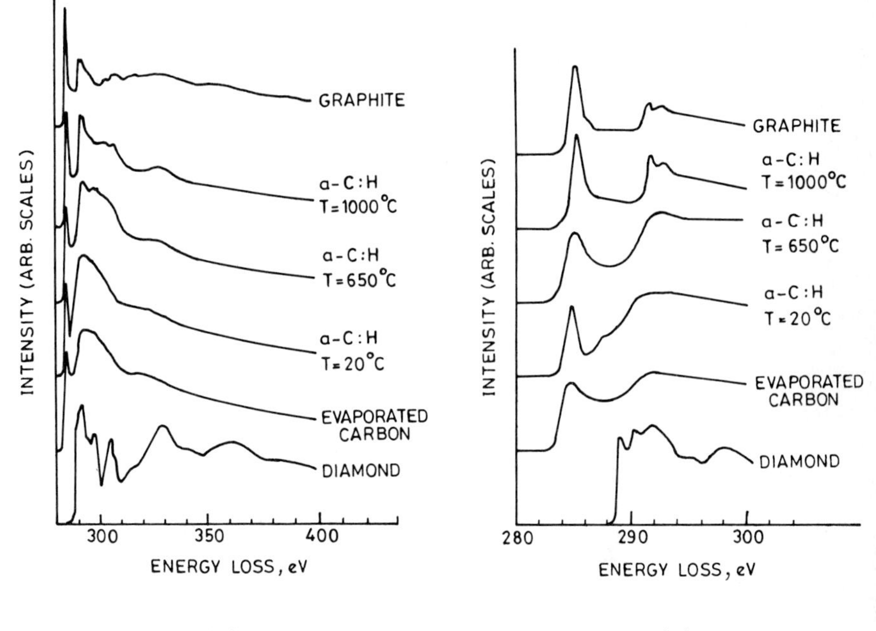

FIGURE 14.62 (*a*) Carbon K-edge. (*b*) Near-edge structure of carbon K-edge of single-crystal diamond, single-crystal graphite, evaporated carbon, and DLC carbon coating as produced by rf-PECVD and after heat treatment at 650 and 1000°C for 1 h (negative self-bias = −400 V, gas pressure = 5.0 Pa or 37.5 mtorr). [*Adapted from Fink et al. (1983).*]

electron beam evaporation) are also shown for comparisons. In the spectra of a-C:H and AC, the intensity oscillation are strongly damped. No oscillations beyond 50 eV above the absorption edge are detectable, whereas in the spectra of graphite and of diamond, oscillations beyond 100 eV above the absorption edge are clearly visible. After heating the a-C:H coating, the oscillations get less damped. At 1000°C, the spectrum shows some similarity to the graphite spectrum. Thus, Fink et al. (1983) concluded that no diamondlike crystallites were detected in the as-deposited coatings, and after heat treatment at 1000°C for 1 h, the coating transformed to a graphitelike structure. A further indication of the nonexistence of diamondlike crystallites can be derived from the near-edge structure shown in Fig. 14.62(*b*). The onset of carbon K-absorption edge for diamond and graphite is near 288.9 and 284 eV, respectively. Thus, the large peak at about 285 eV in the spectra of a-C:H and AC reveals the existence of carbon atoms having a sp^2 electron configuration. At 1000°C, the near edge structure is very similar to that of graphite. Fink et al. (1983) obtained further information on the ratio of sp^3 to sp^2 electron configuration by analyzing the valence excitations.

Raman scattering is a very useful tool to study existence of any diamondlike crystallites. Figure 14.63 shows Raman spectrum of DLC coating produced by rf-PECVD. The spectra for single-crystal diamond and single-crystal graphite are also included for comparisons. We note that Raman spectra of the DLC coating is different from those of diamond and graphite; therefore, we believe that DLC

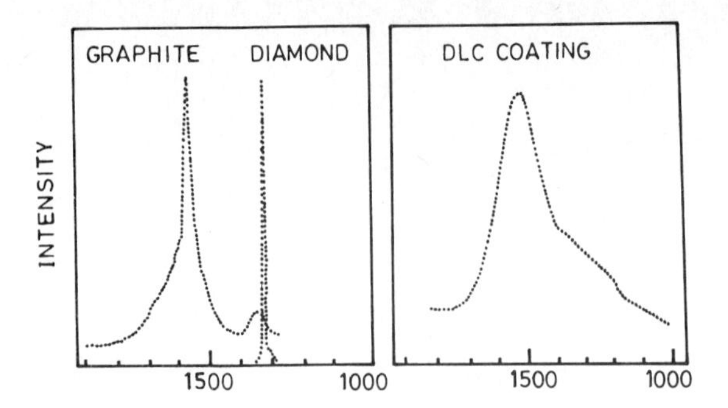

FIGURE 14.63 Raman spectra of single-crystal diamond, single-crystal graphite, and DLC carbon coating on a silicon substrate produced by rf-PECVD. [*Adapted from Kurokawa et al. (1987).*]

coatings have a structure different than that of diamond and graphite. XPS analysis showed that these coatings are amorphous (Berg and Andersson, 1979). TED analysis showed that rf-PECVD coatings consisted of small diamond crystallites in the amorphous matrix (Mori and Namba, 1983). IR spectroscopic analysis showed that coatings produced by rf-PECVD exhibited a polymerlike amorphous structure. In an ellipsometry study, Khan et al. (1983) measured optical constants of rf-PECVD and ion beam sputtered coatings. They found that coatings were essentially amorphous.

The DLC coatings are under significant compressive internal stresses ranging from about 1 to 10 GPa (Dimigen and Hübsch, 1983–84; Turner et al., 1980). For this reason, DLC coatings thicker than about 1 μm have a tendency to delaminate from the substrate (Whitmell and Williamson, 1976). The stresses may be caused by the presence of dissolved hydrogen in the coating. The magnitude of the stresses was found to be closely dependent on the hydrocarbon gas pressure and the rf voltage in a rf plasma system (Dimigen and Hübsch, 1983–84) (Fig. 14.64). Relatively high compressive stress values are observed in the diamondlike region. Relatively low values are obtained with polymeric coatings deposited at high ethane (C_2H_4) pressure and low electrode potential, and with graphitic coatings deposited at a low ethane pressure and high electrode potential. The low compressive stress usually found in polymeric coatings is not so surprising in view of their softness. The lower compressive stress of graphitic coatings compared to that of DLC coatings is probably connected with reduced incorporation of the hydrogen contained in the plasma. Zelez (1983) has reported the stress-free (very low stress < 10 MPa) coatings produced by a hybrid process involving rf-bias sputtering with ultrapure C electrodes and plasma decomposition of normal butane. Nir (1984) has studied the shape of stress-relief features.

The adhesion of the coatings varies widely depending on the substrate and the method of preparation. Best adhesion is obtained on substrates that form carbides, e.g., Si, Fe, and Ti. Good adhesion has also been obtained on Ge and quartz (Angus et al., 1986; Sander et al., 1987). Sander et al. (1987) conducted depth profile analyses using Auger and XPS of a-C:H coatings deposited on silicon substrate. They reported that substantial amounts of silicon carbide (roughly 10 nm thick) was present in the a-C:H (Si) interface for the coatings with good adhesion and high hardness.

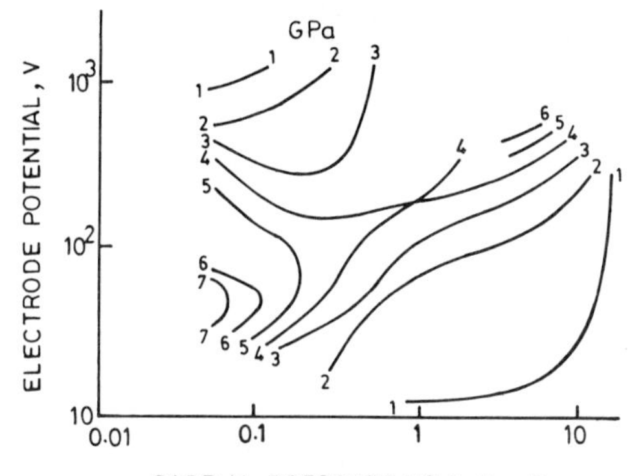

FIGURE 14.64 Lines of equal compressive internal stress in DLC carbon coatings, as a function of the partial pressure of ethane ($p_{C_2H_4}$) and the electrode potential V_E during rf-plasma deposition. [*Adapted from Dimigen and Hubsch (1983–84).*]

The dense a-C:H coatings are inert to organic solvents and inorganic acids, including HF. Of special interest is the fact that the coatings are not attacked by a solution of three parts H_2SO_4 and one part HNO_3 at 80°C (Angus et al., 1986). This reagent will dissolve all hydrocarbon polymers and graphitic carbons but will not attack diamond. These coatings would presumably be oxidized by molecular oxygen and water vapor at elevated temperatures.

Friction and wear performance of DLC coatings were found to be strongly dependent on the water vapor content and partial gas pressure in the test environment. Holland and Ojha (1976) reported a coefficient of friction of 0.2 to 0.28 in a steel pin sliding on the carbon coating. Dimigen and Hübsch (1983–84), Enke et al. (1980), and Memming et al. (1986*a*) measured the coefficient of friction and wear rate in a pin-on-disk apparatus with a steel pin sliding on the carbon-coated silicon substrate. The kinetic friction coefficient as a function of partial pressure of water vapor (or relative humidity) in nitrogen atmosphere is shown in Fig. 14.65. The load of the pin was 1 N, and the speed of disk was 1600 r/min. The hardness and electrical resistivity of the coating were 2000 kg mm^{-2} and 10^{12} Ωcm, respectively. In the range of p_{H_2O} = 10^{-6} to 10 Pa (10^{-8} to 10^{-1} torr), the coefficient of friction had the extremely low value of 0.01 to 0.02. Above 10 Pa (10^{-1} torr), friction rises as long as the relative humidity is increased attaining a final value of 0.19 at a relative humidity of nearly 100 percent in nitrogen atmosphere. The coefficient of friction in dry air was 0.6. High friction is related to material transfer during sliding. Doan and MacKintosh (1988) and Bhushan (1990) have reported similar trends for sputtered DLC coatings. This behavior is contrary to graphite, which seems to need a certain amount of water to show low friction (Savage, 1948; Savage and Schaefer, 1956; Rowe, 1960). This behavior is also in contrast to those of Bowden and Young (1951) who reported that the friction of bulk graphite, carbon, and diamond decreases with increasing water vapor pressure.

The durability of the rf-PECVD carbon was measured at various contents of water vapor by Enke et al. (1980). They found that the durability of the coating at low relative humidity was lower than at high humidity.

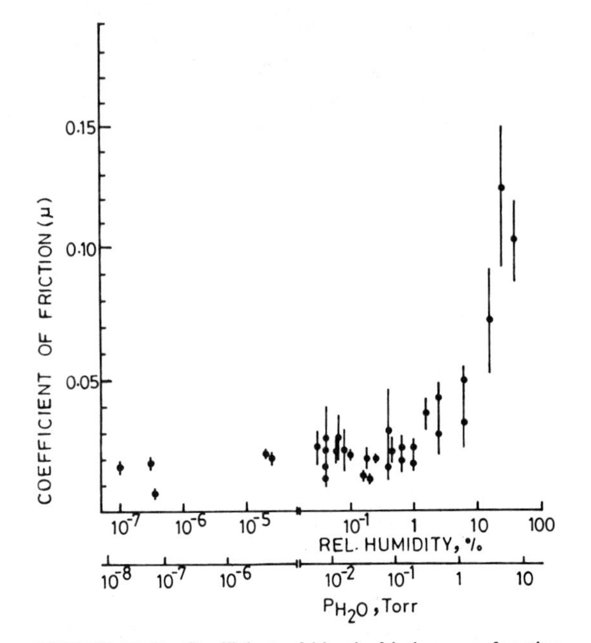

FIGURE 14.65 Coefficient of kinetic friction as a function of water vapor partial pressure (p_{H_2O}) in N_2 atmosphere of a steel pin sliding on an rf-plasma CVD (DLC) carbon coating on a silicon substrate. [*Adapted from Enke et al. (1980).*]

14.4.6.2 Diamond Coatings. Many studies have been conducted to produce cubic diamond coatings at subambient pressures (~10 kPa or 75 torr) and high substrate temperatures (~900 to 1000°C) from (~99 percent) hydrogen-methane mixtures where it is in the metastable form of carbon and usually contains only small amounts of hydrogen impurity. Systematic studies of diamond vapor deposition techniques began primarily in the 1950s in the USSR and the United States (Angus and Hayman, 1988; Badzian et al., 1987; DeVries, 1987; Spear, 1989). In the first two decades of research on chemical vapor-deposited diamond in the USSR, the reported diamond growth rates were low (few nanometers per hour) and the simultaneous codeposition of graphitic carbon was always a problem (Deryagin and Fedoseev, 1975). The early research primarily involved the thermal decomposition of hydrocarbons and hydrogen-hydrocarbon gas mixtures, with no additional activation of the gas.

Similar research was being conducted in the United States. Angus and coworkers (Angus et al., 1968; Chauhan et al., 1976; Poferl et al., 1973) continued to pursue these techniques during the 1960s and early to mid 1970s, and they obtained similar results as Deryagin and coworkers in the USSR. As an example, Chauhan et al. (1976) deposited carbon coatings on 0- to 1-μm diamond-seed crystals in the temperature range of 1140 to 1475°C, methane partial pressures from 10^{-1} to 5×10^3 Pa (10^{-3} to 44 torr) and hydrogen partial pressures from 0 to 2 kPa (0 to 16 torr). A relatively rapid rate of diamond growth decayed exponentially to a final slow constant rate as graphite carbon was codeposited. The maximum initial growth rate was about 0.5 μm/day. Dilution of methane with hydrogen decreased the nucleation rate of graphitic carbon relative to that of diamond resulting in an increase in the diamond yield. These coatings had densities very close to that of diamond and were essentially pure carbon. The crystalline quality, however, could not be established. In these early studies, when hydrocarbons were cracked in the absence of a diamond substrate, pyrolitic carbons were formed. These carbons were soft, black, and conductive and were based on a local graphitic structure.

In the early 1980s, Deryagin's group reported the use of gas activation techniques which resulted in dramatic increases in diamond growth rates. While eliminating much of the graphite codeposition (Spitsyn et al., 1981; Fedoseev et al., 1984), these studies illustrated the importance of: (1) using excess hydrogen in the deposition gas and (2) activating the gas before deposition to produce super equilibrium concentrations of atomic hydrogen. Hydrogen is required for efficient growth. The addition of excess hydrogen (>98 percent) to the hydrocarbon precursor gas leads to less graphite codeposition. It is atomic hydrogen that has to be present in the gas phase in super equilibrium concentration. Activated hydrogen etches graphite at a much higher rate than it does diamond. In addition, activating the gas prior to deposition increases the diamond growth rates to few micrometers per hour. Three methods of obtaining hydrogen-atom super equilibrium were used in a diamond vapor growth system: (1) catalytic, such as heated Pt or dissociating H_2, (2) electric discharge; and (3) a heated tungsten filament located just in front of the substrate.

Spitsyn et al. (1981) showed photographs of deposits that exhibited many faceted crystals with dimensions up to about 30 μm. They also showed an electron diffraction pattern characteristic of crystalline cubic diamond and listed a number of physical property measurements on grown coatings which were characteristic of diamond. Growth rates were on the order of few micrometers per hour.

Following the paper by Spitsyn, a number of papers originated from the United States and Japan. Matsumoto et al. (1982) used the Deryagin's technique of activating the precursor deposition gases with a filament heated to approximately 2000°C about 1 cm from the 1000°C growth surface. This hot filament method (HFCVD) has since been used and studied extensively by many researchers. Matsumoto (1985) reported on diamond growth in a rf glow-discharge system (rf-PECVD). This method also has been combined with a heated filament to improve on the diamond growth rates (Komatsu et al., 1987). Sawabe and Inuzuka (1986) later developed an electron-enhanced method (EECVD) for growing diamond with a mixture of CH_4 + H_2 gas at growth rates of 3 to 5 $\mu m\ h^{-1}$, as opposed to many of the earlier rates of about 1 $\mu m\ h^{-1}$ and less. The successful use of a microwave plasma-enhanced CVD (MPECVD) method to activate the growing crystalline diamond was reported by Kamo et al. (1983). Kawarada et al. (1987) developed a magnetomicrowave system so that large areas could be coated. Fujimori et al. (1986) deposited conducting (1 to 10^{-3} Ωcm) boron-doped diamond coatings by MPECVD method and using B_2H_6/CH_4 and H_2 reactant gas mixtures.

These methods typically utilize total pressures of about 10^4 Pa (76 torr) or less, pressures at which "cold plasmas" operate. Recently Matsumoto et al. (1987) grew diamond particles and coatings on Mo substrates using an rf plasma with a 10^5 Pa (760 torr) Ar-H_2-CH_4 mixture. Their growth rates were on the order of 60 $\mu m\ h^{-1}$, much higher than typical growth rates of few $\mu m\ h^{-1}$ or less. The molybdenum substrates were placed on a water-cooled holder, and held at 700 to 1200°C. Sato et al. (1987) reported that the nature of precursor hydrocarbon gas had little effect on the deposition behavior. Spear (1989) reported that diamond coatings grown from hydrogen 1 mole % methane mixtures usually contain only small amounts of hydrogen impurity (<500 ppm). In general, he found that greater amounts of hydrogen in the feed gas leads to less high hydrogen impurity in the diamond deposit. Spear (1989) reported that small amounts of oxygen added to the precursor gas accelerate the growth rate of diamond coatings. The temperature dependence of the rate of diamond growth exhibits a maximum that is initially increases with temperature and then decreases. This behavior is a manifestation of the competition between the growth of diamond and graphite.

Spear (1989) studied the mechanism of formation of diamond coatings. Graphite is the stable form of carbon under the conditions used in vapor-deposition of crystalline diamond. Why is it then possible to grow diamond at less than atmosphere pressure in the temperature range of 750 to 1100°C? According to Spear, the "metastable" diamond growth rests on the gas-solid interface in the carbon-hydrogen system. The vapor-growth process does not involve just elementary carbon, the one component which is represented on the phase diagram, but it also involves hydrogen. A diamond carbon surface saturated with sp^3 C-H bonds is more stable than a carbon surface free of hydrogen. Once a surface carbon is covered by another diamond growth layer, then that covered carbon possessing sp^3 C-C bonds is metastable with respect to graphitic carbon. This can be understood by close examination of diamond and graphite structure. A diamond structure with puckered {111} planes stacked in their ABCA...sequence is shown above the stacking of the hexagonal planes of graphite in Fig. 14.66. The diamond {111} planes are shaded to emphasize the relationship of the diamond structure to that of graphite. Hydrogen atoms are shown as satisfying the "dangling sp^3 bonds" of the carbons on the top diamond plane. Lander and Morrison (1966) were the first to hypothesize that hydrogen can stabilize a diamond surface by forming sp^3 C-H bonds with the surface carbons. Without the hydrogens maintaining the sp^3 character of these surface carbons, it is easy to imagine the {111} diamond planes collapsing into the more stable planar graphite structure during the growth process.

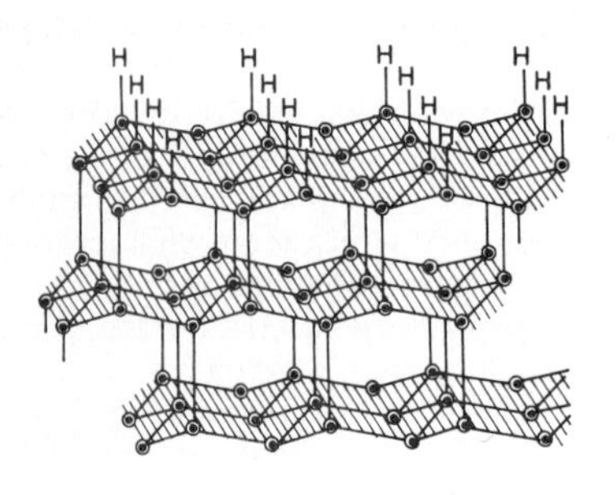

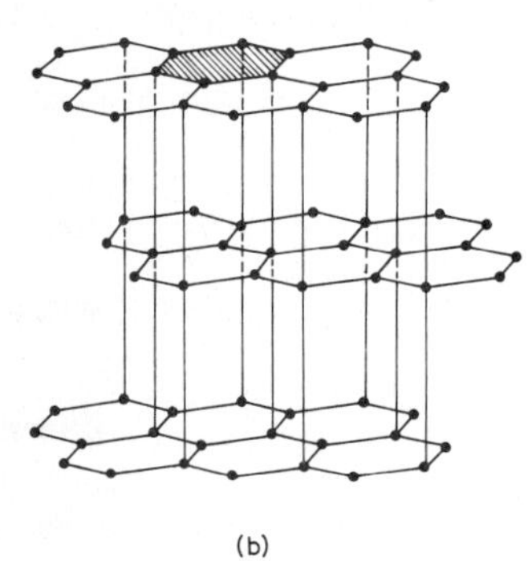

FIGURE 14.66 Schematic diagrams showing the similarities in the crystal structures of: (*a*) Diamond. (*b*) Graphite. The hydrogen atoms bonded to the surface carbons depict their role in stabilizing the diamond surface structure.

It should be pointed out that since all vapor-deposited carbon coatings are in a metastable state, they decompose if used at elevated temperatures in contact with reactive materials. Nadler et al. (1984*a*, 1984*b*) observed a loss of hydrogen and a decomposition started at 400°C. Also Fink et al. (1983) have reported that a decomposition or graphitization occurs upon heating. The coatings were completely transformed to graphite at 1000°C.

The diamond coatings that are currently being produced evolve from single crystals, like those shown in the SEM images in Fig. 14.67. When these crystallites grow together, dense diamond coatings, such as the one shown in Fig. 14.67, can be prepared (Spear, 1989). Raman-scattering spectroscopy is the most sensitive technique for identifying the presence of crystalline diamond in mixtures of various forms of carbon. The Raman-scattering efficiency for the sp^2 bonded graphite is more than 50 times greater than that for the sp^3 bonded diamond; therefore, small amounts of sp^2 bonded carbon in the diamond coatings can be readily detected. Figure 14.68 shows a Raman spectra for a diamond coating deposited on a silicon substrate by MPECVD at the Diamond Materials Institute at Pennsylvania State University, State College, Pennsylvania. The peak at 1332 ± 1 wave number is characteristic of diamond. The broad peaks occurring in the 1500 to 1600 wave number range are characteristic of disordered sp^2 type carbons in the deposit (Knight and White, 1989).

Diamond coatings have been vapor-deposited on a wide variety of substrate materials, though the dominant substrate has been single-crystal silicon. Examples of substrates include Cu, Au, Si, Ta, Mo, W, Ni, SiO_2, SiC, WC, diamond, and graphite. Studies on coating various oxide, carbide, and nitride cutting-tool materials are also underway. The diamond nucleation rates and adhesions vary with the tendency to form intermediate carbide layers such as SiC or Mo_2C (Spear, 1989).

Very little data is available on the physical properties of the diamond coatings. Matsumoto (1985) reported that polycrystalline diamond coatings prepared by rf-PECVD method on Mo substrates, had electrical resistivity in the range of 10^8 to

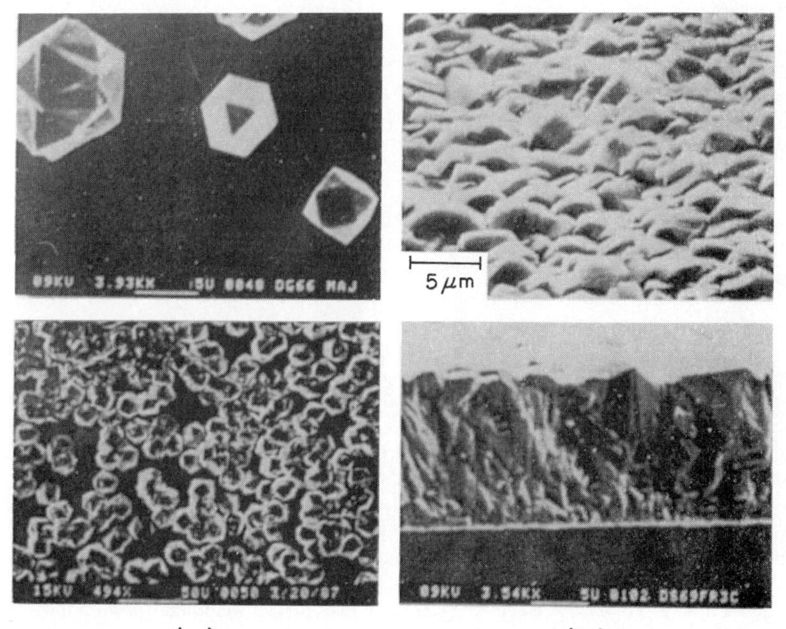

FIGURE 14.67 SEM images of microwave plasma-enhanced chemical-vapor-deposited: (*a*) Diamond crystals in the early phase of deposition. (*b*) Surface and a cross section of a diamond coating on a silicon substrate. [*Adapted from Spear (1989).*]

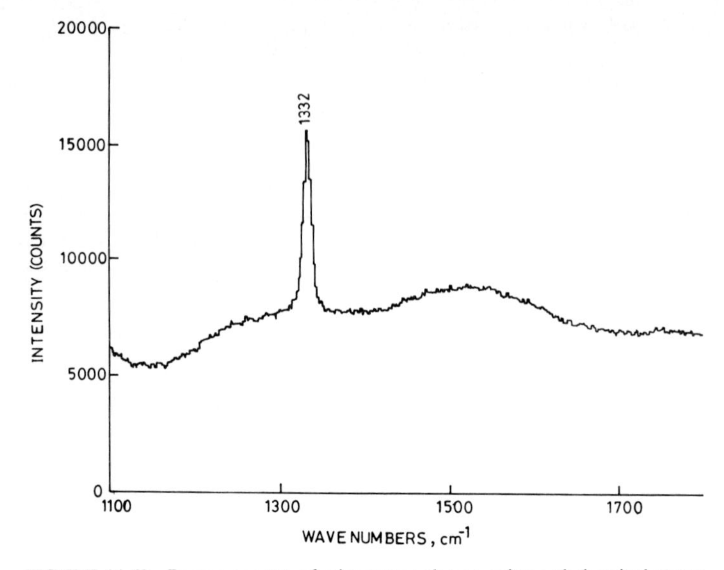

FIGURE 14.68 Raman spectra of microwave plasma-enhanced chemical-vapor-deposited diamond coating on single-crystal silicon. Raman peak coincides with that of diamond crystal.

10^{10} Ωcm and microhardness ranged from 7000 to 12,000 kg mm^{-2}. Sawabe and Inuzuka (1986) reported that electron-assisted CVD coatings had resistivity greater than 10^{13} Ωcm and microhardness of about 10,000 HV. Thermal conductivity of these coatings was very high, roughly 1100 W m^{-1} K^{-1} which is closer to that of diamond solid.

14.5 REFERENCES

Aisenberg, S., and Chabot, R. (1971), "Ion Beam Deposition of Thin Films of Diamond Like Carbon," *J. Appl. Phys.*, Vol. 49, pp. 2953–2958.

Alper, A. M. (1970a), *High Temperature Oxides*, Part I: *Magnesia, Lime, and Chrome Refractories*, Academic, New York.

———, (1970b), *High Temperature Oxides*, Part II: *Oxides of Rare Earths, Titanium, Zirconium, Hafnium, Niobium, and Tantalum*, Academic, New York.

———, (1970c), *High Temperature Oxides*, Part III: *Magnesia, Alumina, Beryllia Ceramics: Fabrication, Characterization, and Properties*, Academic, New York.

Andersson, D. A. (1977), "The Electrical and Optical Properties of Amorphous Carbon Prepared by the Glow Discharge Technique," *Phil. Mag.*, Vol. 35, pp. 17–26.

Andersson, L. P. (1981), "A Review of Recent Work on Hard i-C Films," *Thin Solid Films*, Vol. 86, pp. 193–200.

———, and Berg, S. (1978), "Initial Etching in an RF Butane Plasma," *Vacuum*, Vol. 28, pp. 449–451.

———, ———, Norström, H., Olaison, R., and Towta, S. (1979), "Properties and Coating Rates of Diamond-like-Carbon Films Produced by rf Glow Discharge by Hydrocarbon Gases," *Thin Solid Films*, Vol. 63, pp. 155–160.

Angus, J. C., Will, H. A., and Stanko, W. S. (1968), "Growth of Diamond Seed Crystals by Vapor Deposition," *J. Appl. Phys.*, Vol. 39, pp. 2915–2922.

———, and Hayman, C. C. (1988), "Low-Pressure Metastable Growth of Diamond and 'Diamondlike' Phases," *Science*, Vol. 241, pp. 913–921.

———, and Jansen, F. (1988), "Dense Diamond like Hydrocarbons as Random Covalent Networks," *J. Vac. Sci. Technol.*, Vol. A6, pp. 1778–1782.

———, Koidl, P., and Domitz, S. (1986), "Carbon Thin Films," in *Plasma Deposited Thin Films* (J. Mort and F. Jansen, eds.), pp. 89–127, CRC Press, Boca Raton, Fla.

Anonymous (1975), Stellite Hard-facing Products, Stellite Division, Cabot Corporation, Kokomo, Ind. (Presently manufactured by Deloro Stellite, Inc., Delleville, Ontario, Canada.)

——— (1976), *Engineering Property Data on Selected Ceramics, Vol. I, Nitrides*, Battelle Columbus Laboratory, Columbus, Ohio.

——— (1979), *Engineering Property Data on Selected Ceramics, Vol. II, Carbides*, Battelle Columbus Laboratory, Columbus, Ohio.

——— (1981), *Engineering Property Data on Selected Ceramics, Vol. III, Single Oxides*, Battelle Columbus Laboratory, Columbus, Ohio.

——— (1982), *Metals Handbook, Vol. 5: Surface Cleaning, Finishing, and Coating*, 9th ed., American Society for Metals, Metals Park, Ohio.

——— (1987), *Tribology of Ceramics*, Special Publications SP-23 and SP-24, STLE, Park Ridge, Ill.

Antony, K. C., Bhansali, K. J., Messler, R. W., Miller, A. E., Price, M. O., and Tucker, R. C. (1983), "Hardfacing," in *Metals Handbook*, Vol. 6: *Welding, Brazing and Soldering*, 9th ed., pp. 771–803, American Society for Metals, Metals Park, Ohio.

Aoki, I., Fukutome, R., and Enomoto, Y. (1985), "Relation between the Hue and Composition of TiC-N Films Prepared Using Multiple Hollow Cathode Discharge Ion Plating," *Thin Solid Films*, Vol. 130, pp. 253–260.

Archer N. J. (1975), "Tungsten Carbide Coatings on Steels," *Proc. 5th Int. Conf. on Chemical Vapor Deposition*, pp. 556–573, The Electrochemical Society, Pennington, N.J.

——— (1981), "The Plasma-Assisted Chemical Vapor Deposition of TiC, TiN, and TiC-N," *Thin Solid Films*, Vol. 80, pp. 221–225.

———, and Yee, K. K. (1978), "Chemical Vapor Deposited Tungsten Carbide Wear Resistant Coatings Formed at Low Temperatures," *Wear*, Vol. 48, pp. 237–250.

Aron, P. R., and Grill, A. (1982), "Some Properties of R.F. Sputtered Hafnium Nitride Coatings," *Thin Solid Films*, Vol. 96, pp. 87–91.

Aubert, A., Gillet, R., and Terrat, J. P. (1983), "Hard Chrome Coatings Deposited by Physical Vapor Deposition," *Thin Solid Films*, Vol. 108, pp. 165–172.

———, Danroc, J., Gaucher, A., and Terrat, J. P. (1985), "Hard Chrome and Molybdenum Coatings Produced by Physical Vapor Deposition," *Thin Solid Films*, Vol. 126, pp. 61–67.

Avery, H. S. (1952), "Hard Facing for Impact," *Welding J.*, Vol. 31, pp. 116–143.

Badzian, A. R., Bachmann, P. K., Hartnett, T., Badzian, T., and Messier, R. (1987), "Diamond Thin Films Prepared by Plasma Chemical Vapor Deposition Processes," *Proc. European Research Society Meeting*, Vol. 14: *Les Additions de Physique*, Paris, France, pp. 63–67.

Bair, S., Ramalingam, S., and Winer, W. O. (1980), "Tribological Experience with Hard Coats on Soft Metallic Substrates," *Wear*, Vol. 60, pp. 413–419.

Banks, B. A., and Rutledge, S. K. (1982), "Ion Beam Sputter Deposited Diamond Like Films," *J. Vac. Sci. Technol.*, Vol. 21, pp. 807–814.

Bardos, L., Mulis, J., and Taras, P. (1983), "Nitrogen Activation for Plasma Chemical Synthesis of Thin Si_3N_4 Films," *Thin Solid Films*, Vol. 102, pp. 107–110.

Barnhart, R. E. (1968), "The Contribution of Metallic and Ceramic Coating to Gas Turbine Engines," presented at the ASME Gas Turbine Conf. and Products Show, Washington, D.C.

Bell, G. R. (1972), "Surface Coatings 3-Gas Welded Coatings," *Tribol. Int.*, Vol. 5, pp. 215–219.

Berg, S., and Andersson, L. P. (1979), "Diamond-like Carbon Films Produced in a Butane Plasma," *Thin Solid Films*, Vol. 58, pp. 117–120.

Bewilogua, K., Dietrich, D., Pagel, L., Schurer, C., and Weissmantel, C. (1979), "Structure and Properties of Transparent and Hard Carbon Films," *Surface Sci.*, Vol. 86, pp. 308–313.

———, ———, Rau, B., Rother, B., and Weissmantel, C. (1985), *Proc. Int. Conf. on Diamond Crystallization Under Reduced Pressure*, Warsaw, Poland.

Bhansali, K. J., Silence, W. L., and Hickl, A. J. (1980), "New Low-Cost Alloys Cut Hardfacing Costs," *Weld. Design Fabr.*, Vol. 53, May, pp. 75–77.

Bhushan, B. (1979), "Development of R.F. Sputtered Chrome Oxide Coatings for Wear Application," *Thin Solid Films*, Vol. 64, pp. 231–241.

——— (1980a), "Characterization of R.F. Sputter-Deposited Chrome Oxide Films," *Thin Solid Films*, Vol. 73, pp. 255–265.

——— (1980b), "High Temperature Self-lubricating Coatings and Treatments—A Review," *Metal Finish.*, Vol. 78, May, pp. 83–88; June, pp. 71–75.

——— (1980c), *High Temperature Self-lubricating Coatings for Air-lubricated Foil Bearings for the Automotive Gas-Turbine Engine*, Report No. CR-159848, NASA Lewis Research Center, Cleveland, Ohio.

——— (1980d), *Coating Development and Friction Measurements for Solid-Lubricated Rolling-Element Bearings*, MTI-80 TR 30, Mechanical Technology Inc., Latham, N.Y.

——— (1981a), "Structural and Compositional Characterization of RF Sputter Deposited Ni-Cr + Cr_2O_3 Films," *J. Lub. Technol. (Trans ASME)*, Vol. 103, pp. 211–217.

——— (1981b), "Friction and Wear Results from Sputter-Deposited Chrome Oxide with and without Nichrome Metallic Binders and Interlayers," *J. Lub. Technol. (Trans. ASME)*, Vol. 103, pp. 218–227.

——— (1982), "Materials Development for Marginal Lubrication Conditions," *International Yearbook of Tribology* (W. J. Bartz, ed.), Expert Verlag, West Germany.

——— (1987), "Overview of Coating Materials, Surface Treatments, and Screening Techniques for Tribological Applications Part I: Coating Materials and Surface Treatments," in *Testing of Metallic and Inorganic Coatings* (W. B. Harding and G. A. Di Bari, eds.), Special Publication STP 947, pp. 289–309, ASTM, Philadelphia.

——— (1990), *Tribology and Mechanics of Magnetic Storage Devices*, Springer-Verlag, New York.

———, and Gray, S. (1978), "Static Evaluation of Surface Coatings for Compliant Gas Bearing in an Oxidizing Atmosphere to 650°C"; *Thin Solid Films*, Vol. 53, pp. 313–331.

———, and ——— (1980), "Development of Surface Coatings for Air-Lubricated Compliant Journal Bearings to 650°C," *ASLE Trans.*, Vol. 23, pp. 185–196.

———, Ruscitto, D., and ——— (1978), *Hydrodynamic Air-Lubricated Compliant Surface Bearings for an Automotive Gas Turbine Engine* Part II: *Materials and Coatings*, Report No. CR-135402, NASA Lewis Research Center, Cleveland, Ohio.

Binnig, G., Rohrer, H., Gerber, Ch., and Weibel, E. (1982), "Surface Studies by Scanning Tunneling Microscopy," *Phys. Rev. Lett.*, Vol. 49, pp. 57–61.

Blachman, A. G. (1971), "Stress and Resistivity Control in Sputtered Molybdenum Films and Comparison with Sputtered Gold," *Metall. Trans.*, Vol. 2, pp. 699–709.

Blackburn, A. R., Shevlin, T. S., and Lowers, H. R. (1949), "Fundamental Study and Equipment for Sintering and Testing of Cermet Bodies, I–III," *J. Am. Ceramic Soc.*, Vol. 32, pp. 81–98.

Blocher, J. M. (1974), "Structure/Properties/Process Relationships in Chemical Vapor Deposition," *J. Vac. Sci. Technol.*, Vol. 11, pp. 680–686.

Boving, H., Hintermann, H. E., and Stehle, G. (1983), "TiC-coated Cemented-Carbide Balls in Gyro-application Ball Bearings," *Lub. Eng.*, Vol. 39, pp. 209–215.

Bowden, F. P., and Young, J. E. (1951), "Friction of Diamond, Graphite, and Carbon and the Influence of Surface Films," *Proc. R. Soc. (Lond.)*, Vol. 208, pp. 444–455.

Brainard, W. A., and Wheeler, D. R. (1978), "An XPS Study of the Adherence of Refractory Carbide, Silicide, and Boride rf Sputtered Wear Resistant Coatings," *J. Vac. Sci. Technol.*, Vol. 15, pp. 1800–1805.

———, and ——— (1979a), "Adherence of Sputtered Titanium Carbides," *Thin Solid Films*, Vol. 63, pp. 363–368.

———, and ——— (1979b), "Use of Nitrogen-argon Plasma to Improve Adherence of Sputtered Titanium Carbide Coatings on Steel," *J. Vac. Sci. Technol.*, Vol. 16, pp. 31–36.

Brütsch, R. (1985), "Chemical Vapor Deposition of Silicon Carbide and Its Applications," *Thin Solid Films*, Vol. 126, pp. 313–318.

Bubenzer, A., Dischler, B., Brandt, G., and Koidl, P. (1983), "R.F. Plasma Deposited Amorphous Hydrogenated Hard Carbon Thin Films, Preparation, Properties and Applications," *J. Appl. Phys.*, Vol. 54, pp. 4590–4594.

Bucher, J. P., Ackermann, K. P., and Buschor, F. W. (1984), "RF Reactively Sputtered TiN: Characterization and Adhesion to Materials of Technical Interest," *Thin Solid Films*, Vol. 122, pp. 63–71.

Bucklow, I. A. (1983), "Bulk Coatings for Use at Elevated Temperatures," in *Coatings for High Temperature Applications* (E. Lang, ed.), pp. 139–167, Applied Science, New York.

Buhl, R., Pulker, H. K., and Moll, E. (1981), "TiN Coatings on Steel," *Thin Solid Films*, Vol. 80, pp. 265–270.

Bunshah, R. F. (1974), "High-Rate Evaporation/Deposition Processes of Metals, Alloys, and Ceramics for Vacuum Metallurgical Applications," *J. Vac. Sci. Technol.*, Vol. 11, pp. 814–819.

——— (1977), "Mechanical Properties of PVD Films," *Vacuum*, Vol. 27, pp. 353–361.

———, and Schramm, R. J. (1977), "Al_2O_3 Deposition by Activated Reactive Evaporation," *Thin Solid Films*, Vol. 40, pp. 211–216.

———, and Shabaik, A. H. (1975), "Superhard Coatings for High-Speed-Steel Tools," *Res./Devel.*, June, Vol. 26, pp. 46 and 48.

———, Schramm, R. J., Nimmagadda, R., Movchan, B. A., and Borodin, V. P. (1977), "Structure and Properties of Refractory Compounds Deposited by Direct Evaporation," *Thin Solid Films*, Vol. 40, pp. 169–182.

———, Nimmagadda, R., Dunford, W., Movchan, B. A., Demchishin, A. V., and Chursanov, N. A. (1978), "Structure and Properties of Refractory Compounds Deposited by Electron Beam Evaporation," *Thin Solid Films*, Vol. 54, pp. 85–106.

Caddoff, I., Neisen, J. P., and Miller, E. (1971), in *Transition Metal Carbides and Nitrides* (L. E. Toth, ed.), Academic, New York, p. 180.

Caputo, A. J. (1977), "Fabrication and Characterization of Vapor-deposited Niobium and Zirconium Carbides," *Thin Solid Films*, Vol. 40, pp. 49–55.

Carson, W. W. (1975), "Sputter Gas Pressure and dc Substrate Bias Effects on Thick rf-Diode Sputtered Films of Ti Oxycarbides," *J. Vac. Sci. Technol.*, Vol. 12, pp. 845–849.

Cashon, I. P. (1975), "Wear Resistant Coatings Applied by the Detonation Gun," *Tribol. Int.*, Vol. 8, pp. 111–115.

Chauhan, S. P., Angus, J. C., Gardner, N. C. (1976), "Kinetics of Carbon Deposits on Diamond Powder," *J. Appl. Phys.*, Vol. 47, pp. 4746–4754.

Chessin, H., and Fernald, E. H. (1982), "Hard Chromium Plating," in *Metals Handbook*, Vol. 5: *Surface Cleaning, Finishing, and Coating*, 9th ed., pp. 170–187, American Society for Metals, Metals Park, Ohio.

Chevallier, J., Chabert, J. P., and Spitz, J. (1981), "Microhardness of TiN_x Coatings by Reactive Cathode Sputtering," *Thin Solid Films*, Vol. 80, p. 263.

Chin, J., Gantzel, P. K., and Hudson, R. G. (1977), "The Structure of Chemical Vapor Deposited Silicon Carbide," *Thin Solid Films*, Vol. 40, pp. 57–72.

Cho, N. H., Krishnan, K. M., Veirs, D. K., Rubin, M. D., Hopper, C. B., Bhushan, B., and Bogy, D. B. (1990a), "Power Density Effects in Sputtered Diamondlike Carbon Thin Films," *Mater. Res. Soc. Symp. Proc.*, Vol. 164, pp. 309–314.

———, ———, ———, ———, ———, ———, and ——— (1990b), "Chemical Structure and Physical Properties of Diamond-Like Amorphous Carbon Films Prepared by Magnetron Sputtering," *J. Mater. Res.*, Vol. 5, pp. 2543–2554.

Chollet, L., and Perry, A. J. (1985), "The Stress in Ion Plated HfN and TiN Coatings," *Thin Solid Films*, Vol. 123, pp. 223–234.

Cholvy, G., and Derep, J. L. (1985), "Characterization and Wear Resistance of Coatings in the Cr-C-N Ternary System Deposited by Physical Vapor Deposition," *Thin Solid Films*, Vol. 126, pp. 51–60.

Chopra, K. L., Agarwal, V., Vankar, V. D., Despandey, C. V., and Bunshah, R. F. (1985), "Synthesis of Cubic Boron Nitride Films by Activated Reactive Evaporation of H_3BO_4," *Thin Solid Films*, Vol. 126, pp. 307–312.

Chuanxian, D., Zatorski, R. A., Herman, H., and Ott, D. (1984a), "Oxide Powders for Plasma Spraying—The Relationship between Powder Characteristics and Coating Properties," *Thin Solid Films*, Vol. 118, pp. 467–475.

———, Bingtang, H., and Huiling, L. (1984b), "Plasma-Sprayed Wear-Resistant Ceramic and Cermet Coating Materials," *Thin Solid Films*, Vol. 118, pp. 485–493.

Clarke, P. J. (1977), "Magnetron dc Reactive Sputtering of Titanium Nitride and Indium-tin Oxide," *J. Vac. Sci. Technol.*, Vol. 14, pp. 141–142.

Craig, S., and Harding, G. L. (1982), "Structure, Optical Properties and Decompositions Kinetics of Sputtered Hydrogenated Carbon," *Thin Solid Films*, Vol. 97, pp. 345–361.

Dennis, J. K., and Such, T. E. (1972), *Nickel, and Chromium Plating*, Wiley, New York.

Deryagin, B. V., and Fedoseev, D. B. (1975), "The Synthesis of Diamond at Low Pressure," *Sci. Am.*, Vol. 233, pp. 102–109.

DeVries, R. C. (1987), "Synthesis of Diamond under Metastable Conditions," *Ann. Rev. Mater. Sci.*, Vol. 17, pp. 161–187.

Dimigen, H., and Hübsch, H. (1983–84), "Applying Low Friction Wear Resistant Thin Solid Films by Physical Vapor Deposition," *Philips Tech. Review*, Vol. 41, pp. 186–197.

Doan, T. O., and Mackintosh, N. D. (1988), "The Frictional Behavior of Rigid Disk Carbon Overcoats," in *Tribology and Mechanics of Magnetic Storage Systems* (B. Bhushan and N. S. Eiss, eds.), Vol. 5, STLE, Park Ridge, Ill., pp. 6–11.

Dragoo, A. L., and Diamond, J. J. (1967), "Transitions in Vapor-Deposited Alumina from 300°C to 1200°C," *J. Am. Ceramic Soc.*, Vol. 50, pp. 568–574.

Dreger, D. R. (1977), "Diamond-Studded Coatings Improve Part Life," *Machine Design*, November 24, pp. 95–97.

Dua, A. K., George, V. C., Agarwala, R. P., and Krishnan, R. (1984), "Preparation and Characterization of Carbon and Titanium Carbide Coatings," *Thin Solid Films*, Vol. 121, pp. 35–42.

Duckworth, R. G. (1981), "High Purity Sputtered Tribological Coatings," *Thin Solid Films*, Vol. 86, pp. 213–218.

Dugger, M. T., Chung, Y. W., Bhushan, B., and Rothschild, W. (1990), "Friction, Wear and Interfacial Chemistry in Thin-film Magnetic Rigid Disk Files," *J. Tribol. (Trans. ASME)*, Vol. 112, pp. 238–245.

Dun, H., Pan, P., White, F. R., and Douse, R. W. (1981), "Mechanism of Plasma-enhanced Silicon Nitride Deposition Using SiH_4/N_2 Mixture," *J. Electrochem. Soc.*, Vol. 128, pp. 1555–1563.

Durmann, G., and Longo, F. N. (1969), "Plasma-Sprayed Alumina-Titania Composite," *Am. Ceramic Soc. Bull.*, Vol. 48, No. 2, pp. 221–224.

Eizenberg, M., and Murarka, S. P. (1983), "Reactively Sputtered Titanium Carbide Thin Films: Preparation and Properties," *J. Appl. Phys.*, Vol. 54, pp. 3190–3194.

Enke, K., Dimigen, H., and Hübsch, H. (1980), "Frictional Properties of Diamond-like Carbon Layers," *Appl. Phys. Lett.*, Vol. 36, pp. 291–292.

Eschnauer, H., and Lugscheider, E. (1984), "Metallic and Ceramic Powders for Vacuum Plasma Spraying," *Thin Solid Films*, Vol. 118, pp. 421–435.

Eser, E., Ogilvie, R. E., and Taylor, K. A. (1978a), "Structural and Compositional Characterization of Sputter-deposited WC + Co Films," *J. Vac. Sci. Technol.*, Vol. 15, pp. 396–400.

———, ———, and ——— (1978b), "Friction and Wear Results from WC + Co Coatings by dc-biased rf Sputtering in a Helium Atmosphere," *J. Vac. Sci. Technol.*, Vol. 15, pp. 401–405.

Fabis, P. M., Cooke, R. A., and McDonough, S. (1990a), "Stress State of Chromium Nitride Films Deposited by Reactive Direct Current Planar Magnetron Sputtering," *J. Vac. Sci. Technol.*, Vol. A8, No. 5, pp. 3809–3818.

———, ———, and ——— (1990b), "The Influence of Microstructure on Stress State of Sputter Deposited Chromium Nitride Films," *J. Vac. Sci. Technol.*, Vol. A8, No. 5, pp. 3819–3826.

Fark, H., Chevallier, J., Reichelt, K., Dimigen, H., and Hübsch, H. (1983), "The Microhardness, Electrical Conductivity, and Temperature Coefficient of Resistance of Reactively Sputtered $TiC_xO_yN_z$ Films," *Thin Solid Films*, Vol. 100, pp. 193–201.

Fedoseev, D. V., Deryagin, B. V., Varshavskaja, Y. G., and Semienova-Tjan-Shanskaja, A. C. (1984), *Crystallization of Diamond* (in Russian) Izd. Nauka, Moscow.

Fenske, G. R., Kaufherr, N., and Sproul, W. D. (1987), "Solid Solution Hardening in High Rate Reactively Sputtered (Hf, Ti)N Coatings," *Thin Solid Films*, Vol. 153, pp. 159–168.

Fields, W. D., Duncan, R. N., Zickgraf, J. R., Baudman, D. W., and Henry, R. A. (1982), "Electroless Nickel Plating," in *Metals Handbook*, Vol. 5: *Surface Cleaning, Finishings and Coating*, 9th ed., pp. 219–243, American Society for Metals, Metals Park, Ohio.

Fink, J., Muller-Heinzerling, T., Pfluger, J., Bubenzer, A., Koidl, P., and Crecelius, G. (1983), "Structure and Bonding of Hydrocarbon Plasma Generated Carbon Films: An Electron Energy Loss Study," *Solid State Commun.*, Vol. 47, pp. 687–691.

———, ———, ———, Scheerer, B., Dischler, B., Koidl, P., Bubenzer, A., and Sah, R. E. (1984), "Investigation of Hydrocarbon-Plasma-Generated Carbon Films by Electron-Energy-Loss Spectroscopy," *Phys. Rev.*, Vol. B30, pp. 4713–4718.

Fishman, G. S., Chen, C. H., Rigsbee, J. M., and Brown, S. D. (1985), "Character of Laser-Glazed, Plasma-Sprayed Zirconia Coatings on Stainless Steel Substrate," in *Ceramics Engineering and Science Proceedings*, Vol. 6, pp. 908–915, American Ceramic Society, New York.

Fleischer, W., Schulze, D., Wilberg, R., Lunk, A., and Schrade, F. (1979), "Reactive Ion Plating (RIP) with Auxiliary Discharge and the Influence of the Deposition Conditions on the Formation and Properties of TiN Films," *Thin Solid Films*, Vol. 63, pp. 347–356.

Foroulis, Z. A. (1984), "Guidelines for the Selection of Hardfacing Alloys for Sliding Wear Resistant Applications," *Wear*, Vol. 96, pp. 203–218.

Freller, H., and Haessler, H. (1987), "$Ti_xAl_{1-x}N$ Films Deposited by Ion Plating with an Arc Evaporator," *Thin Solid Films*, Vol. 153, pp. 67–74.

Fujimori, S., and Nagai, K. (1981), "The Effect of Ion Bombardment on Carbon Films Prepared by Laser Evaporation," *Jap. J. Appl. Phys.*, Vol. 20, pp. L194–L196.

———, Kasai, T., and Inamura, T. (1982), "Carbon Film Formation by Laser Evaporation and Ion Beam Sputtering," *Thin Solid Films*, Vol. 92, pp. 71–80.

———, Imai, T., and Doi, A. (1986), "Characterization of Conducting Diamond Films," *Vacuum*, Vol. 36, pp. 99–102.

Fukutomi, M., Fujitsuka, M., Shinno, H., Kitajima, M., Shikama, T., and Okada, M. (1982), "Thermal Stability of Vapor-plated TiC Coating on Molybdenum and Graphite," *Proc. 7th Int. Conf. on Vacuum Metallurgy*, Vol. 1, pp. 711–724, The Iron and Steel Institute of Japan, Tokyo.

———, Kitajima, M., Okada, M., and Watanabe, R. (1984), "Preparation of Silicon Carbide Films on Molybdenum by R. F. Reactive Ion-Plating," *Trans. Nat. Res. Inst. Metals (Japan)*, Vol. 26, pp. 263–269.

Gabriel, H. M. (1983), "Morphology and Structure of Ion Plated TiN, TiC and Ti(C, N) Coatings," in *Proc. Int. Ion Engineering Congr.* Kyoto (T. Takagi, ed.), Vol. 2, p. 1311, Institute of Electrical Engineers of Japan, Tokyo.

———, and Kloos, K. H. (1984), "Morphology and Structure of Ion-Plated TiN, TiC and Ti (C, N) Coatings," *Thin Solid Films*, Vol. 118, pp. 243–254.

———, Schmidt, F., Broszeit, E., and Kloos, K. H. (1983), "Improved Component Performance of Vane Pumps by Ion Plated TiN Coatings," *Thin Solid Films*, Vol. 108, pp. 189–197.

Gambino, R. J., and Thompson, J. A. (1980), "Spin Resonance Spectroscopy of Amorphous Carbon Films," *Solid State Commun.*, Vol. 34, pp. 15–18.

Gawrilov, G. G. (1979), *Chemical (Electrochemical) Nickel Plating*, Portcullis Press, Redhill, England.

Gebhardt, J. J., and Cree, R. F. (1965), "Vapor-Deposited Borides of Group IV A Metals," *J. Am. Ceramic Soc.*, Vol. 48, pp. 262–267.

———, Tanzilli, R. A., and Harris, T. A. (1975), "Chemical Vapor Deposition of Silicon Nitride," *Proc. 5th Int. Conf. on Chemical Vapor Deposition* (J. M. Blocher, H. E. Hintermann, and L. H. Hall, eds.), The Electrochemical Society, Pennington, N.J.

Gianelos, L. (1982), "Decorative Chromium Plating," in *Metals Handbook*, Vol. 5: *Surface Cleaning, Finishing and Coating*, 9th ed., pp. 188–198, American Society for Metals, Metals Park, Ohio.

———, and McMullen, W. H. (1982), "Nickel Plating," in *Metals Handbook*, Vol. 5; *Surface Cleaning, Finishing and Coating*, 9th ed., pp. 199–218, American Society for Metals, Metals Park, Ohio.

Glang, R., Holmwood, R. A., and Furois, P. C. (1965), "Bias Sputtering of Molybdenum Films," in *Proc. Third Int. Vac. Cong.*, Vol. 2, pp. 643–650, Pergamon, New York.

Gobel, J. (1984), "Anwendung der Oberflachenanalyse in der Dunnschichttechnik an ausgewahlten Beispielen, Fresenius," *Z. Anal. Chem.*, Vol. 319, pp. 771–776.

Goldschmidt, H. J. (1967), *Interstitial Alloys*, Plenum, New York.

Gologan, V. F., and Eyre, T. S. (1974), "Friction and Wear of Some Engineering Materials against Hard Chromium Plating," *Wear*, Vol. 28, pp. 49–57.

Goward, G. W. (1979), "Protective Coatings for High Temperature Alloys-state of Technology," in *Source Book on Materials for Elevated Temperature Applications*, pp. 369–386, American Society for Metals, Metals Park, Ohio.

Gray, S., Heshmat, H., and Bhushan, B. (1981), "Technology Progress on Compliant Foil Air Bearing Systems for Commercial Applications," *Proc. 8th Int. Gas Bearing Symp.*, pp. 69–97, BHRA Fluid Engineering, Cranfield, Bradford, England.

Greene, J. E., and Pestes, M. (1976), "Adhesion of Sputtered Deposited Carbide Films to Steel Substrates," *Thin Solid Films*, Vol. 37, pp. 373–385.

Gregory, E. N. (1978), "Hardfacing," *Tribol. Int.*, Vol. 11, pp. 129–134.

——— (1980), "Surfacing by Welding-Alloys, Processes, Coatings and Materials Selection," *Metal Constr.*, Vol. 12, pp. 685–690.

Grill, A., and Aron, P. R. (1983), "RF Sputtered Silicon and Hafnium Nitrides: Properties and Adhesion to 440C Stainless Steel," *Thin Solid Films*, Vol. 108, pp. 173–180.

Grossklaus, W., and Bunshah, R. F. (1975), "Synthesis and Morphology of Various Carbides in the Ta-C System," *J. Vac. Sci. Technol.*, Vol. 12, pp. 811–814.

Guzman, L., Marchetti, F., Calliari, L., Scotoni, I., and Ferrari, F. (1984), "Formation of BN by Nitrogen Ion Implantation of Boron Deposits," *Thin Solid Films*, Vol. 117, pp. L63–L66.

Hänni, W., and Hintermann, H. E. (1981), "Coated Tools for the Machining of Cu- and Al-Alloys," *Proc. of the 8th Int. Conf. on Chemical Vapor Deposition* (J. M. Blocher, G. E. Vuillard, and G. Wahl, eds.), The Electrochemical Society, Pennington, N.J., pp. 597–605.

Hatto, P. (1985), "Ion Bond and the Arc Evaporation Process," *Proc. of Course on Nitride and Carbide Coatings, Manufacturing Technology: Science and Applications*. Centre Suisse D'Electronique Et Du Microtechnique, S.A. Nêuchatel, Switzerland.

Hauser, J. J. (1977), "Electrical, Structural and Optical Properties of Amorphous Carbon," *J. Noncrystal. Solids*, Vol. 23, pp. 21–41.

Hazzard, R. (1972), "Surface Coatings 2-Arc-Welded Coatings," *Tribol. Int.*, Vol. 5, pp. 207–214.

Hebsur, M. G., and Miner, R. V. (1987), "Stress Rupture and Creep Behavior of Low Pressure Plasma-Sprayed NiCoCrAlY Coating Alloys in Air and Vacuum," *Thin Solid Films*, Vol. 147, pp. 143–152.

Hess, D. W. (1984), "Plasma-Enhanced CVD: Oxides, Nitrides, Transition Metals, and Transition Metal Silicides," *J. Vac. Sci. Technol.*, Vol. A2, pp. 244–252.

Hibbs, M. K., Sundgren, J.-E., Jacobson, B. E., and Johansson, B. O. (1983), "The Microstructure of Reactively Sputtered Ti-N Films," *Thin Solid Films*, Vol. 107, pp. 149–157.

———, Johansson, B. O., Sundgren, J.-E., and Helmersson, U. (1984), "Effects of Substrate Temperature and Substrate Material on the Structure of Reactively Sputtered TiN Films," *Thin Solid Films*, Vol. 122, pp. 115–129.

Hintermann, H. E. (1980), "Exploitation of Wear- and Corrosion-Resistant CVD-Coatings," *Tribol. Int.*, Vol. 13, pp. 267–277.

——— (1981), "Tribological and Protective Coatings by Chemical Vapor Deposition," *Thin Solid Films*, Vol. 84, pp. 215–243.

——— (1983), "Surface Treatments," *Proc. Int. Conf. on Sciences of Hard Materials* (R. K. Viswanadham, D. J. Rowcliffe, and J. Gurland, eds.), pp. 357–394. Plenum, New York.

——— (1984), "Adhesion, Friction and Wear of Thin Hard Coatings," *Wear*, Vol. 100, pp. 381–397.

Hiraki, A., Kawano, T., Kawakami, Y., Hayashi, M., and Miyasato, T. (1984), "Tetrahedral Carbon Film by Hydrogen Gas Reactive rf-Sputtering of Graphite onto Low Temperature Substrate," *Solid State Commun.*, Vol. 50, pp. 713–716.

Hochman, R. F., Erdemir, A., Dolan, F. J., and Thom, R. L. (1985), "Rolling Contact Fatigue Behavior of Cu and TiN Coating on Bearing Steel Substrate," *J. Vac. Sci. Technol.*, Vol. A3, pp. 2348–2353.

Hoffman, D. W., and Thornton, J. A. (1977), "The Compressive Stress Transition in Al, V, Zr, Nb, and W Metal Films Sputtered at Low Working Pressures," *Thin Solid Films*, Vol. 45, pp. 387–396.

———, and Gaerttner, M. R. (1980), "Modification of Evaporated Chromium by Concurrent Ion Bombardment," *J. Vac. Sci. Technol.*, Vol. 17, pp. 425–428.

Holland, L., and Ojha, S. M. (1976), "Deposition of Hard and Insulating Carbonaceous Films of an rf Target in Butane Plasma," *Thin Solid Films*, Vol. 38, pp. L17–L19.

———, and ——— (1979), "The Growth of Carbon Films with Random Atomic Structure from Ion Impact Damage in Hydrocarbon Plasma," *Thin Solid Films*, Vol. 58, pp. 107–116.

Holleck, H. (1986), "Material Selection for Hard Coatings," *J. Vac. Sci. Technol.*, Vol. A4, pp. 2661–2669.

Horvath, E., and Perry, A. J. (1980), "Vanadium Coating of Carbon Steels by Chemical Vapor Deposition," *Thin Solid Films*, Vol. 65, pp. 309–314.

Hosokawa, N., Konishi, A., Hiratsuka, H., and Annoh, K. (1981), "Preparation and Properties of Carbon Films with Magnetron Sputtering," *Thin Solid Films*, Vol. 73, pp. 115–116.

Hove, J. E., and Riley, W. C. (1965), *Ceramics for Advanced Technologies*, Wiley, New York.

Hummer, E., and Perry, A. J. (1983), "Adhesion and Hardness of Ion-Plated TiC and TiN Coatings," *Thin Solid Films*, Vol. 101, pp. 243–251.

Hyder, S. B., and Yep, T. O. (1976), "Structure and Properties of BN Films Grown by High Temperature Reactive Plasma Deposition," *J. Electrochem. Soc.*, Vol. 123, pp. 1721–1724.

Hynecek, J. (1979), "Plasma Deposition of Transparent Conductive Layers," U.S. Patent 4,140,814, February 20.

Ikawa, K. (1972), "Vapor Deposition of Zirconium Carbide-Carbon Composites by the Iodide Process," *J. Less-Common Metals*, Vol. 27, pp. 325–332.

Inagawa, K., Watanabe, K., Ohsone, H., Saitoh, K., and Itoh, A. (1987), "Preparation of Cubic Boron Nitride Film by Activated Reactive Evaporation with a Gas Activation Nozzle," *J. Vac. Sci. Technol.*, Vol. A5, pp. 2696–2700.

Ingham, H. S. (1967), *Plasma Flame Spraying Materials, Techniques, and Applications*, Metal Spraying and Plastic Coating Division Conference, London.

——— (1969), "Flame Sprayed Coatings," in *Composite Engineering Laminates* (G. H. Dietz, eds.), pp. 304–316, M.I.T. Press, Cambridge, Mass.

Inwood, B. C., and Garwood, A. E. (1978), "Electroplated Coatings for Wear Resistance," *Tribol. Int.*, Vol. 11, pp. 113–118.

Itoh, H., Kato, K., and Sugiyama, K. (1986), "Chemical Vapor Deposition of Corrosion Resistance TiN Film to the Inner Walls of Long Steel Tubes," *J. Mater. Sci.*, Vol. 21, pp. 751–756.

Iwamoto, N., Makino, Y., and Kamai, M. (1987), "Characterization of rf-Sputtered Zirconia Coating," *Thin Solid Films*, Vol. 153, pp. 233–242.

Jacobson, B. E., Bunshah, R. F., and Nimmagadda, R. (1978), "Transmission Electron Microscopy Studies in TiC and VC-TiC Deposits Prepared by Reactive Evaporation," *Thin Solid Films*, Vol. 54, pp. 107–118.

———, Nimmagadda, R., and Bunshah, R. F. (1979), "Microstructure of TiN and Ti_2N Deposits Prepared by Activated Reactive Evaporation," *Thin Solid Films*, Vol. 63, pp. 333–339.

———, Deshpandey, C. V., Doerr, H. J., Karim, A. A., and Bunshah, R. F. (1984), "Microstructure and Hardness of Ti (C, N) Coatings on Steel Prepared by the Activated Reactive Evaporation Technique," *Thin Solid Films*, Vol. 118, pp. 285–292.

Jaeger, R., and Cohen, S. (1972), *Proc. of the Third Int. Conf. on Chemical Vapor Deposition* (F. A. Glaski, ed.), p. 500, American Nuclear Society, La Grange, Ill.

Jamal, T., Nimmagadda, R., and Bunshah, R. F. (1980), "Friction and Adhesive Wear of Titanium Carbide and Titanium Nitride Overlay Coatings," *Thin Solid Films*, Vol. 73, pp. 245–254.

Jansen, F., Machonkin, M., Kaplan, S., and Hark, S. (1985), "The Effects of Hydrogenation on the Properties of Ion Beam Sputter Deposited Amorphous Carbon," *J. Vac. Sci. Technol.*, Vol. A3, pp. 605–609.

Jansson, U., and Carlsson, J. (1985), "Chemical Vapor Deposition of Boron Carbide in the Temperature Range 1300–1500 K and at a Reduced Pressure," *Thin Solid Films*, Vol. 124, pp. 101–107.

Je, J. H., Gyarmati, E., and Naoumidis, A. (1986), "Scratch Adhesion Test of Reactively Sputtered TiN Coatings on Soft Substrate," *Thin Solid Films*, Vol. 136, pp. 57–67.

Jenkins, G. M., and Kawamura, K. (1976), *Polymeric Carbon-Carbon Fibre, Glass and Char*, Cambridge University Press, London and New York.

Johannesson, T. R., and Lindstrom, J. N. (1975), "Factors Affecting the Initial Nucleation of Alumina on Cemented Carbide Substrate in the CVD Process," *J. Vac. Sci. Technol.*, Vol. 12, pp. 854–857.

Johansson, B. O., Sundgren, J. E., Greene, J. E., Rockett, A., and Barnett, S. A. (1985), "Growth and Properties of Single Crystal TiN Films Deposited by Reactive Magnetron Sputtering," *J. Vac. Sci. Technol.*, Vol. A3, pp. 303–307.

Johansen, O. A., Dontje, J. H., and Zenner, R. L. D. (1987), "Reactive Arc Vapor Ion Deposition of TiN, ZrN and HfN," *Thin Solid Films*, Vol. 153, pp. 75–82.

Johnson, P. C., and Randhawa, H. (1987), "Zirconium Nitride Films Prepared by Cathodic Arc Plasma Deposition Process," *Surf. Coat. Technol.*, Vol. 33, pp. 53–62.

Jones, D. I., and Stewart, A. D. (1982), "Properties of Hydrogenated Amorphous Carbon Films and the Effect of Doping," *Phil. Mag.*, Vol. B46, pp. 423–434.

Kamo, M., Sato, Y., Matsumoto, S., and Setaka, N. (1983), "Diamond Synthesis from Gas Phase in Microwave Plasma," *J. Crystal. Growth*, Vol. 62, pp. 642–644.

Kaplan, M. L., Schmidt, P. H., Chen, C. H., and Walsh, W. M. (1980), "Carbon Films with Relatively High Conductivity," *Appl. Phys. Lett.*, Vol. 36, pp. 867–869.

Kaplan, S., Jansen, F., and Machonkin, M. (1985), "Characterization of Amorphous Carbon-Hydrogen Films by Solid-State Nuclear Magnetic Resonance," *Appl. Phys. Lett.*, Vol. 47, pp. 750–753.

Karlsson, B., Sundgren, J.-E., Johansson, B. O. (1982), "Optical Constants and Spectral Selectivity of Titanium Carbonitride," *Thin Solid Films*, Vol. 87, pp. 181–187.

Katz, G. (1976), "Adhesion of Copper Films to Aluminum Oxide Using a Spinel Structure Interface," *Thin Solid Films*, Vol. 33, pp. 99–105.

Kawarada, H., Mar, K., and Kiraki, A. (1987), "Large Area Chemical Vapor Deposition of Diamond Particles and Films Using Magneto-Microwave Plasma," *Jap. J. Appl. Phys.*, Vol. 26, L1032–L1034.

Kedward, E. C., Addison, C. A., and Tennent, A. A. B. (1976), "Development of a Wear-Resistant Electrodeposited Composite Coating for Use on Aero Engines," *Trans. Inst. Met. Finish.*, Vol. 54, pp. 8–16.

Keenan, W. F., and Harrison, D. C. (1985), "A Tin Oxide Transparent-Gate Buried-Channel Virtual-Phase CCD Imager," *IEEE Trans. Electron Devices*, Vol. ED-32, pp. 1531–1533.

Kelly, B. T. (1981), *Physics of Graphite*, Applied Science Publishers, London.

Kenton, D. J. (1983), "Particle Enhanced Electroless Coatings," *Int. J. Powder Metall. and Powder Tech.*, Vol. 19, pp. 185–195.

Kessler, G., Bauer, H. D., Pompe, W., and Scheibe, H. J. (1987), "Laser Pulse Vapor Deposition of Polycrystalline Wurtzite-type BN," *Thin Solid Films*, Vol. 147, pp. L45–L50.

Khan, A. A., Mathine, D., Woollam, J. A., and Chung, Y. (1983), "Optical Properties of 'Diamondlike' Carbon Films: An Ellipsometric Study," *Phys. Rev.*, Vol. B28, pp. 7229–7235.

Khan, M. R., Heiman, N., Fisher, R. D., Smith, S., Smallen, M., Hughes, G. F., Veirs, K., Marchon, B., Ogletree, D. F., and Salmeron, M. (1988), "Carbon Overcoat and the Process Dependence on Its Microstructure and Wear Characteristics," *IEEE Trans. Magn.*, Vol. MAG.-24, pp. 2647–2649.

Kim, M. S., and Chun, J. S. (1983), "Effects of the Experimental Conditions of Chemical Vapor Deposition on a TiC/TiN Double-Layer Coating," *Thin Solid Films*, Vol. 107, pp. 129–139.

Kirk, J. A., Flinn, D. R., and Lynch, M. J. (1981), "Wear of TiB_2 Coatings," *Wear*, Vol. 72, pp. 315–323.

Klokholm, E., and Berry, B. S. (1968), "Intrinsic Stresses in Evaporated Metal Films," *J. Electrochem. Soc. Solid State Sci.*, Vol. 115, pp. 823–826.

Kloos, K. H., Broszeit, E., Gabriel, H. M., and Schroder, H. J. (1982), "Thin TiN Coatings Deposited onto Nodular Cast Irons by Ion- and Plasma-Assisted Coating Techniques," *Thin Solid Films*, Vol. 96, pp. 67–77.

Knight, D. S., and White, W. B. (1989), "Characterization of Diamond Films by Raman Spectroscopy," *J. Mater. Res.*, Vol. 4, No. 2, pp. 385–393.

Knoll, R. W., and Bradley, E. R. (1984a), "Correlation between the Stress and Microstructure in Bias-Sputtered ZrO_2-Y_2O_3 Films," *Thin Solid Films*, Vol. 117, pp. 201–210.

———, and ——— (1984b), "Microstructure and Phase Composition of Sputter-Deposited zirconia-yttria Films," in *Plasma Processing and Synthesis of Materials* (J. Szekely and

D. Apelian, eds.), *Mat. Res. Soc. Symp. Proc.*, Vol. 30, pp. 235–243, Materials Research Society, Pittsburgh, Pa.

Knotek, O., Lugscheider, E., and Reimann, H. (1979), "Wear-Resistant and Corrosion-Resistant Nickel-Base Alloys for Coating by Furnace Melting," *Thin Solid Films*, Vol. 64, pp. 365–369.

———, Münz, W. D., and Leyendecker, T. (1987), "Industrial Deposition of Binary, Ternary, and Quaternary Nitrides of Titanium, Zirconium, and Aluminum," *J. Vac. Sci. Technol.*, Vol. A5, pp. 2173–2179.

Kobayashi, M., and Doi, Y. (1978), "TiN and TiC Coatings on Cemented Carbides by Ion-Plating," *Thin Solid Films*, Vol. 54, pp. 67–74.

Kodama, M., Shabaik, A. H., and Bunshah, R. F. (1978), "Machining Evaluation of Cemented Carbide Tools Coated with HfN and TiC by the Activated Evaporation Process," *Thin Solid Films*, Vol. 54, pp. 353–357.

Komatsu, T., Yamashida, H., Tamou, Y., and Kikuchi, N. (1987), "Diamond Synthesis Using Hot Filament Thermal CVD Assisted by rf Plasma," in *Proc. of the 8th Int. Symp. on Plasma Chemistry* (K. Akashi and A. Kinbara, eds.), Vol. 1, pp. 2487–2492, International Union of Pure and Applied Chemistry, Oxford, England.

Komiya, S. (1982), "Ion Plating by Hot Hollow Cathode Discharge and Its Applications to the Tool Industry" (private communication), ULVAC Corp., Japan.

———, Ono, S., Umezu, N., and Narusawa, T. (1977a), "Characterization of Thick Chromium-Carbon and Chromium-Nitrogen Films Deposited by Hollow Cathode Discharge," *Thin Solid Films*, Vol. 45, pp. 433–445.

———, ———, and ——— (1977b), "Hardness and Grain Size Relations for Thick Chromium Films Deposited by Hollow Cathode Discharge," *Thin Solid Films*, Vol. 45, pp. 474–479.

———, Umezu, N., and Hayashi, C. (1979), "Titanium Nitride Films as a Protective Coating for a Vacuum Deposition Chamber," *Thin Solid Films*, Vol. 63, pp. 341–346.

Kramer, B. M. (1983), "Requirements for Wear Resistant Coatings," *Thin Solid Films*, Vol. 108, pp. 117–125.

———, and Suh, N. P. (1980), "Tool Wear by Solution: A Quantitative Understanding," *J. Eng. Indus. (Trans. ASME)*, Vol. 102, pp. 303–309.

Krutenat, R. C., Wielonski, R. F., and Lawrence, F. M. (1968), "The Chemistry and Structure of Several Sputtered Coatings of Multicomponent Alloys for Gas Turbine Engine," *Proc. Vacuum Metallurgy Conf.* (R. B. Barrow and A. Simkovich, eds.), pp. 235–250, American Vacuum Society, New York.

Kurokawa, H., Mitani, T., and Yonezawa, T. (1987), "Application of Diamond Like Carbon Films to Metallic Thin Film Magnetic Recording Media," *IEEE Trans. Magn.*, Vol. MAG 23, pp. 2410–2412.

Kuwahara, K., Sumomogi, T., and Kondo, M. (1981), "Internal Stress in Aluminum Oxide, Titanium Carbide, and Copper Films Obtained by Planar Magnetron Sputtering," *Thin Solid Films*, Vol. 78, pp. 41–47.

Lai, G. Y. (1978), "Evaluation of Sprayed Chromium Carbide Coatings for Gas-Cooled Reactor Applications," *Thin Solid Films*, Vol. 53, pp. 343–351.

——— (1979), "Factors Affecting the Performance of Sprayed Chromium Carbide Coatings for Gas-Cooled Reactor Heat Exchangers," *Thin Solid Films*, Vol. 64, pp. 271–280.

Lander, J. J., and Morrison, J. (1966), "Low-Energy Electron Diffraction Study of the (111) Diamond Surface," *Surf. Sci.*, Vol. 4, pp. 241–246.

Laugier, M. T. (1983), "Load Bearing Capacity of TiN Coated WC-Co Cemented Carbides," *J. Mater. Sci. Lett.*, Vol. 2, pp. 419–421.

Lee, E. H., and Poppa, H. (1977), "Effects of High Temperature and E-Beam Irradiation on the Stability of Refractory Thin Films," *J. Vac. Sci. Technol.*, Vol. 14, January/February, pp. 223–226.

Lee, M., and Richman, H. (1974), "Some Properties of TiC-Coated Cemented Tungsten Carbides," *Metal. Technol.*, Vol. 1, pp. 538–546.

Levy, M., Sklover, G. N., and Sellers, D. J. (1968), *Adhesion and Thermal Properties of Refractory Coating-Metal Substrate System*, U.S. Army Material Research Agency, AMRA TR 66–01, Watertown, Mass.

Li, C. C., and Lai, G. Y. (1979), "Sliding Wear Studies of Sprayed Chromium Carbide-Nichrome Coatings for Gas-Cooled Reactor Applications," *Proc. Int. Conf. on Wear of Material* (K. C. Ludema, W. A. Glaeser, and S. K. Rhee, eds.), pp. 427–436, ASME, New York.

Lin, P., Despandey, C. V., Doerr, H. J., Bunshah, R. F., Chopra, K. L., and Vankar, V. D. (1987), "Preparation and Properties of Cubic Boron Nitride Coatings," *Thin Solid Films*, Vol. 153, pp. 487–496.

Lindstrom, J. N., and Johannesson, T. R. (1975), "Nucleation of Al_2O_3 Layers on Cemented Carbide Tools," *Proc. 5th Int. Conf. on Chemical Vapor Deposition* (J. M. Blocher, H. E. Hintermann, and L. H. Hall, eds.), pp. 453–468, The Electrochemical Society, Pennington, N.J.

Lux, B., Colombier, C., and Altena, H. (1986), "Preparation of Alumina Coatings by Chemical Vapor Deposition," *Thin Solid Films*, Vol. 138, pp. 49–64.

Lynch, J. F., Ruderer, C. G., Duckworth, W. H. (1966), *Engineering Properties of Selected Ceramic Materials*, American Ceramic Society, Columbus, Ohio.

Mann, B. S., Krishnamoorthy, P. R., and Vivekananda, P. (1985), "Cavitation Erosion Characteristics of Nickel Based Alloy-Composite Coatings Obtained by Plasma Spraying," *Wear*, Vol. 103, pp. 43–55.

Mantyla, T., Vuoristo, P., and Kettunen, P. (1984), "Chemical Vapor Deposition Densification of Plasma Sprayed Oxide Coatings," *Thin Solid Films*, Vol. 118, pp. 437–444.

Marchon, B., Salmeron, M., and Siekhaus, W. (1989), "Observation of Graphitic and Amorphous Structures on the Surface of Hard Carbon Films by Scanning Tunneling Microscopy," *Phys. Rev.*, Vol. B39, No. 17, pp. 12907–12910.

Martin, P. J., Netterfield, R. P., and Sainty, W. G. (1982), "Optical Properties of TiN Produced by Reactive Evaporation and Reactive Ion-Beam Sputtering," *Vacuum*, Vol. 32, pp. 359–362.

———, McKenzie, D. R., Netterfield, R. P., Swift, P., Filipczuk, S. W., Muller, K. H., Pacey, C. G., and James, B. (1987a), "Characteristics of Titanium Arc Evaporation Process," *Thin Solid Films*, Vol. 153, pp. 91–102.

———, Netterfield, R. P., McKenzie, D. R., Falconer, I. S., Pacey, C. G., Tomas, P., and Sainty, W. G. (1987b), "Characterization of a Ti Vacuum Arc and the Structure of Deposited Ti and TiN Films," *J. Vac. Sci. Technol.*, Vol. A5, pp. 22–28.

Matsumoto, S. (1985), "Chemical Vapor Deposition of Diamond in rf Glow Discharge," *J. Mater. Sci. Lett.*, Vol. 4, pp. 600–602.

———, Sato, Y., Tsutsumi, M., and Setaka, N. (1982), "Growth of Diamond Particles from Methane-Hydrogen Gas," *J. Mater. Sci.*, Vol. 17, pp. 3106–3112.

———, Lobayashi, T., Hino, M., Ishigaki, T., and Moriyoshi, Y. (1987), "Deposition of Diamond in an rf Induction Plasma," in *Proc. of Int. Symp. of Plasma Chemistry* (K. Akashi and A. Kinbara, eds.), pp. 2458–2462, International Union of Pure and Applied Chemistry, Oxford, England.

Matthews, A. (1985), "Titanium Nitride PVD Coatings Technology," *Surf. Eng.*, Vol. 1, pp. 93–104.

———, and Lefkow, A. R. (1985), "Problems in the Physical Vapor Deposition of Titanium Nitride," *Thin Solid Films*, Vol. 126, pp. 283–291.

———, and Sundquist, H. A. (1983), "Tribological Studies of Ti-N Films Formed by Reactive Ion Plating," in *Proc. Int. Ion Engineering Congress* (T. Takagi, ed.), Vol. 2, pp. 1325–1330, Kyoto Institute of Electrical Engineers, Japan.

———, and Teer, D. G. (1980), "Deposition of Ti-N Compounds by Thermionically Assisted Triode Reactive Ion-Plating," *Thin Solid Films*, Vol. 72, pp. 541–549.

McKenzie, D. R., McPhedran, R. C., Botten, L. C., Savvides, N., and Netterfield, R. P. (1982), "Hydrogenated Carbon Films Produced by Sputtering in Argon-Hydrogen Mixtures," *Appl. Opt.*, Vol. 21, pp. 3615–3616.

McLintock, I. S., and Orr, J. C. (1973), "Evaporated Carbon Films," *Adv. Chemistry and Physics of Carbon* (P. L. Walker, Jr., and P. A. Thrower, eds.), Vol. 11, pp. 243–312, Marcel Dekker, New York.

Memming, R. (1986), "Properties of Polymeric Layers of Amorphous Hydrogenated Carbon Produced by a Plasma-Activated Chemical Vapor Deposition Process, I: Spectroscopic Investigations," *Thin Solid Films*, Vol. 143, pp. 279–289.

———, Tolle, H. J., and Wierenga, P. E. (1986), "Properties of Polymeric Layers of Amorphous Hydrogenated Carbon Produced by a Plasma-Activated Chemical Vapor Deposition Process, II: Tribological and Mechanical Properties," *Thin Solid Films*, Vol. 143, pp. 31–41.

Messier, R., Badzian, A. R., Badzian, T., Spear, K. E., Bachmann, P., and Roy, R. (1987), "From Diamond-like Carbon to Diamond Coatings," *Thin Solid Films*, Vol. 153, pp. 1–9.

Meyerson, B., and Smith, F. W. (1980), "Electrical and Optical Properties of Hydrogenated Amorphous Carbon Films," *J. Noncrystal. Solids*, Vol. 35/36, pp. 435–440.

———, and ——— (1981), "Electrical and Optical Properties of Hydrogenated Amorphous Carbon Films," *J. Noncrystal. Solids*, Vol. 35/36, pp. 435–440.

Milewski, W. (1973), "Sonic Phenomena Occurring during Plasma Spraying WC-Co Compositions," *Proc. 7th Int. Metal Spraying Conf., London, U.K.*

Miller, R. A., and Berndt, C. C. (1984), "Performance of Thermal Barrier Coatings in High Heat Flux Environments," *Thin Solid Films*, Vol. 119, pp. 195–202.

Mirtich, M. J., Swec, D. M., and Angus, J. C. (1985), "Dual-Ion-Beam Deposition of Carbon Films with Diamond Like Carbon," *Thin Solid Films*, Vol. 131, pp. 245–254.

Miyamoto, M., Hirose, M., and Osaka, Y. (1983), "Structural and Electronic Characterization of Discharge-Produced BN," *Jap. J. Appl. Phys.*, Vol. 22, pp. L216–L218.

Miyasato, T., Kawakami, Y., Kawano, T., and Hiraki, A. (1984), "Preparation of sp^3-rich Amorphous Carbon Film by Hydrogen Gas Reactive rf-Sputtering of Graphite and Its Properties," *Jap. J. Appl. Phys.*, Vol. 23, pp. L234–L237.

Miyazawa, T., Misawa, S., Yoshida, S., and Gonda, S. (1984), "Preparation and Structure of Carbon Film Deposited by a Mass-Separated C^+ Ion Beam," *J. Appl. Phys.*, Vol. 55, pp. 188–193.

Miyoshi, K., Buckley, D. H., and Spalvins, T. (1985), "Tribological Properties of Boron Nitride Synthesized by Ion Beam Deposition," *J. Vac. Sci. Technol.*, Vol. A3, pp. 2340–2344.

Mock, J. A. (1974), "Ceramic and Refractory Coatings," *Mater. Eng.*, November, Vol. 80, pp. 101–108.

Mogab, C. J., and Lugujjo, E. (1976), "Backscattering Analysis of the Composition of Silicon-Nitride Films Deposited by rf Reactive Sputtering," *J. Appl. Phys.*, Vol. 47, pp. 1302–1309.

Moravec, T. J., and Orent, T. W. (1981), "Electron Spectroscopy of Ion Beam and Hydrocarbon Plasma Generated Diamondlike Carbon Films," *J. Vac. Sci. Technol.*, Vol. 18, pp. 226–228.

Mori, T., and Namba, Y. (1983), "Hard Diamond Like Carbon Films Deposited by Ionized Deposition of Methane Gas," *J. Vac. Sci. Technol.*, Vol. A1, pp. 23–27.

Most, C. R. (1969), "Surface Coatings Available to Industry Today," Paper No. 690481, presented at Society of Automotive Engineers Meeting, Chicago, Ill.

Movchan, B. A., and Demchishin, A. V. (1969), "Study of the Structure and Properties of Thick Vacuum Condensates of Nickel, Titanium, Tungsten, Aluminum Oxide, and Zirconium Dioxide," *Phys. Met. Metallogr.*, Vol. 28, pp. 83–90.

Mullendore, A. W., and Pope, L. F. (1987), "Wear Resistance of Metastable Ni-B Alloys Produced by Chemical Vapor Deposition," *Thin Solid Films*, Vol. 153, pp. 267–279.

Mumtaz, A., and Class, W. H. (1982), "Color of Titanium Prepared by Reactive dc Magnetron Sputtering," *J. Vac. Sci. Technol.*, Vol. 20, pp. 345–348.

Münz, W. D. (1985), "Reactive Sputtering of Nitrides and Carbides," *Proc. of Course on Nitride and Carbides Coatings*, publication no. 11—SO 7.2, Centre Suisse D'Electronique Et Du Microtechnique, S.A., Nêuchatel, Switzerland.

———, and Hessberger, G. (1981), "Coating of Preformed Pieces with TiN by High Rate Sputtering," *Vak. Tech.*, Vol. 30, pp. 78–86.

———, Hofmann, D., Hartig, K. (1982), "A High Rate Sputtering Process for the Formation of Hard Friction-Reducing TiN Coating on Tools," *Thin Solid Films*, Vol. 96, pp. 79–86.

Murray, S. F. (1984), "Wear Resistant Coatings and Surface Treatments," in *CRC Handbook of Lubrication: Theory and Practice of Tribology*, Vol. 2: *Theory and Design* (E. R. Booser, ed.), pp. 623–644, CRC Press, Boca Raton, Fla.

Nadler, M. P., Donovan, T. M., and Green, A. K. (1984a), "Structure of Carbon Films Formed by the Plasma Decomposition of Hydrocarbons," *Applications Surf. Sci.*, Vol. 18, pp. 10–17.

———, ———, and ——— (1984b), "Thermal Annealing Study of Carbon Film Formed by Plasma Decomposition of Hydrocarbons," *Thin Solid Films*, Vol. 116, pp. 241–247.

Nakamura, K., Inagawa, K., Tsuruoka, K., and Komiya, S. (1977), "Applications of Wear-Resistant Thick Films Formed by Physical Vapor Deposition Process," *Thin Solid Films*, Vol. 40, pp. 155–167.

Nelson, R. L., Ramsay, J. D. F., Woodhead, J. L., Cairns, J. A., and Crossley, J. A. A. (1981), "The Coating of Metals with Ceramic Oxides via Colloidal Intermediates," *Thin Solid Films*, Vol. 81, pp. 329–337.

Nilsson, H. T. G., Andersson, K. A. B., and Nordlander, P. J. (1977), "The Influence of the Condensation Surface Temperature on the Microstructure and Intrinsic Stress in Evaporated Chromium Films," *Thin Solid Films*, Vol. 45, pp. 463–477.

Nimmagadda, R., and Bunshah, R. F. (1976), "Synthesis of TiC-Ni Cermets by the Activated Reactive Evaporation Process," *J. Vac. Sci. Technol.*, Vol. 13, pp. 532–535.

———, and ——— (1979), "High Rate Deposition of Hafnium Nitride by Activated Reactive Evaporation," *Thin Solid Films*, Vol. 63, pp. 327–331.

Nir, D. (1984), "Stress Relief Forms of Diamond-like Carbon Thin Films under Internal Compressive Stress," *Thin Solid Films*, Vol. 112, pp. 41–49.

Noda, T., Inagaki, M., and Yamada, S. (1969), "Glass-like Carbons," *J. Noncrystal. Solids*, Vol. 1, pp. 285–302.

Nyaiesh, A., and Holland, L. (1984), "The Growth of Amorphous and Graphitic Carbon Layers under Ion Bombardment in an rf Plasma," *Vacuum*, Vol. 34, pp. 519–522.

Oakes, J. J. (1983), "A Comparative Evaluation of HfN, Al_2O_3, TiC and TiN Coatings on Cemented Carbide Tools," *Thin Solid Films*, Vol. 107, pp. 159–165.

Obata, T., Aida, H., Hirohata, Y., Mohri, M., and Yasashina, T. (1981), "Titanium Carbide Coatings on Molybdenum by Means of Reactive Sputtering and Electron Beam Techniques," *J. Nucl. Mater.*, Vol. 103/104, pp. 283–288.

O'Brien, T. E., and Chaklades, A. C. D. (1974), "Effect of Oxygen on the Reaction between Copper and Sapphire," *J. Am. Ceramic Soc.*, Vol. 57, pp. 329–332.

Ogawa, T., Ikawa, K., and Iwamoto, K. (1979), "Effect of Gas Composition on the Deposition of ZrC-C Mixtures: The Bromide Process," *J. Mater. Sci.*, Vol. 14, pp. 125–132.

Ojha, S. M. (1982), "Plasma Enhanced Chemical Vapor Deposition of Thin Films," *Physics of Thin Films* (G. Hass, M. H. Francombe, and J. L. Vossen, eds.), Vol. 12, pp. 237–296, Academic, New York.

———, and Holland, L. (1977a), "The Growth Characteristics and Properties of Hard Carbon Films Grown under Ion Impact," *Proc. Conf. on Ion Plating and Allied Techniques*, Edinburgh, pp. 238–242, CEP Consultants, Edinburgh.

———, and ——— (1977b), "Preparation and Properties of Amorphous Carbon Films Grown under Ion Impact in an RF Discharge," *Proc. 7th Int. Vacuum Congr. and 3rd Int. Conf. on Solid Surfaces*, Vienna, Vol. 2, pp. 1667–1669.

———, and ——— (1977c), "Some Characteristics of Hard Carbonaceous Films," *Thin Solid Films*, Vol. 40, pp. L31–L32.

———, Norstrom, H., and McCoulluch, D. (1979), "The Growth Kinetics and Properties of Hard and Insulating Carbonaceous Films Grown in an r.f. Discharge," *Thin Solid Films*, Vol. 60, pp. 213–225.

Packer, C. M., and Clark, P. M. (1978), Report No. FE-2592-6 Lockheed Palo Alto Research Lab., Sunnyvale, Calif.

Padmanabhan, K. R., Chevallier, J., and Sørensen, G. (1986), "The Influence of Ion Bombardment on Deposition of Carbon Films," *Nucl. Instrum. Meth. Phys. Res.*, Vol. B16, pp. 369–372.

Pan, A., and Greene, J. E. (1981), "Residual Compressive Stress in Sputter Deposited TiC Films on Steel Substrates," *Thin Solid Films*, Vol. 78, pp. 25–34.

Parker, K. (1974), "Hardness and Wear Resistance Tests of Electroless Nickel Deposits," *Plating*, Vol. 61, pp. 834–841.

Pawlewicz, W. T., and Hays, D. D. (1982), "Microstructure Control for Sputter-Deposited ZrO_2, $ZrO_2 \cdot CaO$ and $ZrO_2 \cdot Y_2O_3$," *Thin Solid Films*, Vol. 94, pp. 31–45.

Perry, A. J. (1986), "The Structure and Colour of Some Nitride Coatings," *Thin Solid Films*, Vol. 135, pp. 73–85.

———, and Chollet, L. (1986), "States of Residual Stress both in Films and in Their Substrates," *J. Vac. Sci. Technol.*, Vol. A4, pp. 2801–2808.

———, Hammer, B., and Grossel, M. (1983), "The Mode of Detachment of HfN Coatings," *Proc. 4th Eur. Conf. on Chemical Vapor Deposition*, Eindhoven, Netherlands, pp. 70–75.

———, Simmen, L., and Chollet, L. (1984), "Ion Plated HfN Coatings," *Thin Solid Films*, Vol. 118, pp. 271–277.

———, Grossl, M., and Hammer, B. (1985), "Some Characteristics of HfN Coatings on Cemented Carbide Substrates," *Thin Solid Films*, Vol. 129, pp. 263–279.

———, Strandberg, C., Sproul, W. D., Hofmann, S., Ernsberger, C., Nicherson, J., and Chollet, L. (1987), "The Chemical Analysis of TiN Films: A Round Robin Experiment," *Thin Solid Films*, Vol. 153, pp. 169–183.

Peterson, J. R. (1974), "Partial Pressure of $TiCl_4$ in CVD of TiN," *J. Vac. Sci. Technol.*, Vol. 11, pp. 715–718.

Peterson, M. B., and Winor, W. O. (eds.) (1980), *Wear Control Handbook*, ASME, New York.

Pethica, J. B., Koidl, P., Gobrecht, J., and Schuler, C. (1985), "Micromechanical Investigations of Amorphous Hydrogenated Carbon Films on Silicon," *J. Vac. Sci. Technol.*, Vol. A3, pp. 2391–2392.

Pettit, F. S., and Goward, G. W. (1983), "Gas Turbine Applications," in *Coatings for High Temperature Applications* (E. Lang, ed.), pp. 341–359, Applied Science Publishers, London.

Pierson, H. O. (1977), "Titanium Carbonitride Obtained by Chemical Vapor Deposition," *Thin Solid Films*, Vol. 40, pp. 41–47.

———, and Mullendore, A. W. (1980), "The Chemical Vapor Deposition of TiB_2 from Diborane," *Thin Solid Films*, Vol. 72, pp. 511–516.

———, and ——— (1982), "Thick Boride Coatings by Chemical Vapor Deposition," *Thin Solid Films*, Vol. 95, pp. 99–104.

———, and Randich, E. (1978), "Titanium Diboride Coatings and Their Interaction with the Substrates," *Thin Solid Films*, Vol. 54, pp. 119–128.

Poferl, D. J., Gardner, N. C., and Angus, J. C. (1973), "Growth of Boron-Doped Diamond Seed Crystals by Vapor Deposition," *J. Appl. Phys.*, Vol. 44, pp. 1428–1434.

Poitevin, J. M., Lemperiere, G., and Tardy, J. (1982), "Influence of Substrate Bias on the Composition, Structure, and Electrical Properties of Reactively dc Sputtered TiN Films," *Thin Solid Films*, Vol. 97, pp. 69–77.

Pope, L. E., Fehrenbacher, L., and Winer, W. O. (Eds.) (1989), "New Materials Approaches to Tribology: Theory and Applications," *Mat. Res. Soc. Symp. Proc.*, No. 140, Materials Research Society, Pittsburgh, Pa.

Quinto, D. T., Wolfe, G. J., and Jindal, P. C. (1987), "High Temperature Microhardness of

Hard Coating Produced by Physical and Chemical Vapor Deposition," *Thin Solid Films*, Vol. 153, pp. 19–36.

Rabinowicz, E. (1967), "Lubrication of Metal Surfaces by Oxide Films," *ASLE Trans.*, Vol. 10, pp. 400–407.

Raghuram, A. C., and Bunshah, R. F. (1972), "The Effect of Substrate Temperature on the Structure of Titanium Carbide Deposited by Activated Reactive Evaporation," *J. Vac. Sci. Technol.*, Vol. 9, pp. 1389–1394.

Ramalingam, S., and Winer, W. O. (1980), "Reactively Sputtered TiN Coatings for Tribological Applications," *Thin Solid Films*, Vol. 73, pp. 267–274.

Randhawa, H., Johnson, P. C., and Cunningham, R. (1988), "Deposition and Characterization of Ternary Nitrides," *J. Vac. Sci. Technol.*, Vol. A6, pp. 2136–2139.

Randich, E. (1980), "Low Temperature Chemical Vapor Deposition of TaB_2," *Thin Solid Films*, Vol. 72, pp. 517–522.

Randin, J. P., and Hintermann, H. E. (1967), "Electroless Nickel Deposited at Controlled pH: Mechanical Properties as a Function of Phosphorous Content," *Plating*, Vol. 54, No. 5, pp. 523–532.

Reardon, J. D., Mignogna, R., Longo, F. N. (1981), "Plasma- and Vacuum-Plasma Sprayed Cr_3C_2 Composite Coatings," *Thin Solid Films*, Vol. 83, pp. 345–351.

Reinberg, A. R. (1979), "Plasma Deposition of Inorganic Silicon Containing Films," *J. Electron. Mater.*, Vol. 8, pp. 345–375.

Retajczyk, T. F., and Sinha, A. K. (1980), "Elastic Stiffness and Thermal Expansion Coefficients of Various Refractory Silicides and Silicon Nitride Films," *Thin Solid Films*, Vol. 70, pp. 241–247.

Rickerby, D. S., Bellamy, B. A., and Jones, A. M. (1987), "Internal Stress and Microstructure of Titanium Nitride Coatings," *Surf. Eng.*, Vol. 3, pp. 138–146.

Robertson, J. (1986), "Amorphous Carbon," *Adv. Phys.*, Vol. 35, pp. 317–374.

Ron, Y., Raveh, A., Carmi, U., Inspektor, A., and Avni, R. (1983), "Deposition of Silicon Nitride from $SiCl_4$ and NH_3 in a Low Pressure R.F. Plasma," *Thin Solid Films*, Vol. 107, pp. 181–189.

Röser, K. (1981), "TiC Growth Studies on Different Standard Steels," *Proc. 8th Int. Conf. on Chemical Vapor Deposition* (J. M. Blocher, G. E. Vuillard, and G. Wahl, eds.), pp. 586–596, The Electrochemical Society, Pennington, N.J.

Rother, B. (1985), Thesis, Technische Hochscule Karl-Max-Stadt, Germany.

———, Zscheile, H. D., Weissmantel, C., Heiser, C., Holzhuter, G., Leonhardt, G., and Reich, P. (1986), "Preparation and Characterization of Ion-Plated Boron Nitride," *Thin Solid Films*, Vol. 142, pp. 83–99.

Rowe, G. W. (1960), "Some Obsērvations on the Frictional Behavior of Boron Nitride and of Graphite," *Wear*, Vol. 3, pp. 274–285.

Rowe, H. H. (1968), "An Investigation of Methods to Improve the Wear Resistance of Gas-Bearing Ceramic Materials," *J. Lub. Technol. (Trans. ASME)*, Vol. 90, pp. 829–840.

Rubin, M., Hopper, C. B., Cho, N. H., and Bhushan, B. (1990), "Optical and Mechanical Properties of DC Sputtered Carbon Films," *J. Mater. Res.*, Vol. 5, pp. 2538–2542.

Rujkorakarn, R., and Sites, J. R. (1986), "Crystalization of Zirconia Films by Thermal Annealing," *J. Vac. Sci. Technol.*, Vol. A4, pp. 568–571.

Sale, J. M. (1979), "Wear Resistance of Silicon Carbide Composite Coatings," *Metal Prog.*, Vol. 115, pp. 44–55.

Samsonov, G. V., and Vinitskii, I. M. (1980), *Handbook of Refractory Compounds*, IFI/Plenum, New York.

Sander, P., Kaiser, U., Altebockwinkel, M., Wiedmann, L., Benninghoven, A., Sah, R. E., and Koidl, P. (1987), "Depth Profile Analysis of Hydrogenated Carbon Layers on Silicon by X-ray Photoelectron Spectroscopy, Auger Electron Spectroscopy, Electron Energy-Loss Spectroscopy, and Secondary Ion Mass Spectrometry," *J. Vac. Sci. Technol.*, Vol. A5, pp. 1470–1473.

Sangam, M., and Nikitich, J. (1985), "Designing for Optimum Use of Plasma Coatings," *J. Metals*, Vol. 37 (no. 9), September, pp. 55–60.

Sarin, V. K., Bunshah, R. F., and Nimmagadda, R. (1977), "Structure of TiN-Ni Coatings Synthesized by Activated Reactive Evaporation," *Thin Solid Films*, Vol. 40, pp. 183–188.

Sato, K., Mohri, M., and Yamashina, T. (1981), "Preparation of Titanium Nitride onto Molybdenum, Stainless Steel and Carbon by Gas Absorption and Reactive r.f. Sputtering," *J. Nucl. Mater.*, Vols. 103 and 104, pp. 273–278.

Sato, T., Tada, M., Huang, Y. C., and Takei, H. (1978), "Physical Vapor Deposition of Chromium and Titanium Nitrides by the Hollow Cathode Discharge Process," *Thin Solid Films*, Vol. 54, pp. 61–65.

Sato, Y., Kamo, M., and Setaka, N. (1987), "Growth of Diamond and Various Hydrocarbon-Hydrogen Mixtures," *Proc. of 8th Int. Symp. on Plasma Chemistry* (K. Akashi and A. Kinbara, eds.), pp. 2446–2451, International Union of Pure and Applied Chemistry, Oxford, England.

Satou, M., and Fujimoto, F. (1983), "Formation of Cubic Boron Nitride Films by Boron Evaporation and Nitrogen Ion Beam Bombardment," *Jap. J. Appl. Phys.*, Vol. 22, pp. L171–L172.

Savage, R. H. (1948), "Graphite Lubrication," *J. Appl. Phys.*, Vol. 19, pp. 1–10.

———, and Schaefer, D. L. (1956), "Vapor Lubrication of Graphite Sliding Contacts," *J. Appl. Phys.*, Vol. 27, pp. 136–138.

Savvides, N. (1985), "Four-fold to Three-fold Transition in Diamond-like Amorphous Carbon Films: A Study of Optical and Electrical Properties," *J. Appl. Phys.*, Vol. 58, pp. 518–521.

———, and Window, B. (1985), "Diamondlike Amorphous Carbon Films Prepared by Magnetron Sputtering of Graphite," *J. Vac. Sci. Technol.*, Vol. A3, pp. 2386–2390.

Sawabe, A., and Inuzuka, T. (1986), "Growth of Diamond Thin Films by Electron-Assisted Chemical Vapor Deposition and Their Characterization," *Thin Solid Films*, Vol. 137, pp. 89–99.

Schlichting, J. (1980), "Chemical Vapor Deposition of Silicon Carbide," *Powder Metall. Int.*, Vol. 12, pp. 141–147, 196–200.

Schmidt, R. D., and Ferriss, D. P. (1975), "New Materials Resistant to Wear and Corrosion to 1000°C," *Wear*, Vol. 32, pp. 279–289.

Schumacher, W. J. (1981), "A Stainless Steel Alternative to Cobalt Wear Alloys," *Chem. Eng.*, Vol. 21, September, pp. 149–152.

Schwarz, E. (1980), *Proc. 9th Int. Conf. on Thermal Spraying*, Paper 21, The Hague, Netherlands.

Shanfield, S., and Wolfson, R. (1983), "Ion Beam Synthesis of Cubic Boron Nitride," *J. Vac. Sci. Technol.*, Vol. A1, pp. 323–325.

Shikama, T., Araki, H., Fujitsuka, M., Fukutomi, M., Shinno, H., and Okada, M. (1983), "Properties and Structure of Carbon Excess Ti_xC_{1-x} Deposited onto Molybdenum by Magnetron Sputtering," *Thin Solid Films*, Vol. 106, pp. 185–186.

Silence, W. L. (1978), "Effect of Structure on Wear Resistance of Co-, Fe-, and Ni-base Alloy," *J. Lub. Technol. (Trans. ASME)*, Vol. 100, pp. 428–435.

Singer, I. L., Bolster, R. N., Wolf, S. A., Skelton, E. F., and Jeffries, R. A. (1983), "Abrasion Resistance, Microhardness and Microstructures of Single Phase Niobium Nitride Films," *Thin Solid Films*, Vol. 107, pp. 207–215.

Skibo, M., and Gruelich, F. A. (1984), "Characterization of Chemically Vapor Deposited Ni- (0.05–0.2 wt %) B Alloys," *Thin Solid Films*, Vol. 113, pp. 225–234.

Sokolowski, M., Sokolowska, A., Michalski, A., Romanowski, Z., Rusek-Mazurek, A., and Wronikowski, M. (1981), "The Deposition of Thin Films of Materials with High Melting Points on Substrates at Room Temperature Using the Pulse Plasma Method," *Thin Solid Films*, Vol. 80, pp. 249–254.

Spalvins, T. (1977), "Microstructural and Wear Properties of Sputtered Carbides and Silicides," *Proc. Int. Conf. Wear of Materials* (W. A. Glaeser, K. C. Ludema, and S. K. Rhee, eds.), ASME, New York.

———, and Brainard, W. A. (1974), "Nodular Growth in Thick-Sputtered Metallic Coatings," *J. Vac. Sci. Technol.*, Vol. 11, November/December, pp. 1186–1192.

Spear, K. E. (1989), "Diamond-Ceramic Coating of the Future," *J. Amer. Ceramic Soc.*, Vol. 72, pp. 171–191.

Spencer, E. G., Schmidt, P. H., Joy, D. C., and Sansalone, F. J. (1976), "Ion Beam Deposited Polycrystalline Diamond-like Films," *Appl. Phys. Lett.*, Vol. 29, pp. 118–120.

Spitsyn, B. V., Bouilov, L. L., and Deryagin, B. V. (1981), "Vapor Growth of Diamond on Diamond and Other Surfaces," *J. Crystal. Growth*, Vol. 52, pp. 219–226.

Spitz, J., Chevallier, J., and Aubert, A. (1974), "Properties and Structure of Thin Films of Niobium Nitride Prepared by Reactive Cathodic Sputtering, I. Preparation and Properties," *J. Less-Common Met.*, Vol. 35, pp. 181–192.

Sproul, W. D. (1984), "Hafnium Nitride Coatings Prepared by Very High Rate Reactive Sputtering," *Thin Solid Films*, Vol. 118, pp. 279–284.

——— (1985a), "Very High Rate Reactive Sputtering of TiN, ZrN, and HfN," *Thin Solid Films*, Vol. 107, pp. 141–147.

——— (1985b), "Reactively Sputtered TiN, ZrN and HfN," *J. Vac. Sci. Technol.*, Vol. A3, pp. 580–581.

———, and Rothstein, R. (1985), "High Rate Reactively Sputtered TiN Coatings on High Speed Steel Drills," *Thin Solid Films*, Vol. 126, pp. 257–263.

———, and Richman, M. H. (1975), "Effect of the Eta Layer on TiC-Coated, Cemented-Carbide Tool Life," *J. Vac. Sci. Technol.*, Vol. 12, pp. 842–844.

Srivastava, P. K., Rao, T. V., Vankar, V. D., and Chopra, K. L. (1984), "Synthesis of Tungsten Carbide Films by rf Magnetron Sputtering," *J. Vac. Sci. Technol.*, Vol. A2, pp. 1261–1265.

Stiglich, J. J., and Bhat, D. K. (1980), "Friction and Erosive Wear of Controlled Nucleation Thermochemically Deposited W-C Alloys and SiC and Chemically Vapor Deposited Si_3N_4," *Thin Solid Films*, Vol. 72, pp. 503–509.

Stolz, M., Hieber, K., and Wieczorek, C. (1983), "Universal Chemical Vapor Deposition System for Metallurgical Coatings," *Thin Solid Films*, Vol. 100, pp. 209–218.

Storms, E. K. (1967), *The Refractory Carbides*, Academic, New York.

Sundaram, K. B., and Bhagavat, G. K. (1981), "X-Ray and Electron Diffraction Studies of Chemically Vapor-Deposited Tin Oxide Films," *Thin Solid Films*, Vol. 78, pp. 35–40.

Sundgren, J.-E. (1985), "Structure and Properties of TiN Coatings," *Thin Solid Films*, Vol. 128, pp. 21–44.

———, Johansson, B. O., and Karlsson, S. E. (1983a), "Mechanisms of Reactive Sputtering of Titanium Nitride and Titanium Carbide, I: Influence of Process Parameters on Film Composition," *Thin Solid Films*, Vol. 105, pp. 353–366.

———, ———, ———, and Hentzell, H. T. G. (1983b), "Mechanisms of Reactive Sputtering of Titanium Nitride and Titanium Carbide, II: Morphology and Structure," *Thin Solid Films*, Vol. 105, pp. 367–384.

———, ———, Hentzell, H. T. G., and Karlsson, S. E. (1983c), "Mechanisms of Reactive Sputtering of Titanium Nitride and Titanium Carbide, III: Influence of Substrate Bias on Composition and Structure," *Thin Solid Films*, Vol. 105, pp. 385–393.

———, and Hentzell, H. T. G. (1986), "A Review of the Present State of Art in Hard Coatings Grown from the Vapor Phase," *J. Vac. Sci. Technol.*, Vol. A4, September/October pp. 2259–2279.

Suri, A. K., Nimmagadda, R., and Bunshah, R. F. (1980), "Synthesis of Titanium Nitrides by Activated Reactive Evaporation," *Thin Solid Films*, Vol. 72, pp. 529–533.

Takahashi, T., Sugiyama, K., and Tomita, K. (1967), "The Chemical Vapor Deposition of Titanium Carbide Coatings on Iron," *J. Electrochem. Soc.*, Vol. 114, pp. 1230–1235.

———, ———, and Itoh, H. (1970), "Single Crystal Growth of Titanium Carbide by Chemical Vapor Deposition," *J. Electrochem. Soc.*, Vol. 117, pp. 541–545.

———, and ——— (1974), "Fibrous Growth of Tantalum Carbide by AC Discharge Methods," *J. Electrochem. Soc. Solid State Sci. Technol.*, Vol. 121, pp. 714–718.

———, Itoh, H., and Takeuchi, A. (1979), "Chemical Vapor Deposition of Hexagonal Boron Nitride Thick Film on Iron," *J. Crystal. Growth*, Vol. 47, pp. 245–250.

Taylor, K. A. (1977), "High Speed Friction and Wear Characteristics of Vapor Deposited Coatings of Titanium Carbide and Tungsten Carbide Plus Cobalt in a Vacuum Environment," *Thin Solid Films*, Vol. 40, pp. 189–200.

Taylor, B. J., and Eyre, T. S. (1979), "A Review of Piston Rings and Cylinder Liner Materials," *Tribol. Int.*, Vol. 12, pp. 79–89.

Teer, D. G., and Salama, M. (1977), "The Ion Plating of Carbon," *Thin Solid Films*, Vol. 45, pp. 553–561.

Thornton, J. A., and Chin, J. (1977), "Structure and Heat Treatment Characteristics of Sputter-Deposited Alumina," *Ceram. Bull.*, Vol. 56, pp. 504–512.

———, and Ferriss, D. P. (1977), "Structural and Wear Properties of Sputter-Deposited Laves Intermetallic Alloy," *Thin Solid Films*, Vol. 40, pp. 365–374.

———, and Hoffman, D. W. (1981), "Internal Stresses in Amorphous Silicon Films Deposited by Cylindrical Magnetron Sputtering Using Ne, Ar, Kr, Xe, and Ar + H_2," *J. Vac. Sci. Technol.*, Vol. 18, pp. 203–207.

Torres, J., Perio, A., Pantel, R., Campidelli, Y., and D'Avitaya, F. A. (1985), "Growth of Thin Films of Refractory Silicides on Si (100) in Ultrahigh Vacuum," *Thin Solid Films*, Vol. 126, pp. 233–239.

Torok, S., Perry, A. J., Chollet, L., and Sproul, W. D. (1987), "Young's Modulus of TiN, TiC, ZrC, and HfN," *Thin Solid Films*, Vol. 153, pp. 37–43.

Toth, L. E. (1971), *Transition Metal Carbides and Nitrides*, Academic, New York.

Trester, P. W., Hopkins, G. R., Kaae, J. L., and Whitley, J. (1983), "Performance Results on a C-SiC Alloy Coating Chemically Vapor-deposited onto Graphite Substrates," *Thin Solid Films*, Vol. 108, pp. 383–393.

Tsai, H.-C., and Bogy, D. B. (1987), "Characterization of Diamond like Carbon Films and Their Application As Overcoats on Thin Film Media for Magnetic Recording," *J. Vac. Sci. Technol.*, Vol. A5, pp. 3287–3312.

———, ———, Kundman, M. K., Veirs, D. K., Hilton, M. R., and Mayers, S. T. (1988), "Structure and Properties of Sputtered Carbon Overcoats on Rigid Magnetic Media Disks," *J. Vac. Sci. Technol.*, Vol. A6, pp. 2307–2315.

Tucker, R. C. (1974), "Structure Property Relationships in Deposits Produced by Plasma Spray and Detonation Gun Techniques," *J. Vac. Sci. Technol.*, Vol. 11, pp. 725–734.

Turner, P. G., Howson, R. P., and Bishop, C. A. (1980), "Preparation and Optical Properties of Transparent Carbon Films Prepared in Various Hydrogen Atmosphere rf Plasmas," in *Low Energy Ion Beams* (I. H. Wilson and K. G. Stephens, eds.), *Conf. Ser. No. 54*, pp. 229–234, The Institute of Physics, London.

Ulla, M., and Juhani, V. (1984), "Tribological Properties of PVD TiN and TiC Coatings," *Proc. of the Nordtrib-84 Conf. Tempre*, Finland.

Vandentop, G. J., Kawasaki, M., Nix, R. M., Brown, I. G., Salmaron, M., and Somorjai, G. A. (1990), "Formation of Hydrogenated Amorphous Carbon Films of Controlled Hardness from a Methane Plasma," *Phy. Rev.*, Vol. B41 (no. 5), pp. 3200–3210.

Vilok, Y. N. (1978), "Character of Microhardness Variation in the Homogeneity Field of Titanium Carbonitride," *Sov. Powder Metall. Met. Ceram.*, Vol. 17, pp. 467–470.

Vinayo, M. E., Kassabji, F., Guyonnet, J., and Fauchais, P. (1985), "Plasma Sprayed WC-Co Coating: Influence of Spray Conditions (Atmospheric and Low Pressure Plasma Spraying) on the Crystal Structure, Porosity, and Hardness," *J. Vac. Sci. Technol.*, Vol. A3, pp. 2483–2489.

Von Seefeld, H., Cheung, N. W., Manpaa, M., and Nicolet, M.-A. (1980), "Investigation of Titanium-nitride Layers for Solar-Cell Contacts," *IEEE Trans. Electron Devices*, Vol. ED27, pp. 873–876.

Wada, N., Gaczi, P. J., and Solin, S. A. (1980), "Diamond-like 3-fold Coordinated Amorphous Carbon," *J. Noncrystal. Solids*, Vols. 35–36, pp. 543–548.

Weast, R. C. (1983), *CRC Handbook of Chemistry and Physics*, 63rd ed., CRC, Boca Raton, Fla., p. F-186.

Wedge, R. H., and Eaves, A. V. (1980), "Coatings in the Aero Gas Turbine," *Proc. 9th Int. Conf. on Thermal Spraying*, The Hague, Netherlands, paper 19.

Weissmantel, C. (1977), "Trends in Thin Film Deposition Methods," *Proc. 7th Int. Vacuum Congr. and 3rd Int. Conf. on Solid Surfaces*, Vienna, Vol. 2, pp. 1533–1544.

——— (1981), "Ion Beam Deposition of Special Film Structures," *J. Vac. Sci. Technol.*, Vol. 18, pp. 179–185.

———, Erler, H.-J., and Resisse, G. (1979a), "Ion Beam Techniques for Thin and Thick Film Deposition," *Surf. Sci.*, Vol. 86, pp. 207–221.

———, Bewilogua, K., Schurer, C., Breuer, K., and Zscheile, H. (1979b), "Characterization of Hard Carbon Films by Electron Energy Loss Spectroscopy," *Thin Solid Films*, Vol. 62, pp. L1–L4.

———, ———, Dietrich, D., Erler, H.-J., Hinneberg, H. J., Klose, S., Nowick, W., and Reisse, G. (1980), "Structure and Properties of Quasi-amorphous Films Prepared by Ion Beam Techniques," *Thin Solid Films*, Vol. 72, pp. 19–31.

———, ———, Breuer, K., Dietrich, D., Ebersbach, U., Erler, H.-J., Rau, B., and Reisse, G. (1982), "Preparation and Properties of Hard i-C and i-BN Coatings," *Thin Solid Films*, Vol. 96, pp. 31–44.

———, Breuer, K., and Winde, B. (1983), "Hard Films of Unusual Microstructure," *Thin Solid Films*, Vol. 100, pp. 383–389.

Wheeler, D. R., and Brainard, W. A. (1977), "X-ray Photoelectron Spectroscopy Study of Radiofrequency Sputtered Chromium Diboride, Molybdenum Disilicide and Molybdenum Disulfide Coatings and Their Friction Properties," Tech. Note TN D-8482, NASA, Washington, D.C.

———, and ——— (1978), "Composition of rf-sputtered Refractory Compounds Determined by X-ray Photoelectron Spectroscopy," *J. Vac. Sci. Technol.*, Vol. 15, pp. 24–30.

Whitmell, D. S., and Williamson, R. (1976), "The Deposition of Hard Surface Layers by Hydrocarbon Cracking in a Glow Discharge," *Thin Solid Films*, Vol. 35, pp. 255–261.

Wieczorek, C. (1985), "Chemical Vapor Deposition of Tantalum Disilicide," *Thin Solid Films*, Vol. 126, pp. 227–232.

Wiggins, M. D., Aita, C. R., and Hickernell, F. S. (1984), "Radio Frequency Sputter Deposited Boron Nitride Films," *J. Vac. Sci. Technol.*, Vol. A2, pp. 322–325.

Wild, C., Wagner, J., and Koidl, P. (1987), "Process Monitoring of a-C:H Plasma Deposition," *J. Vac. Sci. Technol.*, Vol. A5, pp. 2227–2230.

Wolfla, T. A., and Johnson, R. N. (1975), "Refractory Metal Carbide Coatings for LMFBR Applications—A Systems Approach," *J. Vac. Sci. Technol.*, Vol. 12, pp. 777–783.

———, and Tucker, R. C. (1978), "High Temperature Wear-Resistant Coatings," *Thin Solid Films*, Vol. 53, pp. 353–364.

Yamamoto, K., Ichikawa, Y., Nakayama, T., and Tawada, Y. (1988), "Relationship between Plasma Parameters and Carbon Atom Coordination in a-C:H Film Prepared by rf Glow Discharge Decomposition," *Jap. J. Appl. Phys.*, Vol. 27, pp. 1415–1421.

Yamashina, T., Aida, H., Kawamoto, O., and Suzuki, M. (1983), "TiC and TiN Coatings by Reactive Sputtering for Fusion Applications," *Thin Solid Films*, Vol. 108, pp. 395–400.

Yamashita, T., Chen, G. L., Shir, J., and Chen, T. (1988), "Sputtered ZrO_2 Overcoat with Superior Corrosion Protection and Mechanical Performance in Thin Film Rigid Disk Application," *IEEE Trans. Magn.*, Vol. MAG-24, pp. 2629–2634.

Yokohama, S., Kajihara, N., Hirose, M., Osaka, Y., Yoshihara, T., and Abe, H. (1980), "Characterization of Plasma-Deposited Silicon Nitride Films," *J. Appl. Phys.*, Vol. 51, pp. 5470–5474.

Yoshihara, H., and Mori, H. (1979), "Enhanced ARE Apparatus and TiN Synthesis," *J. Vac. Sci. Technol.*, Vol. 16, pp. 1007–1012.

Zega, B., Kornmann, M., and Amiguet, J. (1979), "Hard Decorative TiN Coatings by Ion Plating," *Thin Solid Films*, Vol. 45, pp. 577–582.

Zelez, J. (1983), "Low-Stress Diamond-like Carbon Films," *J. Vac. Sci. Technol.*, Vol. A1, pp. 305–307.

CHAPTER 15

SCREENING METHODOLOGY FOR MATERIALS, COATINGS, AND SURFACE TREATMENTS

Screening tests have to be conducted during the development of materials, coatings, and surface treatments for a particular application. These screening tests can be divided into two broad categories: physical and chemical examinations and friction and wear tests, including accelerated and functional tests. First, physical and chemical examinations of the candidate materials are conducted to determine if the materials meet the specifications. Next, simulated accelerated friction and wear tests are conducted to rank the candidate materials. Finally, functional (actual component) tests are conducted in order to make the final selection. If the accelerated friction and wear tests are properly simulated, an acceleration factor between the simulated test and the functional test can be empirically determined so that the subsequent functional tests can be minimized, saving considerable test time. This chapter presents a review of physical examinations and friction and wear tests.

15.1 PHYSICAL EXAMINATIONS

The tribological performance of materials, coatings, and surface treatments is strongly influenced by their chemical composition, microstructure, morphology, surface roughness, hardness, and toughness (or ductility), as well as coating thickness, porosity, density, residual stresses, and coating-to-substrate adhesion in the case of coatings. The electrical resistivity of metallic coatings is of importance for applications in the electronics industry. A careful examination of surfaces in sliding or rolling contact with appropriate surface analytical techniques can provide a considerable amount of information and insight into the behavior of the materials and a history of what has transpired on the surface. Excellent reviews of commonly used techniques to conduct physical (microstructure and morphology) and chemical or compositional analyses are presented by several authors (e.g., Bhushan, 1990; Buckley, 1981; Czanderna, 1975; Ferrante, 1981; Kane and Larrabee, 1974; Walker, 1970; Willard et al., 1974). This section presents an overview of methods used to measure coating thickness, coating density, coating porosity, electrical resistivity, surface roughness, hardness, residual stresses, and coating-to-substrate adhesion.

15.1.1 Techniques for Coating-Thickness Measurements

Thickness is undoubtedly the coating characteristic most often measured and specified quantitatively. In addition to dimensional tolerance, thickness is very significant to coating characteristics such as wear, corrosion, porosity, and other mechanical properties and monetary value. Thickness may be measured by in situ monitoring of the rate of deposition or by tests after the coated part is taken out of the deposition chamber.

Coating-thickness measurement techniques based on different principles are listed in Fig. 15.1 (Chopra, 1979; Lowenheim, 1974; Piegari and Masetti, 1985; Pliskin and Zanin, 1970; Schwartz and Schwartz, 1967). Optical, mechanical, electrical, magnetic, electromagnetic, radiation, and miscellaneous techniques have been used for coating-thickness measurements. Various thickness-measurement techniques based on different principles are commercially available. All the techniques are nondestructive, except coulometric and sectioning techniques.

Dial-gauge and gravimetric techniques are commonly used to measure thick coatings (>3 μm) and will not be discussed here. Among the various techniques, the stylus (mechanical) technique has gained the widest acceptance for thickness measurements. Optical interference phenomena have been found to be most useful when the coating thickness is on the order of a wavelength of light. The optical techniques can be used for both opaque and transparent coatings. Ellipsometry is used for thickness measurements of extremely thin (down to 1 nm) transparent coatings. Techniques for in situ monitoring of the rate of vapor deposition include optical interference (photometric, only for transparent coatings on reflective substrates), resistance, capacitance, and ionization and microbalance techniques. The quartz-crystal oscillator (microbalance) technique is the most commonly used in situ vapor deposition-rate monitor.

Commercially available thickness gauges are based on electrical resistance, eddy current, magnetic attractive force, electromagnetic induction, x-ray fluorescence, beta backscattering, and coulometric techniques. UPA Technology, Inc., Syosset, New York, is a major supplier of thickness-measuring instruments. Radiation techniques (x-ray fluorescence and beta backscattering) are commonly used as on-line measuring tools in various deposition processes. The characteristic features of common thickness-measurement techniques are compared in Table 15.1.

15.1.1.1 Optical Techniques. The usual optical techniques for thickness measurements of thin coatings (down to 3 nm) include optical interference for coat-

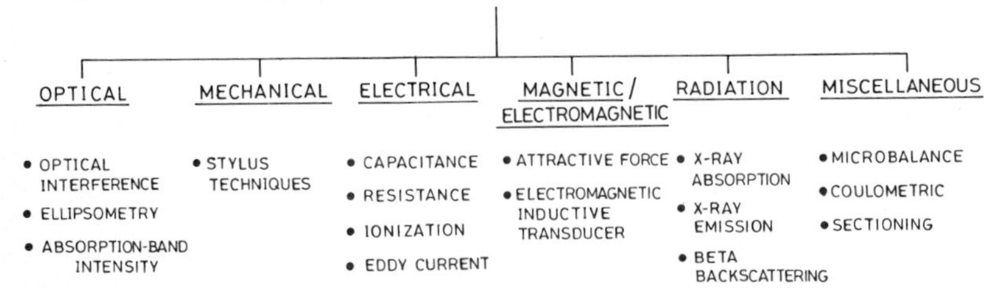

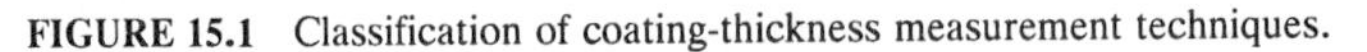
FIGURE 15.1 Classification of coating-thickness measurement techniques.

TABLE 15.1 Comparison of Some Common Thickness-Measurement Techniques

Measurement technique	Coating materials	Thickness range	Accuracy or precision	Remarks
Multiple-beam interference	All	3 nm to 2 μm	~1 nm	Nondestructive; step and highly reflective surface required; most accurate but time consuming
CARIS/VAMFO	Transparent coatings	40 nm to 20 μm	~1 nm to 0.1%	Nondestructive
Ellipsometry	Transparent coatings	1 nm to a few micrometers	~0.1 nm to 0.1%	Nondestructive; mathematics complicated; most accurate, especially with very thin coatings
Stylus	All	3 nm to a few millimeters	5%	Nondestructive; step in the coating required; the coating should be hard enough to resist the penetration of stylus; substrate must be quite smooth and rigid
Resistance	Metals	20 nm to 1 μm	1%	Nondestructive; also used as deposition-rate monitor
Ionization	All	No limit	<0.1 nm s^{-1}	Only used as vapor deposition-rate monitor
Eddy current	Metals on metals provided conductivities are different	1–1000 μm	1%	Nondestructive
Magnetic force/ electromagnetic	Coatings or substrate to be magnetic	3–125 μm	1%	Nondestructive; sensitive to the magnetic conditions and properties of the test specimen
X-ray absorption	All	10 nm to 1000 μm	5%	Nondestructive; substrate must produce the characteristic radiations; simple, rapid scan, used in on-line manufacturing

TABLE 15.1 Comparison of Some Common Thickness-Measurement Techniques (*Continued*)

Measurement technique	Coating materials	Thickness range	Accuracy or precision	Remarks
X-ray emission	All	2 nm to 1000 μm	2%	Nondestructive; substrate must not contain any of the elements in the coating material; multicomponent coatings can be measured; simple, rapid scan, used in on-line manufacturing
Beta backscattering	All	10 nm to 50 μm	5%	Nondestructive; coating and substrate material must have difference in atomic number; simple, rapid scan, used in on-line manufacturing
Microbalance (gravimetric)	All	No limit	$\sim$0.1 nm cm^{-2}	Used as vapor deposition-rate monitor; simple
Microbalance (momentum type)	All	No limit	$\sim$0.1 nm min^{-1} per degree rotation	Used as vapor deposition-rate monitor; simple
Oscillating quartz-crystal	All	No limit	$\ll$0.1 nm cm^{-2}	Most commonly used vapor deposition-rate monitor; simple
Coulometric	Metals (must be dissolved anodically)	1 to 50 μm	10%	Destructive; generally used for plated coatings

ings of any material (with highly reflective surfaces), ellipsometry, and absorption-band intensity for transparent coatings (Bennett and Bennett, 1967; Pliskin and Zanin, 1970). Most optical techniques are extremely accurate, but they are also time consuming.

Optical Interference. Optical interference techniques can be used for both transparent and opaque coatings. Thickness measurements of transparent coatings on reflective substrates are based on the interference between radiations reflected from the dielectric-air interface and those from the dielectric-substrate interface. These interferometric techniques can be divided into two general categories. In the most commonly used technique, the radiation is reflected from the sample coating in a spectrophotometer with fringes being formed as a function of wavelength. This technique has been named *constant-angle reflection interference spectroscopy* (*CARIS*), or simply *interference spectroscopy*. In the other technique, interference fringes are formed by varying the angle of observation, and this technique is therefore called *variable-angle monochromatic fringe observation* (*VAMFO*) (Pliskin and Zanin, 1970). While these arrangements yield relatively broader fringes (and hence less accuracy) compared with the step technique (to be described below), the optical flat required in step technique physically does not touch the coating, so the coating does not get scratched.

The wedge or step technique is generally used for thickness measurements of opaque coatings. A step is made in the coating by masking a portion of the substrate during deposition or by removing part of the coating from the substrate. The interference fringes are formed between an optical flat and the surface of the coated specimen using a monochromatic light that can be observed with commercially available interference microscopes (Fig. 15.2). If there is a slight angle between the two surfaces of the optical flat and the specimen, the fringe pattern will appear as alternating parallel light and dark bands that are perpendicular to the step on the coating. Two adjacent fringes are separated by half the light wavelength ($\lambda/2$). The step in the coated specimen surface will displace the fringe pattern, and this displacement is a measure of the difference in the level of the surface of the specimen. The coating thickness is measured as the displacement of the fringe pattern at the step area. The sharpness of the fringes (resulting in greater accuracy in the measurement) increases markedly if interference occurs between many beams (multiple-beam interference), and this can be accomplished

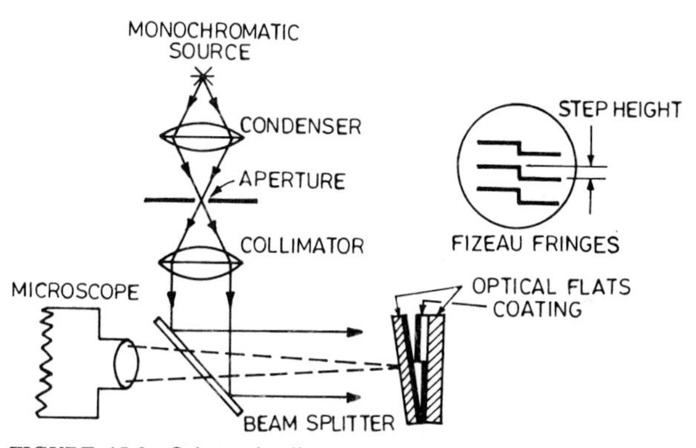

FIGURE 15.2 Schematic diagram for thickness measurement by multiple-beam optical interferometer using Fizeau method of fringes of constant thickness.

if the reflectivity of the two samples at the interfaces is very high. While the displacement of the fringes of a double-beam instrument can be estimated to about one-tenth of the fringe spacing, the fringes of a multiple-beam instrument are much more sharply defined, and their displacement can be estimated to perhaps one-hundredth of the fringe spacing (~3 nm). Two-beam and multiple-beam interference microscopes (Saur 1965; Schwartz and Schwartz, 1967; Thomas and Rouse, 1955; Tolansky, 1960) are commercially available for measuring coating thicknesses. There is no fixed limit to how thick a coating may be measured, but the method is best suited to coatings less than about 2 μm thick.

Ellipsometry. Ellipsometry provides another nondestructive method for measuring both thickness and optical constant (refractive index) of transparent coatings. It is very useful for extremely thin coatings (down to 1 nm), but it also may be used for very accurate measurements of thicker coatings. However, ellipsometry is more complicated than such relatively simple methods as optical interference and stylus techniques. Ellipsometry is used primarily in the electronics industry.

Ellipsometry is based on evaluating the change in the state of polarization of monochromatic light reflected from the substrate of a coated surface. The state of polarization is determined by the relative amplitude of the electric fields, i.e., parallel (E_p) and perpendicular (E_s) to the plane of incidence and the phase difference between the two components ($\Delta_p - \Delta_s$). On reflection from a surface, bare or coated, the ratio of the two amplitudes (E_p/E_s) and the phase difference between the two components ($\Delta_p - \Delta_s$) undergo changes that are dependent on the optical constants of the substrate (n_s, k_s), the angle of incidence, the optical constant of the coating (n_c, k_c), and the coating thickness (t). If the optical constants of the substrate are known, and if the coating is nonabsorbing (that is, $k_c = 0$), then the only unknowns in the equations describing the state of polarization are n_c and t. In principle, then, with a complete knowledge of the state of polarization of the incident and reflected light, n_c and t can be determined.

A schematic diagram of a typical ellipsometer is shown in Fig. 15.3. The collimated, unpolarized light is first linearly polarized by the polarizer. It is then elliptically polarized by the compensator, which can be a mica quarter-wave plate or a Babinet-Soleil compensator. Although the compensator can be set at any azimuth, for simplicity it is usually oriented so that its fast axis is at ±45° to the plane of incidence. After reflection from the sample, the light is transmitted through a second polarizer, which serves as the analyzer. Finally, the light intensity is determined by the eye, or photomultiplier detecter. The polarizer and analyzer are rotated until extinction is obtained. From the extinction readings of the polarizer and analyzer, the phase difference and amplitude ratio can be determined (Azzam and Bashara, 1976).

Absorption-Band Intensity. The thicknesses of transparent coatings also can be determined from characteristic absorption-band intensities. A simple example of this technique presented by Pliskin and Gnall (1964) involves the measurement of SiO_2 coating thickness on silicon wafers from the optical density [log (I_0/I)] of the 9-μm SiO_2 absorption peak in the infrared transmission spectrum. The optical density is not perfectly linear with coating thickness because the shape of the absorption band varies with coating thickness. Thus, in some cases, it is better to use the integrated-band intensity.

15.1.1.2 Mechanical Techniques. Stylus (profiler) instruments are widely used for the measurement of surface roughness and surface finish. If a step is made in a coating by masking a portion of the substrate during deposition or by removing part of the coating from the substrate, then a stylus instrument can be used to

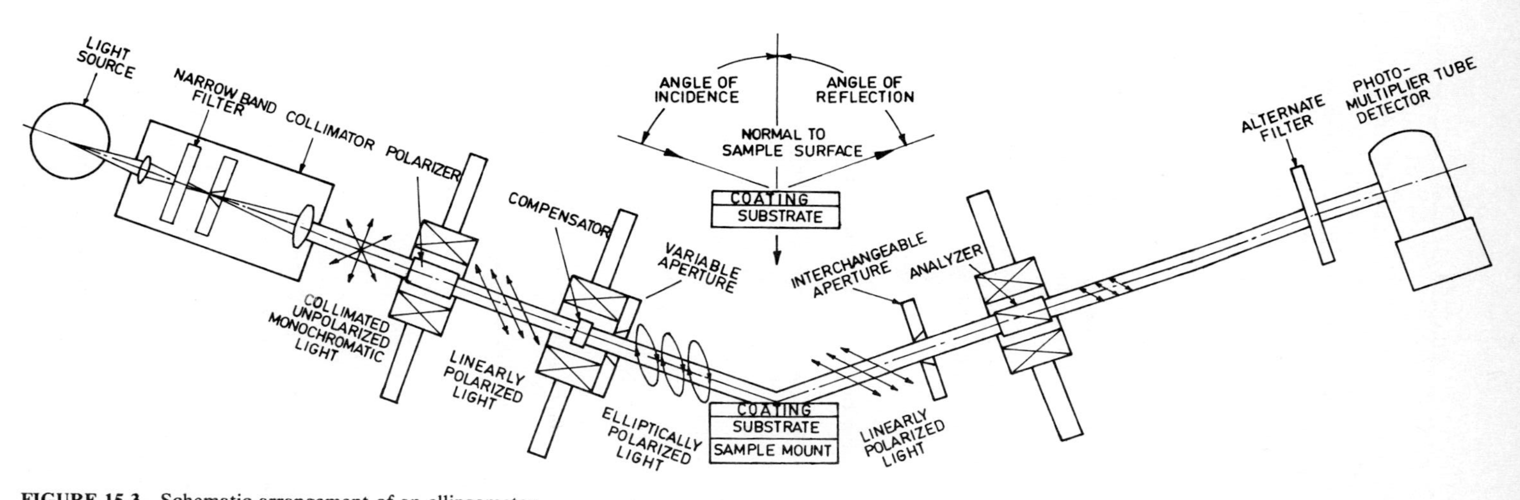

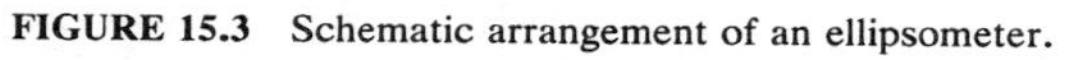

FIGURE 15.3 Schematic arrangement of an ellipsometer.

measure coating thickness ranges from about 100 μm down to 5 nm. In this method, the diamond-tipped stylus is drawn across the step and the horizontal and vertical motions of the stylus are magnified and recorded. The difference in the vertical displacement is converted to electric signals by means of a linear variable differential transformer (LVDT) and represents the coating thickness. The signal is then amplified and recorded on a strip-chart recorder that also amplifies to a lesser extent the horizontal movement of the stylus relative to the sample surface. The penetration of the stylus into the coating can be reduced by increasing the tip radius and reducing the load in order to reduce the contact pressure. For example, King et al. (1972) used a stylus with a radius of 12.5 μm and a load of 10 to 20 μN without observing deformation of soft gold film. The reliability of these instruments is normally not much better than 5 percent for the most sensitive instruments, although a step as small as few nanometers can be measured.

Figure 15.4 shows the trace of a coating step obtained with a stylus profiler (Talystep 1, manufactured by Rank Taylor Hobson, Leicester, England). Silver was deposited onto a glass substrate, but part of the substrate protected by a mask so that a step was formed on the sample (Pliskin and Zanin, 1970). The coating thickness corresponds to the vertical distance between linear extrapolations of the lower and upper portions of the trace. Generally, more accurate measurements are obtained with grooves than with steps.

15.1.1.3 ***Electrical Techniques.*** Electrical techniques are commonly used to measure coating thickness by eddy currents as well as to monitor vapor-deposition rates.

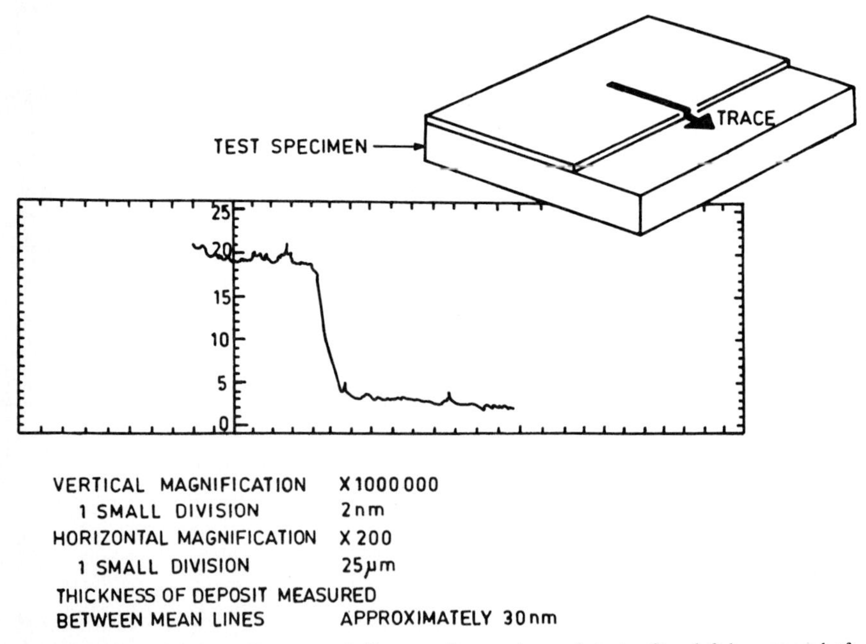

FIGURE 15.4 Stylus profiler trace of silver coating on glass substrate. Step left by removal of mask.

Vapor-Deposition-Rate Monitors. The electrical properties that are useful in measuring coating thicknesses are dielectric strength (breakdown voltage), capacitance, and resistance of the coating. The most commonly used electrical technique is the in situ vapor-deposition-rate monitor. Other methods include ionization (in situ deposition-rate) techniques (Chopra, 1979; Pliskin and Zanin, 1970; and Schwartz and Schwartz, 1967). Because of the unreliability of electrical breakdown voltage as an indicator of coating thickness, this technique is not recommended.

Capacitance measurements are sometimes used for determining the thickness of a dielectric coating deposited on a conducting substrate, employing the parallel-plate capacitor configuration shown in Fig. 15.5(*a*). The rate of evaporation is measured by determining the changes in capacitance of a parallel-plate condenser due to changes in the dielectric constant resulting from the presence of the vapor of the evaporant. The stringent requirements for these measurements are that there should be no pin holes in the coating and that the exact dielectric constant must be known. Usually the dielectric constant of the coating depends on the deposition method, purity of the coating material, etc. Capacitance measurements for thickness evaluation are not used frequently because of the abovementioned constraints.

The measurement of coating resistance is a simple operation that can be used to determine the thickness of conducting coatings on nonconducting substrates and of semiconductor epitaxial coating layers. However, this technique requires accurate knowledge of the coating resistivity and that it is not variable with coating thickness. These conditions are rarely met because the resistivity of a coating is strongly dependent on deposition conditions and often is quite different from that of the bulk coating material. The resistance of metal coatings, however, can be measured very easily by coating one arm of a Wheatstone bridge [Fig. 15.5(*b*)]. Coating resistance can be measured automatically by recording the potential across the bridge. The most commonly used instrument for measuring coating resistance is the collinear four-point probe (to be discussed later). In some cases, it can be used as a process-control technique, provided the coatings are not so thin that the drastic resistivity changes associated with the nucleation and island-growth stages are encountered. Coating thicknesses of such materials as gold, silver, and copper have been measured with good accuracy from 20 nm to about 1 μm (Leonard and Ramey, 1966).

In ionization deposition-rate monitors, the deposition rate is monitored by ion-

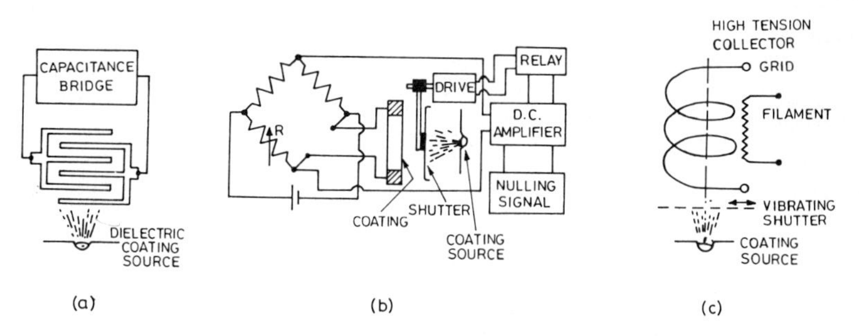

FIGURE 15.5 Schematic diagrams of in situ vapor-deposition-rate monitors: (*a*) parallel-plate capacitance monitor for insulator coatings, (*b*) Wheatstone-bridge resistance monitor for metal coatings, (*c*) a triode-type ionization monitor.

izing the coating material in the vapor phase and measuring the resulting ion current. The ion current is proportional to the total number of vapor atoms and their ionization probability. Basically, an ionization monitor is a triode-type ionization gauge used for measuring pressures. It consists of a filament (thermal source) for electrons that are accelerated by the spiral-wire grid (anode) at a positive potential and injected to ionize the vapor in the region between the anode and the collector wire [Fig. 15.5(*c*)]. The collector has a negative potential that attracts the ions. The small currents (typically fractions of a microampere) can be measured by a variety of instruments. Ionization monitors can be used in ultrahigh-vacuum conditions at high temperatures.

Eddy-Current Gauge. Brenner and Garcia-Rivera (1953) and Brodell and Brennar (1957) described a coating-thickness gauge (called a *Dermitron gauge*) that utilizes the principle of eddy currents to measure coating thickness. The instrument measures the thickness of metallic coatings on metals provided the conductivities of the coating and substrate differ sufficiently. Measurements can be made on combinations of metals regardless of whether they are magnetic or nonmagnetic. The operation of the gauge depends on the *skin effect*, which is the tendency of alternating currents to travel closer and closer to the surface as the frequency is increased. For a given frequency, the magnitude of the eddy current induced in a surface layer will depend on the conductivity of that layer through which the current flows, other factors being constant.

An eddy-current thickness gauge is shown schematically in Fig. 15.6. When high-frequency current is passed through the exciter/sensing coil and the coil is brought near an electrically conductive material, then eddy currents are induced in the material. These currents, in turn, react with the sensing coil, causing back emf energy loss and a change in coil impedance. It is this impedance change that the instrument measures and converts into a thickness-proportional signal. The frequency of the current (typically 10 kHz to 10 MHz) is chosen so that, by virtue of the skin effect, an appreciable proportion of the eddy current flows in both the substrate and the coating. Provided that the coating and substrate have dif-

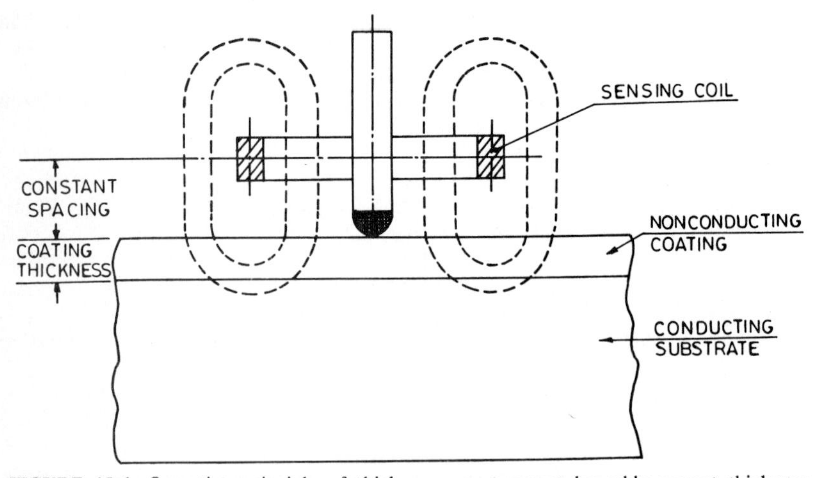

FIGURE 15.6 Operating principle of thickness measurement by eddy-current thickness gauge.

ferent electrical conductivities and/or magnetic properties, the strength of the induced eddy current varies with the thickness of the coating. The strength of the induced current is indirectly measured through the interaction of its magnetic field with the probe coil. The net result is a change in the apparent impedance of the probe coil. The current through the probe coil changes correspondingly, and the amount of the change, or the unbalance of the circuit, is measured with a microammeter.

The thickness of a coating is thus translated into a change in the reading on a microammeter. By calibrating the gauge with standard coatings of known thickness, the meter readings can be directly related to the thickness of the coatings being tested.

15.1.1.4 Magnetic/Electromagnetic Techniques. Magnetic/electromagnetic methods are normally used for determining the thickness of a nonmagnetic coating on a magnetic substrate. Although these techniques are primarily intended for the measurement of nonmagnetic coatings on iron and steel, they can be adapted to measure magnetic coatings on nonmagnetic substrates (Bennett, 1949). Gauges based on magnetism range from simple mechanical devices that rely on the measurement of magnetic forces to electromagnetic induction. These methods are characterized by considerable speed and accuracy of measurement. The magnetic techniques are generally limited to coating thicknesses on the order of few micrometers.

The magnetic-force techniques are based on the attractive force between a magnet and a magnetic substrate that is covered by a nonmagnetic coating. As the coating thickness increases, the force of attraction decreases. In the case of a magnetic coating on a nonmagnetic substrate, the force increases with the increasing thickness of the magnetic coating (ASTM Standards B499-69, B530-70; IS-2178, and IS-2361). The instruments may use either a permanent magnet or a soft iron core surrounded by a solenoid through which passes a direct or alternating current. The magnetic force of attraction is compared with some mechanical forces (e.g., a spring or a weight). A commonly used attractive-force thickness gauge is shown in Fig. 15.7 (Brenner, 1938). A permanent magnet is attached to the tip of a horizontal lever that turns in virtually friction-free bearings. The mass of the magnet is balanced by a counterweight. If the arm with the permanent magnet is brought near a ferromagnetic base, then a torsional moment is produced. The magnitude of the moment depends on the thickness of the intervening nonmagnetic coating.

In electromagnetic inductive-type instruments, the nonmagnetic coating on a magnetic substrate again forms the nonmagnetic gap in a circuit, and the inductance of the coil is diminished as the air gap is increased. However, if a magnetic coating on a nonmagnetic substrate is measured, the inductance increases with thickness. Several designs of these instruments operate on this principle, employing a standard ac bridge circuit or a transformer circuit (Bennett, 1949). Figure 15.8 shows schematically the essential elements of a transformer type of thickness tester, consisting of a single- or two-pole (or yoke) probe which together with the coated substrate forms a magnetic circuit. The principle is similar to that of a transformer; i.e., an alternating current passing through a primary winding produces a magnetic field that induces a voltage in a secondary winding. The magnitude of this voltage depends on the amount of coupling between the primary and secondary winding and hence on the length and nature of the magnetic path, i.e., also on the thickness of the intervening nonmagnetic coating.

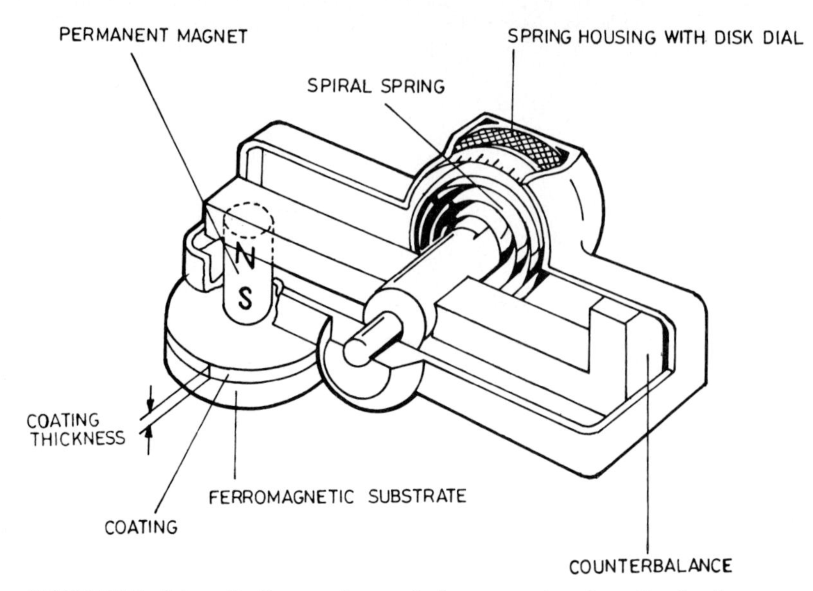

FIGURE 15.7 Schematic diagram of magnetic-force gauge based on attractive forces.

In calibrating all magnetic gauges, it is essential to have a series of coating-thickness standards covering the range of a particular instrument. A series covering 5 to 750 μm includes most requirements.

15.1.1.5 ***Radiation Techniques.*** The thickness dependence of the absorption of light, x-rays, α-rays, and β-rays emitted from a radioactive source is utilized to determine coating thickness in radiation techniques (Zemany, 1960). The various techniques involve measurements of absorption, attenuation of emitted radiation (or x-ray fluorescence), and scattering by the coating (Pliskin and Zanin, 1970). Two radiation techniques—x-ray emission and beta backscattering—are commonly used as on-line measuring tools in various deposition processes. Coating-thickness measurements can be made very quickly with commercial instruments. X-ray absorption, x-ray emission, and beta backscattering can be used for both opaque and transparent coatings. The major disadvantage of these techniques is that thickness measurements are made over a large area (typically a few millimeters).

Absorption. Absorption of x-rays, α-rays, and β-rays has been used to measure coating thicknesses; x-rays are most commonly used. Absorption techniques are based on measuring the attenuation by the coating of the characteristic radiation of the substrate material. The attenuation for a particular wavelength is an exponential function of the coating thickness and is dependent on the mass-absorption coefficient of the coating material (Light, 1969; Weinstein et al., 1968; Zemany, 1960). The thicknesses measurable are dependent on the energy of the substrate radiation and the absorption coefficient of the coating and generally range from less than 10 nm to about 1 mm. The absorption method is generally used to measure the thickness of single-component coatings with an accuracy of about 5 percent.

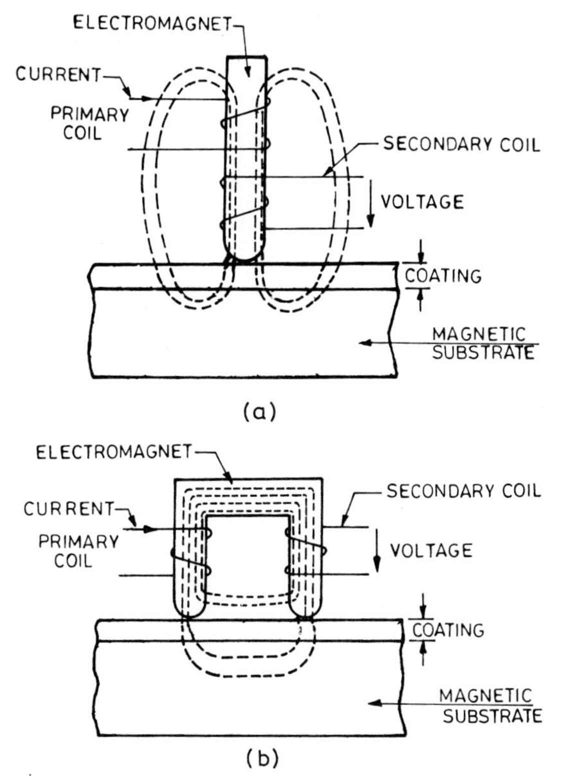

FIGURE 15.8 Operating principle of thickness measurement by electromagnetic inductive probes: (*a*) single-pole probe, (*b*) two-pole or yoke probe.

Emission. X-ray emission methods, also called *x-ray fluorescence methods*, depend on the excitation of x-rays by either the coating or the substrate and the measurement of their intensity after attenuation by the coating (Fig. 15.9). The intensity of the radiation emitted from the coating is linearly proportional to the coating thickness and increases exponentially for thicker coatings up to a maximum value (Schreiber et al., 1963; Heller, 1969; Zemany, 1960). Usually x-rays from x-ray tubes are used to fluoresce the specimens, and measurements are made with an x-ray spectrometer. X-rays and γ-rays from radioisotopes are sometimes used in conjunction with nondispersive measurement techniques. For heavy metals such as gold, γ-rays can excite the K radiation, which permits measurement of thicker gold coatings than is done conventionally with the L radiation excited by x-rays.

X-ray fluorescence techniques are more broadly applicable than absorption techniques because the only requirement is that the substrate material does not contain any of the elements of the coating. Multicomponent coatings also can be measured. Accuracies of less than 2 percent can be obtained.

X-ray techniques require instruments that are expensive but easy to operate. However, standards are required, and these must be calibrated by other thickness-measuring techniques.

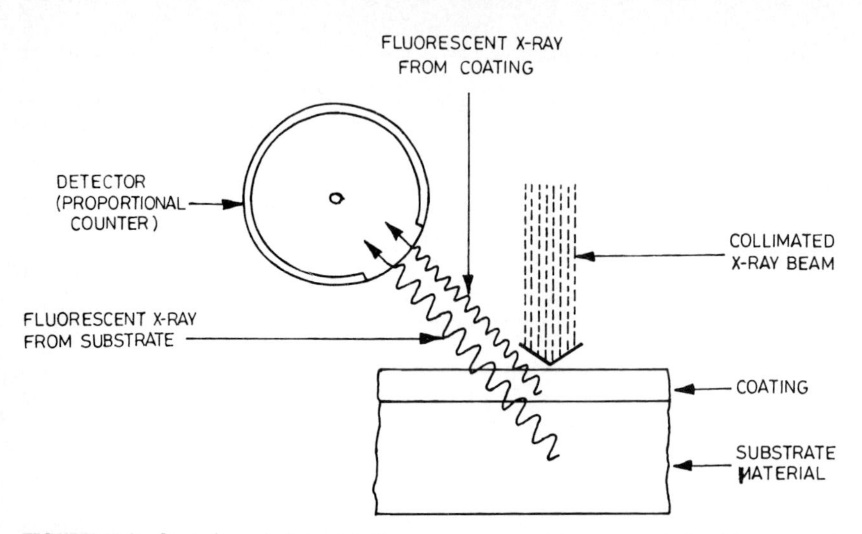

FIGURE 15.9 Operating principle of thickness measurement by x-ray emission (fluorescence).

Beta Backscattering. When radiations are incident on a surface, some scattering takes place. The backscattering of β-rays depends on the atomic number, density, and thickness of the scattering materials, and its measurement therefore allows a determination of coating density and thickness (ASTM Standard B567; Brodsky et al., 1972; Clarke et al., 1951; Cooley and Lemons, 1969). Figure 15.10 illustrates the principle of the beta backscattering technique. An isotope such as promethium-147 emits a collimated beam of beta particles that are directed toward the test material. The particles impinge on the material, and some are scattered back as a diffused beam that passes to a detector mounted below the source. The detector is normally a Geiger-tube counter. The intensity of these backscattered β-rays is a function of mass per unit area (thus coating thickness). As the difference between the atomic number of the coating material and that of

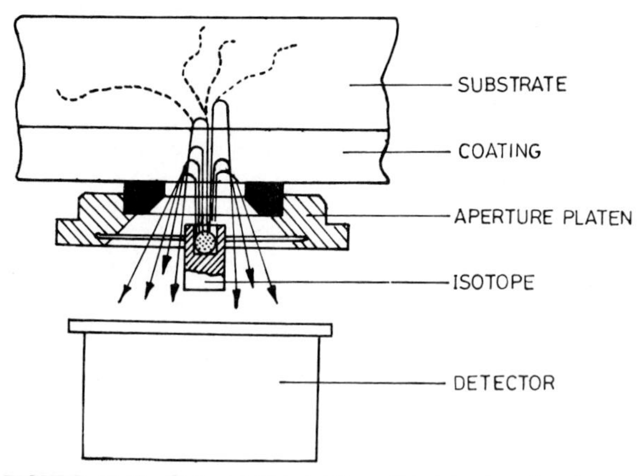

FIGURE 15.10 Cross-sectional view of betabackscatter transducer.

substrate increases, the sensitivity to the weight per unit area of coating increases.

The beta backscattering method is inexpensive, very simple to use, fast, and nondestructive. Standards of the coating material in the thickness range of interest are required. With a suitable choice of sources, thickness measurements from 10 nm to 50 μm on about 1 cm^2 of coating can be made with an accuracy of about 5 percent. Gauges of this type are commercially available.

15.1.1.6 Miscellaneous Techniques

Microbalance Techniques. Microbalance techniques are used for in situ monitoring of the rate of vapor deposition. These techniques are termed *gravimetric* or *momentum* type depending on whether they measure the weight or the momentum of the impinging vapor, respectively. Gravimetric methods are among the earliest developed and most convenient to use. Various types of balances have been used, such as pivotal, torsion fiber, quartz or tungsten helical spiral, and magnetic suspension (all of which employ null-balance principles using mechanical, optical, electromagnetic, or electrostatic deflection methods). The detection sensitivities of these balances range from 1 to 10^{-2} μg. A sensitivity of 0.1 μg, which is conveniently obtained in a commercial pivotal balance, corresponds to about a 0.1-nm-thick Ag coating of a one square centimeter area.

Microbalance monitors that use moving-coil current meters have been used by several researchers (e.g., Hayes and Roberts, 1962). One arrangement, shown in Fig. 15.11(*a*), uses a lightweight large-area vane (made of mica) with a counterweight on the other side of the meter movement for obtaining a high sensitivity. Depending on the relative orientation of the vane and the vapor stream, the arrangement may be used to measure the total weight (average thickness), the force due to the vapor momentum transfer (rate of deposition), or both. For example, if the vane is vertical and the vapor stream impinges horizontally, only rate of deposition can be measured. By impinging vapor onto the vane inclined to the horizontal and vertical positions, the components of force at right angles can be measured so that both rate of deposition and total weight can be determined. Neugebauer (1964) used a simple torsion pendulum [Fig. 15.11(*b*)] to measure the momentum transfer rate. A sensitivity of 0.1 nm min^{-1} per degree rotation for Sn vapor was obtained.

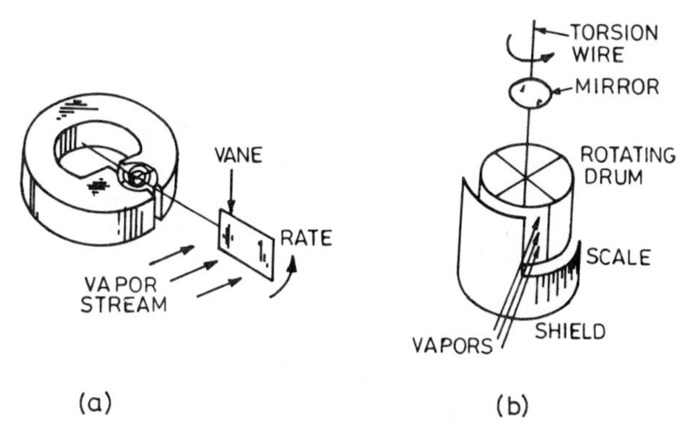

FIGURE 15.11 Schematic diagrams of in situ vapor-deposition-rate microbalance monitors: (*a*) gravimetric type, (*b*) momentum type.

A sensitive and rugged microbalance is based on measuring changes in the resonance frequency of an oscillating piezoelectric quartz crystal with mass loading when operated in a thickness shear mode of vibration. The oscillating quartz crystal for monitoring and controlling deposition rates in vapor deposition has become universally accepted (Hartman, 1965; Warner and Stockbridge, 1963). Here the major crystal surfaces are antinodal, and mass added on either one or both sides shifts the resonance frequency. For a given resonance frequency, the frequency shift is linearly proportional to the mass per unit area of the coating. AT-cut quartz crystals with resonance frequencies of about 5 MHz are commonly used. At 5 MHz, a 1-Hz change corresponds to a weight increase of about 10 ng cm^{-2}, which for a Fe coating corresponds to 0.12 monolayers. The frequency and frequency changes of the quartz-crystal oscillator circuit can be measured by using a suitable pulse-analog or pulse-digital counter with a moderate accuracy of ± 1 Hz for a 5-MHz crystal.

Coulometric Technique. Nondestructive measuring techniques have many advantages but cannot always be employed. For technical and economical reasons, some problems that occur in practice just do not lend themselves to solution by the methods and equipment discussed so far. The coulometric method, a destructive technique, provides an alternative. This is an extremely versatile and easily applicable method which, for example, can be used to measure the ternary coating system Cr-Ni-Cu on Fe and Zn die castings.

The coulometric method is a reversal of the electroplating process. The coating thickness is determined by measuring the time required to dissolve the coating anodically at a constant current density and calculating the thickness from the current density, area, and density of the coating (ASTM Standards B504, IS-2177; Baldwin, 1970; Bennett, 1949; Waite, 1953). This method generally is used only for coatings less than 50 μm thick. The instruments are calibrated at the time of use with coating-thickness standards, and the thickness is read directly. In a commercial thickness tester reported on by Waite (1953), the deplated area is defined by a rubber washer placed on the test surface. The electrolyte is contained in a metal tube that seats against the washer and serves as the cathode. An electronic relay is used in conjunction with an automatic counter to measure the time of deplating. When the deplating current is turned on, the counter is automatically started. When the plated metal is deplated, the rate of change in voltage actuates the relay, which turns off the instrument, including the counter. This counter is calibrated in thickness units (Mathur and Karuppanan, 1961; Narayanan and Venkatachalam, 1961). This method is suitable for most electroplated coatings, including thin decorative coatings of chromium, tin-nickel alloy, and gold. This method also has been used for coatings on wire, as well as for multilayer coatings (Baldwin, 1970, Waite, 1953). In general, accuracy of better than ± 10 percent for a coating of 1 to 50 μm thickness can be achieved.

Sectioning. Various sectioning methods can be used for measuring coating thicknesses. However, these methods are destructive and are not generally recommended. If the coating is thick enough (>1 μm), then microscopic measurements can be made on a normal cross section. With thinner coatings, the sample is beveled at a small angle to amplify the thickness of the coating. Since the coating boundary cannot be detected directly, it is necessary to differentiate the coating from the substrate material by chemical-staining techniques. Sectioning techniques are sometimes used for thickness measurements of epitaxial or diffused semiconductor coatings. Unfortunately, these techniques are slow and completely destructive and require considerable skill. However, since the measurements are necessarily made over localized areas, sectioning is well adapted to checking general thickness variations over the surface of a specimen.

15.1.2 Techniques for Coating-Density Measurements

Coating density is important because it depends on the physical structure and composition of the materials. Hence its measurement can give information about the soundness of a coating and its impurity content. Density ρ_c can be calculated from direct measurements of the mass M_c and volume V_c of the the specimen, that is, $\rho_c = M_c/V_c$. For most coatings, density measurement presents some problems because of the small mass and volume. Also, coating nonuniformity, surface roughness, pores, cracks, and recesses may introduce errors into thickness measurements. Several methods have been used to accurately measure the mass and thickness of the coatings to obtain reliable density measurements. Techniques involving hydrostatic weighing (sink or float technique) have been used for detached coatings.

15.1.2.1 ***Measurement of Coating Mass and Thickness.*** One of the oldest and still commonly used techniques for determining coating mass is to weigh the substrate before and after deposition (or removal) of the coating; this is known as the *gravimetric technique*. The weighings must be very accurate, since relatively small differences in large numbers are being sought. A weighing accuracy of up to ± 1 μg can be attained with the modern piezoelectric quartz-crystal microbalance. This accuracy is sufficient provided the substrates are small (on the order of 10×10 mm) and thin (on the order of 25 to 250 μm thick) and the coatings are fairly thick (on the order of 0.5 μm or thicker). Mass measurements also can be obtained from in situ measurement of the deposition rate using an oscillating piezoelectric quartz crystal (Blois and Rieser, 1954; Hartman, 1965; Pliskin and Zanin, 1970). Beta backscattering (described earlier) is another technique to measure the mass per unit area of coatings (Brodsky et al., 1972; Cooley and Lemons, 1969). Anttila et al. (1986) and Ingram et al. (1986) used Rutherford backscattering (RBS) to measure mass per unit area of diamond-like carbon coatings.

The most commonly used technique to measure coating thickness is the wedge or step technique. The step in the coating is measured by using an interference microscope or a stylus profiler (see previous section). Optical interferometry and ellipsometry are used for thickness measurements of transparent coatings. Other techniques used to measure coating thicknesses were discussed earlier.

15.1.2.2 ***Sink or Float Technique.*** In the sink or float technique, also called the *hydrostatic technique*, the apparent mass of the specimen in two fluids is determined by suspending the specimen in the fluids from one arm of a balance by a fine wire. The density ρ_c can then be calculated from the following two equations:

$$\text{Mass in air} = M_c - \rho_{L1} V_c \tag{15.1a}$$

$$\text{Mass in liquid} = M_c - \rho_{L2} V_c \tag{15.1b}$$

where ρ_{L1} and ρ_{L2} are the densities of two liquids. Measurements with an accuracy of about 0.02 percent can be made with this technique. Bowman and Schoonover (1967) used air and water or a fluorocarbon (density = 1800 kg m^{-3})

for density measurements of electrodeposits. Banks and Rutledge (1982) used fluid mixtures of 1,1,2,2-tetrabromoethane (density = 2970 kg m^{-3}) and 1,1,2-trichloroethane (density = 1460 kg m^{-3}) for density measurements of ion-beam sputter-deposited diamond-like carbon coatings.

Franklin and Spal (1971) described a sink or float technique that used two immiscible liquids and could measure coatings on samples as small as a few milligrams up to a gram or more. For small samples, the reported error is a few percent, and it is much smaller for larger samples. A schematic of the apparatus is shown in Fig. 15.12. Two immiscible liquids of differing density (such as fluorinated hydrocarbon with a density of about 1770 kg m^{-3} as the lower liquid and distilled water containing 0.2 g L^{-1} of sodium lauryl sulfate) are contained in the cylinder. A cup containing the test specimen is suspended in the lower, more dense liquid by a wire from the float in the upper liquid. The three tubes, *A, B*, and *C*, when viewed from above, form an equilateral triangle. The cup is lowered through tube *A* to the vane. By vertical and rotational movement of the rod in

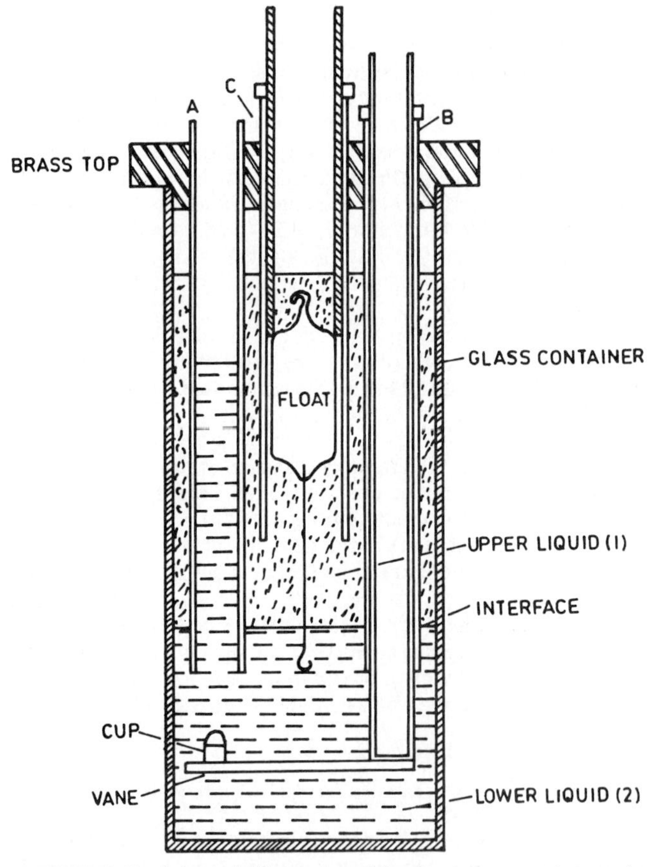

FIGURE 15.12 Schematic illustration of density-measurement apparatus by sink or float technique using two immiscible liquids of differing density. [*Adapted from Franklin and Spal (1971).*]

tube B, the cup is transferred by the vane to the hook suspended from the float. Tube C consists of two concentric tubes. The inner one serves to keep the float submerged at a level convenient for hooking the cup. The outer tube controls any horizontal drift. When the loaded float is approaching its equilibrium position, this outer tube is raised so that it will not interfere with the free vertical movement of the float. The vertical position of the hook at its rest point is measured with a telescope equipped with an eyepiece scale.

The ratio of the density of the test specimen to that of the calibration standards for the two reference specimens used is calculated from the equation

$$\frac{\rho_c}{\rho_s} = \left[\frac{M_s}{M_c} + \frac{\rho_s}{\rho_{LL}}\left(1 - \frac{M_s}{M_c}\right) - \left(\frac{\rho_s}{\rho_{LL}} - 1\right)\frac{\Delta_1}{\Delta_2}\frac{\Delta_M}{M_c}\right]^{-1} \tag{15.2}$$

where ρ_s = density of the reference specimens
M_s = mass of the lightest standard
ρ_{LL} = density of lower liquid
Δ_M = absolute difference in masses of the two reference specimens
Δ_1 = difference in rest positions between the test specimen and lightest reference specimen
Δ_2 = difference in rest positions between the reference specimens

The hydrostatic method has the advantage of being entirely suitable for irregularly shaped objects. In addition, nonuniformity of coating thickness, surface roughness, pores, cracks, and recesses are of little consequence. This method is also useful for detached coatings on samples a few milligrams or more in weight. This technique is commonly used for electroplated coatings.

15.1.3 Techniques for Coating-Porosity Measurements

Thin-coating application processes generally involve a phase transition from vapor to solid state that is accompanied by a large density change, and therefore, it is almost inevitable that nonequilibrium volume defects in the form of vacancies and voids are incorporated into the coatings. Vacancies and voids in coatings are formed when incoming atoms from the vapor fail to cover certain regions below the grown surface; i.e., the presence of vacancies and voids in a coating implies that the coating contains locally unfilled regions inside the lattice. The size of unfilled regions in coatings may range from that of a single vacancy to that of a void with a finite dimension. These vacancies and voids in coatings result into microporosity. If the size of condensing species is considerable (several micrometers), such as molten particles in thermal spraying, electroless, and electrochemical deposition processes, the size of the voids also increases in the same ratio and results into macroporosity or porosity. The porosity can be open (through) pore or closed pore (Nakahara, 1979; Poate et al., 1978; Tu and Lau, 1978).

The techniques used to measure porosity can be divided into direct observation and indirect methods. Direct observation methods allow observation of both isolated voids (closed pores) and through (open) pores. Both macroscopic and microscopic (<5 nm) pores can be detected. Indirect detection methods include measurements of the density and the porosity of the coatings. Porosity measurements generally measure only open pores and do not permit the detection of closed pores. Porosity measurements are more commonly used for rather large pores, typically on the order of a micrometer in size. Density measurements allow assessment of the aggregate of both open and closed pores. Porosity results in a coating density that is lower than that of the corresponding bulk material. However, coating density may vary because of chemical and structural changes that have nothing to do with porosity. Therefore, density measurements can only be used for materials in which no chemical or structural changes are expected. The indirect methods do not provide any information on the geometry of the pores in the coatings. Several commonly used methods to measure coating porosity are listed in Fig. 15.13. These methods will be described in this section. Techniques to measure coating density were described in the preceding section.

15.1.3.1 ***Direct Observation.*** Direct observation techniques are used to detect macroscopic and microscopic (a few nanometers) pores (both open and closed). These include optical microscopy (OM), scanning electron microscopy (SEM), transmission electron microscopy (TEM), and x-ray scattering. These techniques are commonly used for vapor-deposited coatings with microscopic pores. Studies employing OM (Kurov et al., 1975) and SEM (Cooksey and Campbell, 1970; Rehrig, 1974) detect the voids intersecting the coating surface. TEM is more useful in examining both small and large pores that can be isolated inside the coating or located on the coating surface (Donovan and Heinemann, 1971; Fuks and Cheremskoy, 1974; Jacobson et al., 1978; Lloyd and Nakahara, 1977; Tu and Lau, 1978). The most powerful features of TEM are the high spatial resolution capability (< 1 nm) and the variety of operational modes (such as diffraction contrast and defocus contrast) that can be used for the structural analysis. However, TEM requires the sectioning of the coating specimen into very thin wafers (a few nanometers thick).

Figure 15.14 shows how the incident plane wave of electrons or x-rays scatters around a vacancy or void in a thin coating. A vacancy effectively acts as a point scatterer, and thus the incident wave will be scattered to create a nearly hemispherical wave front around the vacancy. This wide-angle scattering makes the imaging of a vacancy by TEM practically impossible, but it contributes to so-called diffuse scattering. X-ray diffuse scattering is used to demonstrate the presence of vacancies. In the case of a void, however, the incident plane wave is scattered at low-angle regions only. Therefore, the incident plane wave and the scattered wave can interfere constructively or destructively below or above the

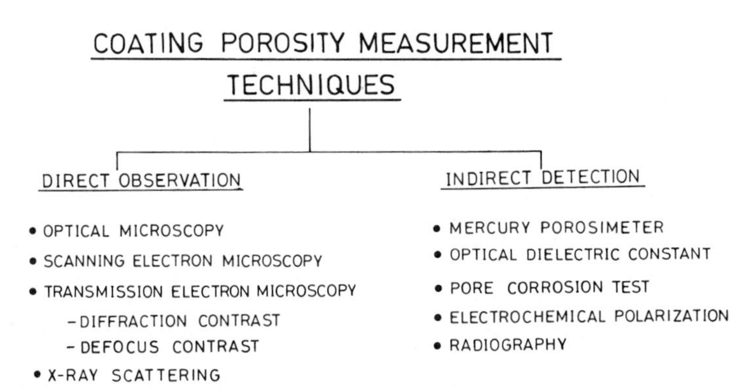

FIGURE 15.13 Classification of coating-porosity measurement techniques.

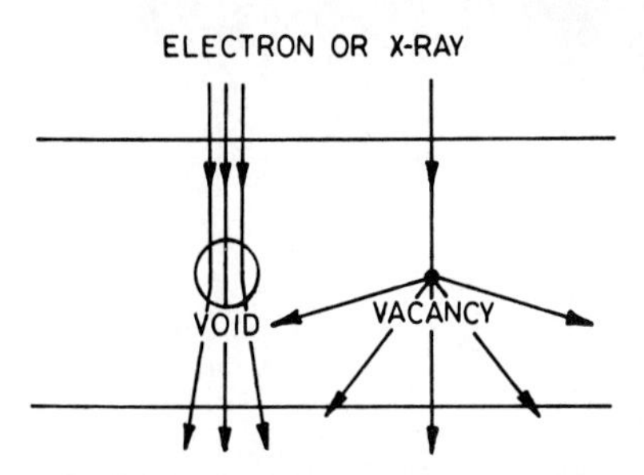

FIGURE 15.14 Schematic diagram showing how the incident plane wave of electrons or x-rays interacts with a vacancy and with a void in a thin coating. Note that a single vacancy acts as a point scatterer, which causes the incident plane wave to scatter at large angles, whereas a void contributes to a scattering of low-angle regions only. [*Adapted from Nakahara (1979).*]

plane containing a void and can produce an interference pattern similar to the so-called Fresnel fringes. The defocus contrast technique is used to observe these interference fringes formed at the boundaries between a void and its matrix at a distance slightly below (overfocus) or above (underfocus) the plane of a void. Observation of the corresponding diffraction pattern rather than the image also can directly yield information about the size, geometry, and distribution of voids. The use of low-angle electron diffraction (Moss and Graczyk, 1969) or small-angle x-ray scattering (Fuks et al., 1971) for detecting voids is based on the same principle. In the powerful defocus contrast technique, small voids appear as white voids surrounded by black rings in the underfocused condition and as black dots surrounded by white rings in the overfocused condition.

Figure 15.15 shows high-density voids in their gold coatings applied by evaporation, sputtering, and electrodeposition. This figure demonstrates that voids are generally formed in thin coatings irrespective of the deposition technique as long as the deposition process involves a phase transition from the vapor phase to the solid state. Nakahara (1977) reported the presence of a high density (about 10^{17} cm^{-3}) of very small (~1 nm) voids in all three coating types. Careful examination of these voids by stereomicroscopy (Nakahara, 1977) indicates a network of interconnecting voids that lies near the coating-substrate interface (Staudinger and Nakahara, 1977).

15.1.3.2 Indirect Detection. Indirect detection methods include mercury porosimetry, optical dielectric constant, pore corrosion tests, electrochemical polarization, and radiography. All these tests except the optical dielectric constant test are more commonly used for rather large pores, typically on the order of 1 μm, such as found in thermal spraying and electrodeposition.

Mercury Porosimetry. Mercury porosimetry is commonly used to measure open pores (Huisman et al., 1983; Orlow et al., 1974). This technique involves immersing a sample in mercury and compressing the system in stages up to a typical pressure of 200 to 1300 MPa while measuring corresponding volumes. These volume changes are calculated from the reduction of the mercury level in the capillary tube of a dilatometer. The volume changes are the result of several processes that take place during the experiment, such as compression of the mercury, compression of the dilatometer, compression of the sample, and filling of the pores with mercury. Compressibility corrections must be made to obtain the coating porosity. Figure 15.16 presents an example of porosity measurement of a particulate γ-Fe_2O_3 magnetic tape sample (curve *A*). Curve *C* is the compressibility correction curve calculated from known volumes and compressibility coefficients. Curve *B* is the corrected pore volume, expressed as a percent of the coating volume.

Optical Dielectric Constant. The introduction of voids into an absorbing host leads both to diminution of absorption characteristic of the host and to the

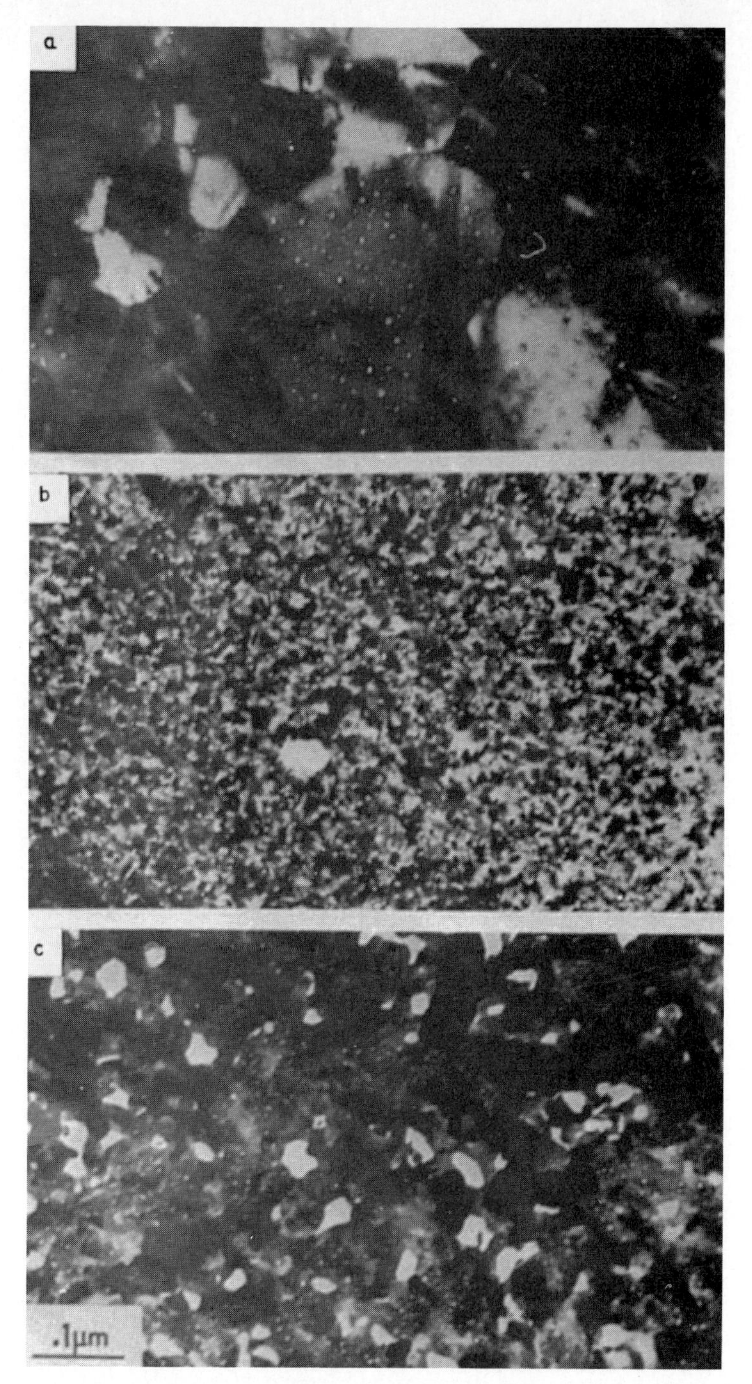

FIGURE 15.15 TEM micrographs (using defocus contrast) showing a high density of voids (white spots) in thin gold coatings prepared by (*a*) evaporation, (*b*) sputtering, and (*c*) electrodeposition. [*Adapted from Nakahara (1979).*]

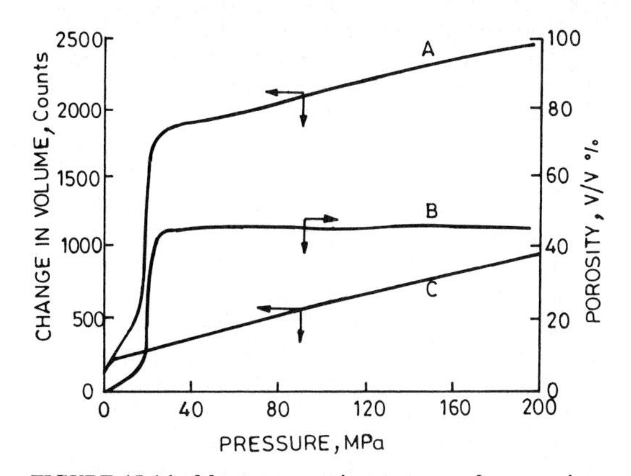

FIGURE 15.16 Mercury porosimetry curve for a particulate γ-Fe_2O_3 magnetic tape. Curve *A* is the actual measured porosity curve; curve *C* is the compressibility correction curve; and curve *B* is the porosity curve after correction for compressibilities. [*Adapted from Huisman et al. (1983).*]

appearance of new resonance peaks in the imaginary part of the effective dielectric constant. These latter peaks are termed *submicroscopic-void resonances* when caused by internal voids too small to be resolved by radiation of the wavelengths involved. Void resonance can reveal defects (a few nanometers in dimension) that are unresolvable by electron microscopy on coatings thicker than about 50 nm (Galeener, 1971; Motulevich and Shubin, 1963). This technique can be used to measure both open and closed pores.

Pore Corrosion Test. In this test, open pores are detected in situ by producing visible substrate corrosion products. This test is commonly used for plated coatings on metallic substrates (Clarke, 1975). Pores about a micrometer or larger in size can be detected with this technique. The coated surface is exposed to a fluid (liquid or gas) that penetrates pores and corrodes the substrate exposed therein. The corrosion product spills out of the pores. In the simplest tests, sufficient product is formed to be visible to the eye. Usually, the substrate is a corroding anode. A cathode reaction occurs on the pore walls (gas tests) or coating surface (liquid tests). A liquid reagent allows the use of an external cathode and applied potential to modify or accelerate the reaction, but this brings with it an increased probability of attacking the coating. A print of the pore pattern can be made by absorbing the product on moist paper pressed against the surface. Pore printing combined with an external cathode is known as *electrography* (Rehrig, 1974).

There are two forms of electrography: (1) paper-print methods and (2) gel and plaster-cast techniques. Prints are made by pressing absorbent paper moistened with electrolyte on flat or gently curved surfaces. A metal cathode placed over the paper brings sufficient substrate dissolution products to the paper to form a colored spot. Gels or casts are applied to complex shapes, and contact is made through a salt bridge. Electrolysis produces visible spots at pore sites. It has been found generally that gels can detect smaller pores.

Pore corrosion tests are used for testing gold coatings deposited on copper-alloy, nickel, or silver substrates for electrical parts. Four standard tests find widespread use: nitric acid vapor tests for copper-alloy substrates (ASTM Standard B583-73), electrography for copper-alloy and nickel substrates, the 48-hour SO_2/H_2S test for copper-alloy or silver substrates (British Standard Spec. 4292: 1968, Appendix B4.2), and the 24- or 2-hour SO_2 test for copper-alloy, nickel, or silver substrates (ASTM Standard B583-73).

Electrochemical Polarization. Electrochemical polarization methods include leakage current and corrosion potential methods for conducting coatings and substrates (Clarke, 1975). Shome and Evans (1951) measured porosities in electrodeposits of nickel and cobalt on steel by measuring the cell currents (leakage currents) passed when the specimens were connected, in an electrolyte of 3% NaCl plus 0.1% Rochelle salt, to a large copper gauze that served as an auxillary cathode. These authors reasoned that if the copper gauze were made sufficiently large, the cell currents would be limited only by the rate of anodic dissolution and would therefore be directly proportional to the exposed area of the substrate.

Morrissey (1970, 1972) measured the porosity of gold electrodeposits on copper by comparing the corrosion potentials of the specimens with those of suitable reference electrodes in an electrolyte that serves as a mild corrodant for the exposed substrate. The corrosion potentials were shown to vary with the exposed area fraction of the substrate metal. The method is simple and offers high sensitivity at low porosities.

Radiography. Radiographic techniques have been used with limited success to measure open and closed pores in plated coatings. Wolff et al. (1955) developed a radiographic method that calls for plating a radioactive metal (such as radioactive iron) over the substrate and then depositing the coating under investigation (such as nickel) over the radioactive layer. By placing a suitable photographic film (x-ray film) over the plated surface, a radiograph of the porosity can be obtained. Pores, which provide an easy path for the radioactive undercoat, produce spots on a radiograph. This method is destructive, however. Ogburn and Hilkert (1956) described a radiographic technique that does away with this intermediate layer of radioactive metal, which can have an effect on the porosity of the final coating. They used external radiation from the rear, which makes this technique nondestructive. The final coating is placed directly on a thin sheet of the substrate metal. An x-ray film is placed over the coating, and the substrate side is exposed to a suitable x-ray source. The substrate must be limited to thicknesses that yield good radiographs in a reasonable time of exposure (on the order of few tens of a micrometer). Radiographic techniques require specialized facilities and are expensive.

15.1.4 Techniques for Electrical-Resistivity Measurements

Electrical-resistivity measurements in thin metallic coatings are of considerable interest for applications in the electronics industry. The electrical resistivity of thin coatings can be measured using either a two-point or a four-point resistivity probe. However, the four-point probe is used most commonly for measuring resistivity.

15.1.4.1 Two-Point Probe Technique. In this method, a long, narrow conductor stripe with a large pad at each end is fabricated over an insulated substrate. The resistance between the two pads is then directly measured using an ohmmeter or by measuring a suitable modification of the Wheatstone bridge (Dunlap, 1959) [Fig. 15.17(*a*)]. Once the resistance R has been measured, the resistivity ρ is obtained from

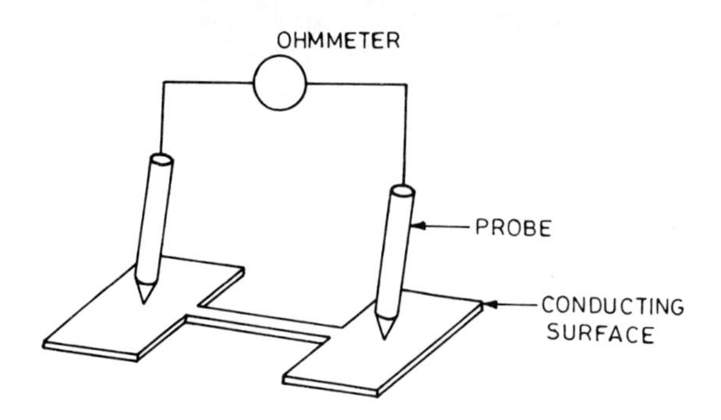

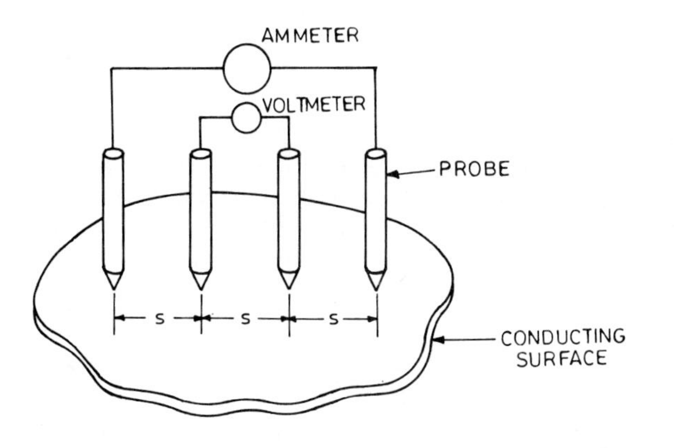

FIGURE 15.17 (*a*) Schematic arrangement for two-point probe resistivity measurement on thin coatings. (*b*) Schematic arrangement for four-point probe resistivity measurement on thin coatings.

$$\rho = R\frac{wt}{l} \tag{15.3}$$

where w, t, and l are the width, thickness, and length of the conductor stripe, respectively.

The basic advantage of two-point probe methods is simplicity, which, however, is gained at the cost of accuracy. There are several sources of errors, the most important of them being that the resistance R includes all contact resistances between the probes and the pads. Two-probe methods are suitable for measuring the sheet resistance with an accuracy of about 5 percent.

15.1.4.2 Four-Point Probe Technique. The four probes are usually arranged in a straight line with equal separation between the points; the two outside probes are current leads, and the two inside probes are used for measuring the voltage drop (Valdes, 1954) [Fig. 15.17(*b*)]. The measuring circuit is usually one of the versions of the Kelvin bridge, and the current is precisely monitored by measuring the voltage drop across a standard resistance connected in series with the current probes.

If I is current passing between the outer points and V is the potential drop between the inner points with a point spacing s, then the resistivity ρ for a specimen of infinite volume is given by

$$\rho = \frac{V}{I} 2\pi s \tag{15.4a}$$

$$\text{and } I = \frac{V_{st}}{R_{st}} \tag{15.4b}$$

where V_{st} is the voltage drop across the standard resistance and R_{st} is the standard resistance. If the specimen dimensions are comparable with the spacing of the points, then the resistivity for various geometries and different types of boundaries is given by

$$\rho = \frac{V}{I} 2\pi s f\left(\frac{L}{s}\right) \tag{15.5}$$

where L is the distance between the nearest probe and the limiting boundary. Valdes (1954) has calculated the function f for several cases.

For a thin specimen ($t \ll s/2$), the resistivity for the collinear four-probe relation is

$$\rho = \frac{V}{I} t f(g) \tag{15.6a}$$

and

$$f(g) = 4.53 \tag{15.6b}$$

for sheet coating of infinite dimensions, where $f(g)$ is a geometric factor depending on the physical dimensions of the thin coating and the probe spacing s. The numerical values of $f(g)$ for regular geometrics are listed by Smits (1958).

The four-point probe technique has two advantages. First, blanket coatings can be used without any further processing. Second, all quantities involved (except the coating thickness t) can be determined with a high degree of accuracy. Consequently, sheet resistance can be determined by the four-point probe techniques with an accuracy of about 0.5 percent or better.

15.1.5 Techniques for Surface-Roughness Measurements

Surface texture is the repetitive or random deviation from the nominal surface that forms the three-dimensional topography of the surface. Surface texture includes (1) roughness (microroughness), (2) waviness (macroroughness) (3) lay, and (4) flaws. Figure 15.18 is a diagrammatic display of surface texture with unidirectional lay.

Microroughness is formed by fluctuations in the surface of short wavelengths characterized by hills (asperities) and valleys of varying amplitudes and spacings, and these are large compared with molecular dimensions. Microroughness includes those features intrinsic to the production process. Waviness is surface irregularities of longer wavelengths and is referred to as *macroroughness*. Wavi-

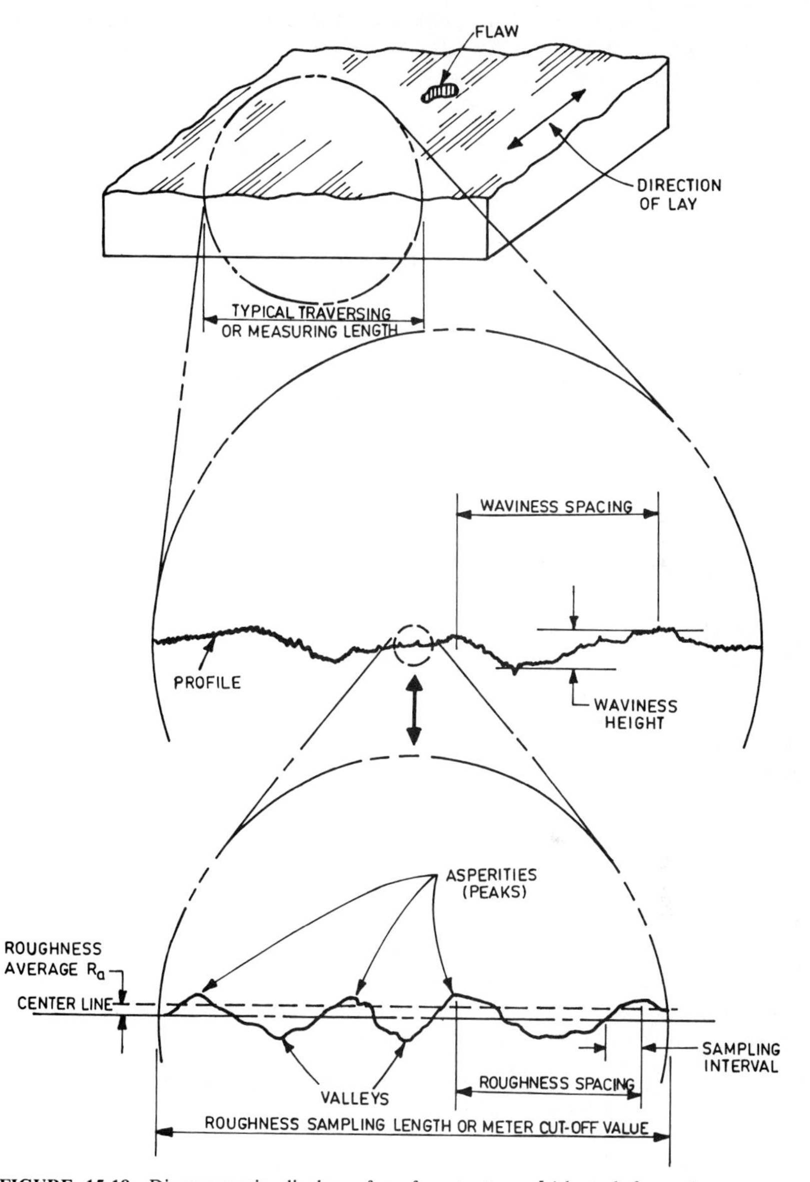

FIGURE 15.18 Diagrammatic display of surface texture. [*Adapted from Anonymous (1985).*]

ness may result from such factors as machine or workpiece deflections, vibration, chatter, heat treatment, or warping strains. Lay is the principal direction of the predominent surface pattern, ordinarily determined by the production method. Flaws are unintentional, unexpected, and unwanted interruptions in the topograph (Anonymous, 1985; Thomas, 1982).

The microroughness is most commonly measured along a single line profile and is usually characterized by one of the two statistical height descriptors. These are center-line average (cla) or arithmetic average (aa) and the standard deviation (σ) or root mean square (rms). We define a *center line* or *mean line* as a line such that the area between the profile and the mean line that is above the line is equal to that below the mean line. The center-line average (cla) is the arithmetic mean of the vertical deviation from the mean:

$$\text{cla} = \frac{1}{L}\int_0^L |z - m|\, dx \tag{15.7}$$

where $z(x)$ = surface profile
L = profile sample length
m = mean

The standard deviation σ is the square root of the arithmetic mean of the square of the vertical deviation from the mean line:

$$\sigma^2 = \frac{1}{L}\int_0^L (z - m)^2\, dx \tag{15.8}$$

Various instruments are available for measuring microroughness. The measurement instruments can be divided into two broad categories: (1) a contact type, in which, during measurement, a component of the instrument actually touches the surface to be measured, and (2) a noncontact type. Contact-type instruments can damage highly polished and soft surfaces; therefore, noncontact-type instruments are preferred for these surfaces (Bhushan, 1990; Bhushan et al., 1988b). A contact-type stylus instrument is the most popular. This instrument, solely recommended by the ISO, is generally used for reference purposes. A noncontact digital optical profiler is used for optically smooth surfaces that will scratch readily with a stylus. A review of other techniques is presented by Bhushan (1990).

15.1.5.1 Stylus Instrument. The stylus instrument amplifies and records the vertical motions of a stylus displaced at a constant velocity by the surface to be measured. The instrument consists of the following components: a pickup, driven by a traverse unit (gear box) that draws the stylus over the surface at a constant speed, roughly 3 mm min^{-1}; an electronic amplifier to boost the signal from the stylus to a useful level; and a chart recorder and a meter for recording the amplified signal (Fig. 15.19). The transducers used in most stylus instruments for sensing vertical motion are linear variable differential transformers (LVDT).

The electric signal from the pickup represents the relative movement between the stylus and the pickup body. For this signal to be truly representative of the surface profile, the pickup body must traverse a path parallel to the general shape of the surface without vertical movement due to surface irregularities. This is achieved by supporting the pickup body on the surface being measured by means of skid or datum that has a large radius (~50 mm) in comparison with the texture spacing. The skid provides a datum only if the crests are close enough together.

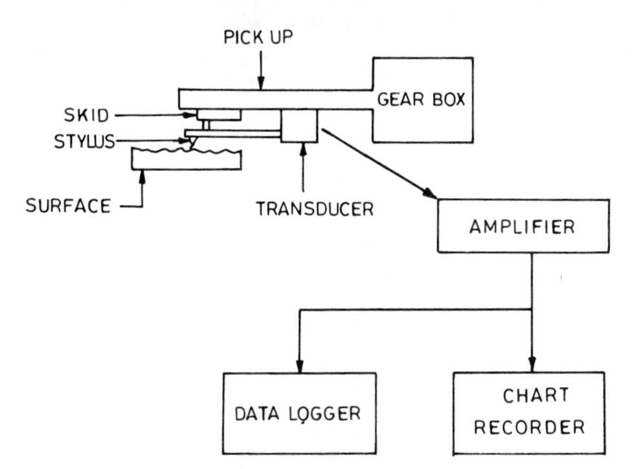

FIGURE 15.19 Schematic diagram of a typical stylus instrument.

If the crests are widely spaced, the measurements are made without the skid, and the traverse shaft provides a straight reference datum.

Diamond styli are universally used. In many instruments they are cones of 90° included angle with a tip radius of 4 to 12 μm (Thomas, 1982).

15.1.5.2 Noncontact Digital Optical Profiler (DOP). The optical profiler consists of a microscope, an interferometer, a detector array, interface electronics, and a computer. The surface-profile measurement is based on two-beam optical interference. Figure 15.20 is a schematic diagram of a digital optical profiler with a Linnik interferometer attached (Bhushan et al., 1988b). The light source is a halogen microscope illuminator. While different wavelengths can be selected using spectral filters, in most instances a spectral filter with a center wavelength of 650 nm and a passband of 40 nm is used. The intensity of the interference fringes is read out using either a 256 × 256 element photodiode array with pixel spacing of about 40 μm. The reference surface in the interferometer is mounted on a piezoelectric transducer (PZT) so that a voltage can be applied to the PZT during measurements to move it at a constant velocity. This allows the use of electronic phase-shifting techniques to measure the phase of the interference pattern. This phase is proportional to the surface height at that location. The instrument produces a two-dimensional array of height elements of the surface under test over a 256 × 256 pixel array. The types of interferometers determine the lateral resolution of the instrument, which can vary from about 0.5 μm to about 25 μm.

15.1.6 Techniques for Hardness Measurements

Hardness implies resistance to deformation; in the case of metals, this characteristic is a measure of their resistance to plastic (permanent) deformation. Hardness measurements usually fall into three main categories: static indentation hardness, rebound or dynamic hardness, and scratch hardness (Tabor, 1951). In static indentation tests, a ball, a diamond cone, or a pyramid is forced into the

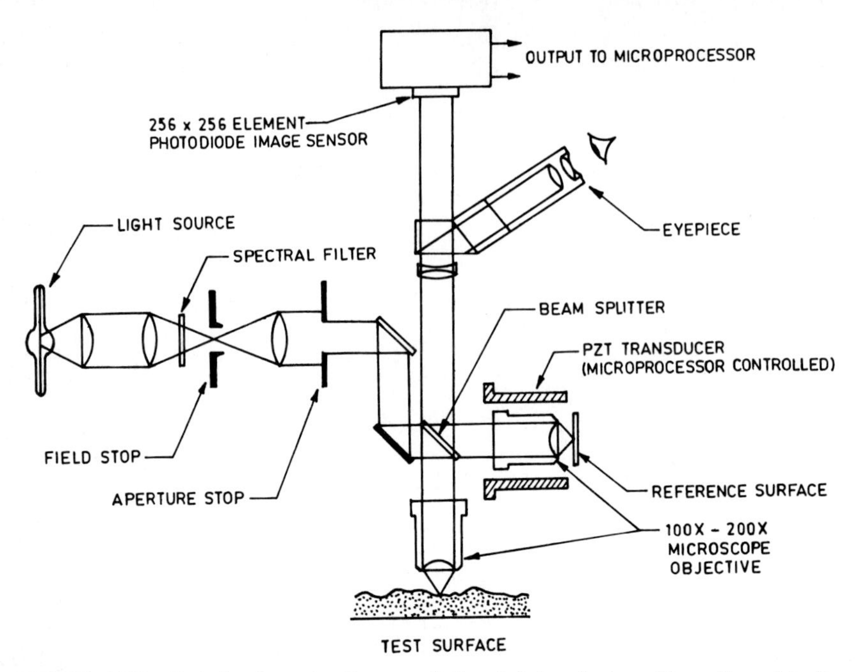

FIGURE 15.20 Optical schematic diagram of the digital optical profiler with a Linnik interferometer. [*Adapted from Bhushan et al. (1988).*]

material being tested. The relationship of total load to the area or depth of indentation provides the measure of hardness. Static indentation tests are most widely used in determining the hardness of materials other than polymers. In rebound or dynamic hardness tests, an object of standard mass and dimension is bounced from the test surface; its height of rebound becomes a measure of hardness. Rebound hardness tests are not as common as static indentation tests. Scratch hardness is the oldest form of hardness measurement. It depends on the ability of one solid to scratch or be scratched by another solid. Scratch tests are at best semiquantitative and are not common at all.

Static identation tests on polymers (rubbers and plastics) are made using special depth-sensing instruments.

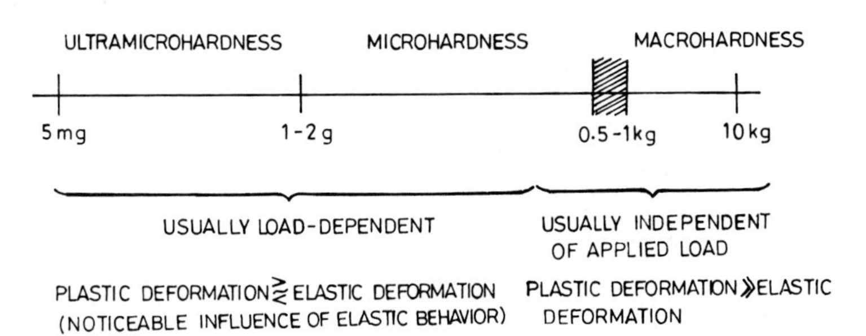

FIGURE 15.21 Extended load range of static indentation hardness testing.

15.1.6.1 Static Indentation Hardness Tests. The methods most widely used in determining the hardness of materials other than polymers are quasi-static indentation methods. These methods involve the formation of permanent (plastic) indentations on the surface of the material to be tested, the hardness number (kg mm^{-2} or GPa) being determined by the normal load divided by either the curved (surface) area (Brinell hardness number and Vickers hardness number) or the projected area (Meyer hardness number, Knoop hardness number, and Berkovich hardness number) of the contact between the indenter and the material being tested under load (Bhushan, 1990; Boyer, 1987; Fee et al., 1985; Tabor, 1951; Westbrook and Conrad, 1973). In a conventional indentation hardness test, the area of contact is determined by measuring the indentation size after the sample is unloaded. For metals at least, there is little change in the size of the indentation of unloading so that the conventional hardness test is essentially a test of hardness under load, although it is subjected to some error due to the varying elastic contraction of the indentation.

Indentation tests differ with regard not only to the shape of the indenter (ball, cone, or pyramid) but also to the load range employed. The extended load range of static indentation hardness testing is shown schematically in Fig. 15.21. Using a large load on the indenter produces a large impression. Large loads are used for the bulk hardness of metals, whereas smaller loads (microhardness tests) allow investigation of the individual constituents of an alloy or coating. It should be noted that only the lower micro- and ultramicrohardness load ranges can be employed successfully for measurements of extremely thin coatings. The intrinsic hardness of thin coatings becomes meaningful only if the influence of the substrate can be eliminated. It is therefore generally accepted that the depth of indentation should never exceed 10 to 20 percent of the coating thickness. One way to overcome this difficulty is to reduce the normal load during measurement.

This section provides an overview of (indentation) macro-, micro-, and ultramicrohardness (nanohardness) measurement techniques. Table 15.2 presents a comparison of various static indentation hardness test methods. The relationship among various hardness methods is not exact, since no two methods measure exactly the same sort of hardness, and a relationship determined on steels of different hardnesses (shown in Fig. 15.22) will be found to be only approximately true with other materials.

Macrohardness Instruments. Macrohardness measurements are generally used to determine bulk hardness. They include Brinell and Rockwell hardness testers (Tabor, 1951).

Brinell hardness is determined by forcing a hardened sphere under a known load into the surface of a material and measuring the diameter of the indentation left after the test. The load is maintained for about 30 seconds, and the diameter of the indentation at the surface is measured with an optical microscope (10×) after the ball has been removed. For production work, however, several testing machines are available that automatically measure the depth of the indentation and from this give a reading of hardness. The Brinell hardness number (BHN), or simply the Brinell number, is defined in kilograms per square millimeter as the ratio of the load used to the actual surface area of the indentation, which is, in turn, given in terms of the imposed load W, ball diameter D, and indentation diameter d or depth of the indentation h:

$$\text{BHN} = \frac{2W}{\pi D\left(D - \sqrt{D^2 - d^2}\right)} = \frac{W}{\pi D h} \tag{15.9}$$

TABLE 15.2 Comparison of Commonly Used Static Indentation Hardness Test Methods

Hardness test	Indenter(s)	Indent: Diagonal or diameter	Indent: Depth	Load(s)	Method of measurement	Surface preparation	Applications	Remarks
Brinell	Ball 2.5 or 10 mm in diameter	1–5 mm	Up to about 1 mm	3000 kg for ferrous materials, down to 100 kg for soft metals	Measure diameter of indentation under microscope; read hardness from tables	Specially ground area for measurements of diameter	Bulk hardness of metals	Damage to specimen minimized by use of lightly loaded ball indenter
Rockwell	120° diamond cone, 1.59-mm-diameter ball	0.1–1.5 mm	25–350 μm	Major, 60, 100, 150 kg; Minor, 10 kg	Read hardness directly from display	No preparation necessary on many surfaces	Bulk hardness of hard materials	Measure depth of penetration; may be used for thinner material than Brinell test
Rockwell superficial	As for Rockwell	0.1–0.7 mm	10–100 μm	Major, 15, 30, 45 kg; Minor, 3 kg	As for Rockwell	Finished surface	Used for thin specimens	Indentation size and load are lower than that of Rockwell
Vickers	Square pyramid with 136° apex angle	10 μm to 1 mm	1–100 μm	1–120 kg; can be as low as 25 g	Measure indent with low-power microscope; read hardness from tables	Smooth, clean surface	Surface layers, thin specimens down to about a micrometer	Less sensitive to surface variations than Knoop
Knoop	Rhombic pyramid with longitudinal edge angles of 172.5° and 130°	10 μm to 1 mm	0.3–30 μm	0.2–4 kg; can be as low as 1 g	Measure indentation with low-power microscope; read hardness from tables	Smooth, clean surface	Surface layers, thin specimens down to about a micrometer	Laboratory test used on brittle materials or microstructural constituents
Berkovich (used in nanoindenter)	Triangular-based pyramid with angle of 65.3° between the face and vertical axis	20 nm to 1 μm	As low as 20 nm	As low as 5 mg	Measure depth in situ by capacitance sensing gauge	Smooth, clean surface	Submicron surface layers	Highly specialized equipment

Note: The minimum material thickness for a test is usually taken to be 5 to 10 times the indentation depth.
Source: Adapted from Fee et al. (1985).

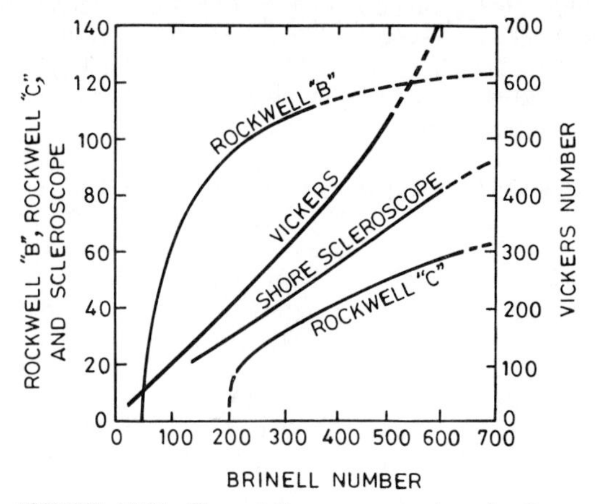

FIGURE 15.22 The relation among various hardness methods.

Hardened steel bearing balls may be used up to 450 BHN, but beyond this hardness, specially treated steel or tungsten carbide balls should be used to avoid flattening of the indenter. The standard size ball is 10 mm, and the standard loads are 3000, 1500, and 500 kg, with 250, 125, and 100 kg sometimes used for softer materials (see *Methods of Brinell Hardness Testing*, ASTM E 10-61T). The load and ball diameter should be adjusted to keep the ratio d/D within the range 0.3 to 0.5. The nearest edge of the specimen should be no closer than 2.5 impression diameters, and the thickness should be more than one diameter. Since d/D is normally less than 0.5, this means that for a 10-mm ball, the uninterrupted width and depth of the specimen may have to be as great as 25 and 10 mm, respectively, to avoid spurious side and bottom effects.

Use of the curved area in the Brinell test was originally introduced to try to compensate for the effects of work-hardening. It is cumbersome to use the surface area of the indentation, though, so to overcome this disadvantage, *Meyer hardness* is defined in kilograms per square millimeter as the load used divided by the projected area of the indentation:

$$\text{Meyer hardness} = \frac{4W}{\pi d^2} \tag{15.10}$$

In the *Rockwell hardness test*, the indenter may be either a steel ball of some specified diameter or a spherical-tipped conical diamond of 120° angle with a 0.2-mm tip radius, called a *Brale*. A minor load (or preload) of 10 kg is first applied, and this causes an initial penetration that holds the indenter in place. At this point, the dial is set to zero and the major load is applied. The standard loads are 150, 100, and 60 kg. Upon removal of the major loads, the reading is taken while the minor load is still on. The hardness number may then be read directly from the scale that measures penetration. The Rockwell hardness number is defined by an arbitrary equation of the following form:

$$\mathrm{R} = C_1 - C_2 \Delta t \tag{15.11}$$

where C_1 and C_2 are constants for a given indenter size, shape, and hardness scale and Δt is the penetration depth in millimeters between the major and minor loads. Although Rockwell hardness increases with Brinell hardness, the two are not proportional, and the dimensions of Rockwell hardness are not force per unit area. In fact, the Rockwell hardness number cannot be assigned any dimensions, since it is defined in an arbitrary equation (15.11).

A variety of combinations of indenter and major load are possible; the most commonly used are HRB or R_B, which uses a 1.59-mm-diameter (1/16 inch) ball as the indenter and a major load of 100 kg; HRC or R_C, which uses a cone as the indenter and a major load of 150 kg; and HRA or R_A, which uses a cone as the indenter and major load of 60 kg. Rockwell B is used for soft metals, and Rockwell C and A are used for hard metals. There is also a *Rockwell superficial hardness test* that is similar to the standard Rockwell test, except that the indentation load, and consequently, the indentation size, is much smaller. The superficial scales are used with thin specimens or to minimize the impression. The superficial scales used are R_{15N}, R_{30N}, R_{45N}, R_{15T}, R_{30T}, and R_{45T}. The N scale uses the 120° diamond cone, and T scale uses a 1.59-mm-diameter (1/16 inch) ball, and the numbers represent the major load. The minor load applied in each case is 3 kg (see *Rockwell Hardness and Rockwell Superficial Hardness of Metallic Materials*, ASTM E 18-61).

Microhardness Instruments. Microhardness measurements allow the indentation to be shallow and of small volume so as to measure the hardness of brittle materials or thin materials or coatings. The instruments include Vickers and Knoop hardness testers (Tabor, 1951). Both testers use highly polished diamond pyramidal indenters. The Vickers indenter is a diamond in the form of a square pyramid with face angles of 136° (corresponding to edge angles of 148.1°) (Fig. 15.23*a*), and relatively low loads varying between 1 and 120 kg are used. The Knoop indenter is a rhombic-based pyramidal diamond with longitudinal edge angles of 172.5° and 130° (Fig. 15.23*b*). In general, the loads used in the Knoop tester vary from about 0.2 to 4 kg. Smaller loads (as low as 1 to 25 g) also may be used in Vickers and Knoop testers when a microhardness examination is needed. In making Vickers or Knoop hardness measurements, the lengths of the diagonals of the indentation are measured using a medium-power compound microscope after the load is removed. If d is the mean value

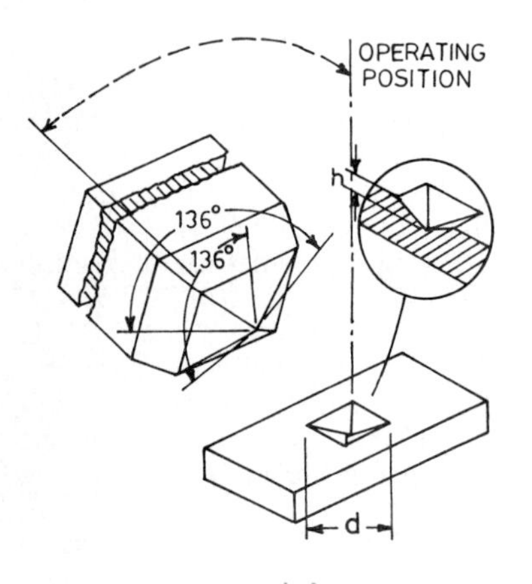

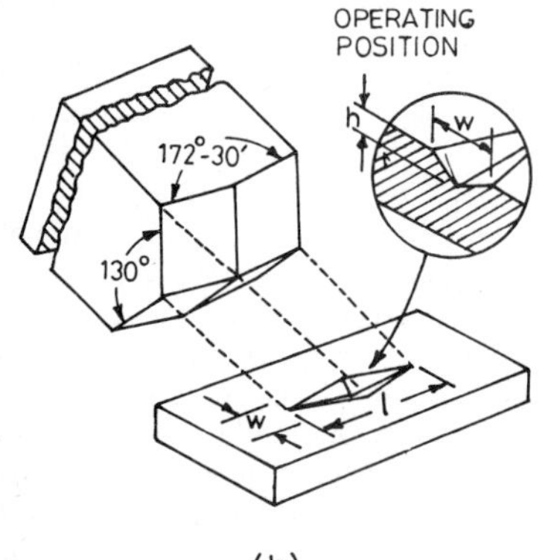

FIGURE 15.23 Geometry and indentation with (*a*) a Vickers indenter and (*b*) a Knoop indenter.

of a diagonal in millimeters and W is the imposed load in kilograms, Vickers hardness number (V or HV), sometimes called *diamond-pyramid hardness* (*DPH*), is given by the load divided by the actual surface area, that is,

$$HV = \frac{1.8544W}{d^2} \tag{15.12}$$

If l is the long diagonal* in millimeters (used for hardness measurement), Knoop hardness number (K or HK) is given by the load divided by the projected area of the indentation, that is,

$$HK = \frac{14.229W}{l^2} \tag{15.13}$$

The depth to the diagonal for the Vickers indenter is about 7, and the depth to the long diagonal (used in the Knoop indenter) is about 30.5. The advantages of the Knoop indenter over the Vickers indenter in microhardness testing lie in the fact that a longer diagonal is obtained for a given depth of indentation or a given volume of material deformed. It can be shown that for the same measured diagonal length, the depth and area of the Knoop indentation are only about 15 percent of those of the Vickers indentation. The Knoop indenter is thus advantageous when shallow specimens or thin, hard layers must be tested. The Knoop indenter is also desirable for brittle materials (such as glass or diamond), in which the tendency for fracture is related to the area of stressed material.

Nanohardness Instruments. The minimum load for commercial microhardness measurements is about 1 g, and this may produce a depth of indentation of a few tenths of a micrometer, exceeding 10 to 20 percent of coating thicknesses on the order of a micrometer. Loads on the order of 5 to 100 mg are desirable if the indentation size approaches the resolution limit of a light microscope, and it is almost impossible to find such a small imprint if the measurement is made with a light microscope after the indentation load has been removed. Therefore, either the indentation instruments are used with a scanning electron microscope (SEM) or in situ depth measurements are made. Several instruments have been developed to measure the nanohardness of extremely thin coatings (submicrometer thickness) and can measure indentation depths as small as 20 nm (Bangert and Wagendristel, 1986; Bangert et al., 1981; Bhushan, 1990; Bhushan et al., 1988a; Oliver et al., 1986; Pulker and Saltzmann, 1986; Williams et al., 1988). Most of these instruments also can be used to measure Young's modulus of elasticity. A review of various instruments is presented by Bhushan (1990).

A nanoindentation apparatus is commercially available from Nano Instruments, Inc., Oak Ridge, Tennessee (Oliver et al., 1986). The tester continuously monitors the load and penetration depth during the indentation process (Fig. 15.24). The position of the indenter is determined by a capacitance displacement gage. A coil and magnet assembly located at the top of the loading column is used to drive the indenter toward the sample. The force imposed on the column is controlled by varying the current in the coil. The loading column is suspended by flexible springs, and the motion is damped by airflow around the central plate of the capacitor, which is attached to the loading column. The capacitance displacement gauge permits one to detect displacement changes of 0.2 to 0.3 nm. The

*The geometry of the Knoop indenter has the shape of a parallelogram in which the longer diagonal is about 7.11 times as large as the shorter diagonal.

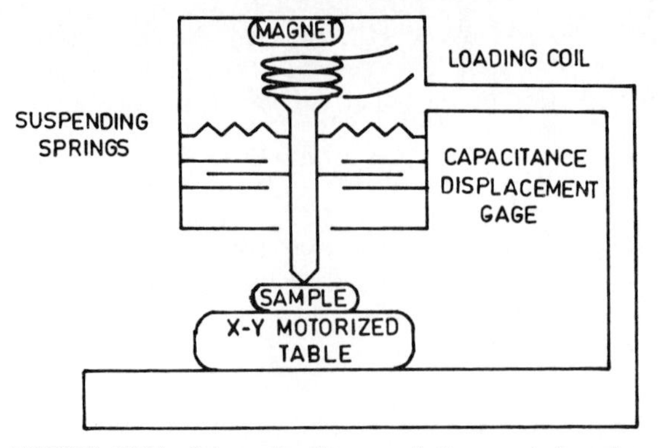

FIGURE 15.24 Schematic diagram of the nanoindentation hardness apparatus manufactured by Nano Instruments, Inc., Oakridge, Tennessee.

force resolution of the system is about 0.5 μN (50 μg). Measurements can be made at a minimum penetration of about 20 nm.

The trends has been to use a Berkovich indenter with depth-sensing instruments (Doerner and Nix, 1986). The Berkovich indenter is a triangular-based, highly polished pyramidal diamond with a nominal angle of 65.3° between the face and the vertical axis and a tip radius of less than 0.1 μm. The Berkovich indenter has a sharply pointed tip in comparison with the Vickers or Knoop indenters, which are four-sided pyramids that have a slight offset (0.5 to 1 μm). A sharper tip is especially desirable for extremely thin coatings that require shallow indentations. The Berkovich hardness (HB or H_B) is defined as the load in kilograms divided by the projected area of the indentation in square millimeters, that is,

$$\mathrm{HB} = 0.04083 \frac{W}{h^2} \tag{15.14}$$

where h is the depth of indentation. Note that the projected area-to-depth relationship of a Berkovich indenter is equivalent to that of a Vickers indenter.

15.1.6.2 Rebound or Dynamic Hardness Tests. The *dynamic hardness* of a metal may be defined, by analogy with static hardness, as the resistance of the material to local indentation when the indentation is produced by a rapidly moving indenter of spherical or conical form. In most practical methods, the indenter is allowed to fall under gravity onto the metal surface. It rebounds to a certain height and leaves an indentation in the surface. The dynamic hardness during impact is equal to the energy expended in producing the indentation after rebound has occurred (= energy of impact minus energy of rebound) divided by the volume of indentation (Tabor, 1951; Fee et al., 1985).

A different approach was adopted by A. F. Shore in 1907. In this method, the height of the rebound itself is used as a measure of the dynamic hardness. It has been found that if the height of the fall is constant, the height of the rebound is roughly proportional to the static hardness of the material concerned (Fig. 15.22). (For most metals at the velocity of impact typically used in dynamic tests, the dynamic yield pressure is not widely different from the static yield pressure.)

Testers of this type are called *Shore rebound Scleroscopes* (a trademark of Shore Instruments and Manufacturing Company, New York).

The Scleroscope test is carried out by allowing a small diamond-tipped (spherically), hardened steel cylinder fall freely through a tube onto the surface to be tested. On the rebound up the tube, the cylinder is caught at its highest point and held there. The height of the rebound is a measure of the Scleroscope hardness. The height of the fall is divided into 140 equal divisions, and a quenched (to maximum hardness) and untempered water-hardening tool steel gives a rebound height about 100 units (~850 HV). Thus materials with hardnesses greater than that of fully hardened tool steel can be measured. Two types of Scleroscope hardness testers are commercially available. In the model D Scleroscope, the hammer is longer and heavier (about 0.36N) than the hammer used in model C Scleroscope (about 23 mN); however, the hammer is dropped a shorter distance (about 18 mm) in model D than in model C (about 250 mm) in order to develop the same striking energy (see *Scleroscope Hardness Testing of Metallic Materials*, ASTM E 448). Scleroscope D values for steels are generally about 2 to 6 percent larger than scleroscope C values, with larger differences at lower values (below 100).

The Scleroscope is used most often when numerous quality checks must be made or for testing very large specimens such as forged steel or wrought alloy steel rolls because the rebound measurements can be carried out in situ. Other advantages of Scleroscope hardness testing include a short test time and simplicity. However, Scleroscope hardness tests are more sensitive to variations in surface conditions (such as roughness) than other hardness tests. In addition, these tests are not very reproducible and are not as common as static indentation tests.

15.1.6.3 Scratch Hardness Tests. Scratch hardness tests represent the oldest type of hardness test. Scratch tests depend on the ability of one solid to scratch or to be scratched by another solid. The three most common techniques for measuring scratch hardness are the Mohs scale, file hardness, and ploughing tests.

The scratch test was first put on a semiquantitative basis by Mohs in 1822, who selected ten minerals as standards, beginning with talc (scratch hardness of 1) and ending with diamond (scratch hardness of 10) (Table 15.3). The Mohs scale is widely used by mineralogists and lapidaries. It is not well suited for metals, however, because the intervals are not well spaced in the higher ranges of hardness and most harder metals in fact have a Mohs hardness ranging between 4 and 8.

TABLE 15.3 Mohs Scratch Hardness Scale

Material (minerals)	Mohs hardness
Talc	1
Gypsum	2
Calcite	3
Fluorite	4
Apatite	5
Orthoclase	6
Quartz	7
Topaz	8
Corundum	9
Diamond	10

Further, the actual values obtained may depend on the experimental procedure, in particular on the inclination and orientation of the scratching edge.

The file hardness test was one of the first scratch tests used for evaluating the hardness of metallic materials. Standard test files are heat treated to about 67 to 70 HRC. The flat face of the file is pressed firmly against and is slowly drawn across the surface to be tested. If the file does not bite, the material is designated as *file hard*. The results of the test are influenced by a number of factors, such as pressure, speed, angle of contact, and surface roughness. Consequently, its ability to give reproducible hardness values is rather limited, and reasonable accuracy is obtained only at the highest hardness levels. The file test is used in estimating the hardness of steels in the high hardness ranges. It provides information on soft spots and decarburization quickly and easily and is readily adaptable to odd shapes and sizes that are difficult to test by other methods.

The ploughing test consists of drawing a diamond stylus under a definite load (typically ranging from 10 to 250 mN) across the surface to be examined. The hardness is determined by the width or depth of the resulting scratch; the harder the material, the smaller is the scratch. This is a useful tool for studying the relative hardness of microconstituents, but it does not lend itself to high reproducibility or extreme accuracy.

15.1.6.4 Static Indentation Hardness Tests on Polymers. In polymers, the indentation diminishes when the force is removed, and therefore, hardness is measured with the force applied. Hardness of rubber is related to its elastic stiffness or Young's modulus rather than to flow stress. On the contrary, in plastics, the elastic moduli are relatively high, and the range over which plastics deform elastically is relatively small. Consequently, the deformation in plastics takes place predominantly outside the elastic range, with considerable plastic deformation. Therefore, the hardness of plastics is bound up primarily with plastic properties and only secondarily with elastic properties.

There are two general classes of hardness-measuring devices for polymers. One class is comprised of the durometer (pocket size, widely used). In this device, the indenter point projects from the flat bottom of the case, held there in zero position by a spring. When pressed against a sample, the indenter point is pushed back into the case against the spring, and this motion is read directly on a dial. The harder the sample, the farther it will push back the indenter point and the higher will be the numerical reading on the dial. All durometers are calibrated in the same arbitrary scale from 0 (soft) to 100 (hard). Several types of durometers are available, as shown in Table 15.4. Of these, only two (type A and type D) conform to ASTM Standard D 2240, *Indentation Hardness of Rubber and Plastic by Means of a Durometer*. Normally, a type A durometer is used for testing soft rubbers, and a type D durometer is used for hard rubber and harder grades of plastics. Durometers are sold under proprietary names such as Shore, Rex, and Wallace.

The other class of hardness testers for polymers consists of tabletop instruments, such as the ASTM hardness tester, the Pusey and Jones Plastometer, the Firestone Penetrometer, and the Admiralty and the Wallace hardness testers. As a class, they are not as convenient or as widely used as durometers, but their measurements are more reproducible. Indenters are loaded under dead weight, and the dial readings are a direct measure of the depth of indentation. Testers of this class differ from each other in the shape of the indenter point, the amount of dead weight, and other details of construction (ASTM Standards D676, D314, and D531).

TABLE 15.4 Specifications of Durometers Used for Hardness Testing of Polymers

Durometer type	Load, N	Indenter	For use on
A (ASTM D2240)	8.22	Frustum cone	Soft vulcanized rubber and all elastomeric materials, natural rubber, nitrile rubber
B	8.22	Sharp 30° included angle	Moderately hard rubber
C	45.4	Frustum cone	Medium hard rubber and plastics
D (ASTM D2240)	45.4	Sharp 30° included angle	Hard rubber and the harder grade of plastics
0	8.22	2.4-mm sphere	Soft polymers, e.g., Nylon
00	1.14	2.4-mm sphere	Sponge rubber and plastics
000	1.14	12.7-mm sphere	Ultrasoft sponge rubber and plastics

15.1.7 Techniques for Measuring Residual Stresses

Nearly all coatings, by whatever means they are produced, and surface layers of treated parts are found to be in a state of residual (intrinsic or internal) stress. These are elastic stresses that exist in the absence of external forces and are produced through the differential action of plastic flow, thermal contraction, and/or changes in volume created by phase transformations. Differential plastic flow can occur during the forming of a part. Differential thermal contraction often occurs during the nonuniform cooling of a large part. Differential volume changes occur during the precipitation of second phases, i.e., the atomic volume of the precipitating phase generally differs from that of the host matrix. The residual stresses can be of two basic types: macrostress and microstress. Residual macrostresses are long-range relative to the scale of the microstructure—that is, they extend continuously across a part. Residual microstresses are short-range relative to the scale of the sample—that is, they vary from grain to grain or between phases (Noyan and Cohen, 1987).

The presence of residual stresses can affect the mechanical behavior of the material. The residual stress may be compressive (i.e., the coating would like to expand parallel to the surface) or tensile (i.e., the coating would like to contract). In the extreme cases, the interfacial shear stresses may exceed the adhesion strength of the coating-substrate interface and lead to cracking and delamination of the coating. However, compressive stresses are believed to increase coating hardness (e.g., see Sundgren and Hentzell, 1986).

Techniques used to measure residual stresses involve determination of the stresses by measuring the physical properties of the body that are affected by the presence of these stresses. The methods used to measure residual stresses are the deformation (or curvature) method, x-ray, electron-, and neutron-diffraction methods, and other miscellaneous methods (Bhushan, 1990; Campbell, 1970; Masubuchi, 1983; Nix, 1989; Noyan and Cohen, 1987). It should be noted that x-ray, electron, and neutron-diffraction methods will give the strain and hence the stress in a crystalline lattice. This is not necessarily the same as that measured by substrate bending, since the stress on the grain boundaries may not be the same as that in the crystallites. Deformation methods are most commonly used for measuring residual stresses.

15.1.7.1 ***Deformation Methods.*** If a coating is deposited in a stressed state on a thin substrate, the substrate will be bent by a measurable degree. A tensile stress will bend it so that the coated surface is concave, and a compressive stress will bend it so that the coating is convex. The deformation of the substrate is measured either by observing the displacement of the center of a circular disk or by using a thin beam as a substrate and calculating the radius of curvature of the bent beam. The deformation can then be related to the residual stresses.

Disk Methods. In disk methods, the stress of a coating is measured by observing the deflection of the center of a circular disk when the coating is deposited on one side. A stylus profiler can be used to measure the disk deflection (e.g., Pan and Greene, 1981). In another variation, the interference fringes (Newton's rings) between the disk and an optical flat are used to measure the deflection of the disk, which typically may be a glass, quartz, or silicon crystal wafer that is 25 to 250 μm in thickness. Figure 15.25 shows the apparatus used by Hoffman (1966). Because of the limited flatness of available substrates, the coating is usually dissolved from the substrate after the test and the remaining profile is used as a reference. Chapman (1971) reported on an instrument known as a *stresometer*, in which metal disks are used and the bowing of the disk is observed hydrostatically by having a fluid in contact with one side of the disk and observing the movement of the fluid in a capillary as the disk bows. Disk methods generally have been applied to electroplated coatings.

Disks strained by the presence of a coating will bend into a paraboloid. The stress σ can then be related to the deflection δ by the disk at a distance r from the center of the disk by the following equation (Glang et al., 1965):

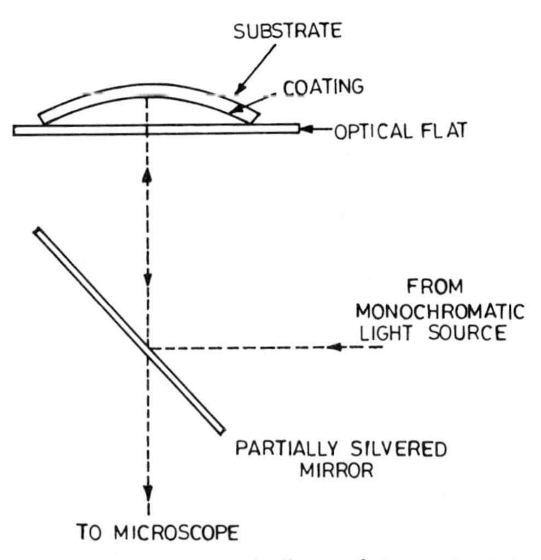

FIGURE 15.25 Schematic diagram for measurement of deflection of the center of a disk plate by optical interference.

$$\sigma = \frac{\delta}{r^2} \frac{E_s}{3(1 - \nu_s)} \frac{t_s^2}{t_c} \tag{15.15}$$

where E_s = Young's modulus of the substrate
ν_s = Poisson's ratio for the substrate
t_s = substrate thickness
t_c = coating thickness

This equation is valid provided the deflection is much less than the thickness of the substrate. It should be noted that the bulge surface is not actually spherical, as has been assumed in the derivation of Equation (15.15).

Beam Methods. In beam methods, the substrate is usually between 2 and 50 times longer than its width and usually 25 to 250 μm thick. Beam curvature is found by measuring the deflection of one end of the beam while other end is clamped or by measuring the deflection of the middle of the beam if it is supported at both ends. Various methods have been used to measure the curvature of the bent beam (Bhushan, 1990; Campbell, 1970). Many of the methods can be used in situ during the deposition process. Microscopic observation of the free end of a cantilevered beam during coating deposition is the simplest method of measuring beam deflection in situ (Fig. 15.26). Other in situ methods for measuring beam deflection include electromechanical, capacitive, inductive, and optical interference methods (Campbell, 1970).

Postdeposition methods include optical interference, laser, strain-gauge indentation, and stylus methods. Letner and Snyder (1954) laid an optical flat on top of a reference surface and used an optical interferometer with a monochromatic light source to measure the curvatures in two orthogonal directions in the surface plane. Their apparatus was similar to that in Figure 15.25. McInerney and Flinn (1982) and Flinn et al. (1987) used a laser method. The substrate, illuminated by a laser beam, is translated, and the beam reflected by the bent substrate is monitored by a position-sensitive photodetector (Fig. 15.27). The reflected beam is displaced by an amount inversely proportional to the substrate radius, that is,

$$R = 2L \frac{\Delta S}{\Delta D} \tag{15.16}$$

where L is the distance between the diode and the substrate, and ΔS and ΔD are the translation distances of the substrate and detector, respectively. The laser

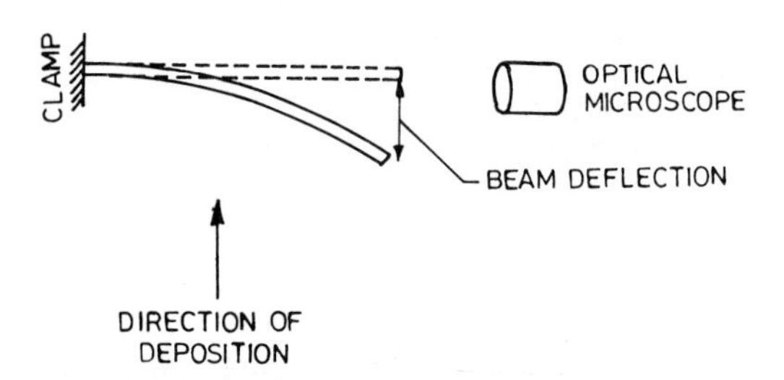

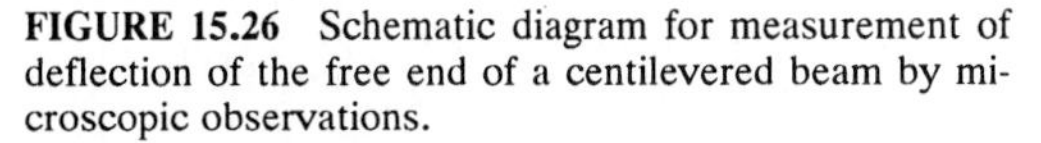

FIGURE 15.26 Schematic diagram for measurement of deflection of the free end of a centilevered beam by microscopic observations.

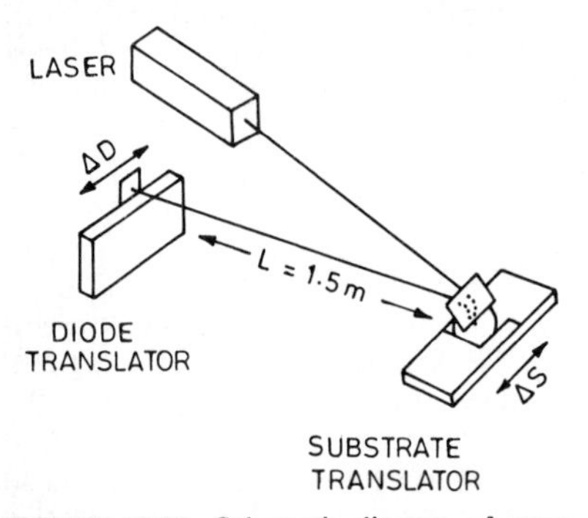

FIGURE 15.27 Schematic diagram of curvature measurement setup using a laser source.

method is very sensitive and accurate for measuring the bending of a flat substrate to a radius of up to about 10 km. For typical coating dimensions, this radius of curvature corresponds to a residual stress of about 0.2 MPa.

Brinksmeier et al. (1982) used strain gauges bonded to the specimen to measure strain in two orthogonal directions. Tsukamoto et al. (1987) measured the deflection at the center of the bent beam by pressing the beam flat with a nanoindenter. The bent beam is placed on a flat glass surface supported by two fulcrums, and a load-deflection curve is generated (Fig. 15.28). The distance h_a can be estimated from the inflection point in the curve. Because of the limited flatness of most substrates, the coating is removed from the substrate, and then the initial deflection is measured. The true deflection resulting from residual stresses in the coating is equal to $h_a - h_b$. The curvature of the substrate can then be calculated by the geometric relationship

$$R = \frac{L^2}{8(h_a - h_b)} \tag{15.17}$$

where L is the span.

Treuting and Reed (1951) and Chandrasekar et al. (1987) determined the curvature by measuring the deflection at the center of the bent beam with diamond stylus tracing on the beam surface.

The biaxial stresses (σ_x and σ_y) in a coating on a rectangular substrate can be deduced from the curvature of a thin strip of substrate by the well-known Stoney equation (Stoney, 1909; Brenner and Senderoff, 1949; Jaccodine and Schlegel, 1976):

$$\sigma_x = \frac{E_s t_s^2}{6t_c(1 - \nu_s^2)}\left(\frac{1}{R_x} + \nu\frac{1}{R_y}\right) \tag{15.18b}$$

$$\sigma_y = \frac{E_s t_s^2}{6t_c(1 - \nu_s^2)}\left(\frac{1}{R_y} + \nu\frac{1}{R_x}\right) \tag{15.18b}$$

where t_s and t_c are the thicknesses of the substrate and coating, respectively, R_x and R_y are the radii of curvature in the x and y directions, E_s is Young's modulus, and ν_s is Poisson's ratio of the substrate. For multiple coatings on a square substrate, the stress in each coating contributes independently to the total substrate curvature by

$$\sigma_1 t_{c1} + \sigma_2 t_{c2} + \sigma_3 t_{c3} = \frac{E_s}{1 - \nu_s}\left(\frac{t_s^2}{6R}\right) \tag{15.19}$$

The distribution of residual stresses as a function of depth of a coated or treated part can be obtained by removing the surface layers and measuring the resulting deformations (curvatures). This method employs a rectangular speci-

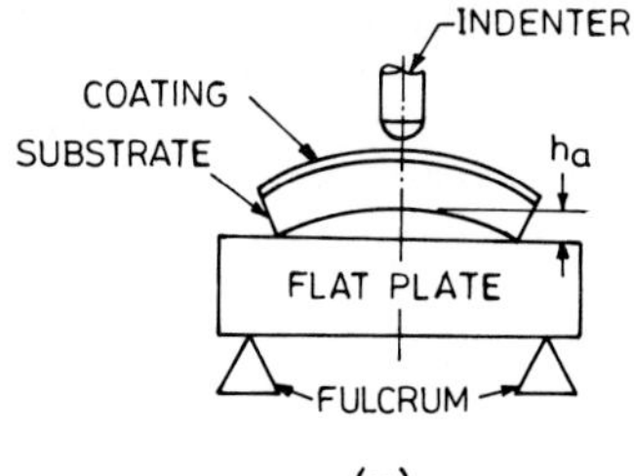

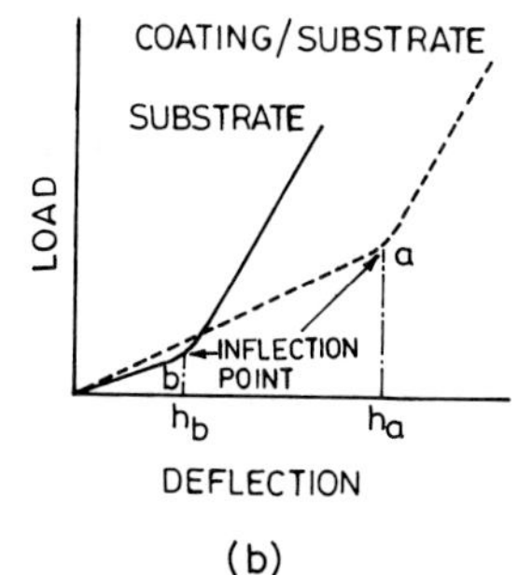

FIGURE 15.28 (*a*) Schematic diagram of the deflection measurement of a bent beam using a nanoindenter. (*b*) Load-deflection curve for a warped composite-beam substrate.

men that is coated or treated on one surface (test surface) and lapped to a high degree of flatness on the other parallel surface (reference surface). The test surface is progressively etched away while the polished reference surface is protected from etching by deposition of a coating. After each etching step, the amount removed is determined, and the reference surface is cleaned and its curvature is measured in two orthogonal directions—in the directions of the principal stresses on the surface. The stress formulas for this technique are derived based on the fact that in a free specimen the internal stresses are self-equilibrating. Removing a part of the stress distribution, as in removing a layer of material by etching, leaves an unbalanced distribution of internal stresses so that external restraints are required to prevent deformation. When the restraints are released, the specimen deforms to restore internal equilibrium. The forces and moments associated with the deformation are equal and opposite to those of the residual stresses in the layer of the material removed by etching.

Treuting and Reed (1951) have derived equations for the principal stresses in terms of curvature measurements, dimensions, and elastic constants of the specimen:

$$\sigma_1(z, w_0) = \frac{E_s}{6(1 - \nu_s^2)} \left\{ z^2\left(\frac{dR_1}{dz} + \nu \frac{dR_2}{dz}\right) + 4z[R_1(z) + \nu R_2(z)] + 2(w_0 - 3z) \right.$$

$$\left. [R_1(w_0) + \nu R_2(w_0)] - 2\int_z^{w_0} [R_1(z) + \nu R_2(z)]dz \right\} \quad (15.20)$$

and a similar equation with subscripts 1 and 2 interchanged for $\sigma_2(z, w_0)$, where w_0 is the initial thickness before etching, w is the thickness after etching, z is coordinate measuring distance below the test surface, E_s is Young's modulus of the substrate, ν_s is Poisson's ratio of the substrate, $R_1(w_0)$ and $R_2(w_0)$ are the equilibrium curvatures of the specimen in the two principal directions when thickness is w_0, and $\sigma_1(z, w_0)$ and $\sigma_2(z, w_0)$ are the residual stresses in the two principal directions existing at any level z before any layer is removed from a specimen of thickness w_0.

15.1.7.2 X-Ray, Electron-, and Neutron-Diffraction Methods. In the x-ray, electron-, and neutron-diffraction methods, changes in lattice spacing are measured, and hence lattice strains or stresses are determined (Anonymous, 1971; Dölle, 1979; Krawitz and Holden, 1990; Rickerby et al., 1987, 1989). These methods are capable of measuring very localized stresses, and most other methods measure macroscopic stresses. In the x-ray technique, if the dimensions of the crystallites of the coating are less than 100 nm, then the diffraction lines will be broadened. The broadening effect due to crystallite sizes can be distinguished by using glancing-angle techniques. Electron-diffraction techniques can be applied without the line-broadening effects due to crystallite size down to 10-nm crystallite dimensions. The application of neutron diffraction complements and extends x-ray diffraction methods. The greater penetration depth of neutrons enables the determination of macrostress gradients in bulk materials and microstress states in composites and multiphase alloys.

Microstresses give rise to peak shift, and if they vary from point to point, line broadening occurs in x-ray diffraction patterns as well. When microstresses range over dimensions of 0.1 to 100 nm they give rise only to x-ray broadening. The x-ray technique is based on the measurement of lattice spacing $d_{\phi\psi}$ at various tilts ϕ and ψ to the x-ray beam (Fig. 15.29). In a polycrystalline coating, only those grains properly oriented to diffract at each tilt contribute to the diffraction pattern. This selectivity implies that the elastic constants relating the measured strains (or change in lattice spacing) to the stresses will vary with a particular set of planes (hkl) chosen for measurements.

From the theory of elasticity, the strains $\epsilon_{\phi\psi}$ measured along the L_3 axis for a biaxial state of stress (see Fig. 15.29) is given by

$$\epsilon_{\phi\psi} = \frac{d_{\phi\psi} - d_0}{d_0} = \frac{1}{2} S_2(\text{hkl})(\sigma_{11} \cos^2 \phi + \sigma_{22} \sin^2 \phi) \sin^2 \psi + S_1(\text{hkl})(\sigma_{11} + \sigma_{22}) \quad (15.21)$$

where $d_{\phi\psi}$ is the lattice planar spacing of the crystalline planes which lie perpen-

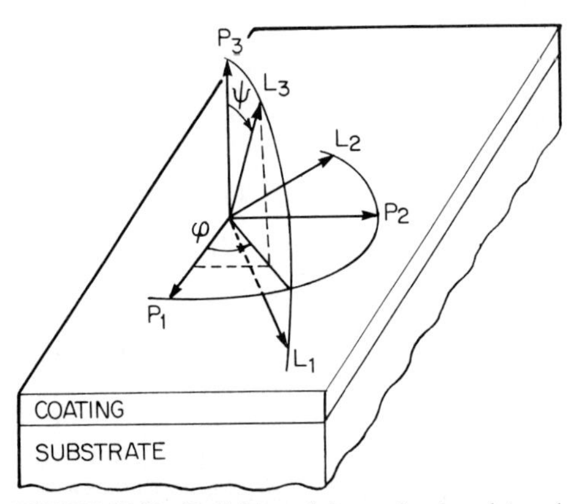

FIGURE 15.29 Definition of the angles ϕ and ψ and orientation of the laboratory system L_i with respect to the sample systems P_i for measurement of residual stresses by x-ray diffraction method. The measurement direction is L_3 or the diffracting planes are normal to L_3. [*Adapted from Dölle (1979).*]

dicular to the ψ direction in the stressed situation; d_0 is the lattice planar spacing in the unstressed situation; S_1(hkl) = $[-(\nu/E)]$ is the x-ray elastic constant for a particular (hkl) plane; ½ S_2 (hkl) = $[(1 + \nu)/E]$ is the other elastic constant for a particular (hkl) plane; ϵ is Young's modulus of the coating; ν is Poisson's ratio of the coating; and σ_{11}, σ_{22} are principal normal stresses.

The variations in lattice strain $\epsilon_{\phi\psi}$ with $\sin^2 \psi$, according to Eq. (15.21), should give a straight line whose slope gives $\sigma_\phi = \sigma_{11} \cos^2 \phi + \sigma_{22} \sin^2 \phi$, the required stress component. However, the existence of a stress gradient normal to the coating surface gives rise to the curvature effects, and the existence of shear stresses in the coated specimen give rise to ψ splitting effects. The details of further analysis of these effects are reported by Rickerby et al. (1987, 1989). Equation (15.21) also suggests that the lattice parameter d(hkl) is a function of internal stress and the peak displacements vary with the crystallographic directions since both S_1 and ½ S_2 vary with (hkl). This change in S_1 and ½ S_2 with (hkl) will have a direct effect on the peak position in the diffraction pattern. Perry (1989) made an attempt to correlate the lattice parameters, measured by conventional x-ray diffraction (2 θ method) to the x-ray elastic constants, i.e., S_1 and ½ S_2 and calculated the stresses in TiN coatings.

The line broadening in the x-ray-diffraction pattern of a stressed coating results because of lattice strain and crystallite size. These effects can be separated by the integral breadth method which depends on the fact that strain is a function of the order of reflection (hkl) while the crystallite size is not.

Rickerby et al. (1987, 1989) have used this $\sin^2 \psi$ method for the residual stress analysis of TiN coatings deposited by PVD on various substrates, such as stainless steel, tool steel, and cemented carbide, and correlated the influence of process parameters, such as substrate biasing, deposition rate, and coating microstructure to the type and level of residual strain and stress in the coating.

15.1.7.3 Miscellaneous Methods. Raman spectroscopy provides an optical method for determining residual stresses in the near-surface region of crystalline materials. Stresses in the crystal lattice cause a shift in the peak spectral frequency of the light scattered from an incident laser beam. The direction of the spectral peak shift is an indication of the sign of the residual stress (i.e., tensile or compressive), and its magnitude scales directly with the magnitude of the stress (Bifano, 1988). Electromagnetic methods are based on the stress dependence of electrical conductivity and magnetic properties of ferromagnetic materials. Ultrasonic methods depend on the fact that propagation of ultrasonic waves in a solid medium is influenced by the state of strain in the medium (Bhushan, 1990).

15.1.8 Techniques for Adhesion Measurements

Adhesion describes the sticking together of two materials (Good, 1976). Adhesion strength, in a practical sense, is the stress required to remove a coating from a substrate. The value of the adhesive strength can lie somewhere between the van der Waals forces at one extreme and the cohesive strength of either component at the other extreme, i.e., typically between the limits of about 1 MPa and 3 GPa. One important factor that governs such functional characteristics as durability and corrosion resistance of thin coatings (<1 μm), thick coatings (>1 μm), and bulk coatings (>25 μm) is the adhesion between the coating and the substrate.

An ideal adhesion test should be (1) quantitative, (2) reproducible, (3) not very time consuming, (4) easily adaptable to routine testing, (5) relatively simple to perform, (6) nondestructive, (7) independent of the coating thickness, (8) applicable to all combinations of coatings and substrates, and (9) valid over a wide range of sample sizes. Indeed, no method has been found so far that can be adapted universally. It is important, therefore, to select the most suitable method for each coating material and the application under investigation.

Many techniques have been developed to measure coating adhesion (Benjamin and Weaver, 1960; Bhushan, 1987, 1990; Campbell, 1970; Chalker et al., 1990; Chapman, 1971; Mittal, 1976, 1978), and the various techniques can be divided in three categories: nucleation methods, mechanical methods, and miscellaneous methods (Fig. 15.30). Studies of the nucleation and initial growth of the coating allow determination of values of the adhesion energy under certain circumstances. These methods are fairly complicated and find very limited application. However, the nucleation methods are useful in understanding coating formation. Unlike mechanical tests, nucleation studies yield the adsorption energy of a single adatom, but the validity of using such adsorption energy values to calculate adhesion energy per unit area is questionable because bond energy in a continuous coating will not in general be the same as that for a single atom. Mechanical methods are more obvious and direct and are used most commonly. Most techniques can be used where the adhesion is low, i.e., less than 100 MPa. [Most hard coatings deposited by vapor deposition would have adhesive strength in excess of 100 MPa (Bhushan, 1979, 1980).] Ultracentrifugal and ultrasonic methods can be used for relatively thick coatings with high adhesion values; however, these require specialized equipment and have not become popular. The Pulsed-laser or electron-beam method also can be used for coatings with high adhesion values, but again, they require large, expensive, and complicated equipment. Bend, indentation, and scratch methods are most common for thin or thick coatings with high adhesion. The bend test is a qualitative technique. The scratch method is the most commonly used technique for testing thin, hard, and well-adhered coatings. Various techniques to measure adhesion will be described in this section.

15.1.8.1 Nucleation Methods. On an atomic scale, the removal of a coating involves breaking the bonds between individual atoms of the coating and the substrate; thus macroscopic adhesion can be considered as the sum of individual atomic forces. In principle, therefore, it should be possible to relate the adsorption energy E_a of a single atom on the substrate to the total adhesion of a coating.

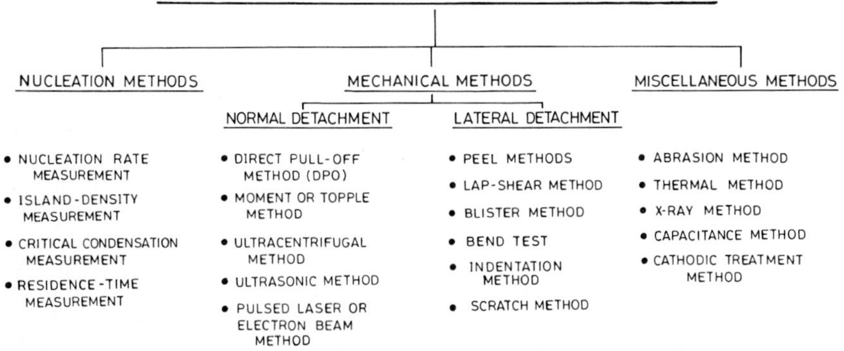

FIGURE 15.30 Classification of coating-to-substrate adhesion measurement techniques.

The adsorption energy of a single atom also helps to govern the behavior of condensing atoms on a surface. It controls the lifetime before an adsorbed atom reevaporates and thus it controls the nucleation of the coating on the substrate. Electron microscopic observations of the nucleation and initial stages of growth of a coating can therefore give measurements from which E_a can be derived (Campbell, 1970).

The nucleation methods are based on the measurement of (1) nucleation rate, (2) island density, (3) critical condensation, and (4) residence time of the atoms comprising the coating. Nucleation density and hence rate measurements are made by using an electron microscope. The concept behind critical condensation measurements is that if the rate of arrival of atoms is very low or if the substrate temperature is high, the lifetime of single atoms on the substrate is too short for collisions with other atoms to occur and thus for stable nuclei to be formed. Experimentally, this means that the incidence rate must be raised at a constant substrate temperature or the substrate temperature must be lowered at a constant incident rate until condensation is observed. The concept behind residence-time measurements is that for materials in which the atoms on the substrate are at such a temperature that stable cluster formation is not immediate, it is possible to measure the mean adsorption lifetime directly and so arrive at a figure for the adsorption energy (Campbell, 1970).

Nucleation methods of measuring adsorption energy are fairly complicated and generally of limited applicability. It is necessary to have access to an electron microscope in order to count island densities, and this, in turn, means that it must be possible to remove the islands from the substrate without moving them relative to one another. Even if this can be done, it may not always be possible to make enough observations to separate the cohesion-energy term from the adsorption-energy term.

Benjamin and Weaver (1959) calculated the adsorption energies of aluminum, silver, and cadmium coatings on glass substrates, as well as of aluminum and silver coatings on single-crystal cleavage surfaces of sodium chloride and potasium chloride. They concluded that the adhesion energy is due to physical adsorption and can be explained in terms of van der Waals forces alone. Chapman (1971) has compared mechanical stripping measurements with nucleation results for gold on glass. He found that with the lowest stripping speeds (the value of adhesion depends on the speed of stripping) used (50 nm s^{-1}), the average adhesion energy was double that obtained by nucleation methods, but he mentioned that it is necessary to work at even lower stripping speeds to obtain better agreement.

*15.1.8.2 **Mechanical Methods.*** All mechanical methods use some means of removing the coating from the substrate. These methods can be broadly classified into two categories depending on the mode of detachment of the coating: (1) methods involving detachment normal to the interface and (2) methods involving lateral detachment of the coating from the interface.

Methods Involving Detachment Normal to the Interface (Tensile Methods). In tensile methods, a force of increasing magnitude in a direction normal to the interface is applied to the coating-substrate interface until the interface is disrupted. Either the maximum force applied is measured, or some other criterion, such as area of detachment, is used to designate adhesion. The tensile methods include the direct pull-off (DPO) method, the moment or topple method, the ultracentrifugal method, the ultrasonic method, and the pulsed-laser or electron-beam method. The direct pull-off, moment or topple, and lap-shear (to be described later) methods suffer from the drawback that some sort of adhesive or solder is required to attach some device to pull the coating, and this is sometimes highly undesirable. Such tests are limited by the strength of the adhesive. In the ultracentrifugal, ultrasonic, and pulsed-laser or electron-beam methods, no adhesive or solder is used. However, these methods required specialized equipment, and the ultracentrifugal and ultrasonic methods are generally used for very thick coatings (>100 μm).

(a) Direct Pull-Off (DPO) Method: The basic principle of this method is to attach some kind of pulling device (such as flat-headed brass pins) to the back of the coating by means of solder or high-strength adhesives and then pull the coating in a direction normal to the interface using a tensile tester (e.g., Instron) (Fig. 15.31) (Bhushan, 1987). The direct pull-off method suffers from the following difficulties: (1) alignment must be perfect to ensure uniform loading across the interface; (2) most of the time these simple tensile tests involve a complex mixture of tensile and shear forces, which renders the interpretation of the results more difficult; (3) such tests are limited by the strengths of available adhesives (maximum strength of about 100 MPa) or solders; and (4) there is always the possibility that adhesive or solvent will penetrate into and affect the coating-substrate interface and that stresses produced during setting of the cement or adhesive will have an effect. All these factors affect the final adhesion strength measured.

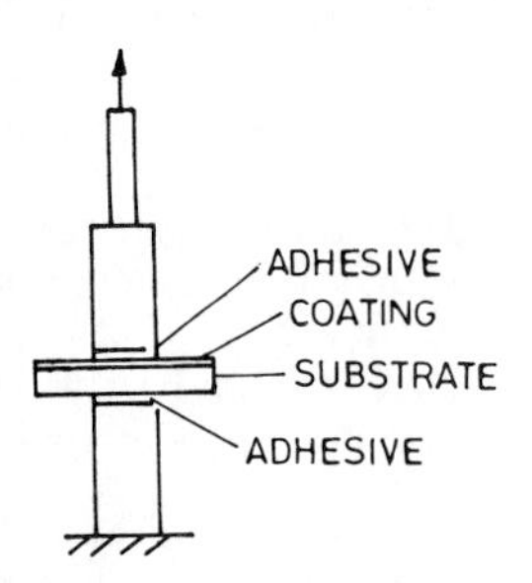

FIGURE 15.31 Schematic illustration of direct pull-off method for adhesion measurements.

Belser and Hicklin (1956) and Jacobsson and Krause (1973) measured the adhesion of evaporated and sputtered metal coatings on glass, quartz, and glazed-tile surfaces. They soldered flat-headed brass pins to the metal coating using Cerroseal (50% tin, 50% indium; mp 118°C) or tin-lead solder, and then they measured the force needed to pull the coating covered by the pinhead perpendicularly from the substrate surface with a spring balance and winch system. The area of the pinhead was about 5 mm^2.

(b) Moment or Topple Method: This is a slight variation of the direct pull-off method. Instead of applying the force vertical to the interface, it is applied in a direction horizontal to a rod bonded to the coating. The moment of the force required to break the coating from the substrate is a measure of adhesion (Fig. 15.32) (Bhushan, 1987). This method has the following advantages over the direct pull-off method: (1) it offers less substrate distortion because there is no overall force normal to the plane of the substrate, and (2) the apparatus does not require critical alignment, as in the case of normal pull.

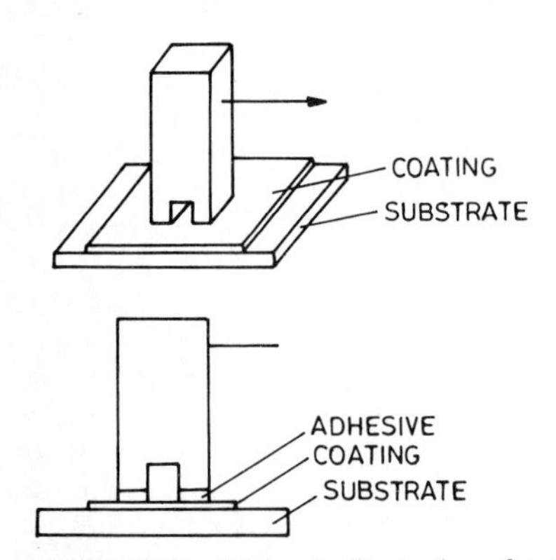

FIGURE 15.32 Schematic illustration of moment or topple method for adhesion measurements.

In this apparatus, one end of the rod is cut to form two short legs, and adhesive strength is calculated assuming pure tension under one leg and compression below the other leg. An adhesive, such as Eastman 910, can be used. Butler (1970) has described a simple coating adhesion comparator based on this principle.

(c) ULTRACENTRIFUGAL METHOD: As the name suggests, the specimen is in the form of a rotor that is spun at extremely high speeds to provide the requisite centrifugal force, as shown schematically in Figure 15.33 (Campbell, 1970). At a critical speed of rotation, the coating can no longer withstand the centrifugal stress and becomes detached. Beam (1956) suspended the rotor in a vacuum and applied a rotating magnetic field to spin it. Dancy (1970) designed a solid-state ultracentrifuge with angular speeds in excess of 80,000 rps for small rotors with diameters on the order of 2.5 mm. Coatings on nonmagnetic substrates also can be investigated with the magnetic support system. The nonmagnetic specimen must be rendered magnetic, however, and this can be accomplished by including a ferromagnetic rod that is aligned along the axis of rotation or by adding a ferromagnetic disk that is cemented to one end of the test rotor.

In the ultracentrifugal system, the forces on the coating are given by

$$\sigma_h + \frac{\sigma_a r}{t} = \frac{4\pi^2 n^2 r^2 \rho}{g} \tag{15.22}$$

where σ_h = hoop stress (in g cm^{-2})
σ_a = adhesion stress (in g cm^{-2})
r = radius of the rotor (in cm)
n = number of revolutions per second
ρ = density of the coating (in g cm^{-3})
t = thickness of the coating (in cm)
g = acceleration due to gravity (981 cm s^{-2})

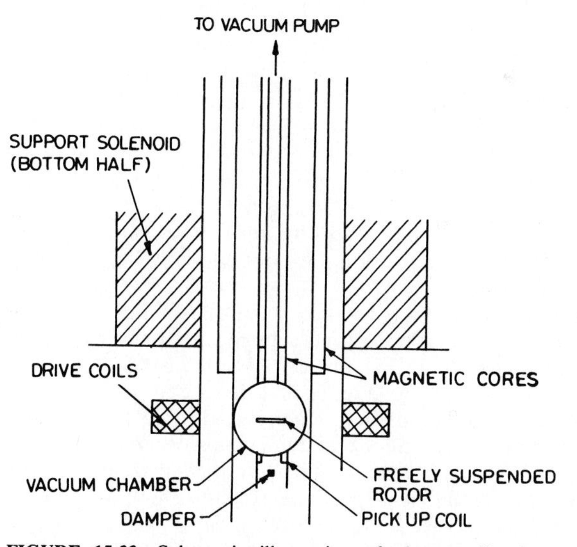

FIGURE 15.33 Schematic illustration of ultracentrifugal arrangement for adhesion measurements. [*Adapted from Maissel and Glang (1970).*]

The hoop stress σ_h can be eliminated by making slits in the coating that are parallel to the axis of rotation. When $\sigma_h = 0$, the preceding equation reduces to

$$\sigma_a = \frac{4\pi^2 n^2 r \rho t}{g} \tag{15.23}$$

From this equation, it is clear that the rotor size and rotational speed will be dictated by the degree of adhesion to be measured. For example, for $n = 30{,}000$ rps, the maximum measurable adhesion using a 0.47-cm-diameter rotor with a 125-μm-thick coating having a density of 8.3 g cm^{-3} would be 90 MPa.

(d) ULTRASONIC METHOD: Usually, this method is used for relatively thick coatings. Here, the normal force is supplied by the inertia of the coating as it is subjected to rapid reversals of motion at ultrasonic frequencies. Moses and Witt (1949) studied the adhesion of resin coatings on the ends of metal cylinders. The apparatus utilizes an electrodynamic system that produces longitudinal ultrasonic vibrations in a metal cylinder. A coating attached to the free end of the vibrating cylinder separates from the metal when the force due to acceleration exceeds the adhesion force at the interface. The accelerating force is determined by the frequency and amplitude of the vibrations and by the mass and area of the coating. The amplitude of the rod or cylinder vibrations is calculated from the input voltage. The maximum acceleration is given by $4\pi^2 f_0^2 a$, where f_0 is the fundamental frequency of vibration, given by $(l/2)\sqrt{E/\rho}$, in which E is Young's modulus, l is the rod length, and ρ is the density. By knowing the dimensions and density of the coating, one can determine the force at each reversal. At 10 MHz, accelerations on the order of 10^9 g can be obtained.

Moses (1949) measured the adhesion of polystyrene coatings deposited from toluene solution onto a dural cylinder at a frequency of 23.6 kHz, and the adhesion was on the order of 410 kPa.

(e) PULSED-LASER OR ELECTRON-BEAM METHOD: This method, also called the *shock-wave method*, is a potentially powerful method of stressing the coating-substrate interface. A compressive-wave test is initiated in the substrate such that the wave propagates perpendicular to the surface. The wave is next transferred by reflection into a tensile wave, which then stresses the interface to be tested. If the peak stress value of the tensile wave exceeds the adhesion, then the coating is lost when the reflected wave traverses the interface (Anderson and Goodman, 1972; Vossen, 1978). The important parameter in this test is the ratio of wavelength to coating thickness; this must be greater than about 0.5 or destructive interference on reflection does not allow the maximum tensile stress to be attained at the interface.

In practice, the coating is patterned into dots smaller in diameter than the incident beam on the substrate. A high-energy laser pulse or electron-beam pulse impinges on the backside of the substrate (Fig. 15.34). In the case of transparent substrates, a thick, absorbing layer is first deposited on the backside of the substrate. The incident laser radiation is converted rapidly to thermal energy in the absorbing substrate or coating, and the explosive evaporation of the material sends a compressive shock wave through the substrate toward the coating-substrate interface; then the substrate snaps back under tension, and this leads to removal of the coating. By increasing the incident laser power, one can find a threshold at which the coating is torn loose from the substrate. This threshold can be related directly to the bond strength of the coating-substrate interface (Vossen, 1978).

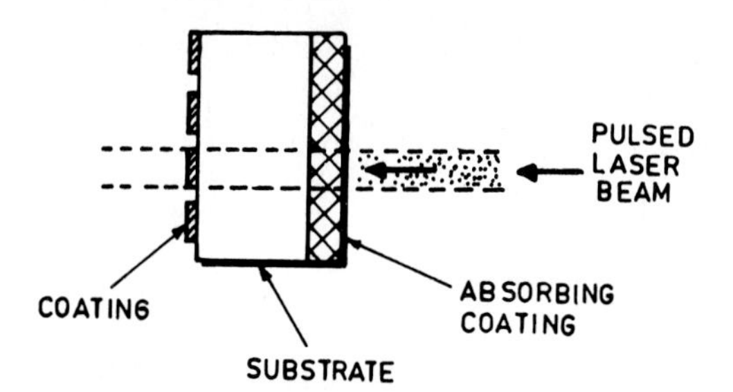

FIGURE 15.34 Schematic illustration of pulsed-laser method. The laser beam typically used is about 10 times larger than the dot so that a plane shock wave is approximated. [*Adapted from Vossen (1978).*]

The main advantage of this method is the ability to detach a coating in a controlled manner without disturbing the coating surface.

Methods Based on the Application of Lateral Stresses for Detachment (Shear Methods). In shear methods, lateral stresses are applied so as to cause detachment of the coating from the substrate. The shear methods include peel methods, the lap-shear method, the blister method, the bend test, the indentation-method, and the scratch method. Peel methods are generally used for coatings with relatively low adhesion; however, their speed and simplicity ensure that they will continue to be used in the rejection of poor coatings. The lap-shear method simulates the tribological conditions. However, an adhesive is used in this test, and therefore, this technique suffers from same weaknesses as the direct pull-off and moment or topple methods. The blister method is also used for coatings with relatively low adhesion (such as polymeric coatings). The bend test, the indentation method, and the scratch method measure coating performance, which depends on other properties such as hardness and fracture toughness in addition to adhesion. The bend test is a qualitative technique commonly used to screen thin, hard, and well-adhered coatings. The scratch method is the most commonly used semiquantitative technique for testing thin, hard, and well-adhered coatings.

(a) PEEL METHODS: In these methods, the coating can be peeled from the substrate in two ways: (1) by applying some sort of backing material to the coating and then holding onto the backing and (2) by directly holding onto the coating.

BY PULLING THE BACKING MATERIAL One of the oldest and most widely used peel methods is the Scotch-tape method, shown in Fig. 15.35(*a*). Using a strip from a fresh roll of cellophane-backed adhesive tape, a strip is firmly pressed onto the coating to be tested. If the coating-substrate adhesion is less than the adhesion between the tape and the coating, the coating will be removed by stripping the tape. Whether the coating comes off or not is then taken as a criterion for poor or good adhesion. Briskly lifting the tape applies a force of about 1 ± 0.1 N mm^{-1} (100 ± 10 g mm^{-1}) of width. The threshold level of the test is too low to allow it to serve as a generally useful test for coatings with adhesion strengths greater than 0.5 MPa, but its speed and simplicity ensure that it will continue to be used in the rejection of poorly adherent coatings. This test is commonly used to measure the adhesion of paints and air-sprayed solid-lubricant coatings such as plastics, molybdenum disulfide, and graphite.

The qualitative Scotch-tape method has been improved to get quantitative information. In this variation, adhesive tape is attached to the back of a coating and

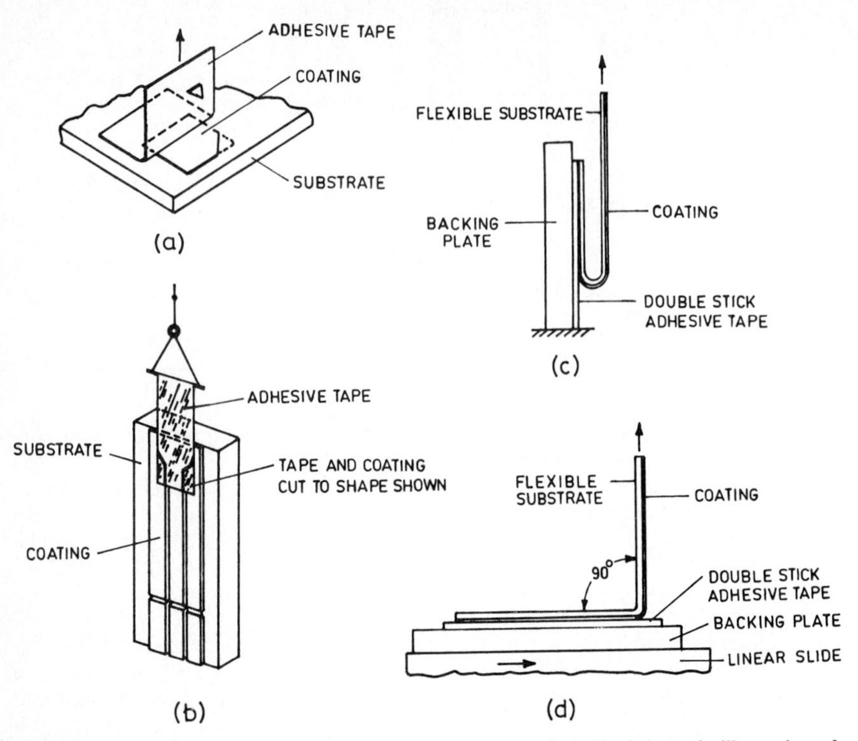

FIGURE 15.35 (*a*) Schematic illustration of Scotch-tape method. (*b*) Schematic illustration of quantitative version of Scotch-tape method. (*c*) Schematic illustration of 180° peel method for flexible, coated substrates. (*d*) Schematic illustration of 90° peel method for flexible, coated substrates.

the tape and the coating are scribed with a razor blade, as shown in Fig. 15.35(*b*). The tape is then attached by means of a thin trapeze to a winding string. As the tape is pulled, the tape and coating are removed from the substrate and the load required to remove the coating is measured.

If the coated substrate is very flexible (e.g., 25- to 40-μm-thick particulate magnetic tapes), the coated substrate is bonded to a smooth backing plate using double-sided adhesive tape. The coated substrate is pulled away from the backing plate at an angle of 90° or 180°, as shown in Fig. 15.35(*c*) and (*d*). In a 90° peel test, the sample is mounted on a linear slide so that the peel angle remains at 90°. Then the load required to remove the coating from the substrate is measured (Bhushan, 1987).

BY DIRECTLY HOLDING ON TO THE COATING If a coating is removable, it can be peeled, as shown in Fig. 15.36. The variation of the peel test involves peeling a specified width of coating at a specified angle θ, 90° being the most common. To make a useful measurement, the coating must be completely removed from the substrate, which limits the applicability of this technique to interfaces that exhibit relatively poor adhesion.

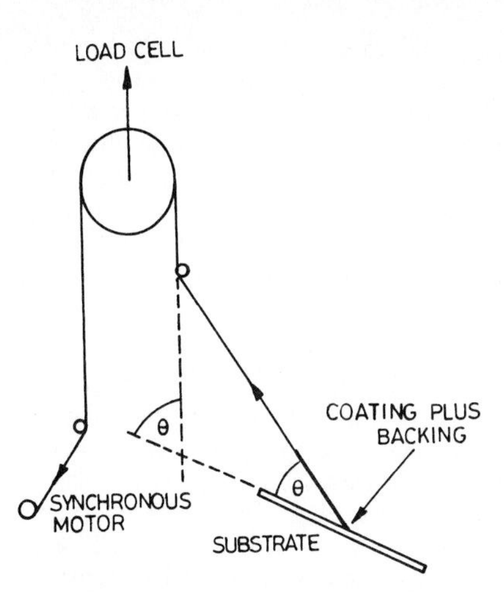

FIGURE 15.36 Schematic illustration of peel method by holding onto the coating for adhesion measurement.

(b) LAP-SHEAR METHOD: The lap-shear method is not as popular as the direct pull-off method, but in tribological applications, lap shear (or tangential shear) simulates the actual application conditions of coatings better. Therefore, the results using this test may be more meaningful. A piece of a glass microscope slide is bonded with an adhesive, such as Eastman 910, to the upper surface of the coating, which has been deposited on the desired substrate. The coating and the substrate are then clamped rigidly in position, and a shear force is applied parallel to the coating by weights attached to a thin stainless steel wire passing over a low-friction pulley (Fig. 15.37). Shear-stress values are calculated from the force required to break the bond between the coating and the substrate divided by the area of contact of the adhesive (Lin, 1971). An adhesive is used in this test, as in direct pull-off and moment or topple methods, and thus this technique suffers from same weaknesses as these other methods. The maximum adhesive strength that can be measured with the lap-shear method is limited by the strength of the adhesive, which is usually less than 100 MPa.

Dini et al. (1972) and Dini and Johnson (1974) have described a ring-shear test that measures quantitatively the adhesion of electrodeposits. In this method, a cylindrical rod is coated with separate rings of electrodeposits of predetermined width. This rod is forced through a hardened-steel die that has a hole diameter larger than the diameter of the rod but smaller than that of the rod plus the coating. The coating is detached, and the adhesion strength σ_a is determined by the formula $\sigma_a = W/\pi dt$, where d is the diameter of the rod, t is the width of the deposit, and W is the force required for detaching.

(c) BLISTER METHOD: In this method, a fluid-gas (Hoffmann and Georgoussis, 1959) or liquid (Dannenberg, 1961) is injected beneath the coating at the coating-substrate interface, and the hydrostatic pressure is increased until the coating begins to detach (peel away) from the interface. The beginning of the peel is usually indicated by discontinuity on the pressure-volume plot. Most of the work using this technique has been done on relatively thick (25 μm) paint coatings. Since the work required to detach metallic coatings would be considerably more than that for polymeric coatings, only less adherent metallic coatings may be amenable to this test.

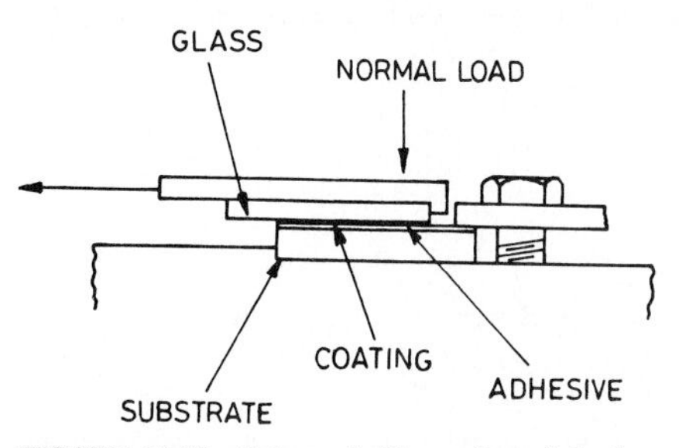

FIGURE 15.37 Schematic illustration of the lap-shear method for adhesion measurement. [*Adapted from Lin (1971).*]

(d) BEND TEST: In a bend test, the coating is deposited on a flexible substrate (50 to 500 μm thick), and the coated substrate is bent on itself (generally 180°) with the coating on the outside of the bend. The bent sample is then photographed in a direction normal to the outside of the bend. If the coating is poorly adhered, it will spall. Hard coatings also may craze because of material ductility rather than poor adhesion. This is a severe test and is used for coatings that have adhesion strengths high enough (>100 MPa) that they cannot be measured by some of the methods discussed earlier. This technique was developed by Bhushan (1979, 1980) for qualitatively screening sputtered hard coatings that have adhesion strengths in excess of 70 MPa.

(e) INDENTATION METHOD: In this test method, the coating sample is indented at various loads. At low loads, the coating deforms with the substrate. However, if the load is sufficiently high, a lateral crack is initiated and propagated along the coating-substrate interface. The lateral crack length increases with the indentation load. The minimum load at which the coating fracture is observed is called the *critical load* and is employed as the measure of coating adhesion (Fig. 15.38). For relatively thick films, the indentation can be created with a Rockwell hardness tester with a Brale diamond indenter (Mehrotra and Quinto, 1985; Mehrotra et al., 1984) or a Vickers pyramidal indenter (Chiang et al., 1981). However, for extremely thin films, the nanoindenters described earlier should be used.

It should be noted that the measured critical load W_{cr} is a function of hardness and fracture toughness in addition to the adhesion of coatings. Chiang et al. (1981) have related the measured crack length during indentation and the applied load to the fracture toughness of the substrate-coating interface. A semianalytical relationship is derived between the measured crack length c and the applied load W:

$$c = \alpha\left(1 - \frac{W_{cr}}{W}\right)^{1/2} W^{1/4} \tag{15.24}$$

where

$$\alpha^2 = \frac{\alpha_1 t_c^{3/2} H^{1/2}}{(K_{Ic})_{\text{interface}}}$$

α_1 is a numerical constant, t_c is the coating thickness, H is the mean hardness, and $(K_{Ic})_{\text{interface}}$ is the fracture toughness of the substrate-coating interface.

(f) SCRATCH METHOD: Scratching the surface with a fingernail or a knife is probably one of the oldest methods for determining the adhesion of paints and other coatings. In the modern version of this method, a smoothly rounded chrome-steel stylus with a tungsten carbide or Rockwell C diamond tip (in the form of a 120° cone with a hemispherical tip of 0.2 mm radius) is drawn across the

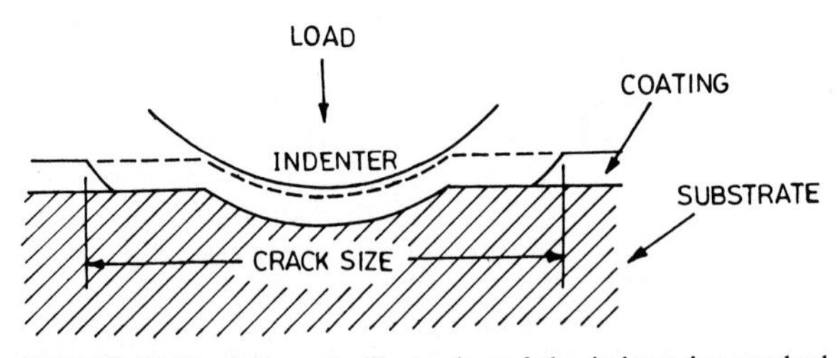

FIGURE 15.38 Schematic illustration of the indentation method for adhesion measurement.

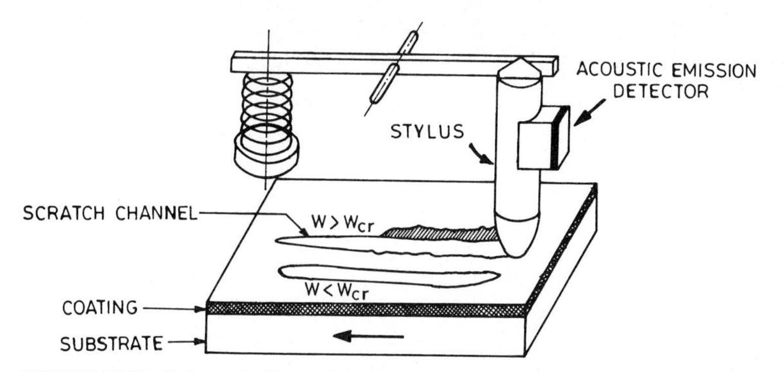

FIGURE 15.39 Schematic illustration of the scratch method for adhesion measurement.

coating surface, and a vertical load is applied to the point and is gradually increased until the coating is completely removed, as shown in Fig. 15.39. The minimum or critical load at which the coating is detached or completely removed is used as a measure of adhesion (Benjamin and Weaver, 1960; Heavens, 1950; Laugier, 1981; Perry, 1981, 1983; Valli et al., 1985).

Figure 15.40 shows the geometry of the scratch. Substrate hardness H is given by

$$H = \frac{W_{cr}}{\pi a^2} \tag{15.25}$$

and adhesion strength τ is given by (Benjamin and Weaver, 1960)

$$\begin{aligned} \tau &= H \tan \theta \\ &= \frac{W_{cr}}{\pi a^2} \left[\frac{a}{(R^2 - a^2)^{1/2}} \right] \end{aligned} \tag{15.26a}$$

$$\text{or } \tau = \frac{W_{cr}}{\pi a R} \quad \text{if } R \gg a \tag{15.26b}$$

where W_{cr} is the critical normal load, a is the contact radius, and R is the stylus radius. The preceding equation (15.26*a*) is generally not used. Instead, adhesion is measured on the basis of critical-load values.

An accurate determination of critical load W_{cr} sometimes is difficult. Several techniques, such as (1) microscopic observation (optical or scanning electron microscope) during the test, (2) chemical analysis of the bottom of the scratch channel (with electron microprobes) and (3) acoustic emission, have been used to obtain the critical load. The acoustic-emission technique has been reported to be very sensitive in determining critical load. Acoustic emission starts to increase as soon as cracks begin to form perpendicular to the direction of the moving stylus.

Several scratch testers are available commercially, such as the Taber shear/scratch tester model 502 with a no. 139-58 diamond cutting tool (manufactured by Teledyne Taber, North Tonawanda, N.Y.) for thick films and a Revetest automatic scratch tester (manufactured by Centre Suisse d'Electronique et de Microtechnique, CH-2007, Neuchatel, Switzerland) for thin films (Steinmann et

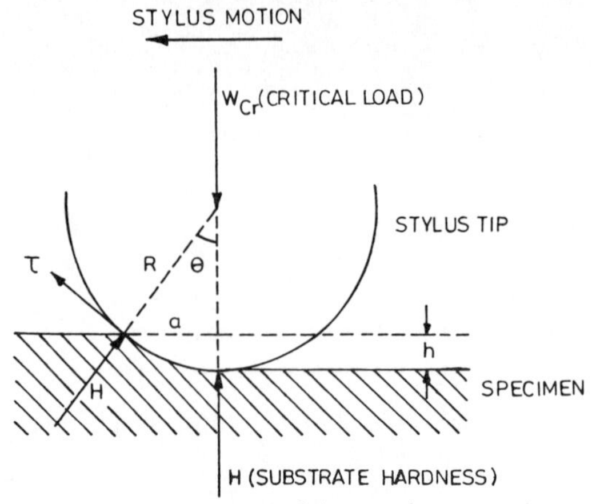

FIGURE 15.40 Geometry of the scratch.

al., 1984, 1987). The latter unit uses an acoustic transducer in the form of an accelerometer mounted just above the diamond stylus. It registers signals emitted in the range of 0 to 50 kHz. The acoustic signal resulting from the coating fracture event identifies the critical load (Je et al., 1986; Mehrotra and Quinto, 1985; Pan and Greene, 1982; Perry, 1981, 1983; Sekler et al., 1988). The application of an acoustic transducer obviates the use of microscopy (although it is still extremely useful to correlate the characteristics of the acoustic signal with microscopic observations of the scratch channel). In some instruments, friction force is also measured during the scratching.

As an example of a scratch test, optical micrographs of the typical failure mode of a sputtered TiN coating on a steel substrate are presented in Fig. 15.41*a*. The acoustic emission curves showing the maximum and minimum signals are plotted as a function of stylus load in Fig. 15.41(*b*). As the stylus load was increased to 7.8 N, the coating started to crack perpendicular to the steel surface at the edge of the channel. The onset of coating loss was accompanied by a sudden increase in acoustic emmission at loads greater than 7.8 N [region *b* in Fig. 15.41(*b*)]. The very high normal load can cause the detached coating particles either to be completely pressed back into the substrate or be partially removed, thus leaving a smooth surface within the channel.

15.1.8.3 ***Miscellaneous Methods.*** Under this heading we describe miscellaneous testing methods that have been used. However, these have not proven to be practical. They require specialized equipment, and some are only qualitative.

Abrasion Method. The abrasion resistance of coatings has received much attention, and the literature can be traced back to 1930, but in such studies, the main objective was to determine the durability of the coating. However, careful consideration will reveal that the abrasion resistance not only depends on the hardness of the coating but also is influenced by the adhesion of the coating to the substrate. Holland (1956) determined abrasion resistance by rubbing the surface with emery-loaded rubber.

The abrasion method undoubtedly involves burnishing as well as stripping action, and these additional processes can certainly affect the values of adhesion

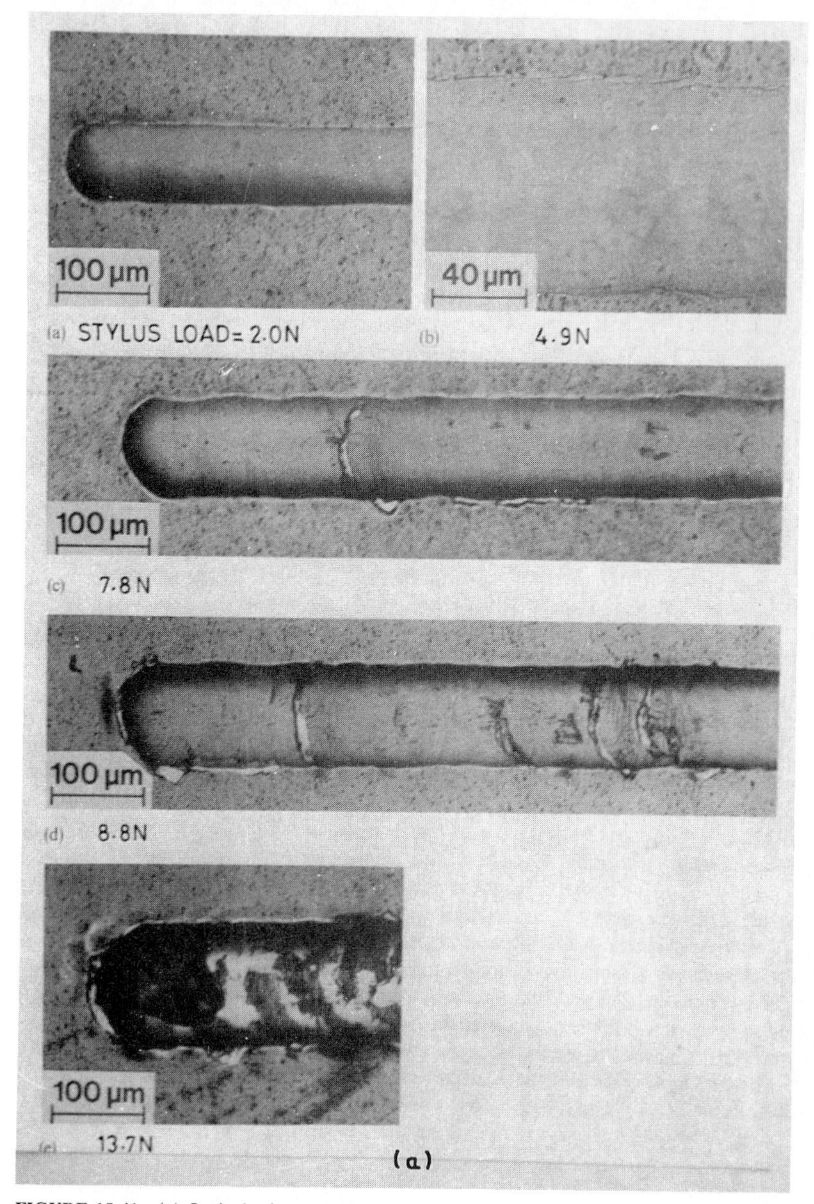

FIGURE 15.41 (*a*) Optical micrographs of channels produced during scribing from right to left under various stylus loads in Crofer 1700 coated with sputtered TiN about 1.5 μm thick. (*b*) Acoustic emission signal maxima and minima taken from the curves recorded within the scratch channel length between 0.5 and 2.5 mm. [*Adapted from Je et al. (1986).*]

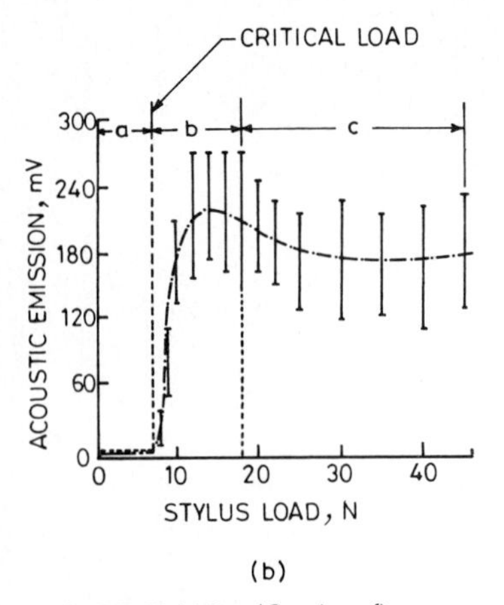

FIGURE 15.41(*b*) (*Continued*)

obtained. This method has not found wide use as a practical method of obtaining quantitative adhesion values.

Thermal Method. The principle of the thermal method is as follows: When a coating is chemically dissolved from its substrate, the energy liberated is equal to the heat of solution of the coating minus the energy of adhesion between the coating and the substrate. A microcalorimeter (Calvet and Prat, 1963) is used to measure small heat changes. Since the liberated heat is very small, the microcalorimeter in its present state may not offer viable results. Chapman (1971) employed this technique in the study of adhesion of vacuum-deposited metallic films on an NaCl substrate.

X-Ray Method. This is one of the nondestructive methods of studying adhesion at the coating-substrate interface. This method is based on the observation that the strains at the coating-substrate interface and the poor adhesion of the coating modify the diffraction contrast of the substrate and thus provide qualitative information on strain and adhesion. This method can be used to compare the adhesions of epitaxial coatings, but it is applicable only in certain situations and cannot be generalized.

Capacitance Method. In this method, an elastic electrode is placed on a coated metal panel, and the absolute capacitance of the electrode so formed is measured at very low and very high frequencies. The ratio of the differences in the capacitance to the capacitance at high frequency is claimed to give a coefficient that is characteristic of the adhesion of the coating. This coefficient is an inverse measure of the adhesion of the coating (Bullett and Prosser, 1972).

Cathodic Treatment Method. Davies and Whittaker (1967) have made mention of this relatively old technique. In a sense, this method is similar to the blister method, with the difference that here hydrogen is used as the fluid to cause blistering. The method is based on the following protocol: The coated part is made the cathode in an electochemical cell, and the hydrogen that evolves diffuses through the coating to collect at the interface and cause blistering. It is obvious that this method has limited applicability because (1) the substrate should be metal, (2) it is qualitative, and (3) specific interactions between the coating material and the hydrogen evolved will affect the results.

15.2 FRICTION AND WEAR TESTS

Accelerated friction and wear tests are conducted to rank the friction and wear resistance of materials in order to optimize materials selection or development for specific applications. After the surfacing materials have been ranked by accelerated friction and wear tests, the most promising candidates (typically from 1 to 3) should be tested in the actual machine under actual operating conditions (functional tests). In order to reduce test duration in functional tests, the tests can be conducted for times shorter than the end-of-life. By collecting friction and wear data at intermediate intervals, end-of-life can be predicted.

Accelerated friction and wear tests should accurately simulate the operating conditions to which the material pair will be subjected. If these tests are properly simulated, an acceleration factor between the simulated test and the functional test can be empirically determined so that the subsequent functional tests can be minimized, saving considerable test time. Standardization, repeatability, short testing time, and simple measuring and ranking techniques are desirable in these accelerated tests. This section describes the design methodology and typical test geometries for friction and wear tests.

15.2.1 Design Methodology

Design methodology of a friction and wear test consists of four basic elements: simulation, acceleration, specimen preparation, and, friction and wear measurements. Simulation is the most critical, but no other elements should be overlooked.

15.2.1.1 Simulation. Proper simulation ensures that the behavior experienced in the test is identical to that of the actual system. Given the complexity of wear and the current incomplete understanding of wear mechanisms, test development is subject to trial and error and is dependent on the capabilities of the developer. The starting point in simulation is the collection of available data on the actual system and the test system. A successful simulation requires the similarity between the functions of actual system and those of test system, i.e., similarity of inputs and outputs and of the functional input-output relations.

To obtain this similarity, first the mating materials, the lubricant, and the operating conditions of the test should be the same as the actual system requirements. Selection of test geometery is a critical factor in simulating wear conditions. Generally, in laboratory testing for sliding contacts, three types of contacts are employed: point contact (such as ball-on-disk), line contact (such as cylinder-on-disk), and conforming contact (such as flat-on-flat). Selection of the geometery depends on the geometery of the function to be simulated. Each of these contact geometeries has its advantages and disadvantages. Point-contact geometery eliminates alignment problems and allows wear to be studied from the initial stages of the test. However, the stress level changes as the mating surfaces wear out. Conforming-contact tests generally allow the mating parts to wear in to establish a uniform and stable contact geometry before taking data. As a result, it is difficult to identify wear-in phenomena, because there is no elaborate regular monitoring of wear behavior.

Other factors besides contact type that significantly influence the success of a simulation include type of motion, load, speed, and operating environment (contamination, lubrication, temperature, and humidity). The type of motion that exists in actual system is one of four basic types: sliding, rolling, spin, and impact. These motions can be simulated by performing wear tests under unidirectional, reciprocating, and oscillating (reciprocating with a high frequency and low amplitude) motions and combinations thereof. Load conditions are simulated by applying static or dynamic load by dead weight, spring, hydraulic means, or electromagnetic means.

The lubrication (or lack thereof), temperature, and humidity also considerably influence the friction and wear characteristics of certain materials. The ambient temperature and contact temperature determine the thermal state of the system and should be precisely simulated/controlled during the test. Measurements at the surface of a specimen can be made with thermocouples, thermistors, or infrared (IR) methods. Thermocouples are the cheapest and simplest to use, and IR methods are the most accurate but complicated.

15.2.1.2 ***Acceleration.*** Accelerated tests are extremely inexpensive and fast. However, if the acceleration is not done properly, the wear mechanism to be simulated may change. Accelerated wear is normally achieved by increasing load, speed, temperature, and continuous operation.

15.2.1.3 ***Specimen Preparation.*** Specimen preparation plays a key role in obtaining repeatable/reproducible results. However, specimen preparation may vary depending on the type of material tested. For metals, surface roughness, geometery of the specimen, microstructure, homogeneity, hardness, and the presence of surface layers must be controlled carefully for both the mating materials. Similar controls are necessary for the wear-causing medium. For instance, in an abrasive wear test, purity, particle size, particle shape, and moisture content of the sand must be controlled.

15.2.1.4 ***Friction and Wear Measurements.*** The coefficient of friction is calculated from the ratio of friction force to applied normal force. The stationary member of the material pair is mounted on a flexible member, and the frictional force (force required to restrain the stationary member) is measured using the strain gauges (known as *strain-gauge transducers*) or displacement gauges (based on capacitance or optical methods) (Bhushan, 1990). Under certain conditions, piezoelectric force transducers (mostly for dynamic measurements) are also used for friction-force measurements (Bhushan, 1980).

Common wear measurements are weight loss, volume loss or displacement scar width or depth, or other geometric measures and indirect measurements such as time required to wear through a coating or load required to cause severe wear or a change in surface finish. Scanning electron microscopy (SEM) of worn surfaces is commonly used to measure microscopic wear. Other less commonly used techniques include radioactive decay and the scanning tunneling microscopy (STM). The resolutions of several techniques are presented in Table 15.5 (Bhushan, 1990).

TABLE 15.5 Resolutions of Several Wear-Measurement Techniques

Measurement technique	Resolution
Weight loss	10–100 μg
Radioactive decay	~1 pg
Stylus profiler	25–50 nm
Microhardness indentation	25–50 nm
Optical profiler	0.5–2 nm
Scanning electron microscope	0.1 nm
Scanning tunneling microscope	0.02–0.05 nm

Weight-loss measurements are suitable for large amounts of wear. However, weight-loss measurements have two major limitations. First, wear is related primarily to the volume of material removed or displaced. Thus such methods may furnish different results if materials to be compared differ in density. Second, this measurement does not account for wear by material displacement; that is, a specimen may gain weight by transfer. Thus weight-loss measurements are valid only when density remains constant and transfer does not occur during the wear process. This technique is not sensitive enough in the case of thin wear-resistant coatings, where wear is very small.

A stylus or noncontact optical profiler and Vickers or Knoop microhardness indentation techniques are easy to use and are commonly used to measure depth of wear with a resolution of up to a fraction of a nanometer (Bhushan, 1990). An example of a wear-track profile obtained with a stylus profiler is shown in Fig. 15.42. Three-dimensional worn-surface profiles also can be obtained with fully automated profilers (Bhushan et al., 1988b; Sayles and Thomas, 1976). In the microhardness indentation technique, Vickers or Knoop indentations are made on the wear surface. By measuring the width of the indentations before and after wear test under a microscope, the wear depth can be calculated (Bhushan and Martin, 1988).

For measurements of microscopic wear, scanning electron microscopy of worn surfaces is commonly used (Bhushan, 1990). For wear on an atomic scale, scanning tunneling microscopy can be used (Dugger et al., 1990). Radioactive decay (also called *autoradiography*) is very sensitive, but it requires facilities to irradiate one of the members and to measure the changes in radiation (Bhushan et al., 1986).

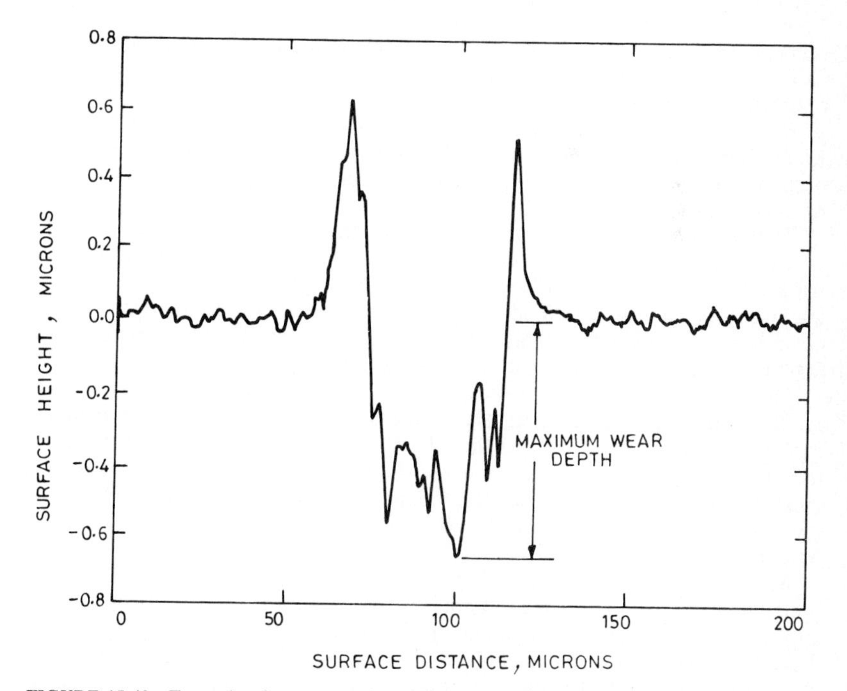

FIGURE 15.42 Example of a wear-track profile obtained by a stylus profiler.

15.2.2 Typical Test Geometries

15.2.2.1 ***Sliding Friction and Wear Tests.*** Many accelerated test apparatuses are commercially available that allow control of such factors as sample geometry, applied load, sliding velocity, ambient temperature, and humidity. Benzing et al. (1976), Bayer (1976, 1979, 1982), Bhushan (1987), Clauss (1972), Nicoll (1983), and Yust and Bayer (1988) have reviewed the various friction and wear testers that have been used in various tribological applications. The most commonly used interface geometries for screening materials, coatings, and surface treatments are shown in Fig. 15.43 and are compared in Table 15.6. Many of the test configurations are one-of-a-kind machines; others are available as commercial units from such companies as Falex–Le Valley Corporation, Downers Grove, Ill., Cameron Plint Tribology, Berkshire, U.K., Swansea Tribology Center, Swansea, U.K., and Optimol Instruments GmbH, Munchen, Germany. However, testing is not limited to such equipment; tests often are performed with replicas and facsimiles of actual devices. Brief descriptions of the typical test geometries, illustrated in Fig. 15.42 and compared in Table 15.6, are presented in the following subsections. Static or dynamic loading can be applied in any of the test geometries.

Pin-on-Disk (Face Loaded). In the pin-on-disk test apparatus [Fig. 15.43(*a*)], the pin is held stationary and the disk rotates or oscillates. The pin can be a nonrotating ball, a hemispherically tipped rider, a flat-ended cylinder, or even a rectangular parallelepiped. This test apparatus is probably the most commonly used during the development of materials for tribological applications.

Pin-on-Flat (Reciprocating). In the pin-on-flat test apparatus [Fig. 15.43(*b*)], a flat moves relative to a stationary pin in reciprocating motion, such as in a Bowden Leben apparatus. In some cases, the flat is stationary and the pin reciprocates. The pin can be a ball, a hemispherically tipped pin, or a flat-ended cylinder. By using a small oscillation amplitude at high frequency, fretting wear experiments can be conducted.

Pin-on-Cylinder (Edge Loaded). The pin-on-cylinder test apparatus [Fig. 15.43(*c*)] is similar to the pin-on-disk apparatus, except that loading of the pin is perpendicular to the axis of rotation or oscillation. The pin can be flat or hemispherically tipped.

Thrust Washer (Face Loaded). In the thrust-washer test apparatus [Fig. 15.43(*d*)], the flat surface of a washer (disk or cylinder) rotates or oscillates on the flat surface of a stationary washer, such as in the Alpha model LFW-3. The testers are face loaded because the load is applied parallel to the axis of rotation. The washers may be solid or annular. This configuration is most common for testing materials for low-stress applications, such as journal bearings and face seals.

Pin-into-Bushing (Edge Loaded). In the pin-into-bushing test apparatus [Fig. 15.43(*e*)], the axial force necessary to press an oversized pin into a bushing is measured, such as in the Alpha model LFW-4. The normal (axial) force acts in the radial direction and tends to expand the bushing; this radial force can be calculated from the material properties, the interference, and the change in the bushing's outer diameter. Dividing the axial force by the radial force gives the coefficient of friction.

Rectangular Flats on a Rotating Cylinder (Edge Loaded). In the rectangular-flats-on-a-rotating-cylinder test apparatus [Fig. 15.43(*f*)], two rectangular flats are loaded perpendicular to the axis of rotation or oscillation of the disk. This apparatus includes some of the most widely used configurations, such as the Hohman A-6 tester shown in Fig. 15.43(*e*). In the Alpha model LFW-1 or

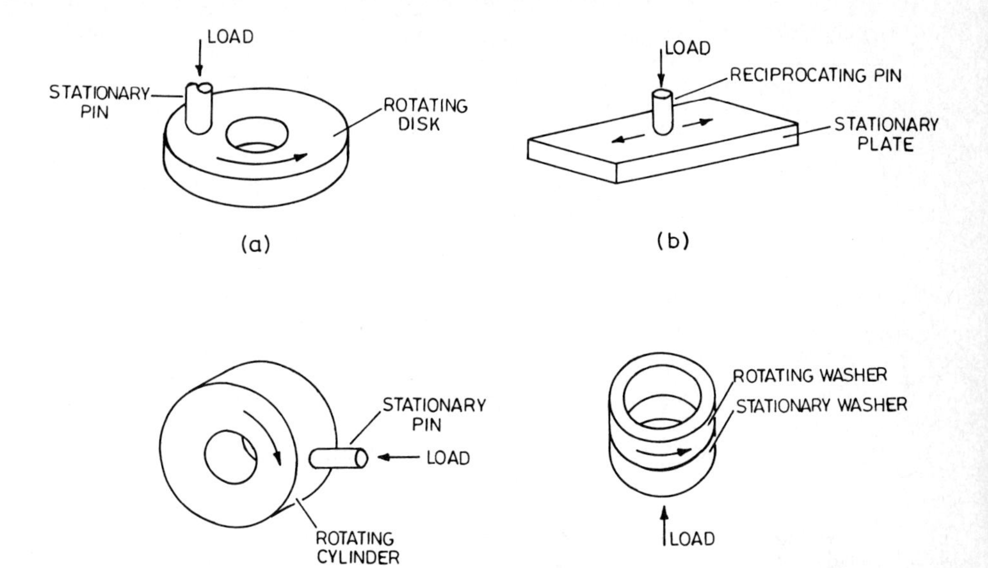

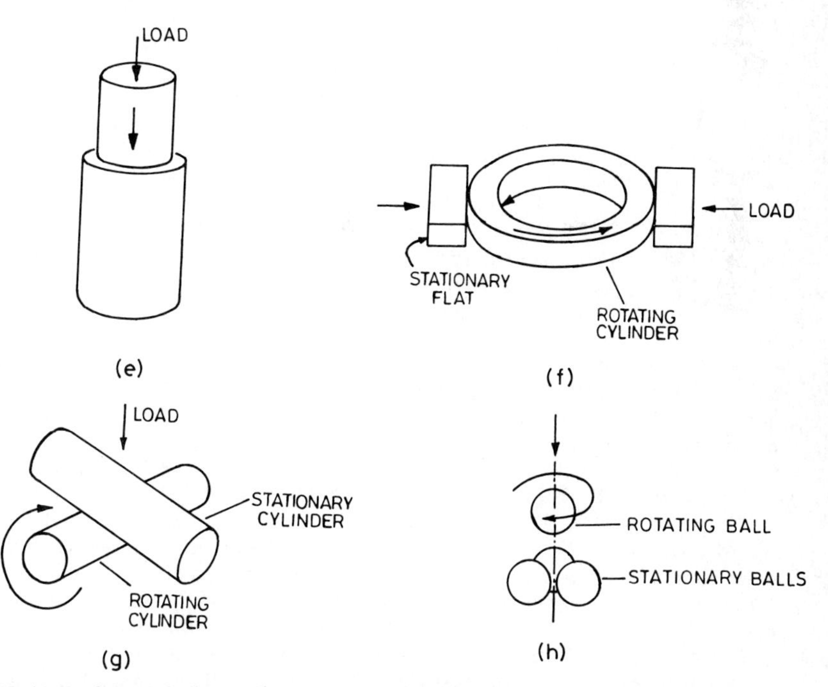

FIGURE 15.43 Schematic illustrations of typical interface geometries used for sliding friction and wear tests: (*a*) pin-on-disk, (*b*) pin-on-flat, (*c*) pin-on-cylinder, (*d*) thrust washer, (*e*) pin-into-bushing, (*f*) rectangular flats on rotating cylinder, (*g*) crossed cylinders, and (*h*) four ball.

TABLE 15.6 Some Details of Typical Test Geometries for Friction and Wear Testing

Geometery*	Type of contact	Type of loading	Type of motion
1. Pin-on-disk (face loaded)	Point/conformal	Static, dynamic	Unidirectional sliding, oscillating
2. Pin-on-flat (reciprocating)	Point/conformal	Static, dynamic	Reciprocating sliding
3. Pin-on-cylinder (edge loaded)	Point/conformal	Static, dynamic	Unidirectional sliding, oscillating
4. Thrust washers (face loaded)	Conformal	Static, dynamic	Unidirectional sliding, oscillating
5. Pin-into-bushing	Conformal	Static, dynamic	Unidirectional sliding, oscillating
6. Flat-on-cylinder (edge loaded)	Line	Static, dynamic	Unidirectional sliding, oscillating
7. Crossed cylinders	Elliptical	Static, dynamic	Unidirectional sliding, oscillating
8. Four balls	Point	Static, dynamic	Unidirectional sliding

*See Fig. 15.43.

the Timken tester, only one flat is pressed against the cylinder. The major difference between Alpha and Timken testers is in the loading system. In the Falex tester, a rotating pin is sandwiched between two V-shaped (instead of flat) blocks so that there are four lines of contact with the pin. In the Almon-Wieland tester, a rotating pin is sandwiched between two conforming bearing shells.

Crossed Cylinders. The crossed-cylinders test apparatus [Fig. 15.43(*g*)] consists of a hollow (water-cooled) or solid cylinder as the stationary wear member and a solid cylinder as the rotating or oscillating wear member that operates at 90° to the stationary member, such as in the Reichert wear tester.

Four Ball. The four-ball test apparatus [Fig. 15.43(*h*)], also called the Shell four-ball tester, consists of four balls in the configuration of an equilateral tetrahedron. The upper ball rotates and rubs against the lower three balls, which are held in a fixed position.

15.2.2.2 ***Abrasion Tests.*** Abrasion tests include two-body and three-body tests. In a two-body abrasion test, one of the moving members is abrasive. In a three-body abrasion test, abrasive particles are introduced at the interface. Abrasion tests can be conducted using any of the so-called conventional test geometries just described, with one of the surfaces being made of abrasive material or in the presence of abrasive particles. A few commonly used specialized tests are described here.

Taber Abrasion Test. The Taber tester (manufactured by Teledyne Taber, North Tonawanda, N.Y.) is widely used for determining the abrasion resistance of various materials and coatings. Test specimens (typically 100 mm square or 110 mm in diameter) are placed on the abrader turntable [Fig. 15.44(*a*)] and are subjected to the rubbing action of a pair of rotating abrasive wheels (resilient calibrade, nonresilient calibrade, wool felt, plain rubber, and tungsten carbide) at known weights (250, 500, or 1000 g). Wear action results when a pair of abrasive wheels is rotated in opposite directions by a turntable on which the specimen material is mounted. The abrading wheels travel on the material about a horizontal axis displaced tangentially from the axis of the test material, which results in a

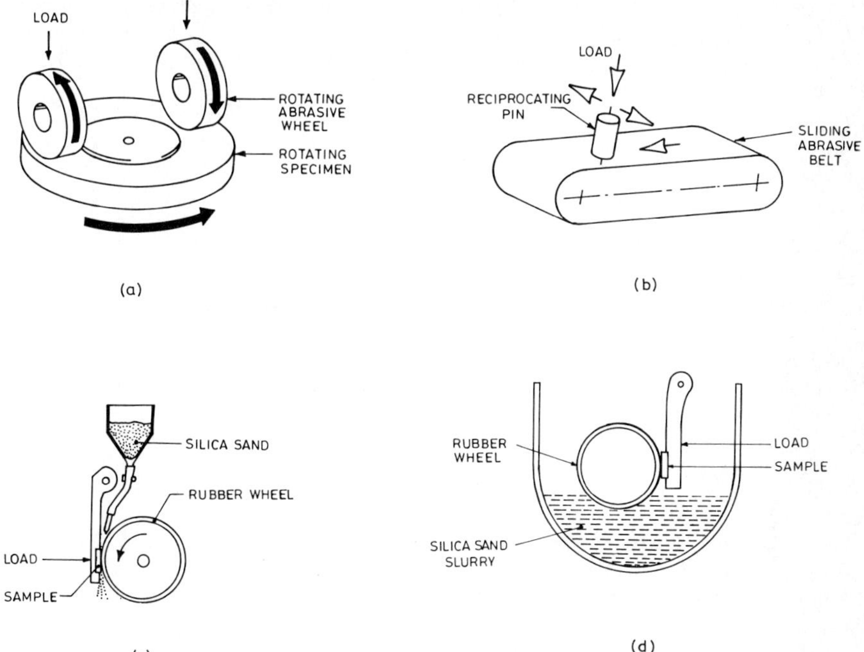

FIGURE 15.44 Schematic illustrations of abrasion test apparatuses: (*a*) two abradent wheels weighted on test specimen driven in opposite directions in the Taber abrasion test apparatus, (*b*) abrasive belt test apparatus, (*c*) dry-sand abrasion test apparatus, (*d*) wet-sand abrasion test apparatus.

sliding action. Results are evaluated by four different methods: end point or general breakdown of the material, comparison of weight loss between materials of the same specific gravity, volume loss in materials of different specific gravities, and measuring the depth of wear.

The Taber abrasion test provides a technique for conducting comparative wear performance evaluations with an intralaboratory precision of ±15 percent. These test are currently used by industry, government agencies, and research institutions for product development, testing, and evaluation.

Abrasive Belt Test. A flat-ended block or cylindrical specimen is abraded by sliding against an abrasive belt [Fig. 15.44(*b*)]. The belt runs horizontally, while the specimen runs transversely across the belt. The specimen also can be rotated during this abrasion test (Benzing et al., 1976).

Dry-Sand Abrasion Test. The dry-sand abrasion or dry-sand rubber-wheel abrasion test apparatus (ASTM G65) is shown in Fig. 15.44(*c*). The specimen is loaded against the rotating rubber wheel. The load is applied along the horizontal diametral line of the wheel. The abrasive is gravity-fed into the vee formed at the contact between the sample block and the wheel. The abrasive is typically 50 to 70 mesh (200 to 300 μm) dry AFS (American Foundary Society) test sand. Specimen weight loss is used as a measure of abrasive wear (Bayer, 1982).

Wet-Sand Abrasion Test. The wet-sand abrasion test apparatus, also known as the SAE wet-sand rubber-wheel test apparatus, is shown in Fig. 15.44(*d*). The specimen is pressed against the rubber wheel. It consists of a neoprene rubber rim on a steel hub that rotates through a silica-sand slurry. The wheel has stirring paddles on each side to agitate the slurry as it rotates. Sand is carried by the rubber wheel to the interface between it and the test specimen. The slurry consists of 940 g of deionized water and 1500 g of AFS 50 to 70 mesh silica test sand. Specimen weight loss is used as a measure of abrasive wear (Bayer, 1982).

Mar-Resistance Abrasion Test. The mar-resistance abrasion test apparatus, also called the falling silicon carbide test apparatus (ASTM D673), simulates abrasive wear resulting from the impingement or impact of coarse, hard silicon carbide particles. The test involves allowing a weighed amount of no. 80 silicon carbide grit to fall through a glass tube and strike the surface of a test specimen at a 45° angle. The abrasion resistance is determined by measuring the percent change in haze of the abraded test specimen by ASTM D1003 (Bayer, 1982).

15.2.2.3 Rolling-Contact Fatigue Tests. A number of rolling-contact fatigue (RCF) tests are used for testing materials and lubricants for rolling-contact applications such as antifriction bearings and gears.

Disk-on-Disk. The disk-on-disk test apparatus [Fig. 15.45(*a*)] uses two disks or a ball-on-disk rotating against each other on their outer surfaces (edge loaded). The disk samples may be crowned or flat. Usually, the samples rotate at different sliding speeds to produce some relative sliding (slip) at the interface (Benzing et al., 1976).

Rotating Four Ball. The rotating four-ball test apparatus [Fig. 15.45(*b*)] consists of four balls in the configuration of an equilateral tetrahedron. The rotating upper ball is dead weight–loaded against the three support balls (positioned 120° apart), which orbit the upper ball in rotating contact. In some tests, five balls instead of four balls are used. Also, in some studies, the lower balls are clamped (Bhushan and Sibley, 1982).

Rolling-Element-on-Flat. The rolling-element-on-flat test apparatus consists of three balls or rollers equispaced by a retainer that are loaded between a stationary flat washer and a rotating grooved washer [Fig. 15.45(*c*)]. The rotating washer produces ball motion and serves to transmit load to the balls and the flat washer (Bhushan and Sibley, 1982).

15.2.2.4 Solid-Particle Erosion Test. Erosion testing is generally conducted at room temperature using an air-blast test apparatus, shown in Fig. 15.46. The tester is operated by feeding the eroding particles from a vibrating hopper into a stream of gas. A known amount of eroding particles is directed onto one or more test specimens. The weight loss of the test specimens is used as a measure of erosive wear (Bayer, 1976).

15.2.2.5 Corrosion Tests. Corrosion can occur because of electrochemical or chemical interactions with the environment. Electrochemical corrosion, also called *electrolytic corrosion* (*EC*), and accelerated business environment (ABE) tests are used to test specimens. The EC test, where the test time is few minutes, is used to rank specimens that corrode by electrochemical means. These tests are useful during early development. In the ABE test, the test time can be from a fraction of a day to several days depending on the environmental conditions. A properly simulated ABE test is generally used to predict the component life under actual conditions (Bhushan, 1990).

Electrochemical (EC) Test. In this test, two- or three-electrode cells are normally used to measure (1) corrosion potential to determine practical nobility of a material, (2) the corrosion-current density to determine the corrosion rate of a material, (3) the corrosion current–potential relationships under defined experimental conditions, which can unravel the corrosion mechanism, and (4) the gal-

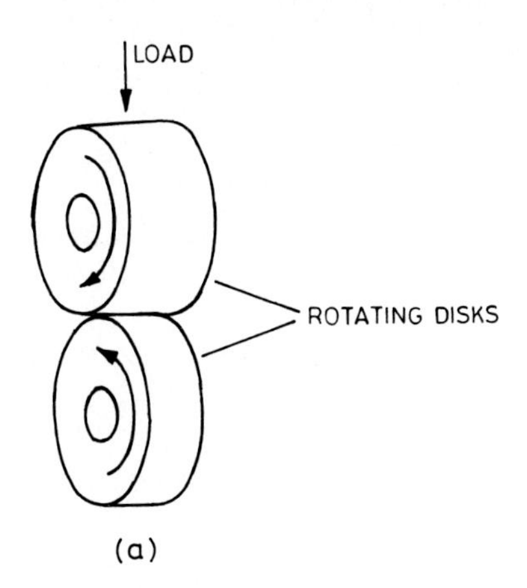

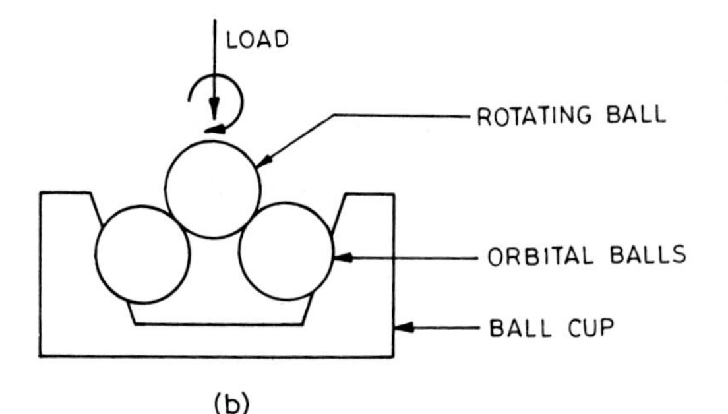

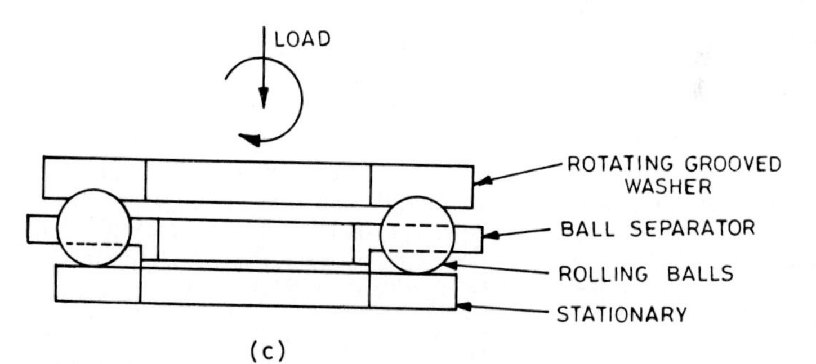

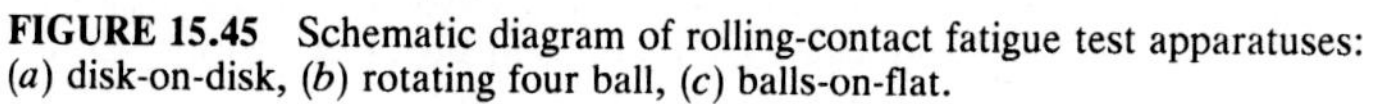

FIGURE 15.45 Schematic diagram of rolling-contact fatigue test apparatuses: (*a*) disk-on-disk, (*b*) rotating four ball, (*c*) balls-on-flat.

vanic potential of two dissimilar materials to determine the corrosion rate of a material couple. A number of electrochemical instruments are commercially available for EC tests (Dean et al., 1970).

Accelerated Business Environment (ABE) Test. The test samples are exposed to a controlled accelerated corrosive environment representative of the business environment. Degradation of the test specimen in the corrosive environ-

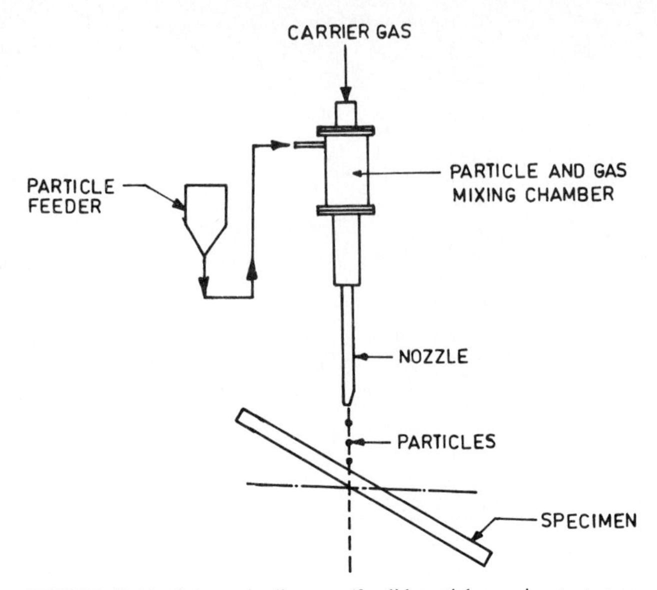

FIGURE 15.46 Schematic diagram of solid-particle erosion test apparatus.

mental test is measured by various methods, such as measuring weight loss, quantifying dimensional changes, noting changes in physical or chemical properties, measuring the size and number of defects on the surface using optical or scanning electron microscopes (visual methods), determining total defect density by light-scattering techniques, determining the atomic concentration of substrate material on a coated surface (chemical analysis), and quantifying performance degradation (if measurable). Commonly used corrosion tests that approximate the corrosion produced in service are discussed here. Frequently, combinations of corrosive environments are used.

(a) Salt Spray (Fog): The neutral salt-spray (fog) test utilizes a box of suitable size, from about 2 m^3 to walk-in size, into which a 5% NaCl solution is aspirated with air. A common testing time is 72 hours, although exposure duration can vary considerably (ASTM B117-73). This test is commonly used for zinc coatings (Saur, 1975). For corrosion tests of gas-turbine components, salts such as Na_2SO_4 and NaCl are added in air (Nicoll, 1983).

(b) Seawater: The test samples are partially or completely submerged in natural or synthetic (ASTM D1141-52) seawater for a fraction of a day to several months. This method is commonly used for marine applications (Bhushan and Dashnaw, 1981; Bhushan and Winn, 1981).

(c) Corrosive Gases: In this test, the test specimen is exposed to an accelerated corrosive gas environment (with constituents representative of the business environment). The corrosive gases may consist of small fractions of Cl_2, NO_2, H_2S and SO_2 (such as air with 5 ppb Cl_2, 500 ppb NO_2, 35 ppb H_2S, and 275 ppb SO_2 at 70 percent relative humidity and 25°C). An exposure of a fraction of a day to few days is sufficient (Bhushan, 1990).

(d) TEMPERATURE/HUMIDITY: In this test, test specimens are exposed to high temperature (*T*) and/or high humidity (*H*) for a fraction of a day to several days (Nicoll, 1983; Bhushan, 1990).

15.3 REFERENCES

Anderson, N. C., and Goodman, A. (1972), "A Technique for Measuring the Adhesion of Thin Films and the Dynamic Tensile Strength of Bonds, with Commercial Applications," Technical Report, Sandia Laboratories, Albuquerque, N.M. Available from NTIS under number SC-DR-69-320.

Anonymous (1971), "Residual Stress Measurement by X-Ray Diffraction," SAE J784a, SAE Handbook Supplement, Society of Automotive Engineers, New York.

——— (1985), *Surface Texture (Surface Roughness, Waviness, and Lay)*, ANSI/ASME B46.1, ASME, New York.

Anttila, A., Koskinen, J., Bister, M., and Hirvonen, J. (1986), "Density Measurements of Diamond-Like Coatings Using a Low Energy Accelerator," *Thin Solid Films*, Vol. 136, pp. 129–134.

Azzam, R. M., and Bashara, N. M. (1976), *Ellipsometry and Polarized Light*, North-Holland, Amsterdam.

Baldwin, P. C. (1970), "Measurement of Electroplate Thickness by Microcoulometric Methods," *Plating*, Vol. 57, pp. 927–932.

Bangert, H., and Wagendristel, A. (1986), "Ultralow Load Hardness Testing of Coatings in a Scanning Electron Microscope," *J. Vac. Sci. Technol.*, Vol. A4, pp. 2956–2958.

———, ———, and Aschinger, H. (1981), "Ultramicrohardness Tester for Use in Scanning Electron Microscope," *Colloid Poly. Sci.*, Vol. 259, pp. 238–240.

Banks, B. A., and Rutledge, S. K. (1982), "Ion Beam Sputter-Deposited Diamondlike Films," *J. Vac. Sci. Technol.*, Vol. 21, pp. 807–814.

Bayer, R. G. (1976), "Selection and Use of Wear Tests for Metals," STP-615, ASTM, Philadelphia.

——— (1979), "Wear Tests for Plastics: Selection and Use," STP-701, ASTM, Philadelphia.

——— (1982), "Selection and Use of Wear Tests for Coatings," STP-769, ASTM, Philadelphia.

Beam, J. W. (1956), "Study of Adhesion by the High-Speed Rotor Technique." *Tech. Proc. Am. Electroplat. Soc.*, Vol. 43, pp. 211–214.

Benjamin, P., and Weaver, C. (1959), "Condensation Energies for Metals on Glass and Other Substrates," *Proc. R. Soc. Lond.*, Vol. A252, pp. 418–430.

——— and ——— (1960), "Measurement of Adhesion of Thin Films," *Proc. R. Soc. Lond.*, Vol. A254, pp. 163–176.

Bennett, R. S. (1949), "A Review of Methods for Coating-Thickness Determinations," *J. Sci. Instr.*, Vol. 26, pp. 209–216.

Bennett, H. E., and Bennett, J. M. (1967), "Precision Measurements in Thin-Film Optics," *Phys. Thin Films*, Vol. 4, pp. 1–96.

Benzing, R. J., Goldblatt, I., Hopkins, V., Jamison, W., Mecklenburg, K., and Peterson, M. B. (1976), *Friction and Wear Devices*, 2d Ed., ASLE, Park Ridge, Ill.

Belser, R. B., and Hicklin, W. H. (1956), "Simple Rapid Sputtering Apparatus," *Rev. Sci. Instrum.*, Vol. 27, pp. 293–296.

Bhushan, B. (1978), "Surface Pretreatment of Thin Inconel X-750 Foils for Improved Coating Adherence," *Thin Solid Films*, Vol. 53, pp. 99–107.

——— (1979), "Development of RF-Sputtered Chromium Oxide Coating for Wear Application," *Thin Solid Films*, Vol. 64, pp. 231–241.

——— (1980), "Stick-Slip Induced Noise Generation in Water-Lubricated Compliant Rubber Bearings," *J. Tribol.* (*Trans. ASME*), Vol. 102, pp. 201–212.

——— (1987), "Overview of Coating Materials, Surface Treatments, and Screening Techniques for Tribological Applications Part 2: Screening Techniques," in *Testing of Metallic and Inorganic Coatings* (W. B. Harding and G. A. DiBari, eds.), STP-947, pp. 310–319, ASTM, Philadelphia.

——— (1990), *Tribology and Mechanics of Magnetic Storage Devices*, Spinger-Verlag, New York.

——— and Dashnaw, F. (1981), "Material Study for Advanced Stern-tube Bearings and Face Seals," *ASLE Trans.*, Vol. 24, pp. 398–409.

——— and Martin, R. J. (1988), "Accelerated Wear Test Using Magnetic-Particle Slurries," *Tribol. Trans.*, Vol. 31, pp. 228–238.

——— and Sibley, L. B. (1982), "Silicon Nitride Rolling Bearings for Extreme Operating Conditions," *ASLE Trans.*, Vol. 25, pp. 417–428.

——— and Winn, L. W. (1981), "Material Study for Advanced Stern-tube Lip Seals," *ASLE Trans.*, Vol. 24, pp. 410–422.

———, Nelson, G. W., and Wacks, M. E. (1986), "Head-Wear Measurements by Autoradiography of Worn Magnetic Tapes," *J. Tribol.* (*Trans. ASME*), Vol. 108, pp. 241–255.

———, Williams, V. S., and Shack, R. V. (1988a), "In-situ Nanoindentation Hardness Apparatus for Mechanical Characterization of Extremely Thin Films," *J. Tribol.* (*Trans. ASME*), Vol. 110, pp. 563–571.

———, Wyant, J. C., and Meiling, J. (1988b), "A New Three-Dimensional Digital Optical Profiler," *Wear*, Vol. 122, pp. 301–312.

Bifano, T. G. (1988), "Ductile-Regime Grinding of Brittle Materials," Ph.D thesis, Dept. of Mech. Eng., North Carolina State University, Raleigh, N.C.

Blois, M. S., Jr., and Rieser, L. M. (1954), "Apparent Density of Thin Evaporated Films," *J. Appl. Phys.*, Vol. 25, pp. 338–340.

Bowman, H. A., and Schoonover, R. M. (1967), "Procedure for High Precision Density Determinations by Hydrostatic Weighing," *J. Res. Nat. Bur. Stand.*, Vol. 71C, pp. 179–198.

Boyer, H. E. (1987) *Hardness Testing*, American Society for Metals, Metals Park, Ohio.

Brenner, A. (1938), "Magnetic Method for Measuring the Thickness of Nickel Coatings on Non-Magnetic Base Metals," *Mon. Rev. Am. Electroplat. Soc.*, Vol. 25, pp. 252–260.

——— and Garcia-Rivera, J. (1953), "An Electronic Thickness Gauge," *Plating*, Vol. 40, pp. 1238–1244.

——— and Senderoff, S. (1949), "Calculation of Stress in Electrodeposits from the Curvature of a Plated Strip," *J. Res. Nat. Bur. Stand.*, Vol. 42, pp. 105–123.

Brinksmeier, D., Cammett, J. T., Koenig, W., Leskovar, P., Peters, J., and Tonshoff, H. K. (1982), "Residual Stresses: Measurement and Causes in Machining Processes," *Ann. CIRP*, Vol. 31–2, pp. 491–510.

Brodell, F. P., and Brenner, A. (1957), "Further Studies of an Electronic Thickness Gauge," *Plating*, Vol. 44, pp. 591–601.

Brodsky, M. H., Kaplan, D., Ziegler, J. F. (1972), "Densities of Amorphous Si Films by Nuclear Backscattering," *Appl. Phys. Lett.*, Vol. 21, pp. 305–307.

Buckley, D. H. (1981), *Surface Effects in Adhesion, Friction, Wear, and Lubrication*, Elsevier, Amsterdam.

Bullett, T. R., and Prosser, J. L. (1972), "The Measurement of Adhesion." *Prog. Org. Coatings*, Vol. 1, pp. 45–73.

Butler, D. W. (1970), "A Simple Film Adhesion Comparator," *J. Phys.* [*E*] *Sci. Instrum.*, Vol. 8, pp. 979–980.

Calvet, E., and Prat, H. (1963), *Recent Advances in Microcalorimetry*, Pergamon, New York.

Campbell, D. S. (1970), "Mechanical Properties of Thin Films," in *Handbook of Thin Film Technology* (L. I. Maissel and R. Glang, eds.), Chap. 12, McGraw-Hill, New York.

Chalker, P. R., Bull, S. J., and Rickerby, D. S. (1990), "A Review of the Methods for the Evaluation of Coating—Substrate Adhesion," in *Proc. 2d Int. Conf. on Plasma Surface Engineering,*, Garmisch-Partenkirchen, Germany.

Chandrasekar, S., Shaw, M. C., and Bhushan, B. (1987), "Comparison of Grinding and Lapping of Ferrites and Metals," *J. Eng. Ind.* (*Trans. ASME*), Vol. 109, pp. 76–82.

Chapman, B. N. (1971), "Adhesion of Thin Magnetic Films," in *Aspects of Adhesion* (D. J. Alner, ed.), pp. 43–54, CRC Press, Boca Raton, Fla.

Chiang, S. S., Marshall, D. B., and Evans, A. G. (1981), "Surfaces and Interfaces," in *Ceramics and Ceramic-Metal Systems* (J. Pask and A. G. Evans, eds.), p. 603, Plenum, New York.

Chopra, K. L. (1979), *Thin Film Phenomenon*, Robert E. Krieger Publishing Co., Huntington, N.Y.

Clarke, E., Carlin, J. R., and Barbour, W. E. (1951), "Measuring the Thickness of Thin Coatings with Radiation Backscattering," *Elect. Eng.*, Vol. 70, pp. 35–37.

Clarke, M. (1975), "Porosity and Porosity Tests," in *Properties of Electrodeposits: Their Measurements and Significance* (R. Sard, H. Leidheiser, and F. Ogburn, eds.), pp. 122–141, The Electrochemical Society, Princeton, N.J.

Clauss, F. J. (1972), *Solid Lubrication and Self-Lubricated Solids*, Academic, New York.

Cooksey, G. L., and Campbell, H. S. (1970), "Scanning Electron Microscope Studies of Porosity in Gold Electrodeposits," *Trans. Inst. Met. Finish*, Vol. 48, pp. 93–98.

Cooley, R. L., and Lemons, K. E. (1969), "Densities of Electroplated Gold," *Plating*, Vol. 56, pp. 511–515.

Czanderna, A. W. (1975), *Methods of Surface Analysis*, Elsevier, Amsterdam.

Dancy, W. H., Jr. (1970), "A Solid State Ultracentrifuge for Adhesion Testing of Electrodeposits," NTIS Tech. Rep. RE 701-138 AD 752460, NTIS, Springfield, Va.

Dannenberg, H. (1961), "Measurement of Adhesion by a Blister Method," *J. Appl. Poly. Sci.*, Vol. 5, pp. 125–134.

Davies, D., and Whittaker, J. A. (1967), "Methods of Testing the Adhesion of Metal Coatings to Metals," *Metallurg. Rev.*, Vol. 12, pp. 15–26.

Dean, S. W., France, W. D., and Ketcham, S. J. (1970), "Electrochemical Methods of Testing," paper presented at the symposium on State of the Art in Corrosion Testing Methods, ASTM Annual Meeting, Toronto, Canada.

Dini, J. W., Helms, J. R., and Johnson, H. R. (1972), "Ring Shear Test for Quantitatively Measuring Adhesion of Metal Deposits," *Electroplat. Metal Finish.*, Vol. 25, pp. 5–11.

——— and Johnson, H. R. (1974), "Ring Shear Adhesion Tests of Various Deposit-Substrate Combinations," *Metal Finish.*, Vol. 72 (Aug.), pp. 44–48.

Doerner, M. F., and Nix, W. D. (1986), "A Method of Interpreting the Data from Depth Sensing Indentation Instruments," *J. Mater. Res.*, Vol. 1, pp. 601–609.

Dölle, H. (1979), "The Influence of Multiaxial Stress States, Stress Gradients, and Elastic Anisotropy on the Evaluation of (Residual) Stresses by X-Rays," *J. Appl. Cryst.*, Vol. 12, pp. 489–501.

Donovan, T. M., and Heinemann, K. (1971), "High Resolution Electron Microscope Observation of Voids in Amorphous Ge," *Phys. Rev. Lett.*, Vol. 27, pp. 1794–1796.

Dugger, M. T., Chung, Y. W., Bhushan, B., and Rothschild, W. (1990), "Friction, Wear and Interfacial Chemistry in Thin-Film Magnetic Rigid Disk Files," *J. Tribol.* (*Trans. ASME*), Vol. 112, pp. 238–245.

Dunlap, W. C. (1959), "Conductivity Measurements on Solids," *Methods of Experimental Physics* (K. Larks-Horovitz and V. A. Johnson, eds.) Vol. 6B, pp. 32–70. Academic, New York.

Fee, A. R., Segabache, R., and Tobolski, E. L. (1985), "Hardness Testing," in *Metals Handbook*, Vol. 8: *Mechanical Testing*, 9th ed., pp. 69–108, American Society for Metals, Metals Park, Ohio.

Ferrante, J. (1981), "Practical Applications of Surface Analytical Tools in Tribology," *Lub. Eng.*, Vol. 38, pp. 223–236.

Flinn, P. A., Gardner, D. S., and Nix, W. D. (1987), "Measurement and Interpretation of Stress in Aluminum-Based Metallization as a Function of Thermal History," *IEEE Trans. on Electron Devices*, Vol. ED-34, pp. 689–699.

Franklin, A. D., and Spal, R. (1971), "A Method for the Precision Comparison of the Densities of Small Specimens," *Rev. Sci. Instrum.*, Vol. 42, pp. 1827–1833.

Fuks, M. Y., and Cheremskoy, P. G. (1974), "Oriented Porosity, Shape of Structure Elements and Anisotropy of the Microstresses in Vacuum Condensates," *Phys. Metals Metallogr.*, Vol. 37, No. 4, pp. 117–124.

———, Palatnik, L. S., Cheremskoy, P. G., and Tu, V. Z. (1971), "Crystallization Submicroporosity in Condensed Polycrystalline Films," *Sov. Phys. Solid State*, Vol. 13, pp. 1467–1473.

Galeener, F. L. (1971), "Submicroscopic Void Resonance: The Effect of Internal Roughness on Optical Absorbtion," *Phys. Rev. Lett.*, Vol. 27, pp. 421–423.

Glang, R., Holmwood, R. A., and Rosenfeld, R. L. (1965), "Determination of Stress in Films on Single Crystalline Silicon Substrates," *Rev. Sci. Instrum.*, Vol. 36, pp. 7–10.

Good, R. J. (1976), "On the Definition of Adhesion," *J. Adhes.*, Vol. 8, pp. 1–9.

Hartman, T. E. (1965), "Density of Thin Evaporated Aluminum Films," *J. Vac. Sci. Technol.*, Vol. 2, pp. 239–242.

Hayes, R. E., and Roberts, A. R. V. (1962), "A Control System for the Evaporation of Silicon Monoxide Insulating Films," *J. Sci. Instrum.*, Vol. 39, pp. 428–431.

Heavens, O. S. (1950), "Some Factors in Influencing the Adhesion of Films Produced by Vacuum Evaporation," *J. Phys. Radium*, Vol. 11, pp. 355–360.

Heller, H. A. (1969), "Analytical Control in Electroplating Technology," *Plating*, Vol. 56, pp. 277–284.

Hoffmann, E., and Georgoussis, O. (1959), "Measurement of Adhesion of Paint Films," *J. Oil Color. Chem. Assoc.*, Vol. 42, pp. 267–269.

Hoffman, R. W. (1966), "The Mechanical Properties of Thin Condensed Films," in *Physics of Thin Films* (G. Hass and R. E. Thun, eds.), Vol. 3, pp. 211–273, Academic, New York.

Holland, L. (1956), *Vacuum Deposition of Thin Films*, Chapman and Hall, London.

Huisman, H. F., Rosenberg, C. J. F. M., and Winsum, J. A. V. (1983), "An Improved Mercury Porosimetry Apparatus: Some Magnetic Tape Applications," *Powder Technol.*, Vol. 36, pp. 203–213.

Ingram, D. C., Woollam, J. A., and Bu-Abbud, G. (1986), "Mass Density and Hydrogen Concentration in Diamond-Like Carbon Films: Proton Recoil, Rutherford Backscattering and Ellipsometric Analysis," *Thin Solid Films*, Vol. 137, pp. 225–230.

Jaccodine, R. J., and Schlegel, W. A. (1966), "Measurement of Strains at Si-SiO_2 Interface," *J. Appl. Phys.*, Vol. 37, pp. 2429–2434.

Jacobson, B. E., Hammond, R. H., Geballe, T. H., and Salem, J. R. (1978), "Transmission Electron Microscopy Studies of Electron Beam Coevaporated Nb_3Sn-Cu Superconducting Composites," *Thin Solid Films*, Vol. 54, pp. 243–258.

Jacobsson, R., and Krause, B. (1973), "Measurement of Adhesion of Thin Evaporated Films on Glass Substrates by Means of Direct Pull-Off Method," *Thin Solid Films*, Vol. 15, pp. 71–77.

Je, J. H., Gyarmati, E., and Naoumidis, A. (1986), "Scratch Adhesion Test of Reactively Sputtered TiN Coatings on a Soft Substrate," *Thin Solid Films*, Vol. 136, pp. 57–67.

Kane, P. F., and Larrabee, G. R. (1974), *Characterization of Solid Surfaces*, Plenum, New York.

King, R. J., Downs, M. J., Clapham, P. B., Raine, K. W., and Talim, S. P. (1972), "A Comparison of Methods for Accurate Film Thickness Measurements," *J. Phys.* [*E*] *Sci. Instrum.*, Vol. 5, pp. 445–449.

Krawitz, A. D., and Holden, T. M. (1990), "The Measurement of Residual Stresses Using Neutron Diffraction," *MRS Bulletin*, Vol. 15, Nov., pp. 57–64.

Kurov, G. A., Zhil'kov, E. A., Dubodel, V. M., Toroptsev, B. A., and Tselibeeva, V. N. (1975), "Microscopic Defects in Thin Metal Films," *Sov. Phys. Dokl.*, Vol. 19, pp. 772–774.

Laugier, M. (1981), "The Development of Scratch Test Technique for the Determination of the Adhesion of Coatings," *Thin Solid Films*, Vol. 76, pp. 289–294.

Leonard, W. F., and Ramey, R. L. (1966), "Thickness of Au, Ag, and Cu Films from Theoretical Expressions for Conductivity and Isothermal Hall Effect," *J. Appl. Phys.*, Vol. 37, pp. 2190–2191.

Letner, H. R., and Snyder, J. H. (1954), "Grinding and Lapping Stresses in Manganese Oil Hardening Tool Steel," *Trans. ASME*, Vol. 76, pp. 317–323.

Light, T. B. (1969), "Density of "Amorphous" Ge," *Phys. Rev. Lett.*, Vol. 22, pp. 999–1000.

Lin, D. S. (1971), "The Adhesion of Metal Films to Glass and Magnesium Oxide in Tangential Shear," *J. Phys.* [*D*] *Appl. Phys., Vol. 4, pp. 1977–1990.*

Lloyd, J. R., and Nakahara, S. (1977), "Voids in Thin as Deposited Gold Films Prepared by Vapor Deposition," *J. Vac. Sci. Technol.*, Vol. 14, pp. 655–659.

Lowenheim, F. A. (ed.) (1974), *Modern Electroplating*, 3d Ed., Wiley, New York.

Maissel, L. I., and Glang, R. (1970), *Handbook of Thin Film Technology*, McGraw-Hill, New York.

Masubuchi, K. (1983), "Residual Stresses and Distortion," in *Metals Handbook*, Vol. 6: *Welding, Brazing and Soldering*, 9th ed., pp. 856–895, American Society for Metals, Metals Park, Ohio.

Mathur, P. B., and Karuppanan, N. (1961), "Measurement of the Thickness of Electroplates by Electrolytic Stripping Method: Stripping Solutions for Silver over Copper, Brass, Mild Steel and Nickel," *Plating*, Vol. 48, pp. 170–172.

McInerney, E. J., and Flinn, P. A. (1982), "Diffusivity of Moisture in Thin Films," *Proc. 20th Annual Int. Relia. Phys. Symp.*, pp. 264–267, IEEE Electron Devices Society and IEEE Reliability Society, New York.

Mehrotra, P. K., and Quinto, D. T. (1985), "Techniques for Evaluating Mechanical Properties of Hard Coatings," *J. Vac. Sci. Technol.*, Vol. A3, pp. 2401–2405.

———, ———, and Wolfe, G. J. (1984), "Composition and Microhardness Profiles in CVD Coatings with a Ball Wear Scar Technique," *Proc. 9th Int. Conf. on CVD* (Mc. D. Robinsen, ed.), pp. 757–771, The American Electrochemical Society, Pennington, N.J.

Mittal, K. L. (1976), "Adhesion Measurement of Thin Films," *Electrocomponent Sci. Technol.*, Vol. 3, pp. 21–42.

——— (ed.) (1978), *Adhesion Measurement of Thin Films, Thick Films, and Bulk Coatings*, STP-640, ASTM, Philadelphia.

Morrissey, R. J. (1970), "Electrolytic Determination of Porosity in Gold Electroplates I. Corrosion Potential Measurements," *J. Electrochem. Soc.*, Vol. 117, pp. 742–747.

——— (1972), "Electrolytic Determination of Porosity in Gold Electroplates II. Controlled Potential Techniques," *J. Electrochem. Soc.*, Vol. 119, pp. 446–450.

Moses, S. (1949), "The Nature of Adhesion," *Ind. Eng. Chem.*, Vol. 41, pp. 2338–2342.

——— and Witt, R. K. (1949), "Evaluation of Adhesion by Ultrasonic Vibrations," *Ind. Eng. Chem.*, Vol. 41, pp. 2334–2338.

Moss, S. C., and Graczyk, J. F. (1969), "Evidence of Voids Within the As-Deposited Structure of Glassy Silicon," *Phys. Rev. Lett.*, Vol. 23, pp. 1167–1171.

Motulevich, G. P., and Shubin, A. A. (1963), "Determination of the Microscopic Characteristics of Indium from Its Infrared Optical Constants and Electrical Conductivity," *Sov. Phys. JETP*, Vol. 17, pp. 33–36.

Nakahara, S. (1977), "Microporosity Induced by Nucleation and Growth Process in Crystalline and Noncrystalline Films," *Thin Solid Films*, Vol. 45, pp. 421–432.

——— (1979), "Microporosity in Thin Films," *Thin Solid Films*, Vol. 64, pp. 149–161.

Narayanan, U. H., and Venkatachalam, K. R. (1961), "A Simple Transistorized Thickness Meter for Electroplates," *Plating*, Vol. 48, pp. 1211–1212.

Neugebauer, C. A. (1964), "Thin Films of Niobium Tin by Codeposition," *J. Appl. Phys.*, Vol. 35, pp. 3599–3603.

Nicoll, A. R. (1983), "A Survey of Methods Used for the Performance Evaluation of High Temperature Coatings," in *Coatings for High Temperature Applications* (E. Lang, ed.), pp. 269–339, Applied Science Publishers, London.

Nix, W. D. (1989), "Mechanical Properties of Thin Films," *Metall. Trans.*, Vol. 20A, pp. 2217–2245.

Noyan, I. C., and Cohen, J. B. (1987), *Residual Stresses*, Springer Verlag, New York.

Ogburn, F., and Hilkert, M. (1956), "The Nature, Cause and Effect of Porosity in Electrodeposits," *Proc. Am. Electroplaters Soc.*, Vol. 43, pp. 256–259.

Oliver, W. C., Hutchings, R., and Pethica, J. B. (1986), "Measurement of Hardness at Indentation Depth as Low as 20 Nanometers," in *Microindentation Techniques in Materials Science and Engineering* (P. J. Blau and B. R. Lawn, eds.), STP-889, pp. 90–108, ASTM, Philadelphia.

Orlow, Y. F., Sakseev, D. A., Timofeeva, L. G., Karel'skaya, V. F., and Kononova, D. V. (1974), "Study of the Porous Structure of Particulate Condensates of Antimony Sulfide," *J. Appl. Chem. USSR*, Vol. 47, pp. 2021–2024.

Pan, A., and Greene, J. E. (1981), "Residual Compressive Stress in Sputter-Deposited TiC Films on Steel Substrates," *Thin Solid Films*, Vol. 78, pp. 25–34.

——— and ——— (1982), "Interfacial Chemistry Effects on the Adhesion of Sputter-Deposited TiC Films to Steel Substrates," *Thin Solid Films*, Vol. 97, pp. 79–89.

Perry, A. J. (1981), "The Adhesion of Chemically Vapor Deposited Hard Coatings to Steel: The Scratch Test," *Thin Solid Films*, Vol. 78, pp. 77–93.

——— (1983), "Scratch Adhesion Testing of Hard Coatings," *Thin Solid Films*, Vol. 107, pp. 167–180.

——— (1989), "The Relationship Between Residual Stress, X-Ray Elastic Constants and Lattice Parameters in TiN Films Made by Physical Vapor Deposition," *Thin Solid Films*, Vol. 170, pp. 67–70.

Piegari, A., and Masetti, E. (1985), "Thin Film Thickness Measurement: A Comparison of Various Techniques," *Thin Solid Films*, Vol. 124, pp. 249–257.

Pliskin, W. A., and Gnall, R. P. (1964), "Evidence for Oxidation Growth at the Oxide-Silicon Interface from Controlled Etch Studies," *J. Electrochem. Soc.*, Vol. 111, pp. 872–873.

——— and Zanin, S. J. (1970), "Film Thickness and Composition," in *Handbook of Thin Film Technology* (L. I. Maissel and R. Glang, eds.), Chap. 11, McGraw-Hill, New York.

Poate, J. M., Tu, K. N., and Mayer, J. W. (1978), *Thin Films: Interdiffusion and Reactions*, Wiley Interscience, New York.

Pulker, H. K., and Salzmann, K. (1986), "Micro/Ultramicro Hardness Measurements with Insulating Films," *SPIE Thin Film Technol.*, Vol. 652, pp. 139–144.

Rehrig, D. L. (1974), "Effect of Deposition Method on Porosity in Au Thin Films," *Plating*, Vol. 61, pp. 43–46.

Rickerby, D. S., Bellamy, B. A., and Jones, A. M. (1987), "Internal Stress and Microstructure of Titanium Nitride Coatings," *Surf. Eng.*, Vol. 3, No. 2, pp. 138–146.

———, Jones, A. M., and Bellamy, B. A. (1989) "X-Ray Diffraction Studies of Physically Vapour-Deposited Coatings," *Surf. Coat. Technol.*, Vol. 37, pp. 111–137.

Saur, R. L. (1965), "New Interference Microscope Techniques for Microtopographic Measurements in the Electroplating Laboratory," *Plating*, Vol. 52, pp. 663–666.

——— (1975), "Corrosion Testing: Protective and Decorative Coatings," in *Properties of Electrodeposits: Their Measurements and Significance* (R. Sard, H. Leidheiser, and F. Ogburn, eds.), pp. 170–186, The American Electrochemical Society, Princeton, N.J.

Sayles, R. S., and Thomas, T. R. (1976), "Mapping a Small Area of a Surface," *J. Phys. [E].*, Vol. 9, pp. 855–861.

Schreiber, T. P., Ottolin, A. C., and Johnson, J. L. (1963), "The X-Ray Emission Analysis of Thin Films Produced by Lubricative Oil Additives," *Appl. Spectroscopy*, Vol. 17, pp. 17–19.

Schwartz, B., and Schwartz, N. (1967), *Measurement Techniques for Thin Films*, Electrochemical Society, Princeton, N.J.

Sekler, J., Steinmann, P. A., and Hintermann, H. E. (1988), "The Scratch Test: Different Critical Load Determination Techniques," *Surface Coat. Technol.*, Vol. 36, pp. 519–529.

Shome, S. C., and Evans, U. R. (1951), "Studies in the Discontinuities of Electrodeposited Metallic Coatings Part II," *J. Electrodepositors Tech. Soc.*, Vol. 27, pp. 45–64.

Smits, F. M. (1958), "Measurements of Sheet Resistivities with the Four-Point Probe," *Bell Systems Tech. J.*, Vol. 37, pp. 711–718.

Staudinger, A., and Nakahara, S. (1977), "The Structure of the Crack Network in Amorphous Films," *Thin Solid Films*, Vol. 45, pp. 125–133.

Steinmann, P. A., Laeng, P., and Hintermann, H. E. (1984), "Determination of Adherence by Scratch Test Applied to Hard, Thin Film Metallurgical Surfaces," *Le Vide les Couches Minces*, Vol. 39, pp. 87–94.

———, Tardy, Y., and Hintermann, H. E. (1987), "Adhesion Testing by the Scratch Test Method: The Influence of Intrinsic and Extrinsic Parameters on the Critical Load," *Thin Solid Films*, Vol. 154, pp. 333–349.

Stoney, G. G. (1909), "The Tension of Metallic Films Deposited by Electrolysis," *Proc. R. Soc. (Lond.)*, Vol. A82, pp. 172–175.

Sundgren, J.-E., and Hentzell, H. T. G. (1986), "A Review of the Present State of the Art in Hard Coatings Grown from the Vapor Phase," *J. Vac. Sci. Technol.*, Vol. A4, pp. 2259–2279.

Tabor, D. (1951), *The Hardness of Metals*, Clarendon Press, Oxford, England.

Thomas, J. D., and Rouse, R. (1955), "Using the Interference Microscope for Thickness Measurements of Decorative Chromium Plating," *Plating*, Vol. 42, pp. 55–57.

Thomas, T. R. (1982), *Rough Surfaces*, Longman, London.

Tolansky, S. (1960), *Surface Microtopography*, Interscience Publishers, London.

Treuting, R. G., and Reed, W. T. (1951), "A Mechanical Determination of Biaxial Residual Stress in Sheet Materials," *J. Appl. Phys.*, Vol. 22, pp. 130–134.

Tsukamoto, Y., Yamaguchi, H., and Yanagisawa, M. (1987), "Mechanical Properties of Thin Films: Measurements of Ultramicroindentation Hardness, Young's Modulus and Internal Stresses," *Thin Solid Films*, Vol. 154, pp. 171–181.

Tu, K. N., and Lau, S. S. (1978), *Thin Films: Interdiffusion and Reactions* (J. M. Poate, K. N. Tu, and J. W. Mayer, eds.), p. 81, Wiley Interscience, New York.

Valdes, L. B. (1954), "Resistivity Measurements on Germanium for Transistors," *Proc. IRE*, Vol. 42, pp. 420–427.

Valli, J., Makela, U., Matthews, A., and Murawa, V. (1985), "TiN Coating Adhesion Studies Using the Scratch Test Method," *J. Vac. Sci. Technol.*, Vol. A3, pp. 2411–2414.

Vossen, J. L. (1978), "Measurements of Film-Substrate Bond Strength by Laser Spallation," in *Adhesion Measurement of Thin Film, Thick Film and Bulk Coatings* (K. L. Mittal, ed.), STP-640, pp. 122–133, ASTM, Philadelphia.

Waite, C. F. (1953), "Thickness of Electrodeposited Coatings by the Anodic Solution Method," *Plating*, Vol. 40, pp. 1245–1254.

Walker, G. A. (1970), "Structural Investigations of Thin Films," *J. Vac. Sci. Technol.*, Vol. 7, pp. 465–473.

Warner, A. W., and Stockbridge, C. D. (1963), "Quartz Resonators: Reduction of Transient Frequency Excursions Due to Temperature Change," *J. Appl. Phys.*, Vol. 34, pp. 437–438.

Weinstein, A. M. D., Antonio, C., and Mukherjee, K. (1968), "Density of Copper Thin Films Measured by X-Ray Absorption," *Nature*, Vol. 220, pp. 777–778.

Westbrook, J. H., and Conrad, H. (eds.) (1973), *The Science of Hardness and Its Research Applications*, American Society for Metals, Metals Park, Ohio.

Willard, H. H., Merritt, L. L., and Dean, J. A. (1974), *Instrumental Methods of Analysis*, 5th Ed., Van Nostrand, New York.

Williams, V. S., Landesman, A. L., Shack, R. V., Vukobratovich, D., and Bhushan, B. (1988), "In Situ Microviscoelastic Measurements by Polarization-Interferometric Monitoring of Indentation Depth," *Appl. Optics*, Vol. 27, pp. 541–546.

Wolff, R. H., Henderson, M. A., and Eisler, S. L. (1955), "Porosity of Nickel Deposits by Autoradiographic Techniques," *Plating*, Vol. 42, pp. 537–544.

Yust, C. S., and Bayer, R. G. (1988), "Selection and Use of Wear Tests for Ceramics," STP-1010, ASTM, Philadelphia.

Zemany, P. D. (1960), *X-Ray Absorption and Emission in Analytical Chemistry*, Wiley, New York.

CHAPTER 16

TRIBOLOGICAL APPLICATIONS OF MATERIALS, COATINGS, AND SURFACE TREATMENTS

The performance and life of a component that is subjected to friction, wear, and corrosion failures during its use can be considerably improved by selection of suitable materials and/or application of a suitable coating and surface treatment. Such coatings and surface treatments provide a greater flexibility in the design and selection of materials for a particular component. Table 16.1 summarizes the typical applications of common coatings and surface treatments (Bhushan, 1987). This chapter discusses the bulk materials, coatings, and treatments used to improve the performance and life of some common tribological components, such as sliding-contact bearings, rolling-contact bearings, seals, gears, cams and tappets, piston rings, electrical brushes, and cutting and forming tools, and some common industrial applications, such as gas turbines, fusion reactors, and magnetic storage devices.

16.1 SLIDING-CONTACT BEARINGS

The machine elements that support a moving shaft against a stationary housing are called *bearings*. In general, we can classify bearings as either sliding-contact or rolling-contact bearings. In sliding-contact bearings, also known as *sliding* or *plain bearings* or *bushings*, the load is transmitted between moving parts by sliding contact. The motion can be a planar motion (e.g., plane, curved, step, and composite sliders and pivoted-pad sliders) or a rotational motion (e.g., full and partial journal bearings, foil bearings, floating-ring bearings) (Bisson and Anderson, 1964; Fuller, 1984; Neale, 1973). Sliding bearings can be lubricated with a film of air, water, oil, grease, or the process fluid. The journal bearing is perhaps the most familiar and most widely used of all bearing types. It consists fundamentally of a sleeve of bearing material wrapped partially or completely around a rotating shaft or journal, and it is designed to support a radial load (Fig. 16.1).

Selection of materials for sliding bearings is a multifunctional optimization problem. In general, the standard requirements are as follows: score resistance, conformability, embeddability, compressive strength, fatigue strength, thermal conductivity, wear resistance, corrosion resistance, and cost. Bearing materials fall into two major categories: metals and nonmetals (Bhushan and Dashnaw,

TABLE 16.1 Some Examples of Coatings and Surface Treatments for Tribological Applications

Coatings/treatments	Typical applications
Soft coatings	
Layered solid coatings (MoS_2 and graphite)	Bearings and seals, bearings operating in vacuum and/or high temperatures (aerospace), piston rings, valves
Nonlayered coatings	Bearings and seals at high temperatures
Polymers	Bearings and seals, pumps, impellers
Soft metals	Bearings, electrical contacts
Hard coatings	
Hard metallic	Bearings and seals, piston rings, cylinder liners, and cross-head pins in internal combustion engines and compressors, cannon and gun tubes, metal-working tools, tape-path components of computer tape drives
Ceramic	Bearings and seals operative in extreme environments, slurry pump seals, acid pump seals, valves, knife sharpeners, cutting tools, metal-working tools, gas turbine blades of aircraft engines, gun steels, magnetic heads, tape-path components of computer tape drives
Surface treatments	
Surface hardening	Camshafts, crank shafts, sprockets and gears, shear blades, and bearing surfaces
Diffusion	Cams and camshafts, rockets and rocker shafts, bushes, bearings, transmission gears, drills, milling cutters, extrusion punches, slitting saws, taps, die casting dies, pump parts, cylinder liners, and cylinder heads
Ion implantation	Ball bearings, dies, press tools, injection molds

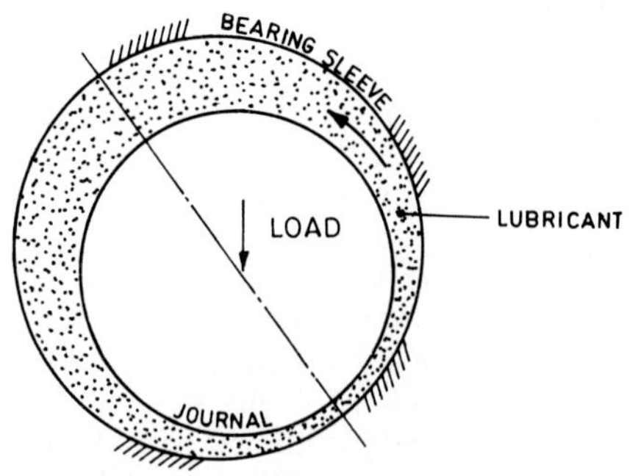

FIGURE 16.1 Schematic diagram of a journal bearing.

1981; Bhushan and Winn, 1979; DeHart, 1984; Fuller, 1984; Glaeser, 1980; Inwood and Garwood, 1978; Neale, 1973; Pratt, 1973). Included in the metals are several types of soft metals—babbitts, copper-based materials, aluminum-based materials, cast iron, silver, and porous metals. The nonmetals include wood, carbon-graphites, plastics, elastomers, ceramics, cermets, and several other proprietary materials. These materials can be used as bulk material or as a lining on a bearing surface. In a lined bearing, the bearing material is bonded to a stronger backing material such as steel. The thickness of liner material usually ranges from 0.25 mm to as high as 10 mm. Many soft metals, carbon-graphites, plastics, and elastomers are used as one of the slider materials against a sliding member such as stainless steel under unlubricated conditions. In many bearing applications, tribological materials are used to coat bearing substrates by various deposition techniques.

Many soft lead- and tin-based babbitts bonded to a harder and stronger shell are used as one of the sliding surfaces in oil-lubricated bearing systems designed for very low unit loads (on the order of 1.5 MPa) with life expectancies of over 20 years. The effective strength (fatigue rating) of babbitts can be improved by reducing their thickness (Pratt, 1973). The highest-strength babbitts are obtained by electrodeposition of lead and tin or lead, tin, and copper onto a bearing substrate. Copper-based bearing materials are widely used both as solid walls and as multilayered materials in which the copper is either cast or sintered onto a steel backing. Copper-based materials are classified as tin, lead, or aluminum bronzes. Bronze bushings are one of the most common types of bearings for dry and liquid-lubricated bearing systems designed for low to moderate loads.

Aluminum-based bearings have superior corrosion resistance and good thermal conductivity. Aluminum has long been used as a solid-wall bearing in heavy- and medium-duty diesel engines as an Al-Sn-Si-Cu (SAE 780) or Al-Sn-Cu-Ni (SAE 770) alloy. Cast iron has been used successfully against steel journals. Electroplated silver bearings are used in specialized applications, such as high-performance diesel engines and turbochargers.

Porous-metal bearings are used where plain-metal bearings are impractical because of lack of space or inaccessability for lubrication. These bearings are made of powdered metals that are pressed in dies. After compression, they are sintered at a high temperature in a reducing atmosphere, which causes the powdered metal to fuse into a strong compact. After sintering, the bearings are submerged in oil for impregnation and finish-sized. Porous bearings are strong and contain from 10 to 35 percent interconnected porosity (voids). Porous bearings also have been impregnated with plastics, such as polytetrafluoroethylene (PTFE). The most common sintered materials are babbitts (e.g., 90% Cu, 10% Sn), leaded bronzes, iron bronzes, iron with copper, and leaded iron.

Carbon-graphites, plastics, and elastomers are self-lubricating materials that are used as bearing liners in both dry and lubricated applications where loads, speeds, and temperatures are low. Carbon-graphite bearings can operate above 200°C under unlubricated conditions. At this temperature, most oils and greases start to decompose in the presence of air. Some plastic materials soften and lose their load-carrying capacity. Carbon, in contrast, actually grows stronger and harder with increasing temperature. Service applications of plastic bearings are limited by low thermal conductivity, low thermal expansion, and thermal degradation. The most commonly used plastics are PTFE, polyimide, Nylon, acetal, polyethylene, polyphenylene sulfide, and phenolics with various fillers, such as MoS_2, graphite, and glass, and various resins in the form of powder or fibers, such as phenolics or polyester with a woven fabric. Phenolics and butadiene-acrylonitrile rubber are commonly used as stave materials in stave-type stern-tube bearings in commercial ships in the presence of oil and water (Bhushan et al., 1982; Bhushan and Dashnaw, 1981; Bhushan and Winn, 1981; Paxton, 1979; Smith, 1967).

A number of ceramics and cermets (ceramics that have been bonded with metals to improve their impact resistance) are used for gas-bearing applications. Ceramics provide low wear, but they have relatively high friction. An alternative technique is burnishing the surface with either MoS_2 or PTFE. Aluminum oxide or silicon nitride against itself is commonly used in instrument gyroscope air bearings.

A large number of coatings are used for sliding bearings. Coatings of MoS_2, graphite, PTFE, polyphenylene sulfide, polyimide, poly(amide-imide), and their combinations applied by atomized liquid spray and other proprietary techniques to a typical thickness range of from 7.5 to 50 μm are commonly used for foil or other types of gas-bearing applications (Bhushan, 1980a, 1982; Bhushan et al., 1978). Sputtered MoS_2 coatings and ion-plated lead coatings with 50- to 500-nm thicknesses exhibit excellent adhesion and are used for high-vacuum applications. Despin bearings for communication satellites, gimbal bearings for reaction motors and momentum wheels of spacecraft, bearings for antennas, gyro bearings, and gears and bearings for spacecraft harmonic drive assemblies are presently MoS_2-coated. Soft-metal coatings of gold, silver, and lead are also used for high-vacuum service.

Hard coatings deposited by hardfacing, vacuum, and other deposition processes are successfully used in various gas-bearing applications. Bhushan (1980b, 1981) developed a sputtered Cr_2O_3 coating (with a thickness of about 1 μm) for sliding against detonation gun–sprayed Cr_3C_2 coating (with a thickness of about 75 μm), for foil air bearings for operation to an ambient temperature of up to 1200°C. Plasma-sprayed Cr_2O_3 coating against itself and tungsten carbide coating applied by chemical vapor deposition against itself are used in several air-bearing systems. Various other coatings of oxides, carbides, and nitrides are also used for gas-bearing applications.

16.2 ROLLING-CONTACT BEARINGS

Rolling-contact or rolling-element or antifriction bearings employ a number of balls or rollers between two surfaces known as *inner* and *outer races* (Fig. 16.2). These elements accommodate relative motion between surfaces primarily by the action of rolling (with a small slip) rather than sliding so that the only frictional forces remaining between the surfaces are due to rolling resistance. Because of the use of balls or rollers, the actual area of contact is reduced to near zero; therefore, contact stresses are very high (hertzian stresses), typically on the order of 500 MPa.

Rolling bearings are generally lubricated with a liquid oil or grease. Some bearings for operation in vacuum and high temperatures must be self-lubricated. The classic rolling-bearing failure mode is fatigue spalling, in which a sizable piece of the contact surface is dislodged during operation by fatigue cracking in the bearing metal under cyclic contact stressing. When the surface motion at rolling-bearing contacts contains substantial sliding, the surface damage and wear can change abruptly from mild to severe adhesive wear, commonly called *scuffing* or *smearing* (Scott, 1977; Tallian, 1967).

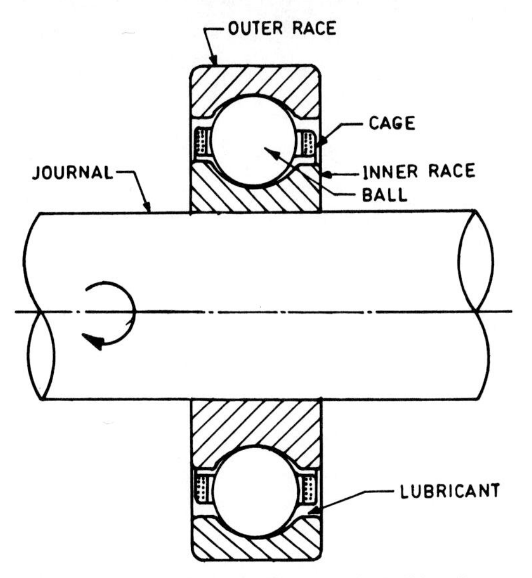

FIGURE 16.2 Schematic diagram of a ball bearing.

16.2.1 Bearings and Races

The bearing industry has used SAE 52100 steel as a standard material since the 1920s. This is a high-carbon chromium steel that also contains small amounts of Mn, Si, Ni, Cu, and Mo. A minimum tolerable hardness for bearing components is about 58 HRC. SAE 52100 steel can be used only up to temperatures of about 200°C. Molybdenum air-hardening steels, known as *high-speed steels* (a class of tool steels), e.g., AISI M-1, M-2, and M-50, can be used to a maximum temperature of about 430°C. For a corrosive environment, AISI 440C stainless steel with a hardness of about 60 HRC is used for temperatures up to 180°C (Bamburger, 1980; Neale, 1973; Zaretsky and Anderson, 1970).

Carburized or case-hardened steels are also used for rolling bearings. These materials are characterized by a hardened surface (greater than 400 μm thick) and a soft core. The room-temperature hardness of most carburized bearing steels is roughly 58 to 63 HRC, with a core hardness of 25 to 40 HRC. Steels such as AISI 4320 and 4620 are commonly used for temperatures up to about 200°C. N^+ ion implanation of SAE 52100 bearing steel has been shown to improve rolling-contact fatigue life (Dearnaley, 1982; Fischer et al., 1983).

Some high-performance applications dictate the need for bearings that operate at high speeds and/or high temperatures (up to 1200°C) with high precision. Since high temperatures are beyond the range in which most ferrous materials are incapable of operating, the more refractory materials and compounds are considered (Baumgartner, 1978; Bhushan and Sibley, 1982; Neale, 1973). Selected physical properties of various high-temperature bearing materials are presented in Table 16.2. Several Co-Cr-W refractory alloys (such as Haynes 6B, Haynes 6K, Haynes 25, Stellite 1000, and Stellite 3) have been used as rolling-bearing materials. Silicon nitride is the favorite ceramic material for high-performance applications. It has two to three times the hardness and one-third the dry friction coefficient of bearing steels, has good fracture toughness compared with other ceramic materials, and maintains

TABLE 16.2 Selected Physical Properties of Rolling-Bearing Materials

Material	Hardness at 20°C, HRC	Maximum useful temperature,* °C	Density, kg m^{-3}	Elastic modulus at 20°C, GPa	Poisson's ratio	Thermal conductivity, W m^{-1} °C^{-1} 20°C	Thermal conductivity, W m^{-1} °C^{-1} 800°C	Coefficient of thermal expansion, × 10^{-6} °C^{-1} (0–800°C)
52100 steel	58	200	7,800	200	0.28	—	—	12.4
440C+Mo steel† (4% Mo)	62	260						
Cr-Mo-V steel† (BG-42)	63	370	7,800	200	0.28	24	—	10.1 (100°C)
Cr-Co-Mo-V-W steel† (Cartech EX-00007)	66	450						
M-50 steel	64	320						
M-1 steel	66	430	7,600	190	0.28	56	—	12.3 (300°C)
M-42 steel	69	540						
Co-Cr-W alloy (Stellite 3)	54	650‡	8,600	210	0.28	9	19	14.1
WC-Co (96-4)	72	500	15,100	655	0.24	84	74	7.6
TiC-Ni-Mo-NbC cermet (K162B)	67	800	6,300	390	0.23	57	28	10.7
Al_2O_3	85	1000	3,900	350	0.25	30	7	8.5
SiC	90	1200	3,200	410	0.25	146	50	5.0
Si_3N_4	78	1200	3,200	310	0.26	30	20	2.9

*Maximum operating temperature when used for rolling bearings based on a minimum hot hardness of 57 HRC after long-term soaking at temperature, compared with a maximum temperature of 180°C for 52100 steel.

†These steels have significant stainless properties, whereas the others do not.

‡Special bearing designs must be used with superalloys to compensate for their low hardness.

its strength and oxidation resistance up to 1200°C, which makes it a promising high-temperature rolling-bearing material (Bhushan and Sibley, 1982). Si_3N_4 has been used for bearing elements as well as for complete bearing configurations. Silicon nitride bearings are used in the aerospace/defense, tool spindle, chemical processing, nuclear, and automotive industries.

Various coatings have been developed for lightly loaded, long-life rolling-element bearings for applications in high vacuum and/or high temperatures, especially in applications with no external lubrication. Plated gold is used on some space bearings. Hard coatings of materials such as TiC or TiN applied by chemical vapor deposition (CVD) onto balls and races provide low friction and significantly longer wear life compared with uncoated bearing (typically as much as eight times) under extreme conditions, i.e., vacuum, low pressure, and high temperature (Boving et al., 1983; Hintermann, 1981, 1984). TiC CVD-coated cemented carbide ball bearings are used for gyro applications. These ball bearings have shown a life of 25,000 hours at journal speeds of 24,000 r/min.

Substrate structural changes accompanying high-temperature CVD coatings can be avoided by sputter coating. Brainard (1978) showed that thin coatings (200 to 300 nm thick) of hard materials, such as TiC, Mo_2C, Mo_2B_5, and $MoSi_2$, deposited by rf sputtering onto a 440C stainless steel substrate showed considerable wear-life improvement. Experiences with both hard and soft coatings suggest that duplex coatings (soft-coating overlays on hard coatings) may be satisfactory for unlubricated applications at high vacuum and/or high temperatures. A soft overlay facilitates run-in and provides low friction in addition to low wear. Bhushan (1980c) reported that sputtered duplex coatings of MoS_2/TiN and MoS_2/Cr_2O_3 combinations onto 440C stainless steel balls, 440C stainless steel races, and phenolic cage surfaces showed considerable wear-life improvement under unlubricated conditions.

16.2.2 Retainers or Cages

Retainers or cages maintain the proper distance between rolling elements. In conventional rolling bearings, both metallic and nonmetallic retainers are used. Under normal temperatures, a large percentage of ball and roller bearings use stamped retainers of low-carbon steel or machined retainers of copper-based alloys such as iron-silicon bronze or a lead brass. In applications where marginal lubrication exists during operation, silver-plated bronze or PTFE-based cage materials are used (Zaretsky and Anderson, 1970).

For high-temperature and marginally lubricated or unlubricated conditions, such as in aerospace applications, potential self-lubricated cage materials are phenolic, polyimide, graphite and composites, Ga-In-WS_2 compact, Ta-Mo-MoS_2 compact, and metals containing MoS_2-graphite (Bhushan and Sibley, 1982; Gardos and McConnell, 1982). A Ga-In-WSe_2 material has been found to be an excellent high-temperature solid-lubricant additive to a carbon fiber–reinforced polyimide composite. This composite has exhibited low friction and wear when tested against an Ni-based alloy (Rene 41) at a normal stress of about 175 MPa and a temperature of 315°C (Gardos and McConnell, 1982).

16.3 SEALS

The seals, commonly called *fluid seals*, are divided into two main classes: static and dynamic seals. Static seals are gaskets, O-ring joints, packed joints, and similar devices used to seal static connections or openings. A dynamic seal is used to restrict fluid flow through an aperture closed by relative moving surfaces. Figure 16.3 is a schematic diagram of the mechanical face seal or end face seal used for sealing the annular space between a rotating shaft and a housing.

The primary function of seals is to limit loss of lubricants or process fluid from systems and to prevent contamination of systems by the operating environment. Solid contact takes place between two surfaces, where one element of the primary sealing interface moves and the other is stationary. Lubrication of the sealing interface varies from hydrodynamic to no lubrication (e.g., in gas-path components such as a turbine or compressor blade tips). Adhesive wear is the dominant type of wear in well-designed seals. Other wear modes are abrasive wear, corrosive wear, fatigue wear, and blistering (Johnson and Schoenherr, 1980; Neale, 1973; Stair, 1984).

Materials used for seals depend on the seal design and the operating requirements. Mechanical face seals employed in oil-lubricated applications range from very hard combinations such as ceramics (e.g., alumina, tungsten carbide, silicon carbide, boron carbide) and cermets (e.g., cemented carbides) to babbitts, bronzes, carbon-graphites, thermoplastic resins, and elastomers. Carbon-graphites and polymers are also used in the presence of process fluids that are poor lubricants or under unlubricated conditions. A large number of mating materials have been used, such as Niresist iron, tool steel, hard-faced steel, nickel-copper base materials (e.g., Monel), nickel-molybdenum alloys (e.g., Hastelloy B or Hastelloy C), Co-based alloys (e.g., Stellite), tungsten carbide, boron carbide, alumina, or plasma-sprayed coatings of various ceramics. For example, many merchant ships employ asbestos-filled phenolic in oil-lubricated face-seal configurations as the stationary-face insert. The rotating mating shaft is made of Niresist iron or chrome-nickel steel. Carbon-graphites and plastics such as PTFE, polyimide, poly(amide-imide), or phenolic composites are used in water-lubricated face seals (Bhushan and Wilcock, 1982; Bhushan and Winn, 1979; Paxton, 1979; Stair, 1984).

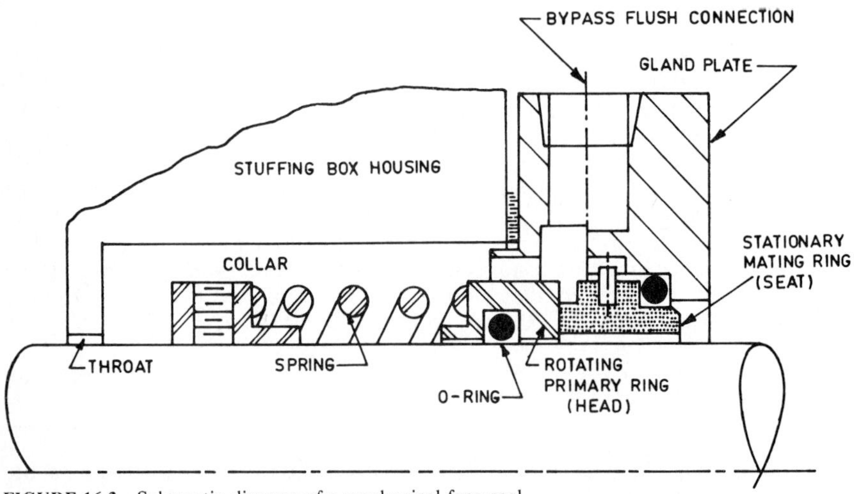

FIGURE 16.3 Schematic diagram of a mechanical face seal.

Lip seals are made of compliant materials—elastomers. The elastomers most commonly used are butadiene-acrylonitrile (Buna N), polyacrylate, and vinylidene fluoride–hexafluoropropylene (Viton) at high temperatures (up to about 170°C). Other lip-seal materials that are used for specialized applications include silicone, fluorosilicone, and perfluoroelastomer (Kalrez). The main prerequisite for the mating surfaces of lip seals are hardness and high abrasion resistance. The liner materials include chrome-nickel steels, case-hardened cast iron, and plasma-sprayed coatings of various ceramics, such as Cr_2O_3, Al_2O_3, or mixtures of oxides with other materials to achieve specific characteristics (Bhushan and Winn, 1979, 1981).

Abradable (rub tolerant) seals are used in the compressor and turbine sections of aircraft gas turbine engines (Meetham, 1981). Abradable seals are used at the rotor-stator interface to maintain the close tolerance without catastrophic failure. One of the sliding members is supposed to abrade against another if there is any interference. Abradable coatings are applied to compressor and turbine castings in some engines. In some cases, wear-resistant coatings are applied to the tip regions of rotating blades to minimize wear. Many of the commonly used abradable materials are sintered ceramics and plasma-sprayed coatings of Ni-Cr–bonded chrome carbide and tungsten carbide sliding against rub-tolerant materials such as felt metal and plasma-sprayed coatings of ceramics such as chrome oxide.

16.4 GEARS

Gears are toothed wheels used for transmission of rotary motion from one shaft to another (Fig. 16.4). Contact occurs on lines or points, resulting in high hertzian contact stresses. The gear motion is associated primarily with rolling and some sliding motions. Gear teeth may operate under boundary, mixed, and fluid-film lubrication regimes. Typical failure modes of gears are scoring, pitting, scuffing, abrasion, and corrosive wear. The essential material requirements for gears are adequate bending fatigue strength and resistance to surface faitgue, adequate toughness to withstand the impact loads, adequate resistance to scuffing, and adequate resistance to abrasive wear (Coleman, 1970; Dudley, 1964).

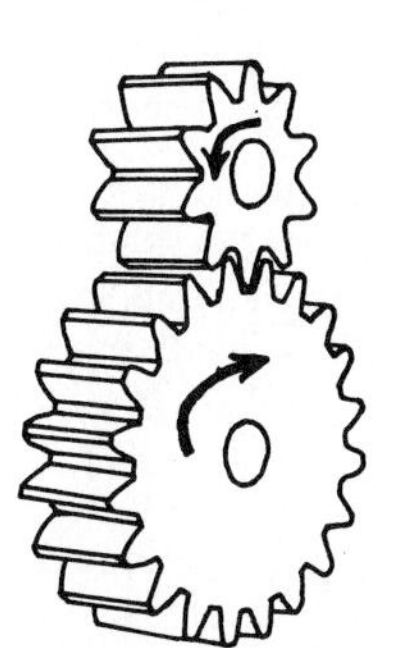

FIGURE 16.4 Schematic diagram of meshing spur gears.

For gears of useful load capacity, hard materials are generally required because high hertzian stresses occur at contact spots. The wear of gears is reduced by heat treatments or thermochemical treatments or by applying coatings. The most commonly used gear material for power transmission is steel. Cast iron, bronze, and some nonmetallic materials are also used. In instrument gearing, toys, and gadgets, such materials as aluminum, brass, zinc, and plastics (such as Nylon) are often used. Nonmetallic gears made of plastics are generally used against steel gears. The through-hardened steels for gears should have about 0.4 percent C. Their hardnesses range from 24 to 43 HRC. Cast-iron gears are generally not used above 250 Brinell (24 HRC).

Cast-iron gears have unusually good wear resistance for their low hardness (Dudley 1964, 1980; Merritt, 1971; Neale, 1973).

Case-hardening (carburizing) methods are used to put a thin, hard case on a medium-hard core for gears. When the hardness is over about 400 Brinell (43 HRC), gear teeth become too brittle when through-hardened. The case-hardening treatments, when properly done, make a gear tooth that is tough and has good to excellent strength. Case-carburized gears at 55 to 63 HRC have the best capability when the surface layer of metal is around 0.8 to 0.9 percent C. Some improvements in wear resistance may occur if the outer layer is 1.0 to 1.1 percent C at the expense of a reduction in tooth-breaking strength. Other surface treatments that are used for steel gears include nitriding, Sulphinuz, and phosphating to reduce friction and wear.

The performance limits of each gear of a pair can be rated in terms of maximum allowable transmitted power, which is proportional to allowable bending stress S_{at} for the gear material for one-way bending. The allowable power rating also should be checked for each gear of a pair in terms of the risk of tooth pitting, which is denoted by allowable contact stress S_{ac} for the gear material. In other words, the performance limits of various gear materials can be compared in terms of allowable bending stress S_{at} and allowable contact stress S_{ac}. Table 16.3 compares these numbers for a variety of surface-treated steels, cast irons, and bronzes.

It is common practice, except where both members of a gear pair are surface-hardened, to have the two members of different strengths and usually of different alloys. This has been found to reduce the likelihood of scuffing. The smaller gear (pinion), which is the high-speed member and has the more arduous and frequent duty, is usually the harder member. Table 16.4 presents common material combinations suitable for the principal types of gears.

16.5 CAMS AND TAPPETS

Cams and tappets (or cam follower systems) are extensively employed in engineering machines to transform rotary motion to reciprocating sliding motion or vice versa, e.g., in automotive valve trains and textile machines. The cam follower can be a flat follower or a roller follower (Fig. 16.5). The contact conditions are nominal points or line contacts which under load lead to elliptical and rectangular contact areas, respectively. There is always a rolling motion through the contact, accompanied by some sliding in the direction of rolling motion. Under heavy duty, cams and tappets suffer from burnishing (due to adhesive/abrasive wear processes), scuffing (due to severe adhesive wear processes), and pitting (due to fatigue wear processes) (Neale, 1973). The wear modes for cams and tappets are very similar to those for gears.

The wear of cams and cam followers can be reduced considerably by selecting hard material combinations or by hardening the cam material by heat treatments or thermochemical treatments or by applying coatings. Tappet materials are usually through-hardened high C, Cr, or Mo types of carburized low-alloy steels. The most common tappet material in automotive applications is gray hardenable cast iron containing Cr, Mo, and Ni or chilled cast iron.

The types of coatings and surface treatments that can be adapted for cam and tappet materials to reduce wear are as follows:

- Running-in coatings: These include phosphate coatings, chemically produced oxide coatings on ferrous metals, and electrochemically deposited Sn and Al.

TABLE 16.3 Hardness, Bending Stress, Contact Stress Allowable in Various Heat-Treated Steels, Cast Iron, and Bronze Materials Used for Gear Applications

Material	AGMA class	Commercial designation	Heat treatment	Minimum hardness surface (HV)	Allowable bending stress number S_{at}, MPa	Allowable contact stress number S_{ac}, MPa
Steel	A-1 thru A-5	—	Through-hardened and tempered	180 BHN (208)	170–230	590–660
				240 BHN (250)	210–280	720–790
				300 BHN (317)	250–320	830–930
				360 BHN (380)	280–360	1000–1100
				400 BHN (423)	290–390	1100–1200
			Flame- or induction-hardened*	50–54 HRC (513–577)	310–380	—
			Flame- or induction-hardened	50–54 HRC (513–517)	150	1200–1300 1200–1300
			Carburized and case-hardened	55 HRC (595)	380–450	1250–1400
				60 HRC (697)	380–480	1400–1550
		AISI 4140*	Nitrided†	48 HRC (484)	230–310	1100–1250
		AISI 4340	Nitrided†	46 HRC (458)	250–325	1050–1200
		Nitralloy 135 M	Nitrided†	60 HRC (697)	260–330	1170–1350
		2.5% Chrome	Nitrided†	54–60 HRC (577–697)	380–450	1100–1500
Cast iron	20		As cast	—	35	340–410
	30		As cast	175 BHN (195)	69	450–520
	40		As cast	200 BHN (228)	90	520–590

TABLE 16.3 Hardness, Bending Stress, Contact Stress Allowable in Various Heat-Treated Steels, Cast Iron, and Bronze Materials Used for Gear Applications (*Continued*)

Material	AGMA class	Commercial designation	Heat treatment	Minimum hardness surface (HV)	Allowable bending stress number S_{at}, MPa	Allowable contact stress number S_{ac}, MPa
Nodular (ductile iron)	A-7-a	60-40-18	Annealed	140 BHN (162)	90 to 100 percent of S_{at} value of steel with same hardness	90 to 100 percent of S_{ac} value of steel with same hardness
	A-7-c	80-55-06	Quenched and tempered	180 BHN (208)		
	A-7-d	100-70-03	Quenched and tempered	230 BHN (242)		
	A-7-e	120-90-02	Quenched and tempered	270 BHN (285)		
Malleable iron (pearlitic)	A-8-c	45007	—	165 BHN (197)	70	500
	A-8-e	50005	—	180 BHN (208)	90	540
	A-8-f	53007	—	195 BHN (225)	110	570
	A-8-i	80002	—	240 BHN (252)	145	650
Bronze	Bronze 2	AGMA 2C	Sand cast	Tensile strength minimum: 275 MPa	40	205
	Al/Br 3	ASTM B-148-52 Alloy 9C	Heat-treated	Tensile strength minimum: 620 MPa	160	450

*Core hardness 300 BHN (317 HV).
†The overload capacity of nitrided gears is low.

TABLE 16.4 Common Material Combinations Suitable for the Principal Types of Gears

Gear duty	Material combination	
Motion only	Plastics, brass, mild steel, stainless steel in any combination	
Light power	Carbon steel	Brass Plastics Cast iron Steel
Worm drives	Alloy steel	Cast iron Phosphor bronze
High duty, industrial and marine	Alloy steel	Carbon steel Alloy steel
	Nitrided alloy steel	Nitrided alloy steel Alloy steel
Automotive	Carburized case-hardened alloy steel	Carburized case-hardened alloy steel
Aircraft and high duty	Carburized case-hardened and ground alloy steel	Carburized case-hardened and ground alloy steel

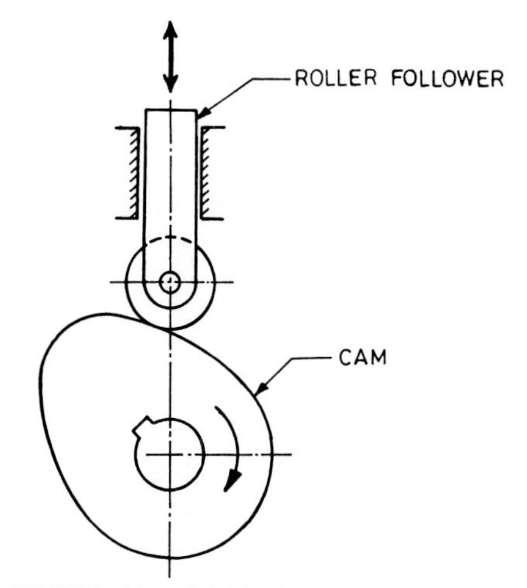

FIGURE 16.5 Schematic diagram of a cam and a translating roller follower (tappet).

- Hard surfaces: Surfaces of cams can be hardened through diffusion treatments such as carburizing, nitriding, and Tufftriding.
- Hard coatings: Several hard coatings such as TiN and TiC, applied by PVD and CVD, also can be applied on cams and tappets to achieve low coefficients of friction and low wear.

16.6 PISTON RINGS

Piston rings are mechanical sealing devices used for sealing pistons, piston plungers, reciprocating rods, etc. In gasoline and diesel engines and lubricated reciprocating-type compressor pumps, the rings are generally split-type compression metal rings. When they are placed in the grooves of the piston and provided with a lubricant, a moving seal is formed between the piston and the cylinder bore. Piston rings are divided into two categories: compression rings and oil-control rings. Compression rings, generally two or more, are located near the top of the piston to block the downward flow of gases from the combustion chamber. Oil rings, generally one or more, are placed below the compression rings to prevent the passage of excessive lubricating oil into the combustion chamber yet provide adequate lubrication for the piston rings. In typically lubricated situations, the piston skirt is in direct contact with the cylinder and acts as a bearing member that supports its own weight and takes thrust loads. In unlubricated arrangements, it is necessary to keep these two surfaces separated because they are not frictionally compatible. This is usually accomplished with a rider ring that supports the piston (Neale, 1973) (Fig. 16.6).

An ideal piston-ring material must meet the following requirements: low friction and wear losses, superior scuffing resistance, tolerances for marginal lubrication and rapidly varying environments, good running-in wear behavior, long-term reliability and consistency of performance, long maintenance-free life, and low production cost.

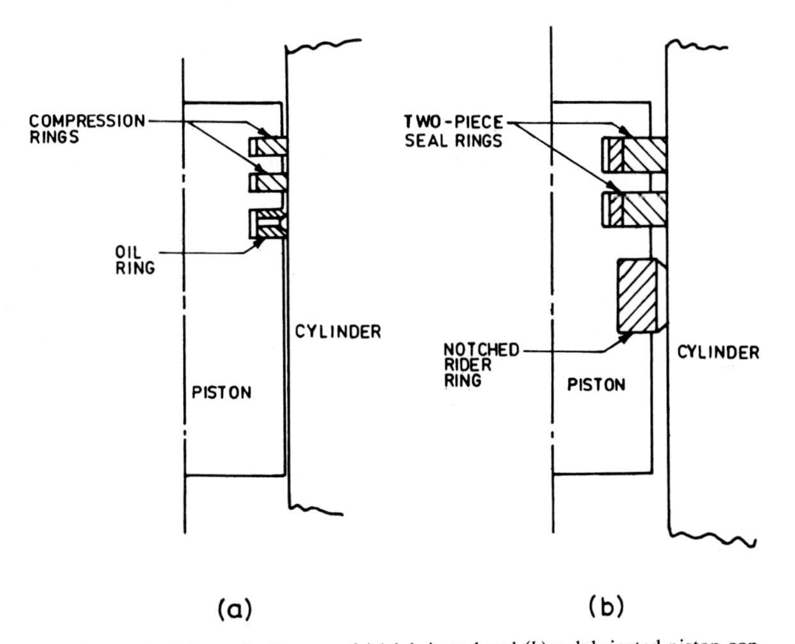

FIGURE 16.6 Schematic diagram of (*a*) lubricated and (*b*) unlubricated piston configurations.

16.6.1 Lubricated Piston Rings

Gray cast iron with a hardness ranging from 200 to 400 Brinell is probably the most commonly used material for compression and oil rings. Table 16.5 shows the various types of cast irons used in gasoline engines. Pearlitic gray iron produced by either centrifugal or sand-cast methods is most widely used. In the heavier-duty engine applications, chromium-molybdenum alloy iron, spheroidal graphite iron, and carbidic malleable iron are used. Harder materials such as carbon steel or even En31 ball-bearing steel are also used.

Relatively thick coatings (up to 0.2 mm) of plated chromium on the ring periphery provide the best compromise between scuffing, wear, and corrosion resistance and low friction and resistance to oxidation at high temperatures; only one mating surface is coated (Inwood and Garwood, 1978). Generally, the use of chromium-plated top rings (with a hardness of 700 to 900 HV) run against cast-iron cylinder liners can reduce the ring and liner wear by a factor of 2 to 3. Rings coated with flame- or plasma-sprayed coatings of molybdenum (in thicknesses up to 0.25 mm) with a hardness of over 1000 HV are believed to have higher scuffing resistance than chromium-plated rings. The main limitation of molybdenum ring coatings is that they are subject to oxidation at 500°C, and at 730°C the oxide volatizes.

Several plasma-sprayed coatings of composites, such as Mo-Cr-Ni alloy, ceramics, such as chromium oxide, have been developed for achieving improved scuffing resistance under conditions of marginal lubrication (Neale, 1973; Scott et al., 1975; Taylor and Eyre, 1979; Ting, 1980). Promising results have been obtained with plasma-sprayed coatings of 88 percent WC/CO, 75 percent Cr_3C_2/Ni-Cr, and 75 percent Cr_3C_2/Mo, which can give better performance for operation in an abrasive environment or where their superior oxidation resistance is required. Numerous other metallic and ceramic coatings and surface treatments have been tested for piston-ring applications (Scott et al., 1975; Taylor and Eyre, 1979).

A wide variety of running-in coatings have been used on piston rings in order to reduce the scuffing resistance. The following are typical examples:

- Iron oxide (Fe_3O_4) coatings, which can be developed just by heating piston rings in steam atmospheres at 500°C
- Phosphating treatments, such as Granodizing and Parco-lubrizing (proprietary processes), which produce soft, porous, and crystalline manganese phosphate coatings that provide better retention of lubricating oil and shelf life
- Sulfinuz treatment, which involves the diffusion of sulfur, nitrogen, and carbon to produce a surface layer of increased hardness that has increased abrasive wear and fatigue resistance. This treatment also increases scuff resistance and reduces the coefficient of friction.
- Tufftriding, which produces a less porous and some what harder surface than Sulfinuz in the absence of sulfur so as to give a reduced scuff resistance
- Soft metal coatings, which are produced by electrochemically depositing coatings of Cd, Sn, or Cu to provide storage protection and to assist running-in. The wear resistance of these coatings is very low, and they probably assist running-in by quickly adapting the necessary profile for good oil-film generation and blow-by control.

Most cylinder liners are made of gray cast iron. To increase the liner mechanical strengths, nickel, chromium, copper, molybdenum, titanium, and vanadium

TABLE 16.5 Typical Piston Ring Materials

			Properties			Range of composition, wt% (balance Fe)								
Type	Approx. spec.	Method of manufacture	UTS, MPa	E, GPa	HB	C	Si	Mn	S	P	Ni	Cr	Mo	Others
Gray cast iron	BSS 4K6	Centrifugally cast	245	117	210 min.	Max. 3.5	1.8–2.5	Max. 1.2	Max. 1.0	0.40–0.65	Max. 0.4	Max. 0.5	Max. 0.4	
	BSS 4K6	Sand cast	245	103	200–245	Max. 3.5	1.0–1.8	0.6–1.2	Max. 0.15	0.25–0.60	Max. 0.4	Max. 0.4	Max. 0.4	
	DTD 233A	Centrifugally cast	310	110	255–296	3.1–3.4	2.1–2.5	0.7–1.0	Max. 0.08	0.40–0.65	—	0.3–0.6	0.5–1.0	
Carbidic/ malleable iron	DTD 485A	Individually cast	585	159	270–320	2.7–3.3	2.0–3.0	0.5–0.9	Max. 0.1	Max. 0.5	—	0.5–0.85	—	V Max. 0.4
	DTD 485A	Centrifugally cast	400	155	259–302	2.75–3.3	1.8–2.5	0.6–1.0	Max. 0.12	0.3–0.5	Max. 0.3	0.65–1.15	0.7–1.0	Al Max. 0.5
Malleable/ nodular iron		Centrifugally cast	540	155	200–400	3.0–3.3	1.0–1.4	0.6–0.9	Max. 0.1	Max. 0.1	Max. 0.3	0.1–0.5	—	
	BS2789 SNG	Sand or centrifugally cast	570	165	240–297	Varied according to mechanical properties required								
Sintered		Sintered	385	117	150 HV	0.9–2.0	Max. 0.3	Max. 0.5	Max. 0.2	Max. 0.2	—	—	0.4–1.7	Cu 2.0–4.5

Key: UTS, ultimate tensile strength; E, modulus of elasticity; HB, Brinell hardness number (kg mm^{-2}).
Source: Adapted from Taylor and Eyre (1979).

are added. Steel cylinder liners also have been used, and their advantage is that the walls can be made much thinner. They have to be hardened to at least 400 Brinell, however, for satisfactory resistance to wear and scuffing, or they are hard-chromium plated. It should be noted that chromium-plated liners should not run against the chromium-plated piston rings, but rather against plain cast-iron rings or rings with other types of coatings. The Sulfinuz and Tufftriding treatments also have been used in cylinder liners and are claimed to be comparable with chromium plating for scuff resistance and as an aid in running-in. Aluminum cylinder liners with high silicon content also have been used. Some engine test results show that the liner wear of cylinders made from this material is lower than that of conventional cast-iron liners (Taylor and Eyre, 1979; Ting, 1980).

16.6.2 Unlubricated Piston Rings

Materials used for unlubricated piston rings are almost without exception nonmetals. This is due to the tendency of metals to weld under dry sliding (Fuchsluger and Vandusen, 1980; Scott et al., 1975). Exceptions are metals impregnated with lubricants or coated with wear-resistant materials. In the nonmetal category, plastics, carbons, and ceramics are the materials most widely used. Among plastics, filled PTFE is used most commonly. Another family of plastics is the filled polyimides. The polyimides, while not as chemically resistant or as low in friction as PTFE, have a high temperature limit (315 to 370°C) and are more rigid. Some other rigid plastics that find use as piston rings are poly(amide-imides), polyphenylene sulfides, and aramids. For lower-temperature (below 150°C), lightly loaded, low-speed applications, rings of Nylon, acetal, ultrahigh-molecular-weight polyethylene, etc. have found use.

Carbon is another commonly used ring material. It is the material most commonly used in the temperature range from 370 to 540°C. Below 370°C, carbon is limited by its reaction to varying degrees of humidity and its fragility. Ceramics have low wear rates, but virtually all ceramics are expensive to fabricate and are subject to thermal shock. Consequently, the use of ceramics for piston ring applications has been limited to ceramic coatings for metal rings.

The most commonly used mating liner materials for nonmetallic piston rings are cast irons and hardened steels. When other materials, such as the 300 and 400 series stainless steels, must be used, a chrome-plated or nitrided wear surface frequently provides the compatibility needed. When neither course of action is open, a solid-lubricant coating is frequently used. Carbon-graphite is reported to have a lower wear rate in combination with Ni-resist or nitrided steel than with chrome plate.

16.7 ELECTRICAL BRUSHES

Machines that utilize electrical brushes can be broadly classified into two groups. In the first group, the machines require a commutator. In these machines, the brushes must be capable of transferring the load current to the external circuit as well as assisting the commutation function. Within this class of machines are dc motors and generators. In the second class of machines, brushes are used only to transmit electric power from a stationary source to a moving component by means of a slip ring (Fig. 16.7). Examples of slip-ring applications are ac generators, motors, and special applications. Brush wear is believed to be due to adhesion and particle transfer, while fatigue has been identified in some circumstance. A further wear mechanism that can occur is fracture caused by mechanical impact between the brush and the slip ring.

The need to transfer electric current efficiently across a sliding interface complicates the situation compared with normal wear. For example, it is not possible to consider reducing wear by using boundary lubrication or low-friction coatings (e.g., PTFE) on the surfaces, since this would cause unacceptably high electrical losses. For all brush applications, the brushes themselves are chosen as the sacrificial elements, so their wear rates exceed those of the machine slip ring or commutator by a factor of at least 10.

Most brushes are made of graphite-based materials such as electrographite, natural graphite, resin-bonded natural graphite, carbon and carbon-graphite, and metal-graphite. Graphite is picked for its low friction and wear and high electrical conductivity. Electrographites are synthetic graphitic materials characterized by a low coefficient of friction but a medium to high contact drop. Brushes of this material are among the most widely used because they exhibit good strength and controlled quality. Natural graphites have low density and low friction but exhibit a relatively high contact drop. They are particularly recommended for brushes operating on high-speed slip rings or commutators. Resin-bonded natural graphite is based on natural graphite that has been bonded with a phenolic or other resin. This produces high electrical resistivity and improved strength and gives good commutating ability for low operating current densities. Carbon-graphites, which are prepared by blending carbon and graphite and bonding them with a pitch or resin binder prior to baking, possess significant abrasive properties.

Metal-graphites are produced either by a PM process or by infilteration of porous graphite with metals. By far the most common metal constituent is copper, although a good range of silver-graphite brushes is also manufactured. Small

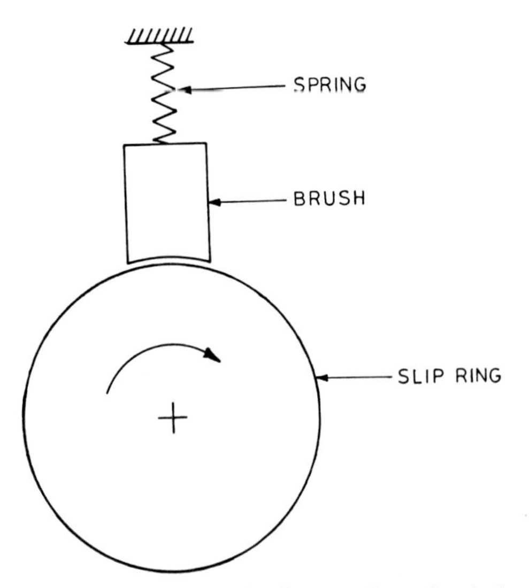

FIGURE 16.7 Schematic diagram of an electrical brush-slip ring.

amounts of other metals (e.g., lead and tin) also may to be added to provide improved bonding or reduced friction. Because of the metal constituent, these brushes have an appreciably lower electrical resistance than pure graphite or carbon-graphite grades. However, this is achieved at the expense of an increase in friction. Copper-graphite brushes are used where a high current-carrying capability is required (Anonymous, 1979; Shobert, 1965).

Copper alloys and steel are most commonly used in slip-ring commutators or countersurfaces against which the brush is operated, although noble and rare metals are sometimes used for small-scale or special applications. Copper is chosen because of its good electrical and thermal properties. Tough-pitch high-conductivity copper is most commonly used for commutators, although silver-bearing copper (<0.1 percent silver), chrome-copper (<1 percent chrome), and zirconium-copper (0.25 percent zirconium) also may be used, especially where higher strength at elevated temperatures is required. Common materials are bronze (copper-tin), phospher bronze (copper-tin-phosphorous), gun metal (copper-tin-zinc), cupronickel (copper, 4 percent nickel), and Monel (nickel, 25 to 40 percent copper). Steel slip rings are used in applications demanding high sliding speeds, which make copper alloys unusable because of their lower mechanical strength and high wear.

Electroplating or other surface treatments can be used to provide an improved current transfer on a higher-strength substrate (e.g., copper on steel or high-strength aluminum) and will give a performance close to that observed for the bulk material of the surface coatings. Good commutating performance can be achieved by using a sintered-copper facing on steel commutator bars (Anonymous, 1979; McNab and Johnson, 1980).

16.8 CUTTING TOOLS

Cutting tools are used to cut, shape, and form bars, plates, sheets, castings, forgings, etc. to produce engineering components. In metal cutting, tool wear occurs by adhesive, abrasive, diffusive, and electrochemical wear. The most important properties of a cutting tool material are hot hardness (i.e., resistance to softening under temperatures generated at the cutting edge of the tool), toughness, and chemical stability and reactivity. Other properties that are also relevant are elastic modulus, rupture strength, compressive strength, and coefficient of thermal expansion. Cutting-tool life is a most important factor in the economics of production.

Tool wear usually takes place on the face or the flank of a cutting tool (Fig. 16.8). Face or crater wear results from a chip moving across the face of the tool, whereas flank wear results from the rubbing action on freshly formed surface of the job. The extent and location of crater wear are considerably affected by the formation of a built-up edge composed of highly strained and hardened fragments of material (Tipnis, 1980). In general, the life of a cutting tool is judged by one of the following criteria: (1) complete failure of the tool, (2) cutting time for material removal to a predetermined crater depth or flank-wear land width, and (3) loss of workpiece dimensional tolerance and degradation of surface finish.

Tools can be made from anything from an elastomer to a diamond, but a few tool materials dominate. Of the different cutting-tool materials in use today, about 40 percent are high-speed steels (HSS) (a class of tool steel), about 10 percent are cemented carbides (cermets), about 30 percent are carbon, alloy, and

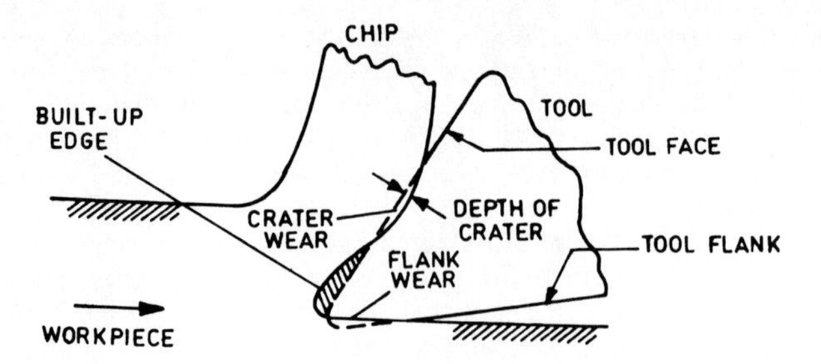

FIGURE 16.8 Crater (or face) and flank wear in cutting tools.

stainless steels, 5 percent are ceramics including diamonds and cubic boron nitride, and the rest are other materials (cast alloys, cast irons, nonferrous metals, and elastomers) (Budinski, 1980). According to some estimates, 60 percent of all carbide tools are coated grades.

A comparison of hot hardness (hardness as a function of temperature) for commonly used tool materials is presented in Fig. 16.9. Carbon steels are confined to hand tools, light-duty woodworking tools, and few minor industrial applications on metals. HSS and cemented-carbide tools are primarily used for most metal cutting and processing of hard materials. Various ceramics are used only for specialized applications. Coated tools exhibit considerable improvement in tool life and permit faster feed rates, their use becoming more and more common. Each machining operation and type of work material place a specific demand on a cutting tool. An approximate grouping of bulk tool materials applicable to different machining operations is presented in Table 16.6, and general uses for various tool material classes are given in Table 16.7.

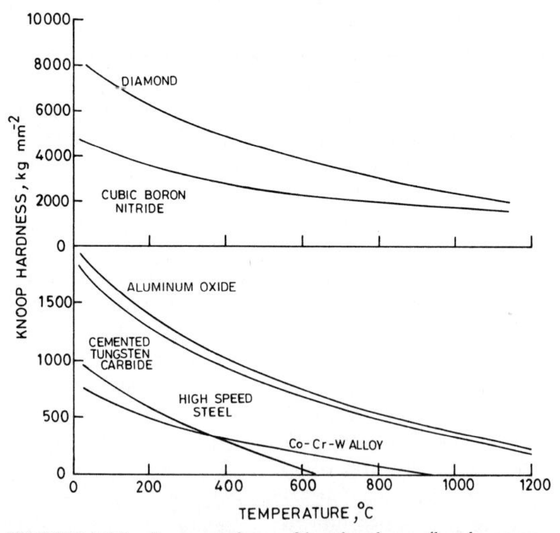

FIGURE 16.9 A comparison of hot hardness (hardness as a function of temperature) of high-speed steel, cemented tungsten carbide, alumina, cast Co-Cr-W alloys, cubic boron nitride, and diamond tool materials.

TABLE 16.6 Approximate Grouping of Tool Materials for Machining Operations

	Machining operation						
Tool class	Turning, facing, boring	Forming, parting, grooving	Planing, shaping, breaching	Milling, hobbing	Drilling, reaming, tapping	Tapping	Sawing
High-speed steel (HSS)	✔	+	+	+	+	+	+
Cast Co-Cr-W alloys	Δ	Δ	Δ	Δ	Δ	0	0
Cemented carbide	+	✔	✔	✔	X	0	Δ
Alumina/composite	X	0	0	Δ	0	0	0
Cubic boron nitride/diamond	Δ	0	0	0	0	0	0

Key: Most commonly utilized, +; frequently utilized, ✔; occasionally utilized, X; rarely utilized, Δ; and not used, 0.

TABLE 16.7 General Uses of Tool Material Classes

Tool class	Use
High-speed steel (HSS)	Major tool material—relative toughness plus retention of hardness at relatively high temperatures allows high cutting speeds, fine cutting edges, and use in rough conditions; relatively low cost
Cast Co-Cr-W alloys	Retains hardness to higher temperatures than HSS; less tough; cuts more difficult material or at higher speed; higher cost than HSS
Cemented carbide	Major tool material—higher hardness plus retention of hardness at much higher temperatures allows much higher cutting speeds and lower wear rates; less tough but tough enough for rough usage if edge not too fine; higher cost but used as tool tip only
Alumina/composite	Retains hardness to higher temperature than cemented carbide and much lower wear rate; this allows much higher cutting speeds; much less tough—use restricted to tools with strong edge and to good cutting conditions; main use turning and facing cast iron at 8 to 10 m s^{-1}; cost similar to cemented carbide
Cubic boron nitride/ diamond	Much higher hardness; less tough; very low wear rate; very high cost; mainly high-speed finishing of nonferrous metals for good surface finish and very close dimensional tolerances

16.8.1 Bulk Tool Materials

Carbon and alloy steels usually contain less than 5 percent total alloy content. This, combined with lower quality-control standards in their manufacture, makes them lower in cost than high-speed tool steels. However, these steels simply cannot equal the hardening and wear characteristics of tool steels. Low-carbon steels (such as AISI 1010 and 1020) can be hardened by cold working to less than 200 HB. Medium-carbon grades (such as AISI 1040 and 1060) can be hardened to 500 to 600 HB by direct-hardening or flame-hardening techniques. Alloy steels can be carburized (such as AISI 4320 and AISI 4620) or direct-hardened (such as AISI 4130 and AISI 4140) to a hardness no more than about 58 HRC.

Nearly 90 percent of the HSS tools are M series (with molybdenum as the major alloying element); the remaining are T series (with tungsten as the major alloying elements). Although T-HSS are somewhat superior in wear resistance, they are more difficult to grind. Cobalt-containing M-HSS can be heat treated to a higher hardness, 900 to 940 HV (67 to 70 HRC) versus 800 to 860 HV (64 to 66 HRC) for the non-cobalt-containing series. When tools are used in a corrosive environment, it is often necessary to use stainless steels, such as in the food industry and the chemical process industry. Of the various stainless steels, Cr-Ni steels with a martensitic structure, such as 440C, are most useful as tool materials. Type 440C is capable of being quench-hardened to 58 to 60 HRC, and its wear characteristics are similar to those of air-hardening cold-worked tool steel such as A2.

Cast Co-Cr-W alloys often contain some molybdenum and boron. In addition, vanadium, tantalum, and columbium as alloying elements and manganese and silicon as deoxidizers are generally present. The hardness of cast alloys ranges from 650 to 800 HV. The cast alloys, however, can withstand higher temperatures than HSS and hence provide properties that are in between those of HSS and sintered carbides.

Sintered or cemented carbide tool materials are made of finely divided carbide particles of tungsten, titanium, tantalum, niobium, and other refractory metals bonded with cobalt, nickel, nichrome, molybdenum, or even steel alloys and are produced by PM processes. The most commonly used cemented carbides are tungsten carbide with a cobalt binder. The most unique property of cemented carbides is their hardness. The carbide particles that make up the major portion of these composites are harder than any metal. For example, the hardness of tungsten carbide, which is the most commonly used carbide, is about 2000 HV. The high hardness makes this class of materials unmachinable with normal techniques.

Ceramic tool materials primarily consist of polycrystalline alumina (Al_2O_3), cubic boron nitride (CBN), and diamonds. The ceramic composite tool materials typically are composites of Al_2O_3 and TiC or WC. Oxide ceramics and cermets are manufactured by either sintering or hot pressing. The principal elevated-temperature properties of the oxide ceramic tool materials are high hardness, chemical inertness, and wear resistance. However, these materials are relatively brittle (low transverse rupture strength) compared with HSS and cemented carbides.

The hardness of CBN is two to three times that of sintered carbides. These materials can withstand very high temperatures without appreciable loss of hardness. This makes it possible to machine high-temperature Ni-based alloy work materials at tenfold the speeds normally employed with carbide tools. However, CBN is relatively brittle. Thin CBN wafers are compacted on carbide substrates to lend strength and to minimize cost. Natural and synthesized polycrystalline diamond compacts are used as tool materials. Diamond tools consist of 0.5-mm-thick layers of sintered polycrystalline diamond bonded on a cemented tungsten carbide substrate. These tools are not suitable for machining ferrous alloys because at high temperatures the diamond tends to react with the carbon in steel and cast irons.

16.8.2 Coatings

Coatings (typically 2 to 25 μm thick) of various ceramic materials, such as TiC, TiN, TiC_xN_y, TiO_xN_y, Al_2O_3, HfC, ZrC, TaC, HfN, and ZrN, on high-speed-steel and cemented-carbide substrates (usually WC-Co or WC-TiC-TaC-Co compacts) have been deposited successfully to increase tool life. The coatings have been deposited by various vapor-deposition techniques: activated reactive evaporation (ARE), ion plating, sputtering, and chemical vapor depositions (CVD). The use of CVD techniques for such coatings (e.g., TiC) on cemented tungsten carbide tools is well established. However, the CVD process is not suited to coat high-speed tools because of the high deposition temperature (1000 to 1200°C) of the process, which results in metallurgical changes and distortion of the tool. The low deposition temperatures of evaporation, ion plating, and sputtering are particularly attractive. Various surface treatments such as carburizing and nitriding of some steel tools and ion implantation are also found to improve the tool life considerably.

Laboratory data show an improvement in tool life typically by a factor of 2 to 10 depending on the type of cutting tool and the coating/treatment. A summary of wear-life improvement of various tool materials and deposition techniques is presented in Table 16.8. Cemented carbide tools coated with TiC, TiN, TiCN, HfN, and Al_2O_3 by CVD to a thickness of 4 to 10 μm and HSS tools and drills coated with TiN by magnetron sputtering to a thickness of 2 to 5 μm are most commonly used commercially.

Almost all CVD-coated carbide tools (98 percent) are used in turning, i.e., in continuous cutting operations. These tools are not very satisfactory for interrupted cutting, e.g., in milling operations. The cemented-carbide and HSS tools coated by sputtering or ion plating exhibit superior toughness and wear life compared with CVD-coated carbide tools (Hintermann, 1980; Kobayashi and Doi, 1978).

16.9 FORMING TOOLS

The performance of various forming or metal-working tools, such as forming punches and dies, used to blank-pierce metallic and nonmetallic sheets is usually dictated by the amount of wear, which is influenced by the kind and thickness of the sheet being drawn, the sharpness of the die radii, die construction and finish, lubrication, and hardness of the die material. The conventional materials used for making dies and punches include alloy cast irons (typical carbon content 2.8 to 3.5 percent C), tool steels (typically D2, W1, O2, and A2 grades), and sintered carbides (typically WC-Co) (Neale, 1973).

The application of hard coatings and surface treatments has resulted in marked improvements in the useful life of forming tools. Komiya (1982) has reported an improvement in useful life up to a factor of 10 for punches, dies, and taps coated with TiN applied by ion plating as compared with uncoated ones (made of tool steel). Hintermann (1980) has reported that application of a TiC, TiN, or TiC/TiN coating by CVD to the chromium steel punches in punch and die assemblies results in an improvement in wear life by a factor of 10 to 200. Plumb and Glaeser (1977) reported that applying a TiB_2 coating (8.7 μm thick) by CVD to A-6 and H-11 steel injection molding dies improved their wear life by two orders of magnitude.

Surface treatments such as carburizing, nitriding, boriding and aluminizing of steels and ion implantation also have been used to improve tool life. Surface treatment by boriding has been found to lead to significant improvements in wear life for deep drawing tools (Fiedler and Sieraski, 1971). Ion nitriding, boriding,

TABLE 16.8 Summary of Tool Wear-Life Improvement by Coating Deposition and Surface Treatments

Tools substrate	Coating material/ treatment	Coating thickness, μm	Deposition method	Improvement in wear life	Reference
Cemented-carbide cutting tool	TiC, TiN	8–10	CVD	Several times	Hintermann (1981, 1984)
Cemented-carbide cutting tool	HfN, TiC/Al_2O_3, Al_2O_3, TiC/TiN	—	CVD	HfN most superior	Oakes (1983)
Cemented-carbide (M15) cutting tool	(Ti, Al)N, TiN/TiC	—	CVD	Three times better with (Ti, Al)N	Knotek et al. (1987)
HSS drill	(Ti, Al)N, TiN	—	IP	Three times better with (Ti, Al)N	Knotek et al. (1987)
HSS (M-7) drill	TiN	1–2	MS	Fifty times	Sproul and Rothstein (1985)
Cemented-carbide cutting tool	TiN	5	MS	Several times	Ramalingam and Winer (1980)
Cemented-carbide cutting tool	TiN, HfN, ZrN	8–15	CVD, IP, MS, ARE	Hardness of IP, MS, ARE coatings at room temperature superior to CVD	Quinto et al. (1987)
Cemented carbide cutting tool	TiN, TiC	5	IP, CVD	IP comparable to CVD*	Kobayashi and Doi (1978)
HSS (M42) cutting tool	TiC	5–8	ARE	Three to eight times	Bunshah et al. (1977)
HSS (M-10) drill	TiC, TiN	2	ARE	Twenty times	Nimmagadda et al. (1981)
HSS tool	N^+ implantation	—	Ion implantation	Four times	Dearnaley (1982)
Cemented-carbide cutting tool	N^+ implantation	—	Ion implantation	Two times	Dearnaley (1982)

Key: CVD, chemical vapor deposition; MS, magnetron sputtering; IP, ion plating; ARE, activated reactive evaporation.

*Toughness obtained by ion plating (arc evaporation) superior to that of CVD coatings.

and aluminizing treatments are helpful in extending the wear life of barrels and extrusion screws in polymer injection-molding machines (Penzera and Saltzmann, 1979). Dearnaley (1982) reported that dies of cemented carbide used for the drawing of copper or steel wire for metal-forming operations exhibited from two to six times longer life as a result of ion implantation with carbon or nitrogen.

16.10 GAS TURBINE ENGINES

Schematics of the major gas turbine engine components are presented in Fig. 16.10. Compression of the air entering the gas turbine is achieved by centrifugal or axial compressors. The combustion system heats the air from the compressor to the required turbine entry temperature. Hot gases from the combustion chamber are accelerated and directed by stationary nozzle guide vanes into the turbine assembly, which provides the power to drive the compressor. For high-performance gas turbine engines, turbine entry temperatures should be as high as possible. Turbine temperatures can be as high as 1100°C for some aerospace applications (Meetham, 1981).

Gas turbine engines are used in a wide variety of applications. The most demanding of these in terms of materials durability and reliability requirements under relatively severe conditions are aircraft propulsion, marine propulsion, and electric power generation. Table 16.9 provides a summary of the relative severities of surface-damage factors that cause turbine airfoil degradation leading to failure (Goward, 1976; Pettit and Goward, 1983). In all the applications identified,

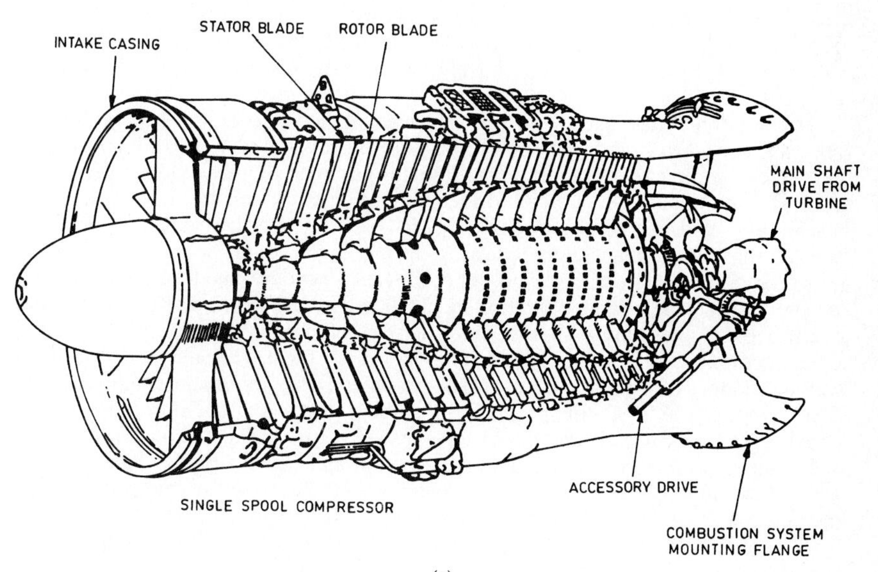

(a)

FIGURE 16.10 Schematic diagrams of gas turbine engine components: (*a*) a typical axial-flow compressor, (*b*) a typical combustion chamber, (*c*) a typical turbine assembly. [*Adapted from Meetham (1981).*]

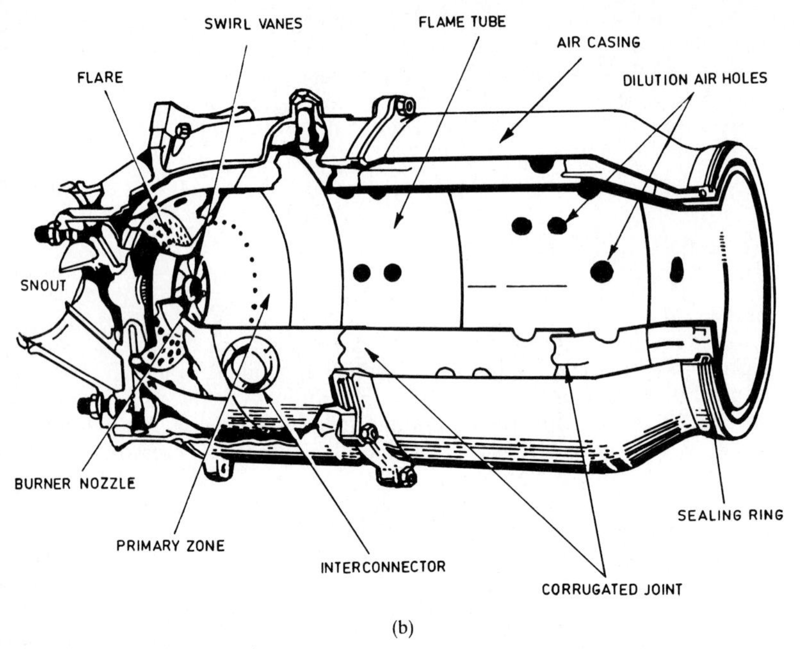

(b)

FIGURE 16.10 (*Continued*)

it is generally accepted that the gas stream constitutes an oxidizing environment. The severity of hot corrosion usually can be correlated with the increased level of salt ingestion associated with high fractions of time near marine or related environments.

Hot-section (turbine end) airfoils in such engines are required to retain mechanical and surface integrity for thousands of hours under conditions of very high stress at elevated temperatures. The hot gases are highly oxidizing and may contain contaminants such as chlorides and sulfates, which can lead to hot corrosion, and also can contain erosive media. Erosion may be caused by ingested sand. Temperature transients occurring during engine operation can cause thermal fatigue. Today, nickel-based superalloys are widely used because of their outstanding strength and oxidation resistance over the temperature range encountered. Nickel-based superalloys such as Nimonic 90, 105, and 115 and Udimet 500 and 700, containing Cr, Co, Ti, Al, and Mo, are commonly used for turbine blades, turbine and compressor casings, and combustion-chamber liners (Anonymous, 1967; Meetham, 1981). A casting process is used to produce turbine blades for economic and design reasons. Sheet materials are used for turbine and compressor casings, combustion chambers, and various pipes. Various coatings are commonly used for protection and will be discussed in this section. The mainline shafts are made of low-alloy steels or maraging steels. High-speed steel (18-4-1) and M-50 steels are the common materials in the mainline rolling-contact bearings. The cages are made of silver-plated En steel. Bearings are normally lubricated with low-viscosity synthetic oils. It should be noted that since early 1970s, prototypes of stationary and rotating components of gas turbine engines

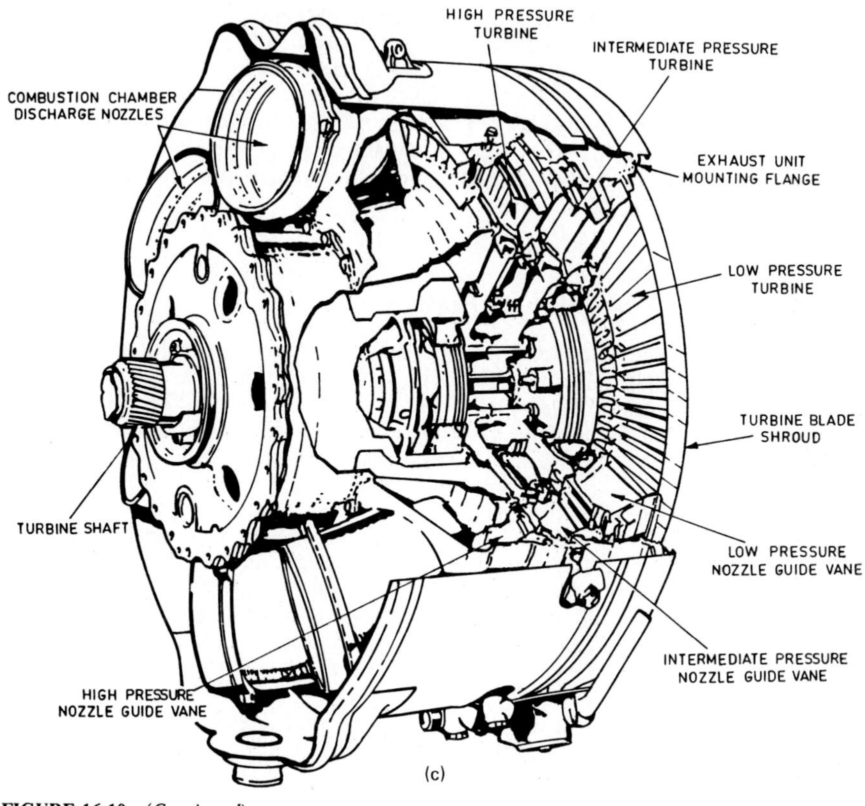

FIGURE 16.10 (*Continued*)

TABLE 16.9 Comparison of Surface-Damage Problems for the Hot Sections of Gas Turbine Applications

	Oxidation	Hot corrosion	Interdiffusion	Thermal fatigue
Aircraft	S	M	S	S
Utility	M	S	M	L
Marine	M	S	L	M

Key: L, light; M, moderate; and S, severe.
Source: Adapted from Pettit and Goward (1983).

for high-temperature (>1000°C) applications have been built with some success of ceramic materials such as SiC (Burke et al., 1978).

With rotational speeds exceeding 10,000 r/min in the high-pressure compressors of large engines, compressor blades experience high tensile stresses. Therefore, specific tensile strength is a basic material requirement. Fatigue strength is also required to resist cyclic stresses. Resistance to erosion and impact by ingested foreign bodies such as sand, stones, and birds is important in early compressor-blade stages. Aluminum alloys were initially used for compressor blades. As exit temperatures exceeded 200°C, the aluminum alloys were superseded by the 12 percent chromium martensitic steels and more recently by titanium alloys (such as Ti-6Al-4V) because of their attractive combination of low density, high specific strength, good fatigue and creep resistance, and excellent corrosion resistance. Coatings are applied to provide corrosion and erosion resistance (Meetham, 1981; Vargas et al., 1980).

16.10.1 Coating Protection Mechanisms

High-temperature (870 to 1200°C) oxidation resistance is generally achieved by aluminum-containing (aluminide) coatings, whether from diffusion or overlays (Goward and Boone, 1971; Goward et al., 1967). The beneficial effect is based on the formation of alumina under oxidizing conditions. The practical durability of the aluminum-containing coatings is dependent on the adherence of the protective oxide. Oxide adherence is enhanced by the presence of oxygen-active elements such as hafnium in the substrate or yttrium added to the overlay coating nominally independent of the substrate alloy composition. The presence of chromium imparts resistance to hot corrosion degradation of alloys and coatings induced by the presence of salt layers. Chromium in coatings or alloys forms protective Cr_2O_3, which is resistant to hot corrosion at certain acidity levels. Similarly, SiO_2 has minimum solubility in highly acidic melts; this suggests that coatings or alloys containing Si should develop significant resistance to corrosion. With regard to thermal fatigue, coatings with moderate (1 to 2 percent) ductility at the temperatures at which maximum strain occurs should have enhanced thermal fatigue resistance compared with elastically brittle coatings.

Ceramic thermal barrier coatings prolong gas turbine component life by insulating the base metal under backside cooling conditions. Steady-state metal temperatures are lowered, and transient thermal strains and the resulting thermal fatigue effects are also reduced by these insulation effects.

16.10.2 Types of Coatings Used

16.10.2.1 Diffusion Coatings. The first widely used coating for the protection of superalloys was based on surface enrichment, e.g., nickel-based alloys with aluminum. The nickel-aluminum intermetallic phase, Ni_2Al_3, or NiAl coatings formed on the Ni-based alloy surface by reaction with aluminum species in gas phase (pack cementation process) provide some protection in applications involving very high temperatures (1100°C) or a severe hot-corrosion environment (Anonymous, 1967). Modest improvements in oxidation and/or hot-corrosion resistance can be achieved by modifying the superalloy surface with elements such as Cr or Pt prior to aluminizing. Similar diffusion coatings based on surface enrichment with chromium or silicon (Anonymous 1967; Bauer et al., 1977, 1979) have found limited applications for protection of utility gas turbine airfoils. Refractory metals and alloys developed for much higher temperatures than superalloys can be protected by fused silicide coatings (Packer and Perkins, 1973; Priceman and Kubick, 1971). These alloys are generally protected by M_5Si_3 and MSi_2 compounds, which primarily form SiO_2 on oxidation (Smialek, 1974).

16.10.2.2 MCrAlY Overlay Coatings. Overlay coatings, typified by the so-called MCrAlY series (M = Fe, Co, Ni, or combinations thereof), are used frequently. MCrAlY alloys allow increased flexibility of design for compositions tailored to a wider variety of applications. Oxidation and hot-corrosion protection and ductility can be varied over wide ranges to meet various requirements for aircraft, utility, and marine propulsion applications. These coatings have been applied by electron-beam evaporation (Rairden, 1978; Restall, 1981), plasma spraying, sputtering, and foil cladding. The most advanced of the MCrAlY series of coatings for high-temperature oxidation resistance is NiCoCrAlY, with a nominal composition (in wt %) of 18 percent Cr, 23 percent Co, 12 percent Al, 0.5 percent Y, and the balance Ni. As applied by electron-beam evaporation on a production scale, this coating has proved useful for the protection of aircraft engine blade airfoils.

The oxidation resistance and ductility of NiCoCrAlY coatings can be improved by the addition of certain metals such as Si, Ta, and Hf (Grisik et al., 1980). Table 16.10 compares the ranking of NiCoCrAlY coatings applied by

TABLE 16.10 Critical Property Ranking* of MCrAlY Coatings Applied by Plasma Spray

Composition, wt % (NiCoCrAlY base)	Oxidation resistance	Ductility
+1.6Si + 8.5Ta	1	—
+1.6 to 2.6 Si	2	4
+1.6Si + 4Ta	3	—
+8Ta	4	5
+0.8Hf	5	2
Base plasma sprayed	6	1
Base electron-beam vapor deposited	7	3

*1 corresponds to the best performance.
Source: Adapted from Pettit and Goward (1983).

plasma spraying on the basis of oxidation resistance, measured at 1120°C by burner cycle rig, and ductility (Pettit and Goward, 1983). Of the compositions tested, NiCoCrAlY coatings having 1.6 to 2.6 percent Si seem to have the best compromise between oxidation resistance and ductility.

As a group, the cobalt-based compositions exhibit much more hot-corrosion resistance than the nickel-based superalloys. The iron-based coatings exhibit relatively poor hot-corrosion resistance. The hot-corrosion rates and thermal expansion coefficients of various Co-, Ni-, and Fe-based coatings applied by electron-beam evaporation on Udimet-500 and IN-738 substrates are compared in Table 16.11. Co-based coatings (with a typical composition of 22 percent Cr, 13 percent Al, 1 percent Y, and the balance Co) show the maximum resistance to hot corrosion.

16.10.2.3 Thermal-Barrier Coatings. Thermal-barrier coatings (TBCs) of zirconia, which have insulating properties and are hot-corrosion resistant, are finding increased utilization in the hot sections of gas turbine engines. All such coatings in practical use are applied by plasma spraying. The major benefit of TBC is the reduction in heat transferred into air-cooled components that they provide. The zirconia ceramic has a thermal conductivity of about 1½ orders of

TABLE 16.11 870°C Burner Rig Hot-Corrosion Rates of Plasma-Sprayed MCrAlY Coatings

	Corrosion rate, h μm^{-1} penetration		Thermal expansion coefficient, × 10^{-6} °C^{-1}	
Coating source composition	Udimet-500 substrate	IN-738 substrate	38–650°C	38–950°C
Uncoated (control)	1.6	1.3	—	—
Co-22Cr-13Al-1Y	44.6	7.1	14.0	15.3
Co-30Cr-9Al-1Y		28.6		
Co-30Cr-9Al-1Y+Al		26.5		
Co-32Cr-3Al-1Y		14.5	16.2	18.1
Co-32Cr-3Al-1Y+Al		37.9		
Co-29Cr-6Al-1Y		33.5	15.8	17.8
Co-29Cr-6Al-1Y+Al		78.0		
Co-26Cr-9Al-1Y		60.8	15.3	17.1
Co-26Cr-9Al-1Y+Al		73.7		
Co-25Cr-4Al-1Y	18.3			
Co-35Cr	33.7	24.8	15.8	17.8
Co-35Cr+Al		41.2		
Ni-25Cr-4Al-1Y	5.3			
Ni-20Cr-15Al	3.1			
Ni-20Cr-5Al	7.8			
Ni-35Cr	30.7			
Ni-35Cr+Al		14.4		
Ni-34.5Cr-1Y	9.5			
Ni-50Cr	0.7			
Fe-25Cr-10Al-1Y	8.4			
Fe-25Cr-4Al-1Y	6.9			
Fe-25Cr-4Al	13.8			
Fe-20Co-30Cr-1Y	8.4			
In-738			15.7	17.5
Udimet-500			14.4	16.7

Note: Substrates, cast pins 4.3 mm in diameter; coating thickness, 75 to 100 μm; test environment, combusted diesel oil plus 1% S +467 ppm sea salt.

Source: Adapted from Rairden (1978).

magnitude lower than that of the superalloys (Restall, 1981; Sink et al., 1979; Wilkes and Lagerdrost, 1973). This means that 50- to 250-μm-thick ZrO_2 coatings can effectively insulate air-cooled combustor and turbine components. The life of such coatings is limited, however, because of the thermal-induced spalling that results from the large thermal expansion mismatch between the coating and the substrate.

One of the approaches used to improve spalling resistance has been based on recognition that failure is caused by flaws that induce cracking propagation in the plane of the coating. The stresses causing such propagation might be reduced by controlling the coating's microstructure such that appropriate discontinuities cause a reduction in the elastic modulus of the coating, thereby reducing local stresses. Such discontinuities include increased porosity, random microcracking, and columnar structures with the column normal to the substrate surface (known as *coating segmentation*). Table 16.12 summarizes the outcomes of several such

TABLE 16.12 Thermal Stress Resistance of Various Modifications of Ceramic Thermal-Barrier Coatings

Coating type	Cycles to failure in thermal stress test*
Unmodified PS† ZrO_2–20 wt % Y_2O_3	200
Microcracked PS ZrO_2–21 wt % MgO	5200
Segmented PS ZrO_2–20 wt % Y_2O_3	5500
Segmented EB‡ ZrO_2–20 wt % Y_2O_3	>1000
Substrate temperature controlled as follows during plasma spraying (ZrO_2–20 wt % Y_2O_3):	
−35°C	3500
20°C	9500
315°C	9000
650°C	200

*Thermal cycled in a petroleum fuel–fired burner with Mach 0.3 gas velocity. Thermal cycle consisted of heating to 1010°C for 4 minutes followed by 2 minutes of cooling to below 315°C.

†Plasma sprayed.

‡Electron-beam vapor deposited.

Source: Adapted from Pettit and Goward (1983).

approaches. The thermal stress-induced spalling, which is measured by cyclic burner testing, is considerably reduced by applying ZrO_2 coatings with microcracked structures and segmented or columnar structures. Microcracked structures can be obtained easily by adding excess MgO (typically 21 percent), and segmented structures can be obtained by optimizing the plasma-spray process parameters. The addition of Y_2O_3 in partially stabilized ZrO_2 coatings applied by electron-beam evaporation also provides segmented structures. The substrate temperature during application of plasma-sprayed coatings is also found to influence the resistance to coating spalling. Application at room temperature shows the best performance.

Another approach is to improve the thermal mismatch between the zirconia coating and the substrate by introducing a metallic interlayer (Fig. 16.11) (Liebert, 1979; Liebert and Levine, 1982; Restall, 1981; Strangman, 1985). MCrAlY interlayers form an adherent oxide scale (Al_2O_3) that inhibits oxidation. The two-layer thermal-barrier coatings with Co- or Fe-based MCrAlY bond coatings with 6.1 percent Y_2O_3– or 8.0 percent Y_2O_3–stabilized zirconia exhibit longer life than MCrAlY coatings alone in cyclic furnace tests between 1120 and 1175°C (Stecura, 1980). Figure 16.12 compares the life of duplex TBC coatings of ZrO_2 and 8 percent Y_2O_3 with Ni-, Co-, and Fe-based bond coatings tested in a cyclic furnace at 1125°C (Stecura, 1986). The longest life was obtained with an FeCrAlY/TBC, followed by the NiCrAlY/TBC and CoCrAlY/TBC. Increasing the Cr in all bond coatings from about 16 to 38 percent is very beneficial, while increasing the Al from about 6 to 20 percent is very detrimental to producing good-quality bond coatings. Stecura (1980) reported that an increase in the bond coating thickness of NiCrAlY/TBC from about 0.25 to 0.75 mm increases the thermal-barrier system life substantially.

Spalling failure of TBC coatings has so far limited their use to less critical gas

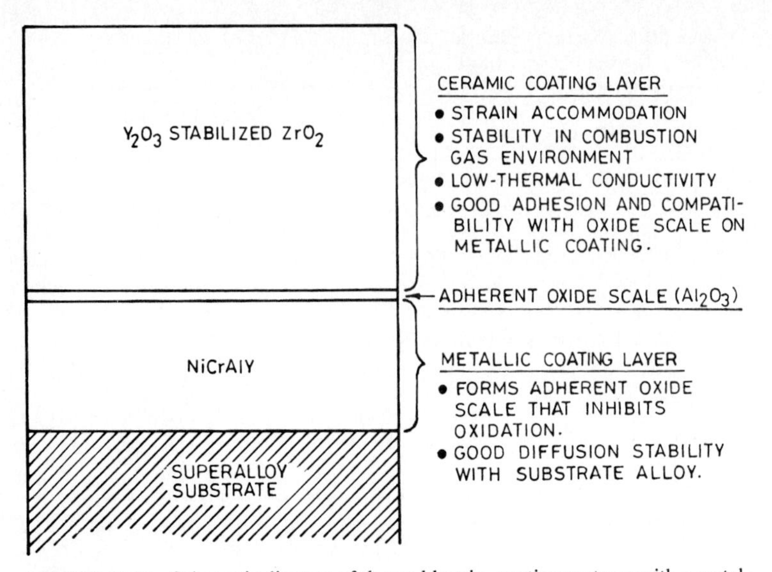

FIGURE 16.11 Schematic diagram of thermal-barrier coating systems with a metallic interlayer.

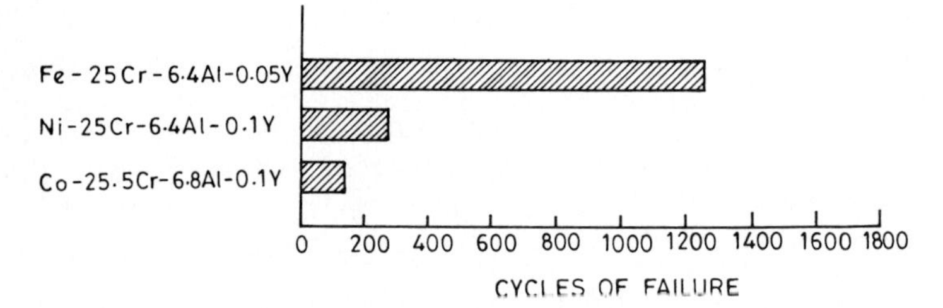

FIGURE 16.12 Life comparison of plasma-sprayed Ni-, Co-, and Fe-based bond coating/ZrO_2–8 wt % Y_2O_3 duplex coating on MAR-M 200 + Hf substrate tested in a cyclic furnace at 1125°C [*Adapted from Stecura (1986)*.]

turbine components such as burner liners and transition ducts fabricated from sheet metal.

16.10.2.4 Coatings for Erosion Resistance. As mentioned earlier, severe erosion of gas turbine components such as compressor blades can occur when sand and dust particles are ingested by gas turbine engine. Chromium steels have reasonable erosion resistance, but titanium alloys require the use of erosion-resistant coatings. Gentner (1974) has identified sputtered SiC as a promising material for antierosion purposes. Hard coatings of TiC-N deposited by CVD have been found to enhance the erosion resistance of PH 17-4 stainless steel blades by a factor of 70 to 80 compared with those of plasma-sprayed with tungsten carbide (Yaws and Wakefield, 1973). Hansen (1979) has reported that SiC deposited by CVD and TiB_2 deposited by electrochemical deposition possess outstanding erosion resistance both at room temperature and at 700°C.

16.11 FUSION REACTORS

Controlled thermonuclear reactors (CTRs) present a unique and demanding environment for materials. Figure 16.13 shows the design of a tokamak (toroidal magnetically confined) reactor with its attendant internal components. The plasma is confined by magnetic fields and is heated by passing large currents through it and injecting high-energy (about 100 keV) neutral hydrogenic particles that give up their energy to the plasma particles by collision. Wall armor opposite the neutral-beam injectors absorbs the "shinethrough" of energetic particles that are not thermalized by the plasma. After the ignition temperature is reached, the plasma is heated by the alpha particles produced by the reaction (Mattox, 1979). The chamber wall is called the *first wall*, and it maintains the vacuum integrity of the system.

There are number of erosion processes operational on a surface in a fusion reactor environment. These include unipolar arcing, physical sputtering, chemical sputtering, blistering, and vaporization. The properties of a surface in recycling hydrogen isotopes are important to the operation of a fusion plasma reactor because, under sustained operations, the hydrogenic fuel will be recycled many times. A hydrogen particle incident on a surface can be reflected, physically or chemically absorbed on the surfaces and desorbed, or become incorporated into the near-surface region and then desorbed. The mechanical properties of a material in a tokamak environment are of concern. Hydrogen and helium embrittlement, surface defects, radiation damage, and extreme thermal and me-

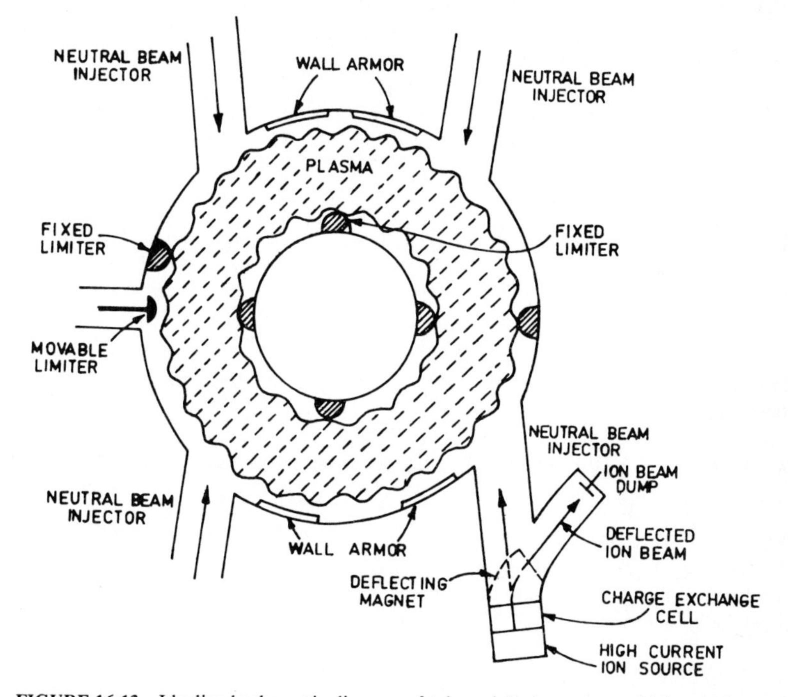

FIGURE 16.13 Idealized schematic diagram of tokamak fusion reactor. [*Adapted from Mattox (1979).*]

chanical environments may lead to premature mechanical failure of structural materials (Mattox, 1979; McCracken and Stott, 1979).

The release of plasma contaminants with large-atomic-number (Z) materials from first-wall components in a plasma reactor may cause significant increases in plasma resistivity and plasma power losses due to electromagnetic radiation. For this reason, first-wall component materials of low to moderate Z that have small impurity release rates and minimal surface erosion rates for typical plasma radiations are preferred. Coating materials with low Z, low erosion rates, good adhesion properties, and high-temperature stability are preferred. To meet these requirements, a variety of hard coatings such as TiB_2, B_4C, SiC, hard carbon, boron, beryllium, and VBe_{12} and methods of deposition such as plasma spraying, vacuum evaporation, sputtering, CVD, PECVD, chemical conversion, and cladding have been evaluated (Kaminsky and Das, 1979, Mattox and Sharp, 1978; Mattox et al., 1979; Veprek and Bruganza, 1979; Wilson and Haggmark, 1979). The coatings have been deposited on graphite, copper, stainless steel, and molybdenum substrates. In addition to the relatively thick coatings normally considered for the first wall, it is also possible to form in situ coatings that are very thin but are renewable by a variety of techniques, including (1) deposition by plasma decomposition of compounds (Braganza et al., 1978; Veprek et al., 1976), (2) vapor deposition from a heated solid (Norem and Bowers, 1978), (3) solute segregation (Johnson and Lam, 1976), and (4) redeposition of eroded material (Cohen et al., 1978).

Several ceramic erosion-resistant and preferably low-atomic-number materials such as TiC, TiB_2, Be, and VBe_{12} have been applied to copper and stainless steel substrates by plasma spraying (Mullendore et al., 1979). These coatings, in general, display good resistance to thermal shock failure. The TiC and TiB_2 coatings exhibit favorable ion-erosion characteristics, and the resistance of the coatings to arc erosion was better than that of stainless steel. Table 16.13 compares some of the important properties such as porosity and erosion yield (atoms per ion) for a few of the most common plasma-sprayed low-Z coatings. TiB_2 coatings exhibited the least erosion yield.

TABLE 16.13 Some Important Properties* of Common Plasma-Sprayed Coatings for Fusion Reactors

Properties	Coatings			
	TiC	TiB_2	Be	VBe_{12}
Total impurities, ppm	3050	2470	940	800
Surface area,† $m^2\ m^{-2}$	340	302	1070	500
Porosity, %	24.8	33.4	29.7	19.3
Porosity below 10 μm, %	90.9	94.7	93.0	74.1
Thickness, mm	0.16	0.16	0.17	0.19
Erosion yield (atoms per ion), $\times 10^{-3}$				
Hydrogen (dose 2.2×10^{21}, energy 250 eV)	0.8	0.9	6	4
Hydrogen (dose 2.1×10^{21}, energy 1000 eV)	8	9.6	16	16
Xenon (dose 7×10^{19}, energy 250 eV)	70	340	90	110
Xenon (dose 2.1×10^{19}, energy 1000 eV)	600	1120	600	810

*Average for copper and stainless steel substrates of 6 mm thickness.
†Coating surface area per unit substrate area.
Source: Adapted from Mullendore et al. (1979).

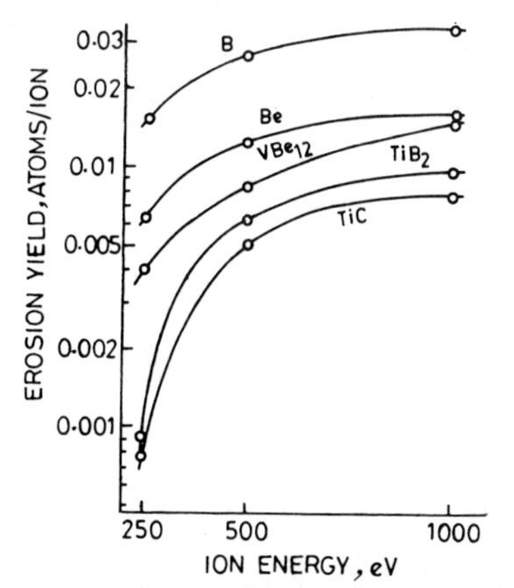

FIGURE 16.14 Yield for hydrogen ion erosion of various coatings applied by CVD on graphite for fusion reactor applications. [*Adapted from Pierson and Mullendore (1979).*]

Boron coatings applied by CVD on graphite were found to be very hard (typically on the order of 2400 HV) and to have pseudoamorphous structures with essentially no porosity and high adhesion. These coatings have demonstrated an excellent resistance to hydrogen ion erosion (Fig. 16.14). These coatings are also good insulators, and they display resistance to the formation of electric arcs. Carbon coatings applied by plasma-enhanced CVD from methane (typically 10 μm thick) have shown negligible damage after 1000 cycles of pulsed thermal heating at 37 MW m^{-2} (Das et al., 1979). Study of surface texture of irradiated surfaces by SEM revealed no significant surface damage for either D^+ or He^+ irradiations with energies of 40 and 60 keV at doses of 4×10^{22} and 8.1×10^{22} ions m^{-2}, respectively. This relative stability under high-energy bombardment favors the potential use of these coatings on various components in fusion reactors.

Several materials have been selected on the basis of the preceding criteria for cladding onto the inner surfaces of fusion reactors (Kaminsky, 1980). Typical materials such as Ti, V, and V–20 percent Ti have been cladded by explosion bonding to copper surfaces. V/Cu-cladded surfaces have been shown to withstand 1000 pulses with a pulse length of 0.5 s for a power density of 3.5 kW cm^{-2}.

16.12 MAGNETIC STORAGE DEVICES

Magnetic storage devices used for audio, video, and data-processing applications are tape, floppy disks, and rigid disk drives. The magnetic recording and playback are accomplished by the relative motion between the magnetic medium and a stationary (audio and data-processing) or rotating (video) read-write magnetic head. Under steady operating conditions, a load-carrying air film is formed. There is a physical contact between the medium and the head during starting and stopping. Figure 16.15 is a schematic drawing of a data-processing IBM 3480 tape drive. After loading the cartridge into the drive, the tape leader is threaded into the drive to the take-up reel by a pentagon threading mechanism. A decoupler column placed near the cartridge entrance decouples any tape vibrations that may occur inside the cartridge. The tension is sensed and controlled by a tension transducer. In the head-disk interface of an IBM 3370 type disk drive, the head slider is mounted on a suspension assembly (Fig. 16.16). The slider-suspension assembly is actuated by a stepping motor or a voice-coil motor for read-write operation (Bhushan, 1990a, 1990b).

16.12.1 Tape Drives

The bearing surfaces of the conventional (inductive-type) magnetic heads are generally coated with wear-resistant coatings such as plasma-sprayed Al_2O_3-TiO_2

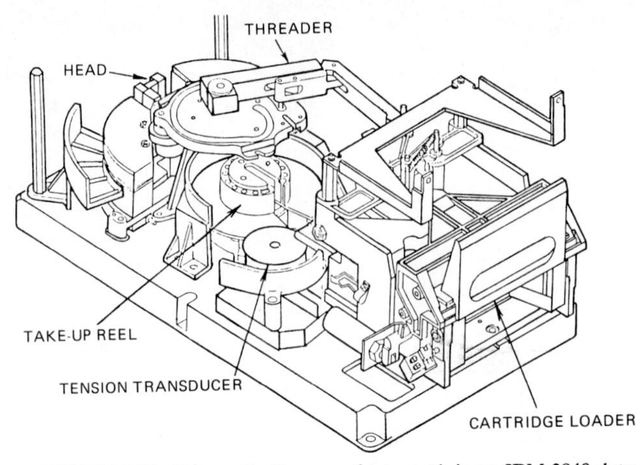

FIGURE 16.15 Schematic diagram of tape path in an IBM 3840 data-processing tape drive.

(87 wt % Al_2O_3, 13 wt % TiO_2) or yttria-stablized zirconia (8 wt % Y_2O_3 and 92 wt % ZrO_2). The bearing surfaces of modern (thin-film magnetoresistive type) heads are made of Ni-Zn ferrite (11 wt % NiO, 22 wt % ZnO, and 67 wt % Fe_2O_3). A sectional view of a particulate magnetic tape is shown in Fig. 16.17. The base film is typically made of poly(ethylene terephthalate) (PET) film. The magnetic coating is composed of magnetic particles [acicular particles such as γ-Fe_2O_3, CrO_2, Fe (metal) or hexagonal platelets of barium ferrite], polymeric binders (such as polyester-polyurethane, polyether-polyurethane, nitrocellulose, poly(vinyl chloride), poly(vinyl alcohol–vinyl acetate), poly(vinylidene chloride), phenoxy, ep-

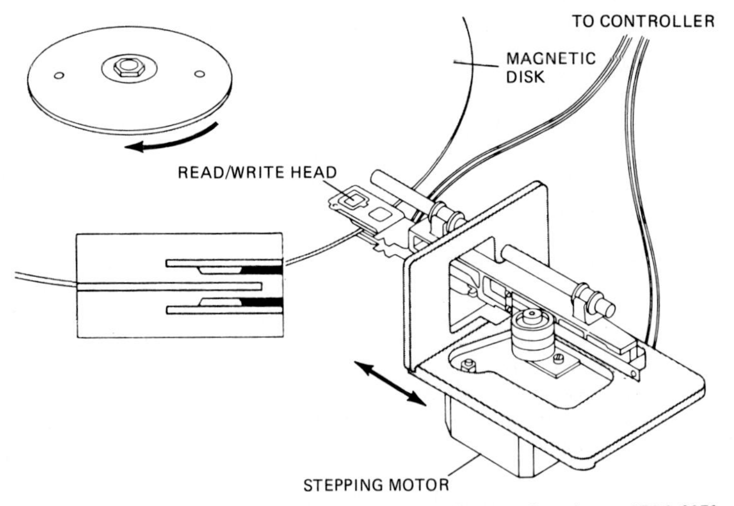

FIGURE 16.16 Schematic diagram of the head-disk interface in an IBM 3370 rigid-disk drive. The inset shows the profile of the head and disk and the spacing between them.

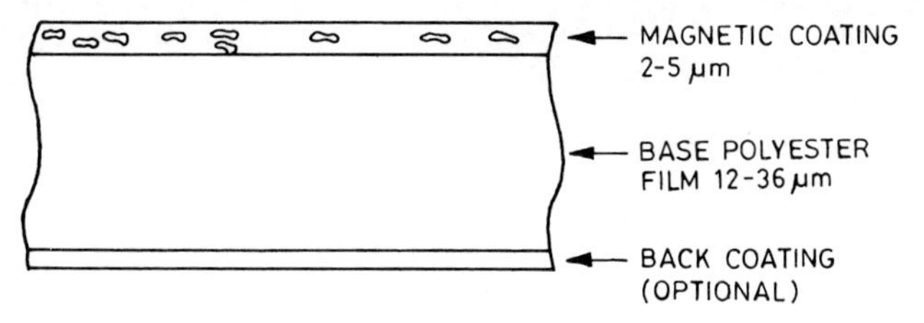

FIGURE 16.17 Sectional view of a particulate magnetic tape.

oxy, or combinations thereof, and lubricants (such as fatty acid esters, e.g., butyl stearate, butyl palmitate, stearic acid) (Bhushan, 1990a, 1990b; Klaus and Bhushan, 1985).

The materials of the tape path are generally selected to be very hard for low wear. Air-bearing surfaces are made of sintered nickel-silver, tungsten carbide, and anodized aluminum, and bearing flange materials are made of tungsten carbide and alumina. The tape-reel hub and flanges are made of glass-filled polycarbonate, phenolic, or polyvinyl chloride. Flanges of high-precision tape reels are made of chemically strengthened glass (Bhushan, 1987, 1990a).

16.12.2 Floppy-Disk Drives

The bearing surfaces of magnetic head materials typically are either Mn-Zn ferrite (17 wt % MnO, 11 wt % ZnO, 72 wt % Fe_2O_3) or calcium titanate. Floppy disks have a similar construction and composition as tapes, except that the substrate is thicker and is magnetically coated on both sides. Floppy disks operate in physical contact with the head; therefore, load-bearing Al_2O_3 or Cr_2O_3 particles are sometimes added to increase the wear resistance. Floppy disks are packaged inside a polyvinyl chloride (PVC) jacket or a hard jacket. Inside the jacket, a soft liner made of a protective fabric is used to minimize wear or abrasion of the media. The liner is made of nonwoven fibers of polyester, poly(ethylene terephthalate), or rayon (Bhushan, 1990a).

16.12.3 Rigid-Disk Drives

The bearing surfaces of conventional (inductive-type) head sliders are made of Mn-Zn ferrite or calcium titanate. The bearing surfaces of modern (thin-film type) head sliders are made of Al_2O_3-TiC (70 wt % Al_2O_3, 30 wt % TiC) or yttria-stablized zirconia/Al_2O_3-TiC composite. Sectional views of particulate and thin-film rigid disks are presented in Fig. 16.18. The substrate for rigid disks is generally non-heat-treatable aluminum-magnesium alloy 5086. For particulate disks, the Al-Mg substrate is sometimes passivated with a very thin (<100 nm) chromate conversion coating based on chromium phosphate. The finished substrate is spin-coated with the magnetic coating. The magnetic coating is composed of acicular magnetic particles of γ-Fe_2O_3 and a binder of a hard copolymer (of epoxy, phenolic, or polyurethane). A small percentage by volume of Al_2O_3 particles is added to improve the wear resistance of the magnetic coating. A 10- to 30-nm-

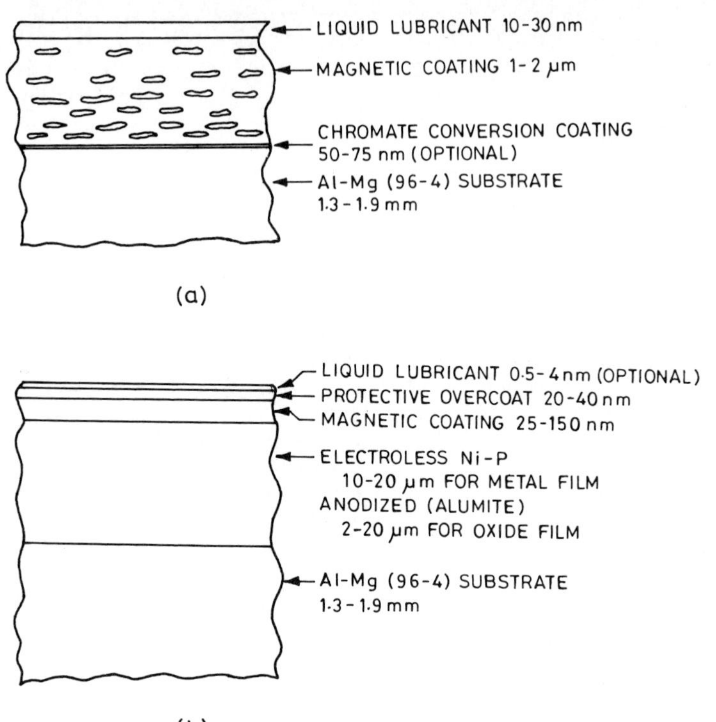

FIGURE 16.18 Sectional views of rigid disks: (*a*) particulate disk, (*b*) thin-film disk.

thick film of perfluoropolyether lubricant (such as Fomblin Z-25, YR, or Krytox 143AD) is applied typically (Bhushan, 1990a, 1990b; Klaus and Bhushan, 1985).

Thin-film disks have either a metallic or an oxide magnetic layer. For thin-film disks with metallic magnetic layers, the Al-Mg substrate is electroless plated with a 10- to 20-μm-thick Ni-P layer (90 wt % Ni, 10 wt % P) to improve its surface hardness to 600 to 800 HV and it smoothness. For thin-film disks with oxide magnetic layers, a 2- to 20-μm-thick alumite layer is formed on the Al-Mg substrate through anodic oxidation in a CrO_3 bath. Recently, chemically strengthened glass has been used as a substrate for thin-film disks with high performance and reliability. The finished substrate is coated with a magnetic film. The magnetic films that have been in development and use are sputtered γ-Fe_2O_3, Co-Ni, Co-Cr, Co-Cr-Ni, Co-Pt-Ni, and other Co-Cr-X alloy films and electroless plated Co-P, Ni-P, and Co-Ni-P (Bhushan, 1990a, 1990b).

Some protective overcoats with or without lubricant overlay are used to improve the corrosion and wear resistance of the thin magnetic film. Protective coatings used to date are plated rhodium and chromium, sputtered or spin-coated SiO_2, sputtered diamond-like carbon; sputtered yttria-stablized zirconia, and plasma-polymerized protective (PPP) films (Harada, 1981; Nagao et al., 1979; Tago et al., 1977; Tsai et al., 1988; Yanagisawa, 1985a). PPP coatings have been made from monomers: toluene-2,4-diisocyanate (TC_y), α-pyrrolidone (α-P_{yr}), tetrafluoroethylene (TFE), and hexafluoropropylene (HFP) (Harada, 1981). The thickness of the protective overcoats for modern disks typically are 20 to 30 nm. In most cases, a thin lubricant coating (0.5 to 4 nm thick) over the protective overcoat is used to further reduce friction and wear. Perfluoropolyether lubricants (such as Fomblin Z-15, Z-25, or Z-dol) are typically used (Bhushan, 1990a, 1990b; Klaus and Bhushan, 1985; Yanagisawa, 1985b).

16.13 REFERENCES

Anonymous (1967), "Hot Corrosion Problems Associated with Gas Turbines," STP-421, ASTM, Philadelphia.

——— (1979), *Proc. Twenty-Fifth Holm Conf.*, Illinois Institute of Technology, Chicago, Ill.

Bamburger, E. N. (1980), "Materials for Rolling Element Bearings," in *Bearing Design: Historical Aspects, Present Technology and Future Problems*, pp. 1–46, ASME, New York.

Bauer, R., Grunling, H. W., and Schneider, K. (1977), "Hot Corrosion Behaviour of Chromium Diffused Ion Coatings," *Proc. Conf. on Materials and Coatings to Resist High Temperature Corrosion*, p. 369, Applied Science Publishers, London.

———, ———, and ——— (1979), "Silicon and Chromium Based Coatings for Stationary Gas Turbines," *Proc. 1st Conf. on Advance Materials for Alternative Fuel Capable Directly Fired Heat Engines,* p. 505, U.S. Department of Energy, Washington, D.C., and Electric Power Research Institute, Palo Alto, Calif.

Baumgartner, H. R. (1978), "Ceramic Bearings for Turbine Applications," in *Ceramics for High Performance Applications*, Vol. 2 (J. J. Burke, E. N. Lenoe, and R. N. Katz, eds.), pp. 423–443, Brook Hill, Chestnut Hill, Mass.

Bhushan, B. (1980a), "High Temperature Self-Lubricating Coatings and Treatments: A Review," *Metal Finish.*, Vol. 78, May, pp. 83–88; June, pp. 71–75.

——— (1980b), "High Temperature Self-Lubricating Coatings for Air-Lubricated Foil Bearings for Automotive Gas Turbine Engine," Tech. Rep. CR-159848, NASA Lewis Research Center, Cleveland, Ohio.

——— (1980c), "Coating Development and Friction Measurements for Solid-Lubricated Rolling-Element Bearings," Rep. No. MTI-80TR30, Mechanical Technology Inc., Latham, N.Y.

——— (1981), "Friction and Wear Results from Sputter-Deposited Chrome Oxide with and without Nichrome Metallic Binders and Interlayers," *J. Lub. Technol.* (*Trans. ASME*), Vol. 103, pp. 218–227.

——— (1982), "Development of CdO-Graphite-Ag for Gas Bearings to 427°C," *Wear*, Vol. 75, pp. 333–356.

——— (1987), "Development of Wear-Test Apparatus for Screening Bearing Flange Materials in Computer Tape Drives," *ASLE Trans.*, Vol. 32, pp. 187–195.

——— (1990a), *Tribology and Mechanics of Magnetic Storage Devices*, Springer-Verlag, New York.

——— (1990b), "Magnetic Media Tribology: State of the Art and Future Challenges," *Wear*, Vol. 137, pp. 41–50.

——— and Dashnaw, F. (1981), "Material Study for Advanced Stern-tube Bearings and Face Seals, *ASLE Trans.*, Vol. 24, pp. 398–409.

——— and Sibley, L. B. (1982), "Silicon Nitride Rolling Bearings for Extreme Operating Conditions," *ASLE Trans.*, Vol. 25, pp. 417–428.

——— and Wilcock, D. F. (1982), "Wear Behavior of Polymeric Compositions in Dry Reciprocating Sliding," *Wear*, Vol. 75, pp. 41–70.

——— and Winn, L. W. (1979), "Material Study for Advanced Sterntube Seals and Bearings for High Performance Merchant Ships," Rep. No. MA-920-79069, U.S. Maritime Administration, Washington, D.C.

——— and ——— (1981), "Material Study for Advanced Stern-tube Lip Seals," *ASLE Trans.*, Vol. 24, pp. 410–422.

——— Ruscitto, D., and Gray, S. (1978), "Hydrodynamic Air-Lubricated Compliant Surface Bearing for an Automotive Gas Turbine Engine Part II: Materials and Coatings," Rep. No. CR-135402, NASA, Cleveland, Ohio.

——— Gray, S., and Graham, R. W. (1982), "Development of Low-Friction Elastomers for Bearings and Seals," *Lub. Eng.*, Vol. 38, pp. 626–634.

Bisson, E. E., and Anderson, W. J. (1964), "Advanced Bearing Technology," Special Publication SP-38, NASA, Washington, D.C.

Boving, H., Hintermann, H. E., and Stehle, G. (1983), "TiC-Coated Cemented-Carbide Balls in Gyro-Application Ball Bearings," *Lub. Eng.*, Vol. 39, pp. 209–215.

Braganza, C., Kordis, J., and Věprek, S. (1978), "In Situ Preparation of Protective Coatings for Controlled Nuclear Fusion Devices," *J. Nucl. Mater.*, Vol. 76–77, pp. 612–613.

Brainard, W. A. (1978), "The Friction and Wear Properties of Sputtered Hard Refractory Compounds," *Proc. Second Int. Conf. Solid Lubrication*, pp. 139–147, ASLE, Park Ridge, Ill.

Budinski, K. G. (1980), "Tool Materials," in *Wear Control Handbook* (M. B. Peterson and W. O. Winer, eds.), pp. 931–985, ASME, New York.

Bunshah, R. F., Shabaik, A. H., Nimmagadda, R., and Covey, J. (1977), "Machining Studies on Coated High Speed Steel Tools," *Thin Solid Films*, Vol. 45, pp. 453–462.

Burke, J. J., Lenoe, E. N., and Katz, R. N. (eds.) (1978), *Ceramics for High Performance Applications*, Brook Hill, Chestnut Hill, Mass.

Cohen, S. A., Dyela, H. F., Rossnagel, S. M. Picraux, S. T., and Borders, J. A. (1978), "Long-Term Changes in the Surface Conditions of PLT," *J. Nucl. Mater.*, Vol. 76–77, pp. 459–471.

Coleman, W. (1970), "Gear Design Considerations," in *Interdisciplinary Approach to the Lubrication of Concentrated Contacts* (P. M. Ku, ed.), pp. 551–589, Special Publication SP-237, NASA, Washington, D.C.

Das, S. K., Kaminsky, M., Rovner, L. H., Chin, J., and Chen, K. Y. (1979), "Carbon Coatings for Fusion Applications," *Thin Solid Films*, Vol. 63, pp. 227–236.

Dearnaley, G. (1982), "Practical Applications of Ion Implantation," *J. Metals*, Vol. 34, pp. 18–28.

DeHart, A. O. (1984), "Sliding Bearing Materials," in *Handbook of Lubrication: Theory and Practice of Tribology*, Vol. 2: *Theory and Design* (E. R. Booser, ed.), pp. 463–476, CRC Press, Boca Raton, Fl.

Dudley, D. W. (1964), *Gear Handbook*, McGraw-Hill, New York.

——— (1980), "Gear Wear," in *Wear Control Handbook* (M. B. Peterson and W. O. Winer, eds.), pp. 755–830, ASME, New York.

Fiedler, H. C., and Sieraski, R. J. (1971), "Boriding of Steels for Wear Resistance," *Metal Prog.*, Vol. 99, pp. 101–107.

Fischer, T. E., Luton, M. J., Williams, J. M., White, C. W., and Appleton, B. R. (1983), "Tribological Properties of Ion-Implanted 52100 Steel," *ASLE Trans.*, Vol. 26, pp. 466–474.

Fuchsluger, J. H., and Vandusen, V. L. (1980), "Unlubricated Piston Rings," in *Wear Control Handbook* (M. B. Peterson and W. O. Winer, eds.), pp. 667–698, ASME, New York.

Fuller, D. D. (1984), *Theory and Practice of Lubrication for Engineers*, 2d Ed., Wiley, New York.

Gardos, M. N., and McConnell, B. D. (1982), "Development of High Speed, High Temperature Self-Lubricating Composites," Special Publication SP-9, ASLE, Park Ridge, Ill.

Gentner, K. (1974), "Thin Erosion Resistant Coatings Deposited by Means of Sputtering Techniques," *Proc. 4th Int. Conf. on Rain Erosion and Associated Phenomena* (A. A. Fyall and R. B. King, eds.), pp. 701–704, Meersburg, Germany.

Glaeser, W. A. (1980), "Bushings," in *Wear Control Handbook* (M. B. Peterson and W. O. Winer, eds.), pp. 583–607, ASME, New York.

Goward, G. W. (1976), "Hot Corrosion of Gas Turbine Airfoils: A Review of Selected Topics," *Tri-Service Conf. on Corrosion*, MCIC-77-3, p. 239.

——— and Boone, D. H. (1971), "Mechanisms of Formation of Diffusion Aluminide Coatings on Nickel-Base Superalloys," *Oxidation of Metals*, Vol. 3, pp. 475–495.

———, ———, and Giggins, C. S. (1967), "Formation and Degradation Mechanisms of Aluminide Coatings on Nickel-Base Superalloys," *Trans. ASM*, Vol. 60, pp. 228–241.

Grisik, J. J., Miner, R. G., and Wortman, J. (1980), "Performance of Second Generation Airfoil Coatings in Marine Service," *Thin Solid Films*, Vol. 73, pp. 397–406.

Hansen, J. S. (1979), "Relative Erosion Resistance of Several Materials," in *Erosion: Prevention and Useful Applications* (W. F. Adler, ed.), SP-664, ASTM, Philadelphia.

Harada, K. (1981), "Plasma Polymerized Protective Films for Plated Magnetic Disks," *J. Appl. Poly. Sci.*, Vol. 26, pp. 3707–3718.

Hintermann, H. E. (1980), "Exploitation of Wear- and Corrosion-Resistant CVD Coatings," *Tribol. Int.*, Vol. 13, pp. 267–277.

——— (1981), "Tribological and Protective Coatings by Chemical Vapor Deposition," *Thin Solid Films*, Vol. 84, pp. 215–243.

——— (1984), "Adhesion, Friction, and Wear of Thin Hard Coatings," *Wear*, Vol. 100, pp. 381–397.

Inwood, B. C., and Garwood, A. E. (1978), "Electroplated Coatings for Wear Resistance," *Tribol. Int.*, Vol. 11, pp. 113–119.

Johnson, R. A., and Lam, N. Q. (1976), "Solute Segregation in Metals Under Irradiation," *Phys. Rev.* [*B*], Vol. 13, pp. 4364–4375.

Johnson, R. L., and Schoenherr, K. (1980), "Seal Wear," in *Wear Control Handbook* (M. B. Peterson and W. O. Winer, eds.), pp. 727–753, ASME, New York.

Kaminsky, M. (1980), "Clad Materials for Fusion Applications," *Thin Solid Films*, Vol. 73, pp. 117–132.

——— and Das, S. K. (1979), "Surface Damage of TiB_2 and C Coatings under Energetic D^+ and $4He^+$ Irradiations," *J. Nucl. Mater.*, Vol. 85–86, pp. 1095–1100.

Klaus, E. E., and Bhushan, B. (1985), "Lubricants in Magnetic Media: A Review," in *Tribology and Mechanics of Magnetic Storage Systems*, Vol. 2 (B. Bhushan and N. S. Eiss, eds.), pp. 7–15, SP-19, ASLE, Park Ridge, Ill.

Knotek, O., Münz, W. D., and Leyendecker, T. (1987), "Industrial Deposition of Binary, Ternary, and Quaternary Nitrides of Titanium, Zirconium, and Aluminum," *J. Vac. Sci. Technol.*, Vol. A5, pp. 2173–2179.

Kobayashi, M., and Doi, Y. (1978), "TiN and TiC Coating on Cemented Carbides by Ion Plating," *Thin Solid Films*, Vol. 54, pp. 67–74.

Komiya, S. (1982), "Ion Plating by Hot Hollow Cathode Discharge: Its Applications to the Tool Industry," unpublished review article, ULVAC Corp., Japan.

Liebert, C. H. (1979), "Tests for NASA Ceramic Thermal Barrier Coating for Gas Turbine Engines," *Thin Solid Films*, Vol. 64, pp. 329–333.

——— and Levine, S. R. (1982), "Further Industrial Tests of Ceramic Thermal-Barrier Coatings," Tech. Rep. 2057, NASA, Washington, D.C.

Mattox, D. M. (1979), "Coatings for Fusion Reactor Environments," *Thin Solid Films*, Vol. 63, pp. 213–226.

——— and Sharp, D. J. (1978), "Low Energy Hydrogen Ion Erosion Yields as Determined with a Kaufman Source," Rep. SAND 78-1029, Sandia Laboratories, Albuquerque, N.M.

———, Mullendore, A. W., Pierson, H. O., and Sharp, D. J. (1979), "Low Z Coatings for Fusion Reactor Applications," *J. Nucl. Mater.*, Vol. 85–86, pp. 1127–1131.

McCracken, G. M., and Stott, P. E. (1979), "Plasma-Surface Interactions in Tokamaks," *Nucl. Fusion*, Vol. 19, pp. 889–981.

McNab, I. R., and Johnson, J. L. (1980), "Brush Wear," in *Wear Control Handbook* (M. B. Peterson and W. O. Winer, eds.), pp. 1053–1101, ASME, New York.

Meetham, G. W. (ed.) (1981), *The Development of Gas Turbine Materials*, Applied Science Publishers, London.

Merritt, H. E. (1971), *Gear Engineering*. Pitman, London.

Mullendore, A. W., Mattox, D. M., Whitley, J. B., and Sharp, D. J. (1979), "Plasma-Sprayed Coatings for Fusion Reactor Applications," *Thin Solid Films*, Vol. 63, pp. 243–249.

Nagao, M., Suganuma, Y., Tanaka, H., Yanagisawa, M., and Goto, F. (1979), "787 BPM 140 TPM Feasibility of a Plated Disk," *IEEE Trans. Magn.*, Vol. MAG-15, pp. 1543–1545.

Neale, M. J. (ed.) (1973), *Tribology Handbook*, Newnes-Butterworth, London.

Nimmagadda, R. R., Doerr, H. J., and Bunshah, R. F. (1981), "Improvement in Tool Life of Coated High Speed Steel Drills Using the Activated Reactive Evaporation Process," *Thin Solid Films*, Vol. 84, pp. 303–306.

Norem, J., and Bowers, D. A. (1978), "Thin Low Z Coatings for Plasma Devices," Tech. Rep. ANL/EPP/TM-108, Argonne National Labs.

Oakes, J. J. (1983), "A Comparative Evaluation of HfN, Al_2O_3, TiC, and TiN Coatings on Cemented Carbide Tools," *Thin Solid Films*, Vol. 107, pp. 159–165.

Packer, C. M., and Perkins, R. A. (1973), "Modified Fused Silicide Coatings for Tantalum (Ta-10W) Reentry Heat Shields," Rep. No. CR-120877, NASA, Washington, D.C.

Paxton, R. R. (1979), *Manufactured Carbon: A Self-Lubricating Material for Mechanical Devices*, CRC Press, Boca Raton, Fl.

Penzera, C., and Saltzmann, G. A. (1979), "Effect of Diffusion Coatings and Work Hardening on Wear Resistance of Commercial Hard-Facing Alloys," in *Proc. Wear of Materials* (K. C. Ludema, W. A. Glaeser, and S. K. Rhee, eds.), pp. 441–448, ASME, New York.

Pettit, F. S., and Goward, G. W. (1983), "Gas Turbine Applications," in *Coatings for High Temperature Applications* (E. Lang, ed.), pp. 341–359, Applied Science Publishers, New York.

Pierson, H. O., and Mullendore, A. W. (1979), "Boron Coatings on Graphite for Fusion Reactor Applications," *Thin Solid Films*, Vol. 63, pp. 257–261.

Plumb, R. S., and Glaeser, W. A. (1977), "Wear Prevention in Mold Used to Mold Boron-Filled Elastomers, in *Proc. Wear of Materials* (W. A. Glaeser, K. C. Ludema, and S. K. Rhee, eds.), pp. 337–343, ASME, New York.

Pratt, G. C. (1973), "Materials for Plain Bearings," *Int. Metallurg. Rev.*, Vol. 18, pp. 62–68.

Priceman, S., and Kubick, R. (1971), "Development of Protective Coatings for Columbium Alloy Gas Turbine Blades, Tech. Rep. AFML-TR-71-172, p. 33, Sylvania Electric Products, Hicksville, N.Y.

Quinto, D. T., Wolfe, G. J., and Jindal, P. C. (1987), "High Temperature Microhardness of Hard Coatings Produced by Physical and Chemical Vapor Deposition," *Thin Solid Films*, Vol. 153, pp. 19–36.

Rairden, J. R. (1978), "Hot-Corrosion-Resistant Duplex Coatings for a Superalloy," *Thin Solid Films*, Vol. 53, pp. 251–258.

Ramalingam, S., and Winer, W. O. (1980), "Reactively Sputtered TiN Coatings for Tribological Applications," *Thin Solid Films*, Vol. 73, pp. 267–274.

Restall, J. E. (1981), "Surface Degradation and Protective Treatments," in *The Development of Gas Turbine Materials* (G. W. Meetham, ed.), pp. 259–290, Applied Science Publishers, London.

Scott, D. (1977), "Lubricant Effects on Rolling Contact Fatigue—A Brief Review, Performance, Testing of Lubricants," *Proc. Symp. on Rolling Contact Fatigue* (R. Tourret and E. P. Wright, eds.), pp. 3–15, 39–44, Heyden, London.

———, Smith, A. I., Tait, J., and Tremain, G. R. (1975), "Metals and Metallurgical Aspects of Piston Ring Scuffing—A Literature Survey," *Wear*, Vol. 33, pp. 293–315.

Shobert, E. I. (1965), *Carbon Brushes*. Chemical Publishing Corp., New York.

Sink, L. W., Hoppin, G. S., and Fujii, M. (1979), "Low-Cost Directionally Solidified Turbine Blades," Rep. No. CR-159464, NASA, Washington, D.C.

Smialek, J. L. (1974), "Fused Silicon-Rich Coatings for Super Alloys," Tech. Memo TM X-3001, NASA, Washington, DC.

Smith, W. V. (1967), "Wear Properties of Plastics as Ships Bearings," in *Testing of Polymers*, Vol. 3, pp. 221–244, Interscience Publishers, New York.

Sproul, W. D., and Rothstein, R. (1985), "High Rate Reactive Sputtered TiN Coatings on High Speed Steel Drills," *Thin Solid Films*, Vol. 126, pp. 257–263.

Stair, W. K. (1984), "Dynamic Seals," in *Handbook of Lubrication: Theory and Practice of Tribology*, Vol. 2: *Theory and Design* (E. R. Booser, ed.), pp. 581–622, CRC Press, Boca Raton, Fla.

Stecura, S. (1980), "Effects of Yttrium, Aluminum, and Chromium Concentrations in Bond Coatings on the Performance of Zirconia-Yttria Thermal Barriers," *Thin Solid Films*, Vol. 73, pp. 481–489.

——— (1986), "Advanced Thermal Barrier System Bond Coatings Used on Nickel-, Cobalt-, and Iron-Based Alloy Substrates," *Thin Solid Films*, Vol. 136, pp. 241–256.

Strangman, T. E. (1985), "Thermal Barrier Coatings for Turbine Airfoils," *Thin Solid Films*, Vol. 127, pp. 93–105.

Tago, A., Masuda, T., and Ando, T. (1977), "Plated Magnetic Disk," *Elect. Comm. Lab. Tech. J. (Japan)*, Vol. 26, pp. 471–498.

Tallian, T. E. (1967), "On Competing Failure Modes in Rolling Contact," *ASLE Trans.*, Vol. 10, pp. 418–439.

Taylor, B. J., and Eyre, T. S. (1979), "A Review of Piston Rings and Cylinder Liner Materials," *Tribol. Int.*, Vol. 12, pp. 79–89.

Ting, L. L. (1980), "Lubricated Piston Rings and Cylinder Bore Wear," in *Wear Control Handbook* (M. B. Peterson and W. O. Winer, eds.), pp. 609–665, ASME, New York.

Tipnis, V. A. (1980), "Cutting Tool Wear," in *Wear Control Handbook* (M. B. Peterson and W. O. Winer, eds.), pp. 891–930, ASME, New York.

Tsai, H.-C., Bogy, D. B., Kundmann, M. K., Veirs, D. K., Hilton, M. R., and Mayer, S. T. (1988), "Structure and Properties of Sputtered Carbon Overcoats on Rigid Magnetic Media Disks," *J. Vac. Sci. Technol.*, Vol. A6, pp. 2307–2315.

Vargas, J. R., Ulion, N. E., and Goebel, J. A. (1980), "Advanced Coating Development for Industrial/Utility Gas Turbine Engines," *Thin Solid Films*, Vol. 73, pp. 407–413.

Vĕprek, S., and Bruganza, C. M. (1979), "Boron Compounds and Metal Nitride Coatings for Fusion Devices Prepared by Means of Low Pressure Discharges," Topical Meeting on Fusion Reactor Materials, Bal Harbour, Fl.

———, Haque, M. R., and Oswald, H. R. (1976), "On the Chemical Erosion of Some Low-Z Materials by Hydrogen Plasma and on the Possibility of Regeneration of the First Wall by Low Pressure Plasma CVD," *J. Nucl. Mater.*, Vol. 63, pp. 405–409.

Wilkes, K. E., and Lagerdrost, J. E. (1973), "Thermophysical Properties of Plasma Sprayed Coatings," Rep. No. CR-121144, NASA Lewis Research Center, Cleveland, Ohio.

Wilson, K. L., and Haggmark, L. G. (1979), "Deuterium Trapping Measurements in Aluminum and Plasma Sprayed Aluminum Coatings," *Thin Solid Films*, Vol. 63, pp. 283–288.

Yanagisawa, M. (1985a), "Tribological Properties of Spin-Coated SiO_2 Film on Plated Magnetic Recording Disks," in *Tribology and Mechanics of Magnetic Storage Systems* Vol. 2 (B. Bhushan and N. S. Eiss, eds.), pp. 21–26, SP-19, ASLE, Park Ridge, Ill.

——— (1985b), "Lubricants on Plated Magnetic Recording Disks," in *Tribology and Mechanics of Magnetic Storage Systems* Vol 2 (B. Bhushan and N. S. Eiss, eds.), pp. 16–20, SP-19, ASLE, Park Ridge, Ill.

Yaws, C. L., and Wakefield, G. F. (1973), "Scale-Up Process for Erosion Resistant Titanium Carbonitride," *Proc. 4th Int. Conf. on Chemical Vapor Deposition*, pp. 577–588, Electrochemical Society, Princeton, N.J.

Zaretsky, E. V., and Anderson, W. J. (1970), "Effect of Materials: General Background," in *Interdisciplinary Approach to the Lubrication of Concentrated Contacts*, Special Publication SP-237, pp. 379–408, NASA, Washington, D.C.

AUTHOR INDEX

SUBJECT INDEX

ABOUT THE AUTHORS

Dr. Bharat Bhushan is the manager of the head/disk interface at IBM's Almaden Research Center and is an internationally recognized expert in tribology and materials science. He received an M.S. in mechanical engineering from the Massachusetts Institute of Technology, an M.S. in mechanics and a Ph.D. in mechanical engineering from the University of Colorado at Boulder, an M.B.A. from Rensselaer Polytechnic Institute, and a D.Sc. (Tech.) from the University of Trondheim in Norway.

Dr. Bhushan has authored or coauthored three books, more than 125 papers in referred journals, more than 60 technical reports, and holds four U.S. patents. He is a recipient of numerous awards from various professional societies, universities, and industries. He is editor in chief of "Advances in Information Storage Systems," an ASME Series, and annually chairs a tribology symposium. He is a fellow of ASME and NY Academy of Sciences, a senior member of IEEE, and a member of STLE.

Dr. B. K. Gupta received a Ph.D. in Physics from the Indian Institute of Technology in New Delhi, and is currently a senior design engineer at that institution. His current interests are in the areas of wear and corrosion-resistant coatings deposited by ion-plating, and the erosion behavior of materials.